HANDBOOK OF SURFACE AND COLLOID CHEMISTRY

FOURTH EDITION

HANDBOOK OF

SURFACE AND COLLOID CHEMISTRY

FOURTH EDITION

Edited by **K.S. Birdi**

CRC Press is an imprint of the
Taylor & Francis Group, an **informa** business

CRC Press
Taylor & Francis Group
6000 Broken Sound Parkway NW, Suite 300
Boca Raton, FL 33487-2742

First issued in paperback 2020

ISBN 13: 978-0-367-57566-3 (pbk)
ISBN 13: 978-1-4665-9667-2 (hbk)

Visit the Taylor & Francis Web site at
http://www.taylorandfrancis.com

and the CRC Press Web site at
http://www.crcpress.com

To Leon, Esma, and David.

Contents

Preface

The theoretical data and application areas of science related to the subject of *surface and colloid chemistry* have been expanding at a rapid pace in the past decade. Especially, this science has been found to be of significant importance in new areas such as energy sources (oil and gas reservoirs), environmental control (pollution), and biotechnology.

Already about half a century ago, the theoretical understanding of surface and colloid systems was found to be of much importance. The amount of information published since then has increased steadily, considering that there are, at present, more than half a dozen different specialty journals related mainly to surface and colloids. The application area of this subject has developed rapidly in both the industrial and biological areas.

During the last few decades, many empirical observations have been found to have basis in the fundamental laws of physics and chemistry. These laws have been extensively applied to the science of surface and colloid chemistry. This development gave rise to investigations based upon molecular description of surfaces and reactions at interfaces. Especially during the last decade, theoretical analyses have added to the understanding of this subject with increasing molecular detail. These developments are moving at a much faster pace with each decade.

The application area of surface and colloid science has increased dramatically during the past decades. For example, the major industrial areas have been soaps and detergents, emulsion technology, colloidal dispersions (suspensions, nanoparticles), wetting and contact angle, paper, cement, oil recovery (enhanced oil recovery [EOR] and shale oil/gas reservoir technology), pollution control, fogs, foams (thin liquid films), food industry, biomembranes, membranes, and pharmaceutical industry.

In the previous editions of this handbook, various important data have been delineated related to theoretical and experimental information on the systems related to surface and colloids. The purpose of the fourth edition of this handbook is to bring the reader up to date with the most recent developments in this area. In this handbook, a team of international experts presents an updated unifying theme of information on surface and colloid chemistry. The subject content is presented in such a manner that the reader can follow through the physical principles, which are needed for application, and extensive references are included for understanding the related phenomena. This edition adds new insights and includes most recent literature data. Further, most important and new areas of research are also included and highlighted.

As the subject area and the quantity of knowledge are immense, there is always a need for a team of experts to join together and compile a handbook. It is therefore an honor for me to be able to arrange and present to the reader chapters written by experts on various subjects pertaining to this science, with bibliographies in excess of a thousand.

It is most impressive to find how theoretical knowledge has led to some fascinating developments in the technology. The purpose of this handbook is also to further this development. The molecular description of liquid surfaces has been obtained from surface tension and adsorption studies. The emulsion (microemulsion) formation and stability are described by the interfacial film structures. The surfaces of solids are characterized by contact angle and adsorption studies. The ultimate in interfaces is an extensive description of chemical physics of colloid systems and interfaces. Contact angle and adhesion is described at a very fundamental level. The thermodynamics of polymer solutions is reviewed. Polymer–surfactant systems are described. The colloidal structures and their stability have been found to be of much interest as described extensively in this handbook.

Finally, with great pleasure, I thank the staff of CRC Press for their patience and endurance in helping me through at every stage in this task as an editor.

I also thank my family for providing the right kind of consideration while working through the material for this handbook.

Editor

Professor K.S. Birdi received his BSc (Hons) from Delhi University, Delhi, India, in 1952. He then traveled to the United States for further studies, majoring in chemistry at the University of California at Berkeley. After graduation in 1957, he joined Standard Oil of California, Richmond. In 1959, he moved to Copenhagen, Denmark, where he joined Lever Brothers as chief chemist in the Development Laboratory. During this period, he became interested in surface chemistry and joined, as assistant professor, the Institute of Physical Chemistry (founded by Professor J. Brønsted), Danish Technical University, Lyngby, Denmark, in 1966. He initially did research on surface science aspects (e.g., detergents, micelle formation and emulsion technology, adsorption and adhesion, and biophysics). During the early exploration and discovery stages of oil and gas in the North Sea, Professor Birdi got involved in Research Science Foundation programs, with other research institutes around Copenhagen, in the oil recovery phenomena and surface science (enhanced oil recovery [EOR]). Later, research grants on the same subject were awarded from the European Union projects. These projects also involved extensive visits to other universities and an exchange of guests from all over the world. Professor Birdi was appointed research professor in 1985 (Nordic Science Foundation) and was then appointed, in 1990 (retired in 1999), to the School of Pharmacy, Copenhagen, as professor in physical chemistry.

There was continuous involvement with various industrial contract research programs throughout these years. These projects have actually been a very important source of information in keeping up with real problems and helped in the guidance of research planning at all levels.

Professor Birdi has been a consultant to various national and international industries. He has been a member of various chemical societies and a member of organizing committees of national and international meetings related to surface science. He has been a member of selection committees for assistant professors and professors and was an advisory member (1985–1987) of the ACS journal *Langmuir*.

Professor Birdi has been an advisor for about 90 advanced student projects and various PhD projects. He is the author of about 100 papers and articles (and a few hundred citations).

In order to describe these research observations and data, he realized that it was essential to delineate these in the form of books related to surface and colloid chemistry. His first book on surface science was published in 1984: *Adsorption and the Gibbs Surface Excess*, Chattoraj, D.K. and Birdi, K.S., Plenum Press, New York. His further publications include *Lipid and Biopolymer Monolayers at Liquid Interfaces*, K. S. Birdi, Plenum Press, New York, 1989; *Fractals, in Chemistry, Geochemistry and Biophysics*, K. S. Birdi, Plenum Press, New York, 1994; *Handbook of Surface and Colloid Chemistry*, K. S. Birdi (editor) (first edition, 1997; second edition, 2003; third edition, 2009; CD-ROM, 1999), CRC Press, Boca Raton, FL; *Self-Assembly Monolayer*, Plenum Press, New York, 1999; *Scanning Probe Microscopes*, CRC Press, Boca Raton, FL, 2002; *Surface and Colloid Chemistry*, CRC Press, Boca Raton, FL, 2010 (translated to Kazakh, 2013); *Introduction to Electrical Interfacial Phenomena*, K. S. Birdi (editor), 2010, CRC Press, Boca Raton, FL, 2013; and *Surface Chemistry Essentials*, CRC Press, Boca Raton, FL, 2014. Surface chemistry (theory and applications) has remained his major area of research interest throughout these years.

Editor

Contributors

Waisudin Badri
University of Lyon
Lyon, France

and

University Lyon-1
Villeurbanne, France

Sanjib Bagchi
Department of Chemical Sciences
Indian Institute of Science Education and Research, Kolkata
Kolkata, India

K.S. Birdi
KSB Consultants
Holte, Denmark

Krassimir D. Danov
Faculty of Chemistry and Pharmacy
Department of Chemical and Pharmaceutical Engineering
Sofia University
Sofia, Bulgaria

Abdelhamid Elaïssari
University of Lyon
Lyon, France

and

University Lyon-1
Villeurbanne, France

V.S. Gevod
Department of Inorganic Chemistry
Ukranian State Chemical Technology University
Dnipropetrovsk, Ukraine

Georgios M. Kontogeorgis
Department of Chemical and Biochemical Engineering
Center for Energy Resources Engineering
Technical University of Denmark
Lyngby, Denmark

Peter A. Kralchevsky
Faculty of Chemistry and Pharmacy
Department of Chemical and Pharmaceutical Engineering
Sofia University
Sofia, Bulgaria

Qiuli Lu
Department of Biomedical, Chemical and Environmental Engineering
University of Cincinnati
Cincinnati, Ohio

Ali A. Mohsenipour
Department of Chemical Engineering
University of Waterloo
Waterloo, Ontario, Canada

Rajinder Pal
Department of Chemical Engineering
University of Waterloo
Waterloo, Ontario, Canada

Costas G. Panayiotou
Department of Chemical Engineering
Aristotle University of Thessaloniki
Thessaloniki, Greece

and

Department of Chemical Engineering
Technical University of Denmark
Lyngby, Denmark

Chittaranjan Ray
Nebraska Water Center
University of Nebraska
Lincoln, Nebraska

I.L. Reshetnyak
Department of Mechanical Engineering
Ukranian State Chemical Technology University
Dnipropetrovsk, Ukraine

Tushar Kanti Sen
Department of Chemical Engineering
Curtin University
Perth, Western Australia, Australia

Nicolas von Solms
Department of Chemical and Biochemical Engineering
Center for Energy Resources Engineering
Technical University of Denmark
Lyngby, Denmark

George A. Sorial
Department of Biomedical, Chemical and Environmental Engineering
University of Cincinnati
Cincinnati, Ohio

Ioannis Tsivintzelis
Department of Chemical Engineering
Aristotle University of Thessaloniki
Thessaloniki, Greece

Ziheng Wang
Department of Chemical Engineering and Applied Chemistry
University of Toronto
Toronto, Ontario, Canada

Liang Yan
Department of Biomedical, Chemical and Environmental Engineering
University of Cincinnati
Cincinnati, Ohio

1 Introduction to Surface and Colloid Chemistry

Recent Advances and General Remarks

K.S. Birdi

CONTENTS

1.1 INTRODUCTION

Surface and colloid chemistry is now recognized to play a very important role in everyday life. Accordingly, the number of research publications related to this subject is also very extensive. Therefore, it is useful to delineate in this chapter some recent advances and general remarks, as related to the fundamental principles of the subject. All natural phenomena are defined by scientific observations of the matter surrounding the earth and heaven. It has been recognized that the science needs information on the structures of the matter (defined as *solids*, *liquids*, *gases*) (Figure 1.1). Research conducted during the last decades has shown that one needs a much more detailed picture of these structures in all kinds of processes (chemical industry and technology and natural biological phenomena). Application of this research has been relevant to many modern industrial innovations (as regards future challenges: drinking water, energy resources [conventional and non-conventional oil/gas reservoirs], food, clean air, transportation, pollution control, housing, health and medicine, paper, printing, etc.).

Matter exists as

Gas
Liquid
Solid

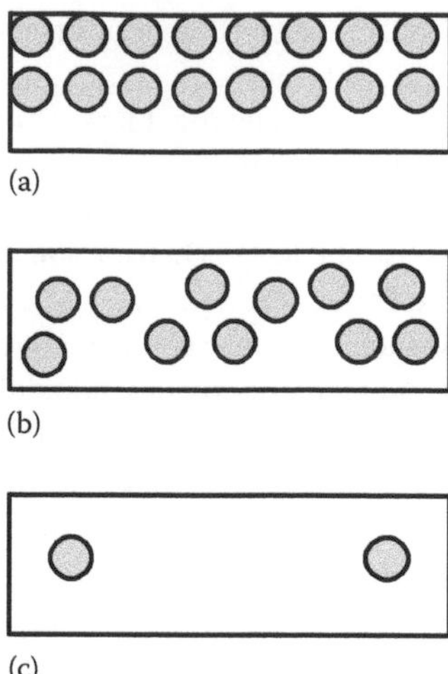

FIGURE 1.1 Molecular structure (schematic) of solid (a), liquid (b), and gas (c).

This has been recognized by classical science (as depicted in the following text):

Solid phase–Liquid phase–Gas phase

Several decades of research has shown that the molecules that are situated at the *interfaces* (e.g., between gas–liquid, gas–solid, liquid–solid, $liquid_1$–$liquid_2$, $solid_1$–$solid_2$) are known to behave differently (Figure 1.2) than those in the bulk phase (Adam, 1930; Adamson and Gast, 1997; Aveyard and Hayden, 1973; Bakker, 1928; Bancroft, 1932; Barnes, 2011; Birdi, 2003a, 2003b, 2009, 2010a, 2014; Birdi and Nikolov, 1979; Biresaw and Mittal, 2008; Butt, 2010; Butt et al., 2013; Castner and Ratner, 2002; Chattoraj and Birdi, 1984; Cini et al., 1972; David and Neumann, 2014; Davies and Rideal, 1963; De Gennes, 2003; Defay et al., 1966; Fanum, 2008; Fendler and Fendler, 1975; Gaines, 1966; Harkins, 1952; Hiemenz and Rajagopalan, 1997; Holmberg, 2002; Israelachvili, 2011; Janaiaud et al., 2014; Kelkar et al., 2014; Kolasinski, 2008; Krungleviclute et al., 2012; Lara et al., 1998; MacDowell et al., 2014; Masel, 1996; Matijevic, 1969; McCash, 2001; Merk et al., 2014; Miller and Neogi, 2008; Mousny et al., 2008; Mun et al., 2014; Nicot and Scanlon, 2012; Oura, 2003; Packham, 2005; Papachristodoulou and Trass, 1987; Partington, 1951; Rao and Geckeler, 2011; Rosen, 2004; Schramm, 2005; Sinfelt, 2002; Soltis et al., 2004; Somasundaran, 2006; Somorjai and Li, 2010, 2011; Taylor, 1984; Yates and Cambell, 2011; Zana, 2008.).

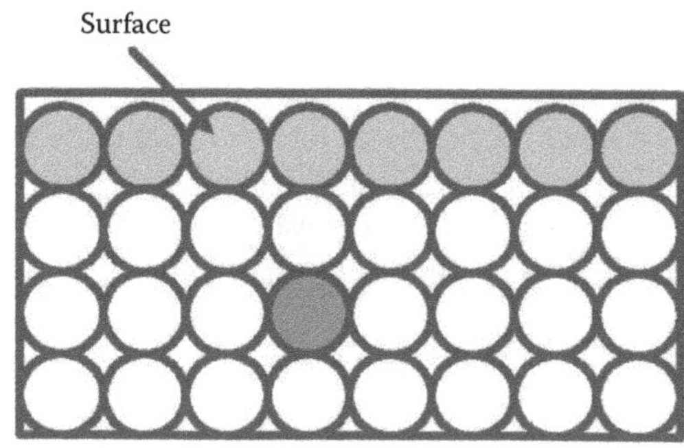

FIGURE 1.2 Molecules in the bulk phase (liquid or solid) and in surface phase (molecular dimension: schematic).

Typical examples are

- Liquid surfaces
  - Surfaces of oceans, lakes, and rivers
  - Lung surface, biological cell surfaces
- Solid surfaces
  - Adhesion, glues, tapes
  - Cement (and building) industry
  - Paper and printing industry
  - Construction industry (dams, tunnels, etc.)
  - Catalysis
  - Road surfaces (car tire, etc.)
  - Paint industry
- Liquid–solid interfaces
  - Washing and cleaning (dry cleaning)
  - Wastewater treatment
  - Air pollution
  - Power plants
- Liquid–liquid interfaces (oil–water systems)
  - Emulsions (cosmetics; pharmaceutical products)
- Diverse industrial applications:
  - Oil and gas (conventional reservoirs and extraction of shale oil and gas fracking technology, methane hydrates)
  - Food and milk products
  - Cleaning and washing
  - Medical applications
  - Electronics and related industries

Modern surface and colloid chemistry science is based on molecular-level understanding and principles. These scientific developments have been useful in providing the information (at molecular level) on various phenomena. Further, in some instances, such as oil spills, one can easily realize the importance of the role of surface of oceans (Figure 1.3). It is found that part of oil evaporates, while

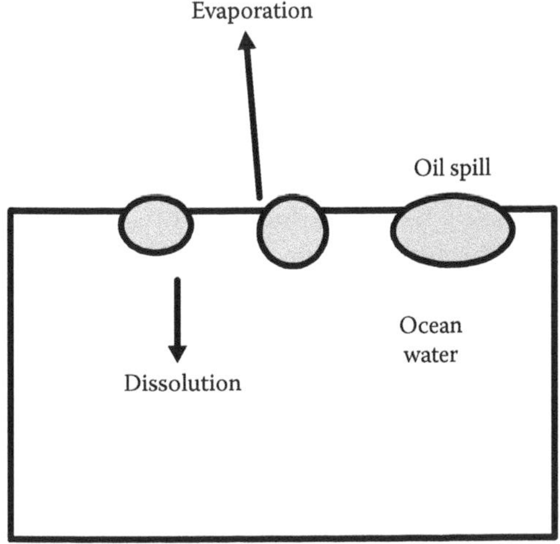

FIGURE 1.3 Ocean surface and oil spill (evaporation; solution; sinking; floating states).

some sinks to the bottom, and main part remains floating on the surface of water. This process is in fact one of the major areas of surface chemistry applications.

Another observation of great importance arises from the fact that oceans cover some 75% of the surface of the earth. Obviously, in this case, the reactions on the *surface of oceans* will have much consequence for life on the earth. For example, carbon dioxide as found in air is distributed in various systems, such as

In air
In plants
In oceans, rivers, and lakes (CO_2 is soluble in water)

Thus, CO_2 (*as present in air* [*concentration around 350 ppm*]) *is in equilibrium* in these different systems, which will have consequences on the pollution control. The adsorption of CO_2 in oceans (lakes, rivers: about 25% of CO_2 in air) thus is primarily a surface phenomenon. In other words, it is not enough to monitor CO_2 in air alone; one must have the knowledge of CO_2 concentrations in the other two states (i.e., plants and oceans). The chemical potentials of CO_2 (μ_{CO_2}) in different phases are at equilibrium:

$$\mu_{CO_2}\text{(in air)} = \mu_{CO_2}\text{(in oceans)} = \mu_{CO_2}\text{(in plants)}$$

At the same time, it is important to mention that CO_2 being the main component of photosynthesis in all plant (i.e., food) growth on the earth, it is *absolutely* the only necessary element for the existence of life on the earth! The global CO_2 equilibrium is thus a complex quantity and an important challenge for scientists.

Experiments have shown that the molecules situated near or at the interface (i.e., liquid–gas) will be interacting *differently* with respect to each other than the molecules in the bulk phase. The intramolecular forces acting would thus be different in these two cases. In other words, all processes occurring near any interface will be dependent on these molecular orientations and interactions. Furthermore, it has been pointed out that, for a dense fluid, the repulsive forces dominate the fluid structure and are of primary importance. The main effect of the repulsive forces is to provide a uniform background potential in which the molecules move as hard spheres. These considerations have shown that the molecules at the interface would be under an *asymmetrical* force field, which gives rise to the so-called surface tension or interfacial tension (IFT) (Adamson and Gast, 1997; Birdi, 1989, 1997, 1999, 2003a; Chattoraj and Birdi, 1984). At a molecular level, when one moves from one phase to another, that is, *across an interface*, this leads to the adhesion forces between liquids and solids, which is a major application area of surface and colloid science. The resultant force on molecules will vary with time because of the movement of the molecules; the molecules at the surface will be pointed downward into the bulk phase. The nearer the molecule is to the surface, the greater is the magnitude of the force due to *asymmetry*. The region of *asymmetry* near the surface plays a very important role. Thus, when the surface area of a liquid is increased, some molecules must move from the interior of the continuous phase to the interface. Surface tension of a liquid is the force acting normal to the surface per unit length of the interface, thus tending to decrease the surface area. Neighboring molecules surround the molecules in the liquid phase, and these interact with each other in a symmetrical way. Further, molecules at the surface in most cases will also be oriented in a different way than in the bulk phase (as found from experiments).

In the gas phase, where the density is *1000* times *lesser* than that in the liquid phase, the interactions between molecules are very weak as compared to those in the dense liquid phase. Thus, when one crosses the line from the liquid phase to the gas phase, there is an abrupt change in the density of factor *1000*. This means that while in liquid phase a molecule occupies a volume, which is *1000* times smaller than when in the gas phase. The interfacial region is found to be of

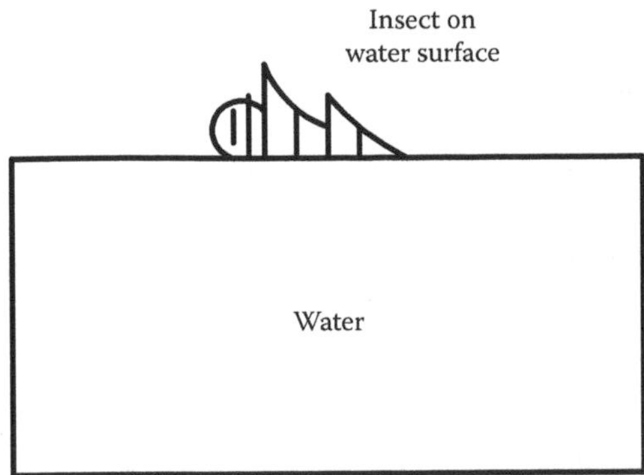

FIGURE 1.4 Insect (many different kinds of insects, such as mosquitoes, etc.) strides on surface of water (such as lakes, rivers, oceans) (schematic).

molecular dimension. Some experiments show it to be of one or few molecules thick. Further, since molecules are evaporating and condensing in this region, the interface is very turbulent!

Surface tension is the differential change of free energy with change of surface area. In any system, an increase in surface area requires that molecules from bulk phase be brought to the surface phase. The same is valid when there are two fluids or solid–liquid, it is usually designated *IFT.* A molecule of a liquid attracts the molecules that surround it, and in its turn they attract it. For the molecules, which are inside a liquid, the resultant of all these forces is neutral and all of them are in equilibrium by reacting with each other. When these molecules are on the surface, they are attracted by the molecules below and by the lateral neighbor molecules, but not toward the outside (i.e., gas phase). The cohesion among the molecules supplies a force tangential to the surface. Thus, a fluid surface behaves like an *elastic* membrane, which wraps and compresses the liquid below the surface molecules. The surface tension expresses the force with which the surface molecules attract each other. It is a common observation that due to the surface tension it takes some effort for some bugs to climb out of the water in lakes. On the contrary, other insects, like marsh treaders and water striders, exploit the surface tension to skate on water without sinking (Figure 1.4).

Insects that move about on the surfaces of lakes are actually also collecting food from the surface of the water. Another well-known example is the floating of a metal needle (or any object heavier than water) on the surface of water. The surface of a liquid being under tension maintains a sort of *skin-like* structure. In other words, energy is required to carry any object from air through the surface of a liquid.

The *surface of a liquid* can thus be regarded as the plane of potential energy. It may be assumed that the surface of a liquid behaves as a membrane (at a molecular scale) that stretches across and needs to be broken in order to penetrate. One observes this *tension* when considering that a heavy iron needle (heavier than water) can be made to float on water surface when carefully placed.

Surface tension and floating of iron needle on water:

Iron needle
Surface of water (*surface tension*)

The reason a heavy object can float on water is due to the fact that in order for the latter to sink, it must overcome the surface forces. Of course, if one merely drops the metal object, it will overcome the surface tension force and sink, which one generally observes. This clearly shows that at any liquid surface, there exists a tension (*surface tension*), which needs to be broken when any contact is made between the liquid surface and the material (here the metal needle). One notices ample of examples on the surfaces of rivers and lakes, where stuff is seen floating about. Based on the same principles, it is found that the smooth hull of a ship exerts less resistance to sailing than a rough bottom, thus saving in energy.

Definition of liquid *interfaces*:

1. Liquid and vapor or gas (e.g., ocean surface and air)
2. $Liquid_1$ and $liquid_2$ immiscible (water–oil; *emulsion*).
3. Liquid and solid interface (water drop resting on a solid, wetting, cleaning of surfaces, adhesion)

Definition of solid surfaces or interfaces:

1. $Solid_1$–$solid_2$ (cement, adhesives).

Surface properties of solids: For example, in the case when a solid sample is crushed and surface area increases per unit gram (Figure 1.5), the surface area per gram of the solid increases. For example, finely divided talcum powder has a surface area of 10 m^2/g. Active charcoal exhibits surface areas corresponding to over 1000 m^2/g. In fact, in cement, the energy input in producing very finely divided particles is much greater than the rest of the production costs. This property has been found to be an appreciable quantity and has important consequences. Qualitatively, one must notice that *work* has to be put into the system when one increases the surface area (both for liquids and solids or any other interface). Cement is mainly based on the mechanical energy used to make the particles as small as possible, such that cost is dependent on this process. The solid surface properties change drastically when one increases the surface area per unit weight.

The surface chemistry of *small particles* is an important part of everyday life (such as dust, talcum powder, sand, rain drops, emissions, etc.). *Thomas Graham already defined colloid chemistry and its relation to surface chemistry a century ago.* A particle having dimensions in the range of 10^{-9} m (= 1 nm = 10 Å) and 10^{-6} m (1 μm) was designated to be a colloidal (Birdi, 2003a, 2010a; Gitis and Sivamani, 2004; Scheludko, 1966). Colloids are an important class of materials, intermediate between bulk and molecularly dispersed systems. Colloidal particles may be spherical, but in some cases, one dimension can be much larger than the other two (as in a needle-like shape). The size of particles also determines whether one can see them by the naked eye. Colloids are not visible to the naked eye or under an ordinary optical microscope. The scattering of light can be suitably used to see such colloidal particles (such as dust particles, etc.). The size of colloidal particles may range from 10^{-4} to 10^{-7} cm. The units used are as follows:

1 μm = 10^{-6} m
1 nm = 10^{-9} m
1 Å (Angstrom) = 10^{-8} cm = 0.1 nm = 10^{-10} m (nm = nano-meter)

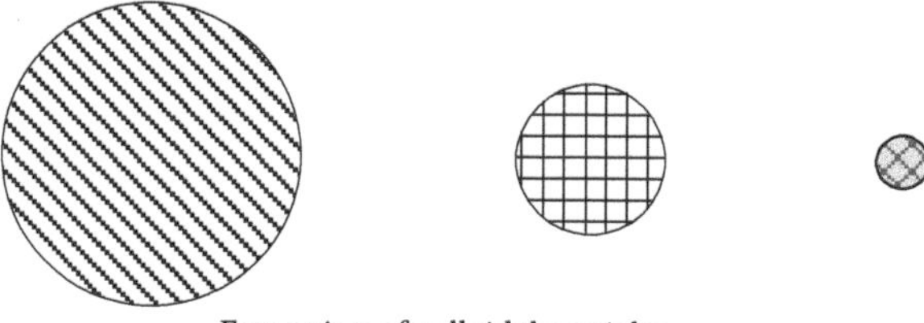

Formation of colloidal particles

Diameter decreases from mm to less than μm

FIGURE 1.5 Formation of fine (colloidal) particles (such as talcum powder, active charcoal, cement) (schematic: size less than μm).

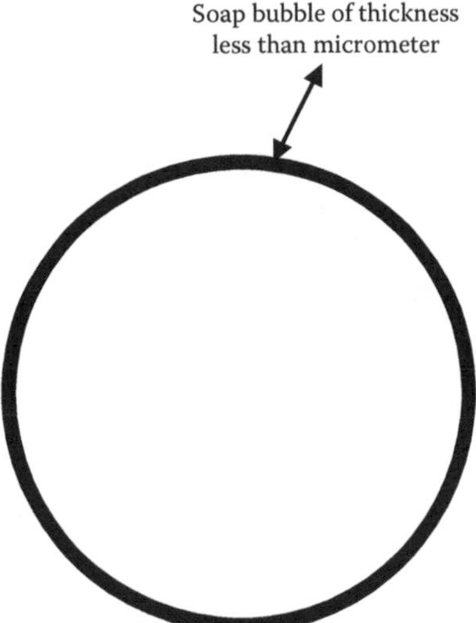

FIGURE 1.6 Soap bubble of thickness of micrometer or less (1000 nm) (schematic).

Considering the following simple examples can provide some information about the nanodimension scale:

Hair: 1/1000th diameter of hair is about nanosize.

Thickness of soap bubbles varies from micro- to nanometer (black rings) (Figure 1.6). Actually, soap bubble is the closest eyesight can see some structures, which is of molecular dimension! In nature, some examples of colors are found, which is related to similar dimensions (such as colors of feathers, etc.).

In surface chemistry, there is a great range of dimensions as needed to describe a variety of systems. As seen here, the range of dimensions is many folded! Accordingly, a unit Angstrom (Å = 10^{-8} cm) was used for systems of molecular dimension (famous Swedish scientist). Most common unit is though nanometer (= 10 Å = 10^{-9} m), which is mainly used for molecular scale features. In recent years, nanosize (nanometer range) particles are of much interest in different applied science systems (Nano = from Greek and means dwarf) (Lin et al., 2005). Nanotechnology is actually strongly getting a boost from the last decade of innovation as reported by the surface and colloid literature (Mun et al., 2014; Rao and Geckeler, 2011). In fact, light-scattering is generally used to study the size and size distribution of such systems. Since colloidal systems consist of two or more phases and components, the interfacial area-to-volume ratio becomes very significant. Colloidal particles have a high surface area-to-volume ratio compared with ordinary bulk materials. A significant proportion of colloidal molecules lie within, or close to, the interfacial region. Hence, the interfacial region has significant control over the properties of colloids. To understand why colloidal dispersions can be either *stable* or *unstable*, we need to consider

1. The effect of the large surface area-to-volume ratio (*e.g.*, 1000 m^2 surface area per gram of solid [active charcoal])
2. The forces operating between the colloidal particles
3. Surface charges (varying from positive or negative or neutral surface charge) are very important characteristics of such systems

There are some very special characteristics, which must be considered as regards colloidal particle behavior: *size and shape, surface area, and surface charge density.*

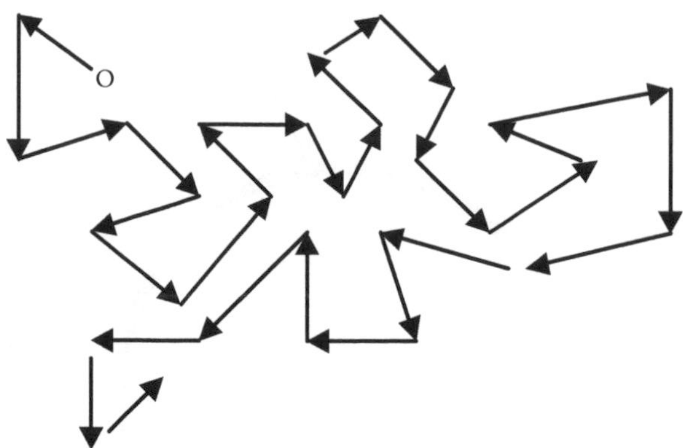

FIGURE 1.7 Brownian motion (schematic movement of small [dust] particles or large molecules [polymers, etc.]) (in air or liquid media).

The *Brownian motion* of particles is a much studied field (Figure 1.7). The motion of colloidal particles is the basis of the physical characteristics of the system. The *fractal* nature of surface roughness has recently been shown to be of importance (Avnir, 1989; Birdi, 1993; Feder, 1988). Recent applications have been reported where *nanocolloids* have been employed. It is thus found that some relevant terms are needed to be defined at this stage. The definitions generally employed are as follows. *Surface* is a term used when one considers the dividing phase between

Gas–*Liquid*
Gas–*Solid*

Interface (Figure 1.8) is the term used when one considers the dividing phase:

Solid–*Liquid*	(*colloids*)
Liquid_1–$\textit{Liquid}_2$	(oil–water *emulsion*)
Solid_1–$\textit{Solid}_2$	(*adhesion:* glue, cement)

It is thus found that surface tension may be considered to arise due to a certain well-defined degree of unsaturation of bonds that occurs when a molecule resides at the surface and not in the bulk. The term *surface tension* is used for solid–vapor or liquid–vapor interfaces. The term *IFT* is more generally used for the interface between two liquids (oil–water), two solids, or a liquid and solid. It is, of course, obvious that in a one-component system, the fluid is uniform from the bulk phase to the surface. However, the *orientation* of the surface molecules will be different from those molecules in the bulk phase in all systems. For instance, in the case of water, the orientation of molecules inside the bulk phase will be different than those at the interface. The hydrogen bonding will orient the oxygen atom toward the interface. The question one may ask, then, is how sharply does the density change from that of being fluid to that of gas (a change by a factor of 1000). Is this a transition region a monolayer deep or many layers deep? The Gibbs adsorption theory (Birdi, 1989, 1999, 2003a, 2009, 2010a; Chattoraj and Birdi, 1984; Defay et al., 1966) considers surface of liquids to be *monolayer.* The surface tension of water decreases appreciably on the addition of very small quantities of *soaps and detergents.* The Gibbs adsorption theory relates the change in surface tension to the change in soap concentration (chemical potential). The experiments, which analyze the spread monolayers, are also based on one molecular layer. The latter data conclusively indeed verify the Gibbs assumption. Detergents and other (soaps, etc.) similar kind of molecules are found to exhibit *self-assembly* characteristics

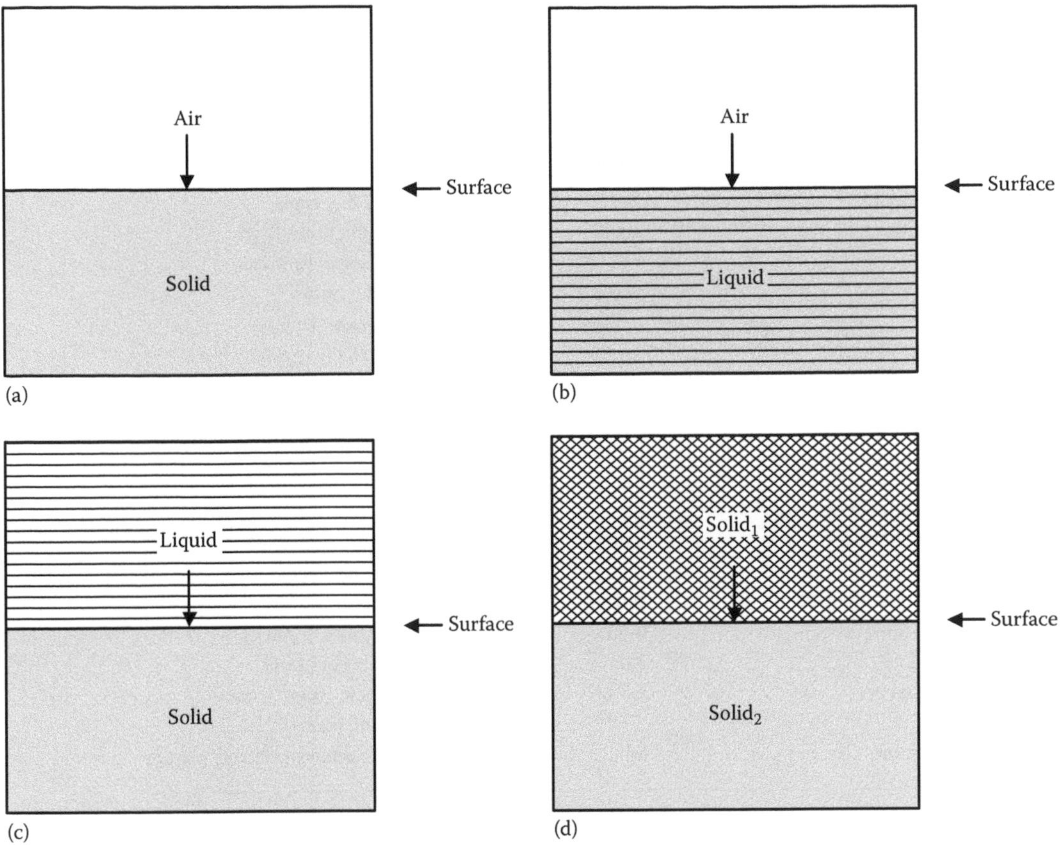

FIGURE 1.8 Characterization of different *interfaces*: (a) solid–gas (air); (b) liquid–gas (air); (c) solid–liquid; and (d) $solid_1$–$solid_2$.

self-assembly monolayer (SAM) (Birdi, 1999). However, there exists no procedure, which can provide information by a direct measurement.

Colloids (Greek word for glue-like) are a wide variety of systems consisting of finely divided particles (or macromolecules [such as glue, gelatine, proteins, etc.]), which are found in everyday life (Table 1.1). In general, one classifies colloidal systems into three distinct types (Adamson and Gast, 1997; Birdi, 2003a, 2009; Dukhin and Goetz, 2002; Lyklema, 2000):

1. In simple colloids, clear distinction can be made between the disperse phase and the disperse medium, for example, simple emulsions of oil-in-water (o/w) or water-in-oil (w/o).
2. Multiple colloids involve the coexistence of three phases of which two are finely divided, for example, multiple emulsions (mayonnaise, milk) of water-in-oil-in-water (w/o/w) or oil-in-water-in-oil (o/w/o).
3. Network colloids have two phases forming an interpenetrating network, for example, polymer matrix.

The stability of *colloidal* systems (in the form as solids or liquid drops) is determined by the free energy (surface free energy or the interfacial free energy) of the system. The main parameter of interest is the large surface area exposed between the dispersed phase and the continuous phase. Since colloidal particles move about constantly, the Brownian motion determines their dispersion energy. The energy imparted by collisions with the surrounding molecules at temperature $\mathbf{T} = 300$ K is $3/2\mathbf{k_B T} = (3/2)(1.38 \times 10^{-23})(300) = 0.6 \times 10^{-20}$ J (where k_B is the Boltzmann constant).

TABLE 1.1
Typical Colloidal Systems

Phases		
Dispersed	**Continuous**	**System Name**
Liquid	Gas	Aerosol fog, spray
Gas	Liquid	Foam, thin films, froth
		Fire extinguisher foam
Liquid	Liquid	Emulsion (milk)
		Mayonnaise, butter
Solid	Liquid	Sols, AgI, photography films
		Suspension wastewater
		Cement
		Oil recovery (shale oil)
		Coal slurry
Bio colloids		
Corpuscles	Serum	Blood
		Blood-coagulants
Hydroxyapatite	Collagen	Bone–teeth
Liquid	Solid	Solid emulsion (toothpaste; paints)
Solid	Gas	Solid aerosol (dust)
Gas	Solid	Solid foam (polystyrene)
		Insulating foam
Solid	Solid	Solid suspension/solids in plastics

This energy and intermolecular forces would thus determine the colloidal stability. In the case of colloid systems (particles or droplets), the kinetic energy transferred on collision will be thus $\mathbf{k_B T} = 10^{-20}$ J. However, at a given moment, there is a high probability that a particle may have a larger or smaller energy in comparison to $\mathbf{k_B T}$. Further, the probability of total energy several times $\mathbf{k_B T}$ (over 10 times $\mathbf{k_B T}$) thus becomes very small. The instability will be observed if the ratio of the barrier height to $\mathbf{k_B T}$ is around 1–2 units.

The idea that two species ($solid_1$–$solid_2$) should interact with each other, so that their mutual potential energy can be represented by some function of the distance between them, has been described in the literature (Israelachvili, 2011). Furthermore, colloidal particles frequently adsorb (and even absorb) ions from their dispersing medium (such as in ground water treatment and purification). Sorption that is much stronger than what would be expected from dispersion forces is called *chemisorption*, a process that is of both chemical and physical interest.

Emulsions: As one knows from experience: oil and water do not mix, which suggests that these systems are dependent on the oil–water interface (Figure 1.9). The $liquid_1$–$liquid_2$ (oil–water) interface is found in many systems, which is most important in the world of *emulsions*.

The trick in using emulsions is based on the fact that one can apply both *water* and *oil* (latter is insoluble in water) simultaneously. Further, one can then include other molecules, which may be soluble in either phase (water or oil). This obviously leads to the common observation where we find thousands of applications of emulsions. It is very important to mention here that actually nature uses this trick in most of the major biological fluids. The most striking example is milk. The emulsion chemistry of milk (and other food emulsion systems) has been found to be the most complex (Kristensen, 1997). Paint consists of polymer molecules dispersed in water phase. After application, water evaporates, leaving behind a glossy layer of paint.

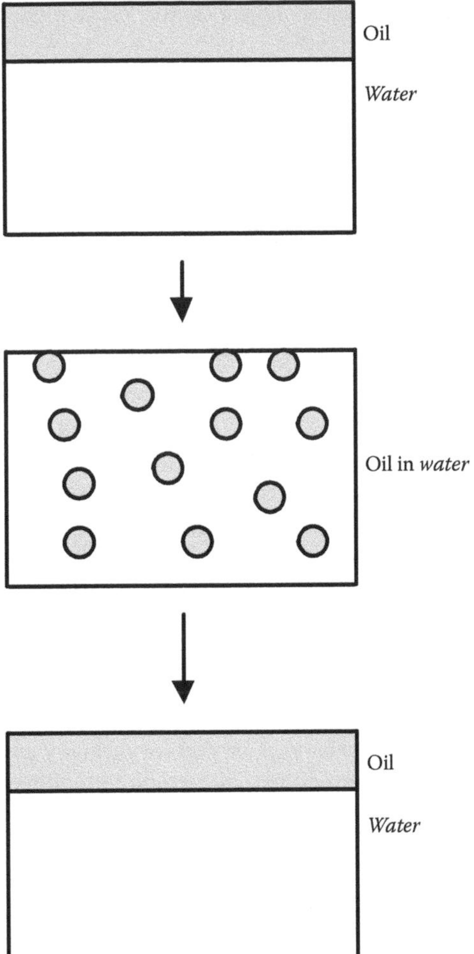

FIGURE 1.9 Mixture of oil–water phases (see text for details).

In fact, the state of mixing oil and water is an important example of interfacial behavior at $liquid_1$–$liquid_2$. Emulsions of oil–water systems are useful in many aspects of daily life: milk, foods, paint, oil recovery, pharmaceutical, and cosmetics.

If one mixes olive oil with water and on shaking, one gets the following (Figure 1.9):

About 1 mm diameter oil drops are formed.
After a few minutes, the oil drops merge together and two layers (oil and water) are again formed.

However, if one adds suitable substances, which change surface forces, then the olive oil drops formed can be very small (micrometer [μm] range). The latter leads to a stable emulsion.

The degree of emulsion stability is basically dependent on the size of the oil drops (as dispersed in water), besides other factors:

Low stability—large oil drops
Long stability—small drops
Very long stability—microsize drops

In addition, one finds that these considerations are important in regard to the different systems as follows:

Paints, cements, adhesives, photographic products, water purification, sewage disposal, emulsions, chromatography, oil recovery, paper and printing industry, microelectronics, soap and detergents, catalysts, and biology (cell, virus). In some oil–surfactant–water–diverse components, liquid crystal (LC) phases (lyotropic LC) are observed. These lyotropic LCs are indeed the basic building blocks in many applications of emulsions in technology. LC structures can be compared with a layer-cake where each layer is molecular thick. It is thus seen that surface science pertains to investigations, which take place between two different phases.

1.2 CAPILLARITY AND SURFACE FORCES (IN LIQUIDS) (CURVED SURFACES)

It is known that liquids and solids are different in many physical characteristics. Liquids take the shape of a container, which surrounds or contains it. The question arises then what happens when the liquid surface is *curved*, in comparison to a flat surface (Figure 1.10).

It is found that the *curvature* at the surface of a liquid imparts different properties to a liquid (or solid) (Adamson and Gast, 1997; Birdi, 1997, 2009; Goodrich et al., 1981). Infact, the extensive surface science research is mainly based on the curvature effect, besides other properties. In the following systems, one can recognize that the liquid surfaces are involved:

1. The most common behavior is bubble and foam formation.
2. Another phenomenon is that when a glass capillary tube is dipped in water, the fluid rises to a given height. It is observed that the narrower the tube, the higher the water rises.
3. The role of liquids and liquid surfaces is important in many everyday natural processes (e.g., oceans, lakes, rivers, and rain drops). In everyday life, the most important liquid one is concerned with is water.

Further, the degree of curvature is found to impart different characteristics to liquid surfaces.

The state of molecules in different phases can be described as follows:

1. *Molecules inside the bulk phase*: molecules are surrounded *symmetrically* in all directions.
2. *Molecules at the surface*: molecules are interacting with molecules in the bulk phase (same as under I), but towards the gas phase, the interaction is weaker due to larger distances between molecules (Figure 1.1).

The molecular forces, which are present between the surface molecules, are found to be different from the forces acting on molecules in the bulk phase or the gas phase. Accordingly, these forces are called *surface forces*. Surface forces make the liquid surface behave like a stretched elastic membrane in that it tends to contract. One cannot see this phenomenon directly, but it is observed through indirect experimental observations (both qualitatively and quantitatively!). The latter arises from the observation that when one empties a beaker with a liquid, the liquid breaks up into spherical drops. This indicates that drops are being created under some forces, which must be present at the surface of the newly formed interface. These surface forces become even more important when a liquid is in contact with a solid (such as ground water, oil reservoir). The flow of liquid (e.g., water or oil) through small pores in the underground is

FIGURE 1.10 Flat and curved surfaces of a liquid.

mainly governed by the *capillary forces*. One can also accept that recovering oil from large depths (10 km!) in underground needs appreciable pressure if pores are very small. It is found that capillary forces play a very dominant role in many of these systems. Thus, the interaction between liquid and any solid (with pores) will form curved surfaces, which being different from planar fluid surface gives rise to capillary forces. Another essential aspect is the mechanical surface tension present at *curved surfaces*. The curved liquid surfaces, such as in drops, or small capillaries, exhibit very special properties than flat liquid surfaces. This is essential, since these principles are the building blocks for the understanding of the subject. These surface forces interact both at $liquid_1$–$liquid_2$ (such as oil–water) and liquid–solid (such as cement–water) interfaces.

1.2.1 Origin of Surface Forces (in Liquids)

A gas phase is converted to a liquid phase when appropriate pressure and temperature are present (which means that molecules become so close that a new phase, liquid, is formed). If temperature is lowered even further, then a solid phase is formed. Liquid water freezes at 0°C to form solid ice (at 1 atm). This kind of transition can be explained by analyzing molecular interactions in liquids, which are responsible for their physicochemical properties (such as *boiling point*, *melting point*, *heat of vaporization*, and *surface tension*). Since molecular structure is the basic parameter involved in these transitions, one needs to analyze the former characteristics. This analysis is the basis of the *quantitative structure activity related* (QSAR) (Barnes, 2011; Birdi, 2003b). QSARs have been used to predict physical properties of liquids in extensive detail (Birdi, 2003a; Gotch, 1974; Livingstone, 1996).

In solids, molecular structures are measured by using x-ray methods. However, one cannot estimate molecular structures of liquids with the same accuracy. Liquid structures have been estimated through indirect methods. In the following, based on simple principles, one can estimate the difference in molecular distances in liquid or gas as follows. In the case of water (for example), following data are known (at room temperature and pressure):

Example: *Water*
Volume per mole *liquid* water = V_{liquid} = 18 mL/mol
Volume per mole water in *gas* state (at STP) (V_{gas}) = 22 L/mol
Ratio $V_{gas}/V_{liquid} \cong 1000$

This illustrates that the approximate ratio of distance between molecules in gas phase and liquid phase will be about $10 \cong (1000)^{1/3}$ (from simple geometrical considerations of volume [proportional to length^3] and length). In other words, the surface chemistry is related to those molecules, which are situated in this *transition region*. Experiments have clearly shown that this transition region is of molecular dimension. The same is true for all liquids, with only minor differences.

In other words, the density of water changes *1000* times as the surface is crossed from liquid phase to the gas phase (air). This means that in gas phase, each molecule occupies 1000 more volume than in the liquid phase. Thus, the molecules in the gas phase move larger distances before interacting with another molecule. This large change means that the surface molecules must be under different environment than in the liquid phase or in the gas phase. Surface phase is the transition region. Distance between gas molecules is approximately 10 times larger than in a liquid. Hence, the forces (*all forces increase when distances between molecules decrease*) between gas molecules are much weaker than in the case of liquid phase. All interaction forces between molecules (solid phase, liquid phase, and gas phase) are related to the distance between molecules. It is also obvious that the surface of a liquid is a very busy place, since molecules are evaporating into the air and some are also returning into the liquid phase. This is a snapshot at a molecular level, so one does not see it by the naked eye.

In the case of water, it is cohesive forces that maintain water, for example, in liquid state at room temperature and pressure (Franks, 1975; Fraxedas, 2014). It is useful as an example to compare cohesive forces in two different molecules, such as H_2O and H_2S. At room temperature and pressure, H_2O is *liquid* while H_2S is a *gas*. This means that H_2O molecules interact with different forces, which are stronger and thus form a liquid phase. On the other hand, H_2S molecules exhibit much lower interactions and thus are in a gas phase at room temperature and pressure. In other words, hydrogen bonds (between H and O) in water are stronger than hydrogen–sulfur bonds.

1.2.1.1 Surface Energy

Molecules interact with each other, and the energy will be dependent on the geometrical packing (thus the magnitude of distances between molecules) in any given phase. The state of *surface energy* has also been described by the following classic example (Adamson and Gast, 1997; Birdi, 1989, 1997, 2003b, 2010a; Chattoraj and Birdi, 1984). Let us consider the area of a liquid film, which is stretched in a wire frame by an increment d**A**, whereby the surface energy changes by (γd**A**) (Figure 1.11). Under this process, the opposing force is **f**. From these data, one finds

Surface tension of a liquid = γ
Change in area = d**A** = ldx
Change in x-direction = dx

$$\mathbf{f}dx = \gamma d\mathbf{A} \tag{1.1}$$

or

$$\gamma = \mathbf{f}\left(\frac{dx}{d\mathbf{A}}\right) = \frac{\mathbf{f}}{21} \tag{1.2}$$

where

dx is the change in displacement
l is the length of the thin film

The quantity γ represents the force per unit length of the surface (mN/m = dyn/cm), and this force is defined as surface tension or IFT. *Surface tension*, γ, is the differential change of free energy with the change of surface area at constant temperature, pressure, and composition.

One may consider another example to describe the surface energy. Let us imagine that a liquid fills a container of the shape of a funnel. In the funnel, if one moves the liquid upward, then there will be an increase in *surface area*. This requires that some molecules from bulk phase have to move into the surface area and create extra surface A_S. The work required to do so will be

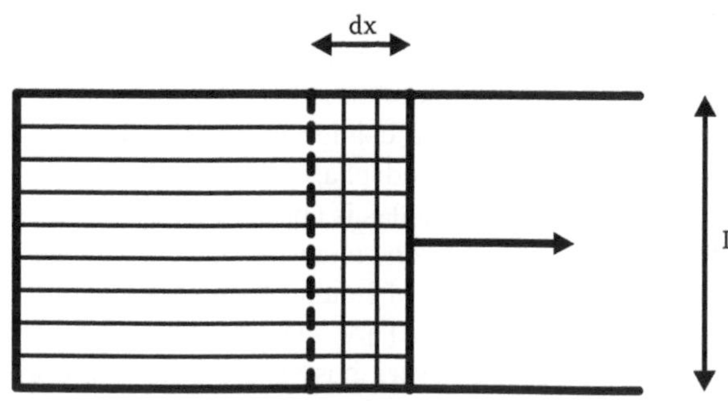

FIGURE 1.11 Surface film of a liquid (schematic) (see text for details).

(force times area) γA_S. This is reversible work at constant temperature and pressure; thus, it gives the increase in free energy of the system:

$$\mathbf{dG} = -\gamma A_S \tag{1.3}$$

Thus, the tension per unit length in a single surface, or surface tension, γ, is numerically equal to the surface energy per unit area. Then G_S, the surface free energy per unit area, is

$$G_S = \gamma = \left(\frac{\mathbf{dG}}{\mathbf{dA}}\right) \tag{1.4}$$

Under reversible conditions, the heat (**q**) associated with it gives the surface entropy, $\mathbf{S_S}$

$$\mathbf{dq = TdS_S} \tag{1.5}$$

Combining these equations, we find that

$$\frac{d\gamma}{\mathbf{dT}} = -\mathbf{S_S} \tag{1.6}$$

Further, one finds

$$H_S = G_S + T_S \tag{1.7}$$

and one can also write for surface energy, E_S,

$$E_S = G_S + \mathbf{TS_S} \tag{1.8}$$

These relations give

$$E_S = \gamma - \mathbf{T}\left(\frac{d\gamma}{\mathbf{dT}}\right) \tag{1.9}$$

The quantity E_S has been found to provide more useful information on surface phenomena than any other quantities (Birdi, 2009).

Here we find that the term S_S is the surface entropy per square centimeter of surface. This shows that to change the surface area of a liquid (or solid: as described later), there exists a *surface energy* (γ: *surface tension*), which one needs to consider.

The quantity γ means that to create a fresh surface of 1 m^2 (=10^{20} $Å^2$) of new surface of water, one will need to use 72 mJ energy. To transfer a molecule of water from the bulk phase (where it is surrounded by about 10 near neighbors by about 7 $\mathbf{k_BT}$) ($\mathbf{k_BT}$ = 4.12 × 10^{-21} J) to the surface, one need to break about half of these hydrogen bonds (i.e., 7/2$\mathbf{k_BT}$ = 3.5$\mathbf{k_BT}$). The free energy of transfer of one molecule of water (with area of 12 $Å^2$) will be thus about 10^{-20} J (or about 3$\mathbf{k_BT}$). This is a reasonable quantity under these simple geometrical assumptions (Adamson and Gast, 1997; Birdi, 2009).

In the case of solid systems, similar consideration is needed if one increases the surface area of a solid (e.g., by *crushing*). In the latter case, one needs to measure and analyze the *surface tension of the solid.* It is found that energy needed to crush a solid is related to the surface forces (i.e., solid surface tension).

1.2.2 Capillary Forces: Laplace Equation (Liquid Curvature and Pressure) (Mechanical Definition)

Liquids show characteristics, which are specific due to the fact that molecules in liquids are able to move inside a container, while a solid cannot exhibit this property. It is found that this property of liquid gives some specific properties, such as curved surfaces in narrow tubings (Adamson and Gast, 1997; Birdi, 2009; Goodrich et al., 1981). It is interesting to consider aspects in the field of the wettability. Surely everybody has noticed that water tends to rise near the walls of a glass container. This happens because the molecules of this liquid have a strong tendency to adhere to the glass. Liquids that wet the walls make concave surfaces (e.g., water/glass), and those that do not wet them make convex surfaces (e.g., mercury/glass). Inside tubes with internal diameter smaller than 2 mm, called capillary tubes, a wettable liquid forms a concave meniscus in its upper surface and tends to go up along the tube. On the contrary, a nonwettable liquid forms a convex meniscus and its level tends to go down. The amount of liquid attracted by the capillary rises until the forces, which attract it, balance the weight of the fluid column. The rising or the lowering of the level of the liquids into thin tubes is named *capillarity* (*capillary force*). Also, the capillarity is driven by the forces of cohesion and adhesion we have already mentioned.

Any liquid inside a large beaker has almost a *flat* surface. However, the same liquid inside a fine tubing will be found to be *curved* at the surface (Figure 1.12). In other words, the curved liquid surface gives rise to some very characteristic properties, which one finds in everyday life (e.g., water flow in earth, oil recovery from reservoirs, blood flow in arteries) (Murrant and Sarelius, 2000).

The surface tension, γ, and the mechanical equilibrium at interfaces have been described in the literature in detail (Adamson and Gast, 1997; Birdi, 1989, 2003a, 2009, 2010a; Chattoraj and Birdi, 1984). The surface has been considered as a hypothetical stretched *membrane*; this is termed as the surface tension. In a real system undergoing an infinitesimal process, it can be written as

$$\mathbf{dW} = \mathbf{pdV} + \mathbf{p'dV'} - \gamma \mathbf{dA} \tag{1.10}$$

where

- $\mathbf{dW}$ is the work done by the systems when a change in volume, $\mathbf{dV}$ and $\mathbf{dV'}$, occurs
- $\mathbf{p}$ and $\mathbf{p'}$ are pressures in the two phases α and ß, respectively, at equilibrium
- $\mathbf{dA}$ is the change in interfacial area

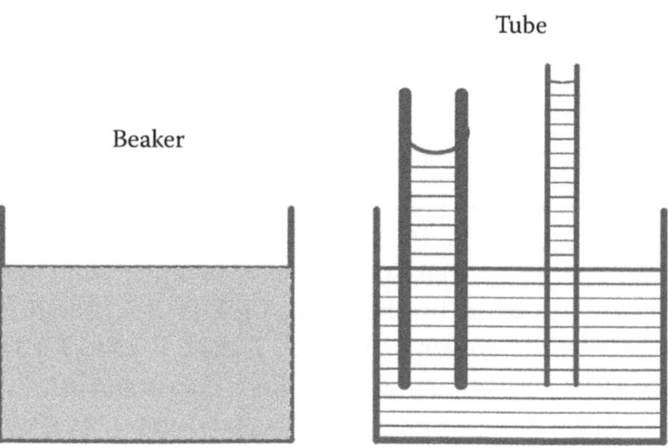

FIGURE 1.12 Surface of water inside a large beaker and in a narrow tubing. The rise of water in the tubing is due to the *capillary force*.

The sign of the interfacial work is designated negative by convention (Adamson and Gast, 1997; Chattoraj and Birdi, 1984).

The fundamental property of liquid surfaces is that they tend to contract to the smallest possible area. This is not observed in the case of solids. This property is observed in the spherical form of small drops of liquid, in the tension exerted by soap films as they tend to become less extended, and in many other properties of liquid surfaces. In the absence of gravity effects, these curved surfaces are described by the Laplace equation, which relates the mechanical forces as (Adamson and Gast, 1997; Birdi, 1997; Chattoraj and Birdi, 1984)

$$\mathbf{p} - \mathbf{p}' = \gamma\left(\frac{1}{r_1} + \frac{1}{r_2}\right) \tag{1.11}$$

$$\mathbf{p} - \mathbf{p}' = 2\left(\frac{\gamma}{r}\right) \tag{1.12}$$

where r_1 and r_2 are the radii of curvature (in the case of an ellipse), while r is the radius of curvature for a spherical-shaped interface. It is a geometric fact that surfaces for which Equation 1.12 holds are surfaces of minimum area. These equations thus give

$$d\mathbf{W} = \mathbf{p}d\left(\mathbf{V} + \mathbf{V}'\right) - \gamma d\mathbf{A} \tag{1.13}$$

$$d\mathbf{W} = \mathbf{p}d\mathbf{V}^t - \gamma d\mathbf{A} \tag{1.14}$$

where

$\mathbf{p} = \mathbf{p}'$ for plane surface

$\mathbf{V}^t$ is the total volume of the system

It is found that there exists a pressure difference across the curved interfaces of liquids (such as drops or bubbles). For example, if one dips a tube into water (or any fluid) and applies a suitable pressure, then a bubble is formed (Figure 1.13). This means that the pressure inside the bubble is greater than the atmosphere pressure. It thus becomes apparent that curved liquid surfaces induce effects, which need special physicochemical analyses in comparison to flat liquid surfaces. It must be noticed that in this system a mechanical force has induced a change on the surface of a liquid. This phenomenon is also called *capillary forces*. Then one may ask, does this also require similar consideration in the case of solids? The answer is yes, and will be discussed later in detail. For example, in order to remove liquid, which is inside a porous media such as a sponge, one would need force equivalent to these capillary forces. Man has been fascinated with bubbles for many centuries. As seen in Figure 1.13, the bubble is produced by applying a suitable pressure, $\Delta\mathbf{P}$, to obtain a bubble of radius $\mathbf{R}$, where the surface tension of the liquid is γ.

It is useful to consider the phenomena where one expands the bubble by applying pressure, P_{inside}. In this process, one has two processes to consider:

1. The surface area of the bubble will increase by d$\mathbf{A}$.
2. The volume will increase by d$\mathbf{V}$.

In other words, there are two opposing actions: expansion of volume and increase of surface area.

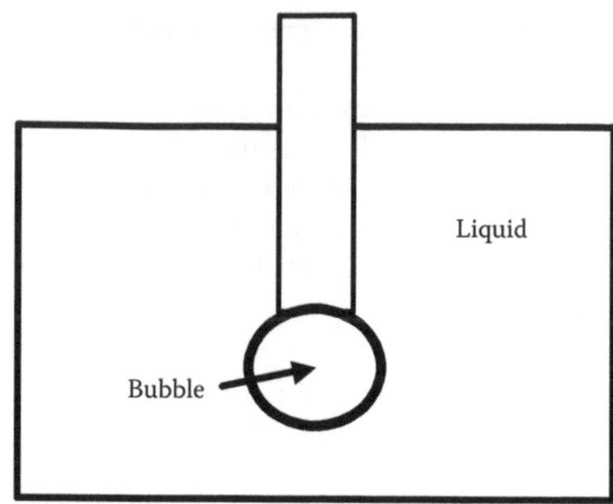

FIGURE 1.13 An air bubble in a liquid. The pressure inside the bubble is greater than the pressure outside the bubble.

The work done can be expressed in terms of that done against the forces of surface tension and that done in increasing the volume. At equilibrium, there will exist following condition between these two kinds of work:

$$\gamma d\mathbf{A} = (P_g - P_{liquid}) d\mathbf{V} \tag{1.15}$$

where

$d\mathbf{A} = 8\pi\mathbf{R}d\mathbf{R}$ ($\mathbf{A} = 4\pi\mathbf{R}^2$)
$d\mathbf{V} = 4\pi\mathbf{R}^2 d\mathbf{R}$ ($\mathbf{V} = 4/3\pi\mathbf{R}^3$)

Combining these relations gives following:

$$\gamma 8\pi\mathbf{R}d\mathbf{R} = \Delta\mathbf{P}4\pi\mathbf{R}^2 d\mathbf{R} \tag{1.16}$$

and

$$\Delta\mathbf{P} = \frac{2\gamma}{\mathbf{R}} \tag{1.17}$$

where $\Delta\mathbf{P} = (P_g - P_{liquid})$. Since the free energy of the system at equilibrium is constant, $dG = 0$, then these two changes in system are equal. If the same consideration is applied to the soap bubble, then the expression for $\Delta\mathbf{P}$ bubble will be

$$\Delta\mathbf{P}_{bubble} = p_{inside} - p_{outside} = \frac{4\gamma}{\mathbf{R}} \tag{1.18}$$

because now there exist *two* surfaces and the factor 2 is needed to consider this state.

The pressure applied gives rise to work on the system, and the creation of the bubble gives rise to the creation of surface area increase in the fluid. The Laplace equation relates the pressure difference across any curved fluid surface to the *curvature*, 1/*radius,* and its surface tension, γ. In those cases where nonspherical curvatures are present (Birdi et al., 1988; Kendoush, 2007), one obtains the more universal equation:

$$\Delta\mathbf{P} = \gamma\left(\frac{1}{R_1} + \frac{1}{R_2}\right) \tag{1.19}$$

It is also seen that in the case of spherical bubbles, since $R_1 = R_2$, this equation becomes identical to Equation 1.18. It is thus seen that in the case of a liquid drop in air (or gas phase), the Laplace pressure would be the difference between the pressure inside the drop, p_L, and the gas pressure, p_G:

$$p_L - p_G = \Delta \mathbf{P} \tag{1.20}$$

$$p_L - p_G = \frac{2\gamma}{\text{Radius}} \tag{1.21}$$

In the case of a water drop with radius 2μ, there will be $\Delta \mathbf{P}$ of magnitude:

$$\Delta \mathbf{P} = \frac{2(72\ \text{mN/m})}{2 \times 10^{-6}\ \text{m}} = 72{,}000\ \text{N/m}^2 = 0.72\,\text{atm} \tag{1.22}$$

The magnitude of vapor pressure is known to be dependent on the pressure. Thus, $\Delta \mathbf{P}$ will effect the vapor pressure and lead to many consequences in different systems. In fact, the capillary (Laplace) pressure determines many industrial and biological systems. The lung alveoli are dependent on the radii during the inhale–exhale process, and the change in the surface tension of the fluid lining the lungs alveoli. In fact, many lung diseases are related to the lack of surface pressure and capillary pressure balance. Blood flow through arteries of different diameter throughout the body is another system where Laplace pressure is of much interest for analytical methods (such as heart function and control).

The Laplace equation is useful for analyses in a variety of systems as described in the following:

1. Bubbles or drops (rain drops or combustion engines, spray, fog).
2. Blood cells (flow of blood cells through arteries)
3. Oil or ground-water movement in rocks (and shale)
4. Lung vesicles.

It is interesting to consider a fe in the following:

From Laplace equation, one may notice that $\Delta \mathbf{P}$ is larger inside a *small* bubble than in a *larger* bubble with the same magnitude of γ. This means that when two bubbles meet, the smaller bubble will enter the larger bubble to create a new bubble, Figure 1.14. This phenomena will have much important consequences in various systems (such as emulsion stability, lung alveoli, oil recovery, bubble characteristics [such as in champagne, beer]).

Let us consider an example as given in Figure 1.15. This is a system, which initially shows two bubbles of different curvature and connected through a regulator (which can be closed or open). After the tap is opened, one finds that the *smaller* bubble shrinks, while the larger bubble (with lower $\Delta \mathbf{P}$) increases in size until equilibrium is reached (when the curvature of the two bubbles becomes equal in magnitude). This kind of equilibrium is the basis of lung alveoli where fluids (containing lipid surfactants) balance out the expanding–contracting cycle (Birdi, 1989).

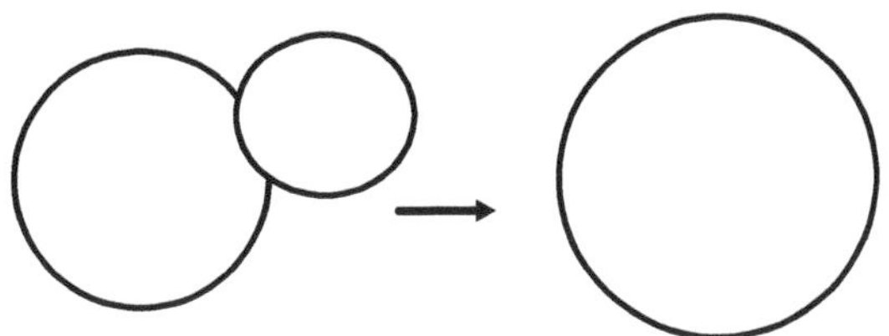

FIGURE 1.14 Coalescence of two bubbles with different radii.

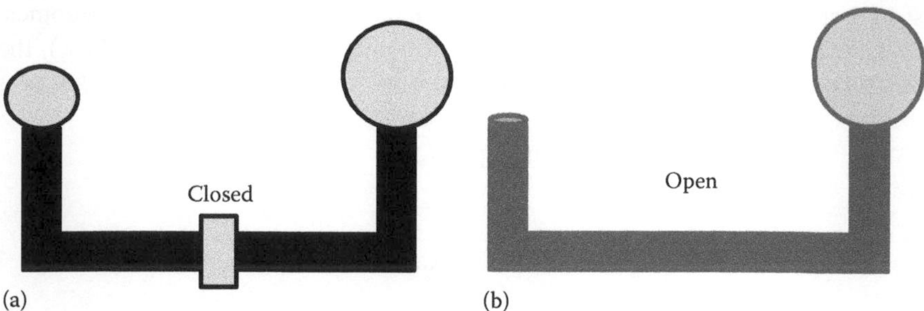

FIGURE 1.15 (a, b) Equilibrium state of two bubbles of different radii (see text).

It has been observed that a system with varying sized bubbles collapses faster than bubbles (or liquid drops) that are of exactly the same size. In other words, in systems with bubbles of similar diameter, there will be a slow coalescence than in systems with varying sized bubbles. Another major consequence is observed in the oil recovery phenomena (see later). Oil production takes place (in general) by applying gas or water injection. When gas or water injection is applied, where there are small pores, the pressure needed will be *higher* than that in the large pore zone. Thus, the gas or water will *bypass* the small pore zone and leave the oil behind (at present more than 30%–50% of oil in place is not recovered under normal production methods). This is obviously a great challenge to the surface and colloid chemists in the future. Enhanced oil recovery (EOR) technology is of enormous interest at this stage in the literature (Birdi, 2003a).

Another aspect is that *vapor pressure* over a curved liquid surface, p_{curve}, will be larger than on a flat surface, p_{flat}. A relation between pressure over *curved* and *flat* liquid surface was derived (Kelvin equation):

$$\ln\left(\frac{p_{curve}}{p_{flat}}\right) = \left(\frac{v_L}{\mathbf{RT}}\right)\left(\frac{2\gamma}{R_{curve}}\right) \tag{1.23}$$

where

p_{curve} and p_{flat} are the vapor pressures over curved and flat surfaces, respectively
R_{curve} is the radius of curvature
v_L is the molar volume

Kelvin equation thus suggests that if liquid is present in a porous material, such as cement, then the difference in vapor pressure exists between two pores of different radii. Infact, in cement industry, one reduces the value of γ, such that the vapor pressure difference between pores of different radii is reduced. Similar consequence of vapor pressure exists when two solid crystals of different size are concerned. The smaller sized crystal will exhibit higher vapor pressure and will also result in faster solubility rate.

Rain drop formation from clouds: In the clouds (water is in a gas phase) where the distance between water molecules is roughly the same as in the gas phase, the transition from water vapor in clouds to rain drops (liquid state) is not as straightforward process as might seem. The formation of a large liquid rain drop requires that a certain number of water molecules in the clouds (as gas phase) have formed a nuclei (which is the formation of first liquid drop). This is observed in some cases where there are many clouds, but one does not get any rain as needed for irrigation (this happens if nuclei are absent). The nuclei or *embryo* will grow, and Kelvin relation will be the determining factor. Artificial rain has been attempted by using fine particles (finely divided silver particles or similar), which leads to nuclei formation and assisting in rain drop formation (Adamson and Gast, 1997). Pollution (such as dust particles) will also give rise to abnormal rain fall due to the latter effect on nucleation.

1.2.3 Capillary Rise (*or Fall*) of Liquids (Capillary Forces)

The surface tension of liquids becomes evident when one observes the following experiment. Capillary forces are the reason that liquids behave different when a narrow capillary tube is dipped into a liquid (Figure 1.1). The curvature inside the tubing is the capillary force, which is related to the surface tension of a liquid. Liquids in narrow tubes are found in many different technical and biological systems:

1. The range of these applications is from the blood flow in the veins to the oil recovery in the reservoir.
2. Fabric's properties are also governed by capillary forces (i.e., wetting, etc.).
3. The sponge absorbs water or other fluids where the capillary forces push the fluid into many pores. This is also called *wicking* process (as in the candle wicks).

The curvature in a system where a narrow capillary circular tube is dipped into a liquid exhibits properties, which are not observed in a large beaker. The liquid is found to *rise* in the capillary, when the fluid wets the capillary (like water and glass or water and metal). The *curvature* of the liquid inside the capillary will lead to pressure difference between this state and the relatively flat surface outside the capillary (Figure 1.16).

The rise or fall of a liquid in a capillary (arising from the capillary force) is dependent on the wetting characteristics. Inside the capillary, the liquid (with surface tension, γ) attains an equilibrium of capillary forces. However, if the fluid is nonwetting (such as Hg in glass), then one finds that the fluid *falls*. This arises from the fact that Hg does not wet the tube. Capillary forces arise from the difference in attraction of the liquid molecules to each other and the attraction of the liquid molecule to those of the capillary tube. The fluid rises inside the narrow tube to a height, **h**, until the surface tension forces balance the weight of the fluid. This equilibrium gives following relation:

$$\gamma 2\pi \mathbf{R} = \text{surface tension force} \tag{1.24}$$

$$= \rho_L g_g \mathbf{h} \pi \mathbf{R}^2 = \text{fluid weight} \tag{1.25}$$

where

γ is the surface tension of the liquid
R is the radius of curvature

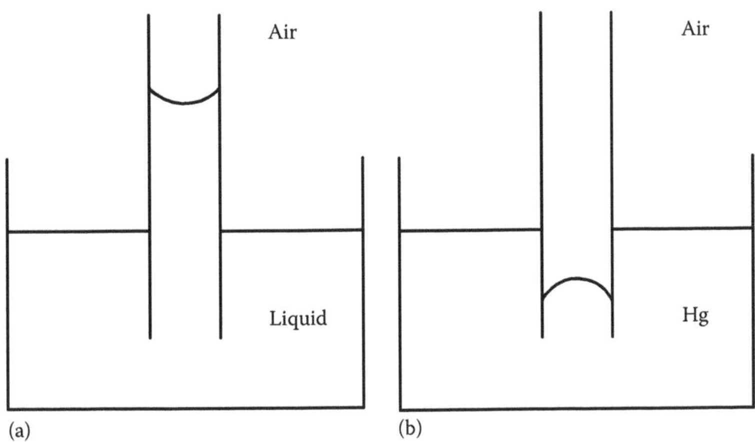

FIGURE 1.16 Rise or fall of a liquid in a glass capillary: (a) rise of water and (b) fall of Hg in a capillary.

In the case of narrow capillary tubes, less than 0.5 mm, the curvature can be safely set equal to the radius of the capillary tubing. The fluid will rise inside the tube to compensate for surface tension force; thus, at equilibrium, we get

$$\gamma = 2\mathbf{R}\rho_L\mathbf{gh} \tag{1.26}$$

where

ρ_L is the density of the fluid
g is the acceleration of gravity
h is the rise in the tube

Example

Liquid = water (25°C)
Water: γ = 72.8 mN/m; ρ_L = 1000 kg/m^3; **g** = 9.8 m/s^2 will rise to a height of
Magnitude of capillary rise in different size capillary:

$$h = \frac{\gamma}{2\mathbf{R}\rho_L\mathbf{g}} = \frac{72.8}{2\mathbf{R}(1000)\mathbf{9.8}}$$

The magnitudes for h for different radii of tubings are the following:

0.015 mm in a capillary of radius 1 m
1.5 mm in a capillary of radius 1 cm
14 cm in a capillary of radius 0.1 mm

One can measure the magnitude of **h** very accurately and thus allows one to measure the value of γ (Adamson and Gast, 1997; Birdi, 2009).

It is found that these assumptions are only precise when the capillary tubing is rather small. In the case of larger sized capillaries, correction tables are found in the literature (Adamson and Gast, 1997). However, in some particular case where the contact angle, θ, is not zero, one will need a correction, and equation will become

$$\gamma = 2\mathbf{R}\rho_L\mathbf{gh}\left(\frac{1}{\cos(\theta)}\right) \tag{1.27}$$

It is seen that when the liquid wets the capillary wall, the magnitude of θ is 0, and cos(0) = 1. In the case of Hg, the contact angle is 180°, since it is nonwetting fluid (see Figure 1.16b). Since cos(180) = −1, then the sign of **h** in equation will be negative. This means that Hg will show a *drop* in height in glass tubing. Hence, the rise or fall of a liquid in a tubing will be governed by the sign of cos(θ).

Thus, capillary forces will play an important factor in all systems where liquids are present in porous environment. Similar result can also be derived by using the Laplace equation (1.28) (1/radius = 1/R):

$$\Delta\mathbf{P} = \frac{2\gamma}{\mathbf{R}} \tag{1.28}$$

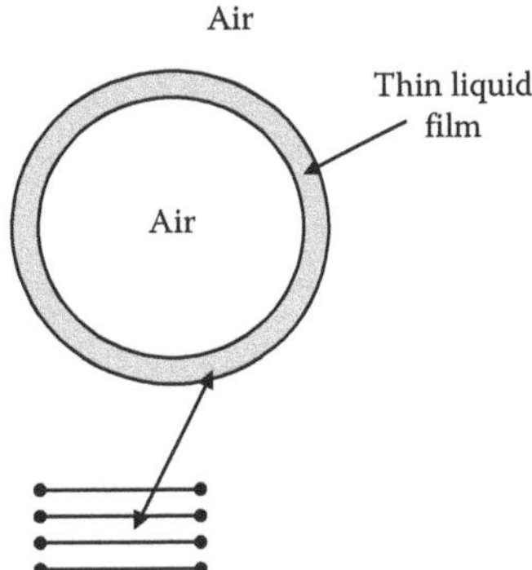

FIGURE 1.17 Bilayer soap film structure (schematic).

The liquid rises to a height **h**, and the systems achieve equilibrium, and the following relation is found:

$$\frac{2\gamma}{\mathbf{R}} = \mathbf{h}g_g\rho_L \tag{1.29}$$

This can be expressed as

$$\gamma = 2\mathbf{R}\rho_L g_g \mathbf{h} \tag{1.30}$$

It is found that the various surface forces are responsible for the capillary rise. The lower the surface tension, the lower is the height of column in the capillary. The magnitude of γ is determined from the measured value of **h** for a fluid with known ρ_L. The magnitude of **h** can be measured directly by using a suitable device (e.g., photograph image).

Further, it is known that in real world, capillaries or pores are not always circular shaped. In fact, one considers that in oil reservoirs (or water seepage), pores are more like triangular or square shaped than circular. In this case, one can measure the rise in other kinds of shaped capillaries, such as rectangular or triangular (Birdi, 1997, 2003a; Birdi et al., 1988). These studies have much importance for oil recovery or water treatment systems. Especially in shale oil/gas technology, one uses large amounts of water to transport chemicals through pores where curved liquid surfaces are present (Nicot and Scanlon, 2012). In any system where fluid flows through porous material (e.g., seepage of water), one would expect that capillary forces will be one of the most dominant factors. Further, it is known that the vegetable world is dependent on capillary pressure (and osmotic pressure) to bring water up to the higher parts of plants. In this way, some trees succeed in bringing this essential liquid (water) up to *120 m* above the ground.

1.2.4 Soap Bubbles (Formation and Stability)

Perhaps the phenomenon of bubble formation is the most common observation mankind experiences since childhood. Bubbles are also commonly observed in many different instances:

1. Beer, champagne, etc.
2. Along the coasts of lakes and oceans
3. Shampoo and detergent solutions

One also knows that soap bubbles are extremely thin (1000 times thinner than the diameter of a hair!) and unstable (Birdi, 1997; Boys, 1959; Li et al., 2014; Lovett, 1994; Scheludko, 1966;

Taylor, 2011). In spite of the latter, under special conditions, one can keep soap bubbles for long lengths of time, which thus allows one to study its physical properties (such as thickness, composition, conductivity, spectral reflection, etc.). The thickness of a bubble is in most cases over hundreds of micrometers in the initial state. The thin liquid film (TLF) consists of bilayer of detergent that contains the solution. The film thickness decreases with time due to following reasons:

Drainage of fluid away from the film
Evaporation

Therefore, the stability and lifetime of such thin films will be dependent on these different characteristics. This is found from the fact as an air bubble is blown under the surface of a soap or detergent solution, air bubble will rise up to the surface. It may remain at the surface, if the speed is slow, or it may escape into the air as a soap bubble. Experiments show that a soap bubble consists of a very TLF with an iridescent surface. But as the fluid drains away and the thickness decreases, the latter approaches to the equivalent of barely *two surfactant molecules* plus a few molecules of water. It is worth noting that the limiting thickness is on the order of *two or more surfactant molecules*. This means that one can see with the naked eye molecular size structures of TLFs. As the air bubble enters the surface region, the soap molecules along with water molecules are pushed up and as the bubble is detached, it leaves as a TLF with following characteristics (as found from various measurements):

A bilayer of soap (approximately 200 Å thick) on the outer region contains the aqueous phase (Figure 1.17).
The thickness of the initial soap bubble is some micrometers.
The thickness decreases with time and one starts to observe rainbow colors, as the reflected light is of the same wavelength as the thickness of the bubble (few hundreds of Angstroms). The thinnest liquid film consists mainly of the bilayer of surface-active substance (SAS) (such as soap = 50 Å) and some layers of water. The light interference and reflection studies show many aspects of these TLFs.

The bilayer soap film may be depicted as arrays of the soap molecule (with a few layers of water) (-----O: length of the soap molecule is about 15 Å):

Thick bubble film (micrometer or more) (shows no colors):

-----OWWWWWO-----
-----OWWWWWO-----
-----OWWWWWO-----

Thin bubble film (shows rainbow colors):

-----OWO-----
-----OWO-----
-----OWO-----

The thickness of the bilayer is thus about 30–50 Å (almost twice the length of the soap molecule). The thickness is approximately the size of wavelength of light.

The iridescent colors of the soap bubble arise from the interference of reflected light waves. The reflected light from the outer surface and the inner layer gives rise to this interference effect (Figure 1.17). The rainbow colors are observed as the bubble thickness decreases (and reaches magnitudes corresponding to that in the light waves) due to the evaporation of water.

Rainbow colors (violet, indigo, blue, green, yellow, orange, red: *VIBGYOR*) are observed when the thickness is on the order of wavelength of light (Adamson and Gast, 1997; Birdi, 2009). The wavelength increases from violet to red color. One observes rainbow colors in all sorts of situations where interference of waves of light takes place. Another interesting observation is that in some natural phenomena, colors are produced by simply structures made similar to the bubble; that is, thickness (of lipid-like molecules) varies in the range of wavelength of light (Orna, 2013). This is, for instance, the case in the colors observed in peacock feathers (and many other colored objects in nature and otherwise).

Soap films that are rather thick (thickness is mainly due to the water) reflect red light and one observes blue–green colors. Lesser thin films cancel out yellow wavelength and blue color is observed. As the thickness approaches the wavelength of light, all colors are cancelled out and a *black (or gray) film* is observed. This corresponds to 25 nm (250 Å).

The transmitted light, I_{tr}, is related to the incident, I_{in}, and the reflected intensity, I_{re}:

$$I_{tr} = I_{in} - I_{re} \tag{1.31}$$

As one finds from the analyses of the daylight, it consists of different wavelengths of colors (violet, indigo, blue, green, yellow, orange, red [VIBGYOR: easy to remember!]):

Red	680 nm
Orange	590 nm
Yellow	580 nm
Green	530 nm
Blue	470 nm
Violet	405 nm

Slightly thicker soap films (ca. 1500 nm) sometimes look *golden*. In the thinning process, the different colors get cut off. Thus, if the blue color gets cut off, the film looks amber to magenta.

The bubble film's (which consists of SAS + water + salts, etc.) stability can be described as follows:

Bubble film:

Evaporation of water (making thin film and unstable)
Flow of water away from the film
Stability of the bubble film (sensitive to vibration or other mechanical action)

It is thus obvious that the rate of water evaporation plays an important role. The evaporation can be reduced by containing the bubble in a closed bottle. One also finds that in such closed system, the bubbles remain stable for a very long time. The drainage of water away from the film is dependent on the viscosity of the fluid. Therefore, additives such as glycerin (or other thickening agents [polymers]) assist in the stability of the bubble films.

1.2.5 Measurement of Surface Tension of Liquids

In the literature, one finds a variety of methods used to measure the magnitude of surface tension of liquids. This arises from the fact that one needs a specific method for each situation, which one may use in the measurement of γ. For example, if the liquid is water (at room temperature), then the method will be different than if the system is molten metal (at very high temperature, ca. 500°C or higher). In the oil reservoirs, one finds oil at high temperature (over 80°C) and pressures (over 200 atm). In many cases, one has developed specific instruments that allow one to measure the magnitude of surface tension under the given situation (Adamson and Gast, 1997; Birdi, 2009).

1.2.5.1 Shape and Weight of Liquid Drop

The formation of liquid drops when flow occurs through thin tubes is a common daily phenomenon. In some cases, such as eye-drop application, the size of drop plays a significant role in the application and dosage of medicine (such as eye-drop solutions). The drop formed when liquid flows through a circular tube is shown in Figure 1.18.

In many preprocesses (such as oil recovery, blood flow, underground water), one encounters liquid flow through thin (micrometer diameter) noncircular-shaped tubes or pores. In the literature, one finds studies that address these latter systems. In other contexts, the liquid drop formation, for example, in an ink jet nozzle, this technique falls under a class of scientifically challenging technology. All combustion engines are controlled by the oil drop formation and evaporation characteristics. The important role of capillary forces is obvious in such systems. As the liquid drop grows larger, it will at some stage break off the tube (due to gravity force being larger than the surface force holding it to the capillary) and will correspond to the maximum weight of the drop that can hang (Birdi, 2003a; Munz and Mills, 2014). The equilibrium state where the weight of the drop is exactly equal to the detachment surface energy is given as

$$m_m\mathbf{g} = 2\pi\mathbf{R}\gamma \tag{1.32}$$

where

m_m is the weight of the detached drop
R is the radius of the tubing

A simple method is to *count* the number of drops (e.g., 10 or more) and measure the weight (with a very high accuracy: less than microgram).

One may also use a more convenient method where a fluid is pumped and the drops are collected and weighed. Since in some systems (solutions) there may be kinetic effects, one must be careful to keep the flow as slow as possible. This system is very useful in studying systems, which one finds in daily life phenomena: oil flow in reservoirs, blood cells flow through arteries. In those cases where the volume of fluid available is limited, one may use this method with advantage. By decreasing the diameters of the tubing, one can work with less than 1 μL fluids. This may be useful in systems such as eye fluids, etc.

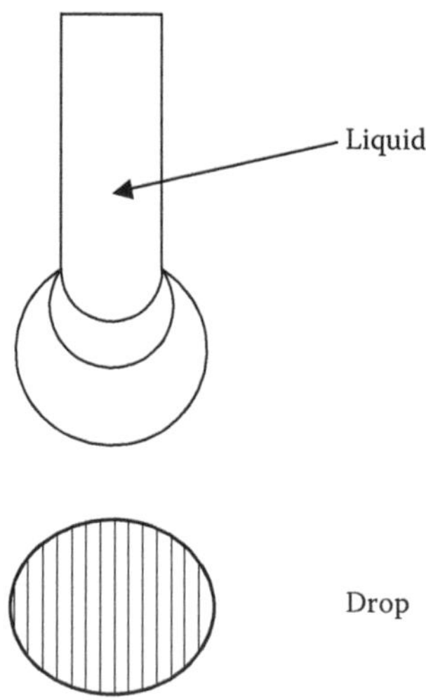

FIGURE 1.18 Liquid drop formation at the end of a tube.

1.2.5.1.1 Maximum Liquid Drop-Weight Method

The "detachment" method is based upon the following: to detach a body from the surface of a liquid that wets the body, it is necessary to overcome the same surface tension forces, which operate when a drop is broken away. The liquid attached to the solid surface on detachment creates following surfaces:

Initial stage: Liquid attached to solid.
Final stage: Liquid separated from solid.

In the process from *initial* to the *final* stage, the liquid molecules that were near the solid surface have been moved away and are now under the influence of their own molecules. This requires energy (interfacial energy), and the force required to make this happen is proportional to the surface area of contact and to the surface tension of the liquid. The methods of determining the surface tension by measuring the force required to detach a body from a liquid are therefore similar to the stalagmometer method described earlier. However, their advantage over the latter method is that is, possible to choose the most convenient form and size of the body (platinum rod, ring or plate) so as to enable the measurement to be carried out rapidly, but without any detriment to its accuracy. The *detachment* method has found an application in the case of liquid whose surface tension changes with time.

1.2.5.1.2 Shape of the Liquid Drop (Pendant Drop Method)

In some cases, the amount of liquid available for surface tension measurement is very small, such as fluid from the eye, etc. Under these conditions, one finds that the following procedure is most suitable for the measurement of γ. The liquid drop forms as it flows through a tubing, Figure 1.19. At a stage just before it breaks off, the shape of the *pendant drop* has been used to estimate γ. The drop shape is photographed, and from the diameters of the shape, one can accurately determine γ. Actually, if one has only a drop of fluid, then one can measure its γ without the loss of sample volume (as in the case of eye fluid, etc.).

The parameters needed are as follows:

A quantity pertaining to the ratio of two significant diameters is

$$S_s = \frac{d_s}{d_e} \tag{1.33}$$

where

d_e is the equatorial diameter
d_s is the diameter at a distance d_e from the tip of the drop, Figure 1.19

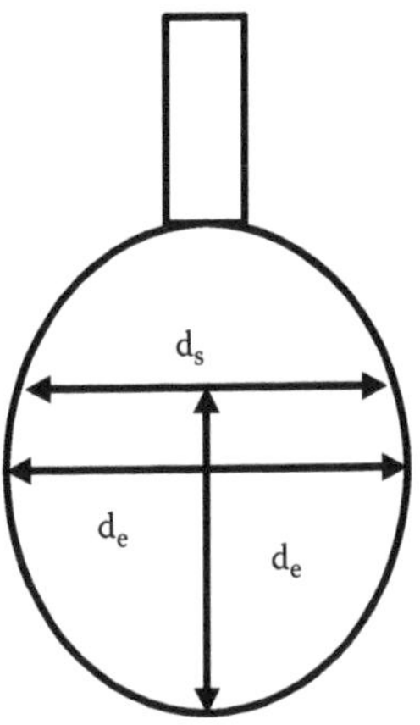

FIGURE 1.19 Pendant drop of liquid (shape analysis).

The relation between γ and d_e and $\mathbf{S_s}$ is found as

$$\gamma = \frac{\rho_L \mathbf{g} d_e^2}{\mathbf{H}} \tag{1.34}$$

where ρ_L is the density of the liquid and $\mathbf{H}$ is related to $\mathbf{S_s}$, but the values of $1/\mathbf{H}$ for varying $\mathbf{S_s}$ were obtained from experimental data. For example, when $\mathbf{S_s} = 0.3$, $1/\mathbf{H} = 7.09837$, while when $\mathbf{S_s} = 0.6$, $1/\mathbf{H} = 1.20399$. The magnitude of γ is estimated from accurate mathematical functions used to estimate $1/\mathbf{H}$ for a given d_e value (Adamson and Gast, 1997; Birdi, 2003a). The accuracy (0.1%) of γ is satisfactory for most of the systems especially when experiments are carried out under extreme conditions (such as high temperatures and pressures).

The pendant drop method is very useful under specific conditions:

1. Technically, only a drop (a few microliters) is required. For example, eye fluid can be studied, since only a drop of microliter is needed.
2. It can be used under very extreme conditions (very high temperature or corrosive fluids).
3. Under very high pressure and temperatures. Oil reservoirs are found typically at 100°C and 300 atm pressure. Surface tension of such systems can be conveniently studied by using *high pressure and temperature* cells with optical clear windows (sapphire windows:1 cm thick; up to 2000 atm). For example, γ of inorganic salts at high temperatures (ca. 1000°C) can be measured by using this method. The variation of surface tension can be studied as a function of various parameters (temperature and pressure, additives [gas, etc.]).

1.2.5.2 Ring Method (Detachment)

In the classical methods used to measure surface tension of liquids, we find that detachment of a solid from a liquid surface provides very accurate results. A method that has been rather widely used involves the determination of the force to detach a ring or loop of wire from the surface of a liquid. This method is based on using a ring (platinum) and measuring the force when it is dipped in the liquid surface.

The method is one of the many detachment methods, of which the drop-weight and Wilhelmy slide methods are also examples. It is based on the principle that within an accuracy of few percents, the detachment force is given by the surface tension multiplied by the periphery of the surface (liquid surface) detached (from a solid surface of a tubing or ring or plate). This assumption is also found to be acceptable for most experimental purposes. Thus, for a ring,

$$W_{total} = W_{ring} + \mathbf{2}(2\pi \mathbf{R}_{ring})\gamma \tag{1.35}$$

$$= W_{ring} + 4\pi R_{ring}\gamma \tag{1.36}$$

where

W_{total} is the total weight of the ring
W_{ring} is the weight of the ring in air
R_{ring} is the radius of the ring

The circumference is $2\pi R_{ring}$, and factor **2** is because of the two sides of contact.

This relation assumes that the contact between the fluid and the ring is geometrically simple. It is also found that this relation is fairly correct (better than 1%) for most working situations. However, it was observed that Equation 1.36 needed correction factor, in much the same way as was done for the drop-weight method. Here, however, there is one additional variable so that the correction factor **f** now depends on two dimensionless ratios (Adamson and Gast, 1997).

1.2.5.3 Plate (Wilhelmy) Method

The methods so far discussed have required more or less tabular solutions, or else correction factors to the respective *ideal* equations. Further, if one needs to make continuous measurements, then it is not easy to use some of these methods (such as capillary rise or bubble method). The most useful method of measuring the surface tension is by the well-known Wilhelmy plate method. If a smooth and flat plate-shaped metal is dipped in a liquid, the surface tension forces will be found to give rise to a tangential force, Figure 1.20. This is because a new contact phase is created between the plate and the liquid.

The total weight measured, W_{total}, would be

$$W_{total} = \text{weight of the plate} + \gamma(\text{perimeter}) \quad (\text{up-drift}) \tag{1.37}$$

Perimeter of a plate is the sum of twice the length + breadth. The surface force will act along the perimeter of the plate (i.e., length [L_p] + width [W_p]). The plate is often very thin (less than 0.1 mm) and made of platinum, but even plates made of glass, quartz, mica, and filter paper can be used. The forces acting on the plate consist of the gravity and surface tension force (downward), and buoyancy due to displaced water upward. For a rectangular plate of dimensions L_p and W_p and of material density ρ, immersed to a depth h_p in a liquid of density ρ_L, the net downward force, **F**, is given by the following equation (i.e., weight of plate + surface force (γ) × perimeter of the plate–upward drift):

$$\mathbf{F} = \rho_p \mathbf{g}(L_p W_p t_p) + 2\gamma(t_p + W_p)(\cos(\theta)) - \rho_L \mathbf{g}(t_p W_p h_p) \tag{1.38}$$

where

γ is the liquid surface tension
θ is the contact angle of the liquid on the solid plate
g is the gravitational constant

The weight of plate is constant and can be tared. If the plate used is very thin (i.e., $t_p \ll W_p$) and the up-drift is negligible (i.e., h_p is almost zero), then γ is found from

$$\gamma = \frac{\mathbf{F}}{2W_p} \tag{1.39}$$

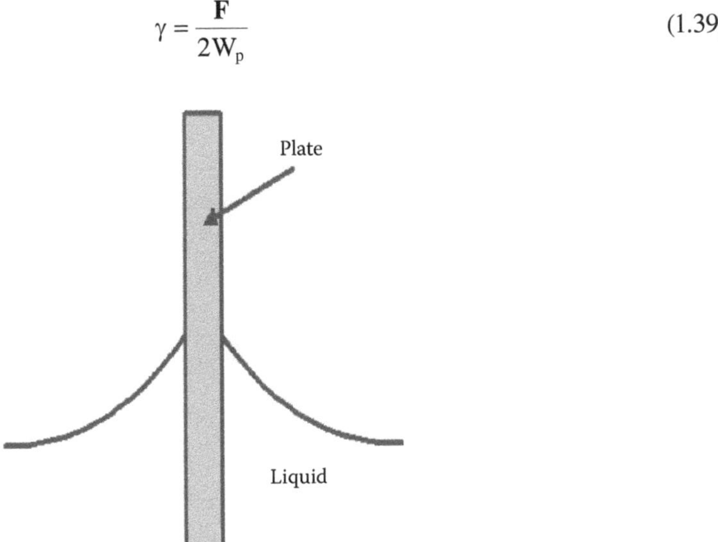

FIGURE 1.20 Wilhelmy plate in a liquid (plate with dimensions: length = L_p, width = W_p).

The sensitivity of γ by using these procedures has been found to be very high (±0.001 dyn/cm [mN/m]) (Birdi, 2009). The change in surface tension (surface pressure = Π) is then determined by measuring the change in **F** for a stationary plate between a clean surface and the same surface with a monolayer present. If the plate is completely wetted by the liquid (i.e., $\cos(\theta) = \cos(0) = 1$), the surface pressure is then obtained from the following equation:

$$\Pi = -\left[\frac{\Delta F}{2(t_p + W_p)}\right] = \frac{-\Delta F}{2W_p}, \quad \text{if } W_p \gg t_p \tag{1.40}$$

In general, by using very thin plates with thickness 0.1–0.002 mm, one can measure surface tension with very high sensitivity. The apparatus is calibrated using pure liquids, such as water and ethanol. The buoyancy correction is made very small (and negligible) by using a very thin plate and dipping the plate as little as possible. The wetting of water on platinum plate is achieved by using commercially available platinum plates that are roughened to increase wettability. The latter property gives rise to almost complete wetting, that is, $\theta = 0$. The force is in this way determined by measuring the changes in the mass of the plate, which is directly coupled to a sensitive electrobalance or some other suitable device (such as pressure transducer, etc.). It must be noticed that in all systems where an object is in contact with a liquid there will exist a force toward the liquid (related to surface tension).

1.3 TYPICAL SURFACE TENSION DATA OF LIQUIDS

At this stage, it is important to consider how the magnitude of surface tension of different molecules changes with respect to the molecular structure (Table 1.2). Extensive studies are found that attempt to correlate surface tension to other physicochemical properties of liquids, such as boiling point, heat of evaporation, etc. This concept has been extensively analyzed in the literature (Birdi, 2010a; Jasper, 1972). In research and other applications where the surface tension of liquids plays an important role, it is necessary to be able to predict the magnitude of γ of different kinds of molecules.

These data need some comments in order to explain the differences in surface tension data and molecular structure. The range of γ is found to vary from ca. 20 to over 1000 mN/m. The surface tension of Hg is high (425 mN/m), for example, as compared to that of water, because it is a liquid metal with a very high boiling point. Latter indicates that it needs much energy to break the bonds between Hg atoms to evaporate. Similarly, γ of NaCl as a liquid (at high temperature) is also very high (Table 1.2). The same is found for metals in liquid state. The other liquids can be considered as under each type, which should help understand the relation between the structure of a molecule and its surface tension.

It is useful to consider some data (Jasper, 1972) that will provide information about structure and surface tension relationship in some simple liquids.

Alkanes (*normal*): It is found that the magnitude of γ (at 25°C) increases by 1.52 mN/m per two $-CH_2$, when alkyl chain length increases from 10 to 12 (*n*-decane = 23.83 mN/m; *n*-dodecane = 25.35 mN/m).

n-Heptane	20.14
n-Hexadecane	27.47
n-Hexane	18.43
n-Octane	21.62

Alcohols: The *magnitude* of γ changes by 23.7 − 22.1 = 1.6 mN/m per $-CH_2-$ group. This is based upon the γ data of ethanol (22.1 mN/m) and propanol (23.7 mN/m).

These observations indicate the molecular correlation between bulk forces and surface forces (tension) (γ) for homologous series of substances.

TABLE 1.2
Surface Tension Values of Some Common Liquids

Liquid	Surface Tension (20°C; mN/m [dyn/cm])
1,2-Dichloro-ethane	33.3
1,2,3-Tribromo propane	45.4
1,3,5-Trimethylbenzene (mesitylene)	28.8
1,4-Dioxane	33.0
1,5-Pentanediol	43.3
1-Chlorobutane	23.1
1-Decanol	28.5
1-Nitro propane	29.4
1-Octanol	27.6
Acetone (2-propanone)	25.2
Aniline	43.4
2-Aminoethanol	48.9
Anthranilic acid ethylester	39.3
Anthranilic acid methylester	43.7
Benzene	28.9
Benzylalcohol	39.0
Benzylbenzoate (BNBZ)	45.9
Bromobenzene	36.5
Bromoform	41.5
Butyronitrile	28.1
Carbon disulfide	32.3
Quinoline	43.1
Chlorobenzene	33.6
Chloroform	27.5
Cyclohexane	24.9
Cyclohexanol	34.4
Cyclopentanol	32.7
p-Cymene	28.1
Decalin	31.5
Dichloromethane	26.5
Diiodomethane (DI)	50.8
1,3-Diiodopropane	46.5
Diethylene glycol	44.8
Dipropylene glycol	33.9
Dipropylene glycol monomethylether	28.4
Dodecyl benzene	30.7
Ethanol	22.1
Ethylbenzene	29.2
Ethylbromide	24.2
Ethylene glycol	47.7
Formamide	58.2
Fumaric acid diethylester	31.4
Furfural (2-furaldehyde)	41.9
Glycerol	64.0
Ethylene glycol monoethyl ether (ethyl cellosolve)	28.6
Hexachlorobutadiene	36.0
Iodobenzene	39.7

(Continued)

TABLE 1.2 (*Continued*)
Surface Tension Values of Some Common Liquids

Liquid	Surface Tension (20°C; mN/m [dyn/cm])
Isoamylchloride	23.5
Isobutylchloride	21.9
Isopropanol	23.0
Isopropylbenzene	28.2
Isovaleronitrile	26.0
m-Nitrotoluene	41.4
Mercury	425.4
Methanol	22.7
Methyl ethyl ketone (MEK)	24.6
Methyl naphthalene	38.6
N,*N*-dimethyl acetamide (DMA)	36.7
N,*N*-dimethyl formamide (DMF)	37.1
n-Decane	23.8
n-Dodecane	25.3
n-Heptane	20.1
n-Hexadecane	27.4
n-Hexane	18.4
n-Octane	21.6
n-Tetradecane	26.5
n-Undecane	24.6
n-Butylbenzene	29.2
n-Propylbenzene	28.9
Nitroethane	31.9
Nitrobenzene	43.9
Nitromethane	36.8
o-Nitrotoluene	41.5
Perfluoroheptane	12.8
Perfluorohexane	11.9
Perfluorooctane	14.0
Phenylisothiocyanate	41.5
Phthalic acid diethylester	37.0
Polyethylene glycol 200 (PEG)	43.5
Polydimethyl siloxane	19.0
Propanol	23.7
Pyridine	38.0
3-Pyridylcarbinol	47.6
Pyrrol	36.6
sym-Tetrabromoethane	49.7
tert-Butylchloride	19.6
sym-Tetrachloromethane	26.9
Tetrahydrofuran (THF)	26.4
Thiodiglycol	54.0
Toluene	28.4
Water	72.8
o-Xylene	30.1
m-Xylene	28.9
a-Bromonaphthalene (BN)	44.4

(*Continued*)

TABLE 1.2 (*Continued*)
Surface Tension Values of Some Common Liquids

Liquid	Surface Tension (20°C; mN/m [dyn/cm])
a-Chloronaphthalene	41.8
Liquid metals (at the melting point) (liquid state)	
Aluminum (650°C)	871
Bismuth (270°C)	382
Copper (1085°C)	1330
Gold (1065°C)	1145
Inorganic salts (at the melting point)	
Potassium chloride (780°C)	100
Sodium chloride (810°C)	113
Oxygen (−184°C),	13
Nitrogen (−183°C)	6

1.3.1 Effect of Temperature and Pressure on Surface Tension of Liquids

All natural processes are found to be dependent on the temperature and pressure effects on any system under consideration. For example, oil reservoirs are generally found under high temperature (ca. 100° C) and pressure (over 200 atm). Actually, mankind is aware of the great variations of both temperature (sun) and pressure (earthquakes, storms, and winds) with which the natural phenomena are surrounding the earth. Even on the surface of the earth itself, one has temperature variations of −50°C to +50°C. On the other hand, inside the central mantle of the earth, one finds that temperature and pressure increase as one goes from its surface to the center of the earth (about 5000 km). In fact, the scientific research interest comprises from temperatures ranging from −273°C (absolute temperature) to over 2000°C (such as inside the earth or the surface of sun). The surface tension is related to the internal forces in the liquid (surface), and one must thus expect it to bear relationship to the internal energy. Further, it is found that surface tension *always* decreases with increasing temperature. Surface tension, γ, is a quantity that can be measured accurately and applied in the analyses of all kinds of surface phenomena. If a new surface is *created*, then in the case of a liquid, molecules from the bulk phase must move to the surface. The work required to create extra surface area, d**A**, will be given as follows:

$$\mathbf{dG} = \gamma \mathbf{dA} \tag{1.41}$$

The surface free energy, G_S, per unit area is given as

$$G_S = \gamma = \left(\frac{\mathbf{dG}}{\mathbf{dA}} \right)_{T,P} \tag{1.42}$$

Hence, the other thermodynamic surface quantities will be

Surface entropy (S_S):

$$\mathbf{S_S} = -\left(\frac{dG_S}{\mathbf{dT}} \right)_P \tag{1.43}$$

$$= -\left(\frac{d\gamma}{\mathbf{dT}} \right) \tag{1.44}$$

We can thus derive for surface enthalpy, H_S,

$$H_S = G_S + \mathbf{T}\mathbf{S_S} \tag{1.45}$$

All natural processes are dependent on the effect of temperature and pressure. For instance, oil reservoirs are found under high temperatures (ca. 80°C) and pressure (around 100–400 atm [depending on the depth]). The thermodynamics of the system is defined by the fundamental equation for the free energy, G, and to the enthalpy, H, and entropy, S, of the system:

$$\mathbf{G} = H - \mathbf{T}S \tag{1.46}$$

It is thus important that in all practical analyses, one should be aware of the effects of temperature and pressure. The molecular forces that stabilize liquids will be expected to decrease as temperature increases. Experiments also show that in all cases, surface tension decreases with increasing temperature. Surface entropy of liquids is given by ($-d\gamma/dT$). This means that the entropy is positive at higher temperatures (because the magnitude of γ always decreases with temperature for all liquids). The rate of decrease of surface tension with temperature is found to be different for different liquids, which supports the aforementioned description of liquids (Jasper, 1972).

For example, the surface tension of water data is given as

At 5°C, γ = 75 mN/m
At 25°C, γ = 72 mN/m
At 90°C, γ = 60 mN/m

Extensive γ data for water were fitted to the following equation (Birdi, 2003a; Cini et al., 1972):

$$\gamma = 75.69 - 0.1413 t_C - 0.0002985 t_C^2 \tag{1.47}$$

where t_C is in °C. This equation gives the value of γ at 0°C as 75.69 mN/m. The value of γ at 50°C is found to be $(75.69 - 0.1413 \times 50 - 0.0002985 \times 50^2) = 67.88$ mN/m, and at 25°C, it is 71.97 mN/m. In the literature, one finds such relationships for other liquids, which allows one to calculate the magnitude of γ at different temperature. This kind of application is important in oil and other industrial phenomena.

Further, these data show that γ of water decreases with temperature from 25°C to 60°C, $(72 - 60)/(90 - 25) = 0.19$ mN/m °C. This is as one should expect from physicochemical theory. The surface tension change with temperature analyses thus provides a very sensitive information as regards the molecular forces in any liquid (surface entropy). In all systems, as the temperature increases, the energy between molecules gets weaker and thus the surface energy, that is, surface tension, will decrease. The difference in the surface entropy gives information on the structures of different liquids. It is also observed that the effect of temperature on γ will be lower on liquids with higher boiling point (such as Hg) than for low boiling liquids (such as *n*-hexane). Actually, there exists a correlation between γ and heat of vaporization (or boiling point). In fact, many systems even show big differences (due to effects of temperature on γ) when comparing winter or summer months (such as rain drops, sea waves, foaming in natural environments). Different thermodynamic relations have been derived, which can be used to estimate the surface tension at different temperatures. Especially, straight chain alkanes have been extensively analyzed. The data show that a simple correlation between surface tension, temperature, and n_C exists. This allows one to estimate the value of γ at any temperature of a given alkane. This observation has many aspects in the applied industry. It allows one to estimate the magnitude of surface tension of an alkane at the required temperature. Further, one has a fairly good quantitative analyses about how surface tension will change for a given alkane under a given experimental condition (e.g., in oil industry).

1.3.2 Heat of Liquid Surface Formation and Evaporation

All matter is stabilized by forces interacting between molecules. Therefore, as one moves inside a liquid phase toward the surface, one needs to consider the consequences of the changing interaction forces, which are related to the surrounding structure. In order to understand which forces stabilize liquid structures, one has suggested a relation between the surface tension of a liquid and the latent heat of evaporation. This is a reasonable argument considering the geometrical packing of molecules. It has been argued (Stefan, 1886) that when a molecule is brought to the surface of a liquid from the interior, the work done in overcoming the attractive force near the surface should be related to the work expended when it escapes into the scarce vapor phase (Adamson and Gast, 1997; Birdi, 1997, 2003a). It was suggested that the first quantity should be approximately half of the second. According to the Laplace theory of capillarity, the attractive force acts only over a small distance equal to the radius of the sphere (see Figure 1.1), and in the interior the molecule is attracted equally in all directions and experiences no resultant force. In the surface, it experiences a force due to the liquid in the hemisphere, and half the total molecular attraction is overcome in bringing it there from the interior. A very useful molecular model was suggested (Stefan, 1886) that the energy necessary to bring a molecule from bulk phase to the surface of a liquid should be *half* the energy necessary to bring it entirely into the gas phase. It is known from geometrical considerations that a sphere can be surrounded by 6 molecules (in two dimensions) (Figure 1.2) and 12 molecules (in three dimensions) of the same size. This corresponds with the most densely packed top (surface) monomolecular layer half filled and the next layer completely filled—to a very dilute gas phase (the distance between gas molecules is approximately *10* times greater than in liquids or solids). This indicates that intermolecular forces in liquids would be weaker than in solids by a few orders of magnitude, as is also found experimentally. The ratio of the enthalpy of surface formation to the enthalpy of vaporization, h_s:h_{vap}, for various substances is given in Table 1.3. Substances with nearly spherical shaped molecules have ratios near 1/2, while substances with a polar group on one end give a much smaller ratio. This difference indicates that the latter molecules are oriented with the nonpolar end toward the gas phase and the polar end toward the liquid. In other words, molecules with dipoles would be expected to be oriented perpendicularly at the gas/liquid interfaces.

A simple analysis was proposed, which was based on purely spherical geometrical packing of molecules (most of which are certainly not spherical) and ideal situation. Hence, any deviation from

TABLE 1.3
Ratio of Enthalpy of Surface Formation, h_s, and the Enthalpies of Vaporization, h_{vap} (See Text for Details)

Molecules (Liquid)	h_s/h_{vap}
Hg	0.64
N_2	0.51
O_2	0.5
CCl_4	0.45
C_6H_6	0.44
Diethyl ether	0.42
$Cl–C_6H_5$	0.42
Methyl formate	0.40
Ethyl acetate	0.4
Acetic acid	0.34
H_2O	0.28
C_2H_5OH	0.19
CH_3OH	0.16

Stefans law is an indication that the surface molecules are oriented differently than in the bulk phase. This observation is useful in order to understand the surface phenomena.

As an example, one may proceed with this theory and estimate the surface tension of a liquid with data on its heat of evaporation. The number of near neighbors of a surface molecule will be about half (6 = 12 [near neighbors in the bulk phase]/2) than those in the bulk phase (12 neighbors). It is now possible to estimate the ratio of the attractive energies in the bulk and in the surface, per molecule. We have the following data for a liquid such as CCl_4:

$$\text{Molar energy of vaporization} = \Delta U_{vap} \tag{1.48}$$

$$= \Delta h_{vap} - \mathbf{RT} = 34{,}000\ \text{J mol}^{-1} - 8.315\ \text{J K}^{-1}\ \text{mol}^{-1}\,(298\ \text{K}) = 31{,}522\ \text{J mol}^{-1} \tag{1.49}$$

$$\text{Energy change per molecule} = \frac{31{,}522\ \text{J mol}^{-1}}{6.023\times 10^{23}\ \text{mol}^{-1}} = 5.23\times 10^{-20}\ \text{J} \tag{1.50}$$

If we assume that about half of energy is gained when a molecule is transferred to the surface, then we get

$$\text{Energy per molecule at surface} = 5.23 \times 10^{-20}\ (2)\ \text{J} = 2.6 \times 10^{-20}\ \text{J} \tag{1.51}$$

Example: Estimation of γ CCl_4

The molecules at the surface of CCl_4 occupy a certain value of area, which can be estimated only roughly as follows.

Density of CCl_4 = 1.59 g/cm^3
Molar mass of CCl_4 = 154 g/mol
Volume per mol = 154/1.59 = 97 cm^3/mol
Volume per molecule = 97×10^{-6} (m^3/mol)/6.023×10^{23} mol^{-1} = 1.6×10^{-28} m^3
The radius of a sphere (volume = $4/3\Pi R^3$) with this magnitude of volume = $[1.6 \times 10^{-28}/(4/3\Pi)]^{1/3}$ = 3.5×10^{-10} m
Area per molecule = $\Pi R^2 = \Pi\ (3.5 \times 10^{-10})^2 = 38 \times 10^{-20}$ m^2
Surface tension (calculated) for CCl_4 = 2.6×10^{-20} J/38×10^{-20} m^2 = 0.068 J/m^2 = 68 mN/m

(Measured value of γ for CCl_4 is 27 mN/m.)

The measured value of γ of CCl_4 is 27 mN/m (Table 1.2). The large difference can be ascribed to the assumption that a Stefan ratio of 2 was used in this example. As expected, the simple ratio with factor 2 may vary for nonspherical molecules (as in the case of CCl_4). Under these assumptions, one may conclude that the estimated value of γ is an acceptable description of the surface molecules. This example is useful for basic considerations about the molecular interactions at the surfaces.

1.3.3 Other Surface Properties of Liquids

There are a variety of other surface chemical properties of liquid surfaces. This arises from the fact that different forces stabilize liquids. Since these are out of scope of this book, only a few important examples will be mentioned.

Surface waves on liquids: Liquid surface, for example, on oceans or lakes, exhibits waves formation when strong winds are blowing over it. It is known that such waves are created by the wind energy being transposed to waves. Hence, mankind has tried to convert wave energy to other useful forms of energy source. Both transverse capillary waves and longitudinal waves can deliver information about the elasticity and viscosity of surfaces, albeit on very different timescales. Rates of adsorption and desorption can also be deduced. Transverse capillary waves are usually generated with frequencies between 100 and 300 Hz. The generator is a hydrophobic knife-edge situated in the surface and oscillating vertically, while the usual detector is a lightweight hydrophobic wire lying in the surface parallel to the generator edge. The generator and detector are usually close together (15–20 mm), so reflections set up a pattern of standing transverse wave. Optical detection, on the other hand, causes no interference to the generated pattern. The damping of capillary ripples arises primarily from the compression and expansion of the surface and the interaction between surface film and water phase (Adamson and Gast, 1997; Birdi, 1997). This leads to compression and expansion of the surface. If a surface film is present, compression tends to lower the surface tension, while expansion raises it. This generates a Marangoni flow, which opposes the wave motion and dampens it (Birdi, 2003a, 2007, 2010a). Furthermore, if the material is soluble, the compression–expansion cycle will be accompanied with the hydrodynamic characteristics of capillary ripples.

1.3.4 Interfacial Tension of Liquid$_1$–Liquid$_2$

Oil and water do not mix; this is an everyday observation. Main reason is oil is insoluble in water, and vice versa. At the oil–water interface, one will thus have interfacial surface forces. In this chapter, the methods in which one can indeed *disperse* oil in water (or vice versa) will be described. The analyses of the IFT, which exists at any oil–water interface, will be described. In the literature, the IFT, γ_{AB}, between two liquids with γ_A and γ_B has been described in much detail (Adamson and Gast, 1997; Chattoraj and Birdi, 1984; Miqueu et al., 2011; Peng et al., 2011; Somasundaran, 2006). An empirical relation was suggested (Antonow's rule) by which one can predict the surface tension γ_{AB}:

$$\gamma_{AB} = \left| \gamma_{A(B)} - \gamma_{B(A)} \right| \tag{1.52}$$

The prediction of γ_{AB} from this rule is approximate but found to be useful in a large number of systems (such as alkanes: water), with some exceptions (such as water: butanol) (Tables 1.3 and 1.4). For example,

$$\begin{aligned}
&\gamma_{\text{water}} = 72 \text{ mN/m} \quad (\text{at } 25^\circ\text{C}),\\
&\gamma_{\text{hexadecane}} = 20 \text{ mN/m} \quad (\text{at } 25^\circ\text{C})\\
&\gamma_{\text{water-hexadecane}} = 72 - 20 = 52 \text{ mN/m} \quad (\text{measured} = 50 \text{ mN/m})
\end{aligned} \tag{1.53}$$

However, for general considerations, one may only use it as a reliable guideline and when exact data are not available. The Antonow rule can be understood in terms of a simple physical picture. There should be an adsorbed film or Gibbs monolayer of substance B (the one of lower surface tension) on the surface of liquid A. If we regard this film as having the properties of bulk liquid B, then $\gamma_{A(B)}$ is effectively the IFT of a duplex surface and would be equal to $[\gamma_{A(B)} + \gamma_{B(A)}]$.

Measurement of IFT (between two immiscible liquids):

IFT can be measured by different methods, depending on the characteristics of the system. Following methods can be applied:

Wilhelmy plate method.
Drop-weight method (can be also used for high pressure and temperature)
Drop shape method (can be also used for high pressure and temperature)

TABLE 1.4
Antonow's Rule and Interfacial Tension Data (mN/m) (See Text for Details)

Oil Phase	w(o)	o(w)	o/w	w(o)–o
Benzene	62	28	34	34
Chloroform	52	27	23	24
Ether	27	17	8	9
Toluene	64	28	36	36
n-Propylbenzene	68	29	39	40
n-Butylbenzene	69	29	41	40
Nitrobenzene	68	43	25	25
i-Pentanol	28	25	5	3
n-Heptanol	29	27	8	2
CS_2	72	52	41	20
Methylene iodide	72	51	46	22

TABLE 1.5
IFT between Water and Organic Liquids (20°C)

Water/Organic Liquid	IFT (mN/m)
n-Hexane	51.0
n-Octane	50.8
CS_2	48.0
CCl_4	45.1
$Br–C_6H_5$	38.1
C_6H_6	35.0
$NO_2–C_6H_5$	26.0
Ethyl ether	10.7
n-Decanol	10
n-Octanol	8.5
n-Hexanol	6.8
Aniline	5.9
n-Pentanol	4.4
Ethyl acetate	2.9
Isobutanol	2.1
n-Butanol	1.6

The Wilhelmy plate is placed at the surface of water, and oil phase is added until the latter covers the whole plate. The apparatus must be calibrated with known IFT data, such as water–hexadecane (52 mN/m; 25°C) (Table 1.5).

The drop-weight method is carried out by using a pump (or a syringe) to deliver liquid phase into the oil phase (or vice versa, as one finds suitable). In the case of water, water drops sink to the bottom of the oil phase. The weight of the drops is measured (by using an electrobalance), and IFT can be calculated. The accuracy can be very high by choosing the right kind of setup. The drop shape (pendant drop) is most convenient if small amounts of fluids are available and if extreme temperature and pressures are involved. Modern digital image analyses also make this method very easy to apply in extreme situations.

1.3.5 Thermodynamics of Liquid Surfaces (Corresponding States Theory of Liquids)

All natural phenomena are dependent on different parameters, such as temperature and pressure. In industry and research, one manipulates with large data of substances that could be systemized in order to predict and understand the system properties. It is thus important to be able to describe the interfacial forces of liquids as a function of temperature and pressure. This is most important in the case of oil recovery from reservoirs where oil is found at high temperatures (ca. 80°C) and pressures (ca. 200 atm). In the following analyses, it is important to notice that in some studies, pressure is constant and therefore not a general model of the equation of state. The magnitude of γ *decreases* almost linearly with temperature (**t**) (for most liquids) within a narrow range (Bahadori, 2011; Birdi, 2003a, 2009; Defay et al., 1966; Ghatee et al., 2010; Kirmse and Morgner, 2006; Kuespert et al., 1995):

$$\gamma_t = k_o(1 - k_1\mathbf{t}) \tag{1.54}$$

where k_o is a constant. It was found that coefficient k_1 is approximately equal to the rate of decrease of density, (ρ), of liquids with the rise of temperature:

$$\rho_t = \rho_o(1 - k_1\mathbf{t}) \tag{1.55}$$

where ρ_o is the value of density at $\mathbf{t}$ = 0°C, and values of constant k_1 were found to be different for different liquids. The magnitude of γ decreases with temperature and vanishes at the critical point (T_c and P_c). Experiments show that if one heats a liquid in a losed tuning, the liquid expands with higher T and P, until at the critical point, the phase boundary between the liquid and the gas disappears. At the critical point, the liquid and the gas have the same density. The following equation relates surface tension of a liquid to the density of liquid, ρ_l, and vapor, ρ_v (Birdi, 1989, 2010a,b; Partington, 1951):

$$\frac{\gamma}{(\rho_l - \rho_v)^4} = C \tag{1.56}$$

where the value of constant C is nonvariable only for organic liquids, while it is not constant for liquid metals. At the critical temperature, T_c, and critical pressure, P_c, a liquid and its vapor are identical, and the surface tension, γ, and total surface energy, like the energy of vaporization, must be zero (Birdi, 1997). At temperatures below the boiling point, which is $2/3T_c$, the total surface energy and the energy of evaporation are nearly constant. The variation in surface tension, γ, with temperature is given in Figure 1.21 for different liquids.

These data clearly show that the variation of γ with temperature is a very characteristic physical property. This observation becomes even more important when it is considered that the sensitivity of γ measurements can be as high as ±0.001 dyn/cm (=mN/m).

The change in γ with temperature in the case of mixtures would thus be dependent on the composition. For example, the variation of γ of the system: CH_4 + hexane is given as follows:

$$\gamma_{(CH_4 + \text{hexane})} = 0.64 + 17.85x_{\text{hexane}} \tag{1.57}$$

This relation shows that addition of hexane to CH_4 increases the magnitude of γ of the mixture. In fact, one can estimate the concentration of hexane in CH_4 by using Equation 1.57. This has much interest in oil reservoir engineering operations where one finds CH_4 in the crude oil. Experiments show that addition of a gas to a liquid will *always decrease* the value of γ of the mixture.

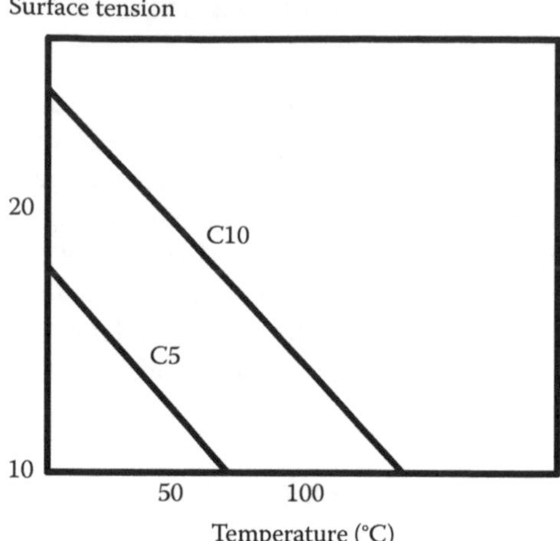

FIGURE 1.21 Variation of γ with temperature of different alkanes (C5: *n*-pentane; C10: *n*-decane).

It is well known that the corresponding states theory can provide much useful information about the thermodynamics and transport properties of fluids. For example, the most useful two-parameter empirical expression, which relates the surface tension, γ, to the critical temperature, T_c, is given as

$$\gamma = k_o \left(\frac{1 - T}{T_c} \right)^{k_1} \tag{1.58}$$

where k_o and k_1 are constants. The magnitude of constant k_1 has been reported as follows:

=3/2, although experiments indicated that $k_1 = 1.23$ or 11/9 (Birdi, 1997, 2003a). Data for some liquids have also reported the value of k_1 to be between 6/5 and 5/4.

It was found that constant k_o was proportional to $T_c^{1/3} P_c^{2/3}$. The relation in Equation 1.58 when fitted to the surface tension, γ, data of liquid CH_4, has been found to give the following correlation:

$$\gamma_{CH_4} = 40.52 \left(\frac{1 - T}{T_c} \right)^{1.287} = 40.52 \left(\frac{1 - T}{190.55} \right)^{1.287} \tag{1.59}$$

where $T_c = 190.55$ K. This equation has been found to fit the γ data for liquid methane from 91 to −190 K, with an accuracy of ±0.5 mN/m. In a recent study, the γ versus T data on *n*-alkanes, from *n*–pentane to *n*-hexadecane, were analyzed (Birdi, 1997, 2010a). The constants k_o (between 52 and 58) and k_1 (between 1.2 and 1.5) were found to be dependent on the number of carbon atoms, n_C, and since T_c is also found to be dependent on n_C (Birdi, 1997). The estimated values of different *n*-alkanes were found to agree with the measured data within a few percent: γ for *n*-$C_{18}H_{38}$, at 100°C, was 21.6 mN/m, from both measured and calculated values. This agreement shows that the surface tension data on *n*-alkanes fits the corresponding state equation very satisfactorily. It is worth mentioning that the equation for the data on γ versus T, for polar (and associating) molecules like water and alcohols, gives magnitudes of k_o and k_1 that are significantly different than those found for nonpolar molecules such as alkanes, etc.

In the following, calculated values of γ for different alkanes are given based upon the analyses using these equations (Birdi, 1997).

Comparison of calculated and measured values of surface tension (γ) (Birdi, 1997)

n-Alkane	Temperature (°C)	Measured	Calculated
C5	0	18.23	18.25
	50	12.91	12.8
C6	0	20.45	20.40
	60	14.31	14.3
C7	30	19.16	19.17
	80	14.31	14.26
C9	0	24.76	24.70
	50	19.97	20.05
	100	15.41	15.4
C14	10	27.47	27.4
	100	19.66	19.60
C16	50	24.90	24.90
C18	30	27.50	27.50
	100	21.58	21.60

The surface entropy (S_S) corresponding to Equation 1.61 is

$$\mathbf{S_S} = \frac{-d\gamma}{\mathbf{dT}} \tag{1.60}$$

$$= k_1 k_o \left(\frac{1-\mathbf{T}}{T_c}\right)^{k_1} - \frac{1}{T_c} \tag{1.61}$$

and the corresponding surface enthalpy, H_s, is

$$H_s = G_s - \mathbf{TS_s} = -\mathbf{T}\left(\frac{d\gamma}{\mathbf{dT}}\right) = k_o\left(\frac{1-\mathbf{T}}{T_c}\right)^{k_1-1}\left(\frac{1+(k_1-1)\mathbf{T}}{T_c}\right) \tag{1.62}$$

The reason heat is absorbed on the expansion of a surface is that the molecules must be transferred from the interior against the inward attractive force to form the new surface. In this process, the motion of the molecules is retarded by this inward attraction so that temperature of the surface layers is lower than that of the interior, unless heat is supplied from outside.

The following relationship relates γ to density (Birdi, 1997, 2003a):

$$\gamma\left(\frac{M}{\rho}\right)^{2/3} = k\left(T_c - \mathbf{T} - 6\right) \tag{1.63}$$

where

M is the molecular weight
ρ is the density (M/ρ = molar volume)

The quantity ($\gamma(M/\rho)^{2/3}$) is called the molecular surface energy. It is important to notice the correction term 6 on the right-hand side. This is the same as found for *n*-alkanes and *n*-alkenes in the estimation of T_c from γ versus temperature data (Birdi, 1997; Jasper, 1972). Surface tension and temperature relationships of alkanes are useful in oil industry.

1.4 SURFACTANTS (SOAPS AND DETERGENTS) AQUEOUS SOLUTIONS (SURFACE-ACTIVE SUBSTANCES)

All kinds of washing processes are probably one of the oldest known systems to mankind, which relate directly to surface chemistry. The physical (thermodynamic) property of any liquid will change when a substance (called *solute*) is dissolved in it. Of course, the change may be small or large, depending on the concentration and other parameters. Accordingly, the magnitude of surface tension of a liquid will change (increase or decrease) when a solute is dissolved in it. It also becomes apparent that if one could manipulate surface tension of water, then many applications areas would be affected drastically. There are some specific substances that are used to change (decrease) the surface tension of water in order to apply this characteristic property for some useful purpose in everyday life. The magnitude of surface tension change (e.g., that of water) will depend on the concentration and on the solute added. In some cases, γ of the solution increases (such as NaCl and other salts are added to water). The change in surface tension may be small (per mole of added solute) (as in the case of inorganic salts) or large (as in the case of such molecules as ethanol or other soap-like molecules) (Figure 1.22). The data for soap solutions are very unique, as explained later.

Surface tension of typical inorganic salt–water solutions (20°C):

Pure water	72.75 mN/m
NaCl (0.1 mol/L)	72.92 mN/m
NaCl (0.93 mol/L)	74.4 mN/m
NaCl (4.43 mol/L)	82.55 mN/m
KCl (0.93 mol/L)	74.15 mN/m
HCl (0.97 mol/L)	72.45 mN/m

KCl gives a larger increase in γ than HCl (per mole of concentration). This indicates that the degree of adsorption of KCl at the surface of water is larger than that of HCl. Further, it is found from experiments that the aqueous solutions of salts exhibit surface charges (Birdi, 2010a; Chattoraj and Birdi, 1984). This arises from the fact that the number of positive ions and negative ions is *not equal* (contrary to the bulk phase!). In other words, it is found that the *surface potential* of KCl solution is lesser than in the case of HCl. From this, it is concluded that K^+ and Cl^- ions adsorb at the surface almost to the same degree.

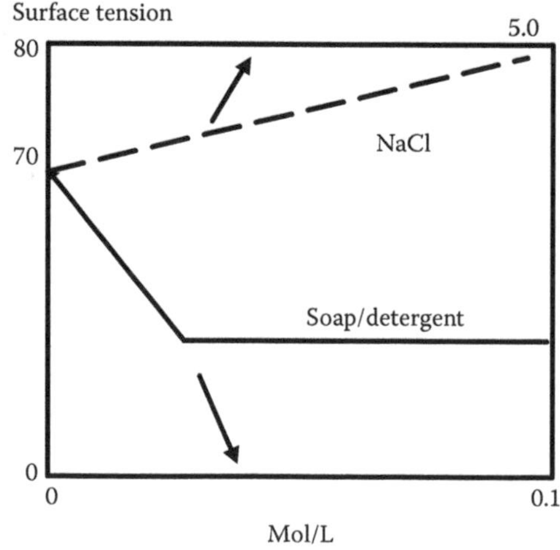

FIGURE 1.22 Change in surface tension of water on addition of inorganic salt (e.g., NaCl or KCl) or a soap.

It is thus seen that even very small differences in surface tension of these solutions indicate some significant characteristics that are unique for the system. These phenomena have very significant consequences in systems where these ions are present in connection with surfaces. One example is that in biological cells, the role of Na^+ and K^+ (and other cations) is significantly different (Birdi, 2010a,b, 2014).

Change in γ with the addition of solute (equal molar concentration):

Inorganic salt	*Minor change* (increase) in γ
Ethanol or similar	*Small change* (decrease) in γ
Soap (similar)	*Large change* (decrease) in γ

Some typical surface tension data of different solutions are given in the following:

Surface tension (mN/m)	72	50	40	30	22
Surfactant (mol/L)					
$C_{12}H_{25}SO_4Na$ (SDS)	0	0.0008	0.003	0.008	—
Ethanol (%)	0	10	20	40	100

This shows that to *reduce* the value of γ of water from 72 to 30 mN/m, one would need 0.005 mol of SDS or 40% ethanol. There are special substances, which are called *soaps* or *detergents* or *surfactants*, which exhibit unique physicochemical properties (Birdi, 2009, 2010a; Romsted, 2014; Rosen, 2004). The most significant structure of these molecules is due to the presence of a *hydrophobic* (alkyl group) and a *hydrophilic* (polar group such as –OH; $-CH_2CH_2O^-$; –COONa; $-SO_3Na$; $-SO_4Na$; $-CH_33N-$; etc.]).

The different polar groups are as follows:

Ionic groups

–COONa, SO_3Na, $-SO_4Na$ (negatively charged—*anionic*)

$-(N)(CH_3)_4Br$ (positively charged—*cationic*)

$-(N)(CH_3)_2-CH_2-COONa$ (amphoteric)

Nonionic groups

$-CH_2CH_2OCH_2CH_2OCH_2CH_2OH$

$-(CH_2CH_2OCH_2CH_2O)_x(CH_2CH_2CH_2O)_yOH$

Accordingly, one also calls these substances *amphiphile* (meaning: *two kinds*) (i.e., alkyl part and the polar group):

CCCCCCCCCCCCCCCCCCCC–**O**

Alkyl group (*CCCCCCC*–)–Polar group (–**O**)=

Amphiphile molecule

$(CH_3CH_2CH_2CH_2CH_2CH_2CH_2CH_2CH_2)$–Polar

For instance, surfactants dissolve in water and give rise to low surface tension (even at very low concentrations [few grams per liter or 1–100 mmol/L]) of the solution; therefore, these substances are called surface-active molecules (*surface-active agents* or substances). On the other hand, most inorganic salts increase the surface tension of water. All surfactant molecules are amphiphilic, which means these molecules exhibit hydrophilic and hydrophobic properties. Ethanol reduces surface tension of water, but one will need over few moles per liter to obtain the same reduction in γ as when using a few millimoles of surface-active agents.

As expected, if one adds ethanol (with γ of 22 mN/m) to water (with γ of 72 mN/m), then the magnitude of γ in the mixed solution will decrease. This is always the case in mixtures. Further, if one dissolves a gas in a liquid, then the magnitude of γ of the mixture will *always* decrease. In fact, one can estimate the amount of gas dissolved by measuring the decrease in γ in such mixtures.

However, the value of γ of surfactant solutions decreases to 30 mN/m with surfactant concentration around mmol/L (range of 1–10 g/L). Soaps have been used by mankind for many centuries. In biology, one finds a whole range of natural amphiphile molecules (bile salts, fatty acids, cholesterol and other related molecules, phospholipids). In fact, many important biological structures and functions are based on amphiphile molecules.

Moreover, one finds that many surfactants as found in nature (such as bile acids in the stomach) behave exactly the same way as the man-made surface-active agent. Proteins (which are large molecules [with molecular weights varying from six thousand to millions]) also decrease γ when dissolved in water (Chattoraj and Birdi, 1984; Tanford, 1980). It is found that aqueous protein solutions exhibit low surface tension values. The decrease in surface tension is related to the amino acid composition of the protein molecule (Chattoraj and Birdi, 1984; Tanford, 1980).

Soaps and surfactants are molecules, which are characterized as *amphiphiles*. Amphiphile is a Greek word, which means likes both kinds. A part of the amphiphile likes oil or hydrophobic (lipophilic = likes fat) (Tanford, 1980), while the other part likes water or hydrophilic (also called lipophobic). The balance between these two parts, hydrophilic–lipophilic, is called hydrophilic–lipophilic balance (HLB). The latter quantity can be estimated by experimental means, and theoretical analyses allow one to estimate its value (Adamson and Gast, 1997; Birdi, 2009). HLB values are applied in the emulsion industry (Birdi, 2009; Hansen, 2007).

Soap molecules are made by reacting fats with strong alkaline solutions (this process is called saponification). In water solution, the soap molecule, $C_nH_{2n+1}COONa$ (with n greater than 12–22), dissociates at high pH into $RCOO^-$ and Na^+ ions. It is found that the magnitude of n must be 12 or more for effective results.

A great variety of surfactants were synthesized from oil by-products, especially, $C_{12}H_{25}C_6H_4SO_3Na$, sodium dodecyl benzene sulfonates, were used in detergents. Later, these were replaced by sodium dodecyl sulfates (SDS) or sulfonates, because sodium dodecyl benzene was found to be not biodegradable (Rosen, 2004). This means that bacteria in the sewage plants were not able to degrade the alkyl group. The alkyl sulfonates were degraded to alkyl hydroxyls and alkyl aldehydes and later to CO_2, etc.

In many applications, one found it necessary to employ surfactants that were nonionic. For example, nonionic detergents as used in washing clothes are much different in structure and properties than those used in dish-washing machines. In washing machines, foam is crucial as it helps in keeping the dirt away from clothes once it has been removed. On the other hand, in machine dish-washing, one does not need any foam. This arises from the fact that foam will hinder mechanical effect of dish-washing process.

1.4.1 Surface Tension Properties of Aqueous Surfactant Solutions

Experiments show that the surface tension (γ) of any pure liquid (water or organic liquid) changes when another substance (solute) is dissolved. The change (increase or decrease) in γ depends on the characteristics of the solute added. The surface tension of water increases (in general) when inorganic salts (such as NaCl, KCl, and Na_2SO_4) are added, while its value decreases when organic substances are dissolved (ethanol, methanol, fatty acids, soaps, and detergents).

The surface tension of water increases from 72 to 73 mN/m when 1 M NaCl is added. On the other hand, the magnitude of surface tension decreases from 72 to 39 mN/m when only 0.008 M (0.008 M × 288 = 2.3 g/L) SDS (molecular weight of SDS = 288) is dissolved. It thus becomes obvious that in all those systems where surface tension plays an important role, the additives will play an important role in these systems.

The magnitude of γ changes slowly in the case of methanol as compared to detergent solutions. The methanol–water mixtures gave following γ data (20°C):

wt% methanol	0	10	25	50	80	90	100
γ (mN/m)	72	59	46	35	27	25	22.7

It is important to have an understanding about the change in the surface tension, γ, of water as a function of molecular structure of solute. The surface tension data in the case of homologous series of alcohols and acids show some simple relation to the alkyl chain length. It is found that each addition of $-CH_2-$ group gives a value of concentration and surface tension such that the value of concentration is lower by about *factor 3*. However, it must be mentioned that such dependence in the case of nonlinear alkyl chains will be different. The effective $-CH_2-$ increase in the case of nonlinear chain will be lesser (ca. 50%) than in the case of a linear alkyl chain. The tertiary $-CH_2-$ group effect would be even lesser. In general, though, one will expect that the change in γ per mole substance will increase with any increase in the hydrocarbon group of the amphiphile. The effect of chain length on surface tension arises from the fact that as the hydrophobicity increases with each $-CH_2-$ group, the amphiphile molecule adsorbs more at the surface. This will thus be a general trend also in more complicated molecules, such as in proteins and other polymers.

In the case of proteins, the amphiphilic property arises from the different kinds of amino acids (25 different amino acids) (Chattoraj and Birdi, 1984; Tanford, 1980). Some amino acids have lipophilic groups (such as phenylalanine, valine, leucine, etc.), while others have hydrophilic groups (such as glycine, aspartic acid, etc.).

In fact, one finds from surface tension measurements that some proteins are considerably more hydrophobic (such as hemoglobin) than others (such as bovine serum albumin, ovalbumin). These properties of proteins have been extensively investigated, and these data have been found to be related to biological functions (Birdi, 1999; Chattoraj and Birdi, 1984; Tanford, 1980).

1.4.2 Surface-Active Substances (*Amphiphiles*)

All molecules that when dissolved in water reduce (significantly) surface tension are called SASs (such as soaps, surfactants, detergents). The adsorption of the SAS at the liquid surface gives rise to large reduction in surface tension. This also indicates that the concentration of SAS in the surface is higher than in the bulk. The same will happen if one added SAS to a system of oil–water system. The IFT of oil–water interface will be reduced accordingly. Inorganic salts, on the other hand, increase (in general) the surface tension of water.

Surfactants exhibit surface activity at different kinds of interfaces:

Air–water
Oil–water
Solid–water

The magnitude of surface tension is reduced, since the hydrophobic (alkyl chain or group) is energetically more attracted to the surface than being surrounded by water molecules inside the bulk aqueous phase. Figure 1.23 shows the monolayer formation of the SAS at high bulk concentration. Since the close-packed SAS at the surface looks like alkane, it would be thus expected that the

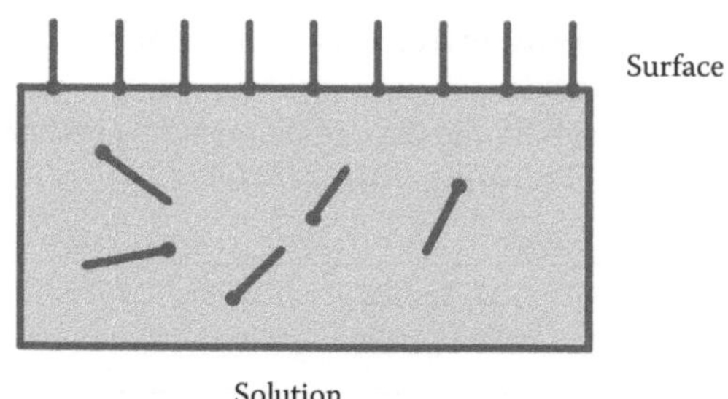

FIGURE 1.23 Orientation of soap (SAS) at the surface of water (alkyl group: ▬, polar group: •).

surface tension of SAS solution would decrease from 72 mN/m (surface tension of pure water) to alkane-like surface tension (close to 25 mN/m).

The orientation of surface molecule at the interface will be dependent on the system. This is shown as follows:

Air–water: Polar part toward water and hydrocarbon part toward air
Oil–water: Polar part toward water and hydrocarbon part toward oil
Solid–water: Polar part toward water and hydrocarbon part toward solid (in general)

1.4.3 Aqueous Solution of Surfactants

The solution properties of the various surfactants in water are very unique and complex in many aspects, as compared to solutes as NaCl or ethanol. This of course is related to the dual nature of the surfactant molecules (one part is hydrophobic [alkyl group] and the other part is polar) (Birdi, 2003a; Rosen 2004; Tanford, 1980). The solubility of charged and noncharged surfactants is very different, especially as regards the effect of temperature and added salts (such as NaCl). These characteristics are important when one needs to apply these substances in diverse systems. For instance, one cannot use the same soap molecule at seas as on the land, the main reason being that higher concentrations of salts (such as Ca^{2+} and Mg^{2+}) as found in sea water affect the foaming and solubility characteristics of major SASs. For similar reasons, one cannot use a nonionic detergent for shampoos (only anionic detergents are used). Therefore, tailor-made surface-active agents have been devised by the industry to meet these specific demands. In fact, the whole soap industry develops detergents designed for each specific system.

1.4.4 Solubility Characteristics of Surfactants in Water

The solubility characteristics (especially in water) of any substance are very important kind of information, which one must investigate. In the present case, one must have the precise information about the solubility and temperature characteristics of the surfactant in water (Birdi, 2009; Rosen, 2004; Tanford, 1980). Even though the molecular structures of surfactants are rather simple, one finds that the solubility in water is rather complex as compared with other amphiphiles, such as long chain alcohols, etc. The solubility in water will be dependent on the alkyl group and as well as on the polar group. This is easily seen from the fact that the alkyl groups will behave mostly as alkanes. However, it is also found that the solubility of surfactants is also dependent on the presence (or absence) of charge on the polar group. Ionic surfactants exhibit different solubility characteristics than nonionic surfactants, with regard to dependence on the temperature. In-fact, in all industrial applications of SAS, the solubility parameter is one of the most important criteria. This characteristic is the determining factor about which SAS to be used in a given system.

1.4.4.1 Ionic Surfactants

The solubility of all ionic surfactants (both anionics [i.e., negatively charged] and cationics [positively charged]) is low at low temperature, but at a specific temperature, the solubility suddenly increase, Figure 1.24. For instance, the solubility of SDS at 15°C is about 2 g/L. This temperature is called Krafft point (KP). KP (or temperature) can be obtained by cooling an anionic surfactant solution (ca. 0.5 M) from a high to a lower temperature until cloudiness appears sharply. The KP is not very sharp in the case of impure surfactants as generally found in industrial chemicals.

It is found that the magnitude of solubility near the KP is almost equal to the critical micelle concentration (CMC). The magnitude of KP is dependent on the chain length of the alkyl chain for a homologous series of nonionic detergents, Figure 1.25.

The linear dependence of KP on the alkyl (linear) chain length is very clear. KP for C12 sulfate is 21°C, and it is 34°C for C14 sulfate.

It may be concluded that KP increases by approximately 10°C per CH_2 group. Since no *micelles* can be formed below the KP, it is important that one keeps this information in mind when using any

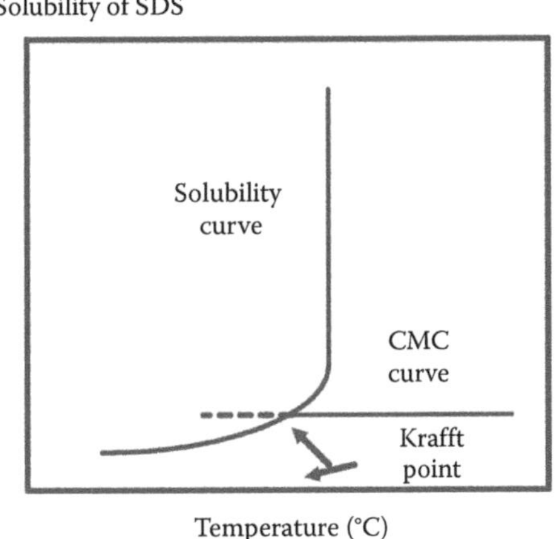

FIGURE 1.24 Solubility (KP) of ionic (anionic or cationic) surfactants in water (as a function of temperature).

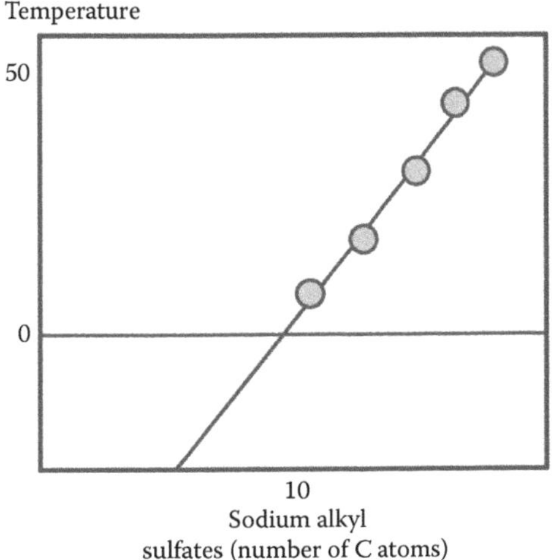

FIGURE 1.25 Variation of KP with chain length of sodium alkyl sulfates.

anionic detergent. Therefore, the effect of various parameters on the KP needs to be considered in the case of ionic surfactants. Some of these are given as follows:

Alkyl chain length (KP increases with alkyl chain length).
KP decreases if lower chain surfactant is mixed with a longer chain surfactant.

1.4.4.2 Nonionic Surfactants

The solubility of nonionic surfactants in water is completely different than those of charged surfactants (especially as regards the effect of temperature). The solubility of nonionic surfactant is high at low temperature, but it decreases abruptly at a specific temperature, called the cloud point (CP) (Figure 1.26). This means that nonionic detergents will not be suitable if used *above* the (CP) temperature. The solubility of such detergent molecules in water arises from the hydrogen bond formation between hydroxyl (–OH) and ethoxy groups (–CH_2CH_2O–) and water molecules. At high temperatures, the degree of hydrogen bonding gets weaker (due to high molecular vibrations) and thus the nonionic detergents become insoluble, at the *t*CP. The name CP is at the temperatures when the solution becomes cloudy (because a new phase with surfactant-rich concentration is formed). Nonionics thus exhibit opposite solubility properties than ionic surfactants, as regards the effect of temperature. The solution separates into two phases: one with rich water phase and another phase with high concentration of nonionic surfactant. The rich nonionic detergent phase is found to consist of low water content. Experiments have shown that there are roughly four molecules of bound water per ethylene oxide group (–CH_2CH_2O–) (Birdi, 2009) in the rich nonionic concentration phase.

It thus must be noted that when a surfactant is needed for any application one must consider the solubility characteristics, besides other properties, which should conform to the experimental conditions. Their area of application characterizes thus the surfactants, which are available in the industry. In fact, the detergent manufacturer is in constant collaboration with washing industry and tailor-made surfactants are commonly developed in collaboration. In detergent industry one, has a whole spectra of molecules, which can be manipulated, depending on the application area.

Anionic surfactants are used in different areas while cationic surfactants are used in completely different systems. For instance, anionic surfactants are used for shampoos and washing, while cationics are used for hair conditioners. Hair has negative (–) charged surface and thus cationic

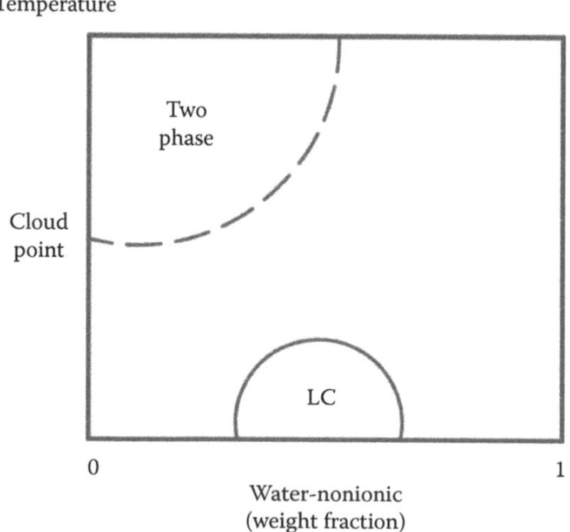

FIGURE 1.26 Solubility of a nonionic surfactant in water (CP) (schematic) (dependent on temperature).

surfactants strongly adsorb at the surface and leave a smooth surface. These detergents are used in the so-called *hair conditioner* formulations. This is because the (positive) charged end is oriented toward the hair (negative) surface and the alkyl group is pointing away (as depicted in the following).

Cationic detergent (positive charge) + hair (negative charge):

Alkyl group (cationic detergent)–polar group/(+)hair(−)

This conditioner process is based on the interaction of a positively charged molecule (cationic detergent) with a negatively charged substance (hair). This imparts hydrophobicity to hair and feels soft and looks fluffy. Fabric softener is based on the same principle (cotton is generally negatively charged).

1.4.5 Micelle Formation of Surfactants (in Aqueous Media) (Critical Micelle Concentration)

The solution properties of ordinary salts, such as NaCl, in water are rather simple. One can dissolve at a given temperature a specific amount of NaCl, giving a saturated solution (approximately 5 mol/L). Similarly, the solution characteristics of methanol or ethanol in water are also simple and straightforward. These alcohols mix with water in all proportions. However, the solution behavior of surfactant molecules in water is much more complex. Besides the effect on surface tension, the solution behavior is found to be dependent on the charge of the surfactant. Surfactant aqueous solutions manifest two major forces that determine the solution behavior. The alkyl part being *hydrophobic* would tend to separate out as a distinct phase, while the *polar part* tends to stay in solution. The difference between these two opposing forces thus determines the solution properties. The factors that one has to consider are the following:

1. The alky group and water
2. The interaction of the alkyl hydrocarbon groups with themselves
3. The solvation (through hydrogen bonding and hydration with water) of the polar groups
4. Interactions between the solvated polar groups

According to the principles of physical chemistry science, the micelle formation is a very intriguing system. One finds a great number of publication in the literature that describes the accurate state of phases in such system (Bai and Lodge, 2010; Birdi, 2003a, 2010a; Birdi and Ben-Naim, 1981; Chen and Ruckenstein, 2014; Danov et al., 2014; Golub and de Keizer, 2004; Raju et al., 2001; Rathman and Scamehorn, 1988; Tanford, 1980; Zana, 1995). Below CMC, detergent molecules are present as single monomers. Above CMC, one will have that monomers, C_{mono}, are in equilibrium with micelles, C_{mice}. The physical chemistry of such an equilibrium is found to be of great interest in the literature. The micelle with aggregation number, N_{ag}, is formed from monomers:

$$N_{ag} \text{ monomer} = \text{Micelle} \tag{1.64}$$

N_{ag} monomers which were surrounded by water aggregate together, above CMC, and form a micelle. In this process, the alkyl chains have transferred from water phase to a alkane-like micelle interior. This occurs because the alkyl part is at a lower energy in micelle than in the water phase. The aggregation process is stepwise (i.e., monomer to dimer to trimer to tetramer and so on). This is based on the fact that some surfactants, such as cholates, form micelles with few aggregation numbers (between 5 and 20), while SDS can form micelles varying from 100 to 1000 aggregation numbers. The micelle formation takes place when the alkyl chain of the molecule as surrounded by water (above the CMC) is transferred to a micelle phase (where alkyl chain is in contact with neighboring alkyl chains). Thus, in the latter case, the repulsion between alkyl chain and water has been removed. Instead, the alkyl chain–alkyl chain attraction (van der Waals forces) is the driving force for the micelle formation. The surfactant molecule forms a micellar aggregate at a concentration higher than CMC, because it moves from water phase to micelle phase (lower energy). The micelle reaches an equilibrium after a certain number of monomers have formed a micelle. This means that there are both *attractive* and *opposing* forces involved in this process. Otherwise, one would expect very large aggregates if there was only attractive force involved. This would mean phase separation, that is, two phases: one water-rich phase and another surfactant-rich phase. Thus, aggregation is a specific property where instead of phase separation molecules are able to form small aggregates, micelles (not visible to naked eye), and very stable micellar solutions.

Thus, one can write the standard free energy of a micelle formation, ΔG^{o}_{mice}, as follows:

$$\Delta G^{o}_{mice} = \text{attractive forces} + \text{opposing forces} \tag{1.65}$$

If there are only attractive forces present (as in alkanes), then one will not observe any significant solubility in water. The attractive forces are associated with the hydrophobic interactions between the alkyl part (alkyl–alkyl chain attraction) of the surfactant molecule, $\Delta G^{o}_{hydrophobic}$. The opposing forces arise from the polar part (charge–charge repulsion, polar group–hydration), ΔG^{o}_{polar}. *These forces are of opposite signs.* The attractive forces would lead to larger aggregates. The opposing forces would hinder the aggregation. A micelle with a definite aggregation number is where the value of ΔG^{o}_{mice} is zero (Figure 1.27). Hence, we can write for ΔG^{o}_{mice}:

$$\Delta G^{o}_{mice} = \Delta G^{o}_{hydrophobic} + \Delta G^{o}_{polar} \tag{1.66}$$

The standard free energy of micelle formation is as follows:

$$\Delta G^{o}_{mice} = \mu^{o}_{mice} - \mu^{o}_{mono} = RT\ln\left(\frac{C_{mice}}{C_{mono}}\right) \tag{1.67}$$

At CMC, one may neglect C_{mice}, which leads to

$$\Delta G^{o}_{mice} \approx RT\ln(CMC) \tag{1.68}$$

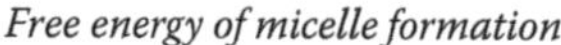

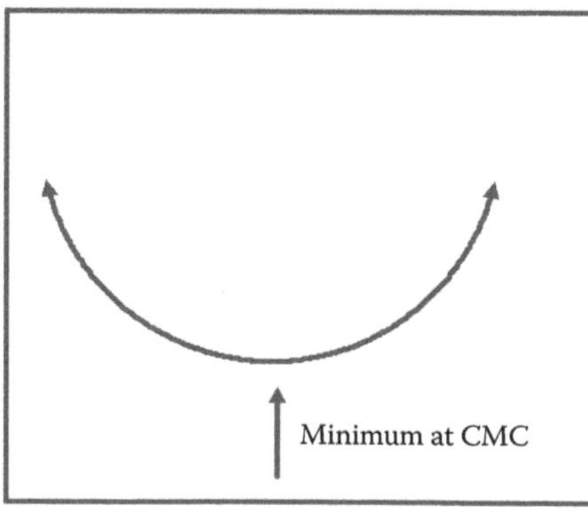

FIGURE 1.27 Attraction forces between alkyl chains and repulsion forces between polar groups give a minimum energy in the system at CMC.

This relation is valid for nonionic surfactants. In the case of ionic surfactants, one needs to modify this equation. This equilibrium shows that if we dilute the system, then micelles will break down to monomers to achieve equilibrium. This is a simple equilibrium for a nonionic surfactant. In the case of ionic surfactants, there will be different charged species present in the solution. In the case of ionic surfactant, such as SDS (with sodium S^+, and alkyl sulfate, SD^-, ions), the micelle with aggregation number, N_{SD^-}, will consist of counter-ions (sodium ions), C_{S^+}:

$$N_{SD^-} \text{ ionic surfactant monomers} + C_{S^+} \text{ counter} = \text{micelle with charge} \left(N_{SD^-} - C_{S^+}\right) \tag{1.69}$$

Since N_{SD^-} is found to be larger than C_{S^+}, all anionic surfactants are negatively charged. Similarly, cationic micelles will be positively charged. For instance, CTAB, we have following equilibria in micellar solutions:

CTAB dissociates into CTA^+ and Br^- ions.

The micelle with N_{CTA^+} monomers will have C_{Br^-} counter-ions. The positive charge of the micelle will be the sum of positive and negative ions $(N_{CTA^+} - C_{Br^-})$. The actual concentration will vary for each species with the total detergent concentration (e.g., SDS, Figure 1.28). The change in surface tension, γ, versus concentration is given in Figure 1.29b.

The surface tension curve, Figure 1.29b, is typical for all kinds of detergents. Below CMC, the SDS molecules in water are found to dissociate into SD^- and Na^+ ions. Conductivity measurements show the following data:

1. SDS behaves as a strong salt, and SD^- and Na^+ ions are formed (same as one observes for NaCl).
2. A break in the plot is observed at SDS concentration equal to the CMC. This clearly shows that the number of ions decreases with concentration. The latter indicates that some ions (in the present case cations, Na^+) are partially bound to the SDS-micelles, which results in change in the slope of the conductivity of the solution. Same behavior is observed for other ionic detergents, such as cationic (CTAB) surfactants. The change in surface tension also shows a break at CMC.

At CMC, micelles (aggregates of SD^- with some counter-ions, Na^+) are formed and some Na^+ ions are bound to these, which is also observed from conductivity data. In fact, an analysis of these data has shown that approximately 70% Na^+ ions are bound to SD^- ions in the micelle. The surface

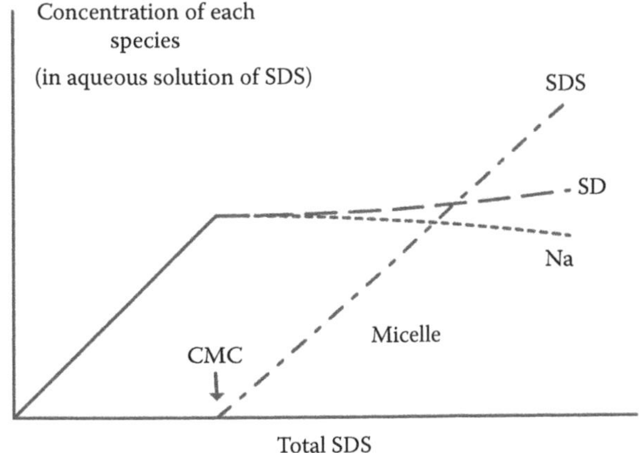

FIGURE 1.28 Variation of concentration of different ionic species for SDS solutions (Na^+, SD^-, $SDS_{micelle}$) and change in surface tension of a detergent solution with concentration.

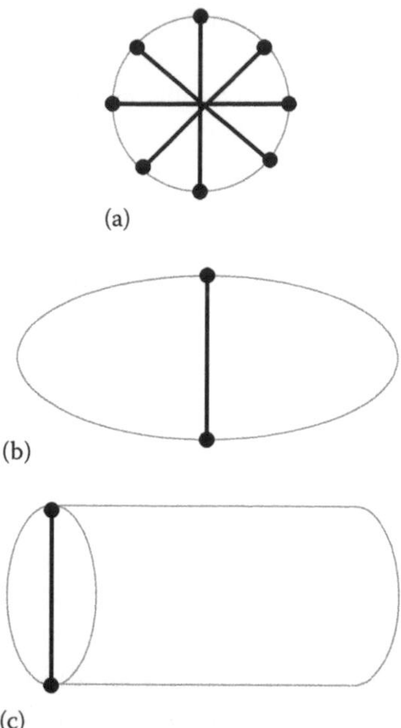

FIGURE 1.29 Different types of micellar aggregates: (a) spherical, (b) disc like, and (c) cylindrical (schematic).

charge (negative charge) was estimated from conductivity measurements (Birdi, 2003a). Therefore, the concentration of Na^+ will be higher than SD^- ions after CMC. The same is true in the case of cationic surfactants. In the case of CTAB solutions, one thus has CTA^+ and Br^- ions below CMC. Above CMC, there are additionally $CTAB^+$ micelles. In these systems, the counter-ion is Br^-. It is important to notice that due to these differences the two systems are completely divergent from each other, as regards the areas of application (such as one cannot use CTAB for washing clothes!).

1.4.5.1 Analyses of CMC of Surfactants

The quantity of CMC is related to the free energy of the system. Experiments show that the magnitude of CMC is dependent on both the alkyl part and the polar part (Mukerjee and Mysels, 1971). It has been found that CMC decreases with increasing alkyl chain length. Highly pure sodium alkyl sulfates were studied as regards CMC (Birdi et al., 1980). This indicates that as the solubility in water of the alkyl part decreases, CMC also decreases. In linear chain, Na-alkyl sulfate detergents following simple relationship has been found:

$$\ln(\mathrm{CMC}) = k_1 - k_2(C_{\mathrm{alkyl}}) \tag{1.70}$$

where

k_1 and k_2 are constants

C_{alkyl} is the number of carbon atoms in the alkyl chain

The magnitude of CMC will change if the additive has an effect on the monomer–micelle equilibrium. It will also change if the additive changes the detergent solubility. The CMC of all ionic surfactants will decrease if co-ions are added. However, nonionic surfactants show very little change in CMC on the addition of salts. This is as one should expect from theoretical considerations. It is important to notice how merely the addition of 0.01 mol/L of NaCl changes the CMC by 65%. This shows that the charge–charge repulsion is very significant and is reduced appreciably by the addition of counter-ions (in this case Na^+).

The change of CMC with NaCl for SDS is as follows (at 25°C):

NaCl (mol/L)	CMC (mol/L)	g/L	N_{ag}
0	0.008	2.3	80
0.01	0.005	1.5	90
0.03	0.003	0.09	100
0.05	0.0023	0.08	104
0.1	0.0015	0.05	110
0.2	0.001	0.02	120
0.4	0.0006	0.015	125

The radius of the spherical micelle is reported as 20 Å, which increases to 23 Å (for the nonspherical).

Experiments have shown that in most cases, such as for SDS, the initial spherical-shaped micelles may grow under some influence into larger aggregates (disk-like, cylindrical, lamellar vesicle) (Figure 1.29). The spherical micelle has a radius of 17 Å. The extended length of the SDS molecule is about 17 Å. However, larger micelles (as found in 0.6 mol/L NaCl solution) have dimensions of 17 and 25 Å, radii of an ellipse.

It is important to notice how CMC changes with even very small addition of NaCl (Birdi et al., 1980; Mukerjee and Mysels, 1971). It has been found in general the change in CMC (in the case of ionic surfactants only) with the addition of ions follows the relation:

$$\ln(\text{CMC}) = \text{Constant}_1 - \text{Constant}_2(\ln(\text{CMC} + C_{ion})) \quad (1.71)$$

The magnitude of Constant_2 was found to be related to the *degree of micelle charge*. Its magnitude varied from 0.6 to 0.7, which means that micelles have 30% (effective) charge. This indicates that if there are 100 monomers per micelle, then ca. 70% counter-ions are bound.

Data were reported for the CMC of cationic surfactants that decreased on the addition of KBr as follows:

Dodecyltrimethylammonium bromide (DTAB) and TrTAB:

$$\ln(\text{CMC}) = -6.85 - 0.64 \ln(\text{CMC} + C_{KBr})$$

TrTAB:

$$\ln(\text{CMC}) = -8.10 - 0.65 \ln(\text{CMC} + C_{KBr})$$

TTAB:

$$\ln(\text{CMC}) = -9.43 - 0.68 \ln(\text{CMC} + C_{KBr})$$

It is noticed that the magnitude of Constant_1 increases with the increase in alkyl chain length. Similar relationship has been reported for Na-alkyl sulfate homologous series (Birdi et al., 1980).

The alkyl chain length has a very significant effect (decrease with increase in n_C) on the CMC. The CMC data for soaps are found to give the following dependence on the alkyl chain length:

Soap	CMC (mol/L) 25°C
C7COOK	0.4
C9COOK	0.1
C11OOK	0.025
C7COOCs	0.4

These data show that CMC decreases by a factor of *4 for each increase in chain length* by $-CH_2CH_2-$. This again indicates that due to lower solubility in water with increasing chain length, CMC is related to the latter characteristic of the molecule. Further, this effect will be valid for all kinds of detergent molecules (both with and without charges) (Mukerjee and Mysels, 1971; Tanford, 1980).

1.4.6 Gibbs Adsorption Equation in Solutions

All liquids (in pure state) when shaken do not form any foam. This merely indicates that the surface layer consists of pure liquid. However, if one adds a very small amount of surface-active agent (soap or detergent [ca. milli-mole concentration: or about ppm by weight]) then if one shakes the aqueous solution there is formed foam at the surface of the solution. This indicates that the surface active-agent has *accumulated* at the surface (i.e., the concentration of surface-active agent is much higher at surface than in the bulk phase: in some cases many thousand times) and thus forms a TLF that constitutes the bubble. In fact, one can use the bubble or foam formation as a useful criterion as regards the purity of the water system. One generally observes at the shores of lakes or ocean that foam bubbles are formed under different conditions. If water in these sites is polluted with SASs, then very stable foams are observed. However, one also finds that there are naturally formed SASs that also contribute to foaming. It must be mentioned that if one adds instead an inorganic salt, NaCl, then no foam is formed. The foam formation indicates that the surface-active agent adsorbs at the surface, and forms a *TLF* (consisting of two layers of amphiphile molecule and with some water).

Analyses of TLFs:

One layer of (SAS)	SSSSSSSSSSSSSSSS
Water (W) molecules in between	wwwwwwwwwwwww
One layer of SAS (S)	SSSSSSSSSSSSSSSS

This may be compared to a *sandwich* kind of structure. It is important to mention that these bubble structures are of nanometer dimensions, but still easily visible to the naked eye. It also indicates the self-assembly properties of such SASs.

The thermodynamics of surface adsorption has been extensively described by Gibbs adsorption theory (Chattoraj and Birdi, 1984; Spaull, 2004; Zhou, 1989). Further, Gibbs adsorption theory has also been applied in the analyses of diverse phenomena (such as solid–liquid or $liquid_1$–$liquid_2$, adsorption of solute on polymers, etc.). In fact, in any system where adsorption takes place at an interface the Gibbs theory will be applicable (such as solid–liquid; protein molecule–solution with solutes that may adsorb).

1.4.7 Gibbs Adsorption Theory at Liquid Interfaces

The magnitude of surface tension of water is sensitive to the addition of different molecules. The surface tension, γ, of water changes with the addition of organic or inorganic solutes, at constant temperature and pressure (Birdi, 1989, 1997; Chattoraj and Birdi, 1984; Defay et al., 1966; Jungwirth and Tobias, 2006; Lu et al., 2000). The extent of surface tension change and the sign of change is determined by the molecules involved (see Figure 3.1). The magnitude of γ of aqueous solutions generally *increases* with different electrolyte concentrations. The magnitude of γ of aqueous solutions containing organic solutes invariably *decreases*. As mentioned earlier, the surface of a liquid is where the density of liquid changes to that of a gas, by a factor 1000, Figure 1.1. As an example, now let us look what happens to surface composition when ethanol is added to water, Figure 1.30.

The reason ethanol concentration in vapor phase is higher than in water is due to its lower boiling point. Next, let us consider the situation when a detergent is added to water whereby the surface tension is lowered appreciably, Figure 1.31.

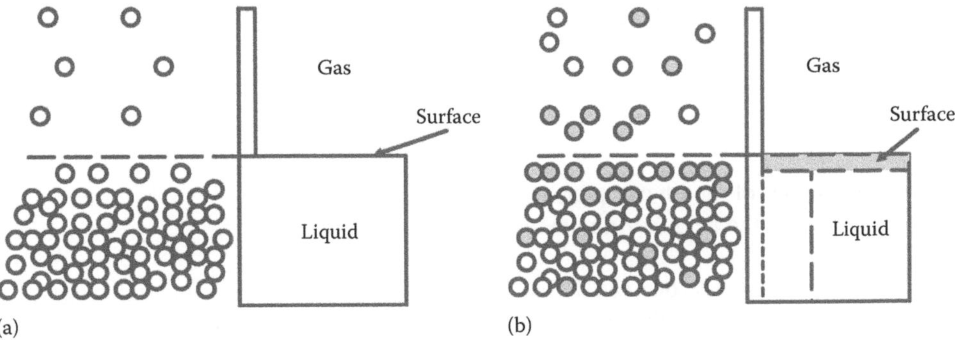

FIGURE 1.30 Surface composition (a) of pure water and (b) of an ethanol–water solution (shaded = ethanol).

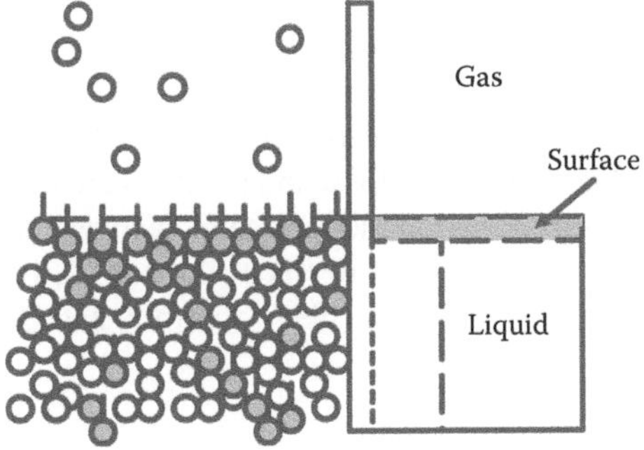

FIGURE 1.31 Concentration of detergent (shaded with tail) in solution and at surface. The shaded area at surface is the excess concentration due to accumulation.

Change in the surface tension of water on the addition of different solutes:

Inorganic salts = increase in γ
Organic substances (such as ethanol) = decrease in γ
Soaps or detergents = appreciable decrease in γ

The schematic concentration profile of detergent molecules is such that the concentration is homogenous up to the surface. At the surface, there is almost only detergent molecule plus the necessary number of water molecules (which are in a bound state to the detergent molecule). The solution thus shows very low surface tension, ca. 30 mN/m. The surface concentration profile of detergent is not easily determined by any direct method. Here it is shown as a rectangle for convenience, but one may also imagine other forms of profiles, such as curved. One observes that surface tension decreases due to ethanol. This suggests that there are more ethanol molecules at the surface than in the bulk. This is also seen in a cognac glass. The ethanol vapors are observed to condense on the edge of the glass. This shows that the concentration of ethanol in the surface of the solution is very high. To analyze these data, one has to use the well-known *Gibbs* adsorption equation (Birdi, 1989, 2009; Chattoraj and Birdi, 1984).

1.4.7.1 Gibbs Adsorption Equation

The Gibbs *adsorption equation* has been a subject of many investigations in the literature (Chattoraj and Birdi, 1984; Fainerman et al., 2002). Gibbs considered that the interfacial region is inhomogeneous and difficult to define, and he therefore also considered a more simplified case in which the interfacial region is assumed to be a mathematical plane.

Gibbs defined a quantity, surface excess Γ_{n_i} of the ith component as follows:

$$n_i^x = n_i^t - n_{i\alpha} - n_{i\beta} \tag{1.72}$$

where
n_i^t is the total moles of the ith component
$n_{i\alpha}$ and $n_{i\beta}$ moles in the neighboring phases, in the real system

In an exactly similar manner, one can define the respective surface excess internal energy, E_x, and entropy, S_x by the following mathematical relationships (Birdi, 1989; Chattoraj and Birdi, 1984):

$$E^x = E^t - E^\alpha - E^\beta \tag{1.73}$$

$$S^x = S^t - S^\alpha - S \tag{1.74}$$

Here E^t and S^t are the total energy and entropy, respectively, of the system as a whole for the actual liquid system. The energy and entropy terms for α and ß phases are denoted by the respective superscripts. The excess (x) quantities thus refer to the surface molecules in an adsorbed state. At constant **T** and **p**, for a two-component system (say water(1) + alcohol(2)), the classical Gibbs adsorption equation has been derived (Adamson and Gast, 1997; Chattoraj and Birdi, 1984):

$$\Gamma_2 = -\left(\frac{d\gamma}{d\mu_2}\right)_{T,p} \tag{1.75}$$

The chemical potential μ_2 is related to the activity of alcohol by the equation

$$\mu_2 = \mu_2^o + \mathbf{RT}\ln(a_2) \tag{1.76}$$

If the activity coefficient can be assumed to be equal to unity, then

$$\mu_2 = \mu_2^o + \mathbf{RT}\ln(C_2) \tag{1.77}$$

where C_2 is the bulk concentration of solute 2.

The Gibbs adsorption then can be written as follows:

$$\Gamma_2 = \frac{-1}{\mathbf{RT}}\left(\frac{d\gamma}{d\ln(C_2)}\right) = \frac{-C_2}{\mathbf{RT}}\left(\frac{d\gamma}{dC_2}\right) \tag{1.78}$$

This shows that the *surface excess* quantity on the left-hand side is proportional to the change in surface tension with the concentration of the solute ($d\gamma/d\ln(C_{surfaceactivesubstance})$). A plot of $\ln(C_2)$ versus γ gives a slope equal to

$$\Gamma_2(\mathbf{RT})$$

From this, one can estimate the value of Γ_2 (mol/area). This indicates that all SASs, will always have a higher concentration at the surface than in the bulk of the solution. This relation has been verified by using radioactive tracers. Further, as will be shown later under spread monolayers, one finds a very convincing support to this relation and the magnitudes of Γ for various systems. The surface tension of water (72 mN/m, at 25°C) decreases to 63 mN/m in a solution of SDS of concentration 1.7 mmol/L. The large decrease in surface tension suggests that SDS molecules are concentrated at the surface, as otherwise there should be very little change in surface tension. This means that the concentration of SDS at the surface is much higher than in the bulk. The molar ratio of SDS:water in the bulk is 0.002:55.5. At the surface, the ratio will be expected to be of a completely different value, as found from the value of Γ (ratio is 1000:1). This is also obvious when considering that foam bubbles form on solutions with very low surface-active agent concentrations. In fact, it is easy to consider the state of surfactant solutions in terms of molecular ratios.

Example: SDS Aqueous Solution

SDS is a strong electrolyte and in water it dissociates almost completely (at concentrations below CMC):

$$\begin{aligned} &C_{12}H_{25}SO_4Na \equiv C_{12}H_{25}SO_4^- + Na^+ \\ &SDS \equiv DS^- + S^+ \end{aligned} \quad (1.79)$$

The appropriate form of the Gibbs equation is as follows:

$$-d\gamma = \Gamma_{DS} - d\mu_{DS^-} + \Gamma_{S^+} d\mu_{S^+} \quad (1.80)$$

where surface excess, Γ_i, terms for each species in the solution, for example, DS^- and S^+, are included.

This equation can be simplified, assuming electrical neutrality is maintained in the interface:

$$\Gamma_{SDS} = \Gamma_{DS^-} = \Gamma_{S^+} \quad (1.81)$$

and

$$C_{SDS} = C_{DS^-} = C_{S^+} \quad (1.82)$$

that on substitution in Equation 1.80 gives

$$-d\gamma = 2\mathbf{RT}\Gamma_{SDS} d(\ln C_{SDS}) \quad (1.83)$$

$$= \left(\frac{2\mathbf{RT}}{C_{SDS}}\right)\Gamma_{SDS} dC_{SDS} \quad (1.84)$$

and one obtains

$$\Gamma_{SDS} = \frac{-1}{2(\mathbf{RT})}\left(\frac{d\gamma}{(d\ln C_{SDS})}\right) \quad (1.85)$$

In the case where the ion strength is kept constant, that is in the presence of added NaCl, then the equation becomes

$$\Gamma_{SDS} = \frac{-1}{(\mathbf{RT})}\left(\frac{d\gamma}{(d\ln C_{SDS})}\right) \quad (1.86)$$

Comparing Equation 1.85 with 1.86, it will be seen that they differ by a factor of 2 and that the appropriate form will need to be used in the experimental test of Gibbs equation (Adamson and Gast, 1997; Chattoraj and Birdi, 1984). It is also quite clear that any partial of ionization would lead to considerable difficulty in applying the Gibbs equation. Further, if SDS were investigated in a solution using a large excess of sodium ions, produced by the addition of, say, sodium chloride, then the sodium ion term in Equation 1.85 will vanish and will arrive back at an equation equivalent to Equation 1.86.

Experimental data of γ versus the $\log(C_{alkylsulfate})]$ give the following:

Concentration (mol/L)	$\Gamma_{S_{alkylsulfate}}$ (10^{-12} mol/cm^2)	A (Area/Molecule) (Å^2)
NaC10 sulfate		
0.03	3.3	50
NaC12 sulfate		
0.008	3.4	50
NaC14 sulfate		
0.002	3.3	50

In the plots of γ versus concentration, the slope is related to the surface excess, $\Gamma_{S_{alkylsulfate}}$.

The Gibbs adsorption equation for a simple 1:1 ionic surfactant, such as NaSDS (SD, Na).

In the absence of any other additives, one gets the following relationship between change in and the change in NaSDS concentration (at a given T):

$$\Gamma_{SDS} = \frac{-1}{4.606\mathbf{RT}}\left(\frac{d\gamma}{(d\log C_{SDS})}\right)$$

The magnitude of Γ thus obtained at the interface provides information about the orientation and packing of the NaSDS molecule. This information is otherwise not available by any other means.

The area/molecule values indicate that the molecules are aligned vertically on the surface, irrespective of the alkyl chain length. If the molecules were oriented flat, then the value of area/mol would be much larger (greater than 100 Å^2). Further, since the alkyl chain length has no effect on the area also proves this assumption is acceptable. These conclusions have been verified from spread monolayer studies. Further, one also finds that the polar group, that is, $-SO^{4-}$, would occupy something like 50 Å^2.

Gibbs adsorption equation is a relation about the solvent and a solute (or many solutes). The solute is present either as excess (if there is an excess surface concentration) if the solute decreases the γ or a deficient solute concentration (if surface tension is increased by the addition of the solute) (Chattoraj and Birdi, 1984).

Let us consider the system: a solution of water with a surfactant (soap, etc.), such as SDS. In an aqueous solution, SDS molecule dissociates into SD^- and S^+ ions. The composition of surface of water in these systems will be as depicted in the following:

Pure water ([w] bulk and surface [**w**] phase):

air air air air

wwwwwwwwwwwwwwwwwwwwwwwwwwwwww

wwwwwwwwwwwwwwwwwwwwwwwwwwwwww

wwwwwwwwwwwwwwwwwwwwwwwwwwwwww

Water plus SDS ([S] in bulk phase and [**S**] in surface phase) (2 g/L):

air air air air
ssswssssssswsssssssssssssswsssswssssssssssws
sssswssswwwwswwwsssswwwwwwwwwww
sssswssswwwwswwwsssswwwwwwwwwww
wwwwwwwwwsswwwwwwwwwsswwwwwwwwww

It is found that the surface tension of pure water of 72 mN/m decreases to 30 mN/m by the addition of 2.3 g/L (8 mmol/L) of SDS. Thus, the surface of SDS solution is mostly a monolayer of SDS plus some bound water. The ratio of water:SDS in the system (1 L solution) is roughly as follows:

In bulk phase: 55 mol water:8 mmol SDS
At the surface: roughly 100 mol SDS:1 mol water

This description is in accordance with the decrease in γ of the system.

Investigations have shown that if one carefully sucked a small amount of surface solution of a surfactant, then one can estimate the magnitude of Γ. Further, this indicates that when there is 8 mmol/L in the bulk of the solution, at the surface the SDS molecules completely cover the surface. The area per molecule at the surface data (as found to be 50 Å^2) indicates that the SDS molecules are oriented with the SO^{4-} groups pointing toward water phase, while the alkyl chains are oriented away from the water phase. This means if one used foam bubbles, the collected foam would continuously remove more and more SAS from the surface. This method of bubble foam separation has been used to purify wastewater of SASs (Birdi, 2009; Boyd, 2008). Latter method is especially useful when very minute amounts of SASs (dyes: in printing industry, pollutants in wastewater) need to be removed. It is economical and free of any chemicals or filters. In fact, if the pollutant is very expensive or poisonous, then this method can have many advantages over the other methods. A simple example is given to understand the useful application of bubbles for wastewater treatment.

Calculation of amount of SDS in each bubble:
Bubble of radius = 1 cm.

Assuming that there is almost no water in the bilayer of the bubble (this is a reasonable assumption in the case of very thin films), then the surface area of the bubble can be used to estimate the amount of SDS.

Surface area of bubble = $(4 \Pi 1^2)2 = 25$ cm^2 = 25×10^{16} Å^2
Area per SDS molecule (as found from other studies) = 50 Å^2/molecule SDS
Number of SDS molecules per bubble = $25 \times 10^{16}/50 = 0.5 \times 10^{16}$ molecules
Amount of SDS per bubble = $0.5 \times 10^{16}/6 \times 10^{23}$ g = 0.01 μg SDS

These data show that it would require 100 million bubbles to remove 1 g of SDS from the solution! However, pollutants generally are found in concentrations less than 1 mg/L. Thus, one would need about *100,000 bubbles* to remove 1 mg of SDS/L of water solution. This is a rather low number of bubbles. This seems to be a very large number. Since bubbles can be easily produced at very fast rates (ca. 100–1000 bubbles/min), this is not a big hindrance. Consequently, any kind of other SAS (such as pollutants in industry) can be thus removed by foaming. For example, in the recycle process in paper industry, ink pollutant molecules are removed by bubble foam technology. It is also important to consider that if an impurity in water was surface-active molecule, then this procedure can be used to purify water.

During the past decades, a few experiments have been reported where verification of Gibbs adsorption has been reported. One of these methods has been carried out by removing using a microtone blade the thin layer of surface of a surfactant solution. Actually, this is almost the same as the procedure of bubble extraction or merely by a careful suction of the surface layer of solution. The surface excess data for a solution of SDS were found to be acceptable. The experimental data were 1.57×10^{-18} mol/cm^2, while from Gibbs adsorption equation one expected it to be 1.44×10^{-18} mol/cm^2.

Example for Surface Excess

Aqueous solution of CTAB shows following data (at 25°C):

$$\gamma = 47 \text{ mN/m}, \quad C_{CTAB} = 0.6 \text{ mmol/L}$$
$$\gamma = 39 \text{ mN/m}, \quad C_{CTAB} = 0.96 \text{ mmol/L}$$

From aforementioned equations, one can calculate

$$\frac{d\gamma}{d\log(C_{CTAB})} = \frac{(47-39)}{(\log(0.6)-\log(0.96))} = \frac{8}{-0.47} = 17$$

From these data, the area/molecule for CTAB is found to be 90 Å^2, which is reasonable.

The Gibbs adsorption equation thus shows that near the CMC the surfactant molecules are oriented horizontally with the alkyl chains pointing up while the polar groups are interacting with the water. Accordingly, if one analyzes data of systems with varying alkyl chain lengths, say, C_8SO_4Na and $C_{12}SO_4Na$, then the area per molecule should be the same. This is indeed the case as found from experimental data.

1.4.7.2 Kinetic Aspects of Surface Tension of Detergent Aqueous Solutions

It goes without exception that one needs the information about kinetic aspects of any phenomena. In the present case, one would like to ask how fast the surface tension of a detergent solution reaches equilibrium after it is freshly created.

It is of interest to examine what happens to the surface tension of a detergent solution if one pours a detergent solution into a container. At almost the instantaneous time concentration of the detergent will be uniform throughout the system, that is, it will be the same in the bulk and at the surface. Since the concentration of SAS is very low, the surface tension of solution will be the same as of pure water (i.e., 72 mN/m, at 25°C). This is due to the fact that the surface excess is zero at time zero. However, it is found that the freshly formed surface of a detergent solution exhibits varying rates of change in surface tension with time. A solution is uniform in solute concentration until a surface is created. At the *surface,* SAS will accumulate dependent on time, and accordingly, surface tension will decrease with time. In some cases, the rate of adsorption at the surface is very fast (less than a second), while in other cases it may take longer time. As expected, one finds that the freshly created aqueous solution shows surface tension of almost pure water, that is, 70 mN/m. However, due to diffusion, the magnitude of surface tension starts to decrease rapidly and reaches an equilibrium value after a given time. The formation of foam bubbles as one pours the solution is indicative of that surface adsorption is indeed very fast (as pure water does not foam on shaking!).

Especially, in the case of high molecular weight SASs (such as proteins), the period of change in γ may be sufficiently prolonged to allow easy observations. This arises from the fact that proteins are surface-active. All *proteins* behave as SAS because of the presence of hydrophilic–lipophilic properties (imparted from the different *polar* [such as glutamine, lysine] and *apolar* [such as alanine, valine, phenylalanine, iso-valine] amino acids). Proteins have been extensively investigated as regards their polar–apolar characteristics as determined from surface activity.

Based on simple diffusion assumptions, the rate of adsorption at the surface,

$$\frac{d\Gamma}{dt} = \left(\frac{\mathbf{D}}{\pi}\right)^2 C_{bulk} t^{-2} \tag{1.87}$$

which on integration gives

$$\Gamma = 2C_{bulk} \left(\frac{\mathbf{D}t}{\pi}\right)^2 \tag{1.88}$$

where

D is the diffusion constant coefficient

C_{bulk} is the bulk concentration of the solute

The following procedure has been used to investigate the surface adsorption kinetics.

Solution surface at equilibrium	Low γ
After suction at the surface	High γ (almost pure water)
After some time	Surface tension as at equilibrium

In the literature, studies using suction at the surface have been used to investigate the surface adsorption kinetics using high-speed measurement techniques (Birdi, 1989).

The magnitude of γ increases right after suction, corresponding to pure water (i.e., 72 mN/m) and decreases with time as Γ increases (from initial value of zero). This experiment actually verifies the various assumptions as made in the Gibbs adsorption equation. Experimental data show good correlation to this equation when the magnitude of t is very small. It is also obvious that different detergents will exhibit different adsorption rates. This property will effect the functional properties of the detergent solution.

1.4.8 Solubilization in Micellar Solutions (of Organic Water Insoluble Molecules) in Micelles

In everyday life, one finds systems that involve *organic-water-insoluble compounds* (both in industry and biology). In many of these systems, one is interested in the mechanism of solubility of such organic compounds in water. One of the most important examples is if one is interested in the solubility of a medical compound for pharmaceutical application. It has been found that micelles (both ionic and nonionic) behave as a micro-phase, where the *inner core* behaves as (liquid) alkane, while the surface area behaves as a polar phase (Birdi and Ben-Naim, 1981; Tanford, 1980; Todorov et al., 2007). This was concluded from the fact that all mixed detergent solutions make mixed micelles (Ben-Naim, 1980; Birdi and Ben-Naim, 1981; Tanford, 1980). The inner core of a micelle is also found to exhibit liquid-alkane-like characteristics. The *inner core* thus has been found to exhibit *alkane-like* properties while being surrounded by a water phase. In fact, micelles are *nanostructures*. What this then suggests is that one can design surfactant solution systems in water that can have both aqueous and alkane-like properties. This unique property is one of the main applications of surfactant micelle solutions in all kinds of systems (especially in washing and cleaning, cosmetics, pharmaceutical, and oil and gas recovery). Further, in ionic surfactant micelles, one can additionally create *nano-reactor* systems. In the latter reactors, the counter-ions are designed to bring two reactants to a very close proximity (due

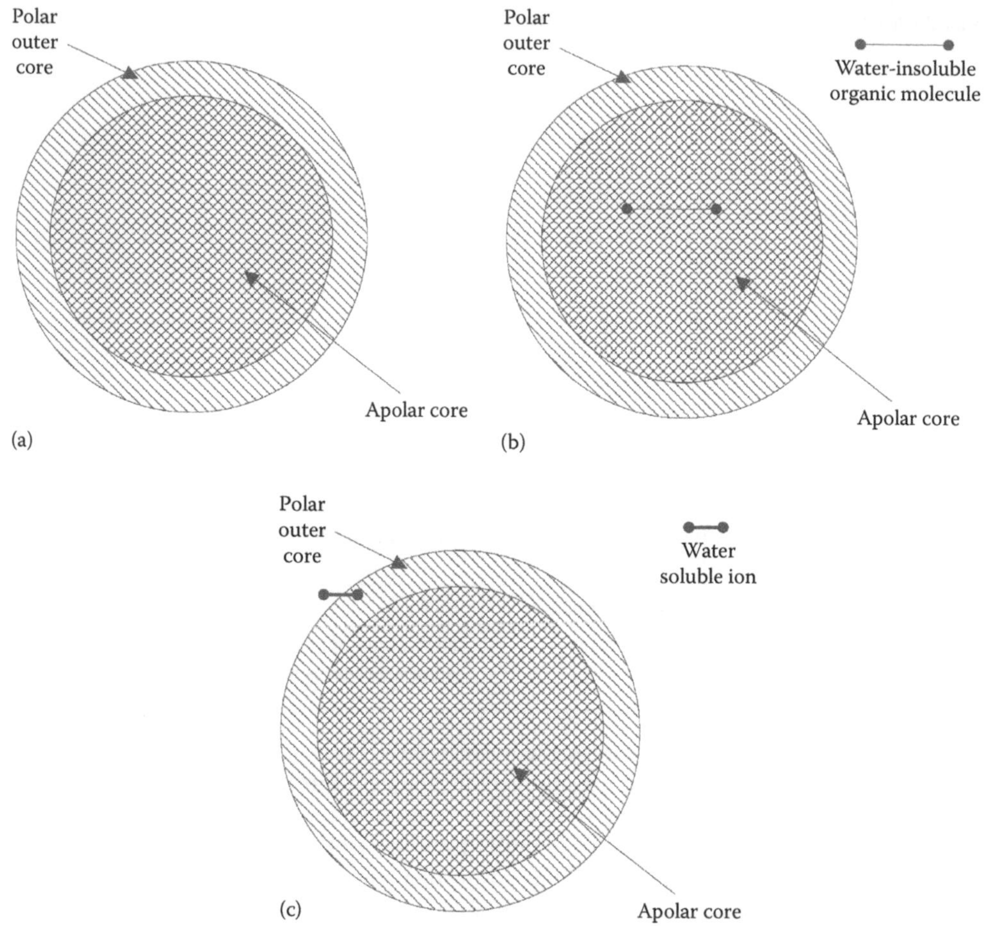

FIGURE 1.32 Micelle structure: (a) inner part = liquid paraffin-like; outer polar part; (b) solubilization of apolar molecule; and (c) binding of counter-ion to the polar part (schematic).

to electrical double layer (EDL)). These reactions would otherwise have been impossible (Birdi, 2009, 2010a). The different characteristics of micelle can be delineated as shown in Figure 1.32.

Alkyl chains attract each other (van der Waals forces); thus, the inner part consists of alkyl groups that are closely packed. It is known (from experimental data) that these clusters behave as *liquid paraffin* ($C_nH_{2n+2)}$. This was concluded from the fact that micelles of different alkyl chain lengths are miscible (Tanford, 1980). The alkyl chains are thus not fully extended. Hence, one would expect that this inner hydrophobic part of micelle should exhibit properties that are common for alkanes, such as ability to solubilize all kinds of water-insoluble organic compounds (as also found from experimental data). The solute enters the alkyl core of the micelle and it swells. The equilibrium is reached when the ratio between moles solute: moles detergent is reached corresponding to the (free energy) thermodynamic value. Size analyses of micelles (by using light-scattering) of some spherical micelles of SDS have indeed shown that the radius of the micelle is almost the same as the length of the SDS molecule. However, if the solute interferes (that is forms aggregates) with the outer polar part of micelle, then the micelle free energy may change (i.e., the CMC and other properties change). This is observed in the case of *n*-dodecanol addition to SDS solutions. In general, the addition of very small amounts of solutes show very little effect on CMC (i.e., the free energy of the micelle formation). However, if a solute makes mixed-micelles, then the free energy of micelle system changes and gives a change in CMC (Birdi, 2010a; Tanford, 1980). The data in

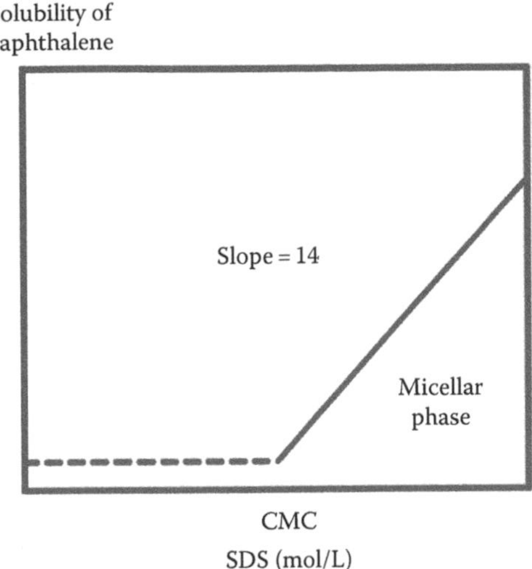

FIGURE 1.33 Solubilization of naphthalene in SDS aqueous solution (at 25°C).

Figure 1.33 show the change in the solubility of naphthalene (a water-insoluble organic molecule) in SDS aqueous solutions. Similar data have been reported for many other water-insoluble organic compounds (such as anthracene, phenanthrene, azobenzene, etc.).

Below CMC, the amount dissolved remains constant, which corresponds to its solubility in pure water. The slope of the plot above CMC corresponds to 14 mol SDS:1 mol naphthalene. It is seen that at the CMC the solubility of naphthalene abruptly increases. This is due to the fact that micelles can solubilize water-insoluble organic compounds. A more useful analysis can be carried out by considering the thermodynamics of this solubilization process.

At equilibrium, the chemical potential of a solute (naphthalene; etc.) will be given as

$$\mu_s^s = \mu_s^{aq} = \mu_s^M \tag{1.89}$$

where μ_s^s, μ_s^{aq}, and μ_s^M are the chemical potentials of the solute in the solid state, aqueous phase, and micellar phase, respectively. This equilibrium state is common in all kinds of physic/chemical systems. It must be noted that in these micellar solutions we will describe the system in terms of *aqueous* phase and *micellar* phase. Before CMC, the solute will only be present in the water phase (as if no detergent is added). The standard free energy change involved in the solubilization, ΔG_{so}^o, is given as follows:

$$\Delta G_{so}^o = -\mathbf{RT}\ln\left(\frac{C_{s,M}}{C_{s,aq}}\right) \tag{1.90}$$

where $C_{s,aq}$ and $C_{s,M}$ are the concentrations of the solute in the aqueous phase and in the micellar phase, respectively. The free energy change is the difference between the energy when solute is transferred from solid state to the micelle interior. It has been found from many systematic studies that ΔG_{so} is dependent on the chain length of the alkyl group of the surfactant. The magnitude of ΔG_{so} changes by −837 J (−200 cal)/mol, with the addition of $-CH_2-$ group. In most cases, the addition of electrolytes to the solution has no effect (Birdi, 1982, 1999, 2003a). The kinetics of solubilization has an effect on its applications (Birdi, 2003a; Yoshida et al., 2002).

Another important aspect is that the slope in Figure 3.20 corresponds to ($1/C_{s,M}$). This allows one to determine moles SDS required to solubilize one mole of solute. This magnitude is useful in understanding the mechanism of solubilization in micellar systems (Birdi, 2009; Chaibundit et al., 2002; Yan et al., 2012).

Analyses of various solutes in SDS micellar systems showed that

Azobenzene	14 mol SDS/mol azobenzene
Naphthalene	14 mol SDS/mol naphthalene
Anthracene	780 mol SDS/mol anthracene
Phenanthrene	47 mol SDS/mol phenanthrene

The ratio of detergent:solute (in the case of naphthalene, etc.) decreases as the chain length of the detergent molecule increases. This kind of study thus allows one to determine (quantitatively) the range of solubilization in any such application. These systems when used to solubilize water-insoluble organic compounds would require such information (in systems such as pharmaceutical, agriculture sprays, paints, etc.). Dosage of any substance is based on the amount of material per volume of a solution. Thus, this also shows that wherever detergents are employed, the major role (besides lower surface tension) would be the solubilization of any water-insoluble organic compounds (e.g., in pharmaceutical products). This process would then assist in the cleaning or washing or any other effect. In some cases, such as bile salts, the solubilization of lipids (especially lecithins) gives rise to some complicated micellar structures (Birdi, 2010a; Tanford, 1980). Due to the formation of *mixed* lipid–bile salt micelles, one observes changes in CMC and aggregation number. This has major consequences in the bile salts in biology. Dietary fat consists essentially of mixed triglycerides. These fatty lipids pass through the stomach into the small intestine, without much change in structure. In the small intestine, the triglycerides are partly hydrolyzed by an enzyme (lipase), which leads to the formation of oil–water emulsion. This shows the importance of surface agents in biological systems, such as stomach.

1.4.9 Biological Micelles (Bile Salt Micelles)

Bile salts are steroids with detergent properties, which are used by nature to emulsify lipids in foodstuff passing through the intestine to enable fat digestion and absorption through the intestinal wall. They are secreted from the liver stored in the gall bladder and passed through the bile duct into the intestine when food is passing through. Bile salts in general form micelles with low aggregation numbers (ca. 10–50) (Tanford, 1980). However, bile salt micelles grow very large in size when it solubilizes lipids (this phenomenon is called mixed-micelle formation).

1.4.10 Washing and Laundry (Dry Cleaning)

The most important application of surface and colloid chemistry principles in everyday life is in the systems where washing (cleaning) and laundry (detergency) are involved. These processes are one of the most important phenomena for mankind (as regards health and welfare and technology), and it has been regarded as such for many centuries. For example, the effect of clean wings of aeroplanes is of utmost concern in the flight security. Mankind has been aware of the role of cleanliness on health and disease for many thousands of years. Many critical diseases, such as AIDS or similar infections, are found to be lesser in incidence in those areas of the world where cleanliness is highest. The term *detergency* is used for processes such as *washing clothes* or *dry cleaning* or cleaning. The substances used are designated as detergents (Tung and Daoud, 2011; Zoller, 2008). In all these processes, the object is to remove dirt from fabrics or solid surfaces (floors or walls or other surfaces of all kinds). The shampoo is used to clean the hair. Hair consists of portentous material and thus requires different kinds of detergents than when washing clothes or cars. Shampoo should not interact strongly with the hair, but it should remove dust particles or other material. Another important requirement is that the ingredients in the shampoo should not damage or irritate the eye or the skin with which it may come in contact. In fact, all shampoos are tested for eye irritation and skin irritation before marketing. In some cases by merely increasing the viscosity, one achieves a great deal of protection. For example, a surfactant solution (ca. 20%), alkyl sulfate with two EO (ethylene oxide groups), gives very high viscosity if a small amount of salt is added. Shampoo industry is highly specialized, and a large industrial state-of-the art research is applied in this product.

1.4.11 Solubility of Organic Molecules in Water (A Surface Tension–Cavity Model Theory)

The phenomenon of solubility of organic molecules in water is very important in everyday life. In pure water, all molecules are surrounded by similar molecules in a symmetrical geometry. Of course, this is also the case for any other liquid. In the present case, we will consider what happens when an organic molecule dissolves (to a varying degree: 10^{-10} to 10 mol/L) in water. The degree of solubility of molecules in water varies a great range. For example, NaCl can dissolve in water up to 10 mol/L, while an alkane (such as hexane) shows a very low solubility (ca. 0.001 mol/L). Let us take a snapshot of what happens when a foreign molecule (water-insoluble organic molecule) is dissolved in water (Figure 1.34).

In the following, one may consider the process of solubility of any substance in water as follows:

A salt (e.g., NaCl) when dissolved in water dissociates into anions and cations. NaCl dissociates into Na^+ and Cl^- ions. These ions are surrounded by water molecules (bound water molecules to each ion). Ions are known to interact through dipoles and hydrogen bonds with water molecules. On the other hand, when a water-insoluble-organic molecule (hydrophobic molecule) is placed in water, this system is found to be different than in the case of NaCl or other similar electrolytes. The organic (nonpolar) molecule will not interact with dipoles of surrounding water molecules. The solubility of such organic molecule thus requires a cavity in water. It is also essential to visualize that the bigger the cavity needed, the lower is the solubility in water (since cavity requires energy). In order to understand in more detail, one may investigate the methane hydrate systems (Tanford, 1980). This has the advantage that ice structures can be analyzed by x-ray analysis that provides molecular details. It has been found that when water crystallizes in the presence of CH_4 or Cl_2, following data are obtained from x-ray analyses of hydrates.

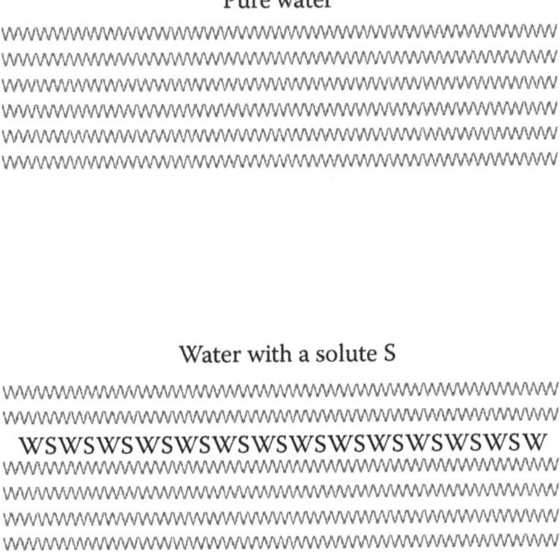

FIGURE 1.34 Pure water (W) and a foreign solute (organic) molecule (S) is dissolved in water. The *cavity* is created for S (see text).

Gas hydrate composition

$CH_4 \cdot 5\frac{3}{4}H_2O$: Methane hydrate
$Cl_2 \cdot 8H_2O$: Chlorine hydrate

This indicates that some kind of structure will be present in liquid water surrounding CH_4 or alkane molecule, which would be indicative of hydrate structure (Tanford, 1980). These data are of current interest in the future (from nonconventional reservoirs) gas recovery from methane hydrate reservoirs.

Inorganic salts, such as NaCl, in aqueous solutions dissociate into Na^+ and Cl^- ions and interact with water through hydrogen bonds. Hexane molecule merely dissolves (although very low solubility) in water by placing inside water structure without taking part in the water bonds (hydrogen bonds). Since water structure is stabilized mainly by hydrogen bonds, hexane molecule will give rise to some rearrangements of these bonds but without breaking (since no heat is evolved in this process).

If a salt exhibits a maximum solubility (called saturation solubility: at a specific temperature and pressure) of 10 mol/L in water, then it corresponds to ca. 10 mol of salt:55 mol of water (ratio of 1:5.5). On the other hand, an alkane molecule may show a maximum solubility of 0.0001 mol/L in water (0.0001:55 mol), or a ratio of 1:550.000. Many decades ago, this model was found to be able to predict the solubilities of both simple organic molecules, and for the case of more complicated. In most simple case, the solubility of heptane is lower than that of hexane, due to the addition of one $-CH_2-$ group. In the case of alkane molecules, a linear relation between the solubility and the number of $-CH_2-$ groups (Acree, 2004; Birdi, 1997; Kyte, 2003; Tanford, 1980).

This model is thus based on the following assumptions when alkane molecule is placed in water: Alkane (CCCCCCCCC) is placed in a cavity in water (ww).

wwwwwwwwwwwwwwwwwwwwwwwwwwwwwwwwww
wwwwwwwwwwwwCCCCCCCCCwwwwwwwwwwwww
wwwwwwwwwwwwwwwwwwwwwwwwwwwwwwwwwwww

The alkane molecule merely makes a *cavity* in water *without breaking hydrogen bonds* of water structure. It has been argued that the structure of water exhibits *voids* where a solute (nonpolar) can reside (Birdi, 2009; Tanford, 1980). Some analogy has been related to the data of methane hydrate structures (Tanford, 1980). In the ice structure, it is found that methane (and other gases, such as chlorine, Cl_2) does fit in voids in ice and is trapped. In water, the ice-like structure persists, and thus, alkane molecules can dissolve in the cavity spaces. The energy needed to create a surface area of the cavity will be proportional to the degree of solubility of the alkane. Thus, the free energy of solubility of any alkane molecule will be given as follows:

$$\begin{aligned}\text{Free energy of solubility} &= \text{Proportional to the product (cavity surface area)}\\ &\quad \times (\text{surface tension of the cavity})\end{aligned} \tag{1.91}$$

By analyzing the solubility data of a whole range of alkane molecules in water, the following relation was found to fit the experimental data:

$$\text{Free energy of solubility} = \Delta \mathbf{G}^{o}_{sol} = \mathbf{RT}\log(\text{solubility}) \tag{1.92}$$

$$= (\gamma_{cavity})(S_{area\,alkane}) \tag{1.93}$$

$$= 25.5(S_{area\,alkane}) \tag{1.94}$$

For the solubility of alkanes in water, the total surface area (TSA) gives the solubility:

$$\ln(\text{sol}) = -0.043(\text{TSA}) + 11.78 \tag{1.95}$$

where solubility (sol) is in molar units and TSA in A^2. For example,

Alkane	**(sol)**	**TSA**	**Predicted (sol)**
n-Butane	0.00234	255	0.00143
n-Pentane	0.00054	287	0.0004
n-Hexane	0.0001	310	0.0001
n-Butanol	1.0	272	0.82
n-Pentanol	0.26	304	0.21
n-Hexanol	0.06	336	0.05

The constant 0.043 is equal to $\gamma_{cavity}/RT = 25.5/600$.

CH_3	CH_2	CH_2	CH_2	CH_2	CH_2	CH_2	CH_2	CH_2	OH
85	43	32	32	32	32	32	40	45	59

The surface areas of each group in *n*-nonanol ($C_9H_{19}OH$) were estimated by different methods. These data are as follows:

The magnitude of the surface area of the CH_3 group is obviously expected to be larger than the CH_2 groups. One can estimate the magnitude of TSA of *n*-decanol: it will be (TSA of nonanol + TSA of CH_2) 431 + 32 = 463 Å^2. From this value, one can estimate its solubility.

The data for solubility of homologous series of *n*-alcohols in water (at 25°C) are of interest, as shown in the following.

Alcohol	**Solubility (mol/L)**	**log(S)**	**Difference per CH_2**
C_4OH	0.97	−0.013	—
C_5OH	0.25	−0.60	0.6
C_6OH	0.06	−1.22	0.62
C_7OH	0.015	−1.83	0.61
C_8OH	0.004	−2.42	0.59
C_9OH	0.001	−3.01	0.59
$C_{10}OH$	0.00023	−3.63	0.62

Accordingly, this algorithm allows one to estimate the solubility of water of any organic substance. The estimated solubility of cholesterol (add) was almost in accord with the experimental data (Birdi, 1997, 2009). The solubility analysis of cholesterol is of much interest in the case of biological phenomena (Tanford, 1980). It is seen that log(S) is a linear function of the number of carbon atoms in the alcohol. Each $-CH_2-$ group reduces log(S) with 0.06 unit.

1.5 MONOMOLECULAR LIPID FILMS ON LIQUID SURFACES (AND LANGMUIR–BLODGETT FILMS)

Ancient Egyptians are known to have poured small amounts of oil (olive oil) over water while their ships were sailing into harbors (as seen from old carvings): small amounts of oil on water surface were known to appreciably calm the waves, thus assisting easy navigation into the harbors. It was later found that some lipid-like substances (almost insoluble in water) formed SAMs

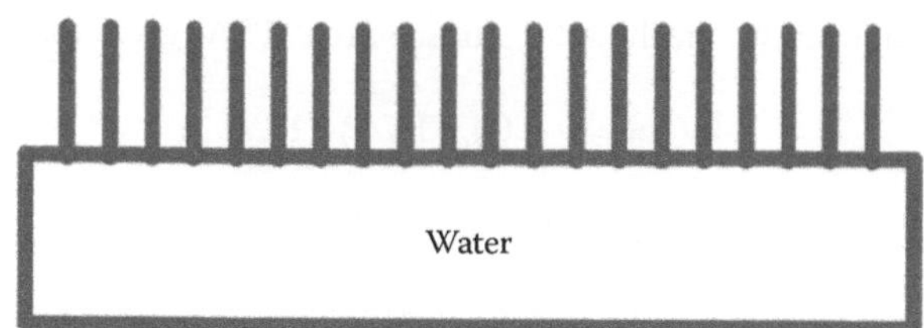

FIGURE 1.35 Lipid monolayer on the surface of water.

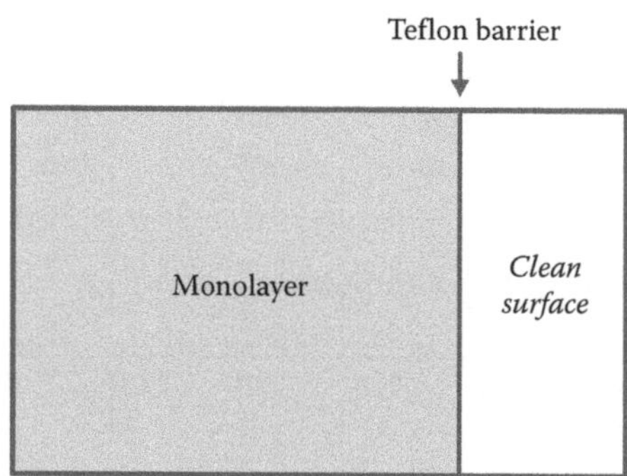

FIGURE 1.36 Monolayer film balance: barrier and lipid film and surface pressure.

(Figure 1.35) on water surface (Adamson and Gast, 1997; Birdi, 1989, 1999, 2003a; Bouvrais et al., 2014; Chattoraj and Birdi, 1984; Gagnon and Meli, 2014; Gaines 1966; Imae, 2007).

A few decades ago, experiments showed that monomolecular films of lipids could be studied by using rather simple experimental methods, Figure 1.36. It is amazing to find that even a monolayer of lipid film (thickness ca. 20 Å) can easily be investigated with a very high precision. In fact, monolayer method is the only technique that can provide detailed information about adsorbed state of molecules in such systems. Furthermore, Langmuir was awarded Nobel Prize (in 1920) for his pioneer monolayer film studies (Birdi, 1999, 2003a; Gaines, 1966).

It was already explained that when a surface-active agent (such as surfactant or a soap) is dissolved in water, it adsorbs preferentially at the surface (surface excess, Γ). This means that the concentration of a surface-active agent at the surface may be as high as 1000 times more than in the bulk. The decrease in surface tension indicates this and also suggests that only a monolayer is present at the surface. For example, in a solution of SDS of concentration 0.008 mol/L, the surface is completely covered with SDS molecules. Let us consider systems where the lipid (almost *insoluble* in water) is present as a monolayer on the surface of water. In these systems, almost all the substance applied to the surface (in the range of few micrograms) is supposed to be present at the interface. This means that one knows (quantitatively) the magnitude of surface concentration (same as the surface excess, Γ).

If one places a very small amount of a lipid on the surface of water, it may affect the surface tension in different ways. It may not show any effect (such as cholesterol). It may also show a drastic decrease in surface tension (such as stearic acid or tetra-decanol). An amphiphile molecule will adsorb at the air–water or oil–water interface, with its alkyl group pointing away from the water phase. The alkyl group is at a lower energy state when pointing toward air that being surrounded by water molecules.

The system can be easily considered as a *two-dimensional* analog to the classical *three-dimensional* systems (such as the gas theory).

It is thus seen that Π of a monolayer is the lowering of surface tension due to the presence of monomolecular film. This arises from the orientation of the amphiphile molecules at the air–water or oil–water interface, where the polar group would be oriented toward the water phase, while the

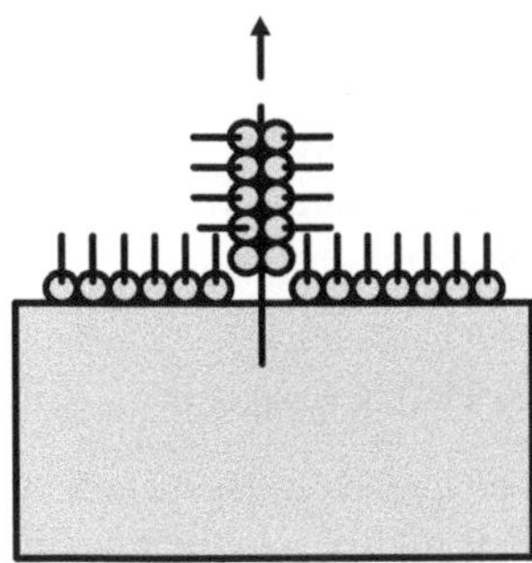

FIGURE 1.37 LB film formation (see text for details).

nonpolar part (hydrocarbon) would be oriented away from the aqueous phase. Later it was observed that if a clean smooth solid is dipped through monolayer, then in most cases a single layer of lipid will be deposited on the solid. This film was called *Langmuir–Blodgett* (LB) film (Figure 1.37). Scanning probe microscopes (Birdi, 2003b) have been used to verify these structures in a very high detail. Further if one repeated this process, one could deposit multilayers. This LB film technique has found much application in the electronic industry. Further, the presence of only one layer of lipid changes the surface properties (such as contact angle-wetting, friction, light reflection, charge, adhesion, etc.) of the solid.

1.5.1 Apparatus for Surface Lipid Film Studies

Since Langmuir reported monolayer studies, a great many instruments (commercial) have been designed about this method. The clean surface of water shows no change in surface tension if one moves a barrier across it. However, if a surface-active agent is present, then the latter molecules will be compressed, and this will give rise to a decrease in surface tension.

Monolayer systems are composed of only a monomolecular lipid film on water surface. The aim in these systems is to study the properties of a monomolecular thin lipid film spread on the surface of water, which is not possible by any other technique. Lately, it has been shown that such lipid films are useful membrane models for biological membranes structure and function studies. The modern methods have allowed one to measure the monolayer properties in much detail than earlier. No other method exists through which one can obtain any direct information about the molecular packing or interactions (forces acting in two dimensions).

The monolayer films were studied by using a teflon trough with a barrier (teflon), which could move across the surface, Figure 1.36. The change in γ was monitored by using a Wilhelmy plate method attached to a sensor. The accuracy could be as high as m mN/m (m dyn/cm).

The film balance (also called Langmuir trough) consists of a teflon (or teflon lined) rectangular trough (typically 20 cm × 10 cm × 1 cm). Teflon poly(tetraflouroethylene) (PTFE) allows one to keep the apparatus clean. A barrier of teflon is placed on one end of the trough, which is used as a barrier to compress the lipid molecules. Currently, there are many commercially available film balances.

Lipid (or protein or other film-forming substance) is applied from its solution to the surface. A solution with a concentration of 1 mg/mL is generally used. Lipids are dissolved in $CHCl_3$ or ethanol or hexane (as found suitable). If the surface area of the trough is 100 cm^2, then one may use 1–100 μL of this solution. After the solvent has evaporated (about 15 min), the barrier is made to compress the lipid film at a rate of 1 cm/s or as suitable. The amount of lipid or protein applied is generally calculated such as to give a compressed film. Most substances cover 1 mg/m^2 to give a solid film. If one has 100 cm^2 in the trough, then 10 μg or less of substance is enough for such an experiment. It is thus obvious that such studies can provide much useful surface chemical information with very minute amounts of material. Most proteins can also be studied as monomolecular films (1 mg of protein spreads to cover about 1 m^2 surface area) (Birdi, 1989, 1999; Gaines, 1966).

Constant area monolayer film method: One can also study Π versus surface concentration (C_s = area/molecule) isotherms by keeping the surface area constant. The surface concentration, C_s, is changed by adding small amounts of a substance to the surface (by using a microliter syring—1 or 5 μL). In general, one obtains a good correlation with the Π versus area isotherms (Adamson and Gast, 1997; Birdi, 1989, 1999; Gaines, 1966).

1.5.2 Monolayer Structures on Water Surfaces

Lipids with suitable HLB are known to spread on the surface of water to form monolayer films. It is obvious that if the lipid-like molecule is highly soluble in water, then it will disappear into the bulk phase (same as is observed for SDS). Thus, the criterion for a monolayer formation is that it exhibits very low solubility in water. The alkyl part of the lipid points away from the water surface. The polar group is attracted to the water molecules and is inside this phase at the surface. This means that the solid crystal when placed on the surface of water is in equilibrium with the film spread on the surface. Detailed analyses of this equilibrium have been given in the literature (Adamson and Gast, 1997; Birdi, 1999; Gaines, 1966). The thermodynamic analysis allows one to obtain extensive physical data of about this system.

1.5.3 Self-Assembly Monolayer Formation

The most fascinating characteristic some amphiphile molecules exhibit is that these when mixed with water form *self-assembly structures*. This was already observed in the case of micelle formation. This is in contrast to simple molecules such as methanol or ethanol in water since most of the biological lipids also exhibit self-assembly structure formation (Birdi, 1989, 1999). The lipid monolayer studies thus provide a very useful method to obtain information about the SAM formation. These studies have also been used as model systems about both technical systems and the cell bilayer structures. Nature has used these lipid molecules throughout the biological world, and the SAM formation has been the basis of all biological cells. Monolayer system thus mimics the cell structure and function.

1.5.4 States of Lipid Monolayers Spread on Water Surface

One is quite familiar with the changes one observes in matter going from solid to liquid or vice versa with temperature or pressure. It is found that even a monolayer of lipid (on water) when compressed can undergo various degrees of packing states. This is somewhat similar to three-dimensional structures (gas–liquid–solid). In the following, the various states of monomolecular film will be described as measured from the surface pressure, Π, versus area, **A**, isotherms, in the case of simple amphiphile molecules. On the other hand, the Π–**A** isotherms of biopolymers will be described separately, since these are found to be of different nature.

But before presenting these analyses, it is necessary to consider some parameters of the two-dimensional states, which should be of interest. We need to start by considering the physical forces acting between the alkyl–alkyl groups (parts) of the amphiphiles, and as well as the interactions between the polar head groups. In the process where two such amphiphiles molecules are brought closer during the Π–**A** measurement, the interaction forces would undergo certain changes that would be related to the packing of the molecules in the two-dimensional plane at the interface in contact with water (subphase).

This change in packing thus is analogous conceptually to the three-dimensional P–V isotherms, as are well known from the classical physical chemistry (Adamson and Gast, 1997; Birdi, 1989; Bouvrais et al., 2014; Gaines, 1966). We know that as pressure, P, is increased on a gas in a container, when $T < T_{cr}$, the molecules approach each other closer and transition to a liquid phase takes place. Further compression of the *liquid state* results in the formation of a *solid phase*.

In the case of alkanes, the distance between the molecules in the solid state phase is ca. 5 Å, while it is 5.5–6 Å in the case of liquid state. The distance between molecules in the gas phase, in general, is ca. $1000^{1/3} = 10$ times larger than in the liquid state (water:volume of 1 mol water = 18 cc; volume of 1 mol gas = 22.4 L). In the *three-dimensional* structural buildup, the molecules are in contact with near neighbors and in contact with molecules that may be 5–10 molecular dimensions apart (as found from x-ray diffraction). This is apparent from the fact that in liquids there is a long range order up to 5–10 molecular dimensions. On the other hand, in the *two-dimensional* films the state is much different. The amphiphile molecules are oriented at the interface such that the polar groups are pointed toward water (subphase), while the alkyl groups are oriented away from the subphase. This orientation gives the minimum surface energy. The structure is stabilized through lateral interaction between

1. Alkyl–alkyl groups: *attraction*
2. Polar group–subphase: *attraction*
3. Polar group–polar group: *repulsion*

The alkyl–alkyl groups attraction arises from the van der Waals forces. The magnitude of van der Waals forces increases with

Increase in alkyl chain length
Decrease in distance between molecules (or when area/molecule [A] decreases)

Experiments show that the alkyl chain length increases the magnitude of Π of films and thus becomes more stable, thus giving higher collapse pressure, Π_{co}. The stable films are thus formed when the attraction forces are stronger than the repulsive forces.

The most convincing results were those as obtained with the normal fatty alcohols and acids. Their monomolecular films were stable and exhibit very high surface pressures (Birdi, 1989, 1999). A steep rise in Π is observed around 20.5 Å, regardless of the number of carbon atoms in the chains. The volume of a ($-CH_2-$) group is 29.4 Å^3. This gives the length of each ($-CH_2-$) group perpendicular to the surface, or the vertical height of each group as ca. 1.42 Å. This compares very satisfactorily with the x-ray data with this value of 1.5 Å. This means that such straight chain lipids are oriented in this compressed state in vertical orientation.

As high pressures lead to transitions from gas to liquid to solid phases in the *three-dimensional* systems, similar state of affairs would be expected in the two-dimensional film compression Π versus A isotherms, Figure 1.38, and as described in the following.

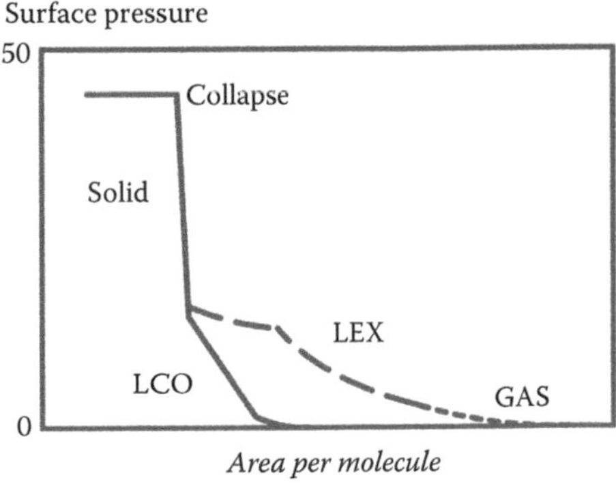

FIGURE 1.38 Lipid monolayer phases (two-dimensional) (monolayer phases are dependent on both molecular structure and temperature).

1.5.4.1 Gaseous Monolayer Films

The most simple type of amphiphile monolayer film or a polymer film would be a *gaseous* state. This film would consist of molecules that are at a sufficient distance apart from each other such that lateral adhesion (van der Waals forces) is negligible. However, there is sufficient interaction between the polar group and the subphase that the film-forming molecules cannot be easily lost into the gas phase, and that the amphiphiles are almost insoluble in water (subphase).

When the area available for each molecule is many times larger than molecular dimension, the gaseous-type film (state 1) would be present. As the area available per molecule is reduced, the other states, for example, liquid-expanded (L_{ex}), liquid-condensed (L_{co}), and finally the solid-like (S or solid-condensed) states, would be present.

The molecules will have an average kinetic energy, that is, $1/2\mathbf{k_B T}$, for each degree of freedom, where k_B is Boltzmann constant (=1.372×10^{-16} ergs/T), and T is the temperature. The surface pressure measured would thus be equal to the collisions between the amphiphiles and the float from the two degrees of freedom of the translational kinetic energy in two dimensions. It is thus seen that the ideal gas film obeys the following relation:

$$\Pi \mathbf{A} = \mathbf{k_B T} \quad \text{(ideal film)} \tag{1.96}$$

$$\Pi \; (\text{mN/m}) \mathbf{A} \; (\text{Å}^{-2} \text{ per molecule}) = 411 (\text{T} = 298 \text{ K}) \quad (\text{ideal film}) \tag{1.97}$$

In general, *ideal gas* behavior is only observed when distances between the amphiphiles are very large, and thus the value of Π is very small, that is, <0.1 mN/m. It is also noticed that from such sensitive data one can estimate the molecular weight of the molecule in the monolayer. This has been extensively reported for protein monolayers (Adamson and Gast, 1997; Birdi, 1989, 1999). The latter observation requires an instrument with very high sensitivity, ±0.001 mN/m. The Π versus A isotherms of *n*-tetradecanol, pentadecanol, pentadecyclic acid and palmitic acid in the low Π region showed data that agreed with the ideal film. Similar data for isotherms were reported for other lipid monolayers by other workers. The various forces that are known to stabilize the monolayers are mentioned as follows:

$$\Pi = \Pi_{kin} + \Pi_{vdW} + \Pi_{electro} \tag{1.98}$$

where

Π_{kin} arises from kinetic forces
Π_{vdW} is related to the van der Waals forces acting between the alkyl chains (or groups)
$\Pi_{electro}$ is related to polar group interactions (polar group–water interaction; polar group–polar group repulsion; charge–charge repulsion)

When the magnitude of A is very large, the distance between molecules is large. If there are no van der Waals or electrostatic interactions, then the film obeys the *ideal* equation. As the area per molecule is decreased, other interactions become significant. The Π versus **A** isotherm can be used to estimate these different interaction forces (Adamson and Gast, 1997; Birdi, 1989, 1999; Bouvrais et al., 2014; Gaines, 1966). The ideal equation has been modified to fit Π versus **A** data, in those films where co-area, $\mathbf{A_o}$, correction is needed (Birdi, 1989) (Table 1.6):

$$\Pi(\mathbf{A} - \mathbf{A_o}) = \mathbf{k_B T} \tag{1.99}$$

In the case of straight chain alcohols or fatty acids, $\mathbf{A_o}$ is almost 20 Å^2, which is the same as found from the x-ray diffraction data of the packing area per molecule of solid alkanes.

TABLE 1.6
Magnitudes of A_o for Different Film-Forming Molecules on the Surface of Water

Compound	A_o (Å²)
Straight chain acid	20.5
Straight chain acid (on dilute HCl)	25.1
N-fatty alcohols	21.6
Cholesterol	40
Lecithins	ca. 50
Proteins	ca. 1 m²/mg
Diverse synthetic polymers (poly-amino acids, etc.)	ca. 1 m²/mg

Source: Birdi, K.S., *Lipid and Biopolymer Monolayers at Liquid Interfaces*, Plenum Press, New York, 1989.

This equation (Equation 1.99) is thus valid when $\mathbf{A} \gg \mathbf{A}_o$. The magnitude of Π is 0.2 mN/m for $\mathbf{A}$ = 2000 Å², for ideal film. However, Π will be about 0.2 mN/m for $\mathbf{A}$ = 20 Å² for a solid-like film of a straight chain alcohol.

Π versus (**A**) for a monolayer of valinomycin (a dodecacyclic peptide) shows that the relation as given in Equation 1.99 is valid. In this equation, it is assumed that the amphiphiles are present as monomers. However, if any association takes place, then the measured values of (Π**A**) would be less than $\mathbf{k_B T} < 411$, as has also been found (Birdi, 1989, 1999). The magnitude of $\mathbf{k_B T}$ = 411 dyn/cm = 4×10^{-21} J, at 25°C. In the case of nonideal films, one will find that the versus data do not fit the relation in equation. This deviation requires that one uses other modified equations-of-state. This procedure is the same as one uses in the case of three-dimensional gas systems.

1.5.4.2 Liquid Expanded and Condensed Films

The Π versus A data are found to provide much detailed information about the state of monolayers at the liquid surface. In Figure 1.38, some typical states are shown. The different states are very extensively analyzed and will be therefore described in the following.

In the case of simple amphiphiles (fatty acids, fatty alcohols, lecithins, etc.), in several cases, transition phenomena have been observed between the gaseous and the coherent states of films, which show a very striking resemblance to the condensation of vapors to liquids in the three-dimensional systems. The liquid films show various states in the case of some amphiphiles, as shown in Figure 1.38 (schematic). In fact, if the Π versus **A** data deviate from the ideal equation, then one may expect following interactions in the film:

Strong van der Waals
Charge–charge repulsions
Strong hydrogen bonding with subphase water

This means that such deviations thus allow one to estimate these interactions.

1. *Liquid expanded films* (L_{exp}): In general, there are two distinguishable types of liquid films. The first state is called the liquid expanded (L_{exp}) (Adamson and Gast, 1997; Chattoraj and Birdi, 1984; Gaines, 1966). If one extrapolates the Π–**A** isotherm to zero Π, the value of **A** obtained is much larger than that obtained for close packed films. This shows that the distance between the molecules is much larger than one will find in the solid film, as will be discussed later. These films exhibit very characteristic elasticity, which will be described further in the following.

2. *Liquid condensed films* (L_{co}): As the area per molecule (or the distance between molecules) is further decreased, there is observed a transition to a so-called liquid condensed (L_{co}) state. These states have also been called "solid expanded" films (Adam, 1930; Adamson and Gast, 1997; Birdi, 1989, 1999; Gaines, 1966), which will be later discussed in further detail. The Π versus **A** isotherms of *n*-pentadecylic acid (amphiphile with a single alkyl chain) have been studied, as a function of temperature. Π–A isotherms for two chain alkyl groups, as lecithins, also showed a similar behavior.

1.5.4.3 Solid Films

As the film is compressed, a transition to a solid film is observed, which collapses at higher surface pressure. The Π versus **A** isotherms, below the transition temperatures, show the liquid to solid phase transition. These solid films have been also called *condensed films*. These films are observed in such systems where the molecules adhere to each other through the van der Waals forces, very strongly, the Π–**A** isotherm shows generally no change in Π at high A, while at a rather low **A** value, one observes a sudden increase in Π, as shown in Figure 1.39. In the case of straight chain molecules, like stearyl alcohol, the sudden increase in Π is found to take place at A = 20–22 Å^2, at room temperature (that is much lower than the phase transition temperature, to be described later). These analyses have shown that the films may under given experimental conditions exhibit three first-order transition states, for example, (1) transition from the gaseous film to the liquid-expanded (L_{ex}), (2) transition from the *liquid-expanded* (L_{ex}) to the *liquid-condensed* (L_{co}), and (3) from L_{ex} or L_{co} to the solid state, if the temperature is below the transition temperature. The temperature above which no expanded state is observed has been found to be related to the melting point of the lipid monolayer.

1.5.4.4 Collapse States of Monolayer Assemblies

The measurements of Π versus **A** isotherms generally exhibit, when compressed, a sharp break in the isotherms that has been connected to the *collapse* of the monolayer under the given experimental conditions. The monolayer of some lipids, such as cholesterol, is found to exhibit an usual isotherm, Figure 1.39. The magnitude of Π increases very little as compression takes place. In fact, the collapse state or point is the most useful molecular information from such studies. The collapse pressure is found to be a very *unique property* of any lipid. It is strongly dependent on the packing state of the lipid monolayer and thus provides important information about molecular interactions. It is found that this is the only direct method, which can provide information about the structure and orientation of amphiphile molecule at the surface of water. However, a steep

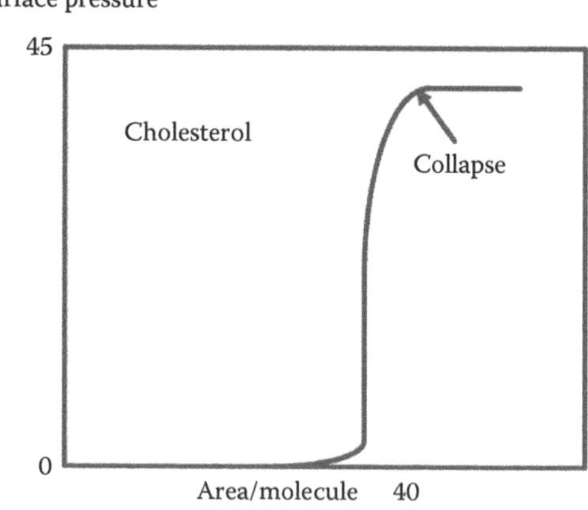

FIGURE 1.39 Surface pressure (Π) versus area/molecule (A) isotherm of cholesterol monolayer on water (at 25°C).

rise in Π is observed and a distinct break in the isotherm is found at the collapse. This occurs at Π = 40 mN/m and **A** = 40 Å^2.

This value of $\mathbf{A}_{co}$ corresponds to the cholesterol molecule oriented with the hydroxyl group pointing toward the water phase. AFM studies cholesterol as LB films have shown that there exist domain structures. This has been found for different collapse lipid monolayers (Birdi, 2003b; Kuo and Chang, 2014). It should be mentioned that monolayer studies are the only procedure, which allows one to estimate the area per molecule of any molecule as situated at the water surface. In general, the collapse pressure, Π_{col}, is the *highest* surface pressure to which a monolayer can be compressed without a detectable movement of the molecules in the films to form a new phase, Figure 1.39. In other words, this surface pressure will be related to the nature of the substance and the interaction between the subphase and the polar part of the lipid or the polymer molecule. The monolayer collapse pressure has been shown to provide much information also in the case of protein monolayers. It was found that a-helical polypeptides exhibited shaper collapse state than b-shaped polypeptides (Birdi and Fasman, 1973).

1.5.5 Other Changes at Water Surfaces due to Lipid Monolayers

The presence of lipid (or similar kind of substance) monolayer at the surface of aqueous phase gives rise to many changes in the properties at the interface. The major effects that have been investigated extensively are

Surface potential (V)
Surface viscosity (η_s)
Surface Viscosity (η_s)

A monomolecular film is resistant to shear stress in the plane of the surface, as also is the case for the bulk phase: A liquid is retarded in its flow by viscous forces. The viscosity of the monolayer may indeed be measured in two dimensions by flow through a canal on a surface or by its drag on a ring in the surface, corresponding to the *Ostwald* and *Couttte* instruments for the study of bulk viscosities. The surface viscosity, η_s, is defined by the relation

$$\{\text{Tangential force per centimeter of surface}\} = \eta_s \ (\text{rate of strain}) \quad (1.100)$$

and is thus expressed in units of (m/t) (surface poises), whereas bulk viscosity, η, is in units of poise (L/(m t)). The relationship between these two kinds of viscosity is

$$\eta_s = \frac{\eta}{d} \quad (1.101)$$

where d is the thickness of the *surface phase*, approximately 10^{-7} cm (= 10 Å = 10^{-9} m = nm) for many films, that the magnitude of η_s is on the order of 0.001–1 surface poise implies that over the thickness of the monolayer, the surface viscosity is about 10^4–10^7 poises. This has been compared to the viscosity comparable to that of butter. The η_s is generally given in surface poise (g/s or kg/s).

Pure water surface consists of only water (W) molecules:

WWWWWWWWWWWWWWWWWWWWWWWWWWWWW: surface

While a system with lipid monolayer (L) gives a surface that is different:

LLLLLLLLLLLLLLLLLLLLLLLLLLLLLLLLLLLLLLLLL: surface
WWWWWWWWWWWWWWWWWWWWWWWWWWWWWWWWWWWW: first layer
WWWWWWWWWWWWWWWWWWWWWWWWWWWWWWWWWW: bulk phase

Thus, the lipid monolayer gives rise to hindrance to any flow or movement at the surface (the layers with *LWLW*). This arises mainly from the fact that water molecules (first layer) bound to the polar part of the lipid are subjected to high viscosity (surface viscosity).

It is easily realized that if a monolayer is moving along the surface under the influence of a gradient of surface pressure, it will carry some of the underlying water with it. In other words, there is no slippage between the monolayer molecules and the adjacent water molecules. The thickness of such regions has been reported to be on the order of 0.003 cm. It has also been asserted that the thickness would be expected to increase as the magnitude of η_s increases. However, analogous to bulk phase, the concept of free volume of fluids should be also considered in these films. As mentioned earlier, the soap films are made of these films when air bubble forms at the liquid surface. Therefore, the soap bubble characteristics are dependent on surface viscosity of the lipid monolayer.

1.5.5.1 Surface Potential (ΔV) of Lipid Monolayers

Any liquid surface, especially aqueous solutions, will exhibit asymmetric dipole or ions distribution at surface as compared to the bulk phase. If SDS is present in the bulk solution, then we will expect that the surface will be covered with SD^- ions. This would impart a negative surface charge (as is also found from experiments).

Due to the surface adsorption of SDS to water not only surface tension changes (reduces) but SDS also imparts *negative* surface potential (due to SD^- at the surface). Similarly, solutions of cationic detergent, such as CTAB, give rise to a *positively* charged potential of water surface.

On the other hand, nonionic surfactants do not change the surface charge of water solutions.

Of course, the surface molecules of methane (in liquid state) obviously will exhibit symmetry in comparison to water molecule. This characteristic can also be associated to the force field resulting from induced dipoles of the adsorbed molecules or spread lipid films (Adamson and Gast, 1997; Birdi, 1989). The surface potential arises from the fact that the lipid molecule orients with polar part toward the aqueous phase. This gives rise to a change in dipole at the surface. There would thus be a change in surface potential when a monolayer is present, as compared to the clean surface. The surface potential, **ΔV**, is thus

$$\Delta \mathbf{V} = \text{potential}_{\text{monolayer}} - \text{potential}_{\text{clean surface}} = \mathrm{V}_{\text{monolayer}} - \mathrm{V}_{\text{clean surface}} \tag{1.102}$$

The magnitude of **ΔV** is measured most conveniently by placing an air electrode (a radiation emitter: e.g., Po^{210} [alpha-emitter]) near the surface (ca. mm in air) connected to a very high impedance electrometer. This is required, since the resistance in air is very high, but it is appreciably reduced by the radiation electrode.

Air electrode
Lipid monolayer IIIIIIIIIIIIIIIIIIIIIIIIIIIII
Water phase electrode

The surface voltage is measured as a change in **V** for pure water surface to that of lipid monolayer system.

Since these monolayers are found to be very useful biological cell membrane structures, it is thus seen that such studies can provide information on many systems where ions are carried actively through cell membranes (Birdi, 1989, 1999; Chattoraj and Birdi, 1984).

The transport of K^+ ions through cell membranes by antibiotics (valinomycin) has been a very important example. Addition of K^+ ions to the subphase of a valinomycin monolayer showed that the surface potential became positive. This clearly indicated the ion-specific binding of K^+ to valinomycin. In biological cells the concentrations of different ions are sometimes 20 times different than in the solution outside the cell. The membrane peptides, such as valinomycin, create a channel in the cell membrane for K^+ ions. This leads to the effect that concentration of K^+ ions becomes the same both inside and outside the cell. This leads to the collapse of the cell. Further, one finds that the theoretical basis of charged lipid monolayers is very well verified from such model monolayer studies.

1.5.6 Charged Lipid Monolayers on Liquid Surfaces

Molecules with charges are found to behave differently than those with no charges (i.e., neutral). In some instances, even it is found that negatively charged molecules behave much differently than positively charged molecules. There is no direct method that can be used to measure these properties. The spread monolayers have provided much useful information about the role of charges at interfaces. In the case of a aqueous solution consisting of fatty acid or SDS, RNa, and NaCl, for example, the Gibbs equation (1.102) may be rewritten as follows (Adamson and Gast, 1997; Birdi, 1989; Chattoraj and Birdi, 1984):

$$-d\gamma = \Gamma_{RNa} d\mu_{RNa} + \Gamma_{NaCl} d\mu_{NaCl} \quad (1.103)$$

Further,

$$\mu_{RNa} = \mu_R + \mu_{Na} \quad (1.104)$$

$$\mu_{NaCl} = \mu_{Na} + \mu_{Cl} \quad (1.105)$$

it can be easily seen that the following will be valid:

$$\Gamma_{NaCl} = \Gamma_{Cl} \quad (1.106)$$

and

$$\Gamma_{RNa} = \Gamma_R \quad (1.107)$$

It is also seen that the following equation will be valid for this system:

$$-d\gamma = \Gamma_{RNa} d\mu_{RNa} + \Gamma d\mu_{Na} + \Gamma d\mu_{Cl} \quad (1.108)$$

This is the form of the Gibbs equation for an aqueous solution containing three different ionic species (e.g., R; Na; Cl). Thus, the more general form would for solutions containing i-number of ionic species as

$$-d\gamma = \Sigma \Gamma_i d\mu_i \quad (1.109)$$

In the case of charged lipid film, the interface will acquire *surface charge*. The surface charge may be positive or negative depending upon the cationic or anionic nature of the lipid or polymer ions. This would lead to the corresponding *surface potential*, ψ, also having a positive or negative potential (Attard, 1996; Birdi, 1989; Chattoraj and Birdi, 1984). The interfacial phase must be electroneutral. This can only be possible if the inorganic counter-ions also are preferentially adsorbed in the interfacial region. If a negatively charged lipid molecule, R^-Na^+, is adsorbed at the interface, the latter will be negatively charged [air–water or oil–water]. According to the Helmholtz model for monolayer, Na^+ ions on the interfacial phase will be arranged in a plane toward the aqueous phase. The charge densities are equal in magnitude, but with opposite signs, Γ (charge per unit surface area), in the adjacent planes. The (negative) charge density of the plane is related to the surface potential [negative], ψ_o, at the Helmholtz charged plane:

$$\psi_o = \frac{(4\pi\sigma\delta)}{\mathbf{D}} \tag{1.110}$$

where **D** is the dielectric constant of the medium (aqueous). According to Helmholtz double layer model, the potential ψ decreases sharply from its maximum value, ψ_o, to zero as δ becomes zero (Birdi, 1989, 1999, 2010a). The Helmholtz model was found not to be able to give a satisfactory analyses of measured data. Later, another theory of the diffuse double layer was proposed by Gouy and Chapman. Due to surface potential, the Na^+ and Cl^- ions will be distributed nonuniformly due to the electrostatic forces. The concentrations of the ions near the surface can be estimated by the Boltzmann distribution. The surface charge, σ, is derived from these assumptions (Adamson and Gast, 1997; Birdi, 2010a; Chattoraj and Birdi, 1984):

$$\sigma = \left[\frac{2\mathbf{DRTC}}{1000\pi}\right]^{1/2}\left[\sinh\left(\frac{\varepsilon\psi^o}{2k\mathbf{T}}\right)\right] \tag{1.111}$$

At 25°C, for uni-univalent electrolytes one gets the following (Adamson and Gast, 1997; Birdi, 2009):

$$k = 3.282 \times 10^7 \text{ cm/C} \tag{1.112}$$

This relates the potential charge of a plane plate condenser to the thickness 1/k. The expression for Π_{el} is derived as

$$\Pi_{el} = 6.1\mathrm{C}^{1/2}\left(\cosh\sinh^{-1}\left(\frac{134}{\mathrm{A}_{el}\mathrm{C}^{1/2}}\right)^{-1}\right) \tag{1.113}$$

The quantity (k**T**) is approximately 4 × 10^{-14} erg at ordinary room temperature (25°C), and (k**T**/ε) = 25 mV. The magnitude of Π_{el} can be estimated from monolayer studies at varying pH. At the iso-electric pH, the magnitude of Π_{el} will be zero (Birdi, 1989). These Π versus A isotherms data at varying pH subphase have been used to estimate Π_{el} in different monolayers.

At the iso-electric pH, the value of Π_{el} was found to be zero, as expected.

1.5.7 Effect of Lipid Monolayers on Evaporation Rates of Liquids

In arid and semi-arid areas, the amount of water lost from reservoirs and lakes by evaporation frequently exceeds the amount beneficially used. As is well known, loss of water by evaporation from lakes and other reservoirs is a very important phenomenon in those parts of world where water is not readily available. Further, from ecological considerations (rain fall and temperature changes), the evaporation phenomenon has much importance for global temperature and other phenomena. Cloud formation depends on the evaporation of water from rivers, lakes, and oceans. The clouds are thus isolators for sunlight reaching on the earth. This cycle is thus related to global heating process. This has been discussed in current literature with regard to CO_2 and global heating of the earth. The reduction of even just a part of the evaporation losses would therefore be of incalculable value for climate control.

The amount of loss of water from surface will be dependent on different parameters:

$$\text{Fall in water level} + \text{rain} - \text{fall} = \text{evaporation} + \text{seepage} + \text{abstractions}$$

Lipid monolayers are found to have extensive effect on the evaporation rates of water.

In one example, one finds the following data:

Surface area of water = 60 cm^2
Rate of evaporation for pure water surface = 0.66 mg/s
Rate of evaporation with a stearic acid monolayer = 0.34 mg/s

The presence of lipid film thus has a significant effect on the evaporation rate of water.

At air–water interface, water molecules are constantly evaporating and condensing in a closed container. In an open container, water molecules at the surface will desorb and diffuse into the gas phase. It thus becomes important to determine the effect of a monomolecular film of amphiphile at the interface. Measurement of the evaporation of water through monolayer films was found to be of considerable interest in the study of methods for controlling evaporation from great lakes. Many important atmospheric reactions involve interfacial interactions of gas molecules (oxygen and different pollutants) with aqueous droplets of clouds and fogs, as well as ocean surfaces. The presence of lipid monolayer films would thus have an appreciable effect on such mass transfer reactions.

In the original procedure, the box containing the desiccant is placed over the water surface, and the amount of water sorbed is determined by simply removing the box and weighing it (Adanson and Gast, 1997; Birdi, 1989; Moroi et al., 2000). The results are generally expressed in terms of specific evaporation resistance, **r**. The methods for calculating r from the water uptake values, together with the assumptions involved, are described in detail in the aforementioned references. The rates of evaporation are measured both without (R_w) and with (R_f) the film. The resistance **r** is given as follows:

$$\mathbf{r} = A(\mathbf{v}_\mathbf{w} - \mathbf{v}_\mathbf{d})(R_f - R_w) \tag{1.114}$$

where
A is the area of the dish
$\mathbf{v}_\mathbf{w}$ and $\mathbf{v}_\mathbf{d}$ are the water vapor concentrations for water and desiccant, respectively

The condensed monolayers gave a much higher **r** values than the expanded films, as expected. Many decades ago, it was found that the evaporation rates of water were reduced by the presence of monomolecular lipid film (Adamson and Gast, 1997). This observation gave rise to important application in reducing the loss of water from great lakes during summer in countries such as Australia and Africa.

The most simple method to investigate is to measure the loss of weight of a water container, with or without the presence of a monomolecular lipid film. La Mer investigated the effect of long chain films ($C_{14}H_{29}OH$, $C_{16}H_{33}OH$, $C_{18}H_{37}OH$, $C_{20}H_{41}OH$, $C_{22}H_{45}OH$) and found that the resistance to evaporation increased with the chain length (Adamson and Gast, 1997). For instance, the resistance increased by a factor 40 for $C_{22}H_{45}OH$ as compared to $C_{14}H_{29}OH$ monolayer. This indicates that evaporation takes place mainly through the alkyl chains films. Since the attraction between alkyl chains increases with the number of carbon atoms (as also observed from the collapse pressure, Π_{col}), the resistance to evaporation increases.

1.5.8 Monolayers of Macromolecules at Water Surface

It is already obvious that monolayers on water are only stable if the hydrophobic part of the molecule is of right magnitude as compared to the polar part. Many macromolecules (such as proteins) form stable monolayers at water surface, if the HLB is of the right balance. Especially, almost all proteins (hemoglobin, ovalbumin, bovine serum albumin, lactoglobulin, etc.) are reported to form stable monolayers at the water surface (Birdi and Fasman, 1973; Birdi, 1999).

1.5.9 Langmuir–Blodgett Films (Transfer of Lipid Monolayers on Solids)

It is obvious that the lipid monolayers as spread on water surface are not easily visualized at molecular scale. Some decades ago, it was reported that when a clean glass plate was dipped into water covered by a monolayer of oleic acid, an area of the monolayer equal to the area of the plate dipped was deposited on withdrawing the plate. Later, it was found that any number of layers could be deposited successively by repeated drippings, later called Langmuir–Blodgett LB method (Birdi, 2003a,b, 2009). However, the films deposited by the LB technique have only recently been used in the electrical applications. The thickness in LB films can be varied from only one monomolecular layer (ca. 25 Å = 25×10^{-10} m), while this is not possible by the evaporation procedures. Monomolecular layers (LB films) of lipids are of interest in a variety of applications, including the preparation of very thin controlled films for interfaces in solid-state electronic devices (Birdi, 1989, 1999; Gaines, 1966). The process of transferring the spread monolayer film to a solid surface by raising the solid surface through the interface has been studied for many decades. This process of transfer has been found that if the monolayer is a closely packed state, then the monolayer is transferred to the solid surface, most likely without any change in the packing density. In recent years, by using the modern AFM techniques, one can determine the molecular orientation and packing of such LB films (Birdi, 1989, 1999, 2003a; Gaines, 1966; Yang et al., 2006).

LB film technology is rapidly developing in areas such as

Micro-lithography
Solid-state polymerization
Light guiding, electron tunneling
Medical diagnostic chips
Photovoltaic effects

In the case of films such as Mg-stearate, if one dips a clean glass slide through the film, a monolayer is adsorbed on the down-stroke. Another layer is adsorbed on the up-stroke. Under careful conditions, one may make LB films with multilayers (varying from a few layers to over thousands). One can monitor the adsorption by measuring the decrease in Π on each stroke. If no adsorption takes place, then one observes no change in surface pressure. There are some lipids, such as cholesterol, which do not form LB films. There are also other methods, such as

Light reflection
IR spectroscopy
STM (scanning tunneling microscope) and AFM (atomic force microscope)
Change in contact angle

which can provide detailed information on these LB films. The LB deposition is traditionally carried out in the solid phase. The surface pressure, Π, is then high enough to ensure sufficient cohesion in the monolayer, for example, the attraction between the molecules in the monolayer is high enough so that the monolayer does not fall apart during transfer to the solid substrate. This also ensures the buildup of homogeneous multilayers. The surface pressure value that gives the best results depends on the nature of the monolayer and is usually established empirically. However, amphiphiles can seldom be successfully deposited at surface pressures lower than 10 mN/m and at surface pressures above 40 mN/m collapse and film rigidity often pose problems. When the solid substrate is hydrophilic (glass, SiO_2, etc.), the first layer is deposited by raising the solid substrate from the subphase through the monolayer, whereas if the solid substrate is hydrophobic (graphite, silanized SiO_2, etc.), the first layer is deposited by lowering the substrate into the subphase through the monolayer. One finds that in some LB systems, the magnitude of Π drops to/by about 1–2 mN/m as each time a plate (with surface area of 5 cm^2) is moved down or up through the monolayer. It is thus a very sensitive method to study the LB deposition phenomena directly. There are several parameters that affect what type of LB film is produced. Depending on the behavior of the molecule, the solid substrate can be dipped through the film until the desired thickness of the film is achieved.

It is obvious that such processes involving monomolecular film transfers will easily be disturbed by defects, arising from various sources. The structural analysis of the molecular ordering within a single LB monolayer is important both to understand how the environment in the immediate vicinity of the surface (i.e., solid) affects the structure of the molecular monolayer and to ascertain how the structure of one layer forms a template for subsequent layers in a multilayer formation.

1.5.9.1 Electrical Behavior of LB Films

Insulating thin films in the thickness range 100–20,000 Å (100–20,000 × 10^{-10} m) have been a subject of varied interest among the scientific community because of the potential applied significance for developing devices, such as optical, magnetic, electronic, etc. Some of the unusual electrical properties possessed by thin LB films, which are unlike those of bulk materials, lead to thinking about their technological applications, and, consequently, interest in thin film studies grew rapidly. In earlier studies one did not prove to be very inspiring, because the LB films obtained always suffered with the presence of pinholes, stacking faults and other imperfections, etc., and hence, the results were not reproducible. It is only in the past few decades that many sophisticated methods have become available for the production and examination in thin films and reproducibility of the results could be controlled to a greater extent. Nevertheless, the unknown nature of inherent defects and a wide variety of thin film systems still complicate the interpretation of many experimental data and thus hinder their use in the devices. It is found that the breakdown conduction in thin films, the major subject of investigation, has been based on the films prepared by thermal evaporation under vacuum or similar techniques. It was realized that the LB films have remained less known among the investigators of this field. The various interesting physical properties of LB films have been investigated in the current literature. As the LB films are very sensitive assemblies, it is necessary that these structures are perfect. There are two crucial factors for making satisfactory electrical measurements on the LB films, which are uniform packing and thickness.

No direct method exists by which one can study the monolayer film molecular structures on water (i.e., in situ). Therefore, one has been using the LB method to study the molecular structures in the past decades. This procedure has been found to provide the most useful information for monolayers. LB deposited film structure is the well-known electron diffraction technique (or scanning probe microscopes (SPMs) [Birdi, 2003b]). The molecular arrangement of deposited monolayer and multilayer films of fatty acids and their salts using this technique has been reported. These analyses showed that the molecules were almost perpendicular to the solid surface in the first monolayer. It was also reported that Ba-stearate molecules have a more precise normal alignment than in the case of stearic acid monolayers. In some investigations, thermal stability of these films has been found to be remarkably stable up to 90°C.

Based on these structural analyses obtained by electron diffraction technique, these deposited films are known to be mono-crystalline in nature; thus, they can be regarded as a special case of a layer-by-layer mechanical growth forming almost *two-dimensional* crystals. There is, however, evidence that Ba-behenate multilayers do in fact show an absence of crystallization which has been demonstrated by electron micrographic studies. Nevertheless, it would be an over-simplification to regard the film transferred at Π high as perfectly uniform, coherent, and defect free. The unidirectional surface conductance of monolayers of stearic acid deposited on a glass support was investigated. The contact angles and adhesional energy changes during the transfer of monolayers from the air–water interface to solid (hydrophobic glass) supports have been analyzed (Birdi, 1989; Gaines, 1966).

Interest in the dielectric studies of deposited LB films of fatty acids and their metal salts was one of the parameters of main investigations in the early stages of research on LB films, for example, capacitance, resistance, and dielectric constant. In early investigations, measurements on impedance of films and related phenomena were carried on Cu-stearate, Ba-stearate, and Ca-stearate films. Initially, Hg droplets were used for small area probe measurements and an AC bridge was used for impedance measurements. The resistance of the films was found to be very low (<1 Ω) with high signal voltages, whereas it was on the order of mega-ohms with signals of 1 or 2 V. In both types of films, the capacity decreased with thickness, as can be expected from the following relation:

$$\text{Capacitance of the deposited LB films} = \mathbf{C_C} = \frac{\mathbf{A}\varepsilon}{4\pi\mathbf{Nd}} \tag{1.115}$$

where

$\mathbf{C_C}$ is the capacity
ε denotes the dielectric constant
N is the number of layers
d is the layer thickness
A is the area of contact between drop and film

On the other hand, the values of the resistance per layer showed a definite increase with the thickness of the film. The specific resistance of the films thus determined from their values of the resistance per layer was ca. 10^{13} Ω. This was based on the results of capacity measurements on some 75 samples. The capacitance measurements thus performed on stearate films (1–10 layers) led to ε values between 2.1 and 4.2, with a bulk value of 2.5.

In many of the measurements reported in the literature, the organic film was sandwiched between an evaporated aluminum electrodes. The fact that an oxide layer grows on the base of aluminum electrode was present and its effect on the capacitance values of the device was neglected considering that the resistivity of oxide film is small compared with resistivity of the organic LB layers. The presence of such thin oxide layer between metal electrodes and fatty acids can be analyzed. The capacitance has been reported to be a linear function of [$1/C_C$] with respect to the number of transferred monolayers (Figure 4.12). LB films of Ba-salts of fatty acids deposited at $\Pi = 50$ mN/m (Birdi, 1999, 2003b) gave following relation between [$1/\mathbf{C_C}$] and **N**:

Ba-stearate:

$$\frac{1}{\mathbf{C_C}} = 15.9\mathbf{N} + 1.13\ (10^6\ \mathrm{F}^{-1}) \tag{1.116}$$

Ba-behenate:

$$\frac{1}{\mathbf{C_C}} = 17.2\mathbf{N} + 8\ (10^6\ \mathrm{F}^{-1}) \tag{1.117}$$

The dielectric anisotropy of long-chain fatty acid monolayers was analyzed. These fatty acids were considered as being oriented in a cylinder cavity with length [L] ≫ diameter [D]. Considering each bond in these molecules as a polarization ellipsoid with axial symmetry about the –C–C– bonds, the mean polarizability of the bonds was calculated.

1.5.9.2 Physical Properties of LB Films

The surface property of a solid changes even after *one layer* of lipid is formed as an LB film: the contact angle decreases after LB deposition (besides other properties). Similarly, many other physical methods have shown that LB films change the surface characteristics of the solid (Adamson and Gast, 1997; Birdi, 2003a,b; Gaines, 1966).

Fourier transform infra red (FTIR)–attenuated total reflection (ATR) spectra have been investigated of LB films of stearic acid deposited on a germanium plate with 1, 2, 3, 5, and 9 monolayers. C=O stretching band at 1702 cm^{-1} was missing for the monolayer. The intensity increased linearly for the multilayer samples. CH scissoring band at 1468 cm^{-1} appeared as a singlet in the case of one-monolayer. A doublet at 1473 and 1465 cm^{-1} was observed for films containing more than three monolayers. Band progression due to $-CH_2-$ wagging vibration of the trans-zigzag hydrocarbon chains is known to appear between 1400 and 1180 cm^{-1}. The intensities increased in this region with the number of layers (Birdi, 1999).

1.5.10 Bilipid Membranes

Scientists have been engaged in determining the molecular structures of biological membranes. However, it was realized that the biological cells were contained by some kind of a thin *lipid membrane*. Some decades ago, one did not have any experimental technique, which allowed one to see the molecular structure of cells. In order to analyze this in more detail, experiments were made (a few decades ago) as follows. Lipids were extracted from the biological cells. These lipids were compressed on the Langmuir balance, and the value of area per molecule was estimated (ca. 45 Å^2/molecule). Knowing the diameter of the cells and from the amounts of lipids (and the area/molecule data), one reached the conclusion that the cell membranes were composed of *bilayer of lipids* (BLM). This was one of the most important results in the history of biological cell membranes. Later, of course, these results were confirmed from x-ray diffraction data and other SPMs (SPMs) (Birdi, 1999, 2003b; Woodson and Liu, 2007).

1.5.11 Vesicles and Liposomes

The most important property all the surface-active molecules have in common is the self-assembly characteristics. It was mentioned that ordinary surfactants (soaps, etc.) when dissolved in water form *self-assembly* micellar structures. The phospholipids are molecules like surfactants; they also have a hydrophilic head and generally have two hydrophobic alkyl chains. These molecules are the main components of the membranes of cells. The lung fluid also consists mainly of lipids of this kind. In fact, usually the membranes of cells are made up of two layers of phospholipids, with the tails turned inward, in an attempt to avoid water.

Phospholipids when dispersed in water may exhibit self-assembly properties (either as micellar self-assembly aggregates or some larger structures). This may lead to aggregates that are called liposomes or vesicles (Birdi, 2009; Harrison et al., 2014; Nair et al., 2014; Reimhult et al., 2003). Liposomes are structures that are essentially empty lipid cells formed due to the self-assembly characteristics. They are microscopic vesicles or containers formed by the lipid membrane alone. They are widely used in the pharmaceutical and cosmetic fields because it is possible to insert chemicals inside them. One may also use liposomes to solubilize (inside the hydrophobic part) hydrophobic chemicals (water-insoluble organic compounds) such as oily substances so that they can be dispersed in an aqueous medium by virtue of the hydrophilic properties of the liposomes (in the alkyl region).

One finds a certain type of lipid (or lipid-like) molecule, which when dispersed in water tends to make self-assembly structures (Figure 1.40). Detergents were shown to aggregate to spherical or large cylindrical-shaped micelles. It is known that if egg phosphatidylethanolamine (egg lecithin)

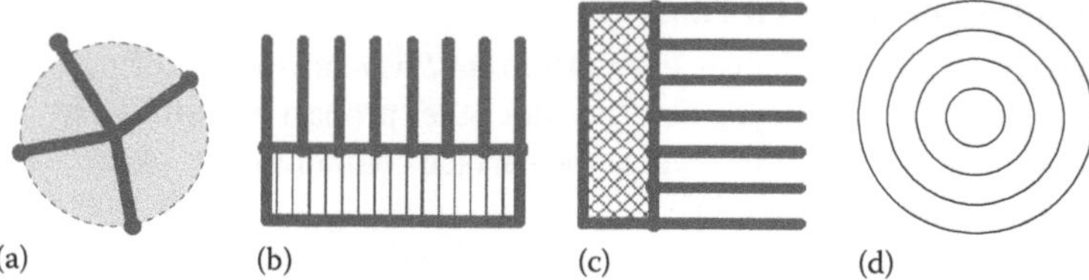

FIGURE 1.40 Different lipid self-assembly monolayer (SAM) structures: (a) micelle; (b) monolayer; (c) LB film; and (d) vesicle.

is dispersed in water at 25°C, it forms self-assembly structure, which is called liposome or vesicle. A liposome is a spherical vesicle with a membrane composed of a phospholipid and cholesterol (less than 50%) bilayer. Liposomes can be composed of naturally derived phospholipids with mixed lipid chains (such as egg phosphatidylethanolamine) or of pure surfactant components like dioleolylphosphatidylethanolamine (DOPE). Liposomes usually contain a core of aqueous solution. Multilayer liposomes are called vesicles. However, one finds a range of mixtures of these structures in mixed lipid systems.

The inherent lipid self-assembly characteristic is the main driving force in these structures. Monolayer studies are the only source of data which provides direct estimation of the stabilizing forces. Hence, it is safe to conclude that many important natural systems are based upon this molecular characteristics of lipids. Vesicles are *unilamellar* phospholipid liposome.

The word *liposome* comprises two words (from Greek-lipid [fat] and *Soma* [body]). The word *liposome* does not in itself denote any size characteristics. Furthermore the term *liposome* does not necessarily mean that it must contain lipophobic contents, such as water, although it usually does. The vesicles may be conceived as microscopic (or *nanosized*) containers of carrying molecules (drugs) from one place to another. The structures are suitable for both transporting water-soluble or water-insoluble drugs. Since the lipids used are biocompatible molecules, this may also enhance their adsorption and penetration into cells.

During the past decades, liposomes have been used for drug delivery due to their unique solubilization characteristics for water-insoluble organic substances. A liposome encapsulates a region on aqueous solution inside a hydrophobic membrane; dissolved hydrophilic solutes cannot readily pass through the lipids. Hydrophobic chemicals can be dissolved into the membrane, and in this way, liposome can carry both hydrophobic molecules and hydrophilic molecules. *Liposomes* can also be designed to deliver drugs in different specific ways. Liposomes that contain low (or high) pH can be constructed such that dissolved aqueous drugs will be charged in solution (i.e., the pH is outside the drug's pI range) (isoelectric pH = pI). As the pH naturally neutralizes within the liposome (protons can pass through a membrane), the drug will also be neutralized, allowing it to freely pass through a membrane. These liposomes work to deliver drug by diffusion rather than by direct cell fusion.

Further, another important property of liposomes is their natural property to target cancer cells. The endothelial walls of all healthy human blood vessels are encapsulated by endothelial cells that are bound together by tight junctions. These tight junctions block large particle in the blood from leaking out of the vessel. It is known that the tumor vessels do not contain the same level of seal between cells and are diagnostically leaky. The size of liposomes can be used to play a specific application. For example, liposomes of certain sizes, typically (i.e., than 400 nm), can rapidly enter tumor sites from the blood but are kept in the bloodstream by the endothelial wall in healthy tissue vasculature. Liposome-based anticancer drugs are now being used as drug delivery systems.

Liposomes can be created by shaking or sonicating phospholipids (dissolved in alcohol) in water. Low shear rates create *multilamellar* liposomes, which have many layers like an onion. Continued high-shear sonication tends to form smaller *unilamellar* liposomes. In this technique, the liposome contents are the same as the contents of the aqueous phase. Sonication is generally considered a "gross" method of preparation, and newer methods such as extrusion are employed to produce materials for human use.

1.6 SOLID SURFACES: ADSORPTION AND DESORPTION (OF DIFFERENT SUBSTANCES)

It was mentioned earlier how the surface molecules in a liquid behave differently than in the bulk phase. It is found that in the case of a large variety of applications where the *surface* of a solid has a specific role and function (e.g., active charcoal, talc, cement, sand, and catalysis). Solids are rigid structures and resist any stress effects. It is thus seen that many such considerations in the case of solid surfaces will be somewhat different than for liquids (Adamson and Gast, 1997; Birdi, 2003a; 2009, 2010a; David and Neumann, 2014; Diebold, 2003; Global CCS, 2013; Kamperman and Synytska, 2012; Neumann, 2010; Zhuravlev, 2000). The mirror polished surfaces of metals, marbles, and plastics (such as in electrical appliances, cars, etc.) are of much importance in technology applications. Further, the *corrosion* process of metals (which initiates at surfaces), thus requiring treatments which are based upon surface properties. Surface treatment technology is constantly developing methods to combat *corrosion*, especially in examples such as cars, bridges, housing, steel structures, etc. (Birdi, 2009; Roberge, 2006). The molecules at the solid surfaces are not under the same force field as in the bulk phase, Figure 1.41.

The differences between *perfect* crystal surfaces and surfaces with *defects* are very obvious in many everyday observations. The solids were the first material that were analyzed at molecular scale. This led to the understanding of the structures of solid substances and the crystal atomic structure. This is because while molecular structures of solids can be investigated by methods such as x-ray diffraction, the same analyses for liquids are not that straightforward. These analyses have also shown that surface defects at molecular level exist.

As pointed out for liquids, one will also need to consider that when surface area of a solid powder is increased by grinding, surface energy is needed. Of course, due to the energy differences between solid and liquid phases, these processes will be many orders of magnitude different from each other. Molecules in the solid are fixed, while in the liquid state, the molecules exchange places. The average distance between molecules in the liquid state is roughly 10% larger than in its solid state. The surface tension of a liquid becomes important when it comes in contact with a solid surface. The interfacial forces are responsible for self-assembly formation and stability on solid surfaces. The interfacial forces that are present between a liquid and solid can be estimated by studying the shape of a drop of liquid placed on any smooth solid surface (Figure 1.42). The balance of forces as indicated (using geometrical considerations) was analyzed very extensively in the last century by Young (1805), who related the different forces at the solid liquid boundary and the contact angle, θ,

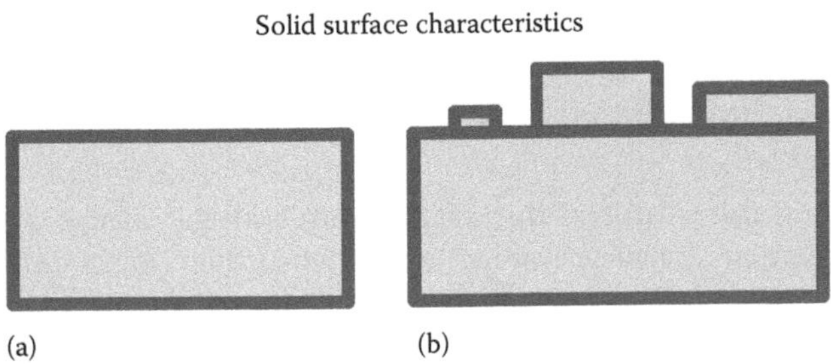

FIGURE 1.41 Solid surface molecules: (a) perfect crystal and (b) surface with defects.

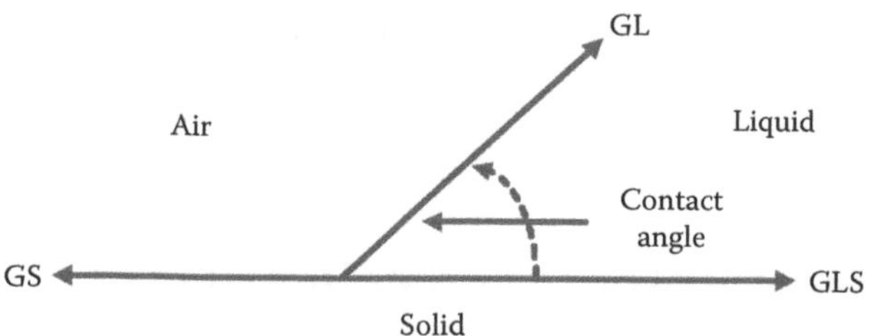

FIGURE 1.42 The surface force equilibrium between surface tensions of liquid (GL)–solid (GS)–liquid–solid (GLS)–contact angle (CA).

as follows (Adamson and Gast, 1997; Birdi, 1997, 2003a; Chattoraj and Birdi, 1984; Coertjens et al., 2014; Karapetsas et al., 2014; Nikolov and Wasan, 2014; Ramiasa et al., 2014):

$$\text{Surface tension of solid } (\gamma_S) = \text{Surface tension of solid/liquid } (\gamma_{SL}) + \text{Surface tension of liquid } (\gamma_L)(\cos(\theta)) \quad (1.118)$$

$$\gamma_S = \gamma_L \cos(\theta) + \gamma_{SL} \quad (1.119)$$

$$\gamma_L \cos(\theta) = \gamma_S - \gamma_{SL} \quad (1.120)$$

where γ is the IFT at the various boundaries between solid, S, liquid, L, and air (or vapor) phases, respectively. The relation of Young's equation is easy to understand as it follows from simple physics laws. At the equilibrium contact angle, all the relevant surface forces come to a stable state, Figure 1.42.

The equilibrium of forces is valid in all kinds of solid–liquid systems. The geometrical force balance is considered only in the X–Y plane. This assumes that the liquid does not affect the solid surface (in any physical sense). This assumption is safe in most cases. However, only in very special cases, if the solid surface is soft (such as contact lens), then one will expect that tangential forces will also need to be included in Equation 1.119. There exist extensive data, which convincingly support the relation in Equation 1.119 for liquids and solids.

1.6.1 Solid Surface Tension (Wetting Properties of Solid Surfaces)

One may ask how can there be surface tension of a solid and how one can measure this force. This question can be answered by the following arguments. Wetting of solid surfaces is well known when considering the difference between teflon and metal surfaces. To understand the degree of *wetting*, between the liquid, L, and the solid, S, it is convenient to rewrite Equation 1.120 as follows:

$$\cos(\theta) = \frac{(\gamma_S - \gamma_{LS})}{\gamma_L} \quad (1.121)$$

which would then allow one to analyze the variation of γ with the change in the other terms. The latter is important, because complete wetting occurs when there is no finite contact angle, and thus, $\gamma_L > \gamma_S - \gamma_{LS}$. However, when $\gamma_L > \gamma_S - \gamma_{LS}$, then $\cos(\theta) < 1$, and a finite contact angle is present. The latter is the case when water, for instance, is placed on hydrophobic solid, such as teflon, polyethylene (PE), or paraffin. The addition of surfactants to water, of course, reduces γ_L; therefore, θ will decrease on the introduction of such SASs (Adamson and Gast, 1997; Birdi, 1997, 2003a; Chattoraj and Birdi, 1984; Eusthopoulos, 1999). The state of a fluid drop under dynamic conditions, such as *evaporation,* becomes more complicated (Birdi et al., 1988). However, we are interested in the spreading behavior of a drop of one liquid when placed on another (immiscible) liquid. The spreading phenomenon by introducing a quantity, spreading coefficient, $\mathbf{S}_{a/b}$, is defined as follows (Adamson and Gast, 1997; Birdi, 2003a; Harkins 1952):

$$\mathbf{S}_{a/b} = \gamma_a - (\gamma_b + \gamma_{ab}) \tag{1.122}$$

where

$\mathbf{S}_{a/b}$ is the spreading coefficient for liquid **b** on liquid **a**
γ_a and γ_b are the respective surface tensions
γ_{ab} is the IFT between the two liquids

If the value of $\mathbf{S}_{b/a}$ is positive, spreading will take place spontaneously, while if it is negative, liquid **b** will rest as a lens on liquid **a**.

However, the value of γ_{ab} needs to be considered as the equilibrium value, and therefore, if one analyzes the system at a nonequilibrium state, then the spreading coefficients would be different. For example, the instantaneous spreading of benzene is observed to give a value of $\mathbf{S}_{a/b}$ as 8.9 dyn/cm, and therefore, benzene spreads on water. On the other hand, as the water phase gets saturated with benzene with time, the value of the spreading coefficient decreases. This leads to the formation of lenses. The short-chain hydrocarbons such as hexane and hexene also have positive initial spreading coefficients and spread to give thicker films. Longer-chain alkanes, on the other hand, do not spread on water, for example, the $\mathbf{S}_{a/b}$ for n-$C_{16}H_{34}$ (n-hexadecane)/water is −1.3 dyn/cm at 25°C. It is also obvious that since impurities can have very drastic effects on the IFTs in Equation 1.122, the value of $\mathbf{S}_{a/b}$ would be expected to vary accordingly (see Table 1.7).

The spreading of a solid substance, for example, cetyl alcohol ($C_{18}H_{38}OH$), on the surface of water has been investigated in some detail (Adamson and Gast, 1997; Birdi, 2003a; Gaines, 1966). Generally, however, the detachment of molecules of the amphiphile into the surface film occurs only at the periphery of the crystal in contact with the air–water surface. In this system, the diffusion of amphiphile through the bulk water phase is expected to be negligible, because the energy barrier now includes not only the formation of a hole in the solid, but also the immersion of the hydrocarbon chain in the water. It is also obvious that the diffusion through the bulk liquid is a rather slow process. Furthermore, the value of $\mathbf{S}_{a/b}$ would be very sensitive to such impurities as regards spreading of one liquid upon another.

Another typical example is as follows: the addition of surfactants (detergents) to a fluid dramatically affects its wetting and spreading properties. Thus, many technologies utilize surfactants for the control of wetting properties (Birdi, 1997). The ability of surfactant molecules to control wetting arises from their *self-assembly* at the liquid–vapor, liquid–liquid, solid–liquid, and solid–air interfaces and the resulting changes in the interfacial energies (Birdi, 1997). These interfacial *self-assemblies* exhibit rich structural detail and variation. In the case of *oil spills* on the seas, these considerations

TABLE 1.7
Calculation of Spreading Coefficients, $S_{a/b}$, for Air–Water Interfaces (20°C)

Oil	$\gamma_{w/a} - \gamma_{o/a} - \gamma_{o/w} = S_{a/b}$	Conclusion
n-$C_{16}H_{34}$	72.8 − 30.0 − 52.1 = −0.3	Will not spread
n-Octane	72.8 − 21.8 − 50.8 = +0.2	Will just spread
n-Octanol	72.8 − 27.5 − 8.5 = +36.8	Will spread

a, air; w, water; o, oil.

become very important. The treatment of such pollutant systems requires the knowledge of the state of the oil. The thickness of the oil layer will be dependent on the spreading characteristics. The effect on ecology (such as birds, plants) will depend on the spreading characteristics. Young's equation at liquid$_1$–solid–liquid$_2$ has been investigated for various systems. This is found in such systems where the liquid$_1$–solid–liquid$_2$ surface tensions meet at a given contact angle. For example, the contact angle of water drop on teflon is 50° in octane (Chattoraj and Birdi, 1984) (Figure 1.43):

water Teflon octane

In this system, the contact angle, θ, is related to the different surface tensions as follows:

$$\gamma_{s-\text{octane}} = \gamma_{\text{water}-s} + \gamma_{\text{octane}-\text{water}} \cos(\theta) \quad (1.123)$$

or

$$\cos(\theta) = \frac{(\gamma_{s-\text{octane}} - \gamma_{\text{water}-s})}{\gamma_{\text{octane}-\text{water}}} \quad (1.124)$$

Experimental value of is found to be 50°. This agrees with the calculated value of θ = 50°, when using the measured values of $(\gamma_{s-\text{octane}} - \gamma_{\text{water}-s})/\gamma_{\text{octane}-\text{water}}$. This analysis showed that the assumptions made in derivation of Young's equation are valid.

The most important property of a surface (solid or liquid) is its capability of interacting with other materials (gases, liquids, or solids). All interactions in nature are governed by different kinds of molecular forces (such as *van der Waals, electrostatic, hydrogen bonds, dipole–dipole interactions*). Based on various molecular models, the surface tension, γ_{12}, between two phases with γ_1 and γ_2, was given as (Adamson and Gast, 1997; Chattoraj and Birdi, 1984):

$$\gamma_{12} = \gamma_1 + \gamma_2 - 2\Phi_{12}(\gamma_1\gamma_2)^2 \quad (1.125)$$

where Φ_{12} is related to the interaction forces across the interface. The latter parameter will depend on the molecular structures of the two phases. In the case of systems such as alkane (or paraffin)–water, there Φ_{12} is found to be unity. Φ_{12} is unity since alkane molecule exhibits no hydrogen bonding property, while water molecules are strongly hydrogen bonded. It is thus found convenient that in all liquid–solid interfaces, there will be present different (apolar (dispersion) forces + polar (hydrogen-bonding; electrostatic forces). Hence, all liquids and solids will exhibit γ composed of different kinds of molecular forces:

Liquid surface tension:

$$\gamma_L = \gamma_{L,D} + \gamma_{L,P} \quad (1.126)$$

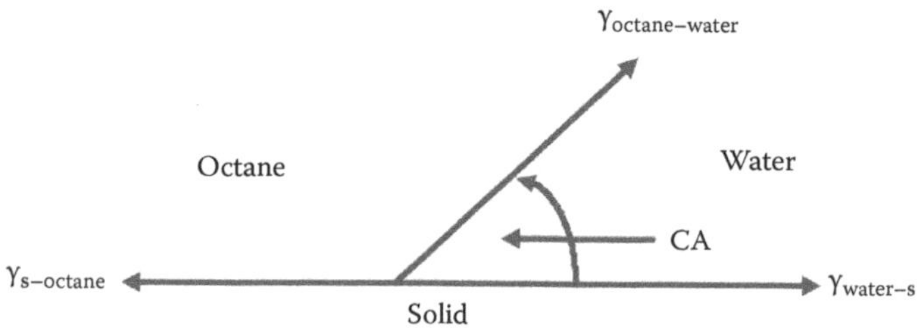

FIGURE 1.43 Contact angle at water–Teflon–octane interface (see Equation 1.123).

Solid surface tension:

$$\gamma_S = \gamma_{S,D} + \gamma_{S,P} \tag{1.127}$$

This means that γ_S for teflon arises only from *dispersion* (γ_{SD}) forces. On the other hand, a glass surface shows γ_S that will be composed of both $\gamma_{S,D}$ and $\gamma_{S,P}$. Hence, the main difference between teflon and glass surface will arise from the $\gamma_{S,P}$ component of glass. This criterion has been found to be of importance in the case of application of adhesives. The adhesive used for glass will need to bind to solid with both polar and apolar forces. This means that one has to design adhesives with varying properties, which depend on the surfaces involved.

The values of γ_{SD} for different solids as determined from these analyses are given in the following.

Solid	γ_S	γ_{SD}	γ_{SP}
Teflon	19	19	0
Polypropylene	28	28	0
Polycarbonate	34	28	6
Nylon 6	41	35	6
Polystyrene	35	34	1
PVC	41	39	2
Kevlar 49	39	25	14
Graphite	44	43	1

In the case of polystyrene surfaces, it was found that the value of sp increased with the treatment of sulfuric acid (due to the formation of sulfonic groups in the surface) (Birdi, 1997). This gave rise to increased adhesion of bacteria cells to the surfaces (Birdi, 1981). The *asymmetrical* forces acting at surfaces of liquids are much shorter than those expected on solid surface. This is due to the high energies, which stabilize the solid structures. Therefore, when one considers solid surface, the *surface roughness* will need to be considered. Although this is a simple analysis of a complicated system, one finds that the model as described here can describe many industrial and biological phenomena.

1.6.2 Definition of Solid Surface Tension (γ_{SOLID})

It was described earlier that the molecules at the surface of a liquid are under tension due to asymmetrical forces, which gives rise to surface tension. However, in the case of solid surfaces, one may not envision this kind of asymmetry as clearly although a simple observation might help one to realize that such surface tension analogy exists.

Let us compare two systems:

Water–teflon
Water–glass

For instance, let us analyze the state of a drop of water (10 μL) as placed on two different smooth solid surfaces, for example, teflon and glass. Experiments show that the magnitudes of the contact angles are different (Figure 1.44).

Since the surface tension of water is the same in the two systems, the difference in contact angle can only arise due to the difference in *surface tension of solids*.

The surface tension of liquids can be measured directly. However, this is not possible in the case of solid surfaces. Experiments show that when a liquid drop is placed on a solid surface, the contact angle, θ, indicates that the molecules interact across the interface. This data of contact angle have been used to estimate the surface tension of solids.

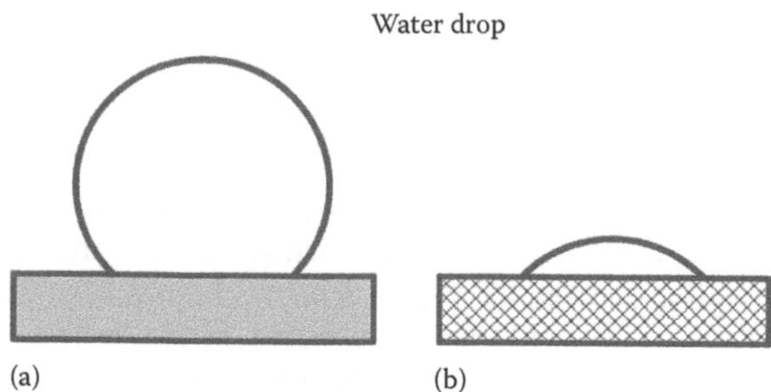

FIGURE 1.44 Drop of water on smooth surface of (a) Teflon and (b) glass.

1.6.3 Contact Angle (Θ) of Liquids on Solid Surfaces

As it is already mentioned, a solid in contact with a liquid leads to interactions related to the surfaces involved (i.e., surface tensions of liquid and solid). The solid surface is being brought in contact with surface forces of the liquid (surface tension of liquid). If a small drop of water is placed on a smooth surface of teflon or glass, Figure 1.47, one finds that these drops are different. The reason is that there are three surface forces (tensions), which at equilibrium give rise to contact angle, θ. The relationship as given by Young's equation describes the interplay of forces (liquid surface tension; solid surface tension; liquid–solid surface tension) at the three-phase boundary line. It is regarded as if these forces interact along a line. Experimental data show that this is indeed true. The magnitude of θ is thus only dependent on the molecules nearest the interface and independent of molecules much further away from the contact line.

Further, one defines that

When θ is less than 90°, the surface is wetting (such as water on glass)
When θ is greater than 90°, the surface is nonwetting (such as water on teflon)

It is also known that by the treatment of glass surface by using suitable chemicals, the surface can be rendered hydrophobic. This is the same technology as used in many utensils, which are treated with teflon or similar items.

1.6.4 Measurements of Contact Angles at Liquid–Solid Interfaces

The magnitude of the contact angle, θ, between a liquid and solid can be determined by various methods. The method to be used depends on the system and on the accuracy required. There are two most common methods: by direct microscope and a goniometer; by photography (digital analyses). It should be mentioned that the liquid drop, which one generally uses in such measurements, is very small, such as 10–100 μL. There are two different systems of interest: liquid–solid and $liquid_1$–solid–$liquid_2$. In the case of some industrial systems (such as oil recovery), one needs to determine θ at high pressures and temperature. In these systems, the value of contact angle can be measured by using photography and suitable high-pressure cells. Recently, digital photography has also been used, since these data can be analyzed by computer programs.

It is useful to consider some general conclusions from these data (Table 1.8). One defines a solid surface as *wetting* if the θ is less than 90°. However, a solid surface is designated as *nonwetting* if θ is greater than 90°. This is a practical and semiquantitative procedure. It is also seen that water, due to its hydrogen bonding properties, exhibits a large θ on nonpolar surfaces (PE, teflon, etc.). On the other hand, one finds lower θ values of water on polar surfaces (glass, mica).

TABLE 1.8
Contact Angles, θ, of Water on Different Solid Surfaces (25°C)

Solid	θ
Teflon (PTE)	108
Paraffin wax	110
Polyethylene	95
Graphite	86
AgI	70
Polystyrene	65
Glass	30
Mica	10

Next, one may consider the effect of *surface charges* on the contact angle of a solid–liquid interface. Metal surfaces will exhibit varying degrees of charges at the surfaces. Biological cells will exhibit charges that will affect adhesion properties to solid surfaces. However, in some applications, one may change the surface properties of solids by chemical modifications of the surface. For instance, polystyrene (PS) has some weak polar groups at the surface. For example, if one treats the PS surface with H_2SO_4, one forms sulfonic groups. This leads to values of θ of water lower than 30° (depending on the time of contact between sulfuric acid and PS surface) (Birdi, 2009). This treatment (or similar) has been used in many applications where the solid surface is modified to achieve a specific property. Since only the surface layer (a few molecules deep) is modified, the solid properties do not change. This analysis shows the significant role of studying the contact angle of surfaces in relation to the application characteristics. The magnitude of contact angle of water (e.g.) is found to vary depending on the nature of the solid surface. The magnitude of θ is almost 100 on a waxed surface of a car paint. The industry strives to create such surfaces to give $\theta > 150$, the so-called superhydrophobic surfaces. Large θ means that water drops do not wet the car polish and are easily blown off by wind.

In many industrial applications, one is both concerned with smooth as well as rough surfaces. The analyses of θ on *rough surfaces* will be somewhat more complicated than on smooth surfaces. The liquid drop on a rough surface, Figure 1.45, may show the *real* θ (solid line) or some lower value (*apparent*) (dotted line), depending on the orientation of the drop.

However, no matter how rough the surface, the forces will be the same as that between a solid and a liquid. In other words, at microscale (i.e., atomic scale), the balance of forces at the liquid–solid and contact angle, surface roughness has no effect. The surface roughness may show contact angle *hysteresis* if one makes the drop move, but this will arise from other parameters (e.g., wetting and dewetting) (Birdi, 1993; Raj et al., 2012). A *fractal* approach has been used to achieve a better understanding of this phenomenon (Barnsley, 1988; Birdi, 1993; Feder, 1988; Koch, 1993; Mandelbrot, 1983).

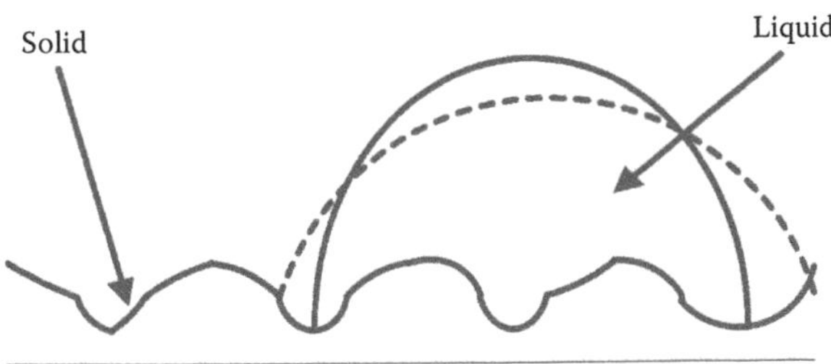

FIGURE 1.45 Analysis of contact angle of a liquid drop on a *rough* solid surface.

TABLE 1.9
Some Typical γ_{cr} Values for Solid Surfaces (Estimated from Zisman Plots)

Surface Group	γ_{cr}
$-CF_2-$	18
$-CH_2-CH_3-$	22
Phenyl–	30
Alkyl chlorine–	35
Alkyl hydroxyl	40

Source: Birdi (2003a).

In spite of its basic assumptions, Young's equation has been found to give useful analyses in a variety of systems. For example, a typical data of cos(θ) versus various liquids on teflon gave an almost straight line plot. The data can be analyzed by the following relation:

$$\cos(\theta) = k_1 - k_2\gamma_L \quad (1.128)$$

This can also be rewritten as

$$\cos(\theta) = 1 - k_3(\gamma_L - \gamma_{cr}) \quad (1.129)$$

where γ_{cr} is the critical value of γ_L at cos(θ) equal to 1 (i.e., =0). The values of γ_{cr} have been reported for different solids using this procedure. The magnitude of γ_{cr} for teflon of 18 mN/m thus suggests that $-CF_2-$ groups exhibit this low surface tension. The value of γ_{cr} for $-CH_2-CH_3-$ alkyl chains gave a higher value of 22 mN/m than for teflon. Indeed, from experience one also finds that teflon is a better water-repellent surface than any other material (Starostina et al., 2014). The magnitudes of γ_{cr} for different surfaces are seen to provide much useful information, Table 1.9.

These data show that the molecular groups of different molecules determine the surface characteristics as related to γ_{cr}. In many cases, the surface of a solid may not behave as desired, and therefore, the surface is treated accordingly, which results in a change of the contact angle of fluids. The low surface energy polymer (PE) is found to change when treated with flame or corona:

Material	Liquid	Contact Angle
PE	Water	87
	Corona	55
PE (corona)	Water	66
	Corona	49

1.6.5 Adsorption of Gases on Solid Surfaces

The most important solid surface property is its interaction with gases or liquids. The adsorption of a gas on a solid surface has been known to be of very much important in various systems (especially in industry involved with catalysis, etc.) (Birdi, 2009; Bonzel, 2014; Do, 1998). The gas–solid surface phenomena can be analyzed as follows.

The molecules in gas are moving very fast, but on adsorption (gas molecules are more or less fixed), there will be thus a large decrease in kinetic energy (thus decrease in entropy, ΔS).

Movement of gas molecules: In gas phase is much larger distances than when adsorbed on a solid surface.

Adsorption takes place spontaneously, which means that ΔG_{ad} is *negative*, which indicates that ΔH_{ad} is negative (exothermic adsorption):

$$\Delta G_{ad} = \Delta H_{ad} - T\Delta S_{ad} \tag{1.130}$$

The adsorption of gas can be of different types. The gas molecule may adsorb on the surface of a solid (as a kind of *condensation* process). It may under other circumstances react with the solid surface (chemical adsorption or chemisorption). In the case of chemoadsorption, one almost expects a chemical bond formation. On carbon while oxygen adsorbs (or chemisorb), one can desorb CO or CO_2. The experimental data can provide information on the type of adsorption. On porous solid surfaces, the adsorption may give rise to *capillary condensation*. This indicates that porous solid surfaces will exhibit some specific properties. The most adsorption process in industry one finds in the case of catalytic reactions (e.g., formation of NH_3 from N_2 and H_2).

Further, it is thus apparent that in gas recovery from shale, the desorption of gas (mainly methane, CH_4) will be determined by the surface forces.

The surface of a solid may differ in many ways from its bulk composition. Especially, solids such as commercial carbon black may contain minor amounts of impurities (such as aromatics, phenol, and carboxylic acid). This would render surface adsorption characteristics different from pure carbon. It is therefore essential that in industrial production, one maintains quality control of surface from different production batches. Otherwise, the surface properties will affect the application. Another example, one may add, arises from the behavior of glass powder and its adsorption character for proteins. It has been found that if glass powder is left exposed to air, then pollutants from air may cover its surface. This leads to lower adsorption of proteins than on a clean surface. Silica surface has been considered to exist as O–Si–O as well as hydroxyl groups formed with water molecules. The orientation of the different groups may also be different at surface. Carbon black has been reported to possess different kinds of chemical groups on their surfaces (Fiueiredo et al., 1999). These different groups are as follows: aromatics, phenol, carboxylic, etc. These different sites can be estimated by comparing the adsorption characteristics of different adsorbents (such as hexane and toluene). When any *clean* solid surface is exposed to a gas, the latter may adsorb on the solid surface to varying degrees (as found from experiments). It has been recognized for many decades that gas adsorption on solid surfaces does not always stop at a monolayer state. Of course, more than one layer (multilayer) adsorption will take place only if the pressure is reasonably high. Experimental data show this when volume of gas adsorbed, v_{gas}, is plotted against p_{gas}.

From experimental analyses, it has been found that five different kinds of adsorption states exist. These adsorption isotherms were classified based on extensive measurements of v_{gas} versus p_{gas} data.

Type I: These are obtained for Langmuir adsorption.
Type II: This is the most common type where multilayer surface adsorption is observed.
Type III: This is somewhat special type with almost only multilayer formation, such as nitrogen adsorption on ice.
Type IV: If the solid surface is porous, then this is found similar to Type II.
Type V: On porous solid surfaces.

The pores in a porous solid are found to vary from 2 to 50 nm (*micropores*).

Macropores are designated for larger than 50 nm. *Mesopores* are used for 2–50 nm range.

1.6.6 Gas Adsorption on Solid Measurement Methods

1.6.6.1 Volumetric Change Methods

The change in volume of gas during adsorption is measured directly in principle, and the apparatus is comparatively simple. One can use a mercury (other suitable liquids) reservoir beneath the manometer, and the burette is used to control the levels of mercury in the apparatus. Calibration involves measuring the volumes of the gas (v_{gas}) lines and of the void space. The apparatus, including the sample, is evacuated, and the sample is heated to remove any previously adsorbed gas. A gas such as helium gas is usually used for the calibration, since it exhibits very low adsorption on the solid surface. After helium gas is pushed into the apparatus, a change in volume is used to calibrate the apparatus and corresponding change in pressure is measured. Different gas (such as nitrogen) is normally used as the adsorbate if one needs to estimate the surface area of a solid. The gas is cooled by liquid nitrogen. The tap to the sample bulb is opened, and the drop in pressure is determined. In the surface area calculations, one uses a value of 0.162 nm^2 for the area of an adsorbed nitrogen molecule.

1.6.6.2 Gravimetric Gas Adsorption Methods

It is obvious that the amount of gas adsorbed on any solid surface will be of a very small magnitude. However, by using a modern sensitive microbalance (or similar procedure), one can measure the adsorption isotherm. The sensitivity is very high since only the difference in weight change is measured. These microbalances can measure weight differences in the range of nanograms to milligrams. It is found possible to measure the weight change caused by the adsorption of a single monolayer on a solid if the surface area is large. The normal procedure is to expose the sample to the adsorbate gas at a certain pressure, allowing sufficient time for equilibrium to be reached and then determining the mass change. This is repeated for a number of different pressures, and the number of moles adsorbed as a function of pressure plotted to give an adsorption isotherm. Microbalances (stainless steel) can be made to handle pressures as high as 120 MPa (120 atm), since gases that adsorb weakly or boil at very low pressures can still be used.

1.6.6.3 Gas Adsorption on Solid Surfaces (Langmuir Theory)

This equation assumes that only *one layer* of gas molecule adsorbs. A monolayer of gas adsorbs in the following case: there are only a given number of adsorption sites for only a monolayer. This is the most simple adsorption model. The amount adsorbed, N_s, is related to the monolayer coverage, N_{sm}, as follows:

$$\frac{N_s}{N_{sm}} = \frac{\mathbf{ap}}{(1+\mathbf{ap})} \tag{1.131}$$

where

p is the pressure
a is dependent on the energy of adsorption

This equation can be rearranged:

$$\frac{\mathbf{p}}{N_s} = \left(\frac{1}{(\mathbf{a}N_{sm})} + \frac{\mathbf{p}}{N_{sm}}\right) \tag{1.132}$$

From the experimental data, one can plot p/N_s versus p. The plot will be linear, and the slope is equal to $1/N_{sm}$. The intersection gives the value of **a**. Charcoal is found to adsorb 15 mg of N_2 as monolayer. Another example is that of adsorption of N_2 on mica surface (at 90 K). The following data were found:

Pressure (Pa)	Volume of Gas Adsorbed (at STP)
0.3	12
0.5	17
1.0	24

Standard temperature and pressure (STP).

In this equation, one assumes

That the molecules adsorb on definite sites
That the adsorbed molecules are stable after adsorption

The surface area of the solid can be estimated from the plot of $\mathbf{p}/N_s$ versus **p**. Most data fit this equation under normal conditions and are therefore widely applied to analyze adsorption process. Langmuir adsorption is found for the data of nitrogen on mica (at 90K). The data were found to be as follows:

$\mathbf{p} = 1\text{–}2/\text{Pa}$
$\mathbf{V}_s = 24\text{–}28\ \text{mm}^3$

This shows that the amount of gas adsorbed increases by a factor 28/24 = 1.2 when the gas pressure increases twofold.

1.6.6.4 Various Gas Adsorption Equations

The gas adsorption on solid surface data has been analyzed by different models. Gas recovery from shale deposits is a very important example of such surface phenomena. Other isotherm equations begin as an alternative approach to the developed equation of state for a two-dimensional ideal gas. As mentioned earlier, the ideal equation of state is found to be as

$$\Pi \mathbf{A} = \mathbf{k_B T} \tag{1.133}$$

In combination with Langmuir equation, one can derive the following relation between N_s and p:

$$N_s = \mathbf{Kp} \tag{1.134}$$

where K is a constant. This is the well-known Henry's law relation, and it is found to be valid for most isotherms at low relative pressures. In those situations where the ideal equation (5.16) does not fit the data, the van der Waals equation type of corrections has been suggested.

The *adsorption–desorption process* is of interest in many systems (such as cement). The water vapor may condense in the pores after adsorption under certain conditions. This may be studied by analyzing the adsorption–desorption data.

Multilayer gas adsorption: As mentioned elsewhere, the fracking process for gas recovery from shale deposits is a process where adsorbed gas is released by the process. In some systems, adsorption of gas molecules proceeds to higher levels where multilayers are observed. From data analyses, one finds that multilayer adsorption takes place, Figure 1.46.

The BET equation has originally been derived for multilayer adsorption only.

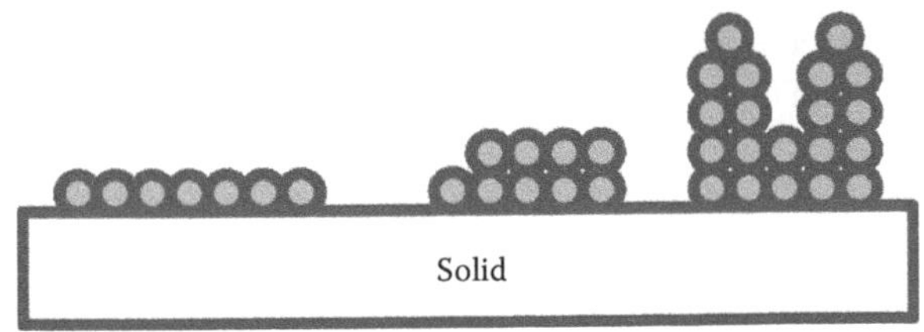

FIGURE 1.46 BET model for multilayer adsorption on solids.

The enthalpy involved in multilayers is related to the differences and was defined by BET theory as

$$E_{BET} = \exp\left[\frac{(E_l - E_v)}{\mathbf{RT}}\right] \quad (1.135)$$

where E_l and E_v are enthalpies of desorption. The BET equation thus after modification of the Langmuir equation becomes

$$\frac{\mathbf{p}}{(N_s(p^o - \mathbf{p}))} = \frac{1}{E_{BET}N_{sm}} + \left[\frac{(E_{BET} - 1)}{(E_{BET}N_{sm})}\frac{\mathbf{p}}{p^o}\right] \quad (1.136)$$

A plot of adsorption data of left-hand side of this equation versus relative pressure (**p**/p°) allows one to estimate N_{sm} and E_{BET}.

1.6.7 Adsorption of Substances (Solutes) from Solution on Solid Surfaces

Any clean solid surface is actually an active center for adsorption from the surroundings, for example, air or liquid. In fact, a clean solid surface can only exist under vacuum. A perfectly cleaned metal surface, when exposed to air, will adsorb a single layer of oxygen or nitrogen (or water) (degree of adsorption will, of course, depend on the system). The most common example of much importance is the process of corrosion (an extensive economic cost) of iron when exposed to air. Or when a completely dry glass surface is exposed to air (with some moisture), the surface will adsorb a monolayer of water. In other words, the solid surface is not as inert as it may seem to the naked eye.

1.6.7.1 Thermodynamics of Adsorption

Activated charcoal or carbon (with the surface area of over 1000 m^2/g) is widely used for vapor adsorption and in the removal of organic solutes from water. These materials are used in industrial processes to purify drinking water and swimming pool water, to decolorize sugar solutions as well as other foods, and to extract organic solvents (especially trace amounts of dangerous substances). Activated charcoal can be made by heat degradation and partial oxidation of almost any carbonaceous matter of animal, vegetable, or mineral origin. For convenience and economic reasons, it is usually produced from bones, wood, lignite, or coconut shells. The complex three-dimensional structure of these materials is determined by their carbon-based polymers (such as cellulose and lignin), and it is this backbone that gives the final carbon structure after thermal degradation. These materials, therefore, produce a very porous high-surface-area carbon in solid form. In addition to a high surface area, the carbon has to be *activated* so that it will interact with and physisorb (i.e., adsorb physically, without forming a chemical bond) a wide range of compounds. This activation process involves controlled oxidation of the surface to produce polar sites. In the present case, the concentration of adsorbate in solution can be monitored. The adsorption process can be analyzed by using the mass action approach.

Solid surface will have molecules arranged at the surface in a very well-defined geometrical arrangement. This will give rise to surface forces, which will determine the adsorption of a particular substance. On any solid surface, one can expect a certain number of possible adsorption *sites* per gram (N_m). This is the number of sites where any adsorbate can freely adsorb. There will be a fraction θ that is filled by one adsorbing solute. It will also be expected that there an *adsorption–desorption* process will exist at the surface. The rate of adsorption will be given as

$$(\text{Concentration of solute})(1-\theta)N_m \tag{1.137}$$

and the rate of desorption a will be given as

$$(\text{Concentration of solute})(\theta)N_m \tag{1.138}$$

It is known that at equilibrium, these rates must be *equal*:

$$k_{ads}C_{bulk}(1-\theta)N_m = k_{des}\theta N_m \tag{1.139}$$

where

k_{ads}, k_{des} are the respective proportionality constants
C_{bulk} is the bulk solution concentration of solute

The equilibrium constant, $K_{eq} = k_{ads}/k_{des}$, gives

$$\frac{C_{bulk}}{\theta} = C_{bulk} + \frac{1}{K_{eq}} \tag{1.140}$$

and since $\theta = N/N_m$, where N is the number of solute molecules adsorbed per gram of solid, one can write

$$\frac{C_{bulk}}{N} = \frac{C_{bulk}}{N_m} + \frac{1}{(K_{eq}N_m)} \tag{1.141}$$

Thus, measurement of N for a range of concentrations (C) should give a linear plot of C_{bulk}/N against C_{bulk}, where the slope gives the value of N_m and the intercept gives the value of the equilibrium constant K_{eq}. This model of adsorption was suggested by Irving Langmuir and is referred to as the *Langmuir adsorption isotherm*.

A typical adsorption experiment is carried out as follows. The solid sample (e.g., activated charcoal) is shaken in contact with a solution with known concentration of acetic acid. After equilibrium is reached (approximately after 24 h), the amount of acetic adsorbed is determined. One can determine the concentration of acetic acid by titration with NaOH solution.

One may also use solutions of dyes (such as methylene blue), and after adsorption, the amount of dye in solution is measured by any convenient spectroscopic method (VIS or UV or flourescence spectroscopy).

1.6.8 Solid Surface Area Determination

As far as surface chemistry is concerned, a solid particle is a very important substance as regards its surface characteristics. In all applications where finely divided powders are used (such as talcum, cement, and charcoal powder), the property of these will depend mainly on the surface area per gram (varying from few m^2 [talcum] to over 1000 m^2/g [charcoal]). For example, if one needs to use charcoal to remove some chemical (such as coloring substances or other pollutants) from wastewater, then it is necessary to know the amount of absorbent needed to fulfill the process.

In other words, if one needs 1000 m^2 area for the adsorption when using charcoal, 1 g of solid will be required. The estimation of surface area of finely divided solid particles from solution adsorption studies is subject to many of the same considerations as in the case of gas adsorption, but with the added complication that larger molecules are involved, whose surface orientation and pore penetrability may be uncertain. The first condition is that a definite adsorption model is obeyed, which in practice means that area determination data are valid within the simple Langmuir equation relation.

In the literature, one has also used fatty acid adsorption for surface area estimation. In these studies it was assumed that fatty acid molecules adsorbed with 20.5 Å^2 area per molecule. This seems to be true for adsorption on such diverse solids as carbon black and for TiO_2. In all of these cases, the adsorption is probably chemisorption in involving hydrogen bonding or actual salt formation with surface oxygen. As another example, the adsorption of surfactants on polycarbonate indicated that depending on the surfactant and concentration, the adsorbed molecules might be lying flat on the surface perpendicular to it, or might form a bilayer. In some cases, one has also used dyes as adsorbates. This method is appealing because of the ease with which analysis may be made calorimetrically. The adsorption generally follows the Langmuir equation. An apparent molecular area of 19.7 Å^2 for methane blue on graphon was found (the actual molecular area is 17.5 Å^2). The fatty acid adsorption method has been used by many investigators. Surface areas of oxide powders were estimated from pyridine adsorption studies. The adsorption data followed the Langmuir equation; the effective molecular area of pyridine is about 24 Å^2 per molecule. In the literature many different approaches have been proposed to estimate the surface area of a solid (Adamson and Gast, 1997; Pereira et al., 2003). Surface areas may be estimated from the exclusion of like-charged ions from a charged interface. This method is intriguing in that no estimation of either site or molecular area is needed. In general, however, surface area determination by means of solution adsorption studies, while convenient experimentally, may not provide the most correct information. Nonetheless, if a solution adsorption procedure has been standardized for a given system, by means of independent checks, it can be very useful for determining relative areas of a series of similar materials. In all cases, it is also more real as it is what happens in real life.

1.6.8.1 Adsorption of a Detergent Molecule

The detergent molecule, such as dodecyl ammoniumchloride, was found to adsorb 0.433 mM per gram of alumina with a surface area of 55 m^2/g. The surface area of alumina as determined from stearic acid adsorption (and using the area/molecule of 21 Å^2 from monolayer) gave the value of 55 m^2/g.

These data can be analyzed in more detail.

$$\text{Surface area of alumina} = 55\ \text{m}^2/\text{g}$$

$$\text{Amount adsorbed} = 0.433\ \text{mM/g} = (0.433\times10^{-3}\ \text{M})(6\times10^{23}\ \text{molecules/mol})$$

$$= 0.433\times10^{20}\ \text{molecules}$$

$$\frac{\text{Area}}{\text{Molecule of detergent}} = \frac{55\times10^4\ \text{cm}^2\ (10^{16}\ \text{Å}^2)}{0.433\times10^{20}\ \text{molecule}} = \frac{55}{0.433\ \text{Å}^2} = 127\ \text{Å}^2$$

The adsorption isotherms obtained for various detergents showed a characteristic feature that an equilibrium value was obtained when the concentration of detergent was over CMC. The adsorption was higher at 40°C than at 20°C. However, the shapes of the adsorption curves were the same (Birdi, 2003a). Detergents adsorb on solids until CMC, after which no more adsorption is observed. This shows that after the solid surface is covered by a monolayer of detergent, the adsorption stops as the system reaches an equilibrium.

Example

One can also calculate the amount of a small molecule, such as pyridine (mol. wt. 100), adsorbed as a monolayer on charcoal with 1000 m²/g. In the following, these data are delineated:

$$\text{Area per pyridine molecule} = 24\ \text{Å}^2 = 24 \times 10^{-16}\ \text{cm}^2$$

$$\text{Surface area of 1 g charcoal} = 1000\ \text{m}^2 = 1000 \times 10^4\ \text{cm}^2$$

$$\text{Molecules pyridine adsorbed} = \frac{1000 \times 10^4\ \text{cm}^2}{24 \times 10^{-16}\ \text{cm}^2/\text{molecule}} = 40 \times 10^{20}\ \text{molecules}$$

$$\frac{\text{Amount of pyridine adsorbed}}{\text{g of charcoal}} = \left(\frac{40 \times 10^{20}\ \text{molecules}}{6 \times 10^{23}}\right) 100 = 0.7\ \text{g}$$

1.6.9 Interaction of Solid with Liquids (Heats of Adsorption)

A solid surface interacts with its surrounding molecules (in gas or liquid phase) with varying degrees. For example, if a solid is immersed into a liquid, the interaction between the two bodies will be interesting. The interaction of a substance with solid surface can be studied by measuring the heat of adsorption (besides other methods). The information one needs is whether the process is exothermic (heat is produced) or endothermic (heat is absorbed). This leads to the understanding of the mechanism of adsorption and helps in the application and design of the system. Calorimetric measurements have provided much useful information (Adamson and Gast, 1997; Chattoraj and Birdi, 1984). When a solid is immersed in a liquid (Figure 1.47), one finds that in most cases, there is a liberation of heat (exotherm):

$$q_{imm} = E_S - E_{SL} \tag{1.142}$$

where E_S and E_{SL} are the surface energies of solid surface and the solid surface in liquid, respectively. The quantity q_{imm} is measured from calorimetry where temperature change is measured after a solid (in finely divided sate) is immersed in a given liquid. One will expect that when a polar solid surface is immersed in a polar liquid, there will be larger q_{imm} than if the liquid was an alkane (nonpolar). Values of q_{imm} for some typical systems are given in Table 1.10.

These data show further that such studies are sensitive to the surface purity of solids.

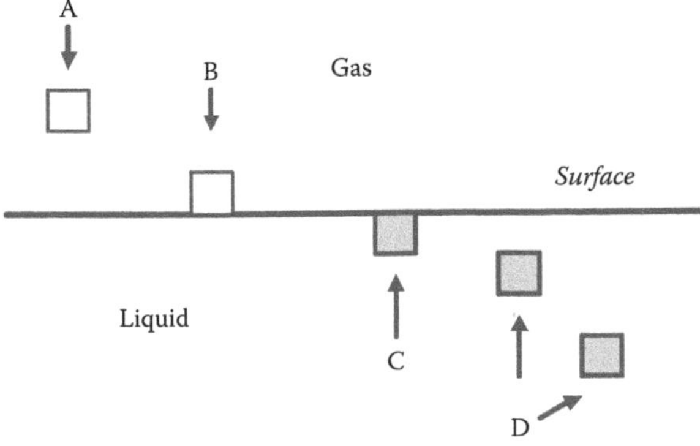

FIGURE 1.47 Solid immersion process in a liquid.

TABLE 1.10
Heats of Immersion (q_{imm}) (erg/cm^2 at 25°C) of Solids in Liquids

	Liquid	
Solid	**Polar (H_2O–C_2H_5OH)**	**Nonpolar (C_6H_{14})**
Polar (TiO_2; Al_2O_3; glass)	400–600	100
Nonpolar (Graphon; Teflon)	6–30	50–100

1.6.10 Particle Flotation Technology (of Solid Particles to Liquid Surface)

In mineralogy, the technique of separating particles of minerals from water media are of great interest. In only rare cases does one find minerals or metals in pure form (such as gold). The earth surface consists of a variety of minerals (major components are iron–silica–oxides–calcium–magnesium–aluminum–chromium–cobalt–titanium). Minerals as found in nature are always mixed with different kinds (e.g., zinc sulfide and felspar minerals). In order to separate zinc sulfide, one suspends the mixture in water and air bubbles are made to achieve separation. This process is called *flotation* (ore [heavier than water] is floated by bubbles).

Flotation is a technical process in which suspended particles are clarified by allowing them to float to the surface of the liquid medium (Fuerstenau et al., 1985; Klimpel, 1995; Rao et al., 1995; Yoon et al., 1990). The material can thus be removed by skimming at the surface flotation is found to be a highly versatile method for physically separating particles based on differences in the ability of air bubbles to selectively adhere to specific mineral surfaces (as determined by surface forces). This is economically much cheaper than any other process. If the suspended particles are heavier than the liquid (such as minerals), then one uses gas (air or CO_2 or other suitable gas) bubbles to enhance the flotation.

Froth flotation commences by grinding the rock, which is used to increase the surface area of the ore for subsequent processing and break the rocks into the desired mineral and gangue (which then has to be separated from the desired mineral); the ore is ground into a fine powder. The desired mineral is rendered hydrophobic by the addition of a surfactant or collector chemical; the particular chemical depends on the mineral that is refined–as an example, pine oil is used to extract copper. This slurry (more properly called the pulp) of hydrophobic mineral-bearing ore and hydrophilic gangue is then introduced to a water bath that is aerated, creating bubbles. The hydrophobic grains of mineral-bearing ore escape the water by attaching to the air bubbles, which rises to the surface, forming foam (more properly called froth). The froth is removed, and the concentrated mineral is further refined.

The *flotation* industry is a very important area in metallurgy and other related processes. The flotation method is based on treating a suspension of minerals (ranging in size from 10 to 50 μm) in water phase to air (or some other gas) bubbles (Figure 1.48).

Flotation leads to separation of ores from the mixtures (add). Especially, in modern mineral industry where rare metals are processed, this technique is being widely applied. It has been suggested that among other surface forces, the contact angle plays an important role. The gas (air or other gas) bubble as attached to the solid particle should have a large contact angle for separation.

Bubbles as needed for flotation are created by various methods. These may be as follows:

Air injection
Electrolytic methods
Vacuum activation

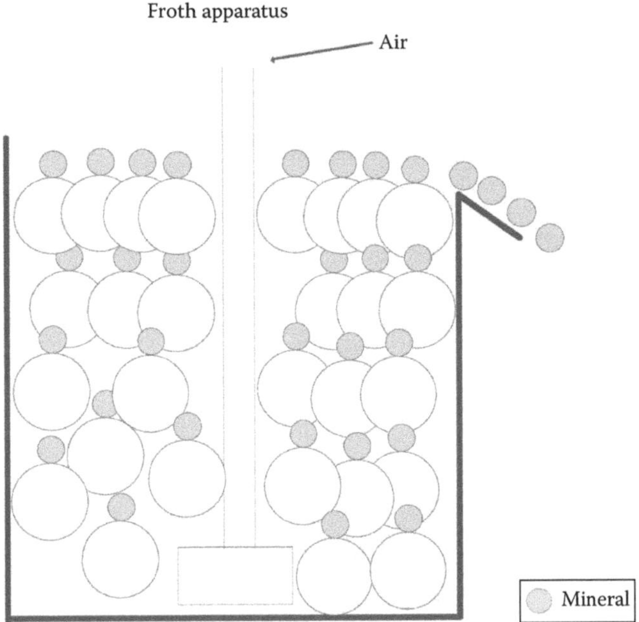

FIGURE 1.48 Flotation of mineral particles as aided by air bubbles.

In a laboratory experiment (Adamson and Gast, 1997), one may use the following recipe. To a 1% sodium bicarbonate solution, one can add a few grams of sand. Then if one adds some acetic acid (or vinegar), the bubbles of CO_2 produced cling to the sand particles and thus make these float on the surface. It must be mentioned that in wastewater treatments, the flotation method is one of the most important procedures. When rocks in crushed state are dispersed in water with suitable surfactants (also called collectors in industry) to give stable bubbles on aeration, hydrophobic minerals will float to the surface due to the attachment of bubbles, while the hydrophilic mineral particles will settle to the bottom. The preferential adsorption of the collector molecules on a mineral makes it hydrophobic. Xanthates (alkyl–O–CS_2) have been used for flotation of lead and copper.

1.6.11 Thermodynamics of Gas Adsorption on Solid Surface

From physicochemical principles, it is known that any pure surface, especially solid, means that it is surrounded by no other foreign molecule (which means it is under vacuum). However, as soon as there is a foreign molecule in the gas phase, the latter will adsorb to some extent depending on the physical conditions (temperature and pressure). On any solid surface, gas will adsorb or desorb under specific conditions (e.g., in gas recovery from shale deposits). Gas molecules will adsorb and desorb at the solid surface as determined by different parameters. At equilibrium, the rates of adsorption (R_{ads}) and desorption (R_{des}) will be equal. The surface can be described as consisting of different kinds of surfaces:

$$TSA = A_t = A_o + A_m$$

$$\text{Area of clean surface} = A_o$$

$$\text{Area covered with gas} = A_m$$

Enthalpy of adsorption = E_{ads} (energy required to adsorb a molecule from gas phase to the solid surface)

One can write following relations:

$$R_{ads} = k_a p A_o \tag{1.143}$$

$$R_{des} = k_b A_m \exp\left(\frac{-E_{ads}}{\mathbf{RT}}\right) \tag{1.144}$$

where k_a and k_b are constants.

At equilibrium,

Rate of adsorption = Rate of desorption

$$R_{ads} = R_{des} \tag{1.145}$$

and the magnitude of A_o is a constant.

Further, we have

Amount of gas adsorbed = N_s
Monolayer capacity of the solid surface = N_{sm}

By combining these relations and

$$\frac{N_s}{N_{sm}} = A_m = A_t \tag{1.146}$$

we get the well-known Langmuir adsorption equation (Birdi, 2008):

$$N_s = \frac{N_{sm}}{((ap)/(1+ap))} \tag{1.147}$$

where p is the pressure, and a is dependent on the energy of adsorption. Additionally, the heat of adsorption has been investigated. For example, the amount of Kr adsorbed on AgI increases when the temperature is decreased from 79 K (0.13 cc/g) to 77 K (0.16 cc/g). This data allow one to estimate the isosteric heat of adsorption (Birdi, 2009; Jaycock and Parfitt, 1981):

$$\frac{d(\ln P)}{dt} = \frac{q_{ads}}{\mathbf{RT}^2} \tag{1.148}$$

The magnitude of q_{ads} was in the range of 10–20 kJ/mol.

1.7 WETTING, ADSORPTION, AND CLEANING PROCESSES

In everyday life, one finds various systems where a liquid comes in contact with the surface of a solid (rain drops on solid surfaces, paint, ink on paper, washing and cleaning, oil reservoirs, etc.). The contact angle studies of liquid–solid systems showed that wetting is dependent on different parameters. It is found that when a liquid comes in contact with a solid, there are a few specific surface-related processes, which one needs to analyze. These processes are

Wetting
Adsorption
Desorption
Cleaning processes: garment industry, etc. (car or aeroplane washing)

One finds that *wetting* characteristics of any solid surface play an important role in all kinds of different systems. The next most important step is the process of *adsorption* of substances on solid surfaces. These phenomena are the crucial steps for all kinds of *cleaning processes.*

These systems may be as follows: washing, coatings, adhesion, lubrication, oil recovery, etc.

The liquid–solid or liquid_1–solid–liquid_2 system is both a contact angle (Young's equation) and capillary phenomena (Laplace equation). These two parameters are

$$\gamma_L \cos(\theta) = (\gamma_S - \gamma_{SL}) \tag{1.149}$$

and

$$\Delta P = \frac{(2\gamma_L \cos(\theta))}{\text{Radius}} \tag{1.150}$$

1.7.1 Oil and Gas Recovery Technology (Oil/Gas Reservoirs and Shale Deposits) (Fracking Technology and Methane Hydrates) and Surface Forces

Current energy demand (approximately 80 million barrels of oil per day + gas + coal + other forms of energy sources) is known to have a very high priority as regards sources, such as oil (and gas). Oil is normally found under high temperatures (80°C) and pressures (200 atm) depending on the depth of the reservoir. The pressure needed for production depends primarily on the porosity of the reservoir rock and the viscosity of the oil, among other factors (Figure 1.49). This creates flow of oil through the rocks, which consists of pores of varying sizes and shapes. Roughly, one may compare oil flow to the squeezing of water out of a sponge. The capillary pressure is lower in larger pores than in pores of smaller diameter. Therefore, in the primary oil production, one recovers mostly from the larger pores of the reservoir. Actually, one may consider it as an advantage that as technology develops, one can anticipate old depleted reservoirs to become productive again in future.

The degree of oil recovery from reservoirs is never 100%, and therefore, some (more than 30%) of the oil in the reservoir remains behind. In order to recover the residual oil, one needs to apply EOR technology (Birdi, 2010a). Major factors arise from capillary forces, as well as adsorption and flow hindrances. The same is valid for nonconventional oil reservoirs recovery from shale deposits. On the other hand, one may consider this as an advantage, since in the long run, as the shortage of

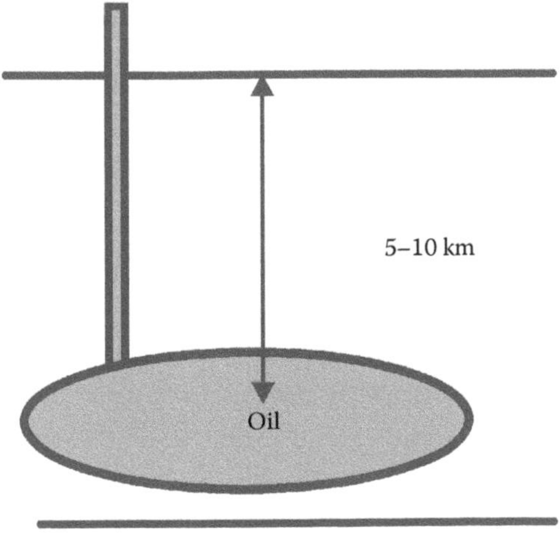

FIGURE 1.49 Oil reservoir is generally found under large depths, ranging from few hundred meters to over 10 km (pressure increases approximately 100 atm/km depth).

oil supplies comes nearer, one may be forced to develop technologies to recover the residual oil EOR (Birdi, 2009; Green and Willhite, 1998; Holmberg, 2002).

EOR is a very important research area in energy supply industry.

Studies have shown that by using additives, one can recover the residual oil in the reservoirs. Especially those additives that lower the IFT give rise to increase in oil production. The latter gives increased production when water flooding (or similar method) is used.

There are various common EOR processes:

Thermal EOR processes: steam-flooding
Gas EOR processes: CO_2 or nitrogen injection
Chemical EOR processes: Micellar or micellar + polymer flooding

The method used is dependent on the type of reservoir

Gas is generally found in conventional reservoirs mostly as methane, CH_4. However, in some cases, gas is also recovered from oil reservoirs. In the last decade, nonconventional reservoirs (both termed as shale gas and *methane hydrate*) have been developed. Natural gas production from shale formations (containing rich contents of hydrocarbons) is one of the most important sources of energy. The shale gas recovery based on creating fractures to enhance gas recovery has added a great potential to U.S. energy resources. Similar potential is believed to exist in other parts of the world (the United Kingdom, Poland, Germany, France, Ukraine, India, China, etc.). In the United States alone, it is believed that there are shale reserves, which could supply gas for over a century. Natural gas is trapped in the pore spaces of the shale. The latter formation requires very special state-of-the art technology to extract the gas.

In the case of conventional reservoirs, one will need to develop oil recovery processes from reservoirs that are depleted. In most oil reservoirs, the primary recovery is based on the natural flow of oil under the gas pressure of the reservoir. In these reservoirs as this gas pressure drops, the water flooding procedure is used. In some reservoirs, one has also used CO_2 to increase the oil recovery. In some cases, one may add substances such as detergents or similar chemicals to enhance the flow of oil through the porous rock structure. The principle is to reduce the Laplace pressure (i.e., $\Delta P = \gamma$/curvature of the pores) and to reduce the contact angle. This process is called *tertiary oil recovery*. The aim is to produce oil that is trapped in capillary-like structure in the porous oil-bearing material. The addition of surface-active agents reduces the oil–water IFT (from ca. 30–50 mN/m to less than 10 mN/m). EOR is becoming a very common procedure since the price of oil is high and recovery expenses are comparable. In the tertiary oil recovery processes, more complicated chemical additives are designed for a particle reservoir. In all these recovery processes, the IFT between oil phase and the water phase is needed. In most reservoirs, one needs methods to improve oil recovery. In most operations, one uses water (with additives) to push the oil to increase the degree of recovery. However, the matter becomes complex due to various reasons. Another important factor is that during the water flooding, the water phase bypasses the oil in the reservoir, Figure 1.50. What this implies is that if one injects water into the reservoir to push the oil, most of the water passes around the oil (bypass phenomena) and comes up without being able to push the oil (Birdi, 2003a, 2009). These observations have convincingly shown that the pores in reservoirs are not circular.

The pressure difference to push the oil drop may be larger than to push the water, thus leading to the so-called bypass phenomena. In other words, as water flooding is performed, due to bypass, there is less oil produced, while more water is pumped back up with oil. The use of surfactants and other SAS leads to the reduction of $\gamma_{oilwater}$. The pressure difference at the oil blob entrapped and the surrounding aqueous phase will be

$$\Delta P_{oilwater} = 2IFT\left(\frac{1}{R_1}\frac{1}{R_2}\right) \tag{1.151}$$

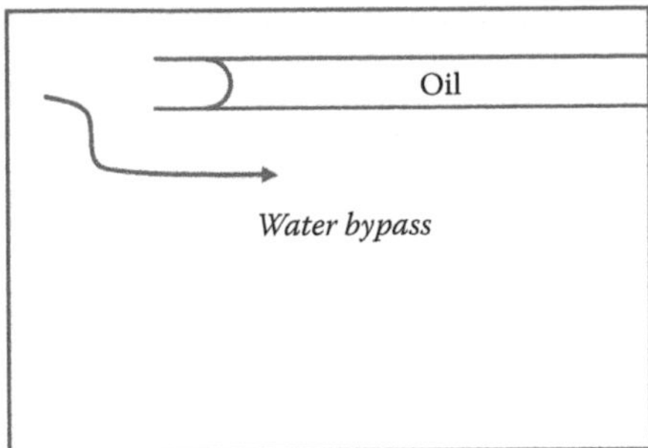

FIGURE 1.50 Water bypass in an oil reservoir. Reservoir pressure is not high enough to push the oil inside the narrow pore.

Thus by decreasing the value of IFT (with the help of surface-active agents) (from 50 to 1 mN/m), the pressure needed for oil recovery would be decreased. In some reservoirs, the addition of CO_2 has also increased the oil production. In water flooding process, one uses mixed emulsifiers. Soluble oils are used in various oil well–treating processes, such as the treatment of water injection wells to improve water injectivity, to remove water blockage in producing wells. The same is useful in different cleaning processes in the oil wells. This is known to be effective since water-in-oil microemulsions are found in these mixtures, and with high viscosity. The micellar solution is composed essentially of hydrocarbon, aqueous phase, and surfactant sufficient to impart micellar solution characteristics to the emulsion. The hydrocarbon is crude oil or gasoline. Surfactants are alkyl aryl sulfonates, more commonly known as petroleum sulfonates. These emulsions may also contain ketones, esters, or alcohols as cosurfactants. In the case of square-shaped pores, one will have to consider bypass in the corners, which are not found in circular-shaped pores (Birdi et al., 1988).

Further, it has been known for decades that gas is found in adsorbed state in shale reserves around the world. Until recently, the recovery of adsorbed gas from shale was very costly as compared to other energy sources (EPA, 2013; Knaus et al., 2010). Shale consists of layered deposits of river beds. The organic matter found in these layered structure over million years has evolved into mostly methane gas. Natural gas is very tightly held within the shale structure and not easily available. In the case of gas and oil recovery from shale, one bores down to the reservoir and pumps water (under high pressure) with sand and additives (surfactants, salts) added. In this process, small explosions are created to make fractures, through which shale releases gas/oil for recovery. The process is specific for each shale reservoir type.

Shale gas fracking technology:

Layer in shale:

= =I = = = =I = = = = = = = = = = =

= = = = = = = = = = = = =I = = = = = =I = = = = = = = = = = = = = = = =

= = = = = = =I = = = = = I =

= = Gas/Oil = = = Gas/Oil = = = Gas/Oil = = = = = = = = = = = = = = = =

= = = = = = =I =

= = = = = = = = = = = = = = = = = = =I = = = = = = = = = = = = = = = = =

The layered salt deposits (depicted as = = = =) contain tightly held gas or oil in shale reservoirs. The recovery of gas is achieved by creating fractures (indicated as I: by inducing controlled explosions) in the reservoir, thus releasing gas (mainly methane: as adsorbed on shale layers) for recovery. High-pressure water has also been found to have an appreciable effect on gas release from shale.

1.7.1.1 Oil Spills and Cleanup Processes on Oceans:

The worldwide concern with oil spills and its treatment is much dependent on the surface chemistry principles. Oil is transported across the seas in very large amounts (over 80 million barrels/day). Oil spill on sea surfaces will be exposed to various parameters:

Oil spill on ocean surface:

1. Loss by evaporation
2. Loss by sinking to the bottom (as such or in conjunction with solids)
3. Emulsification

Oil spills on oceans are treated by various methods, depending on the region (in warmer seas or around cold climate). It is also apparent that the oil spill in the Gulf of Mexico will be of completely different nature than a similar accident near Greenland, for instance. The composition of oil differs from place to place. The light fluids of oil will evaporate into the air. The oil that has adsorbed on solid suspension will sink to the bottom. The remaining oil is skimmed off by suitable machines. In some cases, one also uses surfactants to emulsify the oil and this emulsion sinks to the bottom slowly. However, no two oil spills are the same because of the variation in oil types, locations, and weather conditions involved. However, broadly speaking, there are four main methods of response.

1. Leave the oil alone so that it breaks down by natural means. If there is no possibility of the oil polluting coastal regions or marine industries, the best method is to leave it to disperse by natural means. A combination of wind, sun, current, and wave action will rapidly disperse and evaporate most oils. Light oils will disperse more quickly than heavy oils. Of course, the temperature of the sea water will also have an effect on the evaporation process.
2. Contain the spill with booms and collect it from the water surface using skimmer equipment. Spilt oil floats on water and initially forms a slick that is a few millimeters thick.
3. Use dispersants to break up the oil and speed its natural biodegradation. Dispersants act by reducing the surface tension that inhibits oil and water from mixing. Small droplets of oil are then formed, which helps to promote a rapid dilution of the oil by water movements. The formation of droplets also increases the oil surface area, thus increasing the exposure to natural evaporation and bacterial action. Dispersants are most effective when used within an hour or two of the initial spill. However, they are not appropriate for all oils and all locations. Successful dispersion of oil through the water column can affect marine organisms like deep water corals and sea grass. It can also cause oil to be temporarily accumulated by subtidal seafood. Decisions on whether or not to use dispersants to combat an oil spill must be made in each individual case. The decision will take into account the time since the spill, weather conditions, the particular environment involved, and the type of oil that has been spilt.
4. Introduce biological agents to the spill to hasten biodegradation. Most of the components of oil washed up along a shoreline can be broken down by bacteria and other microorganisms into harmless substances such as fatty acids and carbon dioxide. This action is called biodegradation.

The oil spill control technology is very advanced and can operate under very divergent conditions (from the cold seawaters to the tropic seas). The biggest difference arises from the oil that may contain varying amounts of heavy components (such as tar.).

1.7.2 Detergency and Surface Chemistry Essential Principles

Detergent industry is a very large and important area where surface and colloid chemistry principles have been applied extensively (Rosen, 2004). In fact, some detergent manufacturers are involved in very highly sophisticated research and development for many decades; some of these are protected by patents. Mankind is known to have used soaps for cleaning for many centuries.

The procedure for cleaning fabrics or metal surfaces, etc., is primarily to remove dirt, etc., from surfaces (fabrics: cotton, wool, synthetics, or mixtures). Second, one must make sure that dirt does not redeposit after its removal. *Dry cleaning* is a different process, since here one uses organic solvents. Dirt is adhering to the fabric through different forces (such as van der Waals and electrostatic). Some components of dirt are water soluble, and some are water insoluble. The detergents used are designed specifically for these particular processes by the industry and the environment. The composition of soaps or detergents is mainly tuned to achieve the following effects:

1. Water should be able to wet fibers as completely as possible; that is, θ should be less than 10. This is achieved by lowering the surface tension, γ, of the washing water, which thus lowers the contact angle. The low value of surface tension also makes the washing liquid to be able to penetrate the pores (if present), since from the Laplace equation, the pressure needed would be much low.
   For example, if the pore size of fabric (such as, any modern microcotton, Gortex) is 0.3 μm, then it will require a certain pressure (= $\mathbf{\Delta P} = 2\gamma/\mathbf{R}$) in order for water to penetrate the fibers. In the case of water (γ = 72 mN/m) and using a contact angle of 105°, we obtain

$$\mathbf{\Delta P} = \frac{2(72 \times 10^3)\cos(105)}{0.3 \times 10^6} = 1.4 \text{ bar} \tag{1.152}$$

2. The detergent then interacts with the dirty soil to start the process of removal from the fibers and dispersion into the washing water. In order to be able to inhibit the soil once removed to readsorb on the clean fiber, one uses polyphosphates or similar suitable inorganic salts. These salts also increase the pH (around 10) of the washing water. In some cases, one also uses suitable polymeric anti-redeposition substances (such as carboxymethyl cellulose).
3. After the fabric is cleaned, one uses special brighteners (fluorescent substances), which give a bluish haze to the fabric. This enhances the whiteness (by depressing the yellow tinge). Additionally, these also enhance the color perception. Brighteners as used for cotton are different from those used for synthetic fabrics. Charges on the fabric determine the adsorption characteristics of the brightener. Hence, the washing process is a series of well-designed steps, which the industry has provided with much information and the state-of-the-art technology. Further, all washing technology processes have changed all along as the demands have changed. Washing machines are designed to operate in conjunction with soap industry. The mechanical movement and agitation is coordinated with the soap/detergent characteristics.

Typical compositions of different laundry detergents, shampoos, or dish-washing powders are given as follows:

	Laundry	Shampoo	Dish-Washing
Washing Detergent (Na-alkyl sulfate)	10–20	25	—
Soaps	5	—	—
Nonionics	5–10	—	1
Inorganic salts (polyphosphate, silicates)	30–50	50	70
Optical brighteners	<1	—	—

It is worth noticing that the aim of detergents in these different formulations is different in each case. In other words, detergents are today tailor-made for each specific application (Ruiz, 2008). The detergents in shampoo should give stable foam in order to increase the cleaning effect, at the same time without any adverse effect on the structure of hair. On the other hand, in the case of laundry detergents in the dish-washing should only give a lower surface tension but almost no foaming (because foaming would reduce the cleaning effect). Hence in dish-washing machine formulations, one uses nonionics that are very little soluble in water and thus produce very little (or none) foam. These are sometimes of type EOEOEOPOPOPO (Ethylene oxide [EO]–propylene oxide [PO]). The propylene group behaves as *apolar,* and the oxide group behaves (through hydrogen bonding) as the *polar* part. These EOPO types can be tailor-made by combining various ratios of EO:PO in the surfactant molecule. In some cases, even butylene oxide groups have been used.

Further, soil consists mainly of particulate, greasy matter, etc. Detergents are supposed to keep the soil suspended in the solution and restrict the redeposition. Tests also show that detergents stabilize the suspensions of carbon or other solids such as manganese oxide in water. This suggests that detergents adsorb on the particles. Detergents are necessary also to remove the greasy part of soil. The adsorption of detergents on soil particles is involved in the detergency process.

1.7.3 Evaporation Rates of Liquid Drops

In many natural (raindrops, fogs, river water fall) and industrial systems (sprays, oil combustion engines, cleaning processes), one encounters liquid drops. The *rate of evaporation* of liquid from such drops can be important for the function of these systems. Raindrops contribute also major part to erosion and other related phenomena. Extensive investigations on the evaporation of liquid drops (free hanging drops; drops placed on solid surfaces) have been reported in the current literature (Birdi, 2003a; 2008, 2010a; Pu and Severtson, 2012; Xu et al., 2013; Yu et al., 2004). These drops have been analyzed as a function of

Liquid (water or organic liquids)
Solids (plastics; glass; others)
Contact angle (θ)
Height and diameter and volume
Weight of drops

In these analyses, some assumptions have been made as regards the shape of the drops. The most accurate data were obtained when weight method was used. Different analyses showed that the rate of evaporation was linearly dependent on the radius of the drop. Further, the contact angle of water drop on teflon (i.e., a nonwetting surface) remained constant under evaporation. On the other hand, contact angle decreased as water drop evaporated on glass (i.e., a wetting surface).

1.8 COLLOIDAL DISPERSION SYSTEMS—PHYSICOCHEMICAL PROPERTIES

Solid particles dispersed in water are found in many examples in daily life, such as wastewater treatment. In this chapter, essential aspects in the very large industrial application of *colloid chemistry* will be described. Surprisingly enough, mankind has been aware of colloids for many thousands of years. Old civilizations, such as in old Egypt and Maya, used their knowledge about adhesion (between blocks of stones) when building pyramids thousands of years ago. This was long before the modern-day cement was invented. Even mud houses were built based on the behavior of the colloidal nature of materials used in such constructions, such as clay and cow dung. In everyday life, one comes across solid particles of different sizes, ranging from stones at a beach, sand particles, or dust floating around in the air. There exists a special relation between the size of particles (surface area) and their characteristics. The rather small particles in the range of size from 50 Å to 50 μm are called *colloids*. The most simple difference is when considering sand particles versus dust particles.

It is almost fascinating to observe how dust or other fine particles remain in suspension in air. In fact, once in a while, one observes that a particle gets a collision-like thrust. Already in the nineteenth century (Brown), it was observed under microscope that a small microscopic particle suspended in water made some erratic movements (as if hit by some other neighboring molecules) (Adamson and Gast, 1997). This has since been called *Brownian motion*. The erratic motion arises from the kinetic movement of the surrounding water molecules. Thus, colloidal particles would remain suspended in solution through Brownian motion, only if the gravity forces did not drag these to the bottom (or top). If one throws some sand into air, one finds that the particles fall to the earth rather quickly. On the other hand, in the case of the talcum particles, one finds that these stay floating in the air for a long time.

These characteristics will be described here. The size of particles may be considered from the following data:

Colloidal Dispersions	Size Range (nm–10 μ)
Mist/fog	0.1–10 μ
Pollen/bacteria	0.1–10 μ
Oil drops in smoke/exhaust	1–100 μ
Virus	nm–10 μ
Polymers/macromolecules	0.1–100 nm
Micelles	0.1–10 nm
Vesicles	1–1000 μ

This shows that one has a range of size of particles. Actually, in colloids, it is the size of particles that matters. The stability of such colloidal systems is something that may be compared to whether the system stays as if energetically stable or it will take up a new state of more stable configuration. One may very roughly compare this to a bottle, which is stable when standing up but if tilted beyond a certain angle, it topples and comes to rest (stable state) on its side (Figure 1.51).

A colloidal suspension may be *unstable* and exhibit separation of particles within a very short time. Or it may be *stable* for a very long time, such as over a year or more. And there will thus be found a *metastable* state, which would be in between these two. This is an oversimplified example, but it shows that one should proceed to analyze any colloidal system following these three criteria. In fact, the most remarkable finding one can mention about colloidal suspensions is that these systems can exist at all! Especially, one finds that some solid suspensions can be stable for a very long time. In pharmaceutical applications, one important example is the use of suspension of insulin in pen injections. The insulin suspension is stable for long enough time for its application, which provides very accurate dosage to the patient. In fact, there exists a wide variety of pharmaceutical products that are based on suspended molecules. Nanoparticles have been applied in various pharmaceutical treatments (Sotiriou et al., 2014).

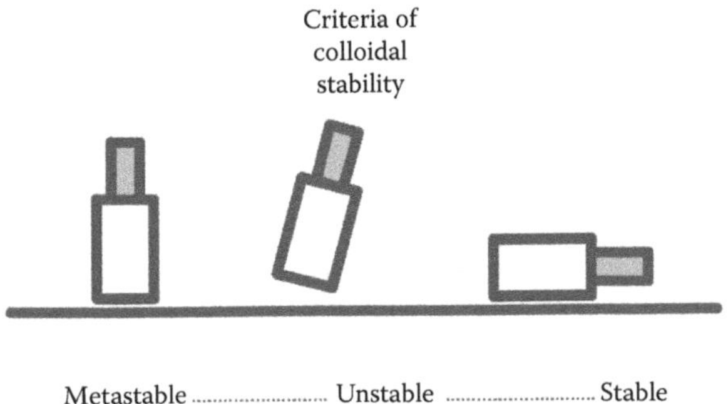

FIGURE 1.51 Stability criteria of any colloidal system: metastable–unstable–stable states.

As an example, one may consider the wastewater treatment process (Cherimisinoff, 2002). The wastewater with colloidal particles is found to be a stable suspension. However, by treating it with some definite methods (such as pH control, electrolyte concentration, etc.), one can change the stability of the system.

The wastewater treatment technology is one of the most important areas of surface chemistry applications. Suspended materials are separated by coagulation and filtered away. Soluble pollutants are removed by other procedures.

van der Waals forces: In colloidal systems, van der Waals forces play an important role (Israelachvili, 2011; Adamson and Gast, 1997). When any two particles (neutral or with charges) come very close to each other, van der Waals forces are strongly dependent on the surrounding medium; in vacuum, two identical particles always exhibit an attractive force. On the other hand, if two different particles are present in a medium (in water), then there may be repulsion forces. This can arise from the particle–solvent adsorption characteristics. One example will be silica particles in water medium and plastics (as in wastewater treatment). It is important to understand under which experimental conditions it is possible that colloidal particles remain suspended for a long time. For example, if paint aggregates in the container, then it is obviously useless. When solid (inorganic) particles are dispersed in aqueous medium, ions are released in the medium. The ions released from the surface of the solid are of opposite charge. For example, when glass powder is mixed in water, one finds that conductivity increases with time.

The presence of same charge on particles in close proximity gives repulsion, which keeps the particles apart, Figure 1.52. The charged positive–positive (or negative–negative) particles will show repulsion. On the other hand, the positive–negative particles will attract each other. The ions distribution will also depend on the concentration of any counterions or coions in the solution. Even glass when dipped in water exchanges ions with its surroundings. Such phenomena can be easily investigated by measuring the change in the conductivity of the water.

The force, F_{12}, acting between these opposite charges is given by Coulomb's law, with charges q_1 and q_2, separated at a distance R_{12}, in a dielectric medium, D_e:

$$F_{12} = \frac{(q_1 q_2)}{(4\pi D_e \varepsilon_o R_{12})} \tag{1.153}$$

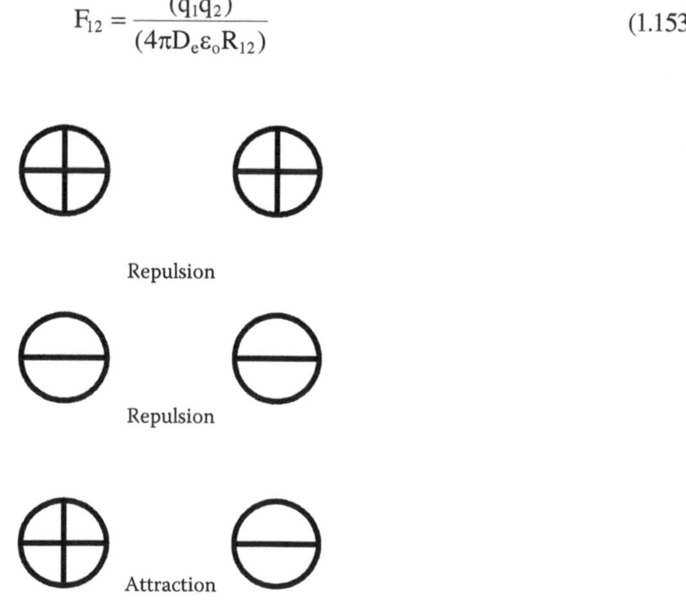

FIGURE 1.52 Solid particles with charges (positive–positive [repulsion] or negative–negative [repulsion] or positive–negative [attraction]).

The force would be attractive between opposite charges while repulsive in the case of similar charges. Since D_e of water is comparatively very high (80 units) as compared to D_e of air (ca. 2), we will expect very high dissociation in water, while there is hardly any dissociation in air or organic liquids. Let us consider the F_{12} for Na^+ and Cl^- ions (with the charge of 1.6×10^{-19} C = 4.8×10^{-10} esu) in water (D_e = 74.2 at 37°C), and at a separation (R_{12}) of 1 nm:

$$F_{12} = \frac{(1.6\times10^{-19})(1.6\times10^{-19})}{[(4\Pi)(8.854\times10^{-12})(10^{-9})(74.2)]} = -1.87 \text{ kJ/mol} \quad (1.154)$$

where ε_o is 8.854×10^{-12} $s^4 A^2/(kg\ m^3)(C^2/(mJ))$.

Another very important physical parameter one must consider is the *size distribution* of the colloids. A system consisting of particles of *same size* is called monodisperse. A system with *different sizes* is called polydisperse. It is also obvious that monodispersed systems will exhibit different properties compared to polydispersed. In many industrial applications such as coating of substances in colloid form on tapes, as used for recording music (coatings on CD or DVD), the size of particles is an important characteristic. The methods used to prepare monodisperse colloids is to achieve a large number of critical nuclei in a short interval of time. This induces all equally sized nuclei to grow simultaneously and thus produces a monodisperse colloidal product.

1.8.1 Colloidal Stability (DLVO Theory)

The question one needs to understand is under which conditions will a colloidal system remain dispersed (and under other conditions become unstable) (Adamson and Gast, 1997; Birdi, 2010a; Israelachvili, 2011; Scheludko, 1966). How colloidal particles interact with each other is one of the important questions that determine the understanding of the experimental results for phase transitions in such system as found in various industrial processes. One also will need to know under which conditions a given dispersion will become unstable (*coagulation*). For example, one needs to apply coagulation in wastewater treatment such that most of the solid particles in suspension can be removed. When any two particles come close to each other, different forces (depending on the distance between the particles) are bound to coexist.

1.8.1.1 Attractive Forces and Repulsive Forces

If attractive forces are larger than repulsive forces, then the two particles will merge together. However, if repulsion forces are larger than attractive forces, then the particles will remain separated. It is important to mention here that the medium in which these particles are present thus will, to some degree, contribute also to the state of equilibrium. Especially, pH and ionic strength (i.e., concentration of ions) are found to exhibit very specific effects.

The different forces of interest are

van der Waals
Electrostatic
Steric
Hydration
Polymer–polymer interactions (if polymers are involved in the system)

In some colloidal systems, one may add large molecules (polymers) that when adsorbed on the solid particles will impart special kind of stability criteria. It is well known that neutral molecules, such as alkanes, attract each other mainly through van der Waals forces. van der Waals forces arise from the rapidly fluctuating dipoles moment (10^{15} s^{-1}) of a neutral atom, which leads to polarization and consequently to attraction. This is also called the London potential between two atoms in a vacuum and is given as

$$V_{vdw} = -\left(\frac{L_{11}}{\mathbf{R}^6}\right) \tag{1.155}$$

where

L_{11} is a constant, which depends on the polarizability and the energy related to the dispersion frequency
R is the distance between the two atoms

Since the London interactions with other atoms may be neglected as an approximation, the total interaction for any macroscopic bodies may be estimated by a simple integration.

When two similarly charged colloidal particles, under the influence of the EDL, come close to each another, they will begin to interact. The potentials will feel each other, and this will lead to consequences. The charged molecules or particles will be under both van der Waals and electrostatic interaction forces. van der Waals forces that operate at short distance between particles will give rise to strong attraction forces. This kind of investigation is important in various industries:

Inorganic materials (ceramics, cements)
Food (milk)
Biomacromolecular systems (proteins and DNA)

Various theoretical models have been presented in the literature as regards the colloidal properties. One of the most accepted theories was named after the scientists who developed the model, the so-called DLVO (*Derjaguin–Landau–Verwey–Overbeek*) theory. The DLVO theory describes that the stability of a colloidal suspension is mainly dependent on the distance between the particles (Adamson and Gast, 1997; Birdi, 2009, 2010a; Grodzka and Pomianowski, 2005; Merk et al., 2014; Scheludko, 1966). DLVO theory has been modified in later years, and different versions are found in the current literature. Electrostatic forces will give rise to repulsion at large distances. This arises from the fact that electrical charge–charge interactions take place at a large distance of separation. The barrier height determines the stability with respect to the quantity kT, the kinetic energy. DLVO theory predicts, in most simple terms, that if the repulsion potential, **W**, exceeds the attraction potential by a value

$$\mathbf{W} \gg kT \tag{1.156}$$

then the suspension will be *stable*. On the other hand, if

$$\mathbf{W} \le kT \tag{1.157}$$

then the suspension will be *unstable* and it will coagulate. It must be stressed that DLVO theory does not provide comprehensive analyses. It is basically a very useful tool for such analyses of complicated systems. Especially, it is a useful guidance theory in any new application or any industrial development.

1.8.2 Charged Colloids (Electrical Charge Distribution at Interfaces)

In everyday life, electrically charged particles or surfaces play a very important role. The interactions between two charged bodies will be dependent on various parameters (e.g., surface charge, electrolyte in the medium, charge distribution). The distribution of ions in an aqueous medium needs to be investigated in such charged colloidal systems. This observation indicates that the presence of charges on surfaces gives rise to a potential (surface potential), which needs to be investigated. On the other hand, in the case of neutral surfaces, one has only the van der Waals forces to be considered. This was clearly seen in the case of micelles, where the addition of small amounts of NaCl to the solution showed

Large decrease in CMC in the case of ionic surfactant
Almost no effect in nonionic micelles (since in these micelles there are no charges or EDL)

The addition of electrolytes produces a different kind of surface potential curve. This is easily verified in applications such as washing clothes, etc. Electrostatic and EDL forces are found to play a very important role in a variety of systems as known in science and engineering (Baimpos et al., 2014; Birdi, 2010b). It would be useful to consider a specific example, in order to understand these phenomena. Let us take a surface with positive charge, which is suspended in a solution containing both positive and negative ions. There will be a definite surface potential, ψ_o, which decreases to a value zero as one moves away into solution. It is obvious that the concentration of positive ions will decrease as one approaches the surface of the positively charged surface (charge–charge repulsion). On the other hand, the oppositely charged ions, negative, will be strongly attracted toward the surface. This gives rise to the so-called Boltzmann distribution:

$$n^- = n_o e^{(z\varepsilon\psi/kT)} \tag{1.158}$$

$$n^+ = n_o e^{-(z\varepsilon\psi/kT)} \tag{1.159}$$

This shows that positive ions are *repelled,* while negative ions are *attracted* to the positively charged surface. At a reasonable distance from the particle, $n^+ = n^-$ (as required by the electroneutrality). In any aqueous solution when an electrolyte, such as NaCl, is present, it dissociates into positive (Na^+) and negative (Cl^-) ions. Due to the requirement of electroneutrality (i.e., there must be same positive and negative ions), each ion is surrounded by an appositively charged ion at some distance. Obviously, this distance will decrease with an increasing concentration of the added electrolyte. The expression $1/\kappa$ is called the Debye length. As expected, the D–H theory tells us that ions tend to cluster around the central ion. A fundamental property of the counterion distribution is the thickness of the ion atmosphere (Birdi, 2010b). This thickness is determined by the quantity Debye length or Debye radius ($1/\kappa$). The magnitude of $1/\kappa$ has dimension in centimeters as follows:

$$\kappa = \left(\frac{\varepsilon_r \varepsilon_o \mathbf{k_B T}}{2 N_A e^2 I} \right)^{1/2} \tag{1.160}$$

The values of $\mathbf{k_B} = 1.38 \times 10^{-23}$ J/molecule K, e = 4.8×10^{-10} esu, ε_r is the dielectric constant, and ε_o is the permittivity of free space. Thus, the quantity $\mathbf{k_B}$T/e = 25.7 mV at 25°C. As an example, with an 1:1 ion (such as NaCl, KBr) with concentration 0.001 M, one gets the value of $1/\kappa$ at 25°C (298 K):

$$\frac{1}{\kappa} = \frac{(78.3)(1.38 \times 10^{-16})(298)}{((2)(4\Pi)(6.023 \times 10^{17})(4.8 \times 10^{-10})^2)^{0.5}} = 9.7 \times 10^{-7}\ \text{cm} = 97\ \text{Å} \tag{1.161}$$

The expression for $\psi(r)$ can be written as

$$\psi(r) = \psi_o(r)\exp(-\kappa r) \tag{1.162}$$

which shows the change in $\psi(r)$ with the distance between particles (r). At a distance $1/\kappa$, the potential has dropped to ψ_o. This is accepted to correspond with the thickness of the double layer. This

TABLE 1.11
Magnitude of the Debye Length ((1/k) nm) in Aqueous Salt Solutions

Salt Concentration (M)	1:1	1:2	2:2
0.0001	30.4	17.6	15.2
0.001	9.6	5.55	4.81
0.01	3.04	1.76	1.52
0.1	0.96	0.55	0.48

is the important analyses, since the particle–particle interaction is dependent on the change in $\psi(r)$. The decrease in $\psi(r)$ at the Debye length is different for different ionic strengths (Table 1.11).

The data in Table 1.11 show values of D–H radius in various salt concentrations. The magnitude of $1/\kappa$ decreases with λ and with the number of charges on the added salt. This means that the thickness of the ion atmosphere around a reference ion will be much compressed with the increasing value of λ and the magnitude of z_{ion}.

A trivalent ion such as Al^{3+} will compress the double layer to a greater extent in comparison with a monovalent ion such as Na^+. Further, inorganic ions can interact with charge surface in one of two distinct ways:

1. Nonspecific ion adsorption where these ions have no effect on the isoelectric point IEP.
2. Specific ion adsorption, which gives rise to change in the value of the IEP.

Under those conditions where the magnitude of $1/\kappa$ is very small (e.g., in high electrolyte solution), one can write

$$\psi = \psi_o \exp - (\kappa x) \tag{1.163}$$

where x is the distance from the charged colloid.

The value of ψ_o is found to be 100 mV (in the case of monovalent ions) (=$4\mathbf{k_B T}/ze$).

Both the experimental data and the theory show that surface potential, ψ, varies with electrolyte concentration and the charge on the ions.

These data show

That the surface potential drops to zero at a faster rate if the ion concentration (C) increases
That the surface potential drops faster if the value of z goes from 1 to 2 or larger

In washing powders, for example, one uses multicharged phosphates, etc., to enable dirt particles staying off the clean fabrics.

1.8.3 Colloidal Electrokinetic Processes

Charged colloids allow one to visually investigate these systems under dynamic conditions.

In the following, let us consider what happens if the charged particle or surface is under dynamic motion of some kind. Further, there are different systems under which electrokinetic phenomena are investigated.

These systems are as follows (Adamson and Gast, 1997; Birdi, 2010a):

1. *Electrophoresis*: This system refers to the movement of the colloidal particle under an applied electric field. In biology, different proteins exhibit different charges and thus can be separated using this property.
   Negatively charged particle moves toward the positive electrode.
   Positively charged particle moves toward the negative electrode.

The speed of movement of a charged particle is dependent on various parameters:

Number of charges
Size and shape of particle

It is thus seen that one can separate particles by electrophoretic technique.

2. *Electro-osmosis*: This system is one where a fluid passes next to a charged material. This is actually the complement of electrophoresis. The pressure needed to make the fluid flow is called the *electro-osmotic pressure.*

Fluid movement through a charged material (like earth) gives rise to electroosmotic pressure. This arises from asymmetrical charge distribution at the liquid–solid interface, which depends on the magnitude of the surface potential.

3. *Streaming potential*: If a fluid is made to flow past a charged surface, then an electric field is created, which is called streaming potential. This system is thus opposite of the electroosmosis.
4. *Sedimentation potential*: A potential is created when charged particles settle out of a suspension. This gives rise to sedimentation potential, which is the opposite of the streaming potential. The reason for investigating electrokinetic properties of a system is to determine the quantity known as the zeta potential.

Electrophoresis is the movement of an electrically charged substance under the influence of an electric field. Experiments have shown that the movement of the charged particle is related to the electric field. F_e is the force, $\mathbf{q}$ is the charge carried by the body, and $\mathbf{E}$ is the electric field (Adamson an Gast, 1997; Birdi, 2010b):

$$F_e = \mathbf{qE} \tag{1.164}$$

The resulting electrophoretic migration is countered by forces of friction, F_f, such that the rate of migration is constant in a constant and homogeneous electric field:

$$F_f = \mathbf{vf_r} \tag{1.165}$$

where
$\mathbf{v}$ is the velocity
$\mathbf{f_r}$ is the frictional coefficient

$$\mathbf{QE} = \mathbf{vf_r} \tag{1.166}$$

The electrophoretic mobility μ is defined as follows:

$$\mu = \frac{\mathbf{v}}{\mathbf{E}} = \frac{\mathbf{q}}{\mathbf{f_r}} \tag{1.167}$$

The aforementioned expression can be applied only to ions at a concentration approaching 0 and in a nonconductive solvent. Polyionic molecules are surrounded by a cloud of counterions that alter the effective electric field applied on the ions to be separated. This renders the previous expression a poor approximation of what really happens in an electrophoretic apparatus.

The mobility depends on both the particle properties (e.g., surface charge density and size) and solution properties (e.g., ionic strength, electric permittivity, and pH). For high ionic strengths, an approximate expression for the *electrophoretic mobility*, μ_e, is given by the Smoluchowski equation:

$$\mu_e = \frac{\varepsilon\varepsilon_o\eta}{\zeta} \tag{1.168}$$

where
 ε is the dielectric constant of the liquid
 ε_o is the permittivity of free space
 η is the viscosity of the liquid
 ζ is the zeta potential (i.e., surface potential) of the particle

1.8.4 Critical Flocculation Concentration: Schultze–Hardy Rule

Suspensions of colloidal particles exhibit different properties that are dependent on various parameters. One of these properties is that the suspension of solid particles will exhibit varying degree of stability. It is thus interesting to determine how these systems are stabilized as regards the various forces interacting between the particles.

Solids in (colloidal) suspension can separate out of solution in various stages and pathways. The two most common pathways are as follows:

1. Stable suspension–flocculation–coagulation–sedimentation–particle separation
2. Stable suspension–partial sedimentation–flocculation–coagulation–particle separation

The most important system for mankind is of course wastewater treatment. However, natural phenomena occurring around as in rivers, lakes, and oceans are of also of prime interest in ecology and future life on the earth.

Particles in all kinds of suspensions or dispersions interact with two different kinds of forces (e.g., attractive forces and repulsive forces). One observes that lyophobic suspensions (sols) must exhibit a maximum in repulsion energy in order to have a stable system. The total interaction energy, $V(\mathbf{h})$, is given as (Adamson and Gast, 1997; Birdi, 2003b, 2009, 2010a; Chattoraj and Birdi, 1984; Chiu and Ducker, 2014; Gisler et al., 1994):

$$V(\mathbf{h}) = V_{el} + V_{vdw} \tag{1.169}$$

where V_{el} and V_{vdw} are electrostatic repulsion and van der Waals attraction components. Dependence of the interaction energy $V(\mathbf{h})$ on the distance $\mathbf{h}$ between particles has been ascribed to coagulation rates as follows:

1. During slow coagulation
2. When fast coagulation sets in

It is found that for large values of h, $V(\mathbf{h})$ is negative (attraction), following the energy of attraction V_{vdw}, which decreases more slowly with increasing distance ($\sim 1/\mathbf{h}^2$). At short distances (small $\mathbf{h}$), the positive component V_{el} (repulsion), which increases exponentially with decreasing $\mathbf{h}$, can overcompensate V_{vdw} and reverse the sign of both $dV(\mathbf{h})/dh$ and $V(\mathbf{h})$ in the direction of repulsion. In accordance with all that has been said before, coagulation will become fast starting from this concentration. This is therefore the *critical concentration*, C_{cc}. In other words, the critical concentration (C_{cc}) can be estimated from simultaneous solution of the following:

$$\frac{dV(\mathbf{h})}{d\mathbf{h}} = 0 \quad \text{and} \quad V(\mathbf{h}) = 0 \tag{1.170}$$

Based on these assumptions, as related to **h** and C, this becomes (Schulze–Hardy Rule) (for solid suspensions in water) (Adamson and Gast, 1997; Birdi, 2010a; Scheludko, 1966),

$$C_{cc} = \frac{8.7 \times 10^{-39}}{Z^6 A^2}$$

$$C_{cc} Z^6 = \text{constant} \tag{1.171}$$

where the constant includes (Hamaker constant is approximately 4.2 × 10^{-19} J) all quantities except **Z**. This shows that critical concentrations of ion to the sixth power of various valencies are inversely proportion to valency

$$\mathbf{Z} = 1 : (2^6)0.016 : (3^6)0.0014 : (4^6)0.000244 \tag{1.172}$$

The flocculation concentrations of mono-, di-, and tri-valent gegen-ions should from this theory expected as (experimental data shows good agreement with theory):

$$1 : (1/2)^6 : (1/3)^6$$
$$1 : 1/64 : 1/729$$

It thus becomes obvious that the colloidal stability of charged particles is dependent on

1. Concentration of electrolyte
2. Charge on the ions
3. Size and shape of colloids
4. Viscosity

The critical concentration (*critical coagulation concentration*) is thus found to depend on the type of electrolyte used and on the valency of the counterion. It is seen that divalent ions are 60 times as effective as monovalent ions. Trivalent ions are several hundred times more effective than monovalent ions. However, ions that specifically adsorb (such as surfactants) will exhibit different behavior. Based on these observations, in the composition of washing powders, one has used multivalent phosphates (or similar kinds of poly-ions), for instance, to keep the charged dirt particles from attaching to the fabrics after having been removed off. Another example is the wastewater treatment, where for coagulation purposes one uses multivalent ions (Cheremisinoff, 2002; Kim and Platt, 2007). Colloidal solutions are characterized by the degree of stability or instability.

It is thus seen that from DLVO considerations the degree of colloidal stability will be dependent on following factors:

1. *Size of particles*: Suspensions of larger particles will be less stable
2. Magnitude of surface potential
3. Hamaker constant (**H**)
4. Ionic strength
5. Temperature
6. Viscosity

The attraction force between two particles is proportional to the distance of separation and a Hamaker constant (specific to the system). The magnitude of **H** is on the order of 10^{-12} erg (Adamson and Gast, 1997; Birdi, 2003b).

The DLVO theory is thus found useful to predict and estimate the colloidal stability behavior. Of course, in such systems with many variables, this simplified theory is to be expected to fit all kinds of systems. In the past decade, much development has taken place as regards measuring the forces involved in these colloidal systems. In one method, the procedure used is to measure the force present between two solid surfaces at very low distances (less than a micrometer). The system can operate under water, and thus, the effect of addictives has been investigated (Israelachvili, 2011). These data have provided a verification of many aspects of DLVO theory. Recently, atomic force microscope (AFM) has been used to measure directly these colloidal forces (Birdi, 2003b). Two particle are brought closer and force (nano-Newton) is measured. In fact, commercially available apparatus are designed to perform such analysis. The measurements can be carried out in fluids and under various experimental conditions (such as added electrolytes, pH, etc.).

1.8.4.1 Flocculation and Coagulation of Colloidal Suspension

It is known from common experience that a colloidal dispersion with smaller particles is more stable than one with larger particles. The phenomena of smaller particles forming aggregates with larger size is called *flocculation* or *coagulation*. For example, to remove insoluble and colloidal metal precipitated, one uses flocculation. This is generally achieved by reducing the surface charges, which gives rise to weaker charge–charge repulsion forces. As soon as attraction forces (van der Waals) become larger than electrostatic forces, then coagulation takes place. Coagulation is initiated by particle charge neutralization (by changing pH or other methods [such as charged polyelectrolytes]), which leads to aggregation of particles to form large size particles.

Coagulation of particles and charge reversal phenomena:

This means the following:

Initial state: Charge–charge repulsion
Final state: Neutral–neutral (attraction)

Coagulation can also be brought by adding suitable substances (coagulants) particularly for a given system. The latter reduces the effective radius of the colloidal particle and leads to coagulation. Flocculation is a secondary process after coagulation, and this leads to very large particle (floccs) formation. Experiments show that coagulation takes place when the zeta potential is around ±0.5 mV. Coagulants such as iron and aluminum inorganic salts are effective in most cases. In wastewater treatment plants, the zeta potential is used to determine the coagulation and flocculation phenomena. The magnitude of zeta potential can be varied by changing the pH. Most of the solid material in wastewater is negatively charged.

1.8.5 Wastewater Treatment: Surface Chemistry Aspects

One of the biggest challenges mankind faces today is the availability of clean drinking water. Waste-water contains different kinds of pollutants (dissolved substances, suspended particles, and colloids). Wastewater is treated in suitable plants before the processed water is released into the surroundings. The substances that are found in wastewater (solutes) are in either molecular (such as benzene, coloring substances, etc.) or ionic form (such as Na^+, Cl^-, Mg^{2+}, K^+, Fe^{2+}, etc.) (Birdi, 2010a; Cheremisinoff, 2002).

The concentrations of pollutants are generally given in various units:

Weight/volume (mg/L; kg/m^3)
Weight/w (mg/kg; parts per million [ppm]; parts per billion [ppb])
Molarity (mmol/L)
Normality (equivalents/L)

Methods needed to treat these pollution systems also depend on the quantitative amounts of substances present. The specific unit used depends on the amounts present. The unit used for trace amounts, such as benzene, is given in ppm or ppb. The hardness of drinking water (mostly Na, Ca, Mg) concentration is given as mg/L. The typical values as found are in the range of less than 10 mg/L (soft water) or hard water (over 20 mg/L).

The solids can be removed by filtration and precipitation methods. The precipitation (of charged particles) is controlled by making the particles flocculate by controlling the pH and ionic strength. The latter gives rise to decrease in charge–charge repulsion and thus can lead to precipitation and removal of finely divided suspended solids. It is thus found that the most important factor that effects *zeta potential* is pH. Imagine a particle in suspension with a negative zeta potential. If more alkali is added to this suspension, then the particle will exhibit an increase in negative charge. On the other hand, if acid is added to the colloidal suspension, then the particle will acquire increasing positive charge. During this process, the particle will undergo a change from negative charge to *zero* charge (where the number of positive charge is equal to negative charge [*point-of-zero-charge*: PZC]). In other words, one can control the magnitude and sign of the surface charge by a potential determining ion. The stability is dependent on the magnitude of electrostatic potential at the surface of the colloid, ψ_o. The magnitude of ψ_o is estimated by using the microelectrophoresis method. When an electric field is applied across an electrolyte, charged particles suspended in the electrolyte are attracted toward the electrode of opposite charge. Viscous forces acting on the particles tend to oppose this movement. When equilibrium is reached between these two opposing forces, the particles move with constant velocity (Adamson and Gast, 1997; Israelachvili, 2011). In this technique, the movement (or rather the speed) of a particle is observed under a microscope when subjected to a given electric field. The field is related to the applied voltage, **V**, divided by the distance between the electrodes (in cm). The velocity is dependent on the strength of electric field or voltage gradient, the dielectric constant of the medium, the viscosity, and zeta potential.

Commercially available electrophoresis instruments are used where the quartz cells designed for any specific system are available. The magnitude of zeta potential, ζ, is obtained from the following relation:

$$\zeta = \frac{\mu\eta}{\varepsilon_o \mathbf{D}} \tag{1.173}$$

where

η is the viscosity of the solution
ε_o is the permittivity of the free space
D is the dielectric constant

The velocity of a particle in a unit electric field is related to its electric mobility.

In another application, the magnitude of zeta potential is measured as a function of added counter-ions. The variation in zeta potential is found to be related to the stability of the colloidal suspension.

Example: Gold Colloidal Suspension

The results of a gold colloidal suspension (gold sol) are reported as follows:

Counter-Ion (Al^{3+})	Velocity	Stability (Flocculation Character)
0 Al^{3+}	3 (−)	Very high stability
20×10^{-6} mol	2 (−)	Flocculates (4 h)
30×10^{-6} mol	0 (*zero*)	Flocculates fast
40×10^{-6} mol	0.2 (+)	Flocculates (4 h)
70×10^{-6} mol	1 (+)	Flocculates slowly

These data show that the charge on the colloidal particles changes from negative to zero (when the particles do not show any movement) to positive, at high counter-ion concentration. This is a very general picture. Therefore, in wastewater treatment plants, one adds counter-ions until the movement of colloids is almost zero and thus one can achieve fast flocculation of pollutant particles. The variation of ζ of silica particles has been investigated as a function of pH (Birdi, 2009). The dissociation of the surface groups –Si(OH) is involved in this characteristic. Under these operations, one constantly monitors the zeta potential by using a suitable instrument.

1.8.6 Application of Scanning Probe Microscopes (STM, AFM, FFM) in Surface and Colloid Chemistry

Microscopes have played a very important role in science, which reveals the structures of the material. The degree of microscopic resolution determines the degree of information. Over many decades, the ultimate aim has been to be able to see single atoms or molecules. During the end of the twentieth century, a big surge in the development of very important techniques has become available for science (*nanoscience*) and technology (self-assembly structures [micelles; monolayers; vesicles]; biomolecules; biosensors; surface and colloid chemistry; nanotechnology). In fact, current literature shows that these developments have no end to this trend as regards the vast expansion in the sensitivity and level of information.

Typical of all humans, *seeing is believing*, so the microscope has attracted much interest for many decades. All these inventions, of course, were basically initiated on the principles laid out by the telescope (as invented by Galileo) and the light-optical microscope (as invented by Hooke). Over the years, the magnification and the resolution of microscopes has improved. However, for the man to understand the nature, the main aim of mankind has been to be able to see atoms or molecules. This goal has been achieved and the subject as described here will explain the latest developments that were invented only few decades ago.

The ultimate aim of scientists has always been to be able to *see molecules* while these are active. In order to achieve this goal, the microscope should be able to operate under ambient conditions. Further, all kinds of molecular interactions between a solid and its environment (gas or liquid or solid), initially, can take place only via the surface molecules of the interface. It is obvious that when a solid or a liquid interacts with another phase, knowledge of the molecular structures at these interfaces is of interest. The term *surface* is generally used in the context of gas–liquid or gas–solid phase boundaries, while the term i*nterface* is used for liquid–liquid or liquid–solid phases. Furthermore, many fundamental properties of surfaces are characterized by morphology scales on the order of 1–20 nm (1 nm = 10^{-9} m = 10 Å [Angstrom = 10^{-8} cm]).

Generally, the basic issues that should be addressed for these different interfaces are as follows:

- What do the molecules of a solid surface look like, and how are the characteristics of these different than the bulk molecules? In the case of crystals, one asks about the kinks and dislocations.
- Adsorption on solid surfaces requires the same information about the structure of the adsorbates and the adsorption site and configurations.
- Solid–adsorbate interaction energy is also required, as is known from the Hamaker theory.
- Molecular recognition in biological systems (active sites on the surfaces of macromolecule; antibody–antigen) and biological sensors (enzyme activity; biosensors).
- Self-assembly structures at interfaces.
- Semiconductors and applications.

Most applications of microscopy are found in the case of surfaces and the study of molecules at the surfaces. Generally, the study of surfaces is dependent on understanding not only the reactivity of the surface but also the underlying structures that determine that reactivity. Understanding the effects of different morphologies may lead to a process for the enhancement of a given morphology and hence to improved reaction selectivities and product yields. Atoms or molecules at the surface of a solid have fewer neighbors as compared with atoms in the bulk phase, which is analogous to the liquid surface. Hence, the surface atoms are characterized by an unsaturated, bond-forming capability and accordingly are quite reactive. Until a decade ago, electron microscopy and some other similarly sensitive methods provided some information about the interfaces. A few decades ago, the best electron microscope images of globular proteins were virtually all little more than shapeless blobs. However, these days, due to relentless technical advances, electron crystallography is capable of producing images at resolutions close to those attained by x-ray crystallography or multidimensional nuclear magnetic resonance (NMR). In order to improve upon some of the limitations of the electron microscope, newer methods were needed. A few decades ago, a new procedure for molecular microscopy was invented. The new *scanning probe microscopes* not only provide new kind of information than hitherto as known from x-ray diffraction, for example, but these also open up a new area of research (e.g., nanoscience and nanotechnology).

The basic method of these SPMs (Birdi, 2003b; Chen, 2008) was essentially to be able to move a tip over the substrate surface with a sensor (probe) with molecular sensitivity (nm) in both longitudinal and vertical directions (Figure 1.53). This may be compared with the act of sensing with a finger over a surface or more akin to the old-fashioned vinyl record player with a metallic needle (a probe for converting mechanical vibrations to music sound).

Scanning probe microscopy was invented by Binnig and Rohrer (Nobel prize-1986) (Birdi, 2003b, 2009; Rabe, 1989). Scanning tunneling microscope (STM) was based on scanning a probe (metallic tip) since it is a sharp tip just above the substrate while monitoring some interactions between the probe and the surface. The tip is controlled to within 0.1 Å (1 nm).

In SPM, various interactions between the tip and the substrate are as follows:

STM: The tunneling current between a metallic tip and a conducting substrate that are at very close proximity but not actually touching in physical contact. This is controlled by piezo-motors in a step-wise method.

AFM: The tip is brought closer to the substrate while van der Waals force is monitored. At a given force, piezo-motor controls this setting while the surface is scanned in x–y direction.

FFM: This is a modification of AFM, where force is measured (Birdi, 2003b).

A schematic description of the SPM with the tip (dimension: 0.2 mm) and sample is shown in Figure 1.54.

The most significant difference between SPM and x-ray diffraction studies is that the former can be carried out both in air and water (or any other fluid). Corrosion and similar systems have been investigated by using STM. The tip is covered by a plastic material and this allows one to operate STM under fluid environment. STM has been used to study the molecules adsorbed on solid surfaces. LB films have been extensively investigated by both STM and AFM.

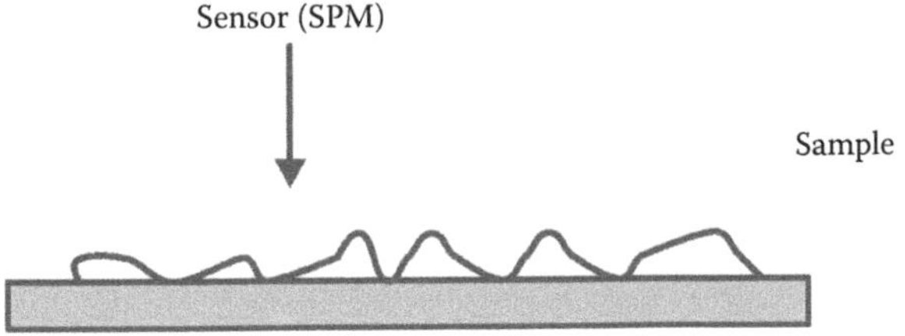

FIGURE 1.53 SPMs. Probe can operate both in air and liquid media.

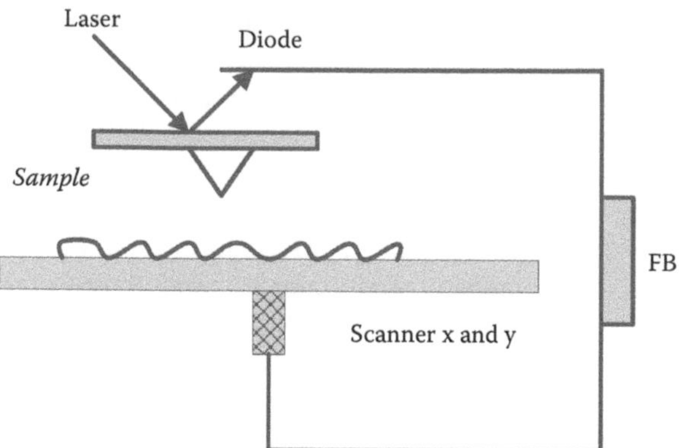

FIGURE 1.54 A schematic drawing of the sensor (tip/cantilever/optical/magnetic device) movement over a substrate in x/y/z direction with nanometer sensitivity (controlled by piezomotor) (at solid–gas or solid–liquid interface).

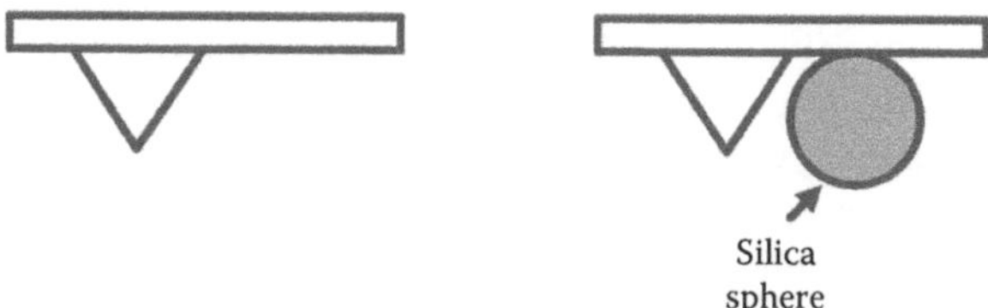

FIGURE 1.55 AFM sensor with SiO_2 sphere (schematic).

Colloid system studies by AFM: AFM has allowed scientists to be able to study molecular forces between molecules at very small (almost molecular size) distances. Further, AFM is a very attractive and sensitive tool for such measurements. In a recent study, the colloidal force as a function of pH of SiO_2 immersed in aqueous phase was reported using AFM. The force between an SiO_2 sphere (ca. 5 mm diameter) and a chromium oxide surface in aqueous phase of sodium phosphate was measured (pH from 3 to 11). The SiO_2 sphere was attached to the AFM sensor as shown in Figure 1.55.

These data showed that the isoelectric point (IEP) of SiO_2 was around pH 2, as expected. The binding of phosphate ions to chrome surface were also estimated as a function of pH and ionic strength (Birdi, 2003b). Further, both STM and AFM have been used to investigate corrosion mechanisms of metals exposed to aqueous phase. Since both STM and AFM can operate under water, this gives rise to a variety of possibilities. It is found that one can use other setups where instead of SiO_2 one can use other molecules to investigate surface phenomena. For example, binding of bacteria to surfaces can be investigated by SPM methods. In the coming decades, one expects much advances in the industrial applications of SPM. As compared to ordinary microscopes, the SPM provides 2D and 3D images. The 3D images allow one to see molecules with different diameters (Birdi, 2003b). SPMs have contributed huge useful information about surface structures in the past decades (Table 1.10).

1.9 GAS BUBBLE FORMATION AND STABILITY (THIN LIQUID FILMS AND FOAMS)

If one shakes pure water, no bubbles are observed at the surface. On the other hand, if one shakes a soap solution, one observes bubbles at the surface after shaking. At the shores of rivers, lakes, and oceans, one also observes bubbles. This shows that some kind of surface TLF is present in the soap solution (or in rivers, lakes, or oceans [mostly due to pollution!]). The formation and structure of TLF such as in foams or bubbles is the most fascinating phenomena that mankind has studied over many decades. TLF is thus the thinnest object one can see without the aid of any kind of microscope. One of the most commonly known liquid thin film structure is the soap bubble or bubbles formed on detergent solutions (as in dish-washing solutions). Everyone has enjoyed the formation of soap bubbles and the view of the rainbow colors. It may look as if the bubble formation and stability have not such a great consequence, but in fact, in everyday life bubbles play an important role (e.g., from lung function to beer and champagne!). It is common observation that ordinary water when shaken does not form any bubbles at the surface. On the other hand, all soap and detergent solutions (shampoo; washing; beer; champagne; sea water) on shaking may form very extensive bubbles. Further, even though one cannot see or observe the surface layer of a liquid directly, the TLFs allow one to make some observations that provide much useful information (Adamson and Gast, 1997; Birdi, 2003a, 2009, 2010a, 2010b; Cox, 2013; Hoher and Addad, 2006; Janaiaud et al., 2014; MacDowell et al., 2014; Mager and Melosh, 2007).

1.9.1 Soap Bubbles and Foams

Let us consider two systems: pure water or a soap (detergent) solution. If pure water is shaken, then no bubbles are observed at the surface. All pure organic fluids exhibit no bubble formation on shaking. This means that as air bubbles rise to the surface of the liquid it merely exits into the air. On the other hand, if an aqueous detergent (SAS) solution is shaken or an air bubble is created under the surface, then a bubble is formed.

This can be described as follows:

Air bubble inside liquid phase—at surface bubble detaches and moves up under gravity.
The detergent molecule forms a bilayer in the bubble film. The water in between is the same as the bulk solution. This may be depicted as follows:
  Surface layer of detergent.
  Bubble with air and a layer of detergent.
  Bubble at the surface forms double layer of detergent with some water in between (TLF) (varying from 10 to 100 μm).

Since even very minute amounts (around ppm) of SAS give rise to stable bubbles, it has been used to detect the presence of the latter. A bubble is composed of a TLF with two surfaces, each with a polar end pointed inward (towards water phase) and the hydrocarbon chains pointing outward. The water inside the films will move away (due to gravity and due to evaporation), giving rise to the thinning of the film. Since the thickness approaches the dimensions of the light wavelength, one observes varying interference colors. The reflected ray will interfere with the incident wavelength. The consequence of this will be that depending on the thickness of the film, one will observe colors. The most amazing observation is found that the thickness of these films is comparable to the wavelength of light. Especially, when the thickness of the film is approximately the same as the wavelength of the light (i.e., between 400 and 1000 nm). The black film is observed when the thickness is the same as the wavelength of the light (approx. 500–700 nm). Thus, this provides the closest visual observation of two molecule thick film by eyesight.

One of the most important roles of bubbles one finds is in the food industry (such as ice-cream; champagne and beer industry). The stability and size of the bubbles determine the taste and the looks of the product. Especially, in the case of champagne, both the size and the stability of bubbles have been found to determine the impact of taste and flavor. It has been estimated that in a bottle of volume 750 mL, there will be about 50 million bubbles (if the average radius of bubbles is 0.1 mm) (Birdi, 2010a). This is a very rough estimate. However, more accurate estimates have been made by using photography of bubbles, etc. In the wine industry, much research has been made on the determination of the factors that control the bubble formation and stability. Another example is ice creams, where air bubbles are trapped in the frozen material.

1.9.2 Foam Formation (Thin Liquid Films)

In general, ordinary foams from detergent solutions are thick initially (micrometer), and as fluid flows away due to gravity or capillary forces or surface evaporation), the film becomes thinner (few hundred Å). Foams are of essential part in many processes, both in industry and biology.

The foam consists of

Air on one side
Outer monolayer of detergent molecule
Some amount of water
Inner monolayer of detergent molecule
Air on outer side

This can be depicted (schematic) as follows:

DETERGENT MOLECULE*WATER***DETERGENT MOLECULE**
DETERGENT MOLECULE*WATER***DETERGENT MOLECULE**
DETERGENT MOLECULE*WATER***DETERGENT MOLECULE**
DETERGENT MOLECULE*WATER***DETERGENT MOLECULE**

The dimension of the detergent molecule is on the order of 15 Å, while the water phase can be many thousands of molecules. It is worth mentioning that a merely one layer of detergent molecule can contain a large volume of water phase. Of course, there is a physical limit to as to how much water can be contained in the film. In thick films, one also observes the movement of water away from the film, this process is called the thinning of the film. The orientation of detergent molecule in TLF is such as that the polar group (OO) is pointing toward the water phase and the apolar alkyl part (CCCCCCCCCC) is pointing toward the air. In the water phase one finds ions in the case of ionic detergents. However, there will be no ions in the case of nonionic detergent films.

AirCCCCCCCCCCOOWATERWATEROOCCCCCCCCCCAir
AirCCCCCCCCCCOOWATERWATEROOCCCCCCCCCCAir
AirCCCCCCCCCCOOWATERWATEROOCCCCCCCCCCAir
AirCCCCCCCCCCOOWATERWATEROOCCCCCCCCCCAir
AirCCCCCCCCCCOOWATERWATEROOCCCCCCCCCCAir
AirCCCCCCCCCCOOWATERWATEROOCCCCCCCCCCAir

The thickness of water phase can vary from over 100 μm to less than 100 nm (Birdi, 2010a; Scheludko, 1966). Foams are thermodynamically unstable, since there is a decrease in total free energy when they collapse. As the thickness of the film decreases to around the wavelength of light (nm), one starts to observe rainbow colors (arising from interference). The TLF at even smaller in thickness (50 Å or 5 nm).

However, certain kinds of foams are known to persist for very long periods of time and many attempts have been made to explain their metastability. The TLF may be regarded as a kind of condenser. The repulsion between the two surfactant layers, Figure 1.17, will be determined by the EDL. The effect of added ions to the solution is to make the EDL contract, and this leads to thin films. It looks *black-grey* and the thickness is around 50 Å (5 nm), which is almost the size of the bilayer structure of the detergent (i.e., twice the length [ca. 25 Å] of a typical detergent molecule plus water). Actually, this is a remarkable fact that one can *see* two molecule thin structure with a naked eye. Rainbow colors are observed since the light is reflected by the varying thickness of the TLF of the bubble.

It may not be obvious at first sight, but in beer industry, the foaming behavior is one of the most important characteristics (Clark et al., 1994). Beer in a bottle is produced under high pressure of CO_2 gas. As soon as one opens a beer bottle, the pressure drops and the gas, CO_2, is released, which gives rise to foaming. In common behavior the foam stays inside the bottle. The foaming is caused due to the presence of different amphiphilic molecules (fatty acids; lipids; proteins). This foam is very rich as the liquid film is very thick and contains lots of aqueous phase (such foams are called *kugelschaum*). The foam fills the empty space in the bottle, and under normal conditions, it barely spills out. However, under some abnormal conditions the foam is highly stable and starts to pour out of the bottle and is considered undesirable. In some reports, one has found that an addition of heavy metal ions could change the foaming characteristics (Birdi, 1989).

As regards foam stability, it was recognized that the surface tension under film deformation must always change in such a way as to resist the deforming forces. Thus, tension in the film where expansion takes place will increase, while it will decrease in the part where contraction takes place. There is, therefore, a force tending to restore the original condition. The film elasticity, E_{film}, has been defined as

$$E_{film} = 2\mathbf{A}\left(\frac{d\gamma}{d\mathbf{A}}\right) \quad (1.174)$$

where

E_{film} and $\mathbf{A}$ are the film elasticity and area of the film
γ is the surface tension of the surface deformed

One of the most important applications is the bubble formation and stability in champagne. The size and the number of bubbles is found to be important for the impact of taste. The stability has also much impact on the looks and taste as well. The taste of impact is related to the size and number of bubbles, as well as how long the bubbles are stable.

Further, the stability of any foam film is related to the kinetics of the thinning of the TLF. As the thickness reaches a critical value, the film becomes very unstable. It was recognized at a very early stage by Gibbs that unstable state will conform when the film diverges from bulk system properties. This thickness was mentioned to be in the range of 50–150 Å. This state is called the *black* film, and the random motion of the molecules may easily give rise to a rupture of the TLF. The flow of liquid is determined by gravity forces. It was found that assuming the fluid in film has the same viscosity and density, the mean velocity will not exceed 1000 $\mathbf{D}^2_{film}$, where latter is distance between the lamella:

D_{film} (mm)	Flow (mm/s)
0.01	0.1
0.001	0.001

1.9.3 Criteria of Foam Stability

As it is known, if one blows air bubbles in pure water, no foam is formed. On the other hand, if a detergent or protein (amphiphile) is present in the system, adsorbed surfactant molecules at the interface give rise to foam or soap bubble formation. Foam can be characterized as a coarse dispersion of a gas in a liquid, where gas is the major phase volume. The foam, or the lamina of liquid, will tend to contract due to its surface tension, and a low surface tension would thus be expected to be a necessary requirement for a good foam-forming property. Furthermore, in order to be able to stabilize the lamina, it should be able to maintain slight differences of tension in its different regions. It is also therefore clear that a pure liquid, which has constant surface tension, cannot meet this requirement. The stability of such foams or bubbles has been related to the monomolecular films structures and stability. For instance, the foam stability has been shown to be related to the surface elasticity or surface viscosity, η_s, besides other interfacial forces. Studies have indicated that foam destabilization is related to the packing and orientation of mixed films, which can be determined from monolayer studies. Since very small (μg = g) amounts are needed for such studies, the monolayer method has been useful in a very large variety of system studies. It is also worth mentioning that foam formation from monolayers of amphiphiles constitutes the most fundamental process in everyday life. The other assemblies such as vesicles and BLM are somewhat more complicated systems, which are also found to be in equilibrium with monolayers. It is important to mention that foam does not form in organic liquids, such as methanol, ethanol, etc.

Although the surface potential, ψ, the electrical potential due to the charge on the monolayers, will clearly affect the actual pressure required to thin the lamella to any given thickness, we shall assume for the purpose of a simple illustration that 1/*k*, the mean *Debye–Huckel* thickness of the ionic double layer, will influence the ultimate thickness when the liquid film is under a relatively low pressure. Let us also assume that each ionic atmosphere extends only to a distance 3/k into the liquid when the film is under a relatively low excess pressure from the gas in the bubbles; this value corresponds to a repulsion potential of only a few millivolts. Thus at about 1 atm pressure,

$$h_{film} = \frac{6}{k} + 2(\text{monolayer thickness}) \tag{1.175}$$

For charged monolayers adsorbed from 10^{-3} *n*-sodium oleate, the final total thickness, $\mathbf{h}_{film}$, of the aqueous layer should thus be on the order of 600 Å (i.e., 6/k or 18 Å). To this value, one needs to add 60 Å (60×10^{-10} m) for the two films of oriented soap molecules, giving a total of 660 Å. The experimental value is 700 Å (Birdi, 2010a; Scheludko, 1966). The thickness decreases on the addition of electrolytes, as also suggested by the preceding equation. For instance, the value of $\mathbf{h}_{film}$ is 120 Å in the case of 0.1 M NaCl. Addition of a small amount of certain nonionic surface-active agents (e.g., *n*-lauryl alcohol, *n*-decyl glycerol ether, laurylethanolamide, laurylsufanoylamide) to anionic detergent solutions has been found to stabilize the foam. Measurements have been carried out on the excess tensions, equilibrium thicknesses and compositions of aqueous foam films stabilized by either *n*-decyl methyl sulfoxide or *n*-decyl trimethyl ammonium-decyl sulfate and containing inorganic electrolytes. It was recognized at a very early stage (Birdi, 2003b, 2009) that the stability of a liquid film must be greatest if the surface pressure strongly resists deforming forces.

It has been shown (Birdi, 2003b, 2009; Friberg et al., 2003) that there exists a correlation between foam stability and the elasticity of the film, that is, monolayer elasticity. In order for elasticity to be large, surface excess must be large. Maximum foam stability has been reported in systems with fatty acid and alcohol concentrations well below the minimum in γ. Similar conclusions have been observed with n-$C_{12}H_{25}SO_4Na$ (SDS) + n-$C_{12}H_{25}OH$ systems, which give minimum in γ versus concentration with maximum foam at the minimum point (Chattoraj and Birdi, 1984), due to mixed monolayer formation. It has been found that SDS + $C_{12}H_{25}OH$ (and some other additives) make *liquid-crystalline* structures at the surface. This leads to a stable foam (and liquid-crystalline structures). In fact, one deliberately uses in technical formulations SDS with some (less than 1%) $C_{12}H_{25}OH$ to enhance the foaming properties. The foam drainage, surface viscosity, and bubble size distributions have been reported for different systems consisting of detergents and proteins. Foam drainage was investigated by using an incident light interference microscope technique. The foaming of protein solution is of theoretical interest and also has a wide application in the food industry (Friberg et al., 2003). Further, in the fermentation industry where foaming is undesirable, the foam is generally caused by proteins. Since mechanical defoaming is expensive due to the high power required, antifoam agents are generally used.

On the other hand, antifoam agents are not desirable in some of these systems, as for instance in food products. Further, antifoam agents deteriorate gas dispersion due to increased coalescence of bubbles. It has been long time that foams are stabilized by proteins, and that these are dependent on pH and electrolyte.

1.9.3.1 Foam Structure

The foam as TLF has a very fascinating structure. If two bubbles of same radius come into contact with each other, this leads to the formation of contact area and subsequently to formation of one large bubble (Adamson and Gast, 1997; Birdi, 2010a; Ghosh, 2004). This leads to following considerations:

I. Two bubbles of same radius.
II. Two bubbles touch each other and form a contact area.
III. Formation of only one bubble.

In stage II, the energy of the system is higher than in I, since the system has formed a contact area (dA_c). The energy difference between II and I is γdA_c. When the final stage is reached, III, there will be a decrease in total area by 41% (i.e., the sum of the area of two bubbles is larger than that of one bubble). This means that system III is at a lower energy state than the initial state I (γdA_{II-III}).

When three bubbles come into contact, the equilibrium angle is 120°. The angle of contact relates to systems equilibrium state, which is 120° from simple geometrical considerations. If four bubbles are attached to each other, then the angle will, at equilibrium, be 109°28′.

1.9.3.2 Foam Formation of Beer and Surface Viscosity

The surface and bulk viscosities not only reduce the draining rate of the lamella but also help in restoration against mechanical, thermal, or chemical shocks. The highest foam stability is associated with appreciable η_s and yield value.

The over-foaming characteristics of beer (*gushing*) has been the subject of many investigations (Birdi, 2003b, 2009). The extreme case of gushing is when a beer on opening starts to foam out of the bottle and in some cases empties the whole bottle (which is indeed a very serious problem). The relationship between surface viscosity, η_s, and gushing was reported by various investigators. Various factors were described for gushing process: pH, temperature, and metal ions, which could lead to protein denaturation.

The stability of a gas (i.e., N_2; CO_2; air) bubble in a solution depends mainly on its dimensions. A bubble with a radius greater than a *critical magnitude* will continue to expand indefinitely and degassing of the solution would take place. Bubbles with a radius equal to the critical value would be in equilibrium, while bubbles with radius less than the critical value would be able to redissolve in the bulk liquid. The magnitude of the *critical radius*, R_{cr}, varies with the degree of saturation of the liquid, that is the higher the level of super-saturation the smaller the R_{cr}. It has been suggested that there is nothing unusual in the stability of the beer and, although carbon dioxide is far from an ideal gas, empirical work supports this conclusion. A possible connection between η_s and gushing has been reported. Nickel ion, a potent inducer of gushing, has been reported to give rise to a large increase in the η_s of beer. Other additives besides Ni, such as Fe or humulinic acid, which cause gushing, have also been reported to give large increase in η_s. On the other hand, additives that are reported to inhibit gushing, such as EDTA (ethylenediamine acetic acid Ba chelating agent), have been reported to decrease η_s of beer. This relation between η_s and gushing has been suggestive that an efficient gushing inhibitor should be very surface active (in order to be able to compete with gushing promoters), but incapable of forming rigid surface layers (i.e., high η_s). Unsaturated fatty acids, such as linoleic acid, is a potent gushing inhibitor, since it destabilizes the surface films. Surface viscosity, η_s (g/s), was investigated by the oscillating-disc method. It was found that low η_s (0.03–0.08 g/s) beer surfaces gave nongushing behavior. Beers with high η_s (2.3–9.0 g/s) were found to give gushing.

1.9.4 Antifoaming Agents (Destabilizing of Foam Bubbles)

In many cases, one finds foaming to be undesirable (such as in machine dish-washing and wastewater treatment). The main criteria for antifoaming molecules is that these exhibit following characteristics (Birdi, 2009):

Do not form mixed monolayers with foaming agents.
Reduce surface viscosity (thus destabilizing the foam films).
Low boiling point liquid additives (such as ethanol).

1.9.5 Bubble Foam Technology (Wastewater Purification Technology)

Foams and bubbles are easily created and require very little energy input. This technology therefore makes it very useful in those applications where bubbles can be of positive benefit.

In the case of pollutants that are present at very low concentrations (around ppm), the bubble foam technology has been used. The biggest challenge mankind is facing today is the need of adequate supply of *pure drinking water* worldwide (current world population being around seven billion). The world population increase (from 1900 to 2000 by a factor of 4) is much faster than the availability of clean drinking water supplies. The increased need for pure water in industrial production also adds further burden to clean water supply. The purification of water for household use has been developed during the past decades. Pollutants as found in wastewaters are of different origin and concentration (Kim and Platt, 2007). Solid particles are mostly removed by filtration, but colloidal particles are not easily removed by this method. Solute compounds are rather difficult to remove, especially toxic substances with very low concentration (around ppm range of concentration). *Flotation* has been used with great advantage in some cases where sedimentation cannot remove all the suspended particles. Following are some examples where flotation is being used with much success:

Paper fiber removal in pulp and paper industry
Oils, greases and other fats in food, oil refinery and laundry wastes
Clarification of chemically treated waters in potable water production
Sewage sludge treatment

Many of the industrial wastewaters amenable to clarification by flotation are colloidal in nature, for example, oil emulsions, pulp and paper wastes, and food processing. For the best results, such wastes must be coagulated prior to flotation. In fact, flotation is always the last step in the treatment. In order to aid the flotation effectivity, one uses surfactants. This leads to lower surface tension and foaming. The latter helps in retaining the particles in the foam under flotation. Further, the effectiveness of flotation is also dependent on what kind of gas is used to make the bubbles (e.g., air and CO_2). This is related to the dipole characteristics of the gas (CO_2 has a stronger dipole than N_2).

1.9.5.1 Bubble Foam Purification of Water

It was found many decades ago that foam or bubbles could be used to purify wastewater. A simple device as used for laboratory froth flotation studies is shown in Figure 1.56.

Bubbles are formed in the sintered glass as air, or other suitable gas (N_2, CO_2, etc.) is bubbled through the solution containing the solid suspension. A flotation agent (a suitable surface-active agent) is added and the air is bubbled. Surface-active pollutants in wastewater have been removed by bubble film separation methods. Especially very minute concentrations are easily removed by this method, which is more economical than more complicated methods (such as active charcoal, filtration, and other chemical methods). This method is now commercially available for such small systems as fish tanks, etc. (Birdi, 2009, 2010a). The principle in this procedure is to create bubbles in wastewater tank and to collect the bubble foam at the top, Figure 1.56.

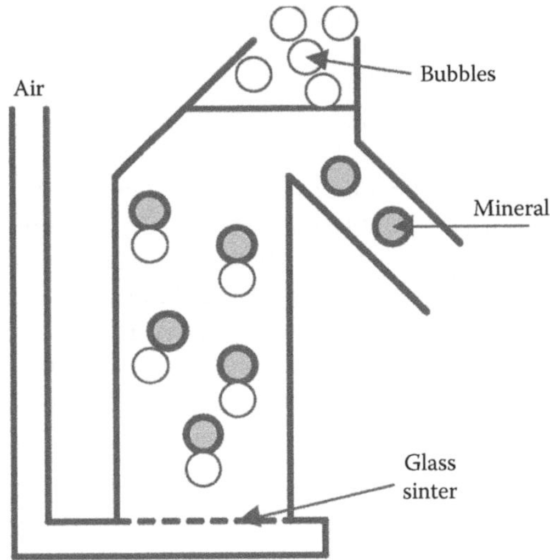

FIGURE 1.56 Bubble foam separation method for wastewater purification.

Bubbles are blown into the inverted funnel. Inside the funnel the bubble film is transported away and collected. Since the bubble film consists of

SAS
Water (and salts, etc.)

it is seen that even very minute amounts (less than milligram per liter) of SAS will accumulate at the bubble surface. As shown previously, it would require a large number of bubbles to remove a gram of substance. However, since one can blow thousands of bubbles in a very short time, the method is found to be very feasible.

In the shale fracking process, the wastewater is treated to remove pollutants (Knaus et al., 2010; Yethiraj and Striolo, 2013). The bubble foam separation can be of importance in this process.

1.10 EMULSIONS–MICROEMULSIONS–LYOTROPIC LIQUID CRYSTALS

One of the most important aspects of surface and colloid chemistry is the subject of oil–water emulsion technology. Oil is sparingly soluble in water, and vice versa. Science of emulsion is also important since many biological systems use the same principles, such as milk. The essential subject of emulsion is based on the fact that oil and water *do not mix* if shaken.

As is well known, if one shakes oil and water, oil breaks up into small drops (about few mm diameters) but these drops join together rather quickly to return to their original state (as shown here).

Step I: *Oil* phase and *water* phase
Step II: Mixing
Step III: *Oil* drops in *water* phase
Step IV: After short time
Step V: *Oil* phase and *water* phase (same as under Step I)

However, one finds that oil and water can be dispersed with the help of suitable *emulsifiers* (surfactants) to give *emulsions* (Becher, 2001; Birdi, 2008, 2010a; Friberg et al., 2003; James et al., 2014; Sjoblom, 2008; Takahashi et al., 2014). This is well known in the emulsions as found in the household, such as milk, mayonnaise, etc. The basic reason is that the IFT between oil and water is around 50 mN/m, which is high and leads to formation of large oil drops. However, addition of suitable emulsifier(s) reduces the value of IFT to very low values (even much less than 1 mN/m). Stable emulsion formation means that oil drops remain dispersed for a given length of time (up to many years). The stability and characteristics of these emulsions are related to the area of applications. Emulsions are a mixture of two (or more) immiscible substances. Some everyday common examples are milk, butter (fats, water, and salts), margarine, mayonnaise, and skin creams. In butter and margarine, the continuous phase consists of lipids. These lipids surround water droplets (water in oil emulsion). All emulsions are prepared by some convenient kind of mechanical agitation or mixing. Remarkably, the natural product, milk (stable oil–water emulsion), is made by the organism without any agitation but inside the glands (Birdi, 2010a; Bouchoux et al., 2014; Kristensen, 1997). Milk is the most life-sustaining emulsion providing nutrition and other health-related support properties.

In general, one finds three different kinds of emulsion types:

Emulsions
Microemulsions
LC (and lyotropic LC)

The main criterion of interest in emulsions is that these systems consist of both water and oil. This may be skin treatment or shoe shine or car polish, etc. In other words, one can apply both these components (water and oil, which do not mix) simultaneously. This also allows one to perform functions that are dependent on properties of water or oil. This, thus, needs information about the IFT and as well as the solubility characteristics of SAS needed to stabilize the emulsions.

Microemulsions are microstructured mixtures of oil–water–emulsifiers–other substances. LCs are substances that exhibit special melting characteristics. Further, some mixtures of surfactant–water–co-surfactant may also exhibit LC (lyotropic crystals) properties. The emulsion technology is basically thus concerned in preparing mixtures of two immiscible substances:

Oil
Water

by adding suitable surface-active agents (emulgators, co-surfactants, and polymers). The emulsion technology is thus very varied, since one finds many simple systems (such as skin creams, etc.) and there are also very complex systems (such as milk, etc.) (Kristensen, 1997).

1.10.1 Formation of Emulsions (Oil and Water)

When a SAS is added to a system of oil–water, the magnitude of IFT decreases from 50 to 30 mN/m (or lower [less than 1 mN/m]). This leads to the observation that on shaking oil–water system (due to low IFT) smaller drops of the dispersed phase (oil or water). The smaller drops also lead to a more stable emulsion. Depending on the surfactant used, one will obtain oil in water (*O/W*) or water in oil (*W/O*) emulsion. These experiments where oil–water or oil–water + surfactant are shaken together are shown in Figure 1.57.

These emulsions are all opaque since these reflect light. Some typical oil–water IFT values are given in Table 1.5.

These data show certain trends. The decrease in IFT is much smaller with decrease in alkyl chain in the case of alkanes than for alcohols.

1.10.2 Types of Oil–Water Emulsions

Emulsions are one of the most important structures that are prepared specifically for a given application. For example, a day cream (skin cream) has different characteristics and ingredients than a

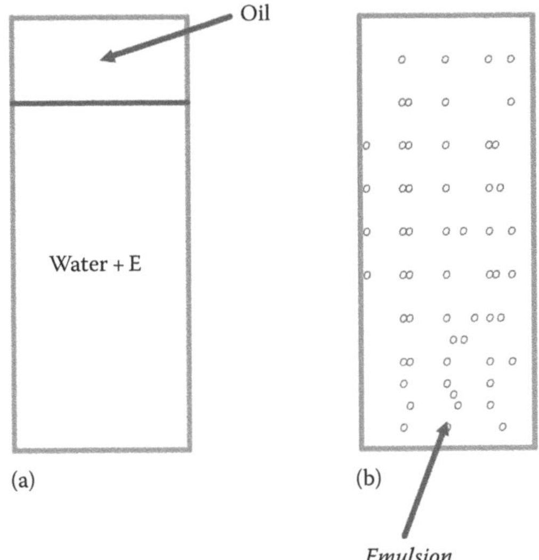

FIGURE 1.57 Mixing of oil–water (a) or oil–water + surfactant (b) by shaking.

night cream. One of the main differences in emulsions is whether oil droplets are dispersed in water phase or water drops are dispersed in oil phase. One can determine this by measuring the conductivity, since it is higher for O/W than for W/O emulsion. Another useful property is that O/W will dissolve water while W/O will not. This, thus, shows that one will choose W/O or O/W depending on the application area. Especially in the case of skin emulsions the type is of much importance.

Oil-in-water emulsions: The main criterion for O/W emulsion is that if one adds water, then it will be miscible with the emulsion. Further, after water evaporates, the oil phase will be left behind. Thus, if one needs oil phase on the substrate (such as skin, metal, and wood), then one shall use an O/W-type emulsion.

Water-in-oil emulsions: The criterion for W/O emulsion is that latter is *miscible* with oil. That means that if one adds the emulsion to some oil, then one obtains a new but diluted W/O emulsion. In some skin creams, one prefers to use W/O-type emulsion (especially if one needs an oil-like feeling after its application). On the other hand, if one needs oil free surface, then O/W emulsion is preferred.

1.10.3 HLB Values of Emulsifiers

The emulsion technology requires a very exact knowledge of the physicochemical properties of the various components. In fact, lipid components used in emulsions are investigated in mixed film monolayer systems. These studies are useful in the determination of the feasibility of components. The emulsifiers used exhibit varying solubility in water (or oil) as related to the HLB value. This will thus have consequences on the emulsion. Let us consider a system where we have oil and water. If we add an emulsifier to this system, then the latter will be distributed in both oil and water phases. The degree of solubility in each phase will of course depend on its structure and HLB character. The emulsifiers as used in making emulsions are characterized with regard to the molecular structure. The amphiphile molecules consist of HLB characteristics. Thus each emulsifier which may need for a given system (e.g., if one needs an O/W or W/O emulsion) will need a specific HLB value. The data in Table 1.12 give a rough estimation of the HLB needed for a given system of emulsion. In general, one may expect that if the emulsifier dissolves in water, then on adding oil, one will obtain an *oil in water emulsion*. Conversely, if the emulsifier is soluble in the oil, then on adding water, one will obtain a *water in oil emulsion*.

TABLE 1.12
Magnitudes of Interfacial Tensions of Different Organic Liquids against Water (20°C)

Oil Phase	IFT (mN/m)
Hexadecane	52
Tetradecane	52
Dodecane	51
Decane	51
Octane	51
Hexane	51
Benzene	35
Toluene	36
CCl_4	45
CCl_3	32
Oleic acid	16
Octanol	9
Hexanol	7
Butanol	2

W/O emulsions are formed by using HLB values between 3.6 and 6. This suggests that one generally uses emulgators that are soluble in oil phase.

O/W emulsions need HLB values around 8–18. This HLB criterion is only a very general observation. However, it must be noticed that HLB values alone do not determine the emulsion type. Other parameters, such as temperature, properties of oil phase, and electrolytes in aqueous phase, also affect the emulsion characteristics. HLB values have no relation to the degree of emulsion stability. HLB values of some surface-active agents are given in Table 1.13.

HLB values decrease as the solubility of surface-active agent *decreases in water.* The solubility of cetyl alcohol in water (at 25°C) is less than 1 mg/L. It is thus obvious that in any emulsion cetyl alcohol will be present mainly in the oil phase, while SDS will be mainly found in the water phase. Empirical HLB values are found to have significant use in applications in emulsion technology. It was shown that the HLB is related, in general, to the distribution coefficient, K_D, of the emulsifier in oil and water phases:

$$K_D = \frac{C(\text{water})}{C(\text{oil})} \tag{9.1}$$

where C(water) and C(oil) are the equilibrium molar concentrations of the emulsifier in water and oil phases, respectively. Based on this definition of K_D, one has found that the magnitude of HLB for emulsifiers can be estimated as follows:

$$(\text{HLB} - 7) = 0.36 \ln(K_D) \tag{9.2}$$

TABLE 1.13
HLB Values of Different Emulsifiers (Commonly Used in Emulsions)

Emulsifier Solubility in Water	HLB	Application
Very low solubility	0–2	W/O
Low solubility	4–8	W/O
Soluble	10–12	Wetting agent
High solubility	14–18	O/W

TABLE 1.14
HLB Values of Some Typical Surface-Active Agents

SAA	HLB
Na-lauryl sulfate	40
Na-oleate	18
Tween 80 (sorbitan monooleate EO20)	15
Tween 81 (sorbitan monooleate EO6)	10
Ca-dodecylbenzene sulfonate	9
Sorbitan monolaurate	9
Soya lecithin	8
Sorbitan monopalmiate	7
Glycerol monolaurate	5
Sorbitan monostearete	5
Span80 (sorbitan monooleate)	4
Glycerol monostearate	4
Glycerol monooleate	3
Sucrose distearate	3
Cetyl alcohol	1
Oleic acid	1

Based on these thermodynamic relations, one could then suggest the relation between HLB and emulsion stability and structure. HLB values can also be estimated from the structural groups of the emulsifier (Table 1.14). This table can be useful in those cases where one needs to estimate HLB values.

There is found an extensive application of food emulsifiers. It is obvious that these emulsifiers must satisfy special requirements (e.g., toxicity) in order to be useful in food industry. One determines the toxicity from animal tests. The test determines the amount of a substance which causes 50% (or more) of the test animals to die (lethal dosage; LD50). It is thus obvious that food emulsions are subject to much strict controls and limitations (Friberg et al., 2003) (Table 1.15).

TABLE 1.15
HLB Group Numbers

Group	Group number
Hydrophilic	
$-SO_4Na$	39
$-COOH$	21
$-COONa$	19
Sulfonate	11
Ester	7
$-OH$	2
Lipophilic	
$-CH-$	0.5
$-CH_2$	0.5
$-CH_3$	0.5
$-CH_2CH_2O-$	0.33

1.10.4 Methods of Emulsion Formation

If one shakes oil and water, the oil breaks up into drops. However, these will quickly coalesce and return to the original state of two different phases. One also observes that the longer one shakes the more the drops reduce in size. In other words, the mechanical energy put into the system makes the drops smaller in size. Emulsions are made based on different procedures. These can be where mechanical agitation is used. There are also used other methods. The emulsion technology is very much of a state-of-the-art type of industry (Birdi, 2003a; Friberg et al., 2003; Sjoblom, 2008). Therefore, there exists a vast literature about methods used for any specific emulsion. In a simple case, an emulsion may be based on three necessary ingredients: water, oil, and emulsifier. In other words one needs to determine in which weight proportions one need to mix these substances in order to obtain an emulsion (at a given temperature) to be stable (or maximum stability). This may be more conveniently (and as an example) carried out in a phase study in the triangle. The micellar region exists on the water–surfactant line (Figure 1.58).

Near the surfactant region, one finds *crystalline* or *lamellar* phase. This is the region that one finds in the hand soaps. The ordinary hand soap is mainly salt of fatty acid (typical composition: coconut oil fatty acids or mixtures) (85%) plus water (15%) and perfume, and some salts, etc. X-ray analyses have shown that the crystalline structure consists of a series of a layer of soap separated by a water layer (with salts). The hand soap is produced by extruding under high pressure. This process aligns the lamellar crystalline structure lengthwise. If one measures the degree of expansion versus temperature, then the expansion is twice along the length than the width. In practice what one is does is as follows. A suitable number (over 50) of test samples are prepared with mixing each component in varying weights to represent a suitable number of the regions. The test samples are mixed under rotation in a thermostat over a few days to reach equilibrium. The test samples are centrifuged and the phases are analyzed. The phase structure is investigated by using a suitable analytical method. It has been found that studies of multicomponent systems as these have shown a very large numbers of phases. However, by analyzing some typical system, one finds that there are some trends (related molecular ratios among the main ingredients) that can be used as guidelines.

A similar conclusion was reached when investigating microemulsions (as described in the following).

The stability of emulsions is dependent on various parameters (size of drops; interactions between drops). In those systems where the emulsifier carries a charge would impart specific characteristics to the emulsion. A double layer will exist around the oil droplets in an O/W emulsion (Birdi, 2010a). If the emulsifier is negatively charged, then it will attract positive counter-ions while repelling negatively charged ions in the water phase. The change in potential at the surface of oil droplets will be dependent on the concentration of ions in the surrounding water phase. The state of stability under

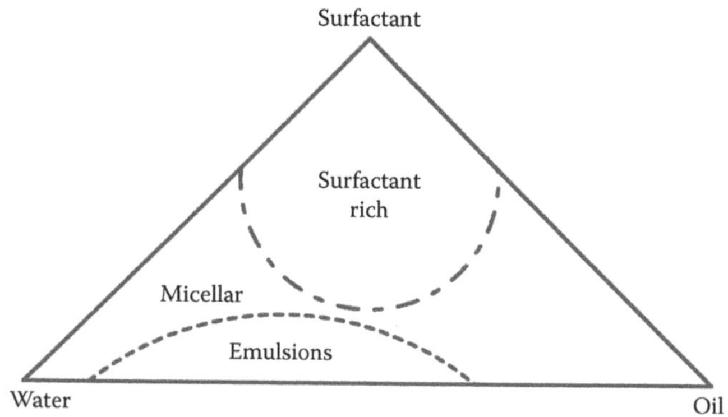

FIGURE 1.58 Different phase equilibria in a water–surfactant (emulsifier)–oil mixture system (schematic).

these conditions can be qualitatively described as follows. As two oil droplets approach each other, the negative charge gives rise to repulsive effect. The repulsion will take place within the EDL region. It can thus be seen that the magnitude of double-layer (EDL) distance will decrease if the concentration of ions in the water phase increases (Birdi, 2010b). This is due to the fact that the EDL region decreases. However, in all such cases where two bodies come closer, there exist two different kinds of forces, which must be considered:

$$\text{Total force between two bodies} = \textit{repulsion forces} + \text{attraction forces}$$

The nature of the total force thus determines whether

The two bodies will stay apart
The two bodies will merge and form a conglomerate

The attraction force arises from van der Waals forces. The kinetic movement will finally determine whether the total force can maintain contact between two (or more) particles.

1.10.5 Orientation of Molecules at Oil–Water Interfaces

At this stage in the literature, there is no method available by which one can directly determine the orientation of molecules of liquids at interfaces. Molecules are situated at interfaces (e.g., air–liquid, liquid–liquid, solid–liquid) under asymmetric forces. Recent studies have been carried out to obtain information about molecular orientation from surface tension studies of fluids (Birdi, 1997; Wu et al., 1993). It has been concluded that interfacial water molecules in the presence of charged amphiphiles are in a tetrahedral arrangement similar to the structure of ice. extensive studies of alkanes near their freezing point had indicated that surface tension changes in abrupt steps. X-ray scattering of liquid surfaces indicated similar behavior (Wu et al., 1993). However, it was found that lower chain alkanes (C16) did not show this behavior. The crystallization of C16 at 18°C shows an abrupt change due to the contact angle change at the liquid–Pt plate interface (Birdi, 1997). It was found that in comparison to C16–air interface one observes super-cooling (to ca. 16.4°C). Each data point corresponded to 1 s; thus, the data showed that crystallization is very abrupt. High speed data ($\ll$1 s) acquisition is needed to determine the kinetics of transition. This kinetic data would thus add more information about molecular dynamics at interfaces. Effect of additives to aqueous phase, such as proteins. The magnitude of IFT is 12.6 and 4 mN/m for BSA and casein, respectively (Birdi, 2003a). These data provide useful information as regards the dynamic interfacial changes as observed directly.

1.10.6 Microemulsions

As mentioned earlier, ordinary emulsions as prepared by mixing oil–water–emulsifier are thermodynamically unstable (Birdi, 2009, 2010a; Guo et al., 1992; Meziani et al., 2003; Ohde et al., 2003). In other words, such an emulsion may be stable over a long length of time, but finally it will separate into two phases (oil phase and aqueous phase). All such emulsions can be separated into two phases, that is oil phase and water phase, by centrifugation. These emulsions are opaque, which means that the dispersed phase (oil or water) is present in the form of large droplets (over μm and thus visible to the naked eye).

A *microemulsion* is defined as a thermodynamically stable and a clear isotropic mixture of water–oil–surfactant–cosurfactant (in most systems it is short-chain alcohol). The co-surfactant is the fourth component that gives rise to the formation of very small aggregates or drops that make the microemulsion almost clear. Microemulsions are also therefore characterized as microstructured,

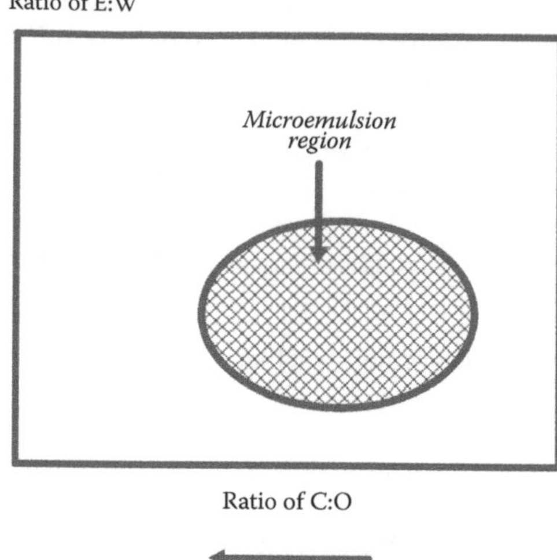

FIGURE 1.59 Four component system: oil (O)–water (W)–emulsifier (E)–co-surfactant (S) (ratio of O:S versus S:W).

thermodynamically stable mixtures of water : oil : surfactant : additional components (such as co-surfactants, etc.). Intensive studies of microemulsions have shown that these are one of the following types:

Microdroplets of oil in water or water in oil
Bicontinuous structure

Emulsifier will be found in both these phases. On the other hand, in systems with four components, Figure 1.59, consisting of oil–water–detergent–co-surfactant, there exists a region where clear phase is found. This is the phase region where microemulsions are found. Microemulsions are thermodynamically *stable* mixtures. The IFT is almost zero. The size of drops is very small, and this makes microemulsions look clear. It has also been suggested that microemulsion may consist of bicontinuous structures. This sounds more plausible in these four-component microemulsion systems. It has been suggested that microemulsion may be compared to swollen micelles (i.e., if one solubilizes oil in micelles). In such isotropic mixtures, there exists short range order between the droplets. Microemulsions have been formed by one of the following procedures (Birdi, 2009, 2010a; Friberg et al., 2003):

Oil–water mixture is added a surfactant. To this emulsion, one keeps adding a short-chain alcohol (with four to six carbon atoms) until a clear mixture (microemulsion) is obtained. It is thus obvious that microemulsion will exhibit very special properties, quite different from those exhibited by the ordinary emulsions. The microdrops may be considered as large micelles.

As an example, a very typical microemulsion which has been extensively investigated consists of a mixture of

$$\text{SDS(detergent)} + C_6H_6\text{(oil)} + \text{water} + \text{co-surfactant } (C_5OH \text{ or } C_6OH)$$

The phase region is determined by mixing various mixtures (approximately 20 samples) and allowing the system to reach equilibrium under controlled temperature. From the literature, one finds the following microemulsion recipe (Birdi, 1982):

Mix 0.0032 mol (0.92 g) SDS (mol. wt. of SDS [$C_{12}H_{25}SO_4Na$] = 288) with 0.08 mol (1.44 g) water and add 40 ml of C_6H_6. This mixture is mixed under vigorous stirring, and one gets a creamy emulsion. Under stirring to this three-component mixture, a co-surfactant ($C_5H_{11}OH$ or $C_6H_{13}OH$) is added slowly until a clear system consisting of a microemulsion is obtained. The stability region is found to be a relation between surfactant:water and surfactant:alcohol. This shows that some kind of structure (at molecular level) is responsible for the microemulsion formation. This shows that LC structure is indeed involved. The size of oil droplets is under a few micrometers and therefore the mixture is clear (Birdi, 1982; 2009, 2010a) as seen by naked eye. These data clearly indicated that the microemulsion phase was formed at certain fixed surfactant: water and co-surfactant:oil ratios.

It is important to consider the different stages when one proceeds to microemulsions from macroemulsions. It was mentioned earlier that surfactant molecules orient with the hydrophobic group inside the oil phase while the polar group orients toward the water phase. The orientation of surfactant at such interfaces cannot be measured by any direct method. Although much useful information can be obtained from monolayer studies of air–water interface or oil–water interface. At present, it is generally accepted that it is not easy to predict microemulsion recipe. However, some suggestions have been extended, which one may summarize as follows:

The HLB value of the SAS need be determined (for deciding the *O/W* or *W/O* type).
The phase diagram of the water–oil–surfactant (and co-surfactant) needs to be determined.
The effect of temperature is found to be very crucial.
The effect of added electrolytes is of additional importance.

In a recent report, the phase equilibria of a microemulsion were reported (Birdi, 2010a). Phase behavior of a microemulsion formed with food grade surfactant sodium bis-(2-ethylhexyl) sulfosuccinate (AOT) was studied. Critical microemulsion concentration was deduced from the dependence of pressure on CP on the concentration of surfactant AOT at constant temperature and water concentration. The results show that there are transition points on the CP curve in a very narrow range of concentration of surfactant AOT. The transition points were changed with the temperature and water concentration. These phenomena show that lower temperature is suitable to forming microemulsion droplet and the microemulsion with high water concentration is likely to absorb more surfactants to structure the interface.

1.10.7 Characteristics and Stability of Emulsions

The stability criteria of any emulsion is dependent on the needs and the application area. In some cases, the emulsion needs to be *stable* for a longer time than in other cases. As in the case of hair-cream, the emulsion should *destabilize* as soon as one applies (mechanical action) to the hair. As otherwise, the hair will be white with emulsion droplets. On the other hand, any emulsion used in spraying on plants needs to be stable for a longer time. Further, if one needs to clean oil spill on oceans, then one needs to *destabalize* the emulsion formation (Nour et al., 2007; Birdi, 2009). There are different processes that are involved in the stability and characteristics.

The various processes are as follows:

1.10.7.1 Creaming or Flocculation of Drops

This process is described in those cases where oil drops (in the case of oil–water) cling to each other and grow in large clusters. The drops do not merge into each other. The density of most oils is lower than that of water. This leads to the fact that instability the oil drop clusters rise to the surface. One can reduce this process by

1. Increasing the viscosity of the water phase and thereby decreasing the rate of movement of the oil drops
2. By decreasing IFT and thus the size of the oil drops

The ionized surfactants will stabilize O/W emulsions by imparting surface EDL.

The curvature (i.e., size) of drops in emulsions is of most importance, as regards the stability and application. The degree of stability of any emulsion is related to the rate of coagulation of two drops (O/W: oil drops; W/O: water drops).

Oil drop + Oil drop—Time—One Oil drop

The length of time is the degree of emulsion stability.

This process indicates that two oil drops in an O/W emulsion come close together, and if the repulsion forces are smaller than the attraction forces, only then the two particles meet and fuse into one larger drop. In the case of charged drops, there will be present an EDL around these drops (Birdi, 2010b). A negatively charged oil drop (charge arising from the emulsifier) will strongly attract positively charged ions in the surrounding bulk aqueous phase. At a close distance from the surface of a drop, the distribution of charges will be very much changing. While at a very large distance there will be electrical neutrality as there will be even number of positive and negative charges. The electrostatic repulsion exists between the two negatively charged drops which would exhibit strong repulsion even at large distances (many times the size of the particle). The shape of the EDL curve will be dependent on the negative and positive charge distribution. It is easily seen that if the concentration of counter-ions increases, then the magnitude of EDL will decrease and this will decrease the maximum of the total potential curve. The stability of emulsions can thus be increased by decreasing the counter-ion concentration. Another important emulsion stabilization is achieved by using polymers. The *large* polymer molecules adsorbed on solid particles, will exhibit repulsion at the surface of particles. The charged polymers will thus also give additional charge–charge repulsion (increased stability). Polymers are used in many pharmaceutical, cosmetics, and other systems (milk).

REFERENCES

Acree, W. E., *J. Phys. Chem. Ref. Data*, 5, 43, 2004.

Adam, N. K., *The Physics and Chemistry of Surfaces*, Clarendon Press, Oxford, U.K., 1930.

Adamson, A. W., Gast, A. P., *Physical Chemistry of Surfaces*, 6th edn., Wiley-Interscience, New York, 1997.

Attard, P., *Adv. Chem. Phys.*, 1, 92, 1996.

Aveyard, R., Hayden, D. A., *An Introduction to Principles of Surface Chemistry*, Cambridge, London, U.K., 1973.

Avnir, D., ed., *The Fractal Approach to Heterogeneous Chemistry*, Wiley, New York, 1989.

Bahadori, A., *Energy Fuels*, 5695, 25, 2011.

Bai, Z., Lodge, T. P., *Langmuir*, 8887, 26, 2010.

Baimpos, T., Shrestha, B. R., Raman, S., Valtiner, M., *Langmuir*, 4322, 30, 2014.

Bakker, G., *Kapillaritat und Uberflachenspannung Handbuch der Eksperimentalphysik*, 3d vi, Leipzig, Germany, 1928.

Bancroft, W. D., *Applied Colloid Chemistry*, McGraw-Hill, New York, 1932.

Barnes, G., *Interfacial Science (An Introduction)*, Oxford University Press, Oxford, U.K., 2011.

Barnsley, M., *Fractals Everywhere*, Academic Press, London, U.K., 1988.

Becher, P., *Emulsions, Theory and Practice*, 3rd edn., Oxford University Press, New York, 2001.

Ben-Naim, A., *The Hydrophobic Interactions*, Plenum Press, New York, 1980.

Birdi, K. S., *J. Theor. Biol.*, 1, 93, 1981.

Birdi, K. S., *J. Colloid Polym. Sci.*, 260, 8, 1982.

Birdi, K. S., *Lipid and Biopolymer Monolayers at Liquid Interfaces*, Plenum Press, New York, 1989.

Birdi, K. S., *Self-Assembly Monolayer (SAM) Structures*, Plenum Press, New York, 1999.

Birdi, K. S., *Fractals (in Chemistry, Geochemistry and Biophysics)*, Plenum Press, New York, 1993.

Birdi, K. S., ed., *Handbook of Surface & Colloid Chemistry*, CRC Press, Boca Raton, FL, 1997.

Birdi, K. S., ed., *Handbook of Surface & Colloid Chemistry—CD ROM*, 2nd edn., CRC Press, Boca Raton, FL, 2003a.

Birdi, K. S., *Scanning Probe Microscopes (SPM)*, CRC Press, Boca Raton, FL, 2003b.

Birdi, K. S., ed., *Handbook of Surface & Colloid Chemistry*, 3rd edn., CRC Press, Boca Raton, FL, 2009.
Birdi, K. S., *Surface & Colloid Chemistry*, CRC Press, Boca Raton, FL, 2010a.
Birdi, K. S., ed., *Interfacial Electrical Phenomena*, CRC Press, Boca Raton, FL, 2010b.
Birdi, K. S., *Surface Chemistry Essentials*, CRC Press, Boca Raton, FL, 2014.
Birdi, K. S., Ben-Naim, A., *J. Chem. Soc., Faraday Trans.*, 741, 77, 1981.
Birdi, K. S., Dalsager, S. U., Backlund, S., *J. Chem. Soc., Faraday Trans.*, 2035, 76(1), 1980.
Birdi, K. S., Fasman, G. D., *J. Polym. Sci.*, 1099, 42, 1973.
Birdi, K. S., Nikolov, A., *J. Phys. Chem.*, 365, 83, 1979.
Birdi, K. S., Vu, D. T., Winter, A., Naargard, A., *J. Colloid Polym. Sci.*, 266, 5, 1988.
Biresaw, G., Mittal, K. L., *Surfactants in Tribology*, CRC Press, New York, 2008.
Bonzel, H. P., ed., *Adsorption on Surfaces & Surface Diffusion of Adsorbates*, Springer Publ., New York, 2014.
Bouchoux, A., Qu, P., Bacchin, P., Gesan-Guiziou, G., *Langmuir*, 22, 30, 2014.
Bouvrais, H., Duelund, L., Ipsen, J. H., *Langmuir*, 13, 30, 2014.
Boyd, D. A., Adleman, J. R., Goodwin, D. G., Psaltis, D., *Anal. Chem.*, 2452, 80, 2008.
Boys, C. V., *Soap Bubbles*, Dover Publ., New York, 1959.
Butt, H. J., *Surface and Interfacial Forces*, Springer Publ., New York, 2010.
Butt, H. J., Graf, K., Kappi, M., *Physics and Chemistry of Interfaces*, Wiley-VCH, Berlin, Germany, 2013.
Castner, D. G., Ratner, B. D., *Surf. Sci.*, 500, 28, 2002.
Chaibundit, C., Ricardo, N. M. P. S., Crothers, M., Booth, C., *Langmuir*, 4277, 18, 2002.
Chattoraj, D. K., Birdi, K. S., *Adsorption and the Gibbs Surface Excess*, Plenum Press, New York, 1984.
Chen, C. J., *Introduction to Scanning Tunneling Microscopy*, Oxford University Press, Oxford, U.K., 2008.
Chen, H., Ruckenstein, E., *Langmuir*, 3723, 30, 2014.
Cheremisinoff, N. P., *Handbook of Water and Wastewater Treatment Technologies*, Butterworth-Heinemann, Oxford, U.K., 2002.
Chiu, C. W., Ducker, W. A., *Langmuir*, 140, 30, 2014.
Cini, R., Loglio, G., Ficalbi, A., *J. Colloid Interf. Sci.*, 41, 287, 1972.
Clark, D. C., Wilde, P. J., Marion, D., *J. Inst. Brewing*, 23, 100, 1994.
Coertjens, S., Moldenaers, P., Vermant, J., Isa, L., *Langmuir*, 4289, 30, 2014.
Complete Guide to Hydraulic Fracturing for Shale Oil and Natural Gas, Environmental Protection Agency (EPA), U.S. Geological Survey, Reston, VI, 2013.
Cox, S., ed., *Foams: Structure and Dynamics*, Oxford University Press, Oxford, U.K., 2013.
Danov, K. D., Kralchevsky, P., Ananthapadmanabhan, K. P., *Adv. Colloid Interf. Sci.*, 17, 206, 2014.
David, R., Neumann, W., *Adv. Colloid Interf. Sci.*, 46, 206, 2014.
Davies, J. T., Rideal, E. K., *Interfacial Phenomena*, Academic Press, New York, 1963.
De Gennes, P. G., *Capillarity and Wetting Phenomena*, Springer Publ., New York, 2003.
Defay, R., Prigogine, I., Bellemans, A., Everett, D. H., *Surface Tension and Adsorption*, Longmans, Green, London, U.K., 1966.
Diebold, U., Surface Science Reports, Vol. 48, p. 53, Elsevier, Amsterdam, the Netherlands, 2003.
Do, D. D., *Adsorption Analysis*, Imperial College Press, London, U.K., 1998.
Dukhin, A. S., Goetz, P. J., *Ultrasound Characterizing Colloids*, Elsevier, New York, 2002
Eusthopoulos, N., *Wettability at High Temperatures*, Pergamon Press, Oxford, U.K., 1999.
Fainerman, V. B., Miller, R., Mohwald, H., *J. Phys. Chem. B*, 809, 106, 2002.
Fanum, M., *Microemulsions*, CRC Press, New York, 2008.
Feder, J., *Fractals: Physics of Solids and Liquids*, Plenum Press, New York, 1988.
Fendler, J. H., Fendler, E. J., *Catalysis in Micellar and Macromolecular Systems*, Academic Press, New York, 1975.
Fiueiredo, J. L., Pereira, M. F. R., Freitas, M. M. A., Orfao, J. J. M., *Carbon*, 37, 1379, 1999.
Franks, F., *Water: A Comprehensive Treatise*, Plenum Press, New York, 1975.
Fraxedas, J., *Water at Interfaces*, CRC Press, New York, 2014.
Friberg, S., Larsson, K., Sjoblom, J., *Food Emulsions*, CRC Press, Boca Raton, FL, 2003.
Frohn, A., *Dynamics of Droplets*, Springer Publ., New York, 2000.
Fuerstenau, M. C., Miller, J. D., Kuhn, M. C., *Chemistry of Flotation*, Society of Mining Engineers, New York, 1985.
Gagnon, B. P., Meli, M. V., *Langmuir*, 179, 30, 2014.
Gaines, G. L., Jr., *Insoluble Monolayers at Liquid-Gas Interfaces*, Wiley-Interscience, New York, 1966.
Ghatee, M. H., Moosavi, F., Zolghadr, A. R., *Ind. Eng. Chem. Res.*, 12696, 49, 2010.
Ghosh, P., *Chem. Eng. Res. Des.*, 849, 82, 2004.
Gisler, T., Schultz, S. F., Borkovec, M., Sticher, H., Schurtenberger, P., D'Aguanno, B., Klein, R., *J. Chem. Phys.*, 101, 1, 1994.

Gitis, N., Sivamani, R., *Tribology Transactions*, Taylor & Francis, New York, 2004.

Global CCS (Carbon Capture and Storage), Institute News-letter, Docklands, VIC, Australia, April 2013.

Golub, T. P., de Keizer, A., *Langmuir*, 9506, 20, 2004.

Goodrich, F. C., Rusanov, A. I., Sonntag, H., *The Modern Theory of Capillarity*, Akademie Verlag, Berlin, Germany, 1981.

Gotch, K., *Ind. Eng. Chem. Fund.*, 287, 13, 1974.

Green, D. W., Willhite, G. P., eds., *Enhanced Oil Recovery*, Society of Petroleum Engineers, 1998.

Grodzka, J., Pomianowski, A., *Physicochem. Probl. Min. Proc.*, 39, 11, 2005.

Guo, J. S., Sudol, E. D., Vanderhoff, J. W., El-Asser, M. S., Polymer latexes, Chapter 7: *ACS Symposium Series*, Vol. 492, p. 99, American Chemical Soc., Washington, DC, 1992.

Hansen, M. C., *Hansen Solubility Parameters*, Taylor & Francis, New York, 2007.

Harkins, W. D., *The Physical Chemistry of Surface Films*, Reinhold, New York, 1952.

Harrison, P., Gardiner, C., Sargent, I. L., *Extracellular Vesicles in Health & Disease*, Pan Standard Publ., New York, 2014.

Hiemenz, P. C., Rajagopalan, R., *Principles of Colloid and Surface Chemistry*, 3rd edn., Marcel Dekker, New York, 1997.

Hoher, S., Addad, S. C., Rheology of liquid foam, *J. Phys. Condens. Matter*, 1041, 17, 2006.

Holmberg, K., ed., *Handbook of Applied Surface & Colloid Chemistry*, John Wiley & Sons Ltd., New York, 2002.

Imae, T., *Advanced Chemistry of Monolayers at Interfaces*, Academic Press, New York, 2007.

Israelachvili, J., *Intermolecular & Surface Forces*, 2nd edn., Academic Press, London, U.K., 2011.

James, C., Hatzopoulos, M. H., Yan, C., Smith, G. N., Alexander, S., Rogers, S. E., Eastoe, J., *Langmuir*, 96, 30, 2014.

Janiaud, E., Bacri, J. C., Andreotti, B., *Phys. Fluids*, 37101, 26, 2014.

Jasper, J. J., *J. Phys. Chem. Ref. Data*, 841, 1, 1972.

Jaycock M. K., Parfitt, G. D., *Chemistry of Interfaces*, John Wiley & Sons, New York, 1981.

Jungwirth, P., Tobias, D. J., *Chem. Rev.*, 1259, 106, 2006.

Kamperman, M., Synytska, A., *J. Mater. Chem.*, 22, 19390, 2012.

Karapetsas, G., Sahu, K. C. Sefiane, K., Matar, O. K., *Langmuir*, 4310, 30, 2014.

Kelkar, A. D., Herr, D. J. C., Ryan, J. G., *Nanoscience & Nanoengineering*, CRC Press, New York, 2014.

Kendoush, A. A., *Ind. Eng. Chem. Res.*, 9232, 46, 2007.

Kim, Y. J., Platt, U., *Advanced Environmental Monitoring*, Springer Publ., The Netherlands, 2007.

Kirmse, K., Morgner, B., *Langmuir*, 2193, 22, 2006.

Klimpel, R. R., in: Kawatra, S. K., ed., *High Efficiency Coal Preparation*, Society for Mining, Metallurgy, and Exploration, Littleton, CO, 1995.

Knaus, E., Killen, J., Bigliarbigi, K., Crawford, P., *Chapter 1: ACS Series*, Vol. 1032, p. 3, American Chemical Soc., Washington, DC, 2010.

Koch, J. P., *Pharmazie*, 48, 643, 1993.

Kolasinski, K. W., *Surface Science*, John Wiley & Sons, New York, 2008.

Kristensen, D., Milk—Fat globule structures, Ph.D. thesis, School of Pharmacy, Copenhagen, Denmark, 1997.

Krungleviclute, V., Migone, A. D., Yudasaka, M., Ijima, S., *J. Phys. Chem. C*, 116, 306, 2012.

Kuespert, D., Donohue, M. D., *J. Phys. Chem.*, 99, 4805, 1995.

Kuo, A. T., Chang, C. H., *Langmuir*, 55, 30, 2014.

Kyte, J., *Biophys. Chem.*, 193, 100, 2003.

Lara, J., Blunt, T., Kotvis, P., Riga, A., Tysoe, W. T., *J. Phys. Chem. B*, 102, 1703, 1998.

Li, J., Chen, H., Zhou, W., Wu, B., Stoyanov, S. D., Pelan, E. G., *Surf. Chem. Colloid*, 30, 4223, 2014.

Lin, Y., Boker, A., Skaff, H., Cookson, D., Dinsmore, A. D., Emrick, T., Russell, T. P., *Langmuir*, 191, 21, 2005.

Livingstone, D., *Data Analysis for Chemists*, Oxford University Press, New York, 1996.

Lovett, D., *Science with Soap Films*, Institute of Physics Publishing, Bristol, U.K., 1994.

Lu, J. R., Thomas, R. K., Penfold, J., *Adv. Colloid Interf. Sci.*, 143, 84, 2000.

Lyklema, J., *Fundamentals of Interfaces and Colloid Science, III, Liquid-Fluid Interfaces*, Academic Press, San Diego, CA, 2000.

MacDowell, L. G., Benet, J., Katcho, N. A., Palanco, M. G., *Adv. Colloid Interf. Sci.*, 150, 206, 2014.

Mager, M. D., Melosh, N. A., *Langmuir*, 9369, 23, 2007.

Mandelbrot, B. B., *The Fractal Geometry of Nature*, W.H. Freeman, New York, 1983.

Masel, R. I., *Principles of Adsorption and Reactions*, John Wiley & Sons, New York, 1996.

Matijevic, E., ed., *Surface and Colloid Science*, Vols. 1–9, Wiley-Interscience, New York, 1969.

McCash, E. M., *Surface Chemistry*, Oxford Press, Oxford, U.K., 2001.
Merk, V., Rehbock, C., Becker, F., Hagemann, U., Nienhaus, H., Barcikowski, S., *Langmuir*, 4213, 30, 2014.
Meziani, M. J., Pathak, P., Allard, L. F., Sun, Y. P., Chapter 20: *ACS Series*, Vol. 860, p. 309, American Chemical Soc., Washington, DC, 2003.
Miller, C. A., Neogi, P., *Interfacial Phenomena*, CRC Press, New York, 2008.
Miqueu, C., Miguez, J. M., Pineiro, M. M., Lafitte, T., Mendiboure, B., *J. Phys. Chem. B*, 9618, 115, 2011.
Moroi, Y., Yamabe, T., Shibata, O., Abe, Y., *Langmuir*, 9697, 16, 2000.
Mousny, M., Omelon, S., Wise, L., Everett, E. T., Dumitriu, M., Holmyar, D. P., *Bone*, 43, 1067, 2008.
Mukerjee, P., Mysels, K. J., *Critical Micelle Concentrations of Aqueous Surfactant Systems*, National Bureau of Standards, Washington, DC, 1971.
Mun, E. A., Hanneli, C., Rogers, S. E., Hole, P., Williams, A. C., Khutoryansky, V., *Langmuir*, 308, 30, 2014.
Munz, M., Mills, T., *Langmuir*, 4243, 30, 2014.
Murrant, C. L., Sarelius, I. H., *Acta Physiol. Scand.*, 531, 168, 2000.
Nair, B. P., Vaikkath, D., Nair, P. D., *Langmuir*, 340, 30, 2014.
Neumann, A. W., *Applied Surface Thermodynamics*, CRC Press, New York, 2010.
Nicot, J. P., Scanlo, B. R., *Environ. Sci. Technol.*, 3580, 46, 2012.
Nikolov, A., Wasan, D., *Adv. Colloid Interf. Sci.*, 207, 206, 2014.
Nour, A. H., Yunus, R. M., Jemaat, Z., *J. Appl. Sci.*, 196, 7, 2007.
Ohde, M., Ohde, H., Wai, C. M., Chapter 27: *ACS Series*, Vol. 860, p. 419, American Chemical Soc., Washington, DC, 2003.
Orna, M. V., *The Chemical History of Color*, Springer, New York, 2013.
Oura, K., *Surface Science*, Springer Publ., New York, 2003.
Packham, D. E., *Handbook of Adhesion*, John Wiley & Sons, New York, 2005.
Papachristodoulou, G., Trass, O., Coal slurry technology, *Can. J. Chem. Eng.*, 65, 177, 1987.
Partington, J. R., *An Advanced Treatise of Physical Chemistry*, Vol. II, Longmans, Green, New York, 1951.
Peng, B. Z., Sun, C. Y., Liu, B., Zhang, Q., Chen, J., Li, W. Z., Chen, G. J., *J. Chem. Eng. Data*, 4623, 56, 2011.
Pereira, M. F. R., Soares, S. F., Orfao, J. J. M., Figueiredo, J. L., *Carbon*, 41, 811, 2003.
Pu, G., Severtson, S. J., *Langmuir*, 10007, 28, 2012.
Rabe, J. P., Surface chemistry with the scanning tunneling microscope, *Adv. Mater.*, 1, 13, 1989.
Raj, R., Enright, R., Zhu, Y., Adera, S., Wang, E. N., *Langmuir*, 28, 45, 2012.
Raju, B. B., Winnik, F. M., Morishima, Y., *Langmuir*, 4416, 17, 2001.
Ramiasa, M., Ralston, J., Fetzer, R., Sedev, R., *Adv. Colloid Interf. Sci.*, 275, 206, 2014.
Rao, J. P., Geckeler, K. E., *Progr. Polym. Sci.*, 36, 887, 2011.
Rao, T. C., Govindarajan, B., Barnwal, J. P., in: Kawatra, M., ed., *High Efficiency Coal Preparations*, Society Mining, Metallurgy, and Exploration, Littleton, CO, 1995.
Rathman, J. F., Scamehorn, J. F., *Langmuir*, 474, 4, 1988.
Reimhult, E., Hook, F., Kasemo, B., *Langmuir*, 19, 1681, 2003.
Roberge, P. R., *Corrosion Basics*, 2nd edn., NACE International, Houston, TX, 2006.
Romsted, L. S., *Surfactant Science & Technology, Retrospects & Prospects*, CRC Press, New York, 2014.
Rosen, M. J., *Surfactants and Interfacial Phenomena*, Wiley-Interscience, New York, 2004.
Ruiz, C. C., *Sugar-Based Surfactants*, CRC Press, New York, 2008.
Scheludko, A., *Colloid Chemistry*, Elsevier, Amsterdam, Holland, 1966.
Schramm, L. L., *Emulsions, Foams and Suspensions*, John Wiley & Sons, New York, 2005.
Sinfelt, J. H., Role of surface science in catalysis, *Surf. Sci.*, 500, 923, 2002.
Sjoblom, J., in: Birdi, K. S., ed., *Handbook of Surface and Colloid Chemistry*, 3rd edn., CRC Press, Boca Raton, FL, 2008.
Soltis, A. N., Chen, J., Atkin, L. Q., Hendy, S., *Curr. Appl. Phys.*, 4, 152, 2004.
Somasundaran, P., *Colloidal and Surfactant Sciences*, CRC Press, New York, 2006.
Somorjai, G. A., Li, Y., *Introduction to Surface Chemistry and Catalysis*, 2nd edn., John Wiley & Sons, New York, 2010.
Somorjai, G. A., Li., Y., *Proc. Natl. Acad. Sci.*, 108, 917, 2011.
Sotiriou, G., Starisch, F., Dasargyn, A., Wurnig, M. C., Krumeich, F., Boss, A., Leroux, J. C., Pratsinis, S. E., *Adv. Funct. Mater.*, 2818, 24, 2014.
Spaull, A. J. B., *J. Chem. Educ.*, 81, 42, 2004.
Starostina, I. A., Stoyanov, O. V., Deberdeev, R. Y., *Polymer Surfaces & Interfaces*, CRC Press, New York, 2014.
Stefan, J., *Ann. Phys.*, 29, 655, 1886
Takahashi, Y., Fukuyasu, K., Horiuchi, T., Kondo, Y., Stroeve, P., *Langmuir*, 41, 30, 2014.
Tanford, C., *The Hydrophobic Effect*, John Wiley & Sons, New York, 1980.

Taylor, K. C., Automobile catalytic converters, in: Anderson, J. R., Boudart, M., eds., *Catalysis (Science and Technology)*, Vol. 5, Springer, Berlin, Germany, p. 120, 1984.
Taylor, R. P., *Non-linear Dyn. Physiol. Life Sci.*, 15, 129, 2011.
Todorov, P. D., Kralchevsky, P. A., Denkov, N. D., Broze, G., Mehreteab, A., *J. Colloid Interf. Sci.*, 371, 245, 2007.
Tung, W. S., Daoud, W. A., *J. Mater. Chem.*, 7858, 21, 2011.
Woodson, M., Liu, C. J., *Phys. Chem. Phys.*, 9, 207, 2007.
Wu, X. Z., Ocko, B. M., Sirota, C. B., Sinha, S. K., Deutsch, M., Cao, B. H., Kim, M. W., *Science*, 261, 1018, 1993.
Xu, W., Leetladhar, R., Kang, Y. T., Coi, C. H., *Langmuir*, 6032, 29, 2013.
Yan, H., Cui, P., Liu, C. B., Yuan, S. L., *Langmuir*, 4931, 28, 2012.
Yang, G., Rao, N., Yin, Z., Zhu, D. M., *J. Colloid Interf. Sci.*, 104, 297, 2006.
Yates, J. T., Cambell, C. T., *PNAS*, 911, 108, 2011.
Yethiraj, A., Striolo, A., *J. Phys. Chem. Lett.*, 687, 4, 2013.
Yoon, R. H., Luttrell, G. H., Adel, G. T., Final Report, Department of Energy, Washington, DC, August 1990.
Yoshida, N., Moroi, Y., Humphry-Baker, R., Gratzel, M., *J. Phys. Chem.*, 3991, 106, 2002.
Yu, H. Z., Soolaman, D. M., Rowe, A. W., Banks, J. T., *J. Chem. Phys. Chem.*, 1035, 5, 2004.
Zana, R., *Giant Micelles*, CRC Press, New York, 2008.
Zana, R., *Langmuir*, 2314, 11, 1995.
Zhou, N. F., *J. Chem. Educ.*, 66, 137, 1989.
Zhuravlev, L. T., *Colloid Surf.*, 1, 173, 2000.
Zoller, U., *Sustainable Development of Detergents*, CRC Press, New York, 2008.

2 Molecular Thermodynamics of Hydrogen-Bonded Systems

Ioannis Tsivintzelis and Costas G. Panayiotou

CONTENTS

2.1 INTRODUCTION

The systems of fluids of practical interest to chemists and chemical engineers are as a rule complex, departing significantly from ideal-solution behavior. The systems at interfaces are, in addition, inhomogeneous exhibiting density gradients and, when multicomponent, they also exhibit composition gradients or peculiar composition profiles across the interfaces. Thus, the development of thermodynamic models for complex fluids applicable over a wide range of conditions remains an active and fascinating research area.

Recent advances in statistical thermodynamics and better understanding of intra- and intermolecular interactions, thanks to accurate experimental measurements and molecular simulations using realistic force fields, have contributed significantly to this end. Many of the recent thermodynamic models based on statistical mechanics are rooted in the pioneering work of Guggenheim [1] and Flory [2] on lattice models for complex fluids, including polymers. The lattice fluid (LF) theory of Sanchez and Lacombe [3,4] is probably one of the most widely used and successful lattice models.

Significant improvement in the performance of these models is obtained by accounting explicitly for the nonrandom distribution of molecular species and free volume and for highly specific forces between neighboring molecules resulting in hydrogen bonding [5–8].

Nonrandomness is essentially omnipresent in fluids as the molecular species are, as a rule, distributed nonrandomly in their mixtures. In other words, the local composition in the immediate neighborhood of a molecule is, in general, different from the overall or bulk composition of the mixture. Even in pure fluids, there is a degree of nonrandomness in the distribution of their functional groups. In fact, modern experimental techniques, such as positron annihilation spectroscopy, reveal significant nonrandomness in the distribution of free volume throughout the volume of the pure fluid, even in nonpolar systems [9,10]. One of the principal causes of nonrandomness is of course hydrogen bonding.

Hydrogen bonding is by itself a subject of remarkable diversity as it is present in and dictates the behavior of an enormous type of systems including aqueous solutions, systems of biological/biomedical interest, pharmaceuticals, colloids and surfactants, physical networks and gels, adhesives and pastes, extractives and binders, and polymer alloys and blends. There are many reviews of the subject in the open literature [11–26] each addressing, usually, one aspect or type of application of hydrogen bonding. Because of its multifaceted character, unified approaches of treatment of hydrogen bonding are particularly useful, especially in areas at the interface of various scientific branches, such as the colloid and interface science.

For the treatment of hydrogen bonding in associated fluids and mixtures, a variety of different approaches are popular. We could divide the overwhelming majority of these approaches into two groups: the association models [5,18,27–31] and the combinatorial models [26,32–35]. Association models invoke the existence of multimers or association complexes and seek expressions for their population. Combinatorial models do not invoke the existence of association complexes, but, instead, they focus on the donor–acceptor contacts and seek combinatorial expressions for the number of ways of forming hydrogen bonds in systems of given proton–donor and proton–acceptor groups. Both types of models imply that the molecules tend to be distributed in the system nonrandomly for more efficient hydrogen-bonding interaction.

Earlier [26], we had compared the two approaches of hydrogen bonding and applied to the description of phase equilibria and mixture properties of systems of fluids. The key conclusion was that in the systems where both approaches apply, they prove to be, essentially, equivalent. However, the combinatorial approach has a much broader field of applications as it can be applied even to systems forming 3D hydrogen-bonding networks. In the second edition of this handbook [36], we had presented an updated review on the thermodynamic aspects of hydrogen bonding in pure fluids and their mixtures by focusing on the combinatorial hydrogen-bonding formalism. A variety of examples were given in applications ranging from phase equilibria of simple aqueous mixtures to (hydro)gel swelling, to intramolecular association, and to hydrogen-bonding cooperativity. In the third edition of this handbook [37], besides hydrogen bonding for the study of bulk phases as well as interfaces, emphasis was also given to the progress in accounting for nonrandomness in solution thermodynamics.

The present review is, in a sense, a continuation of all three aforementioned reviews. The thermodynamic aspects of hydrogen bonding (combinatorial approach) in pure fluids and their mixtures will be presented in a way that can be combined with any thermodynamic model. Here, it should become clear at the outset that, in general, hydrogen bonding makes a *contribution* only and is not sufficient for the complete evaluation of the various thermodynamic properties of fluids and their mixtures. Thus, hydrogen-bonding formalisms are usually combined with thermodynamic models, which account for all other contributions collectively called physical contributions. For the purposes of this chapter, we will use two kinds of thermodynamic frameworks: equation-of-state theories (EoS) and predictive (infinite dilution) activity coefficient models.

The former (EoS approach) is applicable to fluids over an extended range of external conditions encompassing liquids, vapors, gases, supercritical fluids, amorphous and glassy polymers, homogeneous as well as inhomogeneous systems, complex aqueous systems, associated polymer mixtures, rubbers, and gels. Thus, the hydrogen-bonding formalism will be combined with two EoS models, the quasichemical hydrogen-bonding (QCHB) theory [7] and its recent development the nonrandom hydrogen-bonding (NRHB) model [8,38]. Both models take into account the nonrandom distribution of free volume in the system by using Guggenheim's *quasichemical* approach [1].

The latter (predictive approach) is based on the new concept of partial solvation parameters (PSPs) [39–43] and focuses on the prediction of thermodynamic properties of complex systems with minimal or almost no experimental data available.

In what follows, after an exposition of the essentials of the combinatorial hydrogen-bonding formalism, we will present the EoS approach and characteristic applications to mixtures of hydrogen-bonded fluids. Subsequently, we will present the extension of the model to interfaces and to solubility parameters. Finally, we will present the essentials of the novel PSP approach. Throughout the presentation, examples of calculations in systems of practical interest will be given.

2.2 GENERAL APPROACH: THE BASICS OF THERMODYNAMIC MODELS

The main objective of our developments here will be the formulation of the configurational partition function of a multicomponent system of fluids, which consists of N_1, N_2, …, N_t such molecules of components 1, 2, …, t, respectively, at temperature T, total volume V, and external pressure P. Our first key assumption is that we may factorize the partition function into a random or athermal and an energetic or thermal part as follows [1,8,35,44–56]:

$$Q(N_1,N_2,\ldots,N_t,T,V)=\sum_i \Omega_i e^{-\beta E_i}\cong Q_{\text{Athermal}}Q_{\text{Thermal}}=Q_RQ_E \tag{2.1}$$

The summation in the aforementioned equation spans all microstates i in the phase space of our system characterized by a corresponding energy level E_i, while Ω_i is the degeneration or multiplicity factor for all microstates corresponding to the energy level E_i, β is the thermal energy factor, which is equal to $1/kT$, k being the Boltzmann's constant. The random part Q_R represents the limiting value of the partition function when all intermolecular interaction energies vanish and the molecules are distributed randomly throughout the volume of the system. The energetic part, Q_E, is then a correction term for the nonrandom distribution of molecular species, which retains the original form of the partition function and which should become equal to one in the limiting case of zero interaction energies. In highly nonideal systems of molecules interacting with strong specific forces, such as hydrogen-bonding forces, the energetic part may be factored, for convenience, into a nonrandomness term, Q_{NR}, and a specific or hydrogen-bonding term, Q_H. Thus, the configurational partition function can be written alternatively as

$$Q=Q_RQ_E=Q_R\{Q_{NR}Q_H\}=\{Q_RQ_{NR}\}Q_H=Q_PQ_H \tag{2.2}$$

In this way, the partition function can be factored into a "physical," Q_P, and a "chemical" or hydrogen-bonding term, Q_H. This is a general approach.

The total partition function of the system in the P, T ensemble and in its maximum term approximation is given by

$$\Psi(T,P,\{N_k\})=Q_P(T,\{N_k\})Q_H(T,\{N_k\},\{N_{ij}\})\exp\left(\frac{-PV}{RT}\right) \tag{2.3}$$

where

N_{ij} is the number of hydrogen bonds of i–j type

V is the total volume of the system, which is given by

$$V=V_P+V_H=V_P+\sum_i^m\sum_j^n N_{ij}V_{ij}^0 \tag{2.4}$$

V_{ij}^0 is the volume change accompanying a i–j hydrogen bond formation, while m and n are the number of types of proton donors and acceptors, respectively, which are able to hydrogen bond.

The free energy of the system will be given by

$$G = -kT \ln \psi \tag{2.5}$$

A direct consequence of our approach for the factorization of the partition function is the division of the Gibbs free energy, G, into a physical term, and a chemical or hydrogen-bonding term:

$$G = G_P + G_H \tag{2.6}$$

For a system at equilibrium, the free energy is at a minimum. One may then write the following minimization conditions:

$$\left(\frac{\partial G}{\partial \tilde{\upsilon}}\right)_{T,P,\{N_k\},\{N_{ij}\}} = 0 \tag{2.7}$$

$$\left(\frac{\partial G}{\partial N_{ij}}\right)_{T,P,\{N_k\},\{N_{rs}\}} = 0 \tag{2.8}$$

The first one leads to a hydrogen-bonding contribution to the equation of state of the adopted model, while the latter one leads to a system of equations that allow the calculation of the number of hydrogen bonds, N_{ij}.

At equilibrium, the chemical potential of component k is given by

$$\mu_k = \mu_{k,P} + \mu_{k,H} = \left(\frac{\partial G_P}{\partial N_k}\right)_{T,P,N_j,\tilde{\upsilon},\{N_{ij}\}} + \left(\frac{\partial G_H}{\partial N_k}\right)_{T,P,N_j,\tilde{\upsilon},\{N_{ij}\}} \tag{2.9}$$

In the next section, we will present the combinatorial formalism for hydrogen bonding, which is used for estimating the hydrogen-bonding contribution, Q_H, to the partition function. Such approach is not bound to the models for Q_P and can be combined with any other appropriate thermodynamic model able to describe the non-hydrogen-bonding contributions to the thermodynamic properties of the studied systems. However, as it was already mentioned, the *physical* term, Q_P, of the partition function will be estimated using the NRHB [38] and QCHB [7] formalism in the EoS approach, presented in Section 2.4.1, and to the PSP [39–43] formalism, presented in Section 2.5, for the prediction of thermodynamic properties of complex systems.

2.3 COMBINATORIAL FORMALISM FOR HYDROGEN BONDING

Let us consider now the hydrogen-bonding or chemical term of the partition function. Let us assume that there are m different kinds of hydrogen-bonding donors and n kinds of hydrogen-bonding acceptors in the system. Let d_i^k be the number of hydrogen-bonding donors of type i ($i = 1, m$) in each molecule of type k ($k = 1, t$) and α_j^k the number of hydrogen-bonding acceptors of type j ($j = 1, n$) in each molecule of type k. The total number N_d^i of hydrogen-bonding donors i in the system is

$$N_d^i = \sum_k^t d_i^k N_k \tag{2.10}$$

and the total number N_a^j of hydrogen-bonding acceptors j in the system is

$$N_a^j = \sum_k^t \alpha_j^k N_k \tag{2.11}$$

The potential energy of the system due to hydrogen bonding is *in excess* of that due to physical interactions. The total energy E_H of the system due to hydrogen bonding is given by

$$E_H = \sum_i^m \sum_j^n N_{ij} E_{ij}^0 \tag{2.12}$$

where

N_{ij} is the number of hydrogen bonds between donors of type i and acceptors of type j
E_{ij}^0 the corresponding hydrogen-bonding energy of the i–j interaction

The total number of hydrogen bonds in the system is

$$N_H = \sum_i^m \sum_j^n N_{ij} \tag{2.13}$$

What is now required is the number of ways Ω of distributing the N_{ij} bonds among the functional groups of the system. In order to find the different number of isoenergetic configurations of our system (number of the different ways of forming or distributing the hydrogen bonds in the system), we have to find different ways of

1. Selecting the associated donor sites out of the donor population
2. Selecting the associated acceptor sites out of the acceptor population
3. Making hydrogen bonds between the selected donor and acceptor sites

Having the number of ways, Ω, of distributing the N_{ij} bonds among the functional groups of the system, we may write the canonical partition function for hydrogen bonding as following:

$$Q_H(T,\{N_k\},\{N_{ij}\}) = \Omega\left(\prod_i^m \prod_j^n \frac{\tilde{\rho}}{rN} \exp\left(\frac{-N_{ij}F_{ij}^0}{RT}\right)\right) \tag{2.14}$$

where

$$F_{ij}^0 = E_{ij}^0 - TS_{ij}^0 \tag{2.15}$$

and P_{ij} is the mean field probability that a specific acceptor j will be proximate to a given donor i, since in order to form a hydrogen bond, the two interacting groups must be proximate. This term is proportional to the volume of the acceptor group divided by the total system volume; that is, $P_{ij} \sim 1/V$. Even spatial proximity does not guarantee that a bond will form. Bond formation requires that donor and acceptor adopt a unique spatial orientation with respect to one another. Formation of the bond is also accompanied by a loss of rotational degrees of freedom. Steric considerations will also come into play in bond formation. In general and in this framework, for a donor *i*–acceptor *j* pair, this probability is given by [26,35]

$$P_{ij} = e^{S_{ij}^{0}/R} \frac{\tilde{\rho}}{rN} \tag{2.16}$$

where S_{ij}^0 is the entropy loss (intrinsically negative) associated with hydrogen bond formation of an (*i*, *j*) pair. The last term $\tilde{\rho}/rN$ in Equation 2.16 comes from the estimation of the volume *V* by the model framework for the physical term. According to the NRHB model, which will presented next, *r* is the number of molecular segments and $\tilde{\rho}$ the reduced density of the system.

Next we will estimate the number of ways Ω of distributing the N_{ij} bonds among the functional groups of the system for three different cases.

2.3.1 Oligomers and Three-Dimensional Networks of Hydrogen-Bonded Molecules

Let us briefly summarize the rationale for this enumeration process and apply it, first, to the simple case of a system of molecules with *one proton donor* and *one acceptor group*, such as systems with hydroxyl groups, which self-associate. Let us assume that no cyclic *r*-mers are allowed.

Consequently, let us have a system with *N* molecules each having one donor and one acceptor site of type 1 with N_{11} hydrogen bonds among them. The number of ways of selecting the N_{11} associated donors out of the donor population *N* is just the binomial coefficient $N!/(N - N_{11})!N_{11}!$. Similarly, the number of ways of selecting the N_{11} associated acceptors out of the acceptor population *N* is again the binomial coefficient $N!/(N - N_{11})!N_{11}!$. The free donor groups in the system are $N - N_{11} = N_{10}$. This is also the number of free acceptor groups, N_{01}, in this particular system. Now, a specific donor can hydrogen bond with any of the N_{11} acceptors, a second donor can hydrogen bond with any of the remaining $N_{11} - 1$ acceptors, and so on. The number of ways that N_{11} bonds can be formed between N_{11} donors and N_{11} acceptors is just $N_{11}!$. Thus, the total number of ways that N_{11} bonds can form between *N* donors and *N* acceptors is the product of the aforementioned three terms, or

$$\Omega = \frac{N!}{(N-N_{11})!N_{11}!}\frac{N!}{(N-N_{11})!N_{11}!}N_{11}! = \frac{N!}{N_{10}!}\frac{N!}{N_{10}!N_{11}!} \tag{2.17}$$

These arguments, when extended to the general case of multigroup molecules, lead to the following equation [26,34–35]:

$$\Omega = \prod_{i}^{m}\frac{N_d^i!}{N_{i0}!N_{i1}!\ldots N_{in}!}\prod_{j}^{n}\frac{N_a^i!}{N_{0j}!N_{1j}!\ldots N_{mj}!}\prod_{i}^{m}\prod_{j}^{n}N_{ij}! = \prod_{i}^{m}\frac{N_d^i!}{N_{i0}!}\prod_{j}^{n}\frac{N_a^j!}{N_{0j}!}\prod_{i}^{m}\prod_{j}^{n}\frac{1}{N_{ij}!} \tag{2.18}$$

where

N_{i0} is the number of free (non-hydrogen-bonded) donor groups of type *i*

N_{0j} is the respective number of free acceptor groups of type *j*

$$N_{i0} = N_d^i - \sum_{j}^{n} N_{ij} \tag{2.19}$$

$$N_{0j} = N_a^j - \sum_{i}^{m} N_{ij} \tag{2.20}$$

Equation 2.18, for a system with n proton donors and m proton acceptors and if we only account for intramolecular association, leads to the following equations:

$$\frac{\nu_{ij}}{\nu_{i0}\nu_{0j}} = \tilde{\rho}\exp\left(-\frac{G_{ij}^0}{RT}\right) \quad \text{for all } i, j \tag{2.21}$$

or

$$\nu_{ij} = \left[\nu_d^i - \sum_k^n \nu_{ik}\right]\left[\nu_a^j - \sum_k^m \nu_{kj}\right]\tilde{\rho}\exp\left(-\frac{G_{ij}^0}{RT}\right) \tag{2.22}$$

where

$$\nu_{ij} \equiv \frac{N_{ij}}{rN} \quad \nu_{i0} \equiv \frac{N_{i0}}{rN} \quad \nu_d^i \equiv \frac{N_d^i}{rN}\cdots \tag{2.23}$$

and

$$G_{ij}^0 = F_{ij}^0 + PV_{ij}^0 = E_{ij}^0 + PV_{ij}^0 - TS_{ij}^0 \tag{2.24}$$

This is a system of ($m \times n$) quadratic equations for ν_{ij}. This system must be solved in combination with the equation of state, and thus, we are finally left with a system of ($m \times n + 1$) coupled nonlinear equations for the reduced density, $\tilde{\rho}$, and the ν_{ij}.

Furthermore, on the basis of the aforementioned data, we can estimate the hydrogen-bonding contribution to the chemical potential from the following equation:

$$\frac{\mu_{k,H}}{RT} = r_k\nu_H - \sum_i^m d_i^k \ln\frac{\nu_d^i}{\nu_{i0}} - \sum_j^n a_j^k \ln\frac{\nu_a^j}{\nu_{0j}} \tag{2.25}$$

2.3.2 Dimerization of Acids

A second case that will be investigated here is the dimerization of organic acids [57]. In such systems, dimers are the overwhelming majority of the association species. Consequently, for simplicity, we consider dimerization only in order to describe such hydrogen-bonding behavior.

Let N_{dm} be the number of dimers in the system. Then, the number of ways of selecting these dimerized molecules out of the N acid molecules is

$$\frac{N!}{(2N_{dm})!(N - 2N_{dm})!} \tag{2.26}$$

The number of ways of selecting the N_{dm} dimers is

$$\Omega = \frac{N!}{(2N_{dm})!(N - 2N_{dm})!}(2N_{dm} - 1)(2N_{dm} - 3)\cdots 1 = \frac{N!}{(N - 2N_{dm})!N_{dm}!2^{N_{dm}}} \tag{2.27}$$

The free-energy change upon formation of one dimer is

$$G_{dm} = E_{dm} + PV_{dm} - TS_{dm} \tag{2.28}$$

Consequently, the hydrogen-bonding factor in the partition function becomes

$$Q_H = \frac{N!}{(N-2N_{dm})!N_{dm}!2^{N_{dm}}}\left(\frac{\tilde{\rho}}{rN}\right)^{N_{dm}}\exp\left(-\frac{N_{dm}G_{dm}}{RT}\right) \tag{2.29}$$

The equilibrium number of dimers per mol of segments of acid, ν_{dm}, is obtained from the aforementioned equation through the usual free-energy minimization condition, or

$$\nu_{dm} = \frac{2+(1/K_{dm})-\sqrt{\left(1/K_{dm}^2\right)+(4/K_{dm})}}{4r} \tag{2.30}$$

where

$$K_{dm} = \frac{\tilde{\rho}}{r}\exp\left(\frac{-G_{dm}}{RT}\right) \tag{2.31}$$

In this case of dimerization, the hydrogen-bonding contribution to the chemical potential is

$$\frac{\mu_H}{RT} = r\nu_{dm} - \ln\frac{1}{1-2r\nu_{dm}} \tag{2.32}$$

2.3.3 Intramolecular Hydrogen Bonding

Both aforementioned cases referred to intermolecular hydrogen bonding. However, the rationale is easily extended to systems with both inter- and intramolecular hydrogen bonds [58–61].

For example, in the general case of alkoxyalkanols, if x is the number of ether oxygen acceptor sites and the number of proton donors and acceptors of type 1 (–OH) is N_1 then the proton acceptors of type 2 (–O–) is xN_1. As previously mentioned [60,61], x is assumed to be equal to 2. As a consequence, the total number of free proton donors is

$$N_{10} = N_1 - N_{11} - N_{12} - B \tag{2.33}$$

where B refers to the number of intramolecular bonds OH–O– in the system.

In order to find the number of ways, Ω, of distributing the various hydrogen bonds in the system, we have first to select the N_{11}, N_{12}, B, and N_{10} donors out of the N_1 donor population. This can be done in $N_1!/[B!N_{11}!N_{12}!N_{10}!]$ ways. Second, we have to select the hydrogen-bonded N_{11} acceptors of type 1 out of the N_1 acceptor population, which can be done in $N_1!/[N_{11}!\,(N_1 - N_{11})!]$ ways. Next, we have to select the B acceptors of type 2 out of the xN_1 acceptor population. However, once we have selected the B proton donors that participate in intramolecular bonds, we have also selected the molecules with the acceptor sites of type 2 that participate in the B intramolecular bonds. We assume for simplicity that all x acceptor sites are equivalent for the intramolecular bonds. In each of these B molecules, we must now select the acceptor site of type 2 for the intramolecular bond out of the x acceptors of type 2 that exist in such molecule. For each molecule, this can be done in $x!/[1!\,(x-1)!]$ ways and, consequently, for the B molecules, it can be done in

$\{x!/[1!\,(x-1)!]\}^B = x^B$ ways. Having selected the B acceptor sites of type 2, we must now select, out of the remaining $(xN_1 - B)$ acceptor type 2 population, the N_{12}, which will participate in the intermolecular bonds. This can be done in $(xN_1 - B)!/[(xN_1 - B - N_{12})!N_{12}!]$ ways. The N_{11} and N_{12} bonds can be done in $N_{11}!\cdot N_{12}!$ ways, while the B bonds in only one way after having selected, both, the donor and the acceptor sites in each molecule. Thus, the number of configurations in the hydrogen bonded system is

$$\Omega = \frac{N_1!}{B!N_{11}!N_{12}!N_{10}!}\frac{N_1!}{N_{11}!(N_1 - N_{11})!}\left(\frac{x!}{(x-1)!}\right)^B \frac{(xN_1 - B)!}{(xN_1 - B - N_{12})!N_{12}!}N_{11}!N_{12}!$$

$$= \frac{(xN_1 - B)!(N_1!)^2 x^B}{B!N_{11}!N_{12}!N_{10}!(N_1 - N_{11})!(xN_1 - B - N_{12})!} \tag{2.34}$$

The number of the three types of hydrogen bonds can be obtained from the following coupled equations:

$$\frac{B(xN_1 - B)}{(xN_1 - B - N_{12})N_{10}x} = c\exp\left(-\frac{G_B^0}{kT}\right) = K_B \tag{2.35}$$

$$\frac{N_{11}}{(N_1 - N_{11})N_{10}} = \frac{\tilde{\rho}}{rN}\exp\left(-\frac{G_{11}^0}{kT}\right) = \frac{K_{11}}{N} \tag{2.36}$$

$$\frac{N_{12}}{(xN_1 - B - N_{12})N_{10}} = \frac{\tilde{\rho}}{rN}\exp\left(-\frac{G_{12}^0}{kT}\right) = \frac{K_{12}}{N} \tag{2.37}$$

where

- G_{11}^0 and G_{12}^0 are the free-energy changes upon formation of the intermolecular hydrogen bonds of type 1–1 and 1–2, respectively
- G_B^0 is the corresponding free-energy change for the formation of the intramolecular hydrogen bonds

The free-energy change for the i–j bond can be resolved as follows:

$$G_{ij}^0 = E_{ij}^0 + PV_{ij}^0 - TS_{ij}^0 \tag{2.38}$$

where E_{ij}^0, V_{ij}^0, S_{ij}^0, refer to the energy, volume, and entropy change, respectively, upon formation of the same hydrogen bond.

After some algebra, equations (2.35), (2.36) and (2.37) lead to the following equations:

$$N_{12} = \frac{K_{12}}{K_B N x}B(xN_1 - B) \tag{2.39}$$

$$N_{11} = \frac{K_{11}}{K_B N x + B(K_{11} - K_{12})}BN_1 \tag{2.40}$$

$$B = \frac{(NxK_B - K_{12}B)}{N}\left[N_1 - B - B\frac{K_{12}}{xK_B}\frac{xN_1 - B}{N} - N_1\frac{K_{11}B}{xNK_B + B(K_{11} - K_{12})}\right] \tag{2.41}$$

The last equation contains only the unknown B, and it can be solved numerically by successive substitutions. The solution for B can then be replaced in Equations 2.39 and 2.40 in order to obtain N_{12} and N_{11}, respectively.

Having obtained the numbers of hydrogen bonds in the system, the average number of hydrogen bonds (intermolecular) per segment, ν_H, can be calculated by the following equation:

$$\nu_H = \frac{N_H}{rN} = \frac{N_{11} + N_{12}}{rN} \tag{2.42}$$

and the hydrogen-bonding contribution to the chemical potential of alkoxyalkanol is given by [58]

$$\frac{N_1 \mu_{1H}}{RT} = N_H + N_1 \ln\left(1 - \frac{N_H + B}{N_1}\right) + N_1 \ln\left(1 - \frac{N_{11}}{N_1}\right) + xN_1 \ln\left(1 - \frac{N_{12}}{xN_1 - B}\right) \tag{2.43}$$

The corresponding contribution for the inert compound of the mixture is given by

$$\frac{\mu_{2H}}{RT} = r\nu_H \tag{2.44}$$

2.4 EQUATION-OF-STATE APPROACH: THE NRHB MODEL

The expression for the physical contribution to the partition function (see Equation 2.2) depends on the adopted framework. Thus, the hydrogen-bonding part remains the same, as shown in the previous section, and can be implemented in an EoS or an activity coefficient model framework. In this section, the EoS approach is described, and the hydrogen-bonding contribution is combined with an LF model that accounts for the nonrandom distribution of empty space and molecular segments, the NRHB model [8,38].

As mentioned earlier, our molecular system consists of $N_1, N_2, \ldots, N_t$ molecules of components $1,2, \ldots, t$, respectively. Let each component of type i be characterized by r_i segments of segmental volumes v_i^*. The molecules are assumed to be arranged on a quasilattice of coordination number z and of N_r sites, N_0 of which are empty. The total number N_r of lattice sites is given by the expression

$$N_r = N_1 r_1 + N_2 r_2 + \cdots + N_t r_t + N_0 = rN + N_0 = N(x_1 r_1 + x_2 r_2 + \cdots + x_t r_t) + N_0 \tag{2.45}$$

where $N = N_1 + N_2 + \cdots + N_t$ is the total number of molecules in the system and x_i is the mole fraction of component i. The average interaction energy per segment of molecule i is given by

$$\varepsilon_i^* = \left(\frac{z}{2}\right)\varepsilon_{ii} \tag{2.46}$$

where ε_{ii} is the interaction energy per i–i contact. If zq_i is the number of external contacts per molecule i, a geometric characteristic of molecule, i, is its surface-to-volume ratio, $s_i = q_i/r_i$. This ratio can be estimated for a very large number of different compounds by using the widely used group contribution UNIFAC model [53,54]. In this way, s_i is not a fitted parameter but is calculated based on well-established theories.

In a mixture, parameters r and q are calculated through the following simple mixing rules:

$$r = \sum_i^t x_i r_i \tag{2.47}$$

$$q = \sum_{i}^{t} x_i q_i \tag{2.48}$$

and so

$$s = \frac{q}{r} \tag{2.49}$$

Furthermore, segment fractions ϕ_i and surface (contact) fractions θ_i are defined as

$$\phi_i = \frac{r_i N_i}{rN} = \frac{x_i r_i}{r} \quad i = 1,2,\ldots,t \tag{2.50}$$

and

$$\theta_i = \frac{q_i N_i}{\sum_k^t q_k N_k} = \frac{q_i N_i}{qN} = \frac{\phi_i s_i}{\sum_k^t \phi_k s_k} = \frac{\phi_i s_i}{s} \quad i = 1,2,\ldots,t \tag{2.51}$$

The total number of contact sites in the system is

$$zN_q = zqN + zN_0 \tag{2.52}$$

while the total volume of the system is given by the expression

$$V = Nrv^* + N_0 v^* = N_r v^* = V^* + N_0 v^* \tag{2.53}$$

where the same average segmental volume v^* is assigned to an empty site as to an occupied site. Furthermore, it is assumed that two neighboring empty sites on the quasilattice remain discrete and do not coalesce. In earlier versions of the theory, v^* was assumed to be a pure component parameter adjusted to experimental data [7,8]. More often, however, v^* is assumed to be constant for all fluids [38,51] and is set equal to 9.75 $cm^3\ mol^{-1}$.

We will ignore in this section the contributions to the partition function due to hydrogen bonding. In the isothermal–isobaric statistical ensemble, the physical term, Q_P, of the partition function can then be written as

$$Q_P(N,P,T) = Q_R Q_{NR} = \Omega_R \Omega_{NR} \exp\left(-\frac{E+PV}{kT}\right) \tag{2.54}$$

where

- Q_R is the combinatorial term of the partition function for a hypothetical system with a random distribution of the empty and molecular segments
- Q_{NR} is a correction term for the actual nonrandom distribution of the empty sites

For the random combinatorial term, earlier theories used the Flory expression [2], resulting in the LF theory [3,4], or the Guggenheim expression [1], resulting in the PV model [51] and QCHB model [7].

The PV model formed the basis for the development of many other lattice models in recent years. In NRHB [8,38], the generalized Staverman expression [52] is adopted, according to which

$$\Omega_R = \prod_i^t \omega_i^{N_i} \frac{N_r! \prod_i^t N_r^{l_i N_i}}{N_0! \prod_i^t N_i!} \left(\frac{N_q!}{N_r!} \right)^{z/2} \tag{2.55}$$

where ω_i is a characteristic quantity for fluid i that accounts for the flexibility and symmetry of the molecule. In phase equilibrium calculations, this quantity cancels out. Parameter l_i is calculated from the expression

$$l_i = \frac{z}{2}(r_i - q_i) - (r_i - 1) \tag{2.56}$$

For the nonrandom correction, the Guggenheim's quasichemical theory is used [1], as proposed in the original model [51]:

$$Q_{NR} = \frac{N_{rr}^0! N_{00}^0! \left[\left(N_{r0}^0 / 2 \right)! \right]^2}{N_{rr}! N_{00}! \left[\left(N_{r0} / 2 \right)! \right]^2} \tag{2.57}$$

In this equation, N_{rr} is the number of external contacts between the segments belonging to molecules, N_{00} is the number of contacts between the empty sites, and N_{r0} is the number of contacts between a molecular segment and an empty site. Superscript 0 refers to the case of randomly distributed empty sites.

In the random case, N_{rr}^0 takes the form

$$N_{rr}^0 = \frac{1}{2} zqN \frac{qN}{N_0 + qN} = \frac{z}{2} qN\theta_r \tag{2.58}$$

where

$$\theta_r = 1 - \theta_0 = \frac{q/r}{q/r + \tilde{v} - 1} \tag{2.59}$$

and the reduced volume, $\tilde{v}$, is defined as

$$\tilde{v} = \frac{V}{V^*} = \frac{1}{\tilde{\rho}} = \frac{1}{\sum_i f_i} \tag{2.60}$$

where $\tilde{\rho}$ is the reduced density. The site fractions, f_0 and f_i, for the empty sites and molecular segments of component i, respectively, are related by the expression

$$f_0 = \frac{N_0}{N_r} = \frac{N_r - \sum_i r_i N_i}{N_r} = 1 - \sum_i f_i \tag{2.61}$$

In the random case, the number of contacts between empty sites is given by the equation

$$N_{00}^{0} = \frac{1}{2} N_0 z \frac{N_0}{N_q} = \frac{z}{2} N_0 \theta_0 \tag{2.62}$$

while the number of contacts between a segment and an empty site is given by

$$N_{r0}^{0} = zqN \frac{N_0}{N_q} = zN_0 \frac{qN}{N_q} = zqN\theta_0 = zN_0\theta_r \tag{2.63}$$

The total number of intersegmental contacts is calculated as the sum of contributions between like molecules and between unlike molecules:

$$N_{rr}^{0} = \sum_{i}^{t} N_{ii}^{0} + \sum_{i}^{t} \sum_{j>i}^{t} N_{ij}^{0} \tag{2.64}$$

where

$$\begin{aligned} N_{ii}^{0} &= \frac{z}{2} q_i N_i \frac{q_i N_i}{N_q} = \frac{z}{2} q_i N_i \theta_i \theta_r \\ N_{ij}^{0} &= zq_i N_i \frac{q_j N_j}{N_q} = zq_i N_i \theta_j \theta_r = zq_j N_j \theta_i \theta_r \quad i \neq j \end{aligned} \tag{2.65}$$

In order to calculate nonrandom distribution of molecular segments and empty sites, appropriate nonrandom factors Γ are introduced, as explained in the following text. As a result, a total of $2t + t(t-1)/2 + 1$ contact value expressions becomes

$$\begin{aligned} N_{ii} &= N_{ii}^{0}\Gamma_{ii} \quad i = 1,\ldots,t \\ N_{ij} &= N_{ij}^{0}\Gamma_{ij} \quad t > j > i \\ N_{00} &= N_{00}^{0}\Gamma_{00} \\ N_{i0} &= N_{i0}^{0}\Gamma_{i0} \quad i = 1,\ldots,t \end{aligned} \tag{2.66}$$

The nonrandom factors Γ should obey the following material balance expressions:

$$\sum_{i=0}^{t} \theta_i \Gamma_{ij} = 1 \quad j = 0,1,\ldots,t \tag{2.67}$$

Finally, in NRHB model, as well as in previous lattice models, it is assumed that only first-neighbor segment–segment interactions contribute to the potential energy E of the system. Consequently, for a mixture, it is

$$-E = \sum_{i=1}^{t} N_{ii}\varepsilon_{ii} + \sum_{i=1}^{t} \sum_{j>i}^{t} N_{ij}\varepsilon_{ij} \tag{2.68}$$

and

$$\varepsilon_{ij} = \sqrt{\varepsilon_{ii}\varepsilon_{jj}}(1-k_{ij}) = \xi_{ij}\sqrt{\varepsilon_{ii}\varepsilon_{jj}} \tag{2.69}$$

where k_{ij} (or ξ_{ij}) is a binary interaction parameter between species i and j and is fitted to binary experimental data.

By substituting Equation 2.54 to Equation 2.3, all thermodynamic properties can be derived. At equilibrium, neglecting the hydrogen-bonding contribution, the reduced density of the system is obtained from the following minimization condition:

$$\left(\frac{\partial G}{\partial \tilde{\rho}}\right)_{T,P,N,N_{10},\ldots,N_{t0}} = 0 \tag{2.70}$$

which leads to the equation of state

$$\tilde{P} + \tilde{T}\left[\ln(1-\tilde{\rho}) - \tilde{\rho}\sum_{i=1}^{t}\phi_i\frac{l_i}{r_i} - \frac{z}{2}\ln\left[1-\tilde{\rho}+\frac{q}{r}\tilde{\rho}\right] + \frac{z}{2}\ln\Gamma_{00}\right] = 0 \tag{2.71}$$

where

$$\tilde{T} = \frac{T}{T^*} = \frac{RT}{\varepsilon^*} \tag{2.72}$$

$$\tilde{P} = \frac{P}{P^*} = \frac{Pv^*}{RT^*} \tag{2.73}$$

$$\varepsilon^* = \sum_{i=1}^{t}\sum_{j=1}^{t}\theta_i\theta_j\varepsilon_{ij}^* \tag{2.74}$$

and

$$\varepsilon_{ij}^* = \xi_{ij}\sqrt{\varepsilon_i^*\varepsilon_j^*} \quad \text{or} \quad \varepsilon_{ij}^* = \sqrt{\varepsilon_i^*\varepsilon_j^*}(1-k_{ij}) \tag{2.75}$$

The $2t + t(t-1)/2 + 1$ different number of contacts N_{ij} or, equivalently, the nonrandom factors Γ_{ij} are calculated from the following set of minimization conditions:

$$\left(\frac{\partial G}{\partial N_{ij}}\right)_{T,P,N,\tilde{\rho}} = 0 \quad i = 0,1,\ldots,t \quad \text{and} \quad j = i+1,\ldots,t \tag{2.76}$$

which leads to the following set of $t(t+1)/2$ equations:

$$\frac{\Gamma_{ii}\Gamma_{jj}}{\Gamma_{ij}^2} = \exp\left(\frac{\Delta\varepsilon_{ij}}{RT}\right) \quad i = 0,1,\ldots,t \quad \text{and} \quad j = i+1,\ldots,t \tag{2.77}$$

where

$$\Delta\varepsilon_{ij} = \varepsilon_i + \varepsilon_j - 2(1-k_{ij})\sqrt{\varepsilon_i\varepsilon_j} \tag{2.78}$$

and $\varepsilon_0 = 0$. Equations 2.67 and 2.77 form a system of $2t + t(t-1)/2 + 1$ nonlinear algebraic equations, which is solved analytically for pure fluids and numerically for the case of multicomponent mixtures. For this purpose, a robust algorithm was proposed by Abusleme and Vera [62] based on the generalized Newton–Raphson method.

For phase equilibrium calculations, the chemical potential of each component i in the mixture is needed. If we neglect the hydrogen-bonding contribution, it is obtained from the following expression:

$$\mu_i = \left(\frac{\partial G}{\partial N_i}\right)_{T,P,N_{j,j\neq i},N_{10},\ldots,N_{t0},\tilde{\rho}} \tag{2.79}$$

By making the appropriate substitutions, Equation 2.79 results in the following expression for a non-hydrogen-bonding component i (NRHB):

$$\begin{aligned}\frac{\mu_i}{RT} &= \ln\frac{\phi_i}{\omega_i r_i} - r_i\sum_j \frac{\phi_j l_j}{r_j} + \ln\tilde{\rho} + r_i(\tilde{v}-1)\ln(1-\tilde{\rho})\\ &\quad -\frac{z}{2}r_i\left[\tilde{v}-1+\frac{q_i}{r_i}\right]\ln\left[1-\tilde{\rho}+\frac{q}{r}\tilde{\rho}\right]\\ &\quad +\frac{zq_i}{2}\left[\ln\Gamma_{ii}+\frac{r_i}{q_i}(\tilde{v}-1)\ln\Gamma_{00}\right]+r_i\frac{\tilde{P}\tilde{v}}{\tilde{T}}-\frac{q_i}{\tilde{T}_i}\end{aligned} \tag{2.80}$$

The expression for the chemical potential of a pure component, μ_i^o, can be obtained from Equation 2.80 by setting $\phi_i = \theta_i = 1$ and the number of components in the summations equal to 1.

It should be pointed out that the presence of hydrogen bonding influences the equation of state through the minimization of Equation 2.7. As an example, Equation 2.71 becomes

$$\tilde{P}+\tilde{T}\left[\ln(1-\tilde{\rho})-\tilde{\rho}\left(\sum_{i=1}^{t}\phi_i\frac{l_i}{r_i}-\nu_H\right)-\frac{z}{2}\ln\left[1-\tilde{\rho}+\frac{q}{r}\tilde{\rho}\right]+\frac{z}{2}\ln\Gamma_{00}\right]=0 \quad \text{(NRHB)} \tag{2.81}$$

In a similar manner, the QCHB model [7] is obtained by using the Guggenheim combinatorial term [1] instead of the Staverman's one. In this case, the equation of state becomes

$$\tilde{P}+\tilde{T}\left[\ln(1-\tilde{\rho})+\tilde{\rho}\left(1-\frac{1}{\bar{r}}\right)+\frac{z}{2}\ln\Gamma_{00}\right]=0 \quad \text{(QCHB)} \tag{2.82}$$

where

$$\frac{1}{\bar{r}}=\left(\frac{1}{r}-\nu_H\right) \tag{2.83}$$

In the QCHB model, the segmental volume v_i^* was considered a characteristic property of component i. In this case, then, the chemical potential in a binary mixture becomes (QCHB)

$$\frac{\mu_1}{RT} = \ln\phi_1 + \left(1 - \frac{r_1}{r_2}\right)\phi_2 + r_1\tilde{\rho}X_{12}\theta_2^2\Gamma_{rr} + r_1(\tilde{v}-1)\ln(1-\tilde{\rho})$$

$$+\ln\frac{\tilde{\rho}}{\omega_1} - \frac{r_1\tilde{\rho}}{\tilde{T}_1}\Gamma_{rr} + r_1\frac{\tilde{P}\tilde{v}}{\tilde{T}}\frac{v_1^*}{v^*} + \frac{r_1 s_1}{2}[\ln\Gamma_{rr} + (\tilde{v}-1)\ln\Gamma_{00}]$$

$$-r_1\frac{\theta_1}{\phi_1}\frac{1-\tilde{\rho}\Gamma_{rr}}{\tilde{T}} \tag{2.84}$$

2.4.1 Applications of the NRHB Model

2.4.1.1 Pure Fluid Parameters

According to the NRHB model [38], each molecule of type i in the system occupies r_i sites in the quasilattice and is characterized by three pure fluid scaling constants and one geometric or surface-to-volume-ratio factor, s. The mean interaction energy per molecular segment, ε^*, is calculated through the first two scaling parameters, ε_h^* and ε_s^*, according to the following relation:

$$\varepsilon^* = \varepsilon_h^* + (T - 298.15)\varepsilon_s^* \tag{2.85}$$

Subscripts h and s in Equation 2.85 denote an *enthalpic* and an *entropic* contribution to the interaction energy parameter.

The third scaling constant, $v_{sp,0}^*$, is used for the calculation of the close packed density, $\rho^* = 1/v_{sp}^*$, as described by the following relation:

$$v_{sp}^* = v_{sp,0}^* + (T - 298.15)v_{sp,1}^* \tag{2.86}$$

The hard-core volume per segment, v^*, is considered constant and equal to 9.75 $cm^3\ mol^{-1}$ for all fluids. Parameter $v_{sp,1}^*$ in Equation 2.86 is treated as a constant for a given homologous series [63–69], and it is set equal to $-0.412\cdot10^{-3}$ $cm^3\ g^{-1}\ K^{-1}$ for nonaromatic hydrocarbons, $-0.310\cdot10^{-3}$ $cm^3\ g^{-1}\ K^{-1}$ for alcohols, $-0.240\cdot10^{-3}$ $cm^3\ g^{-1}\ K^{-1}$ for acetates, $-0.300\cdot10^{-3}$ $cm^3\ g^{-1}\ K^{-1}$ for water, and $0.150\cdot10^{-3}$ $cm^3\ g^{-1}\ K^{-1}$ for all the other fluids [68,69].

The following relation holds: $r = MW\ v_{sp}^*/v^*$, where MW is the molecular weight. As already mentioned, the shape factor is defined as the ratio of molecular surface to molecular volume, $s = zq/zr = q/r$, and is calculated through the UNIFAC group contribution method [53,54].

Furthermore, in cases of associating compounds, it also required the knowledge of two additional parameters that characterize the specific association, namely, the association energy, $E_{\alpha\beta}^{hb}$, and the association entropy, $S_{\alpha\beta}^{hb}$ (for simplicity, the association volume, $V_{\alpha\beta}^{hb}$, is usually set equal to zero). Pure fluid parameters for some characteristic fluids that were used for the calculations presented in the next section are shown in Table 2.1. Table 2.2 presents the number of the hydrogen-bonding sites assumed on every functional group, while Table 2.3 presents the association energy and entropy for the strong specific (i.e., hydrogen bonding, Lewis acid–base) interactions between two functional groups.

2.4.1.2 Mixture Parameters

Having the scaling parameters for every fluid of interest, the model can predict all thermodynamic properties of pure fluids and mixtures. In case of mixtures, the mixing and combining rules of Equations 2.47, 2.48, 2.74, and 2.75 are used. Furthermore, combining rules are also used for

TABLE 2.1
NRHB Parameters for the Fluids That Were Used in This Study

Fluid	ε_h^* (J mol^{-1})	ε_s^* (J mol^{-1} K^{-1})	$v_{sp,0}^*$ (cm^3 g^{-1})	s	References
n-Hexane	3957.1	1.6580	1.27753	0.857	[66]
n-Heptane	4042.0	1.7596	1.25328	0.850	[66]
n-Octane	4105.3	1.8889	1.23687	0.844	[66]
Benzene	5148.5	−0.2889	1.06697	0.753	[66]
Methanol	4202.3	1.52690	1.15899	0.941	[66]
Ethanol	4378.5	0.75100	1.15867	0.903	[66]
1-Propanol	4425.6	0.87240	1.13923	0.881	[66]
1-Butanol	4463.1	1.19110	1.13403	0.867	[66]
2-Butanol	4125.1	1.45711	1.11477	0.867	[68]
1-Hexanol	4522.7	1.64571	1.11545	0.850	[69]
1-Heptanol	4521.2	1.76223	1.12464	0.844	—
1-Octanol	4532.1	1.86863	1.12094	0.839	[63]
1-Nonanol	4576.7	1.87264	1.10261	0.836	—
Water	5336.5	−6.5057	0.97034	0.861	[66]
Carbon dioxide	3468.4	−4.5855	0.79641	0.909	[64]
Acetone	4909.0	−1.15000	1.14300	0.908	[66]
Acetic acid	6198.8	−2.0598	0.92434	0.941	[66]
Dichloromethane	5163.3	−1.3305	0.688	0.881	[66]
Benzoic acid	7228.5	−3.3473	0.8552	0.774	[64]
Acetanilide	7243.2	1.9411	0.8955	0.780	[63]
Phenacetin	6246.3	−0.5539	0.9333	0.827	[63]
$[C_4mim]Tf_2N$	5465.37	2.11139	0.66240	0.824	[69]
$[C_6mim^+][Tf_2N^-]$	5389.10	2.63354	0.69229		[69]
1,2 Propylene glycol (PG)	5088.9	1.05260	0.9084	0.903	[68]
Poly(ethylene glycol) 200 < MW < 100,000 g mol^{-1}	6273.0	0.71110	$0.86620 - MW \cdot 7.98 \cdot 10^{-08}$	0.829	[68]
Poly(vinyl acetate)	5970.4	2.59194	0.80924	0.825	[68]
Linear polyethylene	7434.9	−7.94296	1.20994	0.800	—

TABLE 2.2
Hydrogen-Bonding (HB) Sites in Each Functional Group

	HB Sites	
Group	**Positive (Proton Donors)**	**Negative (Proton Acceptors)**
–OH	1	1
Benzene	0	1
HOH	2	2
Benzoic acid	1	1
>NH	1	1
>C=O (in pharmaceuticals)	0	1
–O–	0	2
–O–C=O	0	1
ILs	5	5

TABLE 2.3
Parameters for Hydrogen-Bonding (HB) Interactions

HB Groups	E^{hb} (J mol^{-1})	S^{hb} (J mol^{-1} K^{-1})	References
–OH···OH– (Methanol)	−25,100	−26.5	[66]
–OH···OH– (Alkanols)	−24,000	−27.50	[66]
–OH···OH– (Polymers)	−22,500	−27.5	[68]
HOH···HOH	−16,100	−14.70	[66]
Acetic Acid	−14,682	−6.9	[66]
Benzoic Acid	−16,380	−20.34	[64]
>NH···NH<	−8,500	−10.25	[63]
>NH···O=C<	−9,176	−6.94	[63]
>NH···–O–	−10,900	−6.90	[63]
>NH···OH–	−18,100	−16.00	[63]
–OH···O=C<	−16,868	−13.25	[63]
–OH···–O–	−10,943	−13.75	[63]
–OH···–O– (polymers)	−14,900	−27.8	[68]
HOH···HOH	−16,100	−14.70	[66]
HOH···benzene	Combining rule (Equation 2.88)	Combining rule (Equation 2.88)	[67]
–OH···O=C–O–	−22,100	−27.30	[68]
HOH···–O–	−17,000	−27.0	[68]
HOH···OH–	Combining rule (Equation 2.87)	Combining rule (Equation 2.87)	[64]
IL–IL	−19,800	−26.5	[69]
IL–water	−16,100	−20.0	[69]

estimating the cross-hydrogen-bonding parameters in cases where no experimental data exist. The following combining rules were used for the cross-interaction between two self-associating groups:

$$E^{hb}_{\alpha\beta} = \frac{E^{hb}_{aa} + E^{hb}_{\beta\beta}}{2}, \quad S^{hb}_{\alpha\beta} = \left(\frac{S^{hb\,1/3}_{\alpha\alpha} + S^{hb1/3}_{\beta\beta}}{2} \right)^3 \tag{2.87}$$

while, for the cross-interaction between one self-associating and one non-self-associating group, the combining rules were

$$E^{hb}_{\alpha\beta} = \frac{E^{hb}_{aa}}{2}, \quad S^{hb}_{\alpha\beta} = \frac{S^{hb}_{aa}}{2} \tag{2.88}$$

2.4.1.3 Applications

In this section, recent applications of the NRHB model to hydrogen-bonding systems will be reviewed.

NRHB model can be used to predict the monomer fractions (fractions of non-hydrogen-bonded molecules) of pure self-associating fluids and their mixtures with inert solvents. According to the formalism of von Solms et al. [70], the fraction of nonbonded molecules for a pure self-associating fluid is given by the following equation:

$$X_1 = \prod_{i=1}^{k} X^K \tag{2.89}$$

where X^K is the fraction of sites of type K that are not bonded.

In the NRHB framework, alkanols are modeled assuming one proton donor and one proton acceptor site on every molecule. Consequently, the aforementioned relation is transformed to $X_1 = X^d X^a$ for alkanols, where X^a and X^d are the fraction of free acceptor and donor sites, respectively, which can easily be calculated through v_{io} and v_{jo} fractions of Equations 2.21 through 2.23.

According to the aforementioned approach, the fraction of nonbonded molecules for alkanols is obtained from the following equation [67]:

$$X_1 = \frac{2}{1 + 2\rho K + \sqrt{1 + 4\rho K}} \tag{2.90}$$

where ρ is the molar density and

$$K = v^* \exp\left(\frac{-G_{ij}^{hb}}{RT}\right) \tag{2.91}$$

Some characteristic calculations are presented in Figures 2.1 and 2.2 for pure alkanols and 1-hexanol–hexane mixture. From the first one, it is clear that the model captures the trend of monomer fraction change with alkanol's molecular weight and temperature. Relatively higher deviations are observed for ethanol. However, the experimental data for ethanol seem to coincide with the relevant data for methanol, indicating a phenomenon not accounted by the model, or an experimental error. Other EoS models result in similar predictions with the NRHB model [76]. According to Tsivintzelis et al. [76], it is unclear whether it is the ethanol or the methanol data (or none of them) that should be questioned, since these are the only experimental data that can be found in the literature.

On the other hand, the NRHB predictions for the monomer fractions of binary mixtures with alcohols and hydrocarbons seem to be in very good agreement with the experimental data as shown in Figure 2.2 [67].

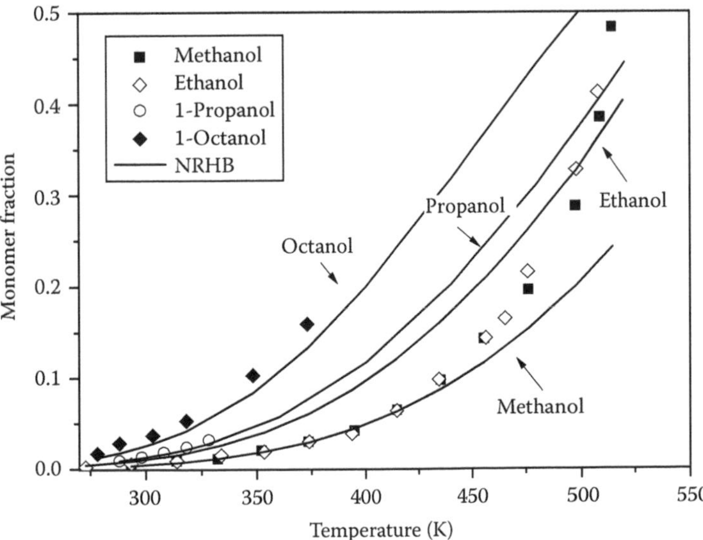

FIGURE 2.1 Experimental monomer fraction data for alcohols and NRHB predictions. Experimental data are from Refs. [71–73] and were used as they were transformed by von Solms et al. [70]. (NRHB pure fluid parameters were adapted from Tsivintzelis, I. et al., *Ind. Eng. Chem. Res.*, 47, 5651, 2008.)

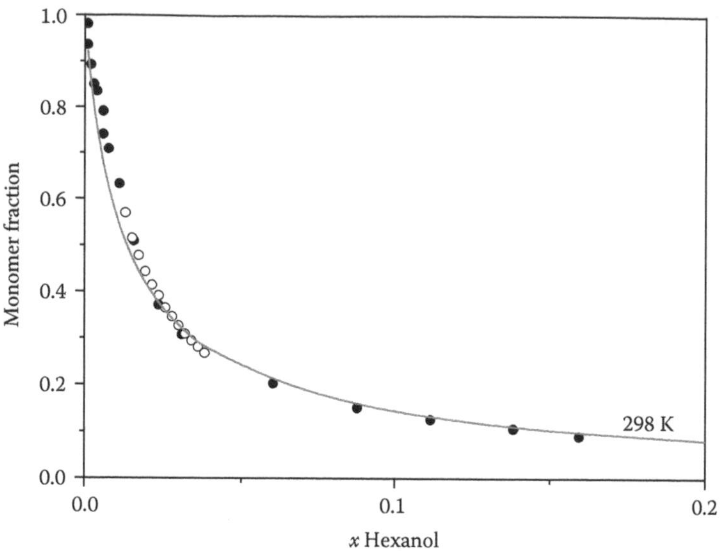

FIGURE 2.2 Fraction of non-hydrogen-bonded molecules in hexane–1-hexanol system. Experimental data [74,75] (points) and NRHB (lines) predictions. (NRHB pure fluid parameters were adapted from Tsivintzelis, I. et al., *Ind. Eng. Chem. Res.*, 47, 5651, 2008.)

The NRHB model has been mainly applied to model the phase equilibrium of hydrogen-bonding mixtures. Grenner et al. [66] used the model to describe the vapor–liquid equilibrium (VLE) of 104 binary mixtures of fluids with different polarity, including hydrogen-bonding systems. Furthermore, Tsivintzelis et al. [67] used the model to describe the liquid–liquid equilibrium (LLE) of carefully selected binary systems, which include water–hydrocarbon, 1-alkanol–hydrocarbon, water–1-alkanol, and glycol–hydrocarbon mixtures. Overall, the model resulted in satisfactory correlations for the investigated systems. In both aforementioned studies [66,67], an extended parameterization of the model was performed and tables with pure fluid and binary interaction NRHB parameters are presented. Some characteristic calculations using such pure fluid and binary parameters are shown in Figures 2.3 through 2.6. The phase behavior, which includes both VLE and LLE, of the methanol–*n*-hexane system, is presented in Figure 2.3. The mutual solubility of water–*n*-hexane and water–benzene is presented in Figure 2.4. Since *n*-hexane and benzene have similar molecular weight, the great difference in their solubility in water (benzene presents almost two orders of magnitude higher solubility than *n*-hexane) originates from the different association behavior. Thus, the higher solubility of benzene is attributed to the relatively strong specific interactions that occur between water proton donors and the π-electrons of the benzene aromatic ring. Accounting for such cross-association, the NRHB model is able to capture the different behavior of such molecules and, as shown in Figure 2.4, results in satisfactory correlations. However, the model is not able to capture the minimum in the hydrocarbon solubility, which is a very difficult task for all the EoS models [67]. Another characteristic calculation is presented in Figure 2.5 for the LLE of water–1,2-propylene glycol. This system also presents a complicated hydrogen-bonding behavior, in which both self- and cross-association interactions occur between the molecules of both fluids. In Figure 2.6, the VLE of the acetic acid–dichloromethane system is presented. Here, the dimerization of acid molecules is taken into account using the analysis of Section 2.3.2 and the hydrogen-bonding parameters of Grenner et al. [66].

Recently, the NRHB model was applied to mixtures with pharmaceuticals [63,64]. Such systems usually present a complicated hydrogen-bonding behavior, which strongly influences drug's solubility. A characteristic example is shown in Figure 2.7, where the solubility of two

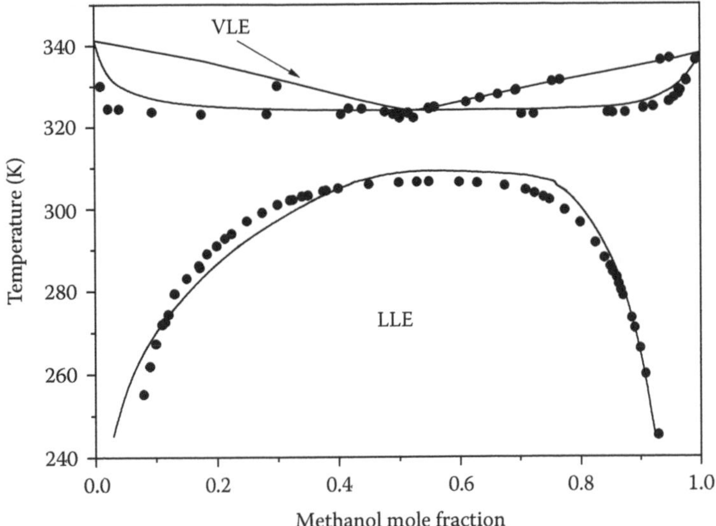

FIGURE 2.3 Methanol–*n*-hexane LLE and VLE. Experimental data [77,78] (points) and model correlations (lines) using a temperature-independent binary interaction parameter, k_{ij} = 0.0304. (Pure fluid parameters adapted from Tsivintzelis, I. et al., *Ind. Eng. Chem. Res.*, 47, 5651, 2008.)

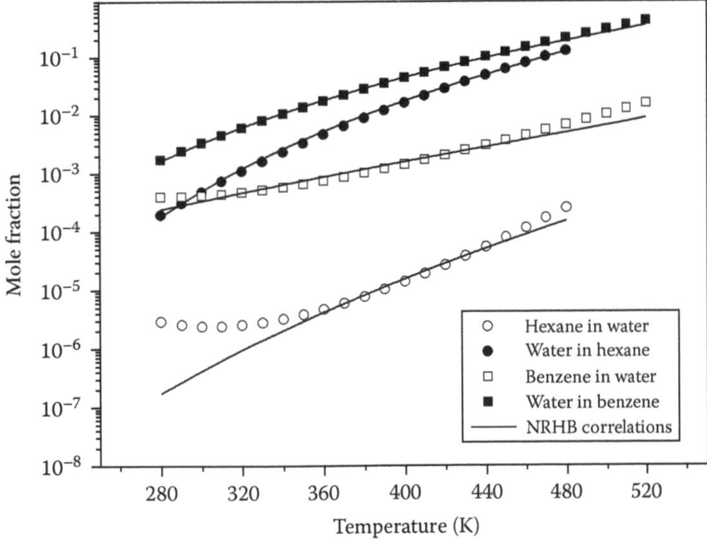

FIGURE 2.4 Hexane–water and benzene–water LLE. Experimental data [79] (points) and model calculations (lines) using pure fluid and binary parameters. (From Tsivintzelis, I. et al., *Ind. Eng. Chem. Res.*, 47, 5651, 2008.)

pharmaceutical molecules in 1-octanol is presented. Both molecules have more than one functional groups that are able to hydrogen bond with the solvent molecules, and their solubility is defined by the crystallization behavior and interplay of self- and cross-association interactions in the solution. Accounting explicitly for all possible association interactions results in very satisfactory calculations with the NRHB model, as shown in Figure 2.7.

However, the advantage of using an EoS model, against the popular activity coefficient models that are usually applied in pharmaceutical systems, is the ability to model systems at high pressures.

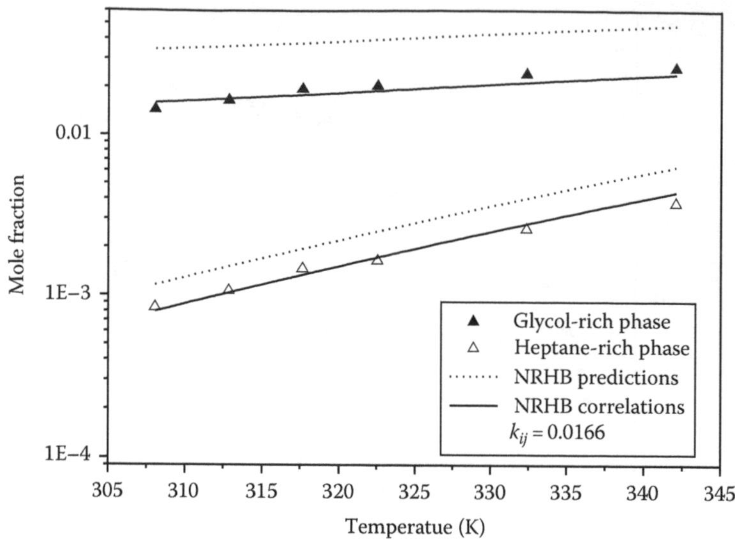

FIGURE 2.5 1,2-Propylene glycol (PG)–heptane LLE. Experimental data [80] (points) and model calculations (lines) using pure fluid and binary parameters. (From Tsivintzelis, I. et al., *Ind. Eng. Chem. Res.*, 47, 5651, 2008.)

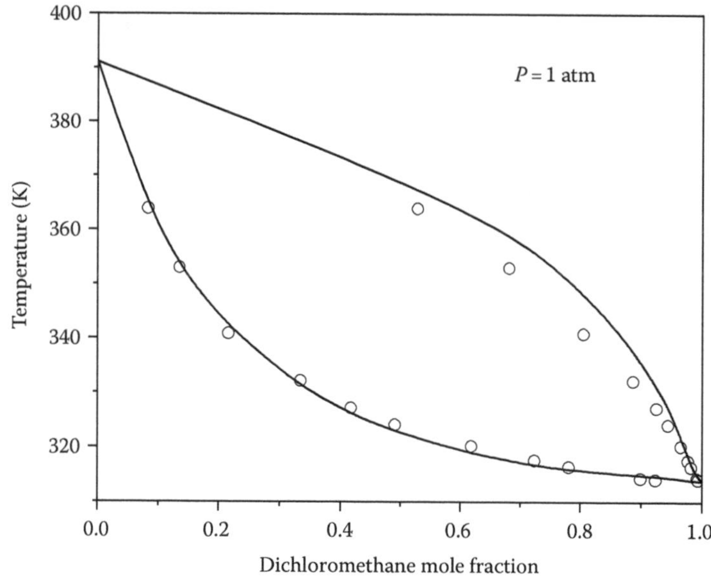

FIGURE 2.6 Acetic acid–dichloromethane VLE. Experimental data [81] (points) and NRHB calculations (lines) using pure fluid and binary parameters. (From Grenner, A. et al., *Ind. Eng. Chem. Res.*, 47, 5636, 2008.)

Such pressures are not used in the great majority of industrial pharmaceutical applications. Recently, however, many attractive alternative processes have been suggested, including the processing of pharmaceuticals with supercritical fluids. The NRHB correlations for the solubility of benzoic acid in supercritical CO_2 and supercritical ethane are presented in Figure 2.8. All NRHB results for systems with pharmaceuticals that are presented in Figures 2.7 and 2.8 were obtained using the pure fluid and binary parameters of Refs. [63,64].

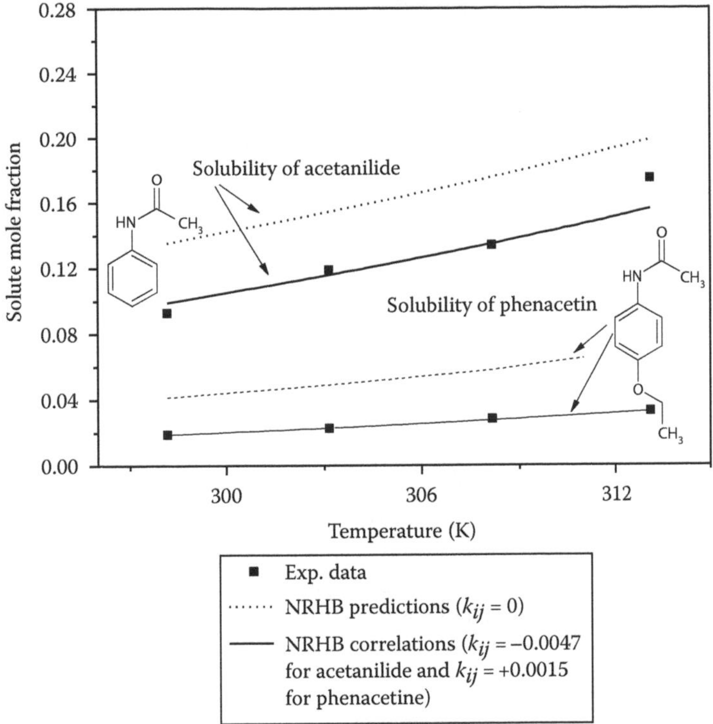

FIGURE 2.7 Solubility of acetanilide and phenacetin in 1-octanol. Experimental data [82] (points) and NRHB calculations (lines) using pure fluid and binary parameters. (From Tsivintzelis, I. et al., *AIChE J.*, 55, 756, 2009.)

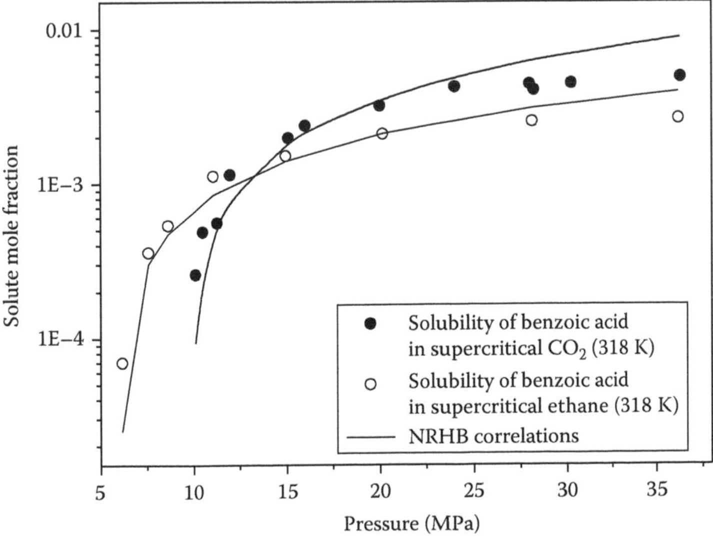

FIGURE 2.8 Solubility of benzoic acid in supercritical CO_2 and supercritical ethane at 318 K. Experimental data [83] (points) and NRHB correlations (lines) with $k_{ij} = -0.04417$ for CO_2 and -0.06974 for ethane. (Pure fluid parameters adapted from Tsivintzelis, I. et al., *AIChE J.*, 55, 756, 2009.)

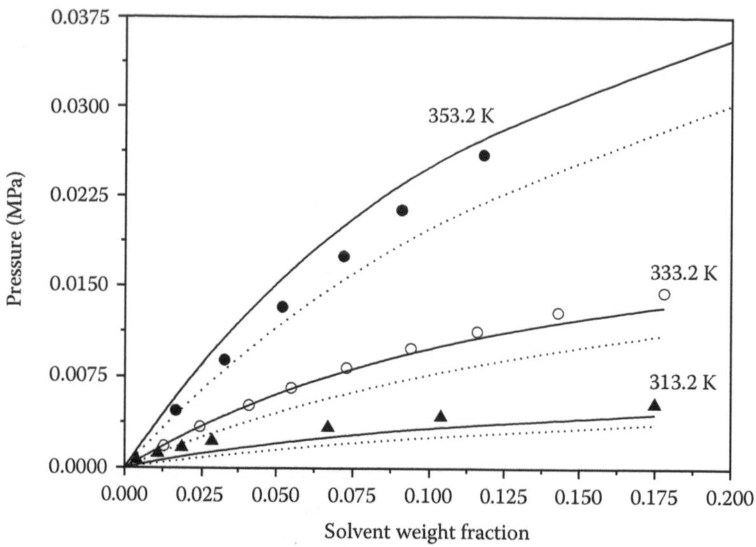

FIGURE 2.9 2-Butanol–poly(vinyl acetate) VLE. Experimental data [84] (points), NRHB predictions (dotted lines, $k_{ij} = 0$), and correlations (solid lines, $k_{ij} = 0.01372$). (Pure fluid parameters were adopted from Tsivintzelis and Kontogeorgis [68]).

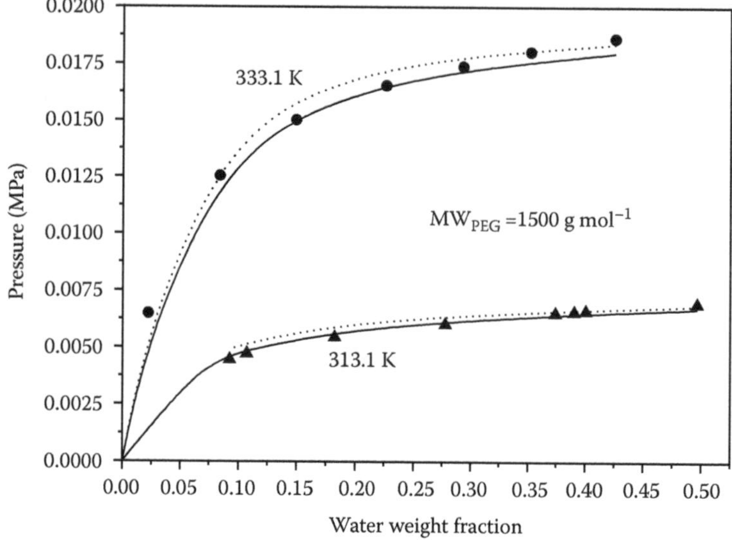

FIGURE 2.10 Water–poly(ethylene glycol) VLE. Experimental data [85] (points), NRHB predictions (dotted lines, $k_{ij} = 0$), and correlations (solid lines, $k_{ij} = 0.01238$). (Pure fluid parameters were adopted from Tsivintzelis and Kontogeorgis [68]).

However, hydrogen bonding can influence the thermodynamic properties of systems with macromolecules as well. Figures 2.9 and 2.10 present the VLE of polymer–solvent systems, in which both self- and cross-association interactions occur between the solvent molecules and between the solvent molecules and the polymer functional groups, respectively. All parameters were adopted by Tsivintzelis and Kontogeorgis [68] who showed that the NRHB model is able to satisfactorily predict (without the use of any binary adjustable parameter) the VLE of such binary mixtures, while using one fitted binary interaction parameter, the model very accurately

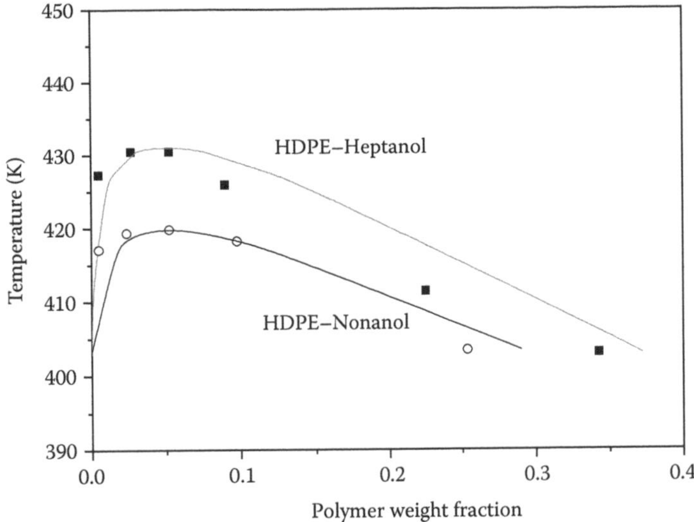

FIGURE 2.11 LLE of high-density polyethylene (HDPE)–1-heptanol and HDPE–1-nonanol. Experimental data [86] (points) and NRHB calculations (lines).

describes the solvent solubility in the liquid phase. The description of the LLE of hydrogen-bonding polymer systems is shown in Figure 2.11.

Over the past decade, an increasing research interest concerning a new class of solvents, known as ionic liquids (ILs), has arisen. Such solvents are considered to be "green," since they present extremely low vapor pressure, which allows for flexible separation processes and for their recycling. Subsequently, the disposal of solvents in the environment can be minimized. Consequently, the description of thermodynamic properties of mixtures with ILs is of great importance.

Recently, the NRHB model was applied to model systems with ILs and a new methodology was suggested for the modeling of relevant systems with a nonelectrolyte model [69]. According to this, the ILs' pure fluid parameters are estimated using PVT data and the Hansen's solubility parameters, while all ionic, polar, and hydrogen-bonding interactions are treated as specific interactions. In this way, the model was able to describe the LLE of aqueous IL systems and the VLE of IL systems with organic solvents or supercritical CO_2.

Some representative calculations, which were performed using the pure fluid and binary parameters of Tsioptsias et al. [69], are illustrated in Figures 2.12 and 2.13, for the LLE of water–1-hexyl-3-methylimidazolium bis(trifluoromethylsulfonyl)imide ($[C_4mim^+][Tf_2N^-]$) system, and the VLE of acetone–1-butyl-3-methylimidazolium bis(trifluoromethylsulfonyl)imide $[C_4mim^+][Tf_2N^-]$ mixture.

Misopolinou et al. [60,61] used a slightly different version of the NRHB, which assumes that the intersegmental energy, ε^*, and the close packed volume, v^*_{sp}, are temperature independent, while the segment volume v^* is an adjustable fluid-specific parameter. They applied the model to systems with alkoxyalkanols, compounds which are capable for both intra- and intermolecular association. Examples of these calculations for the separate contributions from non-hydrogen-bonding (dispersion and polar) interactions, from intermolecular, as well as from intramolecular hydrogen bonds are shown in Figures 2.14 and 2.15. The calculated heats of mixing are compared with the experimental ones [61].

Interestingly, the behavior of the binary with benzene is different in two points: the maximum and the shape of the H^E curve. The maximum value is about half the values of the systems with the n-alkanes. The decreased maximum value for the benzene system could be explained on the basis of the occurrence of σ–π hydroxyl–aromatic ring interactions (exothermic process).

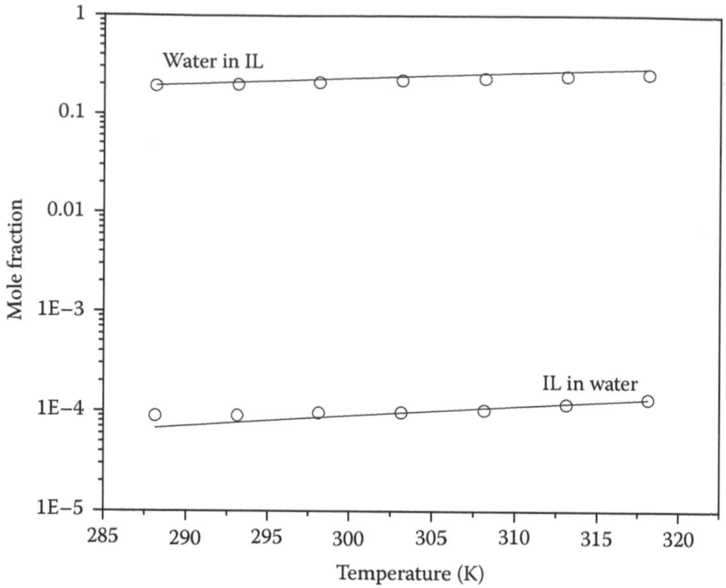

FIGURE 2.12 [C_6mim^+][Tf_2N^-]–water LLE. Experimental data [87] (points) and NRHB calculations (lines).

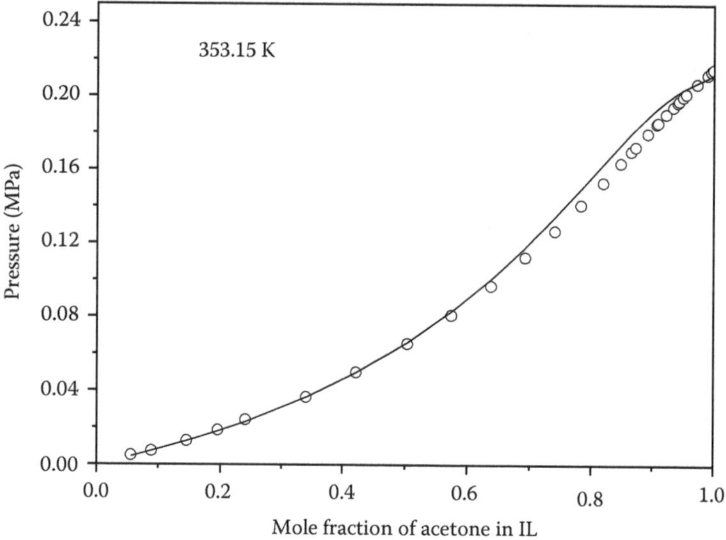

FIGURE 2.13 Acetone–[C_4mim]Tf_2N VLE. Experimental data [87] (points) and NRHB calculations (lines).

The asymmetric shape of the H^E curve could be explained by the difficulty of breaking the interactions in the 2-ethoxyethanol-rich region: In this region, the low presence of benzene fails to break the interactions of the pure 2-ethoxyethanol, while in the benzene-rich region, the breakage of the intermolecular hydrogen bonding is rather extensive.

As observed in Figures 2.14 and 2.15, the contribution of the intramolecular hydrogen bonds to the heats of mixing of the studied systems is by no means negligible. The important point is that its contribution is negative (exothermic). An explanation of this negative contribution comes from our previously published spectroscopic data [59], which indicated that at low concentration, the degree of intramolecular hydrogen bonding is increasing with increasing mole fraction of 2-ethoxyethanol much more than the degree of intermolecular hydrogen bonding. It is also known that, at the

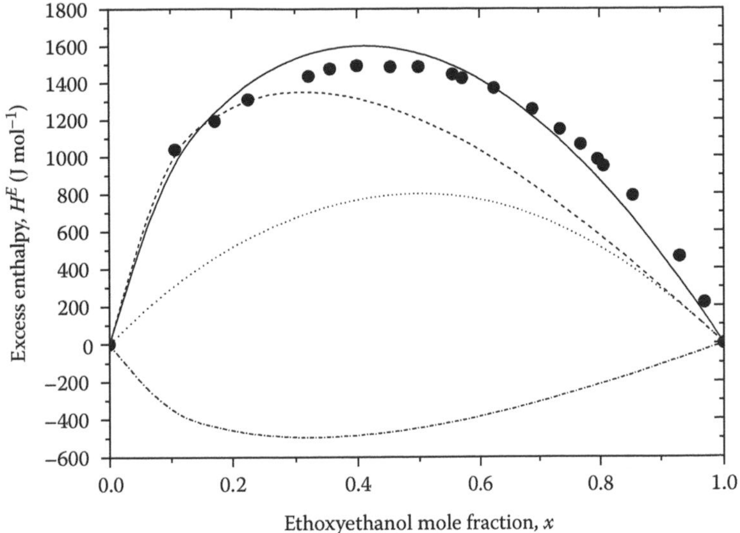

FIGURE 2.14 Experimental (points) [61] and calculated (lines) heats of mixing for the system x 2-ethoxyethanol + $(1 - x)$ n-octane at 318.15 K. The contributions from intermolecular hydrogen bonds (dash, ---), dispersive interactions (dot, ···), and intramolecular hydrogen bonds (dash dot, -·-·-) are shown by separate lines.

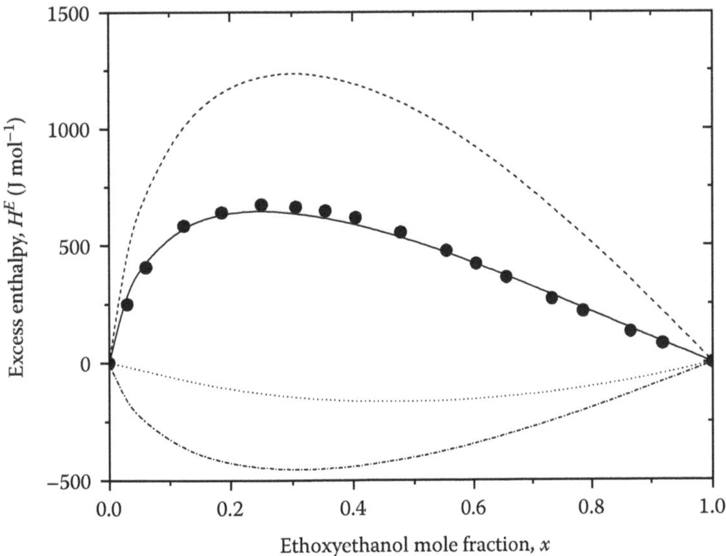

FIGURE 2.15 Experimental [61] (points) and calculated (lines) heats of mixing for the system x 2-ethoxyethanol + $(1 - x)$ benzene at 318.15 K. Symbols as in Figure 2.14.

low-concentration region, the inert solvent does not influence essentially the intramolecular hydrogen bonds, while it causes the destruction of the intermolecular bonds, as the interacting molecules fail to come close together (proximity condition).

The prediction of the NRHB model for the contribution of intermolecular hydrogen bonding is positive for all binaries, while the contribution of the van der Waals interactions is negative (exothermic) for the system with benzene but positive (endothermic) for the other systems with n-alkanes, which corroborates the previous established picture. In binary mixtures of benzene with

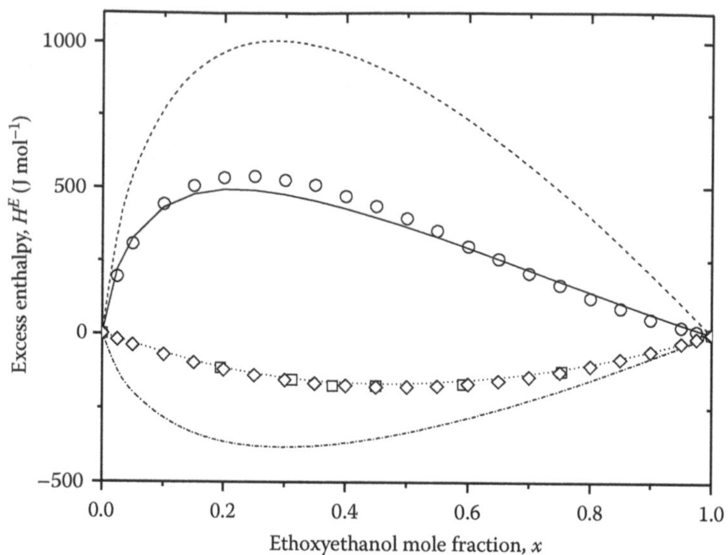

FIGURE 2.16 Comparison of the experimental [92] excess molar enthalpies, H^E, at 298.15 K for the system x 2-ethoxyethanol + $(1 - x)$ benzene, ○, with the predictions of the NRHB model for the contributions from intermolecular hydrogen bonds (---), dispersive interactions (···), intramolecular hydrogen bonds (-·-·-), and the total H^E (—). The contribution of dispersive interactions is compared with the experimental H^E for the system x 1,2-dimethoxyethane + $(1 - x)$ benzene: □, data from Ref. [89]; ◊, data from Ref. [90].

plain ethers, there are no hydrogen-bonding interactions. It is worth pointing out that the enthalpies of mixing for these binaries are negative. Typical examples are the mixtures of 1,4-dioxane with benzene at different temperatures [88] and of 1,2-dimethoxyethane (an isomer to 2-ethoxyethanol) with benzene [89,90]. The latter experimental data are compared with the predictions of the NRHB model for the system x 2-ethoxyethanol + $(1 - x)$ benzene in Figure 2.16 and, as is shown, the experimental values for the binary with 1,2-dimethoxyethane coincide with the calculated contribution of the dispersive (non-hydrogen-bonding) interactions. From literature data, we also see positive enthalpies of mixing in binaries of n-alkanes or cycloalkanes with ethers. Typical examples are the mixtures of n-hexane or cyclohexane with 1,2-dimethoxyethane [91]. This further supports the validity of our model.

2.4.2 Extension to Interfaces

Methods for reliable estimations of interfacial properties in multicomponent systems are essential for a rational design of numerous processes, notably, emulsion and suspension processes in food, pharmaceutical and material technology, in coating and adsorption processes, in industrial separation processes, in microemulsions, and in tertiary oil recovery. In the era of nanotechnology, these properties are also essential for controlling, among others, sizes and shapes of nanostructures. There are many approaches for the study of fluid–fluid interfaces [92–105], but little progress has been made for the calculations of interfacial tensions and interfacial profiles in highly nonideal mixtures, such as the hydrogen-bonded and the azeotropic mixtures.

In a series of papers [7,106,107], we have combined our EoS model with the density gradient approximation of inhomogeneous systems [99–105]. In Refs. [7,106,107], we have addressed in three alternative ways the problem of consistency and equivalence of the various methods of calculating the interfacial tension. In the first case [106], we have simulated the number density profile across the interface with the classical hyperbolic tangent expression [92] (Equation 2.138). In the second case [7], this profile was obtained from the free-energy minimization condition [103,105].

In the latter case, the internal consistency requirement resulted in a universal value of three for the proportionality factor between the interfacial tension and the integral of the Helmholtz's free-energy density difference, $\Delta\psi_0$. In the third case [107], we have also incorporated density gradient contributions into, both, the equation for the chemical potential and the equation of state. The striking result was that a universal value of four was then obtained for the aforementioned proportionality factor. In the following text, we will present the basic formalism, starting from the second case [7], and we will discuss some representative applications.

2.4.2.1 Point-Thermodynamic Approach to Interfaces

Let us consider a multicomponent two-phase system with a plane interface of area A in complete equilibrium, and let us focus on the inhomogeneous interfacial region. Our approach is a point-thermodynamic approach [92–96], and our key assumption is that in an inhomogeneous system, it is possible to define, at least consistently, local values of the thermodynamic fields of pressure P, temperature T, chemical potential μ, number density ρ, and Helmholtz free-energy density ψ. At planar fluid–fluid interfaces, which are the interfaces of our interest here, the aforementioned fields and densities are functions only of the height z across the interface.

Let us first consider the case where density gradient terms do not enter explicitly into the expression for the Helmholtz free energy Ψ. For the free energy Ψ of the entire inhomogeneous system of volume V, entropy S, and of N_i number of molecules of component i ($i = 1, 2, \ldots, t$), we may write [92–96,107]

$$d\Psi = -S\,dT - P^e dV + \gamma\,dA + \sum_{i=1}^{t} \mu_i^e dN_i \tag{2.92}$$

and for the interfacial tension γ

$$\gamma = \left(\frac{\partial \Psi}{\partial A}\right)_{T,V,N} \tag{2.93}$$

P^e and μ_i^e in Equation 2.92 are the equilibrium pressure and chemical potentials, respectively. Applying Euler's theorem for homogeneous functions, we obtain from Equation 2.92

$$\gamma = \frac{\left[P^e V - \left(\sum_{i=1}^{t} \mu_i^e N_i - \Psi\right)\right]}{A} = \frac{\Psi - \Psi^e}{A} \tag{2.94}$$

Let the z axis be normal to the plane interface, $\rho_i(z)$ be the mean segment number density of component i at the height z, and $\psi_i(z)$ be the corresponding mean molecular contribution to Ψ of component i at z. By integrating over the full height H of the system, we have

$$V = \int_H A\,dz \tag{2.95}$$

$$N_i = \int_H A\rho_i(z)dz \tag{2.96}$$

$$\Psi = \sum_{i=1}^{t} \int_H A\rho_i(z)\psi_i(z)dz \tag{2.97}$$

Substituting Equations 2.95 through 2.97 into Equation 2.94, we obtain

$$\gamma = \int_H [P^e - P'(z)]dz \tag{2.98}$$

where

$$P'(z) = \sum_{i=1}^{t} \rho_i(z)\left[\mu_i^e - \psi_i(z)\right] = \sum_{i=1}^{t} \rho_i(z)\mu_i^e - \psi_0(z) \tag{2.99}$$

$\psi_0(z)$ is the local Helmholtz free-energy density of the fluid at T and $\rho(z)$.

For the two bulk phases at equilibrium, we have from classical thermodynamics

$$\mu_i^e = \psi_i^e + P^e\bar{V}_i \tag{2.100}$$

$\bar{V}_i$ is the partial molar volume of component i.

We may rewrite Equation 2.98 in a more convenient form by setting the origin $z = 0$ within the interfacial region:

$$\gamma = \int_{-\infty}^{+\infty} [P^e - P'(z)]dz \tag{2.101}$$

or

$$\gamma = \int_{-\infty}^{+\infty} \Delta\psi(z)dz \tag{2.102}$$

where

$$\Delta\psi = \psi_0(z) - \frac{\Psi^e}{V} = \psi_0(z) - \sum_{i=1}^{t} \rho_i(z)\mu_i^e + P^e \tag{2.103}$$

In a one-component system, the curves $\rho(z)$, $\psi(z)$, $P'(z)$ are completely determined by the temperature, and P^e, μ^e, γ are functions of temperature only. These properties are also functions of composition in multicomponent systems.

Equation 2.101 can be used for the calculation of interfacial tension as long as we know the local value of pressure P' at the height z. In the frame of the local thermodynamic approach, we, now, assume that the aforementioned EoS model(s) can be used to provide with the local quantities $\mu_i(z)$, $\psi_i(z)$, $P'(z)$, and $\bar{V}_i(z)$ connected by the equation

$$\mu_i(z) = \psi_i(z) + P'(z)\bar{V}_i(z) \tag{2.104}$$

Equilibrium, however, requires for the chemical potentials of each component to be equal throughout the total volume of the system. This requirement gives the working form of Equation 2.100 for the local quantities:

$$\mu_i^e = \psi_i(z) + P'(z)\bar{V}_i(z) \tag{2.105}$$

Equation 2.101 can be used now with the understanding that $P'(z)$ is the pressure given by the EoS model for given temperature and number densities (local density and composition). Alternatively, Equation 2.102 can be used, but, now, the EoS model will be used to provide with the local Helmholtz free-energy density $\psi_0(z)$ for given temperature, local density, and composition.

Let us now turn to the case where contributions of density gradients to Helmholtz free energy are explicitly taken into account.

EoS models can also be used in the frame of the gradient approximation, such as the Cahn–Hilliard theory [100] of inhomogeneous systems, for the description of surface properties. In the frame of this theory, the Helmholtz's free-energy density ψ in a one-component inhomogeneous system can be expressed as an expansion of density ρ and its derivatives:

$$\psi = \psi_0(\rho) + c_1\nabla^2\rho + c_2(\nabla\rho)^2 + \cdots \tag{2.106}$$

where the free-energy density, $\psi_0(\rho)$, of a uniform fluid of density ρ is again given by

$$\psi_0(z) = \sum_{i=1}^{t} \rho_i(z)\psi_i(z) \tag{2.107}$$

For sufficiently slow variation of density, the free energy of the system may be approximated by

$$\Psi = A\int_H [\psi_0(\rho) + c(\nabla\rho)^2]dz \tag{2.108}$$

where

$$c = -\frac{dc_1}{d\rho} + c_2 \tag{2.109}$$

With the aforementioned equations, the surface tension for a plane interface can be approximated by

$$\gamma = \int_{-\infty}^{+\infty}\left[\Delta\psi + c\left(\frac{d\rho}{dz}\right)^2\right]dz \tag{2.110}$$

Here again, $\Delta\psi(z)$ is the free-energy density difference given by Equation 2.103. The interaction or *influence* parameter c can be estimated if the intermolecular interaction potential of the fluid is known [103]. Most often, however, it is treated as an adjustable parameter for each fluid.

Minimization of Equation 2.110 yields [100,103,105]

$$\frac{\partial\Delta\psi}{\partial\rho} - 2c\frac{d^2\rho}{dz^2} = 0 \tag{2.111}$$

Multiplying by $d\rho/dz$ and integrating, we obtain

$$\Delta\psi = c\left(\frac{d\rho}{dz}\right)^2 \tag{2.112}$$

or

$$dz = \sqrt{\frac{c}{\Delta\psi}}d\rho \tag{2.113}$$

This is a useful equation, since it may be used to obtain the density profile $\rho(z)$ across the interface. Substituting it into Equation 2.110, we obtain

$$\gamma = 2\int_{\rho_\alpha}^{\rho_\beta} [c\Delta\psi]^{1/2} d\rho \tag{2.114}$$

ρ_α and ρ_β are the densities of the homogeneous phases α and β, respectively, at equilibrium. Substituting Equation 2.112 into Equation 2.110, we also obtain

$$\gamma = 2\int_{-\infty}^{+\infty} \Delta\psi(z)dz \tag{2.115}$$

For a one-component system, Equation 2.103 becomes

$$\Delta\psi = \psi_0(z) - \frac{\Psi^e}{V} = \rho(z)\psi(z) - \rho(z)\mu^e + P^e = \rho(z)(\mu'(\rho(z), P^e) - \mu^e(\rho^e(\infty), P^e)) \tag{2.116}$$

Substituting Equation 2.116 into Equation 2.111, we obtain

$$\mu'(\rho(z), P^e) - 2c\frac{d^2\rho}{dz^2} = \mu^e(\rho^e(\infty), P^e) \tag{2.117}$$

This equation indicates how the chemical potential, calculated at the local density ρ and the equilibrium (normal) pressure of the system, must be corrected in order to make it equal to the equilibrium chemical potential.

In the case of systems of t components with plane interfaces, Equation 2.108 can be generalized as follows [45,48,50]:

$$\Psi = A\int_H \left[\psi_0(\rho) + \sum_i\sum_j c_{ij}\frac{d\rho_i}{dz}\frac{d\rho_j}{dz}\right]dz \tag{2.118}$$

Substituting into Equation 2.94 and recalling that $d\rho_i/dz \to 0$ as $z \to \infty$, we obtain for the interfacial tension

$$\gamma = \int_{-\infty}^{+\infty} \left[\Delta\psi(z) + \sum_i\sum_j c_{ij}\frac{d\rho_i}{dz}\frac{d\rho_j}{dz}\right]dz \tag{2.119}$$

where $\Delta\psi(z)$ is again given by Equation 2.103. The equilibrium interfacial tension can be obtained now by minimizing Equation 2.119. Such minimization results to t coupled differential equations:

$$\frac{\partial \Delta\psi}{\partial \rho_i} - \sum_j c_{ij} \frac{d^2\rho_j}{dz^2} = 0 \quad i, j = 1, 2, \ldots, t \tag{2.120}$$

which are the conditions subject to which the integral in Equation 2.119 must be evaluated. Multiplying Equations 2.120 by $d\rho_i/dz$, summing over all components i, and integrating, we obtain

$$\Delta\psi(z) = \sum_i \sum_j c_{ij} \frac{d\rho_i}{dz} \frac{d\rho_j}{dz} \tag{2.121}$$

Substituting Equation 2.121 into Equation 2.119 results in an equation identical in form to Equation 2.115, or

$$\gamma = 2\int_{-\infty}^{+\infty} \Delta\psi(z)dz = 2\int_{-\infty}^{+\infty} \sum_i \sum_j c_{ij} \frac{d\rho_i}{dz} \frac{d\rho_j}{dz} dz \tag{2.122}$$

An appropriate combining rule for c_{ij} is needed when $i \neq j$. With c_{ij}s available and an EoS model available, the earlier formalism is complete for the calculation of interfacial tensions at fluid–fluid plane interfaces.

Equation 2.120 is not very useful in this form. However, they can turn to a set of most useful equations if we replace the first terms by the derivatives obtained from Equation 2.103, which for convenience is rewritten here in an alternative form:

$$\Delta\psi = \psi_0(z) - \sum_{i=1}^{t} \rho_i(z)\mu_i^e + P^e = \sum_{i=1}^{t} \rho_i(z)\left[\mu_i'(\rho(z), P^e) - \mu_i^e\right] \tag{2.123}$$

Equation 2.120 can, then, be written as

$$\left[\mu_i'(\rho(z), P^e) - \mu_i^e\right] - \sum_j c_{ij} \frac{d^2\rho_j}{dz^2} = 0 \quad i, j = 1, 2, \ldots, t \tag{2.124}$$

Since we have accepted that the equations of the aforementioned EoS model(s) can be used in the inhomogeneous region and in terms of the local variables, we may further simplify Equation 2.124 by using one of the equations for the chemical potential, such as Equation 2.80 or 2.84 (plus Equation 2.25 for the hydrogen-bonding contribution) and the requirement for equilibrium—Equation 2.105:

$$\frac{\partial \Delta\psi}{\partial \rho_i} = \mu_i'(\rho(z), P^e) - \mu_i(\rho(z), P'(z)) = (P^e - P'(z))\tilde{v}v_i^* \quad i = 1, 2, \ldots, t \tag{2.125}$$

or

$$\frac{1}{v_i^*}\frac{\partial \Delta\psi}{\partial \rho_i} = (P^e - P'(z))\tilde{v} \quad i = 1, 2, \ldots, t \tag{2.126}$$

or

$$(P^e - P'(z))\tilde{v}v_i^* = \sum_j c_{ij}\frac{d^2\rho_j}{dz^2} \quad i, j = 1, 2, \ldots, t \tag{2.127}$$

Equation 2.126 is the most useful equation for the evaluation of the composition at the height z. For this purpose, it may be used in the alternative form

$$\frac{1}{v_1^*}\frac{\partial \Delta\psi}{\partial \rho_1} = \frac{1}{v_2^*}\frac{\partial \Delta\psi}{\partial \rho_2} = \frac{1}{v_3^*}\frac{\partial \Delta\psi}{\partial \rho_3} = \cdots \tag{2.128}$$

Equation 2.123 can also be written in a useful form by transforming it from the z space to the ρ space. From Equation 2.121, we may write

$$\frac{d\rho_1}{dz} = \pm\left[\frac{\Delta\psi(z)}{c_{11} + 2\sum_{j>1} c_{1j}(d\rho_j/d\rho_1) + \sum_{j>1} c_{jj}(d\rho_j/d\rho_1)^2}\right] \tag{2.129}$$

This equation can be integrated to give the interfacial profile $z(\rho)$ or $\rho(z)$:

$$z - z_0 = \int_{\rho_1(z_0)}^{\rho_1(z)} \sqrt{\frac{c_{11} + 2\sum_{j>1} c_{1j}(d\rho_j/d\rho_1) + \sum_{j>1} c_{jj}(d\rho_j/d\rho_1)^2}{\Delta\psi}}\, d\rho_1 \tag{2.130}$$

Using Equation 2.129, we may rewrite Equation 2.130 as follows:

$$\gamma = 2\int_{\rho_1^{\alpha}}^{\rho_1^{\beta}} \left[c_{11} + 2\sum_{j>1} c_{1j}\frac{d\rho_j}{d\rho_1} + \sum_{j>1} c_{jj}\left(\frac{d\rho_j}{d\rho_1}\right)^2\right]^{1/2} \Delta\psi^{1/2}\, d\rho_1 \tag{2.131}$$

where α and β refer to the two phases at equilibrium.

2.4.2.2 Calculations of Interfacial Tension and Interfacial Profiles

The aforementioned equations, along with the scaling constants for pure fluids, are sufficient for calculating, both, the interfacial tension as well as the interfacial composition profiles in multicomponent systems. The calculations in the original references [7,106,107] were done with the QCHB formalism. For convenience, these scaling constants are reported in Table 2.4. For the influence parameters c_{ii} we may adopt the rationale of Poser and Sanchez [105] leading to the following equation:

$$c_{ii} = \varepsilon_i^* \left(v_i^*\right)^{5/3} \kappa_{ii} \tag{2.132}$$

TABLE 2.4
QCHB Scaling Constants for Pure Fluids

Fluid	$\varepsilon^* = RT^*$ (J mol^{-1})	$v^* = \varepsilon^* P^{*-1}$ (cm^3 mol^{-1})	$v_{sp}^* = \rho^{*-1}$ (cm^3 g^{-1})	$s = q/r$
Methane	1809	7.100	2.0630	0.980
Ethane	2967	7.824	1.5600	0.965
Propane	3168	9.131	1.4201	0.955
n-Pentane	4043	12.24	1.2614	0.950
n-Hexane	4281	12.80	1.2162	0.947
n-Heptane	4479	13.04	1.1879	0.933
n-Decane	5180	16.92	1.2576	0.900
Carbon monoxide	1378	6.683	1.1250	0.914
Carbon dioxide	2569	4.546	0.6172	1.020
Acetone	4982	9.138	1.1573	0.871
Benzene	5372	12.40	1.0610	0.892
Toluene	5392	12.87	1.0661	0.898
Cyclohexane	5102	13.99	1.1722	0.902
Methanol	5508	15.84	1.3459	0.795
Ethanol	5222	14.69	1.2791	0.795
1-Propanol	5405	15.36	1.2422	0.795
1-Butanol	5700	16.70	1.2347	0.795
1-Pentanol	5878	17.40	1.2249	0.795
1-Hexanol	5911	18.67	1.2131	0.824
1-Octanol	6068	19.25	1.2037	0.835
1-Decanol	6068	19.63	1.1980	0.848
Water	4350	13.92	1.0310	0.830

This equation satisfies the required dimensionality for c_{ii} and lets us treat κ_{ii} as a dimensionless adjustable parameter. Values of this parameter are given in Table 2.5 for some representative fluids. For simplicity, the following geometric mean rule can be adopted for κ_{ij}:

$$\kappa_{ij} = \sqrt{\kappa_{ii}\kappa_{jj}} \tag{2.133}$$

With this combining rule, there are no adjustable parameters for the mixture and, thus, the model can be used for the prediction of interfacial tensions in multicomponent systems. For these systems,

TABLE 2.5
Influence Parameter of Representative Fluids

Fluid	κ
n-Hexane	0.792
Cyclohexane	0.835
Toluene	0.720
Methanol	1.230
Ethanol	0.980
1-Propanol	0.875
Poly(dimethyl siloxane)	0.470
Polystyrene	0.530

however, apart from the density profile $\rho(z)$, we need the profiles $\rho_i(z)$ for each component i. Equivalently, we need the profiles of the compositions $x_i(z)$ or, in the case of multimer molecules, the segment fractions $\varphi_i(z)$. For these profiles, the following equations hold:

$$\sum_{i=1}^{t} \rho_i(z) = \sum_{i=1}^{t} \varphi_i(z)\rho(z) = \rho(z) \tag{2.134}$$

The composition profiles are obtained from Equations 2.128.

A comparison of Equation 2.122 or 2.115 with Equation 2.102 indicates that, by disregarding the density gradient contributions to the Helmholtz free energy of the inhomogeneous system, we underestimate the interfacial tension by a factor of 2. It should be kept in mind, however, that we arrived at this result (the factor of 2 in Equations 2.115 and 2.122) by adopting the truncated form of Equation 2.106 and by disregarding contributions of higher-order derivatives and their higher powers. The key question now is whether this truncation is a good approximation and whether the factor of 2 in Equations 2.115 and 2.122 is sufficient for the correct estimation of interfacial tension. The answer to this question is not independent of the intermolecular interaction potential or the molecular model used for the Helmholtz free energy. The following approach, however, although adapted to the QCHB EoS model, is quite general and can be used with any other mean field model.

Let us return to Equations 2.101, 2.102, and 2.115 or 2.122 and examine the alternative ways for estimating the interfacial tension. Given an EoS model, we may use Equation 2.101 in two ways, which are the two ways we may estimate $P'(z)$ by the model: from the equation of state or from the expression for the chemical potential. In addition, we may use Equation 2.102 for the same purpose. Thus, in total, we may estimate γ in three different ways. By using our EoS model, we observe that Equations 2.101 and 2.102 with $P'(z)$ from the equation for the chemical potential give identical results for γ, while Equation 2.101 with $P'(z)$ from the equation of state gives a much larger value. If we repeat this procedure but instead of Equation 2.102 we use Equation 2.115 or 2.122, we observe that the two estimations from the chemical potential are again identical but, now, closer to the value estimated by the equation of state, though still lower. An example is shown in Figure 2.17.

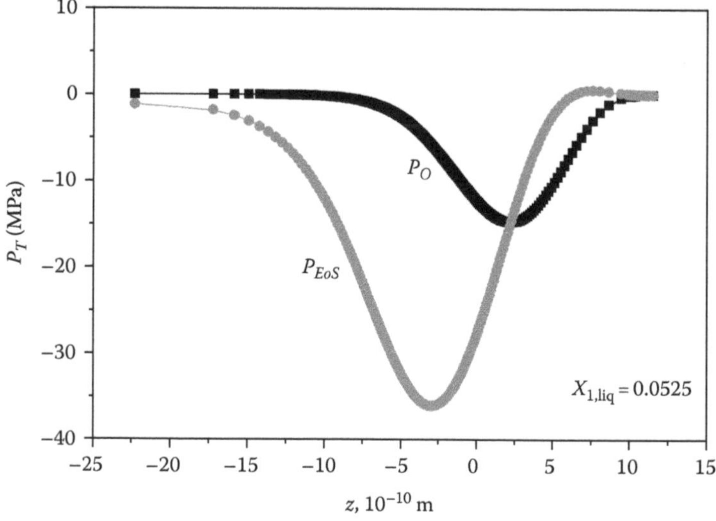

FIGURE 2.17 The pressure profile at the interfacial layer as calculated by the equation of state (P_{EoS}) and by the chemical potential (P_{CP}) for the system 1-propanol (1)–n-hexane (2) at 298.15 K near the azeotropic composition.

TABLE 2.6
Influence Parameter and the Shift Factor of Common Fluids

Fluid	κ	β
Carbon dioxide	0.550	1.000
Propane	0.370	1.000
n-Pentane	0.380	1.000
n-Hexane	0.355	1.000
Cyclohexane	0.375	1.000
Benzene	0.342	1.000
Toluene	0.322	1.000
Acetone	0.255	1.000
Poly(Dimethyl Siloxane)	0.215	1.000
Polyethylene–(linear)	0.615	1.000
Polyisobutylene	0.484	1.000
Polystyrene	1.005	1.000
Methanol	0.570	0.965
Ethanol	0.523	0.980
1-Propanol	0.400	0.987

In Ref. [106], we had considered that this discrepancy is primarily due to the earlier truncation of higher-order contributions to the Helmholtz free energy. Of course, if we had the complete form of Ψ, we could also get the complete forms of the equation of state and the chemical potential in the local thermodynamic approach. In this case, we would have, quite generally [106],

$$\gamma = \int_{-\infty}^{+\infty} [P^e - P'_{EoS}]dz = (2+\beta)\int_{-\infty}^{+\infty} \Delta\psi\, dz = (2+\beta)\int_{-\infty}^{+\infty} [P^e - P'_{CP}]dz \tag{2.135}$$

where β in this equation is a *shift* factor required to make all different estimations of γ equal. This is not an external parameter. It is an internal parameter of the model and, at each set of external conditions, it takes a value that fulfils the consistency requirement, namely, that Equation 2.135 holds true. Neither its value nor its dependence on external conditions is a priori known. We have applied Equation 2.135 to a number of fluids of varying polarity. The striking result is that β not only is independent of external conditions, but it appears that it has a universal value equal to unity—at least for the nonstrongly polar fluids. The parameters κ_{ii}, or simply κ, are different for different fluids, however. These parameters are reported in Table 2.6 for a number of common fluids and should be contrasted with the κ values reported in Table 2.5, where the consistency requirement was ignored. They have been determined at one temperature by fitting Equation 2.135 to the known surface tension at this temperature. Calculations at other temperatures are just predictions. The integrals in Equation 2.135 were evaluated numerically by summation over a finely subdivided interval.

The following equation has been adopted for the influence parameter of alkanols [7,106]:

$$c_{ii} = \left(\varepsilon_i^* + \lambda_H \frac{\nu_H}{2}(E^H - TS^H)\right)\left(v_i^* + \lambda_V \frac{\nu_H}{2} V^H\right)^{5/3} \kappa_{ii} \tag{2.136}$$

where λ_H and λ_V were set equal to one. ν_H is again the number of hydrogen bonds per molecular segment. If needed (as in the case of mixtures with alkanols), the following combining rule is adopted for simplicity:

$$\beta_{ij} = \sqrt{\beta_{ii}\beta_{jj}} \tag{2.137}$$

The values for the consistency parameter β follow the "universality" shown in Table 2.6 if Equations 2.113 and 2.130 are used for the density profile across the interface. The universality is lost and lower values are obtained [106] if a different ad hoc profile equation is adopted, such as the classical equation [92]:

$$\rho(z) = \frac{\rho_\alpha + \rho_\beta}{2} - \frac{\rho_\alpha - \rho_\beta}{2}\tanh\left(\frac{2(z - z_0)}{D}\right) \tag{2.138}$$

where

z_0 is the arbitrary origin

D is a measure of the interfacial layer thickness

Quantities ρ_α and ρ_β are the densities of the two bulk phases at equilibrium

In Figure 2.18, the experimental [108] surface tensions of three representative fluids with the calculated ones are compared. Similar comparison is made in Figure 2.19 for representative high polymers. As observed, the agreement is rather satisfactory over an extended range of temperatures.

In Figure 2.20, the experimental [110,111] surface tensions of the 1-propanol + n-hexane mixture at 298.15 K with the predicted ones by the present combined model are compared. In view of the strong nonideality of this system, the agreement is again rather satisfactory. For the same system, we have calculated the interfacial layer thickness as a function of composition of the liquid phase. The calculations are shown in Figure 2.21. This is an azeotropic system, and it is worth observing that the calculated maximum in Figure 2.21 is close to the azeotropic composition. In Ref. [106], we have shown that the interfacial layer thickness increases with the vapor pressure of the system.

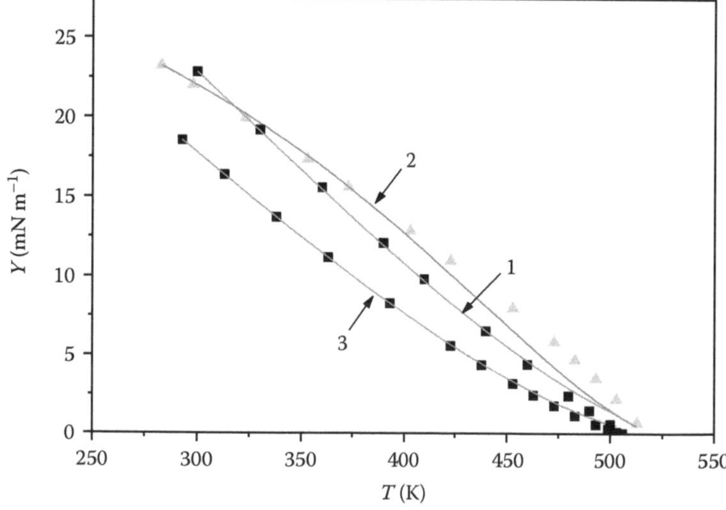

FIGURE 2.18 Experimental [108] (symbols) and calculated (lines) surface tensions for acetone (1), ethanol (2), and n-hexane (3) as a function of temperature.

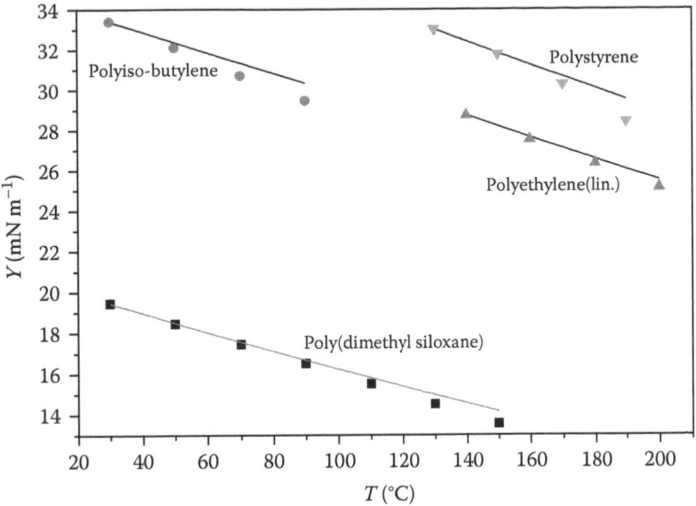

FIGURE 2.19 Experimental [109] (symbols) and calculated (solid lines) surface tensions of pure polymers.

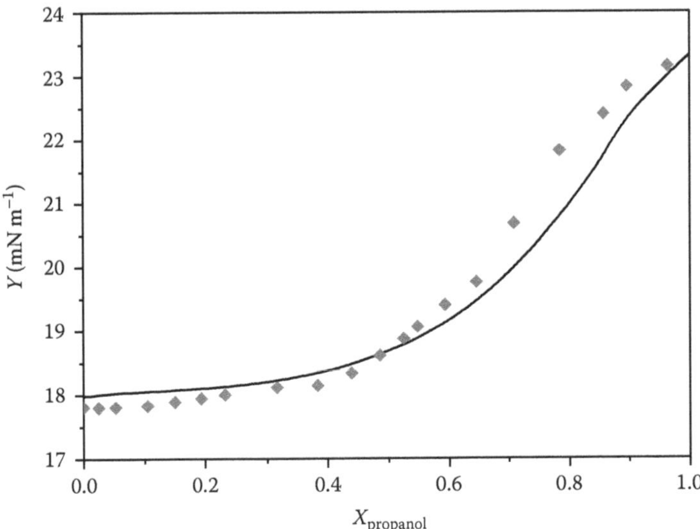

FIGURE 2.20 Experimental [110,111] (symbols) and predicted (lines) surface tensions of 1-propanol(1)–*n*-hexane (2) mixture at 298.15 K.

Since this azeotrope exhibits a maximum in vapor pressure, one may easily explain the occurrence of a maximum in Figure 2.21.

Of interest is the composition profile across the interface for such an azeotropic system. These profiles are shown in Figure 2.22. As shown, near the azeotropic composition, the interfacial composition profile exhibits a deep minimum, indicating that 1-propanol tends to avoid the interface and stay, instead, in the bulk liquid phase where it may form easier its hydrogen bonds or in the bulk vapor phase where it has much freedom of motion. The composition profiles in Figure 2.22 are qualitatively similar to the corresponding profiles calculated in Ref. [106] by simulating the density profiles with Equation 2.138. The density profiles obtained with Equation 2.130 are indeed of the form suggested by Equation 2.138 [7]. In general, the profiles obtained with Equation 2.130 are somewhat broader compared to those obtained with Equation 2.138. Qualitatively, however, we expect similar calculations of composition profiles by the two equations.

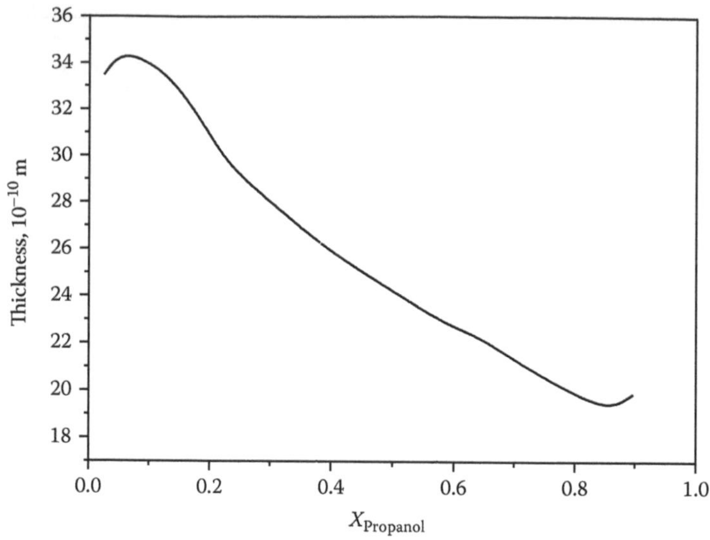

FIGURE 2.21 The calculated interfacial layer thickness as a function of composition for 1-propanol (1)–n-hexane (2) mixture at 298.15 K.

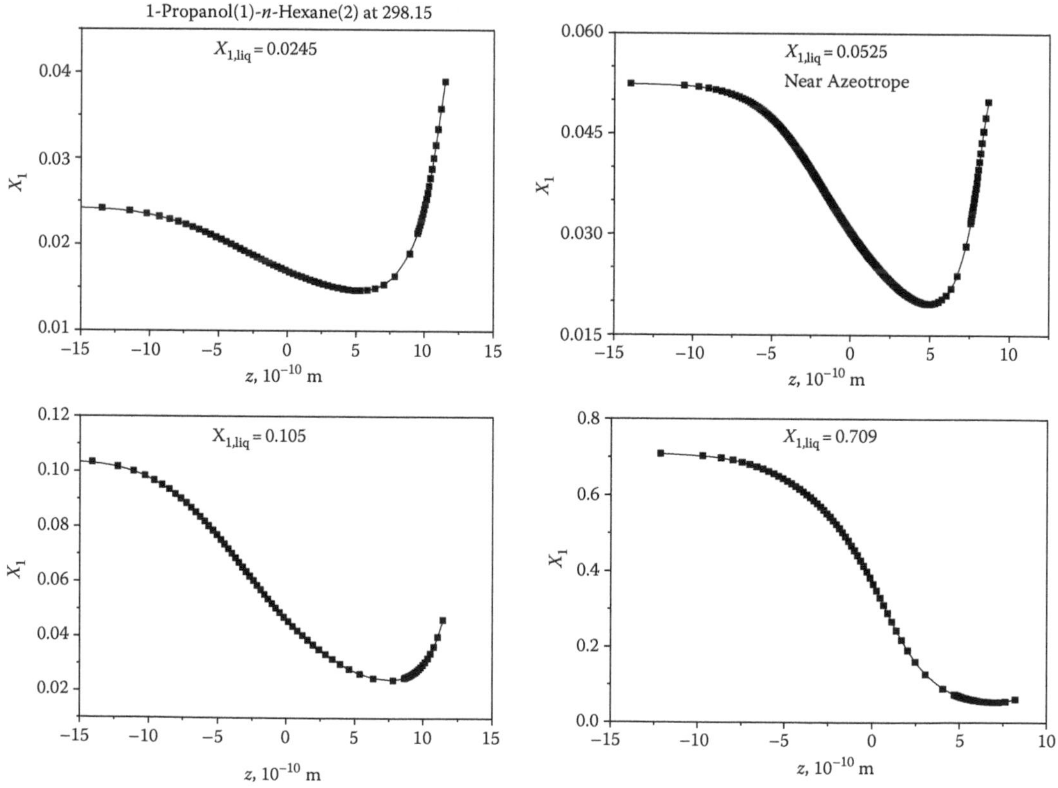

FIGURE 2.22 The evolution of the composition profiles across the interface for the 1-propanol(1)–n-hexane (2) mixture at 298.15 K. The liquid phase composition, x_1, increases from top left to bottom right.

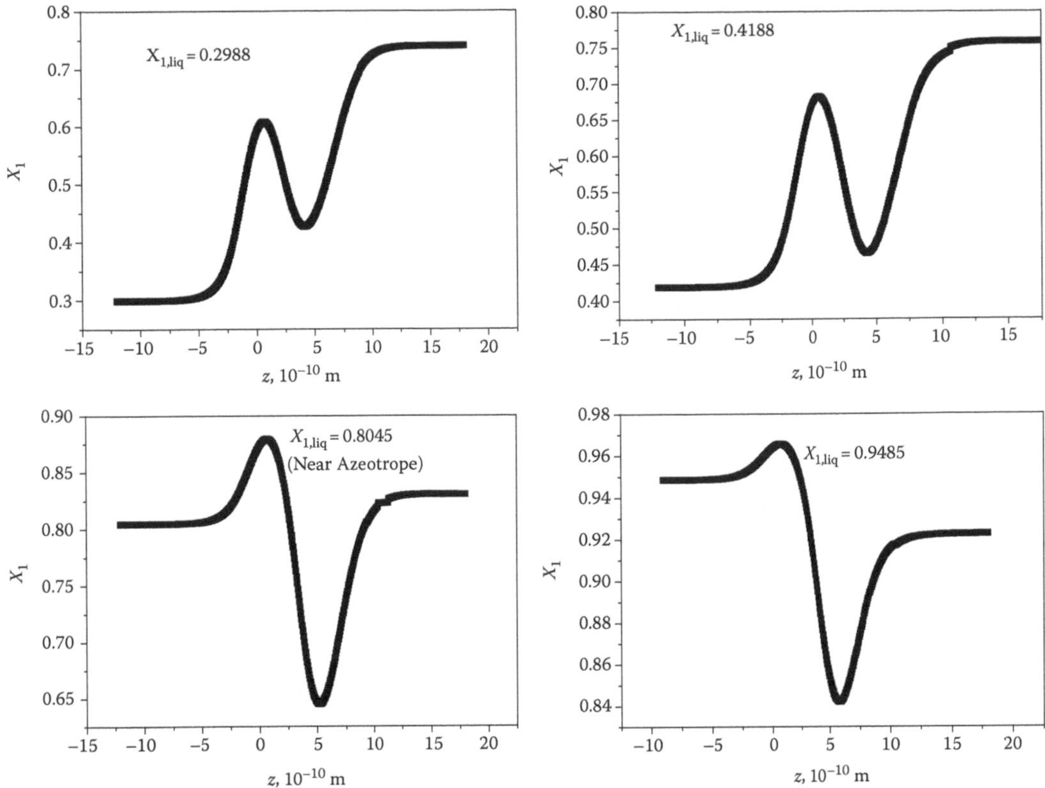

FIGURE 2.23 Methanol (1)–toluene (2) at 308.15 K. The evolution of composition profiles across the interface, as predicted by the equation-of-state model [106].

Very interesting is Figure 2.23 where the evolution of composition profiles is shown for another azeotropic mixture. The calculations were done [106] by adopting Equation 2.138.

In summarizing the results of this section, it is worth focusing on the important aspect that emerged from the extension of the EoS model(s) to interfaces, namely, the unification of three, essentially, different approaches for calculating interfacial properties through the equation $\gamma = \alpha \int_{-\infty}^{+\infty} \Delta\psi(z)dz$. By disregarding density gradient contributions to Helmholtz free energy (and to derived properties), one may use this equation with $\alpha = 1$. By adopting the density gradient approximation and the truncated form of Equation 2.108, this equation should be used with $\alpha = 2$. However, it appears that internal consistency (in the sense that the various ways of calculating γ give the same result) is fulfilled only when $\alpha = 3$, or near 3. It should be pointed out that, in the latter case, the precise value of α is an internal parameter of the model and not an external adjustable parameter. Less than 3 values have been obtained only for the hydrogen-bonded 1-alkanols, but this is not a proof of lack of universality: It should be stressed that this is an approximate model and Equation 2.136 for the evaluation of the influence parameter of hydrogen-bonded fluids is just a simple approximation for the OH–OH interaction.

Work must be done with fluids exhibiting hydrogen bonds between other pairs of proton donors and acceptors. General rules and conclusions for α could then be possible.

As already mentioned, in the earlier approach, we have disregarded density gradient contributions in the chemical potential and in the equation of state. But the alternative approach, of taking into consideration the full density gradient contributions, is by no means an easy task. However, a first approximation has been attempted in Ref. [107] by recognizing that incorporation of these contributions would imply that the total Helmholtz free energy may be written, in general, as

$$\Psi = V^*\left\{\psi_0(\rho) + \frac{1}{\rho}\sum_{n,m,i}\rho_i\lambda_{nm,i}(\rho)(\nabla^n\rho)^m\right\} = V^*\left\{\sum_i \rho_i\left[\psi_{i,0}(\rho) + \sum_{n,m}\delta_{nm,i}(\rho)(\nabla^n\rho)^m\right]\right\} \quad (2.139)$$

This equation, in turn, implies that there may be density gradient contributions to, both, the chemical potential and to the equation of state. As an example, for the QCHB model, we would have

$$\mu_i(z) = \psi_i(z) + P'(z)\bar{V}_i(z) = \mu_{i,0}(z) + \sum_{n,m}\delta_{nm,i}(\rho)(\nabla^n\rho)^m \quad (2.140)$$

$$\tilde{P} + \tilde{T}\left[\ln(1-\tilde{\rho}) + \tilde{\rho}\left(1-\frac{1}{r}\right) + \frac{z}{2}\ln\Gamma_{00}\right] + \sum_{n,m}\omega_{nm,i}(\rho)(\nabla^n\rho)^m = 0 \quad (2.141)$$

where $\mu_{i,0}$ in Equation 2.140 is the chemical potential of component i without the density gradient contribution and is assumed to be obtained from the EoS model as before. By using the last three equations and by following a procedure entirely analogous to the one described in this section, one ends up [107] in the striking result of $\alpha = 4$. In this sense, the truncation used in the density gradient approximation approach [100,103,105] is, rather, severe. It should be repeated, however, that all the aforementioned approaches may give similar estimations of interfacial tension. The key differences appear in the calculated profiles of the various properties across the interface. Since there are no experimental data available for these profiles, we cannot judge safely the various approaches. We hope, however, that such data, though challenging, will be available soon, since they may create new perspectives in the theoretical description of fluid–fluid interfaces.

The approach presented in this work can be extended to mixtures of macromolecules, including systems with strong specific interactions, and to aqueous systems. Of course, entropic corrections must be introduced in the case of polymer systems due to the effects of density and concentration gradients on the spatial conformations of macromolecules in the interfacial regions. The aforementioned interesting results indicate the potential of the EoS approach in studying interfaces. It should be stressed, however, once again that this density-gradient approach is an approximate approach with ad hoc assumptions and simplifications.

2.4.3 Partial Solubility Parameters

Originally introduced by Hildebrand and Scott [112], the solubility parameter, δ, remains today one of the key parameters for selecting solvents in industry, characterizing surfaces, predicting solubility and degree of rubber swelling, polymer compatibility, chemical resistance and permeation rates, and for numerous other applications. There is also much interest in utilizing solubility parameter for rationally designing new processes, such as the supercritical fluid, the coating, and the drug delivery processes [113–119].

The division of δ into its partial components or Hansen solubility parameters [116], δ_d, δ_p, and δ_{hb}, for the dispersion, the polar, and the hydrogen-bonding contribution, respectively, has very much enhanced its usefulness and success in practical applications. Thus, liquids with similar δ_d, δ_p, and δ_{hb}, are very likely to be miscible. The bulk of the developments in solubility parameter reside on this principle of "similarity matching" of properties. Methods for estimating these partial solubility parameters are then particularly useful and much needed.

Recently [57], our previous approach to solubility parameter estimation [120,121] was extended, in an effort to account for all three components of the solubility parameter. This was done by adopting the NRHB EoS framework, modified in order to explicitly account for dipole–dipole interactions and, thus, explicitly calculate the polar component, δ_p. In what follows, we summarize these developments and show some representative applications.

For the purposes of this section, it is convenient to rewrite Equation 2.2 in the following manner:

$$Q(N,T,P) = Q_R Q_{NR} Q_{hb} = \Omega_R \Omega_{NR} \Omega_{hb} \exp\frac{-E_d}{kT} \exp\frac{-E_p}{kT} \exp\frac{-E_{hb}}{kT} \exp\frac{-PV}{kT} \tag{2.142}$$

E_d, E_p, and, E_{hb} are the dispersion, polar, and hydrogen-bonding components, respectively, of the potential energy of the system. From this equation, we may extract in the usual manner the following equation for the chemical potential:

$$\frac{\mu}{RT} = \frac{\mu_{dp}}{RT} + \frac{\mu_H}{RT} \tag{2.143}$$

The first term is the chemical potential for the dispersion and polar interactions, and the second chemical potential for the hydrogen-bonding interactions. The key problem now is the separation of the contributions from dispersion and polar forces. Following the rationale of Twu and Gubbins [122] and Nezbeda and Pavlicek [123–125], it was proposed [57] to replace ε^* by $\varepsilon^*[1+(4/\pi)(m/r)^2(s^2/\tilde{T})]$, where m is the dipole moment of the fluid. In this way, the potential energy of the fluid may be written as

$$-E = \Gamma_{rr} q N \theta_r \varepsilon^* \left[1 + \frac{4}{\pi}\left(\frac{m}{r}\right)^2 \frac{s^2}{\tilde{T}}\right] - N_H E_H \tag{2.144}$$

On the basis of the aforementioned equation, the partial solubility parameters are given by

$$\delta_d = \sqrt{\frac{\Gamma_{rr} q N \theta_r \varepsilon^*}{V}} \tag{2.145}$$

$$\delta_p = \sqrt{\frac{\Gamma_{rr} q N \theta_r \varepsilon^* [(4/\pi)(m/r)^2 (s^2/\tilde{T})]}{V}} \tag{2.146}$$

$$\delta_{hb} = \sqrt{\frac{-N_H E_H}{V}} \tag{2.147}$$

Thus, by knowing the scaling constants of a fluid along with its dipole moment and hydrogen-bonding interaction energies, one may estimate (essentially, predict) the three partial solubility parameters over an extended range of external conditions. On the other hand, if the partial solubility parameters are known, they may be used in order to extract information about the scaling constants of the fluid and its hydrogen-bonding interaction, which in turn could be used for predicting all basic thermodynamic properties of the fluid. Such calculations have been done for a large number of fluids including polymers and extensive tabulations are reported in Ref. [57]. One representative sample from these tabulations is shown in Table 2.7.

TABLE 2.7
Total and Partial Solubility Parameters (MPa$^{1/2}$) of Pure Fluids

Polar/Hydrogen-Bonded Fluids

	δ_{Total}		δ_{hb}		δ_p	
Fluid	**Exp**	**Calc**[a]	**Exp**		**Exp**	**Calc**[a]
Acetone	19.95	21.29(20.04)	6.95	6.95	10.43	10.44(10.14)
Ethyl acetate	18.48	18.66(18.31)	9.20	9.20	5.32	5.70(6.07)
1,4-Dioxane	20.47	19.67(20.08)	7.36	7.36	1.84	1.84(1.91)
$CHCl_3$	18.94	19.46(19.18)	5.73	5.73	3.07	3.34(3.79)
Methanol	29.61	30.86(29.89)	22.30	24.15(24.08)	12.27	12.11(11.34)
Ethanol	26.50	26.26(26.08)	19.43	20.08(19.98)	8.80	8.49(8.24)
1-Butanol	23.35	22.92(22.90)	15.80	15.84(15.80)	5.70	5.70(5.72)
1-Octanol	20.87	20.30(20.27)	11.86	11.97(11.94)	3.27	3.27(3.75)
Phenol	24.63	24.59(24.69)	14.90	14.90	5.90	6.31(5.16)
Ethylene glycol	33.70	33.97(33.64)	25.77	25.93	11.05	16.01(12.2)
Glycerol	34.12	34.14(34.34)	29.25	30.91	12.07	12.10(14.31)
Diethylamine	16.61	16.97(16.80)	6.10	6.10	2.30	2.49(2.89)
n-Butylamine	18.31	18.32(18.48)	8.00	8.00	4.50	4.86(4.75)
Tetrahydrofuran (THF)	19.46	20.05(18.93)	8.00	8.00	5.70	6.04(6.98)
Ammonia	27.40	28.73(26.52)	17.80	17.80	15.70	16.10(15.70)
Water	47.82	47.80(48.68)	42.82	43.15	16.00	18.70(16.00)

[a] Values in parenthesis obtained when ε^* is replaced by $\varepsilon^*\left[1+\pi\left(\frac{m}{r}\right)^2 s^2\right]$.

In Figure 2.24, the calculated components of the solubility parameter of water over an extended range of saturation temperatures are shown. As was expected, the main contribution to δ of water, especially at low temperatures, is the hydrogen bonding. This type of diagrams is most useful for designing applications involving subcritical or supercritical water.

A most useful concept, which quantifies the "similarity" of two substances 1 and 2, especially the similarity of a polymer, 2, and a potential solvent, 1, for it, is the solubility parameter distance, *Ra*, defined by [116]

$$Ra = \sqrt{\left[4(\delta_{d2}-\delta_{d1})^2 + (\delta_{p2}-\delta_{p1})^2 + (\delta_{hb2}-\delta_{hb1})^2\right]} \tag{2.148}$$

The idea is: the smaller the *Ra*, the better is the solvent for the polymer. A sphere with radius *Ra* encompasses the good solvents for this polymer. A refined discussion on *Ra* and the related quantities *Ro* and *RED* = *Ra*/*Ro* is provided by Hansen [117].

In Figure 2.25, the partial solubility parameters for (bisphenol A) polycarbonate as functions of temperature are plotted. As shown, all three components are nonnegligible, and there is a crossover in the polar and hydrogen-bonding components for this polymer. In Figure 2.26, the distances *Ra* of this polymer with three common solvents are compared. According to the calculations, chloroform is the best of the three solvents for this polymer, followed by tetrahydrofuran (THF). Heptane is the worst and, essentially, a nonsolvent for the polymer, and all these findings agree with experiment.

In a similar manner, by comparing the distances *Ra* for polypropylene with three solvents, THF, chloroform, and tetralin, it is shown that the best solvent for polypropylene is tetralin, which is again

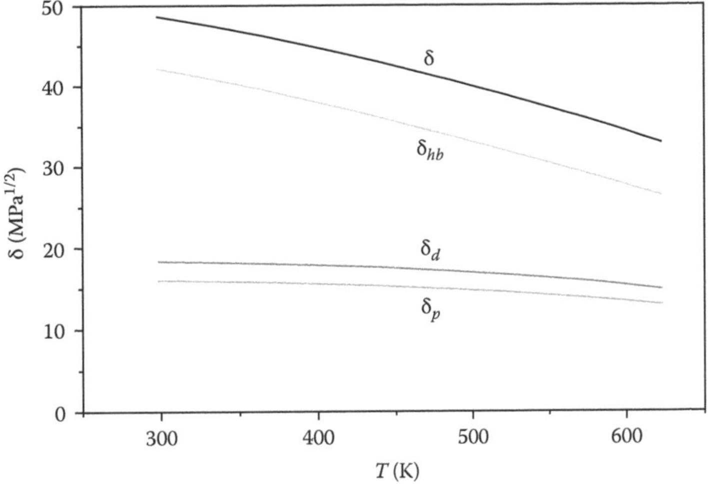

FIGURE 2.24 Fractional solubility parameters for water as calculated when ε^* is replaced by $\varepsilon^*[1 + \pi(m/r)^2 s^2]$.

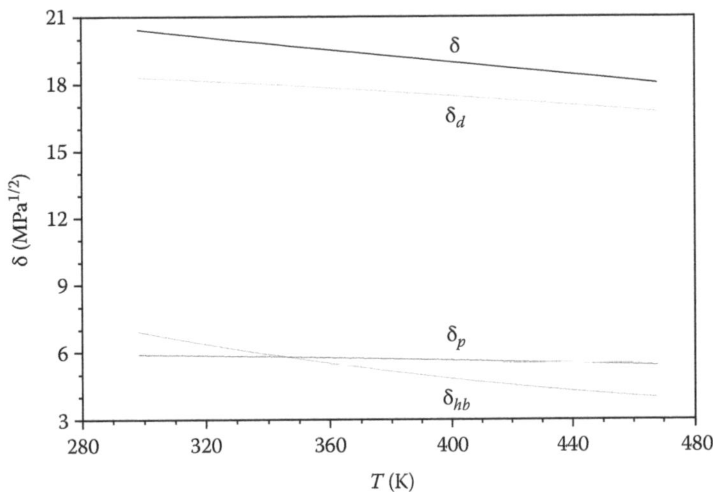

FIGURE 2.25 Partial solubility parameters for (bisphenol A) polycarbonate.

corroborated by the experiment. This type of figures is most useful not only for the mere selection of the solvent, but also for the selection of the external conditions (especially, temperature) for the dissolution of the polymer or any other solute.

2.5 PARTIAL SOLVATION PARAMETER APPROACH

The PSP approach [39–43] is a novel predictive thermodynamic framework, which combines elements from the solubility parameter approach [112,114,116,126] detailed earlier, the solvatochromic/LSER approach [127–133], and the COSMO-RS theory of solutions [134–136]. It retains the simplicity of the solubility parameter approach, it uses molecular descriptors that can be mapped one to one to the Abraham/LSER descriptors, and these descriptors are derived from the moments of the σ-profiles of the quantum mechanics–based COSMO-RS model. Because of this combination, the PSP approach has a broader range of applications compared to each of the earlier three

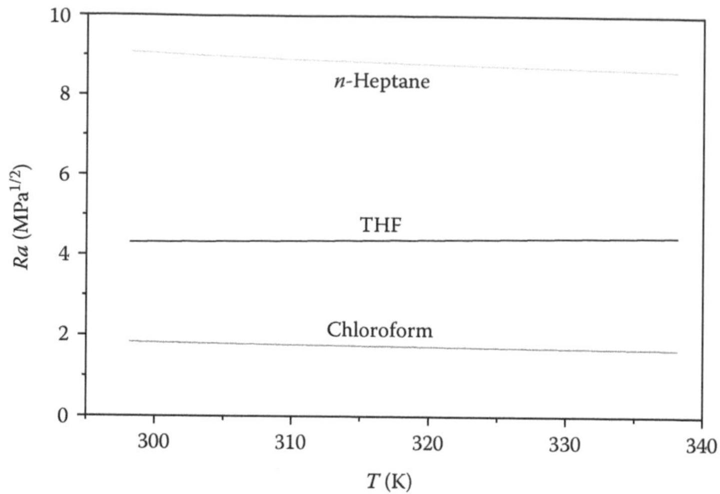

FIGURE 2.26 The estimated solubility parameter distance, *Ra*, of (bisphenol A) polycarbonate (MW = 100,000) with three common solvents, as a function of temperature.

approaches. In what follows, we will briefly present the PSP essentials. Details may be found in the recent relevant literature [39–43].

There are two schemes of PSPs: the *s*-scheme and the σ-scheme. Due to space limitations, we will confine ourselves here to the σ-scheme. According to this scheme, each compound/molecule is characterized by three major σ PSPs, which have the same dimensions as the solubility parameters (MPa$^{1/2}$). The first descriptor, called van der Waals PSP, is defined as follows:

$$\sigma_W = \sqrt{\frac{E_{vdW}}{V_{mol}}} \tag{2.149}$$

E_{vdW} and V_{mol} are the van der Waals interaction energy and the molar volume of the compound, respectively. The van der Waals energy is already available for thousands of compounds [136], but it may be obtained also from alternative straightforward calculation schemes. The hydrogen-bonding PSP, σ_{hb}, is subdivided into a donor or acidic component, σ_a, and an acceptor or basic component, σ_b, through the relation

$$\sigma_{hb}^2 = \sigma_a^2 + \sigma_b^2 = 4\frac{m\text{-}SUM}{V_{\cos m}} \tag{2.150}$$

where V_{cosm} is the COSMO volume [134–136] of the molecule and $m\text{-}SUM$ is obtained from the third hydrogen-bonding σ-COSMOments [134–136] through the equation:

$$m\text{-}SUM = HB\_acc3 + 1.492HB\_don3 \tag{2.151}$$

and

$$\frac{\sigma_b^2}{\sigma_{hb}^2} = 1 - \frac{\sigma_a^2}{\sigma_{hb}^2} = \frac{HB\_acc3}{m\text{-}SUM} \tag{2.152}$$

The third PSP, σ_{pz}, reflects interactions due to polarity/polarizability/refractivity of the molecule, and it may also be subdivided into two components for the polarity, and the refractivity although, for all practical purposes and at this stage, the combined σ_{pz} PSP is sufficient. Its mathematical definition will be better understood after we divert for the introduction of a key concept underlying PSPs.

The cohesive interactions in a pure compound are, in general, different from its solvation interactions with solvents. Some molecules may self-associate in their pure state via hydrogen-bonding interactions, which others can only cross-associate with different molecules. The cohesive energy, then, or the heat of vaporization cannot be used to determine the hydrogen-bonding capacity of these latter molecules. This is why PSPs are focusing on solvation rather than on cohesion. The compounds are divided into two major classes, the homosolvated and the heterosolvated ones. The homosolvated molecules are the molecules in which all types of intermolecular interactions are operational in pure state as well as in solution. In contrast, in heterosolvated molecules, the types of operational intermolecular interactions in solution may be different from those in pure state. Typical representatives of homosolvated and heterosolvated compounds are aspirin and acetone, respectively, as shown graphically in the upper part of Figure 2.27. The surface charge (σ) profile of aspirin contains, both, acidic as well as basic surface areas and may self-associate, while acetone does not possess acidic areas and can only cross-associate with another acidic compound. This distinction leads to the distinction between the well-known cohesive energy density and the solvation energy density.

$$\sigma_W^2 + \sigma_{pz}^2 + \sigma_{hb}^2 = \sigma_{total}^2 = \text{sed} \tag{2.153}$$

In view of the aforementioned text, the σ_{pz} PSP is obtained from the following balance equations:

$$\begin{aligned} &\sigma_W^2 + \sigma_{pz}^2 = \text{ced} - \sigma_{hb}^2 \quad \text{(homosolvation)} \\ &\sigma_W^2 + \sigma_{pz+}^2 = \text{ced} \quad \text{(heterosolvation)} \end{aligned} \tag{2.154}$$

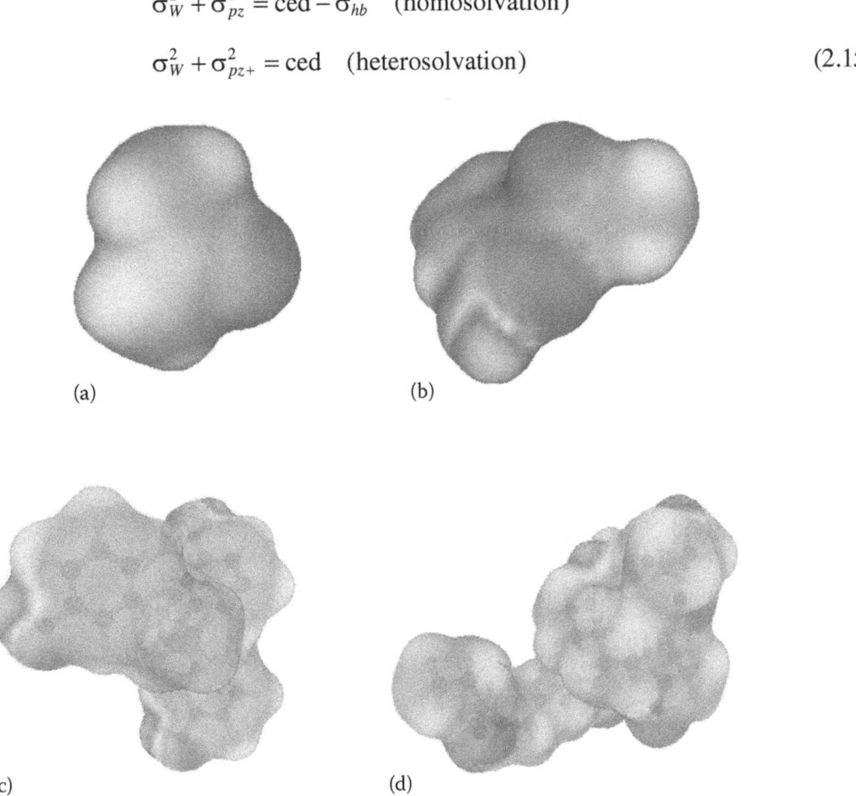

FIGURE 2.27 Surface charge distribution in representative molecules. (a) Acetone, (b) aspirin, (c) vinyl phenol trimer, and (d) nylon-6 pentamer.

Strictly speaking, the σ_a and σ_b hydrogen-bonding PSPs cannot be mapped one to one in the corresponding A and B acidic and basic LSER descriptors since the latter reflect interaction free energies rather than interaction energies. A direct mapping would introduce the *free-energy* hydrogen-bonding PSP, σ_{Ghb}, and its components, σ_{Ga} and σ_{Gb}, satisfying the relations

$$\sigma_{Ga}^2 + \sigma_{Gb}^2 = \sigma_{Ghb}^2 \tag{2.155}$$

$$\frac{\sigma_{Gb}^2}{\sigma_{Ghb}^2} = 1 - \frac{\sigma_{Ga}^2}{\sigma_{Ghb}^2} = \frac{B}{m\text{-}SUMG} \tag{2.156}$$

and

$$\sigma_{Ghb} = 5.5\sqrt{\frac{m\text{-}SUMG}{V_{\mathrm{cosm}}}} \tag{2.157}$$

where

$$m\text{-}SUMG = A + 1.0198B \tag{2.158}$$

Having, however, hydrogen-bonding energy PSPs (from COSMOments) and free-energy PSPs (from LSER), one may calculate energies, entropies, and free energies of hydrogen bond formation of type αβ between molecules of type *i* and *j*, respectively, using the following equation:

$$E_{\alpha\beta,ij}^{H} = -h\sigma_{a,i}^{\alpha}\sigma_{b,j}^{\beta}\sqrt{V_{mol,i}V_{mol,j}} \tag{2.159}$$

$$G_{\alpha\beta,ij}^{H} = -f\sigma_{Ga,i}^{\alpha}\sigma_{Gb,j}^{\beta}\sqrt{V_{mol,i}V_{mol,j}} + g \tag{2.160}$$

and

$$S_{\alpha\beta,ij}^{H} = \frac{\left(E_{\alpha\beta,ij}^{H} - G_{\alpha\beta,ij}^{H}\right)}{T} \tag{2.161}$$

The proposed [40] values for the universal coefficients *h*, *f*, and *g* are $h = 1.86$; $f = 1.70$; $g = 2.50$ kJ/mol. Having, however, these energies and free energies, we may apply directly the aforementioned presented hydrogen-bonding and combinatorial formalism and have a predictive thermodynamic model for hydrogen-bonded systems. In a free-energy expression for a binary mixture, the σ_W and σ_{pz} PSPs contribute via the term [41–43]

$$\frac{\Delta G_m^W + \Delta G_m^{pz}}{RT} = \frac{x_1\varphi_2 V_{mol,1}\Delta\sigma}{RT} \tag{2.162}$$

where

$$\Delta\sigma = (\sigma_{W1} - \sigma_{W2})^2 + \frac{1}{3}(\sigma_{pz1} - \sigma_{pz2})^2 \tag{2.163}$$

This PSP formalism can be used for both concentrated as well as infinitely dilute systems, for small molecules as well as for high polymers and copolymers. It may, in particular, be used for the surface characterization of polymer and other solid surfaces. In fact, it may be used in order to obtain the corresponding surface energy components, γ_w, γ_{pz}, γ_a, and γ_b, of the solid surfaces [42]. The widely used van der Waals–Lifshitz surface tension component is nearly the sum of γ_w and γ_{pz}, while the contact angle θ of a liquid L on a polymer surface P is given by the quadratic equation [42]:

$$\gamma^L(1+\cos\theta) = 2\left(\sqrt{\gamma_W^P\gamma_W^L} + \sqrt{\gamma_{pz}^P\gamma_{pz}^L} + \sqrt{\gamma_a^P\gamma_b^L} + \sqrt{\gamma_b^P\gamma_a^L}\right) \qquad (2.164)$$

The range of applications of the PSP approach is vast. It may be used for the prediction of phase equilibria in highly nonideal systems, for the solubility of solids, such as drugs, in various solvent systems over an extended range of temperatures, or for the sorption of solvent vapors on polymer surfaces. Various examples are reported in Ref. [41]. It may, in particular, be used for the complete thermodynamic characterization of solid surfaces via a novel method exposed in Ref. [42]. With this method, the PSPs of polymers may be obtained graphically, as shown in Figure 2.28. Inverse gas chromatography may now be used for the thermodynamic characterization of polymers or solid powders (drugs, dyes, clays, etc.).

Having the PSPs of a polymer, one may predict its Flory–Huggins interaction parameter, χ_{12}, not only with a solvent but also with another polymer. This then may be used for the prediction of polymer–polymer miscibility—a subject of high technological importance. It may, as an example, predict the effect of copolymer composition on its miscibility with another polymer or copolymer as shown in Figure 2.29. It may also use experimental or theoretical information on oligomers (cf. lower part of Figure 2.27) or low molecular weight analogs in order to characterize high polymers. The detailed formalism for these calculations may be found in Ref. [43].

Due to its simplicity, the PSP approach can be turned into an EoS approach along the lines of the previous paragraphs. This may lead to a very useful predictive model over a broad range of external conditions of temperature, pressure, and composition. Work is underway in our laboratory on these developments.

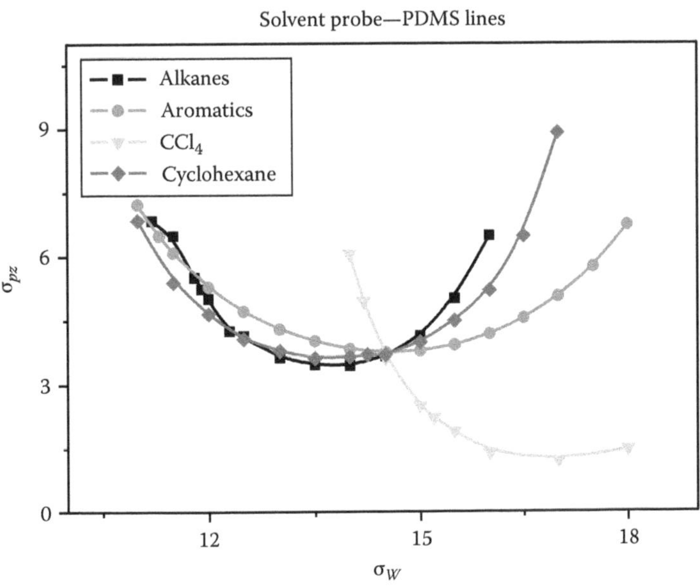

FIGURE 2.28 Determination of the poly(dimethyl siloxane) (PDMS) PSPs. The PSP lines for various non-hydrogen-bonded solute probes intersect at $\sigma_W = 14.5$ and $\sigma_{pz} = 3.7$ MPa$^{1/2}$. Symbols are experimental [137] χ_{12} parameters.

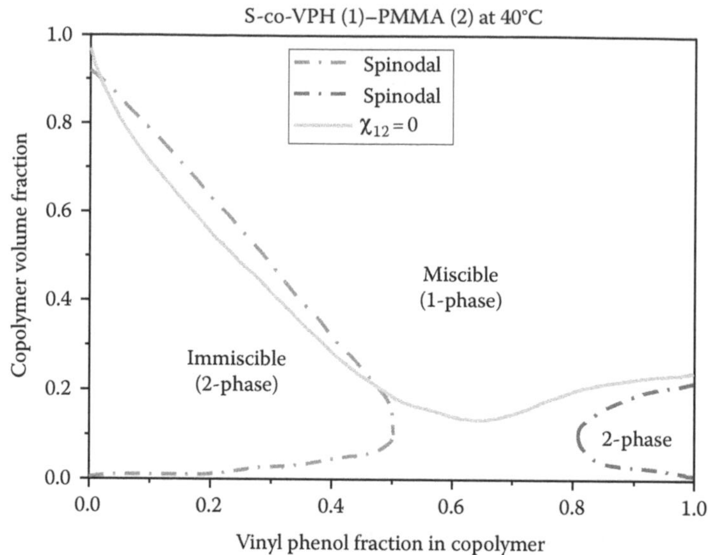

FIGURE 2.29 The miscibility map for the system styrene-co-vinyl phenol (S-co-VPH) (1)–poly(methyl methacrylate) (PMMA) (2) at 40°C as given by the spinodal lines (dash-dotted lines) and the line of zero χ_{12} interaction parameter (solid line) as functions of the vinyl phenol content (F_{VPH}) of the copolymer.

2.6 CONCLUSIONS

In the previous sections, we have exposed the rationale and the formalism of a unified treatment of hydrogen bonding in fluid systems ranging from small gas molecules up to high molecular weight polymeric glasses, in homogeneous as well as heterogeneous systems and interfaces. There is no limit, in principle, in the number and types of different hydrogen bonds that can be treated simultaneously. They can be intermolecular, intramolecular, or cooperative. They are incorporated into an EoS framework—the NRHB, and this facilitates the systematic study of the effect of density or pressure and temperature on the degree of hydrogen bonding or the fraction of hydrogen-bonded donor or acceptor groups in the system. The presented applications are only indicative and by no means exhaustive. Our more recent work on PSPs is a new type of predictive thermodynamic models, which combines elements from quantum mechanics, statistical thermodynamics, and QSPR approaches and has the potential to remove some of the key limitations of the successful COSMO-type thermodynamic models. Its capacity to handle surface characterization of polymers/solids is one of its promising features of particular interest to the colloids and surface science community. Much of its strength resides on the way hydrogen bonding is implemented in both concentrated and infinitely dilute systems.

REFERENCES

1. Guggenheim, E.A., *Mixtures*, Oxford University Press: Oxford, U.K., 1952.
2. Flory, P.J., *Principles of Polymer Chemistry*, Cornell University Press: Ithaca, NY, 1953.
3. Sanchez, I.C. and Lacombe, R.H., *J. Phys. Chem.*, 80, 2352, 1976.
4. Sanchez, I.C. and Lacombe, R.H., *Macromolecules*, 11, 1145, 1978.
5. Panayiotou, C.G., *J. Phys. Chem.*, 92, 2960, 1988.
6. Taimoori, M. and Panayiotou, C., *Fluid Phase Equilib.*, 205, 249, 2003.
7. Panayiotou, C., *J. Chem. Thermodyn.*, 35, 349, 2003.
8. Panayiotou, C., Pantoula, M., Stefanis, E., Tsivintzelis, I., and Economou, I.G., *Ind. Eng. Chem. Res.*, 43, 6592, 2004.
9. Abdel-Hady, E.E., *Polym. Degrad. Stabil.*, 80, 363, 2003.

10. Kilburn, D., Dlubek, G., Pionteck, J., Bamford, D., and Alam, M.A., *Polymer*, 46, 869, 2005.
11. Frank, H.S. and Wen, W.-Y., *Discuss. Faraday Soc.*, 24, 133, 1957.
12. Pimentel, G.C. and McClellan, A.L., *The Hydrogen Bond*, W.H. Freeman: San Francisco, CA, 1960.
13. LaPlanche, L.A., Thompson, H.B., and Rogers, M.T., *J. Phys. Chem.*, 69(5), 1482, 1965.
14. Vinogradov, S. and Linnell, R., *Hydrogen Bonding*, Van Nostrand Reinhold: New York, 1971.
15. Joesten, M.D. and Saad, L.J., *Hydrogen Bonding*, Marcel Dekker: New York, 1974.
16. Huyskens, P.L., *J. Am. Chem. Soc.*, 99, 2578, 1977.
17. Walter, H., Brooks, D., and Fisher, D. (Eds.), *Partitioning in Aqueous Two-Phase Systems*, Academic Press: New York, 1985.
18. Marsh, K. and Kohler, F., *J. Mol. Liquids*, 30, 13, 1985.
19. Kleeberg, H., Klein, D., and Luck, W.A.P., *J. Phys. Chem.*, 91, 3200, 1987.
20. Bourrel, M. and Schechter, R.S. (Eds.), *Microemulsions and Related Systems*, Marcel Dekker: New York, 1988.
21. Hobza, P. and Zahradnik, R., *Intermolecular Complexes*, Academia: Prague, Czech Republic, 1988.
22. Reichardt, C., *Solvent and Solvent Effects in Organic Chemistry*, VCH Verlag GmbH: Weinheim, Germany, 1988.
23. Burchard, W. and Ross-Murphy, S.B. (Eds.), *Physical Networks, Polymers, and Gels*, Elsevier Applied Science: London, U.K., 1990.
24. Coleman, M.M., Graf, J.F., and Painter, P.C., *Specific Interactions and the Miscibility of Polymer Blends*, Technomic: Lancaster, PA, 1991.
25. Maes, G. and Smets, J., *J. Phys. Chem.*, 97, 1818, 1993.
26. Sanchez, I.C. and Panayiotou, C. Equation of State Thermodynamics of Polymer and Related Systems, In *Models for Thermodynamic and Phase Equilibria Calculations*, S. Sandler (Ed.), Marcel Dekker: New York, 1994.
27. Acree, W.E., *Thermodynamic Properties of Nonelectrolyte Solutions*, Academic Press: New York, 1984.
28. Prausnitz, J.M., Lichtenthaler, R.N., and de Azevedo, E.G., *Molecular Thermodynamics of Fluid Phase Equilibria*, 2nd edn., Prentice-Hall: New York, 1986.
29. Heintz, A., *Ber. Bunsenges. Phys. Chem.*, 89, 172, 1985.
30. Panayiotou, C., *J. Solut. Chem.*, 20, 97, 1991.
31. Panayiotou, C. and Sanchez, I.C., *Macromolecules*, 24, 6231, 1991.
32. Levine, S. and Perram, J.W. In *Hydrogen Bonded Solvent Systems*, A.K. Covington and P. Jones (Eds.), Taylor & Francis: London, U.K., 1968.
33. Luck, W.A.P., *Angew. Chem.*, 92, 29, 1980.
34. Veytsman, B.A., *J. Phys. Chem.*, 94, 8499, 1990.
35. Panayiotou, C. and Sanchez, I.C., *J. Phys. Chem.*, 95, 10090, 1991.
36. Panayiotou, C. Hydrogen Bonding in Solutions: The Equation of State Approach, In *Handbook of Surface and Colloid Chemistry*, Birdi, K.S. (Ed.), CRC Press: Boca Raton, FL, 2003.
37. Panayiotou, C. Hydrogen Bonding and Non-randomness in Solution Thermodynamics, In *Handbook of Surface and Colloid Chemistry*, Birdi, K.S. (Ed.), CRC Press: Boca Raton, FL, 2009.
38. Panayiotou, C., Tsivintzelis, I., and Economou, I.G., *Ind. Eng. Chem. Res.*, 46, 2628, 2007.
39. Panayiotou, C., *Phys. Chem. Chem. Phys.*, 14, 3882, 2012.
40. Panayiotou, C., *J. Chem. Thermodynamics*, 51, 172, 2012.
41. Panayiotou, C., *J. Phys. Chem. B*, 116, 7302, 2012.
42. Panayiotou, C.G., *J. Chromatogr. A*, 1251, 194, 2012.
43. Panayiotou, C., *Polymer*, 54, 1621, 2013.
44. Yan, Q., Liu, H., and Hu, Y., *Fluid Phase Equilib.*, 218, 157, 2004.
45. Wilson, G., *J. Am. Chem. Soc.*, 86, 127, 1964.
46. Panayiotou, C. and Vera, J.H., *Can. J. Chem. Eng.*, 59, 501, 1981.
47. Panayiotou, C., *Fluid Phase Equilib.*, 237, 130, 2005.
48. Wang, W. and Vera, J.H., *Fluid Phase Equilib.*, 85, 1, 1993.
49. Vera, J.H., *Fluid Phase Equilib.*, 145, 217, 1998.
50. Hill, T.L., *An Introduction to Statistical Thermodynamics*, Dover Publications: New York, 1986.
51. Panayiotou, C. and Vera, J.H., *Polymer J.*, 14, 681, 1982.
52. Staverman, A.J., *Rec. Trav. Chim. Pays-Bas.*, 69, 163, 1950.
53. Fredenslund, A., Jones, R.L., and Prausnitz, J.M., *AIChE J.*, 21, 1086, 1975.
54. Fredenslund, A. and Sorensen, M.J. Group Contribution Estimation Methods, In *Models for Thermodynamic and Phase Equilibria Calculations*, Sandler, S. (Ed.), Marcel Dekker: New York, 1994.
55. Kemeny, S., Balog, G., Radnai, G., Savinsky, J., and Rezessy, G., *Fluid Phase Equilib.*, 54, 247, 1990.

56. Li, X. and Zhao, D., *J. Chem. Phys.*, 117, 6803, 2002.
57. Stefanis, E., Tsivintzelis, I., and Panayiotou, C., *Fluid Phase Equilib.*, 240, 144, 2006.
58. Missopolinou, D. and Panayiotou, C., *J. Phys. Chem. A*, 102, 3574, 1998.
59. Missopolinou, D., Ioannou, K., Prinos, I., and Panayiotou, C., *Z. Phys. Chem.*, 216, 905, 2002.
60. Missopolinou, D., Tsivintzelis, I., and Panayiotou, C., *Fluid Phase Equilib.*, 238, 204, 2005.
61. Missopolinou, D., Tsivintzelis, I., and Panayiotou, C., *Fluid Phase Equilib.*, 245, 89, 2006.
62. Abusleme, J.A. and Vera, J.H., *Can. J. Chem. Eng.*, 63, 845, 1985.
63. Tsivintzelis, I., Economou, I.G., and Kontogeorgis, G.M., *AIChE J.*, 55, 756, 2009.
64. Tsivintzelis, I., Economou, I.G., and Kontogeorgis, G.M., *J. Phys. Chem. B*, 113, 6446, 2009.
65. Tsivintzelis, I., Spyriouni, T., and Economou, I.G., *Fluid Phase Equilib.*, 253, 19, 2007.
66. Grenner, A., Tsivintzelis, I., Kontogeorgis, G.M., Economou, I.G., and Panayiotou, C., *Ind. Eng. Chem. Res.*, 47, 5636, 2008.
67. Tsivintzelis, I., Grenner, A., Economou, I.G., and Kontogeorgis, G.M., *Ind. Eng. Chem. Res.*, 47, 5651, 2008.
68. Tsivintzelis, I. and Kontogeorgis, G.M., *Fluid Phase Equilib.*, 280, 100, 2009.
69. Tsioptsias, C., Tsivintzelis, I., and Panayiotou, C., *Phys. Chem. Chem. Phys.*, 12, 4843, 2010.
70. von Solms, N., Michelsen, M.L., Passos, C.P., Derawi, S.O., and Kontogeorgis, G.M., *Ind. Eng. Chem. Res.*, 45, 5368, 2006.
71. Luck, W.A.P., *Angew. Chem. (Int. Ed. Engl.)*, 19, 28,1980.
72. Lien, T.R., A study of the thermodynamic excess functions of alcohol solutions by IR spectroscopy. Applications to chemical solution theory. PhD thesis, University of Toronto, Toronto, Ontario, Canada, 1972.
73. Fletcher, A.N. and Heller, C.A., *J. Phys. Chem.*, 71, 3742, 1967.
74. Aspiron, N., Hasse, H., and Maurer, G., *Fluid Phase Equilib.*, 186, 1, 2001.
75. Gupta, R.B. and Brinkley, R.L., *AIChE J.*, 44, 207, 1998.
76. Tsivintzelis, I., Bøgh, D., Karakatsani, E., and Kontogeorgis, G.M., *Fluid Phase Equilibria*, 365, 112, 2014.
77. Sørensen, J.M. and Arlt, W., *Liquid–Liquid Equilibrium Data Collection (Binary Systems)*, Vol. V, Part 1, DECHEMA Chemistry Data Series: Frankfurt/Main, Germany, 1995.
78. Raal, J.D., Best, D.A., Code, R.K., *J. Chem. Eng. Data*, 17, 211, 1972.
79. Tsonopoulos, C. and Wilson, G.M., *AIChE J.*, 29, 990, 1983.
80. Derawi, S.O., Kontogeorgis, G.M., Stenby, E.H., Haugum, T., and Fredheim, A.O., *J. Chem. Eng. Data*, 47, 169, 2002.
81. Dobroserdov, L.L. and Shakhanov, V.D., *Zh. Prikl. Khim. (Leningrad)*, 44, 45, 1971.
82. Baena, Y., Pinzon, J.A., Barbosa, H.J., and Martinez, F., *Phys. Chem. Liquids*, 42, 603, 2004.
83. Schmitt, W.J. and Reid, R.C., *J. Chem. Eng. Data*, 31, 204, 1986.
84. Wibawa, G., Hatano, R., Sato, Y., Takishima, S., and Masuoka, H., *J. Chem. Eng. Data*, 47, 1022, 2002.
85. Herskowltz, M. and Gottlleb, M., *J. Chem. Eng. Data*, 30, 233, 1985.
86. Nakajima, A., Fujiwara, H., and Hamada, F., *J. Polym. Sci. A-2*, 4, 507, 1966.
87. Freire, M.G., Carvalho, P.J., Gardas, R.L., Marrucho, I.M., Santos, L.M.N.B.F., Coutinho, J.A.P., *J. Phys. Chem. B*, 112, 1604, 2008.
88. Andrews, A.W. and Morcon, K.W., *J. Chem. Thermodyn.*, 3, 519, 1971.
89. Kehiaian, H.V., Sosnkowska-Kehiaian, K., and Hryniewicz, R., *J. Chim. Phys. Phys. Chim. Biol.*, 68, 929, 1971.
90. Ohji, H. and Tamura, K., *J. Chem. Thermodyn.*, 35, 1591, 2003.
91. De Torre, A., Velasco, I., Otin, S., and Gutierrez Losa, C., *J. Chem. Thermodyn.*, 12, 87, 1989.
92. Rowlinson, J.S. and Widom, B., *Molecular Theory of Capillarity*, Clarendon Press: Oxford, U.K., 1982.
93. Gibbs, J.W., *Collected Works*, Vol. I, Longmans, Green: New York, 1906.
94. Sprow, F.B. and Prausnitz, J.M., *Can. J. Chem. Eng.*, 45, 25, 1967.
95. Rusanov, A.I., *Pure Appl. Chem.*, 64, 111, 1992.
96. Hill, T.L., *J. Chem. Phys.*, 56, 526, 1952.
97. Toxvaerd, S., *J. Chem. Phys.*, 55, 3116, 1971.
98. Haile, J.M., Gray, C.G., and Gubbins, K.E., *J. Chem. Phys.*, 64, 2569, 1976.
99. van der Waals, J.D., *Z. Phys. Chem.*, 657, 1894.
100. Cahn, J.W. and Hilliard, J.E., *J. Chem. Phys.*, 30, 1121, 1958.
101. Widom, B., *J. Chem. Phys.*, 43, 3892, 1965.
102. Yang, A.J.M., Fleming, P.D., III, and Gibbs, J.H., *J. Chem. Phys.*, 64, 3732, 1976.
103. Bongiorno, V., Scriven, L.E., and Davis, H.T., *J. Colloid Interface Sci.*, 57, 462, 1976.

104. Evans, R., *Adv. Phys.*, 28, 143, 1979.
105. Poser, C.I. and Sanchez, I.C., *Macromolecules*, 14, 361, 1981.
106. Panayiotou, C., *Langmuir*, 18, 8841, 2002.
107. Panayiotou, C., *J. Colloid Interface Sci.*, 267, 418, 2003.
108. Daubert, T.E. and Danner, R.P., *Physical and Thermodynamic Properties of Pure Compounds: Data Compilation*, Hemisphere: New York, 2001.
109. Wu, S., *J. Macromol. Sci.*, C10, 1, 1974.
110. Papaioannou, D. PhD thesis, Department of Chemical Engineering, University of Thessaloniki, Thessaloniki, Greece, 1993.
111. Papaioannou, D. and Panayiotou, C., *J. Chem. Eng. Data*, 39, 457, 1994.
112. Hildebrand, J. and Scott, R.L., *Regular Solutions*, Prentice-Hall: Englewood Cliffs, NJ, 1962.
113. Hansen, C.M., *J. Paint Technol.*, 39, 104, 1967.
114. Barton, A.F.M., *Handbook of Solubility Parameters and Other Cohesion Parameters*, CRC Press: Boca Raton, FL, 1983.
115. Barton, A.F.M., *Polym. Sci. Technol. Pure Appl. Chem.*, 57, 905, 1985.
116. Hansen, C.M., *Hansen Solubility Parameters: A User's Handbook*, CRC Press: Boca Raton, FL, 2000.
117. Hansen, C.M., *Prog. Org. Coat.*, 51, 55, 2004.
118. Tehrani, J., *Am. Labor.*, 40hh–40mm, February 1993.
119. Bustamante, P., Peña, M.A., and Barra, J., *Int. J. Pharm.*, 174, 141, 1998.
120. Panayiotou, C., *Fluid Phase Equilib.*, 131, 21, 1997.
121. Panayiotou, C., *Fluid Phase Equilib.*, 236, 267, 2005.
122. Twu, C.H. and Gubbins, K.E., *Chem. Eng. Sci.*, 33, 863, 1978.
123. Nezbeda, I. and Pavlíček, J., *Fluid Phase Equilib.*, 116, 530, 1996.
124. Nezbeda, I. and Weingerl, U., *Mol. Phys.*, 99, 1595, 2001.
125. Karakatsani, E., Spyriouni, T., and Economou, I., *AIChE J.*, 51, 2328, 2005.
126. Hildebrand, J. and Scott, R.L., *The Solubility of Nonelectrolytes*, 3rd edn., Rheinhold, New York, 1950.
127. Kamlet, M.J. and Taft, R.W., *J. Am. Chem. Soc.*, 98, 377, 1976.
128. Kamlet, M.J. and Taft, R.W., *J. Am. Chem. Soc.*, 98, 2886, 1976.
129. Kamlet, M.J., Abboud, J.-L.M., and Taft, R.W., *Progress in Physical Organic Chemistry*, Wiley: New York, 1981.
130. Taft, R.W., Kamlet, M.J., Abraham, M.H., and Doherty, R.M., *J. Am. Chem. Soc.*, 107, 3105, 1985.
131. Kamlet, M.J., Doherty, R.M., Abboud, J.-L.M., Abraham, M.H., and Taft, R.W., *Chemtech*, 16, 566–574, 1986.
132. Abraham, M.H., *Chem. Soc. Rev.*, 22, 73, 1993.
133. Abraham, M.H., Ibrahim, A., and Zissimos, A.M., *J. Chromatogr. A*, 1037, 29, 2004.
134. Klamt, A. and Eckert, F., *Fluid Phase Equilib.*, 172, 43, 2000.
135. Klamt, A., *COSMO-RS from Quantum Chemistry to Fluid Phase Thermodynamics and Drug Design*, Elsevier: Amsterdam, the Netherlands, 2005.
136. COSMObase Ver. C2.1 Rev. 01.06. COSMOlogic GmbH & Co., K.G., Leverkusen, Germany, 2006.
137. Tian, M. and Munk, P., *J. Chem. Eng. Data*, 39, 742, 1994.

3 Thermodynamics of Polymer Solutions

Georgios M. Kontogeorgis and Nicolas von Solms

CONTENTS

3.1 INTRODUCTION

Knowledge of phase behavior (thermodynamics) of polymer solutions is important for the design of many processes and products, including many specific applications in colloid and surface chemistry.

Among the many applications, we can mention the following:

1. The design of separations for the removal of unreacted monomers, colorants, by-products, and additives after solution or emulsion polymerizations[1]

2. Selection of appropriate mixed solvents for paints and coatings, which can meet strict production and environmental criteria (fewer VOCs, water-based products)[2,3]
3. Numerous applications in the coatings industry, for example, in the control of emissions during production and application, the effect of temperature, and swelling of film or sorption of gases and chemicals from the atmosphere[3–6]
4. Novel recycling methods for polymer waste based on physicochemical methods, the so-called selective dissolution[7,8]
5. Design of advanced materials based on compatible polymer blends[9] or novel structures, for example, star or hyberbranched polymers[10,11]
6. Permeabilities of gases in the flexible polymeric pipelines used, for example, in the North Sea for transporting hydrocarbons from the seabed to the surface[12,13]
7. Use of CO_2 as user-friendly solvent for many polymer-related applications including those involving paints and coatings[14–16]
8. Simultaneous representation of bulk and surface thermodynamic properties[17]
9. Migration of plasticizers from PVC[18–21]
10. Separation of proteins based on the so-called aqueous two-phase systems using polymers like PEG or dextran[1,22,23]
11. Choice of suitable solvents for polymers and especially copolymers used as stabilizers in colloidal dispersions

The aforementioned list shows some of the many applications where polymer thermodynamics plays a key role. Polymer solutions and blends are complex systems: frequent existence of liquid–liquid equilibria (LLE) (UCST, LCST, closed loop, etc.), the significant effect of temperature and polymer molecular weight including polydispersity in phase equilibria, free-volume (FV) effects, and other factors may cause difficulties. For this reason, many different models have been developed for polymer systems and often the situation may seem rather confusing to the practicing engineer.

The choice of a suitable model will depend on the actual problem and depends, specifically, on

- Type of mixture (solution or blend, binary or multicomponent, etc.)
- Type of phase equilibria (VLE, LLE, SLLE, gas solubility, etc.)
- Conditions (temperature, pressure, concentration)
- Type of calculations (accuracy, speed, yes/no answer or complete design, etc.)

This chapter will present key tools in polymer thermodynamics at three different levels from the "simpler" but also "more easy-to-use" methods up to the "more advanced" but also "more complex" and potentially more accurate approaches:

1. Rules of thumb for choosing solvents including a discussion of the Flory–Huggins (FH) approach (Section 3.2)
2. FV activity coefficient models based on UNIFAC (Section 3.3), which are often predictive, and have found widespread applications at low pressures
3. The advanced statistical-associating fluid theory (SAFT) (Section 3.4), which is a theoretically based equation of state, which can be rigorously applied, unlike the activity coefficient models, at both low and high pressures

3.2 CHOICE OF SOLVENTS

A summary of some rules of thumb for predicting polymer–solvent miscibility, with focus on the screening of solvents for polymers, is presented here. These rules are based on well-known concepts of thermodynamics (activity coefficients, solubility parameters) and some specific ones for polymers (the FH parameter). It can be roughly stated that a chemical (1) will be a good solvent for a specific polymer (2), or in other words, the two compounds will be miscible if one (or more) of the following "rules of thumb" are valid:

1. If the polymer and the solvent have similar hydrogen bonding degrees:

$$|\delta_1 - \delta_2| \leq 1.8\,(\text{cal/cm}^3)^{1/2} \tag{3.1}$$

   where δ is the solubility parameter.
2. If the polymer and the solvent have very different hydrogen bonding degrees:

$$\sqrt{4(\delta_{d1} - \delta_{d2})^2 + (\delta_{p1} - \delta_{p2})^2 + (\delta_{h1} - \delta_{h2})^2} \leq R \tag{3.2}$$

   where R is the Hansen-solubility parameter sphere radius and the subscripts d, p, h in the solubility parameter denote the dispersion, polar, and hydrogen bonding contributions, respectively.
3. $\Omega_1^\infty \leq 6$ (the lower the infinite dilution activity coefficient of the solvent, the greater is the miscibility of a chemical). Values of the infinite dilution activity coefficient above 10 indicate nonmiscibility. In the intermediate region, it is difficult to conclude if the specific chemical is a solvent or a nonsolvent.
4. $\chi_{12} \leq 0.5$ (the lower the FH parameter value, the greater is the miscibility). Values much above 0.5 indicate nonmiscibility.

3.2.1 Rules of Thumb Based on Solubility Parameters

They are widely used. The starting point (in their derivation and understanding) is the equation for the Gibbs free energy of mixing:

$$\Delta G^{mix} = \Delta H^{mix} - T\Delta S^{mix} \tag{3.3}$$

A negative value implies that a solvent–polymer system forms a homogeneous solution, that is, the two components are miscible. Since the contribution of the entropic term ($-T\Delta S$) is always negative, it is the heat of mixing term that determines the sign of the Gibbs energy. The heat of mixing can be estimated from various theories, for example, the Hildebrand regular solution theory for nonpolar systems, which is based on the concept of the solubility parameter. For a binary solvent (1)–polymer (2) system, according to the regular solution theory,

$$\Delta H^{mix} = \varphi_1\varphi_2 V(\delta_1 - \delta_2)^2 \tag{3.4}$$

where φ_i is the so-called volume fraction of component i. This is defined via the mole fractions x_i and the molar volumes V_i as (for binary systems):

$$\varphi_i = \frac{x_i V_i}{x_i V_i + x_j V_j} \tag{3.5}$$

According to Equation 3.4, the heat of mixing is always positive. For some systems with specific interactions (hydrogen bonding), the heat of mixing can be negative and Equation 3.4 does not hold.

Thus, the regular solution theory is strictly valid for nonpolar–slightly polar systems, without any specific interactions.

According to Equations 3.3 and 3.4, if solvent and polymer have the same solubility parameters, the heat of mixing is zero and they are therefore miscible at all proportions. The lower the solubility parameter difference, the larger the tendency for miscibility. Many empirical rules of thumb have been proposed based on this observation. Seymour[24] suggests that if the difference of solubility parameters is below 1.8 $(cal/cm^3)^{1/2}$, then polymer and solvent are miscible (Equation 3.1).

Similar rules can be applied for mixed solvent–polymer systems, which are very important in many practical applications, for example, in the paints and coatings industry and for the separation of biomolecules using aqueous two-phase systems.

The solubility parameter of a mixed solvent is given by the equation:

$$\delta = \sum_i \varphi_i \delta_i \tag{3.6}$$

Barton[25,26] provides empirical methods based on solubility parameters for ternary solvent systems.

Charles Hansen introduced the concept of 3D solubility parameters, which offers an extension of the regular solution theory to polar and hydrogen bonding systems. Hansen observed that when the solubility parameter increments of the solvents and polymers are plotted in 3D plots, then the "good" solvents lie approximately within a sphere of radius R (with the polymer being in the center). This can be mathematically expressed as shown in Equation 3.2. The quantity under the square root is the distance between the solvent and the polymer. Hansen found empirically that a universal value 4 should be added as a factor in the dispersion term to approximately attain the shape of a sphere. This universal factor has been confirmed by many experiments. Hansen[27] in his book provides a review of the method together with extensive tables of parameters.

The Hansen method is very valuable. It has found widespread use particularly in the paints and coatings industry, where the choice of solvents to meet economical, ecological, and safety constraints is of critical importance. It can explain some cases in which polymer and solvent solubility parameters are almost perfectly matched and yet the polymer will not dissolve. The Hansen method can also predict cases where two nonsolvents can be mixed to form a solvent. Still, the method is approximate; it lacks the generality of a full thermodynamic model for assessing miscibility and requires some experimental measurements. The determination of R is typically based on visual observation of solubility (or not) of 0.5 g polymer in 5 cm^3 solvent at room temperature. Given the concentration and the temperature dependence of phase boundaries, such determination may seem a bit arbitrary. Still the method works out pretty well in practice, probably because the liquid–liquid boundaries for most polymer–solvent systems are fairly "flat."

3.2.2 Rule of Thumb Based on the Infinite Dilution Activity Coefficient

Since in several practical cases concerning polymer–solvent systems, the "solvent" is only present in very small (trace) amounts, the so-called infinite dilution activity coefficients are important. On a molar and weight basis, they are defined as follows:

$$\begin{aligned} \gamma_i^\infty &= \lim_{x_i \to 0} \gamma_i \\ \Omega_i^\infty &= \lim_{w_i \to 0} \left(\frac{x_i \gamma_i}{w_i} \right) = \gamma_1^\infty \frac{M_{pol}}{M_{solv}} \end{aligned} \tag{3.7}$$

The latter part of the equation is valid for a binary solvent–polymer solution and γ_1^∞ is the infinite dilution activity coefficient of the solvent.

The weight-based infinite dilution activity coefficient, Ω_1^∞, which can be determined experimentally from chromatography, is a very useful quantity for determining good solvents. As mentioned in the previous section, low values (typically below 6) indicate good solvents, while high values (typically above 10) indicate poor solvents according to rules of thumb discussed by several investigators.[28,29] The derivation of this rule of thumb is based on the FH model.

This method for solvent selection is particularly useful, because it avoids the need for direct liquid–liquid measurements and it makes use of the existing databases of solvent infinite dilution activity coefficients, which is quite large (e.g., the DECHEMA and DIPPR databases[30,31]).

Moreover, in the absence of experimental data, existing thermodynamic models (such as the FH, the Entropic-FV, and the UNIFAC-FV discussed later) can be used to predict the infinite dilution activity coefficient. Since, in the typical case today, existing models perform much better for VLE and activity coefficient calculations than directly for LLE calculations, this method is quite valuable and successful, as shown by sample results in Table 3.1.

This rule of thumb makes use of either experimental or predicted, by a model, infinite dilution activity coefficients. However, the results depend not only on the accuracy of the model, but also on the rule of thumb, which in turns depends on the assumptions of the FH approach. A thermodynamically more correct method is to employ the activity–concentration diagram. The maximum indicates phase split, while a monotonic increase of activity with concentration indicates a single liquid phase (homogeneous solutions).

3.2.3 Rule of Thumb Based on the Flory–Huggins Model

The FH model for the activity coefficient, proposed in the early 1940s by Flory and Huggins,[32,33] is a famous Gibbs free energy expression for polymer solutions. For binary solvent–polymer solutions and assuming that the parameter of the model, the so-called FH interaction parameter χ_{12}, is constant, the activity coefficient is given by the equation:

$$\ln \gamma_1 = \ln \frac{\varphi_1}{x_1} + 1 - \frac{\varphi_1}{x_1} + \chi_{12}\varphi_2^2 = \ln \frac{\varphi_1}{x_1} + \left(1 - \frac{1}{r}\right)\varphi_2 + \chi_{12}\varphi_2^2 \tag{3.8}$$

where

φ_i can be volume or segment fractions

r is the ratio of the polymer volume to the solvent volume V_2/V_1 (approximately equal to the degree of polymerization)

Appendix 3.A presents the general expression for the FH model suitable for multicomponent systems.

Using standard thermodynamics and Equation 3.8, it can be shown that for high molecular weight polymer–solvent systems, the polymer critical concentration is close to zero and the interaction parameter has a value equal to 0.5. Thus, a good solvent (polymer soluble in the solvent at all proportions) is obtained if $\chi_{12} \leq 0.5$, while values greater than 0.5 indicate poor miscibility. Since the FH model is only an approximate representation of the physical picture and particularly the FH parameter is often not a constant at all, this empirical rule is certainly subject to some uncertainty. Nevertheless, it has found widespread use and its conclusions are often in good agreement with experiment.

TABLE 3.1
Choice of Suitable Solvents Using the Ω_1^∞-Rule of Thumb for PBMA Systems (See Section 3.2)

Solvent	S/NS	EFV	UFV	GCFL
Hexane	NS	7.1	7.0	10.7
n-Octane	NS	6.7	6.3	10.4
n-Decane	NS	6.5	6.0	10.7
n-Dodecane	NS	6.6	6.0	11.3
n-Hexadecane	NS	6.8	6.1	13.2
Toluene	S	3.2	4.4	4.7
Xylene	S	2.3	3.6	5.7
Methylene dichloride	S	3.3	2.5	3.0
Chloroform	S	1.9	2.1	1.7
Carbon tetrachloride	S	2.2	2.2	2.9
Ethylene dichloride	S	3.5	3.0	—
Trichloroethylene	S	2.5	2.9	33.9
Chlorobenzene	S	2.5	3.0	3.0
o-Dichlorobenzene	S	1.3	2.5	2.7
Acetone	S	10.9	14.2	11.1
MEK	S	8.4	10.5	8.2
MIBK	S	6.3	7.7	5.7
Acetophenone	S	8.1	9.3	8.6
Ethyl acetate	S	6.7	6.7	60.3
Butyl acetate	S	5.3	5.1	31.4
Diethyl ether	S	5.2	5.8	11.6
THF	S	3.8	4.0	—
1,4-Dioxane	S	4.1	4.4	159.4
Methanol	NS	43.7	57.7	35.7
Ethanol	NS	29.2	31.3	17.3
1-Butanol	NS	18.1	17.1	8.1
Cyclohexanol	NS	24.3	20.1	3.0
Ethylene glycol	NS	277.8	—	15,947.0
Propylene glycol	NS	212.6	—	1,879.2
1,3-Butanediol	NS	158.5	—	525.5
Glycerol	NS	294.6	—	2,282.4
Isopropanol	NS	23.4	21.6	10.6
Isobutanol	S	19.0	17.9	7.9
Diethylene glycol	NS	240.1	—	2,470.4
Dipropylene glycol	NS	127.9	945.7	287.8
Nitromethane	NS	16.7	17.2	—
1-Nitropropane	S	4.7	5.2	—
N,N-dimethylformamide	S	3.8	—	—

Source: Modified from Lindvig, Th. et al., *AIChE J.*, 47(11), 2573, 2001.
Notes: EFV, Entropic-FV; UFV, UNIFAC-FV; GCFL, GC-Flory; S, solvent; NS, nonsolvent (experimental observation).

There are several, still rather obscure issues about the FH model, which we summarize here together with some recent developments:

1. The FH parameter is typically not a constant and should be estimated from experimental data. Usually, it varies with both temperature and concentration, which renders the FH model useful only for describing experimental data. It cannot be used for predicting phase equilibria for systems for which no data is available. Moreover, when fitted to the critical solution temperature, the FH model cannot yield a good representation of the whole shape of the miscibility curve with a single parameter.
2. Accurate representation of miscibility curves is possible with the FH model using suitable (rather complex) equations for the temperature and the concentration dependence of the FH parameter.[34]
3. In some cases, a reasonable value of the FH parameter can be estimated using solubility parameters via the equation:

$$\chi_{12} = \chi_s + \chi_h = 0.35 + \frac{V_1}{RT}(\delta_1 - \delta_2)^2 \tag{3.9}$$

   Equation 3.9, without the empirical 0.35 factor, is derived from the regular solution theory. The constant 0.35 is added for correcting for the deficiencies of the FH combinatorial term. These deficiencies become evident when comparing experimental data for athermal polymer and other asymmetric solutions to the results obtained with the FH model. A systematic underestimation of the data is observed, as discussed extensively in the literature,[28,35] which is often attributed to the inability of the FH model in accounting for the FV differences between polymers and solvents or between compounds differing significantly in size such as *n*-alkanes with very different chain lengths. The term, which contains the "0.35 factor," corrects in an empirical way for these FV effects. However, and although satisfactory results are obtained in some cases, we cannot generally recommend Equation 3.9 for estimating the FH parameter. Moreover, for many nonpolar systems with compounds having similar solubility parameters, the empirical factor 0.35 should be dropped.
4. Lindvig et al.[36] proposed an extension of the FH equation using the Hansen solubility parameters for estimating activity coefficients of complex polymer solutions.

$$\ln\gamma_1 = \ln\frac{\varphi_1}{x_1} + 1 - \frac{\varphi_1}{x_1} + \chi_{12}\varphi_2^2$$

$$\chi_{12} = 0.6\frac{V_1}{RT}\left[(\delta_{d1} - \delta_{d2})^2 + 0.25(\delta_{p1} - \delta_{p2})^2 + 0.25(\delta_{h1} - \delta_{h2})^2\right] \tag{3.10}$$

   In order to achieve that, Lindvig et al.[36] have, as shown in Equation 3.10, employed a universal correction parameter, which has been estimated from a large number of polymer–solvent VLE data. Very good results are obtained, especially when the volume-based combinatorial term of FH is employed, also for ternary polymer–solvent systems.[37]
5. Based on the FH model, several techniques have been proposed for interpreting and for correlating experimental data for polymer systems, for example, the so-called Schultz–Flory

(SF) plot. Schultz and Flory[38] have developed, starting from the FH model, the following expression, which relates the critical solution temperature (CST), with the theta temperature and the polymer molecular weight:

$$\frac{1}{\text{CST}} = \frac{1}{\Theta}\left[1 + \frac{1}{\psi}\left(\frac{1}{\sqrt{r}} + \frac{1}{2r}\right)\right] \tag{3.11}$$

where

$\psi = ((1/2) - \chi_s)$ is the entropic parameter of the FH model (Equation 3.9)
r is the ratio of molar volumes of the polymer to the solvent

This parameter is evidently dependent on the polymer's molecular weight. The SF plot can be used for correlating data of critical solution temperatures for the same polymer–solvent system, but at different polymer molecular weights. This can be done, as anticipated from Equation 3.11, because the plot of 1/CST against the quantity in parentheses in Equation 3.11 is linear. The SF plot can also be used for predicting CST for the same system but at different molecular weights than those used for correlation as well as for calculating the theta temperature and the entropic part of the FH parameter. It can be used for correlating CST/molecular weight data for both the UCST and LCST areas. Apparently different coefficients are needed.

3.3 FREE-VOLUME ACTIVITY COEFFICIENT MODELS

3.3.1 The Free-Volume Concept

The FH model provides a first approximation for polymer solutions. Both the combinatorial and the energetic terms require substantial improvement. Many authors have replaced the random van-Laar energetic term by a nonrandom local-composition term such as those of the UNIQUAC, NRTL, and UNIFAC models. The combinatorial term should be modified to account for the FV differences between solvents and polymers.

The improvement of the energetic term of FH equation is important. Local-composition terms like those appearing in NRTL, UNIQUAC, and UNIFAC provide a flexibility, which cannot be accounted for by the single-parameter van Laar term of FH. However, the highly pronounced FV effects should always be accounted for in polymer solutions.

The concept of FV is rather loose, but still very important. Elbro[28] demonstrated, using a simple definition for the FV (Equation 3.12), that the FV percentages of solvents and polymers are different. In the typical case, the FV percentage of solvents is greater (40%–50%) than that of polymers (30%–40%). There are two notable exceptions to this rule; water and PDMS: water has lower FV than other solvents and closer to that of most of the polymers, while PDMS has quite a higher FV percentage, closer to that of most solvents. LCST is, as expected, related to the FV differences between polymers and solvents. As shown by Elbro,[28] the larger the FV differences, the lower is the LCST value (the larger the area of immiscibility). For this reason, PDMS solutions have a LCST, which are located at very high temperatures.

Many mathematical expressions have been proposed for the FV. One of the simplest and successful equations is

$$V_f = V - V^* = V - V_w \tag{3.12}$$

originally proposed by Bondi[39] and later adopted by Elbro et al.[35] and Kontogeorgis et al.[40] in the so-called Entropic-FV model (described later). According to this equation, FV is just the "empty" volume available to the molecule when the molecules' own (hard-core or closed-packed V^*) volume is subtracted.

The FV is not the only concept, which is loosely defined in this discussion. The hard-core volume is also a quantity difficult to define and various approximations are available. Elbro et al.[35] suggested using $V^* = V_w$, that is, equal to the van der Waals volume (V_w), which is obtained from the group increments of Bondi and is tabulated for almost all existing groups in the UNIFAC tables. Other investigators[41] interpreted somewhat differently the physical meaning of the hard-core volume in the development of improved FV expressions for polymer solutions, which employ Equation 3.12 as basis, but with V^* values higher than V_w (about 1.2–1.3V_w).

3.3.2 Entropic-FV Model

The original UNIFAC model does not account for the FV differences between solvents and polymers and, as a consequence of that, it highly underestimates the solvent activities in polymer solutions. One of the most successful and earliest such models for polymers is the UNIFAC-FV by Oishi and Prausnitz.[42] The UNIFAC-FV model was developed for solvent activities in polymers, but it cannot be successfully applied to LLE.

A similar to UNIFAC-FV but somewhat simpler approach, which can be readily extended to multicomponent systems and LLE, is the so-called Entropic-FV model proposed by Elbro et al.[35] and Kontogeorgis et al.[40]:

$$\ln \gamma_i = \ln \gamma_i^{comb-FV} + \ln \gamma_i^{res}$$

$$\ln \gamma_i^{comb-FV} = \ln \frac{\phi_i^{FV}}{x_i} + 1 - \frac{\phi_i^{FV}}{x_i}$$

$$\phi_i^{FV} = \frac{x_i V_{i,FV}}{\sum_i x_j V_{j,FV}} = \frac{x_i (V_i - V_{wi})}{\sum_j x_j (V_j - V_{wj})}$$

$$\ln \gamma_i^{res} \rightarrow \text{UNIFAC} \tag{3.13}$$

As can been seen from Equation 3.13, the FV definition given by Equation 3.12 is employed. The combinatorial term of Equation 3.13 is very similar to that of FH. However, instead of volume or segment fractions, FV fractions are used. In this way, both combinatorial and FV effects are combined into a single expression. The combinatorial–FV expression of the Entropic-FV model is derived from statistical mechanics, using a suitable form of the generalized van der Waals partition function.

The residual term of Entropic-FV is taken by the so-called new or linear UNIFAC model, which uses a linear-dependent parameter table[43]:

$$a_{mn} = a_{mn,1} + a_{mn,2}(T - T_o) \tag{3.14}$$

This parameter table has been developed using the combinatorial term of the original UNIFAC model. As with UNIFAC-FV, no parameter re-estimation has been performed. The same group parameters are used in the "linear UNIFAC" and in the Entropic-FV models.

A common feature for both UNIFAC-FV and Entropic-FV is that they require values for the volumes of solvents and polymers (at the different temperatures where application is required). This can be a problem in those cases where the densities are not available experimentally and have to be estimated using a predictive group-contribution or other method, for example, GCVOL[44,45] or van Krevelen methods. These two estimation methods perform quite well and often similarly even for low molecular weight compounds or oligomers such as plasticizers.

Both UNIFAC-FV and Entropic-FV, especially the former, are rather sensitive to the density values used for the calculations of solvent activities.

As already mentioned, the Entropic-FV model has been derived from the van der Waals partition function. The similarity of the model with the van der Waals equation of state $P = RT/(V - b) - a/V^2$ becomes apparent if the van der Waals equation of state is written (when the classical van der Waals one fluid mixing and classical combining rules are used) as an activity coefficient model:

$$\ln \gamma_i = \ln \gamma_i^{comb-FV} + \ln \gamma_i^{res} = \left(\ln \frac{\varphi_i^{FV}}{x_i} + 1 - \frac{\varphi_i^{FV}}{x_i} \right) + \left(\frac{V_i}{RT} (\delta_i - \delta_i)^2 \varphi_j^2 \right)$$

$$\varphi_i^{FV} = \frac{x_i (V_i - b_i)}{\sum_j x_j (V_j - b_j)} \tag{3.15}$$

$$\delta_i = \frac{\sqrt{a_i}}{V_i}$$

where φ_i is the volume fraction as defined in Equation 3.5. The first term in Equation 3.15 is the same as in Entropic-FV with $V_w = b$, while the latter term is a regular solution theory or van Laar-type term.

3.3.3 Results and Discussion

Table 3.2 presents an overview of the results with Entropic-FV model for different applications together with the corresponding references. Selected results are shown in Tables 3.3 through 3.9 and Figures 3.1 and 3.2.

The most important general conclusions can be summarized as following:

1. Satisfactory predictions are obtained for solvent activities, even at infinite dilution, and for nonpolar, as well as for complex polar and hydrogen bonding systems including solutions of interest to paints and coatings. Rather satisfactory predictions are also achieved when mixed solvents and copolymers are present.

TABLE 3.2
Applications of the Entropic-FV Model

Application	References
VLE binary solutions	[40,41]
VLE complex polymers–solvents	[48]
VLE ternary polymer–solvents	[37]
Paints	[61]
Dendrimers	
VLE copolymers	[47]
VLE athermals systems	[28,35,50,62]
SLE hydrocarbons	[51,52]
SLE polymer–solvents	[49]
Comparison with other models	[40,46,53,60]
LLE polymer–solvents	[58]
LLE ternary polymer–mixed solvents	[8]
Polymer blends	[59]
EFV + UNIQUAC	

TABLE 3.3
Prediction of the Solubility for Characteristic Polymer–Solvent Systems Using the Ω_1^∞ Rule of Thumb and Two FV Models for Solvent Selection

System	Experiment	Entropic-FV	UNIFAC-FV
PBMA/nC_{10}	NS	6.5 (–)	6.1 (–)
PBMA/xylene	S	2.3 (S)	3.6 (S)
PBMA/$CHCl_3$	S	1.9 (S)	9.1 (NS)
PBMA/acetone	S	0.2 (NS)	14.1 (NS)
PBMA/ethyl acetate	S	6.7 (–)	6.7 (–)
PBMA/ethanol	NS	29.2 (NS)	31.3 (NS)
PMMA/acetone	S	10.0 (NS)	16.5 (NS)
PMMA/ethyl acetate	S	6.6 (–)	8.4 (NS)
PMMA/butanol	NS	26.8 (NS)	14.4 (NS)
PEMA/MEK	S	8.1 (NS)	11.7 (NS)
PEMA/diethyl ether	S	5.8 (S)	7.6 (–)
PEMA/nitropropane	NS	4.5 (S)	1.4 (S)
PVAc/hexane	NS	38.7 (NS)	38.6 (NS)
PVAc/methanol	S	18.9 (NS)	19.4 (NS)
PVAc/ethanol	NS	15.2 (NS)	38.9 (NS)
PVAc/nitromethane	S	3.9 (S)	3.8 (S)
PVAc/THF	S	8.4 (NS)	5.6 (S)

Notes: S, good solvent; NS, nonsolvent; –, no answer according to the rule of thumb.

TABLE 3.4
Prediction of Infinite Dilution Activity Coefficients for Polyisoprene (PIP) Systems with Two Predictive Group Contribution Models

PIP Systems	Exper. Value	Entropic-FV	UNIFAC-FV
+Acetonitrile	68.6	47.7 (31%)	52.3 (24%)
+Acetic acid	37.9	33.5 (12%)	17.7 (53%)
+Cyclohexanone	7.32	5.4 (27%)	4.6 (38%)
+Acetone	17.3	15.9 (8%)	13.4 (23%)
+MEK	11.4	12.1 (6%)	10.1 (12%)
+Benzene	4.37	4.5 (2.5%)	4.4 (0%)
+1,2-Dichloroethane	4.25	5.5 (29%)	6.5 (54%)
+CCl_4	1.77	2.1 (20%)	1.8 (0%)
+1,4-Dioxane	6.08	6.3 (4%)	5.9 (2%)
+Tetrahydrofurane	4.38	4.9 (14%)	3.9 (10%)
+Ethylacetate	7.47	7.3 (2%)	6.6 (11%)
+n-Hexane	6.36	5.1 (20%)	4.6 (27%)
+Chloroform	2.13	3.00 (41%)	2.6 (20%)

Experimental values and calculations are at 328.2 K.

TABLE 3.5
Average Absolute Deviations between Experimental and Calculated Activity Coefficients of Paint-Related Polymer Solutions Using the FH/Hansen Method and Two FV Models

Model	% AAD (Systems in Database)	% AAD Araldite 488	% AAD Eponol-55
FH/Hansen, Volume (Equation 3.10)	22	31	28
Entropic-FV	35	34	30
UNIFAC-FV	39	119	62

Source: Adapted from Lindvig, Th. et al., *Fluid Phase Equilib.*, 203, 247, 2002.
The second column presents the systems used for optimization of the universal parameter (solutions containing acrylates and acetates). The last two columns show predictions for two epoxy resins.

TABLE 3.6
Mean Percentage Deviations between Experimental and Calculated Activity Coefficients of Solvents in Various Nearly Athermal Solutions

% AAD Infinite Dilution Conditions γ_1^∞ (Ω_1^∞ for Polymers)	Entropic-FV	UNIFAC-FV	Flory-FV
Short *n*-alkanes/long alkanes	8	15	20
Short branched, cyclic alkanes/long alkanes	10	17	20
Alkanes/polyethylene	9	23	19
Alkanes/polyisobutylene	16	12	38
Organic solvent/PDMS, PS, PVAc	20	29	26
Overall	13	19	25

Source: Kouskoumvekaki, I. et al., *Fluid Phase Equilib.*, 202(2), 325, 2002. With permission.

TABLE 3.7
Mean Percentage Deviations between Experimental and Calculated Activity Coefficients of Heavy Alkanes Solutes in Alkane Solvents

	Entropic-FV	UNIFAC-FV	Flory-FV
% AAD infinite dilution conditions, γ_2^∞			
Symmetric long alkanes/short alkanes	36	47	10
Medium asymmetric long alkanes/short alkanes	34	48	12
Asymmetric long alkanes/short alkanes	44	54	37
Overall	38	50	20
% AAD finite concentrations, γ_2			
Symmetric long alkanes/short alkanes	14	17	6
Medium asymmetric long/short alkanes	23	31	11
Asymmetric long/short alkanes	40	55	16
Overall	26	34	11

Source: Kouskoumvekaki, I. et al., *Fluid Phase Equilib.*, 202(2), 325, 2002. With permission.

TABLE 3.8
Average Absolute Logarithmic Percentage Deviations between Experimental and Predicted Equilibrium Pressures and Average Absolute Deviation (×100) between Calculated and Experimental Vapor Phase Compositions (Mole Fractions) for Various Ternary Polymer–Mixed Solvent Systems

Sys. No.	Variable	SAFT	EFV/UQ	FH	Pa-Ve	EFV/UN	UFV	GCFL	GCLF	FHHa
1	P	11	6	6	6	2	1	21	8	2
	y	4	3	3	3	3	3	3	3	3
2	*P*	4	2	14	8	2	2	—	2	11
	y	5	2	5	1	2	2	—	4	5
3	*P*	—	—	—	—	3	3	12	5	18
	y	—	—	—	—	3	3	2	4	4
4	*P*	—	—	—	—	4	4	13	9	5
	y	—	—	—	—	18	18	19	19	15
5	*P*	14	16	11	4	17	19	93	5	52
	y	17	17	16	11	17	13	2	18	14

Sources: Katayama, T. et al., *Kagaku Kogaku*, 35, 1012, 1971; Matsumara, K. and Katayama, T., *Kagaku Kogaku*, 38, 388, 1974; Tanbonliong, J.O. and Prausnitz, J.M., *Polymer*, 38, 5775, 1997.

Based on the results shown by Lindvig et al.[37]

Notes: 1, PS–toluene–ethylbenzene at 303 K; 2, PS–toluene–cyclohexane at 303 K; 3, PVAc–acetone–ethyl acetate at 303 K; 4, PVAc–acetone–methanol at 303 K; 5, PS–chloroform–carbon tetrachloride at 323.15 K.

TABLE 3.9
Prediction of Liquid–Liquid Equilibria for Polymer–Solvent Systems with Various Thermodynamic Models

Polymer System	Molecular Weight	Entropic-FV	N0065w UNIFAC	GC-Flory
PS/acetone	4,800	84	21	75
PS/acetone	10,300	98	8	42
PS/cyclohexane	20,400	38	62	—
PS/cyclohexane	37,000	26	59	—
PS/cyclohexane	43,600	24	63	—
PS/cyclohexane	89,000	11	60	—
PS/cyclohexane	100,000	15	62	—
PS/cyclopentane	97,200	27	105	—
PS/cyclopentane	200,000	12	103	—
HDPE/*n*-butyl acetate	13,600	10	82	72
HDPE/*n*-butyl acetate	20,000	22	97	70
HDPE/*n*-butyl acetate	61,100	29	107	71
PMMA/1-chloro butane	34,760	53	—	—
PBMA/*n*-pentane	11,600	Hourglass	—	—
PBMA/*n*-octane	11,600	155	—	—

Absolute difference between experimental and predicted UCST for several polymer solutions using various models (based on results from Ref. [58]). All results are predictions. The three last models are based on group contributions. The new UNIFAC model is a combination of FH with the UNIFAC residual term.

Notes: PS, polystyrene; HDPE, high-density polyethylene; PMMA, poly(methyl methacrylate); PBMA, poly(butyl methacrylate).

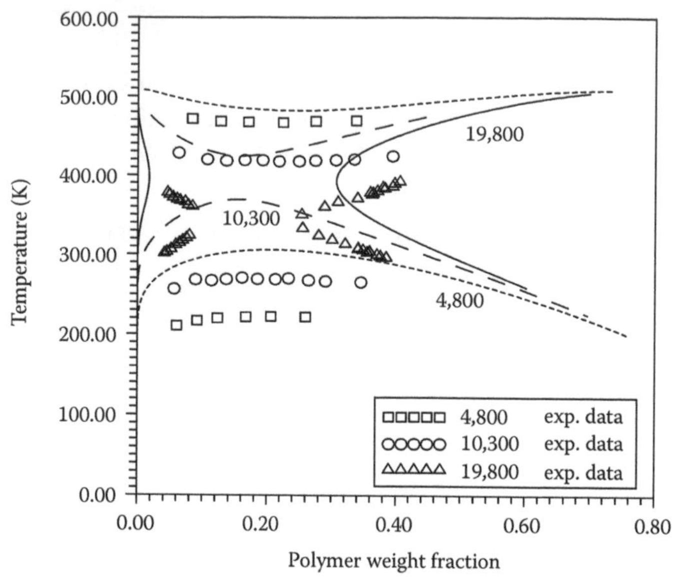

FIGURE 3.1 Experimental and predicted LLE diagram for the system polystyrene/acetone at three polymer molecular weights (4,800, 10,300, 19,800). The points are the experimental data and the lines are the predictions with the Entropic-FV model. (From Kontogeorgis, G.M. et al., *Ind. Eng. Chem. Res.*, 34, 1823, 1995. With permission.)

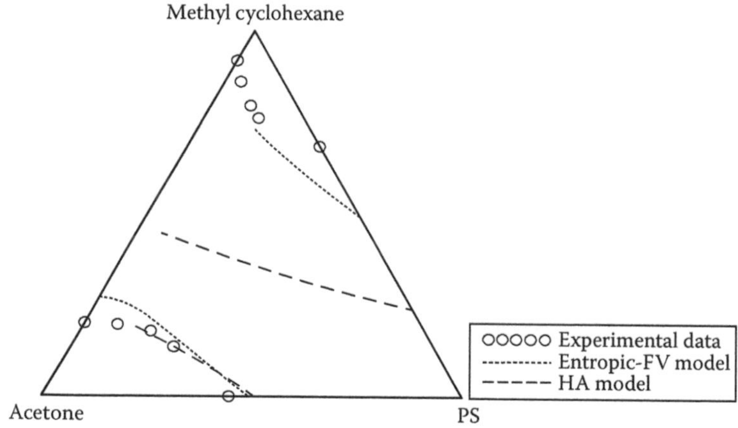

FIGURE 3.2 Ternary LLE for PS(300,000)/methyl cyclohexane/acetone, T = 298.15 K. (Reprinted from Pappa, G.D. et al., *Ind. Eng. Chem. Res.*, 36, 5461, 1997. With permission.)

2. Qualitatively correct and occasionally also quantitatively satisfactory representation of LLE for binary and ternary polymer–solvent systems is achieved, especially for the UCST-type phase behavior. A single investigation for SLLE also shows good results.
3. Less satisfactory results are obtained for polymer blends, where FV effects are not as dominant as for polymer solutions.

Some additional observations for specific cases are hereafter presented:

1. Athermal solutions, for example, polyolefins with alkanes offer a way of testing FV terms and numerous such investigations have been presented. FV models perform generally better than those that do not contain volume-dependent terms.[60] Better FV terms than that of

Entropic-FV (Equation 3.13) have been proposed[41,51,52,57,60] and they may serve as the basis for future developments resulting in even better activity coefficient models for polymer solutions. A rigorous test for newly developed FV expressions is provided by athermal alkane systems, especially the activity coefficients of heavy alkanes in short-chain ones.

2. A difference of 10°–30° should be expected in UCST predictions with Entropic-FV, see Table 3.9.
3. An alternative to Entropic-FV successful approach is the FH/Hansen model presented previously (Equation 3.10). In this case, best results are obtained when the original FH combinatorial rather than a FV term is used together with HSP. It seems that HSP incorporate some FV effects.

3.3.4 Correlative Versions of the Entropic-FV model

Both UNIFAC-FV and Entropic-FV are group contribution models. This renders the models truly predictive, but at the same time with little flexibility if the performance of the models for specific cases is not satisfactory. Two interesting alternative approaches are discussed here, which still maintain the FV terms but use different residual terms.

3.3.4.1 Entropic-FV/UNIQUAC Model

The first approach is to employ the UNIQUAC expression for the residual term. This Entropic-FV/UNIQUAC model has been originally suggested by Elbro et al.[35] and has shown to give very good results for polymer solutions if the parameters are obtained from VLE data between the solvent and the low molecular weight monomer (or the polymer's repeating unit).

The Entropic-FV/UNIQUAC model has been recently further developed and extended independently by two research groups.[55–57] Both VLE and LLE equilibria are considered, but the emphasis is given to LLE. Very satisfactory results are obtained as can be seen for two typical systems in Figures 3.3 and 3.4. It has been demonstrated that the Entropic-FV/UNIQUAC approach can

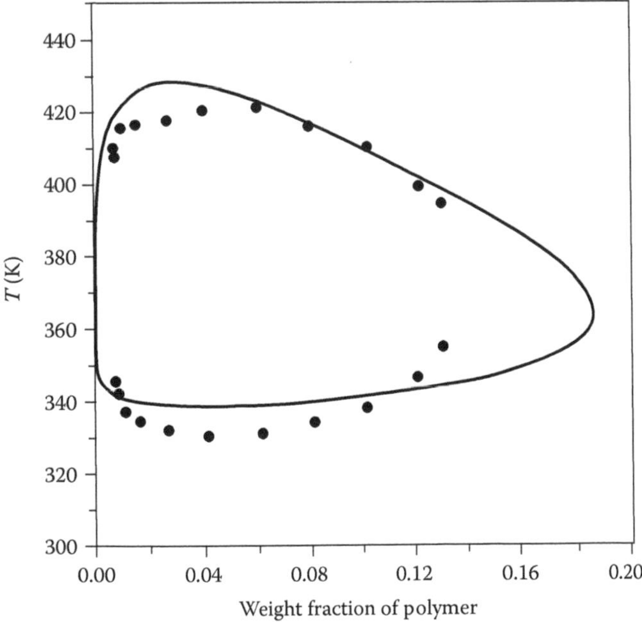

FIGURE 3.3 Correlation of liquid–liquid equilibria for the PVAL/water system with the Entropic-FV/UNIQUAC model. (•) Exp. data (M_n = 140,000 g/mol); (—) correlation. (From Bogdanic, G. and Vidal, J., *Fluid Phase Equilib.*, 173, 241, 2000. With permission.)

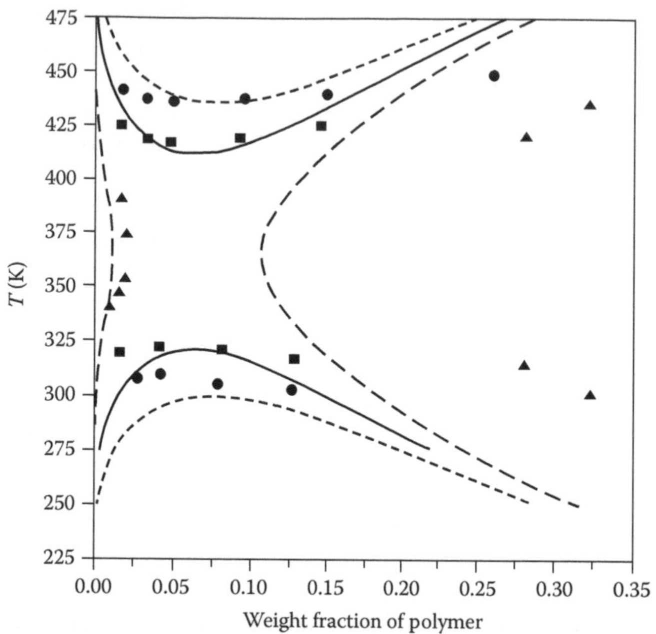

FIGURE 3.4 Correlation and prediction of liquid–liquid equilibria for the PBD/1-octane system with the Entropic-FV/UNIQUAC model. (■) Exp. data (M_v = 65,000 g/mol), (—) correlation; (▲) Exp. data (M_v = 135,000 g/mol), (- -) prediction; and (●) Exp. data (M_w = 44,500 g/mol), (- - - -) prediction. (From Bogdanic, G. and Vidal, J., *Fluid Phase Equilib.*, 173, 241, 2000. With permission.)

correlate both UCST/LCST and closed-loop behavior and even show the pressure dependency of critical solution temperatures (UCST and LCST).

3.3.4.2 Extension of Free-Volume Models to Gas Solubilities in Elastomers

The second approach proposed by Thorlaksen et al.[63] is based on a combination of the Entropic-FV term with Hildebrand's regular solution theory and developed a model for estimating gas solubilities in elastomers. The so-called Hildebrand–Entropic-FV model is given by the equation:

$$\ln \gamma_i = \ln \gamma_i^R + \ln \gamma_i^{C+FV} \tag{3.16}$$

$$\ln \gamma_2^R = \frac{V_2^L \cdot (\delta_1 - \delta_2)^2 \cdot \Phi_1^2}{R \cdot T}$$
$$\ln \gamma_2^{C+FV} = \ln \frac{\Phi_2^{FV}}{x_2} + 1 - \frac{\Phi_2^{FV}}{x_2} \tag{3.17}$$

where

δ_1 is the solvent solubility parameter
δ_2 is the gas solubility parameter
x_2 is the gas mole fraction in liquid/polymer
Φ_2 is the "apparent" volume fraction of solvent, given by

$$\Phi_2 = \frac{x_2 \cdot V_2^L}{x_1 \cdot V_1^L + x_2 \cdot V_2^L}$$

Φ_2^{FV} is the "FV" fraction given by

$$\Phi_2^{FV} = \frac{x_2 \cdot \left(V_2^L - V_2^{VW}\right)}{x_2 \cdot \left(V_2^L - V_2^{VW}\right) + x_1 \cdot \left(V_1^L - V_1^{VW}\right)}$$

V_2^L is a hypothetical liquid volume of the (gaseous) solute.

$$\frac{1}{x_2} = \frac{f_2^l}{\hat{f}_2^g} \cdot \exp \frac{V_2^L \cdot (\delta_1 - \delta_2)^2 \cdot \Phi_1^2}{R \cdot T} \tag{3.18}$$

$\hat{f}_2^g$ is the fugacity of the gas
f_2^l is the fugacity of the hypothetical liquid, which can be estimated from the equation:

$$\ln \frac{f_2^l}{P_c} = 3.54811 - \frac{4.74547}{T_r} + 1.60151 \cdot T_r - 0.87466 \cdot T_r^2 + 0.10971 \cdot T_r^3 \tag{3.19}$$

Finally, the gas solubility in the polymer is estimated from the equation:

$$\frac{1}{x_2} = \frac{f_2^l}{\hat{f}_2^g} \cdot \exp\left(\frac{V_2^L \cdot (\delta_1 - \delta_2)^2 \cdot \Phi_1^2}{R \cdot T} + \ln \frac{\Phi_2^{FV}}{x_2} + 1 - \frac{\Phi_2^{FV}}{x_2} \right) \tag{3.20}$$

Calculations showed that the hypothetical gas "liquid" volumes are largely independent to the polymer used, and moreover, for many gases (H_2O, O_2, N_2, CO_2 and C_2H_2), these are related to the critical volume of the gas by the equation:

$$V_2^L = 1.776 V_c - 86.017 \tag{3.21}$$

Very satisfactory results are obtained as shown in Table 3.10.

A final note for these "classical" activity coefficient models is that, despite the advent of advanced SAFT and other equations of state discussed next (Section 3.4), they are still quite popular and widely used in practical applications. They are also well cited in literature. For example, the historical articles by Flory and Huggins (Refs. [32,33]) are cited 998 (13.5) and 1034 (14) and the citations of the articles by Elbro et al.[35]: 164 (6.6), Lindvig et al.[36]: 44 (3.4), Kontogeorgis et al.[40]: 121 (5.5), and Oishi and Prausnitz[42]: 353 (9.5). The citations are per May 2014 and the numbers in parenthesis are citations per year.

3.4 SAFT FAMILY OF EQUATIONS OF STATE

3.4.1 Introduction

"Statistical mechanics is that branch of physics which studies macroscopic systems from a microscopic or molecular point of view. The goal of statistical mechanics is the understanding and prediction of macroscopic phenomena and the calculation of macroscopic properties from the properties of the individual molecules making up the system."

This is the opening paragraph of "Statistical Mechanics" written by McQuarrie,[64] already in 1976. Attempts to achieve this goal of statistical mechanics have been around for a long time. For example, Wertheim[65] was the first to derive an equation of state for hard-sphere systems. Carnahan and

TABLE 3.10
Summary of the Performance of the Models Tested at $T = 298$ K; $P = 101.3$ kPa

Polymer	Gas	Michaels/Bixler	Tseng/Lloyd	Hildebrand/Scott	Hildebrand Entropic-FV-1	Hildebrand Entropic-FV-2
PIP	N_2	14.7	73	3.9	−7.9	−4.6
	O_2	−16.1	−4	14	10.8	11.8
	Ar	−32.5	−23	—	29.4	−22.2
	CO_2	−3.2	13	4.5	8.7	4.6
PIB	N_2	−2.5	—	6.8	3.1	5.0
	O_2	−6.1	—	−1.7	−8.3	1.7
	Ar	—	—	—	—	—
	CO_2	32.8	—	−1.9	41.1	35.2
PBD	N_2	22.3	—	8.1	8.1	12.6
	O_2	14.9	—	−6	8.7	10.8
	Ar	12.1	—	—	111.1	24.0
	CO_2	−9.7	—	−4.6	0.4	−4.0
PDMB	N_2	—	—	−23	−7.5	−3.1
	O_2	—	—	−32	−16.8	−15.9
	Ar	—	—	—	—	—
	CO_2	—	—	−24	2.3	−2.2
PCP	N_2	58.1	—	49	−7.0	−4.2
	O_2	43.7	—	60	−1.4	−1.4
	Ar	—	—	—	—	—
	CO_2	8.8	—	27	−13.3	−17.1
AAD		19.8	28	18	16.8	10.6

Errors associated with models for predicting gas solubilities in polymers.
Hildebrand Entropic-FV-1: The liquid volume of the gas is determined from its relationship with the critical volume, Equation 3.26. Hildebrand Entropic-FV-2: The average hypothetical liquid volume of a gas is used.

Starling[66] made an empirical modification to Wertheim's solution based on molecular simulation data to arrive at what is by now the famous Carnahan–Starling equation of state. As well as being an early attempt to arrive at an engineering model using results from "hard" science (or what we might cynically call "impractical" science), this work also showed the usefulness of using results from molecular simulation. In a sense, molecular simulation fulfils the goal of statistical mechanics, in that it predicts (some) macroscopic properties using only molecular properties as input. However, molecular simulation is system specific, time consuming, and ultimately can only be as successful as the molecular model it is based upon. Thus, while it is certainly a step forward to be able to predict the properties of a system of hard spheres, the hard sphere as a model is itself an incomplete description of a real molecule. Nevertheless, increasing computer power and ever more detailed knowledge of molecular properties, extending even to the quantum level, means that molecular simulation will continue to be an important tool in engineering thermodynamics. See Economou[67] for a review of current industrial applications of molecular simulation.

However, equations of state, too, will be an essential component of chemical engineering theory and practice for the foreseeable future, and as ever, the balance will need to be struck between rigorous theory and engineering applicability. One equation of state, which seems to have done an admirable job of bridging the gap between molecular theory and engineering application, is statistical associating fluid theory (SAFT) and it is with this equation of state and its spin-offs that the remainder of this discussion is concerned.

3.4.2 SAFT Equation of State—History and Introduction

A series of four seminal papers once again written by Wertheim[68–71] appeared in 1984 and 1986. These papers laid the foundation for the associating theory (or thermodynamic perturbation theory [TPP]), which was to become the key feature in the novel molecular-based equation of state known as SAFT. The theories presented in Wertheim's papers are highly complex and almost intractable—"essentially incomprehensible," as one author has put it.[1] However, in the period 1988–1990 at Cornell University, Chapman and coworkers[72–75] performed the monumental task of transforming the abstruse theory of Wertheim into workable equations, and finally into an engineering equation of state. SAFT is by no means the only molecular-based equation of equation of state out there—PHSC is another[76] but it differs from the vast majority of other similar equations of state in one important respect—it is extensively used. In their review of SAFT published in 2001, Müller and Gubbins[77] estimate that 200 articles dealing with SAFT had appeared. Since that review appeared, a further 70 or so articles have appeared dealing directly with SAFT or one of its variants. Significantly, SAFT is also now available in industrial process simulators such as ASPEN, PRO/II, and ChemCad as well as in the SPECS thermodynamics package of IVC-SEP at the Technical University of Denmark.

One of the drawbacks arising from the extensive use of SAFT is that many versions have appeared, with the result that the literature is complex and can be confusing. Here, we will try to outline the theoretical development in some detail of the original SAFT model (this too is not unambiguous, since "original SAFT" is often used to describe the version due to Huang and Radosz,[78,79] which is slightly different from that of Chapman), as well as some of its modified versions. We will then summarize some of the more interesting results obtained using SAFT and its variants.

3.4.3 Original SAFT[74]

The foundation for what was to become SAFT was laid in two papers[72,75] which appeared in the journal *Molecular Physics* in 1988. The first of these papers developed the theory required for associating fluids, while the second focused on chain formation. However, the first paper to contain an equation of state that can realistically be called SAFT appeared in the August, 1990, issue of *Industrial and Engineering Chemistry Research*.[74] It is interesting that the Huang and Radosz paper appeared in the November issue of the same journal.[78]

To understand exactly what occurs in SAFT, we refer to Figure 3.5, taken from Fu and Sandler.[80] Initially, a pure fluid is assumed to consist of equal-sized hard spheres (b). Next a dispersive potential is added to account for attraction between the spheres (c). Typical potentials are the square-well or Lennard–Jones potential. Next each sphere is given two (or more) "sticky" spots, which enables the formation of chains (d). Finally, specific interaction sites are introduced at certain positions in the chain, which enable the chains to associate through some attractive interaction (hydrogen bonding) (e). This interactive energy is often taken to be a square-well potential. The final single molecule is shown in Figure 3.5a. Each of these steps contributes to the Helmholtz energy. The residual Helmholtz energy is given by

$$a^{res} = a^{seg} + a^{chain} + a^{assoc} \tag{3.22}$$

where

a^{seg} is the Helmholtz energy of the segment, including both hard-sphere reference and dispersion terms

a^{chain} is the contribution from chain formation

a^{assoc} is the contribution from association

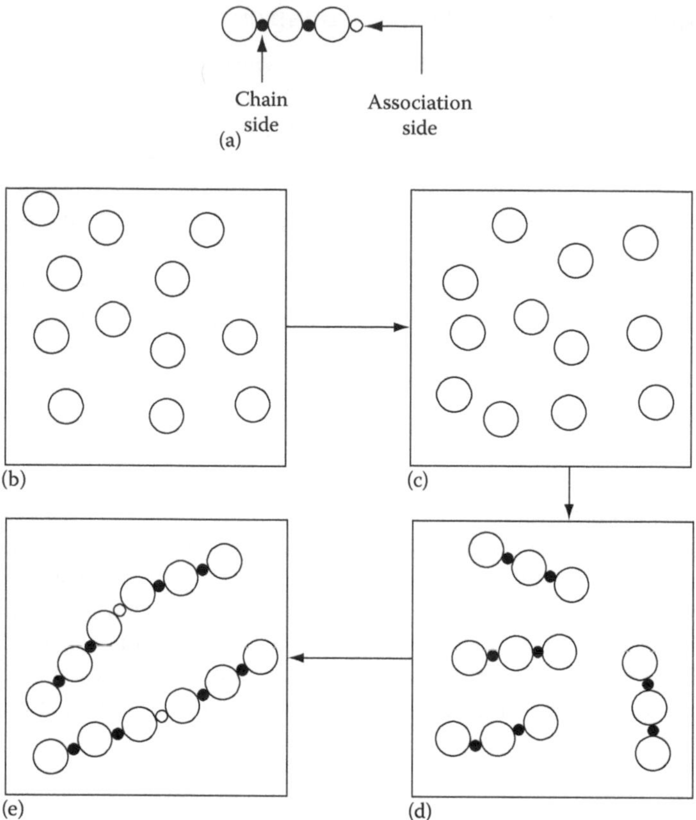

FIGURE 3.5 Procedure to form a molecule in the SAFT model. (a) The proposed molecule. (b) Initially the fluid is a hard-sphere fluid. (c) Attractive forces are added. (d) Chain sites are added and chain molecules appear. (e) Association sites are added and molecules form association complexes through association sites. (From Fu, Y.-H. and Sandler, S.I., *Ind. Eng. Chem. Res.*, 34, 1897, 1995. With permission.)

It is this separation of the Helmholtz energy into additive components that is partly responsible for the fact that SAFT has so many variants—each of the contributions can be considered (and modified) separately, or new terms may be added (such as polar, electrostatic, or other contributions). The individual terms may also be used outside of the context of SAFT. Thus, the term that accounts for association has been combined (with minor modification) with the SRK equation to create CPA, an equation of state, which has had remarkable success in a number of applications.[81]

It is worth noting that both the chain formation and the association term derive from Wertheim's TPT. However, chains (i.e., covalent bonds) are formed in the limit of complete association. It is these two terms that make up the innovative development in SAFT—the first because now we have available a rational method for considering polymer molecules (linear chains with very many bonded segments) and the second because we can now consider associating molecules in a way that more closely resembles the actual physical picture. The calculation of useful thermodynamic properties from a^{assoc} was initially a complex procedure, although Michelsen and Hendriks[82] simplified the computations considerably by recasting the equations in a simpler, although mathematically equivalent form.

We now consider each of the terms individually. We follow the original notation of Chapman et al.[74] Each pure component is characterized by a chain length m, a size-parameter σ, and an energy parameter, ε. If the molecule is self-associating, there are two further parameters, which characterize the volume ($\kappa^{A_iB_i}$) and energy ($\varepsilon^{A_iB_i}$) of association. The association term is given for mixtures by

$$\frac{a^{assoc}}{RT} = \sum_i X_i \left[\sum_{A_i} \left[\ln X^{A_i} - \frac{X^{A_i}}{2} \right] + \frac{1}{2} M_i \right] \tag{3.23}$$

where

X^{A_i} is the mole fraction of molecules i not bonded at site A

M_i is the number of association sites on molecule i

We have

$$X^{A_i} = \left[1 + N_{AV} \sum_j \sum_{B_j} \rho_j X^{B_j} \Delta^{A_iB_j} \right]^{-1} \tag{3.24}$$

where

N_{AV} is the Avogadro's number

ρ_j is the molar density of j

$\Delta^{A_iB_j}$ is the association strength given by

$$\Delta^{A_iB_j} = d_{ij}^3 g_{ij}(d_{ij})^{seg} \kappa^{A_iB_j} \left[\exp\left(\frac{\varepsilon^{A_iB_j}}{kT} \right) - 1 \right] \tag{3.25}$$

Here d_{ij} is a temperature-dependent size-parameter related to σ_{ij} by

$$d = \sigma f\left(\frac{kT}{\varepsilon}, m \right) \tag{3.26}$$

where

$$f\left(\frac{kT}{\varepsilon}, m \right) = \frac{1 + 0.2977kT/\varepsilon}{1 + 0.33163kT/\varepsilon + f(m)(kT/\varepsilon)^2} \tag{3.27}$$

and

$$f(m) = 0.0010477 + 0.025337\frac{m-1}{m} \tag{3.28}$$

This temperature dependence is incorporated to account for the fact that real molecules are not hard spheres, but rather there is some degree of interpenetration between molecules, particularly at high temperatures. Thus, the "effective" hard-sphere diameter of a segment is smaller at higher temperatures. The radial distribution function in Equation 3.25 is given by the mixture version of the Carnahan–Starling equation of state for hard-sphere mixtures:

$$g_{ij}^{seg}(d_{ij}) \approx g_{ij}^{HS}\left(d_{ij}^{+}\right) = \frac{1}{1-\zeta_3} + \left(\frac{d_i d_j}{d_i + d_j} \right) \frac{3\zeta_2}{(1-\zeta_3)^2} + \left(\frac{d_i d_j}{d_i + d_j} \right)^2 \frac{2\zeta_2^2}{(1-\zeta_3)^3} \tag{3.29}$$

where

$$\zeta_k = \frac{\pi N_{AV}}{6}\rho \sum_i X_i m_i d_{ii}^k \tag{3.30}$$

For the chain term in Equation 3.22, we have

$$\frac{a^{chain}}{RT} = \sum_i X_i (1 - m_i) \ln(g_{ii}(d_{ii})^{HS}) \tag{3.31}$$

Finally, for the segment term in Equation 3.22, we have

$$a^{seg} = a_0^{seg} \sum_i X_i m_i \tag{3.32}$$

where the 0 subscript indicates a nonassociated segment. The segment energy consists of a hard-sphere reference and a dispersion contribution:

$$a_0^{seg} = a_0^{HS} + a_0^{disp} \tag{3.33}$$

The Carnahan–Starling equation[66] is used for both pure components and mixtures to give

$$\frac{a_0^{HS}}{RT} = \frac{4\eta - 3\eta^2}{(1-\eta)^2} \tag{3.34}$$

where for mixtures $\eta = \zeta_3$ as defined by Equation 3.30. The dispersion term is given by

$$a_0^{disp} = \frac{\varepsilon R}{k}\left(a_{01}^{disp} + \frac{a_{02}^{disp}}{T_R} \right) \tag{3.35}$$

where

$$a_{01}^{disp} = \rho_R \left[-0.85959 - 4.5424\rho_R - 2.1268\rho_R^2 + 10.285\rho_R^3 \right] \tag{3.36}$$

$$a_{02}^{disp} = \rho_R \left[-1 - 9075 - 9.9724\rho_R - 22.216\rho_R^2 + 15.904\rho_R^3 \right] \tag{3.37}$$

The reduced quantities are given by $T_R = kT/\varepsilon$ and $\rho_R = (6/2^{0.5}\pi)\eta$.

Most of the results of this initial paper are comparisons with simulation data for chains with various parameters, although pure-component parameters for six hydrocarbons and two associating fluids were fitted. No results for mixtures of real fluids are presented.

3.4.4 "Original" SAFT[78]

Probably the main contribution of the Huang and Radosz[78] version of SAFT was the regression of pure-component parameters for over 100 different fluids. There are also some notational differences. Thus, instead of a size parameter σ, they use a volume parameter v^{00}, which is related through the equation

$$v^{00} = \frac{\pi N_{AV}}{6\tau}\sigma^3 \tag{3.38}$$

Here $\tau = 0.74048$ is the highest possible packing fraction for a system of pure hard spheres. They also use the notation u^0 instead of ε for the energy parameter, although these terms are completely equivalent.

The rather complex temperature dependence of the hard-sphere diameter given by Equations 3.26 through 3.28 was simplified by Huang and Radosz, following Chen and Kreglewski[83] to

$$d = \sigma\left[1 - 0.12\exp\left[\frac{-3u^0}{kT}\right]\right] \tag{3.39}$$

The dispersion term is also different from that of Chapman et al.[74] and is given by

$$\frac{a_0^{disp}}{RT} = \sum_i \sum_j D_{ij}\left[\frac{u}{kT}\right]^i \left[\frac{\eta}{\tau}\right]^j \tag{3.40}$$

where

D_{ij} are universal constants

u is the temperature-dependent energy parameter given by $u = u^0(1 + e/kT)$ and e/k is a constant set to −10, with a few exceptions

Another important contribution of Huang and Radosz is the presentation of detailed tables discussing bonding schemes for different associating fluids. These schemes are presented as Tables 3.11 and 3.12 and have been widely adopted in the literature of SAFT and other equations of state for associating fluids.

In their paper on mixture properties,[79] Huang and Radosz also use the full mixture version of the Carnahan–Starling equation for the hard-sphere mixtures reference system:

$$\frac{a^{HS}}{RT} = \frac{1}{\zeta_0}\left[\frac{3\zeta_1\zeta_2}{1-\zeta_3} + \frac{\zeta_2^3}{\zeta_3(1-\zeta_3)^2} + \left(\frac{\zeta_2^3}{\zeta_3^2} - \zeta_0\right)\ln(1-\zeta_3)\right] \tag{3.41}$$

One of the reasons that the Huang and Radosz version of SAFT has been adopted (and is widely referred to as "original" SAFT) is that they undertook an extensive pure-component parameterization for over 100 pure fluids. This meant that their equation of state could be used immediately for real fluids of industrial interest without any intermediate steps. Both Huang and Radosz were employed by Exxon during the development of SAFT. The fact that the model had the backing of a major oil company may also help explain its rapid adoption and use as an engineering tool.

Besides it is interesting to note that the first paper to appear describing PC-SAFT also had pure-component parameters for about 100 species.[84] This fact, coupled with the success of the model, has certainly been partly responsible for the rapid adoption of PC-SAFT in the few years it has been in existence, both in industry and in academia.

TABLE 3.11
Unbonded Site Fractions X^A for Different Bonding Types

Type	Δ Approximations	X^A Approximations	X^A
1	$\Delta^{AA} \neq 0$		$\frac{-1+(1+4\rho\Delta)^{1/2}}{2\rho\Delta}$
2A	$\Delta^{AA} = \Delta^{AB} = \Delta^{BB} \neq 0$	$X^A = X^B$	$\frac{-1+(1+8\rho\Delta)^{1/2}}{4\rho\Delta}$
2B	$\Delta^{AA} = \Delta^{BB} = 0$ $\Delta^{AB} \neq 0$	$X^A = X^B$	$\frac{-1+(1+4\rho\Delta)^{1/2}}{2\rho\Delta}$
3A	$\Delta^{AA} = \Delta^{AB} = \Delta^{BB} = \Delta^{AC} = \Delta^{BC} = \Delta^{CC} \neq 0$	$X^A = X^B = X^C$	$\frac{-1+(1+12\rho\Delta)^{1/2}}{6\rho\Delta}$
3B	$\Delta^{AA} = \Delta^{AB} = \Delta^{BB} = \Delta^{CC} = 0$ $\Delta^{AC} = \Delta^{BC} \neq 0$	$X^A = X^B$ $X^C = 2X^A - 1$	$\frac{-(1-\rho\Delta)+((1+\rho\Delta)^2+4\rho\Delta)^{1/2}}{4\rho\Delta}$
4A	$\Delta^{AA} = \Delta^{AB} = \Delta^{BB} = \Delta^{AC} = \Delta^{BC} = \Delta^{CC} = \Delta^{AD} =$ $\Delta^{BD} = \Delta^{CD} = \Delta^{DD} \neq 0$	$X^A = X^B = X^C = X^D$	$\frac{-1+(1+16\rho\Delta)^{1/2}}{8\rho\Delta}$
4B	$\Delta^{AA} = \Delta^{AB} = \Delta^{BB} = \Delta^{AC} = \Delta^{BC} = \Delta^{CC} = \Delta^{DD} = 0$ $\Delta^{AD} = \Delta^{BD} = \Delta^{CD} \neq 0$	$X^A = X^B = X^C$ $X^D = 3X^A - 2$	$\frac{-(1-2\rho\Delta)+((1+2\rho\Delta)^2+4\rho\Delta)^{1/2}}{6\rho\Delta}$
4C	$\Delta^{AA} = \Delta^{AB} = \Delta^{BB} = \Delta^{CC} = \Delta^{CD} = \Delta^{DD} = 0$ $\Delta^{AC} = \Delta^{AD} = \Delta^{BC} = \Delta^{BD} \neq 0$	$X^A = X^B = X^C = X^D$	$\frac{-1+(1+8\rho\Delta)^{1/2}}{4\rho\Delta}$

Source: Huang, S.H. and Radosz, M., *Ind. Eng. Chem. Res.*, 29, 2284, 1990. With permission.

TABLE 3.12
Types of Bonding in Real Associating Fluids

Species	Formula	Rigorous Type	Assigned Type
Acid	—C(=O---HO)(—OH---O=)C—	1	1
Alkanol	—Ö: (A, B) CH	3B	2B
Water	B:Ö:H C (A) H D	4C	3B
Amines			
Tertiary	—N̈— (A) |	1	Non-self-associating
Secondary	—N̈— (A) H B	2B	2B
Primary	—N̈:H B (C) H A	3B	3B
Ammonia	A H:N̈:H B (D) H C	4B	3B

Source: Huang, S.H. and Radosz, M., *Ind. Eng. Chem. Res.*, 29, 2284, 1990. With permission.

3.4.5 Simplified SAFT[80]

The key idea in the work of Fu and Sandler is the simplification of the dispersion term. All other terms from Huang and Radosz are retained. Since the dispersion term given by Huang and Radosz, Equation 3.40, contains 24 constants, it seems reasonable to attempt to simplify this term. For mixtures, the dispersion Helmholtz free energy is

$$\frac{a^{disp}}{RT} = mZ_M \ln\left(\frac{v_s}{v_s + \langle v^* Y \rangle}\right) \tag{3.42}$$

where

"average" chain length $m = \sum_i x_i m_i$

$Z_M = 36$ is the maximum coordination number

$v_s = 1/\rho m$ is the total molar volume of a segment

$$\langle v^* Y \rangle = \frac{N_{AV} \sum_i \sum_j x_i x_j m_i m_j \left(d_{ij}^3/\sqrt{2}\right)\left[\exp(u_{ij}/kT) - 1\right]}{\sum_i \sum_j x_i x_j m_i m_j} \tag{3.43}$$

The remaining terms have the same meaning as in Huang and Radosz.[78] Generally, simplified SAFT performs as well as Huang and Radosz SAFT, although it requires refitting all the pure-component parameters. Fu and Sandler provide parameters for 10 nonassociating and 8 associating fluids. Table 3.13 is reproduced from Fu and Sandler. It is interesting because it presents different types of cross-association in some detail. This scheme is completely general and applicable to any equation of state incorporating association. Combining rules for cross-association also need to be introduced, however, and are far from self-evident.

3.4.6 LJ-SAFT[85,86]

The main change in the SAFT version of Kraska and Gubbins[85,86] is that they use Lennard–Jones (LJ) spheres for the reference term, rather than hard spheres. The remaining terms are unchanged, except that the radial distribution function used in the calculation of the chain and association contributions in Equations 3.25 and 3.31 is the radial distribution function for LJ spheres rather than hard spheres. Thus, an equation of state for LJ spheres is required. The equation used is that of Kolafa and Nezbeda.[87] The Helmholtz energy for the reference (LJ) system is (for a pure fluid)

$$A_{seg} = m\left(A_{HS} + \exp(-\gamma\rho^{*2})\rho T \Delta B_{2,hBH} + \sum_{i,j} C_{ij} T^{i/2} \rho^{*j}\right) \tag{3.44}$$

where

$$A_{HS} = T\left(\frac{5}{3}\ln(1-\eta) + \frac{\eta(34 - 33\eta + 4\eta^2)}{6(1-\eta)^2}\right) \tag{3.45}$$

$$\Delta B_{2,hBH} = \sum_{i=-7}^{0} C_i T^{i/2} \tag{3.46}$$

TABLE 3.13
Type of Association for Cross-Associating Mixtures

Mixture	Component 1	Component 2	Association Type
Acid–acid	O // C \\ O—H } Site A1	Site A$_2$ { H—O \\ C // O	$\in^{A_1A_1} \neq 0$ $\in^{A_2A_2} \neq 0$ $\in^{A_1A_2} \neq 0$
Alcohol–alcohol	Site B$_1$ C—Ö \\ H Site A$_1$	Site B$_2$ C—Ö \\ H Site A$_2$	$\in^{A_1B_1} \neq 0, \in^{A_2B_2} \neq 0$ $\in^{A_1B_2} = \in^{A_2B_1} \neq 0$ $\in^{A_1A_1} = \in^{A_2A_2} = \in^{A_1A_2} = \in^{A_2A_1} = 0$ $\in^{B_1B_1} = \in^{B_2B_2} = \in^{B_1B_2} = \in^{B_2B_1} = 0$
Acid–alcohol	O // C \\ O—H } Site A$_1$	Site B$_2$ C—Ö \\ H Site A$_2$	$\in^{A_1B_1} \neq 0, \in^{A_2B_2} \neq 0$ $\in^{A_1A_2} = \in^{A_1B_2} \neq 0$ $\in^{A_2A_2} = \in^{B_2B_2} = 0$
Water–acid	Site C$_1$ H—Ö Site B$_1$ \\ H Site A$_1$	Site A$_2$ { H—O \\ C // O	$\in^{A_1C_1} = \in^{B_1C_1} \neq 0, \in^{A_2A_2} \neq 0$ $\in^{A_1A_2} = \in^{B_1A_2} = \in^{C_1A_2} \neq 0$ $\in^{A_1A_1} = \in^{B_1B_1} = \in^{C_1C_1} = \in^{A_1B_1} = 0$
Water–alcohol	Site C$_1$ H—Ö Site B$_1$ \\ H Site A$_1$	Site B$_2$ C—Ö \\ H Site A$_2$	$\in^{A_1C_1} = \in^{B_1C_1} \neq 0, \in^{A_2B_2} \neq 0$ $\in^{A_1B_2} = \in^{B_1B_2} \in^{C_1A_2} \neq 0$ $\in^{A_1A_1} = \in^{B_1B_1} = \in^{C_1C_1} = \in^{A_1B_1} = 0$ $\in^{A_2A_2} = \in^{B_2B_2} = 0$ $\in^{A_1A_2} = \in^{B_1A_2} = \in^{C_1B_2} = 0$

Source: Fu, Y.-H. and Sandler, S.I., *Ind. Eng. Chem. Res.*, 34, 1897, 1995. With permission.

$$\rho^* = \frac{mb}{V_m} \tag{3.47}$$

$$\eta = \frac{\pi}{6} \rho^* \sigma_{BH}^3 \tag{3.48}$$

$$\sigma_{BH} = \sum_{i=-2}^{1} D_i T^{i/2} + D_{\ln} \ln T \tag{3.49}$$

Apart from using a different reference system, the notation in Kraska and Gubbins does not follow the customary SAFT notation, nor are the pure-component parameters defined in the same way (e.g., the energy parameter with units of 1/K is defined inversely as a temperature parameter with units K) and care should be taken in using it. They also incorporate a term to account for dipole–dipole interactions.

3.4.7 SAFT-VR[88]

SAFT-VR is the version of SAFT developed by George Jackson and coworkers first at the University of Sheffield and currently at Imperial College (Gil-Vilegas et al.[88]; McCabe et al.[89]). SAFT-VR is identical to the Huang and Radosz version except in the dispersion contribution. This term incorporated attraction in the form of a square-well potential. Thus, in addition to a segment being characterized by a size and an energy parameter, the square-well width (λ) is also included as a pure-component parameter. Thus, changing the parameter λ changes the range of attraction of the segment (hence the name VR for "variable range"). It is the introduction of this extra term that gives SAFT-VR greater flexibility, since we now have an extra pure component to "play with." Although it is generally desirable to describe pure-component liquid densities and vapor pressure with the minimum number of parameters, the extra variable-range parameter may be necessary for the description of certain anomalous behaviors in systems containing water. The Helmholtz energy for the dispersion energy is given by

$$a^{disp} = \frac{a_1}{kT} + \frac{a_2}{(kT)^2} \tag{3.50}$$

where

$$a_1 = -\rho_s \sum_i \sum_j x_{s,i} x_{s,j} \alpha_{ij}^{VDW} g^{HS}\left[\sigma_x; \zeta_x^{eff}\right] \tag{3.51}$$

The subscript s refers to segment rather than molecule properties. We have

$$\alpha_{ij}^{VDW} = \frac{2\pi\varepsilon_{ij}\sigma_{ij}^3\left(\lambda_{ij}^3 - 1\right)}{3} \tag{3.52}$$

and g^{HS} is the radial distribution function for hard spheres as before except that the arguments are different:

$$\zeta_x^{eff} = c_1\zeta_x + c_2\zeta_x^2 + c_3\zeta_x^3 \tag{3.53}$$

$$\sigma_x^3 = \sum_i \sum_j x_{s,i} x_{s,j} \sigma_{ij}^3 \tag{3.54}$$

$$\zeta_x = \frac{\pi}{6}\rho_s\sigma_x^3 \tag{3.55}$$

The constants c_i in Equation 3.53 are given by

$$\begin{pmatrix} c_1 \\ c_2 \\ c_3 \end{pmatrix} = \begin{pmatrix} 2.25855 & -1.50349 & 0.249434 \\ -0.66927 & 1.40049 & -0.827739 \\ 10.1576 & -15.0427 & 5.30827 \end{pmatrix} \begin{pmatrix} 1 \\ \lambda_{ij} \\ \lambda_{ij}^2 \end{pmatrix} \tag{3.56}$$

The second-order term in Equation 3.50 is given by

$$a_2 = \sum_{i=1}^{n}\sum_{j=1}^{n} x_{s,i}x_{s,j}\frac{1}{2}K_{HS}\varepsilon_{ij}\rho_s\frac{\partial a_1}{\partial \rho_s} \tag{3.57}$$

where

$$K_{HS} = \frac{\zeta_0(1-\zeta_3)^4}{\zeta_0(1-\zeta_3)^2 + 6\zeta_1\zeta_2(1-\zeta_3) + 9\zeta_2^3} \tag{3.58}$$

3.4.8 PC-SAFT[84]

A recent version of SAFT that has appeared is that due to Gross and Sadowski[84] developed at the Technical University of Berlin. Once again, most of the terms in PC-SAFT are the same as those in the Huang and Radosz version. The term that is different is the dispersion term. However, it is not simply a different way of expressing the dispersion attraction between segments, but rather it tries to account for dispersion attraction between whole chains. Referring to Figure 3.5 should make this clear. Instead of adding the dispersion to hard spheres and then forming chains, we first form hard-sphere chains and then add a chain dispersion term, so the route in Figure 3.5 would be (b)–(d)–(c)–(e). To do this, we require interchain rather than intersegment radial distribution functions. These are given by O'Lenick et al.[90] The Helmholtz energy for the dispersion term is given as the sum of a first- and second-order term:

$$\frac{A^{disp}}{kTN} = \frac{A_1}{kTN} + \frac{A_2}{kTN} \tag{3.59}$$

where

$$\frac{A_1}{kTN} = -2\pi\rho m^2\left(\frac{\varepsilon}{kT}\right)\sigma^3\int_1^\infty \tilde{u}(x)g^{hc}\left(m;\frac{x\sigma}{d}\right)x^2dx \tag{3.60}$$

$$\frac{A_2}{kTN} = -\pi\rho m\left(1+Z^{hc}+\rho\frac{\partial Z^{hc}}{\partial\rho}\right)^{-1} m^2\left(\frac{\varepsilon}{kT}\right)^2\sigma^3\frac{\partial}{\partial\rho}\left[\rho\int_1^\infty \tilde{u}(x)^2 g^{hc}\left(m;\frac{x\sigma}{d}\right)x^2dx\right] \tag{3.61}$$

where
$x = r/\sigma$
$\tilde{u}(x) = u(x)/\varepsilon$ is the reduced intermolecular potential

The radial distribution function g^{hc} is now an interchain function rather than a segment function as before. This is a key point in PC-SAFT. The term involving compressibilities is given by

$$\left(1+Z^{hc}+\rho\frac{\partial Z^{hc}}{\partial\rho}\right) = \left(1+m\frac{8\eta-2\eta^2}{(1-\eta)^4}+(1-m)\frac{20\eta-27\eta^2+12\eta^3-2\eta^4}{((1-\eta)(2-\eta))^2}\right) \tag{3.62}$$

We still need to solve the integrals in Equations 3.60 and 3.61. Setting

$$I_1 = \int_1^{\infty} \tilde{u}(x) g^{hc}\left(m; \frac{x\sigma}{d}\right) x^2 dx \tag{3.63}$$

$$I_2 = \frac{\partial}{\partial \rho}\left[\rho \int_1^{\infty} \tilde{u}(x)^2 g^{hc}\left(m; \frac{x\sigma}{d}\right) x^2 dx\right] \tag{3.64}$$

we can substitute the Lennard–Jones potential and the radial distribution function of O'Lenick et al.[90] This was done for the series of n-alkanes and the integrals were fit as a power series:

$$I_1 = \sum_{i=0}^{6} a_i \eta^i \tag{3.65}$$

$$I_2 = \sum_{i=0}^{6} b_i \eta^i \tag{3.66}$$

with

$$a_i = a_{0i} + \frac{m-1}{m} a_{1i} + \frac{m-1}{m}\frac{m-2}{m} a_{2i} \tag{3.67}$$

$$b_i = b_{0i} + \frac{m-1}{m} b_{1i} + \frac{m-1}{m}\frac{m-2}{m} b_{2i} \tag{3.68}$$

Equations 3.67 and 3.68 require a total of 42 constants, which are adjusted to fit experimental pure-component data of n-alkanes. This direct fitting to experimental data to some extent accounts for errors in the reference equation of state, the perturbing potential, and the radial distribution function, which appear in the integrals of Equations 3.63 and 3.64. The dispersion potential given by Equations 3.60 and 3.61 is readily extended to mixtures using the van der Waals one-fluid theory.

Since this first PC-SAFT paper appeared, the authors rapidly published a series of further papers applying PC-SAFT to polymers,[91,92] associating fluids[93] and copolymers.[94]

3.4.9 Simplified PC-SAFT[95]

Two simplifications to PC-SAFT have been proposed, which simplify phase equilibrium calculations substantially for mixtures. For pure components, this simplified PC-SAFT becomes original PC-SAFT, so the simplifications may be considered as a particular set of mixing rules. The advantage of this is that existing pure-component parameters can be used in simplified PC-SAFT—no refitting is required. The targets of the simplifications are Equations 3.29 and 3.41. In other words, this simplified PC-SAFT targets the hard-sphere reference equation of state. The remaining terms are the same, except as mentioned previously, the simpler radial distribution function will affect both the chain and association terms, since the radial distribution function appears in both of these terms.

By setting $\eta \equiv \zeta_3$, Equations 3.29 and 3.40 reduce to

$$g^{HS}(d^+) = \frac{1-\eta/2}{(1-\eta)^3} \tag{3.69}$$

and

$$\tilde{a}^{HS} = \frac{4\eta - 3\eta^2}{(1-\eta)^2} \tag{3.70}$$

respectively. In the initial paper,[95] use of Equation 3.69 only is called modification 1, while use of both Equations 3.69 and 3.70 is called modification 2. This and subsequent work has shown that there is no loss of accuracy using the most simplified version of PC-SAFT, so modification 2 is used throughout and is called "simplified PC-SAFT." Simplified PC-SAFT has since been applied in our group to several polymer systems including: polymer VLE,[96] polymer–solvent binary LLE,[97] ternary polymer–solvent and blend systems,[98] and high-pressure gas solubility in polymers.[99,100]

One of the interesting points about SAFT in general is the ability to extrapolate the properties of higher molecular weight substances from knowledge of similar shorter chain compounds. This is most evident for the *n*-alkane series. Pure-component parameters for polyethylene can in principle be predicted by extrapolating from the properties of the *n*-alkanes. In practice, this is problematic, since for very long chains, effects such as entanglement of the individual polymer chains start to influence the behavior. Figure 3.6 shows PC-SAFT parameters for the alkanes up to eicosane (C20). The parameters m, $m\sigma^3$, and $m\varepsilon/k$ are all linear with molecular weight. Tables 3.14 and 3.15 show computing times for simplified PC-SAFT compared with original PC-SAFT, as well as with other equations of state.

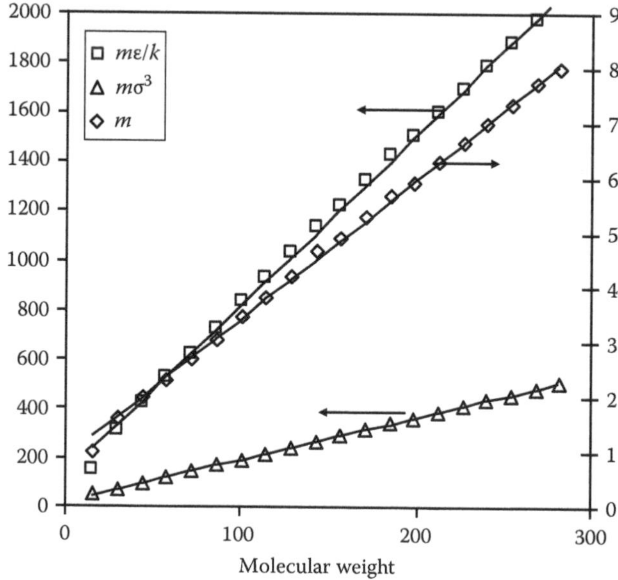

FIGURE 3.6 The groups m, $m\sigma^3$, and $m\varepsilon/k$ vs. molecular weight for linear alkanes up to eicosane. Points are PC-SAFT parameters, lines are linear fits to these points, excluding methane. (From von Solms, N. et al., *Ind. Eng. Chem. Res.*, 42, 1098, 2003. With permission.)

TABLE 3.14
Comparison of Computing Times for 36-Component Phase-Envelope Calculation with PC-SAFT and with Simplified PC-SAFT

Calculation	Computing Time (ms)
Phase-envelope calculation, 36-component mixture, full PC-SAFT	48
Phase-envelope calculation, 36-component mixture, PC-SAFT modification 2	32

Source: von Solms, N. et al., *Ind. Eng. Chem. Res.*, 42, 1098, 2003. With permission.
Computations were performed on a 2.0 GHz Pentium IV machine with DVF compiler.

TABLE 3.15
Comparison of Computing Times for Various Models and Mixtures

	Computing Time (μs)	
Mixture Model	Fugacity Coefficients Only	All Derivatives (T, P, Composition Residual Heat Capacity)
SRK, 6 components	1.9	3.4
SRK, 6 components	3.1	5.6
CPA, 15 components, 0 sites	15	20.1
CPA, 15 components, 2 sites	18.5	24.8
CPA, 15 components, 4 sites	24.4	32.2
CPA, 15 components, 6 sites	32.3	40.9
CPA, 15 components, 8 sites	40.7	50.9
Modification 2, 15 components, 0 sites	15.6	23.1
Modification 2, 15 components, 6 sites	39.1	54.4

Source: von Solms, N. et al., *Ind. Eng. Chem. Res.*, 42, 1098, 2003. With permission.
Number of sites refers to number of association sites on a molecule when employing an equation of state with association.

3.4.10 Extensions to SAFT-Type Equations

The additive contributions within SAFT mean that the equation of state is quite versatile—contributions can be added to account for effects not included in the discussion earlier. Thus, papers have appeared accounting for polar,[101] quadrupolar,[102] and dipolar[103] molecules, as well as polarizable dipoles.[104]

Versions of SAFT have also appeared, which include an electrostatic contribution, the intention being to develop a version of SAFT suitable for modeling electrolytes.[105,106]

Care should be taken when adding very many contributions, since each extra contribution will almost always require one or more pure-component parameters. The actual physical picture should always be borne in mind when considering which contributions to include.

Another recent development is the application of group contribution methods to SAFT (see, e.g., Le Thi et al.[107] and references therein). Rather than the addition of extra terms, which then require more pure-component parameters, the use of group contribution methods is an attempt to generalize the parameters in SAFT (extending even to binary interaction parameters). The hope is that in this way SAFT becomes a more predictive tool, relying less on fitting of parameters to experimental data.

3.4.11 Applications of SAFT

The remainder of this discussion looks at some polymer applications of SAFT. Figure 3.7 shows the results of using parameters extrapolated based on Figure 3.6. The system is methane–tetratetracontane (C44), where the C44 parameters are obtained by extrapolation of the lines in Figure 3.6. While this is not a polymer system, the method of obtaining parameters by extrapolation is applicable to polymer systems. A more sophisticated method for finding polymer parameters based on extrapolation of the monomer properties and polymer density data has been published recently.[108]

Figures 3.8 and 3.9 show a comparison of the various modifications and original PC-SAFT for VLE in the systems polystyrene–propyl acetate (Figure 3.8) and polypropylene–diisopropyl ketone (Figure 3.9). In general, PC-SAFT and simplified PC-SAFT performed similarly, as can be seen from Table 3.16, which gives the errors in prediction for a large number of polymer–solvent VLE systems. Figure 3.10 is a pressure–weight fraction diagram (VLE) for the polymer poly(vinyl acetate) in the associating solvent 2-methyl-1-propanol. A small value of the binary interaction parameter correlates the data well.

Figure 3.11 is an illustration of a novel method developed by von Solms et al.[97] for finding LLE in polymer systems, known as the method of alternating tangents. This figure shows the Gibbs energy of mixing for two binary systems as a function of the mole fraction of component 1. The method will be illustrated with reference to the system methanol(1)–cyclohexane(2), since this curve clearly shows the existence of two phases. The composition of methanol in each phase is found by locating a single line, which is a tangent to the curve in two places (the common tangent). In Figure 3.11, these compositions are given by x_1^{eq1} and x_1^{eq2}. In fact the curve for the system PS(1)–acetone(2) also shows the existence of two phases, although this is not visible. The first step in the procedure is to determine whether a spinodal point exists (this is a necessary condition for phase separation). In the figure, the two spinodal points are given by the compositions x_1^{sp1} and x_1^{sp2}. The spinodal condition is given by $((\partial^2 g^{mix}/RT)/\partial x^2) = 0$, that is, an inflection point on the curve. Once a spinodal point has been found (using a Newton–Raphson method), the next step is to find the point of tangent of a

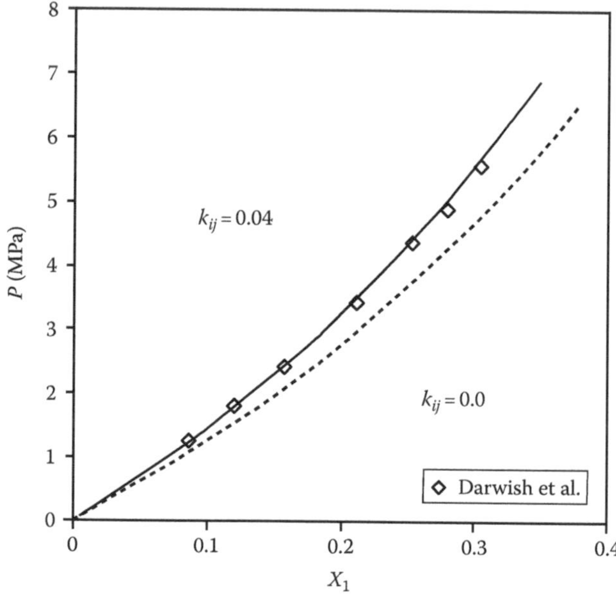

FIGURE 3.7 Vapor pressure curve for the system methane(1)–tetratetracontane(2) at T = 423.2 K. The dashed line is PC-SAFT, the solid line is PC-SAFT with a binary interaction parameter $k_{ij} = 0.04$. The points are experimental data. (From von Solms, N. et al., *Ind. Eng. Chem. Res.*, 42, 1098, 2003. With permission.)

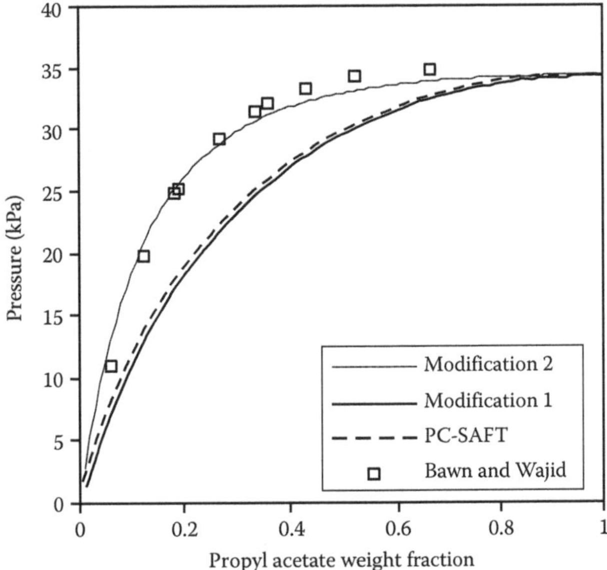

FIGURE 3.8 Vapor pressure curve for the system propyl acetate(1)–polystyrene(2) at T = 343.15 K. The dashed line is PC-SAFT (the lowest line on the plot), the solid line is modification 1, and the dotted line is modification 2. Points are experimental data. All lines are pure predictions ($k_{ij} = 0$). (From von Solms, N. et al., *Ind. Eng. Chem. Res.*, 42, 1098, 2003. With permission.)

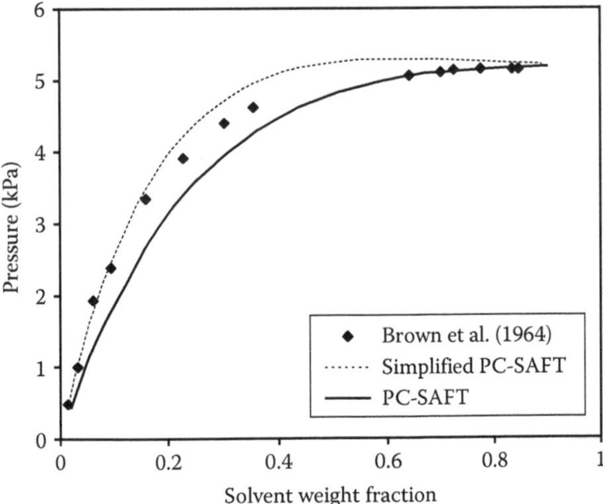

FIGURE 3.9 Pressure–weight fraction plot of polypropylene(1)–diisopropyl ketone(2) at T = 318 K. Polypropylene molecular weight = 20,000. Comparison of experimental data with the predictions of original (solid line) and the simplified version (dotted line) of PC-SAFT. In both curves, the interaction parameter k_{ij} = 0. (From *Fluid Phase Equilib.*, 215, Kouskoumvekaki, I.A., von Solms, N., Michelsen, M.L., and Kontogeorgis, G.M., Application of a simplified perturbed chain SAFT equation of state to complex polymer systems, 71–78, Copyright 2004, with permission from Elsevier.)

TABLE 3.16
Comparison of the Performance of the Simplified against the Original PC-SAFT in Predicting Vapor–Liquid Equilibria of Polymer Solutions (k_{ij} = 0 in All Cases)

	PC-SAFT	
% AAD	Simplified Version	Original Version
Cyclic hydrocarbons		
PS–cyclohexane	13	16
PS–benzene	28	13
PS–ethyl benzene	3	6
PS–*m*-xylene	25	16
PS–toluene	18	7
PVAc–benzene	6	10
Chlorinated hydrocarbons		
PS–carbon tetrachloride	18	12
PS–chloroform	30	11
PP–dichloromethane	59	74
PP–carbon tetrachloride	55	47
Esters		
PS–propyl acetate	5	21
PS–butyl acetate	3	25
PVAc–methyl acetate	3	2
PVAc–propyl acetate	19	18
Ketones		
PS–acetone	6	26
PS–diethyl ketone	7	28
PS–methyl ethyl ketone	14	12
PVAc–acetone	4	7
PP–diethyl ketone	16	27
PP–diisopropyl ketone	4	11
PVAc–methyl ethyl ketone	7	6
Amines		
PVAc–propylamine	4	3
PVAc–isopropyl amine	17	16
Alcohols		
PVAc–1-propanol	56	54
PVAc–2-propanol	84	73
PVAc–1-butanol	59	59
PVAc–2-butanol	39	36
PVAc–2-methyl-1-propanol	29	29
Overall	23	24

Source: Kouskoumvekaki, I.A. et al., *Fluid Phase Equilib.*, 215, 71, 2004. With permission.
Average percentage deviation between experimental and predicted equilibrium pressure curves.

line originating at the spinodal point. In the figure, this is the line connecting x_1^{sp1} and x_1. This point is just to the left of x_1^{eq2} (i.e., we are not yet quite at the equilibrium concentration after one step). Once the first tangent has been found, the point of tangent opposite is then found in a similar way. This process is repeated until the change in the composition at the tangent point is within a specified tolerance. At this point, the equilibrium values have been calculated.

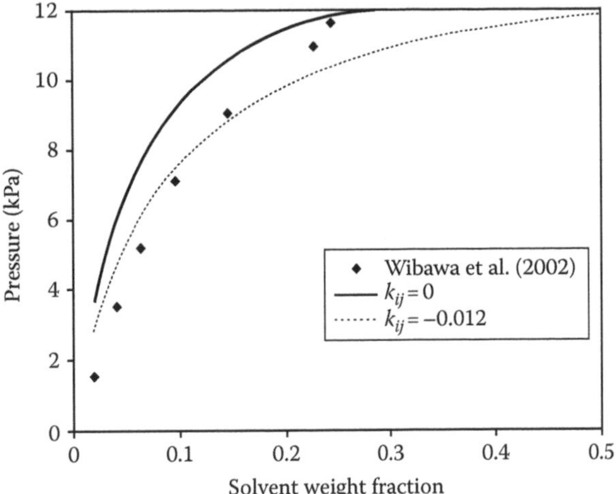

FIGURE 3.10 Pressure–weight fraction plot of poly(vinyl acetate)(1)–2-methyl-1-propanol(2) at T = 313 K. Comparison of experimental data with prediction (k_{ij} = 0) and correlation (k_{ij} = −0.012) results of simplified PC-SAFT. Poly(vinyl acetate) molecular weight = 167,000. (From *Fluid Phase Equilib.*, 215, Kouskoumvekaki, I.A., von Solms, N., Michelsen, M.L., and Kontogeorgis, G.M., Application of a simplified perturbed chain SAFT equation of state to complex polymer systems, 71–78, Copyright 2004, with permission from Elsevier.)

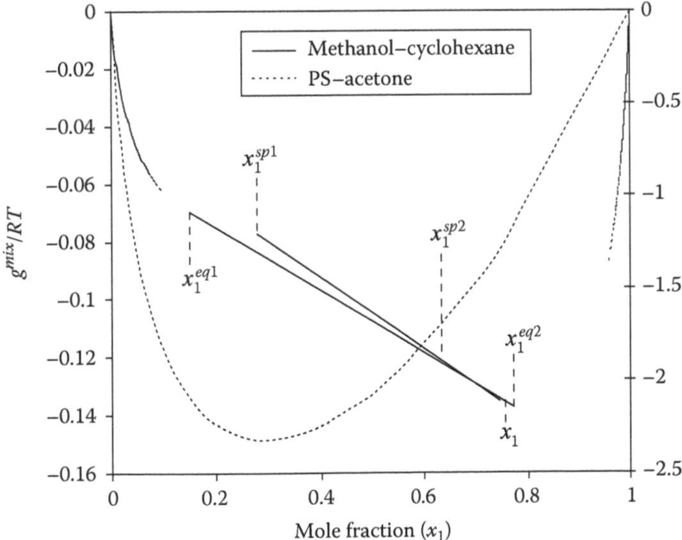

FIGURE 3.11 Illustration of the method of alternating tangents. The solid line is the system methanol(1)–cyclohexane(2). The dotted line is the system PS(1)–acetone. The two spinodal points are indicated by x_1^{sp1} and x_1^{sp2}. The equilibrium (binodal) points are indicated by x_1^{eq1} and x_1^{eq2}. Starting from a spinodal point, the equilibrium values can be calculated by solving for only one point at a time. (From *Fluid Phase Equilib.*, 222–223, von Solms, N., Kouskoumvekaki, I.A., Lindvig, T., Michelsen, M.L., and Kontogeorgis, G.M., A novel approach to liquid-liquid equilibrium in polymer systems with application to simplified PC-SAFT, 87–93, Copyright 2004, with permission from Elsevier.)

Figures 3.12 through 3.15 are examples of binary LLE for polymer systems. Figure 3.12 shows results for the system polystyrene–methylcyclohexane for different molecular weights of polystyrene. The experimental data are from the classic work of Dobashi et al.[109] The lines are simplified PC-SAFT correlations with k_{ij} = 0.0065 for polystyrene molecular weights 10,200, 46,400 and 719,000 in order of increasing critical solution temperature. The data are reasonably well correlated over a very large range of molecular weight with a single value of the binary interaction parameter, k_{ij}. The binary interaction parameter was adjusted to give the correct upper critical solution temperature. However, the correct critical solution concentration is not obtained, although the experimental trends are correctly predicted by the model: The critical solution temperature increases, and the polymer weight fraction at the critical solution temperature decreases with increasing molecular weight.

Figure 3.13 shows results for the system polyisobutylene–diisobutyl ketone at different polymer molecular weights. The experimental data are from Shultz and Flory.[110] The lines are simplified PC-SAFT correlations. A single binary interaction parameter (k_{ij} = 0.0053) was used for all three systems, although it seems that there is a weak dependence of molecular weight on k_{ij}. Incorporating a functional dependence of k_{ij} on molecular weight (e.g., a linear fit) would improve the correlation. It should also be noted that these three systems represent a very large range of molecular weights.

Figure 3.14 shows the results for a single molecular weight of HDPE in five different *n*-alkanol solvents from *n*-pentanol up to *n*-nonanol. The results are well correlated using simplified PC-SAFT with a small value of the binary interaction parameter k_{ij}. A k_{ij} value of around 0.003 gives a good correlation for all the systems, except HDPE–*n*-pentanol. In the figure, a small value (k_{ij} = 0.0006) was used to correlate the data, although the data is also well predicted by simplified PC-SAFT (k_{ij} = 0), giving an error in the upper critical solution temperature of 3 K in the case of HDPE–*n*-pentanol.

Figure 3.15 shows the results for the system HDPE–butyl acetate. This system displays both UCST and LCST behaviors. A single binary interaction parameter (k_{ij} = 0.0156) was used to correlate the data for both molecular weights shown. The binary interaction parameter was adjusted to

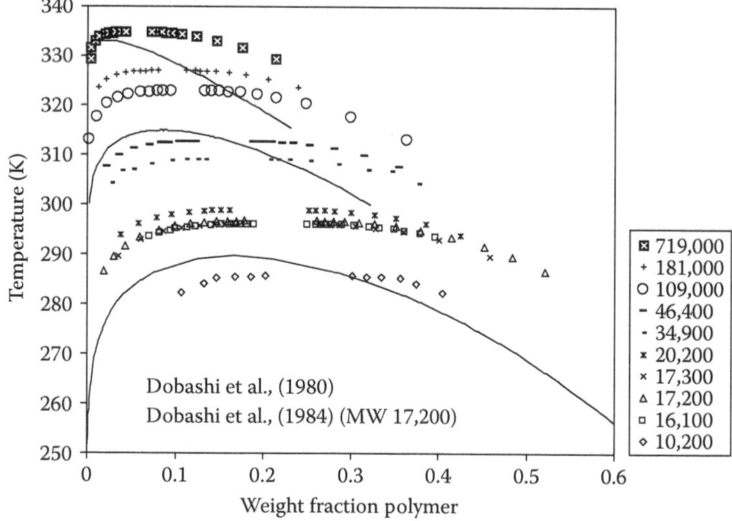

FIGURE 3.12 Liquid–liquid equilibrium in the system polystyrene–methyl cyclohexane for different molecular weights of polystyrene. The experimental data are from Dobashi et al.[109] The lines are simplified PC-SAFT correlations with k_{ij} = 0.0065 for polystyrene molecular weights 10,200, 46,400 and 719,000 in order of increasing critical solution temperature. (From *Fluid Phase Equilib.*, 222–223, von Solms, N., Kouskoumvekaki, I.A., Lindvig, T., Michelsen, M.L., and Kontogeorgis, G.M., A novel approach to liquid-liquid equilibrium in polymer systems with application to simplified PC-SAFT, 87–93, Copyright 2004, with permission from Elsevier.)

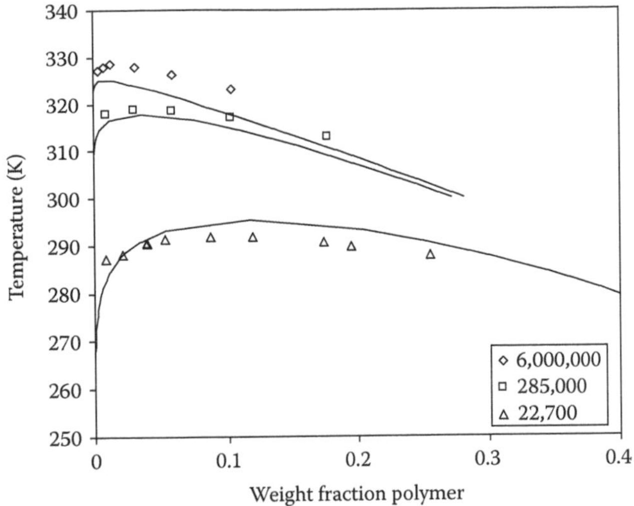

FIGURE 3.13 Liquid–liquid equilibrium in the system polyisobutylene–diisobutyl ketone. PC-SAFT parameters for diisobutyl ketone were obtained by fitting to experimental liquid density and vapor pressure data in the temperature range 260–600 K. This data was taken from the DIPPR database. The parameters were: $m = 4.6179$, $\varepsilon/k = 243.72$ K, and $\sigma = 3.7032$ Å. Average percent deviations were 1.03% for vapor pressure and 0.64% for liquid density. Lines are simplified PC-SAFT correlations with $k_{ij} = 0.0053$, the same at all three molecular weights. (From *Fluid Phase Equilib.*, 222–223, von Solms, N., Kouskoumvekaki, I.A., Lindvig, T., Michelsen, M.L., and Kontogeorgis, G.M., A novel approach to liquid–liquid equilibrium in polymer systems with application to simplified PC-SAFT, 87–93, Copyright 2004, with permission from Elsevier.)

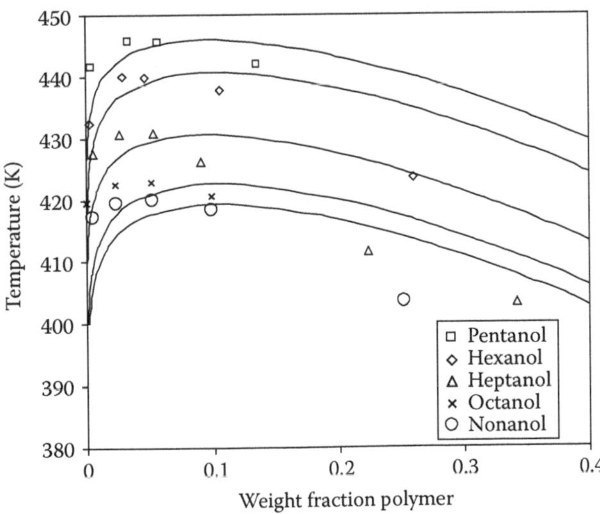

FIGURE 3.14 Liquid–liquid equilibrium for HDPE with *n*-alkanols. Lines are simplified PC-SAFT correlations for each of the five solvents (pentanol highest, nonanol lowest). Polymer molecular weight is 20,000. Binary interaction parameters are as follows: pentanol: 0.0006; hexanol: 0.003; heptanol: 0.0025; octanol: 0.0033; and nonanol: 0.0029. (From *Fluid Phase Equilib.*, 222–223, von Solms, N., Kouskoumvekaki, I.A., Lindvig, T., Michelsen, M.L., and Kontogeorgis, G.M., A novel approach to liquid–liquid equilibrium in polymer systems with application to simplified PC-SAFT, 87–93, Copyright 2004, with permission from Elsevier.)

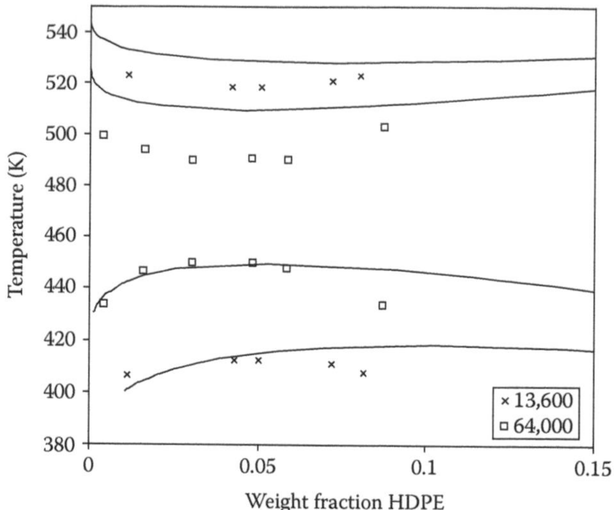

FIGURE 3.15 Liquid–liquid equilibrium in the system HDPE–butyl acetate. The system displays both upper and lower critical solution temperature behaviors. The experimental data for molecular weights 13,600 and 64,000. Lines are simplified PC-SAFT correlations with k_{ij} = 0.0156 for both molecular weights. (From *Fluid Phase Equilib.*, 222–223, von Solms, N., Kouskoumvekaki, I.A., Lindvig, T., Michelsen, M.L., and Kontogeorgis, G.M., A novel approach to liquid–liquid equilibrium in polymer systems with application to simplified PC-SAFT, 87–93, Copyright 2004, with permission from Elsevier.)

give a good correlation for the UCST curve at the higher molecular weight (64,000). As mentioned earlier, the LCST curve is rather insensitive to k_{ij}. Nevertheless, the LCST curve is reasonably well correlated using this value. The prediction (k_{ij} = 0) is almost as good for the LCST curve, although the UCST will then be substantially underpredicted.

Figure 3.16 shows a ternary phase diagram for the system polystyrene–acetone–methylcyclohexane. The binary interaction parameters were obtained by fitting to the individual binary systems.

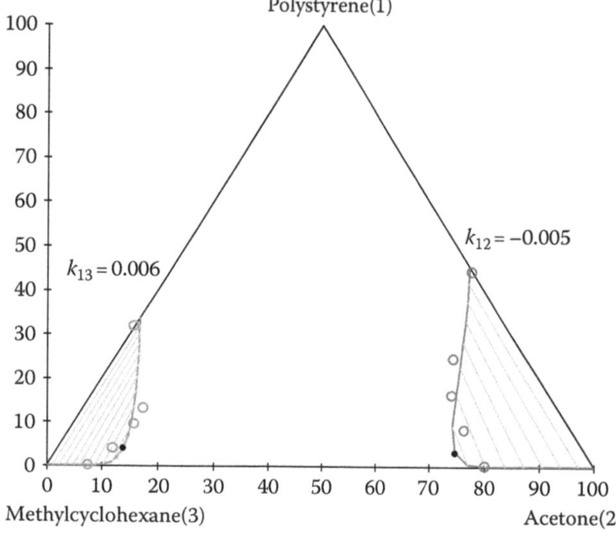

FIGURE 3.16 Ternary phase diagram for the system polystyrene–acetone–methylcyclohexane. The binary interaction parameters were obtained by fitting to the individual binary systems. The ternary coexistence curves are predictions.

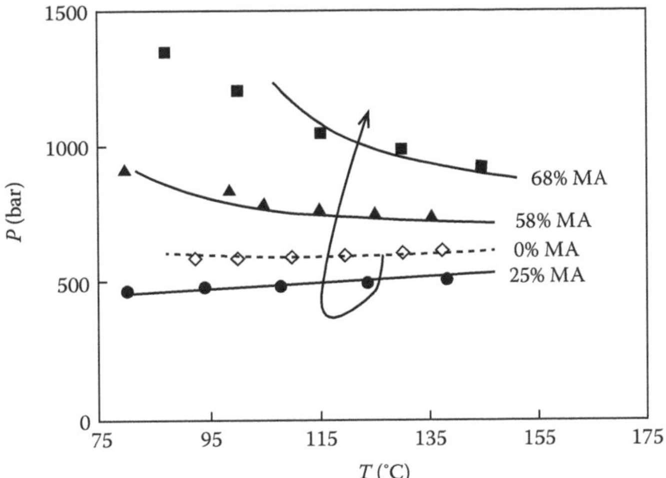

FIGURE 3.17 High-pressure equilibrium for mixtures of poly-(ethylene-*co*-methyl acrylate) (EMA) and propylene for different repeat-unit compositions of the EMA. Comparison of experimental cloud-point measurements to calculation results of the PC-SAFT equation of state. (EMA [0% MA] is equal to LDPE: open diamonds and dashed line.) (From Gross, J. et al., *Ind. Eng. Chem. Res.*, 42, 1266, 2003. With permission.)

The ternary coexistence curves are predictions. The algorithm for finding ternary LLE in systems containing polymers is an extension of the binary algorithm discussed earlier and was developed by Lindvig et al.[98]

Figure 3.17 from Gross et al.[94] shows high pressure equilibrium for mixtures of poly(ethylene-*co*-methyl acrylate) (EMA) and propylene for different repeat unit compositions of EMA. As repeat units of methyl acrylate (MA) are added to the polyethylene chain, the demixing pressure at first declines, but then increases as the composition of MA increases. This effect is correctly predicted by PC-SAFT.

Finally, Figure 3.18 shows a comparison of SAFT-VR and simplified PC-SAFT in a recent study[100] where the two models were compared in their ability to model multicomponent phase equilibrium in systems typical of real polyethylene reactors. The system shown here is polyethylene/nitrogen/1-butene. The results of the simplified PC-SAFT and SAFT-VR calculations for this ternary are consistent. From the figure, one can see that as butene in the vapor is replaced by nitrogen, the calculated absorption of butene decreases (not surprisingly—there is less of it to absorb). However, it is also clear that as nitrogen in the vapor is replaced by butene, absorption of nitrogen increases, even though there is *less* nitrogen to absorb. This suggests that there may be some enhancement/inhibition of absorption effect.

In the time since the previous edition of this work (2008), around 150 articles have appeared applying SAFT to various polymer systems. Certain newer themes become apparent here:

The use of density functional theory combined with equations of state to examine surface and structural properties is a relatively new phenomenon, resulting in equations such as iSAFT.[111,112] For example, Jain et al.[111] used SAFT and density functional theory to probe the structure of tethered polymer chains.

Perhaps the most frequently occurring, relatively new idea is that of group-contribution SAFT, where there has been a great deal of activity.[107,113–122] Part of the motivation for a group-contribution method is the need for an equation that can be readily extended to complex molecules, where pure-component properties may be unavailable, and for which SAFT types of equation are well suited. Examples of complex systems where SAFT has been applied are biopolymers,[123] hyperbranched polymers,[124] and asphaltenes.[125]

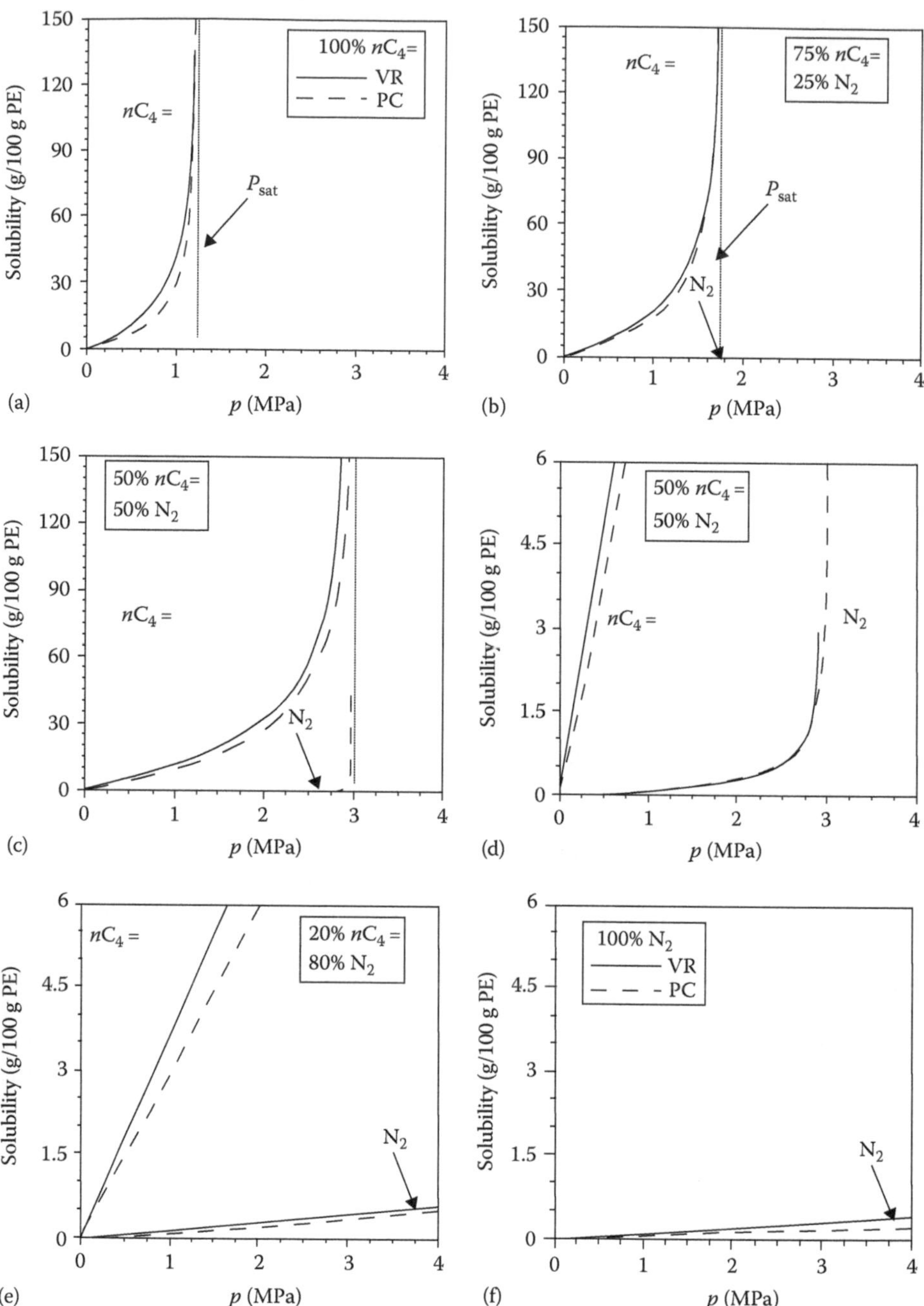

FIGURE 3.18 Gas absorptions in amorphous PE calculated with SAFT-VR and simplified-PC-SAFT for a range of vapor compositions of the ternary mixture of (but-1-ene + nitrogen + the reference PE [MW = 12,000 g mol^{-1}]) at T = 80°C: (a) 100 vapor mol% butene (binary mixture); (b) 75% but-1-ene, 25% nitrogen; (c) 50% but-1-ene, 50% nitrogen (with a vertical scale chosen to highlight butene absorption); (d) 50% but-1-ene, 50% nitrogen (with a vertical scale chosen to highlight nitrogen absorption); (e) 20% butene, 80% nitrogen; and (f) 100% nitrogen (binary mixture). In each case, solid curves represent SAFT-VR calculations and dashed curves represent simplified-PC-SAFT calculations. (From Haslam, A.J. et al., *Fluid Phase Equilib.*, 243, 74, 2006. With permission).

Another encouraging feature is the use of SAFT-type models in industrial applications. In addition to the asphaltene example noted earlier,[125] work has appeared relevant to refrigeration[126,127] and catalytic polymerization processes.[128]

Finally, it can be mentioned that since the review article by Müller and Gubbins,[77] several monographs reviewing various aspects of SAFT have appeared.[129–131]

3.5 CONCLUDING REMARKS AND FUTURE CHALLENGES

Attempting to summarize in a few words the current status in polymer thermodynamics, we could state

1. Many databases (some available in computer form) and reliable group-contribution methods are available for estimating many pure polymer properties and phase equilibria of polymer solutions such as densities, solubility parameters, glass and melting temperatures, and solvent activity coefficients.
2. Simple group-contribution methods based on UNIFAC, containing corrections for the FV effects, satisfactorily predict the solvent activities and vapor–liquid equilibria for binary and ternary polymer solutions. They are less successful for the prediction of liquid–liquid equilibria if the parameters are based on VLE. They are much more successful if the parameters are based on LLE data. The combination of a simple FV expression such as that employed in the Entropic-FV model and a local-composition energetic term such as that of UNIQUAC seems to be a very promising tool for both VLE and LLE in polymer solutions. We expect that such tools may find widespread use in the future for practical applications.
3. The SAFT model will continue to be a very successful tool for polymer systems. The low-pressure and especially the high-pressure results for systems including solvents and nonpolar polymers (with emphasis to those of interest to the polyolefin industry) are very satisfactory. However, the extension to polar systems represents so far a limitation of the model. In many ways, PC-SAFT has fulfilled the early promise of SAFT—it is reasonably simple to implement (compared with many versions of SAFT), it is relatively undemanding computationally, and it has proved successful in predicting and correlating phase equilibria in many systems containing polymers and/or associating compounds. There is still work to be done, however. One area is in finding parameters for pure polymers. Since polymers are nonvolatile, one cannot use vapor-pressure regression and the energy parameter is relatively insensitive within the range of experimental *P–V–T* data. Regressing pure-component parameters from binary systems is problematic as this leads to nonunique pure-component parameters—a situation best avoided. *Another area where research is needed is in the modeling of water. There was been a great deal of effort in this area, but as yet no satisfactory model, fitting within the existing SAFT framework, has been developed for modeling water-containing systems. Examples of such deficiencies are predicting the density maximum for pure water, as well as modeling water–hydrocarbon mutual solubilities.*
4. Most theoretical/modeling studies in polymer thermodynamics are limited to
   a. Organic polymers
   b. Binary systems often involving monodisperse polymers and single solvents
   c. Rather "simple" polymers (polyolefins, polystyrene, PVC, etc.)
   d. Vapor–liquid equilibria and activity coefficients
   e. "Rules of thumb" estimates of miscibility (solubility parameters, theta parameters, etc.)

Some of the future challenges in the area of polymer thermodynamics will involve

1. More emphasis to multicomponent systems including both mixed solvents, blend-solvent systems as well as the effect of polydispersity.

2. Better treatment of condensed phases especially liquid–liquid and liquid–solid equilibria as well as water-soluble polymer systems and polyelectrolytes.
3. Emphasis to high-pressure systems involving both the typical nonpolar and polar polymers as well.
4. New directions such as description of "special materials" including those involving oligomers, copolymers, new structures (star-like and dendrimers), "inorganic" polymers (e.g., tin-based antifouling paints).
5. Proper account for the effects of crystallinity and cross-linking with special attention to swelling phenomena.
6. Closer collaboration with industry, for example, for testing existing theories for polymers having novel structures, for commercial polymers for which so far the structure is not revealed to academic researchers and for many other applications of practical interest. Many industrial systems are much more complex than the ones studied in academia. Closer collaboration in the future between academia and the polymer and paint/adhesives industries may further help the advancements in the area of polymer thermodynamics in the coming years.

3.A APPENDIX

3.A.1 AN EXPRESSION OF THE FH MODEL FOR MULTICOMPONENT MIXTURES

The FH model was originally developed as a model for the entropy of mixing for mixtures containing molecules of different size, but it was soon modified also to account for energetic interactions. The model can be formulated in terms of the excess Gibbs energy as follows (Lindvig et al.[37]):

$$G^E = G^{E,comb} + G^{E,res}$$

$$\frac{G^{E,comb}}{RT} = \sum_{i=1}^{N} n_i \ln \frac{\varphi_i}{x_i}$$

$$\frac{G^{E,res}}{RT} = \sum_{i=1}^{N} \sum_{j=1}^{N} \varphi_i \varphi_j a_{ij}$$

$$\chi_{ij} = 2a_{ij} v_i$$

Using basic thermodynamics, the following expression for the activity coefficient is obtained:

$$\ln \gamma_i = \ln \gamma_i^{comb} + \ln \gamma_i^{res}$$

where the combinatorial term is given by

$$\ln \gamma_i^{comb} = \ln \frac{\varphi_i}{x_i} + 1 - \frac{\varphi_i}{x_i}$$

and the residual term is

$$\ln \gamma_i^{res} = 2v_i \sum_{j=1}^{NC} \varphi_j a_{ij} - v_i \sum_{j=1}^{NC} \sum_{k=1}^{NC} \varphi_j \varphi_k a_{jk}$$

The aforementioned formulation of the FH model is slightly different from the conventionally used formulation using the FH interaction parameter (χ_{12}), although there is an interrelationship based on the simple equation shown above.

For a binary mixture, the multicomponent equation reduces to the traditional FH residual term:

$$\ln \gamma_1^{res} = \chi_{12}\varphi_2^2$$

ABBREVIATIONS

BR	butadiene rubber
CPA	cubic plus association
CST	critical solution temperature
EAC	ethyl acetate
EoS	equation of state
EFV	entropic-free volume
EMA	poly(ethylene-*co*-methyl acrylate)
FH	Flory–Huggins (model/equation/interaction parameter)
FV	free-volume
GC	group contribution (method/principle)
GC Fl(ory)	group contribution Flory equation of state
GCVOL	group contribution volume (method for estimating the density)
HDPE	high density polyethylene
LCST	lower critical solution temperature
LJ-SAFT	Lennard–Jones SAFT
LLE	liquid–liquid equilibria
MA	methyl acrylate
MCSL	Mansoori–Carnahan–Starling–Leland
PBMA	polybutyl methacrylate
PC-SAFT	perturbed chain-SAFT
PDMS	polydimethylsiloxane
PE	polyethylene
PEMA	polyethyl-methacrylate
PEO	polyethylene oxide
PHSC	perturbed hard-sphere chain
PIB	polyisobutylene
PS	polystyrene
PVAC	polyvinyl acetate
PVC	polyvinyl chloride
SAFT	statistical associating fluid theory
SAFT-VR	SAFT-variable range
SLE	solid–liquid equilibria
SLLE	solid–liquid–liquid equilibria
SRK	Soave–Redlich–Kwong
TPT	thermodynamic perturbation theory
UCST	upper critical solution temperature
UNIFAC	UNIQUAC functional activity coefficient (a method for estimating activity coefficients)
UFV	UNIFAC-FV
vdW1f	van der Waals one fluid (mixing rules)
VLE	vapor–liquid equilibria
VOC	volatile organic content

REFERENCES

1. Prausnitz, J.M., Lichtenthaler, R.N., and Azevedo, E.G.D., 1999. *Molecular Thermodynamics of Fluid Phase Equilibria*, 3rd edn. Prentice-Hall International, Upper Saddle River, NJ.
2. Holten-Andersen, J., 1987. Physical models and coatings technology. Report T12-87, Scandinavian Paint and Printing Ink Research Institute, Hørsholm, Denmark. Also Presented at the *International Conference in Organic Coatings Science and Technology*, Athens, Greece, pp. 13–30, 1986.
3. Holten-Andersen, J., 1986. Heat balance and solvent balance in the drying of coatings. *JOCCA*, 69(12): 324–331.
4. Holten-Andersen, J. and Hansen, C.M., 1983. Solvent and Water Evaporation from coatings, *Prog. Org. Coat.*, 11(3): 219.
5. Holten-Andersen, J. and Eng, K., 1988. Activity coefficients in polymer solutions. *Prog. Org. Coat.*, 16: 77.
6. Doong, S.J. and Ho, W.S., 1991. Sorption of organic vapors in polyethylene. *Ind. Eng. Chem. Res.*, 30: 1351–1361.
7. Kampouris, E.M., Diakoulaki, D.C., and Papaspyrides, C.D., 1986. Solvent recycling of rigid poly(vinyl chloride) bottles. *J. Vinyl Technol.*, 8(2): 79.
8. Pappa, G.D., Kontogeorgis, G.M., and Tassios, D.P., 1997. Prediction of ternary liquid-liquid equilibria in polymer-solvent-solvent systems. *Ind. Eng. Chem. Res.*, 36: 5461.
9. Coleman, M.M., Graf, J.F., and Painter, P.C., 1995. *Specific Interactions and the Miscibility of Polymer Blends*. Technomic Publishing Co., Inc., Lancaster, PY.
10. Mio, C., Kiritsov, S., Thio, Y., Brafman, R., Prausnitz, J.M., Hawker, C., and Malmstrom, E.E., 1998. Vapor-liquid equilibria for solutions of dendritic polymers. *J. Chem. Eng. Data*, 43: 541–550.
11. Lieu, J.G., Liu, M., Frechet, J.M.J., and Prausnitz, J.M., 1999. Vapor-liquid equilibria for dendritic-polymer solutions. *J. Chem. Eng. Data*, 44: 613–620.
12. von Solms, N., Nielsen, J.K., Hassager, O., Rubin, A., Dandekar, A.Y. Andersen, S.I., and Stenby, E.H., 2004. Direct measurement of gas solubilities in polymers with a high-pressure microbalance. *J. Appl. Polym. Sci.*, 91: 1476–1488.
13. Patterson, R., Yampolskii, Y.P., Fogg, P.G.T., Bokarev, A., Bondar, V., Illinich, O., and Shishatskii, S., 1999. IUPAC-NIST solubility data series 70. The solubility of gases in glassy polymers. *J. Phys. Chem. Ref. Data*, 28(5): 1255–1452.
14. Leitner, W., May 2000. Designed to dissolve. *Nature*, 405: 129–130.
15. Teja, A.S. and Eckert, C.A., 2000. Commentary on supercritical fluids: Research and applications. *Ind. Eng. Chem. Res.*, 39: 4442–4444.
16. Perrut, M., 2000. Supercritical fluid applications: Industrial developments and economic issues. *Ind. Eng. Chem. Res.*, 39: 4531–4535.
17. Hansen, C.M., 1967. The three-dimensional solubility parameter and solvent diffusion coefficient, PhD Thesis. Technical University of Denmark, Copenhagen, MI.
18. Shashoua, Y., 2000. Permanence of plasticizers in polyvinylchloride objects in the museum environment. *Polym. Preprints*, 41(2): 1796–1797.
19. Su, C.S., Patterson, D., and Schreiber, H.P., 1976. Thermodynamic interactions and the properties of the PVC-plasticiser systems. *J. Appl. Polym. Sci.*, 20: 1025–1034.
20. Demertzis, P.G., Riganakos, K.A., and Akrida-Demertzi, K., 1990. Study of compatibility of PVC and polyester-type plasticizer blends by inverse gas chromatography. *Eur. Polym. J.*, 26(2): 137–140.
21. Demertzis, P.G., Riganakos, K.A., and Akrida-Demertzi, K., 1991. An inverse gas chromatographic study of the compatibility of food grade PVdC copolymer and low volatility plasticisers. *Polym. Int.*, 25: 229–236.
22. Prausnitz, J.M., 1995. Some new frontiers in chemical engineering thermodynamics. *Fluid Phase Equilib.*, 104: 1–20.
23. Prausnitz, J.M., 1989. Biotechnology: A new frontier for molecular thermodynamics. *Fluid Phase Equilib.*, 53: 439–451.
24. Seymour, R.B., 1982. *Plastics vs. Corrosives*. SPE Monograph Series. Wiley, New York.
25. Barton, A.F.M., 1983. *Handbook of Solubility Parameters and Other Cohesion Parameters*. CRC Press, Boca Barton, FL.
26. Barton, A.F.M., 1990. *CRC Handbook of Polymer–Liquid Interaction Parameters and Solubility Parameters*. CRC Press, Boca Raton, FL.
27. Hansen, C.M., 2000. *Hansen Solubility Parameters. A User's Handbook*. CRC Press, Boca Raton, FL.
28. Elbro, H.S., 1992. Phase equilibria of polymer solutions—With special emphasis on free volumes, Ph.D. Thesis. Department of Chemical Engineering, Technical University of Denmark, Copenhagen, MI.
29. Klein, J. and Jeberien, H.E., 1980. Chainlength dependence of thermodynamic properties of polyethylene (glycol). *Makromol. Chem.*, 181: 1237.

30. Wen, H., Elbro, H.S., and Alessi, P., 1991. *Polymer Solution Data Collection*. Dechema Chemistry Data Series, Frankfurt, Germany.
31. High, M.S. and Danner, R.P., 1992. *Polymer Solution Handbook*; DIPPR 881 Project. Design Institute for Physical Property Data, PY.
32. Flory, P.J., 1941. Thermodynamics of high polymer solutions. *J. Chem. Phys.*, 9: 660.
33. Huggins, M.L., 1941. Solutions of long chain compounds. *J. Chem. Phys.*, 9: 440.
34. Qian, C., Mumby, S.J., and Eichinger, B.E., 1990. Application of the theory of phase diagrams to binary polymer solutions and blends. *Polym. Preprints*, 31: 621.
35. Elbro, H.S., Fredenslund, Aa., and Rasmussen, P., 1990. A new simple equation for the prediction of solvent activities in polymer solutions. *Macromolecules*, 23: 4707.
36. Lindvig, Th., Michelsen, M.L., and Kontogeorgis, G.M., 2002. A Flory–Huggins model based on the Hansen solubility parameters. *Fluid Phase Equilib.*, 203: 247.
37. Lindvig, Th., Economou, I.G., Danner, R.P., Michelsen, M.L, and Kontogeorgis, G.M., 2004. Modeling of multicomponent vapour-liquid equilibria for polymer-solvent systems. *Fluid Phase Equilib.*, 220: 11–20.
38. Schultz, A.R. and Flory, P.J., 1953. Phase equilibria in polymer-solvent systems. II. Thermodynamic interaction parameters from critical miscibility data. *J. Am. Chem. Soc.*, 75: 496.
39. Bondi, A., 1968. *Physical Properties of Molecular Crystals, Liquids and Glasses*. John Wiley & Sons, New York.
40. Kontogeorgis, G.M., Fredenslund, Aa., and Tassios, D.P., 1993. Simple activity coefficient model for the prediction of solvent activities in polymer solutions. *Ind. Eng. Chem. Res.*, 32: 362.
41. Kouskoumvekaki, I., Michelsen, M.L., and Kontogeorgis, G.M., 2002. An improved entropic expression for polymer solutions. *Fluid Phase Equilib.*, 202(2): 325.
42. Oishi, T. and Prausnitz, M., 1978. Estimation of solvent activities in polymer solutions using a group-contribution method. *Ind. Eng. Chem. Proc. Des. Dev.*, 17(3): 333.
43. Hansen, H.K., Coto, B., and Kuhlmann, B., 1992. UNIFAC with lineary temperature-dependent group-interaction parameters, IVC-SEP Internal Report 9212, Technical University of Denmark.
44. Elbro, H.S., Fredenslund, Aa., and Rasmussen, P., 1991. Group contribution method for the prediction of liquid densities as a function of temperature for solvents, oligomers, and polymers. *Ind. Eng. Chem. Res.*, 30: 2576.
45. Tsibanogiannis, I.N., Kalospiros, N.S., and Tassios, D.P., 1994. Extension of the GCVOL method and application to some complex compounds. *Ind. Eng. Chem. Res.*, 33: 1641.
46. Bogdanic, G. and Fredenslund, Aa., 1994. Revision of the GC-Flory EoS for phase equilibria in mixtures with polymers. 1. Prediction of VLE for polymer solutions. *Ind. Eng. Chem. Res.*, 33: 1331.
47. Bogdanic, G. and Fredenslund, Aa., 1995. Prediction of VLE for mixtures with co-polymers. *Ind. Eng. Chem. Res.*, 34: 324.
48. Lindvig, Th., Hestkjær, L.L., Hansen, A.F., Michelsen, M.L., and Kontogeorgis, G.M., 2002. Phase equilibria for complex polymer solutions. *Fluid Phase Equilib.*, 663: 194.
49. Harismiadis, V.I. and Tassios, D.P., 1996. Solid-liquid-liquid equilibria in polymer solutions. *Ind. Eng. Chem. Res.*, 35: 4667.
50. Kontogeorgis, G.M., Coutsikos, Ph., Tassios, D.P., and Fredenslund, Aa., 1994. Improved models for the prediction of activity coefficients in nearly athermal mixtures. Part I. Empirical modifications of free-volume models. *Fluid Phase Equilib.*, 92: 35.
51. Coutinho, J.A.P., Andersen, S.I., and Stenby, E.H., 1995. Evaluation of activity coefficient models in prediction of alkane SLE. *Fluid Phase Equilib.*, 103: 23.
52. Polyzou, E.N., Vlamos, P.M., Dimakos, G.M., Yakoumis, I.V., and Kontogeorgis, G.M., 1999. Assessment of activity coefficient models for predicting solid-liquid equilibria of asymmetric binary alkane systems. *Ind. Eng. Chem. Res.*, 38: 316–323.
53. Kontogeorgis, G.M., Fredenslund, Aa., Economou, I.G., and Tassios, D.P., 1994. Equations of state and activity coefficient models for vapor-liquid equilibria of polymer solutions. *AIChE J.*, 40: 1711.
54. Bogdanic, G. and Vidal, J., 2000. A segmental interaction model for liquid-liquid equilibrium calculations for polymer solutions. *Fluid Phase Equilib.*, 173: 241–252.
55. Bogdanic, G., 2001. The FV-UNIQUAC segmental interaction model for liquid-liquid equilibrium calculations for polymer solutions. Part 2. Extension to solutions containing polystyrene. *Fluid Phase Equilib.*, 4791: 1–9.
56. Panagou, E., Vidal, J., and Bogdanic, G., 1998. A segmental interaction model for LLE correlation and prediction—Application to the poly(vinyl alcohol)/water system. *Polym. Bull.*, 40: 117.
57. Pappa, G.D., Voutsas, E.C., and Tassios, D.P., 2001. Liquid-liquid phase equilibrium in polymer-solvent systems: Correlation and prediction of the polymer molecular weight and the pressure effect. *Ind. Eng. Chem. Res.*, 40(21): 4654.

58. Kontogeorgis, G.M., Saraiva, A., Fredenslund, Aa., and Tassios, D.P., 1995. Prediction of liquid-liquid equilibrium for binary polymer solutions with simple activity coefficient models. *Ind. Eng. Chem. Res.*, 34: 1823.
59. Harismiadis, V.I., van Bergen, A.R.D., Saraiva, A., Kontogeorgis, G.M., Fredenslund, Aa., and Tassios, D.P., 1996. Miscibility of polymer blends with engineering models. *AIChE J.*, 42: 3170.
60. Pappa, G.D., Voutsas, E.C., and Tassios, D.P., 1999. Prediction of solvent activities in polymer solutions with simple group-contribution models. *Ind. Eng. Chem. Res.*, 38: 4975.
61. Lindvig, Th., Michelsen, M.L., and Kontogeorgis, G.M., 2001. Thermodynamics of paint related systems with engineering models. *AIChE J.*, 47(11): 2573–2584.
62. Kontogeorgis, G.M., Nikolopoulos, G.I., Tassios, D.P., and Fredenslund, Aa., 1997. Improved models for the prediction of activity coefficients in nearly athermal mixtures. Part II. A theoretically based G^E-model based on the van der Waals partition function. *Fluid Phase Equilib.*, 127: 103.
63. Thorlaksen, P., Abildskov, J., and Kontogeorgis, G.M., 2003. Prediction of gas solubilities in elastomeric polymers for the design of thermopane windows. *Fluid Phase Equilib.*, 211: 17.
64. McQuarrie, D.A., 1976. *Statistical Mechanics*. Harper Collins, New York.
65. Wertheim, M.S., 1963. Exact solution of Percus-Yevick integral equation for hard spheres. *Phys. Rev. Lett.*, 10: 321.
66. Carnahan, N.F. and Starling, K.E., 1970. Thermodynamic properties of a rigid-sphere fluid. *J. Chem. Phys.*, 53: 600.
67. Economou, I.G., 2003. Molecular simulation for industrial applications. In: Kontogeorgis, G. and Gani, R., eds., *Computer Aided Property Estimation for Process and Product Design*, Elsevier, Amsterdam, the Netherlands. pp. 279–308.
68. Wertheim, M.S., 1984. Fluids with highly directional attractive forces. I. Statistical thermodynamics. *J. Stat. Phys.*, 35: 19.
69. Wertheim, M.S., 1984. Fluids with highly directional attractive forces. II. Thermodynamic perturbation theory and integral equations. *J. Stat. Phys.*, 35: 35.
70. Wertheim, M.S., 1986. Fluids with highly directional attractive forces. III. Multiple attraction site. *J. Stat. Phys.*, 42: 459.
71. Wertheim, M.S., 1986. Fluids with highly directional attractive forces. IV. Equilibrium polymerization. *J. Stat. Phys.*, 42: 477.
72. Chapman, W.G., Jackson, G., and Gubbins, K.E., 1988. Phase equilibria of associating fluids: Chain molecules with multiple bonding sites. *Mol. Phys.*, 65: 1057.
73. Chapman, W.G., Gubbins, K.E., Jackson, G., and Radosz, M., 1989. SAFT: Equation-of-state solution model for associating liquids. *Fluid Phase Equilib.*, 52: 31.
74. Chapman, W.G., Gubbins, K.E., Jackson, G., and Radosz, M., 1990. New reference equation of state for associating liquids. *Ind. Eng. Chem. Res.*, 29, 1709.
75. Jackson, G., Chapman, W.G., and Gubbins, K.E., 1988. Phase equilibria of associating fluids: Spherical molecules with multiple bonding sites. *Mol. Phys.*, 65: 1.
76. Donohue, M.D. and Prausnitz, J.M., 1978. Perturbed hard chain theory for fluid mixtures—Thermodynamic properties for mixtures in natural-gas and petroleum technology. *AIChE J.*, 24: 849–860.
77. Müller, E.A. and Gubbins, K.E., 2001. Molecular-based equations of state for associating fluids: A review of SAFT and related approaches. *Ind. Eng. Chem. Res.*, 40: 2193.
78. Huang, S.H. and Radosz, M., 1990. Equation of state for small, large, polydisperse and associating molecules. *Ind. Eng. Chem. Res.*, 29: 2284.
79. Huang, S.H. and Radosz, M., 1991. Equation of state for small, large, polydisperse and associating molecules: Extension to fluid mixtures. *Ind. Eng. Chem. Res.*, 30: 1994.
80. Fu, Y.-H. and Sandler, S.I., 1995. A simplified SAFT equation of state for associating compounds and mixtures. *Ind. Eng. Chem. Res.*, 34: 1897.
81. Kontogeorgis, G.M., Voutsas, E.C., Yakoumis, I.V., and Tassios, D.P., 1996. An equation of state for associating fluids. *Ind. Eng. Chem. Res.*, 35: 4310.
82. Michelsen, M.L. and Hendriks, E.H., 2001. Physical properties of association models. *Fluid Phase Equilib.*, 180: 165.
83. Chen, S.S. and Kreglewski, A., 1977. Applications of augmented van der Waals theory of fluids. 1. Pure fluids. *Ber. Bunsen Phys. Chem.*, 81: 1048.
84. Gross, J. and Sadowski, G., 2001. Perturbed-chain SAFT: An equation of state based on a perturbation theory for chain molecules. *Ind. Eng. Chem. Res.*, 40: 1244.
85. Kraska, T. and Gubbins, K.E., 1996. Phase equilibria calculations with a modified SAFT equation of state. 1. Pure alkanes, alkanols and water. *Ind. Eng. Chem. Res.*, 35: 4727.

86. Kraska, T. and Gubbins, K.E., 1996. Phase equilibria calculations with a modified SAFT equation of state. 2. Binary mixtures of *n*-alkanes, 1-alkanols and water. *Ind. Eng. Chem. Res.*, 35: 4738.
87. Kolafa, J. and Nezbeda, I., 1994. The Lennard-Jones fluid: An accurate analytic and theoretically-based equation of state. *Fluid Phase Equilib.*, 100: 1.
88. Gil-Vilegas, A., Galindo, A., Whitehead, P.J., Mills, S.J., Jackson, G., and Burgess, A.N., 1997. Statistical associating fluid theory for chain molecules with attractive potentials of variable range. *J. Chem. Phys.*, 106: 4168.
89. McCabe, C., Gil-Vilegas, A., and Jackson, G., 1999. Gibbs ensemble computer simulation and SAFT-VR theory of non-conformal square-well monomer-dimer mixtures. *Chem. Phys. Lett.*, 303: 27.
90. O'Lenick, R., Li, X.J., and Chiew, Y.C., 1995. Correlation functions of hard-sphere chain mixtures: Integral equation theory and simulation results. *Mol. Phys.*, 86: 1123.
91. Gross, J. and Sadowski, G., 2002. Modeling polymer systems using the perturbed-chain statistical associating fluid theory equation of state. *Ind. Eng. Chem. Res.*, 41: 1084.
92. Tumakaka, F., Gross, J., and Sadowski, G., 2002. Modeling of polymer phase equilibria using the perturbed-chain SAFT. *Fluid Phase Equilib.*, 194–197: 541.
93. Gross, J. and Sadowski, G., 2002. Application of the perturbed-chain SAFT equation of state to associating systems. *Ind. Eng. Chem. Res.*, 41: 5510.
94. Gross, J., Spuhl, O., Tumakaka, F., and Sadowski, G., 2003. Modeling copolymer systems using the perturbed-chain SAFT equation of state. *Ind. Eng. Chem. Res.*, 42: 1266.
95. von Solms, N., Michelsen, M.L., and Kontogeorgis, G.M., 2003. Computational and physical performance of a modified PC-SAFT equation of state for highly asymmetric and associating mixtures. *Ind. Eng. Chem. Res.*, 42: 1098.
96. Kouskoumvekaki, I.A., von Solms, N., Michelsen, M.L., and Kontogeorgis, G.M., 2004. Application of a simplified perturbed chain SAFT equation of state to complex polymer systems. *Fluid Phase Equilib.*, 215: 71–78.
97. von Solms, N., Kouskoumvekaki, I.A., Lindvig, T., Michelsen, M.L., and Kontogeorgis, G.M., 2004. A novel approach to liquid-liquid equilibrium in polymer systems with application to simplified PC-SAFT. *Fluid Phase Equilib.*, 222–223: 87–93.
98. Lindvig, T., Michelsen, M.L., and Kontogeorgis, G.M., 2004. Liquid-liquid equilibria for binary and ternary polymer solutions with PC-SAFT. *Ind. Eng. Chem. Res.*, 43: 1125–1132.
99. von Solms, N., Michelsen, M.L., and Kontogeorgis, G.M., 2005. Prediction and correlation of high-pressure gas solubility in polymers with simplified PC-SAFT. *Ind. Eng. Chem. Res.*, 44: 3330.
100. Haslam, A.J, von Solms, N., Adjiman, C.S., Galindo, A., Jackson, G., Paricaud, P., Michelsen, M.L., and Kontogeorgis, G.M., 2006. Predicting enhanced absorption of light gases in polyethylene using simplified PC-SAFT and SAFT-VR. *Fluid Phase Equilib.*, 243: 74.
101. Jog, P.K., Sauer, S.G., Blaesing, J., and Chapman, W.G., 2001. Application of dipolar chain theory to the phase behavior of polar fluids and mixtures. *Ind. Eng. Chem. Res.*, 40: 4641.
102. Gross, J., 2005. An equation-of-state contribution for polar components: Quadrupolar molecules. *AIChE J.*, 51: 2556–2568.
103. Gross, J. and Vrabec, J., 2006. An equation-of-state contribution for polar components: Dipolar molecules. *AIChE J.*, 52: 1194–1204.
104. Kleiner, M. and Gross, J., 2006. An equation of state contribution for polar components: Polarizable dipoles. *AIChE J.*, 52: 1951–1961.
105. Behzadi, B., Patel, B.H., Galindo, A., and Ghotbi, C., 2005. Modeling electrolyte solutions with the SAFT-VR equation using Yukawa potentials and the mean-spherical approximation. *Fluid Phase Equilib.*, 236: 241–255.
106. Cameretti, L.F., Sadowski, G., and Mollerup, J.M., 2005. Modeling of aqueous electrolyte solutions with perturbed-chain statistical associated fluid theory. *Ind. Eng. Chem. Res.*, 44, 3355–3362.
107. Le Thi, C., Tamouza, S., Passarello, J.P., Tobaly, P., de Hemptinne, J.-C., 2006. Modeling phase equilibrium of H-2+n-alkane and CO_2+n-alkane binary mixtures using a group contribution statistical association fluid theory equation of state (GC-SAFT-EOS) with a k(ij) group contribution method. *Ind. Eng. Chem. Res.*, 45: 6803–6810.
108. Kouskoumvekaki, I.A., von Solms, N., Lindvig, T., Michelsen, M.L., and Kontogeorgis, G.M., 2004. Novel method for estimating pure-component parameters for polymers: Application to the PC-SAFT equation of state. *Ind. Eng. Chem. Res.*, 43: 2830–2838.
109. Dobashi, T., Nakata, M., and Kaneko, M., 1984. Coexistence curve of polystyrene in methylcyclohexane. 3. Asymptotic-behavior of ternary-system near the Plait point. *J. Chem. Phys.*, 80: 948–953.
110. Shultz, A.R. and Flory, P.J., 1952. Phase equilibria in polymer-solvent systems. *J. Am. Chem. Soc.*, 74: 4760–4767.

111. Jain, S., Jog, P., Weinhold, J., Srivastava, R., and Chapman, W.G., 2008. Modified interfacial statistical associating fluid theory: Application to tethered polymer chains. *J. Chem. Phys.*, 128: 154910.
112. Bymaster, A. and Chapman, W.G., 2010. An iSAFT density functional theory for associating polyatomic molecules. *J. Phys. Chem. B*, 114: 12298.
113. Tamouza, S., Passarello, J.P., Tobaly, P., de Hemptinne, J.C., 2004. Group contribution method with SAFT EOS applied to vapor liquid equilibria of various hydrocarbon series. *Fluid Phase Equilib.*, 222–223: 67.
114. Grandjean, L., de Hemptinne, J.-C., and Lugo, R., 2014. Application of GC-PPC-SAFT EoS to ammonia and its mixtures. *Fluid Phase Equilib.*, 367: 159.
115. Tihic, A., von Solms, N., Michelsen, M.L., Kontogeorgis, G.M., and Constantinou, L., 2009. Application of sPC-SAFT and group contribution sPC-SAFT to polymer systems—Capabilities and limitations. *Fluid Phase Equilib.*, 281: 70.
116. Tihic, A., von Solms, N., Michelsen, M.L., Kontogeorgis, G.M., and Constantinou, L., 2009. Analysis and applications of a group contribution sPC-SAFT equation of state. *Fluid Phase Equilib.*, 281: 60.
117. Tihic, A., Kontogeorgis, G.M., von Solms, N., and Michelsen, M.L., 2008. A predictive group-contribution simplified PC-SAFT equation of state: Application to polymer systems. *Ind. Eng. Chem. Res.*, 47: 5092.
118. Peng, Y., Goff, K.D., dos Ramos, M.C., McCabe, C., 2010. Predicting the phase behavior of polymer systems with the GC-SAFT-VR approach. *Ind. Eng. Chem. Res.*, 49: 1378.
119. Clark, G.N.I., Galindo, A., Jackson, G., Rogers, S., and Burgess, A.N., 2008. Modeling and understanding closed-loop liquid-liquid immiscibility in aqueous solutions of poly(ethylene glycol) using the SAFT-VR approach with transferable parameters. *Macromolecules*, 41: 6582.
120. Sanchez, F.A., Pereda, S., and Brignole, E.A., 2011. GCA-EoS: A SAFT group contribution model-extension to mixtures containing aromatic hydrocarbons and associating compounds. *Fluid Phase Equilib.*, 306: 112.
121. Peters, F.T., Laube, F.S., Sadowski, G., 2012. Development of a group contribution method for polymers within the PC-SAFT model. *Fluid Phase Equilib.*, 324: 70.
122. Papaioannou, V., Lafitte, T., Avendano, C., Adjiman, C.S., Jackson, G., Muller, E.A., and Galindo, A., 2014. Group contribution methodology based on the statistical associating fluid theory for heteronuclear molecules formed from Mie segments. *J. Chem. Phys.*, 140: 054107.
123. Cameretti, L.F. and Sadowski, G., 2008. Modeling of aqueous amino acid and polypeptide solutions with PC-SAFT. *Chem. Eng. Proc.*, 47: 1018.
124. Kozlowska, M.K., Jurgens, B.F., Schacht, C.S., Gross, J., and de Loos, T.W., 2009. Phase behavior of hyperbranched polymer systems: Experiments and application of the perturbed-chain polar SAFT equation of state. *J. Phys. Chem. B*, 113: 1022.
125. Vargas, F.M., Gonzalez, D.L., Hirasaki, G.J., and Chapman, W.G., 2009. Modeling asphaltene phase behavior in crude oil systems using the perturbed chain form of the statistical associating fluid theory (PC-SAFT) equation of state. *Energy Fuels*, 23: 1140.
126. von Solms, N. and Kristensen, J., 2010. Refrigeration plants using carbon dioxide as refrigerant: Measuring and modelling the solubility, diffusivity and permeability of carbon dioxide in polymers used as packing and sealing materials. *Int. J. Refrig.*, 33: 19.
127. Neela, V. and von Solms, N., 2014. Permeability, diffusivity and solubility of carbon dioxide in fluoropolymers: An experimental and modeling study. *J. Polym. Res.*, 21: 401.
128. Krallis, A. and Kanellopoulos, V., 2013. Application of Sanchez-Lacombe and perturbed-chain statistical associating fluid theory equation of state models in catalytic olefins (co)polymerization industrial applications. *Ind. Eng. Chem. Res.*, 52: 9060.
129. Tan, S.P., Adidharma, H., and Radosz, M., 2008. Recent advances and applications of statistical associating fluid theory. *Ind. Eng. Chem. Res.*, 47: 8063.
130. Sadowski, G., 2011. Modeling of polymer phase equilibria using equations of state. In: Enders, S. and Wolf, B.A., eds. *Polymer Thermodynamics: Liquid Polymer-Containing Mixtures*. Advances in Polymer Science, Vol. 238, p. 389, Springer-Verlag, Berlin, Heidelberg.
131. Kleiner, M., Turnakaka, F., and Sadowski, G., 2009. Thermodynamic modeling of complex systems. In: Lu, X. and Hu, Y., eds. *Molecular Thermodynamics of Complex Systems*. Structure and Bonding, Vol. 131, p. 75, Springer-Verlag, Berlin, Heidelberg.
132. Tanbonliong, J.O. and Prausnitz, J.M., 1997. Vapour-liquid equilibria for some binary and ternary polymer solutions. *Polymer*, 38: 5775.
133. Matsumara, K. and Katayama, T., 1974. Katayama, T. Mateumura, K. and Urahama, Y. 1971. Measurement of Vapor-Liquid Equilibria of Binary and, ternary Solutions containing polystyrene as a Component. *Kagaku Kogaku*, 38: 388.
134. Tanbonliong, J.O. and Prausnitz, J.M., 1997. Vapor-Liquid Equilibria of Binary and Ternary Solutions Containing Polyvinylacetate as a Component. *Polymer*, 38: 5775.

4 Chemical Physics of Colloid Systems and Interfaces

Peter A. Kralchevsky and Krassimir D. Danov

CONTENTS

4.1 INTRODUCTION

A *colloidal system* represents a multiphase (heterogeneous) system in which at least one of the phases exists in the form of very small particles: typically smaller than 1 μm but still much larger than the molecules. Such particles are related to phenomena like Brownian motion, diffusion, and osmosis. The terms "microheterogeneous system" and "disperse system" (dispersion) are more general because they also include bicontinuous systems (in which none of the phases is split into separate particles) and systems containing larger, non-Brownian particles. The term dispersion is often used as a synonym of colloidal system.

A classification of the colloids with respect to the state of aggregation of the *disperse* and *continuous* phases is shown in Table 4.1. Some examples are the following.

1. Examples for *gas-in-liquid* dispersions are the foams or the boiling liquids. *Gas-in-solid* dispersions are the various porous media like filtration membranes, sorbents, catalysts, isolation materials, etc.
2. Examples for *liquid-in-gas* dispersions are the mist, the clouds, and other aerosols. *Liquid-in-liquid* dispersions are the emulsions. At room temperature there are only four types of mutually immiscible liquids: water, hydrocarbon oils, fluorocarbon oils, and liquid metals (Hg and Ga). Many raw materials and products in food and petroleum industries exist in the form of oil-in-water or water-in-oil emulsions. The soil and some biological tissues can be considered as examples of *liquid-in-solid* dispersions.

TABLE 4.1
Types of Disperse Systems

	Continuous Phase		
Disperse Phase	**Gas**	**Liquid**	**Solid**
Gas	—	G in L	G in S
Liquid	L in G	L_1 in L_2	L in S
Solid	S in G	S in L	S_1 in S_2

3. Smoke, dust, and some other aerosols are examples for *solid-in-gas* dispersions. The *solid-in-liquid* dispersions are termed *suspensions* or *sols*. The pastes and some glues are highly concentrated suspensions. The *gels* represent *bicontinuous* structures of solid and liquid. The pastes and some glues are highly concentrated suspensions. Examples for *solid-in-solid* dispersions are some metal alloys, many kinds of rocks, some colored glasses, etc.

In the following section, we will consider mostly liquid dispersions, that is, dispersions with liquid continuous phase, like foams, emulsions and suspensions. Sometimes these are called "complex fluids."

In general, the area of the interface between the disperse and continuous phases is rather large. For instance, 1 cm^3 of dispersion phase with particles of radius 100 nm and volume fraction 30% contains interface of area about 10 m^2. This is the reason why the interfacial properties are of crucial importance for the properties and stability of colloids.

The *stabilizing* factors for dispersions are the repulsive surface forces, the particle thermal motion, the hydrodynamic resistance of the medium, and the high surface elasticity of fluid particles and films.

On the contrary, the factors *destabilizing* dispersions are the attractive surface forces, the factors suppressing the repulsive surface forces, the low surface elasticity, gravity, and other external forces tending to separate the phases.

In Sections 4.2 and 4.3 we consider effects related to the surface tension of surfactant solution and capillarity. In Section 4.4 we present a review on the surface forces due to intermolecular interactions. In Section 4.5 we describe the hydrodynamic interparticle forces originating from the effects of bulk and surface viscosity and related to surfactant diffusion. Finally, Section 4.6 is devoted to the kinetics of coagulation in dispersions.

4.2 SURFACE TENSION OF SURFACTANT SOLUTIONS

4.2.1 Static Surface Tension

As a rule, the fluid dispersions (emulsions, foams) are stabilized by adsorption layers of amphiphilic molecules. These can be ionic [1,2] and nonionic [3] surfactants, lipids, proteins, etc. (see Chapter 4 of this Handbook). All of them have the property to lower the value of the surface (or interfacial) tension, σ, in accordance with the *Gibbs adsorption equation* [4–6],

$$d\sigma = -\sum_i \Gamma_i d\mu_i \tag{4.1}$$

where

Γ_i is the surface concentration (adsorption) of the ith component
μ_i is its chemical potential

The summation in Equation 4.1 is carried out over all components. Usually an equimolecular dividing surface with respect to the solvent is introduced for which the adsorption of the solvent is set zero by definition [4,5]. Then the summation is carried out over all other components. Note that Γ_i is an excess surface concentration with respect to the bulk; Γ_i is positive for surfactants, which decreases σ in accordance with Equation 4.1. On the contrary, Γ_i is negative for aqueous solutions of electrolytes, whose ions are repelled from the surface by the electrostatic image forces [5]; consequently, the addition of electrolytes increases the surface tension of water [6]. For surfactant concentrations above the critical micellization concentration (CMC) μ_i = constant and, consequently, σ = constant (see Equation 4.1).

4.2.1.1 Nonionic Surfactants

4.2.1.1.1 Types of Adsorption Isotherms

Consider the boundary between an aqueous solution of a nonionic surfactant and a hydrophobic phase, air or oil. The dividing surface is usually chosen to be the equimolecular surface with respect to water, that is $\Gamma_w = 0$. Then Equation 4.1 reduces to $d\sigma = -\Gamma_1 d\mu_1$, where the subscript "1" denotes the surfactant. Because the bulk surfactant concentration is usually not too high, we can use the expression for the chemical potential of a solute in an ideal solution: $\mu_1 = \mu_1^{(0)} + kT \ln c_1$, where k is the Boltzmann constant, T is the absolute temperature, c_1 is the concentration of nonionic surfactant, and $\mu_1^{(0)}$ is its standard chemical potential, which is independent of c_1. Thus the Gibbs adsorption equation acquires the form

$$d\sigma = -kT\Gamma_1 d\ln c_1 \tag{4.2}$$

The surfactant adsorption isotherms, expressing the connection between Γ_1 and c_1, are usually obtained by means of some molecular model of adsorption. Table 4.2 contains the six most popular surfactant adsorption isotherms, those of Henry, Freundlich [7], Langmuir [8], Volmer [9], Frumkin [10] and van der Waals [11]. For $c_1 \to 0$ all isotherms (except that of Freundlich) reduce to the Henry isotherm: $\Gamma_1/\Gamma_\infty = Kc_1$. The physical difference between the Langmuir and Volmer isotherms is that the former corresponds to a physical model of *localized* adsorption, whereas the latterto *nonlocalized* adsorption. The Frumkin and van der Walls isotherms generalize, respectively, the Langmuir and Volmer isotherms for case, in which the interaction between neighboring adsorbed molecules is not negligible. (If the interaction parameter β is set zero, the Frumkin and van der Walls isotherms reduce to the Langmuir and Volmer isotherms, correspondingly.) The comparison between theory and experiment shows that for air–water interfaces $\beta > 0$, whereas for oil–water interfaces we can set $\beta = 0$ [12,13]. The latter facts lead to the conclusion that for air–water interfaces β takes into account the van der Waals attraction between the hydrocarbon tails of the adsorbed surfactant molecules across air; such attraction is missing when the hydrophobic phase is oil. (Note that in the case of ionic surfactants it is possible to have $\beta < 0$, see the next section.) The adsorption parameter K in Table 4.2 characterizes the surface activity of the surfactant: the greater K, the higher the surface activity. K is related to the standard free energy of adsorption, $\Delta G^\circ = \mu_{1s}^{(0)} - \mu_1^{(0)}$, which is the energy gain for bringing a molecule from the bulk of the aqueous phase to a diluted adsorption layer [14,15]:

$$K = \frac{\delta_1}{\Gamma_\infty} \exp\left(\frac{\mu_1^{(0)} - \mu_{1s}^{(0)}}{kT}\right) \tag{4.3}$$

The parameter δ_1 characterizes the thickness of the adsorption layer; δ_1 can be set (approximately) equal to the length of the amphiphilic molecule. Γ_∞ represents the maximum possible value of the adsorption. In the case of localized adsorption (Langmuir and Frumkin isotherms), $1/\Gamma_\infty$ is the area

TABLE 4.2
Types of Adsorption and Surface Tension Isotherms

Type of Isotherm	Surfactant Adsorption Isotherms (for Nonionic Surfactants: $a_{1s} \equiv c_1$)
Henry	$Ka_{1s} = \frac{\Gamma_1}{\Gamma_\infty}$
Freundlich	$Ka_{1s} = \left(\frac{\Gamma_1}{\Gamma_\infty}\right)^{1/m}$
Langmuir	$Ka_{1s} = \frac{\Gamma_1}{\Gamma_\infty - \Gamma_1}$
Volmer	$Ka_{1s} = \frac{\Gamma_1}{\Gamma_\infty - \Gamma_1} \exp\left(\frac{\Gamma_1}{\Gamma_\infty - \Gamma_1}\right)$
Frumkin	$Ka_{1s} = \frac{\Gamma_1}{\Gamma_\infty - \Gamma_1} \exp\left(-\frac{2\beta\Gamma_1}{kT}\right)$
van der Waals	$Ka_{1s} = \frac{\Gamma_1}{\Gamma_\infty - \Gamma_1} \exp\left(\frac{\Gamma_1}{\Gamma_\infty - \Gamma_1} - \frac{2\beta\Gamma_1}{kT}\right)$
	Surface Tension Isotherm $\sigma = \sigma_0 - kTJ + \sigma_d$ (for Nonionic Surfactants: $\sigma_d \equiv 0$)
Henry	$J = \Gamma_1$
Freundlich	$J = \frac{\Gamma_1}{m}$
Langmuir	$J = -\Gamma_\infty \ln\left(1 - \frac{\Gamma_1}{\Gamma_\infty}\right)$
Volmer	$J = \frac{\Gamma_\infty \Gamma_1}{\Gamma_\infty - \Gamma_1}$
Frumkin	$J = -\Gamma_\infty \ln\left(1 - \frac{\Gamma_1}{\Gamma_\infty}\right) - \frac{\beta\Gamma_1^2}{kT}$
van der Waals	$J = \frac{\Gamma_\infty \Gamma_1}{\Gamma_\infty - \Gamma_1} - \frac{\beta\Gamma_1^2}{kT}$

Note: The surfactant adsorption isotherm and the surface tension isotherm, which are combined to fit experimental data, obligatorily must be of the same type.

per adsorption site. In the case of nonlocalized adsorption (Volmer and van der Waals isotherms), $1/\Gamma_\infty$ is the excluded area per molecule.

The standard free energy of surfactant adsorption, ΔG^o, can be determined by nonlinear fits of surface tension isotherms with the help of a theoretical model of adsorption. The models of Frumkin, van der Waals, and Helfant–Frisch–Lebowitz have been applied, and the results have been compared [16]. Irrespective of the differences between these models, they give close values for the standard free energy because all of them reduce to the Henry isotherm for diluted adsorption layers. The results from the theoretical approach have been compared with those of the most popular empirical approach [17]. The latter gives values of the standard free energy, which are considerably different from the respective true values, with ca. 10 kJ/mol for nonionic surfactants, and 20 kJ/mol for ionic surfactants. These differences are due to contributions from interactions between the molecules in dense adsorption layers. The true values of the standard free energy can be determined with the help of an appropriate theoretical model. The van der Waals model was found to give the best results, especially for the determination of the standard adsorption enthalpy ΔH^o and entropy ΔS^o from the temperature dependence of surface tension [16].

As already mentioned, the Freundlich adsorption isotherm, unlike the other ones in Table 4.2, does not become linear at low concentrations, but remains convex to the concentration axis. Moreover, it does not show saturation or limiting value. Hence, for the Freundlich adsorption isotherm in Table 4.2 Γ_∞ is a parameter scaling the adsorption (rather than saturation adsorption). This isotherm can be derived assuming that the surface (as a rule solid) is heterogeneous [18,19]. Consequently, if the data fit the Freundlich equation, this is an indication, but not a proof, that the surface is heterogeneous [6].

The *adsorption* isotherms in Table 4.2 can be applied to both fluid and solid interfaces. The *surface tension* isotherms in Table 4.2, which relate σ and Γ_1, are usually applied to fluid interfaces, although they could also be used also for solid–liquid interfaces if σ is identified with the Gibbs [4] *superficial* tension. (The latter is defined as the force per unit length which opposes every increase of the wet area without any deformation of the solid.)

The surface tension isotherms in Table 4.2 are deduced from the respective adsorption isotherms in the following way. The integration of Equation 4.2 yields

$$\sigma = \sigma_0 - kTJ \tag{4.4}$$

where σ_0 is the interfacial tension of the pure solvent and

$$J \equiv \int_0^{c_1} \Gamma_1 \frac{dc_1}{c_1} = \int_0^{\Gamma_1} \Gamma_1 \frac{d \ln c_1}{d\Gamma_1} d\Gamma_1 \tag{4.5}$$

The derivative $d \ln c_1/d\Gamma_1$ is calculated for each adsorption isotherm, and then the integration in Equation 4.5 is carried out analytically. The obtained expressions for J are listed in Table 4.2. Each surface tension isotherm, $\sigma(\Gamma_1)$, has the meaning of a *2D equation of state* of the adsorption monolayer, which can be applied to both *soluble* and *insoluble* surfactants [6,20].

An important thermodynamic property of a surfactant adsorption monolayer is its Gibbs (surface) elasticity

$$E_G \equiv -\Gamma_1 \left(\frac{\partial \sigma}{\partial \Gamma_1} \right)_T \tag{4.6}$$

Expressions for E_G, corresponding to various adsorption isotherms, are shown in Table 4.3. The Gibbs elasticity characterizes the lateral fluidity of the surfactant adsorption monolayer. At high values of the Gibbs elasticity the adsorption monolayer behaves as tangentially immobile. In such case, if two emulsion droplets approach each other, the hydrodynamic flow pattern, and the hydrodynamic interaction as well, is almost the same as if the droplets were solid. For lower values of the surfactant adsorption the so-called "Marangoni effect" appears, which is equivalent to appearance of gradients of surface tension due to gradients of surfactant adsorption: $\nabla_s\sigma = -(E_G/\Gamma_1)\nabla_s\Gamma_1$ (here ∇_s denotes surface gradient operator). The Marangoni effect can considerably affect the hydrodynamic interactions of fluid particles (drops, bubbles), (see Section 4.5).

4.2.1.1.2 Derivation from First Principles

Each surfactant adsorption isotherm (that of Langmuir, Volmer, Frumkin, etc.), and the related expressions for the surface tension and surface chemical potential, can be derived from an expression for the surface free energy, F_s, which corresponds to a given physical model. This derivation helps us obtain (or identify) the self-consistent system of equations, referring to a given model,

TABLE 4.3
Elasticity of Adsorption Monolayers at a Fluid Interface

Type of Isotherm (cf. Table 4.2)	Gibbs Elasticity E_G
Henry	$E_G = kT\Gamma_1$
Freundlich	$E_G = kT\dfrac{\Gamma_1}{m}$
Langmuir	$E_G = kT\Gamma_1\dfrac{\Gamma_\infty}{\Gamma_\infty - \Gamma_1}$
Volmer	$E_G = kT\Gamma_1\dfrac{\Gamma_\infty^2}{(\Gamma_\infty - \Gamma_1)^2}$
Frumkin	$E_G = kT\Gamma_1\left(\dfrac{\Gamma_\infty}{\Gamma_\infty - \Gamma_1} - \dfrac{2\beta\Gamma_1}{kT}\right)$
van der Waals	$E_G = kT\Gamma_1\left[\dfrac{\Gamma_\infty^2}{(\Gamma_\infty - \Gamma_1)^2} - \dfrac{2\beta\Gamma_1}{kT}\right]$

Note: The above expressions are valid for both nonionic and ionic surfactants.

which is to be applied to interpret a set of experimental data. Combination of equations corresponding to different models (say Langmuir adsorption isotherm with Frumkin surface tension isotherm) is incorrect and must be avoided.

The general scheme for derivation of the adsorption isotherms is the following:

1. With the help of statistical mechanics an expression is obtained, say, for the canonical ensemble partition function, Q, from which the surface free energy F_s is determined [11]:

$$F_s(T, A, N_1) = -kT \ln Q(T, A, N_1) \tag{4.7}$$

where
A is the interfacial area
N_1 is the number of adsorbed surfactant molecules; see Table 4.4

2. Differentiating the expression for F_s, we derive expressions for the *surface pressure*, π_s, and the *surface chemical potential* of the adsorbed surfactant molecules, μ_{1s} [11]:

$$\pi_s \equiv \sigma_0 - \sigma = -\left(\frac{\partial F_s}{\partial A}\right)_{T,N_1}, \quad \mu_{1s} = \left(\frac{\partial F_s}{\partial N_1}\right)_{T,A} \tag{4.8}$$

Combining the obtained expressions for π_s and μ_{1s}, we can deduce the respective form of the Butler equation [21], see Equation 4.16.

3. The surfactant adsorption isotherm (Table 4.2) can be derived by setting the obtained expression for the surface chemical potential μ_{1s} equal to the bulk chemical potential of the surfactant molecules in the subsurface layer (i.e., equilibrium between surface and subsurface is assumed) [11]:

$$\mu_{1s} = \mu_1^{(0)} + kT \ln\left(\frac{a_{1s}\delta_1}{\Gamma_\infty}\right) \tag{4.9}$$

TABLE 4.4
Free Energy and Chemical Potential for Surfactant Adsorption Layers

Type of Isotherm	Surface Free Energy $F_s(T, A, N_1)$ $(M = \Gamma_\infty A)$
Henry	$F_s = N_1\mu_{1s}^{(0)} + kT\left[N_1 \ln\left(\frac{N_1}{M}\right) - N_1\right]$
Freundlich	$F_s = N_1\mu_{1s}^{(0)} + \frac{kT}{m}\left[N_1 \ln\left(\frac{N_1}{M}\right) - N_1\right]$
Langmuir	$F_s = N_1\mu_{1s}^{(0)} + kT[N_1 \ln N_1 + (M - N_1)\ln(M - N_1) - M\ln M]$
Volmer	$F_s = N_1\mu_{1s}^{(0)} + kT[N_1 \ln N_1 - N_1 - N_1\ln(M - N_1)]$
Frumkin	$F_s = N_1\mu_{1s}^{(0)} + kT[N_1 \ln N_1 + (M - N_1)\ln(M - N_1) - M\ln M] + \frac{\beta\Gamma_\infty N_1^2}{2M}$
van der Waals	$F_s = N_1\mu_{1s}^{(0)} + kT[N_1 \ln N_1 - N_1 - N_1\ln(M - N_1)] + \frac{\beta\Gamma_\infty N_1^2}{2M}$
	Surface Chemical Potential μ_{1s} $(\theta \equiv \Gamma_1/\Gamma_\infty)$
Henry	$\mu_{1s} = \mu_{1s}^{(0)} + kT\ln\theta$
Freundlich	$\mu_{1s} = \mu_{1s}^{(0)} + \frac{kT}{m}\ln\theta$
Langmuir	$\mu_{1s} = \mu_{1s}^{(0)} + kT\ln\frac{\theta}{1-\theta}$
Volmer	$\mu_{1s} = \mu_{1s}^{(0)} + kT\left(\frac{\theta}{1-\theta} + \ln\frac{\theta}{1-\theta}\right)$
Frumkin	$\mu_{1s} = \mu_{1s}^{(0)} + kT\ln\frac{\theta}{1-\theta} - 2\beta\Gamma_1$
van der Waals	$\mu_{1s} = \mu_{1s}^{(0)} + kT\left(\frac{\theta}{1-\theta} + \ln\frac{\theta}{1-\theta}\right) - 2\beta\Gamma_1$

Here a_{1s} is the activity of the surfactant molecule in the subsurface layer; a_{1s} is scaled with the volume per molecule in a dense (saturated) adsorption layer, $v_1 = \delta_1/\Gamma_\infty$, where δ_1 is interpreted as the thickness of the adsorption layer, or the length of an adsorbed molecule. In terms of the subsurface activity, a_{1s}, Equation 4.9 can be applied to ionic surfactants and to dynamic processes. In the simplest case of nonionic surfactants and equilibrium processes we have $a_{1s} \approx c_1$, where c_1 is the bulk surfactant concentration.

First, let us apply the general scheme mentioned earlier to derive the *Frumkin isotherm*, which corresponds to *localized* adsorption of *interacting* molecules. (Expressions corresponding to the Langmuir isotherm can be obtained by setting $\beta = 0$ in the respective expressions for the Frumkin isotherm.) Let us consider the interface as a 2D lattice having M adsorption sites. The corresponding partition function is [11]

$$Q(T,M,N_1) = \frac{M!}{N_1!\,(M-N_1)!}[q(T)]^{N_1}\exp\left(-\frac{n_c w N_1^2}{2kTM}\right) \tag{4.10}$$

The first multiplier in the right-hand side of Equation 4.10 expresses the number of ways N_1 indistinguishable molecules can be distributed among M labeled sites; the partition function for a single adsorbed molecule is $q = q_x q_y q_z$, where q_x, q_y, and q_z are 1D harmonic-oscillator partition functions. The exponent in Equation 4.10 accounts for the interaction between adsorbed molecules in the framework of the Bragg–Williams approximation [11]. w is the nearest-neighbor interaction energy of two molecules and n_c is the number of nearest-neighbor sites to a given site (for example $n_c = 4$ for a square lattice). Next, we substitute Equation 4.10 into Equation 4.7 and using the known Stirling approximation, $\ln M! = M \ln M - M$, we get the expression for the surface free energy corresponding to the Frumkin model:

$$F_s = kT[N_1 \ln N_1 + (M - N_1)\ln(M - N_1) - M \ln M - N_1 \ln q(T)] + \frac{n_c w N_1^2}{2M} \tag{4.11}$$

Note that

$$M = \Gamma_\infty A, \quad N_1 = \Gamma_1 A \tag{4.12}$$

where Γ_∞^{-1} is the area per one adsorption site in the lattice. Differentiating Equation 4.11 in accordance with Equation 4.8, we deduce expressions for the surface pressure and chemical potential [11]:

$$\pi_s = -\Gamma_\infty kT \ln(1 - \theta) - \beta\Gamma_1^2 \tag{4.13}$$

$$\mu_{1s} = \mu_{1s}^{(0)} + kT \ln\frac{\theta}{1-\theta} - 2\beta\Gamma_1 \tag{4.14}$$

where we have introduced the notation

$$\theta = \frac{\Gamma_1}{\Gamma_\infty}, \quad \beta = -\frac{n_c w}{2\Gamma_\infty}, \quad \mu_{1s}^{(0)} = -kT \ln q(T) \tag{4.15}$$

We can check that Equation 4.13 is equivalent to the Frumkin's surface tension isotherm in Table 4.2 for a nonionic surfactant. Furthermore, eliminating $\ln(1 - \theta)$ between Equations 4.13 and 4.14, we obtain the Butler [21] equation in the form

$$\mu_{1s} = \mu_{1s}^{(0)} + \Gamma_\infty^{-1}\pi_s + kT \ln(\gamma_{1s}\theta) \quad \text{Butler equation} \tag{4.16}$$

where we have introduced the surface activity coefficient

$$\gamma_{1s} = \exp\left[-\frac{\beta\Gamma_\infty\theta(2-\theta)}{kT}\right] \quad \text{(for Frumkin isotherm)} \tag{4.17}$$

(In the special case of Langmuir isotherm we have $\beta = 0$, and then $\gamma_{1s} = 1$.) The Butler equation is used by many authors [12,22–24] as a starting point for the development of thermodynamic adsorption models. It should be kept in mind that the specific form of the expressions for π_s and γ_{1s}, which are to be substituted in Equation 4.16, is not arbitrary, but must correspond to the *same* thermodynamic model (to the same expression for F_s—in our case Equation 4.11). At last, substituting Equation 4.16 into Equation 4.9, we derive the Frumkin adsorption isotherm in Table 4.2, where K is defined by Equation 4.3.

Now, let us apply the same general scheme, but this time to the derivation of the *van der Waals isotherm*, which corresponds to *nonlocalized* adsorption of *interacting* molecules. (Expressions corresponding to the Volmer isotherm can be obtained by setting $\beta = 0$ in the respective expressions for the van der Waals isotherm.) Now the adsorbed N_1 molecules are considered as a 2D gas. The corresponding expression for the canonical ensemble partition function is

$$Q(T,M,N_1) = \frac{1}{N_1!} q^{N_1} \exp\left(-\frac{n_c w N_1^2}{2kTM} \right) \tag{4.18}$$

where the exponent accounts for the interaction between adsorbed molecules, again in the framework of the Bragg–Williams approximation. The partition function for a single adsorbed molecule is $q = q_{xy}q_z$, where q_z is 1D (normal to the interface) harmonic-oscillator partition function. On the other hand, the adsorbed molecules have free translational motion in the xy-plane (the interface); therefore we have [11]

$$q_{xy} = \frac{2\pi \tilde{m} kT}{h_p^2} \hat{A} \tag{4.19}$$

where

$\tilde{m}$ is the molecular mass

h_p is the Planck constant

$\hat{A} = A - N_1\Gamma_\infty^{-1}$ is the area accessible to the moving molecules; the parameter Γ_∞^{-1} is the excluded area per molecule, which accounts for the molecular size

Having in mind that $M \equiv \Gamma_\infty A$, we can bring Equation 4.18 into the form

$$Q(T,M,N_1) = \frac{1}{N_1!} q_0^{N_1} (M - N_1)^{N_1} \exp\left(-\frac{n_c w N_1^2}{2kTM} \right) \tag{4.20}$$

where

$$q_0(T) \equiv \frac{2\pi \tilde{m} kT}{h_p^2 \Gamma_\infty} q_z(T) \tag{4.21}$$

Further, we substitute Equation 4.20 into Equation 4.7 and, using the Stirling approximation, we determine the surface free energy corresponding to the van der Waals model [11,20,25]:

$$F_s = kT[N_1 \ln N_1 - N_1 - N_1 \ln q_0(T) - N_1 \ln(M - N_1)] + \frac{n_c w N_1^2}{2M} \tag{4.22}$$

Again, having in mind that $M \equiv \Gamma_\infty A$, we differentiate Equation 4.22 in accordance with Equation 4.8 to deduce expressions for the surface pressure and chemical potential:

$$\pi_s = \Gamma_\infty kT \frac{\theta}{1-\theta} - \beta \Gamma_1^2 \tag{4.23}$$

$$\mu_{1s} = \mu_{1s}^{(0)} + kT\left(\frac{\theta}{1-\theta} + \ln \frac{\theta}{1-\theta} \right) - 2\beta\Gamma_1 \tag{4.24}$$

where

$\mu_{1s}^{(0)} = -kT \ln q_0(T)$

β is defined by Equation 4.14

We can check that Equation 4.23 is equivalent to the van der Waals surface tension isotherm in Table 4.2 for a nonionic surfactant. Furthermore, combining Equations 4.23 and 4.24, we obtain the Butler Equation 4.16, but this time with another expression for the surface activity coefficient

$$\gamma_{1s} = \frac{1}{1-\theta}\exp\left[-\frac{\beta\Gamma_\infty\theta(2-\theta)}{kT}\right] \quad \text{(for van der Waals isotherm)} \tag{4.25}$$

[In the special case of Volmer isotherm we have $\beta = 0$, and then $\gamma_{1s} = 1/(1-\theta)$.] Finally, substituting Equation 4.24 into Equation 4.9, we derive the van der Waals adsorption isotherm in Table 4.2, with K defined by Equation 4.3.

In Table 4.4 we summarize the expressions for the surface free energy, F_s, and chemical potential μ_{1s}, for several thermodynamic models of adsorption. We recall that the parameter Γ_∞ is defined in different ways for the different models. On the other hand, the parameter K is defined in the same way for all models, viz. by Equation 4.3. The expressions in Tables 4.2 through 4.4 can be generalized for multicomponent adsorption layers [20,26].

At the end of this section, let us consider a general expression, which allows us to obtain the surface activity coefficient γ_{1s} directly from the surface pressure isotherm $\pi_s(\theta)$. From the Gibbs adsorption isotherm, $d\pi_s = \Gamma_1 d\mu_{1s}$, it follows that

$$\left(\frac{\partial\mu_{1s}}{\partial\Gamma_1}\right)_T = \frac{1}{\Gamma_1}\left(\frac{\partial\pi_s}{\partial\Gamma_1}\right)_T \tag{4.26}$$

By substituting μ_{1s} from the Butler's Equation 4.16 into Equation 4.26 and integrating, we can derive the sought for expression:

$$\ln\gamma_{1s} = \int_0^\theta \left(\frac{(1-\theta)}{\Gamma_\infty kT}\frac{\partial\pi_s}{\partial\theta} - 1\right)\frac{d\theta}{\theta} \tag{4.27}$$

We can check that a substitution of π_s from Equations 4.13 and 4.23 into Equation 4.27 yields, respectively, the Frumkin and van der Waals expressions for γ_{1s}, viz. Equations 4.17 and 4.24.

4.2.1.2 Ionic Surfactants

4.2.1.2.1 The Gouy Equation

The thermodynamics of adsorption of *ionic* surfactants [13,26–30] is more complicated (in comparison with that of nonionics) because of the presence of long-range electrostatic interactions and, in particular, electric double layer (EDL) in the system, see Figure 4.1. The electro-chemical potential of the ionic species can be expressed in the form [31]

$$\mu_i = \mu_i^{(0)} + kT\ln a_i + Z_i e\psi \tag{4.28}$$

where

- e is the elementary electric charge
- ψ is the electric potential
- Z_i is the valence of the ionic component "i"
- a_i is its activity

In the EDL (Figure 4.1), the electric potential and the activities of the ions are dependent on the distance z from the phase boundary: $\psi = \psi(z)$, $a_i = a_i(z)$. At equilibrium the electrochemical

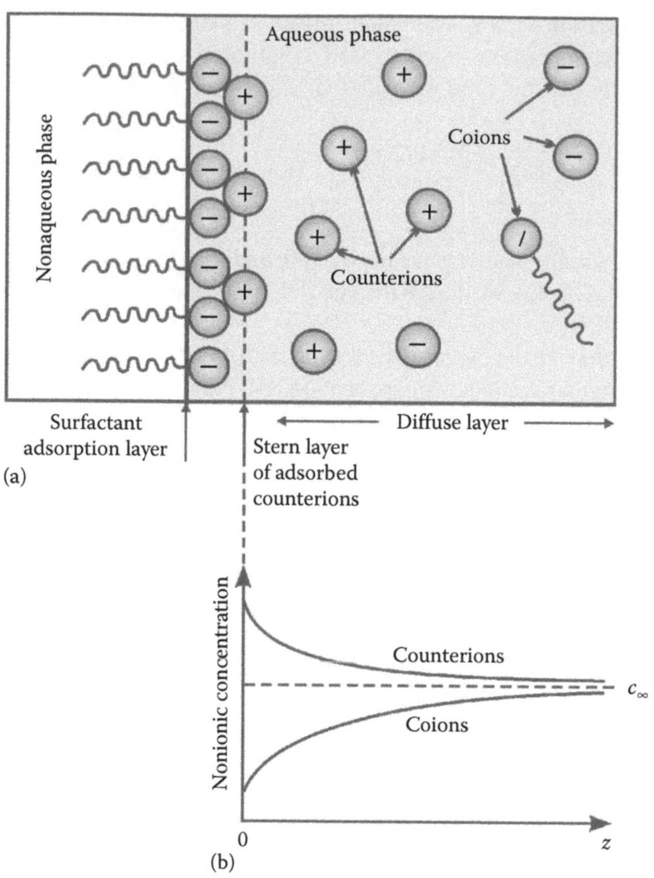

FIGURE 4.1 Electric double layer in the vicinity of an adsorption layer of ionic surfactant. (a) The diffuse layer contains free ions involved in Brownian motion, while the Stern layer consists of adsorbed (bound) counterions. (b) Near the charged surface there is an accumulation of counterions and a depletion of coions.

potential, μ_i, is uniform throughout the whole solution, including the EDL (otherwise diffusion fluxes would appear) [31]. In the bulk of solution ($z \to \infty$) the electric potential tends to a constant value, which is usually set equal to zero, that is $\psi \to 0$ and $\partial\psi/\partial z \to 0$ for $z \to \infty$. If the expression for μ_i at $z \to \infty$ and that for μ_i at some finite z are set equal, from Equation 4.28 we obtain a Boltzmann-type distribution for the activity across the EDL [31]:

$$a_i(z) = a_{i\infty} \exp\left[-\frac{Z_i e\psi(z)}{kT}\right] \tag{4.29}$$

where $a_{i\infty}$ denotes the value of the activity of ion "i" in the bulk of solution. If the activity in the bulk, $a_{i\infty}$, is known, then Equation 4.29 determines the activity $a_i(z)$ in each point of the EDL. A good agreement between theory and experiment can be achieved [12,13,26] using the following expression for $a_{i\infty}$:

$$a_{i\infty} = \gamma_\pm c_{i\infty} \tag{4.30}$$

where

$c_{i\infty}$ is the bulk concentration of the respective ion the activity coefficient $\gamma_\pm$ is calculated from the known formula [32]

$$\log \gamma_\pm = -\frac{A|Z_+Z_-|\sqrt{I}}{1+Bd_i\sqrt{I}} + bI \tag{4.31}$$

which originates from the Debye–Hückel theory; I denotes the ionic strength of the solution:

$$I \equiv \frac{1}{2}\sum_i Z_i^2 c_{i\infty} \tag{4.32}$$

where the summation is carried out over all ionic species in the solution. When the solution contains a mixture of several electrolytes, then Equation 4.31 defines $\gamma_\pm$ for each separate electrolyte, with Z_+ and Z_- being the valences of the cations and anions of *this* electrolyte, but with I being the *total* ionic strength of the solution, accounting for all dissolved electrolytes [32]. The log in Equation 4.31 is decimal, d_i is the ionic diameter, A, B, and b are parameters, whose values can be found in the book by Robinson and Stokes [32]. For example, if I is given in moles per liter (M), the parameters values are $A = 0.5115$ $M^{-1/2}$, $Bd_i = 1.316$ $M^{-1/2}$, and $b = 0.055$ M^{-1} for solutions of NaCl at 25°C.

The theory of EDL provides a connection between surface charge and surface potential (known as the Gouy equation [33,34] of Graham equation [35,36]), which can be presented in the form [26,37]

$$\sum_{i=1}^{N} z_i\Gamma_i = \frac{2}{\kappa_c}\left\{\sum_{i=1}^{N} a_{i\infty}[\exp(-z_i\Phi_s)-1]\right\}^{1/2} \quad \text{(Gouy equation)} \tag{4.33}$$

where

Γ_i ($i = 1, \ldots, N$) are the adsorptions of the ionic species
$z_i = Z_i/Z_1$

the index $i = 1$ corresponds to the surfactant ions

$$\kappa_c^2 \equiv \frac{2Z_1^2e^2}{\varepsilon_0\varepsilon kT}, \quad \Phi_s \equiv \frac{Z_1e\psi_s}{kT} \tag{4.34}$$

ε is the dielectric permittivity of the medium (water)
$\psi_s = \psi(z = 0)$ is the surface potential

Note that the Debye parameter is $\kappa^2 = \kappa_c^2 I$.

For example, let us consider a solution of an ionic surfactant, which is a symmetric 1:1 electrolyte, in the presence of a symmetric, 1:1, inorganic electrolyte (salt). We assume that the counterions due to the surfactant and salt are identical. For example, this can be a solution of sodium dodecyl sulfate (SDS) in the presence of NaCl. We denote by $c_{1\infty}$, $c_{2\infty}$, and $c_{3\infty}$ the bulk concentrations of the surface-active ions, counterions, and coions, respectively (Figure 4.1). For the special system of SDS with NaCl $c_{1\infty}$, $c_{2\infty}$, and $c_{3\infty}$ are the bulk concentrations of the DS^-, Na^+ and Cl^- ions, respectively. The requirement for the bulk solution to be electroneutral implies $c_{2\infty} = c_{1\infty} + c_{3\infty}$. The multiplication of the last equation by $\gamma_\pm$ yields

$$a_{2\infty} = a_{1\infty} + a_{3\infty} \tag{4.35}$$

The adsorption of the coions of the nonamphiphilic salt is expected to be equal to zero, $\Gamma_3 = 0$, because they are repelled by the similarly charged interface [26,38–40]. However, the adsorption of surfactant at the interface, Γ_1, and the binding of counterions in the Stern layer, Γ_2, are different from zero (Figure 4.1). For this system the Gouy Equation 4.33 acquires the form

$$\Gamma_1 - \Gamma_2 = \frac{4}{\kappa_c}\sqrt{a_{2\infty}}\sinh\left(\frac{\Phi_s}{2}\right) \quad (Z_1{:}Z_1 \text{ electrolyte}) \tag{4.36}$$

4.2.1.2.2 *Contributions from the Adsorption and Diffuse Layers*

In general, the *total* adsorption $\tilde{\Gamma}_i$ of an ionic species includes contributions from *both* the adsorption layer (surfactant adsorption layer + adsorbed counterions in the Stern layer), Γ_i, and the diffuse layer, Λ_I [13,26,27,29]:

$$\tilde{\Gamma}_i = \Gamma_i + \Lambda_i, \quad \text{where } \Lambda_i \equiv \int_0^\infty [a_i(z) - a_{i\infty}]dz \tag{4.37}$$

and $\tilde{\Gamma}_i$ represents a surface excess of component "*i*" with respect to the *uniform* bulk solution. Because the solution is electroneutral, we have $\sum_{i=1}^N z_i\tilde{\Gamma}_i = 0$. Note, however, that $\sum_{i=1}^N z_i\Gamma_i \neq 0$, see the Gouy Equation 4.33. Expressions for Λ_i can be obtained by using the theory of EDL. For example, because of the electroneutrality of the solution, the right-hand side of Equation 4.36 is equal to $\Lambda_2 - \Lambda_1 - \Lambda_3$, where

$$\Lambda_2 = 2a_{2\infty}\kappa^{-1}\left[\exp\left(\frac{\Phi_s}{2}\right) - 1\right]; \qquad \Lambda_j = 2a_{j\infty}\kappa^{-1}\left[\exp\left(\frac{-\Phi_s}{2}\right) - 1\right], \quad j = 1,3 \tag{4.38}$$

($\kappa^2 = \kappa_c^2 I$; $Z_1{:}Z_1$ electrolyte). In analogy with Equation 4.37, the interfacial tension of the solution, σ, can be expressed as a sum of contributions from the adsorption and diffuse layers [26,27,34]:

$$\sigma = \sigma_a + \sigma_d \tag{4.39}$$

where

$$\sigma_a = \sigma_0 - kTJ \quad \text{and} \quad \sigma_d = -\varepsilon_0\varepsilon\int_0^\infty \left(\frac{d\psi}{dz}\right)^2 dz \tag{4.40}$$

Expressions for J are given in Table 4.2 for various types of isotherms. Note that Equations 4.39 and 4.40 are valid under both equilibrium and dynamic conditions. In the special case of SDS + NaCl solution (see the text explained earlier), *at equilibrium*, we can use the theory of EDL to express $d\psi/dz$; then from Equation 4.40 we derive [26,27,34]

$$\sigma_d = -\frac{8kT}{\kappa_c}\sqrt{a_{2\infty}}\left[\cosh\left(\frac{\Phi_s}{2}\right) - 1\right] \quad (Z_1{:}Z_1 \text{ electrolyte, at equilibrium}) \tag{4.41}$$

Analytical expressions for σ_d for the cases of 2:1, 1:2, and 2:2 electrolytes can be found in Refs. [26,36].

In the case of ionic surfactants, Equation 4.1 can be presented in two alternative, but equivalent forms [26,37]:

$$d\sigma = -kT\sum_{i=1}^{N}\tilde{\Gamma}_i d\ln a_{i\infty} \quad (T = \text{const.}) \tag{4.42}$$

$$d\sigma_a = -kT\sum_{i=1}^{N}\Gamma_i d\ln a_{is} \quad (T = \text{const.}) \tag{4.43}$$

where $a_{is} = a_i$ ($z = 0$) is the "subsurface" value of activity a_i. From Equations 4.29 and 4.34, we obtain

$$a_{is} = a_{i\infty}\exp(-z_i\Phi_s) \tag{4.44}$$

The comparison between Equations 4.42 and 4.43 shows that the Gibbs adsorption equation can be expressed either in terms of σ, $\tilde{\Gamma}_i$, and $a_{i\infty}$, or in terms of σ_a, Γ_i, and a_{is}. Note that Equations 4.42 and 4.44 are valid under *equilibrium* conditions, whereas Equation 4.43 can also be used for the description of *dynamic* surface tension (Section 4.2.2) in the case of surfactant adsorption under diffusion control, assuming local equilibrium between adsorptions Γ_i and subsurface concentrations of the respective species.

The expression $\sigma_a = \sigma_0 - kTJ$, with J given in Table 4.2, can be used for description of both static and dynamic surface tension of ionic and nonionic surfactant solutions. The surfactant adsorption isotherms in this table can be used for both ionic and nonionic surfactants, with the only difference that in the case of ionic surfactant the adsorption constant K depends on the subsurface concentration of the inorganic counterions [26], see Equation 4.48.

4.2.1.2.3 The Effect of Counterion Binding

As an example, let us consider again the special case of SDS + NaCl solution. In this case, the Gibbs adsorption equation (4.1) takes the form

$$d\sigma_a = -kT(\Gamma_1 d\ln a_{1s} + \Gamma_2 d\ln a_{2s}) \tag{4.45}$$

where as before, the indices "1" and "2" refer to the DS^- and Na^+ ions, respectively. The differentials in the right-hand side of Equation 4.45 are independent (we can vary independently the concentrations of surfactant and salt), and moreover, $d\sigma_a$ is an exact (total) differential. Then, according to the Euler condition, the cross derivatives must be equal [26]:

$$\frac{\partial\Gamma_1}{\partial\ln a_{2s}} = \frac{\partial\Gamma_2}{\partial\ln a_{1s}} \tag{4.46}$$

A surfactant adsorption isotherm, $\Gamma_1 = \Gamma_1(a_{1s}, a_{2s})$, and a counterion adsorption isotherm, $\Gamma_2 = \Gamma_2(a_{1s}, a_{2s})$, are *thermodynamically compatible* only if they satisfy Equation 4.46. The counterion adsorption isotherm is usually taken in the form

$$\frac{\Gamma_2}{\Gamma_1} = \frac{K_2 a_{2s}}{1 + K_2 a_{2s}} \quad \text{(Stern isotherm)} \tag{4.47}$$

where K_2 is a constant parameter. The latter equation, termed the *Stern isotherm* [41], describes Langmuirian adsorption (binding) of counterions in the Stern layer. It can be proven that a sufficient condition Γ_2 form Equation 4.47 to satisfy the Euler's condition 4.46, together with one of the surfactant adsorption isotherms for Γ_1 in Table 4.2, is [26]

$$K = K_1(1 + K_2 a_{2s}) \tag{4.48}$$

where K_1 is another constant parameter. In other words, if K is expressed by Equation 4.48, the Stern isotherm 4.47 is thermodynamically compatible with each of the surfactant adsorption isotherms in Table 4.2. In analogy with Equation 4.3, the parameters K_1 and K_2 are related to the respective standard free energies of adsorption of surfactant ions and counterions $\Delta\mu_i^{(0)}$:

$$K_i = \frac{\delta_i}{\Gamma_\infty}\exp\left(\frac{\Delta\mu_i^{(0)}}{kT}\right) \quad (i = 1,2) \tag{4.49}$$

where δ_i stands for the thickness of the respective adsorption layer.

4.2.1.2.4 Dependence of Adsorption Parameter K on Salt Concentration

The physical meaning of Equation 4.48 can be revealed by chemical-reaction considerations. For simplicity, let us consider *Langmuir*-type adsorption, that is, we treat the interface as a 2D lattice. We will use the notation θ_0 for the fraction of the free sites in the lattice, θ_1 for the fraction of sites containing adsorbed surfactant ion S^-, and θ_2 for the fraction of sites containing the complex of an adsorbed surfactant ion + a bound counterion. Obviously, we can write $\theta_0 + \theta_1 + \theta_2 = 1$. The adsorptions of surfactant ions and counterions can be expressed in the form:

$$\frac{\Gamma_1}{\Gamma_\infty} = \theta_1 + \theta_2; \quad \frac{\Gamma_2}{\Gamma_\infty} = \theta_2 \tag{4.50}$$

Following Kalinin and Radke [28], we consider the "reaction" of adsorption of S^- ions:

$$A_0 + S^- = A_0S^- \tag{4.51}$$

where A_0 symbolizes an empty adsorption site. In accordance with the rules of the chemical kinetics, we can express the rates of adsorption and desorption in the form:

$$r_{1,\text{ads}} = K_{1,\text{ads}}\theta_0 c_{1s}, \quad r_{1,\text{des}} = K_{1,\text{des}}\theta_1 \tag{4.52}$$

where

- c_{1s} is the subsurface concentration of surfactant
- $K_{1,\text{ads}}$ and $K_{1,\text{des}}$ are constants

In view of Equation 4.50, we can write $\theta_0 = (\Gamma_\infty - \Gamma_1)/\Gamma_\infty$ and $\theta_1 = (\Gamma_1 - \Gamma_2)/\Gamma_\infty$. Thus, with the help of Equation 4.52 we obtain the net adsorption flux of surfactant:

$$Q_1 \equiv r_{1,\text{ads}} - r_{1,\text{des}} = K_{1,\text{ads}} c_{1s}\frac{(\Gamma_\infty - \Gamma_1)}{\Gamma_\infty} - K_{1,\text{des}}\frac{(\Gamma_1 - \Gamma_2)}{\Gamma_\infty} \tag{4.53}$$

Next, let us consider the reaction of counterion binding:

$$A_0S^- + M^+ = A_0SM \tag{4.54}$$

The rates of the direct and reverse reactions are, respectively,

$$r_{2,\mathrm{ads}} = K_{2,\mathrm{ads}}\theta_1 c_{2s}, \quad r_{2,\mathrm{des}} = K_{2,\mathrm{des}}\theta_2 \tag{4.55}$$

where

$K_{2,\mathrm{ads}}$ and $K_{2,\mathrm{des}}$ are the respective rate constants
c_{2s} is the subsurface concentration of counterions

Having in mind that $\theta_1 = (\Gamma_1 - \Gamma_2)/\Gamma_\infty$ and $\theta_2 = \Gamma_2/\Gamma_\infty$, with the help of Equation 4.55 we deduce an expression for the adsorption flux of counterions:

$$Q_2 \equiv r_{2,\mathrm{ads}} - r_{2,\mathrm{des}} = K_{2,\mathrm{ads}} c_{2s} \frac{(\Gamma_1 - \Gamma_2)}{\Gamma_\infty} - K_{2,\mathrm{des}} \frac{\Gamma_2}{\Gamma_\infty} \tag{4.56}$$

If we can assume that the reaction of counterion binding is much faster than the surfactant adsorption, then we can set $Q_2 \equiv 0$, and Equation 4.56 reduces to the Stern isotherm, Equation 4.47, with $K_2 \equiv K_{2,\mathrm{ads}}/K_{2,\mathrm{des}}$. Next, a substitution of Γ_2 from Equation 4.47 into Equation 4.53 yields [37]

$$Q_1 \equiv r_{1,\mathrm{ads}} - r_{1,\mathrm{des}} = K_{1,\mathrm{ads}} c_{1s} \frac{(\Gamma_\infty - \Gamma_1)}{\Gamma_\infty} - K_{1,\mathrm{des}} (1 + K_2 c_{2s})^{-1} \frac{\Gamma_1}{\Gamma_\infty} \tag{4.57}$$

Equation 4.57 shows that the adsorption flux of surfactant is influenced by the subsurface concentration of counterions, c_{2s}. At last, if there is equilibrium between surface and subsurface, we have to set $Q_1 \equiv 0$ in Equation 4.57, and thus we obtain the Langmuir isotherm for an ionic surfactant:

$$K c_{1s} = \frac{\Gamma_1}{\Gamma_\infty - \Gamma_1}, \quad \text{with } K \equiv \frac{K_{1,\mathrm{ads}}}{K_{1,\mathrm{des}}}(1 + K_2 c_{2s}) \tag{4.58}$$

Note that $K_1 \equiv K_{1,\mathrm{ads}}/K_{1,\mathrm{des}}$. This result demonstrates that the linear dependence of K on c_{2s} (Equation 4.48) can be deduced from the reactions of surfactant adsorption and counterion binding, Equations 4.51 and 4.54. (For $I < 0.1$ M we have $\gamma_\pm \approx 1$ and then activities and concentrations of the ionic species coincide.)

4.2.1.2.5 Comparison of Theory and Experiment

As illustration, we consider the interpretation of experimental isotherms by Tajima et al. [40,42,43] for the surface tension σ versus SDS concentrations at 11 fixed concentrations of NaCl, see Figure 4.2. Processing the set of data for the interfacial tension $\sigma = \sigma(c_{1\infty}, c_{2\infty})$ as a function of the bulk concentrations of surfactant (DS^-) ions and Na^+ counterions, $c_{1\infty}$ and $c_{2\infty}$, we can determine the surfactant adsorption, $\Gamma_1(c_{1\infty}, c_{2\infty})$, the counterion adsorption, $\Gamma_2(c_{1\infty}, c_{2\infty})$, the surface potential, $\psi_s(c_{1\infty}, c_{2\infty})$, and the Gibbs elasticity $E_G(c_{1\infty}, c_{2\infty})$ for every desirable surfactant and salt concentrations.

The theoretical dependence $\sigma = \sigma(c_{1\infty}, c_{2\infty})$ is determined by the following *full set of equations*: Equation 4.44 for $i = 1, 2$; the Gouy Equation 4.36, Equation 4.39 (with σ_d expressed by Equation 4.41 and J from Table 4.2), the Stern isotherm 4.47, and one surfactant adsorption isotherm from Table 4.2, say the van der Waals one. Thus, we get a set of six equations for determining six unknown variables: σ, Φ_s, a_{1s}, a_{2s}, Γ_1 and Γ_2. (For $I < 0.1$ M the activities of the ions can be replaced by the respective concentrations.) The principles of the numerical procedure are described in Ref. [26].

The theoretical model contains four parameters, β, Γ_∞, K_1, and K_2, whose values are to be obtained from the best fit of the experimental data. Note that all 11 curves in Figure 4.2 are fitted

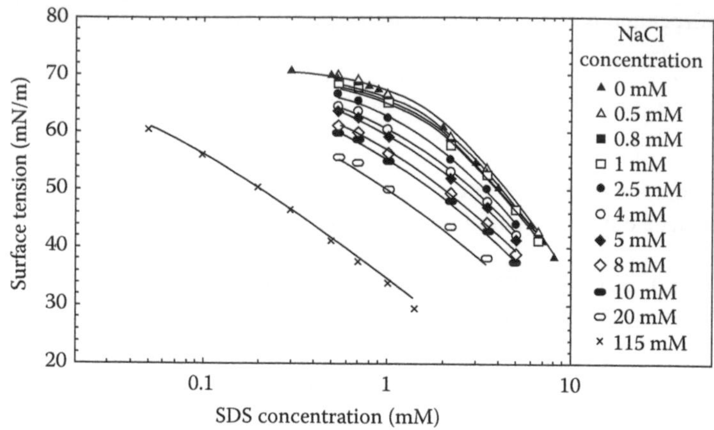

FIGURE 4.2 Plot of the surface tension σ vs. the concentration of SDS, $c_{1\infty}$, for 11 fixed NaCl concentrations. The symbols are experimental data by Tajima et al. [40,42,43]. The lines represent the best fit [42] with the full set of equations specified in the text, involving the van der Waals isotherms of adsorption and surface tension (Table 4.2).

simultaneously [44]. In other words, the parameters β, Γ_∞, K_1, and K_2 are the same for all curves. The value of Γ_∞, obtained from the best fit of the data in Figure 4.2, corresponds to $1/\Gamma_\infty = 29.8$ Å^2. The respective value of K_1 is 99.2 m^3/mol, which in view of Equation 4.49 gives a standard free energy of surfactant adsorption $\Delta\mu_1^{(0)} = 12.53kT$ per DS$^-$ ion, that is 30.6 kJ/mol. The determined value of K_2 is 6.5×10^{-4} m^3/mol, which after substitution in Equation 4.49 yields a standard free energy of counterion binding $\Delta\mu_2^{(0)} = 1.64kT$ per Na$^+$ ion, that is, 4.1 kJ/mol. The value of the parameter β is positive, $2\beta\Gamma_\infty/kT = +2.73$, which indicates attraction between the hydrocarbon tails of the adsorbed surfactant molecules. However, this attraction is too weak to cause 2D phase transition. The van der Waals isotherm predicts such transition for $2\beta\Gamma_\infty/kT > 6.74$.

Figure 4.3 shows calculated curves for the adsorptions of surfactant, Γ_1 (the full lines), and counterions, Γ_2 (the dotted lines), versus the SDS concentration, $c_{1\infty}$. These lines represent the variations

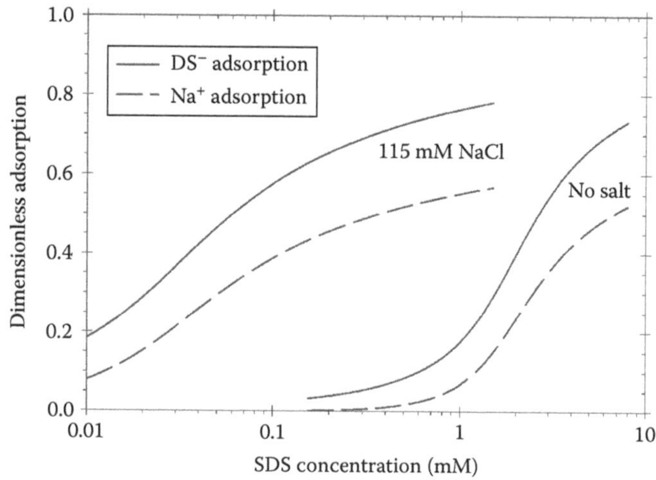

FIGURE 4.3 Plots of the dimensionless adsorptions of surfactant ions Γ^1/Γ_∞ (DS$^-$, the full lines), and counterions Γ^2/Γ_∞ (Na$^+$, the dotted lines), vs. the surfactant (SDS) concentration, $c_{1\infty}$. The lines are calculated [44] for NaCl concentrations 0 and 115 mM using parameter values determined from the best fit of experimental data (Figure 4.2).

of Γ_1 and Γ_2 along the experimental curves, which correspond to the lowest and highest NaCl concentrations in Figure 4.2, viz. $c_{3\infty} = 0$ and 115 mM. We see that both Γ_1 and Γ_2 are markedly greater when NaCl is present in the solution. The highest values of Γ_1 for the curves in Figure 4.3 are 4.2×10^{-6} and 4.0×10^{-6} mol/m^2 for the solutions with and without NaCl, respectively. The latter two values compare well with the saturation adsorptions measured by Tajima et al. [42,43] for the same system by means of the radiotracer method, viz. $\Gamma_1 = 4.3 \times 10^{-6}$ and 3.2×10^{-6} mol/m^2 for the solutions with and without NaCl, respectively.

For the solution *without* NaCl the occupancy of the Stern layer, Γ_2/Γ_1, rises from 0.15 to 0.73 and then exhibits a tendency to level off. The latter value is consonant with data of other authors [45–47], who have obtained values of Γ_2/Γ_1 up to 0.70–0.90 for various ionic surfactants; pronounced evidences for counterion binding have also been obtained also in experiments with solutions containing surfactant micelles [48–52]. As it could be expected, both Γ_1 and Γ_2 are higher for the solution *with* NaCl. These results imply that the counterion adsorption (binding) should be always be taken into account.

The fit of the data in Figure 4.2 also gives the values of the surface electric potential, ψ_s. For the solutions with 115 mM NaCl the model predicts surface potentials varying in the range $|\psi_s| =$ 55–95 mV within the experimental interval of surfactant concentrations, whereas for the solution without salt the calculated surface potential is higher: $|\psi_s| =$ 150–180 mV (for SDS ψ_s has a negative sign). Thus it turns out that measurements of surface tension, interpreted by means of an appropriate theoretical model, provide a method for determining the surface potential ψ_s in a broad range of surfactant and salt concentrations. The described approach could be also applied to solve the inverse problem, viz. to process data for the surface potential. In this way, the adsorption of surfactant on solid particles can be determined from the measured zeta-potential [53].

It is remarkable that the minimal (excluded) area per adsorbed surfactant molecule, $\alpha \equiv 1/\Gamma_\infty$, obtained from the best fit of surface tension data by the *van der Waals isotherm* practically coincides with the value of α estimated by molecular size considerations (i.e., from the maximal cross-sectional area of an amphiphilic molecule in a dense adsorption layer); see for example Figure 7.1 in Ref. [36]. This is illustrated in Table 4.5, which contains data for alkanols, alkanoic acids, (SDS), (DDBS), cocamidopropyl betaine (CAPB), and C_n-trimethyl ammonium bromides (n = 12, 14, and 16). The second column of Table 4.5 gives the group whose cross-sectional area is used to calculate α. For molecules of circular cross section, we can calculate the cross-sectional area from the expression $\alpha = \pi r^2$, where r is the respective radius. For example [54], the radius of the SO_4^{2-} ion is $r =$ 3.09 Å, which yields $\alpha = \pi r^2 = 30.0$ Å^2. In the fits of surface tension data by the van der Waals isotherm, α was treated as an adjustable parameter, and the value $\alpha = 30$ Å^2 was obtained from the best fit. As seen in Table 4.5, excellent agreement between the values of α obtained from molecular size and from surface tension fits is obtained also for many other amphiphilic molecules [54–61].

TABLE 4.5
Excluded Area per Molecule, α, Determined in Two Different Ways

Amphiphile	Group	α from Molecular Size (Å^2)	α from Surface Tension Fits[a] (Å^2)	References
Alkanols	Paraffin chain	21.0	20.9	[54]
Alkanoic acids	COO^-	22–24	22.6	[55,56]
SDS	SO_4^{2-}	30.0	30	[44,57]
DDBS	Benzene ring	35.3	35.6	[58]
CAPB	CH_3–N^+–CH_3	27.8	27.8	[59]
C_nTAB (n = 12, 14, 16)	$N(CH_3)_4^+$	37.8	36.5–39.5	[57,60]

[a] Fit by means of the van der Waals isotherm.

It should be noted the result mentioned earlier holds only for the van der Waals (or Volmer) isotherm. Instead, if the Frumkin (or Langmuir) isotherm is used, the value of α obtained from the surface tension fits is about 33% greater than that obtained from molecular size [44]. A possible explanation of this difference could be the fact that the Frumkin (and Langmuir) isotherm is statistically derived for localized adsorption and is more appropriate to describe adsorption at solid interfaces. In contrast, the van der Waals (and Volmer) isotherm is derived for nonlocalized adsorption, and they provide a more adequate theoretical description of the surfactant adsorption at liquid–fluid interfaces. This conclusion refers also to the calculation of the surface (Gibbs) elasticity by means of the two types of isotherms [44].

The fact that α determined from molecular size coincides with that obtained from surface tension fits (Table 4.5) is very useful for applications. Thus, when fitting experimental data, we can use the value of α from molecular size, and thus to decrease the number of adjustable parameters. This fact is especially helpful when interpreting theoretically data for the surface tension of surfactant mixtures, such as SDS + dodecanol [54]; SDS + CAPB [59], and fluorinated + nonionic surfactant [61]. An additional way to decrease the number of adjustable parameters is to employ the Traube rule, which states that $\Delta\mu_1^{(0)}$ increases with $1.025kT$ when a CH_2 group is added to the paraffin chain; for details see Refs. [54,55,60].

4.2.2 Dynamic Surface Tension

If the surface of an equilibrium surfactant solution is disturbed (expanded, compressed, renewed, etc.), the system will try to restore the equilibrium by exchange of surfactant between the surface and the subsurface layer (adsorption–desorption). The change of the surfactant concentration in the subsurface layer triggers a diffusion flux in the solution. In other words, the process of equilibration (relaxation) of an expanded adsorption monolayer involves two consecutive stages:

1. Diffusion of surfactant molecules from the bulk solution to the subsurface layer;
2. Transfer of surfactant molecules from the subsurface to the adsorption layer; the rate of transfer is determined by the height of the kinetic barrier to adsorption.

(In the case of desorption the processes have the opposite direction.) Such interfacial expansions are typical for foam generation and emulsification. The rate of adsorption relaxation determines whether the formed bubbles/drops will coalesce upon collision, and in final reckoning—how large will be the foam volume and the emulsion drop size [62,63]. In the following section, we focus our attention on the relaxation time of surface tension, τ_σ, which characterizes the interfacial dynamics.

The overall rate of surfactant adsorption is controlled by the slowest stage. If it is stage (1), we deal with *diffusion control*, while if stage (2) is slower, the adsorption occurs under *barrier (kinetic) control*. The next four sections are dedicated to processes under diffusion control (which are the most frequently observed), whereas in Section 4.2.2.5 we consider adsorption under barrier control. Finally, Section 4.2.2.6 is devoted to the dynamics of adsorption from micellar surfactant solutions.

Various experimental methods for dynamic surface tension measurements are available. Their operational time scales cover different time intervals [64,65]. Methods with a *shorter* characteristic operational time are the oscillating jet method [66–68], the oscillating bubble method [69–72], the fast-formed drop technique [73,74], the surface wave techniques [75–78], and the maximum bubble pressure method (MBPM) [57,79–84]. Methods of *longer* characteristic operational time are the inclined plate method [85], the drop-weight/volume techniques [86–90], the funnel [91] and overflowing cylinder [60,92] methods, and the axisymmetric drop shape analysis (ADSA) [93,94]; see Refs. [64,65,95] for a more detailed review.

In this section, devoted to dynamic surface tension, we consider mostly *nonionic* surfactant solutions. In Section 4.2.2.4 we address the more complicated case of *ionic* surfactants. We will restrict our considerations to the simplest case of *relaxation of an initial uniform interfacial dilatation*. The more complex case of simultaneous adsorption and dilatation is considered elsewhere [57,64,80,84,92,95].

4.2.2.1 Adsorption under Diffusion Control

Here, we consider a solution of a *nonionic* surfactant, whose concentration, $c_1 = c_1(z, t)$, depends on the position and time because of the diffusion process. (As before, z denotes the distance to the interface, which is situated in the plane $z = 0$.) Correspondingly, the surface tension, surfactant adsorption, and the subsurface concentration of surfactant vary with time: $\sigma = \sigma(t)$, $\Gamma_1 = \Gamma_1(t)$, $c_{1s} = c_{1s}(t)$. The surfactant concentration obeys the equation of diffusion:

$$\frac{\partial c_1}{\partial t} = D_1 \frac{\partial^2 c_1}{\partial z^2} \quad (z > 0, t > 0) \tag{4.59}$$

where D_1 is the diffusion coefficient of the surfactant molecules. The exchange of surfactant between the solution and its interface is described by the boundary conditions

$$c_1(0,t) = c_{1s}(t), \quad \frac{d\Gamma_1}{dt} = D_1 \frac{\partial c_1}{\partial z} \quad (z = 0, t > 0) \tag{4.60}$$

The latter equation states that the rate of increase of the adsorption Γ_1 is equal to the diffusion influx of surfactant per unit area of the interface. Integrating Equation 4.59, along with 4.60, we can derive the equation of Ward and Tordai [96]:

$$\Gamma_1(t) = \Gamma_1(0) + \sqrt{\frac{D_1}{\pi}} \left[2c_{1\infty}\sqrt{t} - \int_0^t \frac{c_{1s}(\tau)}{\sqrt{t-\tau}} d\tau \right] \tag{4.61}$$

Solving Equation 4.61 together with some of the adsorption isotherms $\Gamma_1 = \Gamma_1(c_{1s})$ in Table 4.2, we can in principle determine the two unknown functions $\Gamma_1(t)$ and $c_{1s}(t)$. Because the relation $\Gamma_1(c_{1s})$ is nonlinear (except for the Henry isotherm), this problem, or its equivalent formulations, can be solved either numerically [97], or by employing appropriate approximations [80,98].

In many cases, it is convenient to use asymptotic expressions for the functions $\Gamma_1(t)$, $c_{1s}(t)$, and $\sigma(t)$ for *short times* ($t \to 0$) and *long times* ($t \to \infty$). A general asymptotic expression for the short times can be derived from Equation 4.61 substituting $c_{1s} \approx c_{1s}(0) = \text{constant}$:

$$\Gamma_1(t) = \Gamma_1(0) + 2\sqrt{\frac{D_1}{\pi}} [c_{1\infty} - c_{1s}(0)]\sqrt{t} \quad (t \to 0) \tag{4.62}$$

Analogous asymptotic expression can also be obtained for long times, although the derivation is not so simple. Hansen [99] derived a useful asymptotics for the subsurface concentration:

$$c_{1s}(t) = c_{1\infty} - \frac{\Gamma_{1e} - \Gamma(0)}{\sqrt{\pi D_1 t}} \quad (t \to \infty) \tag{4.63}$$

where Γ_{1e} is the equilibrium value of the surfactant adsorption. The validity of Hansen's Equation 4.63 was confirmed in subsequent studies by other authors [100,101].

In the following section, we continue our review of the asymptotic expressions considering separately the cases of *small* and *large* initial perturbations.

4.2.2.2 Small Initial Perturbation

When the deviation from equilibrium is small, then the adsorption isotherm can be linearized:

$$\Gamma_1(t)-\Gamma_{1,e} \approx \left(\frac{\partial \Gamma_1}{\partial c_1}\right)_e [c_{1s}(t)-c_e] \quad (4.64)$$

Here and hereafter, the subscript "*e*" means that the respective quantity refers to the equilibrium state. The set of linear Equations 4.59, 4.60, and 4.64 has been solved by Sutherland [102]. The result, which describes the relaxation of a *small* initial interfacial dilatation, reads:

$$\frac{\sigma(t)-\sigma_e}{\sigma(0)-\sigma_e} = \frac{\Gamma_1(t)-\Gamma_{1,e}}{\Gamma_1(0)-\Gamma_{1,e}} = \exp\left(\frac{t}{\tau_\sigma}\right)\operatorname{erfc}\left(\sqrt{\frac{t}{\tau_\sigma}}\right) \quad (4.65)$$

where

$$\tau_\sigma \equiv \frac{1}{D_1}\left(\frac{\partial \Gamma_1}{\partial c_1}\right)_e^2 \quad (4.66)$$

is the characteristic relaxation time of surface tension and adsorption, and

$$\operatorname{erfc}(x) \equiv \frac{2}{\sqrt{\pi}}\int_x^\infty \exp(-x^2)dx \quad (4.67)$$

is the so-called complementary error function [103,104]. The asymptotics of the latter function for small and large values of the argument are [103,104]:

$$\operatorname{erfc}(x) = 1-\frac{2}{\sqrt{\pi}}x + O(x^3) \text{ for } x \ll 1; \quad \operatorname{erfc}(x) = \frac{e^{-x^2}}{\sqrt{\pi}x}\left[1+O\left(\frac{1}{x^2}\right)\right] \text{ for } x \gg 1 \quad (4.68)$$

Combining Equations 4.65 and 4.68, we obtain the short-time and long-time asymptotics of the surface tension relaxation:

$$\frac{\sigma(t)-\sigma_e}{\sigma(0)-\sigma_e} = \frac{\Gamma_1(t)-\Gamma_{1,e}}{\Gamma_1(0)-\Gamma_{1,e}} = 1-\frac{2}{\sqrt{\pi}}\sqrt{\frac{t}{\tau_\sigma}} + O\left[\left(\frac{t}{\tau_\sigma}\right)^{3/2}\right] \quad (t \ll \tau_\sigma) \quad (4.69)$$

$$\frac{\sigma(t)-\sigma_e}{\sigma(0)-\sigma_e} = \frac{\Gamma_1(t)-\Gamma_{1,e}}{\Gamma_1(0)-\Gamma_{1,e}} = \sqrt{\frac{\tau_\sigma}{\pi t}} + O\left[\left(\frac{\tau_\sigma}{t}\right)^{3/2}\right] \quad t \gg \tau_\sigma \quad (4.70)$$

Equation 4.70 is often used as a test to verify whether the adsorption process is under diffusion control: data for $\sigma(t)$ are plotted versus $1/\sqrt{t}$ and it is checked if the plot complies with a straight line; moreover, the intercept of the line gives σ_e. We recall that Equations 4.69 and 4.70 are valid in the case of a *small* initial perturbation; alternative asymptotic expressions for the case of *large* initial perturbation are considered in the next section.

With the help of the thermodynamic Equations 4.2 and 4.6, we derive

$$\frac{\partial\Gamma_1}{\partial c_1}=\frac{\partial\Gamma_1}{\partial\sigma}\frac{\partial\sigma}{\partial c_1}=\frac{\Gamma_1^2 kT}{c_1 E_G} \tag{4.71}$$

Thus Equation 4.66 can be expressed in an alternative form [37]:

$$\tau_\sigma=\frac{1}{D_1}\left(\frac{\Gamma_1^2 kT}{c_1 E_G}\right)_e^2 \tag{4.72}$$

Substituting E_G from Table 4.3 into Equation 4.72, we can obtain expressions for τ_σ corresponding to various adsorption isotherms. In the special case of Langmuir adsorption isotherm, we can present Equation 4.72 in the form [37]

$$\tau_\sigma=\frac{1}{D_1}\frac{(K\Gamma_\infty)^2}{(1+Kc_1)^4}=\frac{1}{D_1}\frac{(K\Gamma_\infty)^2}{(1+E_G/(\Gamma_\infty kT))^4}\quad\text{(for Langmuir isotherm)} \tag{4.73}$$

Equation 4.73 visualizes the very strong dependence of the relaxation time τ_σ on the surfactant concentration c_1; in general, τ_σ can vary with many orders of magnitude as a function of c_1. Equation 4.73 shows also that high Gibbs elasticity corresponds to short relaxation time, and vice versa.

As a quantitative example let us take typical parameter values: K_1 = 15 m^3/mol, $1/\Gamma_\infty$ = 40 Å^2, D_1 = 4.5 × 10^{-6} cm^2/s, and T = 298 K. Then with c_1 = 6.5 × 10^{-6} M, from Table 4.3 (Langmuir isotherm) and Equation 4.73, we calculate $E_G \approx$ 1.0 mN/m and $\tau_\sigma \approx$ 5 s. In the same way, for c_1 = 6.5 × 10^{-4} M we calculate $E_G \approx$ 100 mN/m and $\tau_\sigma \approx$ 5 × 10^{-4} s.

To directly measure the Gibbs elasticity E_G, or to precisely investigate the dynamics of surface tension, we need an experimental method, whose characteristic time is smaller compared to τ_σ. Equation 4.73 and the latter numerical example show that when the surfactant concentration is higher, the experimental method should be faster.

4.2.2.3 Large Initial Perturbation

By definition, we have *large initial perturbation* when at the initial moment the interface is clean of surfactant:

$$\Gamma_1(0)=0,\quad c_{1s}(0)=0 \tag{4.74}$$

In such case, the Hansen Equation 4.63 reduces to

$$c_{1s}(t)=c_{1\infty}-\frac{\Gamma_{1,e}}{\sqrt{\pi D_1 t}}\quad(t\to\infty) \tag{4.75}$$

By substituting $c_{1s}(t)$ for c_1 in the Gibbs adsorption Equation 4.2, and integrating, we obtain the long-time asymptotics of the surface tension of a nonionic surfactant solution after a large initial perturbation:

$$\sigma(t)-\sigma_e=\left(\frac{\Gamma_1^2 kT}{c_1}\right)_e\left(\frac{1}{\pi D_1 t}\right)^{1/2}\quad\text{(large initial perturbation)} \tag{4.76}$$

with the help of Equation 4.72, we can bring Equation 4.76 into another form:

$$\sigma(t) - \sigma_e = E_G \left(\frac{\tau_\sigma}{\pi t} \right)^{1/2} \quad \text{(large initial perturbation)} \tag{4.77}$$

where E_G is given in Table 4.3. It is interesting to note that Equation 4.77 is applicable to both nonionic and ionic surfactants with the only difference that for nonionics τ_σ is given by Equation 4.66, whereas for ionic surfactants the expression for τ_σ is somewhat longer, see Refs. [37,105].

The equations mentioned earlier show that in the case of adsorption under diffusion control the long-lime asymptotics can be expressed in the form

$$\sigma = \sigma_e + St^{-1/2} \tag{4.78}$$

In view of Equations 4.70 and 4.77, the slope S of the dependence σ versus $t^{-1/2}$ is given by the expressions [105]

$$S_s = [\sigma(0) - \sigma_e] \left(\frac{\tau_\sigma}{\pi} \right)^{1/2} \quad \text{(small perturbation)} \tag{4.79}$$

$$S_l = E_G \left(\frac{\tau_\sigma}{\pi} \right)^{1/2} \quad \text{(large perturbation)} \tag{4.80}$$

As known, the surfactant adsorption Γ_1 monotonically increases with the rise of the surfactant concentration, c_1. In contrast, the slope S_l is a nonmonotonic function of c_1: S_l exhibits a maximum at a certain concentration. To demonstrate that we will use the expression

$$S_l = \frac{\Gamma_{1,e}^2 kT}{c_1 \sqrt{\pi D_1}} \tag{4.81}$$

which follows from Equations 4.76 and 4.78. In Equation 4.81 we substitute the expressions for c_1 stemming from the Langmuir and Volmer adsorption isotherms (Table 4.2 with $c_1 = a_{1s}$); the result reads

$$\tilde{S}_l = \theta(1-\theta) \quad \text{(for Langmuir isotherm)} \tag{4.82}$$

$$\tilde{S}_l = \theta(1-\theta) \exp\left(-\frac{\theta}{1-\theta} \right) \quad \text{(for Volmer isotherm)} \tag{4.83}$$

where θ and $\tilde{S}_l$ are the dimensionless adsorption and slope coefficient, respectively:

$$\theta = \frac{\Gamma_{1,e}}{\Gamma_\infty}, \quad \tilde{S}_l = \frac{S_l \sqrt{\pi D_1}}{kT\, K\Gamma_\infty^2} \tag{4.84}$$

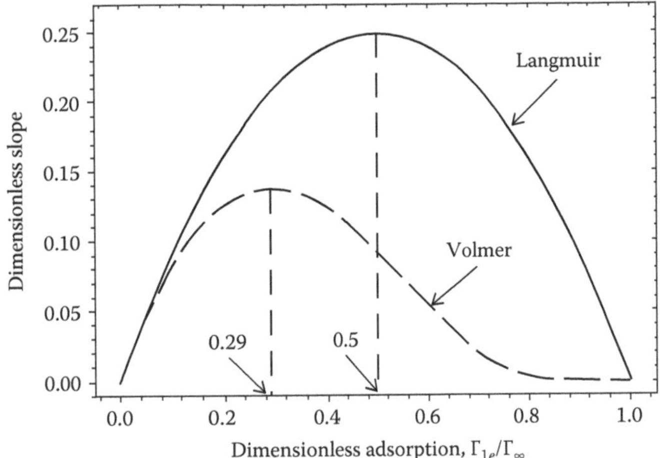

FIGURE 4.4 Plot of the dimensionless slope, $\tilde{S}_l$, vs. the dimensionless equilibrium surfactant adsorption, $\theta = \Gamma_{1e}/\Gamma_\infty$, in accordance with Equations 4.82 and 4.83, corresponding to the cases of localized and nonlocalized adsorption.

Figure 4.4 compares the dependencies $\tilde{S}_l(\theta)$ given by Equations 4.82 and 4.83: we see that the former is symmetric and has a maximum at $\theta = 0.5$, whereas the latter is asymmetric with a maximum at $\theta \approx 0.29$. We recall that the Langmuir and Volmer isotherms correspond to *localized* and *nonlocalized* adsorption, respectively (see Section 4.2.1.1.2). Then Figure 4.4 shows that the symmetry/asymmetry of the plot $\tilde{S}_l$ versus θ provides a *test* for verifying whether the adsorption is localized or nonlocalized. (The practice shows that the fits of *equilibrium* surface tension isotherms do not provide such a test: theoretical isotherms corresponding to localized and nonlocalized adsorption are found to fit surface tension data equally well!)

From another viewpoint, the nonmonotonic behavior of $S_l(\theta)$ can be interpreted as follows. Equation 4.80 shows that $S_l \propto E_G\sqrt{\tau_\sigma}$; then the nonmonotonic behavior stems from the fact that E_G is an increasing function of c_1, whereas τ_σ is a decreasing function of c_1. This qualitative conclusion is also valid for the case of ionic surfactant, as demonstrated in the next section.

4.2.2.4 Generalization for Ionic Surfactants

In the case of ionic surfactants, the dynamics of adsorption is more complicated because of the presence of a dynamic EDL. Indeed, the adsorption of surfactant at the interface creates surface charge, which is increases in the course of the adsorption process. The charged interface repels the new-coming surfactant molecules, but attracts the conversely charged counterions (Figure 4.1); some of them bind to the surfactant headgroups thus decreasing the surface charge density and favoring the adsorption of new surfactant molecules. The theoretical description of the overall adsorption process involves the electro-diffusion equations for the surfactant ions, counterions and coions, and the Poisson equation from electrodynamics. Different analytical and numerical approaches to the solution of this problem have been proposed [13,60,105–114].

In Ref. [114], an approach to the dynamics of ionic surfactant adsorption was developed, which is simpler as both concept and application, but agrees very well with the experiment. Analytical asymptotic expressions for the dynamic surface tension of ionic surfactant solutions are derived in the general case of nonstationary interfacial expansion. Because the diffusion layer is much wider than the EDL, the equations contain a small parameter. The resulting perturbation problem is singular and it is solved by means of the method of matched asymptotic expansions [115]. The derived general expression for the dynamic surface tension is simplified for two important special cases, which are considered in the following section.

The first special case refers to adsorption at an *immobile interface* that has been initially perturbed, and to the *maximum bubble pressure method* (MBPM). The generalization of Equations 4.78 and 4.81 for this case reads [114]:

$$\sigma = \sigma_e + \frac{S_l}{(t_{\text{age}})^{1/2}}, \quad S_l \equiv \frac{kT\Gamma_{1,e}^2 \lambda}{(\pi D_{\text{eff}})^{1/2}\gamma_\pm}\left(\frac{1}{c_{1\infty}} + \frac{1}{c_{2\infty}}\right) \tag{4.85a}$$

As usual, the subscript 'e' denotes equilibrium values; t_{age} is the age of the interface, which is defined as the period of time between the minimum pressure (bubble formation) and the maximum pressure (bubble detachment) in the case of MBPM; λ is a dimensionless parameter; $\lambda = 1$ for immobile interfaces; in the case of MBPM, λ is an apparatus constant that can be determined by calibration experiments [57]; as mentioned earlier, $c_{1\infty}$ and $c_{2\infty}$ are the bulk concentrations of surfactant ions and counterions, respectively; $\gamma_\pm$ is the activity coefficient; D_{eff} is an effective diffusivity that depends on the diffusivities and bulk concentrations of surfactant ions, counterions, and inorganic coions: $D_{\text{eff}} = D_{\text{eff}}(D_1, D_2, D_3, c_{1\infty}, c_{2\infty}, c_{3\infty})$. The latter dependence is described by explicit formulas derived in Ref. [114]; see Equations 6.19 through 6.26 therein.

In the case of the cationic surfactant dodecyl trimethyl ammonium bromide ($C_{12}TAB$), the calculated dependence of D_{eff} on the surfactant and salt concentrations, $c_{1\infty}$ and $c_{3\infty}$, is illustrated in Figure 4.4. Because the range $c_{1\infty} \leq$ CMC is considered, the calculated curves end at the CMC. At very low surfactant concentrations, $c_{1\infty} \to 0$, in the presence of salt ($c_{3\infty} > 0$), the effective diffusivity approaches its limiting value for diluted solutions, $D_{\text{eff}} \to D_1$. We see that D_{eff} increases with the rise of $c_{1\infty}$, except the case without added salt ($c_{3\infty} = 0$), for which D_{eff} is a constant: $1/D_{\text{eff}} = (1/D_1 + 1/D_2)/2$. The curves in Figure 4.5 show that D_{eff} decreases with the rise of salt concentration, $c_{3\infty}$, and becomes $\approx D_1$ for $c_{3\infty} = 100$ mM. Note that the salt concentration affects the dynamic surface tension, σ, also through $\Gamma_{1,e}$ and through the factor $(1/c_{1\infty} + 1/c_{2\infty})$ in Equation 4.85a; see Ref. [114] for details.

The accuracy of Equation 4.85a can be verified in the following way. Each of the dynamic surface tension isotherms for $C_{12}TAB$ in Figure 4.6a are fitted by means of the equation $\sigma = \sigma_e + S_l/[a_\sigma + (t_{\text{age}})^{1/2}]$, and the parameters σ_e, a_σ, and S_l are determined from the best fit. Next, for each

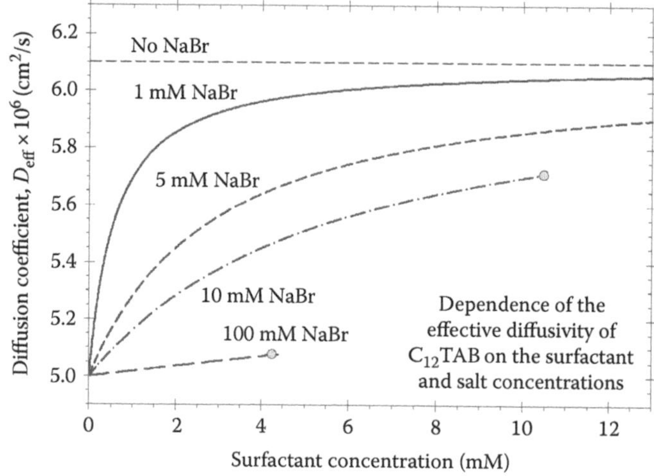

FIGURE 4.5 Dependence of the effective diffusivity, D_{eff}, on the surfactant concentration, $c_{1\infty}$, for various salt concentrations, $c_{3\infty}$, denoted in the figure. The curves are calculated by using the values of D_1, D_2, and D_3 given in Ref. [114] for the cationic surfactant $C_{12}TAB$. The end points of some curves correspond to the CMC.

value of S_l, we calculate the equilibrium surfactant adsorption, $\Gamma_{1,e}$, using Equation 4.85a and the theoretical value of D_{eff} from Figure 4.5; see the points in Figure 4.6b. For the used MBPM set up [57], the apparatus constant is $\lambda = 6.07$. The solid line in the latter figure represents the equilibrium surfactant adsorption independently calculated from the fit of equilibrium surface tension data by means of the van der Waals isotherm [114]. The excellent agreement between the points with the line in Figure 4.6b (no adjustable parameters) confirms the accuracy of Equation 4.85a.

The case of adsorption at an interface that is subjected to stationary expansion needs a special theoretical description. This case is experimentally realized with the strip method [95,116], and the overflowing cylinder method [60,92]. It could be realized also by a Langmuir trough. The interfacial expansion is characterized by the quantity $\dot{\alpha} = dA/(Adt)$, which represents the relative rate of

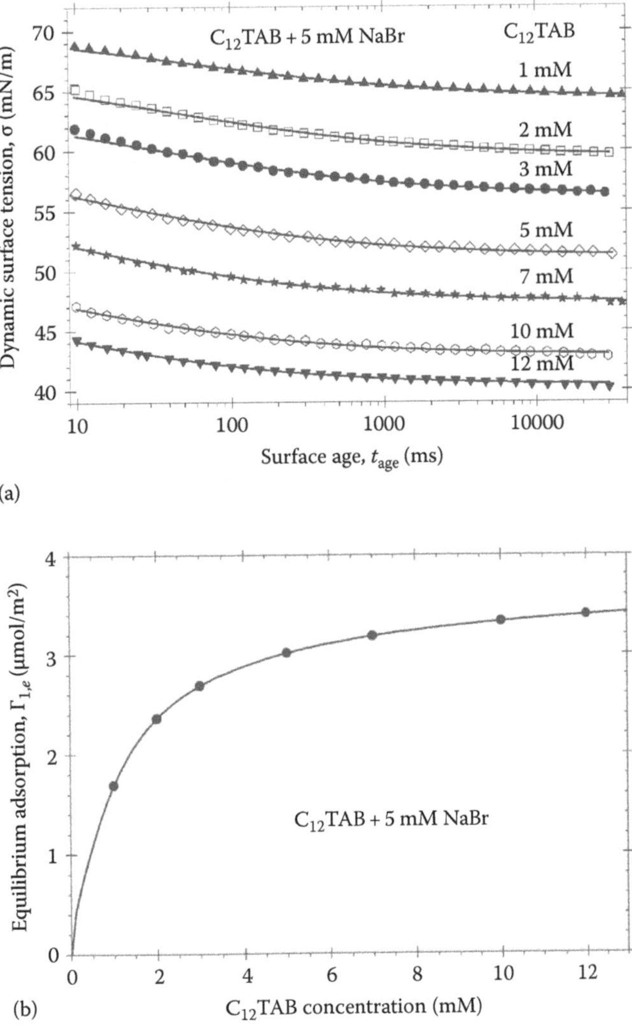

FIGURE 4.6 (a) Data for the dynamic surface tension, σ, vs. the surface age, t_{age}, measured by MBPM [57] at concentrations of C_{12}TAB denoted in the figure; the solid lines are fits (see the text). (b) Dependence of the equilibrium surfactant adsorption, $\Gamma_{1,e}$, on the C_{12}TAB concentration The points are calculated by means of Equation 4.85a for S_l determined from the fits in Figure 4.6a. The solid lines are calculated independently from fits of surface tension data, σ_e vs. $c_{1\infty}$, by means of the van der Waals adsorption model. (From Danov, K.D. et al., *J. Colloid Interface Sci.*, 303, 56, 2006.)

increase of the interfacial area, A. For stationary processes, $\dot{\alpha}$ = const. is a parameter known from the experiment. In this case, the dynamic surface tension is given by the expression [114]:

$$\sigma = \sigma_e + kT\Gamma_{1,e}^2 \left(\frac{\pi\dot{\alpha}}{2D_{\text{eff}}}\right)^{1/2} \frac{1}{\gamma_\pm}\left(\frac{1}{c_{1\infty}} + \frac{1}{c_{2\infty}}\right) \tag{4.85b}$$

where D_{eff} is given by Equations 6.19 through 6.26 in Ref. [114]. Equation 4.85b does not contain the time, t, as it should be for a stationary process. For nonionic surfactants and for ionic surfactants at high salt concentrations the term $1/c_{2\infty}$ in Equation 4.85b disappears and $D_{\text{eff}} = D_1$.

4.2.2.5 Adsorption under Barrier Control

In general, the adsorption is under barrier (kinetic, transfer) control when the stage of surfactant transfer from the subsurface to the surface is much slower than the diffusion stage because of some kinetic barrier. The latter can be due to steric hindrance, spatial reorientation, or conformational changes accompanying the adsorption of molecules, including destruction of the shells of oriented water molecules wrapping the surfactant hydrocarbon tail in water [117]. We will restrict our considerations to the case of pure barrier control, without double layer effects. In such case the surfactant concentration is uniform throughout the solution, c_1 = constant, and the increase of the adsorption $\Gamma_1(t)$ is solely determined by the transitions of surfactant molecules over the adsorption barrier, separating subsurface from surface:

$$\frac{d\Gamma_1}{dt} = Q \equiv r_{\text{ads}}(c_1, \Gamma_1) - r_{\text{des}}(\Gamma_1) \tag{4.86}$$

where r_{ads} and r_{des} are the rates of surfactant adsorption and desorption, respectively. The concept of barrier-limited adsorption originates from the works of Bond and Puls [118], and Doss [119], and has been further developed by other authors [120–127]. Table 4.6 summarizes some expressions for the total rate of adsorption under barrier control, Q. The quantities K_{ads} and K_{des} in Table 4.6 are

TABLE 4.6
Rate of Surfactant Adsorption for Different Kinetic Models

Type of Isotherm	Rate of Reversible Adsorption $Q = r_{\text{ads}}(c_1, \Gamma_1) - r_{\text{des}}(\Gamma_1)$
Henry	$Q = K_{\text{ads}}c_1 - K_{\text{des}}\frac{\Gamma_1}{\Gamma_\infty}$
Freundlich	$Q = K_{\text{ads}}K^{m-1}c_1^m - K_{\text{des}}\frac{\Gamma_1}{\Gamma_\infty}$
Langmuir	$Q = K_{\text{ads}}c_1\left(1-\frac{\Gamma_1}{\Gamma_\infty}\right) - K_{\text{des}}\frac{\Gamma_1}{\Gamma_\infty}$
Frumkin	$Q = K_{\text{ads}}c_1\left(1-\frac{\Gamma_1}{\Gamma_\infty}\right) - K_{\text{des}}\frac{\Gamma_1}{\Gamma_\infty}\exp\left(-\frac{2\beta\Gamma_1}{kT}\right)$
Volmer	$Q = K_{\text{ads}}c_1 - K_{\text{des}}\frac{\Gamma_1}{\Gamma_\infty-\Gamma_1}\exp\left(\frac{\Gamma_1}{\Gamma_\infty-\Gamma_1}\right)$
van der Waals	$Q = K_{\text{ads}}c_1 - K_{\text{des}}\frac{\Gamma_1}{\Gamma_\infty-\Gamma_1}\exp\left(\frac{\Gamma_1}{\Gamma_\infty-\Gamma_1} - \frac{2\beta\Gamma_1}{kT}\right)$

the rate constants of adsorption and desorption, respectively. Their ratio is equal to the equilibrium constant of adsorption

$$\frac{K_{\text{ads}}}{K_{\text{des}}} = K \tag{4.87}$$

The parameters Γ_∞ and K are the same as in Tables 4.2 through 4.4. Setting $Q = 0$ (assuming equilibrium at surface–subsurface), from each expression in Table 4.6 we deduce the respective equilibrium adsorption isotherm in Table 4.2. In addition, for $\beta = 0$ the expressions for Q related to the Frumkin and van der Waals model reduce, respectively, to the expressions for Q in the Langmuir and Volmer models. For $\Gamma_1 \ll \Gamma_\infty$ both the Frumkin and Langmuir expressions in Table 4.6 reduce to the Henry expression.

Substituting Q from Table 4.6 into Equation 4.86, and integrating, we can derive explicit expressions for the relaxation of surfactant adsorption:

$$\frac{\sigma(t)-\sigma_e}{\sigma(0)-\sigma_e} \approx \frac{\Gamma_1(t)-\Gamma_{1,e}}{\Gamma_1(0)-\Gamma_{1,e}} = \exp\left(-\frac{t}{\tau_\sigma}\right) \tag{4.88}$$

Equation 4.88 holds for $\sigma(t)$ only in the case of small deviations from equilibrium, whereas there is not such a restriction concerning $\Gamma_1(t)$; the relaxation time in Equation 4.88 is given by the expressions

$$\tau_\sigma = \left(\frac{K_{\text{des}}}{\Gamma_\infty}\right)^{-1} \quad \text{(Henry and Freundlich)} \tag{4.89}$$

$$\tau_\sigma = \left(\frac{K_{\text{des}}}{\Gamma_\infty} + \frac{K_{\text{ads}}c_1}{\Gamma_\infty}\right)^{-1} \quad \text{(Langmuir)} \tag{4.90}$$

Equation 4.88 predicts that the perturbation of surface tension, $\Delta\sigma(t) = \sigma(t) - \sigma_e$, relaxes exponentially. This is an important difference with the cases of adsorption under diffusion and electro-diffusion control, for which $\Delta\sigma(t) \propto 1/\sqrt{t}$, cf. Equations 4.70, 4.76, and 4.78. Thus, a test to check whether or not the adsorption occurs under purely barrier control is to plot data for $\ln[\Delta\sigma(t)]$ versus t and to check if the plot complies with a straight line.

In the case of *ionic* surfactants, the adsorption of surfactant ions is accompanied by binding of counterions. In addition, the concentrations of the ionic species vary across the EDL (even at equilibrium). These effects are taken into account in Equation 4.57, which can be used as an expression for Q in the case of Langmuir barrier adsorption of an ionic surfactant.

In fact, a pure barrier regime of adsorption is not frequently observed. It is expected that the barrier becomes more important for substances of low surface activity and high concentration in the solution. Such adsorption regime was observed with propanol, pentanol, and 1,6 hexanoic acid [95], as well as with proteins [128].

It may happen that the characteristic times of diffusion and transfer across the barrier are comparable. In such case we deal with *mixed* kinetic regime of adsorption [129]. Insofar as the stages of diffusion and transfer are consecutive, the boundary conditions at the interface are

$$\frac{d\Gamma_1}{dt} = r_{\text{ads}}(c_1,\Gamma_1) - r_{\text{des}}(\Gamma_1) = D_1\left(\frac{\partial c_1}{\partial z}\right)_{z=0} \tag{4.91}$$

The formal transition in Equation 4.91 from mixed to diffusion control of adsorption is not trivial and demands application of scaling and asymptotic expansions. The criterion for occurrence of adsorption under diffusion control (presence of equilibrium between subsurface and surface) is

$$\frac{aK_{\mathrm{des}}}{D_1}\left(\frac{\partial\Gamma_1}{\partial c_1}\right)_e \gg 1 \tag{4.92}$$

where a is the characteristic thickness of the diffusion layer.

An important difference between the regimes of diffusion and barrier control is in the form of the respective initial conditions. In the case of large initial deformations, these are

$$\Gamma_1(0) = 0, \quad c_{1s}(0) = 0 \quad \text{(diffusion control)} \tag{4.93}$$

$$\Gamma_1(0) = 0, \quad c_{1s}(0) = c_{1\infty} \quad \text{(barrier control)} \tag{4.94}$$

Equation 4.93 reflects the fact that in diffusion regime the surface is always assumed to be equilibrated with the subsurface. In particular, if $\Gamma_1 = 0$, then we must have $c_{1s} = 0$. In contrast, Equation 4.94 stems from the presence of barrier: for time intervals shorter than the characteristic time of transfer, the removal of the surfactant from the interface ($\Gamma_1 = 0$) cannot affect the subsurface layer (because of the barrier) and then $c_{1s}(0) = c_{1\infty}$. This purely theoretical consideration implies that the effect of barrier could show up at the short times of adsorption, whereas at the long times the adsorption will occur under diffusion control [129,130]. The existence of barrier-affected adsorption regime at the short adsorption times could be confirmed or rejected by means of the fastest methods for measurement of dynamic surface tension.

4.2.2.6 Dynamics of Adsorption from Micellar Surfactant Solutions

4.2.2.6.1 Dynamic Equilibrium between Micelles and Monomers

At higher concentrations, spherical aggregates of surfactant molecules, called *micelles* [131,132], appear in the aqueous surfactant solutions (Figure 4.7). The number of monomers in a micelle (the aggregation number) is typically between 50 and 100, depending on the size of the surfactant headgroup and the length of its hydrocarbon tail [36]. The micelles appear above a certain surfactant concentration termed the CMC. For concentrations above the CMC, the addition of surfactant to the solutions leads to the formation of more micelles, whereas the concentration of the monomers remains constant and equal to the CMC. In other words, the micelles, irrespective of their concentrations, exist in dynamic equilibrium with a background solution of monomers with concentration equal to the CMC. (Note that at high surfactant concentrations, the spherical micelles could undergo a transition to bigger aggregates, such as rodlike, disclike, and lamellar micelles [36,133–135].)

A detailed physicochemical model of the micelle–monomer equilibria was proposed [136], which is based on a full system of equations that express (1) chemical equilibria between micelles and monomers, (2) mass balances with respect to each component, and (3) the mechanical balance equation by Mitchell and Ninham [137], which states that the electrostatic repulsion between the headgroups of the ionic surfactant is counterbalanced by attractive forces between the surfactant molecules in the micelle. Because of this balance between repulsion and attraction, the equilibrium micelles are in tension free state (relative to the surface of charges), like the phospholipid bilayers [136,138]. The model is applicable to ionic and nonionic surfactants and to their mixtures and agrees very well with the experiment. It predicts various properties of single-component and mixed micellar solutions, such as the compositions of the monomers and the micelles, concentration of counterions, micelle aggregation number, surface electric charge and potential, effect of added salt on the CMC of ionic surfactant solutions, electrolytic conductivity of micellar solutions, etc. [136,139].

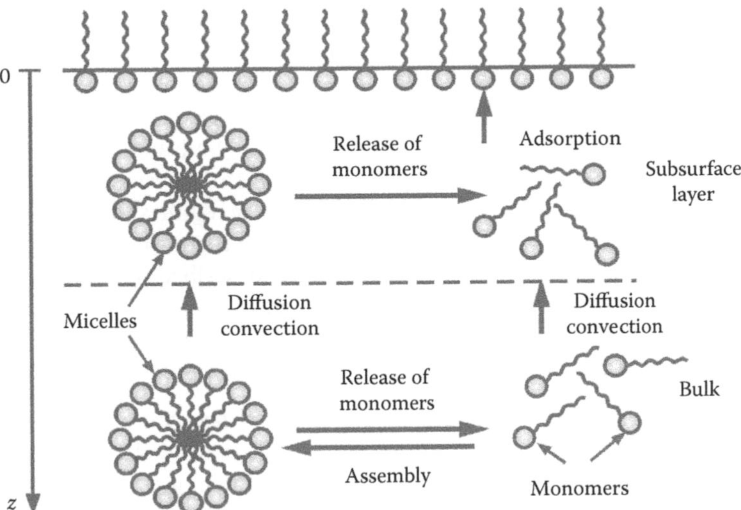

FIGURE 4.7 In the neighborhood of an expanded (nonequilibrium) adsorption monolayer, the micelles (the aggregates) release monomers to restore the equilibrium concentrations of surfactant monomers at the surface and in the bulk. The concentration gradients give rise to diffusion of both monomers and micelles. As a rule, the adsorbing component are the surfactant monomers, whereas the micelles are repelled by the interface and do not adsorb.

When surfactant molecules adsorb at an interface, the concentration of monomers in the subsurface layer decreases, which leads to release of monomers from the neighboring micelles, or to their complete decomposition. The decrease in the concentrations of monomers and micelles gives rise to corresponding diffusion fluxes from the bulk of solution toward the subsurface layer (Figure 4.7). In general, the role of the micelles as sources and carriers of monomers leads to a marked acceleration of surfactant adsorption.

The first models of micellar kinetics in spatially uniform solutions have been developed by Kresheck et al. [140] and Aniansson and Wall [141]. The existence of "fast" and "slow" processes of the micellar dynamics has been established. The *fast* process represents exchange of separate monomers between micelles and the surrounding solution. If the micelle releases monomers, its aggregation number could decrease to a critical value, after which a complete decomposition of the micelle to monomers takes place. This decomposition is known as the *slow* demicellization process [141].

The first theoretical model of surfactant adsorption from micellar solutions, proposed by Lucassen [142], uses the simplifying assumptions that the micelles are monodisperse and that the micellization happens as a single step, which is described as a reversible reaction of order n (the micelle aggregation number). Later, more realistic models, which account for the multi-step character of the micellar process, were developed [143–145]. The assumption for a complete local dynamic equilibrium between monomers and micelles makes possible to use the equilibrium mass action law for the micellization reaction [142,146,147]. In such a case, the surfactant transfer corresponds to a conventional diffusion-limited adsorption characterized by an effective diffusion coefficient, D_{eff}, which depends on the micelle diffusivity, concentration, and aggregation number. D_{eff} is independent of the rate constants of the fast and slow demicellization processes: k_m and k_S. Joos et al. [146,147] confirmed experimentally that in some cases the adsorption from micellar solutions could be actually described as a diffusion-limited process characterized by an apparent diffusivity, D_{eff}. In other experiments, Joos et al. [95,148] established that sometimes the dynamics of adsorption from micellar solutions exhibits a completely different kinetic pattern: the interfacial relaxation is exponential, rather than inverse square root, as it should be for diffusion-limited kinetics.

The theoretical developments [95,129,148] revealed that the exponential relaxation is influenced by the kinetics of micellization, and from the data analysis we could determine the rate constant of the fast process, k_m. The observation of different kinetic regimes for different surfactants and/or experimental methods makes the physical picture rather complicated.

A realistic model of the micellar kinetics was proposed [149] and applied to investigate the dynamics of adsorption at quiescent [150] and expanding [57,151] interfaces. The theoretical analysis reveals the existence of four different consecutive relaxation regimes (stages) for a given micellar solution: two exponential regimes and two inverse-square-root regimes, following one after another in alternating order. The results of these studies are briefly described in the following section, and the agreement between theory and experiment is illustrated.

4.2.2.6.2 *The Four Kinetic Regimes of Adsorption from Micellar Solutions*

In the theoretical model proposed in Refs. [149,150], the use of the quasi-equilibrium approximation (local chemical equilibrium between micelles and monomers) is avoided. The theoretical problem is reduced to a system of four nonlinear differential equations. The model has been applied to the case of surfactant adsorption at a quiescent interface [150], that is, to the relaxation of surface tension and adsorption after a small initial perturbation. The perturbations in the basic parameters of the micellar solution are defined in the following way:

$$\xi_1 \equiv \frac{h_a}{\Gamma_{p,0}} c_{1,p}; \quad \xi_c \equiv \frac{h_a}{\beta \Gamma_{p,0}} C_{m,p}; \quad \xi_m \equiv \frac{h_a c_{1,\text{eq}}}{s_{\text{eq}}^2 \Gamma_{p,0}} m_p \tag{4.95}$$

Here

$c_{1,p}$, $C_{m,p,}$ and m_p are, respectively, the perturbations in the monomer concentration, c_1, micelle concentration, C_m, the micelle mean aggregation number, m, the respective *dimensionless* perturbations are ξ_1, $\xi_{c,}$ and ξ_m;

$\Gamma_{p,0}$ is the perturbation in the surfactant adsorption at the initial moment ($t = 0$);

s_{eq} is the halfwidth of the equilibrium micelle size distribution modeled by a Gaussian bell-like curve;

β and h_a are, respectively, the dimensionless bulk micelle concentration and the characteristic adsorption length, defined as follows:

$$\beta \equiv \frac{C_{\text{tot}} - \text{CMC}}{\text{CMC}}; \quad h_{\text{a}} = \left(\frac{d\Gamma}{dc_1} \right)_{\text{eq}} \tag{4.96}$$

where

C_{tot} is the total surfactant concentration

Γ is the surfactant adsorption

The dimensionless fluxes of the fast and slow demicellization processes, denoted by ϕ_m and ϕ_s, respectively, can be expressed as follows [150]:

$$\varphi_m = \xi_1 - \xi_m \tag{4.97}$$

$$\varphi_s = (m_{\text{eq}} - w s_{\text{eq}})\xi_1 - m_{\text{eq}}\xi_c + s_{\text{eq}} w \xi_m \tag{4.98}$$

(Some small terms are neglected in Equations 4.97 and 4.98.) Here m_{eq} is the equilibrium micelle aggregation number $w = (m_{\text{eq}} - n_r)/s_{\text{eq}}$, where n_r is an aggregation number at the boundary between the regions of the rare aggregates and the abundant micelles [150]

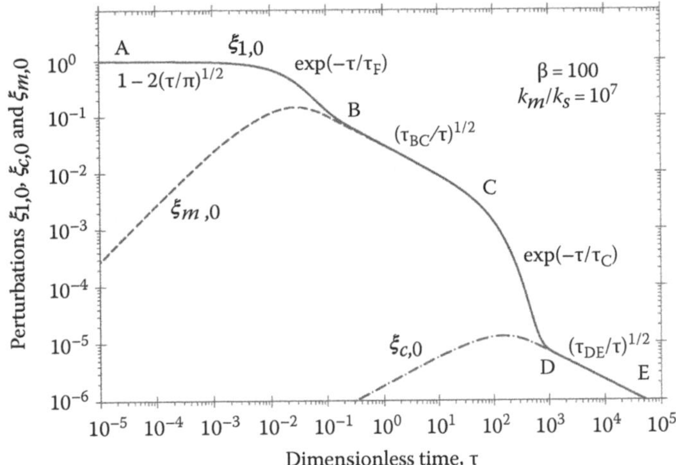

FIGURE 4.8 Time dependence of the perturbations in the subsurface monomer concentration, $\xi_{1,0}$, micelle concentration, $\xi_{c,0}$, and mean aggregation number, $\xi_{m,0}$, for β = 100. The curves are obtained by numerical solution of the general system of equations. (From Danov, K.D. et al., *Adv. Colloid Interface Sci.*, 119, 17, 2006.)

Figure 4.8 shows results obtained by solving numerically the general system of equations in Ref. [150] for a relatively high micelle concentration, β = 100. The calculated curves $\xi_{1,0}(\tau)$, $\xi_{c,0}(\tau)$, and $\xi_{m,0}(\tau)$ represent the subsurface values (at $z = 0$, Figure 4.7) of the perturbations ξ_1, ξ_c, and ξ_m, plotted versus the dimensionless time, $\tau = (D_1/h_a^2)t$, where D_1 is the diffusion coefficient of the surfactant monomers. Note that $\xi_{1,0}$ expresses not only the perturbation in the subsurface monomer concentration, but also the perturbations in the surface tension and adsorption [150]:

$$\frac{\sigma(t)-\sigma_e}{\sigma(0)-\sigma_e} = \frac{\Gamma(t)-\Gamma_e}{\Gamma(0)-\Gamma_e} = \xi_{1,0}(\tau) \tag{4.99}$$

where

- $\sigma(t)$ and $\Gamma(t)$ are the dynamic surface tension and adsorption, respectively
- $\sigma(0)$ and $\Gamma(0)$ are their initial values
- σ_e and Γ_e are their final equilibrium values

A typical value, $k_m/k_S = 10^7$, of the ratio of the rate constants of the fast and slow demicellization processes is used to calculate the curves in Figure 4.8.

The most important feature of the relaxation curves in Figure 4.8, which represents a kinetic diagram, is that $\xi_{m,0}$ merges with $\xi_{1,0}$ at a given point, denoted by B, while $\xi_{c,0}$ merges with $\xi_{1,0}$ (and $\xi_{m,0}$) at another point, denoted by D. The time moments, corresponding to the points B and D, are denoted by τ_B and τ_D, respectively. As seen in Figure 4.8, for $\tau > \tau_B$, we have $\xi_{1,0} = \xi_{m,0}$. In view of Equation 4.97, this means that for $\tau > \tau_B$ the flux of the fast micelle relaxation process, ϕ_m is equal to zero. In other words, for $\tau > \tau_B$ the monomers and micelles are equilibrated with respect to the *fast* micellar process. For a regular relaxation process, the theoretical analysis [150] yields the expression $\tau_B = s_{eq}h_a(2k_m/D_1)^{1/2}$. In addition, for $\tau > \tau_D$ we have $\xi_{c,0} = \xi_{1,0} = \xi_{m,0}$, and then Equation 4.98 indicates that $\phi_s = 0$, that is, the monomers and micelles are equilibrated with respect to the *slow* micellar process.

The computer modeling [150] shows that $\xi_{1,0}(\tau)$ exhibits two exponential (kinetic) regimes, AB and CD, and two inverse-square-root (diffusion) regimes, BC and DE, see Figure 4.8. In particular, the point C corresponds to the moment $\tau_C = (D_1/h_a^2)t_c \approx (\beta D_1\sigma_{eq}^2)/(k_S h_a^2 m_{eq}^3)$, where t_c is the characteristic time of the slow micellar process; see Ref. [149]. τ_C also serves as a characteristic relaxation time of adsorption in the kinetic regime CD. The expressions for the other characteristic times, τ_F, τ_{BC}, and τ_{DE} (Figure 4.8) are [150] the following:

$$\tau_F = \frac{m_{eq}D_1}{\beta k_m h_a^2} \quad \text{(regime AB)} \tag{4.100}$$

$$\frac{1}{\tau_{BC}} = \frac{D_{BC}}{D_1} = (1+u\beta)(1+u\beta B_m) \quad \text{(regime BC)} \tag{4.101}$$

$$\frac{1}{\tau_{DE}} = \frac{D_{DE}}{D_1} = [1+(u+m_{eq})\beta][1+(u+m_{eq})\beta B_m] \quad \text{(regime DE)} \tag{4.102}$$

Here

D_{BC} and D_{DE} are the effective diffusivities of the micellar solutions in the regimes BC and DE, respectively; $u = s_{eq}^2/m_{eq}$

$B_m = D_m/D_1$; D_m is the mean diffusivity of the micelles

Typical parameter values are $u \approx 1$ and $B_m \approx 0.2$.

It should be noted that in addition to the regular kinetic diagrams (Figure 4.8), for low micelle concentrations (β close to 1) we could observe "rudimentary" kinetic diagrams, characterized by merging or disappearance of the stages BC and CD [150,151].

The diffusion regimes BC and DE can be observed not only for adsorption at a quiescent interface, but also in the cases of stationary [151] and nonstationary [57] expansion of an interface. The expressions for the effective diffusivities, D_{BC} and D_{DE}, given by Equations 4.101 and 4.102, are valid in all these cases. In particular, the experimental data by Lucassen [142] correspond to the kinetic regime DE, while the experimental data by Joos et al. [147] correspond to the kinetic regime BC.

As an illustration, in Figure 4.9 we show experimental data for the ionic surfactants (SDS) and $C_{12}TAB$ + 100 mM added inorganic electrolyte. The data are obtained by means of the MBPM described in Ref. [57]. To check whether the kinetic regime is DE, we substitute typical parameter values in Equation 4.102: $m_{eq} = 70$, $\beta = 20$, and $B_m = 0.2$, and as a result we obtain $D_{DE}/D_1 = 3.9 \times 10^5$, which is much greater than the experimental values of D_{eff}/D_1 in Figure 4.9. Consequently, the

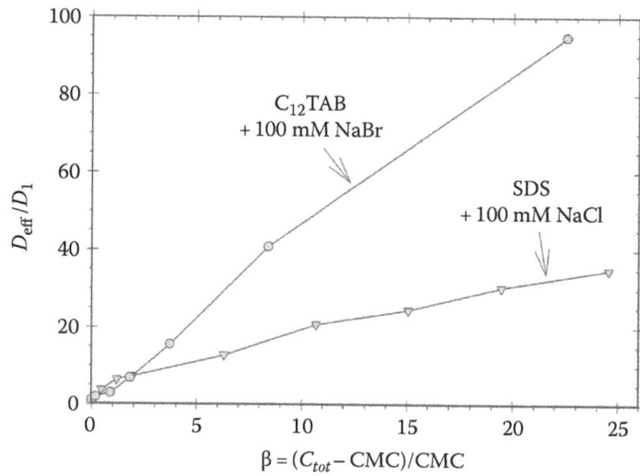

FIGURE 4.9 Plot of the dimensionless effective diffusivity of the micellar solution, D_{eff}/D_1, vs. the dimensionless micelle concentration, β, obtained from dynamic surface tension values measured by the maximum bubble pressure method (MBPM) (From Christov, N.C. et al., *Langmuir*, 22, 7528, 2006); D_1 is the diffusivity of the surfactant monomers. The lines are guides to the eye.

kinetic regime cannot be DE. On the other hand, a similar estimate of D_{BC}/D_1 from Equation 4.101 gives reasonable values. To demonstrate that, from the experimental values of D_{eff}/D_1 in Figure 4.9 we calculated u by means of Equation 4.101, substituting $B_m = 0.2$. For most of the concentrations we obtain values $0.4 < u < 2$, which seem reasonable. Values $u > 2$ are obtained at $\beta < 2$, which indicate that at the lowest micellar concentrations we are dealing with a rudimentary kinetic regime [150,151], rather than with the diffusion regime BC.

4.2.2.6.3 The Case of Stationary Interfacial Expansion

This special case of interfacial dynamics is realized with the strip method [95,147] and the overflowing cylinder method [60,92]. Because the adsorption process is stationary, the time, t, is not a parameter of state of the system. For this reason, in the kinetic diagrams (like Figure 4.10) we plot the perturbations versus the dimensionless rate of surface expansion, $\theta = (h_a^2/D_1)(dA/dt)/A$, where A is the interfacial area, and dA/dt = constant is the interfacial expansion rate. In Figure 4.10, the total perturbations, $\xi_{1,T}$, $\xi_{c,T}$, and $\xi_{m,T}$, are plotted, which represent the local perturbations, $\xi_1(z)$, $\xi_c(z)$, and $\xi_m(z)$, integrated with respect to the normal coordinate z along the whole semiaxis $z > 0$ (Figure 4.7). As seen in Figure 4.10, we observe the same kinetic regimes, as in Figure 4.8, although the diagrams in the two figures look like mirror images: the "young" surface age (the regime AB) corresponds to the left side of Figure 4.8, but to the right side of Figure 4.10. Analytical expressions for the adsorption and surface tension relaxation could be found in Ref. [151]. As mentioned earlier, the expressions for the effective diffusivities, D_{BC} and D_{DE}, given by Equations 4.101 and 4.102, are also valid in the case of stationary interfacial expansion. In particular, it has been found [151] that the kinetic regime of adsorption from the solutions of the nonionic surfactant polyoxyethylene-20 hexadecyl ether (Brij 58), measured by means of the strip method [147], corresponds to the regime BC.

We recall that in the regime BC the rate constants of the fast and slow micellar processes, k_m and k_S, do not affect the surfactant adsorption kinetics, and cannot be determined from the fit of the data. In principle, it is possible to observe the kinetic regime AB (and to determine k_m) with faster methods or with slower surfactants.

In summary, four distinct kinetic regimes of adsorption from micellar solutions exist, called AB, BC, CD, and DE; see Figures 4.8 and 4.10. In regime AB, the fast micellar process governs the adsorption kinetics. In regime BC, the adsorption occurs under diffusion control because the

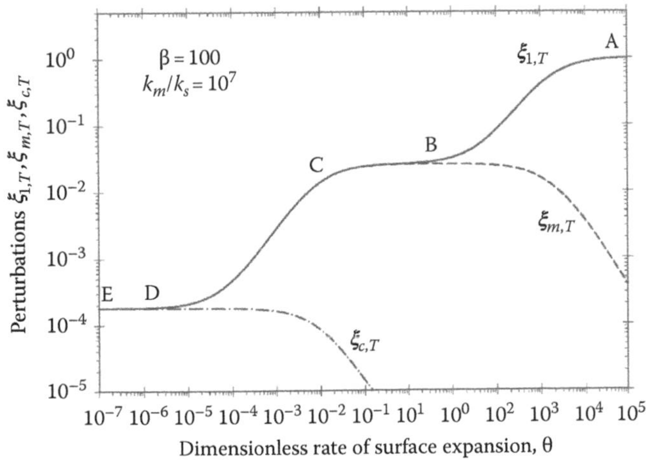

FIGURE 4.10 Total perturbations in monomer concentration, $\xi_{1,T}$, micelle concentration, $\xi_{c,T}$, and mean aggregation number, $\xi_{m,T}$, plotted vs. the dimensionless rate of surface expansion, θ, at micelle dimensionless concentration $\beta = 100$. The curves are obtained by numerical solution of the linear system of equations derived. (From Danov, K.D. et al., *Colloids Surf. A*, 282–283, 143, 2006.)

fast micellar process is equilibrated, while the slow process is negligible. In regime CD, the slow micellar process governs the adsorption kinetics. In regime DE, the adsorption occurs under diffusion control, because both the fast and slow micellar processes are equilibrated. Note that only the regimes BC and DE correspond to purely diffusion processes. For the regimes AB and CD, the rate constants of the fast and slow micellar processes, k_m and k_S, respectively, affect the surfactant adsorption kinetics, and could be in principle determined from the fit of experimental data. For the specific experimental examples considered here, the adsorption kinetics corresponds to the diffusion regime BC.

4.2.2.6.4 Kinetics of Oil Solubilization in Micellar Solutions

The term 'solubilization' was coined by McBain [152] to denote the increased solubility of a given compound, associated with the presence of surfactant micelles or inverted micelles in the solution. The most popular solubilization process is the transfer of oil molecules into the core of surfactant micelles. Thus, oil that has no solubility (or limited solubility) in the aqueous phase becomes water soluble in the form of solubilizate inside the micelles. This process has a central importance for washing of oily deposits from solid surfaces and porous media, and for removal of oily contaminants dispersed in water. The great practical importance of solubilization is related to its application in the everyday life: in the personal care and household detergency, as well as in various industrial processes [153].

The main actors in the solubilization process are the micelles of surfactant and/or copolymer. Their ability to uptake oil is of crucial importance [153,154]. The addition of copolymers, which form mixed micelles with the surfactants [155], is a way to control and improve the micelle solubilization performance. Two main kinetic mechanisms of solubilization have been established whose effectuation depends on the specific system:

1. *Solubilization as a bulk reaction*: Molecular dissolution and diffusion of oil into the aqueous phase takes place, with a subsequent uptake of oil molecules by surfactant micelles [156–161]. This mechanism is operative for oils (like benzene, hexane, etc.), which exhibit a sufficiently high solubility in pure water. Theoretical models have been developed and verified against the experiment [157,159–161]. The bulk solubilization includes the following processes. First, oil molecules are dissolved from the surface of an oil drop into water. Kinetically, this process can be characterized by a mass transfer coefficient. Next, by molecular diffusion, the oil molecules penetrate in the water phase, where they react with the micelles. Thus, the concentration of free oil molecules diminishes with the distance from the oil–water interface. In other words, solubilization takes place in a certain zone around the droplet [159,160].
2. *Solubilization as a surface reaction*: This is the major solubilization mechanism for oils that are practically insoluble in water [156,158,160,162–170]. The uptake of such oils cannot happen in the bulk of the aqueous phase. The solubilization can be realized only at the oil–water interface. The mechanism may include (1) micelle adsorption, (2) uptake of oil, and (3) desorption of the swollen micelles [168–170]. Correspondingly, the theoretical description of the process involves the rate constants of the three consecutive steps. If the empty micelles are long rodlike aggregates, upon solubilization they usually break to smaller spherical aggregates [168,171]. For some systems (mostly solid solubilizates), the intermediate stages in the solubilization process may involve penetration of surfactant solution into the oily phase and formation of a *liquid crystalline phase* at the interface [172–176].

In the case of solubilization as surface reaction, the detailed kinetic mechanism could be multiform. Some authors [156,163] expect that the surfactant arrives at the interface in a monomeric form. Then, at the phase boundary mixed (or swollen) micellar aggregates are formed, which eventually desorb. This version of the model seems appropriate for solid solubilizates because hemimicelles can be formed at their surfaces, even at surfactant concentrations below the bulk CMC [177]. Another concept, presented by Plucinski and Nitsch [165], includes a step of partial fusion of the micelles with the oil–water interface, followed by a step of separation. Such mechanism could take place in the case when microemulsion drops, rather than micelles, are responsible for the occurrence of solubilization.

Experiments with various surfactant systems [166,170,178] showed that the solubilization rates for solutions of *ionic* surfactants are generally much lower than those for nonionic surfactants. This can be attributed to the electrostatic repulsion between the micelles and the similarly charged surfactant adsorption monolayer at the oil–water interface. On the other hand, copolymers have been found to form micelles, which solubilize various hydrophobic compounds well, even in the absence of low-molecular-weight surfactants [179–187]. Moreover, appropriately chosen copolymers can act as very efficient promoters of solubilization [160,168–170].

4.3 CAPILLARY HYDROSTATICS AND THERMODYNAMICS

4.3.1 Shapes of Fluid Interfaces

4.3.1.1 Laplace and Young Equations

A necessary condition for mechanical equilibrium of a fluid interface is the Laplace equation of capillarity [188–191]:

$$2H\sigma = \Delta P \tag{4.103}$$

Here

H is the local mean curvature of the interface
ΔP is the local jump of the pressure across the interface

If $z = z(x, y)$ is the equation of the interface in Cartesian coordinates, then H can be expressed in the form [191]

$$2H = \nabla_s \cdot \left[\frac{\nabla_s z}{\left(1 + |\nabla_s z|^2\right)^{1/2}} \right] \tag{4.104}$$

where ∇_s is the gradient operator in the plane xy. More general expressions for H can be found in the literature on differential geometry [191–193]. Equation 4.103, along with Equation 4.104, represents a second-order partial differential equation which determines the shape of the fluid interface. The interface is bounded by a *three-phase contact line* at which the boundary conditions for the differential equation are formulated. The latter are the respective necessary conditions for mechanical equilibrium at the contact lines. When one of the three phases is solid (Figure 4.11a), the boundary condition takes the form of Young [194] equation:

$$\sigma_{12} \cos \alpha = \sigma_{1s} - \sigma_{2s} \tag{4.105}$$

where

α is the three-phase contact angle
σ_{12} is the tension of the interface between the fluid phases 1 and 2
σ_{1s} and σ_{2s} are the tensions of the two fluid–solid interfaces

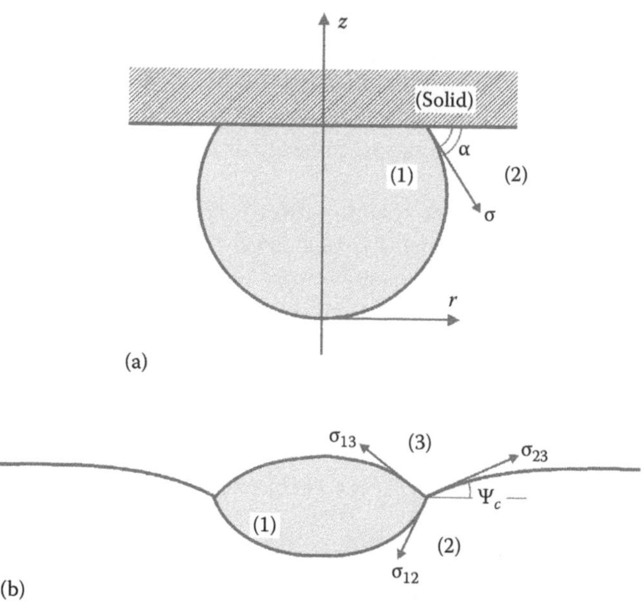

FIGURE 4.11 Sketch of fluid particle (1) attached to the interface between phases (2) and (3). (a) Fluid particle attached to solid interface; α is the contact angle; σ is the interfacial tension of the boundary between the two fluid phases. (b) Fluid particle attached to a fluid interface; σ_{12}, σ_{13}, and σ_{23} are the interfacial tensions between the respective phases; ψ_c is the slope angle of the outer meniscus at the contact line.

Insofar as the values of the three σ's are determined by the intermolecular forces, contact angle α is the material characteristics of a given three-phase system. However, when the solid is not smooth and chemically homogeneous, then the contact angle exhibits hysteresis, that is, α has no defined equilibrium value [6,195]. Contact angle hysteresis can be observed even with molecularly smooth and homogeneous interfaces under dynamic conditions [196].

When all the three neighboring phases are fluids, then the boundary condition takes the form of the Neumann [197] vectorial triangle:

$$\sigma_{12}\boldsymbol{v}_{12} + \sigma_{13}\boldsymbol{v}_{13} + \sigma_{23}\boldsymbol{v}_{23} = 0 \tag{4.106}$$

(see Figure 4.11b); here $\boldsymbol{v}_{ik}$ is a unit vector, which is simultaneously normal to the contact line and tangential to the boundary between phases i and k. The Laplace, Young, and Neumann equations can be derived as conditions for minimum of the free energy of the system [37,191,198]; the effect of the line tension can also be taken into account in Equations 4.105 and 4.106 [198].

In the special case of spherical interface $H = 1/R$, with R being the sphere radius, and Equation 4.103 takes its most popular form, $2\sigma/R = \Delta P$. In the case of axisymmetric meniscus (z-is the axis of symmetry, Figure 4.11), the Laplace equation reduces to either of the following two equivalent forms [190,199]:

$$\frac{1}{r}\frac{d}{dr}\left[\frac{rz'}{(1+z'^2)^{1/2}}\right] = \frac{\Delta P}{\sigma}, \quad z = z(r) \tag{4.107}$$

$$-\frac{r''}{(1+r'^2)^{3/2}} + \frac{1}{r(1+r'^2)^{1/2}} = \frac{\Delta P}{\sigma}, \quad r = r(z) \tag{4.108}$$

Two equivalent parametric forms of Laplace equation are often used for calculations [190,199]:

$$\frac{d\sin\varphi}{dr}+\frac{\sin\varphi}{r}=\frac{\Delta P}{\sigma},\quad \tan\varphi=\frac{dz}{dr} \tag{4.109}$$

or

$$\frac{d\varphi}{ds}=\frac{\Delta P}{\sigma}-\frac{\sin\varphi}{r},\quad \frac{dr}{ds}=\cos\varphi,\quad \frac{dz}{ds}=\sin\varphi \tag{4.110}$$

Here
- φ is the meniscus running slope angle (Figure 4.11a)
- s is the arc length along the generatrix of the meniscus

Equation 4.110 is especially convenient for numerical integration, whereas Equation 4.109 may create numerical problems at the points with $\tan\varphi = \pm\infty$, like the particle equator in Figure 4.11a. A generalized form of Equation 4.109, with account for the interfacial (membrane) bending elastic modulus, k_c,

$$\sigma\left(\frac{d\sin\varphi}{dr}+\frac{\sin\varphi}{r}\right)=\Delta P+\frac{k_c}{r}\cos\varphi\frac{d}{dr}\left\{r\cos\varphi\frac{d}{dr}\left[\frac{1}{r}\frac{d}{dr}(r\sin\varphi)\right]\right\} \tag{4.111}$$

serves for description of the axisymmetric configurations of real and model cell membranes [37,200,201]. The Laplace equation can be generalized to also account for the interfacial bending moment (spontaneous curvature), shear elasticity, etc.; for review, see Refs. [37,200]. The latter effects are physically important for systems or phenomena like capillary waves [202], phospholipid and protein membranes [203–206], emulsions [207], and microemulsions [208].

4.3.1.2 Solutions of Laplace Equations for Menisci of Different Geometry

Very often, the capillary menisci have rotational symmetry. In general, there are three types of axially symmetric menisci corresponding to the three regions denoted in Figure 4.12: (1) meniscus meeting the axis of revolution, (2) meniscus decaying at infinity, and (3) meniscus confined between two cylinders, $0 < R_1 < r < R_2 < \infty$. These three cases are separately considered in the following section.

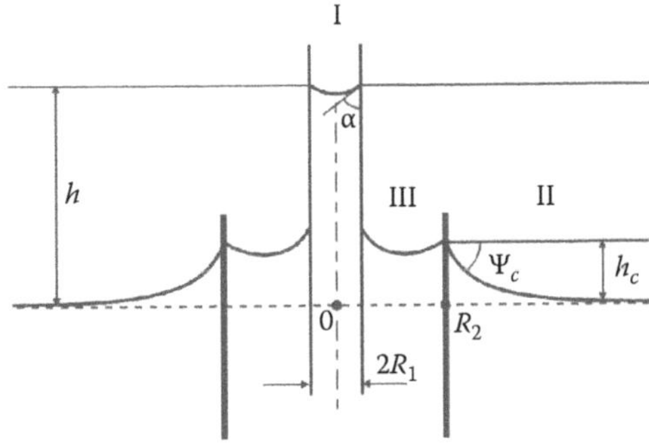

FIGURE 4.12 Capillary menisci formed around two coaxial cylinders of radii R_1 and R_2. (I) Meniscus meeting the axis of revolution; (II) meniscus decaying at infinity; (III) meniscus confined between the two cylinders. h denotes the capillary raise of the liquid in the inner cylinder; h_c is the elevation of meniscus II at the contact line $r = R_2$.

4.3.1.2.1 Meniscus Meeting the Axis of Revolution

This includes the cases of a bubble/droplet under a plate (Figure 4.11a), the two surfaces of a floating lens (Figure 4.11b), and any kind of sessile or pendant droplets/bubbles. Such a meniscus is a part of a sphere when the effect of gravity is negligible, that is when

$$\frac{\Delta\rho g b^2}{\sigma} \ll 1 \tag{4.112}$$

Here

- g is the gravity acceleration
- $\Delta\rho$ is the difference in the mass densities of the lower and the upper fluid
- b is a characteristic radius of the meniscus curvature

For example, if Equation 4.112 is satisfied with $b = R_1$ (see Figure 4.12), the raise, h, of the liquid in the capillary is determined by means of the equation [6]

$$h = \frac{2\sigma\cos\alpha}{\Delta\rho g R_1} \tag{4.113}$$

When the gravity effect is not negligible, the capillary pressure, ΔP, becomes dependent on the z-coordinate:

$$\Delta P = \frac{2\sigma}{b} + \Delta\rho g z \tag{4.114}$$

Here b is the radius of curvature at the particle apex, where the two principal curvatures are equal (e.g., the bottom of the bubble in Figure 4.11a). Unfortunately, Equation 4.107, along with Equation 4.114, has no closed analytical solution. The meniscus shape can be exactly determined by numerical integration of Equation 4.110. Alternatively, various approximate expressions are available [199,209,210]. For example, if the meniscus slope is small, $z'^2 \ll 1$, Equation 4.107 reduces to a linear differential equation of Bessel type, whose solution reads

$$z(r) = \frac{2[I_0(qr) - 1]}{bq^2} \qquad q \equiv \left(\frac{\Delta\rho g}{\sigma}\right)^{1/2} \tag{4.115}$$

where $I_0(x)$ is the modified Bessel function of the first kind and zeroth order [211,212]. Equation 4.115 describes the shape of the lower surface of the lens in Figure 4.11b; similar expression can also be derived for the upper lens surface.

4.3.1.2.2 Meniscus Decaying at Infinity

Examples are the outer menisci in Figures 4.11b and 4.12. In this case the action of gravity cannot be neglected insofar as the gravity keeps the interface flat far from the contact line. The capillary pressure is

$$\Delta P = \Delta\rho g z \tag{4.116}$$

As mentioned earlier, Equation 4.107, along with Equation 4.116, has no closed analytical solution. On the other hand, the region far from the contact line has always a small slope, $z'^2 \ll 1$.

In this region Equation 4.107 can be linearized, and then in analogy with Equation 4.115 we derive

$$z(r) = AK_0(qr) \quad (z'^2 \ll 1) \tag{4.117}$$

where
A is a constant of integration
$K_0(x)$ is the modified Bessel function of the second kind and zeroth order [211,212]

The numerical integration of Equation 4.110 can be carried out by using the boundary condition [199] $z'/z = -qK_1(qr)/K_0(qr)$ for some appropriately fixed $r \ll q^{-1}$ (see Equation 4.117). Alternatively, approximate analytical solutions of the problem are available [199,210,213]. In particular, Derjaguin [214] derived an asymptotic formula for the elevation of the contact line at the outer surface of a thin cylinder,

$$h_c = -R_1 \sin\psi_c \ln\left[\frac{qR_1\gamma_e(1+\cos\psi_c)}{4}\right], \quad (qR_1)^2 \ll 1 \tag{4.118}$$

where
R_1 is the radius of the contact line
ψ_c is the meniscus slope angle at the contact line (Figure 4.12)
q is defined by Equation 4.115
$\gamma_e = 1.781072418\ldots$ is the constant of Euler–Masceroni [212]

4.3.1.2.3 *Meniscus Confined between Two Cylinders ($0 < R_1 < r < R_2 < \infty$)*

This is the case with the Plateau borders in real foams and emulsions, and with the model films in the Scheludko–Exerowa cell [215,216]; such is the configuration of the capillary bridges (Figure 4.13a) and of the fluid particles pressed between two surfaces (Figure 4.13b). When the gravitational deformation of the meniscus cannot be neglected, the interfacial shape can be determined by numerical integration of Equation 4.110, or by iteration procedure [217]. When the meniscus deformation caused by gravity is negligible, analytical solution can be found as described in the following section.

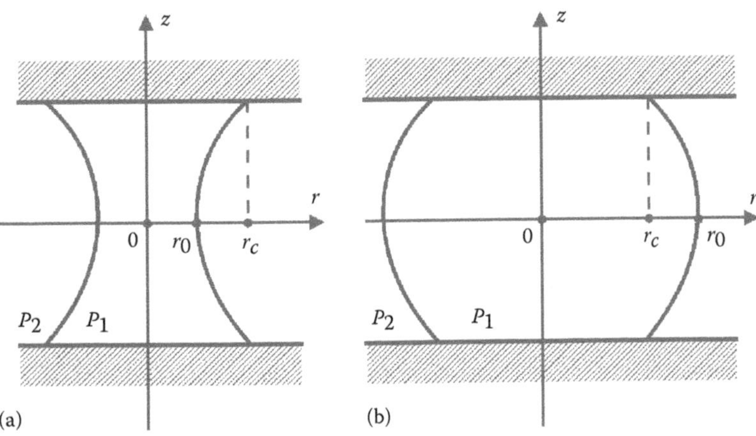

FIGURE 4.13 Concave (a) and convex (b) capillary bridges between two parallel plates. P_1 and P_2 denote the pressures inside and outside the capillary bridge, r_0 is the radius of its section with the midplane; r_c is the radius of the three-phase contact lines.

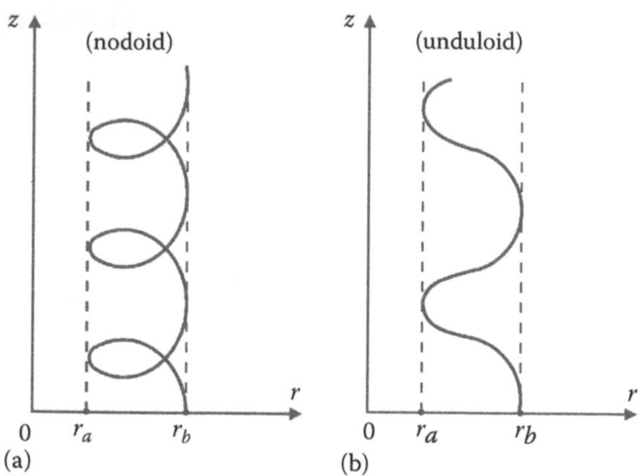

FIGURE 4.14 Typical shape of nodoid (a) and unduloid (b) plateau curves. Note that the curves are confined between two cylinders of radii r_a and r_b.

To determine the shape of the menisci depicted in Figure 4.13a and b, we integrate Equation 4.109 from r_0 to r to derive

$$\frac{dz}{dr} = \frac{k_1\left(r^2 - r_0^2\right) + r_0}{\pm\left[\left(r^2 - r_0^2\right)\left(r_1^2 - r^2\right)\right]^{1/2}|k_1|}, \quad k_1 \equiv \frac{P_1 - P_2}{2\sigma}, \quad r_1 \equiv \left|\frac{1 - k_1 r_0}{k_1}\right| \tag{4.119}$$

The pressures in phases 1 and 2, P_1 and P_2, and r_0 are shown in Figure 4.13. Equation 4.119 describes curves, which after Plateau [189,190,218–220] are called "nodoid" and "unduloid" (see Figure 4.14). The nodoid (unlike the unduloid) has points with horizontal tangent, where $dz/dr = 0$. Then with the help of Equation 4.119, we can deduce that the meniscus generatrix is a part of *nodoid* if $k_1 r_0 \in (-\infty, 0) \cup (1, +\infty)$, while the meniscus generatrix is a part of *unduloid* if $k_1 r_0 \in (0, 1)$.

In the special case, when $k_1 r_0 = 1$, the meniscus is spherical. In the other special case, $k_1 r_0 = 0$, the meniscus has the shape of *catenoid*, that is,

$$z = \pm r_0 \ln\left[\frac{r}{r_0} + \sqrt{\left(\frac{r}{r_0}\right)^2 - 1}\right], \quad (k_1 = 0) \tag{4.120}$$

The meniscus is concave and has a "neck" (Figure 4.13a) when $k_1 r_0 \in (-\infty, 1/2)$; in particular, the generatrix is nodoid for $k_1 r_0 \in (-\infty, 0)$, catenoid for $k_1 r_0 = 0$, and unduloid for $k_1 r_0 \in (0, 1/2)$. For the configuration depicted in Figure 4.13a, we have $r_1 > r_0$ (in Figure 4.14 $r_a = r_0$, $r_b = r_1$) and Equation 4.119 can be integrated to yield (see tables of integrals):

$$z(r) = \pm\left\{r_0 F(\phi_1, q_1) + r_1 \operatorname{sgn} k_1\left[E(\phi_1, q_1) - \frac{1}{r r_1}\sqrt{(r^2 - r_0^2)(r_1^2 - r^2)}\right]\right\} \quad (r_0 \le r \le r_1) \tag{4.121}$$

where

sgn$x$ denotes the sign of x

$q_1 = (1 - r_0^2/r_1^2)^{1/2}$

$\sin\phi_1 = q_1^{-1}(1 - r_0^2/r^2)^{1/2}$

$F(\phi, q)$ and $E(\phi, q)$ are the standard symbols for elliptic integrals of the first and the second kind [211,212]

A convenient method for computation of $F(\phi, q)$ and $E(\phi, q)$ is the method of the arithmetic–geometric mean (see Ref. [211], Chapter 17.6).

The meniscus is convex (Figure 4.13b) when $k_1 r_0 \in (1/2, +\infty)$; in particular, the generatrix is unduloid for $k_1 r_0 \in (1/2, 1)$, circumference for $k_1 r_0 = 1$, and nodoid for $k_1 r_0 \in (1, +\infty)$. For the configuration depicted in Figure 4.13b, we have $r_0 > r_1$ (in Figure 4.14 $r_a = r_1$, $r_b = r_0$) and Equation 4.119 can be integrated to yield (see tables of integrals):

$$z(r) = \mp\left[\left(r_0 - \frac{1}{k_1}\right)F(\phi_2, q_2) - r_0 E(\phi_2, q_2)\right], \quad (r_1 \le r \le r_0) \tag{4.122}$$

where

$q_2 = (1 - r_1^2/r_0^2)^{1/2}$

$\sin\phi_2 = q_2^{-1}(1 - r^2/r_0^2)^{1/2}$

Additional information about the shapes, stability, and nucleation of capillary bridges, and for the capillary-bridge forces between particles, can be found in Chapter 11 of Ref. [37].

Small capillary bridges, called "pendular rings" [221], give rise to cohesion between the particles in the wet sand and to adhesion of particles to a flat plate [222]. In their study on the enhancement of rheology of three-phase (solid–oil–water) dispersions, Koos and Willenbacher [223] identified two states with different structures: (1) the *pendular state*, where the solid particles are interconnected with *concave* capillary bridges, and (2) the *capillary state*, where the particles are interconnected by *convex* capillary bridges; see Figure 4.13. In the former case, the bridging fluid wets the particles well (contact angle $\theta < 90°$), whereas in the latter case the bridging fluid does not wet the particles well (contact angle $\theta > 90°$). However in both cases the capillary bridging phenomenon leads to a considerable enhancement of the rheological response of the dispersion at a minor volume fraction of the second fluid [223].

4.3.1.3 Gibbs–Thomson Equation

The dependence of the capillary pressure on the interfacial curvature leads to a difference between the chemical potentials of the components in small droplets (or bubbles) and in the large bulk phase. This effect is the driving force for phenomena like nucleation [224,225] and Ostwald ripening (see Section 4.3.1.4). Let us consider the general case of a multicomponent two-phase system; we denote the two phases by α and β. Let phase α be a liquid droplet of radius R. The two phases are supposed to coexist at equilibrium. Then we can derive [4,5,226,227]

$$\left(\mu_i^\beta\right)_R - \left(\mu_i^\beta\right)_{R=\infty} = \left(\mu_i^\alpha\right)_R - \left(\mu_i^\alpha\right)_{R=\infty} = V_i^\alpha \frac{2\sigma}{R} \tag{4.123}$$

where

μ is chemical potential

V_i is partial volume

the superscripts denote phase

the subscripts denote component

Equation 4.123 is derived under the following assumptions. When β is a gaseous phase, it is assumed that the partial volume of each component in the gas is much larger than its partial volume in the liquid α; this is fulfilled far enough from the critical point [227]. When phase β is liquid, it is assumed that $P^\beta(R) = P^\beta(R = \infty)$, where P denotes pressure.

When phase β is an ideal gas, Equation 4.123 yields [4,5,226,227]

$$\frac{P_i^{\beta}(R)}{P_i^{\beta}(\infty)} = \exp\left(\frac{2\sigma V_i^{\alpha}}{RkT}\right) \tag{4.124}$$

where $P_i^{\beta}(R)$ and $P_i^{\beta}(\infty)$ denote, respectively, the equilibrium vapor pressure of component i in the droplet of radius R and in a large liquid phase of the same composition. Equation 4.124 shows that the equilibrium vapor pressure of a droplet increases with the decrease of the droplet size. (For a bubble, instead of a droplet, R must be changed to $-R$ in the right-hand side of Equation 4.124 and the tendency becomes the opposite.) Equation 4.124 implies that in an aerosol of polydisperse droplets the larger droplets will grow and the smaller droplets will diminish down to complete disappearance.

The small droplets are "protected" against disappearance when phase α contains a nonvolatile component. Then instead of Equation 4.124 we have

$$\frac{P_i^{\beta}(R)}{P_i^{\beta}(\infty)} = \frac{1-X(R)}{1-X(\infty)}\exp\left(\frac{2\sigma V_i^{\alpha}}{RkT}\right) \tag{4.125}$$

where X denotes the molar fraction of the nonvolatile component in phase α; for $X(R) = X(\infty)$ Equation 4.125 reduces to Equation 4.124. Setting the left-hand side of Equation 4.125 equal to 1, we can determine the value $X(R)$ needed for a liquid droplet of radius R, surrounded by the gas phase β, to coexist at equilibrium with a large ($R = \infty$) liquid phase α of composition $X(\infty)$.

When both phases α and β are liquid, Equation 4.123 yields

$$\frac{X_i^{\beta}(R)}{X_i^{\beta}(\infty)} = \exp\left(\frac{2\sigma V_i^{\alpha}}{RkT}\right) \tag{4.126}$$

where

$X_i^{\beta}(R)$ denotes the equilibrium molar fraction of component i in phase β coexisting with a droplet of radius R

$X_i^{\beta}(\infty)$ denotes the value of $X_i^{\beta}(R)$ for $R \to \infty$, that is, for phase β coexisting with a large phase α of the same composition as the droplet

In the case of oil-in-water emulsion, X_i^{β} can be the concentration of the oil dissolved in the water. In particular, Equation 4.126 predicts that the large emulsion droplets will grow and the small droplets will diminish. This phenomenon is called Ostwald ripening (see Section 4.3.1.4). If the droplets (phase α) contain a component, which is insoluble in phase β, the small droplets will be protected against complete disappearance; a counterpart of Equation 4.125 can be derived:

$$\frac{X_i^{\beta}(R)}{X_i^{\beta}(\infty)} = \frac{1-X(R)}{1-X(\infty)}\exp\left(\frac{2\sigma V_i^{\alpha}}{RkT}\right) \tag{4.127}$$

where X denotes the equilibrium concentration in phase α of the component which is insoluble in phase β. Setting the left-hand side of Equation 4.127 equal to 1, we can determine the value $X(R)$ needed for an emulsion droplet of radius R, surrounded by the continuous phase β, to coexist at equilibrium with a large ($R = \infty$) liquid phase α of composition $X(\infty)$.

4.3.1.4 Kinetics of Ostwald Ripening in Emulsions

The Ostwald ripening is observed when the substance of the emulsion droplets (we will call it component 1) exhibits at least minimal solubility in the continuous phase, β. As discussed above earlier, the chemical potential of this substance in the larger droplets is lower than in the smaller droplets, see Equation 4.123. Then a diffusion transport of component 1 from the smaller toward the larger droplets will take place. Consequently, the size distribution of the droplets in the emulsion will change with time. The kinetic theory of Ostwald ripening was developed by Lifshitz and Slyozov [228] and Wagner [229] and further extended and applied by other authors [230–233]. The basic equations of this theory are the following.

The volume of an emulsion droplet grows (or diminishes) due to the molecules of component 1 supplied (or carried away) by the diffusion flux across the continuous medium. The balance of component 1 can be presented in the form [233]

$$\frac{4\pi}{3}\frac{d}{dt}R^3(t) = 4\pi DRV_1[c_m(t) - c_{\text{eq}}(R)] \tag{4.128}$$

where

t is time
D is the diffusivity of component 1 in the continuous phase
V_1 is the volume per molecule of component 1
c_m is the number-volume concentration of component 1 in the continuous medium far away from the droplets surfaces

$c_{\text{eq}}(R)$ is the respective equilibrium concentration of the same component for a droplet of radius R as predicted by the Gibbs–Thomson equation

Note that Equation 4.128 is rigorous only for a diluted emulsion, in which the concentration of dissolved component 1 levels off at a constant value, $c = c_m$, around the middle of the space between each two droplets. Some authors [231] also add in the right-hand side of Equation 4.128 terms accounting for the convective mass transfer (in the case of moving droplets) and thermal contribution to the growth rate.

Because the theory is usually applied to droplets of diameter not smaller than micrometer (which are observable by optical microscope), the Gibbs–Thomson equation, Equation 4.126, can be linearized to yield [233]

$$c_{\text{eq}}(R) \approx c_\infty\left(1 + \frac{b}{R}\right), \quad b \equiv \frac{2\sigma V_1}{kT} \tag{4.129}$$

with c_∞ being the value of c_{eq} for flat interface. With $\sigma = 50$ mN/m, $V_1 = 100$ Å^3, and $T = 25$°C we estimate $b = 2.5$ nm. The latter value justifies the linearization of Gibbs–Thomson equation for droplets of micrometer size.

Let $f(R, t)$ be the size distribution function of the emulsion droplets such that $f(R, t)dR$ is the number of particles per unit volume in the size range from R to $(R + dR)$. The balance of the number of particles in the system reads

$$df\,dR = (jdt)\,|_R - (jdt)\,|_{R+dR}, \quad \left(j \equiv f\frac{dR}{dt}\right) \tag{4.130}$$

The term in the left-hand side of Equation 4.130 expresses the change of the number of droplets whose radius belongs to the interval $[R, R + dR]$ during a time period dt; the two terms in the right-hand side represent the number of the incoming and outgoing droplets in the size interval $[R, R + dR]$ during time period dt. Dividing both sides of Equation 4.130 by $(dR\,dt)$, we obtain the so-called continuity equation in the space of sizes [229–233]:

$$\frac{\partial f}{\partial t} + \frac{\partial j}{\partial R} = 0 \qquad (4.131)$$

One more equation is needed to determine c_m. In a closed system this can be the total mass balance of component 1:

$$\frac{d}{dt}\left[c_m(t) + \frac{4\pi}{3}\int_0^{\infty} dR\,R^3 f(R,t)\right] = 0 \qquad (4.132)$$

The first and the second terms in the brackets express the amount of component 1 contained in the continuous phase and in the droplets, respectively. This expression is appropriate for diluted emulsions when c_m is not negligible compared to the integral in the brackets.

Alternatively, in opened systems and in concentrated emulsions we can use a mean field approximation based on Equation 4.129 to obtain the following equation for c_m:

$$c_m(t) = c_\infty\left[1 + \frac{b}{R_m(t)}\right], \quad R_m(t) \equiv \frac{\int_{R_0}^{\infty} dR\,Rf(R,t)}{\int_{R_0}^{\infty} dR\,f(R,t)} \qquad (4.133)$$

where R_0 is a lower limit of the experimental distribution, typically $R_0 \approx 1$ μm as smaller droplets cannot be observed optically. The estimates show that neglecting of integrals over the interval $0 < R < R_0$ in Equation 4.133 does not affect the value of R_m significantly. We see that Equation 4.133 treats each emulsion droplet as being surrounded by droplets of average radius R_m, which provide a medium concentration c_m in accordance with the Gibbs–Thomson equation, Equation 4.129. From Equations 4.128 through 4.131 and 4.133 we can derive a simple expression for the flux j:

$$j(R,t) = Q\left(\frac{1}{RR_m} - \frac{1}{R^2}\right)f(R,t), \quad Q \equiv Dbc_\infty V_1 \qquad (4.134)$$

In calculations, we use the set of Equations 4.128, 4.131, and 4.132 or 4.133 to determine the distribution $f(R, t)$ at known distribution $f(R, 0)$ at the initial moment $t = 0$. In other words, the theory predicts the evolution of the system at a given initial state. From a computational viewpoint it is convenient to calculate $f(R, t)$ in a finite interval $R_0 \le R < R_{max}$ (see Figure 4.15). The problem can be solved numerically by discretization: the interval $R_0 \le R < R_{max}$ is subdivided into small portions of length δ, the integrals are transformed into sums, and the problem is reduced to solving a linear set of equations for the unknown functions $f_k(t) \equiv f(R_k, t)$, where $R_k = R_0 + k\delta$, $k = 1, 2, \ldots$.

In practice, the emulsions are formed in the presence of surfactants. At concentrations above the CMC the swollen micelles can serve as carriers of oil between the emulsion droplets of different size. In other words, surfactant micelles can play the role of mediators of the Ostwald ripening. Micelle-mediated Ostwald ripening has been observed in solutions of nonionic surfactants [234–236]. In contrast, it was found that the micelles do not mediate the Ostwald ripening in undecane-in-water emulsions at the presence of an *ionic* surfactant (SDS) [237]. It seems that the surface charge due to

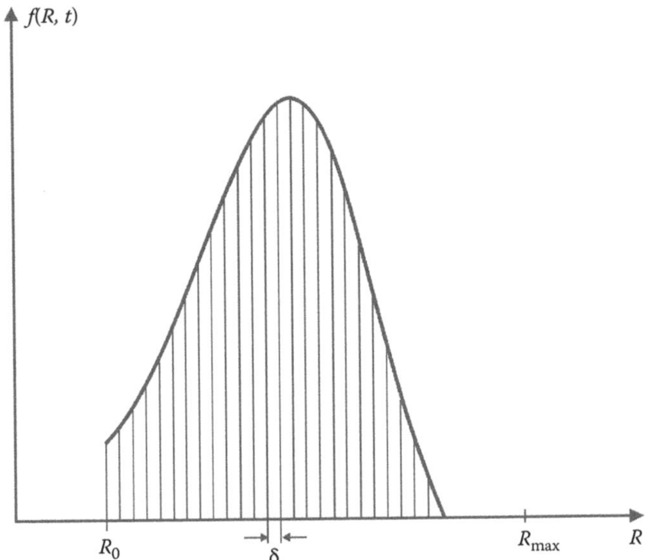

FIGURE 4.15 Sketch of the droplet size distribution function, $f(R, t)$ vs. the droplet radius R at a given moment t. δ is the length of the mesh used when solving the problem by discretization.

the adsorption of ionic surfactant (and the resulting double layer repulsion) prevents the contact of micelles with the oil drops, which is a necessary condition for micelle-mediated Ostwald ripening.

4.3.2 Thin Liquid Films and Plateau Borders

4.3.2.1 Membrane and Detailed Models of a Thin Liquid Film

Thin liquid films can be formed between two colliding emulsion droplets or between the bubbles in foam. Formation of thin films accompanies the particle–particle and particle–wall interactions in colloids. From a *mathematical* viewpoint, a film is thin when its thickness is much smaller than its lateral dimension. From a *physical* viewpoint, a liquid film formed between two macroscopic phases is *thin* when the energy of interaction between the two phases across the film is not negligible. The specific forces causing the interactions in a thin liquid film are called *surface forces*. Repulsive surface forces stabilize thin films and dispersions, whereas attractive surface forces cause film rupture and coagulation. This section is devoted to the *macroscopic* (hydrostatic and thermodynamic) theory of thin films, while the *molecular* theory of surface forces is reviewed in Section 4.4.

In Figure 4.16, a sketch of plane-parallel liquid film of thickness h is presented. The liquid in the film contacts with the bulk liquid in the *Plateau border*. The film is symmetrical, that is, it is formed

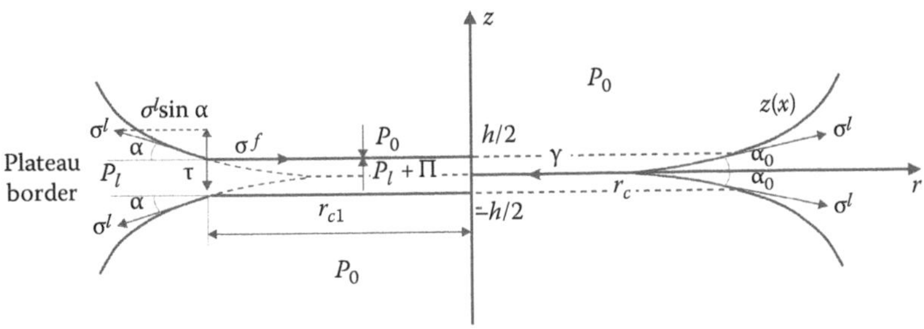

FIGURE 4.16 The detailed and membrane models of a thin liquid film (on the left- and right-hand side, respectively).

between two *identical* fluid particles (drops, bubbles) of internal pressure P_0. The more complex case of nonsymmetrical and curved films is reviewed elsewhere [238–240].

Two different, but supplementary, approaches (models) are used in the macroscopic description of a thin liquid film. The first of them, the "membrane approach," treats the film as a membrane of zero thickness and one tension, γ, acting tangentially to the membrane (see the right-hand side of Figure 4.16). In the "detailed approach", the film is modeled as a homogeneous liquid layer of thickness h and surface tension σ^f. The pressure P_0 in the fluid particles is larger than the pressure, P_l, of the liquid in the Plateau border. The difference

$$P_c = P_0 - P_l \tag{4.135}$$

represents the capillary pressure of the liquid meniscus. By making the balance of the forces acting on a plate of unit width along the y-axis and height h placed normally to the film at $-h/2 < z < h/2$ (Figure 4.16), we derive the Rusanov [241] equation:

$$\gamma = 2\sigma^f + P_c h \tag{4.136}$$

Equation 4.136 expresses a condition for equivalence between the membrane and detailed models with respect to the lateral force. To derive the normal force balance we consider a parcel of unit area from the film surface in the detailed approach. Because the pressure in the outer phase P_0 is larger than the pressure inside the liquid, P_l, the mechanical equilibrium at the film surface is ensured by the action of an additional *disjoining* pressure, $\Pi(h)$, representing the surface force per unit area of the film surfaces [242]

$$\Pi(h) = P_0 - P_l = P_c \tag{4.137}$$

(see Figure 4.16). Note that Equation 4.137 is satisfied only at equilibrium; at nonequilibrium conditions the viscous force can also contribute to the force balance per unit film area. In general, the disjoining pressure, Π, depends on the film thickness, h. A typical $\Pi(h)$-isotherm is depicted in Figure 4.17 (for details see Section 4.4). We see that the equilibrium condition, $\Pi = P_c$, can be satisfied at three points shown in Figure 4.17. Point 1 corresponds to a film, which is stabilized by the double layer repulsion; sometimes such a film is called the "primary film" or "common black film." Point 3 corresponds to unstable equilibrium and cannot be observed experimentally. Point 2 corresponds to a very thin film, which is stabilized by the short range repulsion; such a film is called the "secondary film" or "Newton black film." Transitions from common to Newton black films are often observed with foam films [243–246]. Note that $\Pi > 0$ means repulsion between the film surfaces, whereas $\Pi < 0$ corresponds to attraction.

4.3.2.2 Thermodynamics of Thin Liquid Films

In the framework of the membrane approach the film can be treated as a single surface phase, whose Gibbs–Duhem equation reads [238,247]:

$$d\gamma = -s^f dT - \sum_{i=1}^{k} \Gamma_i d\mu_i \tag{4.138}$$

where

γ is the film tension
T is temperature
s^f is excess entropy per unit area of the film
Γ_i and μ_i are the adsorption and the chemical potential of the ith component, respectively

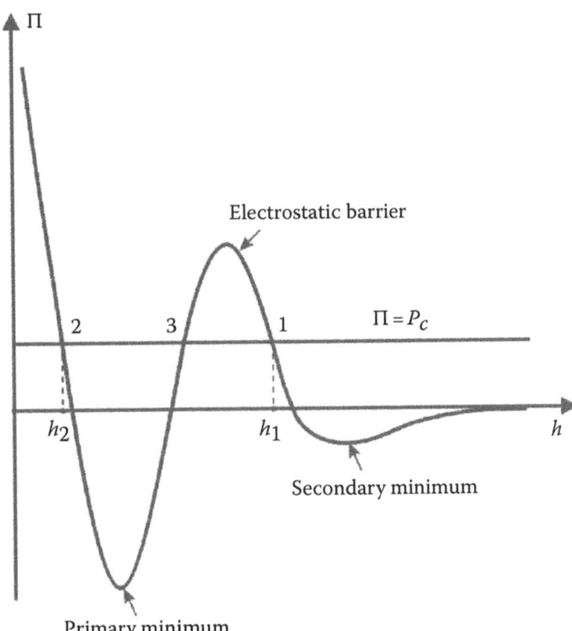

FIGURE 4.17 Sketch of a disjoining pressure isotherm of the DLVO type, Π vs. h. The intersection points of the $\Pi(h)$-isotherm with the line $\Pi = P_c$ correspond to equilibrium films: $h = h_1$ (primary film), $h = h_2$ (secondary film). Point 3 corresponds to unstable equilibrium.

The Gibbs–Duhem equations of the liquid phase (l) and the outer phase (o) read

$$dP_\chi = s_v^\chi dT + \sum_{i=1}^{k} n_i^\chi d\mu_i, \quad \chi = l, o \tag{4.139}$$

where

s_v^χ and n_i^χ are entropy and number of molecules per unit volume
$P\chi$ is pressure ($\chi = l, o$)

The combination of Equations 4.127 and 4.131 provides an expression for dP_c. Let us multiply this expression by h and subtract the result from the Gibbs–Duhem equation of the film, Equation 4.138. The result reads:

$$d\gamma = -\tilde{s}\, dT + h\, dP_c - \sum_{i=1}^{k} \tilde{\Gamma}_i d\mu_i \tag{4.140}$$

where

$$\tilde{s} = s^f + \left(s_v^o - s_v^l\right)h, \qquad \tilde{\Gamma}_i = \Gamma_i + \left(n_i^o - n_i^l\right)h, \quad i = 1,\ldots,k \tag{4.141}$$

An alternative derivation of the same equations is possible [248,249]. Imagine two equidistant planes separated at a distance h. The volume confined between the two planes is thought to be filled with the bulk liquid phase (l). Taking surface excesses with respect to the bulk phases we can derive Equations 4.140 and 4.141 with $\tilde{s}$ and $\tilde{\Gamma}_i$ being the excess surface entropy and adsorption ascribed to the surfaces of this liquid layer [248,249]. A comparison between Equations 4.138 and 4.140 shows that there is one additional differential in Equation 4.140. It corresponds to one supplementary degree of freedom connected with the choice of the parameter h. To specify the model, we need an additional equation to determine h. For example, let this equation be

$$\tilde{\Gamma}_1 = 0 \tag{4.142}$$

Equation 4.142 requires h to be the thickness of a liquid layer from phase (l), containing the same amount of component 1 as the real film. This thickness is called the thermodynamic thickness of the film [249]. It can be on the order of the real film thickness if component 1 is chosen in an appropriate way, say the solvent in the film phase.

From Equations 4.137, 4.140, and 4.142, we obtain [248]

$$d\gamma = -\tilde{s}\,dT + h\,d\Pi - \sum_{i=2}^{k} \tilde{\Gamma}_i d\mu_i \tag{4.143}$$

A corollary of Equation 4.143 is the Frumkin [250] equation

$$\left(\frac{\partial \gamma}{\partial \Pi}\right)_{T,\mu_2,\ldots,\mu_k} = h \tag{4.144}$$

Equation 4.144 predicts a rather weak dependence of the film tension γ on the disjoining pressure, Π, for equilibrium thin films (small h). By means of Equations 4.136 and 4.137, Equation 4.143 can be transformed to read [249]

$$2\,d\sigma^f = -\tilde{s}\,dT - \Pi\,dh - \sum_{i=2}^{k} \tilde{\Gamma}_i d\mu_i \tag{4.145}$$

From Equation 4.145, we can derive the following useful relations [248]

$$2\left(\frac{\partial \sigma^f}{\partial h}\right)_{T,\mu_2,\ldots,\mu_k} = -\Pi \tag{4.146}$$

$$\sigma^f(h) = \sigma^l + \frac{1}{2}\int_h^{\infty} \Pi(h)\,dh \tag{4.147}$$

with σ^l being the surface tension of the bulk liquid. Equation 4.147 allows calculation of the film surface tension when the disjoining pressure isotherm is known.

Note that the thermodynamic equations mentioned earlier are, in fact, corollaries from the Gibbs–Duhem equation of the membrane approach Equation 4.138. There is an equivalent and complementary approach, which treats the two film surfaces as separate surface phases with their own fundamental equations [241,251,252]; thus for a flat symmetric film we postulate

$$dU^f = T\,dS^f + 2\sigma^f\,dA + \sum_{i=1}^{k} \mu_i dN_i^f - \Pi A\,dh \tag{4.148}$$

where

- A is area
- U^f, S^f, and N_i^f are, respectively, excess internal energy, entropy, and number of molecules ascribed to the film surfaces

Compared with the fundamental equation of a simple surface phase [5], Equation 4.148 contains an additional term, $\Pi A dh$, which takes into account the dependence of the film surface energy on the film thickness. Equation 4.148 provides an alternative thermodynamic definition of disjoining pressure:

$$\Pi = -\frac{1}{A}\left(\frac{\partial U^f}{\partial h}\right) \tag{4.149}$$

4.3.2.3 The Transition Zone between Thin Film and Plateau Border

4.3.2.3.1 Macroscopic Description

The thin liquid films formed in foams or emulsions exist in permanent contact with the bulk liquid in the Plateau border, encircling the film. From a macroscopic viewpoint, the boundary between film and Plateau border is treated as a three-phase contact line, the line at which the two surfaces of the Plateau border (the two concave menisci sketched in Figure 4.16) intersect at the plane of the film (see the right-hand side of Figure 4.16). The angle, α_0, subtended between the two meniscus surfaces represents the thin film contact angle. The force balance at each point of the contact line is given by Equation 4.106 with $\sigma_{12} = \gamma$ and $\sigma_{13} = \sigma_{23} = \sigma^l$. The effect of the line tension, κ, can also be taken into account. For example, in the case of symmetrical flat film with circular contact line, like those depicted in Figure 4.16, we can write [252]

$$\gamma + \frac{\kappa}{r_c} = 2\sigma^l \cos\alpha_0 \tag{4.150}$$

where r_c is the radius of the contact line.

There are two film surfaces and two contact lines in the detailed approach (see the left-hand side of Figure 4.16). They can be treated thermodynamically as linear phases and a 1D counterpart of Equation 4.148 can be postulated [252]:

$$dU^L = T\,dS^L + 2\tilde{\kappa}\,dL + \sum_i \mu_i dN_i^L + \tau\,dh \tag{4.151}$$

Here

U^L, S^L, and N_i^L are linear excesses

$\tilde{\kappa}$ is the line tension in the detailed approach

$$\tau = \frac{1}{L}\left(\frac{\partial U^L}{\partial h}\right) \tag{4.152}$$

is a 1D counterpart of the disjoining pressure (see Equation 4.149). The quantity τ, called the *transversal tension*, takes into account the interaction between the two contact lines. The general force balance at each point of the contact line can be presented in the form of the following vectorial sum [238]

$$\sigma_i^f + \sigma_i^l + \sigma_i^\kappa + \tau_i = 0, \quad i = 1,2 \tag{4.153}$$

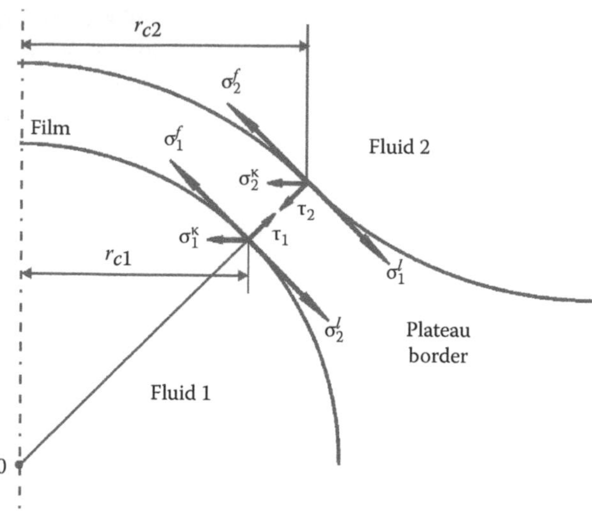

FIGURE 4.18 The force balance in each point of the two contact lines representing the boundary between a spherical film and the Plateau border (see Equation 4.153).

The vectors taking part in Equation 4.153 are depicted in Figure 4.18, where $|\sigma_i^\kappa| = \tilde{\kappa}_i/r_{ci}$. For the case of a flat symmetric film (Figure 4.16) the tangential and normal projections of Equation 4.153, with respect to the plane of the film, read:

$$\sigma^f + \frac{\tilde{\kappa}}{r_{c1}} = \sigma^l \cos\alpha \tag{4.154}$$

$$\tau = \sigma^l \sin\alpha \tag{4.155}$$

Note that, in general $\alpha \neq \alpha_0$ (see Figure 4.16). Besides, both α_0 and α can depend on the radius of the contact line due to line tension effects. In the case of straight contact line from Equations 4.147 and 4.154, we derive [252]

$$\cos\alpha\Big|_{r_{c1}=\infty} = \frac{\sigma^f}{\sigma^l} = 1 + \frac{1}{2\sigma^l}\int_h^\infty \Pi(h)dh \tag{4.156}$$

Because $\cos\alpha \leq 1$, the surface tension of the film must be less than the bulk solution surface tension, $\sigma^f < \sigma^l$, and the integral term in Equation 4.156 must be negative in order for a nonzero contact angle to be formed. Hence, the contact angle, α, and the transversal tension, τ (see Equation 4.155), are integral effects of the long-range *attractive* surface forces acting in the transition zone between the film and Plateau border, where $h > h_1$ (see Figure 4.17).

In the case of a fluid particle attached to a surface (Figure 4.19) the integral of the pressure $P_l = P_0 - \Delta\rho g z$ over the particle surface equals the buoyancy force, F_b, which at equilibrium is counterbalanced by the disjoining pressure and transversal tension forces [238,253]:

$$2\pi r_{c1}\tau = F_b + \pi r_{c1}^2 \Pi \tag{4.157}$$

F_b is negligible for bubbles of diameter smaller than ca 300 μm. Then the forces due to τ and Π counterbalance each other. Hence, at equilibrium the role of the repulsive disjoining pressure is to

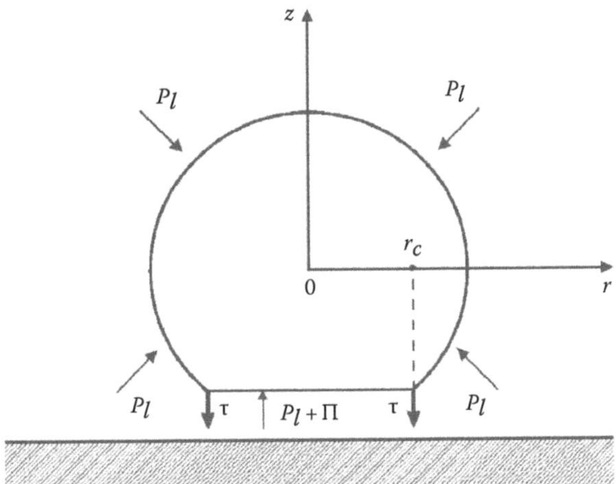

FIGURE 4.19 Sketch of the forces exerted on a fluid particle (bubble, drop, vesicle) attached to a solid surface: Π is disjoining pressure, τ is transversal tension, P_l is the pressure in the outer liquid phase.

keep the film thickness uniform, whereas the role of the attractive transversal tension is to keep the bubble (droplet) attached to the surface. In other words, the particle sticks to the surface at its contact line where the long-range attraction prevails (see Figure 4.17), whereas the repulsion predominates inside the film, where $\Pi = P_c > 0$. Note that this conclusion is valid not only for particle–wall attachment, but also for particle–particle interaction. For zero contact angle τ is also zero (Equation 4.155) and the particle will rebound from the surface (the other particle), unless some additional external force keeps it attached.

4.3.2.3.2 Micromechanical Description

From a microscopic viewpoint, the transition between the film surface and the meniscus is smooth, as depicted in Figure 4.20. As the film thickness increases across the transition zone, the disjoining pressure decreases and tends to zero at the Plateau border (see Figures 4.17 and 4.20).The surface tension varies from σ^f for the film to σ^l for the Plateau border [254,255]. By using local force balance considerations, we can derive the equations governing the shape of the meniscus in the transition zone; in the case of axial symmetry (depicted in Figure 4.20), these equations read [255]:

$$\frac{d}{dr}(\sigma \sin\varphi) + \frac{1}{r}\sigma(r)\sin\varphi(r) = P_c - \Pi(h(r)) \tag{4.158}$$

$$-\frac{d}{dz}(\sigma \cos\varphi) + \frac{1}{r}\sigma(r)\sin\varphi(r) = P_c, \quad \tan\varphi(r) = \frac{dz}{dr} \tag{4.159}$$

where $\varphi(r)$ and $h(r) = 2z(r)$ are the running meniscus slope angle and thickness of the gap, respectively. Equations 4.158 and 4.159 allow calculation of the three unknown functions, $z(r)$, $\varphi(r)$, and $\sigma(r)$, provided that the disjoining pressure, $\Pi(h)$, is known from the microscopic theory. By eliminating P_c between Equations 4.158 and 4.159 we can derive [255]

$$\frac{d\sigma}{dz} = -\Pi(h(r))\cos\varphi(r) \tag{4.160}$$

This result shows that the hydrostatic equilibrium in the transition region is ensured by simultaneous variation of σ and Π. Equation 4.160 represents a generalization of Equation 4.146 for a film of

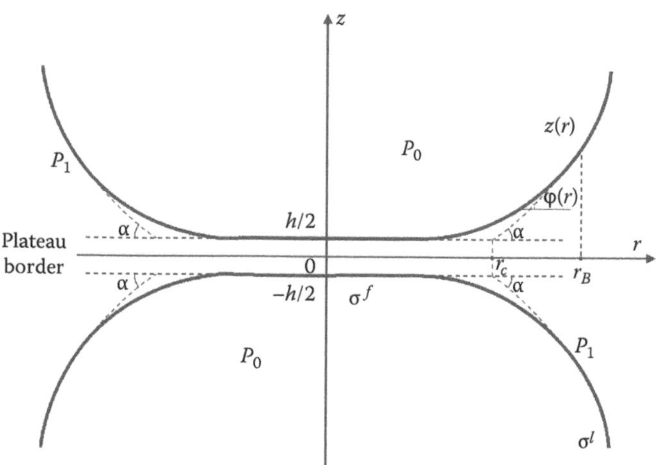

FIGURE 4.20 Liquid film between two attached fluid particles (bubbles, drops, vesicles). The solid lines represent the actual interfaces, whereas the dashed lines show the extrapolated interfaces in the transition zone between the film and the Plateau border.

uneven thickness and axial symmetry. Generalization of Equations 4.158 through 4.160 for the case of more complicated geometry is also available [238,239].

For the Plateau border we have $z \gg h$, $\Pi \to 0$, $\sigma \to \sigma^l$ = constant, and both Equations 4.158 and 4.159 reduce to Equation 4.109 with $\Delta P = P_c$. The macroscopic contact angle, α, is defined as the angle at which the *extrapolated* meniscus, obeying Equation 4.109, meets the *extrapolated* film surface (see the dashed line in Figure 4.20). The real surface, shown by solid line in Figure 4.20, differs from this *extrapolated* (idealized) profile, because of the interactions between the two film surfaces, which is taken into account in Equation 4.158, but not in Equation 4.109. To compensate for the difference between the real and idealized system, the line and transversal tensions are ascribed to the contact line in the macroscopic approach. In particular, the line tension makes up for the differences in surface tension and running slope angle [255]:

$$\frac{\tilde{\kappa}}{r_c} = \int_0^{r_B} \left[\left(\frac{\sigma \sin^2 \varphi}{r \cos \varphi} \right)^{\text{real}} - \left(\frac{\sigma \sin^2 \varphi}{r \cos \varphi} \right)^{\text{idealized}} \right] dr \tag{4.161}$$

whereas τ compensates for the differences in surface forces (disjoining pressure):

$$\tau = \frac{1}{r_c} \int_0^{r_B} [(\Pi)^{\text{id}} - \Pi(r)] r \, dr \tag{4.162}$$

where

$$(\Pi)^{\text{id}} = P_c \quad \text{for } 0 < r < r_c$$

$$(\Pi)^{\text{id}} = 0 \quad \text{for } r > r_c$$

The superscripts "real" and "idealized" in Equation 4.161 mean that the quantities in the respective parentheses must be calculated for the real and idealized meniscus profiles; the latter coincide for

$r > r_B$ (Figure 4.20). Results for $\tilde{\kappa}$ and τ calculated by means of Equations 4.161 and 4.162 can be found in Ref. [256].

In conclusion, it should be noted that the width of the transition region between a thin liquid film and Plateau border is usually very small [257]—below 1 μm. That is why the optical measurements of the meniscus profile give information about the thickness of the Plateau border in the region $r > r_B$ (Figure 4.20). Then if the data are processed by means of the Laplace equation (Equation 4.109), we determine the contact angle, α, as discussed earlier. In spite of being a purely macroscopic quantity, α characterizes the magnitude of the surface forces inside the thin liquid film, as implied by Equation 4.156. This has been pointed out by Derjaguin [257] and Princen and Mason [258].

4.3.2.4 Methods for Measuring Thin Film Contact Angles

Prins [259] and Clint et al. [260] developed a method of contact angle measurement for macroscopic flat foam films formed in a glass frame in contact with a bulk liquid. They measured the jump in the force exerted on the film at the moment, when the contact angle is formed. Similar experimental setup was used by Yamanaka [261] for measurement of the velocity of motion of the three-phase contact line.

An alternative method, which can be used in both equilibrium and dynamic measurements with vertical macroscopic films, was developed by Princen and Frankel [262,263]. They determined the contact angle from the data for diffraction of a laser beam refracted by the Plateau border.

In the case of microscopic films, especially appropriate are the interferometric methods: light beams reflected or refracted from the liquid meniscus interfere and create fringes, which in turn give information about the shape of the liquid surfaces. The fringes are usually formed in the vicinity of the contact line, which provides a high precision of the extrapolation procedure used to determine the contact angle (see Figure 4.20). We can distinguish several interference techniques depending on how the interference pattern is created. In the usual interferometry the fringes are due to interference of beams reflected from the upper and lower meniscus. This technique can be used for contact angle measurements with foam films [217,264–266], emulsion films [267,268], and adherent biological cells [201]. The method is applicable for not-too-large contact angles ($\alpha < 8°-10°$); for larger meniscus slopes the region of fringes shrinks and the measurements are not possible.

The basic principle of the differential interferometry consists of an artificial splitting of the original image into two equivalent and overlapping images (see Françon [269] or Beyer [270]). Thus interferometric measurements are possible with meniscus surfaces of larger slope. The differential interferometry in transmitted light was used by Zorin et al. [271,272] to determine the contact angles of wetting and free liquid films. This method is applicable when the whole system under investigation is transparent to light.

Differential interferometry in reflected light allows for the measurement of the shape of the upper reflecting surface. This method was used by Nikolov et al. [253,273–275] to determine the contact angle, film, and line tension of foam films formed at the top of small bubbles floating at the surface of ionic and nonionic surfactant solutions. An alternative method is the holographic interferometry applied by Picard et al. [276,277] to study the properties of bilayer lipid membranes in solution. Film contact angles can also be determined from the Newton rings of liquid lenses, which spontaneously form in films from micellar surfactant solutions [217].

Contact angles can also be determined by measuring several geometrical parameters characterizing the profile of the liquid meniscus and processing them by using the Laplace equation (Equation 4.109) [278,279]. The computer technique allows processing of many experimental points from meniscus profile and automatic digital image analysis.

Contact angles of microscopic particles against another phase boundary can be determined interferometrically, by means of a *film trapping technique* [280,281]. It consists in capturing of micrometer-sized particles, emulsion drops, and biological cells in thinning free foam films or wetting films. The interference pattern around the entrapped particles allows us to reconstruct the meniscus shape, to determine the contact angles, and to calculate the particle-to-interface adhesion energy [280,281].

A conceptually different method, called *gel trapping technique*, was developed by Paunov [282] for determining the three-phase contact angle of solid colloid particles at an air–water or oil–water interface. The method is applicable for particle diameters ranging from several hundred nanometers to several hundred micrometers. This technique is based on spreading of the particles on a liquid interface with a subsequent gelling of the water phase with a nonadsorbing polysaccharide. The particle monolayer trapped on the surface of the gel is then replicated and lifted up with poly(dimethylsiloxane) (PDMS) elastomer, which allows the particles embedded within the PDMS surface to be imaged with high resolution by using a scanning electron microscope (SEM), which gives information on the particle contact angle at the air–water or the oil–water interface [282]. This method has found applications for determining the contact angles of various inorganic [283,284] and organic [285,286] particles at liquid interfaces.

4.3.3 Lateral Capillary Forces between Particles Attached to Interfaces

4.3.3.1 Particle–Particle Interactions

The origin of the lateral capillary forces between particles captive at a fluid interface leads to deformation of the interface, which is supposed to be flat in the absence of particles. The larger the interfacial deformation, the stronger is the capillary interaction. It is known that two similar particles floating on a liquid interface attract each other [287–289] (see Figure 4.21a). This attraction appears because the liquid meniscus deforms in such a way that the gravitational potential energy of the two particles decreases when they approach each other. Hence the origin of this force is the particle weight (including the Archimedes force).

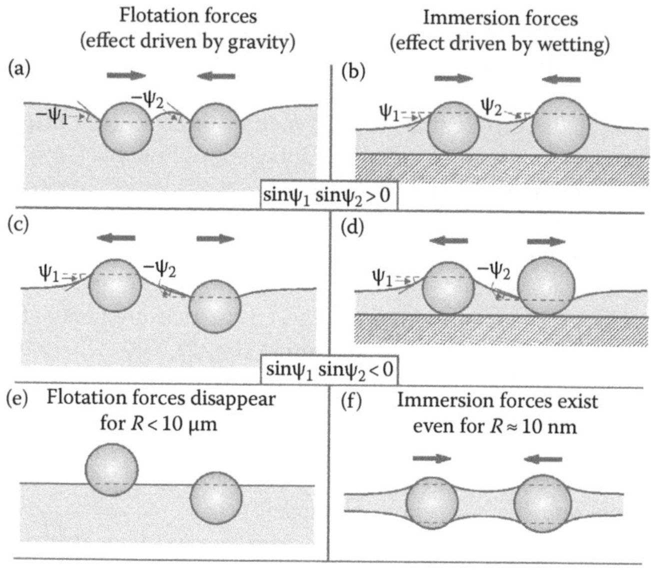

FIGURE 4.21 Flotation (a, c, e) and immersion (b, d, f) lateral capillary forces between two particles attached to fluid interface: (a) and (b) two similar particles; (c) a light and a heavy particle; (d) a hydrophilic and a hydrophobic particle; (e) small floating particles that do not deform the interface; (f) small particles captured in a thin liquid film deforming the interfaces due to the wetting effects.

A force of capillary attraction also appears when the particles (instead of being freely floating) are partially immersed in a liquid layer on a substrate [290–292] (see Figure 4.21b). The deformation of the liquid surface in this case is related to the wetting properties of the particle surface, that is, to the position of the contact line and the magnitude of the contact angle, rather than to gravity.

To distinguish between the capillary forces in the case of floating particles and in the case of partially immersed particles on a substrate, the former are called lateral *flotation* forces and the latter, lateral *immersion* forces [289,292]. These two kinds of forces exhibit similar dependence on the interparticle separation but very different dependencies on the particle radius and surface tension of the liquid (see Refs. [37,293] for comprehensive reviews). The flotation and immersion forces can be both attractive (Figure 4.21a and b) and repulsive (Figure 4.21c and d). This is determined by the signs of the meniscus slope angles ψ_1 and ψ_2 at the two contact lines: the capillary force is attractive when $\sin\psi_1 \sin\psi_2 > 0$ and repulsive when $\sin\psi_1 \sin\psi_2 < 0$. In the case of flotation forces $\psi > 0$ for *light* particles (including bubbles) and $\psi < 0$ for *heavy* particles. In the case of immersion forces between particles protruding from an aqueous layer, $\psi > 0$ for *hydrophilic* particles and $\psi < 0$ for *hydrophobic* particles. When $\psi = 0$ there is no meniscus deformation and, hence, there is no capillary interaction between the particles. This can happen when the weight of the particles is too small to create significant surface deformation (Figure 4.21e).

The immersion force appears not only between particles in wetting films (Figure 4.21b and d), but also in symmetric fluid films (Figure 4.21f). The theory provides the following asymptotic expression for calculating the lateral capillary force between two particles of radii R_1 and R_2 separated by a center-to-center distance L [37,288–293]:

$$F = 2\pi\sigma Q_1 Q_2 q K_1(qL)\left[1 + O\left(q^2 R_k^2\right)\right] \quad r_k \ll L \tag{4.163}$$

where

σ is the liquid–fluid interfacial tension
r_1 and r_2 are the radii of the two contact lines
$Q_k = r_k \sin\psi_k$ ($k = 1, 2$) is the “capillary charge” of the particle [289,292]; in addition

$$q^2 = \frac{\Delta\rho g}{\sigma} \quad \text{(in thick film)}$$

$$q^2 = \frac{\Delta\rho q - \Pi'}{\sigma} \quad \text{(in thick films)} \tag{4.164}$$

Here

$\Delta\rho$ is the difference between the mass densities of the two fluids
Π' is the derivative of the disjoining pressure with respect to the film thickness
$K_1(x)$ is the modified Bessel function of the first order

The asymptotic form of Equation 4.163 for $qL \ll 1$ ($q^{-1} = 2.7$ mm for water),

$$F = \frac{2\pi\sigma Q_1 Q_2}{L} \quad r_k \ll L \ll q^{-1} \tag{4.165}$$

looks like a 2D analogue of Coulomb’s law, which explains the name “capillary charge” of Q_1 and Q_2. Note that the immersion and flotation forces exhibit the same functional dependence on the interparticle distance, see Equations 4.163 and 4.164. On the other hand, their different physical origin results in different magnitudes of the "capillary charges" of these two kinds of capillary force. In this aspect they resemble the electrostatic and gravitational forces, which obey the same power law, but differ in the physical meaning and magnitude of the force constants (charges, masses). In the special case when $R_1 = R_2 = R$ and $r_k \ll L \ll q^{-1}$, we can derive [292,293]

$$F \propto \left(\frac{R^6}{\sigma}\right) K_1(qL) \quad \text{for flotation force}$$

$$F \propto \sigma R^2 K_1(qL) \quad \text{for immersion force} \tag{4.166}$$

Hence, the flotation force decreases, while the immersion force increases, when the interfacial tension σ increases. Besides, the flotation force decreases much more strongly with the decrease of R than the immersion force. Thus $F_{\text{flotation}}$ is negligible for $R < 10$ μm, whereas $F_{\text{immersion}}$ can be significant even when $R = 10$ nm. This is demonstrated in Figure 4.22 where the two types of capillary interactions are compared for a wide range of particle sizes. The values of the parameters used are: particle mass density $\rho_p = 1.05$ g/cm^3, surface tension $\sigma = 72$ mN/m, contact angle $\alpha = 30°$, interparticle distance $L = 2R$, and thickness of the nondisturbed planar film $l_0 = R$. The drastic difference in the magnitudes of the two types of capillary forces is due to the different deformation of the water–air interface. The small floating particles are too light to create substantial deformation of the liquid surface, and the lateral capillary forces are negligible (Figure 4.21e). In the case of immersion forces the particles are restricted in the vertical direction by the solid substrate. Therefore, as the film becomes thinner, the liquid surface deformation increases, thus giving rise to a strong interparticle attraction.

As seen in Figure 4.22, the immersion force can be significant between particles whose radii are larger than few nanometers. It has been found to promote the growth of 2D crystals from colloid particles [294–297], viruses, and globular proteins [298–304]. Such 2D crystals have found various applications: in nanolithography [305], microcontact printing [306], as nanostructured materials in photo-electrochemical cells [307], in photocatalytic films [308], photo- and electroluminescent semiconductor materials [309], as samples for electron microscopy of proteins and viruses [310], as immunosensors [311], etc. (for reviews see Refs. [37,312]).

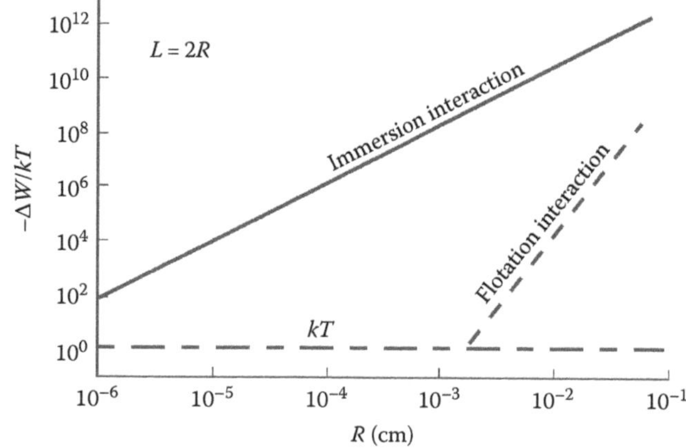

FIGURE 4.22 Plot of the capillary interaction energy in kT units, $\Delta W/kT$, vs. the radius, R, of two similar particles separated at a center-to-center distance $L = 2R$.

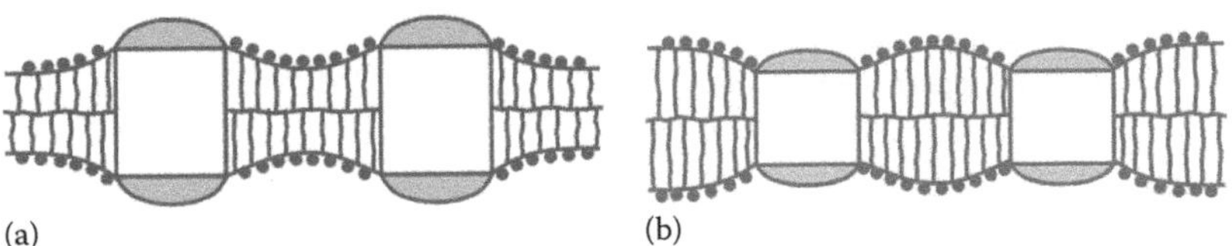

FIGURE 4.23 Inclusions (say, membrane proteins) in a lipid bilayer: the thickness of the inclusion can be greater (a) or smaller (b) than the thickness of the (nondisturbed) lipid bilayer. In both cases, the overlap of the deformations around the inclusions leads to an attraction between them (see Refs. [37,204].)

In the case of interactions between inclusions in lipid bilayers (Figure 4.23), the elasticity of the bilayer interior must also be taken into account. The calculated energy of capillary interaction between integral membrane proteins turns out to be of the order of several kT [204]. Hence, this interaction can be a possible explanation for the observed aggregation of membrane proteins [204,313–316]. The lateral capillary forces have also been calculated for the case of particles captured in a *spherical* (rather than planar) thin liquid film or vesicle [316].

Lateral capillary forces between vertical cylinders or between spherical particles have been measured by means of sensitive electromechanical balance [317], piezo-transducer balance [318], and torsion microbalance [319]. Good agreement between theory and experiment has been established [318,319].

As already mentioned, the weight of micrometer-sized and sub micrometer-sized floating particles is not sufficient to deform the fluid interface and to bring about capillary force between the particles (Figure 4.21e). However, the situation changes if the contact line at the particle surface has *undulated* or *irregular* shape (Figure 4.24a). This may happen when the particle surface is rough, angular, or heterogeneous. In such cases, the contact line sticks to an edge or to the boundary between two domains of the heterogeneous surface. The undulated contact line induces undulations in the surrounding fluid interface [312,320–324]. Let $z = \zeta(x, y)$ be the equation describing the interfacial shape around such isolated particle. Using polar coordinates (r, φ) in the xy-plane, we can express the interfacial shape as a Fourier expansion:

$$\zeta(r,\varphi) = \sum_{m=1}^{\infty} r^{-m}[A_m\cos(m\varphi) + B_m\sin(m\varphi)] \tag{4.167}$$

where

r is the distance from the particle centre

A_m and B_m are coefficients

In analogy with electrical theory, Equation 4.167 can be interpreted as a multipole expansion: the terms with m = 1, 2, 3,…, play the role of capillary "dipoles," "quadrupoles," "hexapoles," and *multipoles* [312,320–324]. The term with m = 0 (capillary "charge") is missing because there is no axisymmetric contribution to the deformation (negligible particle weight). Moreover, the dipolar term with m = 2 is also absent because it is annihilated by a spontaneous rotation of the floating particle around a horizontal axis [321]. Therefore, the leading term becomes the quadrupolar one, with m = 2. The interaction between capillary quadrupoles has been investigated theoretically [321–324]. This interaction is nonmonotonic: attractive at long distances, but repulsive at short distances. Expressions for the rheological properties (surface dilatational and shear elasticity and yield stress) of Langmuir monolayers from angular particles have been derived [37,322,323].

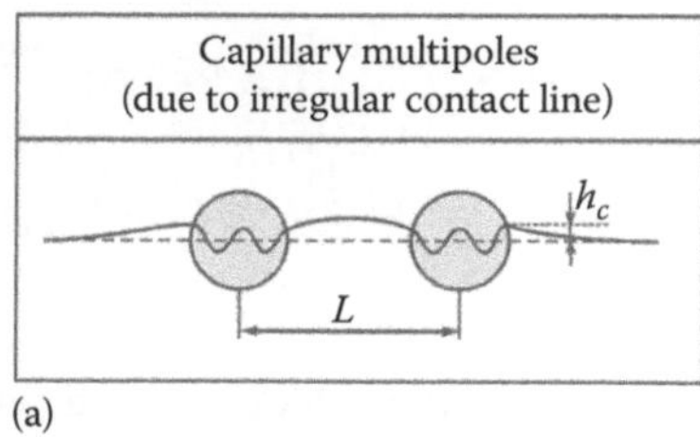

(a)

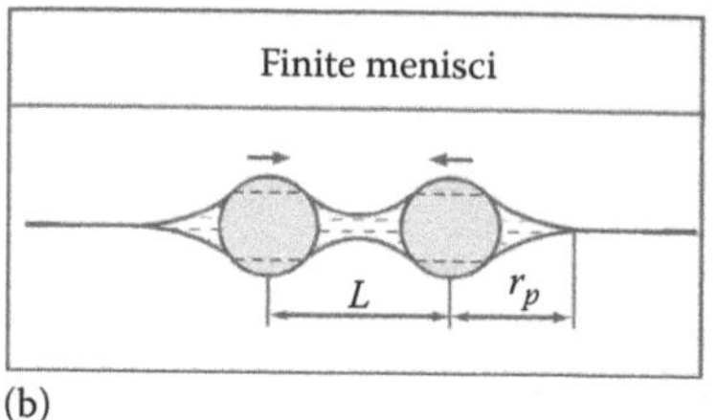

(b)

FIGURE 4.24 Special types of immersion capillary forces: (a) the contact line attachment to an irregular edge on the particle surface produces undulations in the surrounding fluid interface, which give rise to lateral capillary force between the particles. (b) When the size of particles, entrapped in a liquid film, is much greater than the nonperturbed film thickness, the meniscus surfaces meet at a finite distance, r_p; in this case, the capillary interaction begins at $L \leq 2r_p$.

Note that Equation 4.167 is approximate and holds for interparticle distances, which are much smaller than the characteristic capillary length, that is, $qr \ll 1$. The general form of the multipolar expansion, Equation 4.167, for arbitrary interparticle distances reads [321–324]:

$$\zeta(r,\varphi) = A_0K_0(qr) + \sum_{m=1}^{\infty} A_mK_m(qr)\cos[m(\varphi - \varphi_{0,m})] \tag{4.167a}$$

where

A_m and $\varphi_{0,m}$ are constants of integration
K_m is the modified Bessel function of the second kind and mth order

The first term with $m = 0$ in the right-hand side of Equation 4.167a accounts for the contribution of the "capillary charges" (or "capillary monopoles"). Analytical expressions for the force and energy of interaction between two capillary multipoles of arbitrary order have been derived [324].

"Mesoscale" capillary multipoles have been experimentally realized by Bowden et al. [325,326], by appropriate hydrophobization or hydrophilization of the sides of floating plates. Interactions between capillary quadrupoles have been observed between floating particles, which have the shape of curved disks [327]. Loudet et al. [328–330] investigated experimentally and theoretically the capillary forces between adsorbed ellipsoidal particles and found that they behave as capillary quadrupoles. These authors noted that from a purely geometrical viewpoint, the condition of a constant contact angle cannot be met for anisotropic particles if the interface remains flat, which explains the reason for the quadrupolar interfacial deformation. Lateral capillary forces between ellipsoidal, cylindrical (rodlike), and other anisotropic particles have also been investigated by van Nierop et al. [331], Lehle et al. [332], Stebe et al. [333–337], and Yunker et al. [338] Gravitation-like instabilities due to the long-range attractive capillary forces between floating particles have also been studied [339].

At last, let us consider another type of capillary interactions—between particles surrounded by *finite* menisci. Such interactions appear when micrometer-sized or submicrometer-sized particles are captured in a liquid film of much smaller thickness (Figure 4.24b) [340–343]. If such particles are approaching each other, the interaction begins when the menisci around the two particles overlap, $L < 2r_p$ in Figure 4.24b. The capillary force in this case is nonmonotonic: initially the attractive force increases with the increase of interparticle distance, then it reaches a maximum and further decays [343]. In addition, there are hysteresis effects: the force is different on approach and separation at distances around $L = 2r_p$ [343].

4.3.3.2 Particle–Wall Interactions

The overlap of the meniscus around a floating particle with the meniscus on a vertical wall gives rise to a particle–wall interaction, which can be both repulsive and attractive. An example for a controlled meniscus on the wall is shown in Figure 4.25, where the "wall" is a hydrophobic teflon barrier whose position along the vertical wall can be precisely varied and adjusted.

Two types of boundary conditions at the wall are analyzed theoretically [37,344]: fixed contact line (Figure 4.25) or, alternatively, fixed contact angle. In particular, the lateral capillary force exerted on the particle depicted in Figure 4.25 is given by the following asymptotic expression [37,344]:

$$F = -\pi\sigma q\left[2Q_2He^{-qx} + r_2He^{-qx} - 2Q_2^2K_1(qx)\right] \tag{4.168}$$

Here

- Q_2 and r_2 are the particle capillary charge and contact line radius, respectively
- H characterizes the position of the contact line on the wall with respect to the nondisturbed horizontal liquid surface (Figure 4.21)
- x is the particle–wall distance
- q is defined by Equation 4.164 (thick films)

The first term in the right-hand side of Equation 4.168 expresses the gravity force pushing the particle to slide down over the inclined meniscus on the wall; the second term originates from the pressure difference across the meniscus on the wall; the third term expresses the so-called capillary image force, that is, the particle is repelled by its mirror image with respect to the wall surface [37,344].

Static [345] and dynamic [346] measurements with particles near walls have been carried out. In the static measurements the equilibrium distance of the particle from the wall (the distance at which $F = 0$) has been measured and a good agreement with the theory has been established [345].

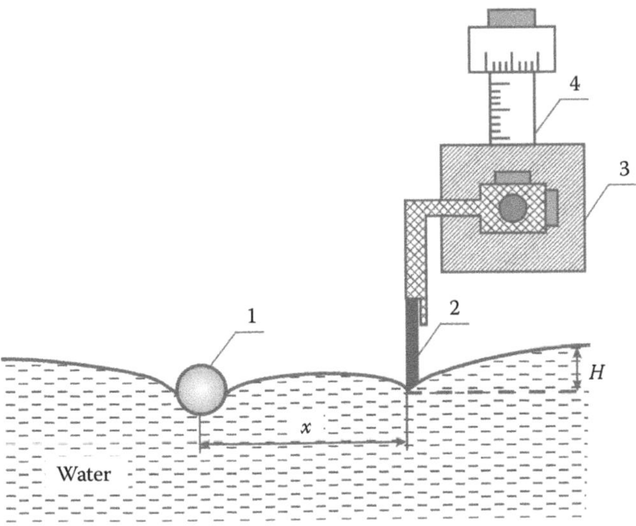

FIGURE 4.25 Experimental setup for studying the capillary interaction between a floating particle (1) and a vertical hydrophobic plate (2) separated at a distance, x. The edge of the plate is at a distance, H, lower than the level of the horizontal liquid surface far from the plate; (3) and (4) are micrometric table and screw; see Refs. [345,346].

In the dynamic experiments [346] knowing the capillary force F (from Equation 4.168) and measuring the particle velocity, $\dot{x}$, we can determine the drag force, F_d:

$$F_d = m\ddot{x} - F, \quad F_d \equiv 6\pi\eta R_2 f_d \dot{x} \tag{4.169}$$

where

R_2, m, and $\ddot{x}$ are the particle radius, mass, and acceleration, respectively
η is the viscosity of the liquid
f_d is the drag coefficient

If the particle were in the bulk liquid, f_d would be equal to 1 and F_d would be given by the Stokes formula. In general, f_d differs from unity because the particle is attached to the interface. The experiment [346] results in f_d varying between 0.68 and 0.54 for particle contact angle varying from 49° to 82°; the data are in good quantitative agreement with the hydrodynamic theory of the drag coefficient [347]. In other words, the less the depth of particle immersion, the less the drag coefficient, as could be expected. However, if the floating particle is heavy enough, it deforms the surrounding liquid surface; the deformation travels together with the particle, thus increasing f_d several times [346]. The addition of surfactant strongly increases f_d. The latter effect can be used to measure the surface viscosity of adsorption monolayers from low molecular weight surfactants [348], which is not accessible to the standard methods for measurement of surface viscosity.

In the case of *protein* adsorption layers, the surface elasticity is so strong that the particle (Figure 4.25) is arrested in the adsorption film. Nevertheless, with heavier particles and at larger meniscus slopes, it is possible to break the protein adsorption layer. Based on such experiments, a method for determining surface elasticity and yield stress has been developed [349].

4.3.3.3 Electrically Charged Particles at Liquid Interfaces

4.3.3.3.1 Particle–Interface Interaction

Let us consider a spherical dielectric particle (phase 1), which is immersed in a nonpolar medium (phase 2), near its boundary with a third dielectric medium (phase 3); see the inset in Figure 4.26. The interaction is due to electric charges at the particle surface. The theoretical problem has been solved exactly, in terms of Legendre polynomials, for arbitrary values of the dielectric constants of the three phases, and expressions for calculating the interaction force, F_z, and energy, W, have been derived [350]:

$$F_z = \frac{\beta_{23}Q^2}{4\varepsilon_2(R+s)^2} f_z, \quad W = \int_s^{\infty} F_z \, ds = \frac{\beta_{23}Q^2}{4\varepsilon_2(R+s)} w \tag{4.170}$$

Here

R is the particle radius
s is the distance between the particle surface and the fluid interface (inset in Figure 4.26)
$Q = 4\pi R^2 \sigma_{pn}$ is the total charge at the boundary particle–nonpolar fluid
f_z and w are dimensionless force and energy coefficients, respectively, which, in general, depend on the parameters s/R, β_{12}, and β_{23}, where $\beta_{ij} = (\varepsilon_i - \varepsilon_j)/(\varepsilon_i + \varepsilon_j)$; $i, j = 1, 2, 3$; ε_1, ε_2, and ε_3 are the dielectric constants of the respective phases

At long distances, $s/R > 1$, we have $f_z \approx w \approx 1$, and then Equation 4.170 reduces to the expressions for the force and energy of interaction between a point charge Q with the interface between phases 2 and 3. This is the known image charge effect. Expressions that allow us to calculate f_z and w for shorter distances ($s/R < 1$) are derived in Ref. [350].

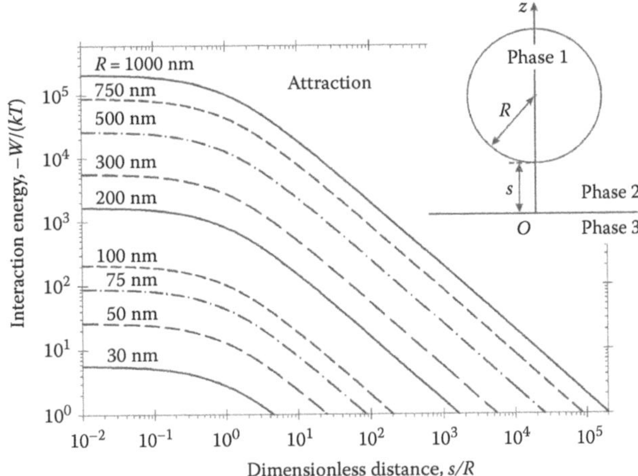

FIGURE 4.26 Plot of the interaction energy W (scaled with kT) vs. the dimensionless distance, s/R, between a charged glass particle (phase 1) and a planar interface; phase 2 is tetradecane; phase 3 is water. The curves correspond to different particle radii, R, denoted in the figure.

In Figure 4.26, numerical results for the particle–interface interaction energy, W, scaled by the thermal energy kT, are plotted versus the relative distance, s/R, for various values of the particle radius, R. The other parameter values correspond to the following choice of the phases: phase 1 (the particle) is glass, phase 2 is tetradecane, and phase 3 is water. The curves in Figure 4.26 describe a strong and long-range attraction between the particle and the interface. The interaction energy, W, becomes comparable, or smaller than the thermal energy kT for particle radius $R < 30$ nm. On the other hand, for $R > 30$ nm W strongly increases with the particle size (in Equation 4.170 $Q^2 \sim R^4$ at fixed surface charge density, σ_{pn}) and reaches $W \approx 10^5\, kT$ for $R = 1$ μm at close contact. In addition, the range of interaction also increases, reaching $s/R \approx 10^5$ for $R = 1$ μm. In general, this is a strong and long-range interaction [350]. For example, water drops could attract charged hydrophobic particles dispersed in the oily phase, which would favor the formation of reverse particle-stabilized emulsions.

4.3.3.3.2 *Forces of Electric Origin between Particles at a Liquid Interface*

Figure 4.27 shows two particles attached to the interface between water and a nonpolar fluid (oil, air). In general, the particles experience three forces of electric origin: F_{ED}—electrodipping force [351]; F_{ER}—direct electric repulsion between the two particles across the oil [352], and F_{EC}—electric field–induced capillary attraction [353], which is termed "electrocapillary force" for brevity. F_{ED} is normal to the oil–water interface and is directed toward the water phase. Physically, F_{ED} is a result of the electrostatic image-charge effect; see the previous section. F_{ED} is acting on each individual particle,

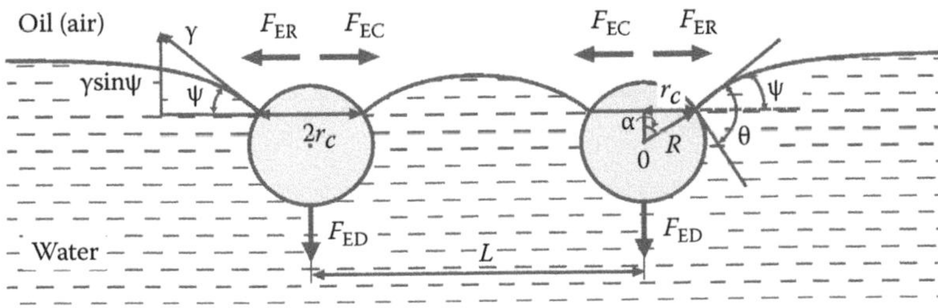

FIGURE 4.27 Sketch of two electrically charged particles attached to an oil–water interface. F_{ED} is the electrodipping force, due to the image-charge effect, that pushes the particles into water and deforms the fluid interface around the particles. F_{ER} is the direct electric repulsion between the two like-charged particles. F_{EC} is the electrocapillary attraction, related to deformations in the fluid interface created by the electric field.

while F_{ER} and F_{EC} are interaction forces between two (or more) particles. The presence of electric field leads to deformations in the fluid interface around the particles, which lead to the appearance of F_{EC}. The three forces, F_{ED}, F_{ER}, and F_{EC}, are separately considered in the following section.

4.3.3.3.3 Electrodipping Force, F_{ED}

At equilibrium, the electrodipping force is counterbalanced by the interfacial tension force: $F_{ED} = 2\pi r_c \gamma \sin \psi$, where γ is the interfacial tension; r_c is the radius of the contact line on the particle surface; and ψ is the meniscus slope angle at the contact line (Figure 4.27) [351,353]. Consequently, F_{ED} can be determined from the experimental values of r_c, γ, and ψ. This approach was used to obtain the values of F_{ED} for silanized glass particles of radii 200–300 μm from photographs of these particles at an oil–water or air–water interface [351]. F_{ED} was found to be much greater than the vertical gravitational force acting on these particles.

As an illustration, Figure 4.28 compares the profiles of the liquid menisci around a noncharged particle and a charged particle. The particles represent hydrophobized glass spheres of density $\rho_p = 2.5$ g/cm^3. The oil phase is purified soybean oil of density $\rho_{oil} = 0.92$ g/cm^3. The oil–water interfacial tension is $\gamma = 30.5$ mN/m. Under these conditions, the calculated surface tension force, $2\pi r_c \gamma \sin \psi$, which counterbalances the gravitational force (particle weight minus the Archimedes force), corresponds to meniscus slope angle $\psi = 1.5°$, and the deformation of the liquid interface caused by the particle is hardly visible (Figure 4.28a). In contrast, for the charged particle (Figure 4.28b), the meniscus slope angle is much greater, $\psi = 26°$. This is due to the fact that the interfacial tension force, $2\pi r_c \gamma \sin \psi$, has to counterbalance the electrodipping force, which pushes the particle toward the water phase. Experimentally, it has been found that the angle ψ is insensitive to the concentration of NaCl in the aqueous phase, which means that (in the investigated case) the electrodipping force is due to charges situated at the particle–oil interface [351,354]. With similar particles, the magnitude of F_{ED} at the air–water interface was found to be about six times smaller than at the oil–water interface [351]. Theoretically, the electrodipping force, F_{ED}, can be calculated from the expression [354,355]:

$$F_{ED} = \left(\frac{4\pi}{\varepsilon_n}\right)(\sigma_{pn} R)^2 (1 - \cos\alpha) f(\theta, \varepsilon_{pn}) \tag{4.171}$$

Here

- ε_n is the dielectric constant of the nonpolar fluid (oil, air)
- σ_{pn} is the surface charge density at the boundary particle–nonpolar fluid
- $\varepsilon_{pn} = \varepsilon_p/\varepsilon_n$ is the ratio of the respective two dielectric constants
- α is a central angle, while $\theta = \alpha + \psi$ is the contact angle (see Figure 4.27)

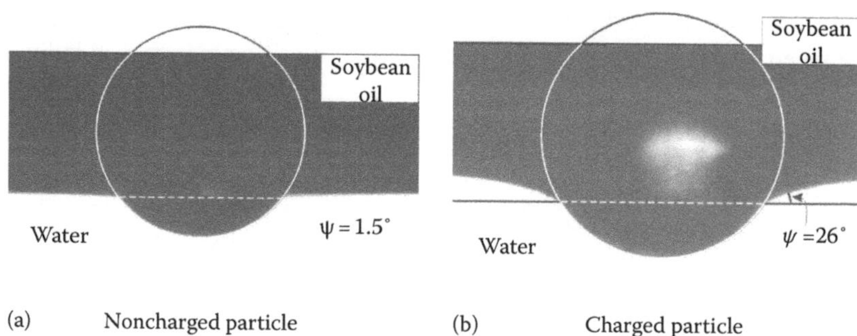

FIGURE 4.28 Side-view photographs of hydrophobized spherical glass particles at the boundary water–soybean oil (no added surfactants). (a) Noncharged particle of radius $R = 235$ μm: the meniscus slope angle due to gravity is relatively small, $\psi = 1.5°$. (b) Charged particle of radius $R = 274$ μm: the experimental meniscus slope angle is $\psi = 26°$ owing to the electrodipping force; if this electric force were missing, the gravitational slope angle would be only $\psi = 1.9°$.

We could accurately calculate the dimensionless function $f(\theta, \varepsilon_{pn})$ by means of the relation $f(\theta, \varepsilon_{pn}) = f_R(\theta, \varepsilon_{pn})/(1 - \cos\theta)$, where the function $f_R(\theta, \varepsilon_{pn})$ is tabulated in Table 4.3 of Ref. [355] on the basis of the solution of the electrostatic boundary problem. The tabulated values can be used for a convenient computer calculation of $f_R(\theta, \varepsilon_{pn})$ with the help of a four-point interpolation formula, Equation D.1 in Ref. [355]. From the experimental F_{ED} and Equation 4.171, we could determine the surface charge density, σ_{pn}, at the particle–oil and particle–air interface. Values of σ_{pn} in the range from 20 to 70 μC/m² have been obtained [351,354].

4.3.3.3.4 *Direct Electric Repulsion, F_{ER}*

Interactions of electrostatic origin were found to essentially influence the type of particle structures at oil–water [352,353,356–358] and air–water [359,360] interfaces. Two-dimensional hexagonal arrays of particles were observed in which the distance between the closest neighbors was markedly greater than the particle diameter [352,353,356–363]. The existence of such structures was explained by the action of direct electrostatic repulsion between like-charged particles. In many cases, the particle arrays are insensitive to the concentration of electrolyte in the aqueous phase [352,356,357]. This fact, and the direct interparticle force measurements by laser tweezers [356], leads to the conclusion that the electrostatic repulsion is due to charges at the particle–oil (or particle–air) interface, which give rise to electric repulsion across the nonpolar phase [352,356–359]. This repulsion is relatively long ranged because of the absence of a strong Debye screening of the electrostatic forces that is typical for the aqueous phase [364]. Evidences about the presence of electric charges on the surface of solid particles dispersed in liquid hydrocarbons could also be found in earlier studies [365,366].

For a particle in isolation, the charges at the particle–nonpolar fluid interface create an electric field in the oil that asymptotically resembles the electric field of a dipole (Figure 4.29). This field practically does not penetrate into the water phase, because it is reflected by the oil–water boundary owing to the relatively large dielectric constant of water. For a single particle, the respective electrostatic problem is solved in Ref. [355]. The asymptotic behavior of the force of electrostatic repulsion between two such particles–dipoles (Figure 4.29) is [355]:

$$F_{ER} = \frac{3p_d^2}{2\varepsilon_n L^4} \quad \left(\frac{R}{L} \ll 1\right) \tag{4.172}$$

L is the center-to-center distance between the two particles; $p_d = 4\pi\sigma_{pn}DR^3 \sin^3\alpha$ is the effective particle dipole moment; as before, R is the particle radius and σ_{pn} is the electric charge density at the particle–nonpolar fluid interface; $D = D(\alpha, \varepsilon_{pn})$ is a known dimensionless function, which can

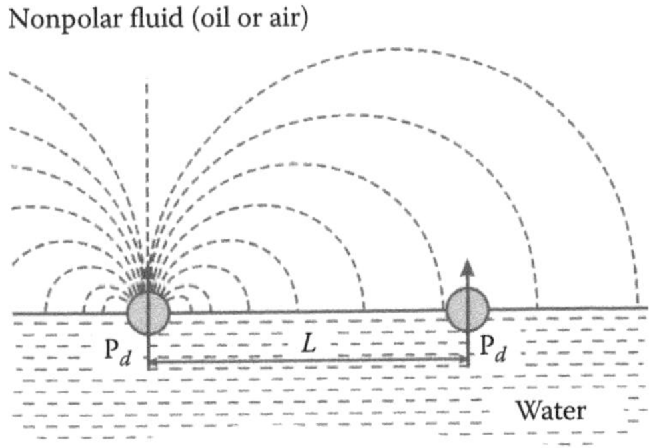

FIGURE 4.29 Two particles attached to the boundary water–nonpolar fluid and separated at a center-to-center distance L. In the nonpolar fluid (oil, air), the electric field of each particle in isolation is asymptotically identical to the field of a dipole of moment p_d. This field is created by charges at the particle–nonpolar fluid interface.

be calculated by means of Table 4.1 and Equation D.1 in Ref. [355]; $\varepsilon_{pn} \equiv \varepsilon_p/\varepsilon_n$ is the ratio of the dielectric constants of the two phases. Equation 4.172 shows that F_{ER} asymptotically decays as $1/L^4$ like the force between two point dipoles. However, at shorter distances, the finite size of the particle is expected to lead to a Coulombic repulsion, $F_{ER} \sim 1/L^2$; see Refs. [356–358].

Monolayers from electrically charged micron-sized silica particles, spread on the air–water interface, were investigated and *surface pressure versus area* isotherms were measured by Langmuir trough and the monolayers' morphology was monitored by microscope [363]. The experiments showed that $\Pi \sim L^{-3}$ at large L, where Π is the surface pressure and L is the mean interparticle distance. A theoretical cell model was developed, which predicts not only the aforementioned asymptotic law but also the whole $\Pi(L)$ dependence. The model presumes a periodic distribution of the surface charge density, which induces a corresponding electric field in the air phase. Then, the Maxwell pressure tensor of the electric field in the air phase was calculated and integrated according to the Bakker's formula [189] to determine the surface pressure. Thus, all *collective effects* from the electrostatic interparticle interactions were taken into account, as well as the effects from the particle finite size.

The effects of applied *vertical external electric field* on the electrostatic forces acting on a colloid particle at a horizontal liquid interface have also been investigated. By varying the strength of the electric field, it is possible to control the distances between the particles in nondensely packed 2D colloid crystals formed at liquid interfaces [367–370]. Theoretical expressions for the forces between floating uncharged [371] and charged [372] dielectric particles in the presence of external electric field were derived. The particles are located on the boundary water–nonpolar fluid (air, oil). The effects of the dielectric constants and contact angle (particle wettability) on the vertical electrodipping force, F_{ED}, acting on each particle, and on the horizontal force between two particles, F_{ER}, were investigated. The external field polarizes the uncharged particles at the fluid interface. The vertical electric force on the particle can be directed upward or downward. The horizontal interparticle repulsion is dipolar and contact angle dependent. At given contact angle (for uncharged particles) and external electric field E_0 (for charged particles), the dipole moment is zero and the repulsion becomes short-range octupolar [371,372]. This minimal electrostatic repulsion could be weaker than the electrocapillary capillary attraction; see the next section.

4.3.3.3.5 Electrocapillary Force, F_{EC}

The electrocapillary forces between particles are due to the overlap of the deformations in the liquid interface created by the particles [353]. The deformations are due not only to the electrodipping force that pushes the particle toward the water (and that determines the value of the angle ψ in Figure 4.28b), but also to the additional electric pressure (Maxwell stress) that is acting per unit area of the oil–water (or air–water) interface owing to the presence of electric field in the nonpolar fluid (see Figure 4.29) [351,353,354,373–375]. The direction of this electric pressure is from the water toward the nonpolar fluid.

The electric field–induced deformation of a liquid interface around charged particles at the interface tetradecane–water has been quantitatively examined in Ref. [354]. An example is given in Figure 4.30. Far from the particle, the interface is flat and horizontal. For particles of radii $R = 200$–300 μm, both gravitational and electric field induced deformations are present. The gravitational deformation is predominant at longer distances, whereas the electric field deformation is significant near the particle. The latter deformation is insensitive to the variation of the concentration of NaCl in the aqueous phase (Figure 4.30), which indicates that this deformation is due to electric charges at the particle–oil interface. Good agreement between experiment (the symbols) and theory (the solid line) has been obtained.

In Ref. [376], the two-particle electrocapillary problem was solved in bipolar coordinates without using any superposition approximations. The following expression (power expansion) was obtained

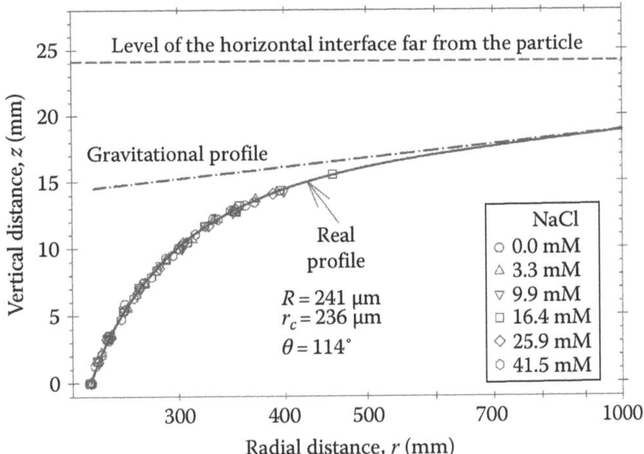

FIGURE 4.30 Profile of the oil (tetradecane)–water interface near the contact line of a charged glass particle, like that in Figure 4.28b: plot of experimental data from Ref. [354]; see Figure 4.27 for the notations. The dash-dot line shows the gravitational; profile calculated under the assumption that the particle is not charged. The difference between the real and the gravitational profiles represents the effect of electric field on the meniscus shape. The fact that the real (experimental) profile is insensitive to the concentration of NaCl in the water phase indicates that the electric charges are located at the particle–oil interface, so that the interfacial deformation is due to electric field in the oily phase.

for two identical floating particles with contact radius r_c, which are separated at a center-to-center distance L (see Figure 4.27):

$$F_x = \frac{3p_d^2}{2\varepsilon_n L^4}\left[1 - \frac{2\delta}{5} + \frac{5\delta}{2}\left(\frac{r_c}{L}\right)^2 - \frac{15\delta}{2}\left(\frac{r_c}{L}\right)^3 + \frac{175\delta}{32}\left(\frac{r_c}{L}\right)^4 + \cdots\right] \tag{4.172a}$$

$\delta = \tan\psi$, where ψ is the meniscus slope angle for each particle in isolation (Figure 4.27). In Equation 4.172a, the first term in the brackets is F_{ER} in Equation 4.172, whereas the next terms, which are proportional to the meniscus deformation angle δ, give F_{EC}. Because for micrometer and submicrometer particles δ is a small quantity, it turns out that for uniform distribution of the surface charges, the electrocapillary attraction is weaker than the electrostatic repulsion at interparticle distances, at which the dipolar approximation is applicable, so that the net force, F_x, is repulsive [376]. The final conclusion from the theoretical analysis is that the direct electrostatic repulsion dominates over the capillary attraction when the surface charge is *uniformly* distributed; no matter whether the surface charge is on the polar–liquid or nonpolar–fluid side of the particle.

Electric field–induced attraction that *prevails* over the electrostatic repulsion was established (both experimentally and theoretically) in the case of not-too-small floating particles, for which the interfacial deformation due to gravity is not negligible [377,378]. If the surface charge is *anisotropically* distributed (this may happen at low surface charge density), the electric field produces a saddle-shaped deformation in the liquid interface near the particle, which is equivalent to a "capillary quadrupole." The interaction of the latter with the axisymmetric gravitational deformation around the other particle (which is equivalent to a "capillary charge") gives rise to a capillary force that decays $\propto 1/L^3$, that is, slower than $F_{ER} \propto 1/L^4$. In such a case, we are dealing with a *hybrid attraction* between a gravity-induced "capillary charge" and an electric field–induced "capillary quadrupole" [378,379].

4.4 SURFACE FORCES

4.4.1 Derjaguin Approximation

The excess surface free energy per unit area of a plane-parallel film of thickness h is [14,380]

$$f(h) = \int_{h}^{\infty} \Pi(h)dh \tag{4.173}$$

where, as before, Π denotes disjoining pressure. Derjaguin [381] derived an approximate formula, which expresses the energy of interaction between two spherical particles of radii R_1 and R_2 through integral of $f(h)$:

$$U(h_0) = \frac{2\pi R_1 R_2}{R_1 + R_2} \int_{h_0}^{\infty} f(h)dh \tag{4.174}$$

Here, h_0 is the shortest distance between the surfaces of the two particles (see Figure 4.31). In the derivation of Equation 4.174 it is assumed that the interaction between two parcels from the particle surfaces, separated at the distance h, is approximately the same as that between two similar parcels in a plane-parallel film. This assumption is correct when the range of action of the surface forces and the distance h_0 are small compared to the curvature radii R_1 and R_2. It has been established, both experimentally [36] and theoretically [382], that Equation 4.174 provides a good approximation in the range of its validity.

Equation 4.174 can be generalized for smooth surfaces of arbitrary shape (not necessarily spheres). For that purpose, the surfaces of the two particles are approximated with paraboloids in the vicinity of the point of closest approach ($h = h_0$). Let the principle curvatures at this point be c_1 and c_1' for the first particle, and c_2 and c_2' for the second particle. Then the generalization of Equation 4.174 reads [380]:

$$U(h_0) = \frac{2\pi}{\sqrt{C}} \int_{h_0}^{\infty} f(h)dh \tag{4.175}$$

$$C \equiv c_1 c_1' + c_2 c_2' + (c_1 c_2 + c_1' c_2')\sin^2 \omega + (c_1 c_2' + c_1' c_2)\cos^2 \omega$$

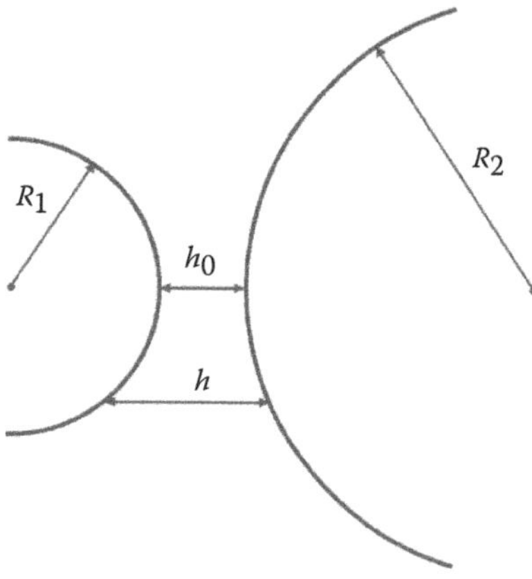

FIGURE 4.31 Two spherical particles of radii R_1 and R_2; the shortest and the running surface-to-surface distances are denoted by h_0 and h, respectively.

where ω is the angle subtended between the directions of the principle curvatures of the two approaching surfaces. For two spheres, we have $c_1 = c_1' = 1/R_1, c_2 = c_2' = 1/R_2$, and Equation 4.175 reduces to Equation 4.174.

For two cylinders of radii r_1 and r_2 crossed at angle ω we have $c_1 = c_2 = 0$; $c_1' = 1/r_1$, $c_2' = 1/r_2$ and Equation 4.175 yields

$$U(h_0) = \frac{2\pi\sqrt{r_1 r_2}}{\sin\omega}\int_{h_0}^{\infty} f(h)dh \tag{4.176}$$

Equation 4.176 is often used in connection to the experiments with the surface force apparatus (SFA) [36,383], in which the interacting surfaces are two crossed cylindrical mica sheets. The divergence in Equation 4.176 for ω = 0 reflects the fact that the axes of the two infinitely long cylinders are parallel for ω = 0 and thus the area of the interaction zone becomes infinite.

The Derjaguin's formula is applicable to any type of force law (attractive, repulsive, oscillatory) if only (1) the range of the forces, and (2) the surface-to-surface distance are much smaller than the surface curvature radii. This formula is applicable to any kind of surface force, irrespective of its physical origin: van der Waals, electrostatic, steric, oscillatory-structural, depletion, etc. It reduces the two-particle interaction problem to the simpler problem for interactions in plane-parallel films.

4.4.2 van der Waals Surface Forces

The van der Waals interaction between molecules *i* and *j* obeys the law:

$$u_{ij}(r) = -\frac{\alpha_{ij}}{r^6} \tag{4.177}$$

where

- u_{ij} is the potential energy of interaction
- r is the distance between the two molecules
- α_{ij} is a constant characterizing the interaction

In fact, the van der Waals forces represent an averaged dipole–dipole interaction, which is a superposition of three main terms: (1) orientation interaction: interaction between two permanent dipoles [384]; (2) induction interaction: interaction between one permanent dipole and one induced dipole [385]; (3) dispersion interaction: interaction between two induced dipoles [386]. The theory yields [36]:

$$\alpha_{ij} = \frac{1}{(4\pi\varepsilon_0)^2}\left[\frac{p_i^2 p_j^2}{3kT} + \left(p_i^2\alpha_{0j} + p_j^2\alpha_{0i}\right) + \frac{3\alpha_{0i}\alpha_{0j}h_p\nu_i\nu_j}{2(\nu_i+\nu_j)}\right] \tag{4.178}$$

where

- p_i and α_{i0} are molecular dipole moment and electronic polarizability, respectively;
- h_p is the Planck constant
- ν_i is the orbiting frequency of the electron in the Bohr atom

For van der Waals interactions between molecules in a gas phase, the orientation interaction can yield from 0% (nonpolar molecules) up to 70% (molecules of large permanent dipole moment, like H_2O) of the value of α_{ij}; the contribution of the induction interaction in α_{ij} is usually low, about 5%–10%; the contribution of the dispersion interaction might be between 24% (water) and 100% (nonpolar hydrocarbons); for numerical data, see Ref. [36].

According to the *microscopic* theory by Hamaker [387], the van der Waals interaction between two macroscopic bodies can be found by integration of Equation 4.177 over all couples of molecules, followed by subtraction of the interaction energy at infinite separation between the bodies. The result depends on the geometry of the system. For a plane-parallel film from component 3 located between two semi-infinite phases composed from components 1 and 2, the van der Waals interaction energy per unit area and the respective disjoining pressure, stemming from Equation 4.166, are [387]

$$f_{vw} = -\frac{A_H}{12\pi h^2}, \quad \Pi_{vw} = -\frac{\partial f_{vw}}{\partial h} = -\frac{A_H}{6\pi h^3} \tag{4.179}$$

where
h is the thickness of the film
A_H is the compound Hamaker constant [14]

$$A_H = A_{33} + A_{12} - A_{13} - A_{23} \quad (A_{ij} = \pi^2 \rho_i \rho_j \alpha_{ij}, i, j = 1,2,3) \tag{4.180}$$

A_{ij} is the Hamaker constant of components i and j
ρ_i and ρ_j are the molecular number densities of phases i and j built up from components i and j, respectively

If A_{ii} and A_{jj} are known, we can calculate A_{ij} by using the Hamaker approximation

$$A_{ij} = (A_{ii} A_{jj})^{1/2} \tag{4.181}$$

In fact, Equation 4.181 is applicable to the dispersion contribution in the van der Waals interaction [36].

When components 1 and 2 are identical, A_H is positive (see Equation 4.180); therefore, the van der Waals interaction between identical bodies, in any medium, is always attractive. Besides, two dense bodies (even if nonidentical) will attract each other when placed in medium 3 of low density (gas, vacuum). When the phase in the middle (component 3) has intermediate Hamaker constant between those of bodies 1 and 2, A_H can be negative and the van der Waals disjoining pressure can be repulsive (positive). Such is the case of an aqueous film between mercury and gas [388].

Lifshitz et al. [389,390] developed an alternative approach to the calculation of the Hamaker constant A_H in condensed phases, called the *macroscopic* theory. The latter is not limited by the assumption for pair-wise additivity of the van der Waals interaction (see also Refs. [36,380,391]). The Lifshitz theory treats each phase as a continuous medium characterized by a given uniform dielectric permittivity, which is dependent on the frequency, ν, of the propagating electromagnetic waves. For the symmetric configuration of two identical phases "i" interacting across a medium "j," the macroscopic theory provides the expression [36]

$$A_H \equiv A_{iji} = A_{iji}^{(\nu=0)} + A_{iji}^{(\nu>0)} = \frac{3}{4} kT \left(\frac{\varepsilon_i - \varepsilon_j}{\varepsilon_i + \varepsilon_j} \right)^2 + \frac{3 h_p \nu_e \left(n_i^2 - n_j^2 \right)^2}{16\sqrt{2} \left(n_i^2 + n_j^2 \right)^{3/2}} \tag{4.182}$$

where
ε_i and ε_j are the dielectric constants of phases i and j, respectively;
n_i and n_j are the respective refractive indices for visible light
h_p is the Planck constant
ν_e is the main electronic absorption frequency which is $\approx 3.0 \times 10^{15}$ Hz for water and the most organic liquids [36]

The first term in the right-hand side of Equation 4.182, $A_{iji}^{(\nu=0)}$, is the so-called zero-frequency term, expressing the contribution of the orientation and induction interactions. Indeed, these two contributions to the van der Waals force represent electrostatic effects. Equation 4.182 shows that the zero-frequency term can never exceed $(3/4)kT \approx 3 \times 10^{-21}$ J. The last term in Equation 4.182, $A_{iji}^{(\nu>0)}$, accounts for the dispersion interaction. If the two phases, i and j, have comparable densities (as for emulsion systems, say oil–water–oil), then $A_{iji}^{(\nu>0)}$ and $A_{iji}^{(\nu=0)}$ are comparable by magnitude. If one of the phases, i or j, has a low density (gas, vacuum), we obtain $A_{iji}^{(\nu>0)} \gg A_{iji}^{(\nu=0)}$. In the latter case, the Hamaker microscopic approach may give comparable $A_{iji}^{(\nu>0)}$ and $A_{iji}^{(\nu=0)}$ in contradiction to the Lifshitz macroscopic theory, which is more accurate for condensed phases.

A geometrical configuration, which is important for disperse systems, is the case of two spheres of radii R_1 and R_2 interacting across a medium (component 3). Hamaker [387] has derived the following expression for the van der Waals interaction energy between two spheres:

$$U(h_0) = -\frac{A_H}{12}\left(\frac{y}{x^2+xy+x} + \frac{y}{x^2+xy+x+y} + 2\ln\frac{x^2+xy+x}{x^2+xy+x+y}\right) \tag{4.183}$$

where

$$x = \frac{h_0}{2R_1}$$

$$y = \frac{R_2}{R_1} \le 1 \tag{4.184}$$

h_0 is the same as in Figure 4.31.

For $x \ll 1$ Equation 4.183 reduces to

$$U(h_0) \approx -\frac{A_H}{12}\frac{y}{(1+y)x} = -\frac{2\pi R_1 R_2}{R_1+R_2}\frac{A_H}{12\pi h_0} \tag{4.185}$$

Equation 4.185 can be also derived by combining Equation 4.179 with the Derjaguin approximation (Equation 4.174). It is worthwhile noting that the logarithmic term in Equation 4.183 can be neglected only if $x \ll 1$. For example, even when $x = 5 \times 10^{-3}$, the contribution of the logarithmic term amounts to about 10% of the result (for $y = 1$); consequently, for larger values of x this term must be retained.

Another geometrical configuration, which corresponds to two colliding deformable emulsion droplets, is sketched in Figure 4.32. In this case the interaction energy is given by the expression [392]

$$U(h,r) = -\frac{A_H}{12}\left[\frac{3}{4} + \frac{R_s}{h} + 2\ln\left(\frac{h}{R_s}\right) + \frac{r^2}{h^2} - \frac{2r^2}{R_s h}\right] \quad (h, r \ll R_s) \tag{4.186}$$

where

h and r are the thickness and the radius of the flat film formed between the two deformed drops, respectively,

R_s is the radius of the spherical part of the drop surface (see Figure 4.32).

Equation 4.186 is a truncated series expansion; the exact formula, which is more voluminous, can be found in Ref. [392]. Expressions for U for other geometrical configurations are also available [37,391].

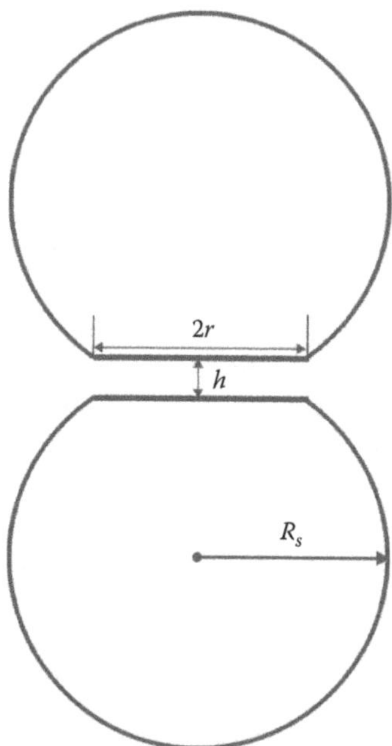

FIGURE 4.32 Thin film of radius r and thickness h formed between two attached fluid particles; the spherical part of the particle surface has radius R_s.

The asymptotic behavior of the dispersion interaction at large intermolecular separations does not obey Equation 4.177; instead $u_{ij} \propto 1/r^7$ due to the electromagnetic retardation effect established by Casimir and Polder [393]. Several different expressions have been proposed to account for this effect in the Hamaker constant [391].

The orientation and induction interactions are electrostatic effects, so they are not subjected to electromagnetic retardation. Instead, they are subject to Debye screening due to the presence of electrolyte ions in the liquid phases. Thus for the interaction across an electrolyte solution, the screened Hamaker constant is given by the expression [36,394]

$$A_H = 2\kappa h A_0 e^{-2\kappa h} + A_d \tag{4.187}$$

where

A_0 denotes the contribution of the (non-screened) orientation and induction interactions to the Hamaker constant

A_d is the contribution of the dispersion interaction

κ is the Debye screening parameter: $\kappa = \kappa_c I^{1/2}$ (see Equation 4.34)

Equation 4.187 is accurate to within 15% for $\kappa h > 2$; see Ref. [36].

4.4.3 Electrostatic Surface Forces

4.4.3.1 Two Identically Charged Planes

First, we consider the electrostatic (double layer) interaction between two identical charged plane parallel surfaces across solution of symmetrical $Z{:}Z$ electrolyte. The charge of a counterion (i.e., ion with charge opposite to that of the surface) is $-Ze$, whereas the charge of a coion is $+Ze$ ($Z = \pm1, \pm2, \ldots$) with e being the elementary charge. If the separation between the two planes is very large, the number concentration of both counterions and coions would be equal to its bulk value, n_0, in the middle of the film. However, at finite separation, h, between the surfaces the two EDLs overlap and the counterion and coion concentrations in the middle of the film, n_{10} and n_{20}, are no longer equal. Because the solution inside the film is supposed to be in electrochemical (Donnan) equilibrium with the bulk electrolyte solution of concentration n_0, we can write [395] $n_{10}n_{20} = n_0^2$, or alternatively

$$n_{10} = \frac{n_0}{\sqrt{m}}, \quad n_{20} = n_0\sqrt{m}, \quad m \equiv \frac{n_{20}}{n_{10}} \tag{4.188}$$

As pointed out by Langmuir [396], the electrostatic disjoining pressure, Π_{el}, can be identified with the excess osmotic pressure in the middle of the film:

$$\Pi_{\mathrm{el}} = kT(n_{10} + n_{20} - 2n_0) = n_0 kT(m^{1/4} - m^{-1/4})^2 \tag{4.189}$$

Equation 4.189 demonstrates that for two identically charged surfaces, Π_{el} is always positive, that is, corresponds to repulsion between the surfaces. In general, we have $0 < m \leq 1$ because the coions are repelled from the film due to the interaction with the film surfaces. To find the exact dependence of Π_{el} on the film thickness, h, we have to solve the Poisson–Boltzmann equation for the distribution of the electrostatic potential inside the film. The solution provides the following connection between Π_{el} and h for symmetric electrolytes [380,397]:

$$\Pi_{\mathrm{el}} = 4n_0 kT \cot^2\theta, \quad \kappa h = 2F(\varphi, \theta)\sin\theta \tag{4.190}$$

where

$F(\varphi, \theta)$ is the elliptic integral of the first kind

φ is related with θ as follows

$$\cos\varphi = \frac{\cot\theta}{\sinh(Z\Phi_s/2)} \quad \text{(fixed surface potential } \Phi_s\text{)} \tag{4.191}$$

$$\tan\varphi = (\tan\theta)\sinh\left(\frac{Z\Phi_\infty}{2}\right) \quad \text{(fixed surface charge } \sigma_s\text{)} \tag{4.192}$$

$$\cosh(Z\Phi_\infty) = 1 + \frac{1}{2}\left(\frac{Ze\sigma_s}{\varepsilon\varepsilon_0 kT\kappa}\right)^2, \quad \Phi_s \equiv \frac{e\psi_s}{kT} \tag{4.193}$$

Here

Φ_s is the dimensionless surface potential

Φ_∞ is the value of Φ_s for $h \to \infty$

Equation 4.190 expresses the dependence $\Pi_{\mathrm{el}}(h)$ in a parametric form: $\Pi_{\mathrm{el}}(\theta)$, $h(\theta)$. Fixed surface potential or charge means that Φ_s or σ_s does not depend on the film thickness h. The latter is important to be specified when integrating $\Pi(h)$ or $f(h)$ (in accordance with Equations 4.173 or 4.176) to calculate the interaction energy.

In principle, it is possible neither the surface potential nor the surface charge to be constant [398]. In such case a condition for *charge regulation* is applied, which represents the condition for dynamic equilibrium with respect to the counterion exchange between the Stern and diffuse parts of the EDL (i.e., condition for constant electrochemical potentials of the ionic species). As discussed in Section 4.2.1.2.3, the Stern layer itself can be considered as a Langmuir adsorption layer of counterions. We can relate the maximum possible surface charge density (due to all the surface ionizable groups) to Γ_1 in Equation 4.47: $\sigma_{max} = Ze\Gamma_1$. Likewise, the effective surface charge density, σ_s, which is smaller by magnitude than σ_{max} (because some ionizable groups are blocked by adsorbed counterions) can be expressed as $\sigma_s = Ze(\Gamma_1 - \Gamma_2)$. Then, with the help of Equation 4.44, the Stern isotherm (Equation 4.47) can be represented in the form

$$\frac{\sigma_{max} - \sigma_s}{\sigma_{max}} = \left[1 + (K_2 I)^{-1} \exp(Z\Phi_s)\right]^{-1} \tag{4.194}$$

The product $Z\Phi_s$ is always positive. At high surface potential, $Z\Phi_s \to \infty$, from Equation 4.194 we obtain $\sigma_s \to \sigma_{max}$, that is, there is no blocking of surface ionizable by adsorbed counterions.

When the film thickness is large enough ($\kappa h \geq 1$) the difference between the regimes of constant potential, constant charge, and charge regulation becomes negligible, that is, the usage of each of them leads to the same results for $\Pi_{el}(h)$ [14].

When the dimensionless electrostatic potential in the middle of the film

$$\Phi_m = \frac{e}{kT}\psi_m = -\frac{1}{2Z}\ln m \tag{4.195}$$

is small enough (the film thickness, h, is large enough), we could use the superposition approximation, that is, we could assume that $\Phi_m \approx 2\Phi_1(h/2)$, where Φ_1 is the dimensionless electric potential at a distance $h/2$ from the surface (of the film) when the other surface is removed at infinity. Because

$$Z\Phi_1\left(\frac{h}{2}\right) = 4e^{-\kappa h/4} \tanh\left(\frac{Z\Phi_s}{4}\right) \tag{4.196}$$

from Equations 4.189, 4.195, and 4.196, we obtain a useful asymptotic formula [399]

$$\Pi_{el} \approx n_0 kTZ^2\Phi_m^2 \approx 64 n_0 kT\left(\tanh\frac{Z\Phi_s}{4}\right)^2 e^{-\kappa h} \tag{4.197}$$

It should be noted that if Φ_s is large enough, the hyperbolic tangent in Equation 4.197 is identically 1, and Π_{el} (as well as f_{el}) becomes independent of the surface potential (or charge). Equation 4.197 can be generalized for the case of 2:1 electrolyte (bivalent counterion) and 1:2 electrolyte (bivalent coion) [400]:

$$\Pi_{el} = 432 n_{(2)} kT\left(\tanh\frac{\nu_{i:j}}{4}\right)^2 e^{-\kappa h} \tag{4.198}$$

where $n_{(2)}$ is the concentration of the bivalent ions, the subscript "$i{:}j$" takes value "2:1" or "1:2,", and

$$\nu_{2:1} = \ln\left[\frac{3}{1+2e^{-\Phi_s}}\right], \quad \nu_{1:2} = \ln\left[\frac{2e^{\Phi_s}+1}{3}\right] \tag{4.199}$$

4.4.3.2 Two Nonidentically Charged Planes

Contrary to the case of two identically charged surfaces, which always repel each other (see Equation 4.189), the electrostatic interaction between two plane-parallel surfaces of different potentials, ψ_{s1} and ψ_{s2}, can be either repulsive or attractive [380,401]. Here, we will restrict our considerations to the case of low surface potentials, when the Poisson–Boltzmann equation can be linearized. Despite that it is not too general quantitatively, this case exhibits qualitatively all features of the electrostatic interaction between different surfaces.

If ψ_{s1} = constant, and ψ_{s2} = constant, then the disjoining pressure at constant surface potential reads [380]:

$$\Pi_{el}^{\psi} = \frac{\varepsilon\varepsilon_0\kappa^2}{2\pi}\frac{2\psi_{s1}\psi_{s2}\cosh\kappa h - \left(\psi_{s1}^2 + \psi_{s2}^2\right)}{\sinh^2\kappa h} \tag{4.200}$$

When the two surface potentials have opposite signs, that is, when $\psi_{s1}\psi_{s2} < 0, \Pi_{el}^{\psi}$ is negative for all h and corresponds to electrostatic attraction (see Figure 4.33a). This result could have been anticipated, because two charges of opposite sign attract each other. More interesting is the case, when $\psi_{s1}\psi_{s2} > 0$, but $\psi_{s1} \neq \psi_{s2}$. In the latter case, the two surfaces repel each other for $h > h_0$, whereas they attract each other for $h < h_0$ (Figure 4.33a); h_0 is determined by the equation $\kappa h_0 = \ln(\psi_{s2}/\psi_{s1})$; $\psi_{s2} > \psi_{s1}$. In addition, the electrostatic repulsion has a maximum value of

$$\Pi_{el}^{\psi}(\max) = \frac{\varepsilon\varepsilon_0\kappa^2}{2\pi}\psi_{s1}^2 \quad \text{at } h_{\max} = \frac{1}{\kappa}\operatorname{arccosh}\frac{\psi_{s2}}{\psi_{s1}},\ \psi_{s2} > \psi_{s1} \tag{4.201}$$

Similar electrostatic disjoining pressure isotherm has been used to interpret the experimental data for aqueous films on mercury [388]. It is worthwhile noting that $\Pi_{el}^{\psi}(\max)$ depends only on ψ_{s1}, that is, the maximum repulsion is determined by the potential of the surface of lower charge.

If σ_{s1} = constant, and σ_{s2} = constant, then instead of Equation 4.200 we have [380]

$$\Pi_{el}^{\sigma}(h) = \frac{1}{2\varepsilon\varepsilon_0}\frac{2\sigma_{s1}\sigma_{s2}\cosh\kappa h + \sigma_{s1}^2 + \sigma_{s2}^2}{\sinh^2\kappa h} \tag{4.202}$$

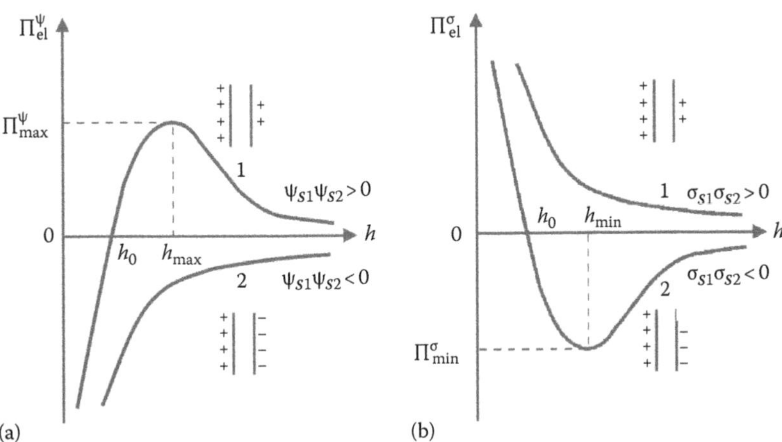

FIGURE 4.33 Electrostatic disjoining pressure at (a) fixed surface potential, Π_{el}^{ψ}, and (b) fixed surface charge density, Π_{el}^{σ}, both of them plotted vs. the film thickness h. ψ_{s1} and ψ_{s2} are the potentials of the two surfaces; σ_{s1} and σ_{s2} are the respective surface charge densities.

When $\sigma_1\sigma_2 > 0$, Equation 4.202 yields $\Pi_{el}^{\sigma} > 0$ for every h (see Figure 4.33b). However, when $\sigma_1\sigma_2 < 0$, Π_{el}^{σ} is repulsive for small thickness, $h < h_0$ and attractive for larger separations, $h > h_0$; h_0 is determined by the equation $\kappa h_0 = \ln(-\sigma_{s2}/\sigma_{s1})$; $|\sigma_{s2}| > |\sigma_{s1}|$. The electrostatic disjoining pressure in this case has a minimum value

$$\Pi_{el}^{\sigma}(\min) = \frac{1}{\varepsilon\varepsilon_0}\sigma_{s1}\sigma_{s2}, \quad \text{at } h_{\min} = \frac{1}{\kappa}\operatorname{arccosh}\left(-\frac{\sigma_{s2}}{\sigma_{s1}}\right) \tag{4.203}$$

Finally, it should be noted that all curves depicted in Figure 4.24 decay exponentially at $h \to \infty$. An asymptotic expression for $Z{:}Z$ electrolytes, which generalizes Equation 4.197, holds [380,399]:

$$\Pi_{el}(h) = 64 n_0 kT \gamma_1\gamma_2 e^{-\kappa h}, \quad \gamma_k \equiv \tanh\left(\frac{Ze\psi_{sk}}{4kT}\right), k = 1,2 \tag{4.204}$$

Equation 4.204 is valid for both low and high surface potentials, only if $\exp(-\kappa h) \ll 1$.

4.4.3.3 Two Charged Spheres

When the EDLs are thin compared with the particle radii ($\kappa^{-1} \ll R_1, R_2$) and the gap between the particles is small ($h_0 \ll R_1, R_2$), we can use Equation 4.204 in conjunction with the Derjaguin approximation, Equations 4.173 and 4.174. The result for thc cnergy of electrostatic interaction between two spheres reads:

$$U_{el}(h_0) = \frac{128\pi R_1 R_2}{\kappa^2(R_1 + R_2)} n_0 kT \gamma_1\gamma_2 e^{-\kappa h} \tag{4.205}$$

Equation 4.205 is valid for any values of the surface potentials ψ_{s1} and ψ_{s2} but only for $\exp(\kappa h) \gg 1$. Complementary expressions, which are valid for every $h \ll R_1, R_2$, but for small surface potentials, can be derived by integrating Equations 4.200 and 4.202, instead of Equation 4.204. In this way, for ψ_{s1} = constant and ψ_{s2} = constant, we can derive [402]:

$$U_{el}^{\psi}(h_0) = \frac{\pi\varepsilon\varepsilon_0 R_1 R_2}{R_1 + R_2}\left[(\psi_{s1} + \psi_{s2})^2 \ln(1 + e^{-\kappa h_0}) + (\psi_{s1} - \psi_{s2})^2 \ln(1 - e^{-\kappa h_0})\right] \tag{4.206}$$

or, alternatively, for σ_{s1} = constant and σ_{s2} = constant we obtain [403]:

$$U_{el}^{\sigma}(h_0) = \frac{-\pi R_1 R_2}{\varepsilon\varepsilon_0\kappa^2(R_1 + R_2)}\left[(\sigma_{s1} + \sigma_{s2})^2 \ln(1 - e^{-\kappa h_0}) + (\sigma_{s1} - \sigma_{s2})^2 \ln(1 + e^{-\kappa h_0})\right] \tag{4.207}$$

The range of validity of the different approximations involved in the derivations of Equations 4.205 through 4.207 is discussed in the book by Russel et al. [404].

As mentioned earlier, Equations 4.205 through 4.207 hold for $h_0 \ll R$. In the opposite case, when h_0 is comparable to or larger than the particle radius R, we can use the equation [14]:

$$U_{el}(h_0) = \frac{4\pi\varepsilon\varepsilon_0\psi_s^2 R^2}{2R + h_0} e^{-\kappa h_0} \tag{4.208}$$

stemming from the theory of Debye and Hückel [405] for two identical particles. Equation 4.208 was derived by using the superposition approximation (valid for weak overlap of the two EDLss) and the linearized Poisson–Boltzmann equation. A simple approximate formula, representing in fact interpolation between Equations 4.208 and 4.206 (the latter for $R_1 = R_2 = R$), has been derived by McCartney and Levine [406]:

$$U_{\text{el}}^{\Psi}(h_0) = 4\pi\varepsilon\varepsilon_0 R\psi_s^2 \frac{R+h_0}{2R+h_0}\ln\left(1+\frac{Re^{-\kappa h_0}}{R+h_0}\right) \tag{4.209}$$

Equation 4.209 has the advantage to give a good approximation for every h_0 provided that the Poisson–Boltzmann equation can be linearized. Similar expressions for the energy of electrostatic interaction between two deformed droplets or bubbles (Figure 4.32) can be derived [392].

4.4.4 DLVO Theory

The first quantitative theory of interactions in thin liquid films and dispersions is the DLVO theory called after the names of the authors Derjaguin and Landau [407] and Verwey and Overbeek [399]. In this theory, the total interaction is supposed to be a superposition of van der Waals and double layer interactions. In other words, the total disjoining pressure and the total interaction energy are presented in the form:

$$\Pi = \Pi_{\text{vw}} + \Pi_{\text{el}}, \quad U = U_{\text{vw}} + U_{\text{el}} \tag{4.210}$$

A typical curve, Π versus h, exhibits a maximum representing a barrier against coagulation and two minima, called primary and secondary minimum (see Figure 4.17); the U versus h curve has a similar shape. The primary minimum appears if strong short-range repulsive forces (e.g., steric forces) are present. With small particles, the depth of the secondary minimum is usually small ($U_{\min} < kT$). If the particles cannot overcome the barrier, coagulation (flocculation) does not take place, and the dispersion is stable due to the electrostatic repulsion, which gives rise to the barrier. With larger colloidal particles ($R > 0.1$ μm), the secondary minimum could be deep enough to cause coagulation and even formation of ordered structures of particles [408].

By addition of electrolyte or by decreasing the surface potential of the particles, we can suppress the electrostatic repulsion and thus decrease the height of the barrier. According to DLVO theory, the critical condition determining the onset of rapid coagulation is

$$U(h_{\max}) = 0, \quad \left.\frac{dU}{dh}\right|_{h_{\max}} = 0 \tag{4.211}$$

where $h = h_{\max}$ denotes the position of the barrier.

By using Equation 4.175 for U_{vw} and Equation 4.205 for U_{el} we derive from Equations 4.210 and 4.211 the following criterion for the threshold of rapid coagulation of identical particles ($R_1 = R_2 = R$; $\gamma_1 = \gamma_2 = \gamma$):

$$\frac{\kappa^6}{n_0^2} = \left[\frac{768\pi}{A_H} kT\, e^{-1} \tanh^2\left(\frac{Ze\psi_s}{4kT}\right)\right]^2 \tag{4.212}$$

For a $Z{:}Z$ electrolyte, substituting $\kappa^2 = (2Z^2e^2n_0)/(\varepsilon_0\varepsilon kT)$ into Equation 4.212, we obtain:

$$n_0(\text{critical}) \propto \frac{1}{Z^6}\tanh^4\left(\frac{Ze\psi_s}{4kT}\right) \tag{4.213}$$

When ψ_s is high enough, the hyperbolic tangent equals 1 and Equation 4.213 yields n_0 (critical) $\propto Z^{-6}$ which is, in fact, the empirical rule established earlier by Schulze [409] and Hardy [410].

4.4.5 Non-DLVO Surface Forces

After 1980, a number of surface forces have been found out which are not taken into account by conventional DLVO theory. They are considered separately in the following section.

4.4.5.1 Ion Correlation Forces

As shown by Debye and Hückel [405], due to the strong electrostatic interaction between the ions in a solution, the positions of the ions are correlated in such a way that a counterion atmosphere appears around each ion, thus screening its Coulomb potential. The energy of formation of the counterion atmospheres gives a contribution to the free energy of the system called correlation energy [25]. The correlation energy also affects a contribution to the osmotic pressure of the electrolyte solution, which can be presented in the form [25]

$$\Pi_{\text{osm}} = kT\sum_{i=1}^{k} n_i - \frac{kT\kappa^2}{24\pi} \tag{4.214}$$

The first term in the right-hand side of the Equation 4.214 corresponds to an ideal solution, whereas the second term takes into account the effect of electrostatic interactions between the ions (the same effect is accounted for thermodynamically by the activity coefficient, see Equation 4.31).

The expression for Π_{el} in the DLVO theory (Equation 4.189) obviously corresponds to an ideal solution, the contribution of the ionic correlations being neglected. Hence, in a more general theory instead of Equation 4.210, we could write:

$$\Pi = \Pi_{\text{vw}} + \Pi_{\text{el}} + \Pi_{\text{cor}} \tag{4.215}$$

where Π_{cor} is the contribution of the ionic correlations to the disjoining pressure. The theory of Π_{cor} takes into account the following effects: (1) the different ionic concentration (and hence the different Debye screening) in the film compared to that in the bulk solution; (2) the energy of deformation of the counterion atmosphere due to the image forces; and (3) the energy of the long-range correlations between charge density fluctuations in the two opposite EDLs. For calculating Π_{cor}, both numerical solutions [411,412] and analytical expressions [413–415] have been obtained. For example, in the case when the electrolyte is symmetrical (Z:Z) and $\exp(-\kappa h) \ll 1$ we can use the asymptotic formula [413]

$$\Pi_{\text{cor}} = \Pi_{\text{el}} \frac{Z^2 e^2 \kappa}{16\pi\varepsilon\varepsilon_0 kT}(\ln 2 + 2I_C) + O(e^{-\kappa h}) \tag{4.216}$$

where Π_{el} is the conventional DLVO electrostatic disjoining pressure,

$$I_C = \frac{1}{2}(1+J)\ln 2 + \frac{2-2z^3+z}{2z(2z^2-1)^2} - \frac{1}{2}(1-J)\ln(z+z^2)$$

$$-\frac{\sqrt{z^2-1}}{z}[1+J+4(2z^2-1)^{-3}]\arctan\sqrt{\frac{z-1}{z+1}}$$

$$J \equiv \frac{2z^2-3}{(2z^2-1)^3}, \quad z \equiv \left[1+\left(\frac{e\sigma_s}{2\varepsilon\varepsilon_0 kT\kappa}\right)^2\right]^{1/2}$$

The results for the case of symmetric electrolytes are the following. Π_{cor} is negative and corresponds to attraction, which can be comparable by magnitude with Π_{vw}. In the case of 1:1 electrolyte, Π_{cor} is usually a small correction to Π_{el}. In the case of 2:2 electrolyte, however, the situation can be quite different: the attractive forces, $\Pi_{cor} + \Pi_{vw}$, prevails over Π_{el} and the total disjoining pressure, Π, becomes negative. The effect of Π_{cor} is even larger in the presence of ions of higher valence. Short-range net attractive ion-correlation forces have been measured by Marra [416,417] and Kjellander et al. [418,419] between highly charged anionic bilayer surfaces in $CaCl_2$ solutions. These forces are believed to be responsible for the strong adhesion of some surfaces (clay and bilayer membranes) in the presence of divalent counterions [36,418,420]. In Ref. [421], the attraction mechanism and the structure of counterionic correlations are discussed in the limit of strong coupling based on numerical and analytical investigations and for various geometries (planar, spherical, and cylindrical) of charged objects.

The theory predicts ion-correlation attraction not only across water films with overlapping EDLs, but also across *oily* films intervening between two water phases. In the latter case, Π_{cor} is not zero because the ions belonging to the two outer double layers interact across the thin dielectric (oil) film. The theory for such a film [422] predicts that Π_{cor} is negative (attractive) and strongly dependent on the dielectric permittivity of the oil film; Π_{cor} can be comparable by magnitude with Π_{vw}; $\Pi_{el} = 0$ in this case.

4.4.5.2 Steric Interaction

4.4.5.2.1 Physical Background

The steric interaction between two surfaces appears when chain molecules, attached at some point(s) to a surface, dangle out into the solution (see Figure 4.34). When two such surfaces approach each other, the following effects take place [36,423–425]: (1) The entropy decreases due to the confining of the dangling chains which results in a repulsive osmotic force known as *steric* or *overlap* repulsion. (2) In a poor solvent, the segments of the chain molecules attract each other; hence the overlap of the two approaching layers of polymer molecules will be accompanied with some *intersegment attraction*; the latter can prevail for small overlap, however at the distance of larger overlap it becomes negligible compared with the osmotic repulsion. (3) Another effect, known as the *bridging attraction*, occurs when two opposite ends of chain molecule can attach (adsorb) to the opposite approaching surfaces, thus forming a bridge between them (see Figure 4.34e).

Steric interaction can be observed in foam or emulsion films stabilized with nonionic surfactants or with various polymers, including proteins. The usual nonionic surfactants molecules are

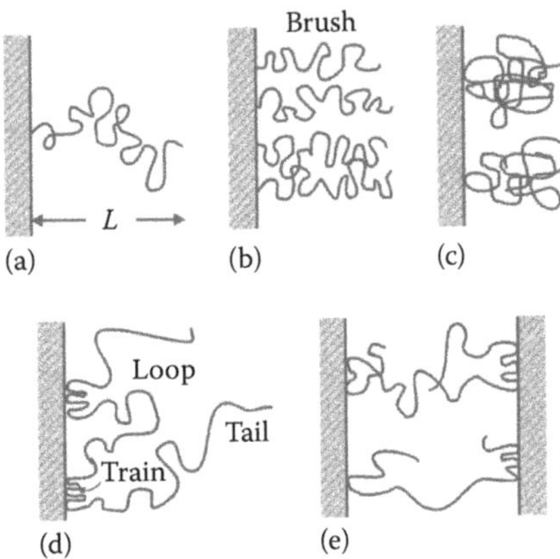

FIGURE 4.34 Polymeric chains adsorbed at an interface: (a) terminally anchored polymer chain of mean end-to-end distance L; (b) a brush of anchored chains; (c) adsorbed (but not anchored) polymer coils; (d) configuration with a loop, trains and tails; (e) bridging of two surfaces by adsorbed polymer chains.

anchored (grafted) to the liquid interface by their hydrophobic moieties. When the surface concentration of adsorbed molecules is high enough, the hydrophilic chains are called to form a brush (Figure 4.34b). The coils of macromolecules, like proteins, can also adsorb at a liquid surface (Figure 4.34c). Sometimes, the configurations of the adsorbed polymers are very different from the statistical coil: loops, trains, and tails can be distinguished (Figure 4.34d).

The osmotic pressure of either dilute or concentrated polymer solutions can be expressed in the form [426]:

$$\frac{P_{osm}}{nkT} = \frac{1}{N} + \frac{1}{2}nv + \frac{1}{3}n^2w + \cdots \tag{4.217}$$

Here

- N is the number of segments in the polymer chain
- n is the number segment density
- v and w account for the pair and triplet interactions, respectively, between segments

In fact, v and w are counterparts of the second and third virial coefficients in the theory of imperfect gases [11]; v and w can be calculated if information about the polymer chain and the solvent is available [404]:

$$w^{1/2} = \frac{\bar{v}m}{N_A}, \quad v = w^{1/2}(1-2\chi) \tag{4.218}$$

where

- $\bar{v}$ (m^3/kg) is the specific volume per segment
- m (kg/mol) is the molecular weight per segment
- N_A is the Avogadro number
- χ is the Flory parameter

The latter depends on both the temperature and the energy of solvent–segment interaction. Then, v can be zero (see Equation 4.218) for some special temperature, called the *theta temperature*. The solvent at the theta temperature is known as the *theta solvent* or *ideal solvent*. The theta temperature for polymer solutions is a counterpart of the Boil temperature for imperfect gases: this is the temperature at which the intermolecular (intersegment) attraction and repulsion are exactly counterbalanced. In a good solvent, however, the repulsion due mainly to the excluded volume effect dominates the attraction and $v > 0$. In contrast, in a poor solvent the intersegment attraction prevails, so $v < 0$.

4.4.5.2.2 *Thickness of the Polymer Adsorption Layer*

The steric interaction between two approaching surfaces appears when the film thickness becomes of the order of, or smaller than, $2L$ where L is the mean-square end-to-end distance of the hydrophilic portion of the chain. If the chain was entirely extended, then L would be equal to Nl with l being the length of a segment; however, due to the Brownian motion $L < Nl$. For an anchored chain, like that depicted in Figure 4.34a, in a theta solvent, L can be estimated as [404]

$$L \approx L_0 \equiv l\sqrt{N} \tag{4.219}$$

In a good solvent $L > L_0$, whereas in a poor solvent $L < L_0$. In addition, L depends on the surface concentration, Γ, of the adsorbed chains, that is, L is different for an isolated molecule and for a brush (see Figure 4.34a and b). The mean field approach [404] applied to polymer solutions provides the following equation for calculating L

$$\tilde{L}^3 - \left(1 + \frac{1}{9}\tilde{\Gamma}^2\right)\tilde{L}^{-1} = \frac{1}{6}\tilde{v} \quad (4.220)$$

where $\tilde{L}$, $\tilde{\Gamma}$, and $\tilde{v}$ are the dimensionless values of L, Γ, and v defined as follows:

$$\tilde{L} = \frac{L}{l\sqrt{N}}, \quad \tilde{\Gamma} = \frac{\Gamma N \sqrt{w}}{l}, \quad \tilde{v} = \frac{v\Gamma N^{3/2}}{l} \quad (4.221)$$

For an isolated adsorbed molecule ($\tilde{\Gamma} = 0$) in an ideal solvent ($\tilde{v} = 0$) Equation 4.220 predicts $\tilde{L} = 1$, that is, $L = L_0$.

4.4.5.2.3 Overlap of Adsorption Layers

We now consider the case of terminally anchored chains, like those depicted in Figure 4.34a and b. Dolan and Edwards [427,428] calculated the steric interaction free energy per unit area, f, as a function on the film thickness, h, in a theta solvent:

$$f(h) = \Gamma kT\left[\frac{\pi^2}{3}\frac{L_0^2}{h^2} - \ln\left(\frac{8\pi}{3}\frac{L_0^2}{h^2}\right)\right] \quad \text{for } h < L_0\sqrt{3} \quad (4.222)$$

$$f(h) = 4\Gamma kT \exp\left(-\frac{3h^2}{2L_0^2}\right) \quad \text{for } h > L_0\sqrt{3} \quad (4.223)$$

where L_0 is the end-to-end distance as defined by Equation 4.219. The boundary between the power-law regime ($f \propto 1/h^2$) and the exponential decay regime is at $h = L_0\sqrt{3} \approx 1.7L_0$, the latter being slightly less than $2L_0$, which is the intuitively expected onset of the steric overlap. The first term in the right-hand side of Equation 4.222 comes from the osmotic repulsion between the brushes, which opposes the approach of the two surfaces; the second term is negative and accounts effectively for the decrease of the elastic energy of the initially extended chains when the thickness of each of the two brushes, pressed against each other, decreases.

In the case of good solvent, the disjoining pressure $\Pi = -df/dh$ can be calculated by means of Alexander–de Gennes theory as [429,430]:

$$\Pi(h) = kT\Gamma^{3/2}\left[\left(\frac{2L_g}{h}\right)^{9/4} - \left(\frac{h}{2L_g}\right)^{3/4}\right] \quad \text{for } h < 2L_g,\ L_g = N(\Gamma l^5)^{1/3} \quad (4.224)$$

where L_g is the thickness of a brush in a good solvent [431]. The positive and the negative terms in the right-hand side of Equation 4.224 correspond to osmotic repulsion and elastic attraction. The validity of Alexander–de Gennes theory was experimentally confirmed by Taunton et al. [432] who measured the forces between two brush layers grafted on the surfaces of two crossed mica cylinders.

In the case of adsorbed molecules, like those in Figure 4.34c, which are not anchored to the surface, the measured surface forces depend significantly on the rate of approaching of the two surfaces [433,434]. The latter effect can be attributed to the comparatively low rate of exchange of polymer between the adsorption layer and the bulk solution. This leads to a hysteresis of the surface force: different interaction on approach and separation of the two surfaces [36]. In addition, we can observe two regimes of steric repulsion: (1) weaker repulsion at larger separations due to the overlap of the tails (Figure 4.34d) and (2) stronger repulsion at smaller separations indicating overlap of the loops [435].

4.4.5.3 Oscillatory Structural Forces

4.4.5.3.1 Origin of the Structural Forces

Oscillatory structural forces appear in two cases: (1) in thin films of pure solvent between two smooth *solid* surfaces and (2) in thin liquid films containing colloidal particles (including macromolecules and surfactant micelles). In the first case, the oscillatory forces are called the *solvation forces* [36,436]. They are important for the short-range interactions between solid particles and dispersions. In the second case, the structural forces affect the stability of foam and emulsion films, as well as the flocculation processes in various colloids. At higher particle concentrations, the structural forces stabilize the liquid films and colloids [437–441]. At lower particle concentrations, the structural forces degenerate into the so-called *depletion attraction*, which is found to destabilize various dispersions [442,443].

In all cases, the oscillatory structural forces appear when monodisperse spherical (in some cases ellipsoidal or cylindrical) particles are confined between the two surfaces of a thin film. Even one "hard wall" can induce ordering among the neighboring molecules. The oscillatory structural force is a result of overlap of the structured zones at two approaching surfaces [444–447]. A simple connection between density distribution and structural force is given by the contact value theorem [36,447,448]:

$$\Pi_{\text{os}}(h) = kT[n_s(h) - n_s(\infty)] \tag{4.225}$$

where

Π_{os} is the disjoining pressure component due to the oscillatory structural forces
$n_s(h)$ is the particle number density in the subsurface layer as a function of the distance between the walls, h

Figure 4.35 illustrates the variation of n_s with h and the resulting disjoining pressure, Π_{os}. We see that in the limit of very small separations, as the last layer of particles is eventually squeezed out, $n_s \to 0$ and

$$\Pi_{\text{os}}(h) \to -kTn_s(\infty) \quad \text{for } h \to 0 \tag{4.226}$$

In other words, at small separations Π_{os} is negative (attractive). Equation 4.226 holds for both solvation forces and colloid structural forces. In the latter case, Equation 4.226 represents the osmotic pressure of the colloid particles and the resulting attractive force is known as the *depletion force* (Section 4.4.5.3.3).

The wall induces structuring in the neighboring fluid only if the magnitude of the surface roughness is negligible in comparison with the particle diameter, *d*. Indeed, when surface irregularities are present, the oscillations are smeared out and oscillatory structural force does not appear. If the film surfaces are fluid, the role of the surface roughness is played by the interfacial fluctuation capillary waves, whose amplitude (between 1 and 5 Å) is comparable with the diameter of the solvent molecules. For this reason, oscillatory solvation forces (due to structuring of solvent molecules) are observed only with liquid films, which are confined between smooth solid surfaces [36]. In order for structural forces to be observed in foam or emulsion films, the diameter of the colloidal particles must be much larger than the amplitude of the surface corrugations. The period of the oscillations is always about the particle diameter [36,441].

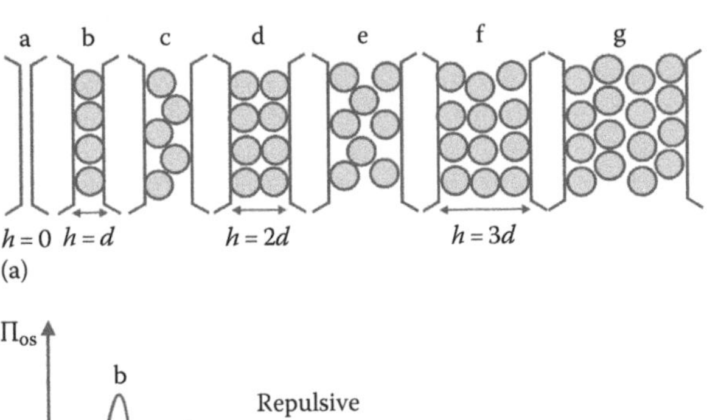

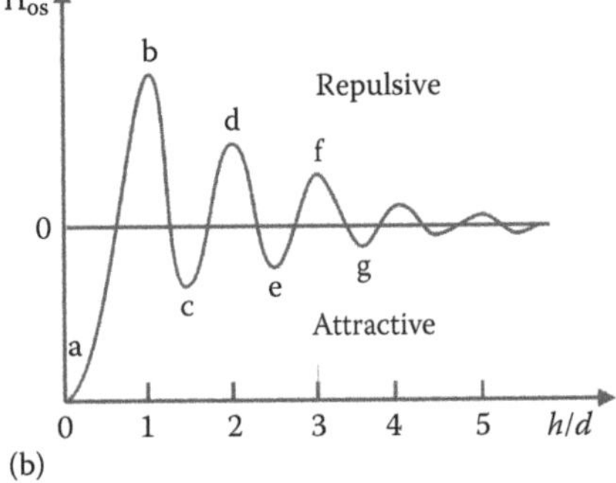

FIGURE 4.35 (a) Sketch of the consecutive stages of the thinning of a liquid film containing spherical particles; (b) plot of the related oscillatory structural component of disjoining pressure, Π_{os}, vs. the film thickness h; see Ref. [36] for details.

The theories developed for calculating the oscillatory force are based on modeling by means of the integral equations of statistical mechanics [449–453] or numerical simulations [454–457]. As a rule, these approaches are related to complicated theoretical expressions or numerical procedures, in contrast with the Derjaguin–Landau–Verwey–Overbeek (DLVO) theory, one of its main advantages being its simplicity [36]. To overcome this difficulty, some relatively simple semiempirical expressions have been proposed [458,459] on the basis of fits of theoretical results for hard-sphere fluids.

The following semiempirical formula for the oscillatory structural component of disjoining pressure reads was proposed in Ref. [458]:

$$\Pi_{os}(h) = P_0 \cos\left(\frac{2\pi h}{d_1}\right)\exp\left(\frac{d^3}{d_1^2 d_2} - \frac{h}{d_2}\right) \quad \text{for } h > d$$
$$= -P_0 \quad \text{for } 0 < h < d \tag{4.227}$$

where

d is the diameter of the hard spheres

d_1 and d_2 are the period and the decay length of the oscillations which are related to the particle volume fraction, ϕ, as follows [458]

$$\frac{d_1}{d} = \sqrt{\frac{2}{3}} + 0.237\Delta\phi + 0.633(\Delta\phi)^2; \quad \frac{d_2}{d} = \frac{0.4866}{\Delta\phi} - 0.420 \tag{4.228}$$

Here

$\Delta\phi = \phi_{max} - \phi$ with $f_{max} = \pi/(3\sqrt{2})$ being the value of ϕ at close packing

P_0 is the particle osmotic pressure determined by means of the Carnahan–Starling formula [460]

$$P_0 = nkT\frac{1+\phi+\phi^2-\phi^3}{(1-\phi)^3}, \quad n = \frac{6\phi}{\pi d^3} \tag{4.229}$$

where n is the particle number density. For $h < d$, when the particles are expelled from the slit into the neighboring bulk suspension, Equation 4.227 describes the depletion attraction. On the other hand, for $h > d$ the structural disjoining pressure oscillates around P_0 as defined by Equation 4.229 in agreement with the results of Kjellander and Sarman [451]. The finite discontinuity of Π_{os} at $h = d$ is not surprising as, at this point, the interaction is switched over from oscillatory to depletion regime. It should be noted that in oscillatory regime, the concentration dependence of Π_{os} is dominated by the decay length d_2 in the exponent (see Equations 4.227 and 4.228). Roughly speaking, for a given distance h, the oscillatory disjoining pressure Π_{os} increases five times when ϕ is increased with 10% [458]. The comparison with available numerical data showed that Equation 4.227 is accurate everywhere except in the region of the first (the highest) oscillatory maximum.

A semiempirical expression for $\Pi_{os}(H)$, which is accurate in the whole region $0 \le H < \infty$, including the region of the first maximum, was proposed by Trokhymchuk et al. [459]:

$$\begin{aligned} \Pi_{os} &= \Pi_0 \cos(\omega h + \varphi_2)e^{-\kappa h} + \Pi_1 e^{(d-h)\delta} && \text{for } h \ge d \\ \Pi_{os} &= -P_0 && \text{for } 0 \le h < d \end{aligned} \tag{4.230}$$

Here, Π_0, Π_1, ω, φ_2, κ, and δ are universal functions of particle volume fraction, ϕ, tabulated in Ref. [459]. Equation 4.230 compares very well with existing computer simulation data [459].

The interactions between the micelles in a nonionic surfactant solution can be adequately described as interactions in a hard-sphere fluid. Experiments with foam films formed from aqueous solutions of two *nonionic* surfactants, Brij 35 and Tween 20, which contain spherical micelles of diameters in the range 7–9 nm, have been carried out [461]. From the measured contact angles, the micelle aggregation number and volume fraction have been determined. In addition, from the measured disjoining pressure isotherms the micelle diameter has been found. In other words, the liquid-film measurements could give information about the micelles, which is analogous to that obtainable by dynamic and static light scattering. As an illustration, Figure 4.36 shows the comparison of theory and experiment for the nonionic surfactant Tween 20. The experimental $\Pi_{os}(h)$ dependence is obtained by using the porous-plate cell by Mysels and Jones, known also as *thin-film pressure balance* [462]. The points on the horizontal axis correspond to the thickness of the metastable states of the film measured by the Scheludko–Exerowa capillary cell [215,216]. The solid line is calculated by means of Equation 4.230 for particle (micelle) diameter determined by light scattering and micelle volume fraction determined from the contact angle of the thin liquid film [461]. The short-range repulsion at $h \approx 10$ nm (Figure 4.36) corresponds to the steric repulsion between the hydrophilic headgroups of the surfactant molecules. Excellent agreement between Equation 4.230 and experimental data obtained by colloidal-probe *atomic force microscopy* (CP-AFM) for micellar solutions of Brij 35 has also been reported [463].

The predictions of different quantitative criteria for stability–instability transitions were investigated [461], having in mind that the oscillatory forces exhibit both maxima, which play the role of barriers to coagulation, and minima that could produce flocculation or coalescence in colloidal dispersions (emulsions, foams, suspensions). The interplay of the oscillatory force with the van der Waals surface force was taken into account. Two different kinetic criteria were considered, which give similar and physically reasonable results about the stability–instability transitions. Diagrams were constructed, which show the values of the micelle volume fraction, for which the oscillatory barriers can prevent the particles from coming into close contact, or for which a strong flocculation in the depletion minimum or a weak flocculation in the first oscillatory minimum could be observed [461].

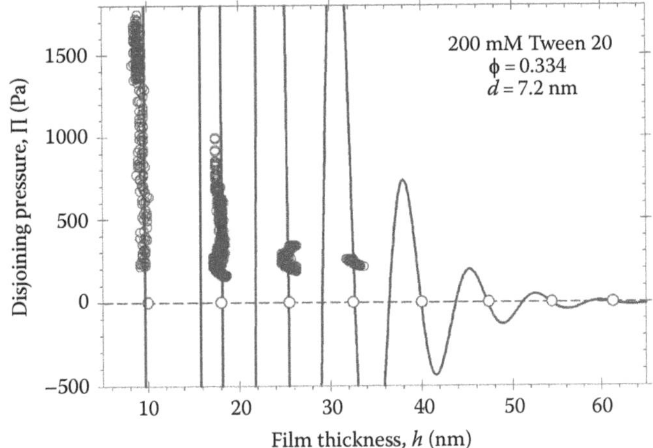

FIGURE 4.36 Plot of disjoining pressure, Π, vs. film thickness, h: comparison of experimental data for a foam film from Ref. [461] (thin-film pressure balance) with the theoretical curve (the solid line) calculated by means of Equation 4.230. The film is formed from 200 mM aqueous solution of the nonionic surfactant Tween 20. The volume fraction of the micelles ($\phi = 0.334$) is determined from the film contact angle; the micelle diameter ($d = 7.2$ nm) is determined by dynamic light scattering. The points on the horizontal axis denote the respective values of h for the stratification steps measured by a thin-film pressure balance.

4.4.5.3.2 *Oscillatory Solvation Forces*

When the role of hard spheres, like those depicted in Figure 4.35, is played by the molecules of solvent, the resulting volume exclusion force is called the *oscillatory solvation force*, or sometimes when the solvent is water, *oscillatory hydration force* [36]. The latter should be distinguished from the *monotonic* hydration force, which has different physical origin and is considered separately in Section 4.4.5.4.

Measurement of the oscillatory solvation force became possible after the precise SFA had been constructed [36]. This apparatus allowed measuring measure the surface forces in thin liquid films confined between molecularly smooth mica surfaces and in this way to check the validity of the DLVO theory down to thickness of about 5 Å, and even smaller. The experimental results with nonaqueous liquids of both spherical (CCl_4) or cylindrical (linear alkanes) molecules showed that at larger separations the DLVO theory is satisfied, whereas at separations on the order of several molecular diameters an oscillatory force is superimposed over the DLVO force law. In aqueous solutions, oscillatory forces were observed at higher electrolyte concentrations with periodicity of 0.22–0.26 nm, about the diameter of the water molecule [36]. As mentioned earlier, the oscillatory solvation forces can be observed only between smooth solid surfaces.

4.4.5.3.3 *Depletion Force*

Bondy [464] observed coagulation of rubber latex in presence of polymer molecules in the disperse medium. Asakura and Oosawa [442] published a theory, which attributed the observed interparticle attraction to the overlap of the depletion layers at the surfaces of two approaching colloidal particles (see Figure 4.37). The centers of the smaller particles, of diameter, d, cannot approach the surface of a bigger particle (of diameter D) at a distance shorter than $d/2$, which is the thickness of the depletion layer. When the two depletion layers overlap (Figure 4.37), some volume between the large particles becomes inaccessible for the smaller particles. This gives rise to an osmotic pressure, which tends to suck out the solvent between the bigger particles, thus forcing them against each other. The total depletion force experienced by one of the bigger particles is [442]

$$F_{\text{dep}} = -kTnS(h_0) \tag{4.231}$$

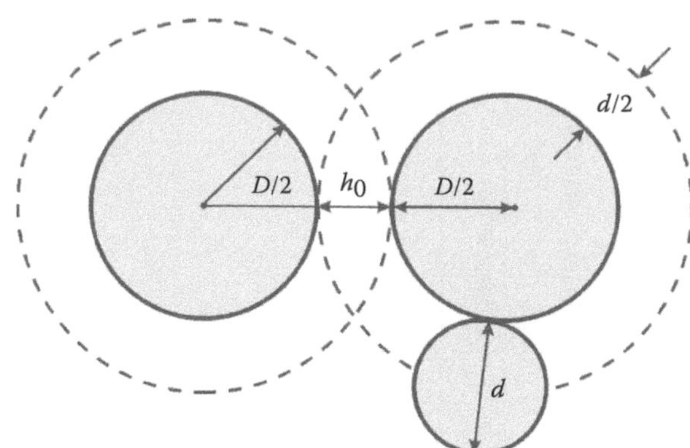

FIGURE 4.37 Overlap of the depletion zones around two particles of diameter D separated at a surface-to-surface distance h_0; the smaller particles have diameter d.

where the effective depletion area is

$$S(h_0) = \frac{\pi}{4}(2D + d + h_0)(d - h_0) \qquad \text{for } 0 \le h_0 \le d$$

$$S(h_0) = 0 \qquad \text{for } d \le h_0 \qquad (4.232)$$

Here

h_0 is the shortest distance between the surfaces of the larger particles
n is the number density of the smaller particles

By integrating Equation 4.233, we can derive an expression for the depletion interaction energy between the two larger particles, $U_{dep}(h_0)$. For $D \gg d$, this expression reads:

$$\frac{U_{dep}(h_0)}{kT} \approx -\frac{3}{2}\phi\frac{D}{d^3}(d - h_0)^2 \quad 0 \le h_0 \le d \qquad (4.233)$$

where $\phi = \pi n d^3/6$ is the volume fraction of the small particles. The maximum value of U_{dep} at $h_0 = 0$ is $U_{dep}(0)/kT \approx -3\phi D/(2d)$. For example, if $D/d = 50$ and $\phi = 0.1$, then $U_{dep}(0) = 7.5kT$. This depletion attraction turns out to be large enough to cause flocculation in dispersions. De Hek and Vrij [443] studied systematically the flocculation of sterically stabilized silica suspensions in cyclohexane by polystyrene molecules. Patel and Russel [465] investigated the phase separation and rheology of aqueous polystyrene latex suspensions in the presence of polymer (Dextran T-500).

The stability of dispersions is often determined by the competition between electrostatic repulsion and depletion attraction [466]. Interplay of steric repulsion and depletion attraction was studied theoretically by van Lent et al. [467] for the case of polymer solution between two surfaces coated with anchored polymer layers. Joanny et al. [468] and Russel et al. [404] re-examined the theory of depletion interaction by taking into account the internal degrees of freedom of the polymer molecules. Their analysis confirmed the earlier results of Asakura and Oosawa [442].

The depletion interaction is always present when a film is formed from micellar surfactant solution; the micelles play the role of the smaller particles. At higher micellar concentrations, the volume exclusion interaction becomes more complicated: it follows the oscillatory curve depicted in Figure 4.34. In this case only, the first minimum (that at $h \to 0$) corresponds to the conventional depletion force.

In the case of plane-parallel films the depletion component of disjoining pressure is

$$\begin{aligned} \Pi_{\text{dep}}(h) &= -nkT \qquad h < d \\ \Pi_{\text{dep}}(h) &= 0 \qquad h > d \end{aligned} \tag{4.234}$$

which is similar to Equation 4.226. This is not surprising because in both cases we are dealing with the excluded volume effect. Evans and Needham [469] succeeded to measure the depletion energy of two interacting bilayer surfaces in a concentrated Dextran solution; their results confirm the validity of Equation 4.236. The effect of polymer polydispersity on the depletion interaction between two plates immersed in a nonadsorbing polymer solution was studied by self-consistent field theory [470]. The results showed that as the two plates approach, the polymers with different chain lengths are excluded from the gap gradually for conformational entropy penalty, and the range of the depletion potential increases and the depth of the potential decreases with increasing polydispersity. Depletion force in a bidisperse granular layer was investigated in experiments and simulations of vertically vibrated mixtures of large and small steel spheres [471].

The interaction between a colloidal hard sphere and a wall or between two spheres in a dilute suspension of infinitely thin rods was calculated numerically [472]. The method allowed to studying the effect of polydispersity on the depletion interaction. It was observed that both the depth and the range of the depletion potential increase drastically if the relative standard deviation of the length distribution is larger than 0.25. In contrast, the potential is virtually indistinguishable from that caused by monodisperse rods, if the standard deviation is ≤ 0.1 [472]. Shear-affected depletion interaction with disc-shaped particles was experimentally investigated [473]. Synergistic effects of polymers and surfactants on depletion forces were also examined. It was established that the formation of relatively large complexes (aggregates) of polymer and surfactant creates a significant depletion force between the particle and plate [474]. A detailed review on depletion surface forces can be found in the book by Lekkerkerker and Tuinier [475].

4.4.5.3.4 Colloid Structural Forces

In the beginning of the twentieth century, Johonnott [476] and Perrin [477] observed that *foam* films decrease their thickness by several stepwise transitions. The phenomenon was called *stratification*. Bruil and Lyklema [478] and Friberg et al. [479] studied systematically the effect of ionic surfactants and electrolytes on the occurrence of the stepwise transitions. Keuskamp and Lyklema [480] anticipated that some oscillatory interaction between the film surfaces must be responsible for the observed phenomenon. Kruglyakov et al. [481,482] reported the existence of stratification with *emulsion* films.

It should be noted that the explanation of the stepwise transitions in the film thickness as a layer-by-layer thinning of an ordered structure of spherical micelles within the film (see Figure 4.35) was first given by Nikolov et al. [437–441]. Before that it was believed that the stepwise transitions are due to the formation of a *lamella*-liquid-crystal structures of surfactant molecules in the films. One of the direct proofs was given by Denkov et al. [483,484], who succeeded in freezing foam films at various stages of stratification. The electron microscope pictures of such vitrified stratifying films containing latex particles (144 nm in diameter) and bacteriorhodopsin vesicles (44 nm in diameter) showed ordered particle arrays of hexagonal packing [484]. The mechanism of stratification was studied experimentally and theoretically in Ref. [485], where the appearance and expansion of black spots in the stratifying films were described as being a process of condensation of vacancies in a colloid crystal of ordered micelles within the film.

The stable branches of the oscillatory curves have been detected by means of a thin-film pressure balance [461,486,487]. Oscillatory forces due to surfactant micelles and microemulsion droplets have also been measured by means of a SFA [488,489]; by atomic force microscopy [463,490]; by a light scattering method [491], in asymmetric films [492], in emulsion films [493], and in films containing solid colloidal spheres [437,438,494–502]. Such forces are also observed in more complex systems like protein solutions, surfactant–polymer mixtures, and ABA amphiphilic block copolymers, where A and B denote, respectively, hydrophilic and hydrophobic parts of the molecule [503–511].

In the case of liquid films that contain *charged* colloid particles (micelles), the *oscillatory period*, Δh, is considerably greater than the particle diameter [437,438]. In this case, the theoretical prediction of Δh demands the use of density functional theory calculations and/or Monte Carlo simulations [456,512]. However, the theory, simulations, and experiments showed that a simple inverse-cubic-root relation, $\Delta h = c_m^{-1/3}$, exists between Δh and the bulk number concentration of micelles (particles), c_m [437,438,498,512–514].

The validity of the semiempirical $\Delta h = c_m^{-1/3}$ law is limited at low and high particle concentrations, characterized by the effective particle volume fraction (particle + counterion atmosphere) [502]. The decrease of the effective particle volume fraction can be experimentally accomplished not only by dilution, but also by addition of electrolyte that leads to shrinking of the counterion atmosphere [500]. The inverse-cubic-root law, $\Delta h = c_m^{-1/3}$, which can be interpreted as an osmotic pressure balance between the film and the bulk [513], is fulfilled in a wide range of particle/micelle concentrations that coincide with the range where stratification (step-wise thinning) of free liquid films formed from particle suspension and micellar solution is observed [139,437–439,493–496,513,514], and where the surface force measured by CP-AFM [463,490,498–502] or SFA [488,489] exhibits oscillations.

Because the validity of the $\Delta h = c_m^{-1/3}$ law has been proven in numerous studies, it can be used for determining the aggregation number, N_{agg}, of ionic surfactant micelles [513]. Indeed, $c_m = (c_s - \text{CMC})/N_{agg}$, where c_s and CMC are the total input surfactant concentration and the CMC expressed as number of molecules per unit volume. The combination of the latter expression with $\Delta h = c_m^{-1/3}$ law yields [513,514]:

$$N_{\text{agg}} = (c_s - \text{CMC})(\Delta h)^3 \tag{4.235}$$

Values of N_{agg} determined from the experimental Δh for foam films containing micelles [513,514] using Equation 4.235 are shown in Table 4.7 for three ionic surfactants, sodium dodecylsulfate (SDS, CMC = 8 mM), cetyl trimethyl ammonium bromide (CTAB, CMC = 0.9 mM), and cetyl pyridinium chloride (CPC, CMC = 0.9 mM) at 25°C. As seen in Table 4.7, the micelle aggregation numbers determined in this way compare very well with data for N_{agg} obtained by other methods.

As mentioned earlier, the experimental Δh is significantly greater than the diameter of the ionic micelle. Δh can be considered as an effective diameter of the charged particle, d_{eff}, which includes its counterion atmosphere. A semiempirical expression for calculating Δh was proposed in [513,514]:

$$d_{\text{eff}} = d_h \left\{ 1 + \frac{3}{d_h^3} \int_{d_h}^{\infty} \left[1 - \exp\left(-\frac{3u_{\text{el}}(r)}{kT} \right) \right] r^2 dr \right\}^{1/3} \tag{4.236}$$

where

$$\frac{u_{\text{el}}(r)}{kT} = \frac{r}{4L_B} \left[\frac{e}{kT} \psi\left(\frac{r}{2} \right) \right]^2 \tag{4.237}$$

Here

d_h is the hydrodynamic diameter of the micelle
k is the Boltzmann constant
T is the absolute temperature
$u_{el}(r)$ is the energy of electrostatic interaction of two micelles in the solution
$\psi(r)$ is the distribution of the electrostatic potential around a given ionic micelle in the solution
r is radial coordinate;
$L_B \equiv e^2/(4\pi\varepsilon_0\varepsilon kT)$ is the Bjerrum length ($L_B = 0.72$ nm for water at 25°C); ε_0 is the permittivity of vacuum; ε is the dielectric constant of the solvent (water); and e is the elementary charge

TABLE 4.7
Measured Period of the Structural Force,[a] Δh, and Micelle Aggregation Number, N_{agg}, Calculated from Equation 4.235

c_s (mM)	Experimental Δh (nm) from Ref. [513]	Aggregation Number N_{agg} from Equation 4.235	N_{agg} from Literature
Sodium dodecyl sulfate (SDS)			
30	15.3	48	50 [515], 55 [516], 59 [517]
40	14.7	61	60 [518], 62 [519], 64 [517]
50	13.7	65	64 [515,517,520], 65 [521]
100	10.6	65	64 [515,517,520], 65 [521]
Cetyl trimethylammonium bromide (CTAB)			
10	25.8	95	92 [522], 95 [523,524], 98 [525]
20	21.8	119	—
30	19.9	137	100 [524]
40	18.0	136	—
50	16.6	135	139 [526], 140 [524]
Cetyl pyridinium chloride (CPC)			
10	21.2	52	45–90 [527], 56 [528]
20	18.7	75	78 [513]
30	16.6	80	82 [525]
40	15.8	93	—
50	14.6	93	87 [529]

[a] Data from Refs. [513,514].

Equation 4.237 reduces the two-particle problem to the single-particle problem; see Refs. [513,514] for details.

It was found [513] that the relationship $d_{eff} = c_m^{-1/3} = \Delta h$ is satisfied in the whole concentration range where stratifying films are observed; d_{eff} is calculated from Equation 4.236, whereas Δh is experimentally determined from the stratification steps of foam films. In contrast, for $d_{eff} < c_m^{-1/3}$ the foam films do not stratify and the oscillations of disjoining pressure vanish. This may happen at low micelle concentrations, or at sufficiently high salt concentrations [513]. Thus, the relation between d_{eff} and c_m can be used as a criterion for the existence of oscillatory structural force with charged colloidal particles.

4.4.5.4 Repulsive Hydration and Attractive Hydrophobic Forces

These two surface forces are observed in thin aqueous films. Their appearance is somehow connected with the unique properties of the water as solvent: small molecular size, large dipole moment, high dielectric constant, formation of an extensive hydrogen-bonding network, and of EDLs near interfaces [36,530].

4.4.5.4.1 Repulsive Hydration Forces

The existence of a short-range ($\leq$4 nm) repulsive pressure was first observed in experiments on the swelling of clays [531,532] and on the stabilization of foam films [533]. This short-range repulsion has been called the "hydration force" [534]. The school of Derjaguin terms this effect "structural component of disjoining pressure" [535]. Indications for its action were found in measurements of interactions between phospholipid bilayers by Parsegian et al. [536,537]. Israelachvili et al. [538–540] and Pashley [541–543] examined the validity of the DLVO theory [399,407] at small film thickness by an SFA in experiments with films from aqueous electrolyte solutions confined between two curved mica surfaces, bare or covered by adsorbed layers. At electrolyte concentrations below 10^{-4} M, they observed the typical DLVO maximum. However, at electrolyte concentrations higher than 10^{-3} M they did not observe the expected DLVO maximum and primary minimum [540]. Instead a strong short-range repulsion was detected, which can be empirically described by exponential law [36]:

$$f_{\text{hydr}}(h) = f_0 e^{-h/\lambda_0} \tag{4.238}$$

where the decay length $\lambda_0 \approx 0.6$–1.1 nm for 1:1 electrolytes and f_0 depends on the hydration of the surfaces but is usually about 3–30 mJ/m^2. Similar repulsion was detected between silica sheets [544,545] and dihexadecyl phosphate monolayers deposited on a solid surface [546].

The conventional electrostatic (double layer) repulsion is suppressed if the solution's ionic strength is increased [399,407]. In contrast, the hydration repulsion is detected at higher ionic strengths [540], at which it is the main stabilizing factor in liquid films and colloidal dispersions. Such strong repulsion at high salt concentrations was observed between apoferritin molecules in solutions [547,548] and between the adsorption layers of this and other proteins on solid surfaces and colloidal particles [549,550]. In general, the hydration force plays an important role for the stability of proteins in physiological media. Hydration forces have also been observed between DNA molecules in aqueous solutions [551,552]. Effects of monovalent anions of the Hofmeister series and other solutes on the hydration repulsion between phospholipid bilayers have been experimentally investigated [553–556]. The hydration repulsion affects the stability of emulsions [557]; the rheology of concentrated suspensions [558]; the interactions of biological cells [559]; and the fusion rate of vesicles in the cellular inter-organelle traffic [560]. Additional information can be found in several review articles [561–566].

The physical importance of the hydration force is that it stabilizes dispersions at high electrolyte concentrations preventing coagulation in the primary minimum of the DLVO curve (Figure 4.17). For example, Healy et al. [567] found that even high electrolyte concentrations cannot cause coagulation of amphoteric latex particles due to binding of strongly hydrated Li^+ ions at the particle surfaces. If the Li^+ ions are replaced by weakly hydrated Cs^+ ions, the hydration repulsion becomes negligible, compared with the van der Waals attraction, and the particles coagulate as predicted by the DLVO theory. Hence, the hydration repulsion can be regulated by ion exchange.

The aforementioned studies indicate that hydration repulsion is observed in (at least) two types of systems. (1) *charged* interfaces at relatively high electrolyte concentrations, where electrostatic and osmotic effects related to the presence of bound and mobile counterions are expected to play an essential role and (2) *electroneutral* surfaces with zwitterionic surface groups, like phospholipid bilayers, where the water structuring near the polar surface and surface charge discreteness could be the main sources of the observed repulsion. Correspondingly, for the theoretical explanation of the hydration repulsion different models have been proposed, which could be adequate for different systems. The most important theoretical models are as follows:

1. *Water-structuring models.* In these models, the short-range repulsive interaction is attributed to alignment of water dipoles in the vicinity of a hydrophilic surface, where the range of the surface force is determined by the orientation correlation length of the solvent molecules [568–570]. Due to the strong orientation of water molecules near polar surfaces, we could expect that there are fewer configurations available to maintain the

bulk water structure, which represents a loss of entropy that leads to a repulsive force [571]. The existence of such effects has been confirmed by molecular dynamics (MD) simulations [572,573].

2. *Image-charge models.* These models take into account the discreteness of surface charges, which induces orientation in the adjacent water dipoles [574–577]. Dipoles due to zwitterionic surface groups, for example, phospholipid headgroups [578], have been also taken into consideration in models of the electrostatic interaction between planar dipole lattices [579–583].
3. *Dielectric-saturation models* attribute the hydration repulsion to the presence of a layer with lower dielectric constant, ε, in the vicinity of the interfaces. Models with a stepwise [584,585] and continuous [586] variation of ε have been proposed.
4. *Excluded-volume models* take into account the fact that the finite size of the ions leads to a lower counterion concentration near a charged surface, and to a weaker Debye screening of the electrostatic field (in comparison with the point-ion model), which results in a stronger repulsion between two charged surfaces at short separations [587,588].
5. *Coion expulsion model.* This model [589] assumes that at sufficiently small thicknesses all coions are pressed out of the film so that it contains only counterions dissociated from the ionized surface groups. Under such conditions, the screening of the electric field of the film surface weakens, which considerably enhances the electrostatic repulsion in comparison with that predicted by the DLVO theory. Such reduced screening of the electric field could exist only in a narrow range of film thicknesses, which practically coincides with the range where the hydration repulsion is observed.

Let us consider in more details the *coion expulsion model*, also called "reduced screening model" [589]. This model was developed to explain the strong short-range repulsion detected in foam films; see Figure 4.38. It was found [589] that the excluded volume model [588] cannot explain the observed large deviations from the DLVO theory. Quantitative data interpretation was obtained by

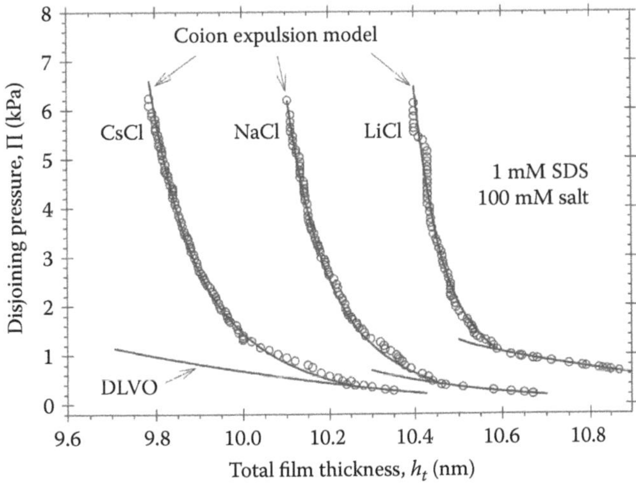

FIGURE 4.38 Plot of the disjoining pressure Π vs. the total thickness h_t of foam films formed from 1 mM aqueous solutions of SDS in the presence of 100 mM electrolyte: LiCl, NaCl and CsCl. At greater thicknesses, the $\Pi(h_t)$ dependence obeys the DLVO theory, whereas at the small thickness the steep parts of the curves are in agreement with the coion-expulsion model. The distances between the experimental curves measured with different electrolytes are due to the different sizes of the hydrated Cs^+, Na^+, and Li^+ counterions. (From Kralchevsky, P.A. et al., *Curr. Opin. Colloid Interface Sci.*, 16, 517, 2011.)

assuming that all *coions* have been pressed out of the thin film (see Figure 4.1). In such case, the Poisson–Boltzmann equation acquires the form

$$\frac{d^2\Phi}{dx^2} = \frac{1}{2}\kappa^2 e^{\Phi} \tag{4.239}$$

where

Φ is the dimensionless electric potential
κ is the Debye screening length

$$\Phi \equiv \frac{e\,|\,\psi\,|}{kT} \qquad \kappa^2 \equiv \frac{2e^2 a_{2\infty}}{\varepsilon\varepsilon_0 kT} \tag{4.240}$$

The notations are the same as in Equation 4.237; in particular, $\psi(x)$ is the dimensional electrostatic potential, x is a coordinate perpendicular to the surfaces of the plane-parallel film, and $a_{2\infty} = \gamma_{\pm} c_{2\infty}$ is the activity of counterions in the bulk solution, which is in contact with the film; see Equation 4.30. The first integral of Equation 4.239 reads:

$$\frac{d\Phi}{dx} = \kappa(e^{\Phi} - e^{\Phi_m})^{1/2} \tag{4.241}$$

Here, Φ_m is the value of Φ in the middle of the film. Integrating Equation 4.241 between the middle of the film and the film surface, one can derive [589]

$$\exp\Phi_s = \frac{\exp\Phi_m}{\cos^2[(\kappa h/4)\exp(\Phi_m/2)]} \tag{4.242}$$

where Φ_s is the value of Φ at the film surface, whereas h is the thickness of the aqueous core of the foam film. The right-hand side of Equation 4.242 has singularities for those h values, for which the cosine in the denominator is equal to zero. For this reason, the region of physical applicability of Equation 4.10 (and of the RS model) corresponds to h values, for which the argument of the cosine is between 0 and $\pi/2$:

$$0 < h < \frac{2\pi}{\kappa}\exp\left(-\frac{\Phi_m}{2}\right) \tag{4.243}$$

In the experiments in Ref. [589], 100 mM electrolyte is present, which leads to $\kappa^{-1} \approx 1$ nm, and the midplane potential is $\Phi_m \approx 0.7$, so that $\exp(-\Phi_m/2) \approx 0.705$. Then, Equation 4.243 reduces to $0 < h < 4.4$ nm. This range of h values includes the range of thicknesses, where the hydration force is operative. In the experiments in Ref. [589], the hydration repulsion appears in the interval $0 < h < 3.71$ nm irrespective of the kind of counterion (Li^+, Na^+, or Cs^+). Note that in Figure 4.38 the data are plotted versus the total film thickness, h_t, which includes not only the water core, but also the two surfactant adsorption layers at the film surfaces.

As already mentioned, the electrostatic component of disjoining pressure can be defined as the excess osmotic pressure in the film midplane with respect to the bulk solution, see Equation 4.189. Hence, if all coions are expelled from the film, the expression for the disjoining pressure acquires the form [589]:

$$\Pi_{el} = kTa_{2\infty}(e^{\Phi_m} - 2) \tag{4.244}$$

For a given surface electric potential, Φ_s, Equations 4.242 and 4.244 determine the $\Pi(h)$ dependence in a parametric form: $h = h(\Phi_m)$ and $\Pi = \Pi(\Phi_m)$. As seen in Figure 4.38, excellent agreement between theory and experiment has been achieved for reasonable parameter values [589].

4.4.5.4.2 Hydrophobic Attraction

The water does not spread spontaneously on hydrocarbons and the aqueous films on hydrophobic surfaces are rather unstable [590]. The cause for these effects is an attractive *hydrophobic* force, which is found to appear in aqueous films in contact with hydrophobic surfaces. The experiments showed that the nature of the hydrophobic surface force is different from the van der Waals and double layer interactions [591–595]. The measurements indicate that the hydrophobic interaction decays exponentially with the increase of the film thickness, h. The hydrophobic free energy per unit area of the film can be described by means of the empirical equation [36]

$$f_{\text{hydrophobic}} = -2\gamma e^{-h/\lambda_0} \tag{4.245}$$

where typically $\gamma = 10$–50 mJ/m^2, and $\lambda_0 = 1$–2 nm in the range $0 < h < 10$ nm. Larger decay length, $\lambda_0 = 12$–16 nm, was reported by Christenson et al. [595] for the range $20 < h < 90$ nm. This long-range attraction could entirely dominate the van der Waals forces. Ducker et al. [596] measured the force between hydrophobic and hydrophilic silica particles and air bubbles by means of an atomic force microscope.

It was found experimentally that 1:1 and 2:2 electrolytes reduce considerably the long-range part of the hydrophobic attraction [594,595]. The results suggest that this reduction could be due to ion adsorption or ion exchange at the surfaces rather than to the presence of electrolyte in the solution itself. Therefore, the physical implication (which might seem trivial) is that the hydrophobic attraction across aqueous films can be suppressed by making the surfaces more hydrophilic. Besides, some special polar solutes are found to suppress the hydrophobic interaction at molecular level in the bulk solution, for example, urea, $(NH_2)_2CO$, dissolved in water can cause proteins to unfold. The polar solutes are believed to destroy the hydrogen-bond structuring in water; therefore they are sometimes called chaotropic agents [36].

There is no generally accepted explanation of the hydrophobic surface force. One of the possible explanations is that the hydrogen bonding in water (and other associated liquids) could be the main underlying factor [36,597]. The related qualitative picture of the hydrophobic interaction is the following. If there were no thermal motion, the water molecules would form an ice-like tetrahedral network with four nearest neighbors per molecule (instead of 12 neighbors at close packing), because this configuration is favored by the formation of hydrogen bonds. However, due to the thermal motion a water molecule forms only about 3–3.5 transient hydrogen bonds with its neighbors in the liquid [598] with lifetime of a hydrogen bond being about 10^{-11} s. When a water molecule is brought in contact with a non-hydrogen-bonding molecule or surface, the number of its possible favorable configurations is decreased. This effect also reduces the number of advantageous configurations of the neighbors of the subsurface water molecules and some ordering propagates in the depth of the liquid. This ordering might be initiated by the orientation of the water dipoles at a water–air or water–hydrocarbon interface with the oxygen atom being oriented toward the hydrophobic phase [599–602]. Such ordering in the vicinity of the hydrophobic wall is entropically unfavorable. When two hydrophobic surfaces approach each other, the entropically unfavored water is ejected into the bulk, thereby reducing the total free energy of the system. The resulting attraction could in principle explain the hydrophobic forces. The existing phenomenological theory [597] has been generalized to the case of asymmetric films [603], and has been applied to interpret experimental data for breakage of emulsion and foam films at low surfactant and high electrolyte concentrations [604,605].

Another hypothesis for the physical origin of the hydrophobic force considers a possible role of formation of gaseous capillary bridges between the two hydrophobic surfaces (see Figure 4.13a) [36,606,607]. In this case, the hydrophobic force would be a kind of capillary-bridge force; see Chapter 11 in Ref. [37]. Such bridges could appear spontaneously, by nucleation (spontaneous dewetting), when the distance between the two surfaces becomes smaller than a certain threshold value, of the order of several hundred nanometers. Gaseous bridges could appear even if there is no dissolved gas in the water phase; the pressure inside a bridge can be as low as the equilibrium vapor pressure of water (e.g., $P_0 = 2337$ Pa at 20°C, which is only 2.3% of the atmospheric pressure) owing to the high interfacial curvature of the nodoid-shaped bridges; see Section 4.3.1.2.3 and Ref. [37].

For example, at air–water–solid contact angle $\theta = 90°–110°$ the maximal length of a nodoid-shaped capillary bridge, h_{max}, can be estimated from the analytical asymptotic formula [37,312]:

$$h_{max} = \frac{-2\sigma\cos\theta}{P - P_0} \quad (90° < \theta < 110°) \tag{4.246}$$

Substituting $P = 1$ atm for the outer pressure, $P_0 = 2337$ Pa for the inner pressure (the equilibrium vapor pressure of water at 20°C), $\sigma = 72.75$ mN/m for the surface tension of water, and $\theta = 94°$, from Equation 4.246 we calculate $h_{max} = 103$ nm for a vapor-filled bridge between two parallel hydrophobic plates in water. This value of h_{max} is close to the distances at which the experimentally observed long-range hydrophobic attraction begins to operate. A number of studies [608–616] provide evidence in support of the capillary-bridge origin of the long-range hydrophobic surface force. In particular, the observation of "steps" in the experimental data was interpreted as an indication for separate acts of bridge nucleation [612].

As discussed in Ref. [617], at present, the accumulated experimental data indicate for the existence of three different force-law regimes of the hydrophobic surface force. At film thickness $h \leq 1–1.5$ nm, a pure *short-range* hydrophobic force is operative, which is probably related to water structuring effects and hydrogen bonds at the water–hydrophobic interface. At intermediate distances, $1.5 < h < 15$ nm, a *long-range* hydrophobic force is acting, which is possibly due to an enhanced Hamaker constant associated with the "proton-hopping" polarizability of water. Finally, at $h > 15$ nm a *super-long-range* attraction is observed, which could be due to gaseous capillary bridges (bridging cavities) or to the electrostatic patch-charge attraction [618].

4.4.5.5 Fluctuation Wave Forces

All fluid interfaces, including liquid membranes and surfactant lamellas, are involved in a thermal fluctuation wave motion. The configurational confinement of such thermally exited modes within the narrow space between two approaching interfaces gives rise to short-range repulsive surface forces, which are considered in the following section.

4.4.5.5.1 Undulation Forces

The undulation force arises from the configurational confinement related to the *bending mode* of deformation of two fluid bilayers. This mode consists in undulation of the bilayer at constant bilayer area and thickness (Figure 4.39a). Helfrich et al. [619,620] established that two such bilayers, apart at a mean distance h, experience a repulsive disjoining pressure given by the expression:

$$\Pi_{und}(h) = \frac{3\pi^2(kT)^2}{64k_t h^3} \tag{4.247}$$

where k_t is the bending elastic modulus of the bilayer as a whole. The experiment [621] and the theory [37,204] show that k_t is of the order of 10^{-19} J for lipid bilayers. The undulation force has been measured, and the dependence $\Pi_{und} \propto h^{-3}$ was confirmed experimentally [622–624].

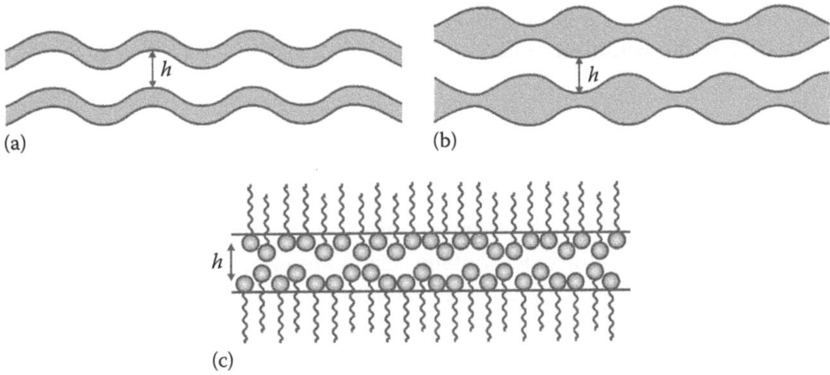

FIGURE 4.39 Surface forces due to configurational confinement of thermally exited modes into a narrow region of space between two approaching interfaces: (a) bending mode of membrane fluctuations giving rise to the undulation force; (b) squeezing mode of membrane fluctuations producing the peristaltic force; (c) fluctuating protrusion of adsorbed amphiphilic molecules engendering the protrusion surface force.

4.4.5.5.2 *Peristaltic Force*

The peristaltic force [625] originates from the configurational confinement related to the peristaltic (squeezing) mode of deformation of a fluid bilayer (Figure 4.39b). This mode of deformation consists in fluctuation of the bilayer thickness at fixed position of the bilayer midsurface. The peristaltic deformation is accompanied with extension of the bilayer surfaces. Israelachvili and Wennerström [625] demonstrated that the peristaltic disjoining pressure is related to the stretching modulus, k_s, of the bilayer:

$$\Pi_{\text{per}}(h) \approx \frac{2(kT)^2}{\pi^2 k_s h^5} \tag{4.248}$$

The experiment [626] gives values of k_s varying between 135 and 500 mN/m, which depend on temperature and composition of the lipid membrane.

4.4.5.5.3 *Protrusion Force*

Due to the thermal motion, the protrusion of an amphiphilic molecule in an adsorption monolayer (or micelle) may fluctuate about the equilibrium position of the molecule (Figure 4.39c). In other words, the adsorbed molecules are involved in a discrete wave motion, which differs from the continuous modes of deformation considered earlier. Aniansson et al. [627,628] analyzed the energy of protrusion in relation to the micelle kinetics. They assumed the energy of molecular protrusion to be of the form $u(z) = \alpha z$, where z is the distance out of the surface ($z > 0$) and determined $\alpha \approx 3 \times 10^{-11}$ J/m for single-chained surfactants. The average length of the Brownian protrusion of the amphiphilic molecules is on the order of $\lambda \equiv kT/\alpha$ [625].

By using a mean-field approach, Israelachvili and Wennerström [625] derived the following expression for the protrusion disjoining pressure which appears when two protrusion zones overlap (Figure 4.39c):

$$\Pi_{\text{protr}}(h) = \frac{\Gamma kT}{\lambda} \frac{(h/\lambda)\exp(-h/\lambda)}{1-(1+h/\lambda)\exp(-h/\lambda)} \tag{4.249}$$

where

λ is the characteristic protrusion length; $\lambda = 0.14$ nm at 25°C for surfactants with paraffin chain
Γ denotes the number of protrusion sites per unit area

Note that Π_{protr} decays exponentially for $h \gg \lambda$, but $\Pi_{protr} \propto h^{-1}$ for $h < \lambda$, that is, Π_{protr} is divergent at $h \to 0$. The respective interaction free energy (per unit film area) is

$$f_{protr} = \int_{h}^{\infty} \Pi_{protr}(\hat{h})d\hat{h} = -\Gamma kT \ln\left[1-\left(1+\frac{h}{\lambda}\right)\exp\left(\frac{-h}{\lambda}\right)\right] \tag{4.250}$$

Equation 4.249 was found to fit well experimental data for the disjoining pressure of liquid films stabilized by adsorbed protein molecules: bovine serum albumin (BSA) [629]. In that case, Γ was identified with the surface density of the loose secondary protein adsorption layer, while λ turned out to be about the size of the BSA molecule. A more detailed statistical approach to the theoretical modeling of protrusion force was proposed [630].

4.5 HYDRODYNAMIC INTERACTIONS IN DISPERSIONS

4.5.1 Basic Equations and Lubrication Approximation

In addition to the surface forces (see Section 4.4), two colliding particles in a liquid medium also experience hydrodynamic interactions due to the viscous friction, which can be rather long range (operative even at distances above 100 nm). The hydrodynamic interaction among particles depends on both the type of fluid motion and the type of interfaces. The quantitative description of this interaction is based on the classical laws of mass conservation and momentum balance for the bulk phases [630–636]:

$$\frac{\partial \rho}{\partial t} + \operatorname{div}(\rho \mathbf{v}) = 0 \tag{4.251}$$

$$\frac{\partial}{\partial t}(\rho \mathbf{v}) + \operatorname{div}(\rho \mathbf{v}\mathbf{v} - \mathbf{P} - \mathbf{P}_b) = 0 \tag{4.252}$$

where

- ρ is the mass density
- **v** is the local mass average velocity
- **P** is the hydrodynamic stress tensor
- $\mathbf{P}_b$ is the body-force tensor which accounts for the action of body forces such as gravity, electrostatic forces (the Maxwell tensor), etc.

In a fluid at rest, and in the absence of body forces, the only contact force given by the hydrodynamic stress tensor is the scalar thermodynamic pressure, p, and **P** can be written as $\mathbf{P} = -p\mathbf{I}$, where **I** is the unit tensor in space. For a fluid in motion, the viscous forces become operative and

$$\mathbf{P} = -p\mathbf{I} + \mathbf{T} \tag{4.253}$$

where **T** is the viscous stress tensor. From the definition of the stress tensor (Equation 4.253), it follows that the resultant hydrodynamic force, **F**, exerted by the surrounding fluid on the particle surface, S, and the torque, **M**, applied to it are given by the expressions [631,633]

$$\mathbf{F} = \int_S \mathbf{P}\cdot\mathbf{n}\,dS, \quad \mathbf{M} = \int_S \mathbf{r}_0 \times \mathbf{P}\cdot\mathbf{n}\,dS \tag{4.254}$$

where

- $\mathbf{r}_0$ is the position vector of a point of S with respect to an arbitrarily chosen coordinate origin
- **n** is the vector of the running unit normal to the surface S

In the presence of body forces, the total force, $\mathbf{F}_{tot}$, and torque, $\mathbf{M}_{tot}$, acting on the particle surface are

$$\mathbf{F}_{tot} = \mathbf{F} + \int_S \mathbf{P}_b \cdot \mathbf{n}\, dS, \quad \mathbf{M}_{tot} = \mathbf{M} + \int_S \mathbf{r}_0 \times \mathbf{P}_b \cdot \mathbf{n}\, dS \tag{4.255}$$

The dependence of the viscous stress on the velocity gradient in the fluid is a constitutive law, which is usually called the bulk rheological equation. The general linear relation between the viscous stress tensor, **T**, and the rate of strain tensor,

$$\mathbf{D} = \frac{1}{2}[\nabla\mathbf{v} + (\nabla\mathbf{v})^T] \tag{4.256}$$

(the superscript T denotes conjugation) reads

$$\mathbf{T} = \zeta(\operatorname{div}\mathbf{v})\mathbf{I} + 2\eta\left[\mathbf{D} - \frac{1}{3}(\operatorname{div}\mathbf{v})\mathbf{I}\right] \tag{4.257}$$

The latter equation is usually referred to as the Newtonian model or Newton's law of viscosity. In Equation 4.257, ζ is the dilatational bulk viscosity and η is the shear bulk viscosity. The usual liquids comply well with the Newtonian model. On the other hand, some concentrated macromolecular solutions, colloidal dispersions, gels, etc., may exhibit non-Newtonian behavior; their properties are considered in detail in some recent review articles and books [636–641]. From Equations 4.252 and 4.257, one obtains the Navier–Stokes equation [642,643]:

$$\rho\frac{d\mathbf{v}}{dt} = -\nabla p + \left(\zeta + \frac{1}{3}\eta\right)\nabla(\nabla\cdot\mathbf{v}) + \eta\nabla^2\mathbf{v} + \mathbf{f}, \quad (\mathbf{f} \equiv \nabla\cdot\mathbf{P}_b) \tag{4.258}$$

for homogeneous Newtonian fluids, for which the dilatational and shear viscosities, ζ and η, do not depend on the spatial coordinates. In Equation 4.258, the material derivative d/dt can be presented as a sum of a local time derivative and a convective term:

$$\frac{d}{dt} = \frac{\partial}{\partial t} + (\mathbf{v}\cdot\nabla) \tag{4.259}$$

If the density, ρ, is constant, the equation of mass conservation (Equation 4.251) and the Navier–Stokes equation 4.258 reduce to

$$\operatorname{div}\mathbf{v} = 0, \quad \rho\frac{d\mathbf{v}}{dt} = -\nabla p + \eta\nabla^2\mathbf{v} + \mathbf{f} \tag{4.260}$$

For low shear stresses in the dispersions, the characteristic velocity, V_z, of the relative particle motion is small enough in order for the Reynolds number, $\text{Re} = \rho V_z L/\eta$, to be a small parameter, where L is a characteristic length scale. In this case, the inertia terms in Equations 4.258 and 4.260 can be neglected. Then, the system of equations becomes linear and the different types of hydrodynamic motion become additive [404,644,645]; for example, the motion in the liquid flow can be presented as a superposition of elementary translation and rotational motions.

The basic equations can be further simplified in the framework of the lubrication approximation, which can be applied to the case when the Reynolds number is small and when the distances

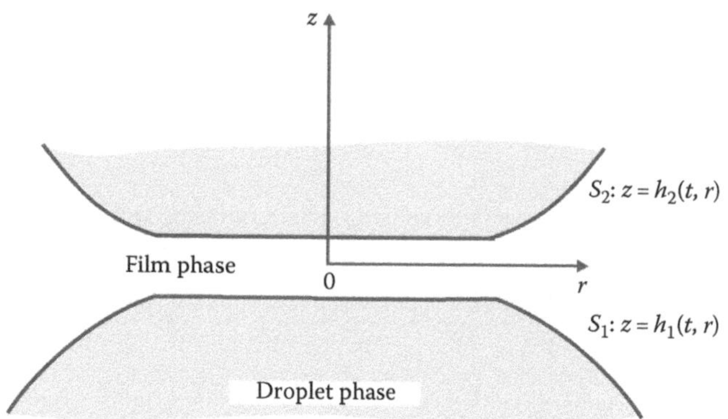

FIGURE 4.40 Sketch of a plane-parallel film formed between two identical fluid particles.

between the particle surfaces are much smaller than their radii of curvature (Figure 4.40) [646,647]. There are two ways to take into account the molecular interactions between the two particles across the liquid film intervening between them: (1) the body force approach and (2) the disjoining pressure approach. The former approach treats the molecular forces as components of the body force, **f** (Equation 4.258); consequently, they give contributions to the normal and tangential stress boundary conditions [648,649]. In the case (2), the molecular interactions are incorporated only in the normal stress boundary conditions at the particle surfaces. When the body force can be expressed as a gradient of potential, $\mathbf{f} = \nabla U$ (that is $\mathbf{P}_b = U\mathbf{I}$), the two approaches are equivalent [650].

If two particles are interacting across an electrolyte solution, the equations of continuity and the momentum balance, Equation 4.260, in lubrication approximation read [651,652]

$$\nabla_{\text{II}} \cdot \mathbf{v}_{\text{II}} + \frac{\partial v_z}{\partial z} = 0, \quad \eta \frac{\partial^2 \mathbf{v}_{\text{II}}}{\partial z^2} = \nabla_{\text{II}} \cdot p + kT \sum_{i=1}^{N} z_i c_i \nabla_{\text{II}} \Phi$$
$$\frac{\partial p}{\partial z} + kT \sum_{i=1}^{N} z_i c_i \frac{\partial \Phi}{\partial z} = 0 \tag{4.261}$$

where

$\mathbf{v}_{\text{II}}$ and ∇_{II} are the projection of the velocity and the gradient operator on the plane xy; the z-axis is (approximately) perpendicular to the film surfaces S_1 and S_2 (see Figure 4.40)
$c_i = c_i(r, z, t)$ is the ion concentration ($i = 1, 2, \ldots, N$)
Φ is the dimensionless electric potential (see Sections 4.2.1.2 and 4.2.2)

It turns out that in lubrication approximation, the dependence of the ionic concentrations on the z coordinate comes through the electric potential $\Phi(r, z, t)$: we obtain a counterpart of the Boltzmann equation $c_i = c_{i,n}(r, z, t)\exp(-z_i\Phi)$, where $c_{i,n}$ refers to an imaginary situation of "switched off" electric charges ($\Phi \equiv 0$). The kinematic boundary condition for the film surfaces has the form:

$$\frac{\partial h_j}{\partial t} + \mathbf{u}_j . \nabla_{\text{II}} h_j = (v_z)_j \quad \text{at } S_j (j = 1,2) \tag{4.262}$$

where

$\mathbf{u}_i$ is the velocity projection in the plane xy at the corresponding film surface S_i, which is close to the interfacial velocity
$(v_z)_i$ is the z component of the velocity at the surface S_i

The general solution of Equations 4.261 and 4.262 could be written as:

$$p = p_n + kT\sum_{i=1}^{N}(c_i - c_{i,n}) \tag{4.263}$$

$$\begin{aligned} \mathbf{v}_{\text{II}} &= \frac{(z-h_1)(z-h_2)}{2\eta}\nabla_{\text{II}}p_n + \frac{h_2 - z}{h}\mathbf{u}_1 + \frac{z-h_1}{h}\mathbf{u}_2 \\ &+ \frac{kTh^2}{4\eta}\sum_{i=1}^{N}\left[m_{2,i}(z) - \frac{h_2 - z}{h}m_{2,i}(h_1) - \frac{z-h_1}{h}m_{2,i}(h_2)\right]\nabla_{\text{II}}c_{i,n} \end{aligned} \tag{4.264}$$

Here $h = h_2 - h_1$ is the local film thickness; the meaning of $p_n(x, y, t)$ is analogous to that of $c_{i,n}(x, y, t)$; the functions, $m_{k,i}(z)$, account for the distribution of the ith ionic species in the EDL:

$$\begin{aligned} m_{0,i} &\equiv \exp(-z_i\Phi) - 1 \\ m_{k,i}(z) &\equiv \frac{2}{h}\int_0^z m_{k-1,i}(\hat{z})d\hat{z} \quad (k = 1,2,3, i = 1,2,\ldots,N) \end{aligned} \tag{4.265}$$

The equation determining the local thickness, h, of a film with fluid surfaces (or, alternatively, determining the pressure distribution at the surfaces of the gap between two solid particles of known shape) is

$$\begin{aligned} &\frac{\partial h}{\partial t} + \nabla_{\text{II}}\cdot\left[\frac{h}{2}(\mathbf{u}_1 + \mathbf{u}_2)\right] = \frac{1}{12\eta}\nabla_{\text{II}}\cdot(h^3\nabla_{\text{II}}p) \\ &+ \frac{kT}{8\eta}\nabla_{\text{II}}\cdot\left\{h^3\sum_{i=1}^{N}[m_{2,i}(h_1) + m_{2,i}(h_2) - m_{3,i}(h_2) + m_{3,i}(h_1)]\nabla_{\text{II}}c_{i,n}\right\} \end{aligned} \tag{4.266}$$

The problem for the interactions upon central collisions of two axisymmetric particles (bubbles, droplets, or solid spheres) at small surface-to-surface distances was first solved by Reynolds [646] and Taylor [653,654] for solid surfaces and by Ivanov et al. [655,656] for films of uneven thickness. Equation 4.266 is referred to as the general equation for films with deformable surfaces [655,656] (see also the more recent reviews [240,657,658]). The asymptotic analysis [659–661] of the dependence of the drag and torque coefficient of a sphere, which is translating and rotating in the neighborhood of a solid plate, is also based on Equation 4.266 applied to the special case of stationary conditions.

Using Equation 4.255, one can obtain expressions for the components of the total force exerted on the particle surface, S, in the lubrication approximation:

$$F_{\text{tot},z} = \int_S\left[p_n + kT\sum_{i=1}^{N}(c_{is} - c_{i,n}) + \Pi_{\text{nel}} - p_\infty\right]dS \tag{4.267}$$

$$\mathbf{F}_{\text{tot,II}} = -\int_S\left(\eta\frac{\partial\mathbf{v}_{\text{II}}}{\partial z} + \frac{2kT}{\kappa_c^2}\frac{\partial\Phi}{\partial z}\nabla_{\text{II}}\Phi\right)dS \tag{4.268}$$

where

- p_∞ is the pressure at infinity in the meniscus region (Figure 4.40)
- $\Pi_{\text{nel}} \equiv \Pi - \Pi_{\text{el}}$ accounts for the contribution of nonelectrostatic (non–double-layer) forces to the disjoining pressure (see Section 4.4)

The normal and the lateral force resultants, F_z and $\mathbf{F}_{II}$, are the hydrodynamic resistance and shear force, respectively.

4.5.2 Interaction between Particles of Tangentially Immobile Surfaces

The surfaces of fluid particles can be treated as tangentially immobile when they are covered by dense surfactant adsorption monolayers that can resist tangential stresses [240,657,658,662,663]. In such a case, the bubbles or droplets behave as flexible balls with immobile surfaces. When the fluid particles are rather small (say, microemulsion droplets), they can behave like hard spheres; therefore, some relations considered in the following section, which were originally derived for solid particles, can also be applied to fluid particles.

4.5.2.1 Taylor and Reynolds Equations, and Influence of the Particle Shape

In the case of two axisymmetric particles moving along the z-axis toward each other with velocity $V_z = -dh/dt$, Equation 4.266 can be integrated; and from Equation 4.267, the resistance force can be calculated. The latter turns out to be proportional to the velocity and bulk viscosity and depends on the shape in a complex way. For particles with tangentially immobile surfaces and without surface electric charge ($\mathbf{u}_1 = \mathbf{u}_2 = 0$, $\Phi = 0$) Charles and Mason [664] have derived

$$F_z = 6\pi\eta V_z \int_0^\infty \frac{r^3}{h^3} dr \tag{4.269}$$

where r is the radial coordinate in a cylindrical coordinate system. In the case of two particles of different radii, R_1 and R_2, film radius R, and uniform film thickness h (see Figure 4.41), from Equation 4.269 the following expression can be derived [665,666]:

$$F_z = \frac{3}{2}\pi\eta V_z \frac{R_*^2}{h}\left(1 + \frac{R^2}{hR_*} + \frac{R^4}{h^2 R_*^2}\right), \quad R_* \equiv \frac{2R_1 R_2}{R_1 + R_2} \tag{4.270}$$

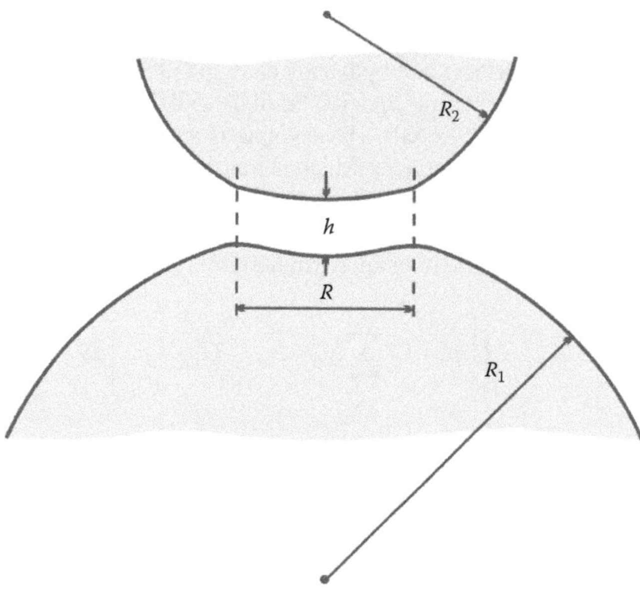

FIGURE 4.41 Sketch of a film between two nonidentical fluid particles of radii R_1 and R_2. The film thickness and radius are denoted by h and R.

This geometrical configuration has proved to be very close to the real one in the presence of electrostatic disjoining pressure [256]. The Charles–Mason formula (Equation 4.269) and Equation 4.267 have been used to calculate the velocity of film thinning for a large number of cases, summarized by Hartland [667] in tables for more than 50 cases (2D and 3D small drops, fully deformed large drops subjected to large forces, 2D hexagonal drops, etc.).

Setting $R = 0$ in Equation 4.270, we can derive a generalized version of the Taylor formula [653,654] for the velocity of approach of two nondeformable spheres under the action of an external (nonviscous) force, F_z [666]:

$$V_{\mathrm{Ta}} = \frac{2hF_z}{3\pi\eta R_*^2} \tag{4.271}$$

When a solid sphere of radius R_c approaches a flat solid surface, we may use the Taylor formula with $R_* = 2R_c$ when the gap between the two surfaces is small compared to R_c. In fact Equation 4.271 does not appear in any of the G.I. Taylor's publications but it was published in the article by Hardy and Bircumshaw [653] (see Ref. [654]).

In the case when two plane-parallel ellipsoidal discs of tangentially immobile surfaces are moving against each other under the action of an external force, $F_{\mathrm{tot},z}$, from Equations 4.266 and 4.267, we can derive the Reynolds equation [646] for the velocity of film thinning:

$$V_{\mathrm{Re}} = \frac{F_z h^3(a^2+b^2)}{3\pi\eta a^3 b^3} \tag{4.272}$$

where a and b are the principal radii of curvature. If there is a contribution of the disjoining pressure, Π, the Reynolds equation for a flat axisymmetrical film ($a = b = R$) between two fluid particles of capillary pressure P_c can be written in the form [216]:

$$V_{\mathrm{Re}} = \frac{2F_z h^3}{3\pi\eta R^4} = \frac{2(P_c - \Pi)h^3}{3\eta R^2} \tag{4.273}$$

From Equations 4.270 and 4.273, the ratio between the Reynolds velocity and the velocity of film thinning for a given force is obtained. In Figure 4.42, this ratio is plotted as a function of the film

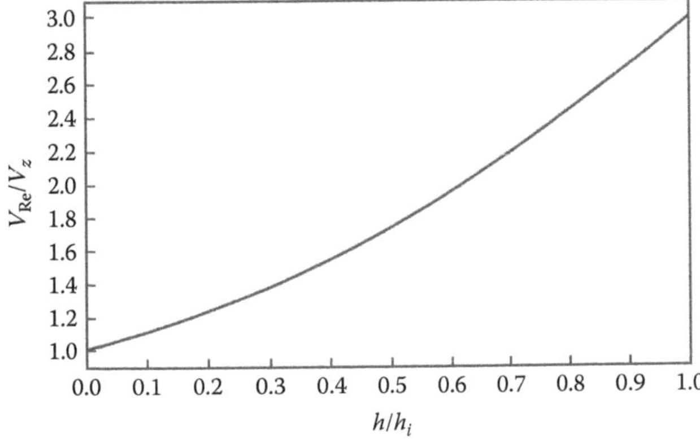

FIGURE 4.42 Plot of V_{Re}/V_z vs. h/h_i for two fluid particles (Equation 4.270) which are deformed because of the viscous friction in the transition zone between the film and the bulk phase (see Figure 4.41).

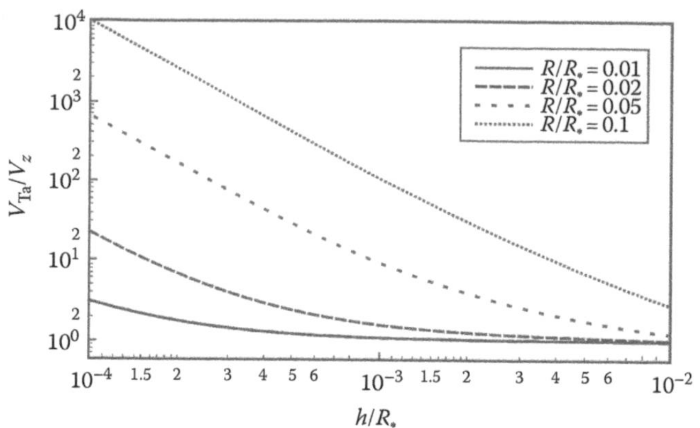

FIGURE 4.43 Plot of V_{Ta}/V_z vs. h/R_* for various values of the dimensionless film radius, R/R_*. V_{Ta} corresponds to two nondeformed (spherical) particles (Equation 4.271), whereas V_z is the velocity of approach of two deformed particles (Equation 4.270).

thickness, h, divided by inversion thickness, $h_i = R^2/R_*$ [657]. We see that the influence of the viscous friction in the zone encircling the film (this influence is not accounted for in Equation 4.273) decreases the velocity of thinning about three times for the larger distances, whereas for the small distances this influence vanishes. From Equations 4.270 and 4.271, the ratio between the Taylor velocity (corresponding to nondeformable spheres) and the approaching velocity of two deformable particles can be calculated. The dependence of this ratio on the distance between the particles for different film radii is illustrated in Figure 4.43. We see that an increase of the film radius, R, and a decrease of the distance, h, lead to a decrease in the velocity. The existence of a film between the particles can decrease the velocity of particle approach, V_z, by several orders of magnitude.

4.5.2.2 Interactions among Nondeformable Particles at Large Distances

The hydrodynamic interaction between members of a group of small particles suspended in a viscous fluid has fundamental importance for the development of adequate models for calculating the particle collective diffusion coefficient and the effective viscosity of suspension [404,644,664,668,669]. The Stokesian resistance is determined for a number of specific particle shapes under the condition that the particles are located so far apart that the hydrodynamic interactions can be ignored [644]. A general theory applicable to a single particle of arbitrary shape has been developed by Brenner [670,671]. This method gives the first-order correction (with respect to the particle volume fraction) of the viscosity and diffusivity. Matrix relations between resistance and velocity for the pure translational and rotational motions of the members of a general multiparticle system involved in a linear shear flow are given by Brenner and O'Neill [672]. In principle, from these relations we can further obtain the higher-order terms in the series expansion of viscosity and diffusivity with respect to the powers of the particle volume fraction.

At present, the only multiparticle system for which exact values of the resistance tensors can be determined is that of two spheres. It turns out that all types of hydrodynamic flows related to the motion of two spherical particles (of radii R_1 and R_2) can be expressed as superpositions of the elementary processes depicted in Figure 4.44 [404,633,644,645,673–682].

The first particle moves toward the second immobile particle and rotates around the line of centers (see Figure 4.44a). This is an axisymmetric rotation problem (a 2D hydrodynamic problem) which was solved by Jeffery [674] and Stimson and Jeffery [675] for two identical spheres moving with equal velocities along their line of centers. Cooley and O'Neill [676,677] calculated the forces for two nonidentical spheres moving with the same speed in the same direction, or alternatively,

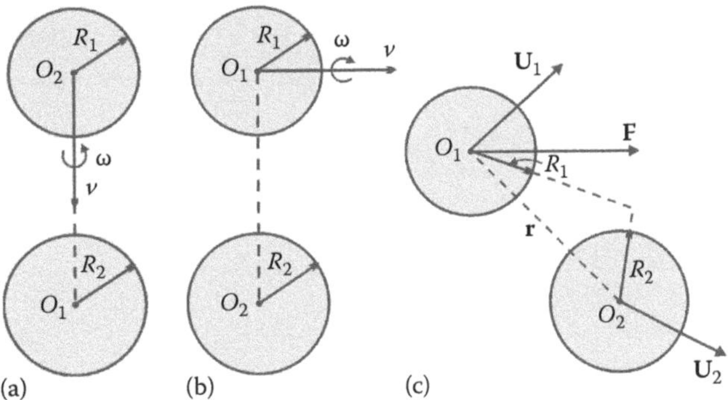

FIGURE 4.44 Types of hydrodynamic interactions between two spherical particles: (a) motion along and rotation around the line of centers; (b) motion along and rotation around an axis perpendicular to the line of centers; (c) the first particle moves under the action of an applied external force, **F**, whereas the second particle is subjected to the hydrodynamic disturbance created by the motion of the first particle.

moving toward each other. A combination of these results permits evaluation of the total forces and torques acting on the particles.

The first particle then moves along an axis perpendicular to the center line and rotates around this axis, whereas the second particle is immobile; see Figure 4.44b (this is a typical 3D hydrodynamic problem). The contribution of this asymmetric motion of the spheres to the resistance tensors was determined by Davis [678] and O'Neill and Majumdar [679].

The first particle moves with linear velocity, $\mathbf{U}_1$, under the action of an applied external force, **F**, whereas the second particle is subjected to the hydrodynamic disturbances (created by the motion of the first particle) and moves with a linear velocity, $\mathbf{U}_2$ (see Figure 4.44c). As a rule, this is a 3D hydrodynamic problem. For this case, Batchelor [683] and Batchelor and Wen [684] have derived the following expressions for the instantaneous translational velocities of the two particles in an otherwise quiescent and unbounded fluid:

$$\mathbf{U}_1 = \frac{\mathbf{F}}{6\pi\eta R_1}\cdot\left[A_{11}(r)\frac{\mathbf{rr}}{r^2} + B_{11}(r)\left(\mathbf{I} - \frac{\mathbf{rr}}{r^2}\right)\right] \tag{4.274}$$

$$\mathbf{U}_2 = \frac{\mathbf{F}}{6\pi\eta(R_1+R_2)}\cdot\left[A_{12}(r)\frac{\mathbf{rr}}{r^2} + B_{12}(r)\left(\mathbf{I} - \frac{\mathbf{rr}}{r^2}\right)\right] \tag{4.275}$$

where $\mathbf{r}$ is the vector connecting the particle centers and $r = |\mathbf{r}|$. Expressions for the mobility functions A_{ij} and B_{ij} $(i, j = 1, 2)$ at large values of the dimensionless distance $s = 2r/(R_1 + R_2)$ and comparable particle radii $\lambda = R_2/R_1 = O(1)$ have been derived by Jeffrey and Onishi [685] and Davis and Hill [682]. The derived far-field expansions arc

$$1 - B_{11} = \frac{68\lambda^5}{(1+\lambda)^6 s^6} + \frac{32\lambda^3(10 - 9\lambda^2 + 9\lambda^4)}{(1+\lambda)^8 s^8} + \frac{192\lambda^5(35 - 18\lambda^2 + 6\lambda^4)}{(1+\lambda)^{10} s^{10}} + O(s^{-12})$$

$$B_{11} - A_{11} = \frac{60\lambda^3}{(1+\lambda)^4 s^4} - \frac{60\lambda^3(8-\lambda^2)}{(1+\lambda)^6 s^6} + \frac{32\lambda^3(20 - 123\lambda^2 + 9\lambda^4)}{(1+\lambda)^8 s^8}$$

$$+ \frac{64\lambda^2(175 + 1500\lambda - 426\lambda^2 + 18\lambda^4)}{(1+\lambda)^{10} s^{10}} + O(s^{-12})$$

$$A_{11} - \frac{2A_{12}}{1+\lambda} = 1 - \frac{3}{(1+\lambda)s} + \frac{4(1+\lambda^2)}{(1+\lambda)^3 s^3} - \frac{60\lambda^3}{(1+\lambda)^4 s^4} + \frac{32\lambda^3(15-4\lambda^2)}{(1+\lambda)^6 s^6} - \frac{2400\lambda^3}{(1+\lambda)^7 s^7}$$
$$- \frac{192\lambda^3(5-22\lambda^2+3\lambda^4)}{(1+\lambda)^8 s^8} + \frac{1920\lambda^3(1+\lambda^2)}{(1+\lambda)^9 s^9} - \frac{256\lambda^5(70-375\lambda-120\lambda^2+9\lambda^3)}{(1+\lambda)^{10} s^{10}}$$
$$- \frac{1536\lambda^3(10-151\lambda^2+10\lambda^4)}{(1+\lambda)^{11} s^{11}} + O(s^{-12}) \tag{4.276}$$

$$B_{11} - \frac{2B_{12}}{1+\lambda} = 1 - \frac{3}{2(1+\lambda)s} - \frac{2(1+\lambda^2)}{(1+\lambda)^3 s^3} - \frac{68\lambda^5}{(1+\lambda)^6 s^6} - \frac{32\lambda^3(10-9\lambda^2+9\lambda^4)}{(1+\lambda)^8 s^8}$$
$$- \frac{192\lambda^5(35-18\lambda^2+6\lambda^4)}{(1+\lambda)^{10} s^{10}} - \frac{16\lambda^3(560-553\lambda^2+560\lambda^4)}{(1+\lambda)^{11} s^{11}} + O(s^{-12})$$

In the case of a small heavy sphere falling through a suspension of large particles (fixed in space), we have $\lambda \gg 1$; the respective expansions, corresponding to Equation 4.276, were obtained by Fuentes et al. [686]. In the opposite case, when $\lambda \ll 1$, the suspension of small background spheres will reduce the mean velocity of a large heavy particle (as compared with its Stokes velocity [687]) because the suspension behaves as an effective fluid of larger viscosity as predicted by the Einstein viscosity formula [683,686].

4.5.2.3 Stages of Thinning of a Liquid Film

Experimental and theoretical investigations [238,246,657,658,663,688,689] show that during the approach of two fluid colloidal particles, a flat liquid film can appear between their closest regions (see Figure 4.32). The hydrodynamic interactions as well as the buoyancy, the Brownian, electrostatic, van der Waals, and steric forces and other interactions can be involved in film formation [207,256,665,690,691]. The formation and the evolution of a foam or emulsion film usually follow the stages shown in Figure 4.45.

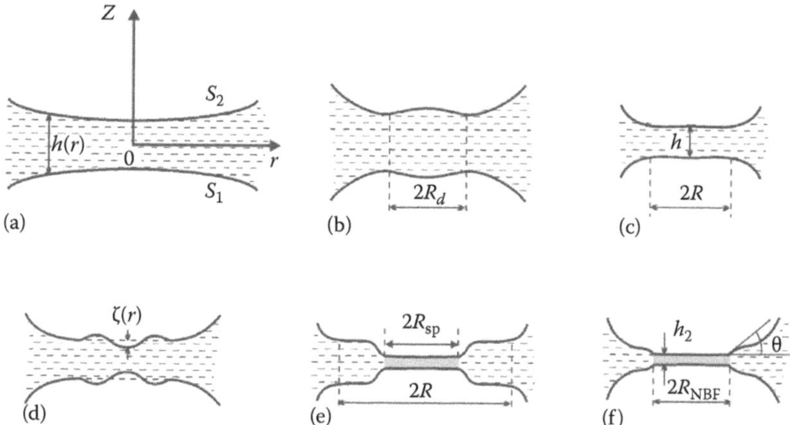

FIGURE 4.45 Main stages of formation and evolution of a thin liquid film between two bubbles or drops: (a) mutual approach of slightly deformed surfaces; (b) at a given separation, the curvature at the center inverts its sign and a dimple arises; (c) the dimple disappears, and eventually an almost plane-parallel film forms; (d) due to thermal fluctuations or other disturbances the film either ruptures or transforms into a thinner Newton black film (e), which expands until reaching the final equilibrium state (f).

Under the action of an outer driving force, the fluid particles approach each other. The hydrodynamic interaction is stronger at the front zones and leads to a weak deformation of the interfaces in this front region. In this case, the usual hydrodynamic capillary number, $Ca = \eta V_z/\sigma$, which is a small parameter for nondeformable surfaces, should be modified to read $Ca = \eta V_z R_*/\sigma h$, where the distance, h, between the interfaces is taken into account. The shape of the gap between two drops for different characteristic times was calculated numerically by many authors [691–711]. Experimental investigation of these effects for symmetric and asymmetric drainage of foam films were carried out by Joye et al. [700,701]. In some special cases, the deformation of the fluid particle can be very fast: for example, the bursting of a small air bubble at an air–water interface is accompanied by a complex motion resulting in the production of a high-speed liquid jet (see Boulton-Stone and Blake [711]).

When a certain small separation, h_i, the inversion thickness, is reached, the sign of the curvature in the contact of the fluid particles (drops, bubbles) changes. A concave lens–shaped formation called a *dimple* is formed (see Frankel and Mysels [712]). This stage is also observed for asymmetric films [701]. A number of theoretical studies have described the development of a dimple at the initial stage of film thinning [691–711,713]. The inversion thickness can be calculated from a simple equation in which the van der Waals interaction is explicitly taken into account (see Section 4.4.2) [240,656,691]

$$h_i = \frac{F_z(\sigma_1+\sigma_2)}{4\pi\sigma_1\sigma_2}\left(1-\frac{A_H R_*}{12F_z h_i}\right) \tag{4.277}$$

where

σ_1 and σ_2 are the interfacial tensions of the phase boundaries S_1 and S_2

in this case F_z is the external force (of nonviscous and non–van der Waals origin) experienced by the approaching particles

A_H is the Hamaker constant

In the case, when the van der Waals force is negligible, Equation 4.277 reduces to $h_i = F_z(\sigma_1 + \sigma_2)/(4\pi\sigma_1\sigma_2)$ [240,656]. Danov et al. [665] have shown that in the case of Brownian flocculation of identical small droplets, h_i obeys the following transcendental equation:

$$h_i = \frac{kT}{2\pi\sigma z_i}\left\{\int_0^{z_i}\left(\frac{z_i}{z}\right)^2\frac{\gamma(z)}{\gamma(z_i)}\exp\left[\frac{U(z)-U(z_i)}{kT}\right]\frac{dz}{z_i}\right\}^{-1} \tag{4.278}$$

where

kT is the thermal energy

$\gamma(z) = F_z/V_z$ is the hydrodynamic resistance given by Equation 4.270

U is the potential energy due to the surface forces (see Equation 4.175)

z is the distance between the droplet mass centers

These authors pointed out that with an increase of droplet size the role of the Brownian force in the film formation decreases, but for micrometer-sized liquid droplets the Brownian force is still by several orders of magnitude greater than the buoyancy force due to gravity. If the driving force is large enough, so that it is able to overcome the energy barrier created by the electrostatic repulsion and/or the increase of the surface area during the droplet deformation, then film with a dimple will be formed. On the contrary, at low electrolyte concentration (i.e., strong electrostatic repulsion) such a dimple might not appear. Parallel experiments [714] on the formation and thinning of emulsion films of macroscopic and microscopic areas, prepared in the Scheludko–Exerowa cell [215,216] and in a miniaturized cell, show that the patterns and the time scales of the film evolution in these two cases are significantly different. There is no dimple formation in the case of thin liquid films of small diameters [714].

In the case of predominant van der Waals attraction, instead of a dimple, a reverse bell-shape deformation, called a *pimple*, appears and the film quickly ruptures [691,698,707,710,713]. The thickness, h_p, at which the pimple appears, can be calculated from the relationship [691]:

$$h_p = \left(\frac{A_H R_*}{12 F_z} \right)^{1/2} \tag{4.279}$$

The pimple formation thickness depends significantly on the radius, R_*. If a drop of tangentially immobile surfaces and radius R_d is driven by the buoyancy force, then we have:

$$F_z = \frac{4}{3} \pi R_d^3 \Delta\rho g \tag{4.280}$$

where

$\Delta\rho$ is the density difference
g is the gravity acceleration

For the collision of this drop with another immobile one, we have $h_p^2 = A_H/(16\pi\Delta\rho g R_d^2)$. We see that h_p is inversely proportional to the drop radius. For typical values of the Hamaker constant $A_H = 4 \times 10^{-20}$ J, density difference $\Delta\rho = 0.12$ g/cm^3, and $R_d = 10$ μm, the thickness of pimple formation is $h_p = 82.3$ nm. Note that this thickness is quite large. The pimple formation can be interpreted as the onset of instability without fluctuations (stability analysis of the film intervening between the drops has been carried out elsewhere [62]).

As already mentioned, if the van der Waals force (or other attractive force) is not predominant, first a dimple forms in the thinning liquid films. Usually the dimple exists for a short period of time; initially it grows, but as a result of the swift outflow of liquid it decreases and eventually disappears. The resulting plane-parallel film thins at almost constant radius R. When the electrostatic repulsion is strong, a thicker primary film forms (see point 1 in Figure 4.17). From the viewpoint of conventional DLVO theory, this film must be metastable. Indeed, the experiments with microscopic foam films, stabilized with sodium octyl sulfate or SDS in the presence of different amount of electrolyte [715], show that a black spot may suddenly form and a transition to secondary (Newton black) film may occur (see point 2 in Figure 4.17). The rate of thinning depends not only on the capillary pressure (the driving force) but also very strongly on the surfactant concentration (for details, see Section 4.5.3.2).

The appearance of a secondary film (or film rupture, if the secondary film is not stable) is preceded by corrugation of the film surfaces due to thermally excited fluctuations or outer disturbances. When the derivative of the disjoining pressure, $\partial\Pi/\partial h$, is positive, the amplitude of the fluctuations (ζ in Figure 4.45d) spontaneously grows. As already mentioned, the instability leads to rupture of the film or to formation of black spots. The theory of film stability was developed by de Vries [716], Vrij [717], Felderhof [648], Sche and Fijnaut [649], Ivanov et al. [718], Gumerman and Homsy [719], Malhotra and Wasan [720], Maldarelli and Jain [650], and Valkovska et al. [721]. On the basis of the lubrication approximation for tangentially immobile surfaces, Ivanov et al. [718] and Valkovska et al. [721] derived a general expression for the critical film thickness, h_{cr}, by using long-waves stability analysis:

$$h_{cr} = h_{tr} \left(\frac{\sigma h_{tr}^2}{kT} \right)^{1/4} \exp\left(-\frac{k_{cr}^2 R^2}{32 h_{cr}^3} \int_{h_{cr}}^{h_{tr}} \frac{h^3 \Pi'}{P_c - \Pi} dh \right) \tag{4.281}$$

where k_{cr} is the wave number of the critical wave defined as

$$k_{cr}^2 = \frac{(1/\sigma)\int_{h_{cr}}^{h_{tr}} (h^3\Pi'/(P_c - \Pi))dh}{\int_{h_{cr}}^{h_{tr}} h^6/(P_c - \Pi)dh} \tag{4.282}$$

In Equation 4.282, h_{tr} is the so-called transitional thickness [717,718,721] at which the increase of free energy due to the increased film area and the decrease of free energy due to the van der Waals interaction in the thinner part (Figure 4.45d) compensate each other. At h_{tr} the most rapidly growing fluctuation (the critical wave) becomes unstable. The transitional thickness obeys the following equation [718,721]:

$$\frac{24h_{cr}^3[P_c - \Pi(h_{tr})]}{R^2 k_{cr}^2 h_{tr}^4} + \frac{\sigma k_{cr}^2 h_{tr}^3}{2h_{cr}^3} = \Pi'(h_{tr}) \tag{4.283}$$

Figures 4.46 and 4.47 show the critical thicknesses of rupture, h_{cr}, for foam and emulsion films, respectively, plotted versus the film radius [722]. In both cases the film phase is the aqueous phase, which contains 4.3×10^{-4} M SDS + added NaCl. The emulsion film is formed between two toluene drops. Curve 1 is the prediction of a simpler theory, which identifies the critical thickness with the transitional one [720]. Curve 2 is the theoretical prediction of Equations 4.281 through 4.283 (no adjustable parameters); in Equation 4.182 for the Hamaker constant the electromagnetic retardation effect has also been taken into account [404]. In addition, Figure 4.48 shows the experimental dependence of the critical thickness versus the concentration of surfactant (dodecanol) for aniline films. Figures 4.46 through 4.48 demonstrate that when the film area increases and/or the electrolyte concentration decreases the critical film thickness becomes larger. Figure 4.49 shows the critical thickness of foam film rupture for three concentrations of SDS in the presence of 0.3 M NaCl [605]. The dashed and dash-dotted lines, for 1 and 10 μM SDS, respectively, are computed assuming only the van der Waals attraction (no adjustable parameter). The deviation of the predicted values of h_{cr}

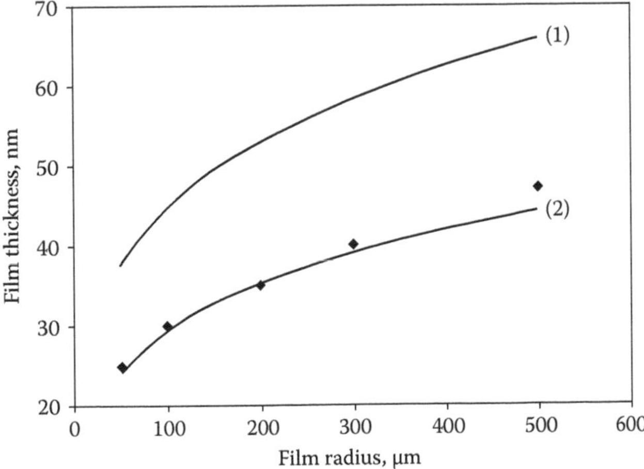

FIGURE 4.46 Dependence of the critical thickness, h_{cr}, on the radius, R, of foam films. (The experimental points are data from Manev, E.D. et al., *J. Colloid Interface Sci.*, 97, 591, 1984.) The films are formed from a solution of 4.3×10^{-4} M SDS + 0.25 M NaCl. Curve 1 is the prediction of the simplified theory [720], whereas Curve 2 is calculated using Equations 4.281 through 4.283; no adjustable parameters.

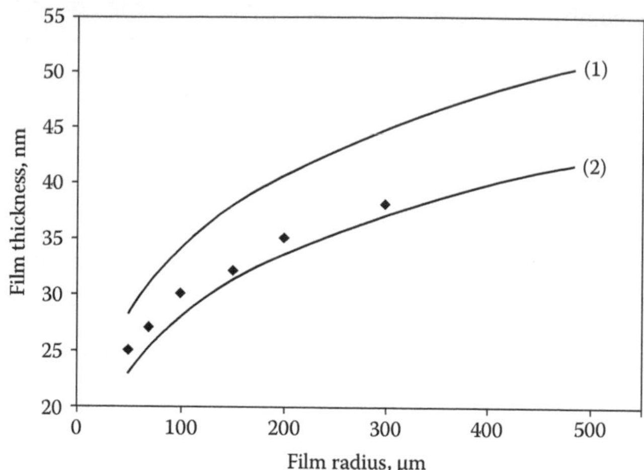

FIGURE 4.47 Critical thickness, h_{cr}, vs. radius, R, of emulsion films, toluene–water–toluene. The experimental points are data from Ref. [722]; the films are formed from a solution of 4.3×10^{-4} M SDS + 0.1 M NaCl. Curve 1 is the prediction of the simplified theory [720], whereas Curve 2 is calculated using Equations 5.281 through 5.283; no adjustable parameters.

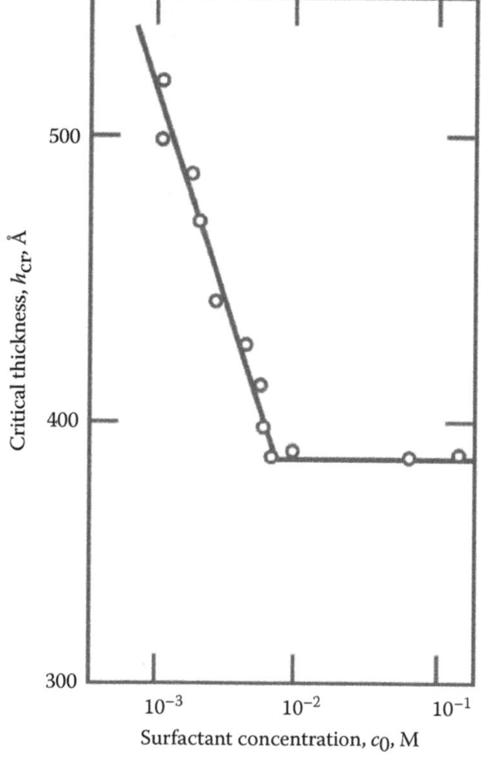

FIGURE 4.48 Dependence of the critical thickness, h_{cr}, of aniline films on the concentration of dodecanol, c_0 [723]. (Modified from Ivanov, I.B., *Pure Appl. Chem.*, 52, 1241, 1980.)

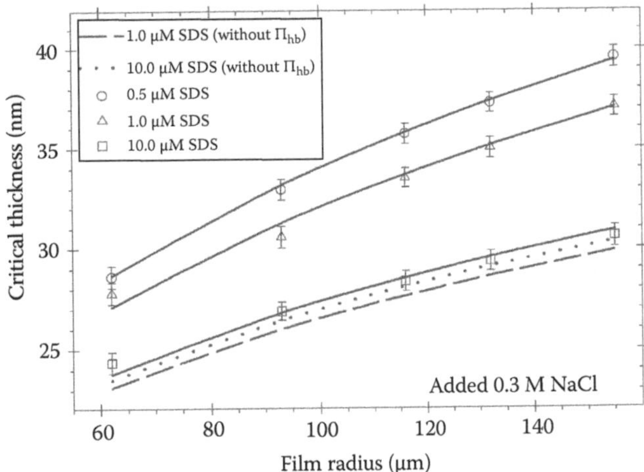

FIGURE 4.49 Critical thickness, h_{cr}, vs the film radius at a 0.3 M fixed concentration of NaCl for three SDS concentrations: 0.5; 1; 10 μM. The dashed and dash-dotted lines, for 1 and 10 μM SDS, respectively, are computed using an absence of the hydrophobic attraction. The solid lines are fits of the experimental points with $\lambda_0 = 15.8$ nm.

from the measured is because of the hydrophobic interaction (Section 4.4.5.4.2). The solid lines represent fits with the decay length of the hydrophobic interactions $\lambda_0 = 15.8$ nm using Equations 4.281 through 4.283.

The surface corrugations do not necessarily lead to film rupture. Instead, black spots (secondary films of very low thickness; h_2 in Figure 4.17) can be formed. The typical thickness of plane-parallel films at stage c (Figure 4.45c) is about 200 nm, while the characteristic thickness h_2 of the Newton black film (Figure 4.45e and f) is about 5–10 nm. The black spots either coalesce or grow in diameter, forming an equilibrium secondary (Newton black) film with a thickness h_2 and radius R_{sp}. These spots grow until they cover the whole film area.

After the entire film area is occupied by the Newton black film, the film radius increases until it reaches its equilibrium value, $R = R_{NBF}$ (Figure 4.45f). Finally, the equilibrium contact angle is established. For more details about this last stage of film thinning, see part IV.C of Ref. [240].

4.5.2.4 Dependence of Emulsion Stability on the Droplet Size

Experimental data [724,725] show that the emulsion stability correlates well with the lifetime of separate thin emulsion films or of drops coalescing with their homophase. To simplify the treatment we will consider here the lifetime of a single drop pressed against its homophase under the action of gravity. To define the *lifetime* (or drainage time) τ, we assume that in the initial and final moments the film has some known thicknesses h_{in} and h_f:

$$\tau = \int_{h_f}^{h_{in}} \frac{dh}{V_z} = \frac{3\pi\eta R_*^2}{2F_z}\left[\ln\left(\frac{h_{in}}{h_f}\right) + \frac{R^2}{h_f R_*}\left(1 - \frac{h_f}{h_{in}}\right) + \frac{R^4}{2h_f^2 R_*^2}\left(1 - \frac{h_f^2}{h_{in}^2}\right)\right] \tag{4.284}$$

The final thickness, h_f, may coincide with the critical thickness of film rupture. Equation 4.284 is derived for tangentially immobile interfaces from Equation 4.270 at a fixed driving force (no disjoining pressure).

In the case of gravity-driven coalescence of a droplet with its homophase, the driving force is given by Equation 4.280 and the mean drop radius is $R_* = 2R_d$. Then from Equations 4.280 and

4.284 we can deduce the droplet lifetime in the so-called Taylor regime, corresponding to nondeformed droplets ($R = 0$):

$$\tau_{\text{Ta}} = \frac{6\pi\eta R_d^2}{F_z}\ln\left(\frac{h_{\text{in}}}{h_f}\right) = \frac{9\eta}{2gR_d\Delta\rho}\ln\left(\frac{h_{\text{in}}}{h_f}\right) \tag{4.285}$$

We see that τ_{Ta} depends logarithmically on the ratio of the initial and final thickness. Moreover, in the Taylor regime the lifetime, τ, decreases with the increase of the driving force, F_z, and the drop radius, R_d. The latter fact is confirmed by the experimental data of Dickinson et al. [726].

In the case of deformed drops ($R \neq 0$), the drainage time, τ, is determined by Equation 4.284, and in such a case the fluid particles approach each other in the Reynolds regime [657,723]. The dependence of τ on R_d in Equation 4.284 is very complex, because the driving force, F_z, and the film radius, R, depend on R_d. The film radius can be estimated from the balance of the driving and capillary force [657,723]:

$$R^2 = \frac{F_z R_d}{2\pi\sigma} \tag{4.286}$$

In this regime, the lifetime, τ, increases with an increase of the driving force, F_z. This is exactly the opposite trend compared to the results for the Taylor regime (see Equation 4.285). The result can be rationalized in view of Reynolds equation (Equation 4.273). In the numerator of this equation, $F_z \propto R_d^3$, whereas in the denominator $R^4 \propto R_d^8$ (see Equation 4.286); as a result, the drainage rate becomes proportional to R_d^{-5}, that is, V_z decreases as the droplet radius increases.

The numerical results from Equations 4.284 through 4.286 for the lifetime or drainage time, τ, versus the droplet radius, R_d, are plotted in Figure 4.50 for parameter values typical for emulsion systems: $\Delta\rho = 0.2$ g/cm^3, $\eta = 1$ cP, $h_f = 5$ nm, and $h_{\text{in}} = R_d/10$. The various curves in Figure 4.50 correspond to different values of the surface tension, σ, shown in the Figure 4.50. The left branches of the curves correspond to the Taylor regime (nondeformed droplets), whereas the right branches correspond to the Reynolds regime (formation of film between the droplets). The presence of a deep minimum on the τ versus R_d curve was first pointed out by Ivanov [727,728]. The theoretical

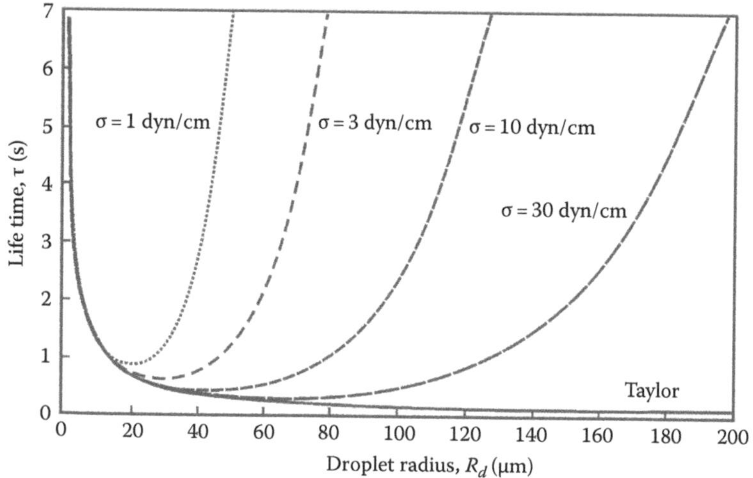

FIGURE 4.50 Calculated lifetime, τ, of drops approaching a fluid interface in Taylor regime (the solid line) and in Reynolds regime (the other lines) as a function of the droplet radius, R_d.

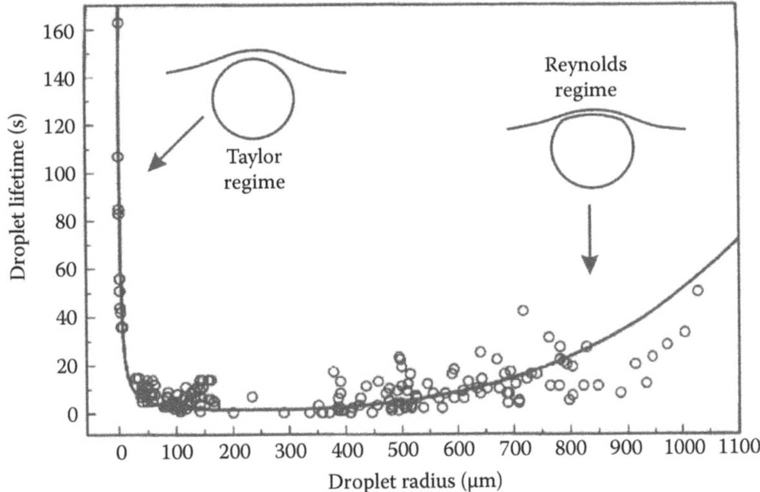

FIGURE 4.51 Stability of oil drops pressed by buoyancy against a large oil–water interface. Measured lifetime, (the points), is plotted vs droplet radius in a system consisting of soybean oil and aqueous solution of $4{\cdot}10^{-4}$ wt% BSA + 0.15 M NaCl (pH = 6.4). The solid line is drawn in accordance to Equation 4.284.

dependencies in Figure 4.50 agree well with experimental data [652,729–731] for the lifetime of oil droplets pressed by the buoyancy force against a large oil–water interface in a system containing protein BSA (Figure 4.51).

4.5.3 Effect of Surface Mobility

The hydrodynamic interactions between fluid particles (drops, bubbles) suspended in a liquid medium depend on the interfacial mobility. In the presence of surfactants, the bulk fluid motion near an interface disturbs the homogeneity of the surfactant adsorption monolayer. The ensuing surface tension gradients act to restore the homogeneous equilibrium state of the monolayer. The resulting transfer of adsorbed surfactant molecules from the regions of lower surface tension toward the regions of higher surface tension constitutes the Marangoni effect. The analogous effect, for which the surface tension gradient is caused by a temperature gradient, is known as the Marangoni effect of thermocapillarity. In addition, the interfaces possess specific surface rheological properties (surface elasticity and dilatational and shear surface viscosities), which give rise to the so-called Boussinesq effect (see the following section) [732].

4.5.3.1 Diffusive and Convective Fluxes at an Interface—Marangoni Effect

To take into account the influence of surfactant adsorption, Equations 4.251 and 4.252 are to be complemented with transport equations for each of the species (k = 1, 2, …, N) in the bulk phases [631,634,641,662,663]

$$\frac{\partial c_k}{\partial t} + \operatorname{div}(c_k \mathbf{v} + \mathbf{j}_k) = r_k \quad (k = 1,2,\ldots,N) \tag{4.287}$$

where

c_k and $\mathbf{j}_k$ are bulk concentration and flux, respectively, of the kth species—note that $\mathbf{j}_k$ includes the molecular diffusive flux, the flux driven by external forces (e.g., electrodiffusion [651,662,663]) and the thermodiffusion flux [662]

r_k is the rate of production due to chemical reactions, including surfactant micellization or micelle decay

The surface mass balance equation for the adsorption, Γ_k, has the form [651,662,663]:

$$\frac{\partial \Gamma_k}{\partial t} + \nabla_s \cdot (\Gamma_k \mathbf{v}_s + \mathbf{j}_k^s) = r_k^s + \mathbf{n} \cdot \langle \mathbf{j}_k \rangle \tag{4.288}$$

where

$\mathbf{n}$ is the unit normal to the interface directed from phase 1 to phase 2
$\langle\ \rangle$ denotes the difference between the values of a given physical quantity at the two sides of the interface
∇_s is the surface gradient operator [733]
$\mathbf{v}_s$ is the local material surface velocity
$\mathbf{j}_k^s$ is the 2D flux of the kth component along the interface
r_k^s accounts for the rate of production of the kth component due to interfacial chemical reactions and could include conformational changes of adsorbed proteins

Equation 4.288 provides a boundary condition for the normally resolved flux, $\mathbf{j}_k$. From another viewpoint, Equation 4.288 represents a 2D analogue of Equation 4.287. The interfacial flux, $\mathbf{j}_k^s$, can also contain contributions from the interfacial molecular diffusion, electrodiffusion, and thermo-diffusion. A simple derivation of the time-dependent convective-diffusion equation for surfactant transport along a deforming interface is given by Brenner and Leal [734–737], Davis et al. [669], and Stone [738]. If the molecules are charged, the bulk and surfaces electrodiffusion fluxes can be expressed in the form [651,739,740]:

$$\mathbf{j}_k = -D_k(\nabla c_k + z_k c_k \nabla \Phi), \quad \mathbf{j}_k^s = -D_k^s(\nabla_s \Gamma_k + z_k \Gamma_k \nabla_s \Phi) \tag{4.289}$$

for the bulk and interfacial phase. Here, D_k and D_k^s are the bulk and surface collective diffusion coefficients, respectively, which are connected with the diffusion coefficients of individual molecules, $D_{k,0}$ and $D_{k,0}^s$, through the relationship [740]

$$D_k = \frac{D_{k,0}}{kT}\frac{K_b(\phi_k)}{(1-\phi_k)}\frac{\partial \mu_k}{\partial \ln \phi_k}, \quad D_k^s = \frac{D_{k,0}^s}{kT} K_s(\Gamma_k)\frac{\partial \mu_k^s}{\partial \ln \Gamma_k} \tag{4.290}$$

where μ_k and μ_k^s are the bulk and surface chemical potentials, respectively. The dimensionless bulk friction coefficient, K_b, accounts for the change in the hydrodynamic friction between the fluid and the particles (created by the hydrodynamic interactions between the particles). The dimensionless surface mobility coefficient, K_s, accounts for the variation of the friction of a molecule in the adsorption layer. Feng [741] has determined the surface diffusion coefficient, the dilatational elasticity, and the viscosity of a surfactant adsorption layer by theoretical analysis of experimental data. Stebe and Maldarelli [742,743] studied theoretically the surface diffusion driven by large adsorption gradients. The determination of bulk and surface diffusion coefficients from experimental data for the drainage of nitrobenzene films stabilized by different concentrations of dodecanol was reported [739].

Note that the adsorption isotherms, relating the surface concentration, Γ_k, with the subsurface value of the bulk concentration, c_k (see Section 4.2.2.1), or the respective kinetic Equation 4.86 for adsorption under barrier control (see Section 4.2.2.5), should also be employed in the computations based on Equations 4.287 through 4.290 in order for a complete set of equations to be obtained.

Another boundary condition is the equation of the interfacial momentum balance [634,635,641,663]:

$$\nabla_s \cdot \sigma = \mathbf{n} \cdot \langle \mathbf{P} + \mathbf{P}_b \rangle \tag{4.291}$$

where $\boldsymbol{\sigma}$ is the interfacial stress tensor, which is a 2D counterpart of the bulk stress tensor, **P**. Moreover, a 2D analogue of Equations 4.253, 4.256, and 4.257, called the Boussinesq–Scriven constitutive law, can be postulated for a fluid interface [240,641,663,732,744–748]:

$$\sigma = \sigma_a \mathbf{I}_s + (\eta_{\mathrm{dl}} - \eta_{\mathrm{sh}})(\nabla_s \cdot \mathbf{v}_s)\mathbf{I}_s + \eta_{\mathrm{sh}}\left[(\nabla_s \mathbf{v}_s)\cdot \mathbf{I}_s + \mathbf{I}_s \cdot (\nabla_s \mathbf{v}_s)^T\right] \quad (4.292)$$

where

η_{dl} and η_{sh} are the interfacial dilatational and shear viscosities, respectively
$\mathbf{I}_s$ is the unit surface idemfactor [733]
σ_a is the scalar adsorption part of the surface tension (see Section 4.2.1.2.2)

In view of the term $\sigma_a \mathbf{I}_s$ in Equation 4.292, the Marangoni effects are hidden in the left-hand side of the boundary condition (Equation 4.291) through the surface gradient of σ_a:

$$\nabla_s \sigma_a = -\sum_{k=1}^{N} \frac{E_k}{\Gamma_k} \nabla_s \Gamma_k - \frac{E_T}{T} \nabla_s T, \quad E_k = -\left(\frac{\partial \sigma_a}{\partial \ln \Gamma_k}\right)_{T,\Gamma_{j \neq k}}, \quad E_T = -\left(\frac{\partial \sigma_a}{\partial \ln T}\right)_{\Gamma_k} \quad (4.293)$$

where

E_k is the Gibbs elasticity for the kth surfactant species (see Equation 4.6)
E_T represents the thermal analogue of the Gibbs elasticity

The thermocapillary migration of liquid drops or bubbles and the influence of E_T on their motion are investigated in a number of works [749–751].

In fact, Equation 4.292 describes an interface as a 2D Newtonian fluid. On the other hand, a number of non-Newtonian interfacial rheological models have been described in the literature [752–773]. Tambe and Sharma [756] modeled the hydrodynamics of thin liquid films bounded by viscoelastic interfaces, which obey a generalized Maxwell model for the interfacial stress tensor. These authors [757,758] also presented a constitutive equation to describe the rheological properties of fluid interfaces containing colloidal particles. A new constitutive equation for the total stress was proposed by Horozov et al. [759], Danov et al. [760], and Ivanov et al. [761] who applied a local approach to the interfacial dilatation of adsorption layers.

The interfacial rheology of protein adsorption layers has been intensively studied in relation to the properties of foams and emulsions stabilized by proteins and their mixtures with lipids or surfactants. Detailed information on the investigated systems, experimental techniques, and theoretical models can be found in Refs. [762–769]. The shear rheology of the adsorption layers of many proteins follows the viscoelastic thixotropic model [770–772], in which the surface shear elasticity and viscosity depend on the surface shear rate. The surface rheology of saponin adsorption layers has been investigated in Ref. [773].

If the temperature is not constant, the bulk heat transfer equation complements the system and involves Equations 4.251, 4.252, and 4.287. The heat transfer equation is a special case of the energy balance equation. It should be noted that more than 20 various forms of the overall differential energy balance for multicomponent systems are available in the literature [631,634]. The corresponding boundary condition can be obtained as an interfacial energy balance [663,748]. Based on the derivation of the bulk [774,775] and interfacial [760,775,776] entropy inequalities (using the Onsager theory), various constitutive equations for the thermodynamic mass, heat, and stress fluxes have been obtained.

4.5.3.2 Fluid Particles and Films of Tangentially Mobile Surfaces

When the surface of an emulsion droplet is mobile, it can transmit the motion of the outer fluid to the fluid within the droplet. This leads to a special pattern of the fluid flow and affects the dissipation of energy in the system. The problem concerning the approach of two nondeformed (spherical) drops or bubbles of pure phases has been investigated by many authors [657,685,686,692,693,777–782]. A number of solutions, generalizing the Taylor equation (Equation 4.271), have been obtained. For example, the velocity of central approach, V_z, of two spherical drops in pure liquid is related to the hydrodynamic resistance force, F_z, by means of a Padé-type expression derived by Davis et al. [692]:

$$V_z = \frac{2hF_z}{3\pi\eta R_*^2}\frac{1+1.711\xi+0.461\xi^2}{1+0.402\xi}, \quad \xi = \frac{\eta}{\eta_d}\sqrt{\frac{R_*}{2h}} \tag{4.294}$$

where

h is the closest surface-to-surface distance between the two drops
η_d is the viscosity of the disperse phase (the liquid in the droplets)

In the limiting case of solid particles, we have $\eta_d \to \infty$, and Equation 4.294 reduces to the Taylor equation (Equation 4.271). Note that in the case of close approach of two drops ($\xi \gg 1$), the velocity V_z is proportional to $\sqrt{h}$. This implies that the two drops can come into contact ($h = 0$) in a finite period of time ($\tau < \infty$) under the action of a given force, F_z, because the integral in Equation 4.284 is convergent for $h_f = 0$. This is in contrast to the case of immobile interfaces ($\xi \ll 1$), when $V_z \propto h$ and $\tau \to \infty$ for $h_f \to 0$.

In the other limiting case, that of two nondeformed gas bubbles ($\eta_d \to 0$) in pure liquid, Equation 4.294 cannot be used; instead, V_z can be calculated from the expression [782,783]

$$V_z = \frac{F_z}{\pi\eta R_d[\ln(R_d/h+1)+2.5407]} \tag{4.295}$$

Note that in this case $V_z \propto (\ln h)^{-1}$, and the integral in Equation 4.284 is convergent for $h_f \to 0$. In other words, the theory predicts that the lifetime, τ, of a doublet of two colliding spherical bubbles in pure liquid is finite. Of course, the real lifetime of a doublet of bubbles or drops is affected by the surface forces for $h < 100$ nm, which should be accounted for in F_z and which may lead to the formation of thin film in the zone of contact [207,392].

In the case of droplets with equal radii, R_d, in a pure liquid (without surfactant), two asymptotic expressions are derived for small interdroplet distances [783]:

1. At not very large droplet viscosity:

$$\frac{F_z}{\pi\eta R_d V_z} = \frac{3\pi^2\eta_d}{16\eta}\sqrt{\frac{R_d}{h}} + \left(1-\frac{\eta_d^2}{3\eta^2}\right)\ln\left(\frac{R_d}{h}+1\right)+2.5407 \quad \text{at } \frac{\eta_d}{\eta}\sqrt{\frac{R_d}{h}}<1 \tag{4.296}$$

2. At very large viscosity of the drop phase:

$$\frac{F_z}{\pi\eta R_d V_z} = \frac{3R_d}{2h} - \frac{9\pi^2\eta}{64\eta_d}\left(\frac{R_d}{h}\right)^{3/2} \quad \text{at } \frac{\eta_d}{\eta}\sqrt{\frac{h}{R_d}} \gg 1 \tag{4.297}$$

Note that for $\eta_d = 0$ Equation 4.296 reduces to Equation 4.295. The second term in the right-hand side of Equation 4.297 represents a correction to the Taylor equation (Equation 4.271).

Next, let us consider the case of deformed fluid particles (Figure 4.32). A number of theoretical studies [784–787] have been devoted to the thinning of plane-parallel liquid films of pure liquid phases (no surfactant additives). Ivanov and Traykov [786] derived the following exact expressions for the velocity of thinning of an emulsion film:

$$V_z = \left(\frac{32\Delta P^2}{\rho_d \eta_d R^4} \right)^{1/3} h^{5/3}, \quad \frac{V_z}{V_{\mathrm{Re}}} = \frac{1}{\varepsilon_e}, \quad \varepsilon_e \equiv \left(\frac{\rho_d \eta_d h^4 F_z}{108\pi\eta^3 R^4} \right)^{1/3} \tag{4.298}$$

where

ρ_d is the density of the disperse phase
V_{Re} is the Reynolds velocity defined by Equation 4.273
ε_e is the so-called emulsion parameter

Substituting typical parameter values in Equations 4.294 and 4.298 we can check that at a given constant force the velocity of thinning of an emulsion film is smaller than the velocity of approach of two nondeformed droplets and much larger than V_{Re}. It is interesting to note that the velocity of thinning as predicted by Equation 4.298 does not depend on the viscosity of the continuous phase, η, and its dependence on the drop viscosity, η_d, is rather weak. There are experimental observations confirming this prediction (see Ref. [34], p. 381).

The presence of surfactant adsorption monolayers decreases the mobility of the droplet (bubble) surfaces. This is due to the Marangoni effect (see Equation 4.293). From a general viewpoint, we may expect that the interfacial mobility will decrease with the increase of surfactant concentration until eventually the interfaces become immobile at high surfactant concentrations (see Section 4.5.2); therefore, a pronounced effect of surfactant concentration on the velocity of film drainage should be expected. This effect really exists (see Equation 4.299), but in the case of emulsions it is present only when the surfactant is predominantly soluble in the continuous phase.

Traykov and Ivanov [787] established (both theoretically and experimentally) the interesting effect that when the surfactant is dissolved in the disperse phase (i.e., in the emulsion droplets), the droplets approach each other just as in the case of pure liquid phases, that is, Equation 4.298 holds. Qualitatively, this effect can be attributed to the fact that the convection-driven surface tension gradients are rapidly damped by the influx of surfactant from the drop interior; in this way, the Marangoni effect is suppressed. Indeed, during the film drainage the surfactant is carried away toward the film border, and a nonequilibrium distribution depicted in Figure 4.52a appears. Because, however, the mass transport is proportional to the perturbation, the larger the deviation from equilibrium, the stronger the flux tending to eliminate the perturbation (the surfactant flux is denoted by thick arrows

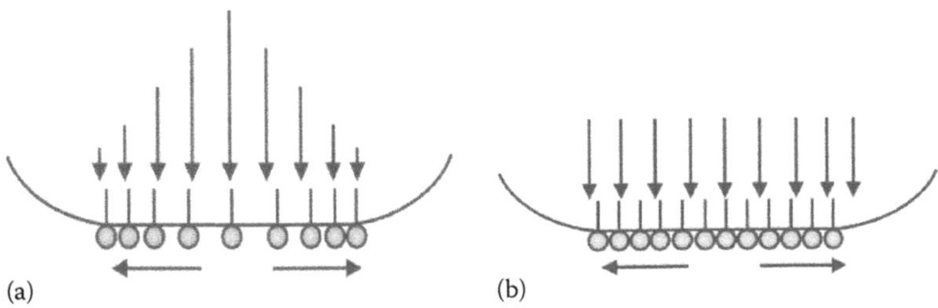

FIGURE 4.52 Damping of convection-driven surface tension gradients by influx of surfactant from the drop interior. (a) Since the mass transport is proportional to the perturbation, the larger the perturbation, the stronger the flux tending to eliminate it. (b) Uniform surfactant distribution is finally reached.

in Figure 4.52b). In this way, any surface concentration gradient (and the related Marangoni effect) disappears. The emulsion films in this case behave as if surfactant is absent.

In the opposite case, when the surfactant is soluble in the continuous phase, the Marangoni effect becomes operative and the rate of film thinning becomes dependent on the surface (Gibbs) elasticity (see Equation 4.293). Moreover, the convection-driven local depletion of the surfactant monolayers in the central area of the film surfaces gives rise to fluxes of bulk and surface diffusion of surfactant molecules. The exact solution of the problem [651,655,689,739,740,787] gives the following expression for the rate of thinning of symmetrical planar films (of both foam and emulsion type):

$$\frac{V_z}{V_{\text{Re}}} = 1 + \frac{1}{\varepsilon_e + \varepsilon_f}, \quad \frac{1}{\varepsilon_f} = \frac{6\eta D_s}{hE_G} + \frac{3\eta D}{\Gamma(\partial\sigma/\partial c)} \tag{4.299}$$

where

as usual, D and D_s are the bulk and interfacial collective diffusion coefficients (see Equation 4.290)
E_G is the Gibbs elasticity
ε_f is the so-called foam parameter [723]

In the special case of foam film, one substitutes $\varepsilon_e = 0$ in Equation 4.299. Note that the diffusive surfactant transport, which tends to restore the uniform adsorption monolayers, damps the surface tension gradients (which oppose the film drainage) and thus accelerates the film thinning. However, at large surfactant concentrations, the surface elasticity, E_G, prevails, ε_f increases, and, consequently, the thinning rate decreases down to the Reynolds velocity, $V_z \to V_{\text{Re}}$ (see Equation 4.299). Similar expressions for the rate of film thinning, which are appropriate for various ranges of values of the interfacial parameters, can be found in the literature [240,655,656,703,788,789]. A table describing the typical ranges of variation of the interfacial properties (Γ, E_G, D, D_s, $\partial\sigma/\partial c$, etc.) for emulsion and foam systems can be found in Ref. [240], Table 4.2 therein. For $h < 100$ nm, the influence of the disjoining pressure should be taken into account (see Equation 4.273). In some studies [240,666,756,790–793], the effect of the interfacial viscosity on the rate of thinning and the lifetime of plane-parallel films is investigated; this effect is found to decrease when the film thickness, h, becomes smaller and/or the film radius, R, becomes larger.

Note that Equation 4.299 does not hold in the limiting case of foam films ($\varepsilon_e = 0$) at low surfactant concentration, $\varepsilon_f \to 0$. The following expression is available for this special case [723]:

$$\frac{V_z}{V_{\text{Re}}} = \frac{(1 + 1/\varepsilon_f)}{[1 + 4h^2/(3R^2\varepsilon_f)]} \tag{4.300}$$

The merit of this equation is that it gives as limiting cases both V_z/V_{Re} for foam films without surfactant, $\varepsilon_f \to 0$, and Equation 4.299 with $\varepsilon_e = 0$ (note that in the framework of the lubrication approximation, used to derive Equation 4.299, the terms $\propto h^2/R^2$ are being neglected). Equation 4.300 has also some shortcomings, which are discussed in Ref. [723].

Another case, which is not described by the above equations mentioned earlier, is the approach of two nondeformed (spherical) bubbles in the presence of surfactant. The velocity of approach in this case can be described by means of the expression [656,666,728,740]:

$$\frac{V_z}{V_{\text{Ta}}} = \frac{h_s}{2h}\left\{\left[\frac{h(1+b)}{h_s} + 1\right]\ln\left[\frac{h_s}{h(1+b)} + 1\right] - 1\right\}^{-1} \tag{4.301}$$

where the parameters b and h_s account for the influence of bulk and surface diffusivity of surfactants, respectively. From Equation 4.290 these parameters are calculated to be [740]

$$b \equiv \frac{3\eta c D_0 K_b(\phi)}{kT\Gamma^2(1-\phi)}, \quad h_s \equiv \frac{6\eta D_0^s K_s(\Gamma)}{kT\Gamma} \tag{4.302}$$

A generalization of Equation 4.301 to the more complicated case of two nondeformed (spherical) emulsion droplets with account for the influence of surface viscosity and the solubility of surfactants in both phases has been published in Ref. [783]. The terms related to the surface viscosities K and E, and the surface elasticity N_{el} are scaled with the drop radius, R_d, as follows [783]:

$$K \equiv \frac{\eta_{dl}}{\eta R_d}, \quad E \equiv \frac{\eta_{sh}}{\eta R_d}, \quad N_{el} \equiv \frac{E_G R_d}{\eta D_s} \tag{4.303}$$

Thus, with the increase of the drop radius the surface viscosity becomes less important and the suppression of surface mobility by the Marangoni effect increases. Figure 4.53 shows the dependence of the drag force coefficient, $f = F_z/(\pi\eta V_z R_d)$, on the dimensionless distance between the droplets, h/R_d. For surfactant concentrations above the CMC, the surfactant relaxation time is so small that the interfacial tension changes insignificantly during the motion of the drops. In this case, the drag force coefficient depends on the bulk and surface viscosities (Figure 4.53a). One sees that with the increase of K and E the drag force changes from the values for two bubbles, f_0 (see Equation 4.295), to those for tangentially immobile drop surfaces, f_{im} (see Equation 4.297). If the effect of bulk and surface viscosities is negligible, then f depends on the surface elasticity, N_{el} (Figure 4.53b). Note that the effect of surface elasticity is inversely proportional to the surface diffusion coefficient (Equation 4.303). A faster surface diffusion suppresses the gradients of the surface tension and decreases N_{el}.

Returning to the parameter values, we note that usually $\varepsilon_e \ll \varepsilon_f$ and $\varepsilon_e \ll 1$. Then, comparing the expressions for V_z/V_{Re} as given by Equations 4.298 and 4.299, we conclude that the rate of thinning is much greater when the surfactant is dissolved in the droplets (the disperse phase) in comparison with the case when the surfactant is dissolved in the continuous phase. This prediction of the theory was verified experimentally by measuring the number of films that rupture during a given period of time [794], as well as the rate of thinning. When the surfactant was dissolved in the drop phase, the average lifetime was the same for all surfactant concentrations (Figure 4.54a), in agreement with Equation 4.298. For the emulsion film with the same, but inverted, liquid phases (the former continuous phase becomes disperse phase, and vice versa), that is, the surfactant is in the film phase, the average lifetime is about 70 times longer—compare curves in Figure 4.54a with curve B in Figure 4.54b. The theoretical conclusions have been also checked and proved in experimental measurements with nitroethane droplets dispersed in an aqueous solution of the cationic surfactant hexadecyl trimethyl ammonium chloride (HTAC) [725].

4.5.3.3 Bancroft Rule for Emulsions

There have been numerous attempts to formulate simple rules connecting the emulsion stability with the surfactant properties. Historically, the first one was the Bancroft rule [795], which states that "to have a *stable* emulsion the surfactant must be soluble in the continuous phase." A more sophisticated criterion was proposed by Griffin [796] who introduced the concept of hydrophilic–lipophilic balance (HLB). As far as emulsification is concerned, surfactants with an HLB number in the range of 3–6 must form water-in-oil (W/O) emulsions, whereas those with HLB numbers from

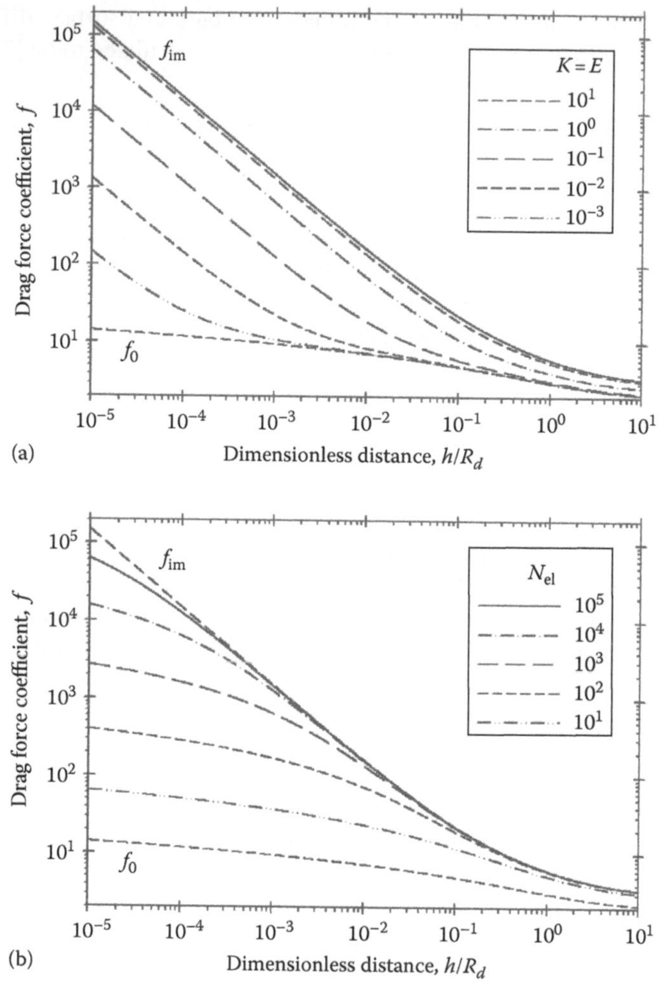

FIGURE 4.53 Dependence of the drag force coefficient, f, on the dimensionless distance, h/R_d. (a) For surfactant concentrations above the CMC at different surface viscosities. (b) For different values of the surface elasticity, the effects of surface viscosities and the viscosity of drop phase are neglected.

8 to 18 are expected to form oil-in-water (O/W) emulsions. Different formulae for calculating the HLB numbers are available; for example, the Davies expression [797] reads

$$\text{HLB} = 7 + (\text{hydrophilic group number}) - 0.475 n_c \tag{4.304}$$

where n_c is the number of $-CH_2-$ groups in the lipophilic part of the molecule. Shinoda and Friberg [798] proved that the HLB number is not a property of the surfactant molecules only, but also depends strongly on the temperature (for nonionic surfactants), on the type and concentration of added electrolytes, on the type of oil phase, etc. They proposed using the phase inversion temperature (PIT) instead of HLB for characterization of the emulsion stability.

Davis [799] summarized the concepts about HLB, PIT, and Windsor's ternary phase diagrams for the case of microemulsions and reported topological ordered models connected with the Helfrich membrane bending energy. Because the curvature of surfactant lamellas plays a major role in determining the patterns of phase behavior in microemulsions, it is important to reveal how the optimal microemulsion state is affected by the surface forces determining the curvature energy

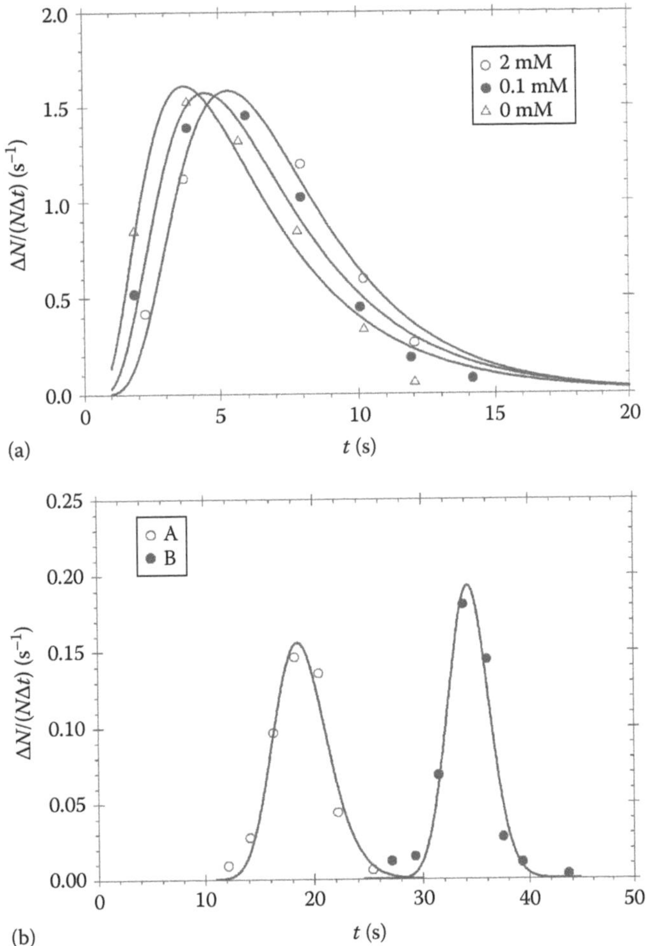

FIGURE 4.54 Histograms for the lifetimes of emulsion films: $\Delta N/N$ is the relative number of films that have ruptured during a time interval Δt. (a) Surfactant in the drops: benzene films between water drops containing surfactant sodium octylsulfonate of concentration: 0 M, 0.1 mM, and 2 mM; (b) Surfactant in the film: (A) benzene film with 0.1 M of lauryl alcohol dissolved in the film, (B) water film with 2 mM of sodium octylsulfonate inside. (From Traykov, T.T. et al., *Int. J. Multiphase Flow*, 3, 485, 1977.)

[239,800,801]. It is hoped that lattice models [802,803] and membrane curvature models [804,805] will lead to predictive formulae for the microemulsion design.

Ivanov et al. [723,727,728,806] have proposed a semiquantitative theoretical approach that provides a straightforward explanation of the Bancroft rule for emulsions. This approach is based on the idea of Davies and Rideal [34] that both types of emulsions are formed during the homogenization process, but only the one with lower coalescence rate survives. If the initial drop concentration for both emulsions is the same, the coalescence rates for the two emulsions—$(\text{Rate})_1$ for emulsion 1 and $(\text{Rate})_2$ for emulsion 2 (Figure 4.55)—will be proportional to the respective coalescence rate constants, $k_{c,1}$ and $k_{c,2}$ (see Section 4.6), and inversely proportional to the film lifetimes, τ_1 and τ_2:

$$\frac{(\text{Rate})_1}{(\text{Rate})_2} \approx \frac{k_{c,1}}{k_{c,2}} \approx \frac{\tau_2}{\tau_1} \approx \frac{V_1}{V_2} \tag{4.305}$$

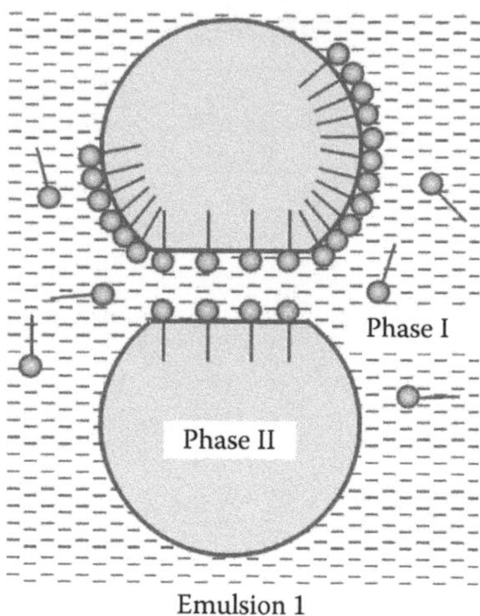

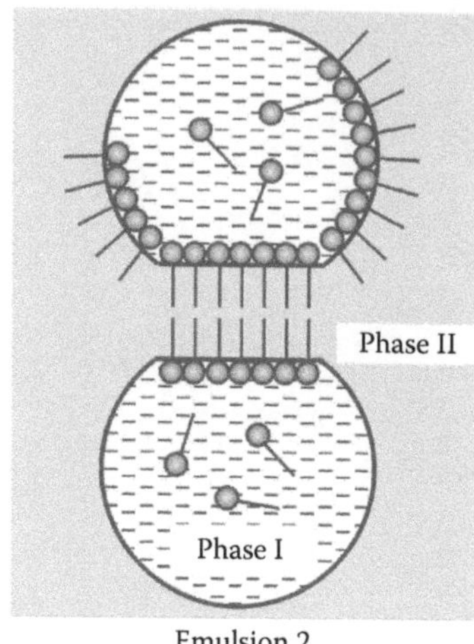

FIGURE 4.55 The two possible types of emulsions obtained just after the homogenization; the surfactant is soluble into Phase I.

Here V_1 and V_2 denote the respective velocities of film thinning. After some estimates based on Equations 4.273, 4.284, 4.298, and 4.299, we can express the ratio in Equation 4.305 in the form:

$$\frac{(\text{Rate})_1}{(\text{Rate})_2} \approx \left(486\rho_d D_s^3\right)^{1/3}\left(\frac{h_{\text{cr},1}^3}{h_{\text{cr},2}^2}\right)^{1/3}\left(\frac{\eta_d}{R^2}\right)^{1/3}\frac{P_c - \Pi_1}{E_G(P_c - \Pi_2)^{2/3}} \tag{4.306}$$

where $h_{\text{cr},1}$ and $h_{\text{cr},2}$ denote the critical thickness of film rupture for the two emulsions in Figure 4.55. Many conclusions can be drawn, regarding the type of emulsion to be formed:

1. If the disjoining pressures, Π_1 and Π_2, are zero, the ratio in Equation 4.306 will be very small. Hence, emulsion 1 (surfactant soluble in the continuous phase) will coalesce much more slowly and it will survive. This underlines the crucial importance of the surfactant location (which is connected with its solubility), thus providing a theoretical foundation for Bancroft's rule. The emulsion behavior in this case will be controlled almost entirely by the hydrodynamic factors (kinetic stability).
2. The disjoining pressure, Π, plays an important role. It can substantially change and even reverse the behavior of the system if it is comparable by magnitude with the capillary pressure, P_c. For example, if $(P_c - \Pi_2) \to 0$ at a finite value of $P_c - \Pi_1$ (which may happen, e.g., for an O/W emulsion with oil-soluble surfactant), the ratio in Equation 4.306 may become much larger than unity, which means that emulsion 2 will become thermodynamically stable. In some cases, the stabilizing disjoining pressure is large enough for emulsions with a very high volume fraction of the disperse phase (above 95% in some cases) to be formed [807].
3. The Gibbs elasticity, E_G, favors the formation of emulsion 1, because it slows down the film thinning. On the other hand, increased surface diffusivity, D_s, decreases this effect, because it helps the interfacial tension gradients to relax, thus facilitating the formation of emulsion 2.
4. The film radius, R, increases and the capillary pressure, P_c, decreases with the drop radius, R_d. Therefore, larger drops will tend to form emulsion 1, although the effect is not very pronounced.

5. The difference in critical thicknesses of the two emulsions only slightly affects the rate ratio in Equation 4.306, although the value of h_{cr} itself is important.
6. The viscosity of the continuous phase, η, has no effect on the rate ratio, which depends only slightly on the viscosity of the drop phase, η_d. This is in agreement with the experimental observations (see Ref. [34]).
7. The interfacial tension, σ, affects the rate ratio directly only through the capillary pressure, $P_c = 2\sigma/R_d$. The electrolyte primarily affects the electrostatic disjoining pressure, Π, which decreases as the salt content increases, thus destabilizing the O/W emulsion. It can also influence the stability by changing the surfactant adsorption (including the case of nonionic surfactants).
8. The temperature strongly affects the solubility and surface activity of nonionic surfactants [3]. It is well known that at higher temperature nonionic surfactants become more oil soluble, which favors the W/O emulsion. Thus, solubility may change the type of emulsion formed at the PIT. The surface activity has numerous implications and the most important is the change of the Gibbs elasticity, E_G, and the interfacial tension, σ.
9. Surface-active additives (cosurfactants, demulsifiers, etc.), such as fatty alcohols in the case of ionic surfactants, may affect the emulsifier partitioning between the phases and its adsorption, thereby changing the Gibbs elasticity and the interfacial tension. The surface-active additives may also change the surface charge (mainly by increasing the spacing among the emulsifier ionic headgroups), thus decreasing the repulsive electrostatic disjoining pressure and favoring the W/O emulsion. Polymeric surfactants and adsorbed proteins increase the steric repulsion between the film surfaces. They may favor either O/W or W/O emulsions, depending on their conformation at the interface and their surface activity.
10. The interfacial bending moment, B_0, can also affect the type of the emulsion, although this is not directly visible from Equation 4.306. (Note that $B_0 = -4k_cH_0$, where H_0 is the so-called spontaneous curvature and k_c is the interfacial curvature elastic modulus [200]. Typically, B_0 is of the order of 5×10^{-11} N.) Usually, for O/W emulsions, B_0 opposes the flattening of the droplet surfaces in the zone of collision (Figure 4.32), but for W/O emulsions favors the flattening [207]. This effect might be quantified by the expression for the curvature contribution in the energy of droplet–droplet interaction [207]:

$$W_c = \frac{-2\pi R^2 B_0}{R_d}, \quad \left(\frac{R}{R_d}\right)^2 \ll 1 \tag{4.307}$$

It turns out that $W_c > 0$ for the droplet collisions in an O/W emulsion, while $W_c < 0$ for a W/O emulsion [207]. Consequently, the interfacial bending moment stabilizes the O/W emulsions but destabilizes the W/O ones. There is supporting experimental evidence [808] for microemulsions, that is, for droplets of rather small size. Moreover, the effect of the bending moment can be important even for micrometer-sized droplets [207]. This is because the bent area increases faster ($R^2 \propto R_d^2$) than the bending energy per unit area which decreases ($W_c/R^2 \propto 1/R_d$) when the droplet radius, R_d, increases (see Equation 4.307).

For micron-sized emulsion droplets the capillary pressure can be so high that a film may not appear between the drops. In such case, instead of Equation 4.306, we can use analogous expression for nondeformed (spherical) drops [783,809]. Figure 4.56 shows the calculated ratio of the coalescence rates of emulsion 1 and emulsion 2 for two water drops (with dissolved sodium alkyl sulfates, C_nSO_4Na) in hexadecane at concentrations close to the CMC. The degree of surface coverage was 0.7 and all parameters are obtained from the fits of surface tension isotherms [783]. The increase of the surfactant chainlength makes the difference between two

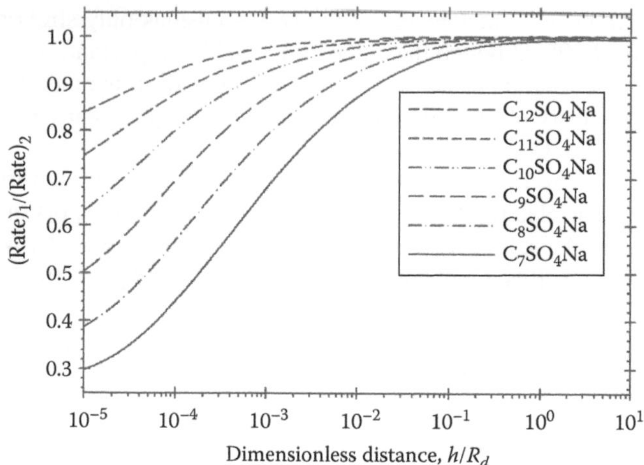

FIGURE 4.56 Ratio of the coalescence rates of emulsion 1 and emulsion 2 vs. the dimensionless distance, h/R_d, for water drops stabilized by sodium alkyl sulfates (C_nSO_4Na) in hexadecane at concentrations close to the CMC.

systems insignificant. In the calculations, the effect of disjoining pressure has not been taken into account. Thus, at concentrations close to the CMC, the two systems have a similar hydrodynamic behavior. The emulsion stability is controlled by the considerable difference between the values of the disjoining pressure in the cases where the surfactants are in the continuous and in the disperse phases.

4.5.3.4 Demulsification

It has been known for a long time [34] that one way to destroy an emulsion is to add a surfactant, which is soluble in the drop phase—this method is termed *chemical demulsification*. To understand the underlying process, let us consider two colliding emulsion droplets with film formed in the zone of collision (see Figures 4.32 and 4.57). As discussed earlier, when the liquid is flowing out of the film, the viscous drag exerted on the film surfaces (from the side of the film interior) carries away the adsorbed emulsifier toward the film periphery. Thus, a nonuniform surface distribution of the emulsifier (shown in Figure 4.57a by empty circles) is established. If demulsifier (the closed circles in Figure 4.57b) is present in the drop phase, it will occupy the interfacial area freed by the emulsifier. The result will be a saturation of the adsorption layer, as shown in Figure 4.57b. If the demulsifier is sufficiently surface active, its molecules will be able to decrease substantially, and even to eliminate completely the interfacial tension gradients, thus changing the emulsion to type 2 (see Figure 4.55 and Section 4.5.3.2). This leads to a strong increase in the

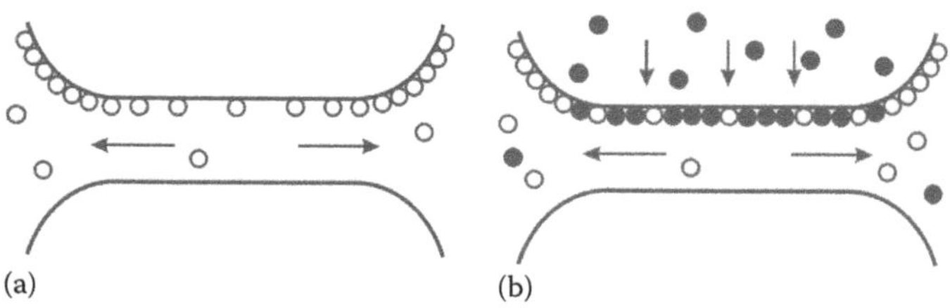

FIGURE 4.57 (a) Nonuniform surface distribution of an emulsifier due to drag from the draining film. (b) Suppression of the surface tension gradients by a demulsifier added in the drop phase.

rate of film thinning, rapid drop coalescence, and emulsion destruction [727,728]. The mechanism mentioned earlier suggests that the demulsifier has to possess the following properties:

1. It must be soluble in the drop phase or in both phases, but in the latter case its solubility in the drop phase must be much higher.
2. Its diffusivity and concentration must be large enough to provide a sufficiently large demulsifier flux toward the surfaces and thus eliminate the gradients of the interfacial tension.
3. Its surface activity must be comparable and even higher than that of the emulsifier; otherwise, even though it may adsorb, it will not be able to suppress the interfacial tension gradients.

4.5.4 Interactions in Nonpreequilibrated Emulsions

The common nonionic surfactants are often soluble in both water and oil phases. In the practice of emulsion preparation, the surfactant (the emulsifier) is initially dissolved in one of the liquid phases and then the emulsion is prepared by homogenization. In such a case, the initial distribution of the surfactant between the two phases of the emulsion is not in equilibrium; therefore, surfactant diffusion fluxes appear across the surfaces of the emulsion droplets. The process of surfactant redistribution usually lasts from many hours to several days, until finally equilibrium distribution is established. The diffusion fluxes across the interfaces, directed either from the continuous phase toward the droplets or the reverse, are found to stabilize both thin films and emulsions. In particular, even films, which are thermodynamically unstable, may exist several days because of the diffusion surfactant transfer; however, they rupture immediately after the diffusive equilibrium has been established. Experimentally, this effect manifests itself in phenomena called *cyclic dimpling* [810] and *osmotic swelling* [811]. These two phenomena, as well as the equilibration of two phases across a film [812,813], are described and interpreted in the following section.

4.5.4.1 Surfactant Transfer from Continuous to Disperse Phase (Cyclic Dimpling)

The phenomenon of cyclic dimpling was first observed [728,810] with xylene films intervening between two water droplets in the presence of the nonionic emulsifier Tween 20 or Tween 80 (initially dissolved in water but also soluble in oil). The same phenomenon also has been observed with other emulsion systems.

After the formation of such an emulsion film, it lowers down to an equilibrium thickness (approximately 100 nm), determined by the electrostatic repulsion between the interfaces. As soon as the film reaches this thickness, a dimple spontaneously forms in the film center and starts growing (Figure 4.58a). When the dimple becomes bigger and approaches the film periphery, a channel forms connecting the dimple with the aqueous phase outside the film (Figure 4.58b). Then, the water contained in the dimple flows out leaving an almost plane-parallel film behind. Just afterward, a new dimple starts to grow and the process repeats again. The period of this cyclic dimpling remains approximately constant for many cycles and could be from a couple of minutes up to more than 10 min. It was established that this process is driven by the depletion of the surfactant concentration on the film surfaces due to the dissolving of surfactant in the adjacent drop phases. The depletion triggers a surface convection flux along the two film surfaces and a bulk diffusion flux in the film interior. Both fluxes are directed toward the center of the film. The surface convection causes a tangential movement of the film surfaces; the latter drag along a convective influx of solution in the film, which feeds the dimple. Thus, the cyclic dimpling appears to be a process leading to stabilization of the emulsion films and emulsions due to the influx of additional liquid in the region between the droplets, which prevents them from a closer approach, and eventually, from coalescence.

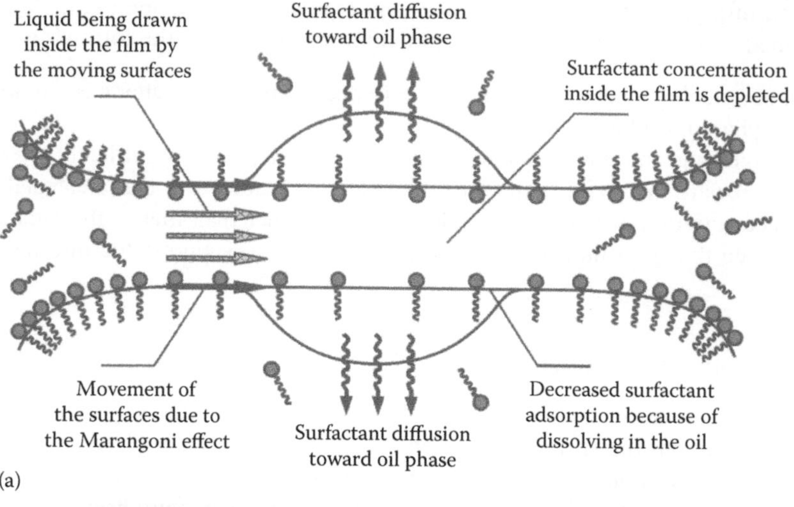

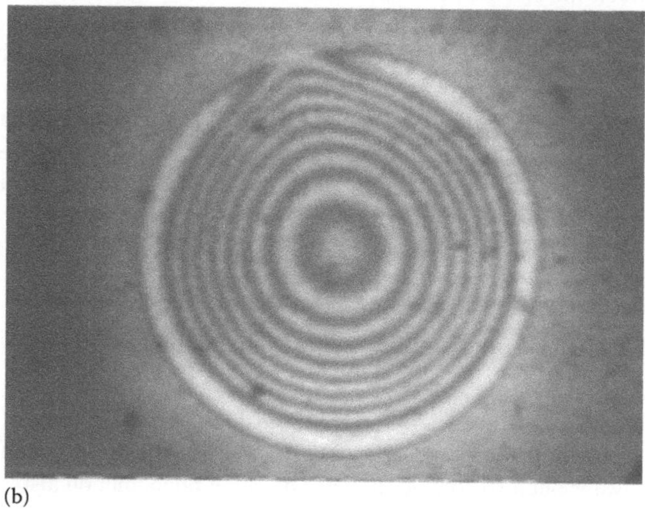

FIGURE 4.58 Spontaneous cyclic dimpling caused by surfactant diffusion from the aqueous film toward the two adjacent oil phases. (a) Schematic presentation of the process. (b) Photograph of a large dimple just before flowing out; the interference fringes in reflected light allow determination of the dimple shape.

Combining the general equation of films with deformable interfaces (Equation 4.266), the mass balance (Equations 4.287 and 4.288), and the boundary condition for the interfacial stresses (Equation 4.292), we can derive [814–817]:

$$\frac{\partial h}{\partial t}+\frac{1}{3\eta r}\frac{\partial}{\partial r}\left\{rh^3\frac{\partial}{\partial r}\left[\frac{\sigma}{r}\frac{\partial}{\partial r}\left(r\frac{\partial h}{\partial r}\right)+\Pi(h)\right]\right\}=\frac{1}{2r}\frac{\partial}{\partial r}\left(\frac{jhr^2}{\Gamma}\right) \tag{4.308}$$

where

- j is the diffusion flux in the drop phase
- r is radial coordinate
- $h(r, t)$ is the film thickness
- σ is surface tension
- Γ is adsorption
- Π is disjoining pressure

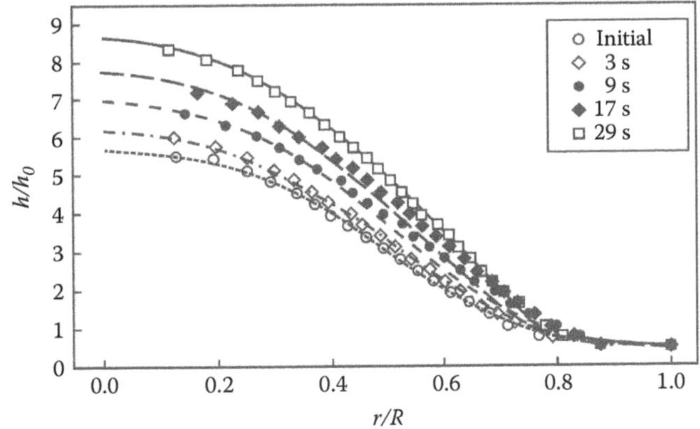

FIGURE 4.59 Comparison between the theory of cyclic dimpling (the lines) and the experimental data (the points) for the dimple shape, $h(r)$, determined from the interference fringes (see Figure 4.58b); emulsifier is anionic surfactant sodium nonylphenol polyoxyethylene-25 sulfate and the oil phase is styrene.

The comparison between the numerical calculations based on Equation 4.308 and the experimental data for the cyclic dimpling with the anionic surfactant sodium nonylphenol polyoxyethylene-25 sulfate shows a very good agreement (Figure 4.59). The experimental points are obtained from the interference fringes (see Figure 4.58). The shape in the initial moment, $t = 0$, serves as an initial condition for determining $h(r, t)$ by solving Equation 4.308. The curves for $t = 3$, 9, 17, and 29 s represent theoretical predictions. The scaling parameters along the h- and r-axes in Figure 4.59 are $h_0 = 350$ nm and $R = 320$ μm, with the latter the film radius; the only adjustable parameter is the diffusion flux, j.

4.5.4.2 Surfactant Transfer from Disperse to Continuous Phase (Osmotic Swelling)

Velev et al. [725] reported that emulsion films, formed from preequilibrated phases containing the nonionic surfactant Tween and 0.1 M NaCl, spontaneously thin to Newton black films (thickness ≈ 10 nm) and then rupture. However, when the nonionic surfactant Tween 20 or Tween 60 is initially dissolved in the xylene drops and the film is formed from the nonpreequilibrated phases, no black film formation and rupture are observed [728,811]. Instead, the films have a thickness above 100 nm, and we observe formation of channels of larger thickness connecting the film periphery with the film center (Figure 4.60). We may observe that the liquid is circulating along the channels for a period from several hours to several days. The phenomenon continues until the redistribution of the surfactant between the phases is accomplished. This phenomenon occurs only when the background surfactant concentration in the continuous (the aqueous) phase is not lower than the CMC. These observations can be interpreted in the following way.

Because the surfactant concentration in the oil phase (the disperse phase) is higher than the equilibrium concentration, surfactant molecules cross the oil–water interface toward the aqueous phase. Thus, surfactant accumulates within the film, because the bulk diffusion throughout the film is not fast enough to transport promptly the excess surfactant into the Plateau border. As the background surfactant concentration in the aqueous phase is not less than CMC, the excess surfactant present in the film is packed in the form of micelles (denoted by black dots in Figure 4.60a). This decreases the chemical potential of the surfactant inside the film. Nevertheless, the film is subjected to osmotic swelling because of the increased concentration of micelles within. The excess osmotic pressure

$$P_{osm} = kTC_{mic} \geq P_c \tag{4.309}$$

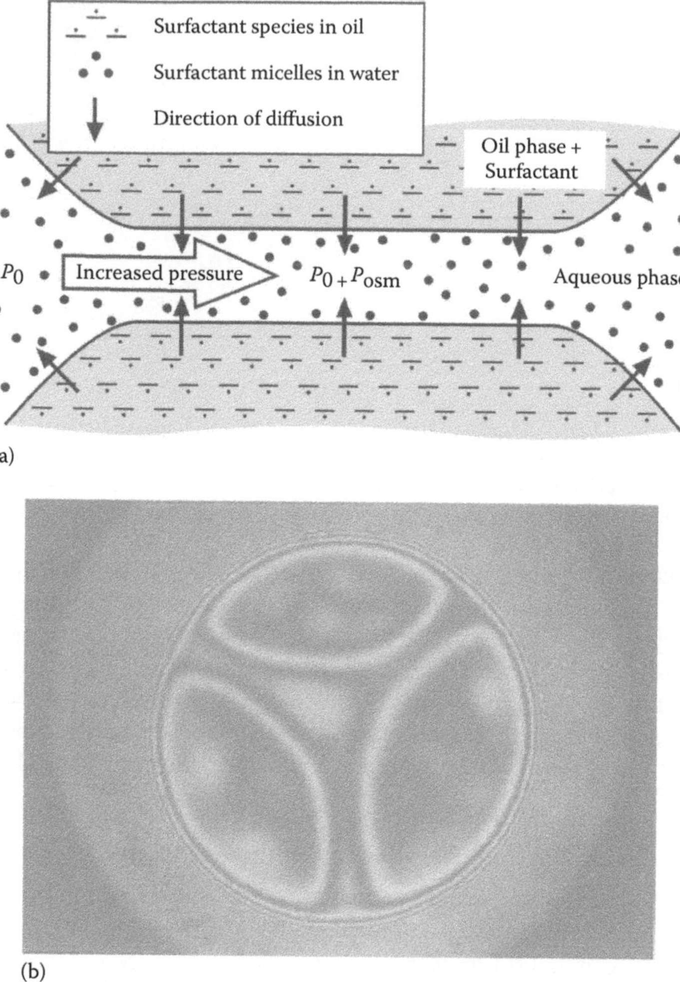

FIGURE 4.60 Osmotic swelling of an aqueous film formed between two oil droplets. (a) The surfactant dissolved in the oil is transferred by diffusion toward the film, where it forms micelles, the osmotic effects of which increase the local pressure. (b) Photograph of a typical pattern from a circular film with channels.

counterbalances the outer capillary pressure and arrests further thinning of the film. Moreover, the excess osmotic pressure in the film gives rise to a convective outflow of solution: this is the physical origin of the observed channels (Figure 4.60b).

Experimental data [728,811] show that the occurrence of the above phenomenon is the same for initial surfactant concentration in the water varying from 1 up to 500 times the CMC, if only some amount of surfactant is also initially dissolved in the oil. This fact implies that the value of the surfactant chemical potential inside the oil phase is much greater than that in the aqueous phase, with the latter closer to its value at the CMC in the investigated range of concentrations.

4.5.4.3 Equilibration of Two Droplets across a Thin Film

In the last two sections, we considered mass transfer from the film toward the droplets and the reverse, from droplets toward the film. In both cases, the diffusion fluxes lead to stabilization of the film. Here we consider the third possible case corresponding to mass transfer from the first droplet toward the second one across the film between them. In contrast with the former two cases, in the last case the mass transfer is found to destabilize the films. Experimentally, the diffusion transfer of alcohols, acetic acid, and acetone was studied [818,819]. The observed destabilization of the films can be attributed to the appearance of Marangoni instability [812], which manifests itself through the growth of capillary waves at the interfaces, which eventually can lead to film rupture.

The Marangoni instabilities can appear not only in thin films, but also in the simpler case of a single interface. In this case, the Marangoni instability may bring about spontaneous emulsification. This effect has been theoretically investigated by Sterling and Scriven [820], whose work stimulated numerous theoretical and experimental studies on spontaneous emulsification. Lin and Brenner [821] examined the role of the heat and mass transfer in an attempt to check the hypothesis of Holly [822] that the Marangoni instability can cause the rupture of tear films. Their analysis was extended by Castillo and Velarde [823], who accounted for the tight coupling of the heat and mass transfer and showed that it drastically reduces the threshold for Marangoni convection. Instability driven by diffusion flux of dissolved oil molecules across an asymmetric liquid film (oil–water–air film) has been theoretically investigated [813]. It was found that even small decrements of the water–air surface tension, caused by the adsorbed oil, are sufficient to trigger the instability.

4.5.5 Hydrodynamic Interaction of a Particle with an Interface

There are various cases of particle–interface interactions, which require separate theoretical treatment. The simpler case is the hydrodynamic interaction of a solid particle with a solid interface. Other cases are the interactions of fluid particles (of tangentially mobile or immobile interfaces) with a solid surface; in these cases, the hydrodynamic interaction is accompanied by deformation of the particle. On the other hand, the colloidal particles (both solid and fluid) may hydrodynamically interact with a fluid interface, which thereby undergoes a deformation. In the case of fluid interfaces, the effects of surfactant adsorption, surface diffusivity, and viscosity affect the hydrodynamic interactions. A special class of problems concerns particles attached to an interface, which are moving throughout the interface. Another class of problems is related to the case when colloidal particles are confined in a restricted space within a narrow cylindrical channel or between two parallel interfaces (solid and/or fluid); in the latter case, the particles interact simultaneously with both film surfaces.

The theoretical contributions are limited to the case of low Reynolds number [644,645,723,824–826,830–832] (mostly for creeping flows, see Section 4.5.1), avoiding the difficulties arising from the nonlinearity of the equations governing the fluid motion at higher velocities. Indeed, for low Reynolds numbers, the term $\mathbf{v}\cdot\nabla\mathbf{v}$ in the Navier–Stokes equation (see Equations 4.258 through 4.260) is negligible, and we may apply the method of superposition to solve the resulting linear set of equations. This means that we may first solve the simpler problems about the particle elementary motions: (1) particle translation (without rotation) in an otherwise immobile liquid, (2) particle rotation (without translation) in an otherwise immobile liquid, and (3) streamlining of an immobile particle by a Couette or Poiseuille flow. Once the problems about the elementary motions have been solved, we may obtain the linear and angular velocity of the real particle motion combining the elementary flows. The principle of combination is based on the fact that for low Reynolds numbers the particle acceleration is negligible, and the net force and torque exerted on the particle must be zero. In other words, the hydrodynamic drag forces and torques originating from the particle translation and rotation are counterbalanced by those originating from the streamlining:

$$\mathbf{F}_{\text{translation}} + \mathbf{F}_{\text{rotation}} + \mathbf{F}_{\text{streamlining}} = \mathbf{0}, \quad \mathbf{M}_{\text{translation}} + \mathbf{M}_{\text{rotation}} + \mathbf{M}_{\text{streamlining}} = \mathbf{0} \tag{4.310}$$

That is the reason why we will now consider expressions for **F** and **M** for various types of elementary motions.

4.5.5.1 Particle of Immobile Surface Interacting with a Solid Wall

The force and torque exerted on a solid particle were obtained in the form of a power series with respect to R_d/l, where R_d is the particle radius and l is the distance from the center of the particle to the wall. Lorentz [827] derived an asymptotic expression for the motion of a sphere along the normal to a planar wall with an accuracy of up to R_d/l. Faxen [828] developed the method of reflection for a sphere moving between two parallel planes in a viscous fluid. Using this method, Wakiya [829] considered the cases of motion in flow of Couette and Poiseuille; however, the method employed by him cannot be applied to small distances to the wall [668]. The next important step was taken by Dean and O'Neil [830] and O'Neil [831], who found an exact solution for the force and the torque acting on a spherical particle moving tangentially to a planar wall at an arbitrary distance from the wall. The limiting case of small distances between the particle and the wall was examined by several authors [550–552,705]. Instead of an exact solution of the problem the authors derived asymptotic formulae for the force and torque. Keh and Tseng [833] presented a combined analytical–numerical study for the slow motion of an arbitrary axisymmetric body along its axis of revolution, with the latter normal to a planar surface. The inertial migration of a small solid sphere in a Poiseuille flow was calculated by Schonberg and Hinch [834] for the case when the Reynolds number for the channel is of the order of unity.

In the following section, we present expressions for the forces and torques for some of the elementary motions. In all cases we assume that the Reynolds number is small, the coordinate plane xy is parallel to the planar wall, and h is the shortest surface-to-surface distance from the particle to the wall.

First, we consider the case of a pure translational motion: a solid spherical particle of radius R_d that translates along the y-axis with a linear velocity U and angular velocity $\omega \equiv 0$ in an otherwise quiescent fluid. In spite of the fact that the particle does not rotate, it experiences a torque, $\mathbf{M}$, directed along the x-axis, due to friction with the viscous fluid. The respective asymptotic expressions [659–661] for the components of the drag force, $\mathbf{F}$, and torque, $\mathbf{M}$, read

$$F_x = 0, \quad F_y = -6\pi\eta U R_d f_y, \quad M_x = -8\pi\eta U R_d^2 m_x, \quad M_y = 0 \tag{4.311}$$

$$f_y = \left(\frac{8}{15} + \frac{16}{375}\frac{h}{R_d}\right)\ln\left(\frac{2R_d}{h}\right) + 0.58461 + O\left(\frac{h}{R_d}\right) \tag{4.312}$$

$$m_x = \left(\frac{1}{10} + \frac{43}{250}\frac{h}{R_d}\right)\ln\left(\frac{2R_d}{h}\right) - 0.26227 + O\left(\frac{h}{R_d}\right) \tag{4.313}$$

where f_y and m_x are dimensionless drag force and torque coefficients, respectively.

Second, we consider the case of pure rotation: a solid spherical particle of radius R_d is situated at a surface-to-surface distance, h, from a planar wall and rotates with angular velocity, ω, around the x-axis in an otherwise quiescent fluid. The corresponding force and torque resultants are [659–661]

$$F_x = 0, \quad F_y = -6\pi\eta\omega R_d^2 f_y, \quad M_x = -8\pi\eta\omega R_d^3 m_x, \quad M_y = 0 \tag{4.314}$$

$$f_y = \frac{2}{15}\ln\left(\frac{R_d}{h}\right) - 0.2526 + O\left(\frac{h}{R_d}\right), \quad m_x = \frac{2}{5}\ln\left(\frac{R_d}{h}\right) + 0.3817 + O\left(\frac{h}{R_d}\right) \tag{4.315}$$

From Equations 4.311 through 4.315, it follows that the force and the torque depend weakly (logarithmically) on the distance, h, as compared to the Taylor or Reynolds laws (Equations 4.271 and 4.272).

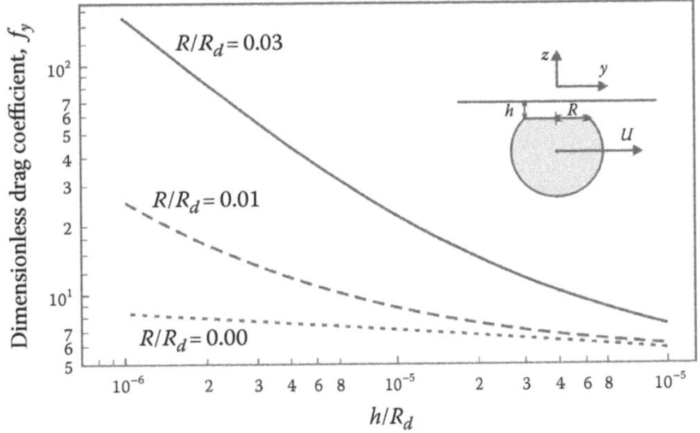

FIGURE 4.61 Deformed fluid particle (the inset) moving tangentially to an immobile solid surface: plot of the dimensionless drag coefficient, f_y, vs. the dimensionless film thickness, h/R_d, for three values of the dimensionless film radius, R/R_d (see Equation 5.317).

As discussed in Sections 4.5.2.1 and 4.5.3.2, a fluid particle in the presence of high surfactant concentration can be treated as a deformable particle of tangentially immobile surfaces. Such a particle deforms when pressed against a solid wall (see the inset in Figure 4.61). To describe the drag due to the film intervening between the deformed particle and the wall, we may use the expression derived by Reynolds [646] for the drag force exerted on a planar solid ellipsoidal disc, which is parallel to a solid wall and is moving along the y-axis at a distance h from the wall:

$$F_x = 0, \quad F_y = -\pi\eta U \frac{h}{ab} \tag{4.316}$$

Here, a and b are the semiaxes of the ellipse; for a circular disc (or film), we have $a = b = R$. By combining Equations 4.311 and 4.312 with Equation 4.316, we can derive an expression for the net drag force experienced by the deformed particle (the inset in Figure 4.61) when it moves along the y-axis with a linear velocity U:

$$F_y = -6\pi\eta U R_d f_y, \quad f_y = \frac{R^2}{6hR_d} + \left(\frac{8}{15} + \frac{16}{375}\frac{h}{R_d}\right)\ln\left(\frac{2R_d}{h}\right) + 0.58461 + O\left(\frac{h}{R_d}\right) \tag{4.317}$$

Here

h and R denote the film thickness and radius

R_d is the curvature radius of the spherical part of the particle surface

The dependence of the dimensionless drag coefficient, f_y, on the distance h for different values of the ratio R/R_d is illustrated in Figure 4.61. The increase of R/R_d and the decrease of h/R_d may lead to an increase of the drag force, f_y, by an order of magnitude. That is the reason the film between a deformed particle and a wall can be responsible for the major part of the energy dissipation. Moreover, the formation of doublets and flocks of droplets separated by liquid films seems to be of major importance for the rheological behavior of emulsions.

4.5.5.2 Fluid Particles of Mobile Surfaces

Let us start with the case of pure phases, when surfactant is missing and the fluid–liquid interfaces are mobile. Under these conditions, the interaction of an emulsion droplet with a planar solid wall was investigated by Ryskin and Leal [835], and numerical solutions were obtained. A new formulation of the same problem was proposed by Liron and Barta [836]. The case of a small droplet moving in the restricted space between two parallel solid surfaces was solved by Shapira and Haber [837,838]. These authors used the Lorentz reflection method to obtain analytical solutions for the drag force and the shape of a small droplet moving in Couette flow or with constant translational velocity.

The more complicated case, corresponding to a viscous fluid particle approaching the boundary between two pure fluid phases (all interfaces deformable), was investigated by Yang and Leal [839,840], who succeeded in obtaining analytical results.

Next, we proceed with the case when surfactant is present and the Marangoni effect becomes operative. Classical experiments carried out by Lebedev [841] and Silvey [842] show that the measured velocity of sedimentation, U, of small fluid droplets in a viscous liquid (pure liquid phases assumed) does not obey the Hadamar [843] and Rybczynski [844] equation:

$$F = 2\pi\eta U R_d \frac{3\eta_d + 2\eta}{\eta_d + \eta} \tag{4.318}$$

where F is the drag force. The limiting case $\eta_d \to 0$ corresponds to bubbles, whereas in the other limit, $\eta_d \to \infty$, Equation 4.318 describes solid particles. Note that Equation 4.318 is derived for the motion of a spherical fluid particle (drop or bubble) of viscosity η_d in a liquid of viscosity η in the absence of any surfactant. The explanation of the contradiction between theory and experiment [841,842] turned out to be very simple: even liquids that are pure from the viewpoint of the spectral analysis may contain some surface-active impurities, whose bulk concentration might be vanishingly low, but which can provide a dense adsorption layer at the restricted area of the fluid particle surface. Then, the effects of Gibbs elasticity and interfacial viscosity substantially affect the drag coefficient of the fluid particle. The role of the latter two effects was investigated by Levich [662], Edwards et al. [663], and He et al. [845] for the motion of an emulsion droplet covered with a monolayer of insoluble surfactant (adsorption and/or desorption not present). These authors used the Boussinesq–Scriven constitutive law of a viscous fluid interface (Equation 4.292), and established that only the dilatational interfacial viscosity, η_{dl}, but not the shear interfacial viscosity, η_{sh}, influences the drag force. If the surfactant is soluble in both phases and the process of adsorption is diffusion controlled (see Section 4.2.2.1), the generalization of Equation 4.318 is [783]

$$F = 2\pi\eta U R_d \left[3 - \left(1 + \frac{\eta_d}{\eta} + \frac{2\eta_{dl}}{\eta R_d} + \frac{R_d E_G}{3\eta D_s} \frac{2}{2 + 2(R_d D/h_a D_s) + (R_d D_d/h_{d,a} D_s)} \right)^{-1} \right] \tag{4.319}$$

where

D_d is the surfactant diffusion coefficient in the drop phase

c and c_d are the concentrations of surfactant in the continuous and drop phases, respectively

$h_a = \partial\Gamma/\partial c$ and $h_{d,a} = \partial\Gamma/\partial c_d$ are the slopes of adsorption isotherms with respect to the surfactant concentration

In the limiting case without surfactant, Equation 4.319 is reduced to the Hadamar [843] and Rybczynski [844] equation (Equation 4.318).

A recently developed experimental technique [846–850] gives the possibility to measure precisely the instantaneous velocity of rising bubbles in a solution as a function of time and the distance to the starting point. The sensitivity of this technique allows one to determine trace amounts of impurities in water.

Danov et al. [347,851–854] investigated theoretically the hydrodynamic interaction of a fluid particle with a fluid interface in the presence of surfactant. The numerical results of these authors reveal that there is a strong influence of both shear and dilatational interfacial viscosities on the motion of the fluid particle when the particle–interface distance, h, is approximately equal to or smaller than the particle radius, R_d. For example, in the presence of an external force acting parallel to the interface (along the y-axis), the stationary motion of the spherical particle close to the viscous interface is a superposition of a translation along the y-axis with velocity V_y and a rotation (around the x-axis) with an angular velocity, ω_x (see the inset in Figure 4.62a). The numerical results of Danov et al. [853,854] for V_y and ω_x normalized by the Stokes velocity, $V_{Stokes} = F/(6\pi\eta R_d)$, are plotted in Figure 4.62a and b versus h/R_d for four different types of interfaces: (1) solid particle and solid wall (see Equations 4.311 through 4.313); (2) fluid particle and fluid interface for $K = E = 100$ (for the definition of K and E see Equation 4.303); (3) the same system as (2) but for $K = E = 10$; (4) the same system as (2) but for $K = E = 1$. (For the definition of the interfacial viscosities, η_{dl} and η_{sh}, see Equation 4.292). As seen in Figure 4.62a, the velocity of the sphere, V_y, is less than V_{Stokes} for the solid (1) and the highly viscous (2) interfaces, and V_y noticeably decreases when the distance h decreases. However, in case (4), corresponding to low surface viscosities, the effect is quite different: V_y/V_{Stokes} is greater than unity (the sphere moves faster near the interface than in the bulk), and its dependence on h is rather weak. The result about the angular velocity, ω_x, is also intriguing (Figure 4.62b). The stationary rotation of a sphere close to a solid (1) or highly viscous (2) interface

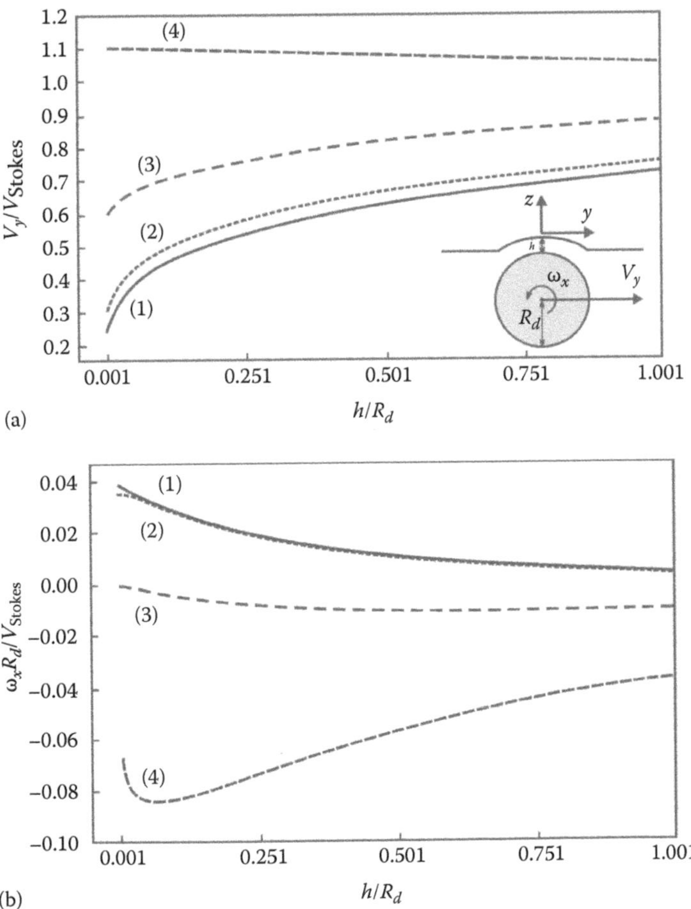

FIGURE 4.62 Spherical particle moving tangentially to a viscous interface: plots of the stationary dimensionless linear (V_y/V_{Stokes}) (a) and angular ($\omega_x R_d/V_{Stokes}$) (b) velocities vs. the dimensionless thickness, h/R_d. The curves correspond to various surface viscosities: (1) $K = E = \infty$ (solid surfaces); (2) $K = E = 100$; (3) $K = E = 10$, and (4) $K = E = 1$ (see Equation 4.303).

is in positive direction, that is, $\omega_x > 0$. For the intermediate interfacial viscosity (3), the sphere practically does not rotate, whereas, for the interfaces of low viscosity (4), the drop rotates in the opposite direction, that is, $\omega_x < 0$. The inversion of the sign of ω_x is due to the fact that the friction of the particle with the bulk fluid below it (see the inset in Figure 4.62a) becomes stronger than the friction with the interface above the particle.

Finally, we consider the case of a solid particle attached to a liquid–fluid interface. This configuration is depicted in Figure 4.21e; note that the position of the particle along the normal to the interface is determined by the value of the three-phase contact angle. Stoos and Leal [855] investigated the case when such an attached particle is subjected to a flow directed normally to the interface. These authors determined the critical capillary number, beyond which the captured particle is removed from the interface by the flow.

Danov et al. [347] examined the case of an attached particle moving along a liquid–gas interface under the action of an applied force directed tangentially to the interface. The effects of the contact angle (the depth of immersion), as well as the effect of adsorbed surfactant on the drag force, were investigated. These authors also calculated the surface diffusion coefficient of a Brownian particle attached to the liquid surface. Let D_p and D_{p0} be the particle surface diffusion coefficient in the

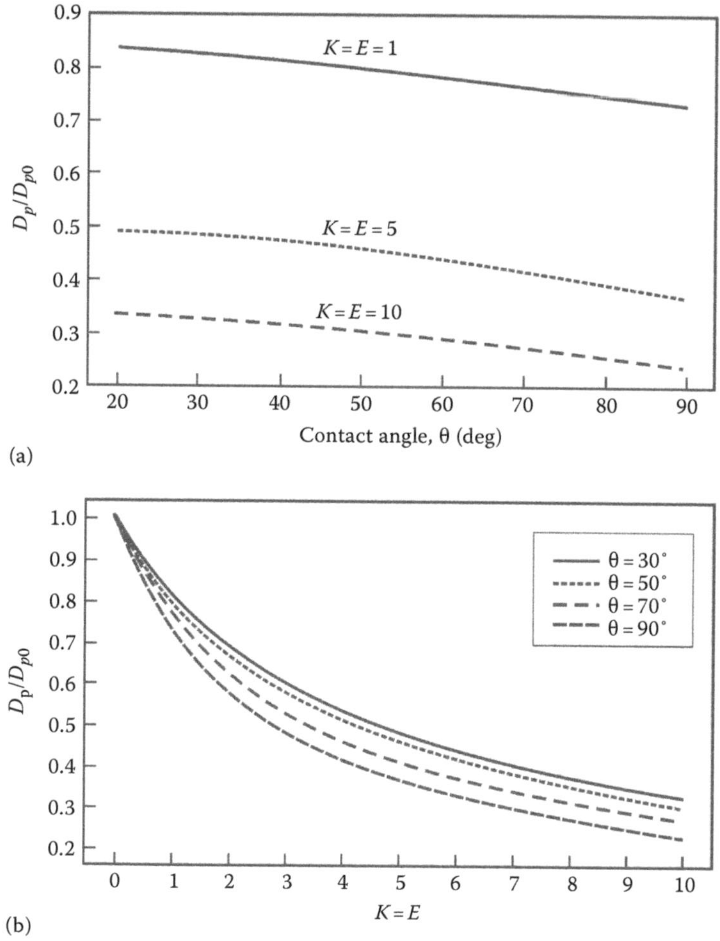

FIGURE 4.63 Effect of adsorbed surfactant on the surface diffusivity, D_p, of a Brownian particle attached to a fluid interface: (a) plot of D_p/D_{p0} vs. particle contact angle, θ, for various surface viscosities (see Equation 4.303); (b) plot of D_p/D_{p0} vs. the dimensionless surface viscosity, $K = E$, for various θ.

presence and absence of surfactant, respectively. In Figure 4.63a, we plot the results for D_p/D_{p0} versus the solid–liquid–gas contact angle, θ, for three different values of the parameters K and E characterizing the surface viscosities (see Equation 4.303): (1) $K = E = 1$; (2) $K = E = 5$, and (3) $K = E = 10$. The relatively small slope of the curves in Figure 4.63a indicates that D_p/D_{p0} depends less significantly on the contact angle, θ, than on the surface viscosity characterized by K and E. Note, however, that D_{p0} itself depends markedly on θ: the absolute value of D_{p0} is smaller for the smaller values of θ (for deeper immersion of the particle in the liquid phase). Figure 4.63b presents the calculated dependence of D_p/D_{p0} on the surface viscosity characterized by K and E ($K = E$ is used in the calculations) for various fixed values of the contact angle, θ. Apparently, the particle mobility decreases faster for the smaller values of K and then tends to zero insofar as the fluid surface "solidifies" for the higher values of the surface viscosities. The experimental data from measurements of the drag coefficient of spherical particles attached to fluid interfaces [346] showed very good agreement with the predictions of the theory [347].

The role of surface viscosity and elasticity on the motion of a solid particle trapped in a thin film, at an interface, or at a membrane of a spherical vesicle has been recently investigated in Refs. [856,857]. The theoretical results [856,857] have been applied to process the experimental data for the drag coefficient of polystyrene latex particles moving throughout the membrane of a giant lipid vesicle [858–864]. Thus, the interfacial viscosity of membranes has been determined. The motion of particles with different shapes trapped in thin liquid films and at Langmuir monolayers is studied intensively both theoretically and experimentally because of biological and medical applications [865–887].

4.5.6 Bulk Rheology of Dispersions

The description of the general rheological behavior of colloidal dispersions requires information regarding the drag forces and torques experienced by the individual particles [404,888,889]. In dilute systems, the hydrodynamic interactions between the particles can be neglected and their motion can be treated independently. In contrast, when the particle concentration is higher, the effect of hydrodynamic interactions between a spherical particle and an interface on the drag force and torque acquires considerable importance. The viscosity and the collective diffusion coefficient of colloidal dispersions can also be strongly affected also by long-range surface forces, like the electrostatic double layer force.

Long ago Einstein [890] obtained a formula for the diffusion coefficient for solid spheres in the dilute limit:

$$D = \frac{kT}{6\pi\eta_m R_p} \tag{4.320}$$

where

R_p is the particle radius

η_m is the viscosity of the liquid medium

This relation was later generalized by Kubo [891] for the cases when the hydrodynamic resistance becomes important. The further development in this field is reviewed by Davis [824].

The particle–particle interactions lead to a dependence of the viscosity, η, of a colloidal dispersion on the particle volume fraction, ϕ. Einstein [892] showed that for a suspension of spherical particles in the dilute limit:

$$\eta = \eta_m[1 + 2.5\phi + O(\phi^2)] \tag{4.321}$$

Later Taylor [893] generalized Equation 4.321 for emulsion systems taking into account the viscous dissipation of energy due to the flow inside the droplets. Oldroyd [894] took into account the effect of surface viscosity and generalized the theory of Taylor [893] to diluted monodisperse emulsions whose droplets have viscous interfaces. Taylor [895], Fröhlich and Sack [896], and Oldroyd [897] applied asymptotic analysis to derive the next term in Equation 4.321 with respect to the capillary number. Thus, the effect of droplet interfacial tension was included. This generalization may be important at high shear rates. Another important generalization is the derivation of appropriate expressions for the viscosity of suspensions containing particles with different shapes [644,645]. A third direction of generalization of Equation 4.321 is to calculate the next term in the series with respect to the volume fraction, ϕ. Batchelor [898] took into account the long-range hydrodynamic interaction between the particles to derive:

$$\eta = \eta_m[1 + 2.5\phi + 6.2\phi^2 + O(\phi^3)] \tag{4.322}$$

From a mathematical viewpoint, Equation 4.322 is an exact result; however, from a physical viewpoint, Equation 4.322 is not entirely adequate to the real dispersions, as not only the long-range hydrodynamic interactions are operative in colloids. A number of empirical expressions have been proposed in which the coefficient multiplying ϕ^2 varies between 5 and 14 [899]. The development of new powerful numerical methods helped for a better understanding of the rheology of emulsions [900–908]. The simple shear and Brownian flow of dispersions of elastic capsules, rough spheres, and liquid droplets were studied in Refs. [901,905,907,908]. The effect of insoluble surfactants and the drop deformation on the hydrodynamic interactions and on the rheology of dilute emulsions are the subject of investigation in Refs. [902,904,906]. Loewenberg and Hinch [900,903] discussed the basic ideas of the numerical simulations of concentrated emulsion flows. These works are aimed at giving a theoretical interpretation of various experimental results for dilute and concentrated dispersions. When the Peclet number is not small, the convective term in the diffusion equation (Equations 4.287 and 4.288) cannot be neglected and the respective problem has no analytical solution. Thus, a complex numerical investigation has to be applied [909,910].

The formulae of Einstein [890,892], Taylor [893], and Oldroyd [894] have been generalized for dilute emulsions of mobile surfaces with account for the Gibbs elasticity and the bulk and surface diffusion and viscosity [911]:

$$\frac{\eta}{\eta_m} = 1 + \left(1 + \frac{3}{2}\langle\varepsilon_m\rangle\right)\phi + O(\phi^2), \quad \langle\varepsilon_m\rangle \equiv \frac{\sum R_d^3\varepsilon_m}{\sum R_d^3} \tag{4.323}$$

where $\langle\varepsilon_m\rangle$ is the average value of the interfacial mobility parameter, ε_m, for all droplets in the control volume. The mobility parameter of individual drops, ε_m, and the effective surfactant diffusion coefficient, D_{eff}, are [911]

$$\varepsilon_m \equiv \frac{(\eta_d/\eta_m) + (2/5)((R_dE_G/2\eta_mD_{\mathrm{eff}}) + (3\eta_{\mathrm{dl}} + 2\eta_{\mathrm{sh}})/R_d\eta_m)}{1 + (\eta_d/\eta_m) + (2/5)((R_dE_G/2\eta_mD_{\mathrm{eff}}) + (3\eta_{\mathrm{dl}} + 2\eta_{\mathrm{sh}})/R_d\eta_m)} \tag{4.324}$$

$$D_{\mathrm{eff}} \equiv D_s + \frac{R_dD}{2h_a} + \frac{R_dD_d}{3h_{d,a}} \tag{4.325}$$

(see Equation 4.319 and the following section). If the droplet size distribution in the emulsion and the interfacial rheological parameters are known, then the average value $\langle\varepsilon_m\rangle$ can be estimated.

For monodisperse emulsions, the average value, $\langle \varepsilon_m \rangle$, and the interfacial mobility parameter, ε_m, are equal. In the special case of completely mobile interfaces, that is, $R_d E_G/(\eta_m D_{\text{eff}}) \to 0$ and $(3\eta_{\text{dl}} + 2\eta_{\text{sh}})/(R_d \eta_m) \to 0$, the mobility parameter, ε_m, does not depend on the droplet size, and from Equation 4.324 and 4.325, the Taylor [891] formula is obtained. It is important to note that the Taylor formula takes into account only the bulk properties of the phases (characterized by η_d/η_m); in such a case ε_m is independent of R_d and the Taylor equation is also applicable to polydisperse emulsions. If only the Marangoni effect is neglected ($E_G \to 0$), then Equations 4.324 and 4.325 become equivalent to the Oldroyd [894] formula, which is originally derived only for monodisperse emulsions.

For higher values of the particle volume fraction, the rheological behavior of the colloidal dispersions becomes rather complex. We will consider qualitatively the observed phenomena, and next we will review available semiempirical expressions.

For a simple shear (Couette) flow, the relation between the applied stress, τ, and the resulting shear rate, $\dot{\gamma}$, can be expressed in the form:

$$\tau = \eta \dot{\gamma} \tag{4.326}$$

(e.g., when a liquid is sheared between two plates parallel to the xy plane, we have $\dot{\gamma} = \partial v_x / \partial z$.) A typical plot of $\dot{\gamma}$ versus τ is shown in Figure 4.64a. For low and high shear rates, we observe Newtonian behavior (η = constant), whereas in the intermediate region a transition from the lower shear rate viscosity, η_0, to the higher shear rate viscosity, η_∞, takes place. This is also visualized in Figure 4.64b, where the viscosity of the colloidal dispersion, η, is plotted versus the shear rate, $\dot{\gamma}$; note that in the intermediate zone η has a minimum value [636,640].

Note also that both η_0 and η_∞ depend on the particle volume fraction, ϕ. De Kruif et al. [899] proposed the semiempirical expansions:

$$\frac{\eta_0}{\eta_m} = 1 + 2.5\phi + (4 \pm 2)\phi^2 + (42 \pm 10)\phi^3 + \cdots \tag{4.327}$$

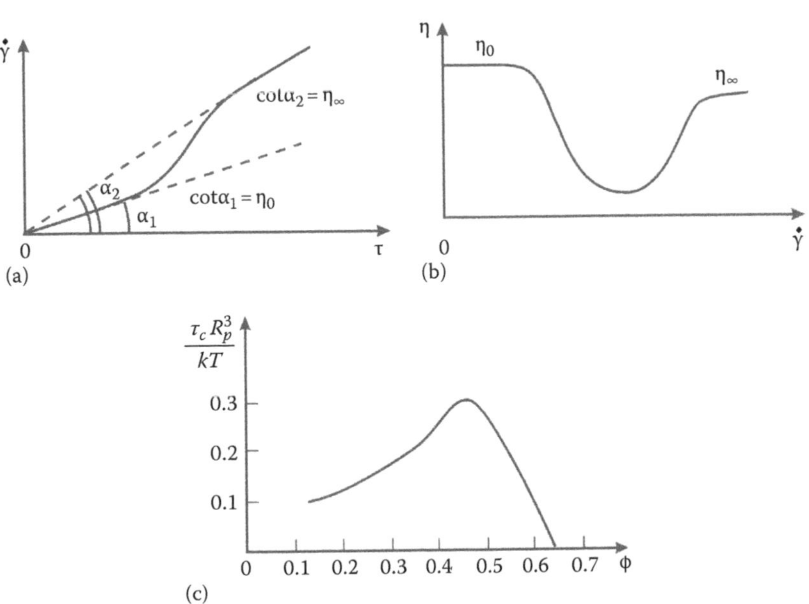

FIGURE 4.64 Qualitative presentation of basic relations in rheology of suspensions: (a) rate of strain, $\dot{\gamma}$, vs. applied stress, τ (see Equation 4.326); (b) average viscosity of a suspension, η, vs. rate of strain, $\dot{\gamma}$; (c) dimensionless parameter τ_c (Equation 4.330) vs. particle volume fraction ϕ.

$$\frac{\eta_\infty}{\eta_m} = 1 + 2.5\phi + (4 \pm 2)\phi^2 + (25 \pm 7)\phi^3 + \cdots \tag{4.328}$$

as well as two empirical expressions which can be used in the whole range of values of ϕ:

$$\frac{\eta_0}{\eta_m} = \left(1 - \frac{\phi}{0.63}\right)^{-2}, \quad \frac{\eta_\infty}{\eta_m} = \left(1 - \frac{\phi}{0.71}\right)^{-2} \tag{4.329}$$

In regard to the dependence of η on the shear stress, τ, Russel et al. [404] reported that for the intermediate values of τ, corresponding to non-Newtonian behavior (Figure 4.64a and b), the experimental data correlate reasonably well with the expression:

$$\frac{\eta - \eta_\infty}{\eta - \eta_0} = \frac{1}{1 + (\tau/\tau_c)^n} \tag{4.330}$$

with $1 \leq n \leq 2$, where τ_c is the value of τ for which $\eta = (\eta_0 + \eta_\infty)/2$. In its own turn, τ_c depends on the particle volume fraction ϕ (see Figure 4.64c). We see that τ_c increases with the volume fraction, ϕ, in dilute dispersions then passes through a maximum and finally decreases down to zero; note that $\tau_c \to 0$ corresponds to $\eta \to \eta_\infty$. The peak at $\phi \approx 0.5$ is the only indication that the hard-sphere disorder–order transition either occurs or is rheologically significant in these systems [404].

The restoring force for a dispersion to return to a random, isotropic situation at rest is either Brownian (thermal fluctuations) or osmotic [912]. The former is most important for submicrometer particles and the latter for larger particles. Changing the flow conditions changes the structure, and this leads to thixotropic effects, which are especially strong in flocculated systems.

Krieger and Dougherty [913] applied the theory of corresponding states to obtain the following expression for the viscosity of hard-sphere dispersions:

$$\frac{\eta}{\eta_m} = \left(1 - \frac{\phi}{\phi_{\max}}\right)^{-[\eta]\phi_{\max}} \tag{4.331}$$

where

$[\eta]$ is the dimensionless intrinsic viscosity, which has a theoretical value of 2.5 for monodisperse rigid spheres

$\phi_{\max}$ is the maximum packing volume fraction for which the viscosity η diverges

The value of $\phi_{\max}$ depends on the type of packing of the particles [636] (Table 4.8). The maximum packing fraction, $\phi_{\max}$, is very sensitive to particle-size distribution and particle shape [914]. Broader particle-size distributions have greater values of $\phi_{\max}$. On the other hand, nonspherical particles lead to poorer space-filling and hence lower $\phi_{\max}$. Table 4.9 presents the values of $[\eta]$ and $\phi_{\max}$ obtained by fitting the results of a number of experiments on dispersions of asymmetric particles using Equation 4.331. The trend of $[\eta]$ to increase and of $\phi_{\max}$ to decrease with increasing asymmetry is clearly seen, but the product, $[\eta]\phi_{\max}$, is almost constant; $[\eta]\phi_{\max}$ is about 2 for spheres and about 1.4 for fibers. This fact can be utilized to estimate the viscosity of a wide variety of dispersions.

A number of rheological experiments with foams and emulsions are summarized in the reviews by Prud'home and Khan [915] and Tadros [916]. These experiments demonstrate the influence

TABLE 4.8
Maximum Packing Volume Fraction, ϕ_{max}, for Various Arrangements of Monodisperse Spheres

Arrangement	ϕ_{max}
Simple cubic	0.52
Minimum thermodynamically stable configuration	0.548
Hexagonally packed sheets just touching	0.605
Random close packing	0.637
Body-centered cubic packing	0.68
Face-centered cubic/hexagonal close packed	0.74

TABLE 4.9
Values of [η] and ϕ_{max} for a Number of Dispersions Obtained by Fitting Experimental Data by Means of Equation 4.331

System	[η]	ϕ_{max}	$[\eta]\phi_{max}$	References
Spheres (submicron)	2.7	0.71	1.92	De Kruif et al. [899]
Spheres (40 μm)	3.28	0.61	2.00	Giesekus [926]
Ground gypsum	3.25	0.69	2.24	Turian and Yuan [927]
Titanium dioxide	4.0	0.55	2.75	Turian and Yuan [927]
Glass rods (30 × 700 μm)	9.25	0.268	2.48	Clarke [928]
Quartz grains (53–76 μm)	4.8	0.371	2.15	Clarke [928]
Glass fibers				
Axial ratio-7	3.8	0.374	1.42	Giesekus [926]
Axial ratio-14	4.03	0.26	1.31	Giesekus [926]
Axial ratio-21	6.0	0.233	1.40	Giesekus [926]

of films between the droplets (or bubbles) on the shear viscosity of the dispersion as a whole. Unfortunately, there is no consistent theoretical explanation of this effect accounting for the different hydrodynamic resistance of the films between the deformed fluid particles as compared to the nondeformed spherical particles (see Sections 4.5.2 and 4.5.3). In the case of emulsions and foams, the deformed droplets or bubbles have a polyhedral shape, and maximum packing fraction can be $\phi_{max} \approx 0.9$ and even higher. For this case, a special geometrical rheological theory has been developed [663,917,918].

Wessel and Ball [919] and Kanai et al. [920] studied in detail the effects of shear rate on the fractal structure of flocculated emulsion drops. They showed that the size of the flocs usually decreases with the increase of the shear stress; often the flocs are split to single particles at high shear rates. As a result, the viscosity decreases rapidly with the increase of shear rate.

Interesting effects are observed when dispersion contains both larger and smaller particles; the latter are usually polymer coils, spherical or cylindrical surfactant micelles, or microemulsion droplets. The presence of the smaller particles may induce clustering of the larger particles due to the depletion attraction (see Section 4.4.5.3.3); such effects are described in the works on surfactant-flocculated and polymer-flocculated emulsions [921–924]. Other effects can be observed in dispersions representing mixtures of liquid and solid particles. Yuhua et al. [925] have established that if the size of the solid particles is larger than three times the size of the emulsion drops, the emulsion can be treated as a continuous medium (of its own average viscosity), in which the solid particles are dispersed; such treatment is not possible when the solid particles are smaller.

Rheological properties of foams (elasticity, plasticity, and viscosity) play an important role in foam production, transportation, and applications. In the absence of external stress, the bubbles in foams are symmetrical and the tensions of the formed foam films are balanced inside the foam and close to the walls of the vessel [929]. At low external shear stresses, the bubbles deform and the deformations of the thin liquid films between them create elastic shear stresses. At a sufficiently large applied shear stress, the foam begins to flow. This stress is called the *yield stress*, τ_0. Then, Equation 4.326 has to be replaced with the Bingham plastic model [930]:

$$\tau = \tau_0 + \eta\dot{\gamma} \tag{4.332}$$

Experiments show that in steadily sheared foams and concentrated emulsions, the viscosity coefficient η depends on the rate of shear strain, and in most cases the Herschel–Bulkley equation [931] is applicable:

$$\tau = \tau_0 + K\dot{\gamma}^n, \quad \eta = K\dot{\gamma}^{n-1} \tag{4.333}$$

Here

K is the consistency

n is the power-law index; $n < 1$ for shear thinning, whereas $n > 1$ for shear thickening

Systematic studies of the foam rheology [932–939] show that the power-law index varies between 0.25 and 0.5 depending on the elasticity of the individual air–solution surfaces. If the elasticity is lower than 10 mN/m, then n is close to 0.25, whereas for large surface elasticity (>100 mN/m) n increases to 0.5.

4.6 KINETICS OF COAGULATION

There are three scenarios for the occurrence of a two-particle collision in a dispersion depending on the type of particle–particle interactions. (1) If the repulsive forces are predominant, the two colliding particles will rebound and the colloidal dispersion will be stable. (2) When at a given separation the attractive and repulsive forces counterbalance each other (the film formed upon particle collision is stable), aggregates or flocs of attached particles can appear. (3) When the particles are fluid and the attractive interaction across the film is predominant, the film is unstable and ruptures; this leads to coalescence of the drops in emulsions or of the bubbles in foams.

To a great extent, the occurrence of coagulation is determined by the energy, U, of particle–particle interaction. U is related to the disjoining pressure, Π, by means of Equations 4.173 and 4.174. Qualitatively, the curves Π versus h (see Figure 4.17) and U versus h are similar. The coagulation is called fast or slow depending on whether the electrostatic barrier (see Figure 4.17) is less than kT or much higher than kT. In addition, the coagulation is termed *reversible* or *irreversible* depending on whether the depth of the primary minimum (see Figure 4.17) is comparable with kT or much greater than kT.

Three types of driving forces can lead to coagulation. (1) The body forces, such as gravity and centrifugal force, cause sedimentation of the heavier particles in suspensions or creaming of the lighter droplets in emulsions. (2) For the particles that are smaller than about 1 μm, the Brownian stochastic force dominates the body forces, and the Brownian collision of two particles becomes a prerequisite for their attachment and coagulation. (3) The temperature gradient in fluid dispersions causes thermocapillary migration of the particles driven by the Marangoni effect. The particles moving with different velocities can collide and form aggregates.

4.6.1 Irreversible Coagulation

The kinetic theory of fast irreversible coagulation was developed by von Smoluchowsky [940,941]. Later, the theory was extended to the case of slow and reversible coagulation. In any case of coagulation (flocculation), the general set of kinetic equations reads:

$$\frac{dn_k}{dt} = \frac{1}{2}\sum_{i=1}^{k-1} a_f^{i,k-i} n_i n_{k-i} - n_k \sum_{i=1}^{\infty} a_f^{k,i} n_i + q_k \quad (k = 1,2,\ldots) \tag{4.334}$$

Here

- t is time
- n_1 denotes the number of single particles per unit volume
- n_k is the number of aggregates of k particles ($k = 2, 3, \ldots$) per unit volume
- $a_f^{i,j}$ ($i, j = 1, 2, 3, \ldots$) are rate constants of flocculation (coagulation; see Figure 4.65)
- q_k denotes the flux of aggregates of size k which are products of other processes, different from the flocculation itself (say, the reverse process of aggregate disassembly or the droplet coalescence in emulsions; see Equations 4.346 and 4.350)

In the special case of irreversible coagulation without coalescence, we have $q_k \equiv 0$. The first term in the right-hand side of Equation 4.334 is the rate of formation of k aggregates by the merging of two smaller aggregates, whereas the second term expresses the rate of loss of k aggregates due to their incorporation into larger aggregates. The total concentration of aggregates (as kinetically independent units), n, and the total concentration of the constituent particles (including those in aggregated form), n_{tot}, can be expressed as

$$n = \sum_{k=1}^{\infty} n_k, \quad n_{tot} = \sum_{k=1}^{\infty} k n_k \tag{4.335}$$

The rate constants can be expressed in the form:

$$a_f^{i,j} = 4\pi D_{i,j}^{(0)} (R_i + R_j) E_{i,j} \tag{4.336}$$

where

- $D_{i,j}^{(0)}$ is the relative diffusion coefficients for two flocks of radii R_i and R_j and aggregation number i and j, respectively
- $E_{i,j}$ is the so-called collision efficiency [696,942]. We give expressions for $D_{i,j}^{(0)}$ and $E_{i,j}$ appropriate for various physical situations in the following section.

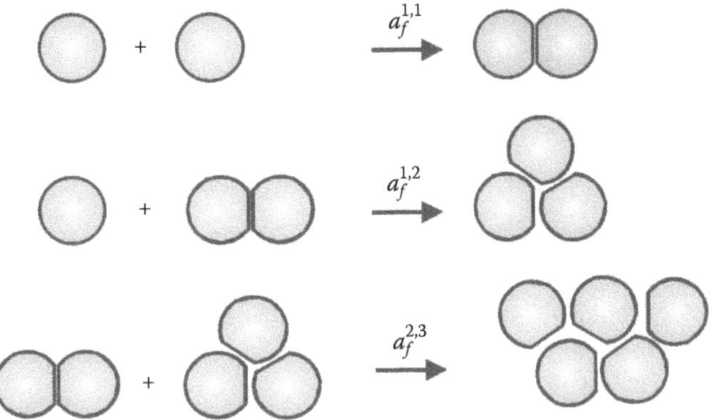

FIGURE 4.65 Elementary acts of flocculation according to the Smoluchowski scheme; $a_f^{i,j}$ ($i, j = 1, 2, 3, \ldots$) are rate constants of flocculation.

The Einstein approach (see Equation 4.320), combined with the Rybczynski–Hadamar equation (Equation 4.318), leads to the following expression for the relative diffusivity of two isolated Brownian droplets:

$$D_{i,j}^{(0)} = \frac{kT}{2\pi\eta}\frac{\eta_d + \eta}{3\eta_d + 2\eta}\left(\frac{1}{R_i} + \frac{1}{R_j}\right) \quad \text{(perikinetic coagulation)} \tag{4.337}$$

The limiting case $\eta_d \to 0$ corresponds to two bubbles, whereas in the other limit ($\eta_d \to \infty$) Equation 4.337 describes two solid particles or two fluid particles of tangentially immobile surfaces.

When the particle relative motion is driven by a body force or by the thermocapillary migration (rather than by self-diffusion), Equation 4.337 is no longer valid. Instead, in Equation 4.336 we have to formally substitute the following expression for $D_{i,j}^{(0)}$ (see Rogers and Davis [943]):

$$D_{i,j}^{(0)} = \frac{1}{4}(R_i + R_j)\left|\mathbf{v}_i - \mathbf{v}_j\right| \quad \text{(orthokinetic coagulation)} \tag{4.338}$$

Here $\mathbf{v}_j$ denotes the velocity of a flock of aggregation number *j*. Physically, Equation 4.338 accounts for the fact that some particle (usually a larger one) moves faster than the remaining particles and can "capture" them upon collision. This type of coagulation is called *orthokinetic* to distinguish it from the self-diffusion-driven perikinetic coagulation described by Equation 4.337. In the case of gravity-driven flocculation, we can identify $\mathbf{v}_j$ with the velocity U in Equation 4.318, where F is to be set equal to the gravitational force exerted on the particle; for a solid particle or a fluid particle of tangentially immobile surface, this yields $\mathbf{v}_j = 2g\Delta\rho R_j^2/(9\eta)$ with g the acceleration due to gravity and $\Delta\rho$ the density difference between the two phases.

In the case of orthokinetic coagulation of liquid drops driven by the thermocapillary migration, the particle velocity $\mathbf{v}_j$ is given by the expression (see Young et al. [944]):

$$\mathbf{v}_j = \frac{2R_j E_T \lambda}{(3\eta_d + 2\eta)(\lambda_d + 2\lambda)}\nabla(\ln T) \quad \text{(thermocapillary velocity)} \tag{4.339}$$

where the thermal conductivity of the continuous and disperse phases are denoted by λ and λ_d, respectively. The interfacial thermal elasticity, E_T, is defined by Equation 4.293.

The collision efficiency, $E_{i,j}$, in Equation 4.336 accounts for the interactions (of both hydrodynamic and intermolecular origin) between two colliding particles. The inverse of $E_{i,j}$ is often called the *stability ratio* or the *Fuchs factor* [945] and can be expressed in the following general form [14,696]:

$$W_{i,j} = \frac{1}{E_{i,j}} = 2\int_0^\infty \frac{\beta(s)}{(s+2)^2}\exp\left[\frac{U_{i,j}(s)}{kT}\right]ds, \quad s \equiv \frac{2h}{R_i + R_j}$$

$$\beta \equiv \left(2\pi\eta R_* \frac{3\eta_d + 2\eta}{\eta_d + \eta}\right)^{-1}\frac{F_z}{V_z} \tag{4.340a}$$

where

h is the closest surface-to-surface distance between the two particles
R_* is defined by Equation 4.270
$U_{i,j}(s)$ is the energy of (nonhydrodynamic) interactions between the particles (see Section 4.4)
$\beta(s)$ accounts for the hydrodynamic interactions
F_z/V_z is the particle friction coefficient

Thus, $\beta \to 1$ for $s \to \infty$, insofar as for large separations the particles obey the Rybczynski–Hadamar equation (Equation 4.318). In the opposite limit, $s \ll 1$, that is, close approach of the two particles,

F_z/V_z can be calculated from Equations 4.271, 4.294 through 4.297, or 4.301, depending on the specific case. In particular, for $s \ll 1$ we have $\beta \propto 1/s$ for two solid particles (or fluid particles of tangentially immobile surfaces), $\beta \propto s^{-1/2}$ for two liquid droplets, and $\beta \propto \ln s$ for two gas bubbles. We see that for two solid particles ($\beta \propto 1/s$), the integral in Equation 4.340a may be divergent. To overcome this problem, one usually accepts that for the smallest separations $U_{i,j}$ is dominated by the van der Waals interaction, as given by Equation 4.185, that is, $U_{i,j} \to -\infty$, and, consequently, the integrand in Equation 4.340a tends to zero for $s \to 0$.

Note that the value of $W_{i,j}$ is determined mainly by the values of the integrand in the vicinity of the electrostatic maximum (barrier) of $U_{i,j}$ (see Figure 4.17), insofar as $U_{i,j}$ enters Equation 4.340a as an exponent. By using the method of the saddle point, Derjaguin [14] estimated the integral in Equation 4.340a:

$$W_{i,j} \equiv \frac{1}{E_{i,j}} \approx \left[\frac{8\pi kT}{-U''_{i,j}(s_m)}\right]^{1/2} \frac{\beta(s_m)}{(s_m+2)^2} \exp\left[\frac{U_{i,j}(s_m)}{kT}\right] \tag{4.340b}$$

where s_m denotes the value of s corresponding to the maximum. We see that the larger the barrier, $U_{i,j}(s_m)$, the smaller the collision efficiency, $E_{i,j}$, and the slower the coagulation.

Note also that for imaginary particles, which experience neither long-range surface forces ($U_{i,j} = 0$) nor hydrodynamic interactions ($\beta = 1$), Equation 4.340a yields a collision efficiency $E_{i,j} = 1$ and Equation 4.336 reduces to the Smoluchowski [940,941] expression for the rate constant of the fast irreversible coagulation. In this particular case, Equation 4.334 represents an infinite set of nonlinear differential equations. If all flocculation rate constants are the same and equal to a_f, the problem has a unique exact solution [940,941]:

$$n = \frac{n_0}{1+a_f n_0 t/2}, \qquad n_k = n_0 \frac{(a_f n_0 t/2)^{k-1}}{(1+a_f n_0 t/2)^{k+1}} \quad (k = 1,2,\ldots) \tag{4.341}$$

It is supposed that the total average concentration of the constituent particles (in both singlet and aggregated form), n_{tot}, does not change and is equal to the initial number of particles, n_0. Unlike n_{tot}, the concentration of the aggregates, n, decreases with time, while their size increases. Differentiating Equation 4.341 we obtain:

$$\frac{dn}{dt} = -\frac{a_f}{2} n^2, \qquad \frac{d\bar{V}}{dt} = \frac{a_f}{2}\phi_0, \qquad \bar{V} \equiv \frac{\phi_0}{n} \tag{4.342}$$

where

$\bar{V}$ is the average volume per aggregate

ϕ_0 is the initial volume fraction of the constituent particles

Combining Equations 4.336 and 4.342, we obtain the following result for perikinetic (Brownian) coagulation:

$$\frac{\bar{V}}{V_0} = 1 + \frac{t}{t_{Br}}, \qquad t_{Br} = \frac{R_0^2}{3\phi_0 D_0 E_0} \tag{4.343}$$

where

$V_0 = 4\pi R_0^3/3$ is the volume of a constituent particle

t_{Br} is the characteristic time of the coagulation process in this case

E_0 is an average collision efficiency

D_0 is an average diffusion coefficient

In contrast, $\bar{V}$ is not a linear function of time for orthokinetic coagulation. When the flocculation is driven by a body force, that is, in case of sedimentation or centrifugation, we obtain [942]:

$$\frac{\bar{V}}{V_0} = \left(1 - \frac{t}{3t_{\text{bf}}}\right)^{-3}, \quad t_{\text{bf}} = \frac{2R_0}{3\phi_0 v_{\text{bf}} E_0} \tag{4.344}$$

where

t_{bf} is the characteristic time in this case

v_{bf} is an average velocity of aggregate motion

As discussed earlier, when the body force is gravitational, we have $v_{\text{bf}} = 2g\Delta\rho R_0^2/(9\eta)$.

When the orthokinetic coagulation is driven by the thermocapillary migration, the counterpart of Equation 4.340 reads [942]

$$\frac{\bar{V}}{V_0} = \exp\left(\frac{t}{t_{\text{tm}}}\right), \quad t_{\text{tm}} = \frac{2R_0}{3\phi_0 v_{\text{tm}} E_0} \tag{4.345}$$

where

v_{tm} is an average velocity of thermocapillary migration

t_{tm} is the respective characteristic time

Note that $D_0 \propto R_0^{-1}, v_{\text{bf}} \propto R_0^2$, and $v_{\text{tm}} \propto R_0$ (see Equations 4.320 and 4.339). Then, from Equations 4.343 through 4.345, it follows that the three different characteristic times exhibit different dependencies on particle radius: $t_{\text{Br}} \propto R_0^3, t_{\text{bf}} \propto R_0^{-1}$, while t_{tm} is independent of R_0. Thus, the Brownian coagulation is faster for the smaller particles, the body force–induced coagulation is more rapid for the larger particles, whereas the thermocapillary driven coagulation is not so sensitive to the particle size [946].

The Smoluchowski scheme based on Equations 4.341 and 4.342 has found numerous applications. An example for biochemical application is the study [947,948] of the kinetics of flocculation of latex particles caused by human gamma globulin in the presence of specific "key-lock" interactions. The infinite set of Smoluchowski equations (Equation 4.334) was solved by Bak and Heilmann [949] in the particular case when the aggregates cannot grow larger than a given size; an explicit analytical solution was obtained by these authors.

4.6.2 Reversible Coagulation

In the case of reversible coagulation, the flocs can disaggregate because the primary minimum (Figure 4.17) is not deep enough [14]. For example, an aggregate composed of $i + j$ particles can be split on two aggregates containing i and j particles. We denote the rate constant of this reverse process by $a_r^{i,j}$ (Figure 4.66a). It is assumed that both the straight process of flocculation (Figure 4.65) and the reverse process (Figure 4.66a) take place. The kinetics of aggregation in this more general case is described by the Smoluchowski set of equations, Equation 4.334, where we have to substitute:

$$q_1 = \sum_{i=1}^{\infty} a_r^{1,i} n_{i+1}, \qquad q_k = \sum_{i=1}^{\infty} a_r^{k,i} n_{i+k} - \frac{1}{2} n_k \sum_{i=1}^{k-1} a_r^{i,k-i} \quad (k = 2,3,\ldots) \tag{4.346}$$

In Equation 4.346 q_k equals the rate of formation of k aggregates in the process of disassembly of larger aggregates minus the rate of decay of the k aggregates. As before, the total number of

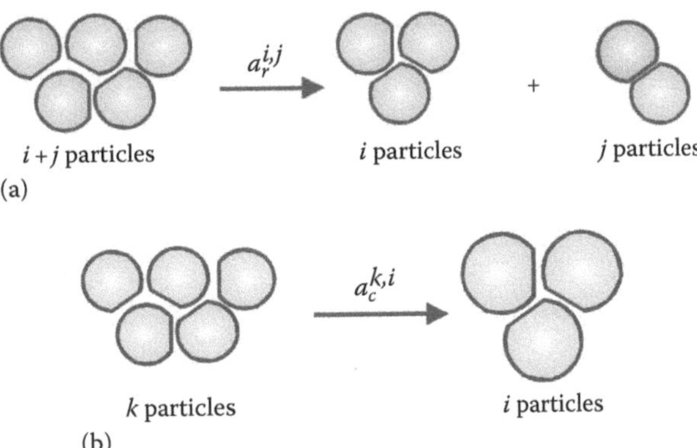

FIGURE 4.66 Elementary acts of aggregate splitting (a) and droplet coalescence within an aggregate (b); $a_r^{i,j}$ and $a_c^{k,i}$ $(i, j, k = 1, 2, 3, \ldots)$ are the rate constants of the respective processes.

constituent particles, n_{tot}, does not change. However, the total number of the aggregates, n, can either increase or decrease depending on whether the straight or the reverse process prevails. Summing up all Equations in 4.334 and using Equation 4.346, we derive the following equation for n:

$$\frac{dn}{dt} = \frac{1}{2}\sum_{i=1}^{\infty}\sum_{j=1}^{\infty}\left(a_r^{i,j}n_{i+j} - a_f^{i,j}n_i n_j\right) \tag{4.347}$$

Martinov and Muller [950] reported a general expression for the rate constants of the reverse process:

$$a_r^{i,j} = \frac{D_{i,j}^{(0)}E_{i,j}}{Z_{i,j}}\frac{1}{(R_i + R_j)^2} \tag{4.348}$$

where $Z_{i,j}$ is the so-called irreversible factor, which can be presented in the form

$$Z_{i,j} = \frac{1}{8}\int_{U_{i,j}<0}(s+2)^2\exp\left[-\frac{U_{i,j}(s)}{kT}\right]ds \tag{4.349}$$

The integration in Equation 4.349 is carried out over the region around the primary minimum, where $U_{i,j}$ takes negative values (see Figure 4.17). In other words, $Z_{i,j}$ is determined by the values of $U_{i,j}$ in the region of the primary minimum, whereas $E_{i,j}$ is determined by the values of $U_{i,j}$ in the region of the electrostatic maximum (see Equations 4.340b and 4.349). When the minimum is deeper, $Z_{i,j}$ is larger and the rate constant in Equation 4.348 is smaller. In addition, as seen from Equations 4.340b and 4.348, the increase of the height of the barrier also decreases the rate of the reverse process. The physical interpretation of this fact is that to detach from an aggregate a particle has to first go out from the well and then to "jump" over the barrier (Figure 4.17).

To illustrate the effect of the reverse process on the rate of flocculation, we solved numerically the set of Equations 4.334, 4.346, and 4.347. To simplify the problem, we used the following assumptions: (1) the von Smoluchowski assumption that all rate constants of the straight process are equal to a_f; (2) aggregates containing more than M particles cannot decay; (3) all rate constants of the reverse process are equal to a_r; and (4) at the initial moment, only single constituent particles of concentration n_0 are available. In Figure 4.67, we plot the calculated curves of n_0/n versus the dimensionless time, $\tau = a_f n_0 t/2$, for a fixed value, $M = 4$, and various values of the ratio of the rate

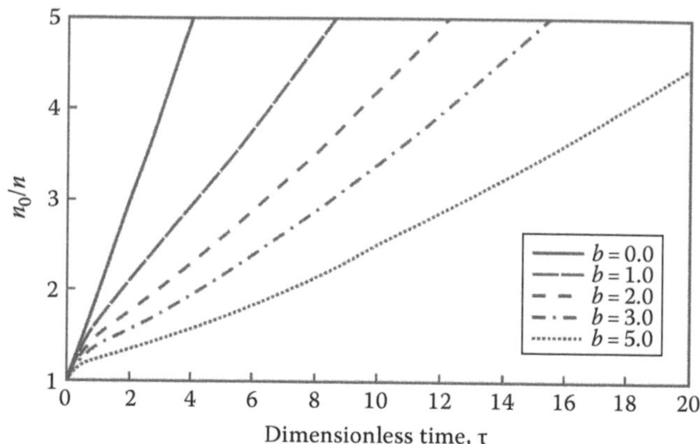

FIGURE 4.67 Reversible coagulation: theoretical plot of the inverse dimensionless aggregate concentration, n_0/n, vs. the dimensionless time, $\tau = a_f n_0 t/2$, in the case of $M = 4$ and various values of the dimensionless ratio, $b = 2a_r/(n_0 a_f)$, of the rate constants of the reverse and straight process, a_r and a_f.

constants of the straight and the reverse process, $b = 2a_r/(n_0 a_f)$. Note that n is defined by Equation 4.335. We see that in an initial time interval all curves in Figure 4.67 touch the von Smoluchowski distribution (corresponding to $b = 0$), but after this period we observe a reduction in the rate of flocculation, which is larger for the curves with larger values of b (larger rate constants of the reverse process). These *S*-shaped curves are typical for the case of reversible coagulation, which is also confirmed by the experiment [14,951].

4.6.3 Kinetics of Simultaneous Flocculation and Coalescence in Emulsions

When coalescence is present, in addition to the flocculation, the total number of constituent drops, n_{tot} (see Equation 4.335), does change, in contrast to the case of pure flocculation considered earlier [34]. Hartland and Gakis [952], and Hartland and Vohra [953] were the first to develop a model of coalescence that relates the lifetime of single films to the rate of phase separation in emulsions of fairly large drops (approximately 1 mm) in the absence of surfactant. Their analysis was further extended by Lobo et al. [954] to quantify the process of coalescence within an already creamed or settled emulsion (or foam) containing drops of size less than 100 μm; these authors also took into account the effect of surfactants, which are commonly used as emulsifiers. Danov et al. [955] generalized the Smoluchowski scheme to account for the fact that the droplets within the flocs can coalesce to give larger droplets, as illustrated in Figure 4.66b. In this case, in the right-hand side of Equation 4.334 we have to substitute [955]

$$q_1 = \sum_{i=2}^{\infty} a_c^{i,1} n_i, \qquad q_k = \sum_{i=k+1}^{\infty} a_c^{i,k} n_i - n_k \sum_{i=1}^{k-1} a_c^{k,i} \quad (k = 2,3,\ldots) \tag{4.350}$$

where $a_c^{k,i}$ is the rate constant of transformation (by coalescence) of an aggregate containing k droplets into an aggregate containing i droplets (see Figure 4.66b). The newly formed aggregate is further involved in the flocculation scheme, which thus accounts for the fact that the flocculation and coalescence processes are interdependent. In this scheme, the total coalescence rate, $a_{c,\text{tot}}^{i}$, and the total number of droplets, n_{tot}, obey the following equation [955]:

$$\frac{dn_{\text{tot}}}{dt} = -\sum_{i=2}^{\infty} a_{c,\text{tot}}^{i} n_i, \qquad a_{c,\text{tot}}^{i} = \sum_{k=1}^{i-1} (i-k) a_c^{i,k} \quad (i = 2,3,\ldots) \tag{4.351}$$

To determine the rate constants of coalescence, $a_c^{k,i}$, Danov et al. [665] examined the effects of droplet interactions and Brownian motion on the coalescence rate in dilute emulsions of micrometer- and submicrometer-sized droplets. The processes of film formation, thinning, and rupture were included as consecutive stages in the scheme of coalescence. Expressions for the interaction energy due to the various DLVO and non-DLVO surface forces between two deformed droplets were obtained [392] (see also Section 4.4).

Average models for the total number of droplets are also available [956,957]. The average model of van den Tempel [956] assumes linear structure of the aggregates. The coalescence rate is supposed to be proportional to the number of contacts within an aggregate. To simplify the problem, van den Tempel has used several assumptions, one of them is that the concentration of the single droplets, n_1, obeys the Smoluchowski distribution (Equation 4.341) for $k = 1$. The average model of Borwankar et al. [957] is similar to that of van den Tempel but is physically more adequate. The assumptions used by the latter authors [957] make their solution more applicable to cases in which the flocculation (rather than the coalescence) is slow and is the rate determining stage. This is confirmed by the curves shown in Figure 4.68 which are calculated for the same rate of coalescence, but for three different rates of flocculation. For relatively high rates of flocculation (Figure 4.68a), the predictions of the three theories differ. For the intermediate rates of flocculation (Figure 4.68b), the prediction of the model by Borwankar et al. [957] is close to that of the more detailed model by Danov et al. [955]. For very low values of the flocculation rate constant, a_f, for which the coalescence is not the rate-determining stage, all three theories [955–957] give numerically close results (Figure 4.68c). Details about the coupling of coalescence and flocculation in dilute oil-in-water emulsions, experimental investigations, and numerical modeling can be found in Refs. [958–966].

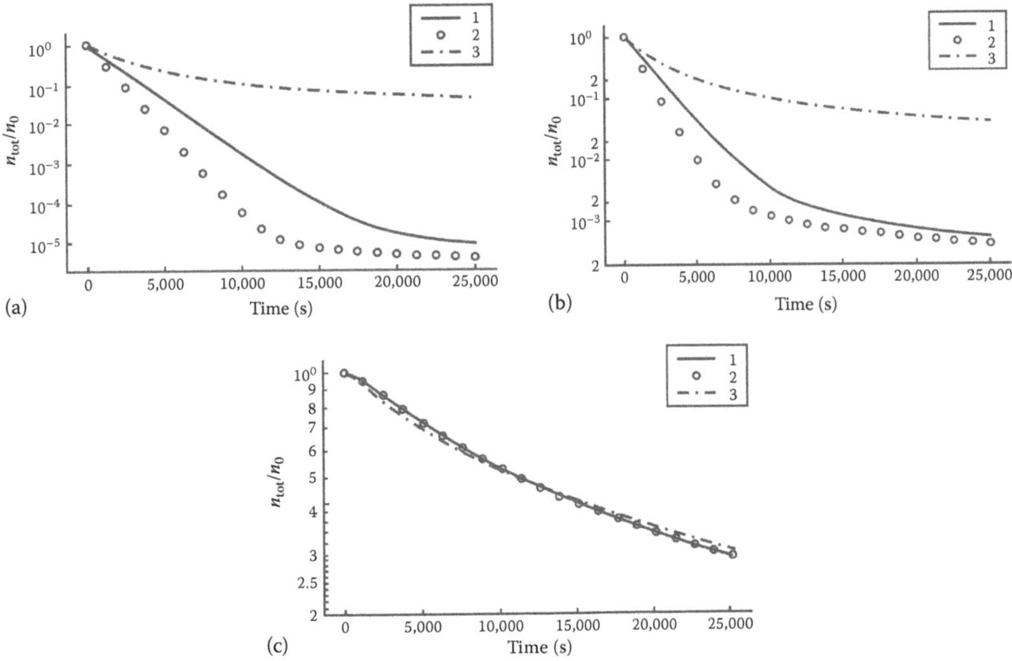

FIGURE 4.68 Relative change in the total number of drops, n_{tot}, vs. time, t; initial number of primary drops $n_0 = 10^{12}$ cm^{-3}; coalescence rate constant $k_c^{2,1} = 10^{-3}$ s^{-1}. Curve 1: numerical solution of Equation 4.351. Curve 2: output of the model of Borwankar et al. [957]. Curve 3: output of the model of van den Tempel [956]. The values of the flocculation rate constant are (a) $a_f = 10^{-11}$ cm^3/s; (b) $a_f = 10^{-13}$ cm^3/s; (c) $a_f = 10^{-16}$ cm^3/s.

ACKNOWLEDGMENTS

The authors gratefully acknowledge the support from the FP7 project Beyond-Everest, and from COST Action CM1101.

REFERENCES

1. Jungermann, E., *Cationic Surfactants*, Marcel Dekker, New York, 1970.
2. Lucassen-Reynders, E.H., *Anionic Surfactants—Physical Chemistry of Surfactant Action*, Marcel Dekker, New York, 1981.
3. Schick, M.J., *Nonionic Surfactants: Physical Chemistry*, Marcel Dekker, New York, 1986.
4. Gibbs, J.W., *The Scientific Papers of J.W. Gibbs*, Vol. 1, Dover, New York, 1961.
5. Ono, S. and Kondo, S., Molecular theory of surface tension in liquids, in *Handbuch der Physik*, Vol. 3/10, Flügge, S. (Ed.), Springer, Berlin, Germany, 1960, pp. 134–280.
6. Adamson, A.W. and Gast, A.P., *Physical Chemistry of Surfaces*, 6th edn., Wiley, New York, 1997.
7. Freundlich, H., *Colloid and Capillary Chemistry*, Methuen, London, U.K., 1926.
8. Langmuir, I., *J. Am. Chem. Soc.*, 40, 1361, 1918.
9. Volmer, M., *Z. Physikal. Chem.*, 115, 253, 1925.
10. Frumkin, A., *Z. Physikal. Chem.*, 116, 466, 1925.
11. Hill, T.L., *An Introduction to Statistical Thermodynamics*, Addison-Wesley, Reading, MA, 1962.
12. Lucassen-Reynders, E.H., *J. Phys. Chem.*, 70, 1777, 1966.
13. Borwankar, R.P. and Wasan, D.T., *Chem. Eng. Sci.*, 43, 1323, 1988.
14. Derjaguin, B.V., *Theory of Stability of Colloids and Thin Liquid Films*, Plenum Press, Consultants Bureau, New York, 1989.
15. Shchukin, E.D., Pertsov, A.V., and Amelina, E.A., *Colloid Chemistry*, Moscow University Press, Moscow, Russia, 1982 (Russian); Elsevier, 2001 (English).
16. Danov, K.D. and Kralchevsky, P.A., *Colloid J.*, 74, 172, 2012.
17. Rosen, M.J. and Aronson, S., *Colloids Surf.*, 3, 201, 1981.
18. Zeldowitch, J., *Acta Physicochim. (USSR)*, 1, 961, 1934.
19. Halsey, G. and Taylor, H.S., *J. Chem. Phys.*, 15, 624, 1947.
20. Gurkov, T.G., Kralchevsky, P.A., and Nagayama, K., *Colloid Polym. Sci.*, 274, 227, 1996.
21. Butler, J.A.V., *Proc. Roy. Soc. Ser. A*, 135, 348, 1932.
22. Fainerman, V.B. and Miller, R., *Langmuir*, 12, 6011, 1996.
23. Vaughn, M.W. and Slattery, J. C., *J. Colloid Interface Sci.*, 195, 1, 1997.
24. Makievski, A.V., Fainerman, V.B., Bree, M., Wüstneck, R., Krägel, J., and Miller, R., *J. Phys. Chem. B*, 102, 417, 1998.
25. Landau, L.D. and Lifshitz, E.M., *Statistical Physics*, Part 1, Pergamon, Oxford, U.K., 1980.
26. Kralchevsky, P.A., Danov, K.D., Broze, G., and Mehreteab, A., *Langmuir*, 15, 2351, 1999.
27. Hachisu, S., *J. Colloid Interface Sci.*, 33, 445, 1970.
28. Kalinin, V.V. and Radke, C.J., *Colloids Surf. A*, 114, 337, 1996.
29. Warszyński, P., Barzyk, W., Lunkenheimer, K., and Fruhner, H., *J. Phys. Chem. B*, 102, 10948, 1998.
30. Prosser, A.J. and Frances, E.I., *Colloids Surf. A*, 178, 1, 2001.
31. Kirkwood, J.G. and Oppenheim, I., *Chemical Thermodynamics*, McGraw-Hill, New York, 1961.
32. Robinson, R.A. and Stokes, R.H., *Electrolyte Solutions*, Butterworths, London, U.K., 1959.
33. Gouy, L.G., *J. Phys.*, 9, 457, 1910.
34. Davies, J. and Rideal, E., *Interfacial Phenomena*, Academic Press, New York, 1963.
35. Grahame, D.C., *Chem. Rev.*, 41, 441, 1947.
36. Israelachvili, J.N., *Intermolecular and Surface Forces*, Academic Press, London, U.K., 2011.
37. Kralchevsky, P.A. and Nagayama, K., *Particles at Fluid Interfaces and Membranes*, Elsevier, Amsterdam, the Netherlands, 2001.
38. Matijevič, E. and Pethica, B.A., *Trans. Faraday Soc.*, 54, 1382, 1958.
39. van Voorst Vader, F., *Trans. Faraday Soc.*, 56, 1067, 1960.
40. Tajima, K., *Bull. Chem. Soc. Jpn.*, 44, 1767, 1971.
41. Stern, O., *Ztschr. Elektrochem.*, 30, 508, 1924.
42. Tajima, K., Muramatsu, M., and Sasaki, T., *Bull. Chem. Soc. Jpn.*, 43, 1991, 1970.
43. Tajima, K., *Bull. Chem. Soc. Jpn.*, 43, 3063, 1970.
44. Kolev, V.L., Danov, K.D., Kralchevsky, P.A., Broze, G., and Mehreteab, A., *Langmuir*, 18, 9106, 2002.

45. Cross, A.W. and Jayson, G.G., *J. Colloid Interface Sci.*, 162, 45, 1994.
46. Johnson, S.B., Drummond, C.J., Scales, P.J., and Nishimura, S., *Langmuir*, 11, 2367, 1995.
47. Alargova, R.G., Danov, K.D., Petkov, J.T., Kralchevsky, P.A., Broze, G., and Mehreteab, A., *Langmuir*, 13, 5544, 1997.
48. Rathman, J.F. and Scamehorn, J.F., *J. Phys. Chem.*, 88, 5807, 1984.
49. Berr, S.S., Coleman, M.J., Marriot, J., and Johnson Jr., J.S., *J. Phys. Chem.*, 90, 6492, 1986.
50. Rosen, M.J., *Surfactants and Interfacial Phenomena*, Wiley, New York, 1989.
51. Clint, J., *Surfactant Aggregation*, Chapman & Hall, London, U.K., 1992.
52. Alargova, R.G., Danov, K.D., Kralchevsky, P.A., Broze, G., and Mehreteab, A., *Langmuir*, 14, 4036, 1998.
53. Dimov, N.K., Kolev, V.L., Kralchevsky, P.A., Lyutov, L.G., Brose, G., and Mehreteab, A., *J. Colloid Interface Sci.*, 256, 23, 2002.
54. Kralchevsky, P.A., Danov, K.D., Kolev, V.L., Broze, G., and Mehreteab, A., *Langmuir*, 19, 5004, 2003.
55. Danov, K.D., Kralchevsky, P.A., Ananthapadmanabhan, K.P., and Lips, A., *J. Colloid Interface Sci.*, 300, 809, 2006.
56. Lunkenheimer, K., Barzyk, W., Hirte, R., and Rudert, R., *Langmuir*, 19, 6140, 2003.
57. Christov, N.C., Danov, K.D., Kralchevsky, P.A., Ananthapadmanabhan, K.P., and Lips, A., *Langmuir*, 22, 7528, 2006.
58. Danov, K.D., Kralchevska, S.D., Kralchevsky, P.A., Broze, G., and Mehreteab, A., *Langmuir*, 19, 5019, 2003.
59. Danov, K.D., Kralchevska, S.D., Kralchevsky, P.A., Ananthapadmanabhan, K.P., and Lips, A., *Langmuir*, 20, 5445, 2004.
60. Valkovska, D.S., Shearman, G.C., Bain, C.D., Darton, R.C., and Eastoe, J., *Langmuir*, 20, 4436, 2004.
61. Day, J.P.R., Campbell, R.A., Russell, O.P., and Bain, C.D., *J. Phys. Chem. C*, 111, 8757, 2007.
62. Valkovska, D.S., Danov, K.D., and Ivanov, I.B., *Colloids Surf. A*, 175, 179, 2000.
63. Danov, K.D., Kralchevsky, P.A., and Ivanov, I.B., Dynamic processes in surfactant stabilized emulsions, in *Encyclopedic Handbook of Emulsion Technology*, Sjöblom, J. (Ed.), Marcel Dekker, New York, 2001, Chapter 26 pp. 621–659.
64. Dukhin, S.S., Kretzschmar, G., and Miller, R., *Dynamics of Adsorption at Liquid Interfaces*, Elsevier, Amsterdam, the Netherlands, 1995.
65. Eastoe, J. and Dalton, J.S., *Adv. Colloid Interface Sci.*, 85, 103, 2000.
66. Rayleigh, L., *Proc. Roy. Soc. (Lond.)*, 29, 71, 1879.
67. Bohr, N., *Philos. Trans. Roy. Soc. (Lond.) A*, 209, 281, 1909.
68. Defay, R. and Pétré, G., Dynamic surface tension, in *Surface and Colloid Science*, Vol. 3, Matijevič, E. (Ed.), Wiley, New York, 1971, p. 27.
69. Miller, R. and Kretzschmar, G., *Adv. Colloid Interface Sci.*, 37, 97, 1991.
70. Wantke, K.-D., Lunkenheimer, K., and Hempt, C., *J. Colloid Interface Sci.*, 159, 28, 1993.
71. Chang, C.-H. and Franses, E.I., *J. Colloid Interface Sci.*, 164, 107, 1994.
72. Johnson, D.O. and Stebe, K.J., *J. Colloid Interface Sci.*, 182, 525, 1996.
73. Horozov, T. and Arnaudov, L., *J. Colloid Interface Sci.*, 219, 99, 1999.
74. Horozov, T. and Arnaudov, L., *J. Colloid Interface Sci.*, 222, 146, 2000.
75. van den Tempel, M. and Lucassen-Reynders, E.H., *Adv. Colloid Interface Sci.*, 18, 281, 1983.
76. Langevin, D., *Colloids Surf.*, 43, 121, 1990.
77. Lemaire, C. and Langevin, D., *Colloids Surf.*, 65, 101, 1992.
78. Grigorev, D.O., Krotov, V.V., and Noskov, B.A., *Colloid J.*, 56, 562, 1994.
79. Mysels, K.J., *Colloids Surf.*, 43, 241, 1990.
80. Kralchevsky, P.A., Radkov, Y.S., and Denkov, N.D., *J. Colloid Interface Sci.*, 161, 361, 1993.
81. Fainerman, V.B., Miller, R., and Joos, P., *Colloid Polym. Sci.*, 272, 731, 1994.
82. Fainerman, V.B. and Miller, R., *J. Colloid Interface Sci.*, 176, 118, 1995.
83. Horozov, T.S., Dushkin, C.D., Danov, K.D., Arnaudov, L.N., Velev, O.D., Mehreteab, A., and Broze, G., *Colloids Surf. A*, 113, 117, 1996.
84. Mishchuk, N.A., Dukhin, S.S., Fainerman, V.B., Kovalchuk, V.I., and Miller, R., *Colloids Surf. A*, 192, 157, 2001.
85. van den Bogaert, R. and Joos, P., *J. Phys. Chem.*, 83, 17, 1979.
86. Möbius, D. and Miller, R. (Eds.), *Drops and Bubbles in Interfacial Research*, Elsevier, Amsterdam, the Netehrlands, 1998.
87. Jho, C. and Burke, R., *J. Colloid Interface Sci.*, 95, 61, 1983.
88. Joos, P. and van Hunsel, J., *Colloid Polym. Sci.*, 267, 1026, 1989.

89. Fainerman, V.B. and Miller, R., *Colloids Surf. A*, 97, 255, 1995.
90. Miller, R., Bree, M., and Fainerman, V.B., *Colloids Surf. A*, 142, 237, 1998.
91. Senkel, O., Miller, R., and Fainerman, V.B., *Colloids Surf. A*, 143, 517, 1998.
92. Bain, C.D., Manning-Benson, S., and Darton, R.C., *J. Colloid Interface Sci.*, 229, 247, 2000.
93. Rotenberg, Y., Boruvka, L., and Neumann, A.W., *J. Colloid Interface Sci.*, 37, 169, 1983.
94. Makievski, A.V., Loglio, G., Krägel, J., Miller, R., Fainerman, V.B., and Neumann, A.W., *J. Phys. Chem.*, 103, 9557, 1999.
95. Joos, P., *Dynamic Surface Phenomena*, VSP BV, AH Zeist, the Netherlands, 1999.
96. Ward, A.F.H. and Tordai, L., *J. Chem. Phys.*, 14, 453, 1946.
97. Miller, R., *Colloid Polym. Sci.*, 259, 375, 1981.
98. McCoy, B.J., *Colloid Polym. Sci.*, 261, 535, 1983.
99. Hansen, R.S., *J. Chem. Phys.*, 64, 637, 1960.
100. Filippov, L.K., *J. Colloid Interface Sci.*, 164, 471, 1994.
101. Daniel, R. and Berg, J.C., *J. Colloid Intrface Sci.*, 237, 294, 2001.
102. Sutherland, K.L., *Aust. J. Sci. Res.*, A5, 683, 1952.
103. Arfken, G.B., Weber, H.J, and Harris, F.E., *Mathematical Methods for Physicists*, Elsevier, Amsterdam, 2013.
104. Korn, G.A. and Korn, T.M., *Mathematical Handbook*, McGraw-Hill, New York, 1968.
105. Danov, K.D., Kolev, V.L., Kralchevsky, P.A., Broze, G., and Mehreteab, A., *Langmuir*, 16, 2942, 2000.
106. Dukhin, S.S., Miller, R., and Kretzschmar, G., *Colloid Polym. Sci.*, 261, 335, 1983.
107. Dukhin, S.S. and Miller, R., *Colloid Polym. Sci.*, 272, 548, 1994.
108. MacLeod, C. and Radke, C.J., *Langmuir*, 10, 3555, 1994.
109. Vlahovska, P.M., Danov, K.D., Mehreteab, A., and Broze, G., *J. Colloid Interface Sci.*, 192, 194, 1997.
110. Danov, K.D., Vlahovska, P.M., Kralchevsky, P.A., Broze, G., and Mehreteab, A., *Colloids Surf. A*, 156, 389, 1999.
111. Diamant, H. and Andelman, D., *J. Phys. Chem.*, 100, 13732, 1996.
112. Diamant, H., Ariel, G., and Andelman, D., *Colloids Surf. A*, 183–185, 259, 2001.
113. Dattwani, S.S. and Stebe, K.J., *J. Colloid Interface Sci.*, 219, 282, 1999.
114. Danov, K.D., Kralchevsky, P.A., Ananthapadmanabhan, K.P., and Lips, A., *J. Colloid Interface Sci.*, 303, 56, 2006.
115. Nayfeh, A.H., *Perturbation Methods*, Wiley, New York, 1973.
116. Rillaerts, E. and Joos, P., *J. Colloid Interface Sci.*, 88, 1, 1982.
117. Durbut, P., Surface activity, in *Handbook of Detergents*, Part A, Broze, G. (Ed.), Marcel Dekker, New York, 1999, Chapter 3 pp. 47–98.
118. Bond, W.N. and Puls, H.O., *Philos. Mag.*, 24, 864, 1937.
119. Doss, K.S.G., *Koll. Z.*, 84, 138, 1938.
120. Blair, C.M., *J. Chem. Phys.*, 16, 113, 1948.
121. Ward, A.F.H., *Surface Chemistry*, Butterworths, London, U.K., 1949.
122. Dervichian, D.G., *Koll. Z.*, 146, 96, 1956.
123. Hansen, R.S. and Wallace, T., *J. Phys. Chem.*, 63, 1085, 1959.
124. Baret, J.F., *J. Phys. Chem.*, 72, 2755, 1968.
125. Baret, J.F., *J. Chem. Phys.*, 65, 895, 1968.
126. Baret, J.F., *J. Colloid Interface Sci.*, 30, 1, 1969.
127. Borwankar, R.P. and Wasan, D.T., *Chem. Eng. Sci.*, 38, 1637, 1983.
128. Alexandrov, N.A., Marinova, K.G., Gurkov, T.D., Danov, K.D., Kralchevsky, P.A., Stoyanov, S.D., Blijdenstein, T.B.J., Arnaudov, L.N., Pelan, E.G., and Lips, A., *J. Colloid Interface Sci.*, 376, 296, 2012.
129. Danov, K.D., Valkovska, D.S., and Kralchevsky, P.A., *J. Colloid Interface Sci.*, 251, 18, 2002.
130. Dong, C., Hsu, C.-T., Chin, C.-Y., and Lin, S.-Y., *Langmuir*, 16, 4573, 2000.
131. McBain, J.W., *Trans. Faraday Soc.*, 9, 99, 1913.
132. Vincent, B., *Adv. Colloid Interface Sci.*, 203, 51, 2014.
133. Missel, P.J., Mazer, N.A., Benedek, G.B., Young, C.Y., and Carey, M.C., *J. Phys. Chem.*, 84, 1044, 1980.
134. Anachkov, S.E., Kralchevsky, P.A., Danov, K.D., Georgieva, G.S., and Ananthapadmanabhan, K.P., *J. Colloid Interface Sci.*, 416, 258, 2014.
135. Kralchevsky, P.A., Danov, K.D., Anachkov, S.E., Georgieva, G.S., and Ananthapadmanabhan, K.P., *Curr. Opin. Colloid Interface Sci.*, 18, 524, 2013.
136. Danov, K.D., Kralchevsky, P.A., and Ananthapadmanabhan, K.P., *Adv. Colloid Interface Sci.*, 206, 17, 2014.
137. Mitchell, D.J. and Ninham, B.W., *J. Phys. Chem.*, 87, 2996, 1983.

138. Evans, E.A. and Skalak, R., *CRC Crit. Rev. Bioengin.*, 3, 181, 1979.
139. Kralchevsky, P.A., Danov, K.D., and Anachkov, S.E., *Colloid J.*, 76, 255, 2014.
140. Kresheck, G.C., Hamory, E., Davenport, G., and Scheraga, H.A., *J. Am. Chem. Soc.*, 88, 246, 1966.
141. Aniansson, E.A.G. and Wall, S.N., *J. Phys. Chem.*, 78, 1024, 1974.
142. Lucassen, J., *Faraday Discuss. Chem. Soc.*, 59, 76, 1975.
143. Noskov, B.A., *Kolloidn. Zh.*, 52, 509, 1990.
144. Johner, A. and Joanny, J.F., *Macromolecules*, 23, 5299, 1990.
145. Dushkin, C.D., Ivanov, I.B., and Kralchevsky, P.A., *Colloids Surf.*, 60, 235, 1991.
146. Joos, P. and van Hunsel, J., *Colloids Surf.*, 33, 99, 1988.
147. Li, B., Joos, P., and van Uffelen, M., *J. Colloid Interface Sci.*, 171, 270, 1995.
148. Geeraerts, G. and Joos, P., *Colloids Surf. A*, 90, 149, 1994.
149. Danov, K.D., Kralchevsky, P.A., Denkov, N.D., Ananthapadmanabhan, K.P., and Lips, A., *Adv. Colloid Interface Sci.*, 119, 1, 2006.
150. Danov, K.D., Kralchevsky, P.A., Denkov, N.D., Ananthapadmanabhan, K.P., and Lips, A., *Adv. Colloid Interface Sci.*, 119, 17, 2006.
151. Danov, K.D., Kralchevsky, P.A., Ananthapadmanabhan, K.P., and Lips, A., *Colloids Surf. A*, 282–283, 143, 2006.
152. McBain, J.W., *Colloidal Science*, D.C. Heat, Lexington, MA, 1950.
153. Christian, S.D. and Scamehorn, J.F., *Solubilization in Surfactant Aggregates*, Marcel Dekker, New York, 1995.
154. Miller, C.A., Micellar systems and microemulsions: solubilization aspects, in *Handbook of Surface and Colloid Chemistry*, 1st edn., Birdi, K.S. (Ed.), CRC Press, Boca Raton, FL, 1997, p. 157.
155. Vasilescu, M., Caragheorgheopol, A., and Caldararu, H., *Adv. Colloid Interface Sci.*, 89–90, 169, 2001.
156. Carroll, B.J., *J. Colloid Interface Sci.*, 79, 126, 1981.
157. Kabalnov, A. and Weers, J., *Langmuir*, 12, 3442, 1996.
158. Weiss, J., Coupland, J.N., Brathwaite, D., and McClements, D.J., *Colloids Surf. A*, 121, 53, 1997.
159. Todorov, P.D., Kralchevsky, P.A., Denkov, N.D., Broze, G., and Mehreteab, A., *J. Colloid Interface Sci.*, 245, 371, 2002.
160. Kralchevsky, P.A. and Denkov, N.D., Triblock copolymers as promoters of solubilization of oils in aqueous surfactant solutions, in *Molecular Interfacial Phenomena of Polymers and Biopolymers*, Chen, P. (Ed.), Woodhead Publishing, Cambridge, U.K., 2005, Chapter 15, p. 538.
161. Sailaja, D., Suhasini, K.L., Kumar, S., and Gandhi, K.S., *Langmuir*, 19, 4014, 2003.
162. Chan, A.F., Fennel Evans, D., and Cussler, E.L., *AIChE J.*, 22, 1006, 1976.
163. Huang, C., Fennel Evans, D., and Cussler, E.L., *J. Colloid Interface Sci.*, 82, 499, 1981.
164. Shaeiwitz, J.A., Chan, A.F.-C., Cussler, E.L. and Fennel Evans, D., *J. Colloid Interface Sci.*, 84, 47, 1981.
165. Plucinski, P. and Nitsch, W., *J. Phys. Chem.*, 97, 8983, 1993.
166. Chen, B.-H., Miller, C.A., and Garrett, P.R., *Colloids Surf. A*, 128, 129, 1997.
167. Chen, B.-H., Miller, C.A., and Garrett, P.R., *Langmuir*, 14, 31, 1998.
168. Christov, N.C., Denkov, N.D., Kralchevsky, P.A., Broze, G., and Mehreteab, A., *Langmuir*, 18, 7880, 2002.
169. Kralchevsky, P.A., Denkov, N.D., Todorov, P.D., Marinov, G.S., Broze, G., and Mehreteab, A., *Langmuir*, 18, 7887, 2002.
170. Todorov, P.D., Marinov, G.S., Kralchevsky, P.A., Denkov, N.D., Durbut, P., Broze, G., and Mehreteab, A., *Langmuir*, 18, 7896, 2002.
171. Granek, R., *Langmuir*, 12, 5022, 1996.
172. Lawrence, A.S.C., *Discuss. Faraday Soc.*, 25, 51, 1958.
173. Lawrence, A.S.C., Bingham, A., Capper, C.B., and Hume, K., *J. Phys. Chem.*, 68, 3470, 1964.
174. Stowe, L.R. and Shaeiwitz, J.A., *J. Colloid Interface Sci.*, 90, 495, 1982.
175. Raterman, K.T. and Shaeiwitz, J.A., *J. Colloid Interface Sci.*, 98, 394, 1984.
176. Lim, J.-C. and Miller, C.A., *Langmuir*, 7, 2021, 1991.
177. Somasundaran, P. and Krishnakumar, S., *Colloids Surf. A*, 123–124, 491, 1997.
178. Ward, A.J., Kinetics of solubilization in surfactant-based systems, in *Solubilization in Surfactant Aggregates*, Christian, S.D. and Scamehorn, J.F. (Eds.), Marcel Dekker, New York, 1995, Chapter 7, pp. 237–272.
179. Nagarajan, R. and Ganesh, K., *J. Colloid Interface Sci.*, 184, 489, 1996.
180. Lebens, P.J.M. and Keurentjes, J.T.F., *Ind. Eng. Chem. Res.*, 35, 3415, 1996.
181. Xing, L. and Mattice, W.L., *Macromolecules*, 30, 1711, 1997.
182. Kříž, J., Masař, B., and Doskočilová, D., *Macromolecules*, 30, 4391, 1997.

183. Marinov, G., Michels, B., and Zana, R., *Langmuir*, 14, 2639, 1998.
184. Kositza, M.J., Bohne, C., Alexandridis, P.T., Hatton, T.A., and Holzwarth, J.F., *Langmuir*, 15, 322, 1999.
185. Walderhaug, H., *J. Phys. Chem. B*, 103, 3352, 1999.
186. Paterson, I.F., Chowdhry, B.Z., Leharne, S.A., *Langmuir*, 15, 6178, 1999.
187. Bromberg, L. and Temchenko, M., *Langmuir*, 15, 8627, 1999.
188. Laplace, P.S., *Traité de mécanique céleste*; Suppléments au Livre X, 1805, 1806.
189. Bakker, G., Kapillatytät und oberflächenspannung, in *Handbuch der Experimentalphysik*, Band 6, Akademische Verlagsgesellschaft, Leipzig, Germany, 1928.
190. Princen, H.M., The equilibrium shape of interfaces, drops, and bubbles, in *Surface and Colloid Science*, Vol. 2, Matijevic, E. (Ed.), Wiley, New York, 1969, p. 1.
191. Finn, R., *Equilibrium Capillary Surfaces*, Springer-Verlag, New York, 1986.
192. Weatherburn, C.E., *Differential Geometry in Three Dimensions*, Cambridge, U.K., 1930.
193. McConnell, A.J., *Application of Tensor Analysis*, Dover, New York, 1957.
194. Young, T., *Philos. Trans. Roy. Soc. (Lond.)*, 95, 55, 1805.
195. Jonson, R.E. and Dettre, Wettability and contact angles, in *Surface and Colloid Science*, Vol. 2, Matijevic, E. (Ed.), Wiley, New York, 1969, p. 85.
196. Starov, V.M., *Adv. Colloid Interface Sci.*, 39, 147, 1992.
197. Neumann, F., *Vorlesungen über die Theorie der Capillarität*, B.G. Teubner, Leipzig, Germany, 1894.
198. Ivanov, I.B., Kralchevsky, P.A., and Nikolov, A.D., *J. Colloid Interface Sci.*, 112, 97, 1986.
199. Hartland, S. and Hartley, R.W., *Axisymmetric Fluid–Liquid Interfaces*, Elsevier, Amsterdam, the Netherlands, 1976.
200. Kralchevsky, P.A., Eriksson, J.C., and Ljunggren, S., *Adv. Colloid Interface Sci.*, 48, 19, 1994.
201. Tachev, K.D., Angarska, J.K., Danov, K.D., and Kralchevsky, P.A., *Colloids Surf. B*, 19, 61, 2000.
202. Meunier, J. and Lee, L.T., *Langmuir*, 7, 1855, 1991.
203. Dan, N., Pincus, P., and Safran, S.A., *Langmuir*, 9, 2768, 1993.
204. Kralchevsky, P.A., Paunov, V.N. Denkov, N.D., and Nagayama, K., *J. Chem. Soc. Faraday Trans.*, 91, 3415, 1995.
205. Danov, K.D., Kralchevsky, P.A., and Stoyanov, S.D., *Langmuir*, 26, 143, 2010.
206. Basheva, E.S., Kralchevsky, P.A., Christov, N.C., Danov, K.D., Stoyanov, S.D., Blijdenstein, T.B.J., Kim, H.-J., Pelan, E.G., and Lips, A., *Langmuir*, 27, 2382, 2011.
207. Petsev, D.N., Denkov, N.D., and Kralchevsky, P.A., *J. Colloid Interface Sci.*, 176, 201, 1995.
208. De Gennes, P.G. and Taupin, C., *J. Phys. Chem.*, 86, 2294, 1982.
209. Concus, P., *J. Fluid Mech.*, 34, 481, 1968.
210. Kralchevsky, P.A., Ivanov, I.B., and Nikolov, A.D., *J. Colloid Interface Sci.*, 112, 108, 1986.
211. Abramowitz, M. and Stegun, I.A., *Handbook of Mathematical Functions*, Dover, New York, 1965.
212. Jahnke, E., Emde, F., and Lösch, F., *Tables of Higher Functions*, McGraw-Hill, New York, 1960.
213. Lo, L.L., *J. Fluid Mech.*, 132, 65, 1983.
214. Derjaguin, B.V., *Dokl. Akad. Nauk USSR*, 51, 517, 1946.
215. Scheludko, A. and Exerowa, D., *Comm. Dept. Chem. Bulg. Acad. Sci.*, 7, 123, 1959.
216. Sheludko, A., *Adv. Colloid Interface Sci.*, 1, 391, 1967.
217. Dimitrov, A.S., Kralchevsky, P.A., Nikolov, A.D., and Wasan, D.T., *Colloids Surf.*, 47, 299, 1990.
218. J. Plateau, Experimental and theoretical researches on the figures of equilibrium of a liquid mass withdrawn from the action of gravity, in The Annual Report of the Smithsonian Institution, Washington, DC, 1863, pp. 207–285.
219. J. Plateau, The figures of equilibrium of a liquid mass, in The Annual Report of the Smithsonian Institution, Washington, DC, 1864, pp. 338–369.
220. J. Plateau, Statique Expérimentale et Théoretique des Liquides Soumis aux Seules Forces Moléculaires, Gauthier-Villars, Paris, France, 1873.
221. Orr, F.M., Scriven, L.E., and Rivas, A.P., *J. Fluid Mech.*, 67, 723, 1975.
222. McFarlane, J.S. and Tabor, D., *Proc. R. Soc. Lond. A*, 202, 224, 1950.
223. Koos, E. and Willenbacher, N., *Science*, 331, 897, 2011.
224. Zettlemoyer, A.C., *Nucleation*, Marcel Dekker, New York, 1969.
225. Abraham, E.F., *Homogeneous Nucleation Theory*, Academic Press, New York, 1974.
226. Thomson, W. (Lord Kelvin), *Proc. Roy. Soc.*, 9, 225, 1858; Thomson, W., *Philos. Mag.*, 17, 61, 1859.
227. Lupis, C.H.P., *Chemical Thermodynamics of Matherials*, North-Holland, New York, 1983.
228. Lifshitz, I.M. and Slyozov, V.V., *Zh. Exp. Teor. Fiz.*, 35, 479, 1958 (in Russian).
229. Wagner, C., *Z. Electrochem.*, 35, 581, 1961.
230. Kalhweit, M., *Faraday Discuss. Chem. Soc.*, 61, 48, 1976.

231. Parbhakar, K., Lewandowski, J., and Dao, L.H., *J. Colloid Interface Sci.*, 174, 142, 1995.
232. Kabalnov, A.S., Pertzov, A.V., and Shchukin, E.D., *Colloids Surf.*, 24, 19, 1987.
233. Kabalnov, A.S. and Shchukin, E.D., *Adv. Colloid Interface Sci.*, 38, 69, 1992.
234. McClements, D.J., Dungan, S.R., German, J.B., and Kinsela, J.E., *Food Hydrocolloids*, 6, 415, 1992.
235. Weiss, J., Coupland, J.N., and McClements, D.J., *J. Phys. Chem.*, 100, 1066, 1996.
236. Weiss, J., Canceliere, C., and McClements, D.J., *Langmuir*, 16, 6833, 2000.
237. Kabalnov, A.S., *Langmuir*, 10, 680, 1994.
238. Ivanov, I.B. and Kralchevsky, P.A., Mechanics and thermodynamics of curved thin liquid films, in *Thin Liquid Films*, Ivanov, I.B. (Ed.), Marcel Dekker, New York, 1988, p. 49.
239. Kralchevsky, P.A. and Ivanov, I.B., *J. Colloid Interface Sci.*, 137, 234, 1990.
240. Kralchevsky, P.A., Danov, K.D., and Ivanov, I.B., Thin liquid film physics, in *Foams: Theory, Measurements and Applications*, Prud'homme, R.K. (Ed.), Marcel Dekker, New York, 1995, p. 1.
241. Rusanov, A.I., *Phase Equilibria and Surface Phenomena*, Khimia, Leningrad, Russia, 1967 (Russian); *Phasengleichgewichte und Grenzflächenerscheinungen*, Akademie Verlag, Berlin, Germany, 1978 (German).
242. Derjaguin, B.V. and Kussakov, M.M., *Acta Physicochem. USSR*, 10, 153, 1939.
243. Exerowa, D. and Scheludko, A., *Bull. Inst. Chim. Phys. Bulg. Acad. Sci.*, 4, 175, 1964.
244. Mysels, K.J., *J. Phys. Chem.*, 68, 3441, 1964.
245. Exerowa, D., *Commun. Dept. Chem. Bulg. Acad. Sci.*, 11, 739, 1978.
246. Kruglyakov, P.M., Equilibrium properties of free films and stability of foams and emulsions, in *Thin Liquid Films*, Ivanov, I.B. (Ed.), Marcel Dekker, New York, 1988, p. 767.
247. Martynov, G.A. and Derjaguin, B.V., *Kolloidn. Zh.*, 24, 480, 1962.
248. Toshev, B.V. and Ivanov, I.B., *Colloid Polym. Sci.*, 253, 558, 1975.
249. Ivanov, I.B. and Toshev, B.V., *Colloid Polym. Sci.*, 253, 593, 1975.
250. Frumkin, A., *Zh. Phys. Khim. USSR*, 12, 337, 1938.
251. de Feijter, J.A., Thermodynamics of thin liquid films, in *Thin Liquid Films*, Ivanov, I.B. (Ed.), Marcel Dekker, New York, 1988, p. 1.
252. Kralchevsky, P.A. and Ivanov, I.B., *Chem. Phys. Lett.*, 121, 111, 1985.
253. Nikolov, A.D., Kralchevsky, P.A., Ivanov, I.B., and Dimitrov, A.S., *AIChE Symp. Ser.*, 82(252), 82, 1986.
254. de Feijter, J.A. and Vrij, A., *J. Electroanal. Chem.*, 47, 9, 1972.
255. Kralchevsky, P.A. and Ivanov, I.B., *Chem. Phys. Lett.*, 121, 116, 1985.
256. Denkov, N.D., Petsev, D.N., and Danov, K.D., *J. Colloid Interface Sci.*, 176, 189, 1995.
257. Derjaguin, B.V., *Acta Physicochim. USSR*, 12, 181, 1940.
258. Princen, H.M. and Mason, S.G., *J. Colloid Sci.*, 20, 156, 1965.
259. Prins, A., *J. Colloid Interface Sci.*, 29, 177, 1969.
260. Clint, J.H., Clunie, J.S., Goodman, J.F., and Tate, J.R., *Nature* (*Lond.*), 223, 291, 1969.
261. Yamanaka, T., *Bull. Chem. Soc. Jpn*, 48, 1755, 1975.
262. Princen, H.M., *J. Phys. Chem.*, 72, 3342, 1968.
263. Princen, H.M. and Frankel, S., *J. Colloid Interface Sci.*, 35, 186, 1971.
264. Scheludko, A., Radoev, B., and Kolarov, T., *Trans. Faraday Soc.*, 64, 2213, 1968.
265. Haydon, D.A. and Taylor, J.L., *Nature* (*Lond.*), 217, 739, 1968.
266. Kolarov, T. and Zorin, Z.M., *Kolloidn. Zh.*, 42, 899, 1980.
267. Kruglyakov, P.M. and Rovin, Y.G., *Physical Chemistry of Black Hydrocarbon Films*, Nauka, Moscow, Russia, 1978 (in Russian).
268. Marinova, K.G., Gurkov, T.D., Velev, O.D., Ivanov, I.B., Campbell, B., and Borwankar, R.P., *Colloids Surf. A*, 123, 155, 1997.
269. Françon, M., *Progress in Microscopy*, Pergamon Press, London, U.K., 1961.
270. Beyer, H., *Theorie und Praxis der Interferenzmicroscopie*, Academische Verlagessellschaft, Leipzig, Germany, 1974.
271. Zorin, Z.M., *Kolloidn. Zh.*, 39, 1158, 1977.
272. Zorin, Z., Platikanov, D., Rangelova, N., and Scheludko, A., Measurements of contact angles between bulk liquids and Newton black films for determining the line tension, in *Surface Forces and Liquid Interfaces*, Derjaguin, B.V. (Ed.), Nauka, Moscow, Russia, 1983, p. 200 (in Russian).
273. Nikolov, A.D., Kralchevsky, P.A., and Ivanov, I.B., *J. Colloid Interface Sci.*, 112, 122, 1986.
274. Lobo, L.A., Nikolov, A.D., Dimitrov, A.S., Kralchevsky, P.A., and Wasan, D.T., *Langmuir*, 6, 995, 1990.
275. Dimitrov, A.S., Nikolov, A.D., Kralchevsky, P.A., and Ivanov, I.B., *J. Colloid Interface Sci.*, 151, 462, 1992.
276. Picard, G., Schneider, J.E., and Fendler, J.H., *J. Phys. Chem.*, 94, 510, 1990.

277. Picard, G., Denicourt, N., and Fendler, J.H., *J. Phys. Chem.*, 95, 3705, 1991.
278. Skinner, F.K., Rotenberg, Y., and Neumann, A.W., *J. Colloid Interface Sci.*, 130, 25, 1989.
279. Dimitrov, A.S., Kralchevsky, P.A., Nikolov, A.D., Noshi, H., and Matsumoto, M., *J. Colloid Interface Sci.*, 145, 279, 1991.
280. Hadjiiski, A., Dimova, R., Denkov, N.D., Ivanov, I.B., and Borwankar, R., *Langmuir*, 12, 6665, 1996.
281. Ivanov, I.B., Hadjiiski, A., Denkov, N.D., Gurkov, T.D., Kralchevsky, P.A., and Koyasu, S., *Biophys. J.*, 75, 545, 1998.
282. Paunov, V.N., *Langmuir*, 19, 7970, 2003.
283. Cayre, O.J. and Paunov, V.N., *Langmuir*, 20, 9594, 2004.
284. Ikeda, S., Takahara, Y.K., Ishino, S., Matsumura, M., and Ohtani, B., *Chem. Lett.*, 34(10), 1386, 2005.
285. Paunov, V.N., Cayre, O.J., Noble, P.F., Stoyanov, S.D., Velikov, K.P., and Golding, M., *J. Colloid Interface Sci.*, 312, 381, 2007.
286. Park, B.J., Pantina, J.P., Furst, E.M., Oettel, M., Reynaert, S., and Vermant, J., *Langmuir*, 24, 1686, 2008.
287. Nicolson, M.M., *Proc. Camb. Philos. Soc.*, 45, 288, 1949.
288. Chan, D.Y.C., Henry, J.D., and White, L.R., *J. Colloid Interface Sci.*, 79, 410, 1981.
289. Paunov, V.N., Kralchevsky, P.A., Denkov, N.D., Ivanov, I.B., and Nagayama, K., *J. Colloid Interface Sci.*, 157, 100, 1993.
290. Kralchevsky, P.A., Paunov, V.N., Ivanov, I.B., and Nagayama, K., *J. Colloid Interface Sci.*, 151, 79, 1992.
291. Kralchevsky, P.A., Paunov, V.N., Denkov, N.D., Ivanov, I.B., and Nagayama, K., *J. Colloid Interface Sci.*, 155, 420, 1993.
292. Kralchevsky, P.A. and Nagayama, K., *Langmuir*, 10, 23, 1994.
293. Kralchevsky, P.A. and Nagayama, K., *Adv. Colloid Interface Sci.*, 85, 145, 2000.
294. Denkov, N.D., Velev, O.D., Kralchevsky, P.A., Ivanov, I.B., Nagayama, K., and Yoshimura, H., *Langmuir*, 8, 3183, 1992.
295. Dimitrov, A.S., Dushkin, C.D., Yoshimura, H., and Nagayama, K., *Langmuir*, 10, 432, 1994.
296. Sasaki, M. and Hane, K., *J. Appl. Phys.*, 80, 5427, 1996.
297. Du, H., Chen, P., Liu, F., Meng, F.-D., Li, T.-J., and Tang, X.-Y., *Mater. Chem. Phys.*, 51, 277, 1977.
298. Price, W.C., Williams, R.C., and Wyckoff, R.W.G., *Science*, 102, 277, 1945.
299. Cosslett, V.E. and Markham, R., *Nature*, 161, 250, 1948.
300. Horne, R.W. and Pasquali-Ronchetti, I., *J. Ultrastruct. Res.*, 47, 361, 1974.
301. Harris, J.R., *Micron Microsc. Acta*, 22, 341, 1991.
302. Yoshimura, H., Matsumoto, M., Endo, S., and Nagayama, K., *Ultramicroscopy*, 32, 265, 1990.
303. Yamaki, M., Higo, J., and Nagayama, K., *Langmuir*, 11, 2975, 1995.
304. Nagayama, K., *Colloids Surf. A*, 109, 363, 1996.
305. Burmeister, F., Schäfle, C., Keilhofer, B., Bechinger, C., Boneberg, J., and Leiderer, P., *Adv. Mater.*, 10, 495, 1998.
306. Xia, Y., Tien, J., Qin, D., and Whitesides, G.M., *Langmuir*, 12, 4033, 1996.
307. Lindström, H., Rensmo, H., Sodergren, S., Solbrand, A., and Lindquist, S.E., *J. Phys. Chem.*, 100, 3084, 1996.
308. Matsushita, S., Miwa, T., and Fujishima, A., *Langmuir*, 13, 2582, 1997.
309. Murray, C.B., Kagan, C.R., and Bawendi, M.G., *Science*, 270, 1335, 1995.
310. Jap, B.K., Zulauf, M., Scheybani, T., Hefti, A., Baumeister, W., Aebi, U., and Engel, A., *Ultramicroscopy*, 46, 45, 1992.
311. De Rossi, D., Ahluwalia, A., and Mulè, M., *IEEE Eng. Med. Biol.*, 13, 103, 1994.
312. Kralchevsky, P.A. and Denkov, N.D., *Curr. Opin. Colloid Interface Sci.*, 6, 383, 2001.
313. Gil, T., Ipsen, J.H., Mouritsen, O.G., Sabra, M.C., Sperotto, M.M., and Zuckermann, M.J., *Biochim. Biophys. Acta*, 1376, 245, 1998.
314. Mansfield, S.L., Gotch, A.J., and Harms, G.S., *J. Phys. Chem., B*, 103, 2262, 1999.
315. Fisher, L.R. and Malloy, A.R., *Annu. Rep. Prog. Chem.*, Sect. C, 95, 373, 1999.
316. Kralchevsky, P.A., Paunov, V.N., and Nagayama, K., *J. Fluid Mech.*, 299, 105, 1995.
317. Camoin, C., Roussel, J.F., Faure, R., and Blanc, R., *Europhys. Lett.*, 3, 449, 1987.
318. Velev, O.D., Denkov, N.D., Paunov, V.N., Kralchevsky, P.A., and Nagayama, K., *Langmuir*, 9, 3702, 1993.
319. Dushkin, C.D., Kralchevsky, P.A., Yoshimura, H., and Nagayama, K., *Phys. Rev. Lett.*, 75, 3454, 1995.
320. Lucassen, J., *Colloids Surf.*, 65, 131, 1992.
321. Stamou, D., Duschl, C., and Johannsmann, D., *Phys. Rev., E*, 62, 5263, 2000.
322. Kralchevsky, P.A., Denkov, N.D., and Danov, K.D., *Langmuir*, 17, 2001, 7694.
323. Danov, K.D., Kralchevsky, P.A., Naydenov, B.N., and Brenn, G., *J. Colloid Interface Sci.*, 287, 121, 2005.

324. Danov, K.D. and Kralchevsky, P.A., *Adv. Colloid Interface Sci.*, 154, 91, 2010.
325. Bowden, N., Terfort, A., Carbeck, J., and Whitesides, G.M., *Science*, 276, 233, 1997.
326. Bowden, N., Choi, I.S., Grzybowski, B.A., and Whitesides, G.M., *J. Am. Chem. Soc.*, 121, 5373, 1999.
327. Brown, A.B.D., Smith, C.G., and Rennie, A.R., *Phys. Rev. E*, 62, 951, 2000.
328. Loudet, J.C., Alsayed, A.M., Zhang, J., and Yodh, A.G., *Phys. Rev. Lett.*, 94, 018301, 2005.
329. Loudet, J.C., Yodh, A.G., and Pouligny, B., *Phys. Rev. Lett.*, 97, 018304, 2006.
330. Loudet, J.C. and Pouligny, B., *Europhys. Lett.*, 85, 28003, 2009.
331. van Nierop, E.A., Stijnman, M.A., and Hilgenfeldt, S., *Europhys. Lett.*, 72, 671, 2005.
332. Lehle, H., Noruzifar, E., and Oettel, M., *Eur. Phys. J. E*, 26, 151, 2008.
333. Lewandowski, E.P., Bernate, J.A., Searson, P.C., and Stebe, K.J., *Langmuir*, 24, 9302, 2008.
334. Lewandowski, E.P., Bernate, J.A., Tseng, A., Searson, P.C., and Stebe, K.J., *Soft Matter*, 5, 886, 2009.
335. Lewandowski, E.P., Cavallaro, M., Botto, L., Bernate, J.C., Garbin, V., and Stebe, K.J., *Langmuir*, 26, 15142, 2010.
336. Cavallaro, M., Botto, L., Lewandowski, E.P., Wang, M., and Stebe, K.J., *Proc. Natl. Acad. Sci. U.S.A.*, 108, 20923, 2011.
337. Botto, L., Lewandowski, E.P., Cavallaro, M., and Stebe, K.J., *Soft Matter*, 8, 9957, 2012.
338. Yunker, P.J., Still, T., Lohr, M.A., and Yodh, A.G., *Nature*, 476, 308, 2011.
339. Bleibel, J., Dominguez, A., and Oettel, M., *Eur. Phys. J.—Special Topics*, 222, 3071, 2013.
340. Velikov, K.P., Durst, F., and Velev, O.D. *Langmuir*, 14, 1148, 1998.
341. Sur, J. and Pak, H.K., *J. Korean Phys. Soc.*, 38, 582, 2001.
342. Danov, K.D., Pouligny, B., Angelova, M.I., and Kralchevsky, P.A., Strong capillary attraction between spherical inclusions in a multilayered lipid membrane, in *Studies in Surface Science and Catalysis*, Vol. 132, Iwasawa, Y., Oyama, N., and Kunieda, H. (Eds.), Elsevier, Amsterdam, the Netehrlands, 2001, p. 519.
343. Danov, K.D., Pouligny, B., and Kralchevsky, P.A., *Langmuir*, 17, 2001, 6599.
344. Kralchevsky, P.A., Paunov, V.N., Denkov, N.D., and Nagayama, K., *J. Colloid Interface Sci.*, 167, 47, 1994.
345. Velev, O.D., Denkov, N.D., Paunov, V.N., Kralchevsky, P.A., and Nagayama, K., *J. Colloid Interface Sci.*, 167, 66, 1994.
346. Petkov, J.T., Denkov, N.D., Danov, K.D., Velev, O.D., Aust, R., and Durst, F., *J. Colloid Interface Sci.*, 172, 147, 1995.
347. Danov, K.D., Aust, R., Durst, F., and Lange, U., *J. Colloid Interface Sci.*, 175, 36, 1995.
348. Petkov, J.T., Danov, K.D., Denkov, N.D., Aust, R., and Durst, F., *Langmuir*, 12, 2650, 1996.
349. Petkov, J.T., Gurkov, T.D., and Campbell, B.E., *Langmuir*, 17, 4556, 2001.
350. Danov, K.D., Kralchevsky, P.A., Ananthapadmanabhan, K.P., and Lips, A., *Langmuir*, 22, 106, 2006.
351. Danov, K.D., Kralchevsky, P.A., and Boneva, M.P., *Langmuir*, 20, 6139, 2004.
352. Aveyard, R., Clint, J.H., Nees, D., and Paunov, V.N., *Langmuir*, 16, 1969, 2000.
353. Nikolaides, M.G., Bausch, A.R., Hsu, M.F., Dinsmore, A.D., Brenner, M.P., Gay, C., and Weitz, D.A., *Nature*, 420, 299, 2002.
354. Danov, K.D., Kralchevsky, P.A., and Boneva, M.P., *Langmuir*, 22, 2653, 2006.
355. Danov, K.D. and Kralchevsky, P.A., *J. Colloid Interface Sci.*, 298, 213, 2006.
356. Aveyard, R., Binks, B.P., Clint, J.H., Fletcher, P.D.I., Horozov, T.S., Neumann, B., Paunov, V.N. et al., *Phys. Rev. Lett.*, 88, 246102, 2002.
357. Horozov, T.S., Aveyard, R., Clint, J.H., and Binks, B.P., *Langmuir*, 19, 2822, 2003.
358. Horozov, T.S., Aveyard, R., Binks, B.P., and Clint, J.H., *Langmuir*, 21, 7407, 2005.
359. Horozov, T.S. and Binks, B.P., *Colloids Surf. A*, 267, 64, 2005.
360. Ray, M.A. and Li, J., *Adv. Mater.*, 19, 2020, 2007.
361. Stancik, E.J., Kouhkan, M., and Fuller, G.G., *Langmuir*, 20, 90, 2004.
362. Reynaert, S., Moldenaers, P., and Vermant, J., *Langmuir*, 22, 4936, 2006.
363. Petkov, P.V., Danov, K.D., and Kralchevsky, P.A., *Langmuir*, 30, 2768, 2014.
364. Domínguez, A., Oettel, M., and Dietrich, S., *J. Chem. Phys.* 127, 204706, 2007.
365. Labib, M.E. and Williams, R., *J. Colloid Interface Sci.*, 115, 330, 1987.
366. Philipse, A.P. and Vrij, A., *J. Colloid Interface Sci.*, 128, 121, 1989.
367. Aubry, N. and Singh, P., *Phys. Rev. E*, 77, 056302, 2008.
368. Aubry, N., Singh, P., Janjua, M., and Nudurupati, S., *Proc. Natl. Acad. Sci. U.S.A.*, 105, 3711, 2008.
369. Janjua, M., Nudurupati, S., Singh, P., and Aubry, N., *Mech. Res. Commun.*, 36, 55, 2009.
370. Singh, P., Joseph, D.D., and Aubry, N., *Soft Matter*, 6, 4310, 2010.
371. Danov, K.D. and Kralchevsky, P.A., *J. Colloid Interface Sci.*, 405, 278, 2013.
372. Danov, K.D. and Kralchevsky, P.A., *J. Colloid Interface Sci.*, 405, 269, 2013.

373. Megens, M. and Aizenberg, J., *Nature*, 424, 1014, 2003.
374. Oettel, M., Domínguez, A., and Dietrich, S., *Phys. Rev. E*, 71, 051401, 2005.
375. Würger, A. and Foret, L., *J. Phys. Chem. B*, 109, 16435, 2005.
376. Danov, K.D. and Kralchevsky, P.A., *J. Colloid Interface Sci.*, 345, 505, 2010.
377. Boneva, M.P., Christov, N.C., Danov, K.D., and Kralchevsky, P.A., *Phys. Chem. Chem. Phys.*, 9, 6371, 2007.
378. Boneva, M.P., Danov, K.D., Christov, N.C., and Kralchevsky, P.A., *Langmuir*, 25, 9129, 2009.
379. Kralchevsky, P.A. and Danov, K.D., Interactions between particles at a fluid interface, in *Nanoscience: Colloidal and Interfacial Aspects*, Starov, V.M. (Ed.), CRC Press, New York, 2010, p. 397.
380. Derjaguin, B.V., Churaev, N.V., and Muller, V.M., *Surface Forces*, Plenum Press, Consultants Bureau, New York, 1987.
381. Derjaguin, B.V., *Kolloid Zeits.*, 69, 155, 1934.
382. Attard, P. and Parker, J.L., *J. Phys. Chem.*, 96, 5086, 1992.
383. Tabor, D. and Winterton, R.H.S., *Nature*, 219, 1120, 1968.
384. Keesom, W.H., *Proc. Amst.*, 15, 850, 1913.
385. Debye, P., *Physik*, 2, 178, 1920.
386. London, F., *Z. Physics*, 63, 245, 1930.
387. Hamaker, H.C., *Physics*, 4, 1058, 1937.
388. Usui, S., Sasaki, H., and Hasegawa, F., *Colloids Surf.*, 18, 53, 1986.
389. Lifshitz, E.M., *Soviet Phys. JETP* (*Engl. Transl.*), 2, 73, 1956.
390. Dzyaloshinskii, I.E., Lifshitz, E.M., and Pitaevskii, L.P., *Adv. Phys.*, 10, 165, 1961.
391. Nir, S. and Vassilieff, C.S., Van der Waals interactions in thin films, in *Thin Liquid Films*, Ivanov, I.B. (Ed.), Marcel Dekker, New York, 1988, p. 207.
392. Danov, K.D., Petsev, D.N., Denkov, N.D., and Borwankar, R., *J. Chem. Phys.*, 99, 7179, 1993.
393. Casimir, H.R. and Polder, D., *Phys. Rev.*, 73, 360, 1948.
394. Mahanty, J. and Ninham, B.W., *Dispersion Forces*, Academic Press, New York, 1976.
395. Moelwyn-Hughes, E.A., *Physical Chemistry*, Pergamon Press, London, U.K., 1961, Chapter 21.
396. Langmuir, I., *J. Chem. Phys.*, 6, 873, 1938.
397. Tenchov, B.G. and Brankov, J.G., *J. Colloid Interface Sci.*, 109, 172, 1986.
398. Vassilieff, C.S., Tenchov, B.G., Grigorov, L.S., and Richmond, P., *J. Colloid Interface Sci.*, 93, 8, 1983.
399. Verwey, E.J.W. and Overbeek, J.Th.G., *The Theory of the Stability of Liophobic Colloids*, Elsevier, Amsterdam, the Netherlands, 1948.
400. Muller, V.M., *Kolloidn. Zh.*, 38, 704, 1976.
401. McCormack, D., Carnie, S.L., and Chan, D.Y.C., *J. Colloid Interface Sci.*, 169, 177, 1995.
402. Hogg, R., Healy, T.W., and Fuerstenau, D.W., *Trans. Faraday Soc.*, 62, 1638, 1966.
403. Usui, S., *J. Colloid Interface Sci.*, 44, 107, 1973.
404. Russel, W.B., Saville, D.A., and Schowalter, W.R., *Colloidal Dispersions*, University Press, Cambridge, U.K., 1989.
405. Debye, P. and Hückel, E., *Z. Phys.*, 24, 185, 1923.
406. McCartney, L.N. and Levine, S., *J. Colloid Interface Sci.*, 30, 345, 1969.
407. Derjaguin, B.V. and Landau, L.D., *Acta Physicochim. USSR*, 14, 633, 1941.
408. Efremov, I.F., Periodic colloidal structures, in *Colloid and Surface Science*, Vol. 8, Matijevic, E. (Ed.), Wiley, New York, 1976, p. 85.
409. Schulze, H., *J. Prakt. Chem.*, 25, 431, 1882.
410. Hardy, W.B., *Proc. Roy. Soc.* (*Lond.*), 66, 110, 1900.
411. Guldbrand, L., Jönsson, B., Wennerström, H., and Linse, P., *J. Chem. Phys.*, 80, 2221, 1984.
412. Kjellander, R. and Marčelja, S., *J. Phys. Chem.*, 90, 1230, 1986.
413. Attard, P., Mitchell, D.J., and Ninham, B.W., *J. Chem. Phys.*, 89, 4358, 1988.
414. Attard, P., Mitchell, D.J., and Ninham, B.W., *J. Chem. Phys.*, 88, 4987, 1988.
415. Kralchevsky, P.A. and Paunov, V.N., *Colloids Surf.*, 64, 245, 1992.
416. Marra, J., *J. Phys. Chem.*, 90, 2145, 1986.
417. Marra, J., *Biophys. J.*, 50, 815, 1986.
418. Kjellander, R., Marčelja, S., Pashley, R.M., and Quirk, J.P., *J. Phys. Chem.*, 92, 6489, 1988.
419. Kjellander, R., Marčelja, S., Pashley, R.M., and Quirk, J.P., *J. Chem. Phys.*, 92, 4399, 1990.
420. Khan, A., Jönsson, B., and Wennerström, H., *J. Phys. Chem.*, 89, 5180, 1985.
421. Naji, A., Jungblut, S., Moreira, A.G., and Netz, R.R., *Phys. A: Stat. Mech. Appl.*, 352, 131, 2005.
422. Paunov, V.N. and Kralchevsky, P.A., *Colloids Surf.*, 64, 265, 1992.

423. Tadros, Th.F., Steric interactions in thin liquid films, in *Thin Liquid Films*, Ivanov, I.B. (Ed.), Marcel Dekker, New York, 1988, p. 331.
424. Patel, S.S. and Tirel, M., *Ann. Rev. Phys. Chem.*, 40, 597, 1989.
425. Ploehn, H.J. and Russel, W.B., *Adv. Chem. Eng.*, 15, 137, 1990.
426. de Gennes, P.G., *Scaling Concepts in Polymer Physics*, Cornell University Press, Ithaca, NY, 1979, Chapter 3.
427. Dolan, A.K. and Edwards, S.F., *Proc. Roy. Soc. (Lond.) A*, 337, 509, 1974.
428. Dolan, A.K. and Edwards, S.F., *Proc. Roy. Soc. (Lond.) A*, 343, 427, 1975.
429. de Gennes, P.G., *C. R. Acad. Sci. (Paris)*, 300, 839, 1985.
430. de Gennes, P.G., *Adv. Colloid Interface Sci.*, 27, 189, 1987.
431. Alexander, S.J., *Physique*, 38, 983, 1977.
432. Taunton, H.J., Toprakcioglu, C., Fetters, L.J., and Klein, J., *Macromolecules*, 23, 571, 1990.
433. Klein, J. and Luckham, P., *Nature*, 300, 429, 1982; Klein, J. and Luckham, P., *Macromolecules*, 17, 1041, 1984.
434. Luckham, P.F. and Klein, J., *J. Chem. Soc. Faraday Trans.*, 86, 1363, 1990.
435. Sonntag, H., Ehmka, B., Miller, R., and Knapschinski, L., *Adv. Colloid Interface Sci.*, 16, 381, 1982.
436. Horn, R.G. and Israelachvili, J.N., *Chem. Phys. Lett.*, 71, 192, 1980.
437. Nikolov, A.D., Wasan, D.T., Kralchevsky, P.A., and Ivanov, I.B., Ordered structures in thinning micellar foam and latex films, in *Ordering and Organisation in Ionic Solutions*, Ise, N. and Sogami, I. (Eds.), World Scientific, Singapore, 1988.
438. Nikolov, A.D. and Wasan, D.T., *J. Colloid Interface Sci.*, 133, 1, 1989.
439. Nikolov, A.D., Kralchevsky, P.A., Ivanov, I.B., and Wasan, D.T., *J. Colloid Interface Sci.*, 133, 13, 1989.
440. Nikolov, A.D., Wasan, D.T., Denkov, N.D., Kralchevsky, P.A., and Ivanov, I.B., *Prog. Colloid Polym. Sci.*, 82, 87, 1990.
441. Wasan, D.T., Nikolov, A.D., Kralchevsky, P.A., and Ivanov, I.B., *Colloids Surf.*, 67, 139, 1992.
442. Asakura, S. and Oosawa, F., *J. Chem. Phys.*, 22, 1255, 1954; Asakura, S. and Oosawa, F., *J. Polym. Sci.*, 33, 183, 1958.
443. de Hek, H. and Vrij, A., *J. Colloid Interface Sci.*, 84, 409, 1981.
444. Snook, I.K. and van Megen, W., *J. Chem. Phys.*, 72, 2907, 1980.
445. Kjellander, R. and Marčelja, S., *Chem. Phys. Lett.*, 120, 393, 1985.
446. Tarazona, P. and Vicente, L., *Mol. Phys.*, 56, 557, 1985.
447. Evans, R., and Parry, A.O., *J. Phys. Condens. Matter*, 2, SA15, 1990.
448. Henderson, J.R., *Mol. Phys.*, 59, 89, 1986.
449. Mitchell, D.J., Ninham, B,W., and Pailthorpe, B.A., *J. Chem. Soc. Faraday Trans. 2*, 74, 1116, 1978.
450. Henderson, D., *J. Colloid Interface Sci.*, 121, 486, 1988.
451. Kjellander, R. and Sarman, S., *Chem. Phys. Lett.*, 149, 102, 1988.
452. Trokhymchuk, A. and Henderson, D., *Curr. Opin. Colloid Interface Sci.*, 20, 33, 2015.
453. Pollard, M.L. and Radke, C.J., *J. Chem. Phys.*, 101, 6979, 1994.
454. Chu, X.L., Nikolov, A.D., and Wasan, D.T., *Langmuir*, 10, 4403, 1994.
455. Chu, X.L., Nikolov, A.D., and Wasan, D.T., *J. Chem. Phys.*, 103, 6653, 1995.
456. Trokhymchuk, A., Henderson, D., Nikolov, A., and Wasan, D.T., *J. Phys. Chem. B*, 107, 3927, 2003.
457. Blawzdziewicz, J. and Wajnryb, E., *Europhys. Lett.*, 71, 269, 2005.
458. Kralchevsky, P.A. and Denkov, N.D., *Chem. Phys. Lett.*, 240, 385, 1995.
459. Trokhymchuk, A., Henderson, D., Nikolov, A., and Wasan, D.T., *Langmuir*, 17, 4940, 2001.
460. Carnahan, N.F. and Starling, K.E., *J. Chem. Phys.*, 51, 635, 1969.
461. Basheva, E.S., Kralchevsky, P.A., Danov, K.D., Ananthapadmanabhan, K.P., and Lips, A., *Phys. Chem. Chem. Phys.*, 9, 5183, 2007.
462. Mysels, K.J. and Jones, M.N., *Discuss. Faraday Soc.*, 42, 42, 1966.
463. Christov, N.C., Danov, K.D., Zeng, Y., Kralchevsky, P.A., and von Klitzing, R., *Langmuir*, 26, 915, 2010.
464. Bondy, C., *Trans. Faraday Soc.*, 35, 1093, 1939.
465. Patel, P.D. and Russel, W.B., *J. Colloid Interface Sci.*, 131, 192, 1989.
466. Aronson, M.P., *Langmuir*, 5, 494, 1989.
467. van Lent, B., Israels, R., Scheutjens, J.M.H.M., and Fleer, G.J., *J. Colloid Interface Sci.*, 137, 380, 1990.
468. Joanny, J.F., Leibler, L., and de Gennes, P.G., *J. Polym. Sci.*, 17, 1073, 1979.
469. Evans, E. and Needham, D., *Macromolecules*, 21, 1822, 1988.
470. Yang, S., Tan, H., Yan, D., Nies, E., and Shi, A.-C., *Phys. Rev. E*, 75, 061803, 2007.
471. Melby, P., Prevost, A., Egolf, D.A., and Urbach, J.S., *Phys. Rev. E*, 76, 051307, 2007.
472. Lang, P.R., *J. Chem. Phys.*, 127, 124906, 2007.

473. July, C., Kleshchanok, D., and Lang, P.R., *Eur. Phys. J. E*, 35, 60, 2012.
474. Tulpar, A., Tilton, R.D., and Walz, J.Y., *Langmuir*, 23, 4351, 2007.
475. Lekkerkerker, H.N.W. and Tuinier, R., *Colloids and the Depletion Interaction*, Springer, Berlin, Germany, 2011.
476. Johonnott, E.S., *Philos. Mag.*, 11, 746, 1906.
477. Perrin, J., *Ann. Phys. (Paris)*, 10, 160, 1918.
478. Bruil, H.G. and Lyklema, J., *Nature*, 233, 19, 1971.
479. Friberg, S., Linden, St.E., and Saito, H., *Nature*, 251, 494, 1974.
480. Keuskamp, J.W. and Lyklema, J., *ACS Symp. Ser.*, 8, 191, 1975.
481. Kruglyakov, P.M., *Kolloidn. Zh.*, 36, 160, 1974.
482. Kruglyakov, P.M. and Rovin, Yu.G., *Physical Chemistry of Black Hydrocarbon Films*, Nauka, Moscow, Russia, 1978 (in Russian).
483. Denkov, N.D., Yoshimura, H., Nagayama, K., and Kouyama, T., *Phys. Rev. Lett.*, 76, 2354, 1996.
484. Denkov, N.D., Yoshimura, H., and Nagayama, K., *Ultramicroscopy*, 65, 147, 1996.
485. Kralchevsky, P.A., Nikolov, A.D., Wasan, D.T., and Ivanov, I.B., *Langmuir*, 6, 1180, 1990.
486. Bergeron, V. and Radke, C.J., *Langmuir*, 8, 3020, 1992.
487. Bergeron, V., Jimenez-Laguna, A.I., and Radke, C.J., *Langmuir*, 8, 3027, 1992.
488. Richetti, P. and Kékicheff, P., *Phys. Rev. Lett.*, 68, 1951, 1992.
489. Parker, J.L., Richetti, P., and Kékicheff, P., *Phys. Rev. Lett.*, 68, 1955, 1992.
490. McNamee, C.E., Tsujii, Y., Ohshima, H., and Matsumoto, M., *Langmuir*, 20, 1953, 2004.
491. Krichevsky, O. and Stavans, J., *Phys. Rev. Lett.*, 74, 2752, 1995.
492. Bergeron, V. and Radke, C.J., *Colloid Polym. Sci.*, 273, 165, 1995.
493. Marinova, K.G., Gurkov, T.D., Dimitrova, T.D., Alargova, R.G., and Smith, D., *Langmuir*, 14, 2011, 1998.
494. Basheva, E.S., Nikolov, A.D., Kralchevsky, P.A., Ivanov, I.B., and Wasan, D.T., Multi-stepwise drainage and viscosity of macroscopic films formed from latex suspensions, in *Surfactants in Solution*, Vol. 12, Mittal, K.L. and Shah, D.O. (Eds.), Plenum Press, New York, 1991, p. 467.
495. Basheva, E.S., Danov, K.D., and Kralchevsky, P.A., *Langmuir*, 13, 4342, 1997.
496. Dushkin, C.D., Nagayama, K., Miwa, T., and Kralchevsky, P.A., *Langmuir*, 9, 3695, 1993.
497. Sethumadhavan, G.N., Nikolov, A.D., and Wasan, D.T., *J. Colloid Interface Sci.*, 240, 105, 2001.
498. Piech, M. and Walz, J.Y., *J. Phys. Chem. B*, 108, 9177, 2004.
499. Klapp, S.H.L., Zeng, Y., Qu, D., and von Klitzing, R., *Phys. Rev. Lett.*, 100, 118303, 2008.
500. Klapp, S.H.L., Grandner, S., Zeng, Y., and von Klitzing, R., *J. Phys.: Condens. Matter*, 20, 494232, 2008.
501. Klapp, S.H.L., Grandner, S., Zeng, Y., and von Klitzing, R., *Soft Matter*, 6, 2330, 2010.
502. Zeng, Y., Grandner, S., Oliveira, C.L.P., Thünemann, A.F., Paris, O., Pedersen, J.S., Klapp, S.H.L., and von Klitzing, R., *Soft Matter*, 7, 10899, 2011.
503. Koczo, K., Nikolov, A.D., Wasan, D.T., Borwankar, R.P., and Gonsalves, A., *J. Colloid Interface Sci.*, 178, 694, 1996.
504. Asnacios, A., Espert, A., Colin, A., and Langevin, D., *Phys. Rev. Lett.*, 78, 4974, 1997.
505. Bergeron, V. and Claesson, P.M., *Adv. Colloid Interface Sci.*, 96, 1, 2002.
506. Kolaric, B., Förster, S., and von Klitzing, R., *Progr. Colloid Polym. Sci.*, 117, 195, 2001.
507. von Klitzing, R. and Kolaric, B., *Tenside Surfactants Deterg.*, 39, 247, 2002.
508. Stubenrauch, C. and von Klitzing, R., *J. Phys.: Condens. Matter*, 15, R1197, 2003.
509. Beltran, C.M., Guillot, S., and Langevin, D., *Macromolecules*, 36, 8506, 2003.
510. Beltran, C.M. and Langevin, D., *Phys. Rev. Lett.*, 94, 217803, 2005.
511. Heinig, P., Beltran, C.M., and Langevin, D., *Phys. Rev. E*, 73, 051607, 2006.
512. Jönsson, B., Broukhno, A., Forsman, J., and Åkesson, T., *Langmuir*, 19, 9914, 2003.
513. Danov, K.D., Basheva, E.S., Kralchevsky, P.A., Ananthapadmanabhan, K.P., and Lips, A., *Adv. Colloid Interface Sci.*, 168, 50, 2011.
514. Anachkov, S.E., Danov, K.D., Basheva, E.S., Kralchevsky, P.A., and Ananthapadmanabhan, K.P., *Adv. Colloid Interface Sci.*, 183–184, 55, 2012.
515. Bales, B.L. and Almgren, M., *J. Phys. Chem.*, 99, 15153, 1995.
516. Turro, N.J. and Yekta, A., *J. Am. Chem. Soc.*, 100, 5951, 1978.
517. Quina, F.H., Nassar, P.M., Bonilha, J.B.S., and Bales, B.L., *J. Phys. Chem.*, 99, 17028, 1995.
518. Gehlen, M.H. and De Schryver, F.C., *J. Phys. Chem.*, 97, 11242, 1993.
519. Mukerjee, P. and Mysels, K.J., Critical micelle concentration of aqueous surfactant systems, National Standard Reference Data Series 36, National Bureau of Standards, Washington, DC, 1971.
520. Shah, S.S., Saeed, A., and Sharif, Q.M., *Colloids Surf. A*, 155, 405, 1999.

521. Benrraou, M., Bales, B.L., and Zana, R., *J. Phys. Chem. B*, 107, 13432, 2003.
522. Weidemaier, K., Tavernier, H.L., and Fayer, M.D., *J. Phys. Chem. B*, 101, 9352, 1997.
523. Hansson, P., Jönsson, B., Ström, C., and Söderman, O., *J. Phys. Chem. B*, 104, 3496, 2000.
524. Van Stam, J., Depaemelaere, S., and De Schryver, F.C., *J. Chem. Educ.*, 75, 93, 1998.
525. Mata, J., Varade, D., and Bahadur, P., *Thermochim. Acta*, 428, 147, 2005.
526. Hassan, P.A., Hodgdon, T.K., Sagasaki, M., Fritz-Popovski, G., and Kaler, E.W., *Comp. Rend. Chim.*, 12, 18, 2009.
527. Mukhim, T. and Ismail, K., *J. Surf. Sci. Technol.*, 21, 113, 2005.
528. Bhat, M.A., Dar, A.A., Amin, A., Rashid, P.I., and Rather, G.M., *J. Chem. Thermodyn.*, 39, 1500, 2007.
529. Amos, D.A., Lynn, S., and Radke, C.J., *Langmuir*, 14, 2297, 1998.
530. Stanley, H.E. and Teixeira, J., *J. Chem. Phys.*, 73, 3404, 1980.
531. Norrish, K., *Discuss. Faraday Soc.*, 18, 120, 1954.
532. Van Olphen, H., *Clays Clay Miner.*, 2, 418, 1954.
533. Clunie, J.S., Goodman, J.F., and Symons, P.C., *Nature* (*Lond.*), 216, 1203, 1967.
534. Jordine, E.St.A., *J. Colloid Interface Sci.*, 45, 435, 1973.
535. Derjaguin, B.V., Churaev, N.V., *Croatica Chem. Acta*, 50, 187, 1977.
536. LeNeveu, D.M., Rand, R.P., Parsegian, V.A., and Gingell, D., *Biophys. J.*, 18, 209, 1977.
537. Lis, L.J., McAlister, M., Fuller, N., Rand, R.P., and Parsegian, V.A., *Biophys. J.*, 37, 657, 1982.
538. Israelachvili, J.N. and Adams, G.E., *J. Chem. Soc. Faraday Trans. 1*, 74, 975, 1978.
539. Pashley, R.M. and Israelachvili, J.N., *Colloids Surf.*, 2, 169, 1981.
540. Pashley, R.M. and Israelachvili, J.N., *J. Colloid Interface Sci.*, 101, 511, 1984.
541. Pashley, R.M., *J. Colloid Interface Sci.*, 80, 153, 1981.
542. Pashley, R.M., *J. Colloid Interface Sci.*, 83, 531, 1981.
543. Pashley, R.M., *Adv. Colloid Interface Sci.*, 16, 57, 1982.
544. Peschel, G., Belouschek, P., Müller, M.M., Müller, M.R., and König, R., *Colloid Polym. Sci.*, 260, 444, 1982.
545. Horn, R.G., Smith, D.T., and Haller, W., *Chem. Phys. Lett.*, 162, 404, 1989.
546. Claesson, P., Carmona-Ribeiro, A.M., Kurihara, K., *J. Phys. Chem.*, 93, 917, 1989.
547. Petsev, D.N. and Vekilov, P.G., *Phys. Rev. Lett.*, 84, 1339, 2000.
548. Petsev, D.N., Thomas, B.R., Yau, S.-T., and Vekilov, P.G., *Biophys. J.*, 78, 2060, 2000.
549. Valle-Delgado, J.J., Molina-Bolívar, J.A., Galisteo-González, F., Gálvez-Ruiz, M.J., Feiler, A., and Rutland, M.W., *Langmuir*, 21, 9544, 2005.
550. Valle-Delgado, J.J., Molina-Bolívar, J.A., Galisteo-González, F., and Gálvez-Ruiz, M.J., *Curr. Opin. Colloid Interface Sci.*, 16, 572, 2011.
551. Stanley, C. and Rau, D.C., *Curr. Opin. Colloid Interface Sci.*, 16, 551, 2011.
552. Zhang, N.H., Tan, Z.Q., Li, J.J., Meng, W.L., and Xu, L.W., *Curr. Opin. Colloid Interface Sci.*, 16, 592, 2011.
553. Aroti, A., Leontidis, E., Dubios, M., and Zemb, T., *Biophys. J.*, 93, 1580, 2007.
554. Leontidis, E., Aroti, A., Belloni, L., Dubios, M., and Zemb, T., *Biophys. J.*, 93, 1591, 2007.
555. Sparr, E. and Wennerström, H., *Curr. Opin. Colloid Interface Sci.*, 16, 561, 2011.
556. Demé, B. and Zemb, T., *Curr. Opin. Colloid Interface Sci.*, 16, 584, 2011.
557. Sanfeld, A. and Steinchen, A., *Adv. Colloid Interface Sci.*, 140, 1, 2008.
558. Kaldasch, J., Senge, B., and Laven, J., *Rheol. Acta*, 48, 665, 2009.
559. Bongrand, P., Intermolecular forces, in *Physical Basis of Cell–Cell Adhesion*, Bongrand, P. (Ed.), CRC Press, Boca Raton, FL, 1987, p. 1.
560. Binder, B., Goede, A., Berndt, N., and Holzhütter, H.-G., *PLoS One*, 4, 1, 2009.
561. Rand, R. and Parsegian, V., *Biochim. Biophys. Acta*, 988, 351, 1989.
562. Cevc, G., *J. Chem. Soc. Faraday Trans.*, 87, 2733, 1991.
563. Leikin, S., Parsegian, V.A., Rau, D.C., Rand, R.P., *Annu. Rev. Phys. Chem.*, 44, 369, 1993.
564. Israelachvili, J. and Wennerström, H., *Nature*, 379, 219, 1996.
565. Butt, H.-J., Capella, B., and Kappl, M., *Surf. Sci. Rep.*, 59, 1, 2005.
566. Leite, F.L., Bueno, C.C., Da Róz, A.L., Ziemath, E.C., and Oliveira Jr., O.N., *Int. J. Mol. Sci.*, 13, 12773, 2012.
567. Healy, T.W., Homola, A., James, R.O., and Hunter, R.J., *Faraday Discuss. Chem. Soc.*, 65, 156, 1978.
568. Marčelja, S. and Radič, N., *Chem. Phys Lett.*, 42, 129, 1976.
569. Besseling, N.A.M., *Langmuir*, 13, 2113, 1997.
570. Forsman, J., Woodward, C.E., and Jönsson, B., *Langmuir*, 13, 5459, 1997.
571. Attard, P. and Batchelor, M.T., *Chem. Phys. Lett.*, 149, 206, 1988.

572. Faraudo, J., *Curr. Opin. Colloid Interface Sci.*, 16, 557, 2011.
573. Schneck, E. and Netz, R.R., *Curr. Opin. Colloid Interface Sci.*, 16, 607, 2011.
574. Jönsson, B. and Wennerström, H., *J. Chem. Soc. Faraday Trans. 2*, 79, 19, 1983.
575. Kjellander, R., *J. Chem. Soc. Faraday Trans. 2*, 80, 1323, 1984.
576. Bratko, D. and Jönsson, B., *Chem. Phys. Lett.*, 128, 449, 1986.
577. Trokhymchuk, A., Henderson, D., and Wasan, D.T., *J. Colloid Interface Sci.*, 210, 320, 1999.
578. Simon, S.A. and McIntosh, T.J., *Proc. Natl. Acad. Sci. U.S.A.*, 86, 9263, 1989.
579. Richmond, P., *J. Chem. Soc. Faraday Trans. 2*, 70, 1066, 1974.
580. Dzhavakhidze, P.G., Kornyshev, A.A., and Levadny, V.G., *Nuovo Cimento*, 10D, 627, 1988.
581. Kornyshev, A.A. and Leikin, S., *Phys. Rev. A*, 40, 6431, 1989.
582. Leikin, S. and Kornyshev, A.A., *J. Chem. Phys.*, 92, 6890, 1990.
583. Attard, P. and Patey, G.N., *Phys. Rev. A*, 43, 2953, 1991.
584. Henderson, D. and Lozada-Cassou, M., *J. Colloid Interface Sci.*, 114, 180, 1986.
585. Henderson, D. and Lozada-Cassou, M., *J. Colloid Interface Sci.*, 162, 508, 1994.
586. Basu, S. and Sharma, M.M., *J. Colloid Interface Sci.*, 165, 355, 1994.
587. Paunov, V.N., Dimova, R.I., Kralchevsky, P.A., Broze, G., and Mehreteab, A., *J. Colloid Interface Sci.*, 182, 239, 1996.
588. Paunov, V.N. and Binks, B.P., *Langmuir*, 15, 2015, 1999.
589. Kralchevsky, P.A., Danov, K.D., and Basheva, E.S., *Curr. Opin. Colloid Interface Sci.*, 16, 517, 2011.
590. Tchaliovska, S., Herder, P., Pugh, R., Stenius, P., and Eriksson, J.C., *Langmuir*, 6, 1535, 1990.
591. Pashley, R.M., McGuiggan, P.M., Ninham, B.W., and Evans, D.F., *Science*, 229, 1088, 1985.
592. Rabinovich, Y.I. and Derjaguin, B.V., *Colloids Surf.*, 30, 243, 1988.
593. Parker, J.L., Cho, D.L., and Claesson, P.M., *J. Phys. Chem.*, 93, 6121, 1989.
594. Christenson, H.K., Claesson, P.M., Berg, J., and Herder, P.C., *J. Phys. Chem.*, 93, 1472, 1989.
595. Christenson, H.K., Fang, J., Ninham, B.W., and Parker, J.L., *J. Phys. Chem.*, 94, 8004, 1990.
596. Ducker, W.A., Xu, Z., and Israelachvili, J.N., *Langmuir*, 10, 3279, 1994.
597. Eriksson, J.C., Ljunggren, S., and Claesson, P.M., *J. Chem. Soc. Faraday Trans. 2*, 85, 163, 1989.
598. Joesten, M.D. and Schaad, L.J., *Hydrogen Bonding*, Marcel Dekker, New York, 1974.
599. Stillinger, F.H. and Ben-Naim, A., *J. Chem. Phys.*, 47, 4431, 1967.
600. Conway, B.E., *Adv. Colloid Interface Sci.*, 8, 91, 1977.
601. Kuzmin, V.L. and Rusanov, A.I., *Kolloidn. Z.*, 39, 455, 1977.
602. Dubrovich, N.A., *Kolloidn. Z.*, 57, 275, 1995.
603. Eriksson, J.C., Henriksson, U., and Kumpulainen, A., *Colloids Surf. A*, 282–283, 79, 2006.
604. Paunov, V.N., Sandler, S.I., and Kaler, E.W., *Langmuir*, 17, 4126, 2001.
605. Angarska, J.K., Dimitrova, B.S., Danov, K.D., Kralchevsky, P.A., Ananthapadmanabhan, K.P., and Lips, A., *Langmuir*, 20, 1799, 2004.
606. Christenson, H.K. and Claesson, P.M., *Science*, 239, 390, 1988.
607. Parker, J.L., Claesson, P.M., and Attard, P., *J. Phys. Chem.*, 98, 8468, 1994.
608. Carambassis, A., Jonker, L.C., Attard, P., and Rutland, M.W., *Phys. Rev. Lett.*, 80, 5357, 1998.
609. Mahnke, J., Stearnes, J., Hayes, R.A., Fornasiero, D., and Ralston, J., *Phys. Chem. Chem. Phys.*, 1, 2793, 1999.
610. Considine, R.F., Hayes, R.A., and Horn, R.G., *Langmuir*, 15, 1657, 1999.
611. Considine, R.F. and Drummond, C., *Langmuir*, 16, 631, 2000.
612. Attard, P., *Langmuir*, 16, 4455, 2000.
613. Yakubov, G.E., Butt, H.-J., and Vinogradova, O., *J. Phys. Chem. B*, 104, 3407, 2000.
614. Ederth, T., *J. Phys. Chem. B*, 104, 9704, 2000.
615. Ishida, N., Sakamoto, M., Miyahara, M., and Higashitani, K., *Langmuir*, 16, 5681, 2000.
616. Ishida, N., Inoue, T., Miyahara, M., and Higashitani, K., *Langmuir*, 16, 6377, 2000.
617. Hammer, M.U., Anderson, T.H., Chaimovich, A., Shell, M.S., and Israelachvili, J., *Faraday Discuss.*, 146, 299, 2010.
618. Popa, I., Gillies, G., Papastavrou, G., and Borkovec, M., *J. Phys. Chem. B*, 113, 8458, 2009.
619. Helfrich, W., *Z. Naturforsch.*, 33a, 305, 1978.
620. Servuss, R.M. and Helfrich, W., *J. Phys. (France)*, 50, 809, 1989.
621. Fernandez-Puente, L., Bivas, I., Mitov, M.D., and Méléard, P., *Europhys. Lett.*, 28, 181, 1994.
622. Safinya, C.R., Roux, D., Smith, G.S., Sinha, S.K., Dimon, P., Clark, N.A., and Bellocq, A.M., *Phys. Rev. Lett.*, 57, 2718, 1986.
623. McIntosh, T.J., Magid, A.D., and Simon, S.A., *Biochemistry*, 28, 7904, 1989.
624. Abillon, O. and Perez, E., *J. Phys. (France)*, 51, 2543, 1990.

625. Israelachvili, J.N. and Wennerström, H., *J. Phys. Chem.*, 96, 520, 1992.
626. Evans, E.A. and Skalak, R., *Mechanics and Thermodynamics of Biomembranes*, CRC Press, Boca Raton, FL, 1980.
627. Aniansson, G.A.E., Wall, S.N., Almgren, M., Hoffman, H., Kielmann, I., Ulbricht, W., Zana, R., Lang. J., and Tondre, C., *J. Phys. Chem.*, 80, 905, 1976.
628. Aniansson, G.A.E., *J. Phys. Chem.*, 82, 2805, 1978.
629. Dimitrova, T.D., Leal-Calderon, F., Gurkov, T.D., and Campbell, B., *Langmuir*, 17, 8069, 2001.
630. Danov, K.D., Ivanov, I.B., Ananthapadmanabhan, K.P., Lips, A., *Adv. Colloid Interface Sci.*, 128–130, 185, 2006.
631. Bird, R.B., Stewart, W.E., and Lightfoot, E.N., *Transport Phenomena*, Wiley, New York, 1960.
632. Germain, P., *Mécanique des Milieux Continus*, Masson et Cie, Paris, France, 1962.
633. Batchelor, G.K., *An Introduction of Fluid Mechanics*, Cambridge University Press, London, U.K., 1967.
634. Slattery, J.C., *Momentum, Energy, and Mass Transfer in Continua*, R.E. Krieger, Huntington, New York, 1978.
635. Landau, L.D. and Lifshitz, E.M., *Fluid Mechanics*, Pergamon Press, Oxford, U.K., 1984.
636. Barnes, H.A., Hutton, J.F., and Walters, K., *An Introduction to Rheology*, Elsevier, Amsterdam, the Netherlands, 1989.
637. Walters, K., Overview of macroscopic viscoelastic flow, in *Viscoelasticity and Rheology*, Lodge, A.S., Renardy, M., and Nohel, J.A. (Eds.), Academic Press, London, U.K., 1985, p. 47.
638. Boger, D.V., *Ann. Rev. Fluid Mech.*, 19, 157, 1987.
639. Barnes, H.A., *J. Rheol.*, 33, 329, 1989.
640. Barnes, H.A., *A Handbook of Elementary Rheology*, University of Wales, Institute of Non-Newtonian Fluid Mechanics, Aberystwyth, U.K., 2000.
641. Sagis, L.M.C., *Rev. Modern Phys.*, 83, 1367, 2011.
642. Navier, M., *Mém. de l'Acad. d. Sci.*, 6, 389, 1827.
643. Stokes, G.G., *Trans. Camb. Philos. Soc.*, 8, 287, 1845.
644. Happel, J. and Brenner, H., *Low Reynolds Number Hydrodynamics with Special Applications to Particulate Media*, Prentice-Hall, Englewood Cliffs, New York, 1965.
645. Kim, S. and Karrila, S.J., *Microhydrodynamics: Principles and Selected Applications*, Butterworth-Heinemann, Boston, MA, 1991.
646. Reynolds, O., *Philos. Trans. Roy. Soc. (Lond.)*, A177, 157, 1886.
647. Lamb, H., *Hydrodynamics*, Cambridge University Press, London, U.K., 1963.
648. Felderhof, B.U., *J. Chem. Phys.*, 49, 44, 1968.
649. Sche, S. and Fijnaut, H.M., *Surf. Sci.*, 76, 186, 1976.
650. Maldarelli, Ch. and Jain, R.K., The hydrodynamic stability of thin films, in *Thin Liquid Films*, Ivanov, I.B. (Ed.), Marcel Dekker, New York, 1988, p. 497.
651. Valkovska, D.S. and Danov, K.D., *J. Colloid Interface Sci.*, 241, 400, 2001.
652. Danov, K.D., Effect of surfactants on drop stability and thin film drainage, in *Fluid Mechanics of Surfactant and Polymer Solutions*, Starov, V. and Ivanov, I.B. (Eds.), Springer, New York, 2004, p. 1.
653. Hardy, W. and Bircumshaw, I., *Proc. Roy. Soc. (Lond.)*, A108, 1, 1925.
654. Horn, R.G., Vinogradova, O.I., Mackay, M.E. and Phan-Thien, N., *J. Chem. Phys.*, 112, 6424, 2000.
655. Dimitrov, D.S. and Ivanov, I.B., *J. Colloid Interface Sci.*, 64, 97, 1978.
656. Ivanov, I.B., Dimitrov, D.S., Somasundaran, P., and Jain, R.K., *Chem. Eng. Sci.*, 40, 137, 1985.
657. Ivanov, I.B. and Dimitrov, D.S., Thin film drainage, in *Thin Liquid Films*, Ivanov, I.B. (Ed.), Marcel Dekker, New York, 1988, p. 379.
658. Danov, K.D., Kralchevsky, P.A. and Ivanov, I.B., in *Handbook of Detergents, Part A: Properties*, Broze, G. (Ed.), Marcel Dekker, New York, 1999, p. 303.
659. O'Neill, M.E. and Stewartson, K., *J. Fluid Mech.*, 27, 705, 1967.
660. Goldman, A.J., Cox, R.G., and Brenner, H., *Chem. Eng. Sci.*, 22, 637, 1967.
661. Goldman, A.J., Cox, R.G., and Brenner, H., *Chem. Eng. Sci.*, 22, 653, 1967.
662. Levich, V.G., *Physicochemical Hydrodynamics*, Prentice-Hall, Englewood Cliffs, NJ, 1962.
663. Edwards, D.A., Brenner, H., and Wasan, D.T., *Interfacial Transport Processes and Rheology*, Butterworth-Heinemann, Boston, MA, 1991.
664. Charles, G.E. and Mason, S.G., *J. Colloid Sci.*, 15, 236, 1960.
665. Danov, K.D., Denkov, N.D., Petsev, D.N., Ivanov, I.B., and Borwankar, R., *Langmuir*, 9, 1731, 1993.
666. Danov, K.D., Valkovska, D.S., and Ivanov, I.B., *J. Colloid Interface Sci.*, 211, 291, 1999.
667. Hartland, S., Coalescence in dense-packed dispersions, in *Thin Liquid Films*, Ivanov, I.B. (Ed.), Marcel Dekker, New York, 1988, p. 663.

668. Hetsroni, G. (Ed.), *Handbook of Multiphase System*, Hemisphere Publishing, New York, 1982, pp. 1–96.
669. Davis, A.M.J., Kezirian, M.T., and Brenner, H., *J. Colloid Interface Sci.*, 165, 129, 1994.
670. Brenner, H., *Chem. Eng. Sci.*, 18, 1, 1963.
671. Brenner, H., *Chem. Eng. Sci.*, 19, 599, 1964; Brenner, H., *Chem. Eng. Sci.*, 19, 631, 1964.
672. Brenner, H. and O'Neill, M.E., *Chem. Eng. Sci.*, 27, 1421, 1972.
673. Van de Ven, T.G.M., *Colloidal Hydrodynamics*, Academic Press, London, U.K., 1988.
674. Jeffery, G.B., *Proc. Lond. Math. Soc.*, 14, 327, 1915.
675. Stimson, M. and Jeffery, G.B., *Proc. Roy. Soc.* (*Lond.*), A111, 110, 1926.
676. Cooley, M.D.A. and O'Neill, M.E., *Mathematika*, 16, 37, 1969.
677. Cooley, M.D.A. and O'Neill, M.E., *Proc. Camb. Philos. Soc.*, 66, 407, 1969.
678. Davis, M.H., *Chem. Eng. Sci.*, 24, 1769, 1969.
679. O'Neill, M.E. and Majumdar, S.R., *Z. Angew. Math. Phys.*, 21, 164, 1970; O'Neill, M.E. and Majumdar, S.R., *Z. Angew. Math. Phys.*, 21, 180, 1970.
680. Davis, M.H., *Two Unequal Spheres in a Slow Linear Shear Flow*, Report NCAR-TN/STR-64, National Center for Atmospheric Research, Bolder, CO, 1971.
681. Batchelor, G.K., *J. Fluid Mech.*, 74, 1, 1976.
682. Davis, R.H. and Hill, N.A., *J. Fluid Mech.*, 236, 513, 1992.
683. Batchelor, G.K., *J. Fluid Mech.*, 119, 379, 1982.
684. Batchelor, G.K. and Wen, C.-S., *J. Fluid Mech.*, 124, 495, 1982.
685. Jeffrey, D.J. and Onishi, Y., *J. Fluid Mech.*, 139, 261, 1984.
686. Fuentes, Y.O., Kim, S., and Jeffrey, D.J., *Phys. Fluids*, 31, 2445, 1988; Fuentes, Y.O., Kim, S., and Jeffrey, D.J., Fuentes, Y.O., Kim, S., and Jeffrey, D.J., *Phys. Fluids*, A1, 61, 1989.
687. Stokes, G.G., *Trans. Camb. Philos. Soc.*, 1, 104, 1851.
688. Exerowa, D. and Kruglyakov, P. M., *Foam and Foam Films: Theory, Experiment, Application*, Elsevier, New York, 1998.
689. Ivanov, I.B., Radoev, B.P., Traykov, T.T., Dimitrov, D.S., Manev, E.D., and Vassilieff, C.S., in *Proceedings of the International Conference on Colloid Surface Science*, Wolfram, E. (Ed.), Akademia Kiado, Budapest, Hungary, 1975, p. 583.
690. Denkov, N.D., Petsev, D.N., and Danov, K.D., *Phys. Rev. Lett.*, 71, 3226, 1993.
691. Valkovska, D.S., Danov, K.D., and Ivanov, I.B., *Colloid Surf. A*, 156, 547, 1999.
692. Davis, R.H., Schonberg, J.A., and Rallison, J.M., *Phys. Fluids*, A1, 77, 1989.
693. Chi, B.K. and Leal, L.G., *J. Fluid Mech.*, 201, 123, 1989.
694. Ascoli, E.P., Dandy, D.S., and Leal, L.G., *J. Fluid Mech.*, 213, 287, 1990.
695. Yiantsios, S.G. and Davis, R.H., *J. Fluid Mech.*, 217, 547, 1990.
696. Zhang, X. and Davis, R.H., *J. Fluid Mech.*, 230, 479, 1991.
697. Chesters, A.K., *Trans. Inst. Chem. Eng. A*, 69, 259, 1991.
698. Yiantsios, S.G. and Davis, R.H., *J. Colloid Interface Sci.*, 144, 412, 1991.
699. Yiantsios, S.G. and Higgins, B.G., *J. Colloid Interface Sci.*, 147, 341, 1991.
700. Joye, J.-L., Hirasaki, G.J., and Miller, C.A., *Langmuir*, 8, 3083, 1992.
701. Joye, J.-L., Hirasaki, G.J., and Miller, C.A., *Langmuir*, 10, 3174, 1994.
702. Abid, S. and Chestrers, A.K., *Int. J. Multiphase Flow*, 20, 613, 1994.
703. Li, D., *J. Colloid Interface Sci.*, 163, 108, 1994.
704. Saboni, A., Gourdon, C., and Chesters, A.K., *J. Colloid Interface Sci.*, 175, 27, 1995.
705. Rother, M.A., Zinchenko, A.Z., and Davis, R.H., *J. Fluid Mech.*, 346, 117, 1997.
706. Singh, G., Miller, C.A., and Hirasaki, G.J., *J. Colloid Interface Sci.*, 187, 334, 1997.
707. Cristini, V., Blawzdziewicz, J., and Loewenberg, M., *J. Fluid Mech.*, 366, 259, 1998.
708. Bazhlekov, I.B., Chesters, A.K., and van de Vosse, F.N., *Int. J. Multiphase Flow*, 26, 445, 2000.
709. Bazhlekov, I.B., van de Vosse, F.N., and Chesters, A.K., *J. Non-Newtonian Fluid Mech.*, 93, 181, 2000.
710. Chesters, A.K. and Bazhlekov, I.B., *J. Colloid Interface Sci.*, 230, 229–243, 2000.
711. Boulton-Stone, J.M. and Blake, J.R., *J. Fluid Mech.*, 254, 437, 1993.
712. Frankel, S. and Mysels, K., *J. Phys. Chem.*, 66, 190, 1962.
713. Tabakova, S.S. and Danov, K.D., *J. Colloid Interface Sci.*, 336, 273, 2009.
714. Velev, O.D., Constantinides, G.N., Avraam, D.G., Payatakes, A.C., and Borwankar, R.P., *J. Colloid Interface Sci.*, 175, 68, 1995.
715. Exerowa, D., Nikolov, A., and Zacharieva, M., *J. Colloid Interface Sci.*, 81, 419, 1981.
716. de Vries, A.J., *Rec. Trav. Chim. Pays-Bas.*, 77, 441, 1958.
717. Vrij, A., *Discuss. Faraday Soc.*, 42, 23, 1966.
718. Ivanov, I.B., Radoev, B.P., Manev, E.D., and Sheludko, A.D., *Trans. Faraday Soc.*, 66, 1262, 1970.

719. Gumerman, R.J. and Homsy, G.M., *Chem. Eng. Commun.*, 2, 27, 1975.
720. Malhotra, A.K. and Wasan, D.T., *Chem. Eng. Commun.*, 48 35, 1986.
721. Valkovska, D. S., Danov, K.D., and Ivanov, I.B., *Adv. Colloid Interface Sci.*, 96, 101, 2002.
722. Manev, E.D., Sazdanova, S.V., and Wasan, D.T., *J. Colloid Interface Sci.*, 97, 591, 1984.
723. Ivanov, I.B., *Pure Appl. Chem.*, 52, 1241, 1980.
724. Aveyard, R., Binks, B.P., Fletcher, P.D.I., and Ye, X., *Prog. Colloid Polymer Sci.*, 89, 114, 1992.
725. Velev, O.D., Gurkov, T.D., Chakarova, Sv.K., Dimitrova, B.I., and Ivanov, I.B., *Colloids Surf. A*, 83, 43, 1994.
726. Dickinson, E., Murray, B.S., and Stainsby, G., *J. Chem. Soc. Faraday Trans.*, 84, 871, 1988.
727. Ivanov, I.B., Lectures at INTEVEP, Petroleos de Venezuela, Caracas, Venezuela, June 1995.
728. Ivanov, I.B. and Kralchevsky, P.A., *Colloid Surf. A*, 128, 155, 1997.
729. Basheva, E.S., Gurkov, T.D., Ivanov, I.B., Bantchev, G.B., Campbell, B., and Borwankar, R.P., *Langmuir*, 15, 6764, 1999.
730. Gurkov, T.D. and Basheva, E.S., Hydrodynamic behavior and stability of approaching deformable drops, in *Encyclopedia of Surface & Colloid Science*, Hubbard, A.T. (Ed.), Marcel Dekker, New York, 2002.
731. Gurkov, T.D., Angarska, J.K., Tahcev, K.D., and Gaschler, W., *Colloids Surf. A*, 382, 174, 2011.
732. Boussinesq, M.J., *Ann. Chim. Phys.*, 29, 349, 1913; Boussinesq, M.J., *Ann. Chim. Phys.*, 29, 357, 1913.
733. Aris, R., *Vectors, Tensors, and the Basic Equations of Fluid Mechanics*, Prentice Hall, Englewood Cliffs, NJ, 1962.
734. Brenner, H. and Leal, L.G., *J. Colloid Interface Sci.*, 62, 238, 1977.
735. Brenner, H. and Leal, L.G., *J. Colloid Interface Sci.*, 65, 191, 1978.
736. Brenner, H. and Leal, L.G., *AIChE J.*, 24, 246, 1978.
737. Brenner, H. and Leal, L.G., *J. Colloid Interface Sci.*, 88, 136, 1982.
738. Stone, H.A., *Phys. Fluids*, A2, 111, 1990.
739. Valkovska, D.S. and Danov, K.D., *J. Colloid Interface Sci.*, 223, 314, 2000.
740. Stoyanov, S.D. and Denkov, N.D., *Langmuir*, 17, 1150, 2001.
741. Feng, S.-S., *J. Colloid Interface Sci.*, 160, 449, 1993.
742. Stebe, K.J. and Maldarelli, Ch., *Phys. Fluids*, A3, 3, 1991.
743. Stebe, K.J. and Maldarelli, Ch., *J. Colloid Interface Sci.*, 163, 177, 1994.
744. Scriven, L.E., *Chem. Eng. Sci.*, 12, 98, 1960.
745. Scriven, L.E. and Sternling, C.V., *J. Fluid Mech.*, 19, 321, 1964.
746. Slattery, J.C., *Chem. Eng. Sci.*, 19, 379, 1964; *Chem. Eng. Sci.*, 19, 453, 1964.
747. Slattery, J.C., *I&EC Fundam.*, 6, 108, 1967.
748. Slattery, J.C., *Interfacial Transport Phenomena*, Springer-Verlag, New York, 1990.
749. Barton, K.D. and Subramanian, R.S., *J. Colloid Interface Sci.*, 133, 214, 1989.
750. Feuillebois, F., *J. Colloid Interface Sci.*, 131, 267, 1989.
751. Merritt, R.M. and Subramanian, R.S., *J. Colloid Interface Sci.*, 131, 514, 1989.
752. Mannheimer, R.J. and Schechter, R.S., *J. Colloid Interface Sci.*, 12, 98, 1969.
753. Pintar, A.J., Israel, A.B., and Wasan, D.T., *J. Colloid Interface Sci.*, 37, 52, 1971.
754. Gardner, J.W. and Schechter, R.S., *Colloid Interface Sci.*, 4, 98, 1976.
755. Li, D. and Slattery, J.C., *J. Colloid Interface Sci.*, 125, 190, 1988.
756. Tambe, D.E. and Sharma, M.M., *J. Colloid Interface Sci.*, 147, 137, 1991.
757. Tambe, D.E. and Sharma, M.M., *J. Colloid Interface Sci.*, 157, 244, 1993.
758. Tambe, D.E. and Sharma, M.M., *J. Colloid Interface Sci.*, 162, 1, 1994.
759. Horozov, T., Danov, K., Kralchevsky, P., Ivanov, I., and Borwankar, R., A local approach in interfacial rheology: Theory and experiment, in *First World Congress on Emulsion*, Paris, France, 3-20, 137, 1993.
760. Danov, K. D., Ivanov, I. B., and Kralchevsky, P. A., Interfacial rheology and emulsion stability, in *Second World Congress on Emulsion*, Paris, France, 2-2, 152, 1997.
761. Ivanov, I.B., Danov, K.D., Ananthapadmanabhan, K.P. and Lips, A. *Adv. Colloid Interface Sci.*, 114–115, 61, 2005.
762. Murray, B.S. and Dickinson, E., *Food Sci. Technol.*, 2, 131, 1996.
763. Miller, R., Wüstneck, R., Kärgel, J., and Kretzschmar, G., *Colloids Surf. A*, 111, 75, 1996.
764. Dickinson, E., *Colloids Surf. B*, 20, 197, 2001.
765. Bos, M.A. and van Vliet, T., *Adv. Colloid Interface Sci.*, 91, 437, 2001.
766. Kärgel, J. and Derkatch, S.R., *Curr. Opin. Colloid Interface Sci.*, 15, 246, 2010.
767. Miller, R., Ferri, J.K., Javadi, A., Kärgel, J., Mucic, N., and Wüstneck, R., *Colloid Polym. Sci.*, 288, 937, 2010.

768. Kärgel, J., Derkatch, S.R., and Miller, R., *Adv. Colloid Interface Sci.*, 144, 38, 2008.
769. Murray, B.S., *Curr. Opin. Colloid Interface Sci.*, 16, 27, 2011.
770. Radulova, G.M., Danov, K.D., Kralchevsky, P.A., Petkov, J.T., and Stoyanov, S.D., *Soft Matter*, 10, 5777, 2014.
771. Danov, K.D., Radulova, G.M., Kralchevsky, P.A., Golemanov, K., and Stoyanov, S.D., *Faraday Discuss.*, 158,195, 2012.
772. Danov, K.D., Kralchevsky, P.A., Radulova, G.M., Basheva, E.S., Stoyanov, S.D., and Pelan, E.G., *Adv. Colloid Interface Sci.*, 2015, doi: 10.1016/j.cis.2014.04.009.
773. Stanimirova, R., Marinova, K., Tcholakova, S., Denkov, N.D., Stoyanov, S., and Pelan, E., *Langmuir*, 27, 12486, 2011.
774. de Groot, S.R. and Mazur, P., *Non-Equilibrium Thermodynamics*, Interscience, New York, 1962.
775. Sagis, L.M.C and Öttinger, H.C., *Phys. Rev. E*, 88, 022149, 2013.
776. Moeckel, G.P., *Arch. Rat. Mech. Anal.*, 57, 255, 1975.
777. Rushton, E. and Davies, G.A., *Appl. Sci. Res.*, 28, 37, 1973.
778. Haber, S., Hetsroni, G., and Solan, A., *Int. J. Multiphase Flow*, 1, 57, 1973.
779. Reed, L.D. and Morrison, F.A., *Int. J. Multiphase Flow*, 1, 573, 1974.
780. Hetsroni, G. and Haber, S., *Int. J. Multiphase Flow*, 4, 1, 1978.
781. Morrison, F.A. and Reed, L.D., *Int. J. Multiphase Flow*, 4, 433, 1978.
782. Beshkov, V.N., Radoev, B.P., and Ivanov, I.B., *Int. J. Multiphase Flow*, 4, 563, 1978.
783. Danov, K.D., Stoyanov, S.D., Vitanov, N.K., and Ivanov, I.B., *J. Colloid Interface Sci.*, 368, 342, 2012.
784. Murdoch, P.G. and Leng, D.E., *Chem. Eng. Sci.*, 26, 1881, 1971.
785. Reed, X.B., Riolo, E., and Hartland, S., *Int. J. Multiphase Flow*, 1, 411, 1974; Reed, X.B., Riolo, E., and Hartland, S., *Int. J. Multiphase Flow*, 1, 437, 1974.
786. Ivanov, I.B. and Traykov, T.T., *Int. J. Multiphase Flow*, 2, 397, 1976.
787. Traykov, T.T. and Ivanov, I.B., *Int. J. Multiphase Flow*, 3, 471, 1977.
788. Lu, C.-Y.D. and Cates, M.E., *Langmuir*, 11, 4225, 1995.
789. Jeelany, S.A.K. and Hartland, S., *J. Colloid Interface Sci.*, 164, 296, 1994.
790. Zapryanov, Z., Malhotra, A.K., Aderangi, N., and Wasan, D.T., *Int. J. Multiphase Flow*, 9, 105, 1983.
791. Malhotra, A.K. and Wasan, D.T., *Chem. Eng. Commun.*, 55, 95, 1987.
792. Malhotra, A.K. and Wasan, D.T., Interfacial rheological properties of adsorbed surfactant films with applications to emulsion and foam stability, in *Thin Liquid Films*, Ivanov, I.B. (Ed.), Marcel Dekker, New York, 1988, p. 829.
793. Singh, G., Hirasaki, G.J., and Miller, C.A., *J. Colloid Interface Sci.*, 184, 92, 1996.
794. Traykov, T.T., Manev, E.D., and Ivanov, I.B., *Int. J. Multiphase Flow*, 3, 485, 1977.
795. Bancroft, W.D., *J. Phys. Chem.*, 17, 514, 1913.
796. Griffin, J., *Soc. Cosmet. Chem.*, 5, 4, 1954.
797. Davies, J.T., in *Proceedings of the Second International Congress on Surface Activity*, Vol. 1, Butterworths, London, U.K., 1957, p. 426.
798. Shinoda, K. and Friberg, S., *Emulsion and Solubilization*, Wiley, New York, 1986.
799. Davis, H.T., Factors determining emulsion type: HLB and beyond, in *Proceedings of First World Congress on Emulsion*, October 19–22, Paris, France, 1993, p. 69.
800. Israelachvili, J., The history, applications and science of emulsion, in *Proceedings of First World Congress on Emulsion*, October 19–22, Paris, France, 1993, p. 53.
801. Kralchevsky, P.A., *J. Colloid Interface Sci.*, 137, 217, 1990.
802. Gompper, G. and Schick, M., *Phys. Rev.*, B41, 9148, 1990.
803. Lerczak, J., Schick, M., and Gompper, G., *Phys. Rev.*, 46, 985, 1992.
804. Andelman, D., Cates, M.E., Roux, D., and Safran, S., *J. Chem. Phys.*, 87, 7229, 1987.
805. Chandra, P. and Safran, S., *Europhys. Lett.*, 17, 691, 1992.
806. Danov, K.D., Velev, O.D., Ivanov, I.B., and Borwankar, R.P., Bancroft rule and hydrodynamic stability of thin films and emulsions, in *First World Congress on Emulsion*, October 19–22, Paris, France, 1993, p. 125.
807. Kunieda, H., Evans, D.F., Solans, C., and Yoshida, *Colloids Surf.*, 47, 35, 1990.
808. Koper, G.J.M., Sager, W.F.C., Smeets, J., and Bedeaux, D., *J. Phys. Chem.*, 99, 13291, 1995.
809. Ivanov, I.B., Danov, K.D., and Kralchevsky, P.A., *Colloids Surf. A*, 152, 161, 1999.
810. Velev, O.D., Gurkov, T.D., and Borwankar, R.P., *J. Colloid Interface Sci.*, 159, 497, 1993.
811. Velev, O.D., Gurkov, T.D., Ivanov, I.B., and Borwankar, R.P., *Phys. Rev. Lett.*, 75, 264, 1995.
812. Danov, K., Ivanov, I., Zapryanov, Z., Nakache, E., and Raharimalala, S., Marginal stability of emulsion thin film, in *Proceedings of the Conference of Synergetics, Order and Chaos*, Velarde, M. (Ed.), World Scientific, Singapore, 1988, p. 178.

813. Valkovska, D.S., Kralchevsky, P.A., Danov, K.D., Broze, G., and Mehreteab, A., *Langmuir*, 16, 8892, 2000.
814. Danov, K.D., Gurkov, T.D., Dimitrova, T.D., and Smith, D., *J. Colloid Interface Sci.*, 188, 313, 1997.
815. Chevaillier, J.P., Klaseboer, E., Masbernat, O., and Gourdon, C., *J. Colloid Interface Sci.*, 299, 472, 2006.
816. Chan, D.Y.C., Klaseboer, E., and Manica, R., *Soft Matter*, 7, 2235, 2011.
817. Chan, D.Y.C., Klaseboer, E., and Manica, R., *Adv. Colloid Interface Sci.*, 165, 70, 2011.
818. Ivanov, I.B., Chakarova, S.K., and Dimitrova, B.I., *Colloids Surf.*, 22, 311, 1987.
819. Dimitrova, B.I., Ivanov, I.B., and Nakache, E., *J. Dispersion Sci. Technol.*, 9, 321, 1988.
820. Sternling, C.V. and Scriven, L.E., *AIChE J.*, 5, 514, 1959.
821. Lin, S.P. and Brenner, H.J., *J. Colloid Interface Sci.*, 85, 59, 1982.
822. Holly, F.J., On the wetting and drying of epithelial surfaces, in *Wetting, Spreading and Adhesion*, Padday, J.F. (Ed.), Academic Press, New York, 1978, p. 439.
823. Castillo, J.L. and Velarde, M.G., *J. Colloid Interface Sci.*, 108, 264, 1985.
824. Davis, R.H., *Adv. Colloid Interface Sci.*, 43, 17, 1993.
825. Uijttewaal, W.S.J., Nijhof, E.-J., and Heethaar, R.M., *Phys. Fluids*, A5, 819, 1993.
826. Zapryanov, Z. and Tabakova, S., *Dynamics of Bubbles, Drops and Rigid Particles*, Kluwer Academic Publishers, London, U.K., 1999.
827. Lorentz, H.A., *Abhandl. Theoret. Phys.*, 1, 23, 1906.
828. Faxen, H., *Arkiv. Mat. Astron. Fys.*, 17, 27, 1923.
829. Wakiya, S., *J. Phys. Soc. Jpn*, 12, 1130, 1957.
830. Dean, W.R. and O'Neill, M.E., *Mathematika*, 10, 13, 1963.
831. O'Neill, M.E., *Mathematika*, 11, 67, 1964.
832. Cooley, M.D.A. and O'Neill, M.E., *J. Inst. Math. Appl.*, 4, 163, 1968.
833. Keh, H.J. and Tseng, C.H., *Int. J. Multiphase Flow*, 1, 185, 1994.
834. Schonberg, J. and Hinch, E.J., *J. Fluid Mech.*, 203, 517, 1989.
835. Ryskin, G. and Leal, L.G., *J. Fluid Mech.*, 148, 1, 1984; Ryskin, G. and Leal, L.G., *J. Fluid Mech.*, 148, 19, 1984; Ryskin, G. and Leal, L.G., *J. Fluid Mech.*, 148, 37, 1984.
836. Liron, N. and Barta, E., *J. Fluid Mech.*, 238, 579, 1992.
837. Shapira, M. and Haber, S., *Int. J. Multiphase Flow*, 14, 483, 1988.
838. Shapira, M. and Haber, S., *Int. J. Multiphase Flow*, 16, 305, 1990.
839. Yang, S.-M. and Leal, L.G., *J. Fluid Mech.*, 149, 275, 1984.
840. Yang, S.-M. and Leal, L.G., *Int. J. Multiphase Flow*, 16, 597, 1990.
841. Lebedev, A.A., *Zhur. Russ. Fiz. Khim.*, 48, 1916.
842. Silvey, A., *Phys. Rev.*, 7, 106, 1916.
843. Hadamar, J.S., *Comp. Rend. Acad. Sci. (Paris)*, 152, 1735, 1911.
844. Rybczynski, W., *Bull. Intern. Acad. Sci. (Cracovie)*, *A*, 1911.
845. He, Z., Dagan, Z., and Maldarelli, Ch., *J. Fluid Mech.*, 222, 1, 1991.
846. Malysa, K., Krasowska, M., and Krzan, M., *Adv. Colloid Interface Sci.*, 114–115, 205, 2005.
847. Krzan, M., Zawala, J., and Malysa, K., *Colloids Surf. A*, 298, 42, 2007.
848. Krzan, M. and Malysa, K., *Physicochem. Problems Mineral Process.*, 43, 43, 2009.
849. Krzan, M. and Malysa, K., *Physicochem. Problems Mineral Process.*, 48, 49, 2012.
850. Zawala, J., Malysa, E., Krzan, M., and Malysa, K., *Physicochem. Problems Mineral Process.*, 50, 143, 2014.
851. Danov, K.D., Aust, R., Durst, F., and Lange, U., *Chem. Eng. Sci.*, 50, 263, 1995.
852. Danov, K.D., Aust, R., Durst, F., and Lange, U., *Chem. Eng. Sci.*, 50, 2943, 1995.
853. Danov, K.D., Aust, R., Durst, F., and Lange, U., *Int. J. Multiphase Flow*, 21, 1169, 1995.
854. Danov, K.D., Gurkov, T.D., Raszillier, H., and Durst, F., *Chem. Eng. Sci.*, 53, 3413, 1998.
855. Stoos, J.A. and Leal, L.G., *J. Fluid Mech.*, 217, 263, 1990.
856. Danov, K.D., Dimova, R.I., and Pouligny, B., *Phys. Fluids*, 12, 2711, 2000.
857. Dimova, R.I., Danov, K.D., Pouligny, B., and Ivanov, I.B., *J. Colloid Interface Sci.*, 226, 35, 2000.
858. Angelova, M.I. and Pouligny, B., *Pure Appl. Optics*, 2, 261, 1993.
859. Pouligny, B., Martinot-Lagarde, G., and Angelova, M.I., *Progr. Colloid Polym. Sci.*, 98, 280, 1995.
860. Dietrich, C., Angelova, M., and Pouligny, B., *J. Phys. II France*, 7, 1651, 1997.
861. Velikov, K., Dietrich, C., Hadjiski, A., Danov, K., and Pouligny, B., *Europhys. Lett.*, 40(4), 405, 1997.
862. Velikov, K., Danov, K., Angelova, M., Dietrich, C., and Pouligny, B., *Colloids Surf. A*, 149, 245, 1999.
863. Dimova, R., Dietrich, C., Hadjiisky, A., Danov, K., and Pouligny, B., *Eur. Phys. J. B*, 12, 589, 1999.
864. Danov, K.D., Pouligny, B., Angelova, M.I., and Kralchevsky, P.A., *Studies in Surface Science and Catalysis*, Vol. 132, Elsevier, Amsterdam, the Netherlands, 2001.

865. Barentin, C., Ybert, C., di Meglio, J.M., and Joanny, J.F., *J. Fluid Mech.*, 397, 331, 1999.
866. Maenosono, S., Dushkin, C.D., and Yamaguchi, Y., *Colloid Polym. Sci.*, 227, 993, 1999.
867. Barentin, C., Muller, P., and Ybert, C., *Eur. Phys. J. E*, 2, 153, 2000.
868. Dimova, R., Pouligny, B., and Dietrich, C., *Biophys. J.*, 79, 340, 2000.
869. Joseph, D.D., Wang, J., Bai, R., Yang, B.H., and Hu, H.H., *J. Fluid Mech.*, 496, 139, 2003.
870. Sickert, M. and Rondelez, F., *Phys. Rev. Lett.*, 90, 126104, 2003.
871. Fischer, T.M., *Phys. Rev. Lett.*, 92, 139603, 2004.
872. Sickert, M. and Rondelez, F., *Phys. Rev. Lett.*, 92, 139604, 2004.
873. Haris, S.S. and Giorgio, T.D., *Gene Therapy*, 12, 512, 2005.
874. Khattari, Z., Ruschel, Y., Wen, H.Z., Fischer, A. and Fischer, T.M., *J. Phys. Chem. B*, 109, 3402, 2005.
875. Singh, P. and Joseph, D.D., *J. Fluid Mech.*, 530, 31, 2005.
876. Fischer, T.M., Dhar, P., and Heinig, P., *J. Fluid Mech.*, 558, 451, 2006.
877. Sickert, M., Rondelez, F., and Stone, H.A., *Europhys. Lett.*, 79, 66005, 2007.
878. Chen, W. and Tong, P., *Europhys. Lett.*, 84, 28003, 2008.
879. Peng, Y., Chen, W., Fisher, T.M., Weitz, D.A., and Tong, P., *J. Fluid Mech.*, 618, 243, 2009.
880. Lee, M.H., Reich, D.H., Stebe, K.J., and Leheny, R.L., *Langmuir*, 26, 2650, 2010.
881. Wilke, N., Vega Merkado, F., and Maggio, B., *Langmuir*, 26, 11050, 2010.
882. Singh, P., Joseph, D.D., and Aubry, N., *Soft Matter*, 6, 4310, 2010.
883. Ally, J. and Amirfazli, A., *AIP Conf. Proc.*, 1311, 307, 2010.
884. Abras, D., Pranami, G., and Abbott, N.L., *Soft Matter*, 8, 2026, 2012.
885. Koynov, K. and Butt, H.-J., *AIP Conf. Proc.*, 1518, 357, 2013.
886. Morse, J., Huang, A., Li, G., Maxey, M.R., and Tang, J.X., *Biophys. J.*, 105, 21, 2013.
887. Memdoza, A.J., Guzman, E., Martines-Pedrero, F., Ritacco, H., Rubio, R.G., Ortega, F., Starov, V.M. and Miller, R., *Adv. Colloid Interface Sci.*, 206, 303, 2014.
888. Hunter, R.J., *Foundation of Colloid Science*, Vol. 1, Clarendon Press, Oxford, U.K., 1987.
889. Hunter, R.J., *Foundation of Colloid Science*, Vol. 2, Clarendon Press, Oxford, U.K., 1989.
890. Einstein, A., *Ann. Phys.*, 19, 289, 1906.
891. Kubo, R., *Rep. Prog. Phys.*, 29, 255, 1966.
892. Einstein, A., *Ann. Phys.*, 34, 591, 1911.
893. Taylor, G.I., *Proc. Roy. Soc. A*, 138, 41, 1932.
894. Oldroyd, J.G., *Proc. Roy. Soc. A*, 232, 567, 1955.
895. Taylor, G.I., *Proc. Roy. Soc. A*, 146, 501, 1934.
896. Fröhlich, H. and Sack, R., *Proc. Roy. Soc. A*, 185, 415, 1946.
897. Oldroyd, J.G., *Proc. Roy. Soc. A*, 218, 122, 1953.
898. Batchelor, G.K., *J. Fluid Mech.*, 83, 97, 1977.
899. De Kruif, C.G., Van Iersel, E.M.F., Vrij, A., and Russel, W.B., *J. Chem. Phys.*, 83, 4717, 1985.
900. Loewenberg, M. and Hinch, E.J., *J. Fluid Mech.*, 321, 395, 1996.
901. Da Cunha, F.R. and Hinch, E.J., *J. Fluid Mech.*, 309, 211, 1996.
902. Li, X. and Pozrikidis, C., *J. Fluid Mech.*, 341, 165, 1997.
903. Loewenberg, M., *J. Fluids Eng.*, 120, 824, 1998.
904. Blawzdziewicz, J., Vajnryb, E., and Loewenberg, M., *J. Fluid Mech.*, 395, 29, 1999.
905. Ramirez, J.A., Zinchenko, A., Loewenberg, M., and Davis, R.H., *Chem. Eng. Sci.*, 54, 149, 1999.
906. Blawzdziewicz, J., Vlahovska, P., and Loewenberg, M., *Physica A*, 276, 50, 2000.
907. Breyannis, G. and Pozrikidis, C., *Theor. Comp. Fluid Dynam.*, 13, 327, 2000.
908. Li, X. and Pozrikidis, C., *Int. J. Multiphase Flow*, 26, 1247, 2000.
909. Rednikov, A.Y., Ryazantsev, Y.S., and Velarde, M.G., *Phys. Fluids*, 6, 451, 1994.
910. Velarde, M.G., *Philos. Trans. Roy. Soc., Math. Phys. Eng. Sci.*, 356, 829, 1998.
911. Danov, K.D., *J. Colloid Interface Sci.*, 235, 144, 2001.
912. Barnes, H.A., Rheology of emulsions—A review, in *Proceedings of First World Congress on Emulsion*, October 19–22, Paris, France, 1993, p. 267.
913. Krieger, L.M. and Dougherty, T.J., *Trans. Soc. Rheol.*, 3, 137, 1959.
914. Wakeman, R., *Powder Tech.*, 11, 297, 1975.
915. Prud'home, R.K. and Khan, S.A., Experimental results on foam rheology, in *Foams: Theory, Measurements, and Applications*, Prud'home, R.K. and Khan, S.A. (Eds.), Marcel Dekker, New York, 1996, p. 217.
916. Tadros, T.F., Fundamental principles of emulsion rheology and their applications, in *Proceedings of First World Congress on Emulsion*, October 19–22, Paris, France, 1993, p. 237.
917. Pal, R., Bhattacharya, S.N., and Rhodes, E., *Can. J. Chem. Eng.*, 64, 3, 1986.

918. Edwards, D.A. and Wasan, D.T., Foam rheology: The theory and role of interfacial rheological properties, in *Foams: Theory, Measurements, and Applications*, Prud'home, R.K. and Khan, S.A. (Eds.), Marcel Dekker, New York, 1996, p. 189.
919. Wessel, R. and Ball, R.C., *Phys. Rev.*, A46, 3009, 1992.
920. Kanai, H., Navarrete, R.C., Macosko, C.W., and Scriven, L.E., *Rheol. Acta*, 31, 333, 1992.
921. Pal, R., *Chem. Eng. Commun.*, 98, 211, 1990.
922. Pal, R., *Colloids Surf. A*, 71, 173, 1993.
923. Prins, A., Liquid flow in foams as affected by rheological surface properties: A contribution to a better understanding of the foaming behaviour of liquids, in *Hydrodynamics of Dispersed Media*, Hulin, J.P., Cazabat, A.M., Guyon, E., and Carmona, F. (Eds.), Elsevier/North-Holland, Amsterdam, the Netherlands, 1990, p. 5.
924. Babak, V.G., *Colloids Surf. A*, 85, 279, 1994.
925. Yuhua, Y., Pal, R., and Masliyah, J., *Chem. Eng. Sci.*, 46, 985, 1991.
926. Giesekus, H., Disperse systems: Dependence of rheological properties on the type of flow with implications for food rheology, in *Physical Properties of Foods*, Jowitt, R. et al. (Eds.), Applied Science Publishers, London, 1983, Chapter 13.
927. Turian, R. and Yuan, T.-F., *AIChE J.*, 23, 232, 1977.
928. Clarke, B., *Trans. Inst. Chem. Eng.*, 45, 251, 1967.
929. Denkov, N.D., Tcholakova, S., Höhler, R., and Cohen-Addad, S., Foam rheology, in *Foam Engineering: Fundamentals and Applications*, Stevenson, P. (Ed.), John Wiley & Sons, Chichester, U.K., 2012, p. 91.
930. Bingham, E.C., *U.S. Bur. Standards Bull.*, 13, 309, 1916.
931. Herschel, W.K. and Bulkley, R., *Kolloid Z.*, 39, 291, 1926.
932. Denkov, N.D., Tcholakova, S., Golemanov, K., Ananthapadmanabhan, K.P., and Lips, A., *Phys. Rev. Lett.*, 100, 138301, 2008.
933. Tcholakova, S., Denkov, N.D., Golemanov, K., Ananthapadmanabhan, K.P., and Lips, A., *Phys. Rev. E*, 78, 011405, 2008.
934. Golemanov, K., Denkov, N.D., Tcholakova, S., Vethamunthu, M., and Lips, A., *Langmuir*, 24, 9956, 2008.
935. Denkov, N.D., Tcholakova, S., Golemanov, K., Hu, T., Lips, A., *Amer. Inst. Physics Conference Proceedings*, 1027, 902, 2008.
936. Denkov, N.D., Tcholakova, S., Golemanov, K., Ananthapadmanabhan, K.P., and Lips, A., *Soft Matter*, 5, 3389, 2009.
937. Politova, N., Tcholakova, S., Golemanov, K., Denkov, N.D., Vethamunthu, M., and Ananthapadmanabhan, K.P., *Langmuir*, 28, 1115, 2012.
938. Mitrinova, Z., Tcholakova, S., Golemanov, K., Denkov, N.D., Vethamunthu, M., and Ananthapadmanabhan, K.P., *Colloids Surf. A*, 438, 186, 2013,
939. Mitrinova, Z., Tcholakova, S., Popova, Z., Denkov, N., Dasgupta, B.R., and Ananthapadmanabhan, K.P., *Langmuir*, 29, 8255, 2013.
940. von Smoluchowsky, M., *Phys. Z.*, 17, 557, 1916.
941. von Smoluchowsky, M., *Z. Phys. Chem.*, 92, 129, 1917.
942. Wang, H. and Davis, R.H., *J. Colloid Interface Sci.*, 159, 108, 1993.
943. Rogers, J.R. and Davis, R.H., *Mettal. Trans.*, A21, 59, 1990.
944. Young, N.O., Goldstein, J.S., and Block, M.J., *J. Fluid Mech.*, 6, 350, 1959.
945. Fuchs, N.A., *Z. Phys.*, 89, 736, 1934.
946. Friedlander, S.K., *Smoke, Dust and Haze: Fundamentals of Aerosol Behaviour*, Wiley-Interscience, New York, 1977.
947. Singer, J.M., Vekemans, F.C.A., Lichtenbelt, J.W.Th., Hesselink, F.Th., and Wiersema, P.H., *J. Colloid Interface Sci.*, 45, 608, 1973.
948. Leckband, D.E., Schmitt, F.-J., Israelachvili, J.N., and Knoll, W., *Biochemistry*, 33, 4611, 1994.
949. Bak, T.A. and Heilmann, O., *J. Phys. A: Math. Gen.*, 24, 4889, 1991.
950. Martinov, G.A. and Muller, V.M., in *Research in Surface Forces*, Vol. 4, Plenum Press, Consultants Bureau, New York, 1975, p. 3.
951. Elminyawi, I.M., Gangopadhyay, S., and Sorensen, C.M., *J. Colloid Interface Sci.*, 144, 315, 1991.
952. Hartland, S. and Gakis, N., *Proc. Roy. Soc. (Lond.)*, A369, 137, 1979.
953. Hartland, S. and Vohra, D.K., *J. Colloid Interface Sci.*, 77, 295, 1980.
954. Lobo, L., Ivanov, I.B., and Wasan, D.T., *AIChE J.*, 39, 322, 1993.
955. Danov, K.D., Ivanov, I.B., Gurkov, T.D., and Borwankar, R.P., *J. Colloid Interface Sci.*, 167, 8, 1994.
956. van den Tempel, M., *Recueil*, 72, 433, 1953.
957. Borwankar, R.P., Lobo, L.A., and Wasan, D.T., *Colloid Surf.*, 69, 135, 1992.

958. Dukhin, S., Sæther, Ø., and Sjöblom, J., Coupling of coalescence and flocculation in dilute O/W emulsions, in *Encycloped Handbook of Emulsion Technology*, Sjöblom, J. (Ed.), Marcel Dekker, New York, 2001, p. 71.
959. Urbina-Villalba, G. and Garcia-Sucre, M., *Mol. Simulat.*, 27, 75, 2001.
960. Madras, G. and McCoy, B.J., *J. Colloid Interface Sci.*, 246, 356, 2002.
961. Barthelmes, G., Pratsinis, S.E., and Buggisch, H., *Chem. Eng. Sci.*, 58, 2893, 2003.
962. Han, B.B., Akeprathumchai, S., Wickramasinghe, S.R., and Qian, X., *AIChE J.*, 49, 1687, 2003.
963. Urbina-Villalba, G., Garcia-Sucre, M., and Toro-Mendoza, J., *Phys. Rev. E*, 68, 061408, 2003.
964. Petsev, D.N. (Ed.), *Emulsions: Structure, Stability and Interactions*, Elsevier, London, U.K., 2004.
965. Urbina-Villalba, G., *Int. J. Mol. Sci.*, 10, 761, 2009.
966. Rahn-Chique, K., Puertas, A.M., Rumero-Cano, M.S., Rojas, C., and Urbina-Villalba, G., *Adv. Colloid Interface Sci.*, 178, 1, 2012.

5 Subsurface Colloidal Fines, Behavior, Characterization, and Their Role in Groundwater Contamination

Tushar Kanti Sen and Chittaranjan Ray

CONTENTS

5.1 INTRODUCTION

There are several classes of subsurface colloids—abiotic and biotic. The mobilization and migration of these subsurface colloidal particles take place under different physical and geochemical conditions (Massoudieh and Ginn, 2010; Sen, 2011; Sen and Khilar, 2006). Therefore, subsurface colloidal fines can enhance or retard the mobility and dispersion of various contaminants in groundwater flows (Sen and Khilar, 2009). There are two categories of colloid-induced subsurface contaminant transport: (a) colloid-associated contaminant transport and (b) transport of biocolloids. General research on subsurface colloid transport originated in the early 1930s (Kretzschmar and Schafer, 2005). First findings on the partitioning of aqueous solution constituents onto colloids appeared in the late 1970s, as summarized by Gustafasson and Gschwend (1987). Subsequently, reports on the transport facilitation of contaminants via association with subsurface colloidal fines emerged in the 1980s (Enfield and Bengtsson, 1988; McCarthy and Zachara, 1989). Our knowledge and understanding of the colloidal fines–associated contaminant transport in subsurface porous media have increased substantially over the last three decades, which is reviewed by various researchers from time to time such as Kretzschmar et al. (1999), Elimelech and Ryan (2002), Sen and Khilar (2006, 2009), and Bin et al. (2011). The focus of this chapter is to review subsurface colloidal fines, sampling methods and characterization, and their role in groundwater contamination. This has been evidenced by various experimental, modeling, and field studies, which have been briefly compiled here and also partially updated in Chapter 4 (Sen and Khilar, 2009). Finally, authors here briefly discussed subsurface inorganic/organic colloid–associated contaminant transport in groundwater and their associated health effects.

5.2 SUBSURFACE COLLOIDAL FINES AND THEIR CLASSIFICATION

Within the subsurface systems such as soils, groundwater aquifers, sediments, or fractured rocks, colloidal particles/fines (particles with an average diameter between 10^{-9} and 10^{-10} m and carrying surface charge) are ubiquitous with a pronounced variability concerning morphology and chemical composition (Borkovec et al., 1993). There are several classes of subsurface colloids, such as abiotic and biotic (Auset and Keller, 2004). Basically, small colloidal particles of inorganic, organic, and microbiological variety exist in natural subsurface systems. These include silicate clays, iron and aluminum oxides, mineral precipitates, humic materials, microemulsions of nonaqueous phase liquids, viruses, and bacteria (Bradford et al., 2002; Sen and Khilar, 2006). In soils, the entire clay fraction is normally considered to be inorganic colloidal, since the clay minerals behave as colloids. The motion of colloids is more dominated by Brownian motions than by gravity. When colloids are stable, the low gravitational settling will render them in suspension for longer periods. Another of the important characteristics of the colloidal fines is that the specific surface is large ($>10^2$ m^2/g) (Kretzschmar et al., 1999). On the other hand, organic colloids can be viruses, bacteria, and protozoa collectively called biocolloids or macromolecular organic matter (OM) (when being smaller than 45 μm often termed dissolved organic matter [DOM]) falls under these colloidal size ranges. Figure 5.1 indicates the size range of colloidal particles including a wide range of organic and inorganic components in the subsurface environment (Kretzschmar et al., 1999).

In general, colloidal fines posses a surface charge, of either permanent or nonpermanent (i.e., variable) nature (Sen and Khilar, 2006). Although a small portion of the colloids in soils are microorganisms and viruses, soil colloids are mostly complex assemblages of clay minerals,

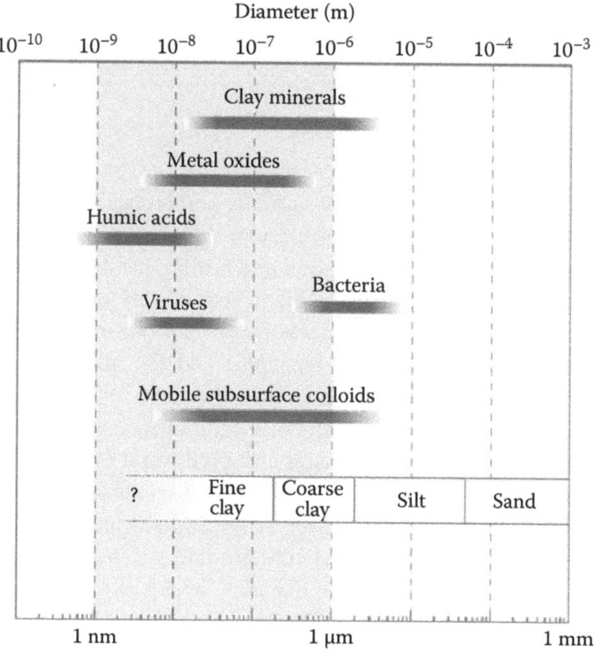

FIGURE 5.1 Diameter range of various types of subsurface colloidal fines. (From Kretzschmar, R. et al., *Adv. Agron.*, 66, 121, 1999.)

oxides and hydroxides, and OM. The clay minerals have fixed charge as well as a variable pH-dependent charge (Khilar and Fogler, 1998; Sen and Khilar, 2009), whereas oxides and hydroxides generally have a positive charge at neutral pH, but the charge is highly pH dependent (Cornell and Schwertmann, 1998). Waterborne viruses include *enteroviruses*, *coxsackievirus*, *echovirus*, *rotavirus*, *norovirus*, and *hepatitis A* and *B*. Bacteria of concern are chiefly enteropathogenic *Escherichia coli*, *Salmonella* and *Shigella* spp., *Campylobacter jejuni*, and *Aeromonas hydrophila*, among others. The main protozoa that have been transmitted by groundwater are *Cryptosporidium parvum* and *Giardia lamblia* (Macler and Merkle, 2000). Bacteria are microscopic (1–6 μm in size) unicellular organisms with a nucleus; viruses are submicroscopic (20–120 nm in size) biological agents consisting of molecules of nucleic acids and a protein envelope (Nevecherya et al., 2005) and negatively charged at high pH (pH 7). With this unique feature of viral structure and colloidal physicochemical properties, the transport of viruses in soil and groundwater can act with a combination of characteristics ranging from solutes, colloids, and microorganisms. The contamination of subsurface water by migrating pathogenic bacteria and viruses has caused large outbreaks of waterborne diseases (Pekdeger et al., 1985; Stevik et al., 2004). The most important pathogenic germs that might possibly be transported on the subsurface water path are the bacteria *Salmonella* sp., *Shigella* sp., *Yersinia enterocolitica*, *Y. pseudotuberculosis*, *Leptospira*, *Dyspepsia coli*, enterotoxigenic *E. coli*, *Vibrio* sp., *Legionella* sp., and the infectious hepatitis virus, polio virus, *coxsackie viruses*, *rotavirus*, and *norwalk-like virus* (Pekdeger et al., 1985). Bradford et al. (2002) mentioned that these inorganic/organic colloid particles can be released into the soil solution and groundwater through a variety of hydrologic, geochemical, and microbiological processes including translocation from the vadose zone (Nyham et al., 1985), dissolution of minerals and surface coatings (Ryan and Gschwend, 1990), precipitation from solution (Gschwend and Reynolds, 1987), deflocculation of aggregates (McCarthy and Zachara, 1989), and microbial-mediated solubilization of humic substances from kerogen and lignitic materials (Quyang et al., 1996). Scientific interest in the way these colloidal particles are transported through the intricate pores of subsurface porous environments has been inspired by two facts affecting public health (Morales et al., 2011): colloids that can be contaminants by themselves (bacteria, viruses, organics) and colloids that can act as carriers for inorganic/organic contaminants (Elimelech and Ryan, 2002; Sen and Khilar, 2006, 2009). Basically, subsurface colloidal particles can be released into soil solution and groundwater through a variety of hydrologic and geochemical processes and also have been implicated in the transport of metals, radionuclide, and certain ionizable organic pesticides in laboratory and field tests, which has been reviewed by Sen and Khilar (2009, 2006), Flurry and Qiu (2008), Elimelech and Ryan (2002), Bin et al. (2011), and Massoudieh and Ginn (2010). It is now generally accepted that in situ release of subsurface soil particles, colloidal in nature, may facilitate the contaminant transport in groundwater (Flurry and Qiu, 2008; Kersting et al., 1999; Ryan and Elimelech, 1996; Sen and Khilar, 2009; Sen et al., 2002).

5.3 CLASSIFICATION OF SUBSURFACE ENVIRONMENT AND IMPORTANCE ON COLLOID-CONTAMINANT TRANSPORT

Major pathways and mechanisms for transport of contaminants may differ in the unsaturated (vadose) and saturated zones, due to different conditions. The vadose zone comprises the subsurface environment, situated between the land surface and the saturated subsurface zone (groundwater). The vadose zone is the first subsurface environment encountered by contaminants released through human activities. In general, the vadose zone is characterized by the presence of oxygen, gas–water interface, and relatively high concentrations of particulate OM and relatively high microbial activity (Figure 5.2). Chemical conditions vary significantly with time and place, due to dilution with rainwater or concentration due to evaporation and due to large horizontal and vertical variations in the composition of the solid phase. Transport of contaminants in the vadose zone will be mainly vertical (Sen and Khilar, 2006). The saturated zone on the other hand is in general characterized by much lower OM contents, much lower oxygen content, and a lower sorption capacity of the solid

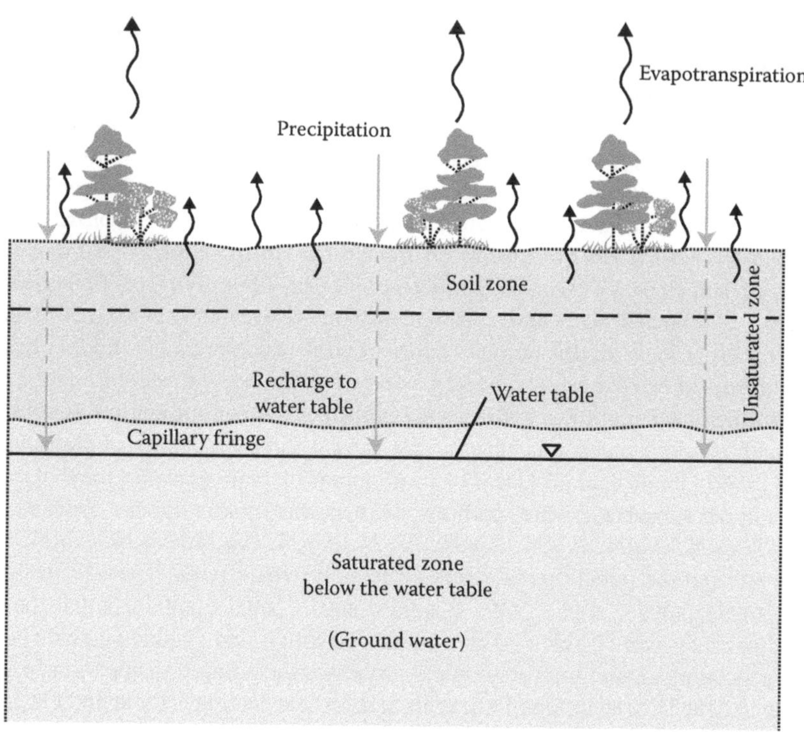

FIGURE 5.2 Classification of subsurface environment and characterization. (From USGS Circular 1186, General facts and concepts about groundwater in sustainability of ground water—Water resources, pubs.usgs.gov/circ/circ1186/html/gen_facts.html.)

phase (Figure 5.2). In comparison with the vadose zone, chemical conditions will be less variable in time and space because, in general, the influence of dilution and evapotranspiration will be negligible and the composition of the solid phase in the horizontal and vertical direction will be less variable. Transport of contaminants in the saturated zone will be mainly horizontal (Sen and Khilar, 2006) and is important for the transport of inorganic contaminants. The transport of colloids/biocolloids through saturated and unsaturated porous media is of significant interest, from the perspective of protecting the groundwater supplies from contamination (e.g., Redman et al., 2001; Surampalli et al., 1997), assessment of risk from pathogens in groundwater (e.g., Bitton and Gerba, 1984; Bruins et al., 2000; Goyal et al., 1989; Morris and Foster, 2000; Nola et al., 2001; Rose et al., 2000; Scandura and Sobsey, 1997; Taylor et al., 2004; Yates and Yates, 1988), and design of better water treatment systems to remove biocolloids from drinking water supplies (e.g., Gerba and Goyal, 1985; Macler and Merkle, 2000; Schijven and Hassanizadeh, 2001). Because biocolloids are living organisms, their transport in the subsurface is more complex than in the case for colloidal solutes transport (Sen and Khilar, 2006, 2009). Not only are they subject to the same physicochemical phenomena as are colloids (Sen et al., 2004), but there are also a number of strictly biological processes that affect their transport. Three key features of the vadose zone play a critical role in colloidal movement: (1) the presence of air–water interfaces, (2) transients in flow and chemistry, and (3) soil structure and heterogeneity.

It has been found that the fate and transport of colloids and colloid-associated contaminants in saturated porous media are controlled by a combination of several basic factors including (1) the presence of colloidal particles in the subsurface environment; (2) their release, dispersion stability, migration, and straining/plugging at pore constrictions; and (3) association of contaminants with colloidal particles/solid matrix (Sen and Khilar, 2006, 2009; Sen et al., 2002, 2004, 2005). Similar observations by several investigators have been compiled by Sen and Khilar (2009).

Chemical colloidal and hydrodynamic forces are important to colloidal particle release, deposition, and straining/plugging in saturated porous media, which are also dependent on geochemical and hydrodynamic conditions of the subsurface medium (Ryan and Elimelech, 1996; Sen and Khilar, 2002, 2006). However, differences in the mobilization mechanisms and deposition behavior of colloids are vastly different for saturated and unsaturated systems: straining, for instance, is anticipated to be even more important in unsaturated than in saturated systems (Bradford et al., 2006; Gao et al., 2006). The retention of colloidal fines in saturated porous media is mainly due to attachment at the solid–liquid interface and the change in surface properties of solid and solution that has been well documented in literature (Kretzschmar et al., 1999; McCarthy and Zachara, 1989; Sen and Khilar, 2006, 2009; Sharma et al., 2008). But mechanisms of colloidal retention in unsaturated porous media may include adsorption (attachment/detachment processes) to the solid–water interface (SWI) (Ryan and Elimelech, 1996; Sen and Khilar, 2006, 2009), the air–water interface (AWI) (Nicole et al., 2004; Torkzaban et al., 2008; Zhang et al., 2010), the solid–water–air (SWA) straining (Bradford et al., 2006; Torkzaban et al., 2008), and film straining (Chin and Flurry, 2005; Kretzschmar and Schafer, 2005; Sen, 2011).

5.4 SAMPLING AND CHARACTERIZATION

As mentioned in earlier sections, a variety of small inorganic and organic materials exists in subsurface environment, termed as colloids. These subsurface colloidal fines include mineral precipitates (such as iron, aluminum, calcium and manganese oxides, hydroxides, carbonates, silicates and phosphates, and oxides and hydroxides of actinide elements), rock and mineral fragments, clay minerals, biocolloids (such as virus, bacteria, and microorganism), and natural organic materials (including humic substances and polymeric materials) (McCarthy and McKay, 2004). Because of their colloidal nature and smaller size, these materials play a significant role as contaminant carriers or blockers in water ecosystems, and these are well documented by experimental, laboratory, and field studies (Sen and Khilar, 2009, 2011).

To access the importance of subsurface colloids, their release, and their role on contaminant transport at a particular site, it is essential to know the information about their existence, the total mobile load of contaminants and colloids, and the distribution of subsurface contaminants between colloid and aqueous phase. In addition, it is also important to know the morphological and surface colloidal characteristics such as size, charge with solution chemistry, sorptive properties, and elemental composition. To know this information, groundwater sampling and characteristics of colloids are essential. Accurate data of colloids and groundwater are essential to develop colloid-associated contaminant transport model. Sampling and characterization of the truly mobile colloidal particles occurring in soils or groundwater aquifers is a challenging task. When collecting colloidal samples from groundwater, special care should be taken to avoid contamination and artifact formation. Factors controlling sampling artifacts are oxygen diffusion that leads to calcite precipitation in groundwater, pumping rates, and filtering techniques. Groundwater samples are routinely obtained by pumping from wells. New protocols should be formulated to minimize artifact formation (Backhus et al., 1993; McCarthy and Degueldre, 1993; McCarthy and McKay, 2004; Sen and Khilar, 2009). Authors are encouraged to go through the article by Takala and Manninen (2006). It is recommended that low pumping rates in the order of 0.2–0.5 L/min should be used (Puls and Barcelona, 1996). Water quality such as solution pH, conductance, dissolved oxygen content, turbidity, and temperature may be monitored during purging. Sometimes passive sample collection may be used for low-permeability formations that are carried out on a *dialysis cell*. Backhus et al. (1993) reported a standard sampling protocol with minimum artifacts (Table 5.1) for in situ mobile colloids, and the associated contaminant concentrations include prolonged slow pumping and groundwater chemistry conditions. They have also reported the case study on sampling of various natures of colloidal fines and their quantitative amounts at different measurement sites. Detailed information on sampling can also be found in Schurtenberger and Newman (1993), Buffle and Leppard (1995), Schulze and Bertsch (1996).

TABLE 5.1
Recommended Protocol for Groundwater Sampling of Colloids and Colloid-Associated Contaminants

1. Measured depth to water calculates standing water volume.
2. (Optional) Make *dipstick* observation of any layering in the well (e.g., floating NAPL).
3. Slowly lower inlet/pump to sampling depth.
4. Inflate packer(s) to reduce well volume and isolate inlet from settled particles at well bottom or floating NAPL at the surface.
5. Initiate pumping at a slow flow rate (100 mL/min). Avoid surging. Observe air bubbles displaced from sample tube to assess progress of steady pumping until water arrives at the surface.
6. Measure volumetric flow rate and adjust pumping rate as necessary.
7. Use a flow-through cell to monitor temperature, specific conductivity, pH, Eh, and dissolved oxygen. Repeatedly collect water and assess turbidity.
8. (Optional) Initiate slow pumping at another well(s) on-site. In addition to sampling wells within the *contaminated* site of the aquifer, *background* wells should be sampled. The background well should be located outside the contaminated zone but constructed and developed in the same manner and screened in the same aquifer as the *contaminated* well.
9. After turbidity has stabilized, collect water samples for mobile contaminant analysis in suitable flow-through vessels. Minimize contact with air. Do not filter for groundwater analyses. Preserve as appropriate (e.g., store in cold and dark for organics).
10. (Optional) Collect colloid samples on membrane filters and store for characterization (SEM/EDX and XRD analysis).

Source: Adapted from Backhus, D.A. et al., *Ground Water*, 31, 466, 1993.

Different techniques are usually used to collect water sample from the vadose zone such as free drainage lysimeters, suction cups or porous plates, and fiberglass wicks or rock wool samplers (Takala and Manninen, 2006). The suitability of fiberglass wicks for colloidal sampling was tested by Czinany (2004). A zero-tension lysimeter is used to sample colloidal particles and to routinely monitor colloidal concentrations in field soils (Thompson and Scharf, 1994). Such zero-tension lysimeters are installed at a certain depth below undisturbed soil cores and these basically consist of a sampling cup with an air inlet tube and a sampling tube. Field flow fractionation (FFF), a rapid and high-resolution fractionation technique, may be used for the characterization of low concentration natural colloids (Beckett and Hart, 1993). This sampling technique is based on colloid interaction with an externally applied force field in a thin, ribbon-like flow channel.

Sampling and analysis of microorganisms in groundwater are especially problematic because traditional analytical methods (APHA, 1995) are based on culturing of viable cells and many pathogens of interest are not readily culturable in the laboratory (Maier et al., 2000; Sen and Khilar, 2009). This problem has been partly solved by the development of nucleic acid-based analysis, which can detect both viable and nonviable microorganisms (Josephson et al., 1993; McCarthy and McKay, 2004).

The nature of colloidal materials is complex and several properties should be considered to characterize the dispersion of colloids. Different techniques for characterization of colloids and nanofines (synthetic or natural) have been presented in different articles (Hassellov et al., 2008; Hennebert et al., 2013; McCarthy and Degueldre, 1993; Schurtenberger and Newman, 1993). When characterizing colloidal particles, one is usually primarily interested in composition (mineralogy, elemental composition, and organic components), concentration (mass concentration, number concentration), size and shape (size distribution), and surface characteristics (specific surface area (SA), surface charge, and sorption capacity). Some important techniques that can be used to analyze subsurface colloids are summarized in Table 5.2, which is updated and taken from our earlier publication (Sen and Khilar, 2009). The elemental and morphological analysis of colloidal particles can be done either from large single particles (<500 nm) or from agglomerate of the colloidal particles or from a filter cake with an energy dispersive x-ray spectroscope, SEM/EDS (Takala and Manninen, 2006).

Mineralogical characteristics of colloidal particles can be found from XRD analysis. Wang and Keller (2009) collected three bulk agricultural soils and one sediment (denoted as Ag#1, Ag#2, and Ag#3, and sediment from the topsoil layer of Santa Barbara, CA) and separated natural colloids and analyzed various properties, which are presented in Table 5.3. From Table 5.3, although natural colloids only vary from 2.7% to 13.8% by weight, they have significantly higher organic carbon (OC), cation exchange capacity (CEC), and BET SA than their bulk soils and sediments and hence have strong implications in the colloid-associated contaminant transport.

There are only few studies dealing with the release and characterization of colloids from solid waste. Hennebert et al. (2013) reported potential colloidal release and their various characterizations

TABLE 5.2
Various Methods Used for the Characterization of Colloidal Particles in Subsurface Systems

Colloid Property	Method	Information
Particle composition	Inductively coupled plasma-atomic emission spectroscopy (ICP-AES) and inductively coupled plasma-mass spectroscopy (ICP-MS)	Bulk elemental composition for most elements with atomic number >10, isotopic analysis with ICP-MS, mass concentration if all constituent elements are analyzed
	Energy dispersive analysis of x-rays (EDAX)	Elemental composition of single particles viewed by SEM or TEM, analysis of most elements with atomic number >10, element mapping
	X-ray diffraction analysis (XRD)	Mineralogical composition
	Fourier transform infrared spectroscopy (FT-IR, FT-IR image)	Mineralogical composition and organic functional groups
	Thermogravimetric analysis (TGA), differential thermal analysis (DTA)	Quantitative analysis of minerals
	Mossbauer spectroscopy	Oxidation state and structure of Fe-containing minerals
Morphology and size distribution	Light microscopy	Particle number, morphology
	Transmission electron microscopy (TEM)	Particle morphology, size distribution
	Atomic force microscopy (AFM)	Particle morphology, size distribution
	Static light scattering (SLS)	Particle size and shape with the small angle option for particle size distribution in the range of 0.1–100 μm
	Dynamic light scattering (DLS)	Average hydrodynamic radius of particles <1 μm, upper size cut-off for polydisperse samples from multiangle measurements
	Sedimentation analysis	Particle size distribution based on sedimentation velocity in a gravitational field, size range 1–50 μm
	Centrifuge techniques	Particle size distribution based on sedimentation velocity in a centrifugal field, analytical and preparative techniques, wide size ranges from 1 to 1000 nm
	Field flow fractionation (FFF)	Size distribution based on sedimentation velocity or diffusion coefficient
Particle surface characteristics	Electrophoretic mobility measurement (EM), zeta analyzer	Electrophoretic mobility, which is an indicator of surface charge (zeta potential)
	Gas adsorption (BET)	Specific surface area, microporosity

Sources: Adapted from Kretzschmar, R. et al., *Adv. Agron.*, 66, 121, 1999; Sen, T.K. and Khilar, K.C., Mobile subsurface colloids and colloid-mediated transport of contaminants in subsurface soils, in: Birdi, K.S. (Ed.), *Handbook of Surface and Colloid Chemistry*, 3rd edn., Taylor & Francis Group/CRC Press, 2009, pp. 107–130.

TABLE 5.3
Measured Properties of Natural Colloids, Bulk Soils, and Sediment

Soil	Weight (%)	OC (%)	CEC (cmol/kg)	Surface Area (m^2/g)	Exchangeable Ca^{2+} + Mg^{2+} (cmol/kg)	Fe Content (%)
Ag#1						
Bulk	100	1.51	6.20	3.5	5.8	0.19
Colloids	5.3	4.95	40.2	30.8	39.0	
Ag#2						
Bulk	100	1.50	15.2	9.4	14.1	0.14
Colloids	6.8	4.36	59.0	53.1	57.5	
Ag#3						
Bulk	100	1.52	15.4	14.1	15.1	0.17
Colloids	13.8	4.50	54.4	63.1	53.6	
Sediment						
Bulk	100	1.12	5.4	2.0	5.2	0.19
Colloids	2.7	6.02	42.2	36.6	40.7	

Source: Adapted from Wang, P. and Keller, A.A., *Langmuir*, 25(12), 6856, 2009.
Abbreviations: CEC, cation exchange capacity; OC, organic carbon

from solid industrial and municipal waste leachates, contaminated soil, contaminated sediments, and landfill leachates. Potentially released colloids and their characterization were carried out with standard leaching, membrane filtration, and elemental determination of filtrate fraction (ICP), and visible morphological characteristics of colloids were done by the TEM-EDS system. They have reported that Mn, As, Co, Pb, Sn, and Zn always had a colloidal form and total organic carbon (TOC), Fe, P, Ba, Cr, Cu, and Ni had a partially colloidal form that existed in various waste leachates and contaminated soil. Koster et al. (2007) show the release of Cu, Pb, Zn, and Cr in colloidal form from municipal solid waste incinerator's bottom ash.

5.4.1 Particle Size

The particle size distribution is one of the most important characteristics of mobile colloidal particles. The size distribution has a large influence on the specific SA of the colloidal particles, which is directly related to the capacity for contaminant sorption and hence in the colloid-associated contaminant transport. The size distribution of colloids is usually determined by light scattering techniques. Colloids are small colloidal fines (10–1000 nm, commonly 20–450 nm) that are present as stable suspension of subsurface environment (Takala and Manninen, 2006). Colloidal fines may be either single solid fine or agglomerates of several smaller solid particles. Figure 5.3 presents SEM of colloidal particles agglomerating from the mine drainage tunnel (Zanker et al., 2000). The size and shape of particles also play a crucial role in determining whether or not plugging will occur and hence colloid-associated contaminant retardation or facilitation takes place. Table 5.4 presents some qualitative results on whether plugging or other possible types of deposition would possibly occur as determined by the ratio of the size of fines to size of the pore constrictions (Khilar and Fogler, 1998; Sen and Khilar, 2006). Typically, the dimensions of the migratory colloidal fines vary from 0.1 to 10 μm (Khilar and Fogler, 1998). The size of the migratory fines in soil mass is usually below 5 μm. Seldom are migratory fines larger than 10 μm.

In case of the last two entries of low size ratios in Table 5.4, there exists a critical particle concentration (CPC) above which the plugging may occur (Pandya et al., 1998).

FIGURE 5.3 SEM micrograph of an agglomerate of the colloidal particles from mine drainage tunnel. (From Zanker, H. et al., *A Separation and Detection Scheme for Environmental Colloids*, Institute of Radiochemistry, Dresden, Germany, 2000, www.fz-rossendorf.de/FWR/COLL/Coll_2.htm; Takala, M. and Manninen, P., Sampling and analysis of groundwater colloids—A literature review, Working report 2006-15, Posiva, Finland, 2006, (29), 43, http://www.posiva.fi/tyoraportit/WR2006-98web.pdf.)

TABLE 5.4
Dependence of Plugging on the Ratio of Size of Fines to Size of Pore Constrictions

Size of Fines/Size of Pore Constrictions	Occurrence
≥1	Plugging due to blocking or size exclusion.
0.1–0.6	Plugging due to bridging and multiparticle blocking.
0.04–0.10	Plugging due to surface deposition, bridging, and multiparticle blocking.
0.01–0.04	Surface deposition. Multiparticle blocking may or may not occur.

Source: Khilar, K.C. and Fogler, H.S., *Migration of Fines in Porous Media*, Kluwer Academic Publishers, Dordrecht, the Netherlands, 1998, Chapters 1, 3, 9.

5.4.2 Surface Charge and Zeta Potential

It is very important to know the surface charge and surface potential (zeta potential) of subsurface colloidal fines in aqueous solution because the adsorption/desorption of ionic species on its surface is determined by the electrical properties at the colloidal fines/aqueous solution interface. Further stability of colloidal fines in subsurface aqueous phase also depends on surface charge of particles. A particle with a zeta potential >30 mV is considered as electrostatically stable (Hassellov and Kaegi, 2009). Solid particles present in water are often charged and depend on particles and geochemical solution chemistry. The mechanisms of charge development have been extensively investigated (Bin et al., 2011; Lopez-Garcia et al., 2007). Fine colloidal particles, in general, carry a surface charge, the origin of which depends on how the fines were formed. A negatively charged surface can exchange its portable cations present on the surface for cations present in the solution. The amount exchanged at equilibrium is known as the CEC. Likewise, we can define anion exchange capacity (AEC) of positively charged colloidal fines. The CEC of several clay minerals has been listed by Grim (1968). For migratory clayey fines such as kaolin and illite, the CEC is of the order of 3–15 meq/g (Khilar and Fogler, 1998). Further soil colloids are heterogeneous in composition and consist of inorganic and organic constituents or a mixture of the two. Therefore, charge development of soil colloids in geochemical environment is a complex process due to heterogeneity in the composition and structures of colloidal particles (Bin et al., 2011).

The surfaces of oxide colloidal particles or silicate minerals in contact with aqueous solutions are charged positively or negatively by adsorption or desorption of H^+ ions. An electrical double layer at solid/liquid interface is formed by adsorbing counterions from the aqueous solution to its surface (Sen and Khilar, 2009). The charge developing process on the surfaces can be represented as follows (Cornell and Schwertmann, 1998):

$$MOH_2{}^+(\text{surf}) = .MOH(\text{surf}) + H^+(\text{aq})K_{a1}^{\text{int}} \tag{5.1}$$

$$MOH(\text{surf}) = MO^-(\text{surf}) + H^+(\text{aq})K_{a2}^{\text{int}} \tag{5.2}$$

where

MOH denotes a surface site

K_{a1}^{int} and K_{a2}^{int} are the intrinsic surface acidity constants

Schindler and Stumm (1987) summarized the intrinsic acidity constants of various metal oxides/minerals under solution chemistry. In this system, H^+ and OH^- are potential determining ions and the surface charge and surface potential are defined by the pH of the solution. There is a particular pH at which the surface charge becomes zero, that is, the point of zero charge (pzc) or the isoelectric point (iep). K_a^{int} is correlated to the pzc, which is as follows (Bin et al., 2011):

$$\text{pzc} = 0.5 \log K_a^{\text{int}} = 0.5 \log K_a \tag{5.3}$$

Here intrinsic acidity constant (K_a^{int}) is equal to the apparent acidity constant (K_a) at a surface potential of zero. Point of zero charge (pzc) can be measured by electrophoresis as the pH at which there is no motion of the particles in suspending medium under an applied external electric field. Table 5.5 presents the point of zero charge of simple hydro(oxides) and other materials (Kosmulski, 2004; Sen and Khilar, 2009).

The more relevant and useful method of surface charge characterization of colloidal fines is, however, the measurement of zeta potential of the particle. The zeta potential of kaolinite has been studied by several investigators (Herrington et al., 1992; Hussain et al., 1996; Khilar and Fogler, 1983; Kia et al., 1987; Williams and Williams, 1978). The kaolinite face zeta potential varies from approximately –35 to –25 mV at a pH of 6.5 in the range of sodium chloride concentration between 10^{-4} and 10^{-2} M (VanOlphen, 1977). Hussain et al. (1996) measured the zeta potential of kaolinite, illite, and chlorite clay minerals against pH and found to have negative zeta potential in the pH range of 2.5–11. The kaolinite is the most negative clay, ranging in zeta potential value from –24 to –49.5 mV. There have been numerous investigations of the electrochemical properties of goethite, hematite, and ferrihydrite (Atkinson et al., 1967; Breeuwsma and Lyklema, 1971, 1973; Crawford et al., 1996; Davis and Leckie, 1978; Onoda and Debruyn, 1966; Park, 1967; Puls and Powell, 1992).

Park (1967) summarized the factors controlling the sign and magnitude of surface charge of oxide and mineral oxides, especially hydrous metal oxides.

TABLE 5.5
Point of Zero Charge (pzc) of Simple Hydro(oxides) and Other Materials

Material	Description	Salt	*T*	Method	pzc	References
Al_2O_3	γ, >99%	0.1 mol/dm³ NaCl	25 60	pH	9.3 9	Fein and Brady (1995)
Al_2O_3	α	0.01 mol/dm³ NaCl	25	IEP	8	Guo et al. (1997)
Al_2O_3	α, A16, Alcoa	None		IEP	8.5	Yang and Troczynski (1999)
Al_2O_3	γ, from *sec*-butoxide	0.001–0.1 mol/dm³ KNO_3	25	CIP	8.6	Ardizzone et al. (2000)
Al_2O_3	Riedel Haen, purity 98%	0.01 mol/dm³ KCl	20	IEP	8	Ramos-Tejada et al. (2002)
Al_2O_3	α, AKP-30	0.01 mol/dm³ $NaNO_3$	22	IEP	9	Hackley et al. (2002)
Al_2O_3	γ, spherical, NanoTek	0.01 mol/dm³ NaCl		IEP	9.6	Tang et al. (2002)
Al_2O_3	γ, Aldrich, mesoporous	0–0.1 mol/dm³ NaCl		CIP	9.1	Wang et al. (2002)
Al_2O_3	Fisher	0–0.1 mol/dm³ NaCl		CIP	8.7	Wang et al. (2002)
Al_2O_3	γ, from nitrate	HNO_3 + KOH	25	IEP CIP pH	8.2 8.6 8–8.4	Vakros et al. (2002), Bourikas et al. (2003)
Al_2O_3	α, sapphire	0.001 mol/dm³ KBr, KNO_3		IEP	5	Franks and Meagher (2003)
Al_2O_3	γ	0.001–0.1 mol/dm³ NaCl	25	IEP	8.4	De Lint et al. (2003)
Al_2O_3	Alcoa			Acousto	7.8	Cordelair and Greil (2003)
Al_2O_3	Aldrich, 99.5%, α			IEP	9	Sun and Berg (2003)
Al_2O_3	Alfa Aesar, 99.9%	0.01 mol/dm³ KCl		IEP	9.1	Hu and Dai (2003)
Al_2O_3	γ, Merck	0.1 mol/dm³ $NaNO_3$	25	CIP	8.2	Lefevre et al. (2004)
AlOOH, Al $(OH)_3$	Six different recipes	HCl + KOH	25	IEP	8.5–9	Katatny et al. (2003)
Co_3O_4	Thermal decomposition of nitrate at 400°C	0.005–0.3 mol/dm³ KNO_3		CIP	7.2	Faria and Trasatti (2003)
Cr_2O_3	From chloride	0.1 mol/dm³ KCl		pH	4	Onija et al. (2003)
Cr_2O_3	Fluka	0.001–0.1 mol/dm³ KCl	22	pH	6.7	Onija and Milonjic (2003)
$Cr(OH)_3$	From chloride	0.1 mol/dm³ KCl		pH	4.7	Onija et al. (2003)
CuO	Tenorite, Merck	0.001 mol/dm³ $KClO_4$		pH	6.9, 7.6	Gonzalez and Laskowski (1974)

(Continued)

TABLE 5.5 (*Continued*)
Point of Zero Charge (pzc) of Simple Hydro(oxides) and Other Materials

Material	Description	Salt	*T*	Method	pzc	References
CuO	Aldrich			IEP	8.5	Rao and Finch (2003)
Fe_3O_4	Magnetite, Sweden	0.01, 0.1 mol/dm^3 KCl		Intersection	6.5	Laskowski and Sobieraj (1969)
Fe_3O_4	Magnetite, synthetic	0.002–1 mol/dm^3 NaCl		IEP CIP	8 7.9	Illes and Tombacz (2003)
Fe_2O_3	From nitrate	0.1 mol/dm^3 $NaNO_3$	25	pH	8.5	Liger et al. (1999)
Fe_2O_3	Baker, washed	0.001–0.1 mol/dm^3 NaCl		IEP CIP	8.5 8.5	Jeon et al. (2001)
Fe_2O_3	Hematite, Laborchemie Apolda	0.01 mol/dm^3 NaCl	25	IEP	7	Chibowski and Wisniewska (2002)
Fe_2O_3	Hematite, Alfa	0.01 mol/dm^3 KNO_3	25	pH	6.3	Preocanin et al. (2002)
Fe_2O_3	Hematite, spherical, from $FeCl_3$	0.01 mol/dm^3 NaCl	25	pH	8.5	Pochard et al. (2002)
Fe_2O_3	Aldrich			IEP	6.8	Rao and Finch (2003)
Fe_2O_3	Aldrich, >99%	0.001, 0.01 mol/dm^3 $NaNO_3$		IEP	8	Ramos-Tejada et al. (2003)
Fe_2O_3	Natural, Clinton, NY	0.01 mol/dm^3 $NaNO_3$ and 0.01 mol/dm^3 KNO_3		IEP	<4	O'Reilly and Hochella (2003)
Fe_2O_3	Natural, Italy	0.01 mol/dm^3 $NaNO_3$ and 0.01 mol/dm^3 KNO_3		IEP	6.8	O'Reilly and Hochella (2003)
Fe_2O_3	Synthetic	0.01 mol/dm^3 $NaNO_3$ and 0.01 mol/dm^3 KNO_3		IEP	7.3	O'Reilly and Hochella (2003)
					7.2	
Fe_2O_3–$n$$H_2O$	Synthetic	0.01–0.1 mol/dm^3 $NaNO_3$	25	CIP	7.9	Trivedi et al. (2003)
Fe_5HO_8–$4H_2O$	From chlorate VII	0.2 mol/dm^3 $NaClO_4$	25	pH	8.3	Spandini et al. (2003)
Fe_5HO_8–$4H_2O$	Synthetic	0.01 mol/dm^3 $NaNO_3$		IEP	6.8	O'Reilly and Hochella (2003)
FeOOH	Geothite, synthetic	0.01 mol/dm^3 NaCl		IEP	9.1	Pozas et al. (2002)
FeOOH	Goethite, from nitrate	0.001–0.1 mol/dm^3 $NaNO_3$	25	IEP	8.7	Kosmulski et al. (2003)
FeOOH	Natural goethite	0.01 mol/dm^3 $NaNO_3$		IEP	7	O'Reilly and Hochella (2003)
FeOOH	Synthetic goethite	0.01 mol/dm^3 $NaNO_3$		IEP	9.2	Luxton and Eick (2003)

(*Continued*)

TABLE 5.5 (*Continued*)
Point of Zero Charge (pzc) of Simple Hydro(oxides) and Other Materials

Material	Description	Salt	*T*	Method	pzc	References
MgO	Merck, 97%	HNO_3 + KOH	25	IEP	12	Vakros et al. (2002)
				CIP	10	
				pH	10.2–10.9	
MgO	Aldrich, >99%, fused			IEP	12.4	Sun and Berg (2003)
MnOOH	Ground natural manganite	0.1 mol/dm^3 $NaNO_3$	23	CIP	5.4	Weaver et al. (2002)
MnO_2	γ, Union Carbide	0.0001–0.01 mol/dm^3 KNO_3		IEP	5.6	Natarajan and Fuerstenau (1983)
MnO_2	γ, electrolytic, Union Carbide	0.0001, 0.001 mol/dm^3 $NaNO_3$		IEP	5.3	Fuerstenau and Shibata (1999)
MnO_2	Ramsdelite, synthetic	0.001–0.1 mol/dm^3 KNO_3	25	CIP	<2.5	Prelot et al. (2003)
MnO_2	Commercial, Sedema	0.001–0.1 mol/dm^3 KNO_3	25	CIP	3.7	Prelot et al. (2003)
MnO_2	Electrochemically synthesized, Delta	0.001–0.1 mol/dm^3 KNO_3	25	CIP	6.7	Prelot et al. (2003)
NIO	Aldrich			IEP	7.8	Rao and Finch (2003)
PbO	In situ precipitated from nitrate	0.01 mol/dm^3 KCl		IEP	11	Liu and Liu (2003)
SiO_2	Merck	0.01 mol/dm^3 NaCl		IEP	<3 if any	Jada et al. (2002)
SiO_2	Davisil, Aldrich	0.001 mol/dm^3 KCl	25	IEP	3.5	Pettersson and Rosenholm (2002)
TiO_2	Baker, 99.9%, anatase			IEP	6.7	Sun and Berg (2003)
ZnO	Aldrich			IEP	7.5	Rao and Finch (2003)
Kaolinite	KGa-2	0.001–0.1 mol/dm^3 NaCl	25 ± 3	pH	2.8	Appel et al. (2003)
Montmorillonite	Fluka		25	IEP	<1 if any	Juang et al. (2002)
$CaCO_3$	Iceland spar		50	IEP	$\zeta \leq 0$	Moulin and Roques (2003)
Kaolin	From China			IEP	2.8–3.8	Hu and Liu (2003)
Activated carbon	CS-1501	0.1 mol/dm^3 NaCl	20	pH	7.5	Faur-Brasquet et al. (2002)

Sources: Taken from Kosmulski, M., *J. Colloid Interface Sci.*, 275, 214, 2004; Sen, T.K. and Khilar, K.C., Mobile subsurface colloids and colloid-mediated transport of contaminants in subsurface soils, in: Birdi, K.S. (Ed.), *Handbook of Surface and Colloid Chemistry*, 3rd edn., Taylor & Francis Group/CRC Press, 2009, pp. 107–130. More data are available in Kosmulski, 2004.

Crawford et al. (1996) measured the zeta potential of amorphous hydrous iron (III) oxide (HFO) and amorphous hydrous chromium (III) oxide (HCO) as a function of pH during adsorption and coprecipitation of Cr^{3+}, Zn^{2+}, and Ni^{2+}. For sand materials, the zeta potential is negative and it increases as the pH increases (Elimelech et al., 2000).

Hennebert et al. (2013) measured the zeta potential of solid industrial and municipal waste leachates under different conditions, and it was found that six leachates had a positive zeta potential (<8 mV) and all others had a negative zeta potential (−15 to −38 mV). A particle with a zeta potential >− 30 mV is considered as electrostatically stable (Hassellov and Kaegi, 2009). The measured zeta potentials for the three types of natural colloids (Ag#1, Ag#2, Ag#3) and engineered zeolite particles under different salt concentrations are presented in Figure 5.4 (Wang and Keller, 2009). Detailed mechanistic explanations are given in Wang and Keller (2009). The zeta potential of all the natural colloids showed similar trends (Figure 5.4).

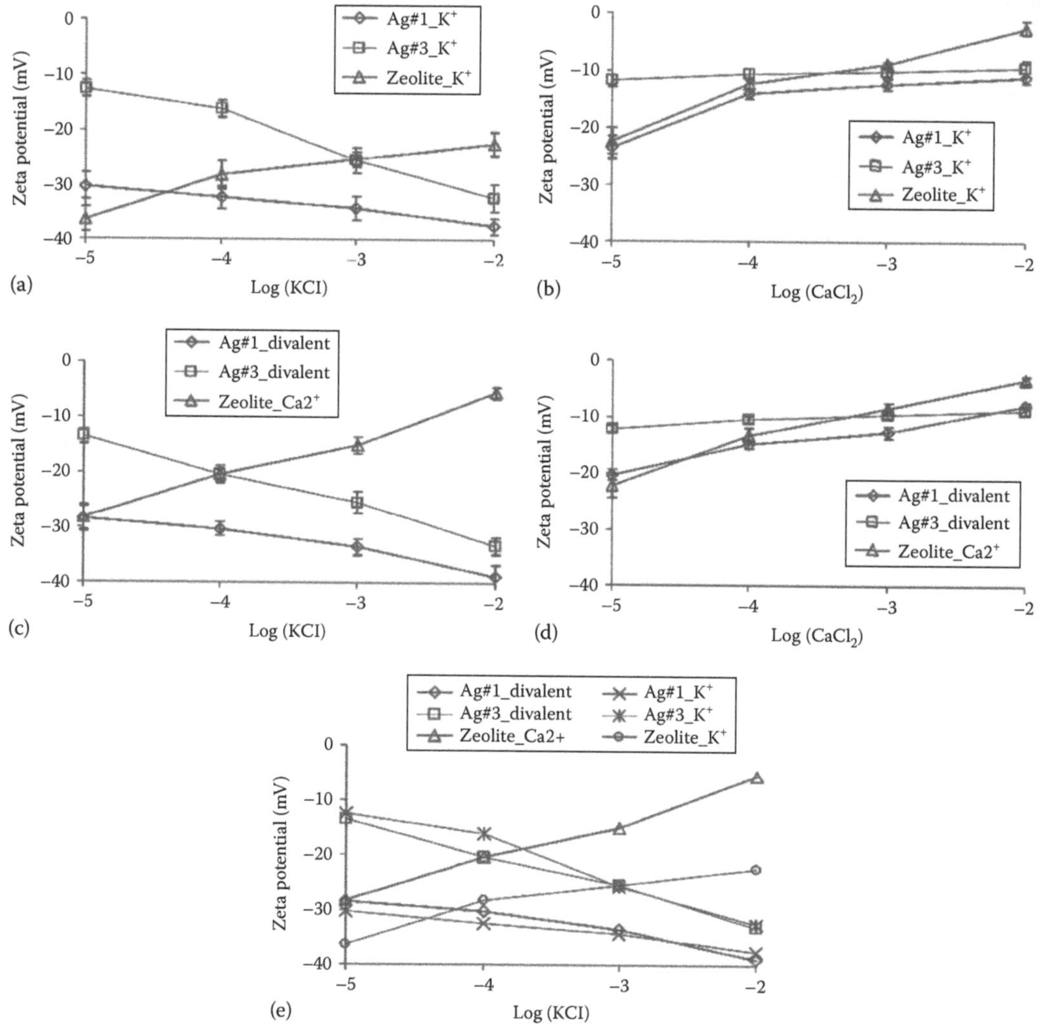

FIGURE 5.4 Zeta potentials of the original and cation-exchanged natural colloids (Ag#1 and Ag#3) and engineered zeolite as a function of log molar concentrations of KCl and $CaCl_2$ (pH 7.0): (a) K^+ saturated colloids with KCl in the bulk solution, (b) K^+ saturated colloids with $CaCl_2$ in the bulk solution, (c) divalent cation saturated colloids with KCl in the bulk solution, (d) divalent cation saturated colloids with $CaCl_2$ in the bulk solution, and (e) comparison of K^+ and divalent cation saturated colloids with KCl in the bulk solution. (From Wang, P. and Keller, A.A., *Langmuir*, 25(12), 6856, 2009.)

TABLE 5.6
Range in Chemical Composition of Migratory Clays

Clay	SiO_2	Al_2O_3	Fe_2O_3	TiO_2	CaO	MgO	K_2O	Na_2O
Kaolin	40–49	35–40	0–13	0–2	0–0.8	0–0.5	0–0.10	0–0.2
Illite	50–56	18–31	2–5	0–0.8	0–0.2	1–4	—	0–1
Chlorite	31–38	18–20	—	—	—	35–38	—	—
Montmorillonite	45–55	19–50	0–3	—	—	0–2	—	—

Source: Khilar, K.C. and Fogler, H.S., *Migration of Fines in Porous Media*, Kluwer Academic Publishers, Dordrecht, the Netherlands, 1998, Chapters 1, 3, 9.

5.4.3 Chemical Characterization of Fines

Chemical characterization of fines implies the elemental and mineral compositional analysis of migratory fines in porous media. Khilar and Fogler (1998) presented the range in chemical composition of migratory clays primarily of kaolinite, illite, montmorillonite, and chlorite particles in Table 5.6. We observe from this table that silica, SiO_2, and alumina are the major minerals.

5.5 BRIEF DESCRIPTION OF THE ROLE OF INORGANIC COLLOIDAL FINES–ASSOCIATED CONTAMINANT TRANSPORT IN GROUNDWATER FLOWS AND ASSOCIATED HEALTH EFFECT

Numerous experimental, theoretical, and field studies have confirmed that subsurface colloidal fines can either facilitate or retardate the transport of inorganic and organic pollutants (Arab et al., 2014; Katzourakis and Chrysikopoulos, 2014; Massoudieh and Ginn, 2011; Sen, 2011; Sen and Khilar, 2009; Wang et al., 2014). There are a couple of review articles published from time to time such as the articles by Ryan and Elimelech (1996, 2002), Kersting et al. (1999), Sen and Khilar (2006, 2009), Sen (2011), and Bin et al. (2011) where a large number of observations on colloid-associated contaminant transports have been compiled. Colloid-facilitated transport has been demonstrated for alkali and alkaline earth cations (Cs^+ and Sr^{2+}), transition metals (Ni^{2+}, Co^{2+}, Cu^{2+}, Pb^{2+}), oxyanions (arsenic and iodate), nonpolar organic compounds (phenanthrene and pyrene), and polar organic compounds (pesticide) (Sen and Khilar, 2009).

Sun et al. (2010) reported the colloidal fines kaolinite facilitated lead (Pb^{2+}) transport through saturated porous media. Similarly soil humic acid (HA)-mediated transport of copper (Cu) was studied by Paradelo et al. (2012). The influence of pH and Cu^{2+} concentration on Cu–HA binding and retention of Cu–HA complexes in the matrix was analyzed with the colloid-surface attachment model. The effect of ionic strength reduction and flow interruption enhanced colloid-facilitated Hg^{2+} transport in contaminated soils, which was reported by Zhu et al. (2014). They have concluded that a large amount of colloidal Hg was associated with colloidal OM, which can be transported over long distances causing environmental pollution. Sen et al. (2002) present experimental studies of colloid-associated contaminant transport in the subsurface. They conducted column experiments to study the effects of the mobilization of colloidal fines, kaolin on nickel transport in sand beds containing kaolin particles under plugged and unplugged conditions. Nickel transport was facilitated when colloidal kaolin present in the sand bed migrated out. Without migration and plugging of kaolin particles, the kaolin-sand bed acted as a poor filter for nickel. However, when migration as well as plugging occurs, the retardation of nickel transport increases and such enhancement in sorption can be attributed to sorption sites that become more accessible as well as higher sweeping a higher pressure drop and higher concentration of kaolin.

Colloids are known to mobilize strongly sorbing compounds (Honeyman, 1999; Kersting et al., 1999; Roy and Dzomback, 1997, 1998; Saiers and Hornberger, 1996). When highly mobile colloids carry with them the attached contaminates, the phenomena have been termed as *colloid-facilitated transport* (McCarthy and Zachara, 1989).

Nanoparticles have found their way for many human use products over the years. They are in the colloidal or subcolloidal size range. Nowack and Bucheli (2007) present a review of classes of nanoparticles relevant to the environment and summarize their formation, emission, and occurrence in the environment. At elevated concentrations, they reported the toxic ecotoxicological effects of certain nanoparticles.

Silver nanoparticles are widely used for their antimicrobial properties. Toxic effects of soluble silver on humans are presented in a review by Panyala et al. (2008). In recent years, the regulating agencies are concerned about the widespread use of inorganic nanoparticles, including the use of carbon nanotubes (CNTs) in various items used by human. Choi et al. (2010) show that nanoparticles that have a hydrodynamic diameter less than 34 nm and have noncationic surface charge can translocate rapidly from the lungs to the mediastinal lymph nodes. Nanoparticles with hydrodynamic radius less than 6 nm can traffic rapidly from lungs to lymph nodes and blood streams. However, the kidney may subsequently clear them. Colvin (2003) is one of the early researchers to alarm about the widespread use of nanoparticles in products for human use. She pointed out that a few studies have investigated the direct or indirect exposure to nanoparticles and no clear guidelines exist to quantify their effects.

Lam et al. (2006) conducted a review of potential CNT toxicity. From rodent studies they summarized that regardless of the synthesis methods or metals used, CNTs caused inflammation, epithelioid granulomas (microscopic nodules), fibrosis, and biochemical/toxicological changes in the lungs. Their review suggests that single-walled CNTs are toxic than quartz (which is considered as a serious occupational hazard when chronically inhaled).

5.6 BRIEF DESCRIPTION OF THE ROLE OF BIOCOLLOIDS IN GROUNDWATER CONTAMINATION AND ASSOCIATED HEALTH EFFECT

Many biocolloids, such as viruses, bacteria, and certain protozoa, being pathogenic can cause disease if they are ingested through drinking waters or recreational waters. Groundwater contamination from biocolloids has been reported over the decades. Crane and Moore (1984) concluded in their review that physical factors such as soil properties, clay content and moisture regime, and pore size distribution along with biological factors such as pH, temperature, nutrient supply, and soil moisture content affect the rate and quantity of bacterial migration through soil. Sen (2011) conducted a review of processes affecting the transport of pathogenic biocolloids in saturated and unsaturated porous media. He suggested that basic processes such as physical, chemical, and biological affect the transport of biocolloids. These factors affecting the transport of biocolloids include advection and dispersion as well as diffusion through water, straining, and physical filtration through the porous media as the biocolloids move, adsorption, and biological processes such as adhesion/detachment, survival, and chemotaxis. For viruses, removal in unsaturated zone seems to be greater than that in the saturated zone and it has been observed by many researchers over three decades (Bitton and Gerba, 1984; Jin et al., 1997; Powelson and Gerba, 1994; Powelson et al., 1990). Chu et al. (2001) suggest that once viruses are adsorbed to the AWI, they are considered to be effectively removed.

Most intestinal (enteric) diseases are infectious and are transmitted through fecal waste. Pathogens—which include virus, bacteria, protozoa, and parasitic worms—are disease-producing agents found in the feces of infected persons. These diseases are more prevalent in areas with poor sanitary conditions. These pathogens travel through water sources and interfuses directly through persons handling food and water. Since these diseases are highly infectious, extreme care and hygiene should be maintained by people looking after an infected patient. Hepatitis, cholera, dysentery, and typhoid are the more common waterborne diseases that affect large populations in the tropical regions (BIONFATE Newsletter, 2013). Tables 5.7 and 5.8 show information about the water-transmitted

TABLE 5.7
Selected Waterborne Bacterial and Protozoan Diseases, Their Agents, Symptoms, and Sources in Water Supply

Disease	Bacterial Species or Genera	Symptoms	Sources in Water Supply
Amoebiasis	*Entamoeba hystolytica*	Diarrhea, bloating, fever	Sewage-contaminated water
Typhoid	*Salmonella typhi*	Very high temperature, sweating, etc.	Fecal-contaminated water
Cryptosporidiosis	*Cryptosporidium parvum*	Watery diarrhea, flu-like symptom	Unfiltered water containing animal manure
Giardiasis	*Giardia lamblia*	Diarrhea, abdominal discomfort	Untreated water, water from beaver and muskrat ponds
Cholera	*Vibrio cholerae*	Watery diarrhea, nausea, vomiting, etc.	Drinking water contaminated with the bacterium
Leptospirosis	Bacteria genus *Leptospira*	Flu-like symptom followed by meningitis, liver damage, etc.	Water contaminated by animal urine carrying the bacterium
Legionellosis	Bacterial genus *Legionella*	Fever, chills, pneumonia, etc.	Contaminated water-cooling towers
Escherichia coli infection	*Escherichia coli*	Diarrhea and can lead to death in people with weak immune response	Water contaminated with the bacterium
Dysentery	Bacterial genera *Shigella* and *Salmonella*	Frequent passing of feces with blood and/or mucus	Water contaminated with the bacteria
Botulism	*Clostridium botulinum*	Dry mouth, blurred vision, slurred speech	Bacterium enters through wounds or gastrointestinal tracts
Campylobacteriosis	*Campylobacter jejuni*	Dysentery with high fever	Drinking water contaminated with feces

Source: Kolwzan, B. et al., *Introduction to Environmental Microbiology*, Oficyna Wydawnicza poletechruki wroclawskieji, Nigeria, 2006, http://www.oficyna.pwr.wroc.pl.

TABLE 5.8
Selected Waterborne Viral Diseases and Symptom

Disease and/or Agent	Symptoms	Sources in Water Supply
Poliomyelitis (polio virus)	Minor or no symptoms to headache, fever, etc.	Feces-contaminated water
Hepatitis A	Fatigue, fever, abdominal pain	Can manifest in water and food
Severe acute respiratory syndrome	Fever, lethargy, cough, sore throat	Manifests in improperly treated water
Coxsackie A	Herpangina, hemorrhagic conjunctivitis, meningitis	Feces contamination
Enteroviruses	Meningitis, encephalitis, hemorrhagic conjunctivitis, fever	Similarity with polio virus
Enteric cytopathic human orphan (ECHO) virus	Meningitis, respiratory system problems, rash, diarrhea, fever	Feces contamination
Norwalk virus	Diarrhea, fever	Feces contamination
Adenoviruses	Respiratory problems, diarrhea, eye infection	Respiratory droplets, fecal route in water
Rotaviruses	Severe diarrhea	Fecal–oral route exposure

Source: Kolwzan, B. et al., *Introduction to Environmental Microbiology*, Oficyna Wydawnicza poletechruki wroclawskieji, Nigeria, 2006, http://www.oficyna.pwr.wroc.pl.

pathogenic microorganisms (bacteria and virus), waterborne bacterial infections, and intestinal viruses that may be transmitted by water and diseases caused by them (Kolwzan et al., 2006).

In rural areas, human and animal waste sources contribute to groundwater contamination. For example, septic tanks and cesspools used for human waste disposal have traditionally been considered as source of fecal contamination of groundwater. Additionally, animal waste lagoons and stockyards are becoming important sources for fecal contamination of shallow groundwater in rural landscapes and the number of these units and the animals housed in the facilities are becoming more and more. Yates (1985) states that septic tanks contribute 800 billion gallons of water per year to subsurface, and this subsurface water is a source of pathogens including bacteria and viruses. She points out that in regions having more than 40 systems per square mile (1 system per 16 acres), the Environmental Protection Agency (EPA) estimates high potential for groundwater contamination. Ahmed et al. (2005) conducted biochemical fingerprinting enterococci and *E. coli* to relate possible migration of these organisms from septic tanks to surface water (a creek).

In 1976, cryptosporidiosis was first recognized to infect humans and the symptoms were diarrhea that generally ends in 2–3 weeks. Among the most well-known public health concerns involving biocolloids is the outbreak of cryptosporidium in the water system of the city of Milwaukee, Wisconsin, in 1993 (MacKenzie et al., 1994). Contamination occurred as the sand filtration system at the Howard Avenue Water treatment Plant on the shore of Lake Michigan failed to remove the oocysts of cryptosporidium during rapid sand filtration. Because of a storm even in early spring season (when cows were out in open areas), the oocysts in manure mixed in runoff water ended up near the intake of the treatment plant. The filtration system was not operating optimally and that led to a bypassing of the oocysts during the backwash of the filters. Over a period of 2 weeks (March 23 to April 8, 1993), over 400,000 people became infected and more than 100 weak or immune-compromised individuals died.

An outbreak of cryptosporidiosis in a disinfected groundwater supply occurred in Warrington, a town in Northwest England, between November 1992 and February 1993 (Bridgeman et al., 1995). A total of 47 cases were recorded over this period, and the bypass flow of surface water containing animal feces to aquifer was attributed to the cause. The bypass flow did not take advantage of filtration in natural sandstone present in subsurface. Once the source (e.g., well) was taken out of service, the outbreak quickly subsided. The fecal samples from infected patients showed the presence of oocysts of cryptosporidium. Chlorination appeared not to have complete disinfection potential in the outbreak.

The town of North Battleford, Saskatchewan, Canada, suffered a cryptosporidium outbreak (Sterling et al., 2001). The mechanism for failure was the defect of the sand filtration units and cryptosporidium oocysts being resistant to chlorine passed through the treatment plant to the distribution system. The outbreak occurred between late March and early May of 2001. Between 5800 and 7100 people were affected by the outbreak that included many visitors. Confirmed 275 cases of cryptosporidiosis were found by May 30, 2001. Again, failure of filtration was attributed to the outbreak. The source water was surface water and the parasites must have entered the river some point upstream from the intake.

Another serious case of contamination of the water supply of Walkerton, Ontario, Canada, with O157:H7 strain of *E. coli* led to the death of 7 people and sickening of 2500 people in year 2000 (Hrudey et al., 2002, 2003). Contaminated surface runoff entered the well possibly due to bypass flow. The operators were held liable for misconduct.

The resort on the island of South Bass in Lake Erie, Ohio, witnessed an outbreak of gastroenteritis in 2004 (Fong et al., 2007; O'Reilly et al., 2007) that exposed the visitors and the local residents. Fong et al. (2007) concluded the following:

> Massive groundwater contamination on the island was likely caused by transport of microbiological contaminants from wastewater treatment facilities and septic tanks to the lake and the subsurface, after extreme precipitation events in May–July 2004. This likely raised the water table, saturated the

subsurface, and along with very strong Lake Erie currents on 24 July, forced a surge in water levels and rapid surface water–groundwater interchange throughout the island. Landsat images showed massive influx of organic material and turbidity surrounding the island before the peak of the outbreak. These combinations of factors and information can be used to examine vulnerabilities in other coastal systems. Both wastewater and drinking water issues are now being addressed by the Ohio Environmental Protection Agency and the Ohio Department of Health.

Understanding the fate and transport of biocolloids in environmental systems is quite important because they pose a serious water quality risk. Biocolloids may remain infectious and may survive for long periods of time in natural environmental systems (Chrysikopoulos et al., 2010). In order to estimate the potential health risk associated with aquifers contaminated by various biocolloids, the prediction of biocolloid fate and transport is necessary.

Virus adsorption has some correlation with soil moisture; whereas its transport is faster when water saturation in pores is high. It has been proved that virus adsorption on AWI in unsaturated porous media is governed by temporal and spatial water saturation fluctuations. Virus adsorption is stronger on AWIs than on liquid–solid interfaces. However, virus inactivation is enhanced under the presence of AWIs. Released viruses in water or wastes can be adsorbed on the sand surface, clays, bacterial cells, natural colloidal dispersions, and sludge particles. Adsorbed viruses survive for longer time periods than viruses suspended in the aqueous phase. The prevailing mechanisms for biocolloid adsorption onto clays are not well defined and they vary for varying kinds of biocolloids. The electrostatic interactions and chemical properties such as ionic strength and pH are dominant parameters for the adsorption of a virus on mineral surfaces (Chrysikopoulos et al., 2010).

5.7 CONCLUSIONS

Within the subsurface systems such as soils, groundwater aquifers, sediments, or fractured rocks, colloidal particles/fines (particles with an average diameter between 10^{-9} and 10^{-10} m and carry surface charge) are ubiquitous with a pronounced variability concerning morphology and chemical composition. There are several classes of subsurface colloids—abiotic and biotic. These include silicate clays, iron and aluminum oxides, mineral precipitates, humic materials, micro emulsions of nonaqueous phase liquids, viruses, and bacteria. The mobilization and migration of these subsurface colloidal particles take place under different physical and geochemical conditions. The contamination of groundwater and soil due to these migrating colloidal particles is well recognized and is classified into two categories: (a) colloid-associated contaminant transport and (b) transport of biocolloids. This chapter has briefly reviewed these subsurface colloidal fines, classifications, sampling methods, and characterization, and finally their role in groundwater contamination. Groundwater contamination due to mobilization and migration of these colloidal fines and associated health effects has also been presented here.

REFERENCES

Ahmed A, Neller N, Katouli M. *J. Appl. Microbiol.*, 2005;98:910.

APHA. *Standard Methods*, 19th edn. American Public Association, Washington, DC, 1995.

Appel C, Ma LQ, Rhue RD, Kennelley E, *Geoderma*, 2003;113:77.

Arab D, Pourafshary P, Ayatollahi AH. *Int. J. Environ. Sci. Technol.*, 2014;11:207.

Ardizzone S, Bianchi CL, Galassi CJ, *Electroanal. Chem.*, 2000;490:48.

Atkinson RJ, Posner AM, Quirk JP, *J. Phys. Chem.*, 1967;71:550.

Auset M, Keller AA, *Water Resour. Res.*, 2004;40:3503.

Backhus DA, Ryan JN, Groher DM, MacFarlane JK, Gschwend PM, *Ground Water*, 1993;31:466.

Beckett R, Hart BT. The use of field-flow fractionation techniques to characterize aquatic particles, colloids and macromolecules. In: Buffle J, Van Leeuwen HP (Eds.), *Environmental Particles*, Vol. 2. Lewis Publisher, London, U.K., 1993, p. 165.

Bin G, Cao X, Dong Y, Luo Y, Ma LQ. *Crit. Rev. Environ. Sci. Technol.*, 2011;41:336.

Bitton G, Gerba CP. *Groundwater Pollution Microbiology: The Emerging Issue*. Wiley, New York, 1984, p. 713.
Borkovec MQ, Wu GD, Laggner P, Sticher H. *Colloids Surf. A*, 1993;73:65.
Bourikas K, Vakros J, Kordulis C, Lycourghiotis A. *J. Phys. Chem. B*, 2003;107:9441.
Bradford SA, Simunik J, Walker SL. *Water Resour. Res.*, 2006;42:W12S12. doi: 10.1029/2005WR004805.
Bradford SA, Yates SR, Bettahar M, Simunek J. *Water Resour. Res.*, 2002;38:1327.
Breeuwsma A, Lyklema J. *Discuss. Faraday Soc.*, 1971;52:324.
Breeuwsma A, Lyklema J. *J. Colloid Interface Sci.*, 1973;43:434.
Bridgeman SA, Robertson RMP, Sayed Q, Speed N, Andrews N, Hunter PR. *Epidemiol. Infect.*, 1995;115:555–566.
Bruins MR, Kapil S, Ochme FW. *Ecotoxicol. Environ. Saf.*, 2000;47(2):105.
Buffle J, Leppard GG. *Environ. Sci. Technol.*, 1995;29:2169.
Chibowski S, Wisniewska M. *Colloids Surf. A*, 2002;208:131.
Chin G, Flurry M. *Colloids Surf. A*, 2005;256:207–216.
Choi HS, Ashitate Y, Lee JH, Kim SH, Matsui A, Insin N, Bawendi MG, Semmler-Behnke M, Frangioni JV, Tsuda A. *Nat. Biotechnol.*, 2010;28:1300.
Chrysikopoulos CV, Masciopinto C, Mantia RL, Manariotis ID. *Environ. Sci. Technol.*, 2010;44(3):971.
Chu Y, Jin Y, Flurry M, Yates MV. *Water Resour. Res.*, 2001;37:253–263.
Colvin VL. *Nat. Biotechnol.*, 2003;21(10):1166.
Cordelair J, Greil P. *J. Colloid Interface Sci.*, 2003;265:359.
Cornell RM, Schwertmann U. *The Iron Oxide*, 1st edn. VCH Publishers, New York, 1998.
Crane SR, Moore JR. *Water Air Soil Pollut.*, 1984;22(1):67–83.
Crawford RJ, Harding IH, Mainwaring DE. *J. Colloid Interface Sci.*, 1996;181:561.
Czinany S. Subsurface colloids: Stability, sampling and transport under gravitational and centrifugal accelerations. Academic dissertation, Department of Crop and Soil Sciences, Washington State University, Pullman, WA, 2004.
Davis JA, Leckie JO. *J. Colloid Interface Sci.*, 1978;67:90.
Degueldre C, Missana T. *Colloids Surf. A: Physicochem. Eng. Aspects*, 2003;217:33.
De Lint WBS, Benes NE, Lyklema J, Bouwmeester HJM, van der Linde AJ, Wessling M. *Langmuir*, 2003;19:5861.
D'Antonio RG et al. *Ann. Intern. Med.*, 1985;103:886–888.
Elimelech M, Nagai M, Ko C-H, Ryan J. *Environ. Sci. Technol.*, 2000;34:2143.
Elimelech M, Ryan JN. The role of mineral colloids in the facilitated transport of contaminants in saturated porous media. In: Huang PM, Bollag J-M, Senesi N (Eds.), *Interactions between Soil Particles and Microorganisms: Impact on the Terrestrial Ecosystem*. John Wiley & Sons, New York, 2002.
Enfield CG, Bengtsson G. *Groundwater*, 1988;26:64.
Faria LAD, Trasatti S. *J. Electroanal. Chem.*, 2003;554:355.
Faur-Brasquet C, Reddad Z, Kadirvelu K, Le Cloirec P. *Appl. Surf. Sci.*, 2002;196:356.
Fein JB, Brady PV. *Chem. Geol.*, 1995;121:11.
Flurry M, Qiu H. *Vadose Zone J.*, 2008;7:682.
Fong T-T, Mansfield LS, Wilson DL, Schwab DJ, Molloy SL, and Joan B. *Environ. Health Perspect.*, 2007;115(6):856.
Franks GV, Meagher L. *Colloids Surf. A*, 2003;214:99.
Fuerstenau DW, Shibata J. *Int. J. Miner. Process.*, 1999;57:205.
Gao B, Saiers JE, Ryan J. *Water Resour. Res.*, 2006;42:W01410.
Gerba CP, Goyal SM. *Pathogen Removal from Wastewater during Groundwater Recharge Artificial Recharge of Groundwater*. Butterworth Publishers, Boston, MA, 1985, p. 317.
Gonzalez G, Laskowski J. *Electronal. Chem. Int. Electrochem.*, 1974;53:452.
Goyal SM, Amundson DA, Robinson RA, Gerba CP. *J. Minn. Acad. Sci.*, 1989;55(1):58.
Grim RE. *Clay Mineralogy*. McGraw Hill, New York, Chapter 6, 1968.
Gschwend PM, Reynolds MD. *J. Contam. Hydrol.*, 1987;1:309.
Guo LC, Zhang Y, Uhida N, Uematsu K. *J. Eur. Ceram. Soc.*, 1997;17:345.
Gustafsson O, Gschwend M. *Limnol. Oceanoger.*, 1997;42(3):519.
Hackley VA, Patton J, Lum LSH, Wasche RJ, Naito M, Abe H, Hotta Y, Pendse H. *J. Dispers. Sci. Technol.*, 2002;23:601.
Hassellov M, Kaegi R. Analysis and characterization of manufactured nanoparticles in aquatic environments. In: Lead JR, Smith E. (Eds.), *Environmental and Human Health Impacts of Nanotechnology*. Wiley, Chichester, U.K., 2009, p. 211.
Hassellov M, Readman JW, Ranville JF, Tiede K. *Ecotoxicology*, 2008;15(5):344.
Hennebert P, Avellan A, Yan J, Aguerre-Chariol O. *Waste Manage.*, 2013;33(9):1870.

Herrington RM, Clarke AQ, Watts JC. *Colloids Surf.*, 1992;68:161.
Honeyman BD. *Nature*, 1999;397:23.
Hoxie NJ, Davis JP, Vergeront JM, Nashold RD, Blair KA, Blair, KA. *Am. J. Public Health*, 1997;12:122.
Hoxie NJ et al. *Am. J. Public Health*, December 1997;87(12):2032–2035, PMC 1381251, PMID 9431298.
Hrudey SE, Payment P, Huck PM, Gillham RW, Hrudey EJ. *Water Sci. Technol.*, 2003;47:7–14.
Hrudey SE et al. *J. Environ. Eng. Sci.*, 2002;1.6:397–407.
Hussain SA, Demirci S, Ozbayoglu G. *J. Colloid Interface Sci.*, 1996;184.
Hu Y, Dai J. *Miner. Eng.*, 2003;16:1167.
Hu Y, Liu X. *Miner. Eng.*, 2003;16:1279.
Illes E, Tombacz E. *11th ICSCS 2003*, Foz de Iguacu, Brazil, 2003.
Jada A, Chaou AA, Bertrand Y, Moreau O. *Fuel*, 2002;81:1227.
Jeon BH, Dempsey BA, Burgos WD, Royer RA. *Colloids Surf. A*, 2001;191:41.
Jin Y, Yates MV, Thompson SS, Jury WA. *Environ. Sci. Technol.*, 1997;31:548.
Josephson KL, Gerba CP, Pepper LI. *Appl. Environ. Microbiol.*, 1993;59:3513.
Juang RS, Lin SH, Tsao KH. *J. Colloid Interface Sci.*, 2002;254:234.
Katatny EA et al. *Powder Technol.*, 2003;132:137.
Katzourakis VE, Chrysikopoulos CV. *Adv. Water Resour.*, 2014;68:62.
Kersting AB, Efurd DW, Finnegan DL, Rokop DJ, Smith DK, Thompson JL. *Nature*, 1999;397:56.
Khilar KC, Fogler HS. *Migration of Fines in Porous Media*. Kluwer Academic Publishers, Dordrecht, the Netherlands, 1998, Chapters 1, 3, 9.
Khilar KC, Fogler HS. *Soc. Pet. Eng. J.*, 1983;23(1):55.
Kia SF, Fogler HS, Reed MG. *J. Colloid Interface Sci.*, 1987;118(1):158.
Kolwzan B, Adamiak W, Grabas K, Welczyk A. *Introduction to Environmental Microbiology*. Oficyna Wydawnicza poletechruki wroclawskieji, Nigeria, 2006, ISBN: 83-7085-880-5, http://www.oficyna.pwr.wroc.pl.
Kosmulski M, Saneluta C, Maczka E. *Colloids Surf. A*, 2003;67:119.
Kosmulski M. *J. Colloid Interface Sci.*, 2004;275:214.
Koster R, Wangner T, Delay M, Frimmel FH. Release of contaminants from bottom ashes-colloid facilitated transport and colloid trace analysis by means of laser-induced breakdown detection (LIBD). In: Frimmel FH, von der Kammer F, Flemming H-C (Eds.), *Colloidal Transport in Porous Media*. Springer, Berlin, Germany, 2007, p. 252.
Kretzschmar R, Borkove M, Grolimund D, Elimelech M. *Adv. Agron.*, 1999;66:121.
Kretzschmar R, Schafer T. *Elements*, 2005;1:205.
Lam C-W, James JT, McCluskey R, Arepalli S, Hunter RL. *CRC Crit. Rev. Toxicol.*, 2006;36(3):189.
Laskowski J, Sobieraj S. *Trans. IMM*, 1969;78:C161.
Lefevre G, Duc M, Fedoroff M. *J. Colloid Interface Sci.*, 2004;269:274.
Liger E, Charlet L, Van Cappellen P. *Geochim. Cosmochim. Acta*, 1999;63:2939.
Liu Q, Liu Y. *J. Colloid Interface Sci.*, 2003;268:266.
Lopez-Garcia JJ, Aranda-Rascon MJ, Horno J. *J. Colloid Interface Sci.*, 2007;316:196.
Luxton T, Eick M. Quoted by O'Reilly and Hochella, 2003.
MacKenzie WR et al. *N. Engl. J. Med.*, 1994;331:161–167. doi: 10.1056/nejm199407213310304.
Macler BA, Merkle JC. *Hydrogeol. J.*, 2000;8(1):29.
Maier RM, Pepper IL, Gerba CP. *Environmental Microbiology*. Academic Press, San Diego, CA, 2000.
Massoudieh, A, Ginn, TR. Colloid-facilitated contaminant transport in unsaturated porous media. In: Hanrahan G (Ed.), *Modelling of Pollutants in Complex Environmental Systems*, Vol. II. ILM Publications, a Trading Division of International Lambbate Limited, Saarbrucken, Germany, 2010.
Massoudieh A, Ginn TR. Colloid-facilitated contaminant transport in unsaturated porous media, Chapter 8. In: Hanrahan G (Ed.), *Modelling of Pollutants in Complex Environmental Systems*, Vol. 2. ILM Publications, a Trading Division of International Labmate Ltd, Konstanz, Germany, 2011, pp. 259–287.
McCarthy J, Zachara J. *Environ. Sci. Technol.*, 1989;23:496.
McCarthy JF, Degueldre C. Sampling and characterization of groundwater colloids for studying their role in the subsurface transport of contaminants. In: Buffle J, Leeuwen HV (Eds.), *Environmental Particles*, Vol. II. Lewis Publishers, Chelsea, MI, 1993, p. 247.
McCarthy JF, McKay LD. *Vadose Zone J.*, 2004;3:326.
Morales, VL et al., *Water Res.*, 2011;45:1691.
Morris BL, Foster SSD. *Water Sci. Technol.*, 2000;41(7):67.
Moulin P, Roques H. *J. Colloid Interface Sci.*, 2003;261:115.
Natarajan R, Fuerstenau DW. *Int. J. Miner. Process.*, 1983;11:139.
Nevecherya IK, Shestakov VM, Mazaev VT, Shlepnina TG, *Water Resour.*, 2005;32(2):209.

Nola M, Njine T, Sikati VF, Djuikom E. *J. Water Sci.*, 2001;14(1):35.

Nowack B, Bucheli TD. *Environ. Pollut.*, 2007;150(1):5–22.

Nyham JW, Brennon BJ, Abeele WV, Wheeler ML, Purtymum WD, Trujillo G Herrera WJ, Booth JW. *J. Environ. Qual.*, 1985;14:501.

Onija AE, Milonjic SK, Cokesa D, Comor M, Miljevic N. *Mater. Res. Bull.*, 2003;38:1329.

Onija AE, Milonjic SK. *Mater. Sci. Forum.*, 2003;413:87.

Onoda GY, Debruyn PL. *Surface Sci.*, 1966; :48.

O'Reilly CE et al. *Clin. Infect. Dis.*, 2007;44(4):506–512.

O'Reilly SE, Hochella MF. *Geochim. Cosmochim. Acta*, 2003;67:4471.

Pandya VB, Bhuniya S, Khilar KC. *AIChE J.*, 1998;44:978.

Panyala NR, Peña-Méndez EM, Havel J. *J. Appl. Biomed.*, 2008;6(3):117–129.

Paradelo M, Rodriguez PR, Calvino DF, Estevez MA, Periago JEL. *Eur. J. Soil Sci.*, October 2012;63:708.

Park GA. Aquous chemistry of oxides and complex minerals. In: Stumm W (Ed.), *Equilibrium Concepts in Natural Water Systems*, Advances in Chemical Series, Vol. 67. American Chemical Society, Washington, DC, 1967, p. 121.

Patelli A, Rigato V. J. *Iberian Geol.*, 2006;32(1):79.

Pekdeger A, Mathess G, Schroter J. Hydrogeology in the service of man. *Memoires of the 18th Congress of the International Association of Hydrogeologists*, Cambridge, U.K., 1985.

Petersen CT, Holm J, Koch CB, Jensen HE, Hansen S. *Pest Manag. Sci.*, 2003;59:85.

Pettersson A, Rosenholm JB. *Langmuir*, 2002;18:8447.

Pochard I, Denoyel R, Couchot P, Foissy A. *J. Colloid Interface Sci.*, 2002;255:27.

Powelson DK, Gerba, CP. *Water Res.*, 1994;28:2175.

Powelson DK, Simpson JR, Gerba CP. *J. Environ. Qual.*, 1990;19:396.

Pozas R, Ocana M, Puerto Morales M, Serna CJ. *J. Colloid Interface Sci.*, 2002;254:87.

Prelot B, Poinsignon C, Thomas F, Schouller E, Villieras F. *J. Colloid Interface Sci.*, 2003;257:77.

Preocanin T, Krehula S, Kallay N. *Appl. Surf. Sci.*, 2002;196:392.

Puls RW, Barcelona MJ. Low-flow (minimal drawdown) ground-water sampling procedures, Ground Water Issue, EPA/540/S-95/504, EPA, Washington, DC, 1996.

Puls WR, Powell RM. *Environ. Sci. Technol.*, 1992;26:614.

Quyang Y, Shinde D, Mansell RS, Harris W. *Crit. Rev. Environ. Sci. Technol.*, 1996;26:189.

Ramos-Tejada MM, Duran JDG, Ontiveros-Ortega A, Espinosa-jimenez M, Perea-Carpio R, Chibowski E. *Colloids Surf. A*, 2002;24:309.

Rao SR, Finch JA. *Int. J. Miner. Process.*, 2003;69:251.

Redman JA, Grant SB, Olson TM, Estes MK. *Environ. Sci. Technol.*, 2001;35(9):1798.

Rose JB, Vasconcelos GJ, Harris SI, Klonicki PT, Sturbaum GD, Moulton-Hancock C. *J. Am. Water Works Assoc.*, 2000;92(9):117.

Roy SB, Dzomback DA. *Environ. Sci. Technol.*, 1997;31:656.

Roy SB, Dzombak DA. *J. Cont. Hydrol.*, 1998;30:179.

Ryan JN, Elimelech M, Harvey RW, Aronhein JS, Bhattacharjee S, Bogatsu Y Loveland JP, Metge DW, Navigato T, Pieper AP. *International Workshop on Colloids and Colloid-Facilitated Transport of Contaminants in Soils and Sediments*, Tjele, Denmark September 19–20, 2002, DIAS Report, Plant Production No. 80, October, 2002, p. 93.

Ryan JN, Elimelech M. *Colloids Surf. A: Physicochem. Eng. Aspects*, 1996;107:1.

Ryan JN, Gschwend PM. *Water Resour. Res.*, 1990;26:307.

Saiers JE, Hornberger GM. *Water. Resour. Res.*, 1996;32:33.

Scandura JE, Sobsey MD. *Water Sci. Technol.*, 1997;38(12):141.

Schijven JF, Hassanizadeh SM. *Water Sci. Technol.*, 2001;46(3):123.

Schindler PW, Stumm W. The surface chemistry of oxides, hydroxides and oxide minerals. In: W. Stumm (Ed.), *Aquatic Surface Chemistry: Chemical Processes at the Particle–Water Interface*. Wiley, New York, 1987, p. 83.

Schulze DG, Bertsch PM. *Adv. Agron.*, 1996;55:1.

Schurtenberger P, Newman ME. *Environ. Particles*, 1993;2:37.

Sen TK, Das D, Khilar KC, Suraish Kumar GK. *Colloids Surf. A: Physicochem. Eng. Aspects*, 2005;260:53.

Sen TK, Khilar KC. *Adv. Colloid Interface Sci.*, 2006;119:71.

Sen TK, Khilar KC. Mobile subsurface colloids and colloid-mediated transport of contaminants in subsurface soils. In: Birdi KS (Ed.), *Handbook of Surface and Colloid Chemistry*, 3rd edn., Taylor & Francis Group/CRC Press, Boca Raton, FL, 2009, pp. 107–130.

Sen TK, Mahajan SP, Khilar KC. *AIChE J.*, 2002a;48:2366.

Sen TK, Nalwaya N, Khilar KC. *AIChE J.*, 2002b;48:2375.
Sen TK, Shanbhag S, Khilar KC. *Colloids Surf. A: Physicochem. Eng. Aspects*, 2004;232:29.
Sen TK. *Water Air Soil Pollut.*, 2011;216(1):239. doi: 10.1007/s11270-010-503-0.
Sharma P, Flury M, Zhou J. *J. Colloid Interface Sci.*, 2008;326:143–150.
Spandini L, Schindler PW, Charlet L, Manceau A, Ragnarsdottir KV. *J. Colloid Interface Sci.*, 2003;266:1.
Sterling D, Reithmeier RA, Casey JR. *J. Biol. Chem.*, 2001;276:47886.
Stevik TK, Aa K, Ausland G, Hanssen JF. *Water Res.*, 2004;38:1355. http://www.bionfate.gr/cms/?p=363&cat=6.
Sun C, Berg JC. *Adv. Colloid Interface Sci.*, 2003;105:151.
Sun H, Gao B, Tian Y, Yin X, Yu C, Wang Y, Ma LQ. *J. Environ. Eng.*, 2010;136(11):1305.
Surampalli RY, Lin KL, Banerji SK, Sievers DM. *J. Environ. Syst.*, 1997;26(3):305.
Takala M, Manninen P. Sampling and analysis of groundwater colloids—A literature review, Working Report 2006-15, Posiva Oy, Finland, 2006 (29), 43, http://www.posiva.fi/tyoraportit/WR2006-98web.pdf.
Tang F, Uchikoshi T, Ozawa K, Sakka Y. *Mater. Res. Bull.*, 2002;37:653.
Taylor R, Cronin A, Pedley S, Barker J, Atkinson T. *FEM Microbial. Ecol.*, 2004;49(1):17–26.
Thompson ML, Scharf RL. *J. Environ. Qual.*, 1994;23:378.
Torkzaban S et al. *J. Contam. Hydrol.*, 2008;96:113.
Trivedi P, Dyer JA, Sparks DL. *Environ. Sci. Technol.*, 2003;37:908.
USGS (Science for a changing world). Environmental Health—Toxic Substances. http://toxics.usgs.gov/definitions/unsaturated_zone.html. Accessed May 3, 2014.
USGS Circular 1186. General facts and concepts about groundwater in sustainability of ground water—Water Resources. http://pubs.usgs.gov/circ/circ1186/pdf/circ1186.pdf. Accessed May 3, 2014.
Vakros J, Kordulis C, Lycourghiotis A. *Chem. Commun.*, 2002; 17: 1980.
VanOlphen H. *Introduction to Clay Colloid Chemistry*. John Wiley & Sons, New York, Appendix II, 1977.
Wang P, Keller AA. *Langmuir*, 2009;25(12):6856.
Wang Y, Bryan C, Xu H, Pohl P, Yang Y, Brinker J. *J. Colloid Interface Sci.*, 2002;254:23.
Weaver RM, Hochella MF, Ilton ES, *Geochim. Cosmochim. Acta*, 2002;66:4119.
Williams DJA, Williams KP. *J. Colloid Interface Sci.*, 1978;65(1):79.
Yang Q, Troczynski T. *J. Am. Cerm. Soc.*, 1999;82:1928.
Yates MV, Yates SR. *Water Sci. Technol.*, 1988;20(11/12):301.
Yates MV. *Groundwater*, 1985;23:586–591.
Zanker H, Hutting G, Richter W, Brendler V. *A Separation and Detection Scheme for Environmental Colloids*. Institute of Radiochemistry, Dresden, Germany, 2000. www.fz-rossendorf.de/FWR/COLL/Coll_2.htm. Accessed 2014.
Zhang, W. et al. *Environ. Sci. Technol.*, 2010;44:4965–4972.
Zhu Y, Lena QM, Dong X, Harris WG, Bonzongo JC, Han F. *J. Hazard. Mater.*, 2014;264:286.

6 Activated Carbon Adsorption in Water Treatment

Liang Yan, Qiuli Lu, and George A. Sorial

CONTENTS

6.1 INTRODUCTION

Activated carbons are widely used as an adsorbent in numerous applications and especially in the phenolics treatment field, due to its excellent adsorption capacity and great flexibility. However, molecular oxygen has been found to be a significant factor in the adsorption of phenolics onto activated carbon (Grant and King 1990; Nakhla et al. 1994; Tessmer et al. 1997). Molecular oxygen was found to be able to promote a significant increase in the adsorptive capacity of activated carbon at the cost of reduction in the recovery of adsorbents. Studies have attributed this phenomenon to the oligomerization of phenolics (oxidative coupling) on the surface of activated carbon (Vidic et al. 1993). Analyzing the solvent extracted products by GC–MS has shown the presence of oligomerized phenols, which are essentially irreversibly bound to carbon surface (Vidic et al. 1997). As a consequence, the regeneration efficiency of activated carbon after the oligomerization of phenolic compounds is low. The cost of adsorbents has been the major concern in activated carbon adsorption usage. In this chapter, the theory as well as factors for controlling oligomerization is outlined. The description of development of novel activated carbon for oligomerization control is also the main subject of this contribution.

6.2 APPLICATION OF ACTIVATED CARBON IN PHENOLICS TREATMENT

6.2.1 Activated Carbon Adsorption

As early as 1500 BC, charcoal has been utilized to purify water in India and Egypt. Meanwhile, the commercial production of powdered activated carbon (PAC) can be traced back to the early nineteenth century. The increased use of activated carbon in the water industry is due to its high internal surface area and pore volume (Walker 1969). Nowadays, different shapes of activated carbon can be fabricated based on different utilizing purposes.

Activated carbon is a porous adsorbent having high capacity for removing various small molecular weight organic compounds from water. The major liquid-phase application of activated carbon is for removing traces of water organic compounds after other conventional treatment processes in order to attain regulatory compliance. In the beginning, activated carbon adsorption was only employed for taste and odor control. But with the great development of analytical technology, many contaminants can be identified and quantified at very low concentration. Therefore, new water quality issues of trace compound contamination were arising. Especially synthetic organic compounds (SOCs) are highly concerned because of their potential toxicity and carcinogenicity. Due to this reason, activated carbon has been used for several decades to treat public water supplies for the removal of SOCs. Activated carbon adsorption is considered to be the best currently available technology for removing SOCs. In addition, due to ubiquity in the environment and risk to human health, volatile organic compounds (VOCs) have received great attention in the field of environmental control. The adsorption of VOCs onto porous adsorbents such as activated carbons has been suggested as an innovative treatment technology.

Activated carbons are used extensively in industrial purification and chemical recovery operations because of their advantage of removing a wide range of organic substances with relatively low cost and no hazardous by-products. The market for activated carbons is indeed vast. The global consumption of activated carbon was 750,000 t in 2002. The estimated growth of worldwide demand is 4%–5% per year, with higher growth rates of 5%–6% per year projected for the United States between 2002 and 2005 (Zhang et al. 2004). Activated carbon is commonly used either in PAC or granular activated carbon (GAC) form. GAC provides continuous removal and is most efficient when organic contaminants are frequently present. It can be regenerated to recover the adsorption capacity and is also a good medium for biological stabilization. PAC has the flexibility of being used whenever it is needed. Therefore, it is best used for occasional taste and odor problems, episodic contaminant spills, and seasonal high level of pesticides and herbicides. Activated carbon adsorption has not been used extensively in wastewater treatment; yet, demands for a better quality of treated wastewater effluent, including toxicity reduction, have led to an intensive examination and use of activated carbon adsorption. In wastewater treatment, activated carbon adsorption is used to remove dissolved organic matter (DOM), the remaining portion of normal biological treatment.

6.2.2 Problems Faced in Activated Carbon Adsorption of Phenols—Oligomerization

Phenolic compounds have been a major environmental concern around the world because they exist widely in industrial effluents such as those from oil refineries and the coal tar, plastics, paper and pulp, leather, paint, pharmaceutical, and steel industries (Singh and Rawat 1994). Phenolic compounds potentially cause long-term contamination of both surface water and groundwater bodies. Among all treatment processes, activated carbon adsorption has been designated as the most widely effective technique in phenolic compounds removal because of its good adsorption capacity and flexibility. However, economic use has been a major concern in GAC usage due to poor regeneration efficiency of activated carbon. The main reason for the drawback is that the presence of molecular oxygen in the aqueous phase promotes chemical transformation, such as oligomerization of the organic compounds absorbed onto the carbon surface. A short review will help the reader to understand the relative theories in this field.

Magne and Walker (1986) first described the existence of oligomerization of phenol adsorption on the surface of activated carbon. They reported that the physisorbed phenol can become chemisorbed in the course of time or by increasing the temperature. Grant and King (1990) clarified the mechanism of the oligomerization of phenol as follows: oligomerization of phenolics involves removing a hydrogen atom from each phenolic molecule and forming intermediate radicals that join, by carbon–carbon bonds, at the ortho- and/or paraposition to the hydroxyl groups. The products of

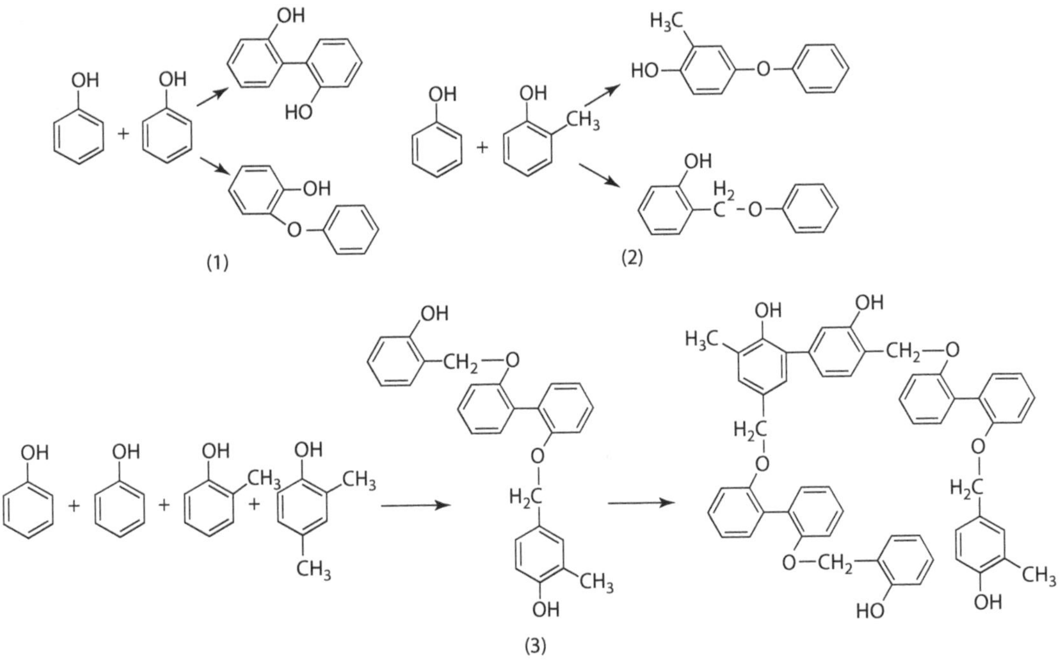

FIGURE 6.1 Possible three (1–3) examples of phenolics oligomerization.

the reaction of phenol oxidation included usually dimers and higher polymers (Figure 6.1). The abundance of the polymer decreased with molecular weight. More importantly, they found the carbon–oxygen linkages are less frequently formed. No minerals in the carbon caused the oxidative coupling reaction. In addition, temperature is always an important factor affecting activated carbon adsorption. The effect of temperature on the dissolved oxygen-induced oligomerization of phenolics was also investigated in their study. They found that higher temperature enhances the polymerization reaction and hence increases the adsorption capacity.

The oligomerization of phenolics depends strongly not only on the presence of molecular oxygen but also on the amount of oxygen (Vidic and Suidan 1991). The variation of molecular oxygen amount can lead to significant differences of adsorption capacity of activated carbon for phenolics. The presence of molecular oxygen increases the adsorption capacity of activated carbon for phenolics by imparting oligomerization reaction. Then the adsorption isotherms can be denoted into two types: anoxic (absence of molecular oxygen) isotherms and oxic (presence of molecular oxygen) isotherms. These suggest that physical adsorption equilibrium on the activated carbon can be achieved by conducting anoxic isotherms with the exclusion of oxygen from the experimental container. In addition, the adsorption capacity including chemisorption like oligomerization can be obtained by conducting oxic isotherms. Therefore, as compared to anoxic isotherm, any increasing amount of adsorbed phenolics predicted under the oxic condition can be attributed to oligomerization. It has been shown that the amount of dissolved oxygen (DO) consumed during the adsorption of phenolic adsorbates is linearly proportional to the amount of irreversibly adsorbed compound (Vidic et al. 1993).

Reviews on studies pertaining to oxygen-containing functional groups on the activated carbon seems to indicate that only basic surface functional groups have little influence on the extent of oligomerization (Tessmer et al. 1997). The presence of basic surface functional groups can slightly promote phenolic compounds adsorption under oxic conditions by increasing its effectiveness in promoting adsorption via oxidative coupling reactions. Tessmer et al. (1997) investigated the modification of activated carbon surface functional groups with outgassing thermal treatment to enhance the catalytic ability of activated carbon. It was found that outgassing at 900°C can effectively eliminate the acidic functional groups and encourage the formation of oxygen basic groups. But outgassing at higher temperature (1200°C) could remove all oxygen complexes. In addition, only oxygenated basic groups can promote the oligomerization on the surface of activated carbon. Hence, outgassing at proper temperature (900°C) can promote the oligomerization by changing the amount and composition of oxygen functional groups.

As the oligomerization of phenolics can only occur in the presence of oxygen, researchers investigated the mechanism of interactions between oxygen and carbons. Zawadzki and Biniak (1988) suggested that super-oxo ions are formed in a first step of oxygen adsorption by the carbon surface. These ions undergo further rearrangement on the carbon surface, forming a number of surface oxygen compounds. This hypothesis was further confirmed by a study conducted by Terzyk (2003) where they found the oligomerization extent was having direct relationship with the concentration of surface lactones at neutral pH. The larger is the concentration of lactones, the less are the oligomerization reactions. They postulated that as lactones have a considerable influence on the temperature dependence of the carbon resistance, it weakens the work function of the electrons. Consequently, lactones can decrease oligomerization by creating strong bonds with adsorbed molecules.

6.2.3 Activated Carbon Fibers

Activated carbon fibers (ACFs) are a fibrous form of activated carbon with carbon content more than 90%. ACFs are relatively new adsorbents for filtration or purification techniques. The unique characteristics of ACFs compared with GAC and PAC could increase the application of activated carbons in various areas. The fiber shape of ACFs can significantly improve the intraparticle adsorption kinetics as compared with PAC and GAC, which are commonly employed in gas-phase and aqueous-phase adsorption. Therefore, ACFs adsorption is a promising technique used for designing adsorption units where intraparticle diffusion resistance is the dominant adsorption factor. As a consequence, the size of adsorption units can be decreased by using ACFs (Yue et al. 2001).

Firstly, in the process of adsorption, the porous structure of adsorbent plays a very important role. Most of the pores in ACFs are micropores with the presence of some mesopores and macropores. The micropores were mainly developed by the evolution of volatile by-products at temperatures as low as 400°C–500°C and its persistence during further heating must be due to the presence of a stable cross-linked structure. The initial pores can be widened by exposing to steam due to diffusion of the etchant into the fiber bulk. It can be noted that pores in the fiber bulk would remain homogeneous and average pore sizes measure between 10 and 20 Å during the activation process (Donnet 1998). However, more heterogeneous and large mesopores and micropores can be observed at the fiber surface. During the adsorption process, although plenty of micropores and mesopores can be found on the surface of GAC and PAC, the large adsorbed molecules in the macropores would easily close the access of adsorbates to smaller pores (micropores and mesopores) where adsorption would occur. However, the homogeneous and a narrower porous structure of ACF may contribute to the less possibility of blockage (Shmidt et al. 1997). The fiber pores can easily reach to the center of fibers because of high microporosity, therefore the diffusion path of adsorbate to reach the adsorption sites is short. Meanwhile, a smaller fiber diameter also results in a larger contact surface area between adsorbents and adsorbates and thus even contact would be achieved. More important, because of unique pore size distribution (PSD), ACFs are very useful in understanding the influence of PSD on the oligomerization of phenolics. Therefore, ACFs have been used in many studies to elucidate the influence of PSD.

Secondly, in the process of adsorption, the surface chemistry of adsorbent always plays an equally critical role in adsorption as compared to physical structure and can be tailored through chemical activation or through post treatment (Donnet 1998). However, elemental analyses of ACFs indicated that they had less than 5% surface oxides and contain very low amounts of inorganic impurities. The low concentration of surface oxides results in limited surface chemistry effects and the low inorganic impurities result in minimal effects on pore development during the activation process (Parker 1995). The advantage of ACFs is that it is very useful in studying the role of functional groups of adsorbate by eliminating the interrupting factors of adsorbent functional groups.

For the adsorption application, ACFs offer the fabrication flexibility that GAC does not have, and thus they are easier to be made into filter media or devices. ACFs can be made in various forms such as fabrics, nonwovens, paper, and composites, which make them suitable for handling in special devices (Suzuki 1994). GAC are widely used in a variety of applications, but a large volume of adsorption devices are always needed in applications due to the relatively slow adsorption rate of GAC. Meanwhile, although the size of PAC is smaller than GAC, PAC may result in a strong flow resistance because of potential compact under flow (Shmidt et al. 1997). ACFs can effectively solve this issue by not only significantly improving the adsorption rate of adsorbates but also have good adsorption capacity. This is probably because of the unique porous structure of ACFs. For ACFs, the surface pores can lead directly into the center of the ACF. But for GAC and PAC, the penetration of adsorbate into the center of PAC and GAC is controlled by diffusion through micropore layers that interconnect individual macropores. In water purification systems, ACFs have advantages of less possibility to produce ash, less pressure drop, and a higher water flow rate. In air purification systems, ACFs show faster adsorption and desorption rate (Shmidt et al. 1997).

6.2.4 Effect of Functional Groups

It is well known that both the adsorbate and the adsorbent properties play a very important role in activated carbon adsorption. Adsorption is a manifestation of complicated interactions among the three components involved, that is, the adsorbent, the adsorbate, and the solvent. Normally, the affinity between the adsorbent and the adsorbate is the main interaction force controlling adsorption. However, the affinity between the adsorbate and the solvent (i.e., solubility) can also play a major role in adsorption. Hydrophobic compounds have low solubility and tend to be pushed to the adsorbent surface and hence are more adsorbable than hydrophilic compounds. Meanwhile, we know that phenolic compounds with different functional groups can lead to different solubility, which may lead to different oligomerization extent. Therefore, the adsorption behavior of phenolic compounds with different functional groups has to be understood. As illustration, we consider the interpretation of experimental isotherms by Lu and Sorial (2007) for the adsorptive capacity of five different phenolics on GAC F400 and two ACFs, ACC-10 and ACC-15, under both anoxic and oxic conditions (Figure 6.2).

The results of anoxic isotherms displayed that the adsorptive capacity was related to the electron withdrawing or electron-donating functional groups. For F400, the adsorptive capacity is in the order 2-nitrophenol > 2-chlorophenol > phenol > 2-ethylphenol > 2-methylphenol. These results can be attributed to the dispersion force between the activated carbon π-electrons of double bonds and π-electrons in phenols. Several studies have confirmed that some phenolics would favor adsorption onto activated carbon due to the existence of these electron-withdrawing substitutes like Cl or NO_2. In contrast, phenolic compounds that have electron-donating substitutes, such as CH_3 and CH_2CH_3, are unfavorable for adsorption (Cooney and Xi 1994; Jung et al. 2001).

In addition, the solubility of phenolics in water is generally regarded as an important factor for their adsorption. It is believed that the lower the solubility of phenolics, the easier the attachment to the activated carbon surface, and as a result, a higher adsorptive capacity can be achieved. However, the adsorptive capacity of phenolics did not follow the order of their water solubility except 2-nitrophenol. They found this phenomenon can be explained by the significant different molecular structures of phenolics. 2-Methylphenol and 2-ethylphenol are three dimensional while 2-chlorophenol and phenol are two dimensional. The adsorbate dimensions played a more important role in the adsorption of phenolics on ACFs due to the narrow PSD of ACFs. As shown in Figure 6.2, the differences of adsorptive capacity on ACC-10 were small as compared to F400.

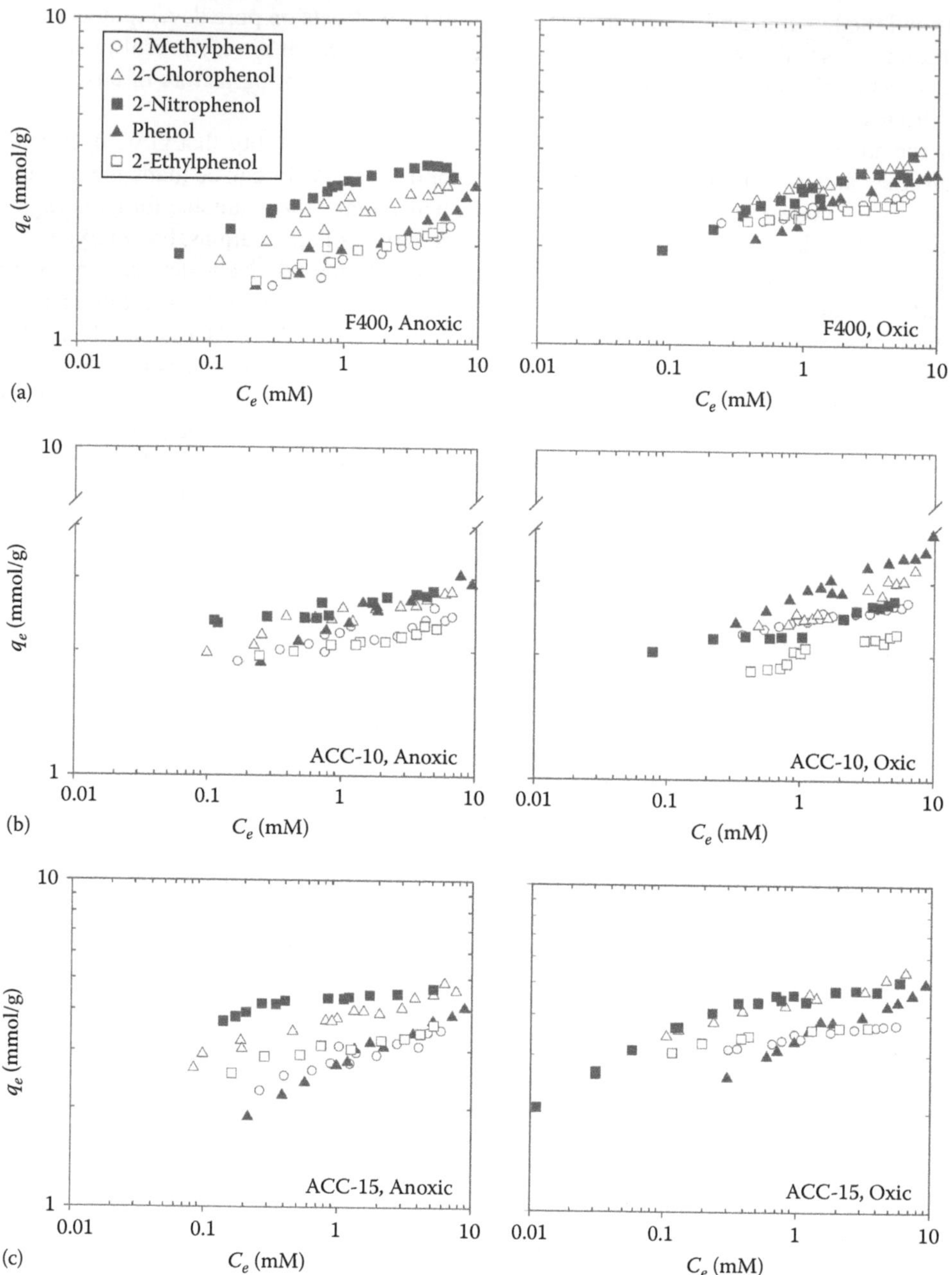

FIGURE 6.2 Adsorption of phenolics on (a) F400, (b) ACC-10, and (c) ACC-15. (Reprinted from Lu, Q. and Sorial, G.A., *J. Hazard. Mater.*, 148, 436, 2007. With permission.)

The trend was similar to F400 with the exception of 2-methylphenol and 2-ethylphenol. Both of these adsorbates are three dimensional, however, 2-ethylphenol (6.06 Å * 5.62 Å * 1.61 Å) is larger than 2-methylphenol (5.68 Å * 4.12 Å * 1.63 Å). The limited PSD of ACC-10 as compared to F400 restricted the adsorption of 2-ethylphenol due to its larger size as compared to 2-methylphenol. In the case of ACC-15 under anoxic conditions, a similar adsorption trend was observed but more differences in adsorption capacities were also displayed as compared to ACC-10. This could be attributed to the larger critical pore diameter of ACC-15 as compared to ACC-10.

More importantly, it can be seen that the oxic adsorptive capacity of phenolic was higher than that of anoxic ones on F400 due to oligomerization of phenolics on the activated carbon surface (Figure 6.2). It is known that the first step in this oligomerization reaction is forming phenoxy radical by losing one proton; the phenoxy radical is highly reactive and can participate in oligomer formation through carbon–oxygen or carbon–carbon bonding. However, they found that the presence of an electron attracting group, like the nitro and the chloro group, inhibited the parent phenol molecule to lose the proton, that is, hydrogen; thus the first step of oligomerization was inhibited to some degree. In contrast, they reported the presence of an electron-donating group, like the methyl and the ethyl group decreased the COP and thus aided the parent phenol molecule to form phenoxy radical.

Figure 6.2 shows that the sequence of phenolic oxic adsorptive capacity was in the same order as the anoxic isotherm while the relative difference under oxic conditions was less. This can be explained by the existence of electron withdrawing groups, like chloro and nitro groups. 2-Chlorophenol and 2-nitrophenol are more adsorbable than other compounds under anoxic conditions, while these existed chloro or nitro groups also make 2-chlorophenol and 2-nitrophenol less oligomerized than the others, which diminished the differences between these compounds. In case of oxic adsorption on ACC-10, the sequence of adsorptive capacity was found to be phenol > 2-chlorophenol > 2-nitrophenol ~ 2-methylphenol > 2-ethylphenol, which is different from the order shown under anoxic conditions. Obviously, the limited pore diameter is more accessible for the 2D molecules (phenol) than the 3D ones (2-methylphenol and 2-ethylphenol). Thus, the large molecular size of phenolics made them less oligomerized on ACC-10.

6.2.5 Pore Size Distribution and Its Effect on Oligomerization

Activated carbon is widely used in drinking water treatment for the removal of organic micropollutants, because of their porous structure and large internal surface area (Martinez et al. 2006). PSD is one of the most important properties that influence the adsorption process. The PSD determines the fraction of the total pore volume that can be accessed by an adsorbate of a given size. Pore size is divided into four types according to the International Union of Pure and Applied Chemistry (IUPAC) classification of pore diameter: macropores (>500 Å), mesopores (20–500 Å), secondary micropores (8–20 Å), and primary micropores (<8 Å). Knowledge of activated carbon characteristics is necessary if we wish to understand carbon's removal of a specific pollutant. The most important activated carbon pore size distribution for a particular water purification process will depend largely upon the size of the organic compounds in that water source (Moore et al. 2001). In summary, at low adsorbate concentration a maximum in adsorption capacity results from the balance between the adsorption potential and the micropore volume of the adsorbent. This means that the factor that determines the adsorption capacity at low concentration is its micropore size distribution and, more specifically, the ratio between the diameter of the molecule to be retained and the micropore size of the adsorbent. This is particularly important for the use of carbons in the adsorption of contaminants at low concentrations since the design of microporous structure of the adsorbent is required to suit the size of the molecules to be removed. On the other hand, at high adsorbate concentrations, the entire micropore system is involved in the adsorption process and the adsorption capacity is directly related to the total pore volume (Centeno et al. 2003).

The structure of activated carbon is known to be very complicated. Microscopically, it is composed of two parts: the macroporous/mesoporous amorphous carbon and microporous graphite crystals. In an adsorption process, the former provides the pore space for intraparticle transport and the latter accommodates the slit-shaped micropores in which most of the adsorption capacities reside. This microporous network of the graphitic crystals dictates the overall adsorption equilibria, and together with the macropore/mesopore network they affect the overall kinetics of adsorbates into activated carbon.

In reported adsorption studies, the activated carbon PSD was shown to be an important parameter affecting the removal of phenolics. Hsieh and Teng (2000) described the adsorption of phenol onto activated carbons developed with different activation levels. They prepared activated carbons with different porosities by carbonizing bituminous coal to different extents of burn-off. Larger surface area and higher surface area were obtained by increasing the extent of carbon burn-off. However, the burn-off level only showed a little influence on the average pore diameter of activated carbons. These indicated that pore deepening, rather than widening, occurred predominantly upon activation in CO_2 atmosphere. Meanwhile, they found that mesopores facilitate the adsorption of the adsorbates in the inner and narrow micropores since the adsorptive capacity was an increasing function of mesopore volume when surface area and micropore volumes were similar. In addition, Juang et al. (2001) reported that the ratio of monolayer capacity and volume of micropores for phenol always decreased with increasing micropore volume, indicating that the adsorption of phenols is not completely restricted to occur within the micropores.

The PSD of activated carbon determines the fraction of its structure that can be accessed by a molecule of given size and shape. For a given size pore, the size of adsorbates relative to the pore size of the adsorbents are important in controlling the adsorptive competitive mechanism. A good understanding of the impact of PSD on oligomerization is required as a basis for developing activated carbon for oligomerization control. However, very few studies have been conducted to study the impact of PSD on oligomerization. Therefore, adsorption isotherms under the control of oxygen can potentially constitute a significant advance in the understanding of the impact of PSD on oligomerization. In addition, activated carbon fiber (ACF) is very suitable for elucidating the impact of PSD on oligomerization due to its perfect surface properties. Firstly, ACFs have unique PSD and all the micropores are directly on the carbon surface (Ryu et al. 2000; Shen et al. 2004). Secondly, elemental analysis of ACFs indicated that they had <5% surface oxides and contain very low amounts of inorganic impurities (Pelekani and Snoeyink 1999, 2000). The low concentration of surface oxides results in limited surface chemistry effects and the low inorganic impurities results in minimal effects on pore development during the activation process (Pelekani and Snoeyink 1999). These parameters plus the controlled pore structure during activation are a key advantage of ACFs over granular activated carbon (GAC) and make them ideal for targeting the effects of PSD on adsorption phenomena (Parker 1995).

Lu and Sorial (2004) conducted an adsorption isotherm of phenolics on two types of ACFs with different pore sizes and GAC F400 to investigate the impact of PSD of activated carbon on the adsorption of phenolics (Figures 6.3 and 6.4). In order to study the impact of molecular oxygen on the adsorptive capacity, two types of isotherms were conducted. One was conducted in the presence of molecular oxygen (oxic isotherm) and the other in the absence of molecular oxygen (anoxic isotherm).

The results clearly indicate that there is a significant difference in adsorptive capacity of 2-methylphenol between oxic and anoxic conditions for the three adsorbents studied (Figure 6.3). The adsorptive capacity of 2-methylphenol under oxic conditions is higher than that under anoxic conditions. The difference in the adsorptive capacity between oxic and anoxic conditions for ACC-10 is smaller as compared to the other adsorbents. A similar behavior is seen for 2-ethylphenol (Figure 6.4). In addition, they compared the adsorptive capacity of phenolics on the three adsorbents under oxic and anoxic conditions at two equilibrium liquid-phase concentrations of 0.5 and 5 mM (Table 6.1). They reported the increase in adsorptive capacity under oxic conditions is in the sequence of F400 > ACC-15 > ACC-10 for all the two adsorbates considered. They found that F400 with a wide range of PSD cannot hamper the oligomerization of phenolics. But ACC-10 with a critical pore diameter of 8.0 Å and narrow PSD (microporosity above 90%) was more effective in hampering the oligomerization of phenolics. It is interesting to note that the molecular size of the adsorbate played another significant role in the increase of adsorptive capacity. The second widest dimension of 2-methylphenol molecules is 4.17 Å. Therefore, it is possible

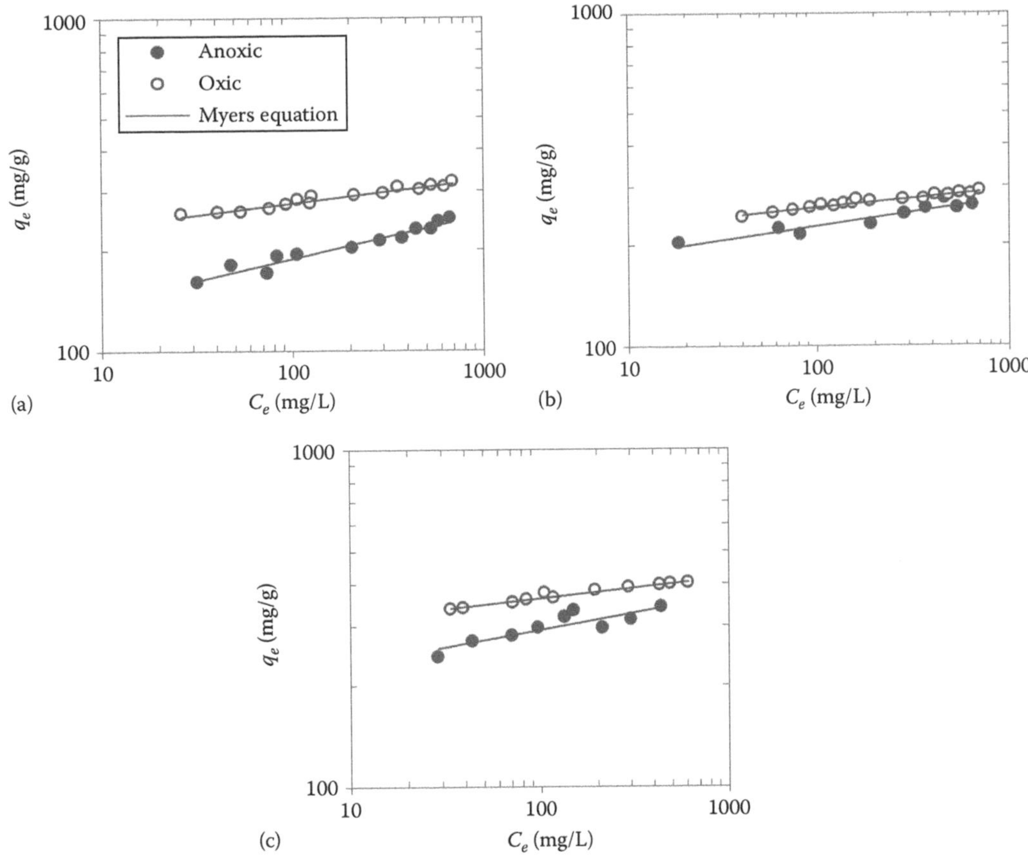

FIGURE 6.3 Adsorption isotherms of o-cresol. (a) F400, (b) ACC-10, and (c) ACC-15. (Reprinted from *Chemosphere*, 55, Lu, Q. and Sorial, G.A., Adsorption of phenolics on activated carbon—Impact of pore size and molecular oxygen, 671–679, Copyright 2002, with permission from Elsevier.)

for 2-methylphenol to form dimers, while the formation of trimers is nearly impossible. In case of 2-ethylphenol, the second widest dimension is 5.62 Å, thus it is impossible for 2-ethylphenol to get oligomerized. This is further confirmed in Table 6.1 where 2-ethylphenol did not show adsorption capacity gain under oxic conditions. In case of ACC-15, forming dimers and trimers of 2-methylphenol is possible, while tetramers or pentamers are very difficult or even impossible. At the same time, they also found that the oligomerization extent is directly related to the critical oxidation potential (COP) value of phenolics. 2-Methylphenol is more amenable to oligomerization than 2-ethylphenol due to its lower COP value (1.04 V). In their study, the effective control of oligomerization on ACF is further confirmed by the high recovery efficiency of ACF after oxic adsorption isotherms. F400 showed the lowest recovery efficiency as compared to ACC-10 and ACC-15, and ACC-15 was lower than ACC-10. ACC-10 showed higher adsorbate extraction efficiency, above 85%, than the other adsorbents. More importantly, the adsorptive capacity of F400 was lower than the two ACFs for both adsorption conditions (oxic and anoxic). Thus, for better use of activated carbon, it is mandatory to use adsorbents with high adsorptive capacity and also high regeneration efficiency. Therefore, activated carbon with similar pore structure as ACC-10 (having their pore volume mostly distributed in micropores and low critical pore diameter) needs to be developed and used for hampering oligomerization in order to make activated carbon more cost effective.

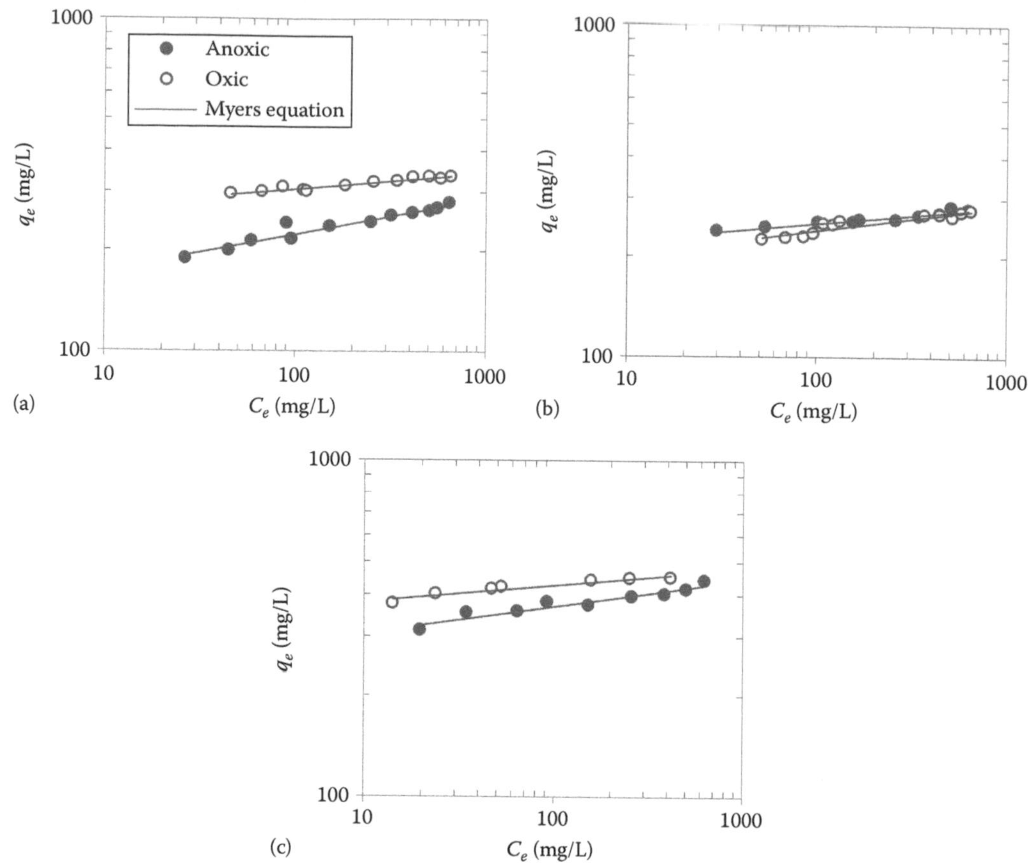

FIGURE 6.4 Adsorption isotherms of 2-ethylphenol. (a) F400, (b) ACC-10, and (c) ACC-15. (Reprinted from *Chemosphere*, 55, Lu, Q. and Sorial, G.A., Adsorption of phenolics on activated carbon—Impact of pore size and molecular oxygen, 671–679, Copyright 2002, with permission from Elsevier.)

TABLE 6.1
Comparison of 2-Methylphenol and 2-Ethylphenol

		F400			ACC-10			ACC-15		
Adsorbate	**Ce (mM)**	**Anoxic q_e (mmol/g)**	**Oxic q_e (mmol/g)**	**Δ q_e (%)**	**Anoxic q_e (mmol/g)**	**Oxic q_e (mmol/g)**	**Δ q_e (%)**	**Anoxic q_e (mmol/g)**	**Oxic q_e (mmol/g)**	**Δ q_e (%)**
2-Methylphenol	0.5	1.61	2.45	52.3	2.00	2.32	15.7	2.54	3.23	27.0
	5	2.17	2.87	32.3	2.42	2.62	8.7	3.20	3.72	16.1
2-Ethylphenol	0.5	1.73	2.44	40.8	2.04	1.93	−5.5	2.90	3.41	17.5
	5	2.23	2.73	22.6	2.28	2.28	0.4	3.49	3.83	9.6

Source: Data from Lu, Q. and Sorial, G.A., *Chemosphere*, 55, 671, 2002.

6.2.6 Sources and Methods of Developing Activated Carbon

Currently, activated carbon is manufactured from a variety of starting materials, that is, precursors, including different types of wood, nutshells, coal, peat, lignite, polymers, and various agricultural by-products. These precursors are usually high in carbon content. The qualities and characteristics of activated carbon depend on the physical and chemical properties of the starting materials and activation methods used (Guo and Lua 2000). It is found that the carbon content of char was significantly higher after carbonization, and the oxygen and hydrogen were lower. In addition, the pore characteristics of activated carbons are important in determining the particular application of the carbon.

Wood is one of the most important materials for preparing activated carbons, which have particular porous characteristics and are appropriate for the adsorption of solutes in liquid phase. For example, quercus agrifolia and eucalyptus woods activated with CO_2 yielded a surface area of 1201 and 1190 m^2/g, respectively, and a fraction of micropore volume of 0.76 and 0.50. In addition, rubber wood sawdust and wood flour activated with 1.5 and 2.4 times of H_3PO_4 obtained an area of 1496 and 1780 m^2/g, respectively, and a fraction of micropore volume of 0.69 and 0.66. These carbons show higher ratios of mass transfer voids (Wu et al. 2005). The general process to produce activated carbon is based on carbonizing and activating the original carbonaceous material. Activated carbons may be produced through physical or chemical activation. For example, physical activation via pyrolysis to produce a char followed by steam gasification, alternatively, the biomass may be treated with a chemical, for example, zinc chloride followed by thermal treatment via pyrolysis to produce an activated carbon. Physical activation involves carbonization of the precursor under inert atmosphere, followed by gasification of the char under oxidizing agents such as steam or carbon dioxide. In industrial practice, physical activation is carried out most frequently by burning off some of the raw carbon in an oxidizing environment to create micropores. The usual commercial choices of activation gas are steam, CO_2, air, or their mixtures. Activation normally takes place at temperatures between 700°C and 1000°C in steam and CO_2, and lower temperature in air (Zhang et al. 2004). Chemical activation is one of the less commonly commercial methods. In this process, the starting materials are impregnated with a solution of dehydrating agent (such as zinc chloride [$ZnCl_2$], phosphoric acid [H_3PO_4], sulfuric acid [H_2SO_4], or potassium hydroxide [KOH]) to retard the formation of tars during the carbonization process. They are then washed, dried, and carbonized in an inert atmosphere and finally activated with CO_2 or steam to produce the final activated carbon (Yamashita and Ouchi 1982; Marsh et al. 1984; Molinasabio et al. 1995). Chemical activation has intrinsic advantages over physical activation, which includes lower temperature, single step, higher product yields, higher surface areas, and better microporosity distribution control (Lozano-Castello et al. 2007). A large number of agricultural by-products such as peach stone, sugar cane bagasse, coconut shell, almond shell, and hazelnut shell have been successfully converted into activated carbons by chemical activation on a laboratory scale. Caturla et al. (1991) studied the preparation of activated carbons from peach stones with $ZnCl_2$ impregnation to determine the effects of various parameters such as starting material size, impregnation method, and activation temperature on the BET surface area, porosity, and density of the resulting carbons. Girgis et al. (1994) prepared activated carbons from sugar cane bagasse impregnated with 50% inorganic acids. The effectiveness of such impregnation was in the following order: $H_3PO_4 > H_2SO_4 > HCl > HNO_3$. They also found out that activated carbons prepared at low temperatures were essentially microporous, but at higher temperatures, products of higher surface areas with developed mesoporosity but low microporosity were obtained. Laine and Calafat prepared a series of activated carbons from coconut shells, impregnated with various potassium solutions (KCl, KNO_3, KOH, K_2CO_3, and K_3PO_4) at different concentrations (Laine et al. 1989; Laine and Calafat 1991). It was found that all the potassium compounds studied except KCl were effective catalysts for activation with CO_2. Balci et al. (1994) used NH_4Cl impregnated and untreated almond and hazelnut shells for the preparation of activated carbons. Experimental results showed that treatment with NH_4Cl increased the total surface area and improved adsorptive capacity. Further, Hussein et al. (1996) prepared activated carbons from oil palm trunk chips by impregnation with $ZnCl_2$. The BET surface area and the porosity of the resulting activated carbons were found to be related to the concentrations of the impregnating solutions. It entails the addition of materials such as zinc salts or phosphoric acid to the carbon precursors could generate secondary environmental pollution during disposal.

The development of micropores and mesopores for the carbon is crucial because the adsorption of low molecular weight constituents from gas or liquid streams is responsible for the existence of these two types of pores. Microporous carbons are generally applied in gas-phase adsorption due to the small gas molecules, whereas mesoporous carbons are often used in liquid-phase adsorption. Recent progress of industrial technologies provides new application fields for activated carbons in super-capacitors and, at the same time, requires the carbons having a desired pore structure. An increased activation temperature and/or time will increase the surface area and the fraction of mesopore volume simultaneously. Hu and Vansant (1995) have shown similar trends for the carbons prepared from eucalyptus wood and activated in the atmosphere containing CO_2 and a small amount of O_2. However, when carbons were activated with KOH, $ZnCl_2$, or H_3PO_4, the increase of such chemical doses did not proportionally enhance the fraction of mesopore volume. Accordingly, the fraction of mesopore volume was usually <0.2, while sometimes it increased up to 0.5. The adsorption energetic heterogeneity commonly observed on activated carbon is induced by the structural heterogeneity, which can be characterized by the distribution of the slit-shaped micropores (MPSD). This MPSD is the intrinsic property of the activated carbon and dictates the adsorption equilibria and kinetics of different adsorbate molecules on the activated carbon through the adsorbate–pore interaction (Wang and Do 1999).

6.2.7 Tailoring of Activated Carbon

Over the years, several different approaches for the control of the pore structure in carbons have been explored, both at the micropore (pore size < 20 Å) and the mesopore (20–500 Å) level. Among them, the so-called template-based technique was found to be particularly suitable for the synthesis of carbons whose porosity is not only uniform in size and shape but also periodically ordered. This approach has been proven extremely successful for the synthesis of ordered mesoporous carbons, which have shown good promise for several uses. However, with regard to the synthesis of microporous carbons with uniform structure, the template-based technique has usually encountered important limitations, which can be mainly attributed to a deficient infiltration of the carbon precursor within the narrow channels of microporous templates. For this reason, very few template-derived microporous carbons with a uniform, controlled structure are documented in the literature. Some studies reported alternative approach for the preparation of microporous carbons, which is centered in the extraction of metals from carbides. Some of the resulting carbons display an outstandingly narrow PSD and are therefore considered to be good candidates for molecular sieving applications. Another possibility for the synthesis of microporous carbons with uniform porosity is based on the well-known and relatively straightforward physical and chemical activation methods, but using highly crystalline organic precursors instead of materials with low or intermediate crystallinity that are typically employed for the preparation of activated carbon (Villar-Rodil et al. 2005).

Some studies have investigated the thermal treatment of carbons in an ammonia gas environment. Economy and Anderson (1966) reported that ACF could be impregnated with NH_3 at elevated temperature to obtain a product with basic characteristic. They also showed that such fibers were effective in adsorbing acidic materials such as HCl and SO_2. Mangun et al. (2001) conducted research on the ammonia treatment of ACF. They reported that ammonia acted as an etchant during contact with the sample. Longer treatment time and higher temperature resulted in an increase of pore size and surface area. The nature of the nitrogen-containing surface differed with the temperature of ammonia treatment. It was believed that amides, aromatic amines, nitriles, and protonated amides dominated at low temperature (400°C–700°C) while pyridine, aromatic amines, and protonated pyridine dominated at higher temperature (>600°C). Their ammonia-tailored samples also showed an enhanced adsorption for HCl. As a generalization, the nitrogen tended to appear in functional groups that were external to the aromatic ring structure at the lower temperatures (with localized charge), whereas nitrogen tended to appear within the aromatic ring structure at higher temperature (with a delocalized charge or no charge). Stohr et al. (1991) observed considerable increase in activated carbon's catalytic activity after the carbon was treated with ammonia at 600°C–900°C for 4 h.

It has long been recognized that physicochemical characteristics affect the performance of activated carbon. Several research groups have studied thermal treatment, apart from other techniques, to enhance the activated carbon properties and improve their sorption capacity for specific contaminants. For instance, Kaneko et al. (1989) tailored activated carbon at high temperature by hydrogen. They concluded that this protocol could remove acidic functional groups from the carbon's surface and enhance the sorption capacity for organic compounds such as benzene. Economy et al. (1996) reported that ACF treated by NH_3 at elevated temperature can demonstrate basic characteristics. These thermally treated carbon fibers showed higher sorption capacity for acidic substances like SO_2 and HCl. Mangun et al. (2001) treated activated carbon with ammonia at higher temperature above 400°C. It was found that a higher temperature and a longer treatment time resulted in an increase of surface area and pore size. They also proposed that the nature of the nitrogen-containing surface groups introduced by this method differed with the treatment temperature. Ammonia-treated samples presented higher HCl sorption capacity. In addition, other researchers have used the chemical vapor deposition of hydrocarbons to modify the physical and chemical properties of activated carbons. Nowack et al. (2004) observed improved 2-methyl isoborneol (MIB) sorption when activated carbon was thermally treated with steam or methane plus steam. Less favorable improvements were achieved with nitrogen or hydrogen. For instance, when GAC was treated with nitrogen between 450°C and 850°C the sorption capacity slightly increased; however, when the modified carbon sample was subsequently exposed to air at ambient temperature it lost its enhanced MIB adsorption capacity and the authors attributed this to the re-incorporation of oxygen. Hydrogen-treated lignite GAC at 1025°C was tested in mini-column experiments and the results showed that this carbon could remove MIB to below detection about 2.5 times more bed volumes longer than the commercial lignite GAC HD4000 could. Conversely, tailored carbon made by applying the methane and steam method at 1000°C presented the highest MIB sorption capacity. This modified carbon sample (13% mass gain above the pyrolysed mass, then 13% net mass loss for the final product) processed about 3.6 times more water than did HD4000 at initial MIB breakthrough. They also found MIB adsorption mainly occurred in the pore size range of 5.5–63 Å. The steam and methane-plus-steam processes have potential to produce a carbon product with high sorption capacity for MIB and other similar organic compounds such as geosmin.

Various methods have been already proposed in order to get a needed porosity by direct elaboration, by pore size reduction, or by pore size widening. The main limitation of those techniques leads in the competition between the simultaneous internal diffusion (heat and mass transfer) and oxidative reactions. This is usually responsible for a radical heterogeneity in textural properties and consequently to a wide PSD. Many researchers have proposed to overcome this limitation by means of the separation of the diffusional and the reaction phenomena by doing the former one at a temperature level at which the second one is ineffective, and the second one at a higher temperature level but under inert atmosphere. Therefore, at the particle scale and during the first step, the gaseous reagent is uniformly chemisorbed at the available intra-particular surface of the material. During the second step, the chemical functional groups previously formed are decomposed withdrawing carbon atoms from the material. In this way, there is no preferential burn-off at the pore mouth but preferential initiation of new micropores. Moreover, this technique presents the advantage of allowing the applicability of a usually unusable activating agent like molecular oxygen, which is known to react with carbon with an exothermic effect leading to combustion of the material. The proposed cycling protocol is based on air as oxidative reagent, which presents an obvious economical interest. The carbon surface concerned by the sorption is heterogeneous; it consists of the faces of graphene sheets and of edges of such layers. The edge sites are much more reactive than the atoms in the interior of the graphene sheets, and chemisorbed foreign elements as oxygen are predominantly located on the edges at the twist and tilt crystallites boundaries. Oxygen can be bound in the form of various functional groups that are similar to those known from organic chemistry. On the other hand, it is well known that essentially all oxygen-containing groups are removed during high temperature treatment in a nitrogen atmosphere. The gasification step consists in the decomposition of the carbon–oxygen complexes into CO and CO_2 and new freshly formed carbon sites that might be viewed as an unsaturated carbon at the edge of the carbon matrix.

6.3 EXPERIMENTAL STUDIES

6.3.1 Development of Novel Activated Carbon for Oligomerization Control

Study has shown that the micropore of carbon is developed by the removal of the carbon atoms during the activation process. However, after the optimum activation, the microporosity evolution will be accompanied with mesopore and macropore development with further activation (Lozano-Castello et al. 2002). Hence, very high surface area carbons with high phenolic compounds adsorption per unit mass may not be effective for controlling oligomerization. Thus, the ideal carbon would be a carbon with the optimum combination of microporosity and surface area. Efforts have to be focused toward getting the optimum degree of activation that will produce an activated carbon with best combination between microporosity and surface area.

Many studies have reported that activated carbon developed by KOH chemical activation from carbonaceous materials, such as charcoal and coal, can provide not only high BET surface area but also narrow PSD and well-developed porosity (Otowa et al. 1997; Lillo-Rodenas et al. 2003, 2004). In the study of Ryu et al. (2002), PAN-based ACFs were developed by using KOH chemical activation and steam physical activation. They found the dominant pores shift from micropore (750°C) to mesopore (860°C) and some macropore (1240°C) with the increasing of activation temperature under steam atmosphere. But KOH chemical activation can supply dominant super micropores (0–7 Å) while some mesopores were also developed during the activation process. Similar results were also confirmed by other studies (Ryu et al. 2000; Martin-Gullon et al. 2001). In addition, Lozano-Castello et al. (2002) reported the novel method to develop activated carbon with optimal PSD for methane storage. In their study, activated carbons were developed based on anthracite raw material by KOH chemical activation method. It was found that the activated carbon with the optimal PSD for methane adsorption is the one developed with KOH/anthracite ratio of 3/1, activation temperature of 750°C, and 2 h activation time. Therefore, KOH chemical activation could be applied to develop activated carbon to control the oligomerization of phenolic compounds. Furthermore, the studies of activated carbon precursors have shown that bituminous coal is a very good raw material for the preparation of activated carbons by chemical activation due to hardness, abrasion resistance, and relatively high density. Studies have shown that activated carbon from bituminous coals allows very high adsorption capacities to be reached with porosity that is mostly micropores (Teng and Yeh 1998; Teng et al. 1998).

Yan and Sorial (2011) developed a series of activated carbon from bituminous coal using KOH as the activation agent (as shown in Table 6.2). The significance of different activation parameters, including KOH/coal mass ratio, activation temperature, activation time, and nitrogen flow rate, was investigated.

The chemical reagent ratio is an essential identifying parameter of activation since this parameter has a significant impact on the micropore size distribution and thus influences the effectiveness of carbon in hampering oligomerization. Therefore, in order to investigate the optimum KOH/coal ratio, the ratio was varied in the range 1/1 to 5/1 while keeping the other factors constant (activation temperature 700°C, activation time 1 h, and nitrogen flow rate 400 mL/min, respectively). They found the PSD of sample prepared with KOH/coal ratio 1/1 and 2/1 was much narrower, which agrees with the micropore volumes percentage calculated from the micropore volume and the total pore volume (Table 6.2). However, it is shown that the BET surface area (354 m^2/g) and micropore volume (0.162 cm^3/g) of the sample using KOH/coal ratio 1/1 are very low, which makes it ineffective for adsorption. In contrast, after activation with a 2/1 ratio, the BET surface area (971 m^2/g) and micropore pore volume (0.405 cm^3/g) of the sample increased significantly. Therefore, they reported a ratio 2/1 is a better condition for adsorption of phenolic compounds. At this ratio, KOH demonstrates spread inside the particle through the interconnected pores and built a well-developed micropore structure inside the coal particle.

The activation temperature was then investigated by varying it from 600°C to 900°C while keeping the rest of the parameters constant (KOH/coal ratio, activation time, and nitrogen flow rate were 2/1, 1 h, and 400 mL/min, respectively). The BET surface area displayed an increase with activation temperature from 650°C to 850°C. Furthermore, the BET surface area of all the

TABLE 6.2
Porous Properties of GAC F400 and Activated Carbon by Using KOH as the Activating Agent

Sample	Activation Parameters	S_{BET}[a] (m²/g)	V_{micro}[b] (cm³/g)	$V_{meso} + V_{macro}$[c] (cm³/g)	V_{Total}[d] (cm³/g)	Micropore Percentage[e] (%)
KOH/coal ratio						
F400	—	993	0.373	0.242	0.615	60.7
BC-11-700-1-400	1:1	354	0.162	0.064	0.226	71.6
BC-21-700-1-400	2:1	971	0.405	0.154	0.559	72.4
BC-31-700-1-400	3:1	1158	0.434	0.235	0.669	64.9
BC-41-700-1-400	4:1	1247	0.389	0.348	0.737	52.8
BC-51-700-1-400	5:1	920	0.295	0.242	0.537	55.0
Temperature						
BC-21-600-1-400	600	845	0.381	0.104	0.485	78.4
BC-21-650-1-400	650	1128	0.507	0.144	0.651	77.9
BC-21-700-1-400	700	971	0.405	0.154	0.559	72.4
BC-21-750-1-400	750	1335	0.529	0.236	0.765	69.2
BC-21-800-1-400	800	1451	0.516	0.319	0.835	61.8
BC-21-850-1-400	850	1285	0.329	0.430	0.759	43.4
Time						
BC-21-650-0.5-400	0.5	1085	0.499	0.122	0.621	80.3
BC-21-650-1-400	1	1128	0.507	0.144	0.651	77.9
BC-21-650-2-400	2	1476	0.684	0.157	0.841	81.3
BC-21-650-3-400	3	1107	0.499	0.139	0.638	78.3
BC-21-650-4-400	4	903	0.415	0.109	0.524	79.2
N_2 flow rate						
BC-21-650-2-100	100	1203	0.534	0.158	0.692	77.1
BC-21-650-2-200	200	1472	0.676	0.179	0.855	79.0
BC-21-650-2-400	400	1476	0.683	0.158	0.841	81.3
BC-21-650-2-600	600	1091	0.480	0.153	0.633	75.8

Source: Modified from Yan, L. and Sorial, G.A., *J. Hazard. Mater.*, 197, 311, 2011.
[a] S_{BET} is BET surface area.
[b] V_{micro} is the volume of micropore.
[c] $V_{meso} + V_{macro} = V_{Total} - V_{micro}$.
[d] V_{Total} is the total pore volume.
[e] Micropore percentage = $V_{micro}/V_{Total} * 100\%$.

samples was larger than F400 (the commercial activated carbon). However, the sample that was made at a temperature below 650°C has a worse texture structure, resulting in much lower total BET surface area. The authors attributed this to the pores in the sample as they are not interconnected because the volatile components need to be released from the uniform structure at higher activation temperature. In addition, volatile components were released much more at higher activation temperature, which cause the development of bigger pores in the activated carbon. On the other hand, the release of volatile components made it easy for the activation agent to penetrate into the coal particle. Consequently, the mesopore and macropore volume increased at higher temperatures. It is possible that during activation, most reactive carbon atoms were eliminated after the activation process, while the carbon microporosity decreased. Therefore, the activation temperature is a very important parameter in developing activated carbon with a suitable structure required for controlling oligomerization. Activated samples with a high level of microporosity can be obtained at 650°C (BC-21-650-1-400, see Table 6.2).

In order to examine the best activation time on the activated carbon, this parameter was varied from 0.5 to 4 h while keeping the other parameters constant. No significant difference of microporosity was found in activated carbon samples, while considerable BET surface area difference could be found between each other (Table 6.2). All the activated carbon samples are mainly microporous (around 80% microporosity). Carbon sample with 2 h activation was found to have highest micropore volume, highest micropore percentage, and highest BET surface area. Therefore, an activation time of 2 h could be considered to be optimum for the preparation of a good activated carbon for oligomerization control.

In the activation process, nitrogen was used as a protection gas in this study. To select the best nitrogen flow rate, the nitrogen flow rate was varied in the range 100–600 mL/min with the values of the other parameters kept constant (KOH/coal ratio of 2:1, activation temperature of 650°C, and activation time of 2 h). The authors found increasing N_2 flow rates favored the development of total BET surface area until a flow rate of 400 mL/min was reached. It can be seen that a further increase in the N_2 flow rate led to a decrease in the surface area (Table 6.2). The highest micropore percentage was obtained with a nitrogen flow rate of 400 mL/min. This finding is consistent with others reported in the literature (Jiang and Zhao 2004; Jimenez et al. 2009). This pore opening behavior could be explained on the basis that the gaseous products (containing CO_2, CO, H_2O, etc.) were obtained during the activation. Previous studies reported that the removal of CO, CO_2, and H_2 would favor the activation process, while the removal of H_2O formed during the pyrolysis reaction work against the activation. In addition, a nitrogen flow rate too high could result in removing a large amount of H_2O from the reaction site and this would be detrimental to the porous properties of carbon, while a nitrogen flow rate too low would probably not remove sufficient amounts of CO, CO_2, which would also be in detrimental to the activation. Therefore, a balance between these two removal processes must be reached under a medium nitrogen flow rate (400 mL/min). Based on these results, it can be seen that the BC-21-650-2-400 sample, which was prepared with activation condition (KOH/coal = 2/1, activation temperature = 650°C, activation time = 2 h, and nitrogen flow rate = 400 mL/min), has the best potential for controlling oligomerization of phenolic compounds. This carbon was referred to as BC-21.

In the study, total acidic and basic groups on carbon surfaces was also performed for the selected BC-21 and F400 using NaOH or HCl uptake methods. Based on the titration test, the BC-21 was found to be an acidic carbon where total acidity and total basicity are 0.54 and 0 meq/g, respectively. In contrast, the total acidity and total basicity for F400 is 0.15 and 0.475 meq/g. Batch equilibrium method was used for determining the pHpzc of BC-21 selected and F400. The pHpzc was determined as the pH of the water sample that did not change after contact with the samples. The values of pHpzc for BC-21 and F400 are 3.4 and 7.1, respectively.

6.3.2 Oligomerization Control of Single Solute Isotherm

The practical consequences of novel activated carbon (BC-21) on hampering oligomerization can be examined by conducting adsorption isotherms of phenolics. Anoxic and oxic adsorption isotherms of phenol, 2-methylphenol, and 2-ethylphenol on BC-21 and GAC F400 are given in Figures 6.5 through 6.7, which are obtained from the study of Yan and Sorial (2011). Myers equation was applied to model the single solute adsorption isotherms. The Myers equation is given by:

$$C_e = \frac{q_e}{H} * \exp(K q_e^p) \tag{6.1}$$

where

C_e is the liquid-phase concentrations, mM
q_e is surface loading, mmol/g
H, K, and p are regression parameters

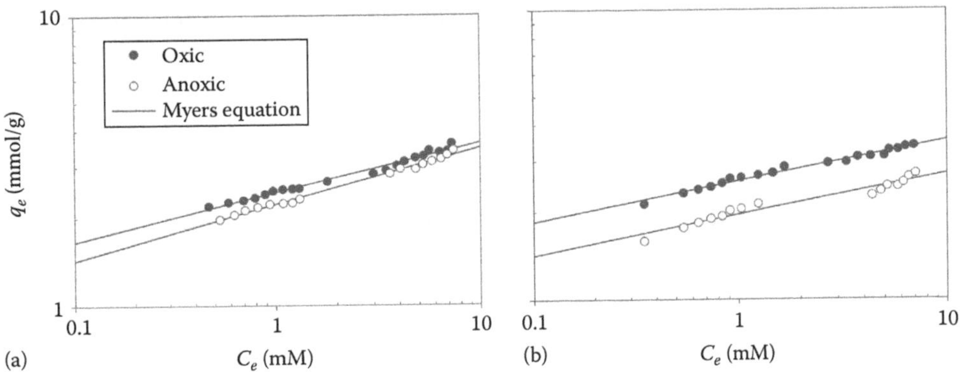

FIGURE 6.5 Single solute adsorption isotherm of phenol. (a) BC-21 and (b) F400. (Reprinted from *J. Hazard. Mater.*, 197, Yan, L. and Sorial, G.A., Chemical activation of bituminous coal for hampering oligomerization of organic contaminants, pp. 311–319, Copyright 2011, with permission from Elsevier.)

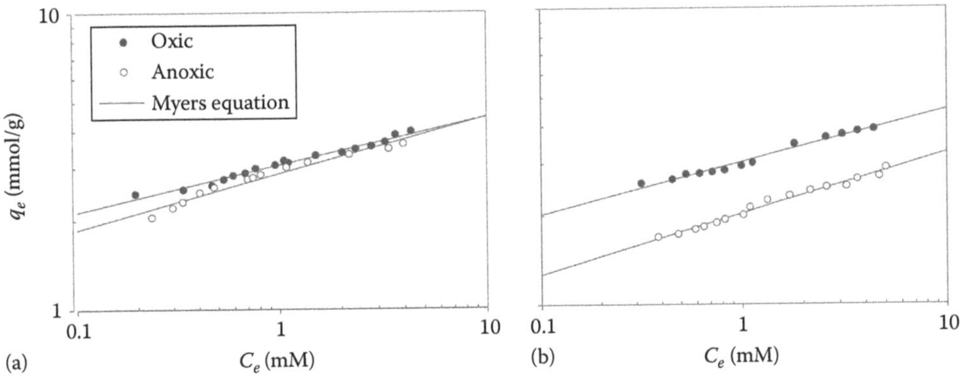

FIGURE 6.6 Single solute adsorption isotherm of 2-methylphenol. (a) BC-21 and (b) F400. (Reprinted from *J. Hazard. Mater.*, 197, Yan, L. and Sorial, G.A., Chemical activation of bituminous coal for hampering oligomerization of organic contaminants, pp. 311–319, Copyright 2011, with permission from Elsevier.)

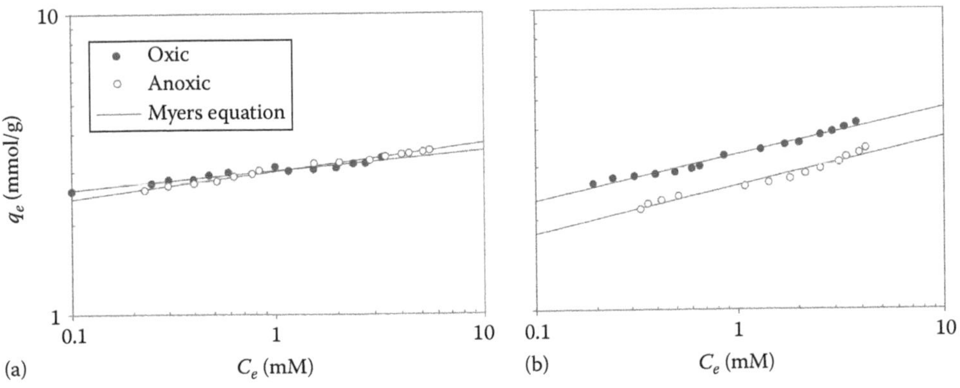

FIGURE 6.7 Single solute adsorption isotherm of 2-ethylphenol. (a) BC-21 and (b) F400. (Reprinted from *J. Hazard. Mater.*, 197, Yan, L. and Sorial, G.A., Chemical activation of bituminous coal for hampering oligomerization of organic contaminants, pp. 311–319, Copyright 2011, with permission from Elsevier.)

At low concentration, the Myers equation becomes a linear equation (6.2), that is, following Henry's law equation,

$$\frac{dC_e}{dq_e} = \frac{1}{H}; \quad q_e \to 0 \tag{6.2}$$

This criterion is very important when predicting multicomponent adsorption systems by the ideal adsorption theory (IAST) because of the integration limits in the Gibbs adsorption equation. Meanwhile, q_e (mmol/g) was calculated from mass balance equation (6.3)

$$q_e = \frac{(C_O - C_e)V}{M} \tag{6.3}$$

where

C_O (mM) is initial liquid-phase concentration
M (g) is the mass of adsorbent
V (L) is the adsorbate volume

The adsorbate volume, V, was 0.125 L for the anoxic isotherms and 0.1 L for the oxic isotherms.

Figure 6.5 clearly indicates that the adsorptive capacity of F400 for phenol under oxic conditions increased significantly as compared to anoxic data due to the presence of molecular oxygen. However, the presence of molecular oxygen in the test environment did not have a tremendous impact on the adsorptive capacity of BC-21 for phenol. A similar behavior is seen for 2-methylphenol and 2-ethylphenol (see Figures 6.6 and 6.7). Furthermore, the difference in the 2-methylphenol adsorptive capacity obtained for F400 under oxic and anoxic conditions is more significant than that for phenol and 2-ethylphenol. However, for BC-21, the adsorption capacity under both oxic and anoxic conditions is nearly identical. The major differences only occurred at low C_e values, that is, at high carbon dosage.

In this experiment by Yan and Sorial, anoxic and oxic adsorptive capacity of phenol, 2-methylphenol, and 2-ethylphenol on the two adsorbents were compared at two equilibrium concentrations of 0.5 and 5 mM (Table 6.3). These two limits were selected in order to represent the low and high carbon dosage. It is clear that the difference between anoxic and oxic condition achieved with F400 is highly significant as compared to BC-21. In addition, they found 2-methylphenol is easier to have oligomerization phenomenon in comparison to phenol and 2-ethylphenol because the difference for 2-methylphenol is larger than 2-ethylphenol and phenol. This distinct performance can be attributed to the differences of critical oxidation potential (COP) of these phenolic compounds. The COP for 2-methylphenol is 1.040 V while that for phenol and 2-ethylphenol is 1.089 and 1.080 V. As shown in Table 6.3, the oxic adsorptive capacity of phenol on F400 compared to the anoxic capacity increased from 32.2% to 32.7% for an equilibrium concentration of 5 and 0.5 mM, respectively. Furthermore, in the case of 2-ethylphenol, for the equilibrium adsorbate concentrations of 5 and 0.5 mM, the adsorptive capacities of F400 exhibited under oxic condition are 25.1% and 33.5% higher than the anoxic adsorptive capacity, respectively. This difference in capacity is much more pronounced for 2-methylphenol. The oxic adsorption capacity for 2-methylphenol compared to the anoxic adsorptive capacity on F400 increases from 42.3% to 52.5% for an equilibrium concentration of 5 and 0.5 mM, respectively.

It is also of interest to notice that the BC-21 exhibited very slight differences between the oxic and anoxic adsorption isotherms for the three phenolic compounds as compared to F400 (see Table 6.3). This indicates that there is no considerable change between both conditions. In other words, it is safe to assume that BC-21 has a significant positive impact on hampering the oligomerization of phenolic compounds on its surface. Especially, in the case of 2-ethylphenol, the capacity difference

TABLE 6.3
Comparison of Anoxic and Oxic Adsorption Capacity

		Phenol			2-Methylphenol			2-Ethylphenol		
	Ce (mM)	Anoxic q_e (mmol/g)	Oxic q_e (mmol/g)	Δ q_e (%)	Anoxic q_e (mmol/g)	Oxic q_e (mmol/g)	Δ q_e (%)	Anoxic q_e (mmol/g)	Oxic q_e (mmol/g)	Δ q_e (%)
F400	0.5	1.854	2.461	32.7	1.760	2.683	52.5	2.146	2.865	33.5
	5	2.401	3.174	32.2	2.828	4.027	42.3	4.001	5.010	25.1
BC-21	0.5	1.943	2.182	12.3	2.552	2.796	9.54	2.844	2.877	1.16
	5	3.195	3.010	6.12	4.185	4.341	3.70	3.537	3.626	2.53

Source: Data from Yan, L. and Sorial, G.A., *J. Hazard. Mater.*, 197, 311, 2011.

on BC-21 under both conditions (low and high carbon dosage) is <2.5%. The main reason for this distinct performance is that BC-21 had a significant higher micropore percentage than F400. In the case of BC-21, forming larger stable oligomerized molecular products is much more difficult than F400. This could be attributed to the molecular dimensions of the adsorbates and the critical pore size of the adsorbent. For example, F400 has a broad PSD from 10 to 500 Å and oligomerization would not be constricted by its PSD. In contrast, BC-21 has a critical pore diameter of 10 Å and a narrow PSD. The second widest dimensions of phenol, 2-methylphenol, and 2-ethylphenol are 4.17, 5.32, and 5.70 Å, respectively. Therefore, it is just possible for phenol to form dimers, while the formation of trimers is impossible. In the case of 2-methylphenol and 2-ethylphenol, it is even impossible for them to get oligomerized. On the other hand, Vidic et al. (1997) have postulated that oxygen-containing basic surface functional groups slightly promote the oligomerization of phenolic compounds. Therefore, the behavior difference between BC-21 and F400 may be partially due to the difference of the total basicity. The total basicity (0 meq/g) for BC-21, as determined by the titration method, was much lower than that of the GAC F400 (0.475 meq/g). This shows that GAC F400 has more basic surface functional groups than BC-21 that could promote the oligomerization of phenolic compounds.

It is worth to note that the equilibrium adsorptive capacity of BC-21 for all three phenolic compounds was higher than F400 under anoxic adsorption condition (see Table 6.3). Some studies have shown that the effect of surface functional group content on phenolic compounds adsorptive capacity in the absence of molecular oxygen can be negligible (Efremenko and Sheintuch 2006). Therefore, the results can be mainly attributed to the following physical reasons. Firstly, these results illustrate that the higher BET surface area of BC-21 can lead to higher adsorption than F400. Secondly, this phenomenon can also be elucidated by the comparison of the microporosity of the two adsorbents. The microporosity of F400 is only 60.7%, while for BC-21 it is 81.8%. Therefore, the adsorption capacity of BC-21 is higher than F400 because micropores are primarily responsible for the adsorption.

6.3.3 Oligomerization Control of Multisolute Adsorption Isotherm

It is worth noting that effluent industrial wastewater will always encounter more than one phenolic compound available. Therefore, competitive adsorption is very common, and capacity reductions of one or two orders of magnitude and also large reductions in the rate of adsorption could occur. Hence, a good understanding of the impact of BC-21 on multicomponent adsorption system is of considerable importance as a basis for applying it in an optimal way. This can be achieved by investigating the equilibrium isotherm of multicomponent systems. Therefore, in the experiment of Yan and Sorial (2013), the effectiveness of BC-21 in hampering the oligomerization of phenolic compounds in the multi-solute system was studied.

Ideal adsorbed solution theory (IAST) was used in this study because it is the most common approach used to predict the multicomponent adsorption isotherms onto activated carbon by using only single solute equilibrium data. The IAST is based on the assumption that the adsorbed mixture forms an ideal solution at a constant spreading pressure. The model can be represented by the following Equation 6.4:

$$C_{e,i} = \frac{C_i^o(\prod m, T) q_{e,i}}{\sum_{i=1}^{i=n} q_{e,i}} \tag{6.4}$$

where $C_i^o(\prod m, T)$ is equilibrium liquid-phase concentration of pure solute i at the same temperature T and spreading pressure Πm of the mixture with N components:

$$\sum_{i=1}^{i=N} \frac{q_{e,i}}{q_i^o} = \sum_{i=1}^{i=N} \frac{C_{e,i}}{C_i^o} = 1 \tag{6.5}$$

Equation 6.5 is the expression of the relationship between mixture equilibrium solid-phase concentration $q_{e,i}$ and single solute equilibrium concentrations q_i^o in the case of an ideal multicomponent system:

$$\frac{A_s \pi_i}{RT} = \int_0^{q_i^o} \frac{q_i^o}{c_i^o} \frac{dc_i^o}{dq_i^o} dq_i^o \tag{6.6}$$

In addition, a further relation is established between the spreading pressure and directly measurable quantities as Equation 6.6, where A_s is the external surface area per unit mass of adsorbent and π_i is the spreading pressure of single component i. This method did not consider the chemical effects during the adsorption process. Therefore, this approach will not accurately predict adsorption equilibrium of trace organic contaminants under oxic condition due to oligomerization of adsorbates on the surface of the activated carbon. However, it is very suitable to examine the effectiveness of BC-21 in hampering oligomerization under oxic conditions in multicomponent adsorption.

In this study, a binary adsorption of 2-methylphenol/2,4-dimethylphenol was performed on F400 and BC-21. Figure 6.8a and b presents the binary adsorption isotherms of phenol/2-methylphenol for F400 and BC-21, respectively. Meanwhile, the Myers parameters obtained from the single solute anoxic isotherm was applied into the IAST model to predict the competitive adsorption behavior. The evaluation of the predictability of IAST was also done by calculating the sum of squares of relative errors (SSREs). All the detailed concentration combination and SSRE values of both runs for both adsorbents are shown in Table 6.4.

Figure 6.8a demonstrates that the presence of molecular oxygen had a significant impact on the binary adsorptive properties of F400. The obvious differences between oxic and anoxic binary adsorption isotherms due to oligomerization on the carbon surface was observed for both 2-methylphenol and 2,4-dimethylphenol. On the other hand, the anoxic adsorption isotherms can be well predicted by IAST. However, the oxic adsorption isotherms for both 2-methylphenol and 2,4-dimethylphenol are obviously higher than oxic adsorption value predicted by IAST. The under prediction of the IAST is due to the oligomerization behavior that is taking place in the presence of molecular oxygen, which is not accounted for in IAST.

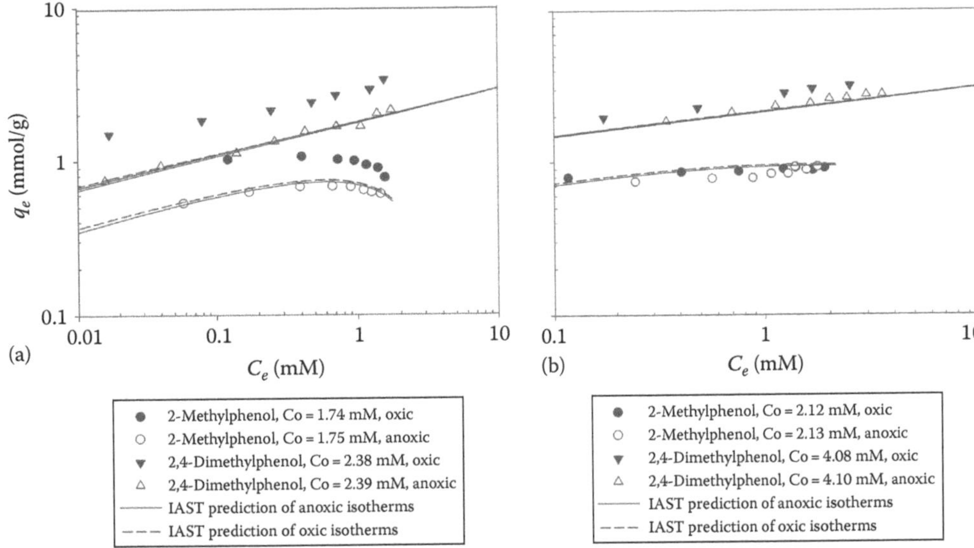

FIGURE 6.8 Binary adsorption isotherms of 2-methylphenol and 2,4-dimethylphenol on (a) F400 and (b) BC-21. (Reprinted from Yan, L. and Sorial, G.A., *Water Air Soil Pollut.*, 224, 1588, 2013. With permission.)

TABLE 6.4
Summary of the Multicomponent Adsorption System

		Binary Adsorption			
		Initial Concentration (mM)		**SSRE**	
Adsorbent	**Conditions**	**2-Methylphenol**	**2,4-Dimethylphenol**	**2-Methylphenol**	**2,4-Dimethylphenol**
F400	Anoxic	1.75	2.39	0.0655	0.0201
	Oxic	1.74	2.38	0.8094	0.4581
$Carbon_{exp}$	Anoxic	2.13	4.10	0.1294	0.029
	Oxic	2.12	4.08	0.0257	0.1810

Source: Date from Yan, L. and Sorial, G.A., *Water Air Soil Pollut.*, 224, 1588, 2013.

In contrast, BC-21 (see Figure 6.8b) showed remarkable different binary adsorption behavior than that documented for F400. For the binary adsorption studied, oligomerization could happen in various ways, such as 2-methylphenol molecules, 2-methylphenol and 2,4-dimethylphenol, or 2,4-dimethylphenol molecules. Therefore, it seems there are more opportunities than single solute adsorption. However, it is apparent from Figure 6.8b that the oxic and anoxic isotherm lines are really close to each other for both adsorbates on $Carbon_{exp}$. This observation can be attributed to three reasons. Firstly, regarding multicomponent adsorption, the possibility for either compound to get oligomerized becomes less as compared to the single solute system because of the competitive effect that would lead to less available sites. Secondly, the narrow PSD also leads to the oligomerization constriction effect on binary adsorption. Thirdly, BC-21 is a more acidic carbon that is unfavorable for the oligomerization of phenolic compounds. It is seen from Figure 6.8b that the binary adsorptive behavior of BC-21 agreed with those predicted by the IAST model for both the oxic and anoxic binary system. Therefore, the good prediction (see Table 6.4) for both the oxic and anoxic binary adsorption on BC-21 by IAST confirmed again the effectiveness of BC-21 in hampering oligomerization for binary solute system adsorption.

REFERENCES

Balci, S., T. Dogu, and H. Yucel. 1994. Characterization of activated carbon produced from almond shell and hazelnut shell. *Journal of Chemical Technology and Biotechnology* 60(4):419–426.

Caturla, F., M. Molinasabio, and F. Rodriguezreinoso. 1991. Preparation of activated carbon by chemical activation with $ZnCl_2$. *Carbon* 29(7):999–1007.

Centeno, T. A., M. Marban, and A. B. Fuertes. 2003. Importance of micropore size distribution on adsorption at low adsorbate concentrations. *Carbon* 41(4):843–846.

Cooney, D. O. and Z. P. Xi. 1994. Activated carbon catalyzes reactions of phenolics during liquid-phase adsorption. *AiChE Journal* 40(2):361–364.

Donnet, J.-B. 1998. *Carbon Fibers*. 3rd edn. New York: Marcel Dekker.

Economy, J. and R. Anderson. 1966. A new route to boron nitride. *Inorganic Chemistry* 5(6):989–992.

Economy, J., M. Daley, and C. Mangun. 1996. Activated carbon fibers—Past, present, and future. *Abstracts of Papers of the American Chemical Society* 211:84-Fuel.

Efremenko, I. and M. Sheintuch. 2006. Predicting solute adsorption on activated carbon: Phenol. *Langmuir* 22(8):3614–3621.

Girgis, B. S., L. B. Khalil, and T. A. M. Tawfik. 1994. Activated carbon from sugar-cane Bagasse by carbonization in the presence of inorganic acids. *Journal of Chemical Technology and Biotechnology* 61(1):87–92.

Grant, T. M. and C. J. King. 1990. Mechanism of irreversible adsorption of phenolic-compounds by activated carbons. *Industrial & Engineering Chemistry Research* 29(2):264–271.

Guo, J. and A. C. Lua. 2000. Textural characterization of activated carbons prepared from oil-palm stones pre-treated with various impregnating agents. *Journal of Porous Materials* 7(4):491–497.

Hsieh, C. T. and H. S. Teng. 2000. Liquid-phase adsorption of phenol onto activated carbons prepared with different activation levels. *Journal of Colloid and Interface Science* 230(1):171–175.

Hu, Z. H. and E. F. Vansant. 1995. Synthesis and characterization of a controlled micropore-size carbonaceous adsorbent produced from walnut shell. *Microporous Materials* 3(6):603–612.

Hussein, M. Z., R. S. H. Tarmizi, Z. Zainal, R. Ibrahim, and M. Badri. 1996. Preparation and characterization of active carbons from oil palm shells. *Carbon* 34(11):1447–1449.

Jiang, Q. and Y. Zhao. 2004. Effects of activation conditions on BET specific surface area of activated carbon nanotubes. *Microporous and Mesoporous Materials* 76(1–3):215–219.

Jimenez, V., J. A. Diaz, P. Sanchez, J. L. Valverde, and A. Romero. 2009. Influence of the activation conditions on the porosity development of herringbone carbon nanofibers. *Chemical Engineering Journal* 155(3):931–940.

Juang, R. S., R. L. Tseng, and F. C. Wu. 2001. Role of microporosity of activated carbons on their adsorption abilities for phenols and dyes. *Adsorption—Journal of the International Adsorption Society* 7(1):65–72.

Jung, M. W., K. H. Ahn, Y. Lee, K. P. Kim, J. S. Rhee, J. T. Park, and K. J. Paeng. 2001. Adsorption characteristics of phenol and chlorophenols on granular activated carbons (GAC). *Microchemical Journal* 70(2):123–131.

Kaneko, Y., M. Abe, and K. Ogino. 1989. Adsorption characteristics of organic-compounds dissolved in water on surface-improved activated carbon-fibers. *Colloids and Surfaces* 37:211–222.

Laine, J. and A. Calafat. 1991. Factors affecting the preparation of activated carbons from coconut shell catalyzed by potassium. *Carbon* 29(7):949–953.

Laine, J., A. Calafat, and M. Labady. 1989. Preparation and characterization of activated carbons from coconut shell impregnated with phosphoric-acid. *Carbon* 27(2):191–195.

Lillo-Rodenas, M. A., D. Cazorla-Amoros, and A. Linares-Solano. 2003. Understanding chemical reactions between carbons and NaOH and KOH—An insight into the chemical activation mechanism. *Carbon* 41(2):267–275.

Lillo-Rodenas, M. A., J. Juan-Juan, D. Cazorla-Amoros, and A. Linares-Solano. 2004. About reactions occurring during chemical activation with hydroxides. *Carbon* 42(7):1371–1375.

Lozano-Castello, D., J. M. Calo, D. Cazorla-Amoros, and A. Linares-Solano. 2007. Carbon activation with KOH as explored by temperature programmed techniques, and the effects of hydrogen. *Carbon* 45(13):2529–2536.

Lozano-Castello, D., D. Cazorla-Amoros, A. Linares-Solano, and D. F. Quinn. 2002. Influence of pore size distribution on methane storage at relatively low pressure: Preparation of activated carbon with optimum pore size. *Carbon* 40(7):989–1002.

Lu, Q. L. and G. A. Sorial. 2004. Adsorption of phenolics on activated carbon—Impact of pore size and molecular oxygen. *Chemosphere* 55(5):671–679.

Lu, Q. L. and G. A. Sorial. 2007. The effect of functional groups on oligomerization of phenolics on activated carbon. *Journal of Hazardous Materials* 148(1–2):436–445.

Magne, P. and P. L. Walker. 1986. Phenol adsorption on activated carbons—Application to the regeneration of activated carbons polluted with phenol. *Carbon* 24(2):101–107.

Mangun, C. L., K. R. Benak, J. Economy, and K. L. Foster. 2001. Surface chemistry, pore sizes and adsorption properties of activated carbon fibers and precursors treated with ammonia. *Carbon* 39(12):1809–1820.

Marsh, H., D. S. Yan, T. M. Ogrady, and A. Wennerberg. 1984. Formation of active carbons from cokes using potassium hydroxide. *Carbon* 22(6):603–611.

Martin-Gullon, I., R. Andrews, M. Jagtoyen, and F. Derbyshire. 2001. PAN-based activated carbon fiber composites for sulfur dioxide conversion: Influence of fiber activation method. *Fuel* 80(7):969–977.

Martinez, M. L., M. M. Torres, C. A. Guzman, and D. M. Maestri. 2006. Preparation and characteristics of activated carbon from olive stones and walnut shells. *Industrial Crops and Products* 23(1):23–28.

Molinasabio, M., F. Rodriguezreinoso, F. Caturla, and M. J. Selles. 1995. Porosity in granular carbons activated with phosphoric-acid. *Carbon* 33(8):1105–1113.

Moore, B. C., F. S. Cannon, J. A. Westrick, D. H. Metz, C. A. Shrive, J. DeMarco, and D. J. Hartman. 2001. Changes in GAC pore structure during full-scale water treatment at Cincinnati: A comparison between virgin and thermally reactivated GAC. *Carbon* 39(6):789–807.

Nakhla, G., N. Abuzaid, and S. Farooq. 1994. Activated carbon adsorption of phenolics in oxic systems effect of Ph and temperature-variations. *Water Environment Research* 66(6):842–850.

Nowack, K. O., F. S. Cannon, and D. W. Mazyck. 2004. Enhancing activated carbon adsorption of 2-methylisoborneol: Methane and steam treatments. *Environmental Science & Technology* 38(1):276–84.

Otowa, T., Y. Nojima, and T. Miyazaki. 1997. Development of KOH activated high surface area carbon and its application to drinking water purification. *Carbon* 35(9):1315–1319.

Parker, G. R. 1995. Optimum isotherm equation and thermodynamic interpretation for aqueous 1,1,2-trichloroethene adsorption isotherms on three adsorbents. *Adsorption—Journal of the International Adsorption Society* 1(2):113–132.

Pelekani, C. and V. L. Snoeyink. 1999. Competitive adsorption in natural water: Role of activated carbon pore size. *Water Research* 33(5):1209–1219.

Pelekani, C. and V. L. Snoeyink. 2000. Competitive adsorption between atrazine and methylene blue on activated carbon: The importance of pore size distribution. *Carbon* 38(10):1423–1436.

Ryu, Z. Y., H. Q. Rong, J. T. Zheng, M. Z. Wang, and B. J. Zhang. 2002. Microstructure and chemical analysis of PAN-based activated carbon fibers prepared by different activation methods. *Carbon* 40(7):1144–1147.

Ryu, Z. Y., J. T. Zheng, M. H. Wang, and B. J. Zhang. 2000. Nitrogen adsorption studies of PAN-based activated carbon fibers prepared by different activation methods. *Journal of Colloid and Interface Science* 230(2):312–319.

Shen, W. Z., J. T. Zheng, Z. F. Qin, J. G. Wang, and Y. H. Liu. 2004. Preparation of mesoporous activated carbon fiber by steam activation in the presence of cerium oxide and its adsorption of Congo red and Vitamin B12 from solution. *Journal of Materials Science* 39(14):4693–4696.

Shmidt, J. L., A. V. Pimenov, A. I. Lieberman, and H. Y. Cheh. 1997. Kinetics of adsorption with granular, powdered, and fibrous activated carbon. *Separation Science and Technology* 32(13):2105–2114.

Singh, B. K. and N. S. Rawat. 1994. Comparative sorption equilibrium studies of toxic phenols on fly-ash and impregnated fly-ash. *Journal of Chemical Technology and Biotechnology* 61(4):307–317.

Stohr, B., H. P. Boehm, and R. Schlogl. 1991. Enhancement of the catalytic activity of activated carbons in oxidation reactions by thermal-treatment with ammonia or hydrogen-cyanide and observation of a superoxide species as a possible intermediate. *Carbon* 29(6):707–720.

Suzuki, M. 1994. Activated carbon-fiber—Fundamentals and applications. *Carbon* 32(4):577–586.

Teng, H. S. and T. S. Yeh. 1998. Preparation of activated carbons from bituminous coals with zinc chloride activation. *Industrial & Engineering Chemistry Research* 37(1):58–65.

Teng, H. S., T. S. Yeh, and L. Y. Hsu. 1998. Preparation of activated carbon from bituminous coal with phosphoric acid activation. *Carbon* 36(9):1387–1395.

Terzyk, A. P. 2003. Further insights into the role of carbon surface functionalities in the mechanism of phenol adsorption. *Journal of Colloid and Interface Science* 268(2):301–329.

Tessmer, C. H., R. D. Vidic, and L. J. Uranowski. 1997. Impact of oxygen-containing surface functional groups on activated carbon adsorption of phenols. *Environmental Science & Technology* 31(7):1872–1878.

Vidic, R. D. and M. T. Suidan. 1991. Role of dissolved-oxygen on the adsorptive capacity of activated carbon for synthetic and natural organic-matter. *Environmental Science & Technology* 25(9):1612–1618.

Vidic, R. D., M. T. Suidan, and R. C. Brenner. 1993. Oxidative coupling of phenols on activated carbon—Impact on adsorption equilibrium. *Environmental Science & Technology* 27(10):2079–2085.

Vidic, R. D., C. H. Tessmer, and L. J. Uranowski. 1997. Impact of surface properties of activated carbons on oxidative coupling of phenolic compounds. *Carbon* 35(9):1349–1359.

Villar-Rodil, S., F. Suarez-Garcia, J. I. Paredes, A. Martinez-Alonso, and J. M. D. Tascon. 2005. Activated carbon materials of uniform porosity from polyaramid fibers. *Chemistry of Materials* 17(24):5893–5908.

Walker, P. L. 1969. Chemical and physical properties of Carbon. In Uoker, F. M. Ed., *Khimicheskie i fizicheskie svoĭstva ugleroda.*

Wang, K., and D. D. Do. 1999. Sorption equilibria and kinetics of hydrocarbons onto activated carbon samples having different micropore size distributions. *Adsorption—Journal of the International Adsorption Society* 5(1):25–37.

Wu, F. C., R. L. Tseng, and R. S. Juang. 2005. Preparation of highly microporous carbons from fir wood by KOH activation for adsorption of dyes and phenols from water. *Separation and Purification Technology* 47(1–2):10–19.

Yamashita, Y. and K. Ouchi. 1982. Influence of alkali on the carbonization process. 1. Carbonization of 3,5-dimethylphenol-formaldehyde resin with NaOH. *Carbon* 20(1):41–45.

Yan, L. and G. A. Sorial. 2011. Chemical activation of bituminous coal for hampering oligomerization of organic contaminants. *Journal of Hazardous Materials* 197:311–319.

Yan, L. and G. A. Sorial. 2013. Carbon activation for hampering oligomerization of phenolics in multicomponent systems. *Water Air and Soil Pollution* 224(6):1–11.

Yue, Z. G., C. Mangun, J. Economy, P. Kemme, D. Cropek, and S. Maloney. 2001. Removal of chemical contaminants from water to below USEPA MCL using fiber glass supported activated carbon filters. *Environmental Science & Technology* 35(13):2844–2848.

Zawadzki, J. and S. Biniak. 1988. IR spectral studies of the basic properties of carbon. *Polish Journal of Chemistry* 62(1–3):195–202.

Zhang, T. Y., W. P. Walawender, L. T. Fan, M. Fan, D. Daugaard, and R. C. Brown. 2004. Preparation of activated carbon from forest and agricultural residues through CO_2 activation. *Chemical Engineering Journal* 105(1–2):53–59.

7 Solvation in Heterogeneous Media

Sanjib Bagchi

CONTENTS

7.1 INTRODUCTION

Solute–solvent interactions play a key role in determining the observed kinetic, equilibrium, and spectroscopic properties of a solute in different media. Since most chemical and biological processes take place in solution, a detailed understanding of molecular interactions involving the solute and its immediate environment is required. It has long been known that physicochemical properties of a molecule (solute) depend considerably on the nature of the interacting molecules in its immediate environment. A solute molecule may be thought as a source of field that induces structural changes in a solvent. Three distinct regions may be distinguished around a solute molecule, namely, a *primary region* with completely oriented solvent molecules, a *secondary region* containing partially oriented solvent molecules, and a *bulk region* where the solvent molecules are not under the influence of the solute and normal distribution of solvent (as in pure solvent) prevails. The primary and the secondary regions where the influence of the solute is felt are commonly referred to as the *cybotactic region* [1]. The net molecular interaction taking place in this region is often referred to as *solvation interaction*. In dilute solutions, the nature of these interactions is mainly solute–solvent and solvent–solvent.

Both theoretical [2,3] and experimental [4] efforts have been made to understand such solvation interactions. Central to the study of this problem of interaction of a solute and its local environment is the question of how the free energy of a solute changes due to the presence of surrounding interacting molecules of the medium. An approach to the problem at a molecular level is fraught with the inherent difficulty that only little information is available about the arrangement of molecules in the cybotactic region. The vast literature of solvent effect on the kinetic, equilibrium, and spectroscopic properties of a solute indicate that one needs to consider only a few modes of solute–solvent interaction [5,6]. One of them is a *nonspecific long-range interaction* due to the collective influence of solvent as a dielectric medium. This involves electrostatic interactions arising due to Coulomb and polarization forces. The continuum dielectric theory provides an isomorphic model for describing nonspecific interaction [5]. It has been found that this mode of interaction is determined by the relative permittivity (ε) and the refractive index (n) or any function thereof. Another is the *specific interaction* involving hydrogen bond donation (HBD) and the acceptance ability of solvents. Since specific solvation interactions are considered to be chemical in nature, the use of a continuum model for describing such interaction is severely limited [7]. Chemists have tried to understand the environmental effect by introducing the concept of *solvent polarity*, which is supposed to represent the overall solvation ability of the solvent and includes both nonspecific and specific modes of interaction [4,8]. For this, a *model process* [1] (e.g., a chemical equilibrium or kinetic process or spectral transition) is chosen, and changes in one of its parameters are recorded when the solvent is changed. Of the various polarity scales, the Z [1,9] and $E_T(30)$ [10] have passed the test of time and are most widely used. In view of the complexity of solute–solvent interaction, the representation of polarity by a single parameter (e.g., Z or $E_T(30)$) is not accurate and one has to use a multiparameter approach [11,12]. In this approach, it is assumed that the change of a physicochemical parameter of a solute (S) due to solvation, $\Delta(S, i)$, can be represented as

$$\Delta(S,i) = P_0(S) + \Sigma a_\alpha(S) P_\alpha(i) \tag{7.1}$$

where

i indicates a solvent

P_0 and a_α terms depend on the solute (S)

P_α's are the solvent properties pertinent to the α mode of interaction of the solute with a solvent

Equation 7.1 is known as the *linear solvation energy relationship* (*LSER*) [4,5,11] and has been found to be of great significance in the theory of solvent effect. Although the single or the multiparameter approach discussed earlier has been applied widely and successfully for describing solvation in pure solvents, their use in the case of heterogeneous media formed by the self-assembly of molecules is rather limited. For example, enhanced solubility of organic molecules in the micellar phase compared to the bulk aqueous medium points to the existence of significant interaction of the molecules with the micellar environment [13]. Successful attempts, however, have been made to find out the descriptors of specific and nonspecific modes of interaction of a solute with the heterogeneous medium [14–18]. Knowledge of the parameters describing specific and nonspecific solvation may be helpful to understand the solubilization of organic molecules like drugs in such media. The objective of this chapter is to discuss the possibility of applying the preceding parametric approach to the case of heterogeneous media and explore the present status of the field. The outline of the chapter is as follows. A brief discussion of the process of self-organization and commonly used microheterogeneous media will be done in Section 7.1. This will be followed by the common procedures for the description of solvation in a homogeneous medium in Section 7.2. The description of solvation in a heterogeneous medium in terms of suitable parameters will be discussed finally in Section 7.3.

7.2 SELF-ORGANIZED ASSEMBLES

Self-assembly is a type of process in which a disordered system of preexisting components forms an organized structure or pattern as a consequence of specific, local interactions among the components themselves, without any external intervention. Self-assembly is either static or dynamic in nature. In static self-assembly, the ordered state is formed as a system approaches equilibrium, reducing its free energy. However in dynamic self-assembly, patterns of preexisting components organized by specific local interactions are commonly described as *self-organized assembles*. Self-assembly can be defined as the spontaneous and reversible organization of component molecular units (building blocks) into ordered structures by *noncovalent interactions*. An important aspect of *self-assembly* is the key role of weak molecular interactions (e.g., van der Waals, $\pi \rightarrow \pi$, hydrogen bonds) with respect to more *traditional* covalent, ionic, or metallic bonds. These weak molecular interactions play an important role especially in biological systems. Examples of *self-organization* in materials science include the formation of molecular crystals, colloids, lipid bilayers, phase-separated polymers, and self-assembled monolayers [19]. A characteristic common to nearly all self-assembled systems is their thermodynamic stability. For self-organization to take place without the intervention of external forces, the process must lead to a lower Gibbs free energy; thus self-assembled structures are thermodynamically more stable than the single, unassembled components. A direct consequence is the general tendency of self-assembled structures to be relatively free of defects. The existence of weak molecular interactions and the condition of thermodynamic stability endows these systems with a special property, namely, the sensitivity to perturbations exerted by the external environment, namely, change of salt concentration, temperature, pH of the medium, and pressure. The weak nature of interactions is responsible for the flexibility of the architecture and allows for rearrangements of the structure in the direction determined by thermodynamics. This gives rise to an important property, namely, reversibility to such type of assemblies. The specific molecular arrangements in self-assembled systems are crucial for a specific property (*functionality*) of the system. Different possible organized assembles are reported in the literature, namely, normal and reverse micelle formed by surfactants, cyclodextrins in aqueous and nonaqueous medium, and vesicles [20–28]. Systems like normal aqueous micelle and vesicles have been dealt with in the following section.

7.2.1 Aqueous Normal Micelles and Vesicles Formed by Surfactants

The term *aqueous* or *normal micelle* is reserved for the organized assemblies formed in water by the aggregation of amphiphilic molecules, commonly known as *surfactants*. Surfactants are characterized by a polar head group and a nonpolar tail. They are of four types, namely, cationic, anionic, nonionic, or zwitterionic depending on the nature of the hydrophilic head group that is bound to the hydrophobic tail. A surfactant, when present at low concentrations in water, adsorbs onto interfaces significantly, thereby changing the interfacial free energy. When surfactant molecules are dissolved in water at concentrations above a critical value, known as the critical micelle concentration (*cmc*), they form aggregates called micelles [29]. The value of *cmc*, however, depends on several factors like surfactant structure, solvent, temperature, and addition of additives. As the concentration of surfactant is increased above the *cmc*, the addition of more monomer molecules results in the formation of more micelle, so that the concentration of the monomer remains practically constant (approximately equal to *cmc*) and the concentration of micelle increases. At surfactant concentration slightly above the *cmc*, the micelles are supposed to be spherical. Shape changes, however, are known to occur as the concentration of surfactant increases. Micelle is not a permanent entity but a dynamic structure that exists in thermodynamic equilibrium with its monomer. In water, the hydrophilic *heads* of surfactant molecules are always in contact with the solvent, regardless of whether the surfactants exist as monomers or as part of a normal micelle. However, the hydrophobic (lipophilic) *tails* of surfactant molecules have less contact with water when they are part of a micelle; this being the basis for the energetic drive for micelle formation, micelles composed of ionic surfactants have an electrostatic attraction to the ions that surround them (*counterions*) in solution. Although the closest counterions partially mask a charged micelle (by up to 90%), the residual of micelle charge affects the structure of the surrounding solvent at appreciable distances from the micelle influencing

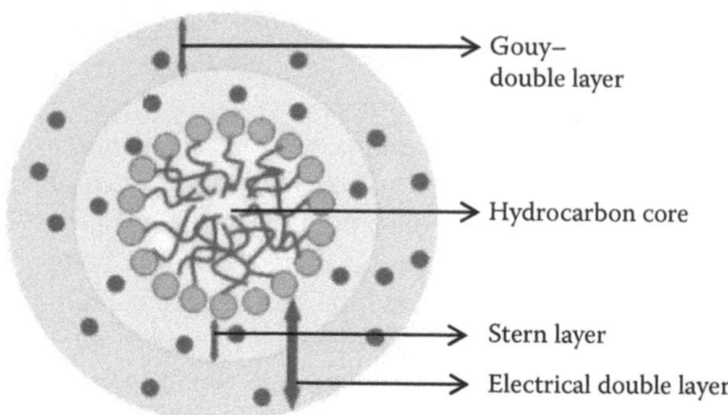

FIGURE 7.1 A model representing the cross section of a spherical ionic micelle.

many properties of the mixture. Adding salts to a colloid containing micelles can decrease the strength of electrostatic interactions and lead to the formation of larger ionic micelles [30]. From the thermodynamic point, the formation of micelles can be understood in terms of a balance between entropy and enthalpy.

Among different possible micellar shapes, the most prevalent is the spherical one as proposed by McBain [31] and supported by Hartley [32]; ellipsoidal and cylindrical shapes are also possible [33–38]. A simplified model for ionic micelles as given by Hartley [32] is useful for understanding features. The cross section of such a micelle is shown schematically in Figure 7.1. The core of a normal micelle is liquid-like, consisting of the hydrocarbon chains of the surfactants. A peripheral shell outside the core contains polar/ionic head groups. This shell, for ionic surfactants, usually has a width up to a few angstroms and is called *Stern layer*; it consists of surfactant head groups, bound counterions, and water molecules. Most of the counterions are dissociated from the micelle and are located in the *Gouy–Chapman double layer*, which is outside the Stern layer and whose width is up to several hundred angstroms [39–41]. The surface potential generated by the net charges of the Stern layer is usually in the range of 50–100 mV, and this acts as an electrostatic barrier to the passage of charged ions to and across the micellar surface. The presence of ionic groups at the micellar interface causes nearby water molecules to solvate the micelles through ion–dipole interaction. For nonionic micelles the shell outside the core is termed as the *Palisade layer* and consists of the polar head groups and hydrogen-bonded water molecules. Recent solvation dynamics studies using time domain optical spectroscopy indicate that the water molecules confined within a small volume of Stern layer/Palisade layer of the micellar interface are fundamentally different from water molecules at the bulk [42,43]. Micelles have particular significance because of their ability to increase the solubility of sparingly soluble substances in water. The spatial position of a solubilized substance in a micelle depends on its polarity. Nonpolar molecules will be solubilized in the micellar core, and substances with intermediate polarity will be distributed along the surfactant molecules in certain intermediate positions.

Apart from the formation of micellar aggregates in aqueous medium, certain naturally occurring or synthetic phospholipids as well as completely synthetic surfactants can form organized assembles. The assemblies formed from phospholipids are typically termed liposomes, and those formed from synthetic surfactants are designated as vesicles [44]. Depending on the method of preparation, the nature of surfactants, and the experimental conditions (pH, ionic strength, concentrations), different vesicular structures can exist. These range from small (8200–500 Å diameter) or large (0.1–10 μ diameter) single-walled bilayer structure to relatively larger (1000–8000 Å diameter) onion-like multicomponent structure. Unlike micelles, vesicles once formed do not break down upon dilution. Moreover, the hydrocarbon part of the vesicle is more ordered [44].

7.3 SOLVENT POLARITY AND SOLVATOCHROMIC PROBES FOR PURE SOLVENTS

Central to the problem of solvation is the question of how the free energy of the solute changes due to interaction with surrounding solvent molecules. But a straightforward solution of the problem is fraught with difficulties since only a little is known about the arrangement of the molecules in the *cybotactic region*. Chemists have introduced empirically the concept of solvent polarity that reflects the complex interplay of all types of solute–solvent interactions in the cybotactic region. Numerous attempts have been made to describe solvent polarity in terms of overall solute–solvent interaction, and several empirical descriptors of polarity have been proposed [1,4,8]. For this, a solvent-sensitive *reference process* or *model process* is introduced [1]. The process may involve a reaction in equilibrium, or a kinetic or a spectroscopic process. A study of free energy change for the process through suitable parameters (e.g., log K in the case of an equilibrium process or log k in the case of a kinetic process, E_T representing the transition energy for a spectroscopic process) in different solvents is supposed to provide an empirical measure of the solvation capability of a particular solvent for the given reference process. It should be emphasized that the polarity defined in this way refers to *microenvironment* around a solute molecule, and often the term *micropolarity* is used to indicate its difference from bulk polarity parameters. Several reference processes have been utilized to give various scales of solvent polarity; the first in this series is the Y-scale of Grunwald and Winstein [45] that used the solvolysis of t-butyl chloride as the model process. UV–visible spectroscopy provides a suitable model process for studying solvation. One may consider a spectroscopic transition in a solution to be represented by the following equation:

$$M_1(\text{Solv}) \pm h\nu \rightarrow M_2(\text{Solv}) \tag{7.2}$$

where M_1 and M_2 are the two different electronic states of a molecule and the positive and negative sign refer to absorption and emission, respectively. The effect of solvation is reflected in the spectroscopic behavior of the solute–solvent system. Interaction of different electronic states with the surrounding solvent molecules is also different, and this phenomenon leads to significant changes in the photophysical properties of the molecule. The observed spectral and photophysical parameters are thus characteristic of the molecule in its environment rather than the isolated one. Thus, a study of a spectral parameter is supposed to provide information regarding solvation interaction. It may be mentioned that while the initial state in a spectroscopic transition is an equilibrium state, the final state is a nonequilibrium Franck–Condon state. A differential solvation interaction between the two electronic states is responsible for the solvent sensitivity of a spectral parameter. In an absorption process, the observed parameter represents the differential solvation interaction of the solute molecule in the S_0 and S_1 state with an environment that is in equilibrium with the S_0 state. On the other hand, the observed emission parameter describes the microenvironment that is in equilibrium with the S_1 state. Moreover, the differential interaction will be large when the electronic transition induces a substantial change in the electronic distribution in the molecule. Thus, the optical response of molecules showing a charge transfer (CT) has been found to depend significantly on solvation interaction. As such, these molecules act as *micropolarity reporters* probing the local environment in a solution. Thus, a study of the parameters associated with the preceding process (e.g., band maximum and bandwidth in the case of absorption; band maximum, bandwidth, quantum yield, lifetime, and anisotropy in the case of fluorescence) provides information regarding solvation interaction. Spectroscopic parameters representing solvent polarity have thus been derived from the solvent-sensitive property of standard solutes (*reporters/probes*) absorbing light in the spectral region corresponding to UV–visible, IR, ESR, and NMR spectra. Most commonly, the parameters are wavelength (frequency) maximum

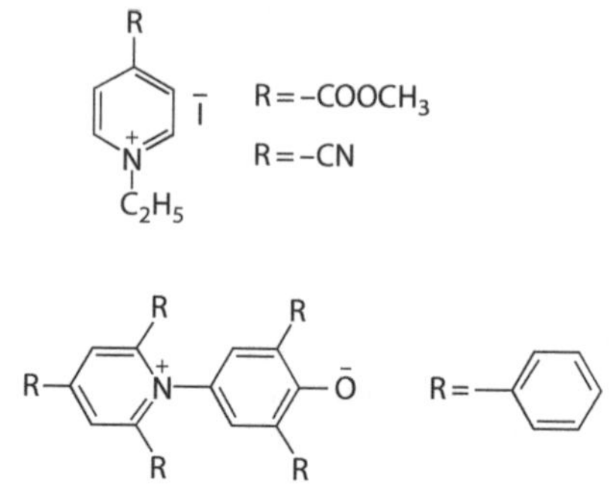

FIGURE 7.2 Some polarity probes (absorption probes).

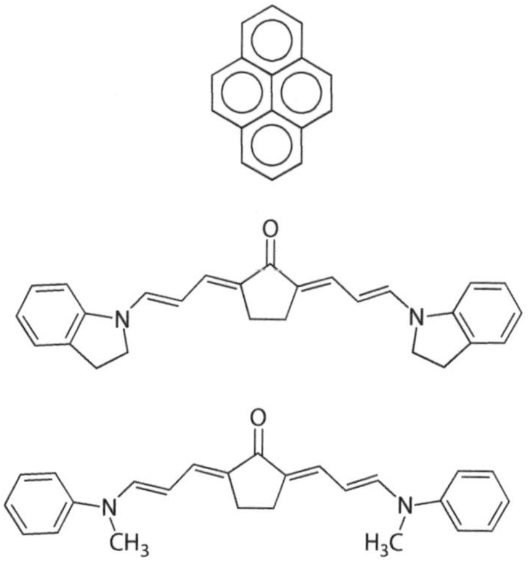

FIGURE 7.3 Some polarity probes (fluorescence probes).

of absorption/fluorescence band, intensity of band, etc. Sometimes the ratio of intensity of various spectroscopic bands has been used (e.g., I_1/I_3 ratio for pyrene emission). In general, several modes of solute–solvent interactions have been identified, namely, nonspecific (dipolarity and polarizability) and specific (acidity, basicity) modes. If one carefully selects an appropriate, sufficiently solvent-sensitive reference process, one may assume that this process would reflect faithfully all possible solute–solvent interactions. It should, therefore, give an empirical measure of the solvation capability of a particular solvent for the given reference process. Several compounds having solvent-sensitive absorption/emission bands have been utilized as *micropolarity reporters*. Probes may be classified as absorption probes or fluorescence probes depending on whether absorption or fluorescence spectral characteristics are utilized. A list of some important reporter molecules (probes) is given in Figures 7.2 and 7.3. Two main approaches for the description of micropolarity have been discussed here.

7.3.1 Single Parameter Approach

7.3.1.1 *Z*-Scale

In this scale, established by Kosower [9], 1-ethyl-4-carbomethoxypyridinium iodide (Figure 7.2, Ia) was chosen as the indicator solute. The position of the band shifts to blue as the solvating ability of the solvent increases. The Z-value was defined as the maximum transition energy in kilo calories per mole corresponding to the longest wavelength CT band of this solute. Thus we have

$$Z\left(\text{kcal mol}^{-1}\right)=\frac{28{,}590}{\lambda_{max}\left(\text{nm}\right)} \tag{7.3}$$

1-Ethyl-4-cyanopyridinium iodide (Figure 7.2, Ib) has also been used to establish a parallel scale [46].

7.3.1.2 $E_T(30)$-Scale

The indicator solute used by Kosower for the Z-scale is not quite soluble in certain nonpolar solvents. To overcome this difficulty, Dimroth and Reichardt suggested the use of pyridinium-N-phenol betaines as the indicator solute [10]. The $E_T(30)$ value of a solvent is the maximum transition energy in kilo calories per mole of the longest wavelength band in that solvent of the solute shown in Figure 7.2, II. The $E_T(30)$ value is obtained from λ_{max} values using an equation similar to Equation 7.3. The $E_T(30)$ values of various neat solvents have been listed in Table 7.1. Owing to the large displacement of the solvatochromic absorption band, the $E_T(30)$ values provide an

TABLE 7.1
List of Values of the Parameters $E_T(30)$, α, β, π^*

Sl. No.	Solvent	$E_T(30)^a$ (k cal mol^{-1})	α^b	β^b	π^{*b}
1	Water	63.1	1.17	0.47	1.09
2	Methanol	56.3	0.93	0.66	0.60
3	Ethanol	51.9	0.83	0.75	0.60
4	1-Propanol	50.7	0.78	0.80	0.52
5	2-Propanol	48.6	0.76	0.84	0.48
6	1-Butanol	50.2	0.79	0.82	0.47
7	1-Hexanol	48.8	—	—	—
8	1-Octanol	48.5	—	—	—
9	Cyclohexanol	46.9	—	—	—
10	Acetone	42.2	0.08	0.48	0.71
11	2-Pentanone	41.1	—	0.50	—
12	3-Pentanone	39.3	—	0.45	0.72
13	Acetonitrile	46.0	0.19	0.40	0.75
14	Butyronitrile	42.5	0.00	0.44	0.71
15	Chloroform	39.1	0.44	0.00	0.58
16	Dichloromethane	41.1	0.30	0.00	0.82
17	Formamide	55.8	0.71	0.70	0.97
18	Dimethylformamide	43.8	0.00	0.69	0.88
19	Dimethylsulfoxide	45.1	0.00	0.76	1.00
20	Dioxane	36.0	0.00	0.37	0.55
21	Tetrahydrofuran	37.4	0.00	0.55	0.58
22	Ethylacetate	38.1	0.00	0.45	0.55
23	Benzene	34.5	0.00	0.10	0.59
24	Toluene	33.9	0.00	0.11	0.54
25	Hexane	31.0	0.00	0.00	−0.08
26	Cyclohexane	31.2	0.00	0.00	0.00
27	Heptane	31.1	0.00	0.00	−0.08

[a] Refs. [8,10].
[b] Kamlet–Taft solvatochromic parameters, Refs. [45,67,70].

excellent and very sensitive characterization of the polarity of the solvents [4,8]. The compound can also act as polarity probe for the investigation of aqueous micellar solutions, microemulsions, and phospholipid bilayers as well as for investigation of interfacial properties of lipid membranes. The disadvantage of $E_T(30)$ probe is, however, that it has a low molar absorptivity and it is generally not fluorescent except under certain circumstances (dye embedded in polymer film or at a low temperature [47]). However, intense fluorescence is essential when probes are used in biological systems since they are mostly applied in highly diluted conditions. An important and widely used scale of solvent polarity has been proposed using a fluorescence probe *pyrene* (Figure 7.3, III). It has been known that the ratio of intensities of the first to third vibronic bands (I_1/I_3) of the $S_1 \rightarrow S_0$ emission of pyrene depends on the medium, and as such I_1/I_3 is often used as a measure of solvent polarity [48–54]. The ratio I_1/I_3 ranges from ~0.6 in hydrocarbon media to ~1.67 in water. A number of other fluorescence probes are known. Kessler and Wolfbeis introduced a group of ketocyanine dyes (Figure 7.3, IVa and IVb) that act as good solvent-sensitive fluorophores [55,56]. Depending on the solvent, these probes display a bright yellow to purple violet color and a strong green, yellow, and red fluorescence. Both the absorption and emission maxima are red shifted in going from nonpolar to polar solvent.

Polarity probes in general should exhibit the general features mentioned in the following text [56,57]: (1) High spectral sensitivity to polarity changes (a spectral shift of both excitation and emission maxima is advantageous because of additional sensitivity and selectivity), (2) high fluorescence quantum yield, (3) long wavelength absorption and fluorescence so that they can be conveniently used in biological systems, and (4) photostability and thermal stability. Although a large number of polarity probes have been reported, none of them appreciably satisfies all the four criteria. It may be mentioned that the chemical structure of an ideal indicator compound should be such that all important nonspecific (dipolar, dispersion) and specific (HBD, hydrogen bond acceptance [HBA], and electron donor–acceptor) solute–solvent interactions are possible. Considering the wide variation in the chemical structure of the probe molecules used for characterization of solvent polarity, the establishment of a universal, generally valid solvent polarity scale seems unattainable.

7.3.2 Multiparameter Approach

The use of a single parameter (e.g., *E*(*A*) or *E*(*F*) or any other photophysical parameter) to describe solvent polarity is based on the assumption that it is necessary to take only one mechanism of solute–solvent interaction into account. The inadequacy of the dielectric model of solvent to represent the solvent effect on the various properties of solutes and proliferation of empirical polarity scales point to the existence of specific solute–solvent interaction. According to Equation 7.1 any solvent-dependent property (*A*) of a solute "*S*" in a solvent "*i*" can be represented as

$$A(S,i) = A(S) + B(S,i) \tag{7.4}$$

where *A*(*S*) is a solvent-independent part. The term *B*(*S*,*i*) is in general a complex function of both solvent and solute comprising several modes of solute–solvent interaction. Experimental data indicate that solvent effect on various kinetic, equilibrium, and spectroscopic parameters is essentially similar in their very nature. The similarity is considered to show that there are comparatively few mechanisms of physical interaction between solvent and solute [6,58–61]. In essence there are three types of interactions: (1) nonspecific long-range solvent–solute interaction, (2) specific short-range solute–solvent interaction, and (3) solvent–solvent interaction from the cavity effect. The most effective nonspecific interactions are considered to be determined by macroscopic physical parameters of the solvent, that is, the relative permittivity or dielectric constant (ε) and refractive index (*n*). The specific solvation is mainly determined by the acidity and basicity of the solvent, in terms of the Lewis concept, which are measures of the solvent hydrogen bond ability to donate (*HBD*) and to accept (*HBA*) a proton, respectively. Disruption and reorganization of solvent–solvent interaction are measured by the work necessary to separate solvent molecules to create a suitable cavity, large enough to accommodate the solute.

It has been shown that under certain simplifying assumptions, the solute–solvent interaction term $B(S,i)$ may be factorized as [62]

$$B(S,i) = \sum a_\alpha(S)P_\alpha(i) \tag{7.5}$$

where suffix α represents various modes of interaction, the parameters $a_\alpha(S)$ and $P_\alpha(i)$ depending on the solute and solvent, respectively. This type of expression, often called *LSER*, has been found to be of great significance in the theory of solvent effect [11,62,63]. In general terms, *LSER* [8,64] is expressed by the following equation:

$$XYZ = XYZ_0 + \text{Cavity formation energy} + \sum \text{Solute} - \text{solvent interaction energy} \tag{7.6}$$

In the preceding equation, *XYZ* is a property linearly related to the free energy of the system. The term XYZ_0 denotes a constant and depends solely on the solute. Summation in the preceding equation extends over all the possible modes of solute–solvent interaction [65,66]. Two approaches in the use of the *LSER* equation may be distinguished, namely, the approach suggested by Koppel and Palm (*KP*) [58] and that by Abraham, Kamlet, and Taft (*AKT*) [11]. In the *KP* approach, functions of dielectric constant (ε) and refractive index (n) were used to describe the nonspecific interaction. Thus Onsager reaction field parameter $(\varepsilon - 1)/(2\varepsilon + 1)$ was used to describe the nonspecific dipolar interaction, while $(n^2 - 1)/(n^2 + 2)$ described the polarizability term. In the *AKT* approach, the dipolarity and polarizability were described by the experimentally determined parameter π^* [62]. The specific interactions were described by the parameter "*E*" (electrophilic solvation ability) and "*B*" (nucleophilic solvation ability) in the *KP* procedure [58]. But *AKT* have preferred the use of *HBD* and *HBA* ability of solvent represented by the empirical parameter α and β, respectively [67]. The endothermic cavity formation term was taken in the *AKT* approach as equal to the solute molar volume times the Hildebrand cohesive energy density (δ_H^2) defined as the enthalpy of vaporization per unit volume [68]. The original *KP* approach did not take this factor into account although this term was introduced later by Makitra and Pirig [69]. Thus, the *KP* equation is

$$A = A_0 + yY + pP + eE + bB \tag{7.7a}$$

$$Y = \frac{2(\varepsilon - 1)}{(2\varepsilon + 1)}; \quad P = \frac{(n^2 - 1)}{(n^2 + 2)} \tag{7.7b}$$

while the *AKT* equation is

$$A = A_0 + \rho\pi^* + a\alpha + b\beta + m\delta_H^2 \tag{7.8a}$$

$$\delta_H^2 = \frac{(\Delta H - RT)}{V} \tag{7.8b}$$

where ΔH is the molar enthalpy of vaporization of the solvent whose molar volume is *V*. It has been found that the various empirical polarity parameters, for example, $E_T(30)$ and *Z*, depend linearly on α-, β-, and π^*-values [47,70]. The Kamlet–Taft solvatochromic parameters π^*, α, and β are usually determined by measuring the maximum transition energy of certain probes using the solvatochromic comparison method [71–73]. For the determination of π^*, the commonly used probes are *4-nitroanisole*, *2-nitroanisole*, *4-ethyl-nitrobenzene*, and *N,N-diethyl-4-nitroaniline*. For these probes, the energy of maximum absorption depends only on the *dipolarity–polarizability* modes of interaction with solvent. For setting up the β scale, a spectroscopic or thermodynamic property of two indicator solutes of differing H-bonding ability (e.g., *4-nitroaniline* and *N,N-diethyl-4-nitroaniline*) in a series of solvents was compared. The value of enhanced solvatochromic displacement normalized with respect to a reference solvent hexamethylphosphoramide (HMPA) gave a dimensionless parameter. To eliminate the specific role of solute, the procedure was repeated for multiple indicators and the average of the dimensionless parameter gave β. A similar procedure using multiple linear regression analysis (MLRA) was adopted for setting up the α-scale. Here due to various complications, the requirements for the indicator solute were considered to be relatively stringent. Sixteen diverse properties involving 13 indicators were used to calculate the final α-values [62]. Table 7.1 lists the relevant polarity parameters of some pure solvent. In fact, the *LSER* equations with these parameters have been used in hundreds of studies involving the effect of solvent on kinetic spectroscopic and equilibrium properties of a solute. Catalan has introduced the parameters *SPP*, *SA*, and *SB* for representing the solvent *dipolarity–polarizability* (nonspecific mode), solvent acidity, and solvent basicity (the specific modes) of a solvent. These parameters are obtained from spectroscopic measurements and their success in accounting for medium effects on reactivity and a host of physiochemical properties is well documented and widely perceived [74–77].

7.4 SOLVATION IN MICROHETEROGENEOUS MEDIA

Micellar phases provide unique environments, and as such they are used in various areas of chemistry. They are used as media for enhancing the solubility of hydrophobic drug molecules [78,79] and as catalysts [80] and detergents [22]. They are also used to alter the selectivity of separation in liquid chromatography and micellar electrokinetic chromatography (MEKC) [81–84]. The utility of micelles in these and other applications depends largely on the interaction of micelle with organic molecules (solutes). As discussed earlier, it is customary to represent the totality of interaction of solute with its environment in terms of polarity parameters. Spectroscopic procedures have been used extensively to characterize the polarity of microheterogeneous media. The use of an effective relative permittivity to describe the polarity of the location of the solute has often been done. The $E_T(30)$ probe (Figure 7.2, II) has also been extensively used to measure the polarity of such media [84]. As discussed earlier, the use of a single parameter to describe solvation has often been criticized. It is desirable to characterize solvation in terms of several independent modes of solute–solvent interaction. According to this procedure, any property of a solute in a medium is described in terms of linear dependence through *LSER* on the solvent parameters representing the independent modes of solute–solvent interaction (Equations 7.7 and 7.8). Three independent modes, namely, dipolarity/polarizability, *HBD*, and *HBA*, represented conveniently by the three parameters π^*, α, and β, respectively, have been recognized.

A problem arises in applying the preceding methodology to micellar solutions. A micellar medium is essentially microheterogeneous. The micellar pseudo-phase always remains in equilibrium with the aqueous phase containing surfactant monomers. The indicator molecules are distributed between the aqueous and micellar phases present in a micellar solution. Since the spectroscopic properties of an indicator depend on its immediate environment, the spectrum (absorption/fluorescence) of the indicator in the micellar phase will differ from that in the aqueous phase. The observed spectrum of an indicator in the solution is a concentration-weighted sum of the indicator's spectra in the aqueous and micellar phases, rendering the determination of the spectrum of the indicator in the micellar phase impossible. In order to obtain the value of a spectroscopic parameter (e.g., maximum energy of absorption/fluorescence, steady-state anisotropy, and lifetime) in the two phases, a knowledge of the distribution coefficient of the solute between the two phases is necessary. This introduces a complication in the data analysis, which is absent when using the Kamlet–Taft methodology to characterize bulk liquids [14].

7.4.1 Estimation of Micellar Phase Property Free from Aqueous Phase Contribution

Two procedures have been adopted to resolve the preceding problem. Vitha et al. have used a modified curve resolution method [14,15]. This method is based on singular value decomposition (SVD) of a matrix of spectra recorded at several concentrations of surfactant. The method assumes that only two components, namely, the indicator in the aqueous and micellar phase, contribute to the observed spectra. A knowledge of the molar volume of the surfactant and the aggregation number is also required. The curve resolution method gives three parameters: the distribution coefficient and the spectra of the indicator solute in the two phases. The value of the spectroscopic parameter of interest (e.g., the position of band maximum) was then determined from the spectrum in the micellar phase.

The method used by Fuguet et al. [16,17] and Shannigrahi and Bagchi [18] also assumed that the observed value of a spectroscopic parameter is a concentration-weighted sum of the values of the parameter in the two phases. The outline of their procedure is given in the following text. Micellization is essentially a dynamic process involving equilibrium between the surfactant molecules (*S*) in the aqueous phase and those in the micelles. For nonionic surfactants, the process can be represented by the following equation:

$$nS\ (\text{aqueous phase}) \rightleftharpoons Sn\ (\text{micellar phase}) \tag{7.9}$$

For ionic micelles, the cation or anion of the amphiphilic molecule remains in equilibrium between the two phases. An indicator solute distributes itself between the two phases. Any property (*P*) of the solute that has dependence on its local environment can be used to reveal the equilibrium process. Due to difference in polarity, the value of *P* is expected to be different in the two phases. The value of *P* for such systems will be determined by the time average location of the probe in the two phases [85]. The measured value of *P* in a micellar media can be assumed to be represented by a mole fraction average of the aqueous phase property (P_{aq}) and micellar phase property (P_m), as given by the following equation:

$$P = \frac{\left(n_{aq}P_{aq} + n_m P_m\right)}{\left(n_{aq} + n_m\right)} \tag{7.10}$$

where

n refers to the number of moles of the solute

subscripts *m* and *aq* denote the micellar phase and aqueous phase, respectively

The calculation of the value of P_m and P_{aq} requires knowledge of n_m and n_{aq}, which in turn are related to the *distribution coefficient* of the indicator solute between the two phases. It has been shown that the parameters are related as follows [17,18]:

$$P = \frac{\left[Kv C_s P_m + \left(1 - vC_s\right)P_{aq}\right]}{\left[KvC_s + \left(1 - vC_s\right)\right]} \tag{7.11}$$

where

v is the molar volume of the surfactant

C_s is the concentration of the surfactant in the aggregated phase, as given by $C_s = C_T - cmc$, where C_T is the total surfactant concentration

K is the molar-based distribution coefficient of the indicator between the two phases

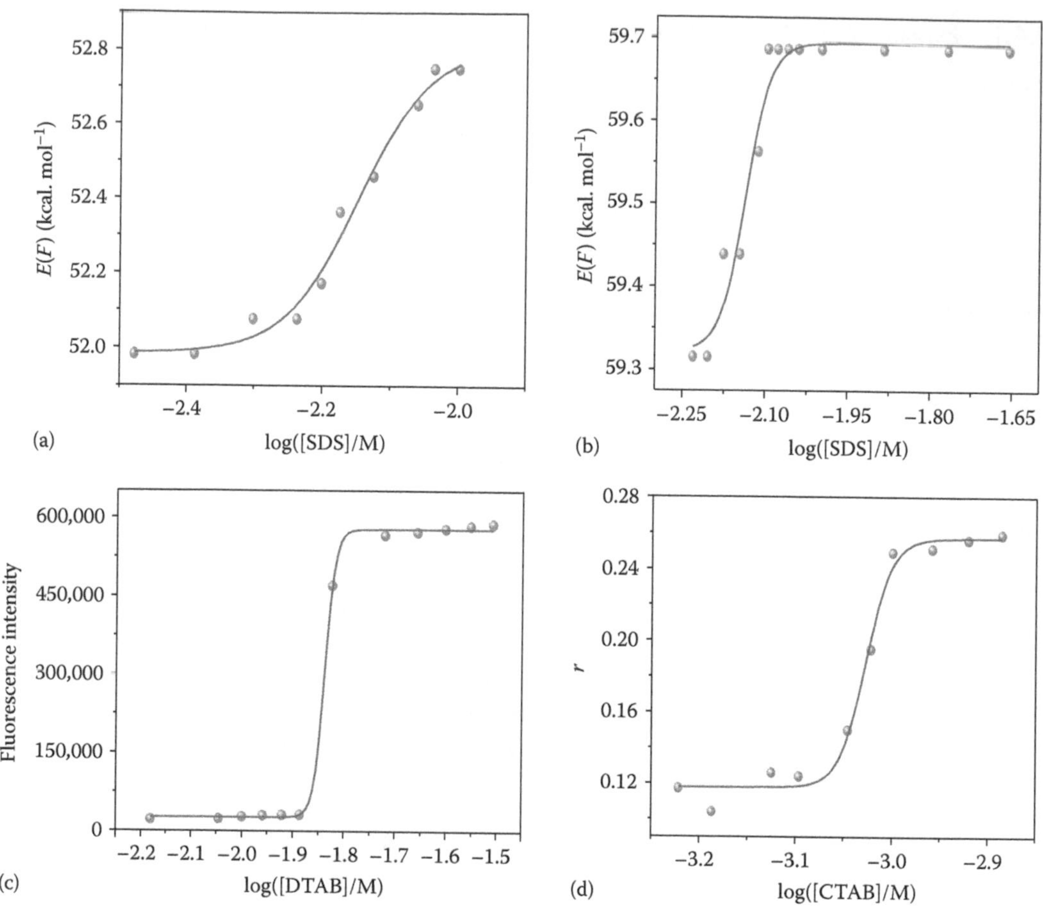

FIGURE 7.4 Plot of the parameters *energy of* (a) *maximum absorption*/ (b) *fluorescence*, (c) *fluorescence intensity* and (d) steady-state anisotropy (r) of (Va) in the aqueous solution of surfactants as a function of log C_T. C_T for the surfactants have been denoted by surfactant. (From Kedia, N. et al., *Spectrochim. Acta A*, 131, 398, 2014.)

The variation of these parameters as a function of log C_T can be studied. Figure 7.4 shows representative plots describing such variation. Values of these parameters change abruptly after a certain concentration. Data points can be fitted to Boltzmann sigmoid equation. It has been proposed that the value of *cmc* can be obtained from the inflexion point of the Boltzmann sigmoid curve obtained by plotting a solute property as a function of C_T [86,87]. Truly speaking, P is related to log C_s by Boltzmann equation as can be seen from Equation 7.11. Thus, the fitting of the experimental data point to the Boltzmann sigmoid plot involving the parameter P and log C_T is only approximate, and hence the assignment of concentration at the inflexion point as *cmc* is likely to be in error. Equation 7.11 can be used for the determination of the *cmc* value [88]. An iterative procedure can be adopted as follows. Equation 7.11 can be rearranged as

$$P = P_{aq} + (KP_m - P_{aq})v(C_T - cmc) + (1 - K)vP(C_T - cmc) \tag{7.12}$$

The value of concentration of surfactant at the inflexion point of P versus log C_T plot is taken as the guess value of *cmc*. The parameter P_{aq} is then determined by linear regression analysis using Equation 7.12. It follows from Equation 7.11 or 7.12 that $P = P_{aq}$ at $C_T = cmc$. Thus, the value of the concentration for which $P = P_a$ was found out from the P versus log C_T curve. This gives the output value of *cmc*. The preceding procedure can be repeated by using the output *cmc* value as the input to get a new output *cmc*. The iteration can be continued until the input and output *cmc* values are within a specified range (e.g., within ±2%). Once *cmc* is determined, values of P_m, P_{aq}, and K can be found out by a linear regression analysis using Equation 7.12. The procedure also requires the value of molar volume of the surfactant. It is apparent from Equation 7.9 that values of parameters in micellar solution, as found out from the iterative procedure, are close to the value of the parameter at a concentration of surfactants much above the *cmc* value. But in view of the possibility of a change in the shape of the micelle, the use of too high a concentration is not advisable.

7.4.2 Determination of Polarity Parameters

The Kamlet–Taft version of *LSER* is expressed by the following equation:

$$A = A_0 + m\delta_H^2 + p\pi^* + a\alpha + b\beta \tag{7.8a}$$

In this equation, the cavity term is written as a product of the coefficient m (depending on the molar volume of the solute) and the Hildebrand solubility parameter, δ_H^2, of the solvent [68], and the solute–solvent interaction is written in terms of nonspecific (π^*) and specific (α and β) interaction parameters. Since entropy does not change during a spectroscopic transition [89], the transition energy can be regarded as related to a change in free energy. Again, during a spectroscopic transition, the volume of the solute remains constant and as a result the cavity term drops out [12]. Thus, one can write the maximum energy of transition (E) of the solute (S)–solvent (i) system as follows:

$$E(S,i) = E_0(S) + p(S)\pi^*(i) + a(S)\alpha(i) + b(S)\beta(i) \tag{7.13}$$

The Kamlet–Taft solvatochromic parameters π^*, α, and β for a micellar medium are usually determined by measuring the spectroscopic parameters of certain probes using known correlation equations. The correlation equations are, however, obtained from studies in pure solvents. It is assumed that the correlation equation obtained for pure solvents is also valid for micellar media. Both absorption [14–17] and fluorescence [18,90] probes have been used for estimating values of α, β, and π^* for the micellar media.

7.4.2.1 Method Using Absorption Probes

For the determination of π^*, the commonly used probes are *4-nitroanisole*, *2-nitroanisole*, *4-ethyl-nitrobenzene*, and *N,N-diethyl-4-nitroaniline*. For these probes, the energy of maximum absorption depends only on the *dipolarity–polarizability* modes of interaction of solvent. Expressing the wavenumbers of transition, σ, in kilokaisers (1 kK = 1000 cm^{-1}), the following relations are known [17]:

$$\text{For 4-nitroanisole,} \quad \pi^* = \frac{(34.12 - \sigma)}{2.343} \tag{7.14}$$

$$\text{For 2-nitroanisole,} \quad \pi^* = \frac{(32.56 - \sigma)}{2.428} \tag{7.15}$$

$$\text{For 2-ethyl-nitrobenzene,} \quad \pi^* = \frac{(37.67 - \sigma)}{2.259} \tag{7.16}$$

$$\text{For } N,N\text{-diethyl-4-nitroaniline,} \quad \pi^* = \frac{(27.52 - \sigma)}{3.182} \tag{7.17}$$

For determining β-values, the commonly used probes are 4-*nitrophenol* and 4-*nitroaniline*. The transition energy of these probes depends on both β and π*. For example, for 4-*nitroaniline* the wavenumber of transition is related to β as follows [17]:

$$\beta = \frac{(31.10 - 3.14\pi^* - \sigma)}{2.79} \tag{7.18}$$

Similarly the α-values are determined by using $E_T(30)$, $E_T(33)$, and $Fe(LL)_2(CN)_2$ as α and π*-probes (transition energy depends on α- and π*-values) [91–93]. The $E_T(33)$ probe is a bis chlorosubstituted derivative of $E_T(30)$ probe. It has the advantage over the $E_T(30)$ probe because of its water solubility.

Thus, relations (7.19) and (7.20) are valid for Dimroth–Reichardt $E_T(30)$ and $E_T(33)$–betaine, respectively:

$$\alpha = 0.197\sigma - 2.091 - 0.899[\pi^* - 0.211\delta] - 0.148\beta \tag{7.19}$$

$$\alpha = 0.203\sigma - 2.809 - 0.899[\pi^* - 0.211\delta] - 0.148\beta \tag{7.20}$$

In the preceding equations, δ is a correction factor to account for the polarizability of the medium [4,67]. Its value is zero for nonaromatic and nonchlorinated pure organic solvents. For the *Fe complex*, the transition energy (*E*) in kcal mol^{-1} is given by the following equation:

$$E = 39.71 + 3.31\pi^* + 4.50\alpha \tag{7.21}$$

The value of α can be obtained using a known value of π* determined by using other probes. This relies on the assumption that these α-probes reside in the same region in the micellar phase as do the π*-probes.

7.4.2.2 Method Using Fluorescence Probes

Dong and Winnik [54] have found that the I_1/I_3 ratio of pyrene fluorescence for homogeneous media is correlated with the π*-values of a series of protic solvents as

$$\left(\frac{I_1}{I_3}\right) = 1.304\pi^* + 0.46 \tag{7.22}$$

This equation can be used to estimate π* for heterogeneous media and the values agree well with those obtained by using other fluorescence probes [94]. The preceding empirical relation has often been used to calculate π* employing the micellar phase value of I_1/I_3 for mixed micelle of varying composition [90].

It is known that the position of the maximum fluorescence of ketocyanine dyes in a solution is highly solvent sensitive and the energy of maximum fluorescence, *E*(*F*), is linearly dependent on the solvatochromic parameters, α and π*. The dyes are solubilized in the micelle–water interface [95]. It is believed that the polarity characteristics of the micelle–water interfacial region resemble those of aqueous alkanols [49]. The values of *E*(*F*) of ketocyanine dyes have been determined in various pure *n*-alkanols and mixed binary aqueous alkanol systems for which the values of solvatochromic parameters (α and π*) are known [96]. An MLRA of *E*(*F*) with the

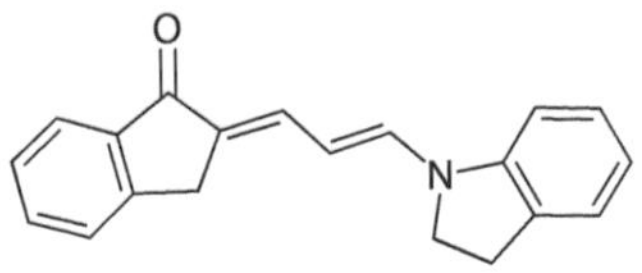

FIGURE 7.5 Fluorescence probes.

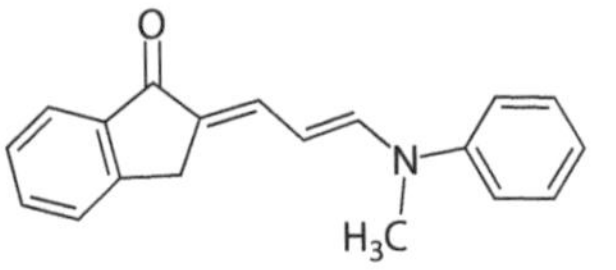

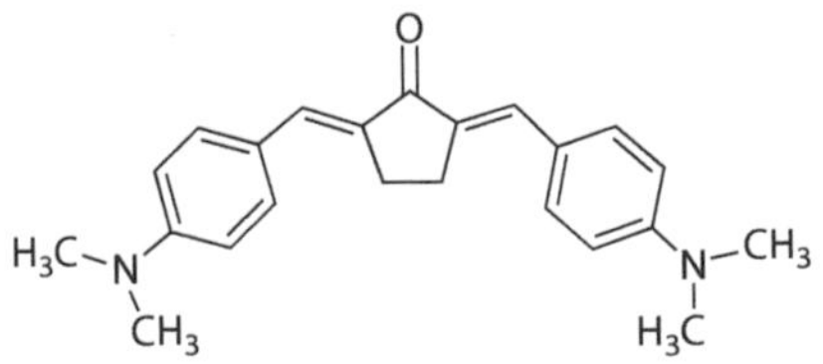

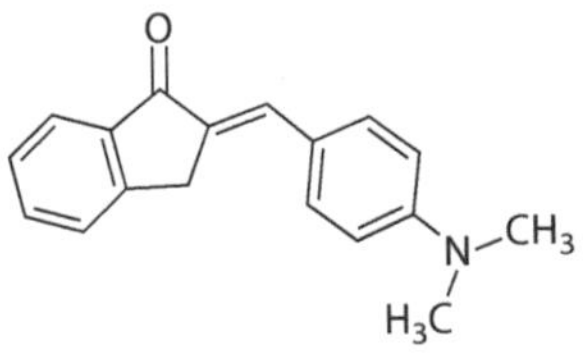

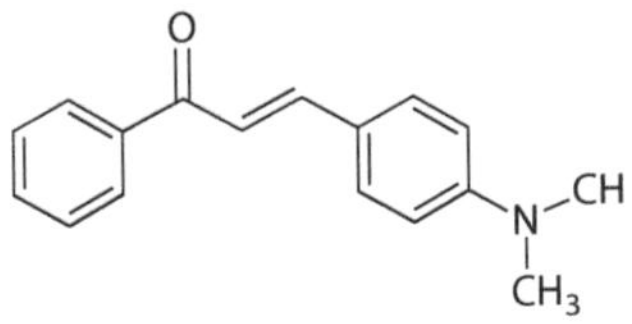

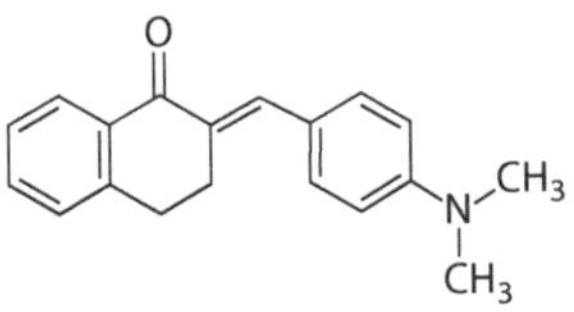

solvatochromic parameters has been performed. It has been found that the $E(F)$ of the ketocyanine dyes, (Va) and (Vb) (Figure 7.5), correlates well with the α- and π*-parameters for a series of pure alkanols and water–alkanol mixtures. Thus, Equations 7.23 and 7.24 are found to be obeyed for dyes (Va) and (Vb), respectively:

$$E(F)\ (\text{kcal mol}^{-1}) = 58.65 - 4.05\alpha - 1.71\pi^* \tag{7.23}$$

$$E(F)\ (\text{kcal mol}^{-1}) = 59.80 - 3.60\alpha - 0.60\pi^* \tag{7.24}$$

One can assume that these relationships are also valid for heterogeneous media. The value of α for the micellar phase can be obtained using Equations 7.22 and 7.23 from the known values of $E(F)$ and π* for the micellar phase.

The coefficients $p(s)$, $a(s)$, and $b(s)$ in Equation 7.13 depend on the solute and are related, respectively, to dipolarity/polarizability and the *HBA* and *HBD* ability of the solute [12]. Equation 7.13 can be used to treat the solvatochromic transition energy (E) for a *fixed solute* for a series of solvents whose π*-, α-, and β-values are known. MLRA of E with π*, α, and β will then provide the p, a, and b terms. This procedure is used to get the *LSER* equation for a particular solute [8]. Conversely, the equation can be used to analyze the transition energy of a series of solutes of known p, a, and b values in a fixed solvent (medium) to get the π*-, α-, and β-values of the medium. Ketocyanine dyes provide an interesting series of solutes, which are characterized by solvatochromic fluorescence [55]. The spectroscopic transition leading to the fluorescence in these molecules has been established as due to intramolecular charge transfer (ICT) involving the carbonyl oxygen and nitrogen atom of the amino group [55,97–99]. Shannigrahi et al. studied the solvatochromic fluorescence energy of the ketocyanine (Figure 7.3, IVa and IVb) and other structurally related dyes (Figure 7.5, Va, Vb, VI, VII, VIII) extensively in various pure solvents [97–99]. MLRA of maximum fluorescence energy in pure solvents has been done with solvent parameters π*, α, and β to get p, a, and b coefficients for the solutes. MLRA of the energy of maximum fluorescence, $E(F)$, of different solutes in the micellar phase has been sought with the known solute parameters p, a, and b to get the values of π*, α, and β for the micellar phase [18].

For a particular micelle, (m), the energy of maximum transition for a particular solute, (s), is given by the following equation:

$$\left[E_m(s)-E_0(s)\right]=a(s)\alpha_m+b(s)\beta_m+p(s)\pi_m^* \tag{7.25}$$

The values of α_m, β_m, and π_m^* that give the best correlation in the least-squares sense of $[E_m(s) - E_0(s)]$ with $a(s)$, $b(s)$, and $p(s)$ for the seven solutes were then determined. Alternatively, one has for the micellar and aqueous phase:

$$\left[E_m(s)-E_{aq}(s)\right]=a(s)\left[\alpha_m-\alpha_{aq}\right]+b(s)\left[\beta_m-\beta_{aq}\right]+p(s)\left[\pi_m^*-\pi_{aq}^*\right] \tag{7.26}$$

A correlation analysis involving $[E_m(s) - E_{aq}(s)]$ in place of $[E_m(s) - E_0(s)]$ provides $\Delta\alpha = [\alpha_m - \alpha_{aq}]$, $\Delta\beta = [\beta_m - \beta_{aq}]$, and $\Delta\pi^* = [\pi_m^* - \pi_{aq}^*]$, that is, the change of the parameters as one goes from an aqueous phase to a micellar phase.

7.4.3 Polarity of the Micellar Phase

Values of the solvatochromic parameters for micellar phases formed by cationic, anionic, and neutral surfactants, as obtained by different workers, have been listed in Table 7.2.

Most of the indicator solutes used in the literature are polar and they probe the polar region of a micelle. It has been found that the π^*-, α-, and β-values for the aqueous phase, as determined from the value of the parameter in the aqueous phase (P_{aq} in Equation 7.12), are almost equal to the values in pure water. Thus, the addition of surfactant molecules at concentrations below *cmc* does not have any significant influence on the solvation properties of water. On the contrary, values of the solvatochromic parameters in the micellar phase differ significantly from those in the aqueous phase. It appears that a micellar phase is characterized by a high polarity. This is rationalizable in view of the fact that the probes reside in the polar part of the micelle and report the values accordingly.

It appears that the *HBD* strength of the micellar region is, in general, less than that of the aqueous phase. The change from the aqueous phase follows the order CTAB > SDS = TX100. Note that the anionic micelle formed by SDS has better *HBD* strength than the other micelles. Thus, the values of α for the micellar phase of SDS and DTAB as reported by Vitha et al. are 0.873 and 0.700,

TABLE 7.2
Solvatochromic Parameters for Various Micellar Phases

Micellar Phase	α	β	π^*
SDS	1.14[a], 1.10[b], 0.873[c], 1.18[f], 1.2[g]	0.70[a], 0.42[b], 0.401[d]	0.52[a,f], 1.14[b], 1.06[d], 0.58[g]
CTAB	0.87[a], 0.62[c], 0.94[g]	0.53[a], 0.47[b]	0.68[a], 0.96[c], 0.69[g]
DTAB	0.700[e], 1.03[g]	0.486[e]	1.02[a], 0.70[g]
TX100	1.06[a], 0.92[f]	0.71[a]	0.56[a], 0.72[f]
Water	1.17	0.47	1.09

[a] Ref. [18].
[b] Ref. [17].
[c] Ref. [16].
[d] Ref. [14].
[e] Ref. [15].
[f] Ref. [90].
[g] Ref. [88].

respectively [14,15]. Similarly, Fuguet et al. reported $\alpha = 0.82$ and 0.62 for SDS and CTAB, respectively [16,17]. The values of β indicate that ionic micelles have relatively more *HBA* basic than the aqueous phase. The neutral TX100 micelle on the other hand has *HBA* ability comparable to that of the aqueous phase. π^*-values for the micellar phase of the surfactant are less than that of the aqueous phase for all the micelles. The micellar phase consists of a hydrocarbon core and a polar region. The polar head groups, the bound counterions (in the case of ionic micelles), and water molecules bound to the head groups mainly constitute the polar region. The polarity monitored by the dye molecules used in the present study corresponds to this region of the micelles. The solvation properties in the micellar phase arise partly due to the presence of bound water molecules and counterions. The head group in an SDS micelle ($-O-SO_3^-$) can bind more water molecules than a head group in CTAB ($-N(CH_3)_3^+$). In the former case, the negative charge is distributed over the oxygen atoms, while in the case of ($-N(CH_3)_3^+$), the intervening methyl groups restrict the water molecules from coming closer to the head group. Moreover, the counterion (Na^+) in the case of SDS can bind more water molecules by solvation than the corresponding counterion (Br^-) in CTAB. The greater *HBD* ability of the anionic SDS micelle relative to the cationic CTAB micelle can be rationalized due to the presence of a larger number of bound water molecules in the Stern layer of the micelle. On the other hand, the TX100 monomer surfactant molecule possesses some *HBD* strength due to the presence of –OH groups. Thus, a higher value of R is expected for TX100 micelles. The higher *HBA* ability of SDS relative to CTAB micelles can also be rationalized in terms of the presence of water molecules as explained earlier. Moreover, in the case of SDS, the presence of ($-O-SO_3^-$) head groups adds to the *HBA* basicity of the micelle formed. The head group of the neutral TX100 surfactant contains a phenyl group followed by a polyoxyethylene chain terminating in a hydroxyl group. Thus, the polar region is expected to resemble an alcohol environment. Also, the determined solvatochromic parameters are close to that of ethanol ($\alpha = 0.86$, $\beta = 0.75$, $\pi^* = 0.54$), which closely resembles the chain end of the surfactant.

7.5 CONCLUDING REMARKS

The difficulty with the use of absorption probes is that a higher probe concentration, as required for some of the probes for the determination of maximum absorption, can lead to a change in micellar structure. The use of fluorescent probes is thus better. Moreover, the use of different indicator solutes for determining different parameters rests on the assumption that all indicator molecules probe the same region of the micelle. It is instructive to determine the parameters using different absorption/fluorescence probes and to examine whether convergent results are obtained. A recent study by Kedia et al. indicates that convergent values of a parameter are obtained when different probes are used. Moreover, for the same probe the use of different spectroscopic properties gives almost the same result. For example, ketocyanine dyes have solvent-sensitive absorption and fluorescence bands. The analysis of maximum energy of absorption or fluorescence gives similar values of a solvatochromic parameter [88]. It has been observed that the probe distributes itself in two different regions of a catanionic vesicle formed by DTAB and SDS, and the solvatochromic parameters (α and π^*) of the regions have been estimated. It is important to mention that a spectroscopic procedure involving indicator solutes can provide only the solvatochromic parameters (α, β, and π^*). No information can be obtained regarding the δ_H^2 parameter. To determine this parameter, one has to study a thermodynamic property. Maitra et al. [100] have determined the δ_H^2 value of SDS micellar phase from solubility studies using a ketocyanine dye (Figure 7.5, Vb).

ACKNOWLEDGMENTS

The author thanks coworkers Dr. Mrinmoy Shannigrahi, Dr. Angahuman Maitra, Dr. Nipamanjari Deb, Dr. Niraja Kedia, and Dr. Amrita Sarkar. The author is also grateful to Dr. Sanjib Sardar and Dr. Amrita Sarkar for their sincere effort in preparing the manuscript.

REFERENCES

1. E. M. Kosower, *An Introduction to Physical Organic Chemistry*, Wiley, New York (1968).
2. R. R. Doganadze, E. Kalman, A. A. Kornyshev, J. Ulstrup, eds., *The Chemical Physics of Solvation*, Elsevier, Amsterdam, the Netherlands (1985).
3. A. Eilmes, Ab initio molecular dynamics simulations of ketocyanine dyes in organic solvents. In: Bubak, M., Szepieniec, T., Wiatr, K. eds., *Building a National Distributed e-Infrastructure—PL-Grid*, Springer, Berlin, Germany (2012), p. 276.
4. C. Reichardt, T. Welton, *Solvents and Solvent Effect in Organic Chemistry*, 4th edn., Wiley-VCH, Weinheim, Germany (2011).
5. I. A. Koppel, V. A. Palm. In: N. B. Chapman, J. Shorter, eds., *Advances in Linear Free Energy Relationships*, Chapter 5: Influence of Solvents on Chemical Reactivity, Plenum Press, London, U.K. (1972).
6. R. M. C. Goncalves, A. M. N. Simoes, L. M. P. C. Albuquerque, *J. Chem. Soc., Perkin Trans.* 2 (1990) 1379.
7. G. Rauhut, T. Clark, T. Steinke, *J. Am. Chem. Soc.* 115 (1993) 9174.
8. C. Reichardt, *Chem. Rev.* 94 (1994) 2319.
9. E. M. Kosower, *J. Am. Chem. Soc.* 80 (1958) 3253.
10. K. Dimroth, C. Reichardt, T. Siepmann, F. Bohlmann, *Liebigs Ann. Chem.* 661 (1963) 1.
11. M. H. Abraham, P. L. Grellier, J. L. M. Abboud, R. M. Doherty, R. W. Taft, *Can. J. Chem.* 66 (1988) 2673.
12. R. W. Taft, J. L. M. Abboud, M. J. Kamlet, M. H. Abraham, *J. Soln. Chem.* 14 (1985) 153.
13. C. O. Rangel-Yagui, A. Pessoa, Jr., L. C. Tavares, *J. Pharm. Pharm. Sci.* 8(2) (2005) 147.
14. M. F. Vitha, J. D. Weckwerth, K. Odland, V. Dema, P. W. Carr, *J. Phys. Chem.* 100 (1996) 18823.
15. M. F. Vitha, P. W. Carr, *J. Phys. Chem. B* 102 (1998) 1888.
16. E. Fuguet, C. Rafols, M. Roses, *Langmuir* 19 (2003) 6685.
17. E. Fuguet, C. Rafols, E. Bosch, M. Roses, *Langmuir* 19 (2003) 55.
18. M. Shannigrahi, S. Bagchi, *J. Phys. Chem. B* 108 (2004) 17703.
19. D. Berillo et al., *J. Colloid Interface Sci.* 368 (2012) 226.
20. Y. Moroi, *Micelles: Theoretical and Applied Aspects*, Plenum, New York (1992).
21. M. M. Breuer, I. D. Robb, *Chem. Ind.* 13 (1972) 530.
22. M. J. Rosen, *Surfactants and Interfacial Phenomena*, Wiley Interscience, New York (1978).
23. D. Attwood, A. T. Florence, *Surfactant Systems*, Chapman & Hall, New York (1983).
24. E. G. Cockbain, *Trans. Faraday Soc.* 49 (1953) 104.
25. M. J. Schwuger, *J. Colloid Interface Sci.* 43 (1973) 491.
26. E. D. Goddard, K. P. Anantapadmanabhan, *Application of Polymer Surfactant Systems*, Marcel Dekker, New York (1998).
27. E. D. Goddard, *Colloids Surf.* 19 (1986) 255.
28. M. N. Jones, *Chem. Soc. Rev.* 21 (1992) 127.
29. E. Dickinson. In: E. D. Goddard, K. P. Anantapadmanabhan, eds., *Interactions of Surfactants with Polymers and Proteins*, CRC Press, Boca Raton, FL, Chapter 7: Proteins in solution and interfaces (1993).
30. (a) N. J. Turro, A. Yekta, *J. Am. Chem. Soc.* 100 (1978) 5951; (b) D. Varadea, T. Joshi, V. K. Aswal, P. S. Goyal, P. A. Hassand, P. Bahadur, *Colloid Surf. A: Physicochem. Eng. Aspects* 259 (2005) 95.
31. J. W. McBain, *Colloid Science*, D. C. Heath and Co., Boston, MA (1950).
32. G. S. Hartley, *Aqueous Solution of Paraffin Chain Salts*, Hermann and Cie, Paris, France (1936).
33. A. D. Abbot, H. V. Tartar, *J. Phys. Chem.* 59 (1955) 1193.
34. H. Schott, *J. Pharm. Sci.* 62 (1973) 162.
35. C. Tanford, *J. Phys. Chem.* 76 (1972) 3020.
36. R. G. Alargova, K. D. Danov, P. A. Kralchevsky, G. Broze, A. Mehreteab, *Langmuir* 14 (1998) 4036.
37. W. Philippoff, *J. Colloid Sci.* 5 (1950) 169.
38. W. D. Harkins, R. Mittelmann, *J. Colloid Sci.* 4 (1949) 367.
39. S. S. Berr, *J. Phys. Chem.* 91 (1987) 4760.
40. S. S. Berr, E. Caponetti, R. R. M. Jones, J. S. Johnson, L. J. Magid, *J. Phys. Chem.* 90 (1986) 5766.
41. G. D. J. Phillies, J. E. Yambert, *Langmuir* 12 (1996) 3431.
42. K. Bhattacharyya, in: V. Ramamurthy, K. S. Schanze, eds., *Organic Molecular Photochemistry*, Molecular and Supramolecular Photochemistry Series, Vol. III, Marcel Dekker, New York (1999).
43. K. Bhattacharyya, *Acc. Chem. Res.* 36 (2003) 95.

44. W. Hinze, N. Srinivasan, T. K. Smith, S. Igarashi, H. Hoshino, eds., *Advances in Multidimensional Luminescence*, Vol. I, JAI Press Inc. Greenwich, CT (1991), p. 149.
45. E. Grunwald, S. Winstein, *J. Am. Chem. Soc.* 70 (1948) 846.
46. K. Medda, P. Chatterjee, A. K. Chandra, S. Bagchi, *J. Chem. Soc., Perkin Trans.* 2 (92) 343.
47. S. Nishiyama, M. Tajima, Y. Yoshida, *Mol. Cryst. Liq. Cryst.* 492 (2008) 130; V. Kharlanov, W. Rettig, *J. Phys. Chem. A* 113 (2009) 10693; J. Catalán, J. L. Garcia de Paz, C. Reichardt, *J. Phys. Chem. B* 114 (2010) 6226.
48. A. Nakajima, *J. Lumin.* 11 (1976) 429.
49. K. Kalyanasundaran, J. K. Thomas, *J. Am. Chem. Soc.* 99 (1977) 2039.
50. A. Nakajima, *Bull. Chem. Soc. Jpn.* 44 (1977) 3272.
51. A. Nakajima, *J. Mol. Spectrosc.* 61 (1976) 467.
52. D. C. Dong, M. A. Winnik, *Photochem. Photobiol. A: Chem.* 35 (1982) 17.
53. A. Nakajima, *Spectrochim. Acta A* 39 (1983) 913.
54. D. C. Dong, M. A. Winnik, *Can. J. Chem.* 62 (1984) 2560.
55. M. A. Kessler, O. S. Wolfbeis, *Spectrochim. Acta A* 47 (1991) 187.
56. M. A. Kessler, O. S. Wolfbeis, *Appl. Fl. Tech.* 2 (1990) 11.
57. B. Valeur, in: G. Schulman, ed., *Modern Luminescence Spectroscopy*, John Wiley & Sons Inc., New York (1993).
58. I. A. Koppel, V. A. Palm, in: N. B. Chapmen, J. Shorter, eds., *Advances in Linear Free Energy Relationships*, Plenum Press, New York (1972).
59. D. Buckingham, P. W. Fowler, J. M. Hutson, *Chem. Rev.* 88 (1988) 963.
60. P. L. Huyskens, W. A. P. Luck. In: T. Zeegers-Huyskenseds, ed., *Intermolecular Forces*, Springer, Berlin, Germany (1991).
61. J. N. Israelachvili, *Intermolecular and Surface Forces*, 2nd edn., Academic Press, New York (1992).
62. M. J Kamlet, J. L. M. Abboud, R. W. Taft, *Prog. Phys. Org. Chem.* 13 (1980) 485.
63. R. W. Taft, J. L. M. Abboud, M. J. Kamlet, M. H. Abraham, *J. Soln. Chem.* 14 (1985) 2877.
64. P. W. Carr, *Microchem. J.* 48 (1993) 4.
65. M. J. Kamlet, R. Doherty, M. H. Abraham, Y. Marcus, R. W. Taft, *J. Phys. Chem.* 92 (1988) 5244.
66. M. H. Abraham, M. Roses, C. F. Poole, S. K. Poole, *J. Phys. Org. Chem.* 10 (1997) 358.
67. M. J. Kamlet, J. L. M. Abboud, M. H. Abraham, R. W. Taft, *J. Org. Chem.* 48 (1983) 2877.
68. J. H. Hildebrand, J. M. Prausnitz, R. L. Scott, *Regular and Related Solutions*, Van Nostrand-Reinhold, New York (1970).
69. R. G. Makitra, Ya. N. Pirig, *Org. React. USSR* 16 (1979) 535.
70. Y. Marcus, *J. Soln. Chem.* 20 (1991) 929.
71. M. J. Kamlet, R. W. Taft, *J. Am. Chem. Soc.* 98 (1976) 377.
72. R. W. Taft, D. Gurka, L. Joris, P. V. R. Schleyer, J. W. Rakshys, *J. Am. Chem. Soc.* 91 (1969) 4801.
73. T. Yokoyama, R. W. Taft, M. J. Kamlet, *J. Am. Chem. Soc.* 98 (1976) 3233.
74. J. Catalan, C. Diaz, F. Garcia-Blanco, *J. Org. Chem.* 65 (2000) 9226.
75. J. Catalan, *J. Org. Chem.* 62 (1997) 8231.
76. J. Catalan, C. Diaz, F. Garcia-Blanco, *J. Org. Chem.* 65 (2000) 3409.
77. J. Catalan, C. Diaz, V. Lopez, P. Perez, J. L. G. de Paz, J. G. Rodriguez, *Liebigs Ann.* 30 (1996) 1758.
78. D. Attwood *Microemulsions*. In: J. Kreuter, ed., *Colloidal Drug Delivery Systems*, Marcel Dekker, New York (1994).
79. H. Alkan-Onyuksel, S. Ramakrishnan, H. B. Chai, J. M. Pezzuto, *Pharm. Res.* 11 (1994) 206.
80. J. H. Fendler, E. J. Fendler, *Catalysis in Micellar and Macromolecular Systems*, Academic Press, New York (1995).
81. S. Terabe, K. Otsuka, K. Ichikawa, A. Ysuchiya, T. Ando, *Anal. Chem.* 56 (1984) 111.
82. S. Terabe, O. Koji, T. Ando, *Anal. Chem.* 57 (1985) 834.
83. Y. Isihhama, T. Oda, K. Uchikawa, N. Asakawa, *Chem. Pharm. Bull.* 42 (1994) 1525.
84. J. D. Baily, G. J. Dorsey, *Chromatography A* 919 (2001) 181.
85. N. J. Turro, B. H. Baretz, P. L. Kuo, *Macromolecules* 17 (1984) 1321.
86. J. Aguir, P. Carpana, J. A. Molina-Boliver, C. C. Ruiz, *J. Colloid Interface Sci.* 258 (2003) 116.
87. G. Basu Ray, I. Chakraborty, S. P. Moulik, *J. Colloid Interface Sci.* 294 (2006) 248.
88. N. Kedia, A. Sarkar, P. Purkayastha, S. Bagchi, *Spectrochim. Acta A* 131 (2014) 398.
89. R. A. Marcus, *J. Chem. Phys.* 43 (1965) 1261.
90. N. Deb, M. Shannigrahi, S. Bagchi, *J. Phys. Chem. B* 112 (2008) 2868.
91. M. A. Kessler, O. S. Wolfbeis, *Chem. Phys. Lipids* 50 (1989) 51.
92. J. Burgess, *Spectrochim. Acta A* 26 (1970) 1957.

93. J. H. Park, A. J. Dallas, P. Chau, P. W. Carr, *J. Phys. Org. Chem.* 7 (1994) 757.
94. M. Shannigrahi, S. Bagchi, *J. Phys. Chem. B* 109 (2005) 14567.
95. D. Banerjee, P. K. Das, S. Mondal, S. Ghosh, S. Bagchi, *J. Photochem. Photobiol. A* 98 (1996) 183.
96. Y. Marcus, *J. Chem. Soc., Perkin Trans.* 2 (1994) 1751.
97. D. Banerjee, A. K. Laha, S. Bagchi, *Indian J. Chem.* 34A (1995) 94.
98. D. Banerjee, A. K. Laha, S. Bagchi, *J. Photochem. Photobiol. A* 85 (1995) 153.
99. P. K. Das, R. Pramanik, D. Banerjee, S. Bagchi, *Spectrochim. Acta A* 56 (2000) 2763.
100. A. Maitra, N. Deb, S. Bagchi, *J. Mol. Liq.* 139 (2008) 104.

8 Water Purification Devices
State-of-the-Art Review

V.S. Gevod and I.L. Reshetnyak

CONTENTS

NOMENCLATURE

Letters

a, b	Filtering parameter, m^{-1}
B'	Correlation coefficient, m^3/kg
C	Concentration, kg/m^3
dG	The quantity of sediment, formed in elementary filter's cell, kg/m^3
D	Diffusion coefficient, cm^2/s
F	Volumetric flow rate, m^3/s
F_1	Airflow rate for bubbling, m^3/s
G	Surfactant mass flow, kg/s
i and j	Anion/cation concentration ratio of fatty acid, m^{-3}
j	The flow of ions, mol/s
J	Volatile substances flow, g/s
K	Boltzmann constant, J/mol·K
$\tilde{K}$	Integrated rate constant of the process, m^3/s
K_A and K_{AS}	Equilibrium dissociation constants of fatty acid, m^3/mol
K_c	The rate constant of the process, s^{-1}
K_{eq} (B)	Adsorption equilibrium constant, kg/m^3
K_i	Rate constant of separate stages of water treatment, s^{-1}
l	The height of the filter, m
m	Mass of the adsorbed molecules, kg
M	Mass of bioconsumable impurities, kg
n	The number of filters in the system
n_s	The adsorption capacity of i-type impurities
n_s^*	Number of adsorbed moles of i-type impurities
N	Avogadro's number, mol^{-1}
N_α	Total number of molecules adsorbed on the surface
N_S	Maximum sites that can be occupied by adsorbed molecules
P	Partial pressure of gas, N/m^2
Q	Heat of adsorption, J/mol
r	Air bubble's radius, m
S	Cross-sections area, m^2
S_{bio}	Total area of biofouling, m^2
S_i	Specific area in monolayer space, m^2
$\tilde{S}_i$	The surface area of one grain of sand, m^2
S_p	The surface area of porous media, m^2
T	Temperature, K
V	Volume, m^3

$\tilde{V}$	Sand volume, m^3
$\tilde{V}_i$	Volume of one grain of sand, m^3
V_{ad}	The rate of adsorption, $n \cdot s^{-1}$
V_d	The rate of desorption, $n \cdot s^{-1}$
w	Superficial velocity of a fluid, defined as the volumetric flow rate divided by the cross-sectional area of the filter, m/s
w_1	Floating velocity of air bubbles, m/s
x	Finite mass of bacteria, kg
x_0	Initial mass of bacteria, kg

Greeks

α	The portion of molecules that can be adsorbed
Γ	Amount adsorbed surfactant, kg/m^2
Γ_∞	Maximum surfactant amount adsorbed at gas–liquid interface, kg/m^2
ΔK_i	Efficiency increment of the separate stage at the expense of feedbacks, s^{-1}
δ	The thickness of adsorbed monolayer, m
δ^1	The thickness of the diffusion layer, m
η, ϕ	Empirical constants with dimensions, $s \cdot m^{-1}$ and s, accordingly
θ	Part of the occupied sites
θ_p	The degree of adsorbent surface filling by adsorbed molecules
μ	Specific rate of biomass growth, s^{-1}
ν_0	The number of adsorption sites per unit surface area of adsorbent
ν_1	Number of collisions between molecules of adsorbate with a unit of adsorbent surface
ν_2	Number of molecules that are desorbed from adsorptive layer per unit time
σ	The surface charge density, C/m^2
τ	Time, s
τ_1	Duration of the time of one cycle, s
ϕ	The electrostatic potential, V
Ω	Total area of adsorption surface, m^2

8.1 BUBBLE-FILM EXTRACTION AS THE FUNCTIONAL CONSTITUENT OF INNOVATIVE WATER PURIFIERS

Bubble-film extraction is an advanced flotation method by means of air bubbles. It provides much better removal from treated water bulk of surface-active substances (SASs) and surface-inactive impurities, which have complementary structure in relation to removed SAS. The improvement consists in a way of surfactants separating from the space of bubble flotator. In this connection, let us note at first the main features of flotation with air bubbles.

Flotation is the process by which dispersed solid or liquid particles, colloidal species, and solutes are floated to the solution surface with the aid of a rising stream of air bubbles. In general, the substance to be floated should be hydrophobic and therefore attachable to the bubbles. The bubbles deliver surface-active material to solution surface and leave it there when they break. This stipulates a concentration gradient up the solution phase. As a result, the SAS concentration inside the surface layer is increased. Different devices, such as scrapers, ventilation suction, and air-blowers, remove this layer. The disadvantage of bubble flotation in the classic version is the significant hydrodynamic circulation. As a result, concentrated surface solution is redispersed back to the water bulk.

In the flotation method, which is named bubble-film extraction, the rising air bubbles do not reach the plane mirror subphase. Instead of that, they penetrate into a special discharge channel as was described in [1–3]. In the discharge channel, the air bubble flow is transformed into air-film flow, and thus, adsorbed substances evacuated more efficiently out of the treated water. The driving force of this process is the compressed air, which is released from air bubbles inside the discharge channel. If necessary, one evacuation channel can be replaced by several channels.

Innovation was applied to prevent redispersion of floated concentrated solution from the surface into the bulk water. As a result, the advanced flotation method was developed. The surfactant concentration in water could be reduced to the level of hundredths of milligrams per liter. This concentration is significantly less than that was achieved by other evacuation methods.

Bubble-film extractors were designed to remove dissolved endogenous and exogenous surfactants from water. These devices have found practical application for quality improvement of tap water at the points of its consumption. They were particularly useful when operating at peripheral points of long pipelines.

The tests have shown that the use of bubble-film extractors is the best way to purify tap water from pipelines' corrosion products and other mineral admixtures, as well as from trihalomethanes (chlorinated derivatives) and different kinds of surface-active contaminants present in tap water. But water purification degree from mineral impurities, especially from colloid rust, by means of bubble-film extraction depends on concentration in the water of endogenous surfactants. These substances are always present in tap water as a result of biofouling inside pipelines. When purifying water by bubble-film extraction, the endogenous surfactants act as collectors and carriers of finely dispersed hydroxides, oxides, and heavy metal ions. They also show the flocculation properties. Concentration of endogenous surfactants in tap water varies with seasons: their maximum concentration was observed in the summer months, and the minimum in winter [4]. To improve purification of tap water in the case of lack of impurities complementary to endogenous surfactants (during the cold times of year), a special synthetic surface-active cationic polyelectrolyte—polyhexamethylene guanidine hydrochloride (PHMGH)—can be added [3,5,6]. Its recommended dose is 2.5 mg/1 dm^3 of water to be purified, when the molecular mass of oligomers ranges from 1000 to 5000. PHMGH is the exogenous SAS.

Another approach consists in the compensation of endogenous surfactant deficiency in treated water by means of biofouling incubation inside the filtration load. As was shown in [4], the surface-active metabolic products of different bacteria cells act as cofactors of bubble-film water treatment process. Rapid accumulation of endogenous surfactants in aqueous media occurs under favorable ecological conditions. An appropriate niche for aerobic reproduction of heterotrophic bacteria is any well-developed nontoxic granular material through which contaminated and oxygen-enriched water passes. Here, the water contaminants are a source of feed for bacteria cells.

Hence, an implementation option of water treatment process on the principle of *rotating wheel* has been substantiated. The main principle of such water treatment process is the recirculation of treated water through cartridge packed with fine grain sand or charcoal and then through bubble-film extraction module. Two units (sand or charcoal cartridge and bubble-film extractor), assembled together and used for water treating in recirculation mode, form an emergent water purification device. Functioning of this device as a whole has intensive influence upon the functionality of the device's parts, what makes the device as a whole more sustained.

8.2 SEPARATION OF IMPURITIES IN KNOWN WATER PURIFICATION DEVICES

8.2.1 Functioning of a Sand Filter

Sand filters are the most known things among the filtering devices with grain packing materials. They represent filtration systems with tiny porous structure of variable cross section in all directions. Porous size depends on the caliber and shape of used sand grains. The parameters of packing material stipulate water filtration quality and working life of the filter.

Sand filters are mainly used to remove suspended and colloidal particles from filtered water.

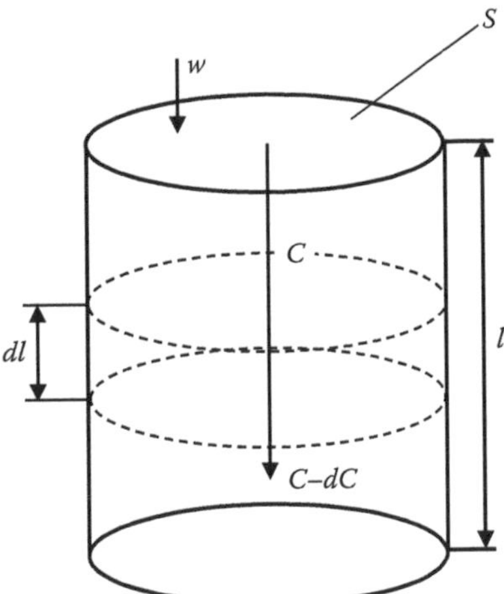

FIGURE 8.1 The model of sand filter.

Sorption of suspended particles occurs by mechanisms of gravitational sedimentation, and the narrow transition bands of the sand filter are clogged with adsorbed particles.

Sorption of colloidal particles occurs inside the filter bed by mechanisms of heterocoagulation. But in time, the list of adsorbed impurities is expanded, as the filter's grain biofouling is developed. So, the contribution of bio-precipitation and bio-oxidation of organic impurities by heterotrophic bacterial colonies becomes essential.

To study the macrokinetics of adsorption processes in the sand filter, let us analyze the elementary filter's layer between two parallel cross sections of area S at a distance dl from each other, as shown in Figure 8.1.

If we take the concentration of dispersed impurities entering this layer as C, then the effluent concentration from this layer will be $(C - dC)$. For a short time interval $d\tau$, at linear rate of water filtration w, the inflow of mass of dispersed particles entering across the depicted boundaries will be equal to $(S \cdot w \cdot C \cdot d\tau)$. And the outflow of mass passing across the boundaries will be equal to $(S \cdot w \cdot (C - dC)\, d\tau)$. The difference of mass inside the flow through elementary cell layer indicates the mass of sediment, formed inside this layer, that is,

$$S \cdot w \cdot C \cdot d\tau - S \cdot w \cdot (C - dC) \cdot d\tau = S \cdot dl \cdot dG, \tag{8.1}$$

where dG is the quantity of sediment, formed in elementary filter's cell.

Dividing both parts of Equation 8.1 on $(S \cdot w \cdot d\tau)$, one will get

$$C - (C - dC) = \frac{dl \cdot dG}{w \cdot d\tau}, \tag{8.2}$$

or

$$\frac{dG}{d\tau} = w \cdot \frac{dC}{dl}. \tag{8.3}$$

As appears from (8.3), at a given filtration rate of turbid water through the sand filter, the effectiveness of water clarification is determined by dC/dl. The high rate of water clarification is obtained at high magnitudes of dC/dl, and at $dC/dl \to 0$, the water clarification process has stopped.

According to Mints's theory, two different processes take place in each functioning layer of the sand filter: (1) adhesion of sediment and colloidal particles onto filter grains and previously entrapped impurities; (2) detachment of previously adhered particles by hydrodynamic forces and their redispersion into filtered water. Mathematically, these phenomena are described as the following:

$$dC = dC_1 - dC_2, \tag{8.4}$$

where

- dC_1 is the decrease in dispersed impurity concentration in the filtering flow at the expense of their adhesion to the grains of filtering material
- dC_2 is the increase in dispersed impurity concentration in the filtering flow at the expense of particle detachment

The precipitation of dispersed particles onto grains of filter material is controlled by the gravitational and hydrodynamic clamping forces, diffusion limitation, and Coulomb and dispersion interactions [7,8]. Mechanisms of particle adhesion onto filter grains are described by theories of Deryagin, Landau, Fairway, Overbek, and Dukhin [9]. To make the adherence of the dispersed particles to the filter grains possible, the total sum of vector forces in regard to interacting objects should result in the reduction of free energy of filtration system. Detachment of particles of dispersed phase from formed sediment increases the free energy of this system. The detachment rate is increased as treated water is pumped faster through the pore space of filter grains. Thus, the increase in the dispersed particle concentration, adherent to the filter grains in elementary layer with thickness dl, is directly proportional to these particles' concentration in filtered water and inversely proportional to their filtration rate. It also depends upon the properties of the filter packing grains and material impurities. To take these factors into account, the proportionality coefficient ($1/b$) is added to the first component of the right-hand part of Equation 8.4, where b is the filtering parameter. It is equal to the ratio of sedimentation rate constant and the surface water flow w, which is defined as the volumetric flow rate divided by the cross-sectional area of filtration bed. So, we shall obtain

$$dC_1 = b \cdot C \cdot dl. \tag{8.5}$$

And the amount of particles leaving the coagulated sediment in the filtering layer, whose thickness is equals to dl, is proportional to the sediment mass G, the water flow rate, and the magnitude of the adhesion forces between the particles in the sediment. All these factors are taken into account by coefficient a, where a is the filtering parameter equal to the ratio of constant of liftoff speed of sedimenting particles to the overall filtration rate, that is,

$$dC_2 = a \cdot G \cdot dl. \tag{8.6}$$

Then, summing up Equations 8.5 and 8.6, we obtain the following:

$$\frac{dC}{dl} = b \cdot C - a \cdot G. \tag{8.7}$$

Partial differentiation of Equation 8.7 with respect to τ leads to

$$\frac{\partial^2 C}{\partial l \cdot \partial \tau} = b\frac{\partial C}{\partial \tau} - a\frac{\partial C}{\partial l}. \tag{8.8}$$

And taking into account Equation 8.3, one can rewrite (8.8) as the following:

$$\frac{\partial^2 C}{\partial l \cdot \partial \tau} + a \cdot w \cdot \frac{\partial C}{\partial l} - b\frac{\partial C}{\partial \tau} = 0. \tag{8.9}$$

Equation 8.9 is the main differential equation of filtration processes of dilute aqueous suspension through granular packing. Integration of Equation 8.9 gives the solution as an infinite series [10], and it is not useful for practical application. Therefore, it is expedient to use Equation 8.7 and get its analytical solution for two extreme cases.

For the case of filter-fresh operating, we can consider that aG is negligible as compared with bC, and when *dead filter* operating, the filtering conditions become such that aG is significantly greater than bC. As a result, one can monitor the clarification efficiency of flow water, depending on the filtering layer thickness at different hydrodynamic conditions.

Integrating Equation 8.7 between $l = 0$ and l, and between $C = C_0$ to C for the case when bC is significantly greater than aG, we shall get

$$\ln\frac{C}{C_0} = -b \cdot l = -\frac{K_c}{w} \cdot l, \tag{8.10}$$

where K_c is the constant of settling process for filtered particles in the filter space.
Or

$$C = C_0 \exp(-b \cdot l) = C_0 \exp\left(-\frac{K_c}{w} \cdot l\right). \tag{8.11}$$

And integrating Equation 8.7 in the same limits but for the case when bC is substantially less than aG, we shall get

$$\ln\frac{C}{C_0} = a \cdot G, \tag{8.12}$$

or

$$C = C_0 \exp(a \cdot G). \tag{8.13}$$

Comparison of Equations 8.11 and 8.13 reveals that the concentration of dispersed phase in filtrate is inversely proportional to the height of filtering layer in the case when adhesion of dispersed particles to the filters grains is the dominated process. And when the settled particles pass into filtrate, then the degree of water flow turbidity depends exponentially upon the amount of accumulated sediment. This is the basic phenomenon, which occurs at filtration using sand and other granular packing. Granular packing adsorbs dispersed impurities according to gravitational deposition mechanisms, inertial and non-inertial heterocoagulation at low and medium filtration rates. And accumulated impurities turn to the filtrate flow at very high filtration rate, or when intensive water backflow is pushed up through the filter material forcing the sand particles into suspension.

According to most of the scientific studies and practical use of sand filters, the optimal height of filtering layer ranges from 0.8 to 1.5 m. Particles of sand range in size from 0.3 to 1.4 mm in diameter, and filtration rate should not exceed 10–20 m/h.

Backwashing of packing material is usually realized by increasing the velocity of reverse water flow at which water passes back through the filter. This, in effect, blasts the clogged particles off of the filter. But in so doing, filtration velocity should not reach the free fall rate of sand particles in filtrated flow in order to avoid the sand loss.

8.2.2 Filtration through Activated Carbon

This type of filtrating material combines two functions: it acts as *inert* filter toward dispersed water impurities, and at the same time, it acts as adsorbent with respect to dissolved gases and some organic substances. The function of contact coagulator is on the outer surface of activated carbon grains at its washing by contaminated water. The function of adsorbent is on the well-developed porous surface area inside the activated carbon granules. We can distinguish three categories of pores with regard to its dimensions: micropores, mesopores (transient pores), and macropores with average curvature radius of their surface of <2 nm, between 2 and 100 nm, and more than 50 nm, respectively.

The specific volume of macroporous active carbons ranges from 0.2 to 0.8 cm^3/g, and its specific surface area is about 2 m^2/g. The specific volume of mesoporous active carbons ranges from 0.5 to 0.7 cm^3/g, and its specific surface area amounts to 20–70 m^2/g. The specific volume of microporous active carbons ranges from 0.6 to 0.8 cm^3/g, and its specific surface area reaches 1000 m^2/g.

In the granules of activated carbon, the macropores act as transport channels through which adsorptive molecules penetrate into the labyrinth of mesopores and into the space of micropores. The most of large organic molecules (proteins, polypeptides, etc.) are adsorbed in mesopores. The small organic molecules and molecules of dissolved gases, such as chloroform, hydrogen sulfide, methane, ammonia, phenol, and benzene, are adsorbed in micropores, which are most prominent in the adsorption of these substances.

The equilibrium adsorption on activated carbon is described by the following equation:

$$\frac{n_s^*}{n_s} = \frac{K_{eq} \cdot C}{1 + K_{eq} \cdot C}, \tag{8.14}$$

where

n_s^* is the number of moles of *i*-type impurities adsorbed per unit weight of activated carbon
n_s is the adsorption capacity of activated carbon with respect to *i*-type impurities
K_{eq} is the adsorption equilibrium constant

Then, the degree of adsorbent surface filling by adsorbed molecules is

$$\theta_p = \frac{K_{eq} \cdot C}{1 + K_{eq} \cdot C}. \tag{8.15}$$

The value of K_{eq}, which appeared in Equations 8.14 and 8.15, is an exponential function of the ratio of heat of adsorption Q of *i*-type impurities to the thermal energy RT, that is,

$$K_{eq} = B' \cdot \exp\left(\frac{Q}{RT}\right), \tag{8.16}$$

where B' is the correlation coefficient.

As follows from (8.15) and (8.16), the maximum value of equilibrium adsorption on activated carbon is achieved if K_{eq} is significantly greater than 1, and if K_{eq} leads to zero, the adsorption is negligible.

The technological practice of the use of coal sorbents is based on these fundamentals. The adsorption process is readily effected at room temperatures, and desorption is realized at high temperatures. So, the adsorbent regeneration is carried out at high temperatures in vacuum or by superheated steam treatment.

8.2.3 Adsorption Kinetics on Activated Carbon

Adsorption kinetics on activated carbon, as well as on the surface of any other adsorbent, is controlled by the capture velocity V_{ad} of substance by adsorbent surface and by the velocity V_d of the reverse process—desorption. The value V_{ad} is proportional to the number of collisions between molecules of adsorbate ν_1 with a unit of adsorbent surface, the free area fraction of this surface $(1 - \theta)$, and the portion of molecules α, which are able to be adsorbed on this surface. The value V_d is proportional to the number of molecules ν_2, which are desorbed from adsorptive layer with filling degree θ. As a result, the rate of adsorption is given by

$$\frac{\nu_0 \cdot d\theta}{d\tau} = V_{ad} - V_d = \alpha \cdot \nu_1 \cdot (1-\theta) - \nu_2 \cdot \theta, \tag{8.17}$$

where

τ is the adsorption process time

ν_0 is the number of adsorption sites per unit surface area of adsorbent

or

$$\frac{d\theta}{d\tau} = \frac{\alpha \cdot \nu_1 - \alpha \cdot \nu_1 \cdot \theta - \nu_2 \cdot \theta}{\nu_0} = x. \tag{8.18}$$

The result of differentiation (8.18) with respect to θ is given by

$$\frac{dx}{d\theta} = -\frac{\alpha \cdot \nu_1 + \nu_2}{\nu_0} = K_c, \tag{8.19}$$

where K_c is the rate constant of adsorption process.

Since x characterizes the rate of adsorption, then at $x = 0$, the degree of adsorbent surface filling by adsorptive molecules must meet the equilibrium condition, that is, $\theta = \theta_p$.

With this in mind, Equation 8.19 can be written in the following integral form:

$$\int_0^x dx = -\int_{\theta_p}^{\theta} K_c \, d\theta. \tag{8.20}$$

Solving this equation for x yields

$$x = -K_c \cdot (\theta_p - \theta). \tag{8.21}$$

Substituting the value x from (8.21) into (8.18), we shall get the required differential equation for adsorption rate in the terms θ, K_c, τ:

$$\frac{d\theta}{d\tau} = -K_c \cdot (\theta_p - \theta), \tag{8.22}$$

or

$$\frac{d\theta}{\theta_p - \theta} = -K_c \cdot d\tau. \tag{8.23}$$

Integration of (8.23) from θ_p to $(\theta_p - \theta)$ and from 0 to τ yields

$$\ln\frac{\theta_p - \theta}{\theta_p} = -K_c \cdot \tau, \tag{8.24}$$

or

$$\theta = \theta_p \cdot [1 - \exp(-K_c \cdot \tau)]. \tag{8.25}$$

In (8.25), the value θ_p is subject to (8.15), and therefore, θ depends on the product of two exponential functions. Namely, it depends on the exponential member characterizing the gain of free energy in the transition of adsorptive molecules from the environment to the adsorbent and from the exponential function that consists of constant of transfer rate of adsorptive molecules to adsorbent. This constant depends on the coefficients of surface and bulk diffusion of substances.

Thus, when in the medium with high concentration of adsorbate, which is characterized by high adsorption heat, the adsorption process occurs quickly, and additional accumulation of adsorptive molecules on the adsorbent surface is impossible. In practice, water is treated using adsorbents in flowing column taking into account the adsorption rate and the stoichiometry of the saturation adsorbent capacity.

8.2.4 Adsorption Wave in Carbon Filter

The adsorptive efficiency of carbon filter is characterized by evolution of concentration profile in the filtered flow. This profile is called the breakthrough curve (adsorption wave). In fixed-bed continuous adsorption, the flow of contaminated water creates a wave front as it flows through the adsorption bed. This wave front is better known as the mass transfer zone (MTZ). This is where active adsorption happens in a packed-bed filter. Within the MTZ, the degree of saturation with adsorbate varies from 100% to effectively zero. As the adsorption bed reaches its equilibrium capacity or in other words becomes exhausted, the MTZ will displace further through the adsorption bed. Figure 8.2a through d shows the MTZ motion across the adsorption bed. Black color corresponds to sections that are completely exhausted, and the white sections correspond with fresh adsorbent. The MTZ will continue to travel through the adsorption bed until it achieves its breakthrough point.

As is seen, at the step d, the adsorbate concentration in the filtrated liquid becomes the same as in the entering stream. The fixed bed of certain dimension has a definite adsorption capacity. The shape of breakthrough curve depends on the inlet flow rate, adsorbate concentration, and other properties, such as column diameter and bed height. The slope of the curve indicates the behavior of equilibrium isotherm. It is the function of activity of the adsorbent with respect to the adsorbate. At high values of adsorption heat (more than 20–40 kJ/mol), the wave front has a well-defined bounding face, which moves parallel to itself as shown in Figure 8.3a. At small values of adsorption heat, the breakthrough curves show tailing in time as shown in Figure 8.3b. And without adsorbate's affinity for the adsorbent, the adsorption wave disappears completely. In this case, the concentration of the adsorptive in filtered flow is unaffected by the adsorbent. Adsorption of chlorine and its organic derivatives on activated carbons is characterized by steady-state shape of the breakthrough curves

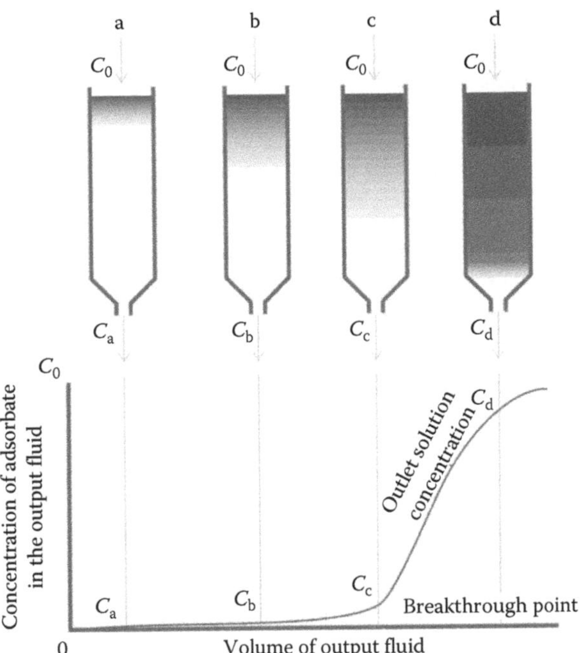

FIGURE 8.2 The picture of MTZ motion across the adsorption bed. (a through d) The steps of the MTZ motion across the adsorption bed (black—completely exhausted section, white—section with *fresh* adsorbent).

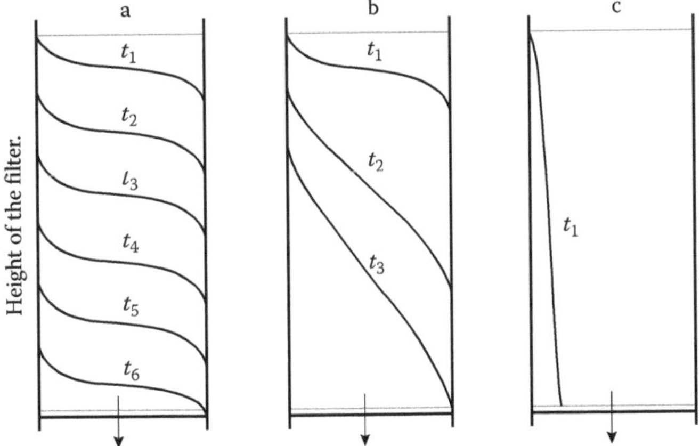

FIGURE 8.3 The shapes of breakthrough curve on activated carbon for the substances with different adsorption heat. The values of adsorption heat: (a) high, (b) average, and (c) small.

(Figure 8.3a). Adsorption of proteins, lipids, polypeptides, and many other organic substances is characterized by *proportional* tailing in time breakthrough curves (Figure 8.3b). Glucose, alcohol, and sucrose are practically unadsorbed on charcoal, and the adsorption wave is not observed (Figure 8.3c).

The time required for the wave to reach the end of the filter bed characterizes the adsorption capacity of the filter. This time depends on the height of packed bed, the quality of adsorption material, concentration of adsorbate in the fluid phase, and the filtration rate. Sometimes the filtration rate is substituted by total amount of solution that will pass through the filter with the requirement

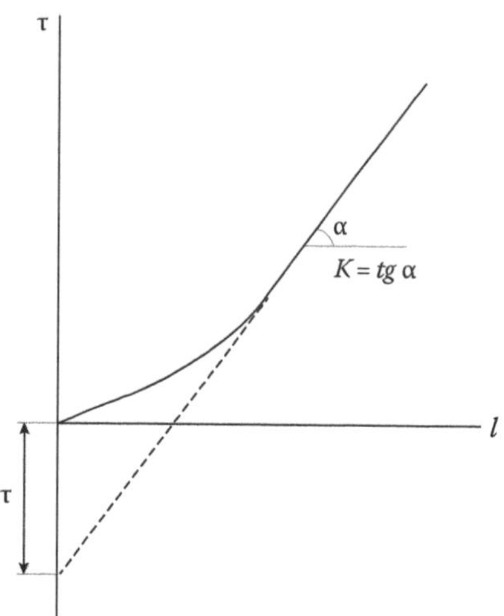

FIGURE 8.4 The filter protective effect depicted as a function of the height of packed bed.

that the concentration of adsorbate in the fluid will be decreased to the given level. Shilov equation gives the relationship between adsorption capacity of the filter and height of packing:

$$\tau = \eta \cdot l - \phi, \tag{8.26}$$

where

τ is the exploitation time for filter (from new filtering material up to breakthrough point)
l is the height of packed bed
η, ϕ are the empirical constants depending upon the adsorbent type and its particle size, the type and concentration of the adsorbate, and the filtration rate

Duration of filter protective effect as a function of the height of packed bed is shown in Figure 8.4. When adsorption capacity of the filter reaches its limit (see Equations 8.14 and 8.25), no more adsorption can take place in the bed.

8.2.5 Microbiological Activity in Filtration System

The information presented in Sections 8.2.1 through 8.2.4 concerns the mechanisms and kinetics of basic physicochemical processes occurring during filtration of contaminated water through sand or carbon filters. But the numerous studies have shown that along with these mechanisms described, biofiltration plays an important role in the removal of impurities [11]. This is for the reason that the porous spaces of any kind of nontoxic filters are acceptable ecological niches for various bacteria. And each bacterial cell is a unique *chemical plant*, which is equipped by a lot of catalytic systems. Bacterial cells can utilize many inorganic and almost all organic substances of natural and synthetic origin.

Heterotrophic bacteria appear in great quantity in the course of water filtration through carbon filters as carbon is an ideal growth medium for these bacteria, which attach to the porous surface of the filter and obtain food by consuming adsorbed water contaminants. Under suitable conditions, they adapt to environment and grow as biofilm. The most suitable conditions for the growth of bacterial colonies occur in the mesoporous activated carbon filters. The substances for microbial consumption are effectively fixed and accumulated within these filters.

At the presence of dissolved oxygen in filtered water, aerobic heterotrophic bacteria grow fast, and as a result, the biofilm is rapidly formed on the grain surface of activated carbon. The phenomenological equations are as follows:

$$C_6H_{12}O_6 + 6O_2 \rightarrow 6CO_2 + 6H_2O, \tag{8.27}$$

$$\frac{5}{6}C_6H_{12}O_6 + NH_3 \rightarrow C_5H_7NO_2 + 3H_2O, \tag{8.28}$$

$$C_5H_7NO_2 + 5O_2 \rightarrow 5CO_2 + NH_3 + 2H_2O, \tag{8.29}$$

where $C_5H_7NO_2$ is the averaged composition of element for bacterial cell.

Equation 8.27 shows the process of substrate oxidation (as exemplified by glucose). Equation 8.28 describes the synthesis of bacterium cell material, and Equation 8.29 describes the auto-oxidation of bacterial cell. Analyzing all these equations, it is easy to see that at the optimal concentration of organic substrate, each milligram of absorbed oxygen gives rise to about 0.6 mg of bacterial mass, and approximately the same amount of bacterial mass disappears in the process of bacterial autolysis.

As biofilm grows, it becomes thicker, it fills and occludes the pores of activated carbon, and thus it reduces the total surface area available for organic substance adsorption. In the extreme case, the flow will be restricted to a few channels, adsorption will be very slow, and the filter will fail. As a result, the delivery of oxygen to biofilm will be slow, too. But the microbial activity in the adsorption–filtration system does not stop by this. Following the aerobic heterotrophic bacteria, the obligate and facultative anaerobes join in the biocatalytic processes. Areas occluded with aerobic biofilm become the suitable environments for them. Due to the functioning of anaerobic bacteria, different organic substances of aerobic bacterial cells (polysaccharides, protein molecules, protein–lipid complexes) split into simpler organic substances (amino acids). In this case, the integrity of biofilm structure is disturbed, and the fission products are rapidly diffused into filtered water. Aerobic heterotrophic bacteria in the presence of dissolved oxygen transfer these substances to the sites with biofilm where they are consumed.

Thus, the adsorption–filtration system is a self-regulating system. It realizes the removal of different organic contaminants from water. As a result, the amounts of removed contaminants can exceed many times the limit adsorption capacity of used adsorbent. At direct-flow filtration, the biocatalytic activity of bacterial colonies in the adsorption-filtering device is proportional to the concentration of organic contaminants in filtered water, the specific area of biofouling in the pore space of the filter, and the height of packed bed. The concentration of bio-decomposed adsorbates in the fluid as it enters and as it leaves the filter bed is defined by the following equation:

$$C = C_0 \cdot \exp\left(-\frac{K_c \cdot l \cdot S}{F}\right), \tag{8.30}$$

where

K_c is the rate constant of biofiltration
F is the filtered water flow
l is the height of packed filter
S is the cross-sectional area of the biofilter

The first systematic tests of granular activated carbon as a biofilter were carried out in 1975 in Germany. Mechanically pretreated water flow was directed through the activated carbon filter.

Filtration velocity was 10 m/h, height of packed bed was 6 m, and filter diameter was 1 m. After a while, a steady-state velocity of organic contaminant decomposition was reached. It was observed that 1 m^2 of surface area of activated carbon decomposes about 60–100 g of organic substances with respect to dissolved organic carbon at the daily oxygen consumption 200–250 g. Comparison of water quality before and after treatment has shown that the activated carbon not only is a vital medium for microorganisms, but also acts as a buffer at sharply increasing concentration of organic impurities in water.

8.2.6 Flotation Treatment of Polluted Water

Flotation processes are an important part of water treatment technologies in modern water treatment plants. Flotation is based on the principle of adhesion of insoluble particles to air bubbles and adsorption of dissolved surfactants at the surface of air bubbles. Flotation allows for different kinds of admixtures to be removed from water bulk in a physical and chemical manner. In this way, suspended and colloidal particles, emulsions of oils and fats, the separate surfactant molecules and their micelles, complexes of surfactants with colloid rust, and multivalent ions of heavy metals can be removed. At present, the flotation processes and equipment for their realization are widely described in the literature [12]. Flotation involves the injection of small bubbles of air or other gas into the water bulk. Surface-active impurities are adsorbed at the bubble surface and transferred through the water bulk to its surface. As a result, the foam concentrate is formed on the surface of bubbling water. It contains surfactants, suspended solid particles (water impurities), emulsified substances, bacterial cells, etc. This foam is evacuated from the surface by means of special scrapers and other devices.

The serious disadvantage of classic flotation process is a significant axial mixing. For that reason, the floated material is not supported on the surface and tends to redisperse into the treated water bulk. So, the efficiency of flotation is significantly decreased.

Because of this phenomenon, the classic flotation does not allow the concentration of surface-active impurities to be decreased <1–5 mg/L. This disadvantage was eliminated in the devices containing special separating channels. Due to the improvements, redispersion of floated products is decreased, and the floated concentrate is transferred directly outside the flotation apparatus [1,3]. These improved apparatuses were named as bubble-film extractors [13,14]. They are able to remove almost 100% of floated product beyond the treated water bulk.

The flotation kinetics depends on many factors. In particular, it depends upon the height of flotation column, the amounts of generated air bubbles, the air bubble size, the probability of air bubble collision with the particles of suspended matter, the diffusion velocity of the dissolved surfactants, and water flow rate in the flotation device.

The concentration of impurities in the fluid as it enters and leaves the flotation column is defined by the following equation:

$$C = C_0 \cdot \exp\left(-\frac{K_c \cdot l \cdot S}{F}\right), \tag{8.31}$$

where

C is the concentration of impurities in the outlet of the flotation column
C_0 is the concentration of impurities in the inlet of the flotation column
K_c is the flotation constant
l is the height of flotation column
F is the flow rate of treated water
S is the cross-sectional area of the flotation unit

This equation is identical to Equation 8.11, which describes the work of a sand filter.

8.3 MODERN TECHNIQUE OF WATER TREATMENT

Water treatment technologies consist of many steps, such as sedimentation, oxidation, coagulation, flocculation, filtration, flotation, ion exchange, adsorption, and reverse osmosis. They are intended for the destruction of water contaminants and their removal from treated water bulk. At that, water is disinfected by different methods, namely, chlorination, ozonation, ultraviolet light treatment, and ultrasound treatment.

For example, in the Netherlands, the multibarrier technology of water treatment is used to obtain drinking water of high quality [15]. According to this technology, the quality of water in open sources is continuously monitored. When quality deteriorates, surface water from the river is mixed with underground water, which is pumped from a depth of about 120 m. If the quality of surface water is reduced drastically and becomes unacceptable, the water is supplied only at the expense of groundwater.

The principal technological scheme of this process is shown in Figure 8.5. According to this scheme, at the first stage, the surface water is treated with coagulant $FeCl_3$. As a result, phosphates and many other contaminants are removed from water. After coagulation, water is collected in two basins. Then, it is transported to the sand dunes and infiltrated. The infiltration of pretreated surface water enables complete removal (more than 99%) of fecal bacteria and viruses, and destruction of nitrates and some organic compounds.

Then, water is pumped to a pool through a special well system. The main amount of water is pumped to the pool from collector and placed before the waterproof layer. At high concentrations of mineral salts in the water, it is diluted with water, pumped from a deep well, and placed under the waterproof layer. Next, water is enriched with oxygen in a special cascade-type aerator, then filtered, and undergoes ozonation. In so doing, the resistant organic substances such as pesticides are destroyed, and the rest of pathogenic bacteria and viruses are inactivated. Then, water is softened, if necessary, and its pH should be adjusted to the desired level.

The next step of water treatment is two-stage adsorption process, whereby the value of the residual concentration of organic substances is strongly reduced due to their physical adsorption on activated carbon and bio-oxidation by heterotrophic bacteria. The water is then drawn through fine filters, made of sand, to remove any remaining contaminants. After slow filter, the water is clear, colorless, without odor and taste. At that stage, the water is ready to be consumed. The water

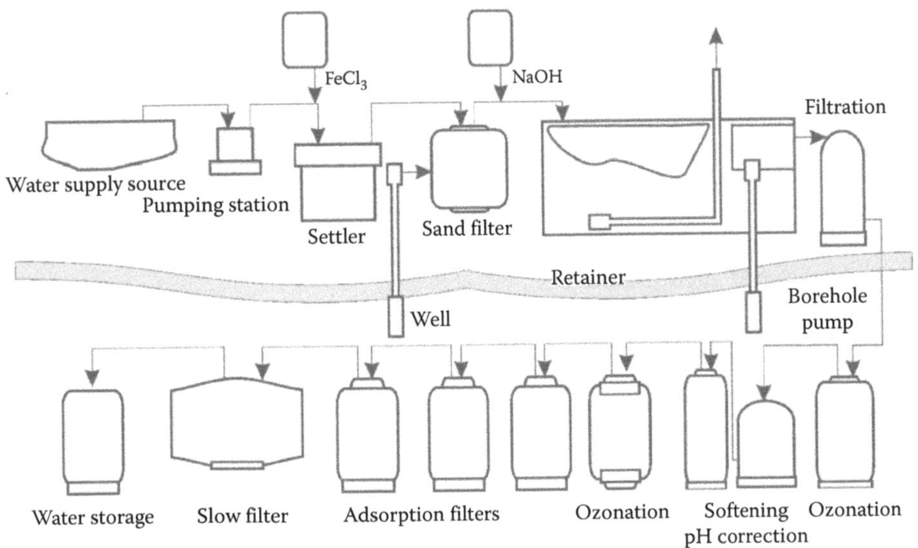

FIGURE 8.5 Technological scheme of drinking water preparation in the Netherlands.

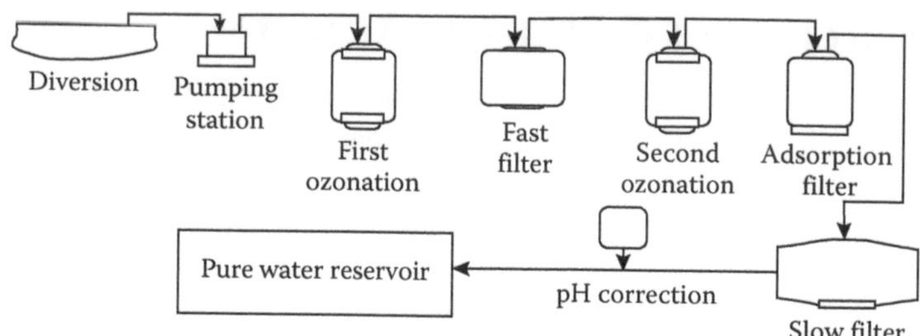

FIGURE 8.6 Technological scheme of drinking water preparation from Lake Zurich.

treatment technology described earlier is the approach that allows the production of drinking water without the use of chlorine.

In modern water treatment technologies, the slow filters should preferably be located at the final stages of the multibarrier process. This allows to essentially reduce the contaminant load on biofilm thereby ensuring effective operation of the sand filters minimizing the numbers of backwashings.

The slow filtration, if properly carried out, provides the high-quality drinking water. But the use of slow filters is possible only in the presence of the required areas at the water treatment plant.

In Switzerland, the water from Lake Zurich is treated by a multibarrier technology, which is shown in Figure 8.6. According to this technology, the water is treated first with ozone, then filtered through a fast filter, after which the water is reozonated, and then filtered through charcoal. After these operations, the water flows through the final slow filter.

The first step of ozone treatment (1.1 mg $O_3/dm^3 \pm 25\%$) is used for disinfection and flocculation of water contaminants before its fast filtering. Fast filters remove more than 95% suspended solids. The next step of ozonation is carried out using 0.5 mg O_3/dm^3, and then water is treated by slow filters with activated carbon.

The filtration zone consists of four layers of gravel and sand of different sizes, with the thickest gravel forming the bottom layer and the finest gravel the top layer: a layer of fine sand (50–85 cm), a layer of coarse sand (5 cm), a layer of fine gravel (5 cm), and a layer of coarse gravel (5 cm).

The main problem of sand filter operation is that these are prone to clogging. In the considered technology, the filters were tested under two operating conditions: at flow rates 0.5 and 4–7 m/h. In the first case, all tested filters have operated more than 9 years with periodic backwashing one time every 2 or 3 years. In the second case, the filters have operated without the need for cleaning at a period of 8 years. In operation, the hydraulic permeability of the filters was gradually reduced to 0.5 m/h, as a result of an increase in clogging. Therefore, the filters were washed, and the first 5 cm of the sand layer were removed and replaced by new sand. This has made it possible to increase the hydraulic permeability of filters by 1–2 m/h. However, the filtration capacity cannot be regenerated completely by this method due to algae clogging of the sand filter bed. At least 10% of the pore space in the top layer of filter bed was filled by biomass.

The total height of the sand bed reaches 100 cm. After 10–15 years of operation, the sand in the filter must be regenerated. In Zurich, at the water treatment plant, there are 14 slow filters. The area of the filter unit is 1.12 m^2, the maximum filtering capacity of each unit is 16 m^3/m^2/day, and the operating conditions are 4°C–8°C in the dark. Whole system provides about 250 m^3 of clean water per day.

An example of the drinking water production from highly polluted surface waters is the technology patented in the United Kingdom. Its scheme is shown in Figure 8.7.

First, contaminated surface water is pretreated by coagulation and collected in a storage tank. Then, water is filtered through the coarse filter. Next, the water passes through a two-stage filtration

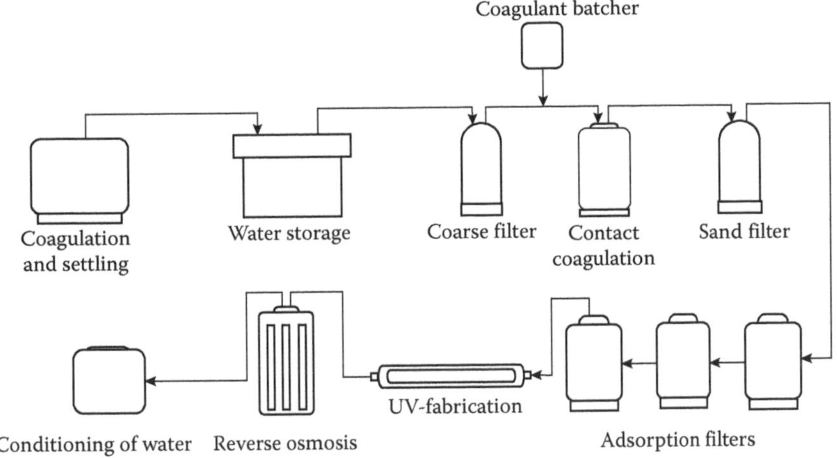

FIGURE 8.7 Scheme of drinking water preparation in the United Kingdom.

unit, using coagulation as a pretreatment process. The several steps of adsorption follow this on activated carbon, and then the water is disinfected with ultraviolet rays. In the final stage, the reverse osmosis system consisting of two units is used to purify water (one unit in service when one is regenerating). Naturally, the end step of the process is the additional mineralization of treated water.

In Namibia, the surface water is mainly used for drinking water preparation. The technological scheme is shown in Figure 8.8. It includes the following processes: ozonation at the first step, then flocculation, flotation, and filtration through a double-layer filter. Next, the water is ozonated again, and then it is filtered through the two-stage carbon filter and treated to safety standards with ultraviolet.

Figure 8.9 describes the basic steps of water treatment process, which is briefly described in [16].

The water enters the plant and goes into accumulation tank. The treatment of water includes mechanical filtration, coagulation and flocculation, flotation, and then water is transported to biological treatment unit, where the organic impurities are decomposed due to heterotrophic

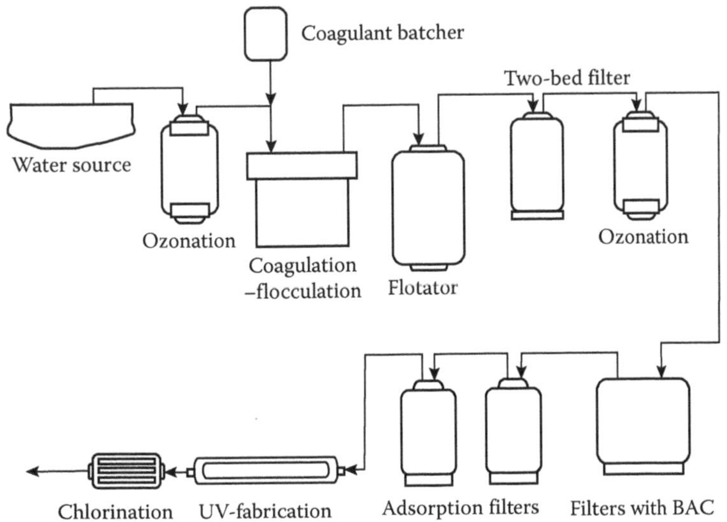

FIGURE 8.8 Scheme of drinking water preparation in Namibia.

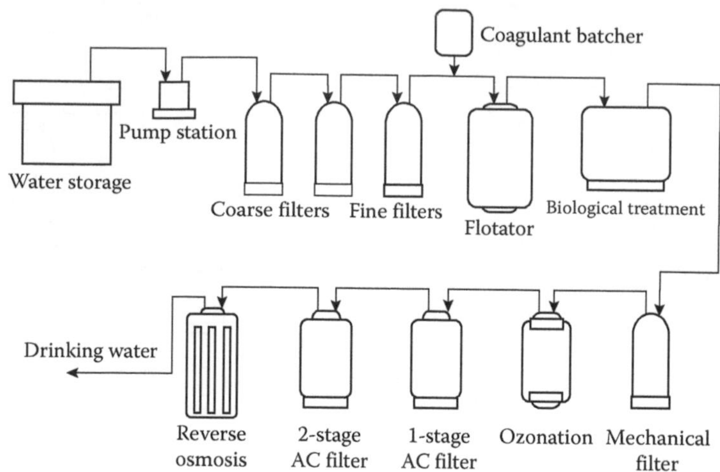

FIGURE 8.9 Scheme of drinking water preparation with the use of biofilters.

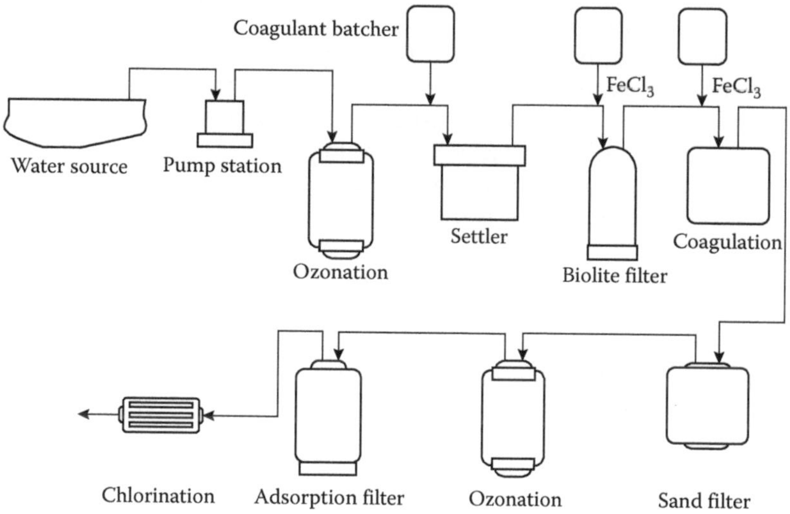

FIGURE 8.10 Scheme of drinking water preparation in France.

bacterial activity. Then, the treated water is filtered again to remove the trace amounts of sludge, and after that, it is ozonated. Then, water flows through activated carbon filters (I and II steps). At the first step, the organic compounds within a certain range of molecular weights and, at the second step, the trace amounts of pesticides and other toxic substances are removed. The final step is water processing by reverse osmosis.

At the drinking water treatment plant in the city of Ivry-sur-Seine (France), a multistage complex technology is also used (Figure 8.10).

The pretreatment process comprises ozonation and then coagulation. Next treating processes are primary filtration stage through biolith, and a secondary filtration stage through sand, with the possibility of coagulation. The finishing treatment includes ozonation, biological filtration through granular activated carbon, and disinfection with the use of chlorine.

As is seen from the earlier examples, the highest degree of purity and safety of drinking water requires complex water treatment technologies.

8.4 TAP WATER QUALITY PROBLEMS STILL REMAIN

As shown earlier, the water treatment plants produce high-quality drinking water, which meets all necessary requirements. But it does not mean that the quality of tap water being delivered to consumers by the distribution networks remains the same.

Drinking water entering the distribution network may contain various bacterial species, such as aerobic and anaerobic heterotrophic bacterial cells. Under favorable conditions, bacterial cells colonize a distribution system and form biofouling inside the water supplying pipes. Chlorination of water at dosages about 1.5–3 mg/dm^3 is not enough to remove arisen colonies. On the other hand, the injection of chlorine at higher concentrations is very harmful for water consumers. Chlorination of water, and especially chlorination at high dosages, leads to water saturation by harmful by-products, including chlorites, haloacetic acids, and total trihalomethanes.

Over the last century, the use of chlorine has led to the appearance of chlorine-resistant bacteria. These bacteria have occupied the water pipelines all over the world. Dissolved chlorine at a concentration about 6 mg/L has only a little effect on biofilms formed from bacteria of this kind. At present, it is impossible to prevent their growth even on cleaned pipe surfaces [17]. In such a way, the intensive biofouling within the distribution networks has the objective reason. And this problem needs to be solved. Later, it will be shown that the phenomenon of *natural* biofouling in tap water may be used with a great profit in innovation water purifiers, which combine flotation and filtration methods at water treatment in a single apparatus.

8.5 WATER TREATMENT BY SAND FILTRATION IN RECIRCULATION MODE AND ITS FEATURES

Modern water purification technologies involve processing water through a direct-flow multibarrier treatment system to produce high-quality drinking water. The constituent steps of this system are the following: sedimentation, oxidation, coagulation, flocculation, filtration, flotation, ion exchange, adsorption, reverse osmosis, etc., as shown in Figure 8.11.

The scheme, as shown earlier, provides consumers with high-quality purified water in the required quantity. But at the same time, this scheme limits the potential ability of used equipment, since the feedback path between the used stages is absent at sequential circuit. By other words, this water treatment technology allows the concentration of water contaminants to be reduced to a certain level. This level is predicted by purification ability of each treatment unit K_i, as shown in Figure 8.11, and does not affect the efficiency of the other units. Therefore, the depicted technological scheme requires an adequate increase in the reaction space areas of each technological unit for its productivity increase. And the limiting factors of water treatment process are the following: thickness and cross-sectional area of filter layers, height and cross-sectional area of flotation columns, dimensions of water clarifier, etc.

Therefore, the multibarrier direct-flow water treatment systems have a large size and require a very heavy investment and a large area of land for their construction and operation. Maintenance costs are also high. Analogues to such systems are available for home use and have reasonable dimensions. But they are expensive in operation and maintenance, because they require frequent replacement of their filters. Otherwise, household water treatment systems do not perform their functions.

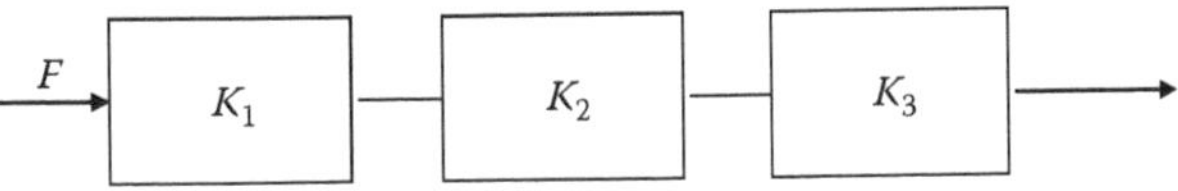

FIGURE 8.11 Direct-flow scheme of water treatment. *F*: flow of treated water; K_1, K_2, K_3: adsorption coefficients of separate stages.

The rational way out at water treatment is to apply water recirculation system instead of the direct-flow system. This enables the use of compact sand filters and biofilters with the gravitational or floating packing, small flotators, aerators, etc. At the same time, the efficiency of water treatment is greatly increased due to arisen positive feedbacks between functional units in the created system.

The features and benefits of recirculation method at water processing can be demonstrated by taking as an example the sand filter. Traditionally, to be effective, sand filter should have the height of sand bed to be 0.7–2.0 m, flow rates in slow sand filters should be between 0.1 and 0.2 m/h, and that in quick sand filters should be between 5 and 20 m/h.

According to [10], the output/input ratio of impurity concentrations C/C_0 for the granular sand filter is described by the following equation:

$$\frac{C}{C_0} = \exp\left(-\frac{K_c \cdot l}{w}\right) \tag{8.32}$$

where

K_c is the rate constant of contaminant adsorption in the porous space of the filter bed
l is the height of the filter bed
w is the face filtration velocity

If we consider the filtration rate w as the water flow related to the cross-sectional area of the filter, then we will obtain

$$\frac{C}{C_0} = \exp\left(-\frac{K_c \cdot l \cdot S}{F}\right), \tag{8.33}$$

where F is the volume flow rate.

It follows from (8.33) that the degree of impurity adsorption in the sand filter bed is an inverse exponential function of the ratio of the filter height l to water flow rate F. The values of two other members of the exponential function are constants by definition. Thus, using a sand filter with a height l and water flow rate F, we will obtain the output/input ratio of impurity concentrations in accordance with (8.33). And we will find volume V of filtered water by the following equation:

$$V = F \cdot \tau, \tag{8.34}$$

where τ is the duration of the filtration process.

Therefore, in direct-flow filtration, we use such height of filter bed l and the cross-sectional area of filter S at a given flow rate F, which gives an opportunity to get the required volume of the filtered water V with its clarification degree C/C_0. But the sand filter washing is complicated with an increase in filter's height. Therefore, sand filters must be connected in a parallel or a series configuration. The filters connected in parallel give productivity benefit, and a series filter configuration has the advantage of filtration quality.

For n filters connected in series, Equation 8.32 is transformed to the following form:

$$\frac{C}{C_0} = \exp\left(-\frac{K_c \cdot l}{w}\right)^n, \tag{8.35}$$

or

$$\frac{C}{C_0} = \exp\left(-\frac{n \cdot K_c \cdot l}{w}\right), \tag{8.36}$$

where n is the number of filters connected in series.

Equation 8.36 shows that the concentration of impurities at the outlet of the filtration system of n filters connected in series is an inverse exponential function of the number of connected filters. But the same equation describes the efficiency in a closed-loop filtration system with a single filter, which has the following parameters: height of filter bed l, surface filtration rate w, and the ratio C/C_0 determined by the nth filtration cycle. Therefore, for water volume V, filtered under recirculation mode through the sand filter with bed height l and cross-section area S during cycle time τ_1, the ratio C/C_0 is determined by the following equation [18,19]:

$$\frac{C}{C_0} = \exp\left(-\frac{K_c \cdot l \cdot S \cdot \tau_1}{V}\right), \tag{8.37}$$

where τ_1 is the cycle time required for the realization of one filtration cycle.

And if the number of filtration cycles will be n, one shall obtain

$$\frac{C}{C_0} = \exp\left(-\frac{n \cdot K_c \cdot l \cdot S \cdot \tau_1}{V}\right). \tag{8.38}$$

As the product of n and τ_1 is the real duration of the recirculation filtration process, we'll have

$$\frac{C}{C_0} = \exp\left(-\frac{K_c \cdot l \cdot S \cdot \tau}{V}\right), \tag{8.39}$$

where $(K_c \cdot S \cdot l)$ is the general rate constant of filtration process.

Equation 8.39 is identical to (8.33) by its form, since the ratio V/τ is the analogue of the flow rate F from Equation 8.33. From Equations 8.33 and 8.39 it follows that at equal heights and cross-section areas of sand filters, equal rate constant of adsorption, and equal volume flow rate of filtered water, the ratio C/C_0 will be the same for direct-flow filtration as well as for recirculation filtration, in the case of $V/\tau = F$. And when V/τ becomes smaller than F, the adsorption degree will be higher in recirculation filtration. Therefore, to obtain the equal degree of water purification using the direct-flow sand filters (the height of filtering layer is equal l), and in recirculation filtration (the height of filtering layer $l_1 < l$) with respect to (8.39) and (8.33), one will get

$$\frac{K_c \cdot l \cdot S}{F} = \frac{K_c \cdot l_1 \cdot S \cdot \tau}{V}. \tag{8.40}$$

If the values K_c and S are equal, then

$$\frac{C}{C_0} \approx \frac{l}{F} = \frac{l_1 \cdot \tau}{V}. \tag{8.41}$$

Thus, the quality of water treatment at direct-flow filtration is directly proportional to the ratio of height of the filter l to flow rate F. And the quality of water treatment under recirculation filtration is directly proportional to the time of filtration τ, to the height of the filter l_1, and inversely proportional to the volume V of filtered water. The dependence of the water clarification degrees upon the filtering mode and filter dimensions is shown in Figure 8.12.

Equation 8.41 indicates that the small sand filters can be used for effective clarification of certain volumes of contaminated water. And the decrease in the filter height is compensated by the recirculation mode of its operation. This statement is true relative to the other units of multibarrier

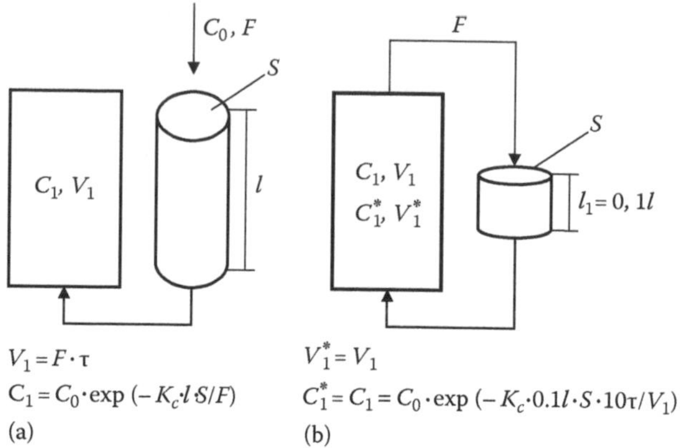

FIGURE 8.12 The scheme of water treatment processes: (a) direct-flow filtration; (b) recirculation filtration.

technological scheme, that is, to floaters, bioreactors, and so on. The earlier described mechanism shows that direct-flow multibarrier water treatment system, which has a large size, can be replaced by the compact prototype, operating under recirculation condition. The recirculation technology provides a new quality as compared with once-through filtration, biofiltration, flotation, etc., where each unit in technological scheme performs only its own function.

8.6 SYNERGISTIC EFFECT OF FILTRATION AND FLOTATION COMBINATION AT WATER TREATING IN RECIRCULATION MODE

As was described in Section 8.3, at direct-flow water treatment technologies, the contaminants are destroyed and removed step by step and with fixed efficiency. For example, the water treatment scheme, represented in Figure 8.9, includes six filters connected in series. Three of them are installed before the flotation unit and three others after biological water treatment unit. According to this scheme, initially, water flows through a coarse mechanical filter, designed to remove the particles larger than the filter pore size. The next two filters collect the suspended impurities. After that, organic and mineral contaminants are removed by water treatment in coagulation–flotation unit. Then, water is directed to the biological treatment. As a result, the organic contaminants are removed due to their oxidation by heterotrophic bacteria. But at the same time, their surface-active degradation products are entered into treated water. Therefore, after biological treatment, the water is fed into the special mechanical filter. It adsorbs bacteria and their degradation products. The remaining contaminants are destroyed completely using ozonation. Then, water is passed through two stages of filtration through activated carbon. And the reverse osmosis treatment is used as the final purification stage to reach constant quality of the treated water.

The disadvantage of the technological scheme of water treatment described earlier is the following: each next step of linear technological chain is unaffected by the work of previous steps. But when the water is treated under recirculation or flow-recirculation conditions, the situation is changed as shown in Figure 8.13.

In particular, if the biofilter, aerator, and bubble-film extractor are connected in series and operate in the recirculation mode of water treatment, then the water flow is saturated by air oxygen in aerator. This stimulates the biofiltration activity of microflora in the pore space of filter. As a result, the magnitude of rate constant K_1 that characterizes the speed of removal of impurities from the water flow in the body of filter should be summarized with the value ΔK_2, which causes a corresponding increase K_1 due to the intensification of biofiltration at saturation of filtered water by atmospheric oxygen. Simultaneously, the concentration of surface-active products of bacterial metabolism at

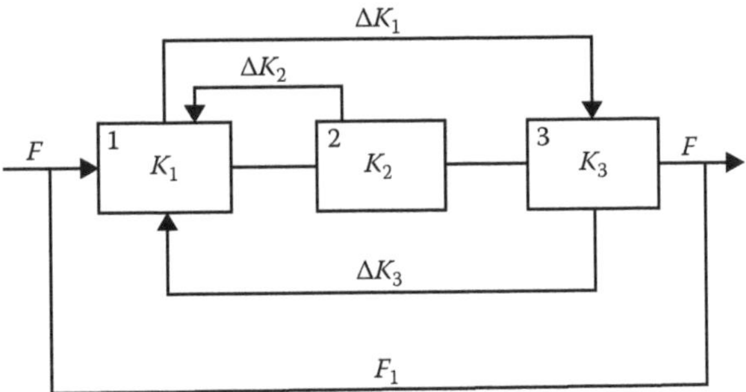

FIGURE 8.13 The scheme of flotation–filtration water purification process under recirculation conditions. 1: bio sand filter; 2: aerator; 3: bubble-film extraction unit; F: flow of the treated water; F_1: recirculation flow; K_1, K_2, K_3: rate constants of separate stages of water treatment; ΔK_1, ΔK_2, ΔK_3: efficiency increments of the stages at the expense of feedbacks in recirculation system.

the output of the filter is also increasing. But these products are natural flocculants and gatherers (collectors) of ionic impurities of water. They provide additional aggregation of colloidal particles and heavy ions in the apparatus of flotation and thereby ensure reduction of the required quantity of, or even refuse from the use of, synthetic surface-active flocculating agents and gatherers. Therefore, the index K_3, which reflects the quality of work of the flotation apparatus, is increased by the value ΔK_1, and we come to the result ($K_3 + \Delta K_1$).

But due to flotation (bubble-film extraction), the products of bacterial metabolism and the products of bacterial degradation together with other water contaminants are removed continuously from recirculation flow through the bubble-film extractor. As a result, another positive feedback is realized in filtration–flotation system. The essence of this effect lies in the inhibition of vital functions of bacteria with increasing microbial metabolite concentration according to the law of chemical kinetics. And in accordance with the same law, bacterial activity is increased as the products of bacterial metabolism are removed from treated water [20]. Thus, we are able to add one more component, namely, ΔK_3, to the magnitude ($K_1 + \Delta K_2$). The component ΔK_3 represents the increase in biofiltration efficiency due to bacterial inhibitors removed by means of bubble-film extraction. In such a way, the resulting rate constant of the process takes the form ($K_1 + \Delta K_2 + \Delta K_3$).

So, there is a great synergy between the biofilter and the flotation unit (bubble-film extractor) in closed-loop flotation and filtration system, and they provide each other with additional water treatment possibilities when operated in the recirculation mode. In other words, the faster biofouling growth of the granular filter leads to increasing natural flocculant concentration in treated water, thereby improving the effectiveness of bubble-film extraction. (And the bubble-film extractions, which saturate the treated water by atmospheric oxygen and remove degradation bacterial products, were enhancing biofilter operation.)

Thus, the mathematical description of the direct-flow sand filtration is the following:

$$\frac{C}{C_0} = \exp\left(-\frac{K_1 \cdot l}{w}\right) = \exp\left(-\frac{K_1 \cdot l \cdot S}{F}\right). \tag{8.42}$$

The operation of the direct-flow flotation unit (bubble-film extractor) is as follows:

$$\frac{C}{C_0} = \exp\left(-\frac{K_3 \cdot l \cdot S}{F}\right). \tag{8.43}$$

And in the case of their combination in series, we shall get

$$C = C_0 \cdot \exp\left(-\frac{K_1 \cdot l \cdot S}{F}\right) \cdot \exp\left(-\frac{K_3 \cdot l \cdot S}{F}\right) = C_0 \cdot \exp\left(-\frac{(K_1 \cdot l \cdot S + K_3 \cdot l \cdot S)}{F}\right). \quad (8.44)$$

And in recirculation mode, this combination is described as

$$\frac{C}{C_0} = \exp\left(-\frac{(K_1 + \Delta K_2 + \Delta K_3) \cdot l \cdot S \cdot \tau}{V}\right) = \exp\left(-\frac{(K_1 + \Delta K_2 + \Delta K_3) \cdot l \cdot S}{F}\right), \quad (8.45)$$

and

$$C = C_0 \exp\left(-\frac{(K_3 + \Delta K_1) \cdot l \cdot S}{F}\right). \quad (8.46)$$

Hence, we shall get the following equation:

$$C = C_0 \exp\left[-\frac{(K_1 + \Delta K_2 + \Delta K_3) \cdot l \cdot S + (K_3 + \Delta K_1) \cdot l \cdot S}{F}\right]. \quad (8.47)$$

Comparing Equations 8.44 and 8.47, one can see the positive effect of combination of filter and bubble-film extractor into recirculating system. The reason is the positive feedbacks arising in this system due to recirculation mode of water treatment.

It is very important to note that in the recirculation filtration–flotation system, there is no need to use more than one filter. One filtration module can operate as well as several filters in series. When the total surface area of the grains in the filter bed is sufficient, the adsorption of water contaminants takes place simultaneously through the following mechanisms: by gravitational sedimentation, by pore-blocking, by heterocoagulation, and by bioprecipitation [7,8,12,21]. In this case, the total adsorption efficiency is subjected to Equation 8.47, and the residual concentration of contaminants at water treating depends upon the working capacities of the filtering and flotation modules.

Thus, there are three positive feedbacks between the filter and the bubble-film extractor in the described water treatment system. Two of them are directed from the bubble-film extractor to the filter and the third from the filter to the bubble-film extractor. Due to these feedbacks, the adsorption rate of both filtration and bubble-film extraction stages is increased significantly. In such a way, the water treatment under recirculation mode shows a new quality, which can be named as *emergence*.

The magnitude of feedback contribution depends upon the amount of bacterial biomass in the filter bed and its lifetime as well as the presence of impurities in water, which support bacterial growth [11,20]. The organic compounds, easily digested by bacterial enzyme systems, are first adsorbed from contaminated water. And then, bacteria digest the more complex organic substances. The recirculation filtration–flotation system allows different impurities (with respect to dispersion degree and nature) to be removed simultaneously. These impurities are the organic and inorganic compounds, inorganic sediments, colloid and true solution of dissolved surfactants, dissolved gases, salts, etc. [1,2].

The other important feature of the recirculation filtration–flotation water treatment system is the possibility to minimize the overall dimensions of equipment maintaining the required water quality.

So, if the direct-flow sand filter of height l enables 10-fold decrease in the concentration of impurities in treated water, then the same result can be obtained after 10 cycles of recirculation through the filter with a height 0.1l, at the same filtering rate. And flotation device or any other water treatment unit is functioned similarly under recirculation mode. In this case, the residual concentration of any kind of water impurity can be reduced to acceptable level, considering that kinetics of withdrawal of i-kind component obeys the inverse exponent law.

8.7 STEP-BY-STEP DESCRIPTION OF DESIGNED SYSTEM

The effective unification of the bubble-film extractor and sand filter (or other filtration device) into joint water treatment system, which is able to act as *emergent water purifier*, must be based on the corresponding relationships in order to achieve the desirable parameters. These relationships are the functions of driving forces, rate constants of the removal of admixtures, dimensions of reaction spaces in the water treatment equipment, etc. They allow to predict the efficiency of water purification by direct-flow and recirculation modes of water treatment.

When using filters with granular load as a filtration medium, the suspended solids and colloids are adsorbed on the pore surface following the mechanisms of gravitational settling and hetero-coagulation. Organic impurities are removed from the water flow according to mechanisms of bio-precipitation and bio-oxidation. In this case, the process is mainly influenced by such parameters as the type of removed contaminants, the type and parameters of filter packing, and the flow rate through the filter.

In the bubble-film extraction device, the surface-active impurities, dissolved gases, and some ions are removed from the treated water flow. In this case, the physical and chemical properties of impurities, their concentration, the intensity of air bubbling, dimensions of flotation device, sizes of gaseous bubbles, and water flow rate have an influence on the process.

In the case of combined device (i.e., filtration–flotation water treatment under recirculation conditions), it is important to know the values of integrated constants of the filtration process, biofiltration process, bubble-film extraction process, gas exchange absorption process, and ion separation process in the functional units of combined device, that is, in the filter and in the bubble-film extractor, operating separately. The initial parameters for the device analysis are the required purification degree and treating time. And the calculated parameters are the integrated rate constants of such processes as filtration, biofiltration, flotation, bubble absorption, and dimensions of reaction spaces.

8.7.1 Constant Rate of Water Clarification at Direct-Flow Sand Filtration and Calculation of Recirculation Filtration Unit Parameters

The graded quartz sand with a size between 0.3 and 2.0 mm is used in direct-flow sand filters. The height of sand bed in these filters is between 0.5 and 2 m. The process of water clarification is realized as quick sand filtration or slow sand filtration depending on the fraction of used sand. In the first case, the filtration rate is about 5.5–14 m/h, and in the second case, it is about 0.1–0.2 m/h. The parameters of these processes were taken from [22–24]. They are represented in Table 8.1. The depicted data allow to determine the rate constants of water clarification at quick and slow sand filtration using Equation 8.32. For slow sand filtration with the height of packed bed equal to 0.5 m, the calculation gives the value of rate constant of water clarification equal to $7.7{\cdot}10^{-4}$ s^{-1}, if the initial turbidity level is 1500 mg/L, finish turbidity level is 1.5 mg/L, and filtration rate is 0.2 m/h. And for fast sand filtration, the value of rate constant of water clarification is equal to $5.26{\cdot}10^{-4}$ s^{-1}, if water turbidity is reduced by 50%, the height of filter bed is 2 m, and filtration rate is 5.5 m/h.

Comparing the obtained results, one can see that the difference between the rate constants of water clarification for both types of filtration is insignificant. These constants depend upon the nature of filter material and contaminants of water but not upon the *quick* or *slow* filtering conditions.

TABLE 8.1
Water Clarification Using Standard Direct-Flow Sand Filters

Type of Filter	Height of Packed Bed, mm	Grain Size, mm	Degree of Water Clarification	Filtration Rate, m/h	Filter Backwash Period
Fast	700–800	0.5–1.2	The water quality	5.5–14	8–12 h
	1300–1800	0.7–1.6			
	1800–2000	0.8–2.0	Initial turbidity 150 mg/L		
			Final turbidity 45–75 mg/L		
Slow	500	0.3–1	Initial turbidity 1500 mg/L	0.1–0.2	10–30 days
	50	1–2	Final turbidity 1.5 mg/L		

TABLE 8.2
Calculated Parameters of Water Clarification in Recirculation Water Treatment System

Volume of Treated Water, V, dm³	Degree of Clarification, %	Height of Filter Bed, l, dm	Cross Section of the Filter, S, dm²	Real Constant of Water Clarification, K_c, s⁻¹	Integrated Constant of Water Clarification under Recirculation, $\tilde{K}$, dm³/s	Water Treatment Period under Recirculation Conditions, τ, h
1	2	3	4	5	6	7
20	90	0.5	6	$6.48{\cdot}10^{-4}$	$1.94{\cdot}10^{-3}$	6.5
200	90	5	30	$6.48{\cdot}10^{-4}$	$9.72{\cdot}10^{-2}$	1.3
2000	90	5	30	$6.48{\cdot}10^{-4}$	$9.72{\cdot}10^{-2}$	13

Known value of the rate constant of water clarification at the direct-flow water filtration allows one to calculate the filter dimensions due to (8.39) and use the result for design recirculation filtration module. The results of these calculations are represented in Table 8.2.

It is seen that the use of recirculation filtration allows one to use the sand filter with smaller dimensions and to obtain the required degree of water clarification. In particular, to obtain 90% of water clarification ($C/C_0 = 0.9$) when treating from 20 to 2000 L of water at the rate constant of water clarification equal to $6.48{\cdot}10^{-4}$ s^{-1}, the calculated dimensions of the sand filter and required times of water treating are shown in the third, fourth, and seventh columns of Table 8.2. So, if one has enough time to implement the water treatment process in recirculation mode, then one can greatly reduce the dimensions of required devices, material consumption, and correspondingly, the cost of the designed equipment.

8.7.2 Constant Rate of Water Biofiltration at Direct-Flow Sand Filtration and Calculation of Recirculation Biofiltration Unit Parameters

Adsorption of bacterial cells from the filtered water to the grain surface in the filter material leads to bacterial growth under favorable ecological conditions. As a result, the biofilms appear at each grain surface. During the time, the biofilms gradually fill the entire surface of porosity space within the filter body. Thickness of biofilms on the surface of filter grains ranges from 50 to 10,000 nm [20].

The vital activity of bacterial cells inside the space of biofouling (see Equations 8.27 through 8.29) leads to removal of impurities from filtered water, which is the substrate of microbial metabolism. The products of bacterial metabolism and dead bacterial cells diffuse back into the filtered water bulk.

The process of biofilm growth obeys the law of chemical auto-synthesis [20]. Its basic equation is

$$x = x_0 \cdot \exp(\mu \cdot \tau), \tag{8.48}$$

where
- x is the finite mass of bacteria
- x_0 is the initial mass of bacteria
- μ is the specific rate of biomass growth
- τ is the current time

The magnitude of μ is a constant value under favorable conditions for microbial vital activity, and thus, the biofilm can growth exponentially.

But the sign and the magnitude of μ are changed at the deterioration of the factors influencing the activity of bacteria, such as decreasing concentration of nutrient substrates, fluctuation in temperature, pH, and redox potential of the filtered water. In this case, the magnitude of μ may decrease to zero. It is due to the lack of food for bacterial cells and their autolysis. So, there is a quick response of bacterial cells to the change of external conditions. Those bacteria and their colonies are able to survive, which have the largest amount of vital nutrients in the filtered water. Therefore, biofiltration takes an important part in modern water treatment technologies that is equivalent to water self-purification.

The technical characteristics of some selected examples of industrial biofilters are shown in Table 8.3. Using the data represented in this table, one can determine the rate constant of biofiltration and the productivity of biofiltration devices. For example, the rate constant of direct-flow biofiltration through sand filter is equal to $1.91 \cdot 10^{-3}$ s^{-1} (see the first row of Table 8.3). This result was obtained from Equation 8.30 at $l = 2.5$ m, $V = 10$ m/h, and $C/C_0 = 25/140$.

TABLE 8.3
Dimensions and Technical Characteristics of some Direct-Flow Biofilters

Name of the Device	Dimensions	Filtration Rate	Productivity of Biofiltration	Notice
Direct-flow biofilter with microorganisms in the sand bed	Height of the filter about 2.5 m	10 m/h—through coarse-grained packing, 20–30 m/h—through fine-grained packing	The nitrate ion concentration decreases—from 140 to 25 mg/L.	
Gravity biofilter with activated coal	Height of the filter bed—6 m	Surface filtration rate—10 m/h	60–100 g of organic substances (with respect to dissolved organic carbon) are decomposed on 1 m² of filtering surface at oxygen daily consumption 200–250 g.	
Membrane bioreactor			Denitrification rate: 5.8 g NO_3^-—N/m²/day. 4 g NO_3^-—N/m²/day. 6.1 g NO_3^-—N/m²/day.	Initial NO_3^- concentration was 200 mg/dm³; the degree of removal of NO_3^- was 90%.

Sources: Refs. [25–27].

If the average diameter of the biofilter grains is ~1 mm, and its cross-sectional area is equal to 1 m^2, then the total area of biofouling in the filter bed reaches 11,250 m^2 or about 4,500 m^2/1 m^3 of sand. The result was obtained using the following formula:

$$S_p = \frac{0.75 \cdot \tilde{V} \cdot \tilde{S}_i}{\tilde{V}_i} = \frac{0.75 \cdot \tilde{V} \cdot 4\pi \cdot r^2}{4/3 \cdot \pi \cdot r^3} = \frac{0.75 \cdot 3 \cdot \tilde{V}}{r}, \tag{8.49}$$

where

S_p is the surface area of porous media of the filter bed
0.75 is the void ratio of packing material
$\tilde{V}$ is the sand volume
$\tilde{V}_i$ is the volume of one grain of sand
$\tilde{S}_i$ is the surface area of one grain of sand

When the surface rate of biofiltration is equal to 10 m/h and conversion biofiltration efficiency is equal to 0.178 ($C/C_0 = 25/140 = 0.178$), 1 m^2 of biofouling surface will convert daily 2.45 g of nitrates. This result was obtained by using the equation

$$M = \frac{w \cdot S \cdot \Delta C \cdot \tau}{S_{bio}}, \tag{8.50}$$

where

M is the mass of bioconsumable impurities accounting to 1 m^2 of biofouling
w is the surface rate of biofiltration
ΔC is the difference of input–output concentrations of bioconsumable impurities
S is the cross-sectional area of the biofilter
τ is the process duration
S_{bio} is the total area of biofouling in the filter bed

In small-size sand filter, which is characterized in the first row of Table 8.2, the surface area of porous space of the filter bed is equal to 18 m^2. On this surface, covered by biofilm, 11 g of nitrates can be destructed during 6 h filtration time (or 0.1 g of nitrates by 1 m^2 of biofouling surface per 1 h) provided having the same hydrodynamic conditions and nitrate concentration as indicated in Table 8.3.

The activated carbon filters are more effective for biofiltration. As is seen from the second row of Table 8.3, when a filter with a volume of filter bed equal to 6 m^3 is used, 60–100 g of organic substances (with respect to dissolved organic carbon) can be decomposed at oxygen daily consumption equal to 200–250 g. This is due to the fact that the porous surface of grained activated carbon available for biofouling is 20–70 m^2/g (see Section 8.2.2). Therefore, total biofilm area in such filter could reach $4{\cdot}10^8$ m^2. And using only 150 g of charcoal grained packing, the surface area reaches a value from 3,000 to 10,500 m^2. It is about 166–583 times as higher as the surface available for biofouling in the sand filter, for which packing material mass is 6 kg and which consists of grains with a radius 0.5 mm (see the first row of Table 8.2). The foregoing information shows that combining sand and charcoal packing for water filtration at recirculation mode, one can achieve a significant improvement in the water quality due to the microfloral activity in the filter bed.

The biofiltration process under recirculation conditions does not require additional equipment. The only need is to get a developed nontoxic porous surface in the filter bed and to saturate treated water with the oxygen of air to create required living conditions for aerobic microorganisms. But in any case, to maintain this fouling at acceptable levels, one need to effectively remove the products of microbial metabolism from the treated water.

8.7.3 Generalized Constant Rate for Water Flotation and Calculation of Recirculation Bubble-Film Extraction

The rate of removal of surface-active contaminants from treated water using flotation depends upon many factors. In particular, it depends on the surfactant concentration in the water bulk, the magnitude of equilibrium constant of surfactant adsorption at the air–water interface, the adsorption rate of surfactants, air flow, and air bubble dimensions. The analytical equation that relates the volume and the surface concentration of the surfactant in adsorption process is the Langmuir adsorption equation (see Appendix 8.A):

$$\Gamma = \Gamma_{\infty}\left(\frac{C}{K_{eq}+C}\right)\cdot\{1-\exp[-(K\uparrow + K\downarrow)\cdot\tau]\}, \tag{8.51}$$

where

- Γ is the amount of surfactant adsorbed at time τ (its surface concentration)
- Γ_{∞} is the maximum surfactant amount adsorbed at the interface
- C is the surfactant concentration in bulk solution
- K_{eq} is the adsorption equilibrium constant of surfactant
- $K\uparrow$ is the constant rate of surfactant transfer to the interface from the subphase
- $K\downarrow$ is the constant rate of the reverse surfactant transition into the subphase
- τ is the duration of adsorption process

The sum of the quantities $K\uparrow$ and $K\downarrow$ in Equation 8.51 is nothing else than the total adsorption rate of surfactant at the interface. It is measured in s^{-1} and is noted by K_c. This, considering Equation 8.51, is transformed into the following form:

$$\Gamma = \Gamma_{\infty}\left(\frac{C}{K_{eq}+C}\right)\cdot\{1-\exp(-K_c\cdot\tau)\}, \tag{8.52}$$

In (8.51) and (8.52), the value of ($\Gamma_{\infty}\cdot(C/K_{eq} + C)$) represents the equilibrium value of adsorption at the air–water interface according to Langmuir model. And the component represented in the curved brackets is the correction factor, which takes into account the surfactant adsorption reducing in time τ, as compared to its equilibrium value. The phenomenon is arisen from the real time of surfactant molecule diffusion from subphase to the interface under the influence of the concentration gradient. The difference between these values is the greater, the smaller the diffusion coefficient of the surfactant molecules in the aqueous medium. The mechanism of this phenomenon is described in Appendices 8.A and 8.B.

If at a small value of τ, the magnitude ($-K_c\tau$) of Equation 8.52 is much greater than 1, the component represented in the curved brackets can be assumed to be 1. This means that the surfactant adsorption occurs without any potential barriers overcoming and diffusion restrictions. And the time factor does not have noticeable influence on the adsorption process, that is, adsorption of substances immediately reaches its equilibrium value at any volume concentration of surfactant. As a result, Equation 8.52 transforms to the classical equation, which describes the equilibrium adsorption following the Langmuir model:

$$\Gamma = \Gamma_{eq} = \Gamma_{\infty}\left(\frac{C}{K_{eq}+C}\right), \tag{8.53}$$

But if the exponent in Equation 8.52 is much less than 1, that is, $(-K_c \cdot \tau) \to 0$, the analysis of adsorption process in this case must be carried out by Equation 8.52 or by the following equation, which is obtained from (8.52)

$$\Gamma_\tau = \Gamma_\infty \left(\frac{C}{K_{eq} + C} \right) \cdot K_c \cdot \tau, \tag{8.54}$$

And if the value of C is much less than K_{eq}, that is, conformed by real conditions of contaminated water treatment when $C \ll K_{eq}$, then Equation 8.54 is transformed into the equation known as Henry's adsorption isotherm:

$$\Gamma = \Gamma_\infty \cdot \frac{C}{K_{eq}} \cdot K_c \cdot \tau, \tag{8.55}$$

where
- Γ_∞ is the maximum surfactant concentration at the interface
- K_{eq} is the adsorption equilibrium constant for the Langmuir model
- K_c is the adsorption rate constant
- τ is the duration of the adsorption process (in the case of flotation, it is equal to the time of air bubbles floating through the water volume)

Thus, taking into account that the magnitude of τ is the ratio of time to air bubbles passing through water bulk to their floating velocity, then Equations 8.52 and 8.55 will be written as follows:

$$\Gamma = \Gamma_\infty \left(\frac{C}{K_{eq} + C} \right) \cdot \left\{ 1 - \exp\left[-K_c \cdot \frac{l_1}{w_1} \right] \right\}, \tag{8.56}$$

and

$$\Gamma = \Gamma_\infty \cdot \frac{C}{K_{eq}} \cdot K_c \cdot \frac{l_1}{w_1}, \tag{8.57}$$

where
- l_1 is the distance of air bubbles passing through water bulk
- w_1 is the floating velocity of air bubbles

At given values of Γ, Γ_∞, C, K_{eq}, l_1, and w_1, we have no problem at calculating the magnitude of K_c following Equations 8.56 and 8.57 (see Appendices 8.A and 8.B).

In order to check the influence of K_c, the adsorption of sodium dodecyl sulfate (SDS), which is a typical detergent, was investigated in detail [3]. The diameter of air bubbles was 0.004 m, the floating velocity of the bubbles was about 0.20 m/s, and the distance of air bubbles passing through water bulk was 0.2 m. The results are shown in Table 8.4.

If one knows the adsorption rate and degree of surfactant adsorption at the surface of air bubbles at their floating, then one can calculate the surfactant mass flow from treated water depending upon the intensity of bubbling, air bubble's radius, and time of bubble floating. The surfactant mass flow change can be expressed by the following equation [3]:

$$\frac{dG}{d\tau} = -\Gamma \cdot 3\frac{F_1}{r} = -\Gamma_\infty \cdot \frac{C}{K_{eq}} \cdot K_c \cdot \frac{l_1}{w_1} \cdot 3\frac{F_1}{r}, \tag{8.58}$$

TABLE 8.4
Adsorption of SDS at the Surface of Air Bubbles of Radius 2·10⁻³ m Rising from a Depth of 0.2 m with a Velocity of 0.2 m/s

Surfactant Concentration in Water Bulk, C, mg/L	Adsorption of Surfactant on Air Bubble, Γ, mol/cm²	Adsorption of Surfactant, Γ, mg/cm²	Degree of Surfactant Adsorption on Air Bubble Surface during Their Floating, $\theta = S_0/S_1$, at S_0 = 30 A²
0.1	$6.66 \cdot 10^{-13}$	$1.92 \cdot 10^{-7}$	0.0012
0.2	$1.33 \cdot 10^{-12}$	$3.84 \cdot 10^{-7}$	0.0024
0.3	$2 \cdot 10^{-12}$	$5.76 \cdot 10^{-7}$	0.0036

and

$$\frac{dG}{d\tau} = -\Gamma \cdot 3\frac{F_1}{r} = \Gamma_\infty \cdot \left(\frac{C}{K_{eq}+C}\right) \cdot \left\{1 - \exp\left(-K_c \cdot \frac{l_1}{w_1}\right)\right\} \cdot 3\frac{F_1}{r}, \tag{8.59}$$

where
F_1 is the air flow rate for bubbling
r is the air bubble's radius

However, in the case when constant water volume is treated under recirculation conditions, the surfactant mass flow is expressed as

$$\frac{dG}{d\tau} = V\frac{dC}{d\tau}, \tag{8.60}$$

where V is the treated volume of contaminated water.

Then, Equations 8.58 and 8.59 will take the form

$$\frac{dG}{d\tau} = V\frac{dC}{d\tau} = -\Gamma \cdot 3\frac{F_1}{r} = -\Gamma_\infty \cdot \frac{C}{K_{eq}} \cdot K_c \cdot \frac{l_1}{w_1} \cdot 3\frac{F_1}{r}, \tag{8.61}$$

$$\frac{dG}{d\tau} = V\frac{dC}{d\tau} = -\Gamma \cdot 3\frac{F_1}{r} = -\Gamma_\infty \cdot \left(\frac{C}{K_{eq}+C}\right) \cdot \left\{1 - \exp\left(-K_c \cdot \frac{l_1}{w_1}\right)\right\} \cdot 3\frac{F_1}{r}, \tag{8.62}$$

$$\frac{dG}{d\tau} = V\frac{dC}{d\tau} = -\Gamma \cdot 3\frac{F_1}{r} = -\Gamma_\infty \cdot \frac{C}{K_{eq}} \cdot \left\{1 - \exp\left(-K_c \cdot \frac{l_1}{w_1}\right)\right\} \cdot 3\frac{F_1}{r}. \tag{8.63}$$

Equation 8.61 is valid in the case when the surfactant is characterized by very low rate constant of its adsorption at the interface and the concentration of flotation solution is very low. Equation 8.62 describes the situation when the rate constant of surfactant adsorption and the concentration of flotation solution are comparable. And Equation 8.63 corresponds to the condition when the surfactant concentration in water is low, but the magnitudes of rate constant of surfactant adsorption are quite different.

Selecting (8.63), one simulates the natural process of tap water purification from its surface-active contaminants with different molecular mass, ranging from simple detergents to finishing proteins.

TABLE 8.5
Parameters of the Flotation (Bubble-Film Treatment) under Recirculation Conditions

No	Treated Water Volume, V, L	Cycle Time of Water Treatment, τ, h	Extraction Degree of Surfactant, C/C_0	Adsorption Equilibrium Constant, K_{eq}, mol/cm³	Adsorption Velocity Constant of Surfactant at Air Bubble, K_c, s⁻¹	Distance of Air Bubbles Passing through Water Bulk, l, cm	Air Flow for Bubbling, F_1, L/min	Integrated Velocity Constant of Flotation, $\tilde{K}$, cm³/s
1	20	5.07	0.1	$0.269{\cdot}10^{-6}$	4.46	30	4.4	2.52
2	200	12.63	0.1	$0.269{\cdot}10^{-6}$	4.46	30	17.6	10.095
3	2000	12.5	0.1	$0.269{\cdot}10^{-6}$	4.46	150	176	101.4

By separating variables in (8.63) and then integrating this equation, one shall get

$$\int_{C_0}^{C}\frac{dC}{C}=-\Gamma_\infty\cdot\frac{1}{V\cdot K_{eq}}\cdot\left\{1-\exp\left(-K_c\cdot\frac{l_1}{w_1}\right)\right\}\cdot 3\frac{F_1}{r}\int_0^{\tau}d\tau, \tag{8.64}$$

$$\ln\frac{C}{C_0}=-\Gamma_\infty\cdot\frac{1}{V\cdot K_{eq}}\cdot\left\{1-\exp\left(-K_c\cdot\frac{l_1}{w_1}\right)\right\}\cdot 3\frac{F_1}{r}\cdot\tau. \tag{8.65}$$

Using Equation 8.65, one can calculate the basic parameters of the flotation process, that is, the magnitude of C/C_0 depending upon Q, K_c, K_{eq}, r, l_1, τ, and V, for the case of ideal (100%) removal of surfactants from the treated water under recirculation conditions. The data were obtained as is shown in Appendix 8.A. These devices were designed to transform the flow of air bubbles with adsorbed surfactant on their surfaces into the flow of surfactant-contained thin liquid films.

At constant intensity of air bubbling Q and known values of Γ_∞, K_c, K_{eq}, r, l_1, and w_1, the right-hand side of Equation 8.65 can be written as

$$\frac{C}{C_0}=\exp\left(-\frac{\tilde{K}\cdot\tau}{V}\right), \tag{8.66}$$

where $\tilde{K}$ is the integrated rate constant of the flotation process.

Parameters of realized flotation process under recirculation conditions are shown in Table 8.5.

As is seen from the data of Table 8.5, the airflow for bubbling from 4.4 up to 176 L/min is needed for a 10-fold decrease in SDS concentration in treated water volumes (20, 200, and 2000 L). The radius of floating air bubbles must be equal to 1 mm.

8.7.4 Water Degassing While Bubbling and Its Saturation by Atmospheric Oxygen

At the course of bubble-film extraction alongside with adsorption of surfactants at the air bubble surface, the volatile substances dissolved in water penetrate to the bubbles' internal surface. Due to this process, the rapid exchange absorption takes place in the bubbling water. As a result, the concentration of volatile substances in water is decreased. And the concentration of air is increased [28].

The difference in the concentration of volatile substances at the air–water interface is the driving force of their diffusion. According to Fick's law,

$$J = -D\frac{dC}{dx}, \tag{8.67}$$

where
- J is the volatile substance flow through the interface
- D is the diffusion coefficient of volatile substances
- dC/dx is the concentration gradient

Most commonly, the mass transfer process in the absorption system is described by two-film model [29]. In this case, the diffusion of volatile substances through gaseous and liquid films is considered individually. But the diffusion rate through water is decreased by about four orders of magnitude as in atmosphere. Therefore, the main diffusion resistance exists in the aqueous phase, exactly in the thin layer around the floating bubbles. So, the concentration gradient in the studied system is represented by the following equation:

$$\frac{dC}{dx} = \pm\frac{C - C_g}{\delta^1}, \tag{8.68}$$

where
- C is the volatile substance concentration in the aqueous phase
- C_g is the volatile substance concentration in the bubbles
- δ^1 is the thickness of the diffusion layer around the surface of floating air bubbles

In Equation 8.68, the plus sign concerns the case when volatile substance is transferred from aqueous phase into the bubble volume and the minus sign concerns the case when the diffusion of volatile substance occurs in opposite direction. And the concentration gradient has the plus sign when the process of water degassing is analyzed (i.e., the evolution of chlorine, chloroform, ammonia, methane, hydrogen sulfide, and other gases analyzed, if their concentration in air is considerably lower than that in aqueous phase).

Equations 8.67 and 8.68 describe the degassing process of water by means of air bubbling. The gas flow J through the air–water interface is described by the following equation:

$$J = \frac{V}{S}\cdot\frac{dC}{d\tau}, \tag{8.69}$$

where
- V is the degassing water volume
- S is the total area of the air–water interface in two-phase system
- C is the concentration of dissolved gas in water
- τ is the time

Combining (8.69), (8.67), and (8.68), we shall get

$$V\frac{dC}{d\tau} = \pm\left(-D\cdot S\cdot\frac{C - C_g}{\delta^1}\right), \tag{8.70}$$

or

$$\frac{dC}{C - C_g} = \pm\left(-\frac{D}{\delta^1}\cdot\frac{S}{V}d\tau\right). \quad (8.71)$$

Integration of differential equation (8.71) gives

$$\ln\frac{C - C_g}{C_0 - C_g} = \pm\left(-\frac{D}{\delta^1}\cdot\frac{S}{V}\cdot\tau\right), \quad (8.72)$$

$$C = \pm(C_0 - C_g)\cdot\exp\left(-\frac{D}{\delta^1}\cdot\frac{S}{V}\cdot\tau\right) + C_g, \quad (8.73)$$

where

C_0 is the initial gas concentration in aqueous phase
C is the final gas concentration in aqueous phase
C_g is the initial gas concentration in the space of bubble

Gaseous concentration can also be expressed in terms of partial pressure:

$$P = \pm(P_0 - P_g)\exp\left(-\frac{D}{\delta^1}\cdot\frac{S}{V}\cdot\tau\right) + P_g, \quad (8.74)$$

where

P is the current partial pressure of gas in aqueous phase
P_0 is the initial partial pressure of gas aqueous phase
P_g is the partial pressure of gas in air bubbles

Equations 8.73 and 8.74 show that the concentration of gas in aqueous phase at bubbling is the exponential function of τ, area of air–water interface, and airflow. Since the total area of the air–water interface in the system is

$$S = \frac{l_1}{w_1}\cdot 3\frac{F_1}{r}, \quad (8.75)$$

then finally, we obtain

$$C = \pm(C_0 - C_g)\exp\left(-\frac{D}{\delta^1}\cdot\frac{l_1}{V}\cdot 3\frac{F_1}{r}\cdot\frac{\tau}{w_1}\right) + C_g, \quad (8.76)$$

where

l_1 is the distance of air bubbles passing through water bulk
w_1 is the floating velocity of air bubbles
F_1 is the air flow
r is the mean radius of air bubbles

Equation 8.76 allows us to estimate the effectiveness of bubble aeration (i.e., oxygen saturation in water) and degassing of water depending upon the airflow, the distance of air bubbles passing

TABLE 8.6
Parameters of Absorption Component in the Circulation System

Treated Water Volume, V, l	Time of Water Treating, τ, h	Degree of Degasation, C/C_0	Specific Rate Constant of Mass Transfer, D/δ, m/s	Distance of Air Bubbles Passing through Water Bulk, l, m	Air Flow for Bubbling, F_1, L/min	Integrated Velocity Constant of Absorption, $\tilde{K}$, cm³/s
20	1.93	0.1	$2\cdot10^{-5}$	0.3	4.4	$3.3\cdot10^{-6}$
200	5.75	0.1	$2\cdot10^{-5}$	0.3	14.8	$1.1\cdot10^{-5}$
2000	5.8	0.1	$2\cdot10^{-5}$	1.5	29.58	$1.1\cdot10^{-4}$

through water bulk, their radius, and the process duration. The values of the diffusion coefficient D and boundary layer thickness δ^1 can be assumed as 10^{-9} m²/s and $0.5\cdot10^{-4}$ m, respectively [30,31]. The results of calculations for the bubbling systems with different capacities are shown in Table 8.6. It is shown that the surface-active impurities are removed from treated water alongside with water degassing, and water also is saturated with oxygen of air.

8.7.5 Water Desalination through the Bubble-Film Extraction Process

In the created system, due to combination of filtration and bubble-film extraction, the surface-active and many other impurities are removed from treated water. Alongside with colloidal matter, surfactants, and dissolved gases, some ions are also evacuated from treated water [18,32,33]. The ion withdrawal is predetermined by formation of double electric layers on the surface of air bubbles.

The double electric layer formation occurs at the air–water interfaces owing to different reasons. First of all, the double electric layer may arise as a result of specific orientation of water molecules inside the boundary layer. So, it may be formed due to dipole–ion interactions. Second, the double electric layer may arise as a result of adsorption to the surface of hydroxyl ions, halide ions, surfactants, and others. So, it may be formed due to ion–ion interactions [34]. For example, the hydroxyl ions, which are present in water due to its dissociation, are hydrophobic when compared with protons. And their concentration at the interface is much greater than that in the water bulk.

The chloride, bromide, iodide, and some other anions offer the advantage of being surface active, too. Their surface excess at the interface can be considered as one of the plates of the electric double layer. And the diffuse part of the double electric layer can be considered as the second plate, which consists of hydrated protons and salt cations.

In desalted and free-from-surfactant water, the electric double layer on the surface of the air bubbles may also include the charges and dipoles. In this case, the double electric layer is formed at the cost of dipoles of water molecules and charges of hydroxyls and hydrated protons. These ions are always present in water as a result of their dissociation. But if the aqueous medium contains salts, then the dissolved anions and cations compete with protons and hydroxyls. The law of mass action predetermines the result of their competition.

The data of many studies show the excess of cations in the diffuse parts of the electrical double layers arisen at the air bubble surface [34]. The electric potential difference between deionized water and air varies in a wide range, and zeta potential of air bubbles in desalinated water reaches the value −110 mV, and it reduced to −75 mV in the presence of $5\cdot10^{-4}$ M NaCl. Some of the results, obtained by different authors, are represented in Table 8.7.

It is reasonable to note that the electric double layer, spontaneously occurring at the air–water interface due to reorientation of the dipoles of water molecules and adsorption of hydroxyl ions, cannot accumulate essential density of electric charges, because there is no sufficient amount in aqueous phase. This can be seen in Appendix 8.C.

TABLE 8.7
Magnitudes of Potential Surface of Water

χ H_2O, V	References
+0.50	Eley D.D., Evans M.G.//*Trans. Faraday Soc.* 1938. V.34. P.1093.
+0.29	Passoth G.//*Z. Phys. Chem.* 1954. Bd.203. S.275.
+0.20	Stokes R.H.//*J. Amer. Chem. Soc.* 1964. V.86. P.979.
+0.06	Krishtalik L.I., Alpatova N.M., Ovsyannikova E.V.//*J. Electroanal. Chem.* 1992. V.329. P.1.
−0.08	Noyes R.M.//*J.Amer. Chem. Soc.* 1964. V.86. P.971.
−0.17	Измайлов Н.А.//*Докл. АН СССР*. 1963. Т.149. Р.1364.
−0.30	De Lidny C.L., Alfenaar M., Veen V.D. 1968. V.87. P.585.
−0.30	Hash N.S.//*Austr. J. Sci. Res.* 1948. V.A1. P.480.
−0.30	Мищенко К.П., Квят Э.И.//Журн. физ. хим. 1954. Р.28. Р.1451.
−0.34	Latimer W.M., Pitzer K.S., Slansky C.M./*J. Chem. Phys.* 1939. V.7. P.108.
−0.36	Strelov H.//*Z. Electrochem.* 1952. Bd.56. P.119.
−0.39	Halliwell H.F., Nyburg S.C.//*Trans. Faraday Soc.* 1963. V.59. P.1126.
−0.48	Vervey E.//*Rec. trav. Chim.* 1942. V.61. P.127.
−1.10	Solomon M.//*J. Phys. Chem.* 1970. V.74. P.2519.

But the situation will change if the air bubble surface will be filled with densely packed adsorption monolayer of ionized surfactants. In this case, the charge density of monolayer can reach 1 C/m^2. And the equal quantity of charges of opposite sign forms the diffuse part of the double electric layer. Due to this phenomenon, the flotation device allows to remove some ionic impurities from treated water, which contain sufficient amount of surfactants. And the exhaustion of surfactant at air bubble surface leads to the end of ion evacuation with bubble-film flow. The dissolved ions are removed due to the action of surfactants as cofactors of flotation.

Thus, the formation of *powerful* electric double layer on the surfaces of floating air bubbles, covered by adsorbed surfactants, makes the removal of ions released from dissociated mineral impurities possible. The selected surfactants should be strong fatty acids, strong fatty bases, or other ionizable suitable compounds with high affinity to the air–water interface.

We have examined the behavior of the adsorption fatty acid monolayer at the air bubble surface in respect to coexistence of ionized and neutral molecules and surface-inactive ions. We can write the dissociation equation for fatty acid, which is the object of the study, as follows:

$$ROH = RO^- + H^+ \tag{8.77}$$

where

ROH is the fatty acid nonionized molecule
RO^- is the surface-active fatty acid anion
H^+ is the surface-inactive cation

To evaluate the ion-transporting ability of air bubble coated with a monolayer of fatty acid, we shall write the expressions for the dissociation constant of fatty acid in the bulk of water and in the state of adsorbed monolayer. They are as follows:

$$K_A = \frac{\left[RO_v^-\right]\cdot\left[H_v^+\right]}{[ROH_v]}, \tag{8.78}$$

$$K_{AS} = \frac{\left[RO_S^-\right]\cdot\left[H_S^+\right]}{[ROH_S]}, \tag{8.79}$$

where K_A and K_{AS} are the equilibrium dissociation constants of fatty acid in aqueous bulk and in adsorption monolayer, respectively.

The indexes S and v represent the surface and the bulk concentration of fatty acids and their dissociation products. The relationships between these concentrations are as follows:

$$[ROH_S] = [ROH_v]\cdot\delta, \tag{8.80}$$

$$\left[RO_S^-\right] = \left[RO_v^-\right]\cdot\delta\cdot i, \tag{8.81}$$

$$\left[H_S^+\right] = \left[H_v^+\right]\cdot\delta\cdot j. \tag{8.82}$$

where

i and j are the anion/cation concentration ratio of fatty acids in the space of adsorbed monolayer
δ is the thickness of adsorbed monolayer

And then,

$$K_{AS} = \frac{\left[RO_v^-\right]\cdot\delta\cdot i\cdot\left[H_v^+\right]\cdot\delta\cdot j}{[ROH_v]\cdot\delta}. \tag{8.83}$$

At the air–water interface, the molecules of fatty acid and its anions form the negatively charged plate of the electric double layer (for the case when water does not contain any additional electrolytes). And hydrated protons (i.e., H_3O^+) form the diffuse part of the electrical double layer, that is, these ions forming a second plate of the electrical condenser. And the local proton density depends exponentially on the electrostatic potential according to the following equation:

$$\left[H_Z^+\right] = \left[H_v^+\right]\cdot\exp\left(-\frac{e\cdot\varphi_Z}{KT}\right), \tag{8.84}$$

where

$[H_Z^+]$ is the local concentration of protons in the diffuse part of the electrical double layer
φ_z is the electrostatic potential at the point of determination of the local proton concentration

So

$$\left[H_S^+\right] = \left[H_v^+\right]\cdot j\cdot\delta = \left[H_v^+\right]\exp\left(-\frac{e\cdot\varphi_S}{KT}\right)\cdot\delta, \tag{8.85}$$

where φ_S is the electrostatic potential at the interface.

Taking into account (8.85), we shall write the balance of concentrations of fatty acid molecules and surfactant anions in the monolayer space:

$$\left[RO_S^-\right] + \frac{\left[RO_S^-\right]\cdot\left[H_v^+\right]\exp(-e\cdot\varphi_S/KT)\cdot\delta}{K_{AS}} = n, \tag{8.86}$$

where n is the total concentration of fatty acid and surfactant anions at the air–water interface, expressed in moles per unit area of the monolayer.

Regarding the surface concentration of fatty acid anions, solution of (8.86) yields

$$\left[RO_S^-\right] = \frac{n}{1+\dfrac{\left[H_v^+\right]\exp(-e\cdot\varphi_S/KT)\cdot\delta}{K_{AS}}}. \qquad (8.87)$$

But when $[H_v^+] = 10^{-\mathrm{pH}}$, $n = 1/S_i \cdot N$, where S_i is the area per molecule of fatty acid in adsorbed monolayer, and N is Avogadro's number, then (8.87) is transformed to the following form:

$$\left[RO_S^-\right] = \frac{1}{S_i\cdot N}\left[1+\frac{10^{-\mathrm{pH}}\cdot\exp(-e\cdot\varphi_S/KT)\cdot\delta}{K_{AS}}\right]^{-1}. \qquad (8.88)$$

Multiplication of both sides of Equation 8.88 by Avogadro's number N and electron charge e describes the relationship between the surface charge density σ and the boundary potential jump φ_S in the monolayer of fatty acid:

$$\left[RO_S^-\right]\cdot N\cdot e = \sigma = -\frac{1}{S_i}\cdot e\cdot\left[1+\frac{10^{-\mathrm{pH}}\cdot\exp(-e\cdot\varphi_S/KT)\cdot\delta}{K_{AS}}\right]^{-1}. \qquad (8.89)$$

And substitution of (8.89) into the Gouy–Chapman equation gives the function of the surface potential φ_S upon the surface charge density σ in fatty acid monolayer and the packing density of molecules at different ionic concentration in subphase:

$$\varphi_S = \frac{2KT}{e} Arsh \frac{(1/S_i)\cdot e\cdot\left[1+10^{-\mathrm{pH}}\cdot\exp(-e\cdot\varphi_S/KT)\cdot(\delta/K_{AS})\right]^{-1}}{A\cdot\sqrt{C}}, \qquad (8.90)$$

where $A = \sqrt{8N\cdot K\cdot T\cdot\varepsilon_0\cdot\varepsilon_1}$.

The transcendental equation (8.90) allows to calculate the surface charge density σ and surface potential jump φ_S at the air bubble surface covered with the monolayer of fatty acid or any other ionizable surfactant, depending upon the values of pH, S_i, K_{AS}, δ, and C.

The results of the calculation of the charge density σ at bubble surface, obtained by Equation 8.90 at $C = 10^{-6}$–10^{-1} M, $S_i = 26\cdot10^{-20}$ m^2, $K_{AS} = 8\cdot10^{-14}$ M/dm^2, and $\delta = 3\cdot10^{-9}$ dm, have showed calculated magnitudes in the range $1\cdot10^{-2}$–$6\cdot10^{-1}$ K/m^2.

The surface charge density of adsorbed fatty acid monolayer increases proportionally with the increase in their ionization constants, ionic strength of the subphase, pH, and packing density of the surface-active ions in the space of monolayer. Maximum value of the surface charge density corresponds to the case when $K_{AS}/\delta = 1\cdot10^{-1}$ M and the minimum when $K_{AS}/\delta = 1\cdot10^{-5}$ M.

The strong fatty acid leads to the formation of monolayer with strong ionization negative charge whose value under investigated conditions is virtually independent of ionic strength and pH of subphase. The weak fatty acid leads to the formation of monolayer with negative charge of smaller magnitude and strongly dependent upon ionic strength and pH of subphase. The same results, but with charges of opposite sign, could be observed in the case of study of monolayer consisting of surface-active bases.

The results obtained by Equation 8.90 allow one to conclude that close-packed adsorption monolayer of strong fatty acid or strong fatty base, localized at the air bubble surface, generates *powerful* surface charges. These surface charges are compensated by counterions of the diffuse part of the double electric layer in the subphase on the mechanisms of ion exchange and trivial shielding.

In the flotation device, the flow of *charged* air bubbles acts as collector and transporter of ionic impurities of treated water. Through the bubble-film extractor, this flow is removed from treated water. So, the mineral composition of treated water can be corrected, and the *target* mineral impurities can be extracted. The efficiency of demineralization of treated water can be evaluated if the relationship of bubble charge density upon pH and ionic concentration of water is known. An appropriate empiric equation is the following:

$$\sigma = \sigma_0 + B \cdot C, \tag{8.91}$$

where

σ is the surface charge density in the ionized monolayer at a given ionic strength of subphase
σ_0 is the hypothetical surface charge density in the monolayer at *zero* ionic concentration of the subphase
B is the correction factor

Equation 8.91 is valid under the following conditions: C = 1·10^{-2}–2.5·10^{-2} M, pH = 6–8, S_i = 25·10^{-20}–50·10^{-20} m^2. K_{AS}/δ = 1·10^{-1}–1·10^{-5} M, δ = 30·10^{-10} m.

With respect to (8.91), the flow of ions initiated by charged air bubbles floating to the air–water interface can be expressed by the following equation:

$$j = \frac{\sigma}{e \cdot N} \cdot \frac{3F_1}{r} = (\sigma_0 + B \cdot C) \cdot \frac{3}{e \cdot N} \cdot \frac{F_1}{r}, \tag{8.92}$$

where

j is the flow of ions collected by air bubbles
F_1 is the air volumetric flow rate
r is the radius of air bubbles
e is the elementary charge
N is Avogadro's number

On the other hand, the flow of ions at air bubble surface is the reason of decrease in their concentration in treated water volume:

$$j = \frac{V \cdot dC}{d\tau}. \tag{8.93}$$

In this case, the combination of (8.92) and (8.93) gives

$$\frac{V \cdot dC}{d\tau} = (\sigma_0 + B \cdot C) \cdot \frac{3}{e \cdot N} \cdot \frac{F_1}{r}. \tag{8.94}$$

We may designate the value of (σ_0 + $B \cdot C$) as X, then

$$\int \frac{dC}{X} = \int \frac{3F_1 \cdot dr}{e \cdot N \cdot r \cdot V}. \tag{8.95}$$

The integration of (8.95) from C_1 to C_2 and from 0 to τ gives

$$\frac{1}{B}\cdot\ln(\sigma_0+B\cdot C_1)-\frac{1}{B}\cdot\ln(\sigma_0+B\cdot C_2)=\frac{3F_1\cdot\tau}{e\cdot N\cdot r\cdot V}, \quad (8.96)$$

or

$$\ln\left[\left(\frac{\sigma_0+B\cdot C_1}{\sigma_0+B\cdot C_2}\right)\right]=B\cdot\frac{3F_1\cdot\tau}{e\cdot N\cdot r\cdot V}. \quad (8.97)$$

Equation 8.97 allows one to determine the time required to reduce the ion concentration from C_1 to C_2 in water volume V (without return of extracted substances to the treated water through the bubble-film extractor), depending on the air volumetric flow rate F_1 and the radius of floating air bubbles r, if these bubbles are filled with monolayers of fatty acid or fatty base with known parameters σ_0 and B.

Maximum ion-transporting efficiency of bubble flow depending upon the volumetric airflow rate and the radius of floating air bubbles can be found from the following equation:

$$j=\frac{V\cdot dC}{d\tau}=\frac{\sigma}{e\cdot N}\cdot\frac{3F_1}{r}, \quad (8.98)$$

or

$$\Delta\tau=\frac{e\cdot N\cdot r\cdot V}{3F_1\cdot\sigma}\cdot\Delta C. \quad (8.99)$$

Here σ is the ultimately realized surface charge density in a monolayer of *strong* fatty acid or a *strong* fat base. It will be about 1.0 K/m^2, if the specific area of the monolayer will be equal to $16{\cdot}10^{-20}$ m^2. In this case, 1 m^3 of the air sprayed to form bubbles with radius 1 mm allows us to transfer about $8{\cdot}10^{-2}$ g of salt impurities. And when the radius of air bubbles will be 0.1 mm, then 1 m^3 of air will be able to remove 0.8 g of salt impurities. It will be cations or anions depending on the nature of surfactants adsorbed at the bubble surfaces. Monolayers of ionized fatty acids will entrain the cations, and monolayers of ionized fatty bases will entrain the anions.

In conclusion, it is reasonable to note that when the aerosol consisting of concentrated solution of suitable fatty acid (or of suitable fatty base) is injected into the ejector's channel (which creates a stream of air bubbles in the space of the flotation unit), the ionized close-packed adsorption monolayer of fatty acid or fatty base at the surface of air bubbles is generated automatically.

8.8 SCHEMES OF DESIGNED DEVICES AND THEIR OPERATION MODE

As followed from the results, represented in Sections 8.7.1 through 8.7.5, the combination of filtration and flotation processes of water treatment under recirculation conditions has a great potential to reduce the number and size of functional units of required equipment in comparison with direct-flow devices. This is the first essential advantage of the created system.

The second advantage is the use of positive feedbacks that are arisen in the system at the recirculation mode of action due to saturation of filtered water with oxygen of air and effective removal of products of microbial metabolism, as described in Sections 8.5 and 8.6. The scheme of combined filtration–flotation device that operates continuously, under recirculating conditions (*rotating wheel*), is shown in Figure 8.14. In this device, the processes of filtration, biofiltration, bubble aeration (exchange absorption), and bubble-film extraction are realized in a common reaction space, separated by the unit case 1. The centrifugal pump 2 is used to ensure the circulation of treated

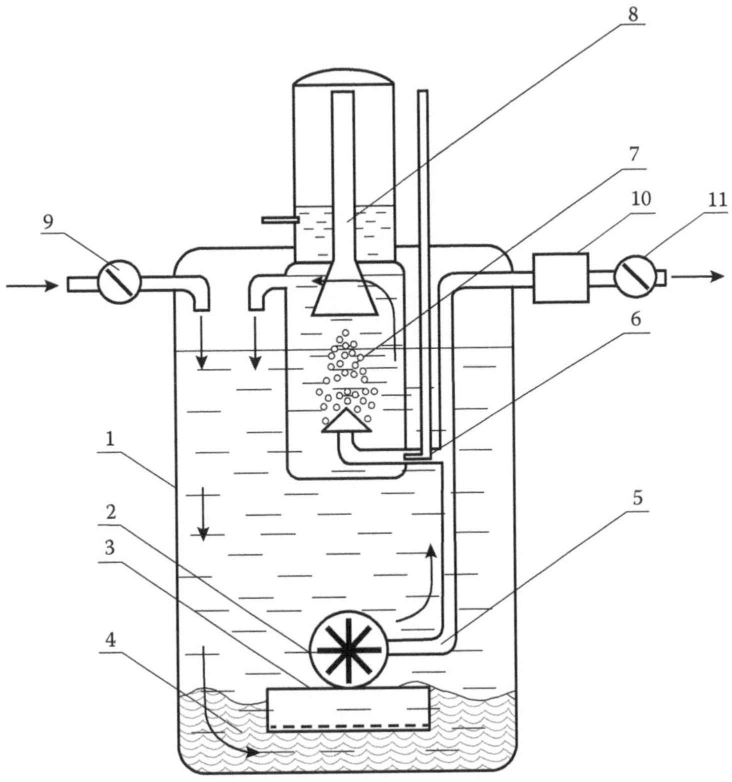

FIGURE 8.14 The scheme of filtration–flotation device. 1: unit case; 2: centrifugal pump; 3: drainage device; 4: sand; 5: pump inlet line; 6: ejector; 7: bubbling compartment; 8: bubble-film extractor; 9: water inlet; 10: UV sterilizer; 11: water outlet.

water. The pump 2 is connected with a drainage unit 3 and generates the water flow through the space of granular filter material 4, which consists of quartz sand or mixture of sand and granular mesoporous activated carbon.

At first, the packing material functions as the adsorption filter and hetero-coagulator. But as the biofouling grows on the filter grains, the sand filter functions as biofilter that combines the functions of bio-precipitator and bio-catalytic reactor.

Discharge line 5 directs the water flow after the pump through ejector 6 into the bubble-film extraction module. This module consists of bubbling compartment 7 and the real bubble-film extractor 8. The water flow is directed through Laval nozzle to form a high velocity flow, and in ejector 6, the air is injected to form the water–air mixture, which is supplied into bubbling compartment 7.

During floating time, the air bubbles adsorb the surfactants and inorganic impurities, which are complementary to adsorbed surfactants, and absorb dissolved gases. The treated water is also saturated with the oxygen of air bubbles.

The impurities collected at air bubble surface are removed from the flotation space through the bubble-film extractor. And treated water enriched with oxygen in the bubbling compartment 7 then pumped through the filter. The processes of filtration and flotation are repeated as long as the concentration of impurities in treated water is reduced to the required level. The time needed to purify the given water volumes are determined by the integral rate constant of withdrawal of i-kind of impurities, and by their initial and residual concentration. Time is calculated according to (8.47).

At the functioning of designed water purifier, the residual concentration of i-type contaminants in the treated water can be reduced to the required level, taking into account that the kinetics of impurities removal obeys the inverse exponential function. Thus, the basic mechanisms of impurities removal from water volume through the sand are the same as in the case of any other granular filter. In particular, the coarse impurities are adsorbed by filter's surface to form loose sediment and block partially the pores of filter media [21]. Colloidal particles form coagulum inside the filter media according to the hetero-coagulation mechanisms [8,12,35].

But the distinguished feature of the use of granular filter in conjunction with flotator (bubble-film extractor) under recirculation conditions is the fact that the water is enriched with oxygen of air and then is fed continuously through the porous space of the filter material. And at the same time, the surface-active products of microbial metabolism are removed from that water flow due to the bubble-film extraction.

Saturation of filtered water with oxygen of air and withdrawal of surfactant inhibitors of microbial metabolism lead to growth of biomass within the filter sand, and thus the sand filter is transformed into the biofilter. Thus, the positive feedback is realized between the functional units of the complex filtration–flotation device. The contribution of these components to the efficiency of water purification under recirculation conditions depends upon the age and form of the species of biofouling and also upon the presence of impurities in the filtered water flow, which are the substrate supply for bacteria [19].

First of all the biofouling absorbs from the water flow the organic substances that are easily decomposed by enzymes of bacterial cells. As the concentration of these substances is reduced, biofouling digests the more complex organic compounds. Due to this process, the water becomes much more transparent at the filter outlet. And the waste products of biofouling, such as carbon dioxide, endogenous surfactants, and the hardly decomposed fragments of dead bacterial cells, are entered into the filter output. These substances are removed from treated water volume through the bubble-film extractor. At the same time, the mineralization level of the treated water is also decreased.

Thus, the created devices allow us to remove impurities from treated water, which belong to different classes in accordance with their dispersion degree, origin, and nature. Finite sterilization of water can be carried out by ultraviolet using bactericidal bulbs of suitable capacity. The different processes of impurities removal through the created devices are illustrated in Figure 8.15.

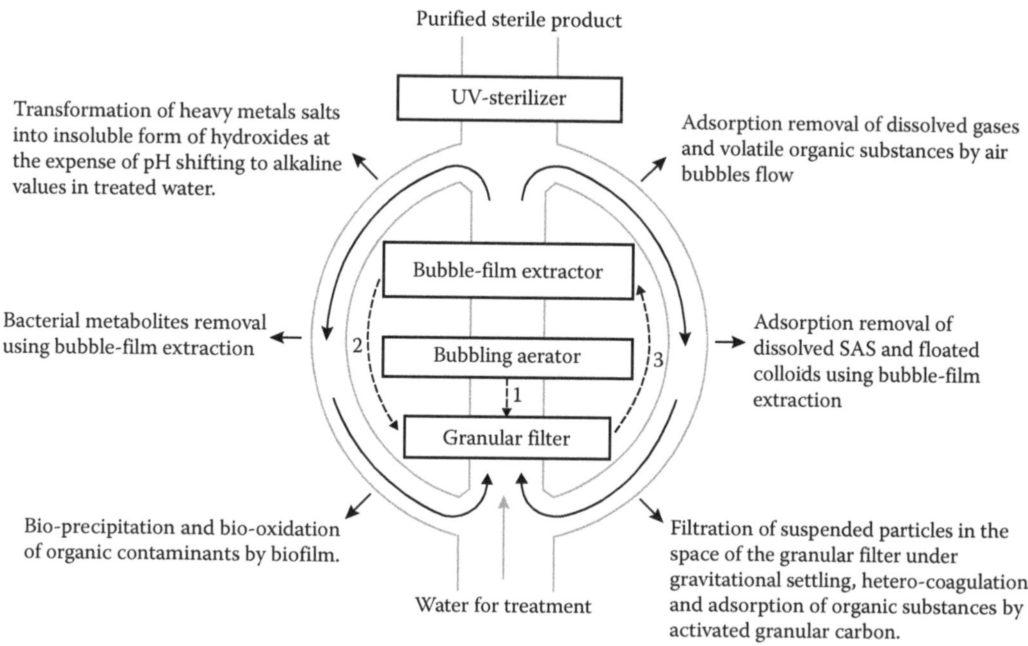

FIGURE 8.15 Topography of the processes of impurities removal through the created devices (1, 2, 3: positive feedbacks in the system).

8.A APPENDIX A

8.A.1 EQUILIBRIUM AND DYNAMIC ADSORPTION OF SURFACTANTS AT THE AIR–WATER INTERFACE

This section concerns the equilibrium adsorption of surfactants and kinetics of this process in aqueous systems. Flotation of SASs is based on this phenomenon. Equilibrium adsorption of surfactants at the air–water interface is described by Gibbs's theory, which shows the relationship between the surface excess of surfactants (SAS), surface tension of SAS solution, and its concentration. Many authors studied the adsorption phenomenon that occurs at the interfaces. Numerous mathematical models have been proposed.

In the following, we consider the equilibrium and kinetics of adsorption of surfactants at the air–water interface on the basis of Langmuir's theory of adsorption of gases on solids. According to Langmuir's theory, it is assumed that the adsorption surface consists of N_S sites, which can be occupied by adsorbed molecules. These sites correspond to the minimum of surface free energy. At achieving the balance between the adsorbed molecules and molecules of gas, only some parts of the potentially available adsorption sites are occupied by gas molecules. This part is equal to θ, and the total number of molecules $N\alpha$ adsorbed on the surface obeys the ratio

$$N_a = N_S \cdot \theta. \tag{8.A.1}$$

The surface sites occupied by adsorbed molecules are not available for other molecules. In this case, only free sites of the surface are available for further adsorption. This part is equal to (1 − θ). And if we assume that the adsorption sites are equivalent and have unit fixed occupancy, then in accordance with the Hertz–Knudsen equation [9], the adsorption rate of molecules from the gaseous phase is determined by the following equation:

$$\left(\frac{dN_a}{d\tau}\right)_{ads} = \Omega \cdot n \cdot (1-\theta) \cdot \left(\frac{KT}{2\pi \cdot m}\right)^{1/2}, \tag{8.A.2}$$

where

Ω is the total area of adsorption surface

n is the number of adsorbed molecules per volume unit of the subphase (air space)

K is the Boltzmann constant

T is the temperature

m is the mass of the adsorbed molecules

And if no other molecules can be adsorbed, then those which velocity components V exceed threshold value V_0, then Equation 8.A.2 is transformed into the following form:

$$\left(\frac{dN_a}{d\tau}\right)_{ads} = \Omega \cdot n \cdot (1-\theta) \cdot \left(\frac{kT}{2\pi \cdot m}\right)^{1/2} \cdot \exp\left(-\frac{1}{2} \cdot \frac{m \cdot V_0^2}{kT}\right). \tag{8.A.3}$$

According to Langmuir, the rate of desorption, which is proportional to value N_S, can be written as

$$\left(\frac{dN_a}{d\tau}\right)_{des} = \theta \cdot N_S \cdot \nu \cdot \exp\left(-\frac{\Psi}{kT}\right), \tag{8.A.4}$$

where $\nu \exp(-\Psi/kT)$ is the rate constant of desorption per one molecule, or the average probability of the desorption of molecule into subphase bulk per 1 s.

The sum of Equations 8.A.3 and 8.A.4 describes the resulting adsorption rate with respect to the total number of molecules adsorbed on the surface and molecules desorbed from the surface, that is,

$$\frac{dN_a}{d\tau} = \Omega \cdot n \cdot (1-\theta) \cdot \left(\frac{kT}{2\pi \cdot m}\right)^{1/2} \cdot \exp\left(\frac{m \cdot V_0^2}{2kT}\right) - \theta \cdot N_S \cdot \nu \cdot \exp\left(-\frac{\Psi}{kT}\right). \quad (8.A.5)$$

In accordance with (8.A.5), this velocity has the dimension of molecules per second, and the first component of the right-hand side of Equation 8.A.5 describes the flow of molecules from the subphase to the adsorption surface, and the second component of this equation describes the reverse flow of molecules (i.e., from the surface into gaseous space). Taking into account that the pressure P and the concentration n of adsorbed molecules in the gaseous phase are related by ratio,

$$P = n \cdot k \cdot T, \quad (8.A.6)$$

the following general expression of adsorption rate for the adsorbed gas in the terms of pressure was obtained by Langmuir:

$$\frac{dN_a}{d\tau} = \Omega \cdot P \cdot (1-\theta) \cdot \frac{1}{\sqrt{2\pi \cdot m \cdot k \cdot T}} \cdot \exp\left(-\frac{m \cdot V_0^2}{2kT}\right) - \theta \cdot N_S \cdot \nu \cdot \exp\left(-\frac{\Psi}{kT}\right). \quad (8.A.7)$$

And for the state when the flow of molecules is equal in both directions (the right-hand side of (8.A.7) is equal to zero), Langmuir has derived the equation of equilibrium adsorption. It relates the degree of adsorption surface filling θ and pressure P of the adsorbed gas, and it is derived from (8.A.7):

$$\theta = \frac{\Omega \cdot P \cdot (1/\sqrt{2\pi \cdot m \cdot k \cdot T}) \cdot \exp\left(-m \cdot V_0^2/2kT\right)}{\Omega \cdot P \cdot (1/\sqrt{2\pi \cdot m \cdot k \cdot T}) \cdot \exp\left(-m \cdot V_0^2/2kT\right) + N_S \cdot \nu \cdot \exp(-\Psi/kT)}, \quad (8.A.8)$$

or

$$\theta = \frac{P}{P + \dfrac{N_S \cdot \nu \cdot \exp(-\Psi/kT)}{\Omega \cdot (1/\sqrt{2\pi \cdot m \cdot k \cdot T}) \cdot \exp\left(-m \cdot V_0^2/2kT\right)}}, \quad (8.A.9)$$

or

$$\theta = \frac{1}{1 + \dfrac{N_S \cdot \nu \cdot \exp(-\Psi/kT)}{\Omega \cdot n\sqrt{kT/2\pi \cdot m} \cdot \exp\left(-m \cdot V_0^2/2kT\right)}}. \quad (8.A.10)$$

And after mathematical transformation,

$$\theta = \frac{1}{1 + \dfrac{N_S \cdot N \cdot \nu \cdot \exp(-\Psi/kT)}{\Omega \cdot h \cdot N \cdot n \cdot (1/h)\sqrt{kT/2\pi \cdot m} \cdot \exp\left(-m \cdot V_0^2/2kT\right)}}. \quad (8.A.11)$$

Here

$N_S/\Omega \cdot N \cdot h = C_S^\infty$ is the maximal adsorption capacity, mol/m^3

$n/N = C$ is the adsorptive concentration, mol/m^3

Thus,

$$\theta = \frac{C}{C + C_S^{\infty} \dfrac{\nu \cdot \exp(-\Psi/kT)}{\sqrt{kT/2\pi \cdot m \cdot h^2} \cdot \exp\left(-m \cdot V_0^2/2kT\right)}}, \tag{8.A.12}$$

or

$$\theta = \frac{C}{C + C_S^{\infty}(K\downarrow/K\uparrow)}. \tag{8.A.13}$$

Here

$K\downarrow = \nu \cdot \exp(-\Psi/kT)$, s^{-1}

$K\uparrow = \sqrt{kT/2\pi \cdot m \cdot h^2} \cdot \exp(-m \cdot V_0^2/2kT)$, s^{-1}

The symbol $K\uparrow$ in Equation 8.A.13 corresponds to the rate constant of adsorptive molecules transferred from the subphase to the adsorption surface, and the symbol $K\downarrow$ corresponds to rate constant for the desorption process. The ratio $K\downarrow/K\uparrow$ is nothing else than the dimensionless equilibrium constant of the adsorption, and the value $C_S^{\infty} \cdot (K\downarrow/K\uparrow)$ is the equilibrium adsorption constant that has the dimension of concentration.

Similarly, the derivation of the Langmuir equation of gas adsorption allows us to consider the kinetics and the equilibrium adsorption of surfactants in the water–air system. In this system, the diffusion resistance layer controls the transfer of surfactant molecules from aqueous phase to its surface. The diffusion flux density J obeys Fick's equation, that is,

$$J = -D\frac{dC}{dx}, \tag{8.A.14}$$

where

D is the diffusion coefficient

dC/dx is the adsorptive concentration gradient in the diffusion layer

At the linear profile of surfactant concentration in the diffusion layer of thickness δ, the total number of surfactant molecules with sufficient energy to overcome this layer through area Ω per time unit is subjected to the following equation:

$$\left(\frac{dN_a}{d\tau}\right)_{ads} = D \cdot \frac{C}{\delta} \cdot \Omega \cdot N \cdot (1-\theta), \tag{8.A.15}$$

where

C is the surfactant concentration in subphase

$(dN_a/d\tau)_{ads}$ is the number of surfactant molecules that overcome the diffusion layer

If the additional energy for adsorption of surfactant molecules is required due to the potential barriers in the adsorption system, then, similarly with (8.A.3), the exponential factor is introduced into Equation 8.A.10, that is,

$$\left(\frac{dN_a}{d\tau}\right)_{ads} = D \cdot \frac{C}{\delta} \cdot \Omega \cdot N \cdot (1-\theta) \cdot \exp\left(-\frac{U_a}{kT}\right), \tag{8.A.16}$$

where U_a is the factor that characterized the activation energy of adsorption.

And the number of surfactant molecules in desorption flow $(dN_a/d\tau)_{des}$ is still described by Equation 8.A.4. Thus, the equilibrium state in the system is described by the following equation:

$$\frac{dN_a}{d\tau}=\left(\frac{dN_a}{d\tau}\right)_{ads}-\left(\frac{dN_a}{d\tau}\right)_{des}=D\cdot\frac{C}{\delta}\cdot\Omega\cdot N\cdot(1-\theta)\cdot\exp\left(-\frac{U_a}{kT}\right)-N_S\cdot\nu\cdot\theta\cdot\exp\left(-\frac{\Psi}{kT}\right)=0. \tag{8.A.17}$$

The solution of this equation with respect to θ is as follows:

$$\theta=\frac{D\cdot(C/\delta)\cdot\Omega\cdot N\cdot\exp(-U_a/kT)}{D\cdot(C/\delta)\cdot\Omega\cdot N\cdot\exp(-U_a/kT)+N_S\cdot\nu\cdot\exp(-\Psi/kT)}, \tag{8.A.18}$$

or

$$\theta=\frac{C}{C+\dfrac{N_S\cdot\nu\cdot\delta\cdot\exp(-\Psi/kT)}{D\cdot\Omega\cdot N\cdot\exp(-U_a/kT)}}. \tag{8.A.19}$$

Then, if the thickness of the adsorbed surfactant layer is designated by h, and when the numerator and the denominator of the second term of divider of Equation 8.A.19 are multiplied by h, we obtain

$$\theta=\frac{C}{C+(\delta\cdot h/D)\cdot C_S^{\infty}\cdot\nu\cdot\exp\left(-(U_a-\Psi)/kT\right)}, \tag{8.A.20}$$

where $C_S^{\infty}=N_S/\Omega\cdot N\cdot h$ is the adsorption capacity of the layer with thickness h.

As is followed from (8.A.20), the equilibrium constant of surfactant adsorption in the water–air system is the product of the capacity of the adsorbed layer upon the ratio of rate constant of SAS molecules transferred from the surface into the subphase to the rate constant of SAS molecules transferred in the opposite direction, that is,

$$K_p=C_S^{\infty}\frac{K\downarrow}{K\uparrow}, \tag{8.A.21}$$

where

$K\downarrow=\nu\cdot\exp(-\Psi/kT)$

$K\uparrow=(D/\delta\cdot h)\exp(-U_a/kT)$

When dividing the numerator and the denominator of (8.A.20) by C_S^{∞}, then the expression for equilibrium constant of the adsorption process becomes dimensionless:

$$K_p=\frac{K\downarrow}{K\uparrow}. \tag{8.A.22}$$

In this case, (8.A.20) can be written as

$$\theta = \frac{C/C_S^\infty}{C/C_S^\infty + (K\downarrow/K\uparrow)}. \quad (8.A.23)$$

And in terms of adsorption Γ, the equilibrium isotherm becomes as follows:

$$\Gamma = \Gamma_\infty \cdot \frac{C/C_S^\infty}{C/C_S^\infty + (K\downarrow/K\uparrow)}, \quad (8.A.24)$$

where

$$\Gamma_\infty = \frac{N_S}{\Omega \cdot N} = \frac{1}{S_i \cdot N}. \quad (8.A.25)$$

In (8.A.25), Γ_∞ value corresponds to the maximum adsorption capacity and has dimensionality of moles per unit area. And S_i is the area per molecule of surfactant in the adsorption layer when the maximum adsorption capacity is reached. The solution of (8.A.24) in relation to $K\downarrow/K\uparrow$ is the following:

$$\frac{K\downarrow}{K\uparrow} = \frac{C}{C_S^\infty}\left(\frac{\Gamma_\infty}{\Gamma} - 1\right). \quad (8.A.26)$$

Equation 8.A.26 yields the value $K\downarrow/K\uparrow$, which is based on the values Γ, Γ_∞, and C_S^∞. These parameters are easily determined from the equilibrium adsorption isotherm by Gibbs equation:

$$\Gamma = \frac{C}{n \cdot R \cdot T} \cdot \frac{d\gamma}{dC} = \frac{1}{n \cdot R \cdot T} \cdot \frac{d\gamma}{d \cdot \ln C}, \quad (8.A.27)$$

where
- γ is the surface tension of the surfactant solution with concentration C
- R is the universal gas constant
- T is the temperature
- n is the factor depending on the type of the adsorbed substance

At adsorption of neutral surfactant molecules, the value of n is assumed to be 1, and at adsorption of ionized surfactant molecules, the value of n is assumed to be 2.

With regard to the kinetics of equilibrium state reaching at surfactant adsorption, the corresponding equation is obtained from (8.A.17) if one will write it in a short form as the equation of reversible first-order reaction, that is,

$$\frac{d[A]}{d\tau} = -K\uparrow \cdot [A] + K\downarrow \cdot [B], \quad (8.A.28)$$

where $[A]$ and $[B]$ correspond to the concentrations of the adsorbed molecules in subphase $[A]$ and in the adsorption layer $[B]$.

At the beginning of the adsorption process, all surfactant molecules are located in the subphase, and their initial concentration is A_0. In the course of adsorption, the surfactant molecules are accumulated at the interface, and their concentration in subphase is decreased:

$$[B]=[A_0]-[A]. \tag{8.A.29}$$

Then, Equation 8.A.28 can be written as

$$\begin{aligned}\frac{d[A]}{d\tau} &= -K\uparrow\cdot[A]+K\downarrow([A_0]-[A]) \\ &= K\downarrow\cdot[A_0]-(K\uparrow+K\downarrow)\cdot[A] \\ &= -(K\uparrow+K\downarrow)\cdot\left([A]-\frac{K\downarrow}{K\uparrow+K\downarrow}\cdot[A_0]\right) \\ &= -(K\uparrow+K\downarrow)\cdot([A]-[A_{eq}]).\end{aligned} \tag{8.A.30}$$

And A_{eq} is expressed as follows:

$$\frac{[B_{eq}]}{[A_{eq}]}=\frac{[A_0]-[A_{eq}]}{[A_{eq}]}=\frac{K\uparrow}{K\downarrow}, \tag{8.A.31}$$

and:

$$[A_{eq}]=\frac{K\downarrow}{K\uparrow+K\downarrow}\cdot[A_0]. \tag{8.A.32}$$

Integration of Equation 8.A.30 gives

$$-\int_{[A_0]}^{[A]}\frac{d[A]}{[A]-[A_{eq}]}=(K\uparrow+K\downarrow)\cdot\int_0^{\tau}d\tau, \tag{8.A.33}$$

and

$$\ln\frac{[A_0]-[A_{eq}]}{[A]-[A_{eq}]}=(K\uparrow+K\downarrow)\cdot\tau. \tag{8.A.34}$$

The value $[A_0]-[A_{eq}]$ in Equation 8.A.34 is none other than the Γ_{eq}, that is, it is the adsorption of the surfactant at the interface, and the value $[A]-[A_{eq}]$ is the difference between the values of the equilibrium adsorption and the current value of adsorption. Therefore, (8.A.34) can be written as

$$\ln\frac{\Gamma_{eq}}{\Gamma_{eq}-\Gamma}=(K\uparrow+K\downarrow)\cdot\tau, \tag{8.A.35}$$

or

$$\frac{\Gamma_{eq}-\Gamma}{\Gamma_{eq}} = \exp[-(K\uparrow + K\downarrow)\cdot\tau], \tag{8.A.36}$$

and

$$\Gamma = \Gamma_{eq}\cdot\{1-\exp[-(K\uparrow + K\downarrow)\cdot\tau]\}. \tag{8.A.37}$$

This equation shows the kinetics of surfactant accumulation at the interface. Taking into account (8.A.37), we shall finally obtain

$$\Gamma_\tau = \frac{1}{S_i\cdot N}\cdot\left(\frac{C}{C_S^\infty(K\downarrow/K\uparrow)+C}\right)\cdot\{1-\exp[-(K\uparrow + K\downarrow)\cdot\tau]\}. \tag{8.A.38}$$

As is followed from (8.A.38), if $C_S^\infty\cdot(K\downarrow/K\uparrow) \gg C$, and $(K\uparrow + K\downarrow)\cdot\tau \gg 1$, the following expression is valid:

$$\Gamma_\tau \approx C\cdot\frac{K\uparrow}{K\downarrow}\cdot h. \tag{8.A.39}$$

This means that the adsorption process is completed almost immediately under indicated conditions. It does not require any length of time to achieve the equilibrium state in the adsorption system. This dynamic is typical for the adsorption of small surfactant molecules and inorganic ions or gases.

As is also followed from (8.A.38), when $C_S^\infty\cdot(K\downarrow/K\uparrow) \gg C$ and $(K\uparrow + K\downarrow)\cdot\tau$ is small, but not so small, that the inverse exponent turns into a real unit, then

$$\Gamma_\tau \approx C\cdot\frac{K\uparrow}{K\downarrow}\cdot h\cdot(K\uparrow + K\downarrow)\cdot\tau = C\cdot K_\tau\cdot\tau, \tag{8.A.40}$$

where $K_\tau \approx (K\uparrow/K\downarrow)\cdot h\cdot(K\uparrow + K\downarrow)$, since $[1-\exp(-a)] \approx a$, if a is the small value.

Equation 8.A.40 shows the dynamics of adsorption according to Henry's law. It is applicable for the analysis of adsorption phenomena in solutions of the most of monomeric surfactants (colloid type of surfactant).

The equilibrium adsorption is decreased, and the time to achieve the equilibrium is increased, when the value $C_S^\infty(K\downarrow/K\uparrow)$ becomes large, and the value of exponent $(K\uparrow + K\downarrow)\cdot\tau$ approaches zero. This corresponds to the adsorption of high-molecular SASs from their solutions. Such substances have low diffusion coefficients, and the equilibrium state is achieved after a very long period of time.

The foregoing information is true in the cases of moderately concentrated solutions, that is, when the adsorption equilibrium is described by Langmuir's equation, that is,

$$\Gamma_p = \Gamma_\infty\cdot\frac{C}{C_S^\infty(K\downarrow/K\uparrow)+C}. \tag{8.A.41}$$

In this case, the kinetics of equilibrium achieving is predetermined only by the exponential term in Equation 8.A.38. So, the sets of Gibbs and Langmuir adsorption equations are the physical–chemical basis for the calculation of any flotation process.

8.B APPENDIX B

8.B.1 FAST AND SLOW SURFACTANT ADSORPTION AT THE AIR BUBBLE SURFACE AND TRANSFER OF ADSORBED MATTER OUTSIDE THE FLOATED WATER BULK

The different surfactants can exist at air–water interface in a form of ions and neutral molecules. When the SASs are adsorbed in the ionized state with the charge of their surface-active ions +1 or −1, then the adsorption Γ_i is calculated vs. concentration C using Equation 8.A.27 from Appendix 8.A at $n = 2$, that is,

$$\Gamma_i = \frac{C}{2RT} \cdot \frac{d\gamma}{dC}. \tag{8.B.1}$$

And if the SAS is adsorbed in unionized state, Equation 8.A.27 is used at $n = 1$, that is,

$$\Gamma_i = \frac{C}{RT} \cdot \frac{d\gamma}{dC}. \tag{8.B.2}$$

In the case, when the maximum (Γ_∞) and equilibrium (Γ_i) adsorption at certain surfactant concentration C is known, the value of adsorption equilibrium constant is determined by following the equation:

$$K_{eq} = C \cdot \left(\frac{\Gamma_\infty}{\Gamma} - 1 \right). \tag{8.B.3}$$

where K_{eq} has the dimensionality of volume concentration.

And the dependence of equilibrium adsorption of surfactant vs. its volume concentration is calculated as

$$\Gamma = \Gamma_\infty \cdot \frac{C}{K_{eq} + C}. \tag{8.B.4}$$

The transport of surfactant molecules with the bubble flows from the water bulk to air–water interface is the dynamic process. Its rate depends upon the SAS adsorption velocity at each air bubble and upon the total area of all floating bubbles, that is,

$$dG = \Gamma \cdot \frac{3F_1}{r} d\tau. \tag{8.B.5}$$

Here

- G is the mass transfer of surfactant with air bubble flow
- Γ is the adsorption of surfactant at the bubble surface during their floating
- F_1 is the volumetric air flow rate
- r is the air bubble radius
- τ is the time

In accordance with (8.B.4), Equation 8.B.5 may be transformed to the following form:

$$dG = \Gamma_\infty \cdot C \cdot \frac{1}{K_{eq}} \left(1 - \exp\left[-K\uparrow \frac{l}{\omega} \right] \right) \cdot \frac{3F_1}{r} d\tau. \tag{8.B.6}$$

The expression contained in the parentheses in (8.B.6) makes correction for the process rate because of the restricted diffusion and potential barriers to the penetration of surfactant molecules to the interface.

For the case of low surfactant concentration in aqueous solution, the large diffusivity of surfactant molecules, and the absence of potential barriers to these molecules to be adsorbed at the air–water interface, the expression can be simplified to the following form:

$$dG = \Gamma_\infty \frac{1}{K_{eq}} \cdot C \frac{3F_1}{r} d\tau. \tag{8.B.7}$$

And taking into account that

$$dG = V \cdot dC, \tag{8.B.8}$$

one shall get the following equation:

$$\int_{C_0}^{C} \frac{dC}{C} = \int_0^\tau \frac{\Gamma_\infty}{K_{eq}} \cdot \frac{3F_1}{r} \cdot \frac{d\tau}{V}. \tag{8.B.9}$$

Here

$$\frac{\Gamma_\infty}{K_{eq}} \cdot \frac{3F_1}{r} = \tilde{K} \tag{8.B.10}$$

and

$$\frac{C}{C_0} = \exp\left(-\frac{\tilde{K} \cdot \tau}{V}\right). \tag{8.B.11}$$

Equation 8.B.11 shows the reduction of surfactant concentration in the given volume V of floated solution under experimental conditions when the extraction efficiency is equal to 100%. Equation 8.B.11 is identical to 8.A.41. Equation 8.B.7, which leads to (8.B.10) and (8.B.11), describes the adsorption process for small molecules with large diffusivity when the essential potential barriers are absent.

8.B.2 SLOW SURFACTANT ADSORPTION AT THE AIR BUBBLES AND TRANSPORT OF ADSORBED MOLECULES OUT OF THE FLOATED BULK

In the case when the diffusion coefficient of surfactant molecules is $<10^{-6}$ cm/s, and there is essential potential barrier for these molecules to penetrate to the interface, the mass-transfer equation of surfactant with the floating bubbles can be written as follows:

$$\frac{dG}{d\tau} = \Gamma_\infty \cdot \frac{1}{K_{eq}} \cdot C \cdot \left\{1 - \exp\left[-(K\uparrow + K\downarrow) \cdot \frac{l}{\omega}\right]\right\} \frac{3F_1}{r}, \tag{8.B.12}$$

where

Γ_∞ is the maximum surfactant adsorption at the interface
K_{eq} is the adsorption equilibrium constant of SAS
C is the surfactant concentration in the water bulk
$K\uparrow$, $K\downarrow$ are the adsorption/desorption velocity constants for surfactant
l is the path length of floating air bubbles
ω is the velocity of floating air bubbles
F_1 is the volume air flow for bubbling
r is the radius of air bubbles

The values of Γ_∞ and K_{eq} are calculated from experimental data of equilibrium surface tension of surfactant solution upon its concentration according to (8.A.27), of Appendix 8.A. The values of r, F_1, and C are obtained from experimental conditions. And the values of $K\uparrow$ and $K\downarrow$ are calculated using (8.B.12), if it is written as the following equation:

$$\Gamma_\infty \cdot \frac{1}{K_{eq}} \cdot C \cdot \left\{1-\exp\left[-(K\uparrow + K\downarrow)\cdot \frac{l}{\omega}\right]\right\} \frac{3F_1}{r} = \frac{dA}{d\tau} \cdot \frac{m}{S_i \cdot N}. \tag{8.B.13}$$

Here

$dA/d\tau$ is the rate of formation of adsorption monolayer at the interface
S_i is the specific area in the adsorption monolayer at a given surface pressure π
m is the molecular mass of adsorbed surfactant
N is Avogadro's number

The left-hand side of Equation 8.B.13 is used to characterize the surfactant flow arisen as a result of their adsorption from the water bulk to the bubble surface floating up to the interface of the aqueous solution. The right-hand side of this equation is used to characterize the increase in the mass of adsorption monolayer arisen at the water surface.

The air bubbles with adsorbed surfactants are destroyed when reaching the air–water interface, and the surfactants from air bubble surfaces are accumulated at the interface. Figure 8.16 shows the experimental assembly used for the study of the SAS adsorption process described by (8.B.13). The process of SAS adsorption at the air bubble surfaces occurs in flotation cylinder, which is connected with Langmuir trough, as it is shown in Figure 8.16a and b. And after that, the floated SASs are accumulated as a monomolecular layer at the air–water interface in Langmuir trough. The experimental assembly allows one to realize the flotation process using both recirculating and continuous-flow conditions. The increase in SAS concentration in adsorption monolayer arisen at the air–water interface in Langmuir trough is controlled by movable barrier and Wilhelmy balance. At low magnitudes of surface pressure in arisen monolayers, the specific area is proportional to the surface pressure [3].

Figure 8.17 shows the $dA/d\tau$ plots concentration of different surfactant solutions, studied in the experimental assembly. The charts represent the rate of subphase's surface filling with adsorption SAS monolayers, compresses to surface pressures of 2.5–5.0 mN/m.

In the case of the adsorption of SDS at π = 4 mN/m, the specific area is about $100 \cdot 10^{-16}$ cm^2. The following values of parameters were determined by experiment and calculation to obtain the

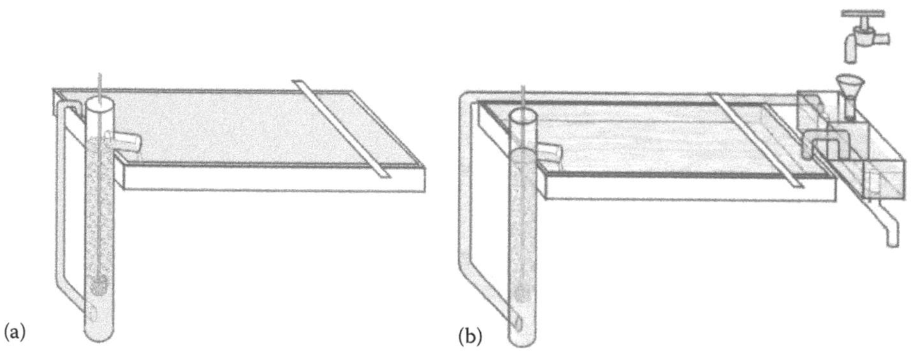

FIGURE 8.16 The schema of experimental assembly used for the realization of the adsorption process. (a) Recirculation operating conditions; (b) continuous flow operating conditions.

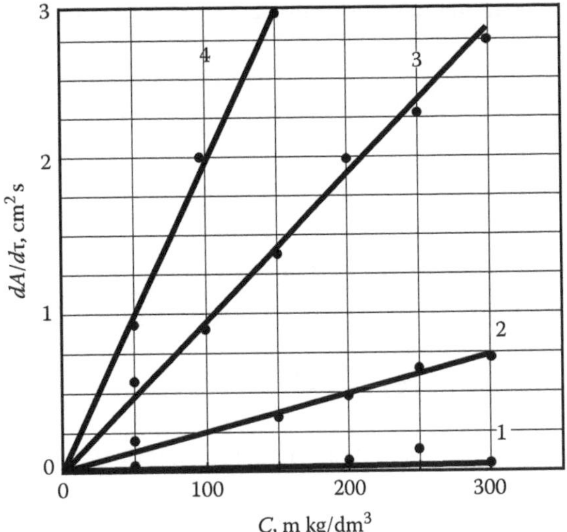

FIGURE 8.17 The dependence of *dA*/*dt* upon *C* for the following solutions: 1: sodium decyl sulfate; 2: sodium dodecyl sulfate; 3: alkyl sulfonate; 4: chemical DC–10.

adsorption rate constant of SDS at the surface of air bubbles in flotation system (Figure 8.16): Γ_∞ = 3.1·10^{-10} mol/cm^2; K_{eq} = 0.269·10^{-6} mol/cm^3; C = 0.3 mg/dm^3 = 3·10^{-4} mg/cm^3; $dA/d\tau$ = 0.75 cm^2/s; π = 4 mN/m; s_i = 100 A^2 = 100·10^{-16} cm^2; m = 288·10^3 mg/mol; l = 10 cm; ω = 25 cm/s; $W_в$ = 250 cm^3/min; r = 0.1 cm; N = 6.02·10^{23} mol^{-1}.

To a first approximation, the sum of the direct and reverse velocity constants of adsorbed molecules transferred from (8.B.13) was expressed as the resultant adsorption rate constant K_τ, that is,

$$K\uparrow + K\downarrow = K_\tau. \tag{8.B.14}$$

Then, transforming Equation 8.B.13, one can obtain

$$1-\exp\left(-K_\tau\frac{l}{\omega}\right)=\frac{dA}{d\tau}\cdot\frac{m}{S_i\cdot N}\cdot\frac{K_{eq}}{\Gamma_\infty\cdot C}\cdot\frac{r}{3F_1}, \tag{8.B.15}$$

$$-K_\tau\frac{l}{\omega}=\ln\left[1-\frac{dA}{d\tau}\cdot\frac{m}{S_i\cdot N}\cdot\frac{K_{eq}\cdot r}{\Gamma_\infty\cdot C\cdot 3F_1}\right], \tag{8.B.16}$$

$$-K_\tau\frac{l}{\varpi}=\ln\left(1-\frac{0.75\cdot 288\cdot 10^3\cdot 0.269\cdot 10^{-6}\cdot 0.1\cdot 60}{100\cdot 10^{-16}\cdot 6.02\cdot 10^{23}\cdot 3.1\cdot 10^{-10}\cdot 3\cdot 10^{-4}\cdot 3\cdot 250}\right)$$

$$=\ln\left(1-\frac{0.348}{0.419}\right)$$

$$=\ln(1-0.832)$$

$$=\ln 0.168.$$

So, $-K_\tau(l/\omega) = -1.78$, and K_τ = 1.78·25/10 = 4.46.

On the other hand, the resultant rate adsorption constant K_τ can be expressed as

$$K\uparrow + K\downarrow = K_\tau = \frac{D}{\delta \cdot h}\exp\left(-\frac{U_a}{KT}\right). \tag{8.B.17}$$

Then,

$$\exp\left(-\frac{U_a}{KT}\right) = \frac{K_\tau \cdot \delta \cdot h}{D} = \frac{4.46 \cdot 0.01 \cdot 18 \cdot 10^{-8}}{1 \cdot 10^{-6}} = 8.028 \cdot 10^{-3},$$

and $-(U_a/KT) = \ln 8.028 \cdot 10^{-3}$, or $(U_a/KT) = 4.82$.

So, $U_a = 4.82 \cdot 1.38 \cdot 6.02 \cdot 293 = 11.73$ kJ/mol.

And, using (8.A.21), that is,

$$K_{eq} = C_s^\infty \frac{K\downarrow}{K\uparrow},$$

where
$K_{eq} = 0.269 \cdot 10^{-6}$ mol/cm^3
$C_S^\infty = N_S/\Omega \cdot N \cdot h = 1.722 \cdot 10^{-3}$ mol/cm^3
$K\uparrow = 4.46$

one will obtain,

$$K\downarrow = \frac{K_{eq}}{C_S^\infty} \cdot K\uparrow = \frac{0.269 \cdot 10^{-6} \cdot 4.46}{1.722 \cdot 10^{-3}} = 6.96 \cdot 10^{-4}.$$

So,

$$K\downarrow = 6.96 \cdot 10^{-4}\ \text{s}^{-1},$$

and

$$K\uparrow = 4.44\ \text{s}^{-1}.$$

$$\text{But } \frac{K\downarrow}{K\uparrow} = \exp\left(-\frac{\Delta F}{RT}\right); \quad \frac{K\downarrow}{K\uparrow} = \frac{6.96 \cdot 10^{-4}}{4.46} = 1.56 \cdot 10^{-4}.$$

Then,

$$-\frac{\Delta F}{RT} = \ln 1.56 \cdot 10^{-4}; \quad \frac{\Delta F}{RT} = 8.76; \quad \Delta F = 8.76 \cdot 8.31 \cdot 293 = 21.32\ \text{kJ/mol}.$$

So, we have got the following characteristics of the adsorption system: $K_{eq} = 0.269 \cdot 10^{-6}$ mol/cm^3; $\Gamma_\infty = 3.1 \cdot 10^{-10}$ mol/cm^2; $C_S^\infty = 1.722 \cdot 10^{-3}$ mol/cm^3; $K_\tau = 4.46$ s^{-1}; $K\downarrow = 6.96 \cdot 10^{-4}$ s^{-1}; $K\uparrow = 4.46$ s^{-1}; $\Delta F = 21.32$ kJ/mol; $U_a = 11.73$ kJ/mol.

In such a way, the kinetics of SDS adsorption in the water–air system is fast. The degree of interface filling with detergent molecules, calculated according to (8.B.12), reaches 67% of its equilibrium value at $\tau = 0.4$ s, $l = 10$ cm, and $V = 25$ cm/s. And it reaches 97% and 99% of its equilibrium value at $l = 20$ cm, $l = 40$ cm, and $V = 25$ cm/s (adsorption time 0.8 and 1.6 s, respectively). Therefore, if the flotation device is used to extract the defined surfactant from its aqueous solution, the value of $\tilde{K}$ (8.A.41), (8.B.1) through (8.B.11) at $l \geq 40$ cm, $r \geq 0.1$ cm, and $V = 25$ cm/s can be calculated according to

$$\tilde{K} = \frac{\Gamma_\infty}{K_{eq}} \cdot \frac{3F_1}{r}. \tag{8.B.18}$$

Here the value $\{1-\exp[-(K\uparrow + K\downarrow)(l/\omega)]\}$, which takes into account the contribution of the track length of floating bubbles, is little affected upon the adsorption, and it may be taken as 1. And if the flotation device is used for the treatment of solutions of high-molecular-weight surfactants that have a small diffusion coefficient, the generalized rate constant of flotation process is calculated according to the following formula:

$$\tilde{K} = \Gamma_\infty \cdot \frac{1}{K_{eq}} \left\{1 - \exp\left[-(K\uparrow + K\downarrow) \cdot \frac{l}{\omega}\right]\right\} \cdot \frac{3F_1}{r}. \tag{8.B.19}$$

The expression represented in square brackets of (8.B.19) corresponds to the product of rate constant of SAS adsorption and duration of air bubbles floating through the water bulk. When the absolute value of this product is small, Equation 8.B.19 is transformed to the following form:

$$\tilde{K} = \Gamma_\infty \cdot \frac{K_\tau \cdot l}{K_{eq} \cdot \omega} \cdot \frac{3F_1}{r}, \tag{8.B.20}$$

or

$$\tilde{K} = \Gamma_\infty \cdot \frac{K\uparrow \cdot l}{K_{eq} \cdot \omega V} \cdot \frac{3F_1}{r}. \tag{8.B.21}$$

So, to calculate the value of generalized rate constants $\tilde{K}$ (8.A.41), (8.B.1) through (8.B.11) of the removal of the defined surfactants from water bulk, it is essential to know the value of the following parameters: constants K_{eq} and K_τ, the track length l of floating bubbles, their floating velocity ω, the value of maximum adsorption Γ_∞, air flow F_1, and the radius of the bubbles r. If it is necessary to remove SASs of the known molecular weight, the concentration of n-type impurities can be calculated according to (8.A.41), (8.B.1) through (8.B.11), using the values of $\tilde{K}$, K_{eq}, K_τ, and Γ_∞ for n-type impurities. And as a whole, the value of $\tilde{K}$ must be calculated as in the case of difficult-to-remove contaminants.

8.C APPENDIX C

8.C.1 CALCULATION OF THE DEGREE OF SURFACE FILLING BY SAS MOLECULES ADSORBED FROM THE SIDE OF AIR

If one is able to get very fine surfactant dispersion in the space of air, then the degree of air–water interface filling with surfactant molecules from air side is described by the following equation:

$$\theta = \frac{C}{C + C_s^{\infty} \cdot \dfrac{\nu \cdot \exp(-\psi/KT)}{\sqrt{KT/2\pi \cdot m \cdot h^2} \cdot \exp\left(-m \cdot V_0^2/2KT\right)}}. \tag{8.C.1}$$

Here,

$C_s^{\infty} = 1/S_i \cdot N \cdot h$

S_i is the specific equilibrium area in adsorbed monolayer

h is the thickness of adsorbed monolayer

N is Avogadro's number

m is the mass of adsorbed molecules

At the adsorption of long-chain surfactants, the value of S_i is varied from $20{\cdot}10^{-20}$ to $60{\cdot}10^{-20}$ m^2. For SDS, octyl sulfate, and tetradecyl benzenesulfonate, the value of S_i must start with $53.6{\cdot}10^{-20}$ m^2 per molecule [36].

The magnitude of h depends upon the molecular structure of adsorbed surfactant: for palmitic acid ($C_{15}H_{31}COOH$), h = 23 Å; and for stearic acid ($C_{17}H_{35}COOH$), h = 26 Å. Therefore, in the case of study of the surfactants with different lengths of its hydrocarbon chains, one can use the following formula:

$$h_i = \frac{26}{17} i,$$

where i is the number of carbon atoms in the hydrocarbon chain of adsorbed SAS molecule.

The value of m is calculated from the relationship $m = M/N$, where M is the molecular mass of adsorbed surfactant. The value of the second component of denominator in (8.C.1) is the equilibrium adsorption constant K_{eq}, which has dimensions mol/cm^3, mol/dm^3, and mol/m^3. So, we can write

$$K_{eq} = C_s^{\infty} \cdot \frac{\nu \exp(-\psi/KT)}{\sqrt{KT/2\pi \cdot m \cdot h^2} \cdot \exp\left(-m \cdot V_0^2/2KT\right)}, \tag{8.C.2}$$

or

$$K_{eq} = C_s^{\infty} \cdot \frac{1}{\sqrt{KT/2\pi \cdot m \cdot h^2}} \cdot \frac{\nu \exp\left((m \cdot V_0^2 - 2\psi)/2KT\right)}{1}, \tag{8.C.3}$$

or

$$K_{eq} = C_s^{\infty} \cdot \frac{1}{\sqrt{KT/2\pi \cdot m \cdot h^2}} \cdot X, \tag{8.C.4}$$

where

$$X = \nu \exp\left(\frac{m \cdot V_0^2 - 2\psi}{2KT}\right). \tag{8.C.5}$$

The expression (8.C.5) describes the energy–frequency behavior of desorption of adsorbed molecules from the interface into the water bulk.

8.C.2 CALCULATION OF THE DEGREE OF THE WATER–AIR INTERFACE FILLING WITH SAS MOLECULES ADSORBED FROM THE SIDE OF WATER

In this case, the adsorption isotherm has the following form:

$$\theta = \frac{C}{C + C_s^\infty \cdot \dfrac{1}{D/\delta h} \cdot \dfrac{\nu \exp(-\psi/KT)}{\exp(-U_a/KT)}}, \tag{8.C.6}$$

or

$$\theta = \frac{C}{C + K_{eq}^1}, \tag{8.C.7}$$

where

$$K_{eq}^1 = C_s^\infty \cdot \frac{1}{(D/\delta h)} \cdot \frac{\nu \exp(-\psi/KT)}{\exp(-U_a/KT)}, \tag{8.C.8}$$

or

$$K_{eq}^1 = C_s^\infty \cdot \frac{1}{(D/\delta h)} \cdot \nu \exp\left(\frac{U_a - \psi}{KT}\right). \tag{8.C.9}$$

And

$$K_{eq}^1 = C_s^\infty \cdot \frac{1}{(D/\delta h)} \cdot X. \tag{8.C.10}$$

$$\text{Here } X = \nu \exp\left(\frac{U_a - \psi}{KT}\right). \tag{8.C.11}$$

Analogous to (8.C.5), the expression (8.C.11) describes the energy–frequency behavior of desorption of adsorbed molecules from adsorption layer into the water bulk.

8.C.3 RATIO OF EQUILIBRIUM ADSORPTION CONSTANTS AS AN INDICATOR OF SURFACTANT ADSORPTION EFFICIENCY FROM WATER BULK AND AIR

At surfactant adsorption from air, the equilibrium adsorption constant is described by Equation 8.C.45. And at surfactant adsorption from water, the equilibrium adsorption constant is described by Equation 8.C.10. In these equations, the values of $1/(D/\delta h)$ and $1/\sqrt{KT/2\pi mh^2}$ are essentially different magnitudes. On the other hand, the values of $\nu \exp((mV_0^2 - 2\psi)/2KT)$ and $\nu \exp((U_a - \psi)/KT)$ have the same or nearly the same magnitudes. These magnitudes reflect the energy–frequency parameters of the desorption stage of adsorbed molecules (i.e., molecules that come back to the water bulk from adsorption layer). Both these members can be depicted as X. Therefore,

$$\frac{K_p}{K_p^1} = \frac{C_s^\infty}{C_s^\infty} \frac{(D/\delta h)(X)}{\sqrt{KT/2\pi mh^2} \cdot (X)}. \tag{8.C.12}$$

Here
D is the diffusion coefficient of surfactant
δ is the thickness of diffusion layer at the interface
h is the thickness of adsorption layer at $\Gamma \to \infty$

$$X \approx \nu \exp\left(\frac{mV_0^2 - 2\psi}{2KT}\right) \quad \text{and} \quad X \approx \nu \exp\left(\frac{U_a - \psi}{KT}\right). \tag{8.C.13}$$

Equation 8.C.12 allows one to evaluate the ratio of adsorption constants for any surfactant. For example, for the adsorption of SDS at the following experimental conditions: $D = 1{\cdot}10^{-6}$ cm²/s; $\delta = 0.01$ cm (at radius of air bubble ~1 mm and its floating velocity 0.25 m/s); $h = 18{\cdot}10^{-8}$ cm; $K = 1.38{\cdot}10^{-23}$ J/K; $T = 293$ K; $m = 0.288$ kg/mol; $N = 6.02{\cdot}10^{-23}$ mol^{-1}, one can obtain

$$\frac{K_{eq}}{K_{eq}^1} = \frac{1 \cdot 10^{-6}/0.01 \cdot 18 \cdot 10^{-8}}{\sqrt{(1.38 \cdot 293 \cdot 6.02/2 \cdot 3.14 \cdot 0.288) \cdot (1/18 \cdot 10^{-10})^{-2}}} = \frac{555.5}{2.03 \cdot 10^{10}} = 2.73 \cdot 10^{-8}, \tag{8.C.14}$$

that is,

$$\frac{K_{eq}}{K_{eq}^1} = 2.73 \cdot 10^{-8} \quad \text{and} \quad \frac{K_{eq}^1}{K_{eq}} = 3.6 \cdot 10^7.$$

It means that in the case when the equilibrium adsorption constant of SDS is equal to $1.16{\cdot}10^{-3}$ mol/dm³ (at adsorption of SDS from bulk water), the value of equilibrium adsorption constant of SDS at adsorption from air is $3.6{\cdot}10^7$ as smaller as at adsorption from water bulk, and it is equal to

$$K_{eq} = \frac{K_{eq}^1}{3.6 \cdot 10^7} = 3.22 \cdot 10^{-11} \text{ mol/dm}^3.$$

So, in the case of essential SAS concentration in air, the adsorption of SDS at the air–water interface will result in fast formation of close-packed adsorption monolayer (8.B.13).

8.C.4 FILLING OF AIR–WATER INTERFACE BY SURFACTANT MOLECULES RELEASED FROM SAS AEROSOL DISPERSED IN AIR

Let us suppose that SAS aerosol is formed in air space by means of suitable device, and its species represent the spheres with radius $1{\cdot}10^{-5}$, $1{\cdot}10^{-6}$, and $1{\cdot}10^{-7}$ m, respectively. If aerosol consists of concentrated SAS solution, in particular SDS solution with a density 1 kg/dm^3, then,

$$1.\ K_a\uparrow_{10\text{mcm}} = \sqrt{\frac{KT}{2\pi m h^2}} = \sqrt{\frac{1.38\cdot 10^{-23}\cdot 293}{2\cdot 3.14\cdot 4/3\cdot 3.14\cdot\left(1\cdot 10^{-5}\right)^3 1000\cdot\left(18\cdot 10^{-10}\right)^2}} = 6.889\cdot 10^3\ \text{s}^{-1},$$

$$2.\ K_a\uparrow_{1\text{mcm}} = \sqrt{\frac{KT}{2\pi m h^2}} = \sqrt{\frac{1,38\cdot 10^{-23}\cdot 293}{2\cdot 3,14\cdot 4/3\cdot 3,14\cdot(1\cdot 10^{-6})^3 1000\cdot(18\cdot 10^{-10})^2}} = 2,17\cdot 10^5\ \text{s}^{-1},$$

$$3.\ K_a\uparrow_{0,1\text{mcm}} = \sqrt{\frac{KT}{2\pi m h^2}} = \sqrt{\frac{1,38\cdot 10^{-23}\cdot 293}{2\cdot 3,14\cdot 4/3\cdot 3,14\cdot(1\cdot 10^{-7})^3 1000\cdot(18\cdot 10^{-10})^2}} = 6,889\cdot 10^6\ \text{s}^{-1}.$$

Here, $K_a\uparrow_i$ is the rate transfer constant of aerosol particles with the mentioned sizes from air to the interface.

And,

$$1.\ \frac{K\uparrow}{K_a\uparrow} = \frac{555,5}{6,889\cdot 10^3} = 0,0806. \quad \text{At } a = 1\cdot 10^{-5}\ \text{m},\ \frac{K\uparrow a}{K\uparrow} = 12,4.$$

$$2.\ \frac{K\uparrow}{K\uparrow a} = \frac{555,5}{2,17\cdot 10^5} = 2,55\cdot 10^{-3}. \quad \text{At } a = 1\cdot 10^{-6}\ \text{m},\ \frac{K\uparrow a}{K\uparrow} = 390,63.$$

$$3.\ \frac{K\uparrow}{K\uparrow a} = \frac{555,5}{6,889\cdot 10^6} = 8,06\cdot 10^{-5}. \quad \text{At } a = 1\cdot 10^{-7}\ \text{m},\ \frac{K\uparrow a}{K\uparrow} = 12401.$$

where $K\uparrow = (D/\delta h) = 555{,}5\ \text{s}^{-1}$. (This value characterizes the diffusion component of rate constant of SDS adsorption from water bulk at the surface of floating air bubble with radius 1 mm, when $\delta = 0{,}01$ cm.)

The comparison of obtained data shows that at a given radius of aerosol particles of SAS solution, the rate adsorption constants of this aerosol at its adsorption from air are essentially higher than the rate adsorption constants from water bulk. This allows one to realize technically the flotation process using the close-pack ionized adsorption monolayer of surfactants, specially obtained at air bubble surface by aerosol dispersion and, thus, to control this process efficiently.

8.D APPENDIX D

8.D.1 PROPERTY OF THE DOUBLE ELECTRIC LAYER FORMED AT THE AIR–WATER INTERFACE BY DIPOLES OF WATER MOLECULES, HYDROXYLS, AND PROTONS

In the system of desalted water–air, when water doesn't contain any surfactants, the double electric layer at the air bubble surface is formed due to the structure and charge distribution in water molecules and due to the presence of charged hydroxyls and protons arisen as a result of dissociation of water molecules. In order to estimate the maximum charge density at the bubbles owing to water molecule dissociation, let's use the expression for the dissociation constant of water in its bulk and at the interface, respectively:

$$K_A = \frac{\left[OH_v^-\right]\cdot\left[H_v^+\right]}{\left[HOH_v\right]}, \tag{8.D.1}$$

$$K_{AS} = \frac{\left[OH_s^-\right]\cdot\left[H_s^+\right]}{\left[HOH_s\right]}, \tag{8.D.2}$$

where K_A and K_{AS} are the dissociation constants of water in the water bulk and at the interface, respectively. The indexes s and v designate the surface and volume concentrations:

$$\left[HOH_s\right] = \left[HOH_v\right]\delta, \tag{8.D.3}$$

$$\left[OH_s^-\right] = \left[OH_v^-\right]\delta\cdot i, \tag{8.D.4}$$

$$\left[H_s^+\right] = \left[H_v^+\right]\delta\cdot j. \tag{8.D.5}$$

In this case,

$$K_{AS} = \frac{\left[OH_v^-\right]\delta\cdot i\left[H_v^+\right]\delta\cdot j}{\left[HOH_v\right]\delta}, \tag{8.D.6}$$

where

i and j are coefficients of hydroxyl and proton concentration at the interface
δ is the thickness of the surface layer

The outside plate of the double electric layer is formed from water molecules and hydroxyl ions, which are located at the interface. At that, the hydrated protons (H_3O^+) form the diffusion part of the double electric layer in the subphase. Their local concentration in this part of double electric layer depends exponentially upon electrostatic potential:

$$\left[H_z^+\right] = \left[H_v^+\right]\cdot\exp\left(-\frac{e\varphi_z}{KT}\right), \tag{8.D.7}$$

where

$[H_z^+]$ is the local concentration of hydrated protons (H_3O^+)
φ_z is the electrostatic potential in the point of local concentration determination

So, the surface concentration of protons obeys the following equation:

$$\left[H_s^+\right] = \left[H_v^+\right] \cdot j \cdot \delta = \left[H_v^+\right] \exp\left(-\frac{e\varphi_s}{KT}\right)\delta, \tag{8.D.8}$$

where φ_s is the electrostatic potential at the interface.

The balance of water molecule concentration and hydroxyl ion concentration in the surface layer obeys the following equation:

$$\left[OH_s^-\right] + \frac{\left[OH_s^-\right] \cdot \left[H_v^+\right] \exp(-e\varphi_s/KT)\delta}{K_{AS}} = n. \tag{8.D.9}$$

Here, n is the total concentration of water molecules and hydroxyl ion concentration at the interface.

Solution of (8.C.9) with respect to the hydroxyl ion surface concentration gives

$$\left[OH_s^-\right] = \frac{n}{1 + \frac{\left[H_v^+\right] \exp(-e\varphi_s/KT)\delta}{K_{AS}}}. \tag{8.D.10}$$

But when $[H_v^+] = 10^{-\mathrm{pH}}$, and $n = 1/(S_i \cdot N)$, where S_i is the area per water molecule at the interface, and N is Avogadro's number, Equation 8.C.10 transforms to the following form:

$$\left[OH_s^-\right] = \frac{1}{S_i N}\left[1 + \frac{10^{-\mathrm{pH}} \exp(-e\varphi_s/KT)\delta}{K_{AS}}\right]^{-1}. \tag{8.D.11}$$

Multiplication of both the parts of (8.C.11) by Avogadro's number N and electron charge N shows the relationship between the surface charge density σ, the boundary potential jump φ_s at air bubble, and pH of desalted water:

$$\left[OH_s^-\right] N \cdot e = \sigma = -\frac{1}{S_i} e\left[1 + \frac{10^{-\mathrm{pH}} \exp(-e\varphi_s/KT)\delta}{K_{AS}}\right]^{-1}. \tag{8.D.12}$$

And the substitution of (8.D.12) into Gui–Chapman equation gives the analytical dependence of surface potential φ_s and surface charge density σ upon packed density of water molecules in the surface layer $1/S_i$, dissociation constant of water K_{AS}, and concentration of dissolved salt C:

$$\varphi_s = \frac{2KT}{e} Arsh \frac{(1/S_i)e\left[1 + 10^{-\mathrm{pH}} \cdot \exp(-e\varphi_s/KT)(\delta/K_{AS})\right]^{-1}}{A\sqrt{C}}, \tag{8.D.13}$$

where $A = \sqrt{8NKT\varepsilon_0\varepsilon_1}$.

Equation 8.D.13 allows one to calculate the boundary potential jump φ_S and surface charge density σ at the air bubble depending upon pH, S_i, K_{AS}, δ, and C. The results of the calculation of σ at pH 1–14, $C = 10^{-6}$–10^{-1} M, $S_i = 6 \cdot 10^{-20}$ m², $K_A = 1{,}8 \cdot 10^{-16}$ mol/dm³, $K_{AS} = 5{,}4.10^{-25}$ mol/dm², and $\delta = 3 \cdot 10^{-9}$ dm show that the surface charge density at the air bubble in this case will be vanishingly small.

REFERENCES

1. Birdi K.S. 2009. *Handbook of Surface and Colloid Chemistry*. CRC Press, Taylor & Francis Group, Boca Raton, FL.
2. Gevod V.S. 2003. NATO Science Series. Modern tools and methods of water treatment for improving standards. IV. *Earth Environ. Sci.*, 48, 137.
3. Гевод В.С., Решетняк И.Л., Шклярова И.Г., Хохлов А.С., Гевод С.В. 2002. Поверхностно-активные и другие загрязнения в водопроводной питьевой воде: свойства, мониторинг, причины накоплений и экономичное удаление. Изд-во УГХТУ, Днепропетровск.
4. Гевод В.С., Решетняк И.Л., Гевод С.В., Шклярова И.Г., Нефедов В.Г. 2004. *Вопросы химии и хим. технологии*, 4, 213.
5. Гевод В.С., Решетняк И.Л., Гевод С.В., Хохлов А.С., Шклярова И.Г. 2001. *Вопросы химии и хим. технологии*, 6, 152.
6. Гевод В.С., Решетняк И.Л., Гевод С.В., Хохлов А.С., Шклярова И.Г. 2005. *Вопросы химии и хим. технологии*, 1, 209.
7. Русанов А.И. 1981. Поверхностное разделение веществ: Теория и методы. Химия, Ленинград.
8. Гевод В.С., Решетняк И.Л., Гевод С.В., Шклярова И.Г. 2007. *Вопросы химии и хим. технологии*, 1, 178.
9. Мелвин-Хьюз Э.А. 1962. *Физическая химия*. Иностранная литература, Москва.
10. Касаткин А.Г. 1971. Основные процессы и аппараты химической технологии. Химия, Москва.
11. Голубовская Э.К. 1978. *Биологические основы очистки воды*. Высшая школа, Москва.
12. Гончарук В.В., Дешко И.И., Герасименок Н.Г. и др. 1998. *Химия и технология воды*, 1, 19.
13. Гевод В.С., Решетняк И.Л., Хохлов А.С., Гевод С.В., К.С. Бирди. 1998. *Вопросы химии и хим. технологии*, 3, 55.
14. Гевод В.С., Решетняк И.Л., Хохлов А.С., Гевод С.В., К.С. Бирди. 1998. *Вопросы химии и хим. Технологии,* 3, 58.
15. Craun G.F. 2001. *Microbial Pathogens and Disinfection by Products in Drinking Water: Health Effects and Management of Risks*. ILSI Press, Washington, DC.
16. Graham H.J., Colloins R.D. 1996. *Advances in Slow Sand Alternative Biological Filtration*. John Wiley & Sons Inc., Chichester, U.K.
17. Woolschlager J., Rittmann B., Piriou P., Kiene L., Schwartz B. 2000. A comprehensive disinfection and water quality model for drinking water distribution systems. *First World Water Congress of the IWA*, Paris, France. ISBN: 2-9515416-0-0. EAN: 9782951541603. NP-119.
18. Гевод В.С., Решетняк И.Л., Гевод С.В., Шклярова И.Г. 2007. *Вопросы химии и хим. технологии*, 5, 222.
19. Гевод В.С., Решетняк И.Л., Гевод С.В., Шклярова И.Г., Нефедов В.Г. 2009. *Вопросы химии и хим. технологии*, 3, 162.
20. Перт С.Дж. 1978. Основы культивирования микроорганизмов и клеток. Мир, Москва.
21. Жужиков В.А. 1980. Фильтрование: теория и практика разделения суспензий. Химия, Москва.
22. Запольский А.К. 2005. *Водопостачання, водовідведення та якість води*. Вища школа, Київ.
23. Кастальский А.К., Минц Д.М. 1962. *Подготовка воды для питьевого и промышленного водоснабжения*. Высшая школа, Москва.
24. Кульский Л.А. 1986. *Технология очистки природных вод*. Вища школа, Киев.
25. Кинеле, Хартмут. 1984. *Активные угли и их промышленное применение*. Химия, Ленинград.
26. Слипченко В.А., Малицкая Т.Н. 1992. *Химия и технология воды*. 1, 35.
27. Гончарук В.В., Клименко Н.А., Савчина Л.А. и др. 2006. *Химия и технология воды*. 1, 44.
28. Гевод В.С., Решетняк И.Л., Гевод С.В., Шклярова И.Г., Хохлов А.С., Губенко И.И. 2002. *Вопросы химии и хим. технологии*, 1, 121.
29. Perry R.H., Green D.W. 2008. *Perry's Chemical Engineers' Handbook*. McGraw-Hill Book Company, New York.
30. Гороновский И.Т. 1987. *Краткий справочник по химии: справочник*. Наукова думка, Киев.
31. Кошкин Н.И. 1966. *Справочник по элементарной физике*. Наука, Москва.
32. Гевод В.С., Решетняк И.Л., Гевод С.В., Шклярова И.Г., Хохлов А.С. 2002. *Вопросы химии и хим. технологии*, 6, 161.
33. Гевод В.С., Решетняк И.Л., Гевод С.В., Шклярова И.Г., Гринев В.М., Нефедов В.Г. 2004. *Вопросы химии и хим. технологии*, 3, 215.
34. Birdi K.S. 2010. *Introduction to Electrical Interfacial Phenomena*. CRC Press, Taylor & Francis Group, Boca Raton, FL.
35. Гевод В.С., Решетняк И.Л., Гевод С.В., Шклярова И.Г. 2007. *Вопросы химии и хим. технологии*, 2, 227.
36. Gaines G.L., Jr. 1966. *Insoluble Monolayers at Liquid–Gas Interface*. Interscience Publishers, New York.

9 Thermally Sensitive Particles

Preparation, Characterization, and Application in the Biomedical Field

Abdelhamid Elaïssari and Waisudin Badri

CONTENTS

9.1 INTRODUCTION

Polymer latex particles are widely used as a solid support in numerous applications and especially in the biomedical field, due to the existence of various polymerization processes (emulsion, dispersion, microemulsion, etc.) for preparing latex that is well-defined in terms of particle size, reactive groups, surface charge density, colloidal stability, etc. Since 1986, precipitation polymerization of alkylacrylamide and alkylmethacrylamide derivatives (water-soluble monomers) has been found to be a convenient method for producing submicronic, functionalized thermally sensitive latex hydrogel particles, as reported by Pelton and Chibante [1]. Since then, thermally responsive microgel latex particles have played a particular and considerable role in academic research and industrial applications. In academic research, studies are mainly focused on the polymerization mechanism and colloidal characterization in dispersed media. From an application point of view, stimuli-responsive microgels have been principally explored in drug delivery as a carrier in therapy.

The implication of such stimuli-responsive particles as a solid polymer support of biomolecules in the biomedical field is probably due to various factors: (1) easiest to prepare via precipitation polymerization (hydrogel particles) or a combination of emulsion and precipitation polymerizations (core–shell particles); (2) the colloidal properties are related to the temperature and to the medium composition (i.e., pH, salinity, surfactant, etc.); (3) the adsorption and the desorption of antibodies and proteins are principally related to the incubation temperature; (4) the covalent binding of proteins onto such hydrophilic and stimuli-responsive particles can be controlled easily by temperature; and finally (5) the hydrophilic character of the microgel particles is an undeniably suitable environment for immobilized biomolecules.

The main objective of this chapter is to report on the preparation, characterization of thermally sensitive particles, and the pertinent aspects that should be considered before their utilization as a polymer support in the biomedical field. This will be followed by an examination of the preparation of such hydrophilic thermally sensitive latex particles bearing reactive groups. Subsequently, the colloidal characterizations that are to be taken into consideration will be presented. Finally, the chapter will be closed by presenting and illustrating recent applications of thermally sensitive polymer colloids as solid supports in the biomedical field.

9.2 SYNTHESIS OF REACTIVE THERMALLY SENSITIVE LATEX PARTICLES

Over the last 20 years, precipitation polymerization leading to the preparation of thermally sensitive hydrogel latexes has been widely reported on and discussed. The first thermally sensitive linear polymer base, using *N*-isopropylacrylamide (NIPAM), was reported by Heskins and Guillet [2] in 1968. The linear homopolymer obtained exhibits a low critical solubility temperature (LCST) of 32°C, corresponding to a dramatic change in the solubility parameters. In fact, below the LCST the polymer is totally soluble in the aqueous phase, whereas above the LCST the solution exhibits phase separation induced by polymer precipitation (coil to globule transition, see Figure 9.1).

Polymer	LCST (°C)
Poly(*N*-isopropylacrylamide), PNIPAM	~32
Poly(*N*-isopropylmethacrylamide), PNIPMAM	~45
Poly(*N*-vinylisobutyramide), PNVIBA	~39
Poly(*N*-ethylacrylamide)	~78
Poly(*N*-acryloylpyrrolidine)	~50
Poly(*N*-acryloylpiperidine)	~5
Poly(*N*-vinyl methyl ether) PVME	~40
Poly(ethylene glycol), PEG	~120
Poly(propylene glycol) PPG	~50

(Continued)

Polymer	LCST (°C)
Poly(methacrylic acid), PMAA	~75
Poly(vinyl alcohol), PVA	~125
Poly(vinyl methyl oxazolidone), PVMO	~65
Poly(vinyl pyrrolidone), PVP	~160
Methylcellulose, MC	~80
Hydroxypropyl cellulose, HPC	~55
Poly(*N*-vinylcaprolactam), PVCL	~30
Polyphosphazene derivates	33–100
Poly(silamine)	~37
Poly(siloxyethylene glycol)	10–60

FIGURE 9.1 Scheme of thermally sensitive linear polymer as a function of temperature. (From Pelton, R.H. and Chibante, P., *Colloid Surf.*, 20, 247, 1986.)

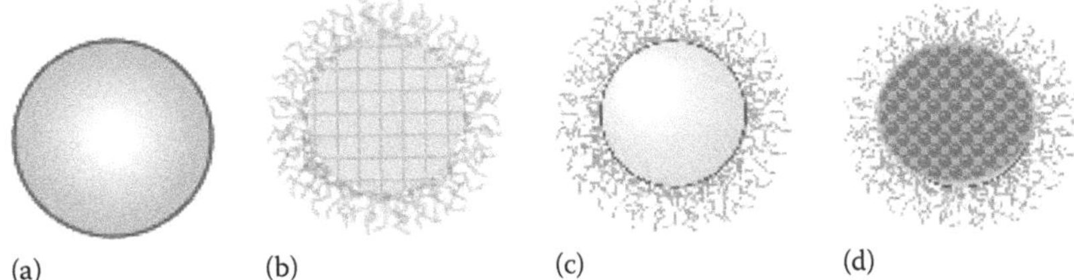

FIGURE 9.2 Illustration of particles morphology: (a) hard sphere (i.e., polystyrene), (b) microgel structure (polyNIPAM), (c) core–shell like (i.e., polystyrene core-polyNIPAM shell), and (d) composite particles (i.e., hard magnetic polymer core and water soluble polymer shell).

In addition, the LCST of thermosensitive polymer has been widely studied using different physical methods—including turbidity, fluorescence, dynamic light scattering, and calorimetric measurements—in order to demonstrate the relationship between the polymer properties and solvent conditions, as reported by Schild [3] in a very thorough review concentrated on poly(NIPAM).

The elaboration of thermally sensitive colloidal particles has been largely studied as can be evidenced by the numerous reported publications. In fact, three kinds of thermally sensitive colloidal particles have been reported: hydrogel and core–shell particles and some composite thermally structured core–shell particles (Figure 9.2).

The most examined microgels are poly(*N*-isopropylacrylamide) [1,4], poly(*N*-isopropylmethacrylamide) [5], poly(*N*-ethylmethacrylamide) [6], and more recently, poly(*N*-vinyl caprolactam)-[7–10] based hydrogel particles.

This section is principally devoted to the preparation of thermally sensitive hydrogel particles (microgels and core–shells) using batch polymerization process. The effect of each reactant and parameter (initiator, temperature, cross-linker agent) on the polymerization process (polymerization kinetic, conversion, final particle size, morphology, water-soluble polymer, etc.) is presented and discussed. Special attention will be focused on the functionality of the elaborated thermally sensitive particles.

9.2.1 Rapid State of the Art

Radical-initiated polymerization has appeared as a suitable process for producing valuable materials such as thermally sensitive hydrogel particles as reported and discussed by several authors. The first work related to the preparation of thermally sensitive microgel has been reported in 1986 [1,11,12] by investigating the precipitation polymerization of *N*-isopropylacrylamide (NIPAM), methylenebisacrylamide (MBA) as a cross-linker, and potassium persulfate (KPS) as an initiator. Using such process, monodisperse microgel particles are easily and rapidly obtained (Figure 9.3). The effect of charged surfactant (i.e., Sodium dodecyl sulfate [SDS]) has been briefly examined on the precipitation polymerization mechanism [11].

The basic principle of such polymerization (in which all reactants are for instance water soluble) is to perform the reaction under the precipitation condition of the formed polymer chains. Then, the formed clusters of chains lead to the particle nucleation step. The colloidal stability of the mature particles originated from the nucleation step is ensured by the use of charged initiator or the use of charged comonomers in the polymerization recipe. In order to maintain the microgel structure rather than resolubilized polymer chains, a cross-linker is absolutely needed to favor the cross-linking efficiency of the precipitated polymer chains under microgel state (Figure 9.4).

Using such process and such batch radical polymerization process, various thermally sensitive hydrogels have been elaborated using NIPAM, NIPMAM, NEMAM, and more recently *N*-vinyl caprolactam (VCL)–based microgels [7–10].

In addition to the use of NIPAM as the main monomer, few works have been dedicated to the use of NIPMAM monomer. Poly(*N*-isopropylmethacrylamide) (PNIPMAM)–based microgel latexes have been prepared at 80°C using methylene-bisacrylamide (MBA) as the cross-linking agent and potassium persulfate (KPS) as the initiator [5]. The polymerization kinetic was found to be rapid and complete but with high water-soluble polymer formation compared to NIPAM batch polymerization [4].

In order to obtain microgels bearing a high volume phase transition temperature compared to PNIPAM and to PNIMAM, batch radical polymerization of NEMAM/MBA/KPS has been

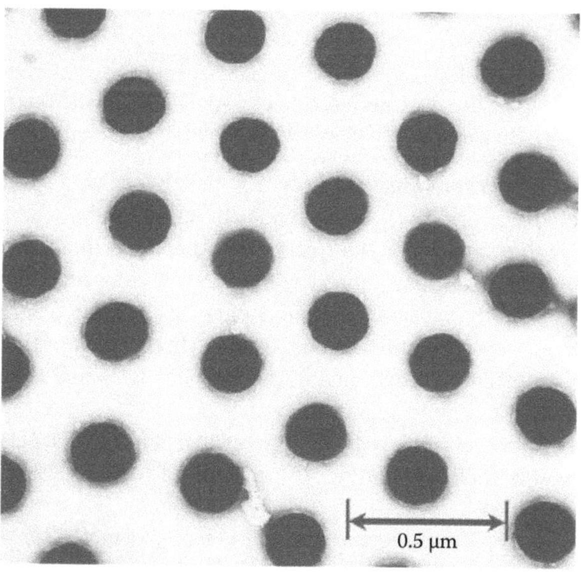

FIGURE 9.3 Transmission electron microscopy of cationic poly(NIPAM) microgel particles deposited from water at a temperature below the T_{VPT}. Due to the hydrophilic character of such particles, hexagonal arrangement is generally observed when the particles are dried at a temperature below the T_{VPT}. (From López-León, T. et al., *J. Phys. Chem. B*, 110, 4629, 2006.)

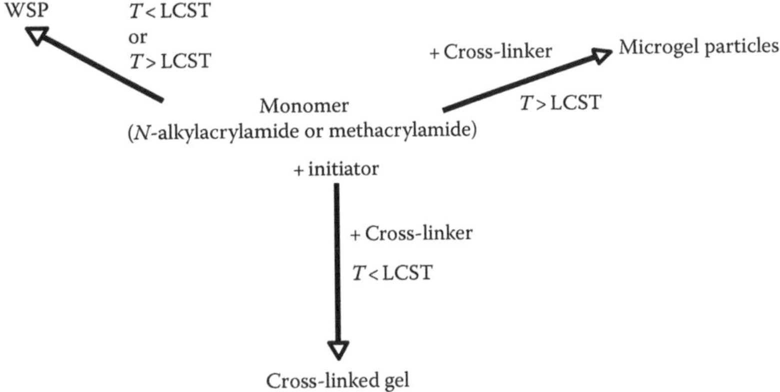

FIGURE 9.4 Polymerization in the aqueous phase of amphiphilic-like monomers, leading to thermally sensitive material preparation (*T* is the polymerization temperature).

explored [6]. Due to the high reactivity of MBA compared to NEMAM, high water soluble polymer formation has been observed and with almost no detectable particles in medium. Consequently, to prepare well-defined microgel particles, it is of paramount importance to take into account not only the reactivity of the principal monomer but also the reactivity of the used cross-linker agent.

More recently, special attention has been focused on the elaboration of vinyl caprolactam (VCL)–based microgels. Poly(*N*-vinylcaprolactam) is thermally sensitive in nature (in the same LCST range as for polyNIPAM) and biocompatible as reported [14]. To prepare VCL-based particles, cross-linker agents such as MBA and charged initiator are used. To some extent, all the study reported on the preparation of polyNIPAM-based microgels (the most studied system) has been extended to VCL monomer. The use of classical water-soluble cross-linker agent (i.e., MBA) to elaborate poly-VCL microgel may lead to non-biocompatible microgels for in vivo applications.

9.2.2 Kinetic Study

To prepare stable microgel particles, the polymerization should be absolutely higher than the LCST of the corresponding homopolymer (or copolymer). In fact, without the precipitation of the formed chains (or lowly cross-linked chains) during the polymerization process, no real particles can be obtained. For instance, the lower critical polymerization temperatures (*LCPT*) in the case of NIPAM, NIPMAM, and NEMAM monomers are 65°C, 70°C, and 80°C, respectively. The observed *LCPT* is related to the properties of the formed oligomers (chain light, molecular weight, composition, and charge distribution). This point is clearly illustrated in the polymerization mechanism described in the following text, but not really established experimentally.

Water-soluble alkylacrylamide or alkylmethacrylamide monomers conversion during the polymerization process can be easily determined or followed using both gas chromatography and ^{1}H-NMR techniques. According to the high reactivity of the generally used water-soluble cross-linkers (i.e., MBA) and the low concentration used in the polymerization recipe, only the main monomer conversion can be easily followed as a function of polymerization time. For most alkyl(metha)acrylamide monomers studied, the polymerization kinetic was generally found to be rapid. In fact, only 30 min were needed to reach total conversion in the case of NIPAM/MBA/V50 or KPS (polymerization temperature from 60°C to 80°C) [7] and 120 min in the case of NIPMAM/MBA/KPS (polymerization temperature from 60°C to 80°C) [5]. The difference in polymerization rate between NIPAM and NIPMAM is related to the high propagation rate constant (k_p) for NIPAM compared to NIPMAM [4,15], as well as for the known reactivity of alkylacrylamides compared

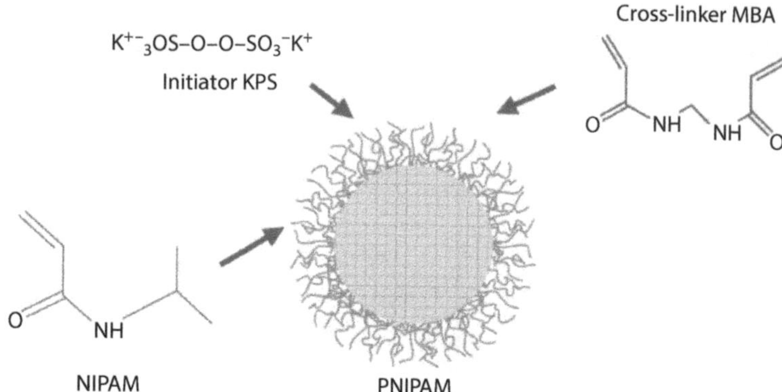

FIGURE 9.5 Schematic representation of batch preparation of functionalized thermally sensitive poly(NIPAM) microgel particles. Basically, to perform such polymerization by a non-polymer chemistry scientist, this recipe based on 1 g of NIPAM, 0.12 g of MBA, 0.012 g of KPS, and 50 mL water in closed battle and placed at 70°C leads to monodisperse microgel particles.

to alkylmethacrylamides derivatives (k_p ~ 18,000 L mol^{-1} s^{-1} for acrylamide [16,17] and k_p ~ 800 L mol^{-1} s^{-1} for methacrylamide [18] at 25°C) (Figure 9.5).

The detailed analysis of particle size versus polymerization time and conversion is one of the best methodologies generally used in polymerization in dispersed media to point out the particle evolution during the polymerization process. The diameter of hydrogel particles (NIPMAM/MBA/KPS at 70°C) [5] versus (conversion)$^{1/3}$ reported in Figure 9.6 was linear relationship, reflecting the fact that the number of particles remained constant and no new particles were formed during the polymerization process. In fact, the particle size (R) is related to the particle number (N_p) and the polymer mass (m, i.e., polymer conversion) by assuming the constancy of the particle density during the polymerization process:

$$R_h \approx N_p^{-1/3} \cdot m^{1/3} \tag{9.1}$$

The observed behavior confirmed by scanning electron microscopy (SEM) analysis of the particle size distribution versus polymerization time is illustrated in Figure 9.7. The utilization of the previously described investigations of both particle size versus conversion$^{1/3}$ and size distribution versus time using TEM or SEM are related to the difficulty of particle number concentration determination (or estimation) induced by the swollen property of the microgels.

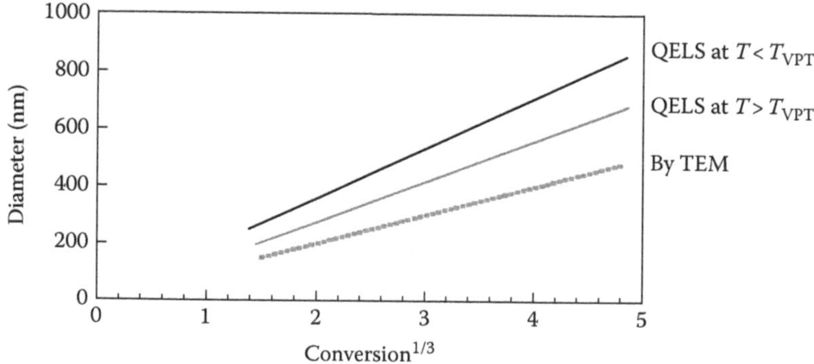

FIGURE 9.6 Illustration of particle size of hydrogel versus conversion in the case of classical alkylacrylamide monomers. See Duracher's work for experimental data with NIPMAM monomer. (From Duracher, D. et al., *J. Polym. Sci. A: Polym. Chem.*, 37, 1823, 1999.)

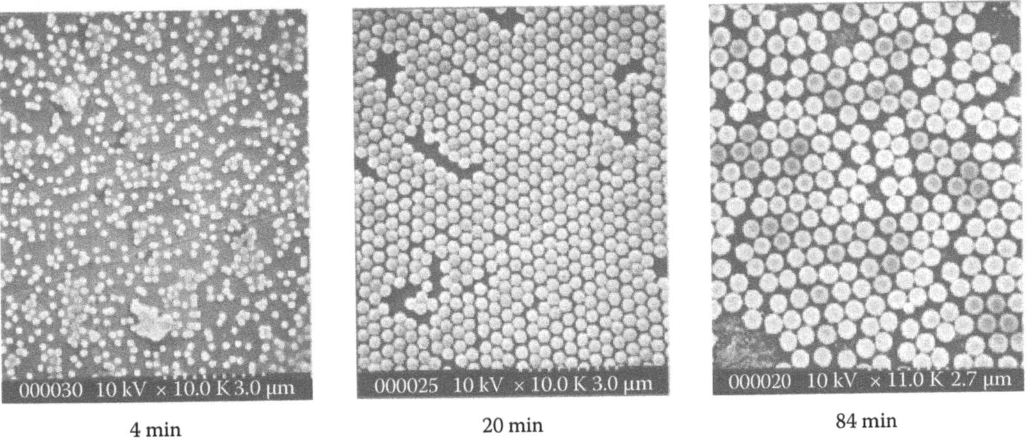

FIGURE 9.7 Scanning electron microscopy (SEM) images of poly(NIPMAM) particles at different polymerization times. (From Duracher, D. et al., *J. Polym. Sci. A: Polym. Chem.*, 37, 1823, 1999.)

9.2.2.1 Effect of Initiator

There are few reports on the influence of initiator on the precipitation polymerization but as expected, the polymerization rate (Figure 9.8) increases together with the initiator concentration, as is often the case in emulsion polymerization. This behavior is attributed to an increase in the polymerization loci. It is interesting to notice that an increase in the initiator concentration leads to an increase in the water-soluble polymer formation (oligomers bearing low molecular weight) and to a decrease in the final hydrodynamic particle size (for low initiator concentrations). The polymerization rate (R_p) is related to the initiator concentration [I] using the low-scale representation:

$$R_p \approx [I]^a \tag{9.2}$$

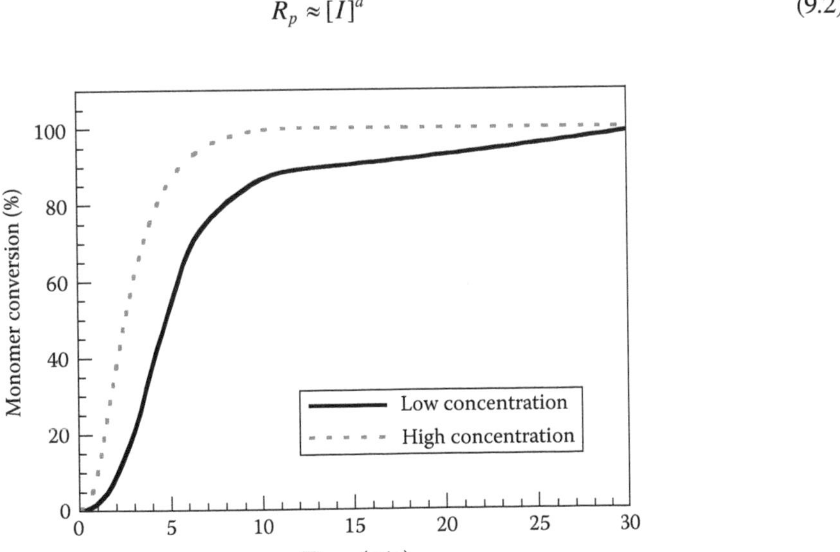

FIGURE 9.8 Monomer conversion versus polymerization time for two initiator concentrations. Example for poly(NIPAM) particles: polymerization temperature T = 70°C, [NIPAM] = 48.51 mmol, [MBA] = 3 mmol, [V50] = 0.1 (—) to 1 mmol (- - - -) for total volume = 250 mL. (From Meunier, F., Synthèse et caractérisation de support polymères particulaires hydrophiles à base de *N*-isopropylacrylamide, Elaboration de conjugués particules/ODN et leur utilisation dans le diagnostic médical, Thèse, University of Lyon-1, France, 1996.)

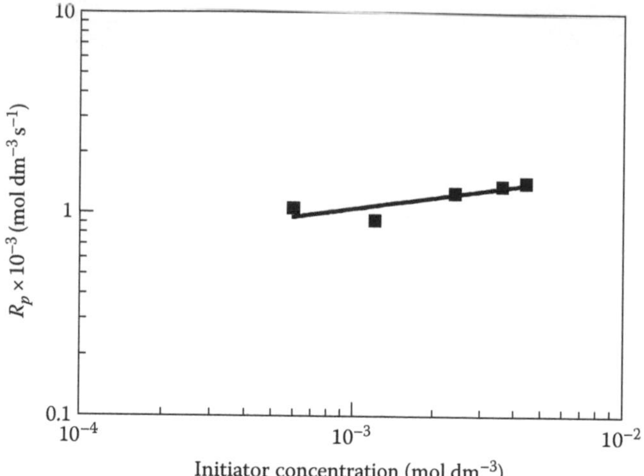

FIGURE 9.9 Dependence of polymerization rate on the initiator concentration in log–log scale. Temperature = 70°C, [NIPAM] = 48.51 mmol, [MBA] = 3 mmol, and total volume = 250 mL ($R_p \approx [I]^{0.18}$, I: V50 initiator concentration). (From Meunier, F., Synthèse et caractérisation de support polymères particulaires hydrophiles à base de *N*-isopropylacrylamide, Elaboration de conjugués particules/ODN et leur utilisation dans le diagnostic médical, Thèse, 1996.)

For instance $a \approx 0.18$ for NIPAM/MBA/KPS at 70°C (Figure 9.9), and $a \approx 0.4$ in the case of emulsion polymerization. With such a complex system, it is more appropriate to consider the relationship between the number of particles (N_p) and the reactant concentration, then:

$$N_p \approx [M]^a[I]^b[CL]^c \tag{9.3}$$

where

- $[M]$ is the monomer
- $[CL]$ is the cross-linker agent
- a, b, and c are the scaling exponents

The particle number cannot be easily determined without a drastic approximation.

9.2.2.2 Effect of Temperature

The influence of temperature on precipitation polymerization was also studied [5,7], and found to be similar to the effect of initiator. In fact, the increase in temperature leads to an increase in the decomposition rate on the initiator, which enhances the polymerization loci, as discussed earlier (Figure 9.10) and described by the following rate decomposition equation:

$$N_{dec} = I(1 - e^{-k_d t}) \tag{9.4}$$

where

- k_d is the decomposition rate constant of the initiator at a given temperature
- I is the initial concentration of initiator
- N_{dec} is the amount of decomposed initiator after a given time and at a given temperature

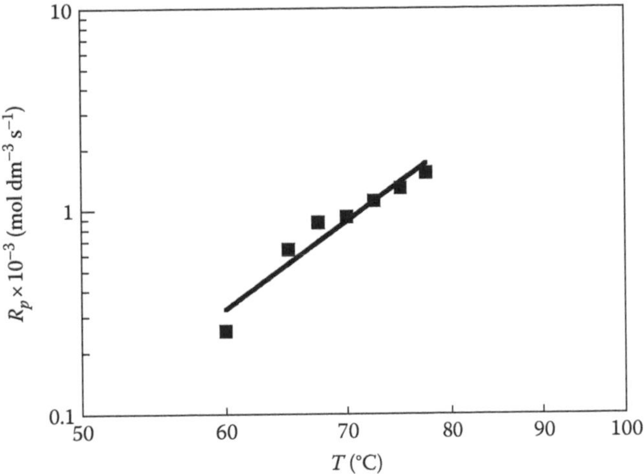

FIGURE 9.10 Polymerization rate versus polymerization temperature in semilog scale. [NIPAM] = 48.51 mmol, [*I*] = 0.3 mmol, [MBA] = 3 mmol, and total volume = 250 mL. (From Meunier, F., Synthèse et caractérisation de support polymères particulaires hydrophiles à base de *N*-isopropylacrylamide, Elaboration de conjugués particules/ODN et leur utilisation dans le diagnostic médical, Thèse, 1996.)

It is important to notice that the polymerization temperature of alkyl(metha)acrylamide in aqueous media should be above the LCST of the corresponding linear polymer, and high enough to favor the precipitation process of the formed water-soluble soluble oligomers. The temperature also affects the final particle size: A low temperature results in a large particle size. Furthermore, the temperature affects the reactivity of each reactant, including the main monomer, initiator, and cross-linker agent. In fact, in the case of emulsion polymerization, the instantaneous rate of polymerization R_p is related to both propagation rate coefficient, monomer concentration (M), the average number of radicals per particle ($\tilde{n}$), the number of latex particles N_p (which is connected to the rate radical generation $N_p \approx \rho^{0.4}$) as expressed by the following equation:

$$R_p = k_p[M] \cdot \tilde{n} N_p \tag{9.5}$$

where k_p is the rate coefficient for propagation of the monomer (M).

9.2.2.3 Effect of the Cross-Linker Agent

The cross-linker agent has a marked and drastic effect on particle formation. It is necessary in the polymerization recipe in order to favor particle formation by cross-linking the precipitated poly(*N*-alkylacrylamide) chains leading to generate the nucleation step. The increase in the cross-linker agent concentration in the batch polymerization recipe leads to a reduction in the final amount of water-soluble polymer (Figure 9.11). In addition, the water-soluble cross-linker agent was found to affect the polymerization rate and the final particle size [5] to a small extent, and the swelling ability of the particles to a greater extent. In accordance with the higher reactivity of the water-soluble cross-linkers (as MBA) generally used in such precipitation polymerization, the final structure of hydrogel particles was found to have a gradient composition (looser and looser from the core to the shell), as evidenced by Guillermo et al. [20] during an investigation of the transverse relaxation of protons using the NMR technique and the internal structure of the particle versus cross-linker concentration can be schematically illustrated as given in Figure 9.12.

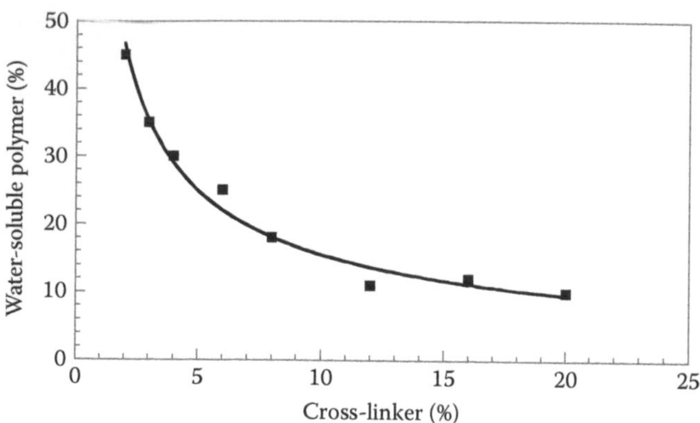

FIGURE 9.11 Effect of MBA (2%–20% w/w) on water-soluble polymer formation using NIPMAM (1 g) monomer and KPS (2% w/w) initiator, polymerization temperature 70°C. (From Duracher, D. et al., *J. Polym. Sci. A: Polym. Chem.*, 37, 1823, 1999.)

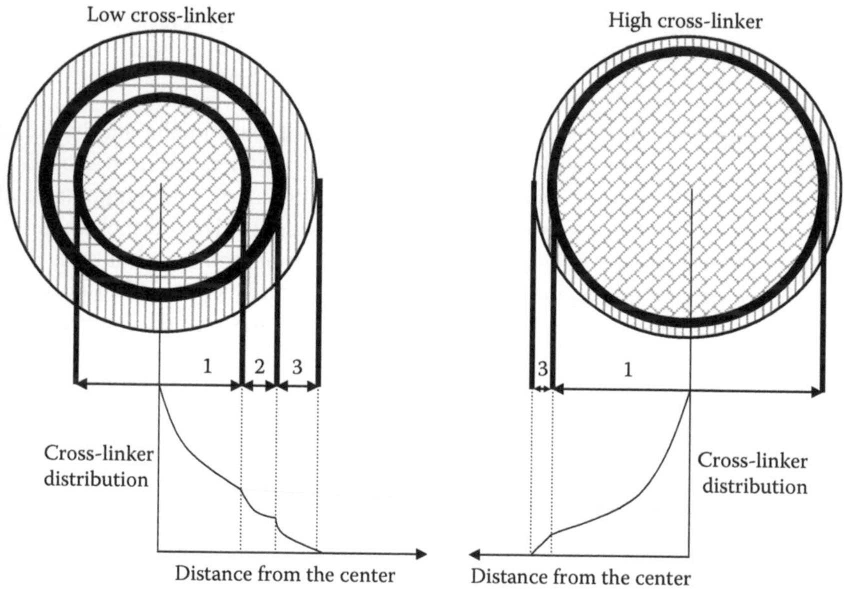

FIGURE 9.12 Schematic illustrations of hydrogel particles as a function of cross-linker concentration. (1) Highly cross-linked part, (2) medium cross-linked phase, and (3) low cross-linked shell.

9.2.2.4 Functionalization Studies and Operating Methods

A large number of processes have been developed during the last decade, permitting the synthesis of reactive latexes and the functionalization of prepared polymer particles with specific properties:

- *Batch polymerization*: Polymerizations performed in a closed reactor, with all the ingredients being introduced at the beginning of a single step. This method, apart from certain exceptions, is of little interest, since a large part of the functional monomer is consumed, providing substantial quantities of water-soluble polymers that disturb nucleation and the final stabilization of the particles.

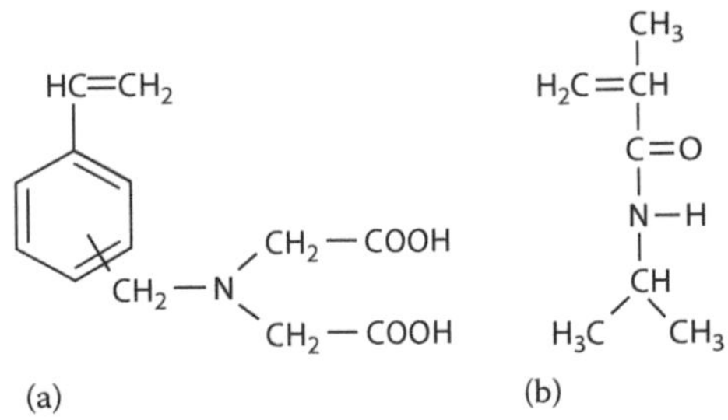

(a) (b)

FIGURE 9.13 Structure of *N*-(vinylbenzylimino)-diacetic acid (IDA) (a) and *N*-isopropylmethacrylamide (b).

- *Semicontinuous addition methods*: These are very useful for copolymerizations requiring the control of homogeneous compositions at the level of the chain and the particle. Variable composition gradient processes permit the introduction of the functional monomer to a suitable conversion, favoring incorporation on the surface or inside the particles.
- Multistage polymerizations, including the deferred addition of an ionic comonomer (constituting the basic latex), favoring a highly efficient surface functionalization.
- *Seed functionalization*: These methods consist of the functionalization of a latex seed by a monomer or monomer mixture. This often permits surface incorporation to be increased and is well adapted for formulating controlled charge density model colloids.
- *Postreaction on reactive latexes*: This process is very useful for modifying the functionality of a given latex if it cannot be obtained directly.

The preparation of functionalized thermally sensitive microgel particles has not been sufficiently investigated, since only a few works have been reported. The first systematic work on the preparation of functionalized poly(NIPAM) hydrogel particles was studied using NIPAM/MBA/AEMH (aminoethylmethacrylate hydrochloride) in a batch polymerization process, and first reported by Meunier et al. [4] and then by Duracher et al. [21] during an investigation of the effect of *N*-(vinylbenzylimino)-diacetic acid (IDA) on the polymerization of NIPMAM/MBA/KPS (Figure 9.13). The effect of the charged functional monomer on the precipitation polymerization of such alkylacrylamide and alkylmethacrylamide monomers (i.e., NIPAM and NIPMAM) was found to resemble the effect of the ionic (or water-soluble) monomer on the emulsion polymerization of styrene, for instance. In fact, the increase in the functional monomer concentration leads to rapid polymerization, high polymerization conversion (>95%), low particle size (Figure 9.14), and high water-soluble polymer formation.

In addition to batch functionalization, the shot-grow process was also performed in order to prepare amino-containing thermally sensitive poly(NIPAM) hydrogel [4] and core–shell (polystyrene core and poly(NIPAM) shell) particles [22]. The obtained results revealed a good functionalization yield with a non-negligible amount of water-soluble polymer formation.

As in the case of emulsion polymerization, an increase in the functional monomer concentration leads to a reduction in the final hydrodynamic particle size and enhanced water-soluble polymer formation, as illustrated in Figures 9.15 and 9.16 for the batch polymerization of NIPMAM/MBA/IDA/KPS. The reduction in particle size versus functional monomer has been attributed to the enhancement of precursor formation and the number of stable particles, which rapidly become the polymerization loci.

9.2.2.5 Polymerization Mechanism

After examination of the role of each reactant implied in the polymerization of water-soluble *N*-alkylacrylamide and *N*-alkylmethacrylamide monomer in the presence of the water-soluble cross-linker agent and radical initiators, the polymerization mechanism of this system in the preparation of thermally sensitive microgel submicron particles can be presented and detailed as follows (Figure 9.17):

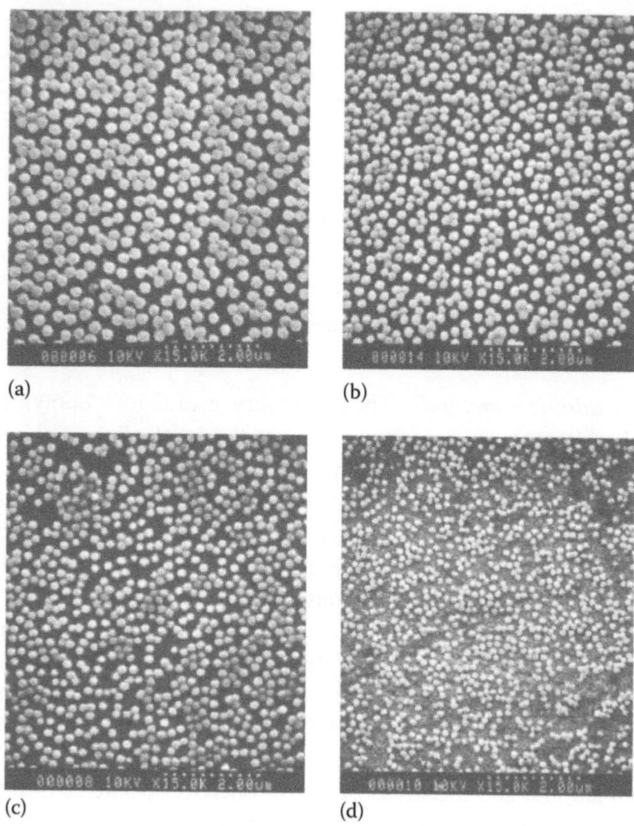

FIGURE 9.14 The influence of IDA (charger monomer) on final particle size and size distribution as examined by scanning electron microscopy (SEM) [21] (the photos are in the same scale): (a) 0%, (b) 0.6%, (c) 1.2%, and (d) 2%.

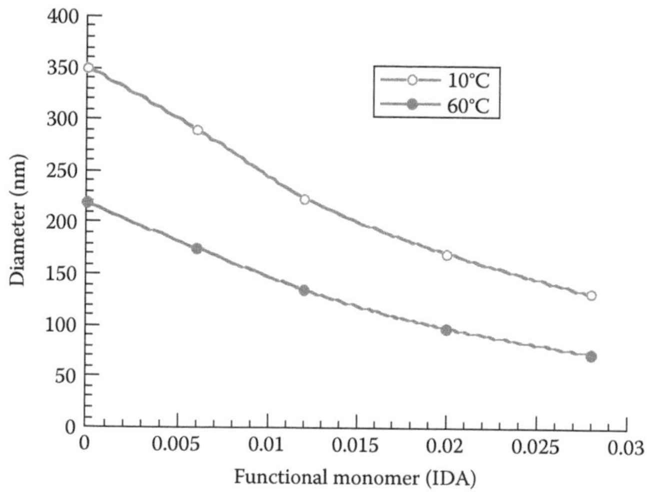

FIGURE 9.15 Effect of functional monomer (IDA in g) on hydrodynamic particle diameter measured by quasi elastic light scattering (QELS) at 10°C and 60°C. (From Duracher, D. et al., *Macromol. Symp.*, 150, 297, 2000.)

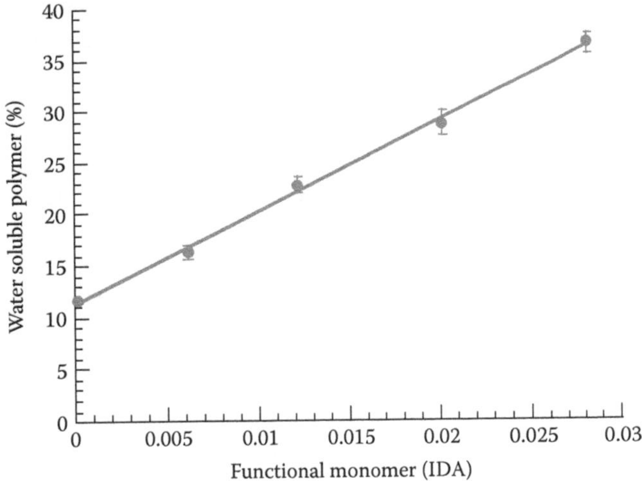

FIGURE 9.16 The influence of IDA (in g) on water-soluble polymer (WSP) formation. (From Duracher, D. et al., *Macromol. Symp.*, 150, 297, 2000.)

- *Water phase polymerization*: Before adding the water-soluble initiator, the medium was totally homogeneous and relatively limpid, even at the polymerization temperature. After adding the initiator, the polymerization requires a low induction period (t^*) (generally less than 3 min in the case of NIPAM/MBA/KPS or V50 system using recrystallized reactants), probably due to the presence of oxygen traces in the aqueous phase. Above the t^* period, the decomposed initiator leads to rapid oligomer formation. The oligomer concentration is principally governed by the initiator decomposition rate and efficiency.

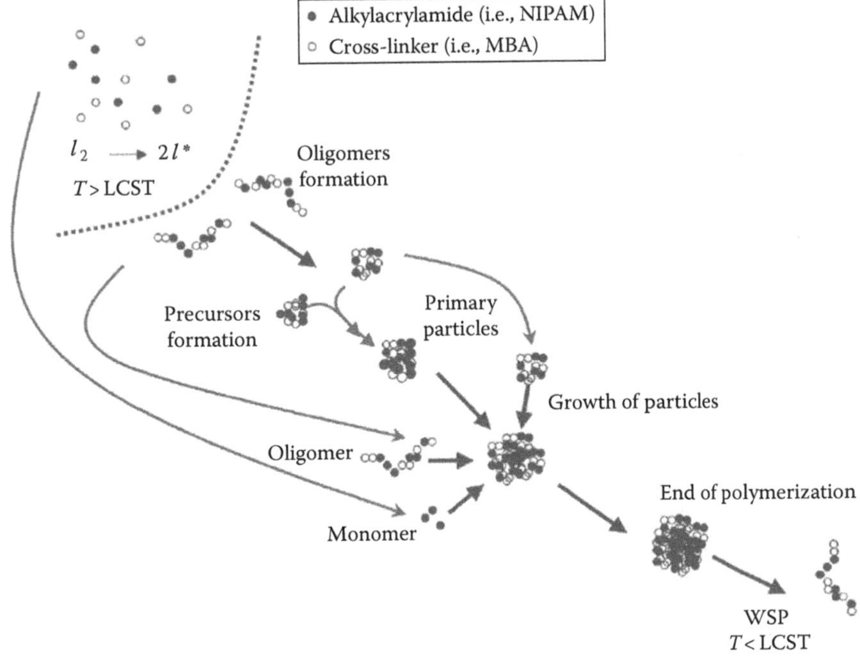

FIGURE 9.17 Schematic illustration of precipitation polymerization mechanism of water-soluble *N*-alkyacrylamide and *N*-alkylmethacrylamide monomer derivatives.

- *Nucleation step*: The nucleation period is mainly due to the precipitation of the oligomers formed when the critical chain lengths (or molecular weight and composition) are reached or when the cross-linked chains start to precipitate. In fact, the scanning electron microscopy (SEM) (Figure 9.7) analysis of the particle size and size distribution revealed rapid and narrow particle distribution formation (i.e., generally in less than 5 min in the case of NIPMAM/MBA/KPS and NIPAM/MBA/V50 systems). In accordance with highly reactive water-soluble cross-linkers such as MBA, it may be assumed that the first oligomers contain a significant amount of the cross-linker agent, and of primary particles.
- *Growth of particles*: After the rapid nucleation period, the polymerization of residual monomer may take place at various domains (i.e., water phase, on the surface of the formed particle, and in the swelled parts of the particles). In fact, the partition of the monomer between the water phase and polymer particles should be considered: (1) polymerization in the water phase leads to the formation of small oligomers that can be cross-linked on the formed particles when there is a sufficient residual cross-linker agent concentration, (2) and (3) the possible polymerization of the monomer in the swollen state. When the cross-linker agent is totally consumed, the oligomers formed contribute to water-soluble polymer formation. The water-soluble polymer formed may originate from the desorption process of adsorbed chains onto the particles when the dispersion is cooled (temperature < T_{VPT} or < LCST), or from low molecular weight chains that are highly water-soluble even above the LCST of the considered homopolymer.

9.2.2.6 Thermally Sensitive Core–Shell Particles

Regarding core–shell like particles, only few works have been reported. The most examined system is polystyrene core/thermally sensitive poly(NIPAM) shell. The first work in this field has been reported by Kawaguchi et al. [23] and then this approach has been adequately examined and explored in order to prepare a well-defined, functionalized, and cross-linked shell. The preparation of such core–shell latexes have been performed via batch combined radical emulsion and precipitation polymerizations and via shot-grow polymerization process [22,24].

Batch emulsion polymerization of styrene in the presence of NIPAM monomer shows rapid consumption of NIPAM monomer [22]. In fact, around 80% of NIPAM monomer is reacted before the styrene monomer starts to polymerize. Then, the formed poly(NIPAM) chains bearing surface active properties acts as a surfactant during the styrene polymerization. Consequently, the initial amount of NIPAM controls the polymerization rate, the particle size, the size distribution, and also the thickness layer of the formed shell. Thus, polymerization should be considered as emulsion polymerization with in situ surfactant production. Examination of the particle size of latexes as a function of polymer conversion by transmission electron microscopy shows that the polymerization of a heterogeneous system such as this occurs in numerous steps that will not be depicted in this chapter, but can be consulted in Refs. [22,24].

In order to prepare a well-defined and functionalized shell, the shot grow polymerization process has been examined. This process leads to thickness layer enhancement and the introduction of surface reactive groups. Obtaining core–shell particles with higher densities of amine groups (in cationic form) can be controlled by using a shot process consisting in polymerizing a mixture of NIPAM, AEMH (amino ethyl methacrylate hydrochloride), and a cross-linking agent (methylene bisacrylamide, MBA) on an initial seed (under formation) at its high conversion rate (>50%).

Besides the previously mentioned complex system, monodisperse thermosensitive poly(*N*-ethylmethacrylamide) microgel particles have been prepared by precipitation polymerization of *N*-ethylmethacrylamide (NEMAM) using hydrophobic cross-linker ethylene glycol dimethacrylate (EGDMA) and potassium persulfate as an initiator. In such batch radical combined (emulsion and precipitation) polymerization system, water-soluble polymer formation amount is still relatively high compared to pure precipitation (NIPAM/MBA)-based radical polymerization. Using hydrophobic cross-linker agent leads to well-defined core–shell like particles. This is attributed to high

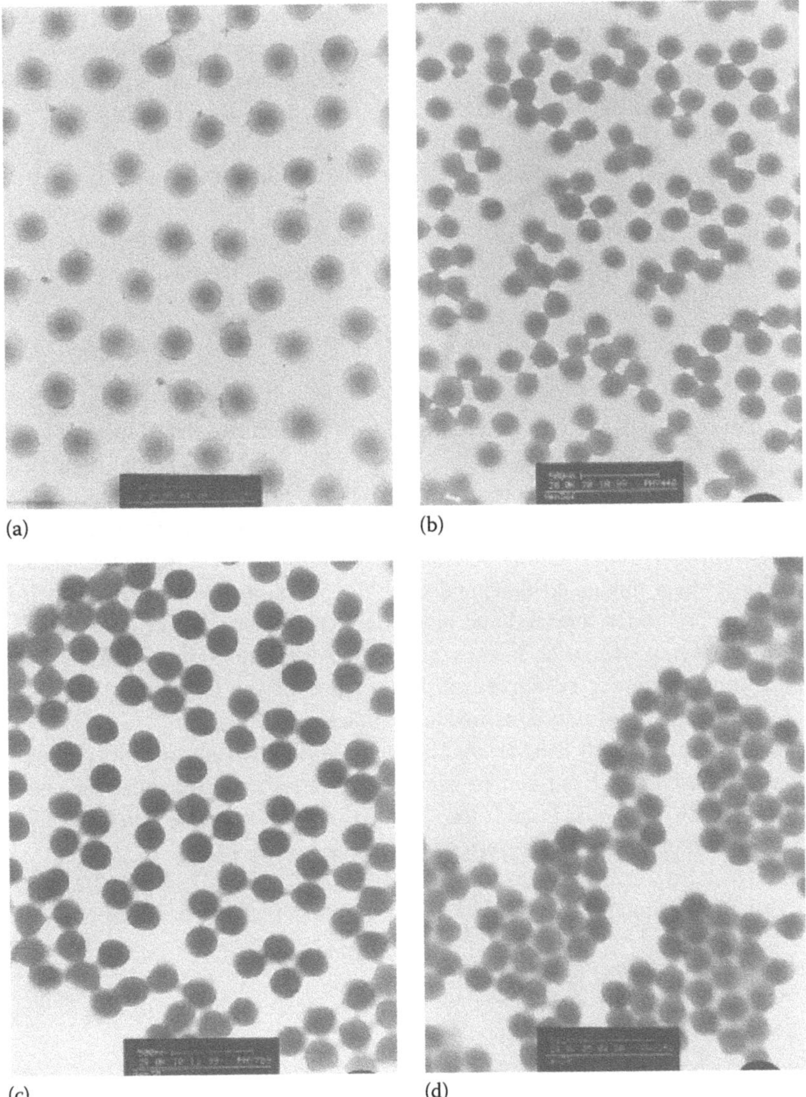

FIGURE 9.18 TEM micrographs of the final PNEMAM microgel particles as a function of the cross-linker concentration (wt% EGDMA) [25]. For low cross-linked microgels, the hexagonal arrangement was clearly observed as for polyNIPAM microgel: (a) 8% EGDMA (D_{TEM} = 20 nm), (b) 15% EGDMA (D_{TEM} = 180 nm), (c) 30% EGDMA (D_{TEM} = 193 nm), and (d) 45% EGDMA (D_{TEM} = 211 nm).

reactivity of the used hydrophobic cross-linker and then constitute the core, whereas, the shell is formed composed of poly(NEMAM) network [6,25,26]. As unexpected, spherical and monodisperse particles are obtained irrespective of the amount of hydrophobic cross-linker as illustrated in Figure 9.18 via the TEM analysis of crude particles.

9.3 COLLOIDAL CHARACTERIZATION

The aim of this analysis stage is to obtain qualitative and quantitative information on particle morphology, particle size and size distribution, surface polarity, assessment of the localization of the functional monomer introduced into the reaction system, and colloidal stability before any implication in the biomedical field. In addition, it is vital to purify polymer particles before colloidal characterization, by separating the particles from impurities originating from polymerization reactants (these are mostly traces of the initiator, residual monomers, and water-soluble polymer). There is now a whole arsenal of techniques whose multiple applications are well referenced. As for separation methods, the most important and frequently used are centrifugation, serum replacement, magnetic separation in the case of magnetic latexes, and ultrafiltration. Dialysis and ionic exchanges on mixtures of cationic and anionic resins deserve mention among the purification methods. In the case of hydrophilic thermally sensitive microgel particles, the centrifugation can be used without any irreversible aggregation risk. In fact, the aggregated particles during the centrifugation step (at high speed and long period) are easily dispersible in aqueous medium by simple stirrer.

9.3.1 Morphology

The morphological characterization of structured latexes is a fundamental aspect of their study, since (1) it provides very useful information on the nature of the mechanisms that regulate the particle's constitution, and (2) knowledge of the organization of the polymer within the particle is the essential foundation for the theoretical interpretation of the behavior of the resulting latex films (mechanical properties, permeability, etc.). From this perspective, there are a great many techniques that require examination in order to eliminate artifacts and wrong conclusions deduced from their use.

Various methods have been reported and published. Of these methods, electronic microscopy is still preferred for studying structured latexes, since improvements in observation techniques (sample preparation, introduction of selective marking of one of the components) permit directly an increasingly refined analysis of polymers inside particles.

Scanning electron microscopy (SEM) can also be used for both surface morphology and particle size and size distribution, as illustrated in Figure 9.19, for thermally sensitive polystyrene core-cross-linked poly(NIPAM) shells. In addition, atomic force microscopy can be used for investigating colloidal particle morphology, as reported by Duracher et al. [22].

In the case of batch radical polymerization of NEMAM/EGDMA/KPS, the thickness layer is mainly governed by the amount of EGDMA in the polymerization recipe as examined by atomic

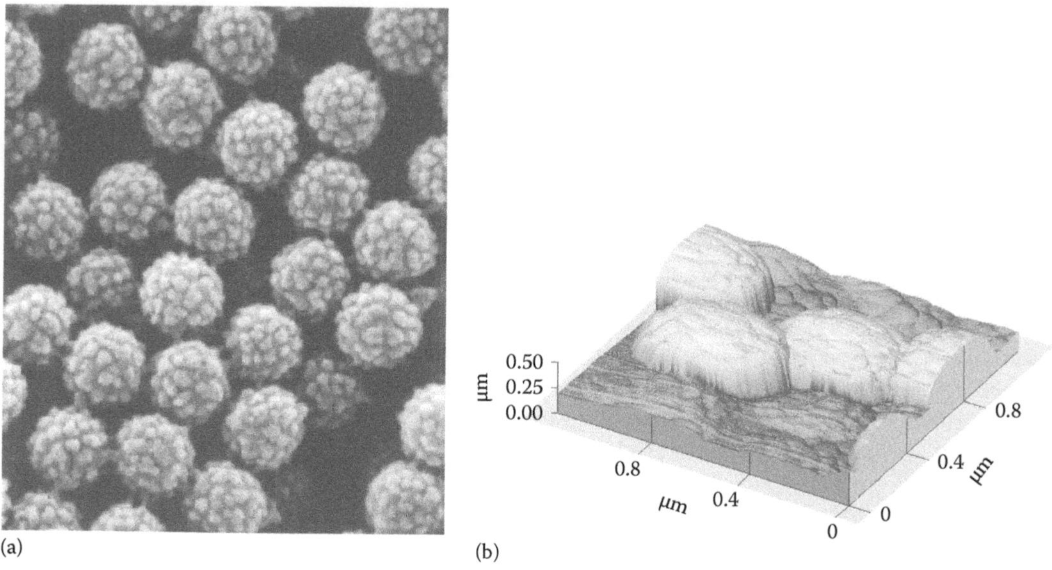

FIGURE 9.19 (a) Scanning electron microscopy (SEM) and (b) atomic force microscopy (AFM) of polystyrene (core)-cross-linked poly(NIPAM) shell microspheres. (From Duracher, D. et al., *Colloid Polym. Sci.*, 276, 219, 1998.)

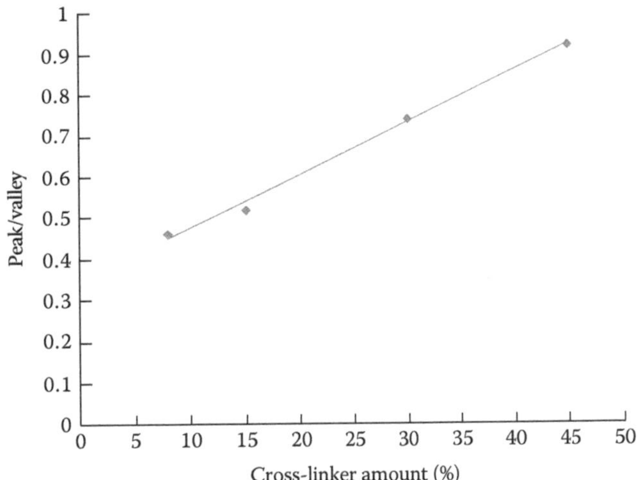

FIGURE 9.20 Peak to valley ratio versus EGDMA concentration in NEMAM/EGDMA/KPS system (1% KPS and 15% wt/wt EGDMA). (From Hazot, P. et al., *J. Polym. Sci. A: Polym. Chem.*, 40, 1808, 2002.)

force microscopy (AFM) via the determination of pick to valley ratio (Figure 9.20). The measured ratio was found to increase (from 0.4 to 0.9) with increasing the cross-linker amount. Pick to valley ratio is close to 1 for 45% cross-linker amount. Consequently, the increase in the EGDMA concentration leads to more rigid particles.

9.3.2 Effect of Temperature on Hydrodynamic Particle Size

The mean hydrodynamic diameter was calculated from the diffusion coefficient measurement, which, in the high dilution limit of negligible particle–particle interactions, is calculated by using the Stokes–Einstein equation:

$$D = \frac{kT}{3\pi\eta D_h} \tag{9.6}$$

where

D is the diffusion coefficient
k is the Boltzmann constant
T is the absolute temperature
η is the viscosity of the medium

Light scattering technology (QELS) is generally suitable for low particle sizes (i.e., diameter < 1 μm). For polystyrene latex particles bearing low charges, density, and hydrophobic surface, the particle size does not generally depend on temperature and salinity. However, in the case of hydrophilic and poly(*N*-alkylacrylamide) derivatives, particle size and swelling ability are affected by the solvent quality of the dispersion medium, temperature, pH, and salinity. By way of illustration, the particle size for various types of particle poly(NIPAM) as a function of temperature is reported in Figure 9.21.

This decrease (in particle size vs. T) is more marked in the case of large particles (with low cross-linking properties), due to the pronounced size difference below and above the volume phase transition temperature (T_{VPT}) between the shrunken and the expanded state. In addition,

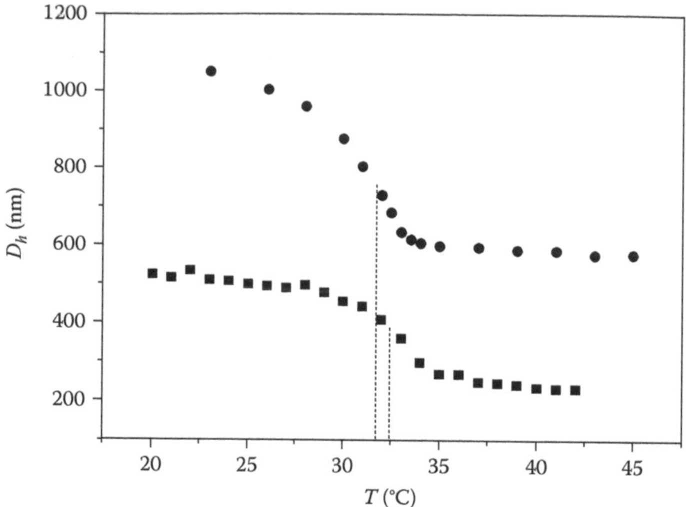

FIGURE 9.21 Hydrodynamic size of negative (•) and positive (■) poly(NIPAM)-based microgels as a function of temperature in a 10^{-3} M NaCl solution. Vertical lines have been drawn to guide the eye toward the T_{VPT} values. Below the T_{VPT} $\chi_{12} = 0$ in good solvent (extended microgel) and above the T_{VPT} $\chi_{12} > 0.5$ in poor solvent (Shrunken microgel) [9.3].

the effect of temperature is more marked in the case of total hydrogel microspheres compared to the core–shell particles, as illustrated earlier (Figure 9.21). The behavior observed has already been reported by many authors with regard to poly(NIPAM) microgel particles, and attributed to the breaking of hydrogen bonds between polymer network and water molecules during the heating process, which induces a decrease in hydrodynamic particle size. Such light scattering experiments can provide two major pieces of information concerning the dispersion: (1) the effect of temperature on particle size (i.e., volume phase transition domain) and (2) the swelling ability of the considered colloidal particles. The swelling ratio calculated from the hydrodynamic particle size above and below the volume phase transition temperature reflects the cross-linking density of the particle but not the cross-linker distribution, as recently reported by Guillermo et al. [20] using the NMR technique as described in Section 9.2.2.3 (Figure 9.12). To compare the swelling capacity of the microgel particles (or core–shell latexes) with different functional monomer concentrations, charge densities, and internal structures, the swelling ratio (S_w) was introduced and defined by the following equation:

$$S_w = \frac{V}{V_c} \tag{9.7}$$

where

V and V_c represent the particle volume calculated from the hydrodynamic radius as determined by QELS

V_c represents the collapsed volume

9.3.3 Electrokinetic Study

The investigation of electrophoretic mobility versus pH and temperature can also be considered as a key point for the analysis of surface charge density variation. In fact, the electrophoretic mobility of thermally sensitive microgel particles is drastically affected by the medium temperature.

The decrease in particle size, and subsequent increase in surface charge density, results in an increase in electrophoretic mobility, as expressed by the following relationship between (μ_e) and particle size (R):

$$\mu_e \approx \frac{N_e}{4\pi\eta R^2\kappa} \tag{9.8}$$

where

N is the number of charged groups per particle of hydrodynamic radius R
e is the electron charge
κ is the reciprocal of Debye length thickness
η is the viscosity of the medium.

The effect of temperature on the electrophoretic mobility of polyNIPAM microgel particles has been investigated as first reported by Pelton and Chibante [1], Kawaguchi et al. [12], and then by Meunier et al. [4].

As illustrated in Figure 9.22, the electrophoretic mobility increased (in the absolute value) together with the temperature, irrespective of the nature of the surface charge. Such behavior is related to the surface charge density versus temperature. The amplitude of the measured transition in the electrokinetic property was more marked for low cross-linked thermally sensitive particles (i.e., high swelling ability). The electrokinetic study of such thermally sensitive particles needs particular attention in order to demonstrate the location of charges implicated in electrophoretic mobility and what is the relationship between the volume phase transition temperature and the electrokinetic transition temperature.

To point out the relationship between the electrophoretic mobility (i.e., charge density) and the reduction of the particle size, the electrophoretic mobility has been examined as a function of (r_h^{-2}). When the linearity (μ_e vs. r_h^{-2}) is observed, it reflects the homogeneous charge distribution on the microgel "surface" and constant reduction of the particles size. When the nonlinearity is observed, it may reflect the complexity of the microgel structure and also the charge distribution in the vicinity of the immobile water molecules [13].

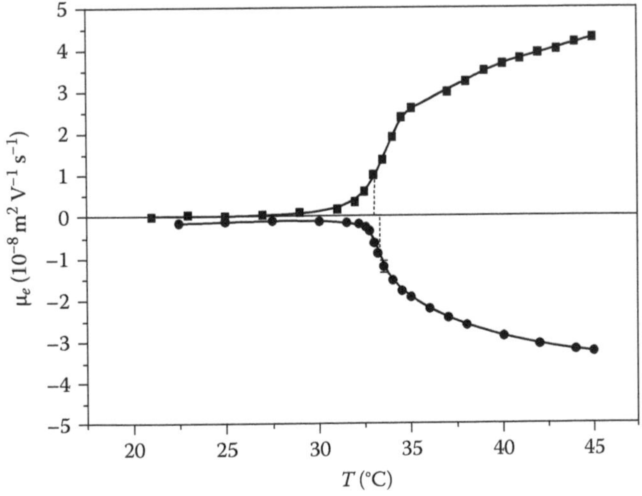

FIGURE 9.22 Electrophoretic mobility of the negative (λ) and positive (ν) poly(NIPAM) microgels as a function of temperature (10^{-3} M NaCl). (From López-León, T. et al., *J. Phys. Chem. B*, 110, 4629, 2006.)

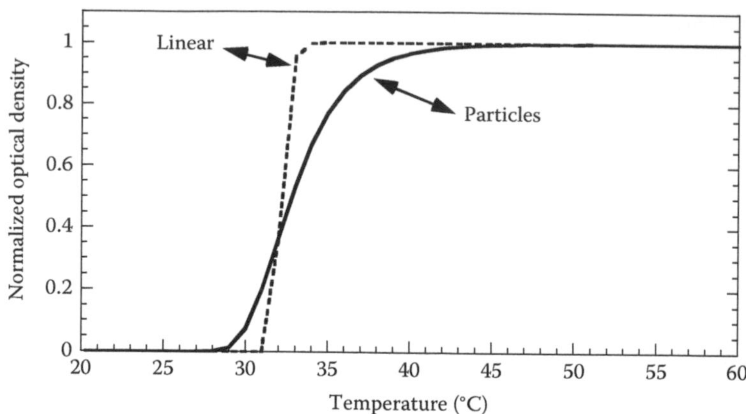

FIGURE 9.23 Normalized optical density versus temperature for both linear and microgel poly(NIPAM) polymer (for highly diluted polymer and low salinity concentration).

9.3.4 Volume Phase Transition Temperature (T_{VPT})

The volume phase transition temperature of thermally sensitive particles can be determined using various methods and techniques: fluorescence study [27], light scattering, differential scanning calorimetry, and turbidity measurement.

The easiest way to determine the LCST in the case of thermally sensitive linear polymer is to investigate the turbidity (τ) of the medium or the optical density (*OD*) variation as a function of temperature ($\tau = 2.303OD/L$, where L is the length of the sample in centimeters). This turbidity method has been adapted to the hydrogel particles and defined as the maximum of the $\delta OD/\delta T$ curve versus temperature.

The normalized optical density variation as a function of the temperature of linear and microgel poly(NIPAM) is illustrated in Figure 9.23. The optical density increases with increasing the temperature for both linear thermally sensitive polymer and microgel particles. Such behavior is related to change in the refractive index of the polymer (ε_p). In fact, below the volume phase transition temperature, the polymer is highly hydrated (i.e., $\varepsilon \sim \Phi{\cdot}\varepsilon_{water} + (1 - \Phi)\varepsilon_p$, Φ is the water fraction in the polymer, here $\Phi \sim 1$), when the temperature increases, the polymer refractive index increases (decrease in $\Phi \sim 0.3$) leading to a rise in the turbidity of the medium.

All the determined volume phase transition temperatures of hydrogel or core–shell bearing thermally sensitive polymer particles are in a broad range compared to the LCSTs. The volume phase transition temperature of thermally sensitive particles is also internal-structure dependent (i.e., polymer composition, cross-linker density, and distribution in the particle).

9.3.5 Colloidal Stability

The colloidal stability has been generally investigated via the stability factor (i.e., Fush factor, W) determination as a function of salinity and temperature using the turbidity-based method or any classical spectrophotometer equipment. The dispersion was highly diluted in water at given pH and temperature. The aggregation rate constants (dOD/dt) were determined by measuring the optical density (*OD*) variation at 600 nm wavelength as a function of time, after adding NaCl solution. The critical coagulation concentration (CCC) of the dispersion was deduced from the stability factor (W) variation as a function of ionic strength (below and above the LCST) plotted in log–log scale. The stability factor was determined using the following classical equation relating W to aggregation rate constants:

$$W = \frac{(dOD/dt)_f}{(dOD/dt)_s} \tag{9.9}$$

where $(dOD/dt)_f$ and $(dOD/dt)_s$ are the initial slopes of the optical density (*OD*) variation as a function of time (*t*), for fast and slow aggregation processes, respectively. The same methodology can be also used in order to investigate the effect of pH or any additive on the colloidal stability.

The colloidal stability of thermally sensitive colloidal microgels has been largely examined in terms of experimental studies. In this direction, interesting and complete work has been recently published by López-León et al. [13]. Interestingly, the salinity affects not only the colloidal stability but also the swelling ability of the microgel. In fact, the increase in the salinity reduced the solubility of thermally sensitive polymer (i.e., polyNIPAM, polyNIPMAM, etc.) and consequently induced reduction of the LCST and then the colloidal stability of the particles. Since the salinity affects the volume phase transition temperature of microgels, it is interesting to take into account not only the effect of salinity on the colloidal stability but the effect of indiscernible temperature and salinity couple on the colloidal stability as illustrated by the stability diagram (Figure 9.24a).

As a general tendency, the colloidal stability of thermally sensitive latex particles is related to both temperature and salinity. In fact, in the case of linear *N*-alkylacrylamide or methacrylamide-based

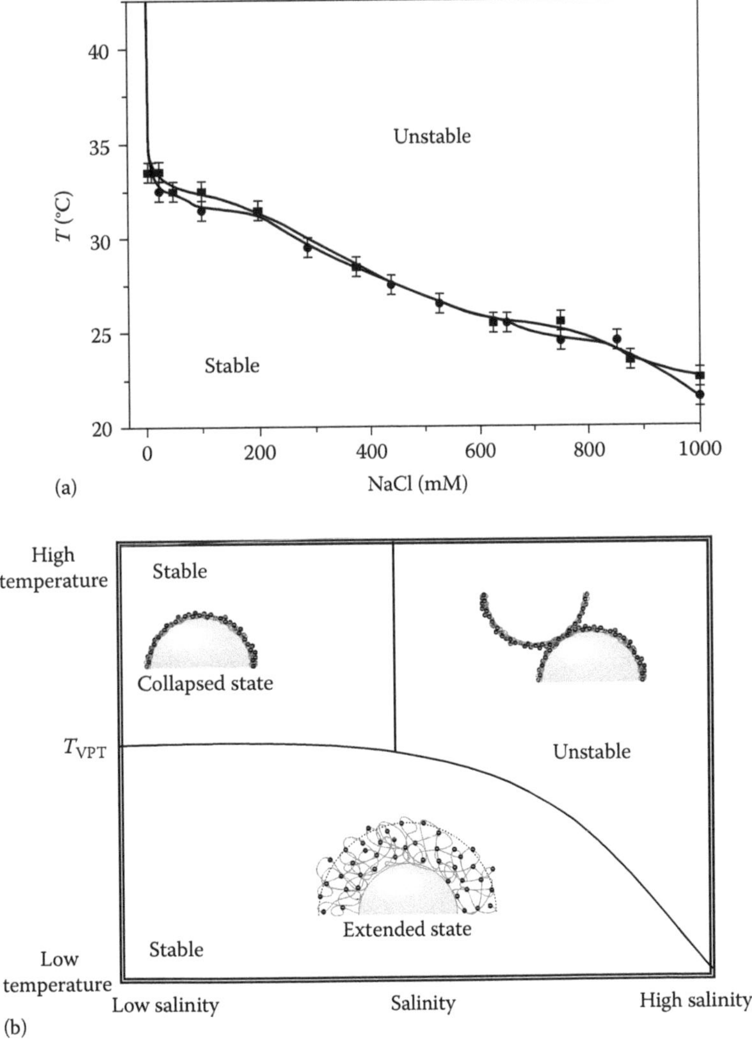

FIGURE 9.24 (a) Stability diagram for the negative (λ) and positive (ν) microgels as a function of salt concentration and temperature. (From López-León, T. et al., *J. Phys. Chem. B*, 110, 4629, 2006.) (b) Schematic illustration of diagram stability of thermally sensitive particles as a function of temperature and salinity.

polymers, the increase in temperature leads to the precipitation polymer chains as discussed earlier. In addition, the increase in salinity reduces the lower critical solution temperature (LCST) of considered polymer in water. The earlier presented effects (temperature and salinity) for linear polymer are also available for microgel and core–shell thermally sensitive particles. The increase in salinity reduces the volume phase temperature (T_{VPT}) and for high salinity medium, the colloidal stability may be affected and the aggregation process may take place as illustrated in Figure 9.24b, in which the colloidal stability domains are depicted versus salinity and temperature. The critical coagulation concentration (CCC) of such a system is high below the volume phase transition temperature in which the particles are in extended state and sterically stabilized. Whereas above the volume phase transition temperature, the CCC is low revealing that the particles are electrostatically stabilized. In addition, to some extent such behavior is reversible by cooling the dispersion and reducing the salt concentration.

9.4 IMMOBILIZATION OF BIOMOLECULES

Polymer colloids have received an increasing interest in various applications and also biomedical areas in which they are mainly used as solid-phase supports of biomolecules. This is due to the versatility of the many heterophase elaboration processes (emulsion, dispersion, precipitation, physical processes) for making well-defined microspheres of appropriate particle sizes and surface reactive groups. In this direction, special attention has been dedicated to the preparation of smart colloids. The principal interest carried to colloidal particles based on alkyl(metha)acrylamide derivative is mainly related to their thermally sensitive colloidal properties.

Before dealing with the adsorption of biomolecules (proteins and nucleic acids) onto latex particles, we should recall certain aspects (which will not be defined here) of the adsorption of macromolecules on solid–liquid interfaces. Biomolecules are complex macromolecules in nature, which, in a polar solvent (usually water), bear a large number of ionized or ionizable functions. The adsorption of macromolecules onto solid surfaces should be well defined before presenting and discussing results. A macromolecule is adsorbed when, after a given contact period with the solid surface, at least one of its sites (or units) is fixed to the support surface. In the biomedical field, the adsorption of biomolecules onto solid phases is undesirable in most cases, but desirable in some. It is of paramount importance to understand the driving forces involved in this interaction process.

In biomedical diagnostics, the adsorption of biomolecules was generally investigated in order to control their covalent binding onto reactive supports. In fact, when the affinity between a given macromolecule and solid support is poor (i.e., low adsorbed amount), the chemically grafted amount is generally negligible. The adsorption process was also studied using biological molecules as a theoretical model for establishing or verifying new theories. Recently, the adsorption of biomolecules (nucleic acids, proteins, etc.) has been investigated using both thermally sensitive magnetic [28] and non-magnetic [29] colloids to concentrate molecules (DNA, RNA, and proteins) and thereby increase the sensitivity of biomedical diagnostics. In fact, the non-specific concentration of biomolecules before the specific detection of the target is one of the most promising technologies and methodologies for the enhancement of biomedical diagnostics.

The utilization of classical polystyrene particles or hydrophobic latexes for protein concentrations can induce undesirable phenomena such as protein denaturation and low concentration yields, on account of the high adsorption affinity between both species, which may lead to low desorbed amount. In addition, the use of such hydrophobic colloids in the polymerase chain reaction (PCR) nucleic-acid amplification step generally leads to total inhibition of the enzymatic reaction. The inhibition phenomena can be attributed to the denaturation of enzymes adsorbed in large numbers onto hydrophobic colloids. The utilization of hydrophilic and highly hydrated latex particles (irrespective of temperature) is the key to solving this problem by suppressing the inhibition of enzyme activity. The purpose of this stage is then to focus on the potential application of thermally responsive poly(NIPAM) particles for both protein and nucleic acid concentrations.

9.4.1 Protein Adsorption

Various kinds of colloidal particles have been used in biomedical domains. In analytical chemistry, they are used as solid supports for sample preparation. In the drug delivery field, nanocolloids, and particularly stimuli responsive polymer-based nanogels, have been intensively explored as protein carriers for in vivo applications. The reported studies in this direction are mainly focused on the release efficiency rather than on the driven forces involved in the loading and the release of the loaded proteins.

Due to its low critical solubility temperature (LCST) around 32°C, poly(*N*-isopropylacrylamide)-based material has long been studied in regard to life science such as versatile tools for the separation and purification of proteins [30], drug delivery [31,32], control of enzyme activity [33–36], therapeutics, and diagnostics [29,37,38]. Pioneering work of Kawaguchi et al. [30] pointed out the temperature dependency of protein loading onto anionic polyNIPAM microgel particles, the results being interpreted in terms of hydrophobic interactions caused by the dehydration of the colloidal microgels. However, Elaïssari et al. [39] reported that the loading of proteins onto charged thermally sensitive similar microgels was mainly governed by electrostatic interactions.

9.4.1.1 Effect of Temperature on the Adsorption of Protein onto Poly[NIPAM] Particles

The effect of temperature on the adsorption of proteins and antibodies onto classical polymer latexes (i.e., polystyrene, poly(MMA), etc.) was found to be negligible at low temperatures ($T < 50°C$). However, the effect of temperature on the adsorption process was principally investigated in the case of stimuli-responsive polymer particles such as poly(NIPAM) microgels and core–shell particles with thermally sensitive shells. In this area, the effect of temperature on poly(NIPAM) microgel particles was found to be negligible below the T_{VPT} when the poly(NIPAM) particles were highly hydrated (amount of water: ~80 wt%). But, when the temperature was increased, the amount of protein adsorbed increased dramatically. The observed behavior was discussed with regard to both hydrophobic and electrostatic interaction. The hydrophobic interaction was attributed to the dehydration of the poly(NIPAM) microgel particles and to the hydrophobic property of linear homopoly(NIPAM) above the LCST (low critical solution temperature). The electrostatic interaction was explained by the increases in surface charge density caused by the shrinkage of the particles at higher temperatures.

The adsorption process can more likely be attributed to electrostatic interaction. In fact, the increase in temperature raises the surface charge density on the thermally sensitive particle, as evidenced by electrophoretic mobility versus temperature. In addition, the amount of water is at least close to 30% above the volume phase transition temperature. This adsorption profile (adsorbed amount versus temperature, as reported in Figure 9.25) is generally observed when the adsorption

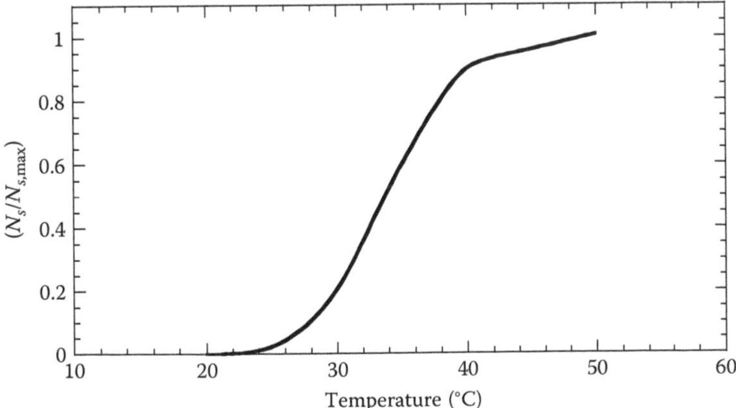

FIGURE 9.25 Reduced amount of HIV-1 capsid P24 protein adsorbed onto thermally sensitive polystyrene core-cationic cross-linked poly(NIPAM) shell as a function of temperature (at pH 6.1, 10 mM phosphate buffer).

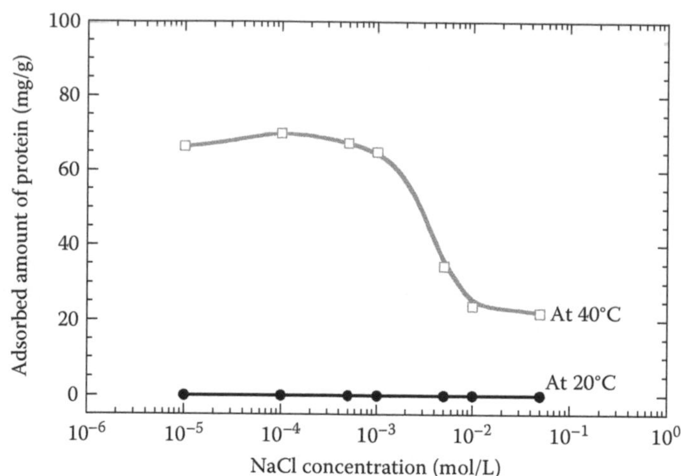

FIGURE 9.26 Effects of electrolyte concentration [NaCl] on modified HIV-1 protein (P24) adsorption onto cationic thermally sensitive core–shell microspheres at 40°C, pH 6.1 at 20°C (●) and 40°C (□). (From Duracher, D. et al., *Langmuir*, 13(23), 9002, 2000.)

temperature is well controlled in the case of attractive electrostatic interactions and only the plateau is drastically affected by the pH, salinity, and surface charge density.

9.4.1.2 Effect of Ionic Strength on the Adsorption of Protein onto Poly[NIPAM]

The effect of ionic strength on the adsorption of protein onto poly[NIPAM] is more complex than was expected. In fact, salinity affects not only electrostatic interactions but also the colloidal properties of such thermally sensitive particles: (1) the increase in ionic strength leads to a reduction in particle size induced by lowering the volume phase transition temperature (i.e., the LCST of linear thermally sensitive polymer decreases as the salinity of the medium increases) and (2) salinity affects the degree of attractive and repulsive electrostatic interactions. As a result, the adsorption of proteins onto thermally sensitive microgel particles is generally and dramatically reduced as salinity increases, irrespective of temperature (as illustrated for P24 [Figure 9.26] adsorption onto poly(NIPAM) particles).

In view of these results, the driving forces in the adsorption of proteins onto thermally sensitive hydrogel particles are debatable, and further research is necessary in order to demonstrate the driving forces involved in the adsorption process.

9.4.1.3 Desorption Study

In accordance with the reversibility of the colloidal properties of thermally sensitive particles, the adsorption of proteins is also found to be reversible in the same cases. In fact, 90% of adsorbed protein can be desorbed just by lowering the temperature (i.e., from above to below the volume phase transition temperature). The hydration processes of the particles lead to a reduction in adsorption affinity, which favors the desorption process (see Figure 9.27). Furthermore, the desorbed amount of protein can be increased by reducing the adsorption affinity through changing the pH and salinity levels. The residual adsorbed (or the non-total desorption) amount is closely related to the adsorption time and to the protein nature. In fact, the more the incubation time (above the T_{VPT}) is increased, the more the desorbed amount (below the T_{VPT}) is reduced.

Such behavior can be explained as follows when batch adsorption is performed above the volume phase transition temperature: (1) the mechanical entrapment of protein molecules in the interfacial shell layer due to the poly(NIPAM) tentacles (octopus-like adsorption process) and (2) the possible reconformation of adsorbed protein occurring during the incubation phase. Consequently, the

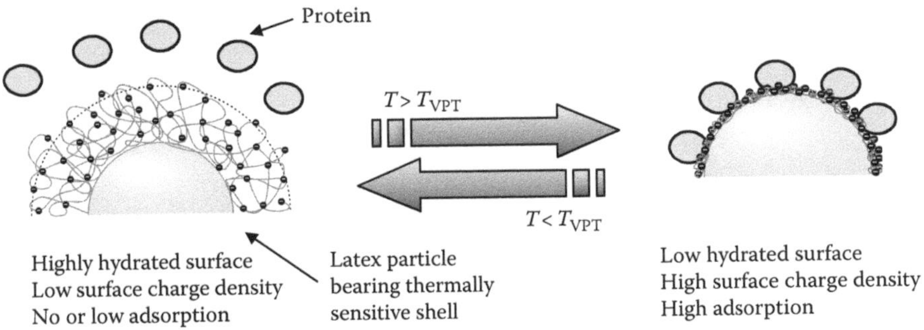

FIGURE 9.27 Schematic representation of protein adsorption desorption as a function of temperature.

tangible interpretation of the protein adsorption and desorption processes should take account the colloidal particle properties (i.e., cross-link density, charge distribution, and hydrophilic–hydrophobic balance) and the protein characteristics (flexibility, charge density, distribution of hydrophobic domains, etc.).

9.4.2 Adsorption of Nucleic Acids

In recent years, numerous studies have been performed on the adsorption of nucleic acids onto colloidal particles. The adsorption study of such polyelectrolytes has mainly been investigated using oligodeoxynucleotides (single-stranded DNA fragments [ssDNA]), and there are only a few works dedicated to the adsorption of DNA or RNA macromolecules [40]. In the biomedical field, much attention has been focused on the extraction, purification, and concentration of nucleic acid molecules (DNA and RNA) from any microbial lysate or biological sample containing a complex mixture of proteins, nucleic acids, lipids, and membrane fragments, using appropriate colloidal particles. To achieve this, various colloids have been used, including macroporous silica beads, polystyrene magnetic latexes, and, more recently, thermally sensitive (magnetic and non-magnetic) particles.

9.4.2.1 Adsorption Kinetic

As for highly charged polyelectrolytes, the adsorption of nucleic acids onto oppositely charged poly(NIPAM) microgel particles is pH, salinity, and charge density dependent. In fact, adsorption is rapid, with the attractive electrostatic forces increased by decreasing the pH (in the case of cationic particles), increasing the surface charge density [41], or lowering the ionic strength of the adsorption medium [40]. As a general tendency, the adsorption kinetic profile of nucleic acids onto highly charged thermally sensitive poly(*N*-isopropylacrylamide) microgel particles bearing cationic groups (amines and amidines) can be illustrated as in the following text.

9.4.2.2 Influence of pH and Ionic Strength

As expected for charged systems, the adsorption of nucleic acids onto latexes is drastically influenced by both salt concentration and pH. The adsorbed amount decreases when the pH value of the incubation medium is increased. In fact, increases in the pH value mainly affect the concentration of the charges involved in the interaction process between negatively charged nucleic acids and cationic charges of the latex particles.

Meanwhile, an increase in the salinity of the dispersed medium leads to a reduction in the attractive electrostatic interactions. In addition, salinity drastically affects the solvency of the thermally sensitive polymers, as mentioned earlier. An increase in the electrolyte concentration leads to an increase in the Flory–Huggins [42] interactions parameter between the polymer and water, resulting in reduced poly(*N*-alkylacrylamide) solvency. Consequently, the amount of nucleic acids adsorbed onto the cationic poly(NIPAM) microgel particles was reduced, as has been widely reported for the

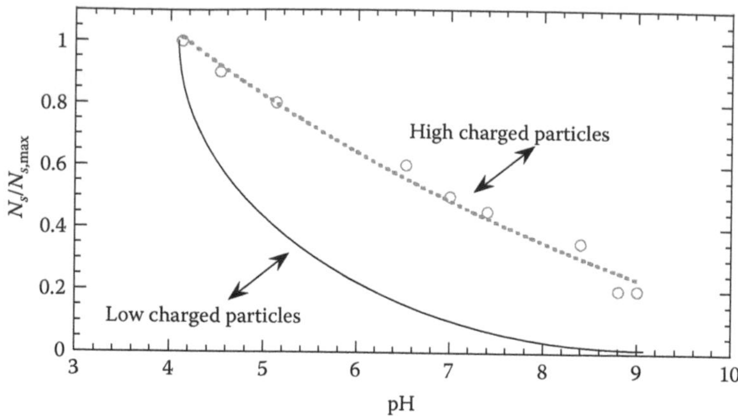

FIGURE 9.28 Reduction in adsorption of nucleic acids onto high cationic polystyrene latexes and low cationic poly(NIPAM) microgel particles as a function of pH at 20°C and 10^{-3} M ionic strength. (From Elaissari, A. et al., *J. Biomater. Sci. Polym. Edn.*, 10, 403, 1999.)

adsorption of polyelectrolytes onto oppositely charged solid supports. The attractive electrostatic interactions are the driving forces in the adsorption process of DNA, RNA, and ssDNA [42,43] onto oppositely charged polymer supports. The variation of the quantity of nucleic acids adsorbed onto cationic thermally sensitive poly(NIPAM) latex particles as a function of both pH and ionic strength are shown in Figures 9.28 and 9.29, respectively.

As for classical polyelectrolytes, the adsorption of oligodeoxyribonucleotides (ssDNA) is basically related to the ssDNA and the adsorption energy as described by the following equation:

$$N_s = k \cdot C_{eq} \cdot e^{(-n \cdot \Delta G)} \tag{9.10}$$

where

k is a constant characterizing the studied system
n is the polymerization degree (i.e., number of bases, for instance, $dT[35]$, $n = 35$)
ΔG is the adsorption energy per monomer (per base)

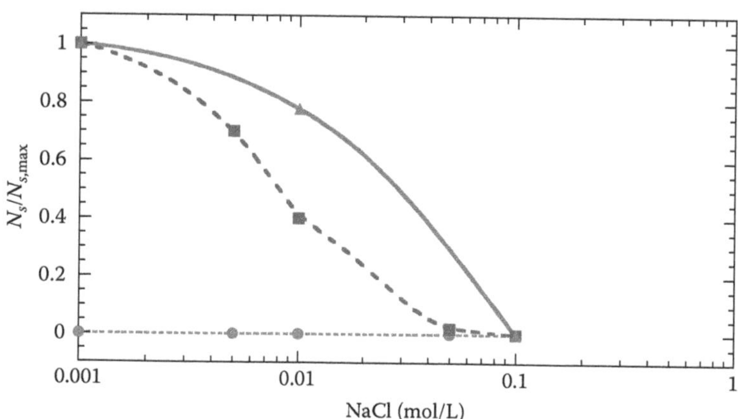

FIGURE 9.29 Reduction in amount of nucleic acids adsorbed onto (amidine groups 5 μmol/g, ●) and amine (amine and amidine groups, 14 μmol/g, ■) poly(NIPAM) microgel particles, and (amine and amidine groups 5) thermally sensitive magnetic bearing poly(NIPAM) shells [28] as a function of NaCl concentration (at pH 4.5 and at 20°C). (From Ganachaud, F. et al., *Langmuir*, 13, 701, 1997.)

The adsorption energy in the case of nucleic acids/polymer particles is the sum of hydrophobic adsorption energy ($\Psi_{hydrophobic}$) attributed to the staking adsorption process (ssDNA/negatively charged polystyrene latex) and electrostatic adsorption energy ($\Psi_{electrostatic}$) related to the charge–charge interaction:

$$n\Delta G = \Psi_{\text{hydrophobic}} + \Psi_{\text{electrostatic}} \tag{9.11}$$

In the case of such hydrophilic thermally sensitive microgel particles, the hydrophobic adsorption can be totally neglected and the electrostatic term can be described by the following relationship:

$$n\Delta G = \Psi_{\text{electrostatic}} \approx \sigma_{\text{ssDNA}} \cdot \sigma_{\text{latex}} \tag{9.12}$$

where σ_{ssDNA} and σ_{latex} are the charge density of ssDNA and latex particles, respectively. The charge density of ssDNA fragment and the surface charge density of latex particles can be approximately expressed as follows:

$$\sigma_{\text{ssDNA}} \approx \frac{n|e|}{L} \tag{9.13}$$

$$\sigma_{\text{latex}} \approx \frac{\varepsilon\kappa\varsigma}{4\pi} \tag{9.14}$$

where

L is the chain length of a given ssDNA
ς is the zeta potential of latex particles
κ is the inverse of double layer thickness
ε is the dielectric constant

In the pH range generally investigated in the adsorption study, the σ_{ssDNA} is negatively charged, whereas, the σ_{latex} is pH dependent as evidenced from any electrokinetic study (zeta potential versus pH). The electrostatic adsorption energy is thus expected to vary linearly with respect to the latex surface charge density leading to linear variation of log(N_s) versus latex surface charge density or zeta potential as well evidenced by Elaissari et al. [29].

9.4.2.3 Desorption Study of Pre-adsorbed Nucleic Acids

In practice, there is a thermodynamic balance in the adsorption process between the macromolecules adsorbed and those free in the solution. This balance can be shifted in one direction or the other; adsorption is favored by changing the nature of the solvent (pH, ionic strength, temperature, etc.) or by increasing the number of adsorption sites, while desorption is generally favored by diluting the free macromolecules or by introducing competitive species.

Desorption is often considered as a slow phenomenon, though its rate can be significant. If the molecule is adsorbed at several sites, there is little chance of it being desorbed. On the other hand, if adsorption occurs via a single contact point, there is competition with neighboring molecules. As a general tendency, experimental results showed that the higher the number of adsorption sites on the surface, the greater the free adsorption of energy and the higher the probability of low exchange rates.

The release of pre-adsorbed nucleic acid onto cationic thermally sensitive latexes was generally investigated in order to purify and concentrate such biomolecules (i.e., adsorption in a few ml from a large volume and desorption in a few μL). Desorption can be performed by changing the pH or salinity level, or by adding a chaotropic agent. According to the adsorption process of such

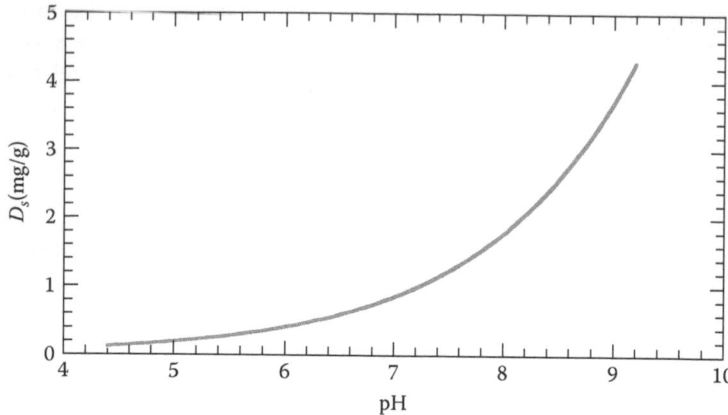

FIGURE 9.30 The effect of pH on the desorption of preadsorbed nucleic acid molecules onto cationic thermally sensitive magnetic latex particles. Adsorption was performed using 10 mM phosphate buffer, pH ~ 5.2, 10^{-3} M NaCl at 20°C, with an incubation time of 180 min. (From Elaissari, A. et al., *J. Magn. Magn. Mater.*, 225, 127, 2001.)

highly charged polyelectrolytes, which is mainly governed by electrostatic interaction, desorption can be intuitively favored by increasing the incubation pH of the medium in order to reduce the attractive forces, as shown in Figure 9.30 (i.e., reduction in the surface charge density of the polymer support), or by increasing salinity in order to screen the intensity of the attractive electrostatic interactions.

9.4.2.4 Specific Extraction of Nucleic Acids

The concentration of biomolecules is of paramount interest in the field of biomedical diagnosis where the sensitivity is very much required. If the enrichment of the sample is necessary, magnetic particles can represent an invaluable tool to concentrate these biological molecules. The generic capture of the nucleic acids negatively charged in nature is obtained, for example, via electrostatic interaction between the negative charges of the nucleic acid molecules and the positive charges available on the cationic magnetic particles (mainly ammonium). This electrostatic interaction depends mainly on physicochemical parameters (pH, salinity, temperature, competitive agents, incubation time). The specific capture [40,45] is obtained by hybridization between the nucleic acid targets and the capture probe (oligonucleotide of well-established sequence) and this capture probe must be fixed in a covalent way on the surface of the magnetic carrier. In this case, the physicochemical characteristics of the medium of incubation play a determining role in the effectiveness of capture efficiency. This explains the great interest to work out magnetic particles as a carrier, compatible with the biological environments and in particular the medium of amplification of the nucleic acids.

The first work concerning the generic capture of the nucleic acids was completed by using the affinity chromatography's columns (silica-based columns) and recently polymer particles such as stimuli-responsive latexes (sensitive to the pH, salinity, and the temperature).

The principle of nucleic acid capture based on the use of silica and magnetic silica is due to the precipitation process of the nucleic acids on the solid phase induced by the high salinity of the medium and the use of chaotropic agents. The use of latex particles for the non-specific capture and concentration of the nucleic acids is based on the attractive electrostatic interactions between the nucleic acids (negatively charged) and the latex particles charged positively. The capture of the total nucleic acids requires the use of particles answering a relatively drastic schedule of conditions. Indeed, the particles owe beings compatible with the enzymes used in the investigated amplification process of the nucleic acids [29].

The extraction of the RNA is a real stake in the biomedical diagnosis where the sensitivity, specificity, and rapidity (fast analysis) are required. There are techniques allowing the specific capture of nucleic acid molecules, but in this case, it is necessary the use of solid phase on which well-defined oligonucleotides sequence are fixed, for instance, the use of reactive colloidal particles [40] bearing oligonucleotides (polyT) for the capture of mRNA (eukaryotes having polyadenyl [polyA] tail). The extraction of the nucleic acids (DNA or RNA) is also performed by using the differential precipitation based on centrifugation on cesium chloride. In this case, only the RNA is recovered. Lastly, the most widespread technique called mini-chromatography based the use of a mini-column material, which fixes RNA. These extracted nucleic acids thereafter are purified and eluted in a low volume, which led to concentrated nucleic acid molecules.

The principal problem generally resides in the specific extraction of RNA from any biological sample containing DNA. The single exit in this case is the selective extraction of ARN that requires examining the chemical difference between DNA and RNA molecules. The ribose in 3′ position of RNA presents a cis-diol function, which led to a complex formation with boronic acid, compound, this in a specific way, even in the presence of DNA.

The specific recognition of cis-diol function by the boronic acid derivatives was the subject of many researches. In fact, certain authors used this affinity for controlling the reversible immobilization of proteins, enzymes, or all biomolecules bearing glucose site. This interaction is sensitive to the pH of the medium due to the pK_a of the boronic acid involved in such affinity. This is due to the effect of the pH on the boronic stereochemistry of the acid. Indeed, it is trigonal in form at low pH and tetragonal at basic pH (i.e., $pH > pK_a = 8.8$). Thus, the complexation reaction is favored at basic pH because of the stability of boronat form. The shift to acid medium lead to the decomplexation and thereafter the release of the cis-diol-containing molecules. It should be noted that the yield of the complexation reaction between the boronic acid and the polyols compound depends on two major factors: (1) the availability of the cis-diol functions and (2) the nature of the charged site close to the boronic acid.

The boronic acid (or phenyl boronic acids) derivatives can be introduced onto particles surface via various processes starting from simple chemical grafting onto preformed reactive magnetic particles to direct incorporation into the used recipes for particle elaboration.

Tuncel et al. [46–48] have reported more recent work. The polymeric microspheres prepared by this group were used for the immobilization of the RNA. The principle of such extraction is illustrated in Figures 9.31 and 9.32.

The specificity and efficiency have been examined by Tuncel et al. [46–48] as a function of various parameters such as temperature, salinity, pH, and also the degree of particle functionalization (Figure 9.33). The immobilization of RNA onto phenyl boronic acid containing polyNIPAM microgel was found to be high compared to nonreactive polyNIPAM microgels. The desorption of pre-adsorbed RNA was examined as a function of pH, temperature, and also the salinity and nature of the salt used. As a general tendency, the desorption of RNA increases with increasing the pH and the salinity of the incubation medium [49].

9.4.3 Amplification of Nucleic Acids

In biomedical diagnostics, the amplification of captured or adsorbed nucleic acids using the classical polymerase chain reaction (PCR) method is one of the aims targeted in various biological applications. The enzymatic amplification of desired nucleic acids is often performed after the desorption or release step. Thanks to hydrophilic, highly hydrated magnetic and non-magnetic latex, direct amplification of adsorbed nucleic acid molecules onto the particles [17] is now possible.

The inhibition of adsorbed nucleic acids after the desorption process can be attributed to the following factors: (1) the possible release of undesirable impurities originating from the particles, such as bare iron oxide nanoparticles, ferric or ferrous ions, surfactant, etc., and (2) the desorption of adsorbed inhibitor initially present in the biological sample being studied.

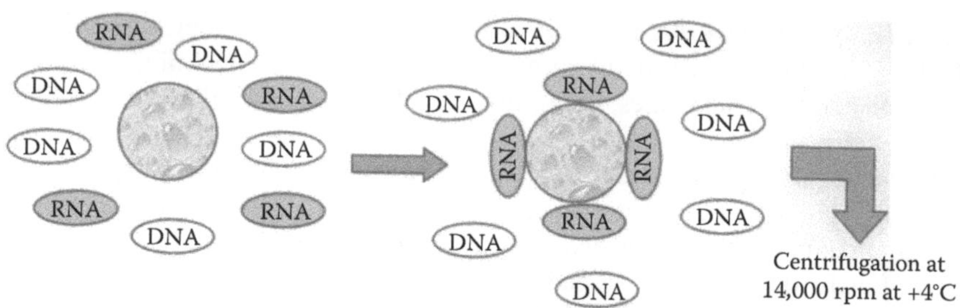

Selective adsorption of RNA from RNA–DNA mixture
+4°C, HEPES buffer, pH:8.5

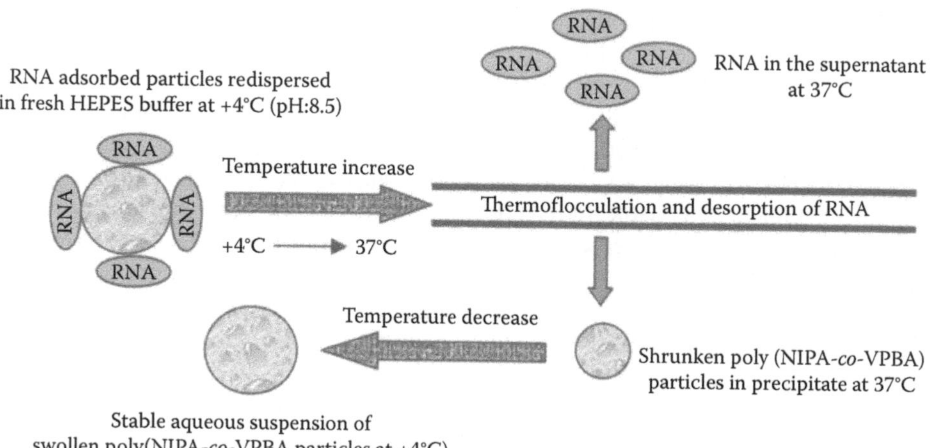

Stable aqueous suspension of
swollen poly(NIPA-*co*-VPBA particles at +4°C)

FIGURE 9.31 Illustration of specific extraction of RNA using phenyl boronic acid containing thermally sensitive microgel particles. (From Elmas, B. et al., *Colloid Surf. A: Physicochem. Eng. Aspects*, 232, 253, 2004.)

The inhibition of direct amplification of adsorbed nucleic acids on the colloidal particles could be due to the factors mentioned earlier, and also to (1) the high affinity between the enzymes and the particles, and (2) the nature of the support (high hydrophobicity, non-coated iron oxide, or denaturizing domains), as well as the high concentration of colloidal particles in the PCR medium. The potential application of hydrophilic, cationic thermally sensitive particles is schematically summarized in Figure 9.34.

9.5 THERMALLY SENSITIVE PARTICLES OR NANOGELS FOR IN VIVO DRUG RELEASE

Stimuli-sensitive or *smart* polymeric systems are defined as the polymers that might rise above dramatic property changes replying to minute changes in the environment [50]. The effect of different stimuli on environmental sensitive polymers (Figure 9.35) has been widely studied, and among these temperature is the most investigated stimulus since it is directly related to the human body [51].

Hydrogels have been employed widely in the development of the smart drug delivery systems. They are the network of hydrophilic polymers that might swell in the water [53]. Hydrogels denote the main class of biomaterials in biotechnology and medicine because several hydrogels reveal exceptional biocompatibility, producing least inflammatory reactions, thrombosis, and tissue damage [54].

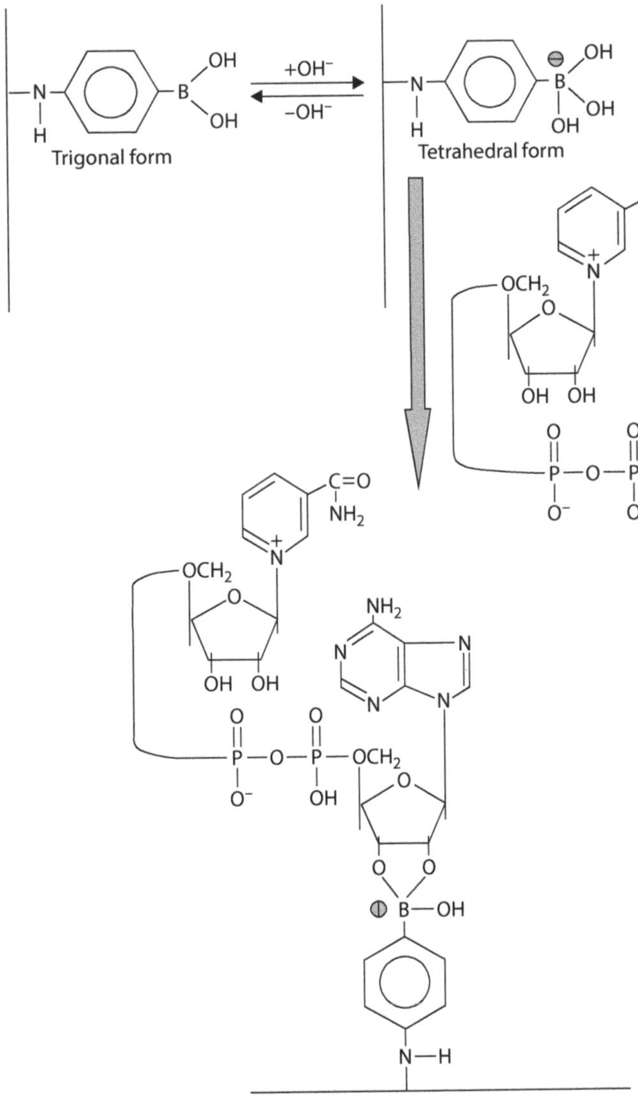

FIGURE 9.32 Schematic illustration of specific immobilization of RNA onto phenyl boronic acid containing polyNIPAM microgels. (From Uguzdogan, E. et al., *Macromol. Biosci.*, 2(5), 214, 2002.)

Nanogels are the nanosized particles produced via physically or chemically cross-linked polymer networks, which are swell in a good solvent. The materials that have the base of Nanogels possess elevated drug loading capacity, biocompatibility, and biodegradability, which are the main issues to design a drug delivery system effectively. It should keep in mind that they can be administered through oral, pulmonary, nasal, parenteral, intra-ocular, and topical routes [55].

Due to following reasons, nanogels have been known as more advanced drug delivery system than others:

1. To avoid fast clearance by phagocytic cells, allowing both passive and active drug targeting their size and surface properties can be manipulated.
2. Drug release control at the target site, therapeutic efficacy enhancement, and decreasing side effects. Drug loading is relatively high and might be achieved exclusive of chemical reactions, which is an important factor for maintaining the drug activity.

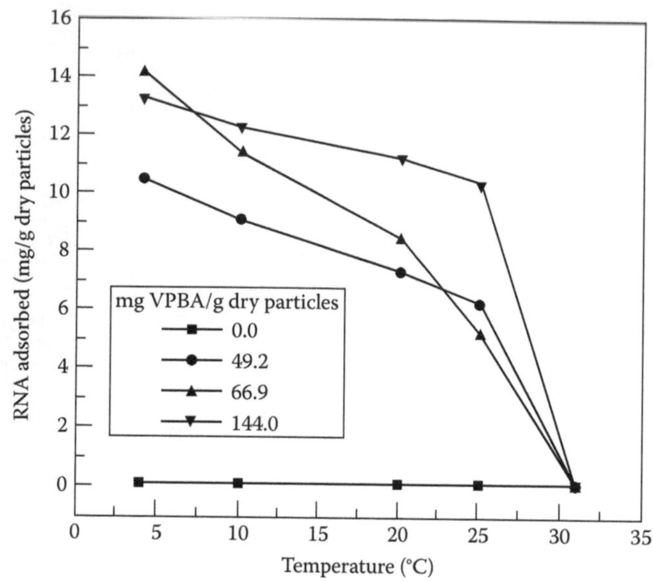

FIGURE 9.33 The influence of VPBA amount (in polyNIPAM microgels) on the fixation of RNA as a function of temperature.

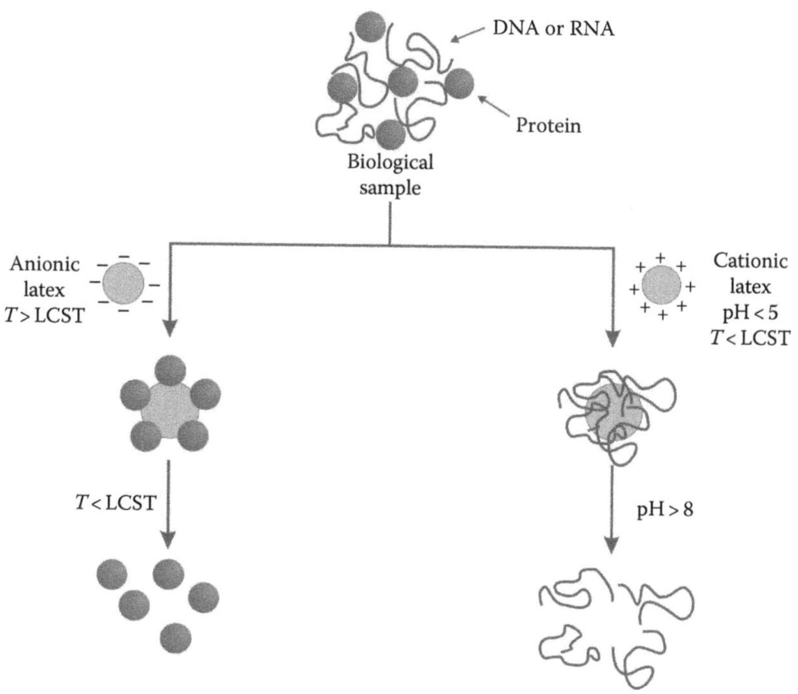

FIGURE 9.34 Schematic illustration of nucleic acids and proteins extraction, purification, and concentration using thermally sensitive microgels.

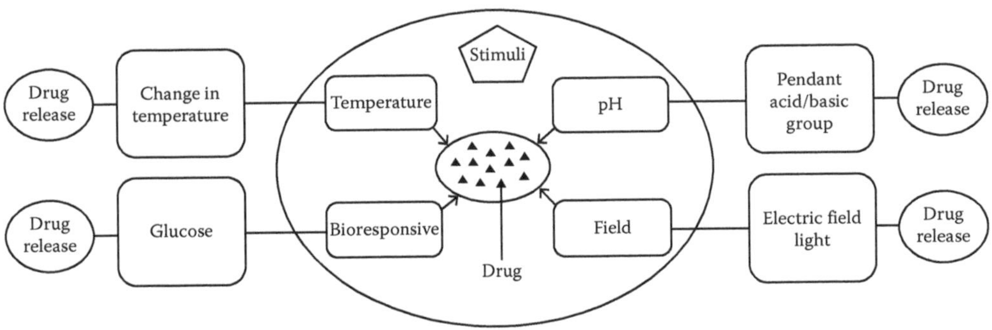

FIGURE 9.35 Various stimuli responsible for controlling drug release from smart polymeric drug delivery systems. (From Honey Priya, J. et al., *Acta Pharm. Sin. B*, 2, 120, 2014.)

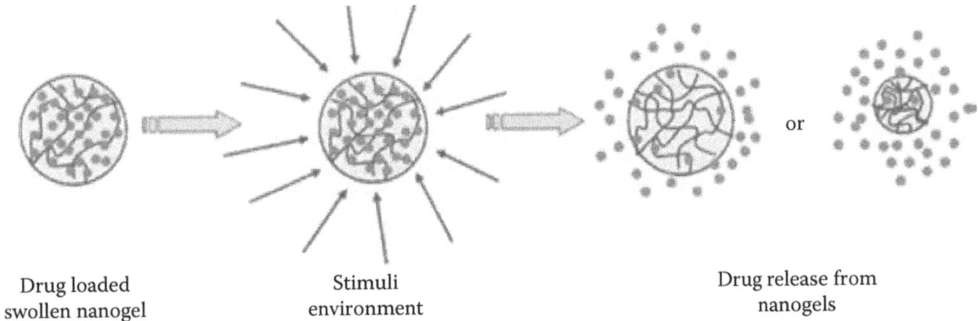

FIGURE 9.36 Drug release model from nanogel. (From Sultana, F. et al., *Appl. Pharm. Sci.*, 3, S95, 2013.)

Ability of these nanogels to reach the smallest capillary vessels and to penetrate the tissues either through the paracellular or the transcellular pathways [56] is mainly due to their tiny volume. In addition, these particles are also biodegradable. A model of drug discharge from nanogel is shown in Figure 9.36.

In general, the size of nanogels are ranged between 20 and 200 nm in diameter and therefore are useful in avoiding the quick renal exclusion but are small enough to prevent the uptake by the reticuloendothelial system. The have good penetration capabilities because of their extremely small size. More particularly, it can cross the blood brain barrier (BBB). Furthermore, nanogels are capable to solubilize hydrophobic molecules and diagnostic agents in their core or gel networks. Hydrophilic and hydrophobic drugs and charged solutes might be given via nanogel. These properties of nanogel are considerably influenced by temperature, existence of hydrophilic/hydrophobic groups in the polymeric networks, the cross-linking density of the gels, surfactant concentration, and type of cross-links present in the polymer networks.

Frequently nanogels are classified into two major ways. The primary classification is according to their reactive manners that might be either stimuli-responsive or nonresponsive:

1. Non-responsive microgels, which have the enlarging characteristic because of absorbing the water.
2. Stimuli-responsive microgels that swell or de-swell upon contact to environmental alterations like temperature, magnetic field, pH, and ionic strength. Multi-responsive microgels are responsive to more than one environmental stimulus [55]. The criteria for the subsequent classification are the type of linkages present in the network chains of gel structure, and polymeric gels (include nanogel) are subdivided into two main categories (Figure 9.37):
   a. Physical cross-linked gels
   b. Chemically cross-linked gels

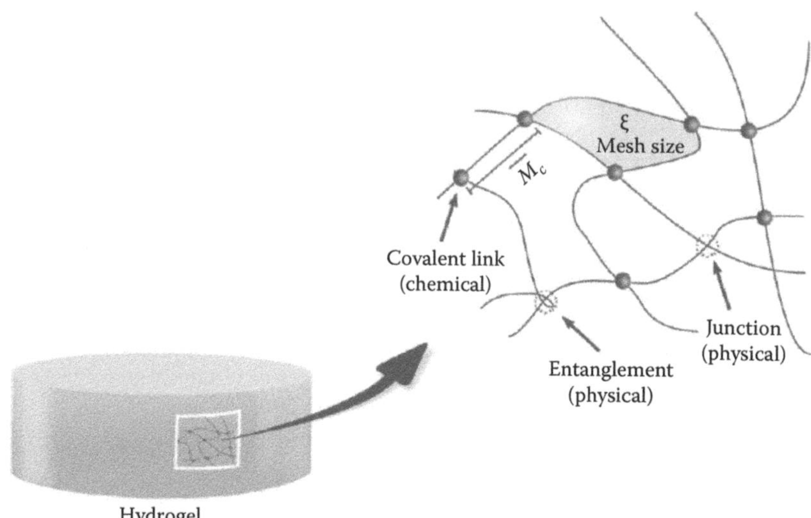

FIGURE 9.37 A cross-linked hydrogel structure with the mesh size ξ and the average molecular weight between the cross-linking points MC, respectively. (From Buenger, D. et al., *Prog. Polym. Sci.*, 37, 1678, 2012.)

The mechanisms of drug release from nanogels are encompassed diffusion, nanogel degradation, and displacement through ions present in the environment [58]. Nanogels, sub-micron hydrogel particles with colloidal properties, are motivating candidates as drug delivery carrier because the hydrogel nanostructure of nanogels can be designed to achieve controlled opening structures, chemical topologies, and swelling responses to environmental incitements, whereas the colloidal nanoparticle macrostructure of nanogels suggests the advantages of high specific surface areas and ready injectability.

Besides photo cross-linking hydrogels, externally "smart" sensitive systems of drug delivery have been studied as a new controlled delivery method to control the release of drugs in answer to alters in the environment [59,60].

Nanogels, which formed exceptional nanovehicles in pharmaceutics are mainly used to encapsulate water soluble active molecules. Discharged nanogels in a swollen state enclose significant amount of water. Loading of biological agents is often attained by mechanisms of self-assembly including electrostatic, van der Waals, and/or hydrophobic interactions among the agent and the polymer matrix. Different nanogels have been demonstrated to deliver their payload inside cells and cross biological barriers. By reason of rise up stability inside cells, nanogels shown good potential for improving of drug bioavailability in digestive tube and brain of low molecular drugs and biomacromolecules [61].

Nowadays, research is focused on preparing smart biocompatible and biodegradable polymers possessing certain desirable characteristics like stimuli-sensitive polymers suffering phase transitions in response to changes in pH, ionic strength, light, electric field, irradiation, or temperature. Particularly, among the temperature-sensitive hydrogels reported to date, poly(*N*-isopropylacrylamide) (PNIPAM) and its copolymers have been commonly used for cell separation also for pharmaceutical and tissue engineering applications because of their distinctive thermal properties [59,62].

Short half-lives, physical and chemical instability, and poor bioavailability are frequently the reasons that can limit the pharmaceutical and biological therapeutics.

Usually it is useful to employ the polymers, which might react with the stimuli that exist intrinsically in the natural systems. In Table 9.1 diverse smart polymeric drug delivery systems have been summarized [52].

TABLE 9.1
Advantages and Disadvantages of the Thermally Sensitive Polymers

Advantages	Disadvantages
Toxic organic solvents avoidance	High burst release of drug
Delivery of hydrophilic and lipophilic drugs	Low mechanical strength of the gel leading to potential dose-dumping
Reduced systemic side effects	Lack of biocompatibility of the polymeric system and gradual lowering of pH of the system due to acidic degradation
Site-specific delivery of the drugs	
Sustained release properties	

Sources: Schmaljohann, D., *Adv. Drug Deliv. Rev.*, 58, 1655, 2006; Ruel-Gariépy, E. and Leroux, J.-C., *Eur. J. Pharm. Biopharm.*, 58, 409, 2004.

The requirement for the new approaches to control the delivery of the compounds such as peptides, proteins, plasmid DNA, antisense oligodeoxynucleotides, and immunotoxins has been created through the development of different advanced medications over the previous decade. Their ability to get the targeted regions is important for the activity of these molecules, nevertheless, once they enter into the body system proteases or DNA-degrading enzymes in vivo simply degraded the mentioned polymers [65].

Because of their ability to employ as controlled drug delivery systems, poly(*N*-isopropylacrylamide) (PNIPAM) is the most popular temperature-sensitive polymer among microgels. Additionally, these polymers shows a sharp lower critical solution temperature at 32°C, which might be shifted to body temperature via formulating with surface active agents or additives. Being cytotoxic because of quaternary ammonium existence in its structure, non-biodegradability and its ability of activation platelets upon contact with body fluids have limited the utilization of the polyNIPAAM [52]. Newly, another thermo-sensible polymer, poly(*N*-vinylcaprolactam) (PVCL), which demonstrates superior biocompatibility in comparison with PNIPAM, is being widely studied, additionally poly(*N*-vinylcaprolactam) (PVCL) is particularly interesting because of the reality that is too stable against hydrolysis [58,66]. Thermally responsive polymers are large molecules that dissolve in cold water but collapse and precipitate upon heating of the aqueous solution over its lower critical solution temperature (LCST) [58]. Prevention of organic toxic solvents, delivery ability of both hydrophilic and lipophilic drugs, decreased systemic side effects, targeting delivery, and sustained release properties are the main advantages of the thermosensitive polymeric systems. Despite these advantages, certain disadvantages such as high-burst release of drug, small mechanical strength of the gel leading to potential dose-dumping, and lack of biocompatibility of the acidic degradation are accompanied with these systems [63,64].

Hydrogels are achieved through cross-linking of these polymers that after heating above the critical temperature reversibly shrink. In a dispersed medium, responsive nanoparticles could be prepared through conducting the polymerization of the suitable monomer and a cross-linker. Latex is the product, in this instance nanosized gel particles dispersed in water [58]. The several responsive characteristics of nanoparticles, like those composed of poly(*N*-isopropylacrylamide) (PNIPAM) have led to them becoming the source of significant interest in the area of drug delivery. They are considered as "smart" polymers because of the conformational changes they suffer in response to changes in environmental conditions such as temperature and pH [67].

Thermosensitive nanogels based on poly(*N*-isopropylacrylamide) (PNIPAM) offer the more potential advantage of undergoing a volume phase transition as they are heated above a critical temperature, resulting in significant gel de-swelling and (in some cases) thermally triggered aggregation of the nanogel particles to form aggregates and/or physically cross-linked hydrogels. Since thermosensitive nanogels can exhibit triggerable changes in pore size and colloidal stability, they have been widely investigated for the "on-demand" delivery of drugs or model drugs, including acetylsalicylic acid, fluorescein-labeled dextran, insulin, and bovine serum albumin, among others [68].

Nanogels of *N*-isopropylacrylamide (NIPAM) are extremely responsive to temperature nearby the physiological temperature. These nanogels not only mimic biological systems but they respond to small environmental variations of body pathological state as well, because of their unique lower critical solution temperature (LCST) in the vicinity of human body temperature. Both properties of liquids and solids describe through the superb swelling and de-swelling comportment of *N*-isopropylacrylamide (NIPAM)-based materials.

In an entirely swollen state, these gels act as soft materials and behave like liquids as the polymer network is solvated by a large amount of trapped solvent and the polymer chains exhibit great mobility. Additionally, the nature of the cross-linked structure contributes to keeping their solid form. The significant interest of *N*-isopropylacrylamide (NIPAM)-based gels in controlled drug release uses has been created by their swelling and de-swelling comportment [61].

The inclusion of the thermoresponsive poly(*N*-isopropylacrylamide) (PNIPAM) constituent suggests the advantages of: (1) degradation mechanism modulation of biodegradable polymers, (2) cytotoxicity reduction of polycationic polymers, (3) thermally localizing medication to target sites after systemic administrations while their lower critical solution temperature (LCST) is tailored to temperatures between 37°C (body temperature) and 42°C (used habitually in clinical hyperthermia), and (4) drugs released at diverse profiles in answer to stimuli such as temperature, pH, etc. [69].

In Figure 9.38, an example of a drug delivery process is illustrated, collapsed microgel particles are positioned in the aqueous drug solution at temperature beneath the volume phase transition temperature (VPTT), and the microgel particles will swell and/or, due to the polymer–drug interaction, small drug molecules will freely penetrate the pores of the polymer network. Subsequently, the drug-loaded microgel particles can be easily removed by centrifugation leaving a less concentrated drug solution behind. The microgel particle will squeeze out the encapsulated drug if the solution is heated up to a temperature above volume phase transition temperature (VPTT).

One research that has been done by Abu Samah and Heard in 2013 showed that in vitro permeation data of caffeine-loaded poly(*N*-isopropylacrylamide) copolymerized with 5% (w/v) of acrylic

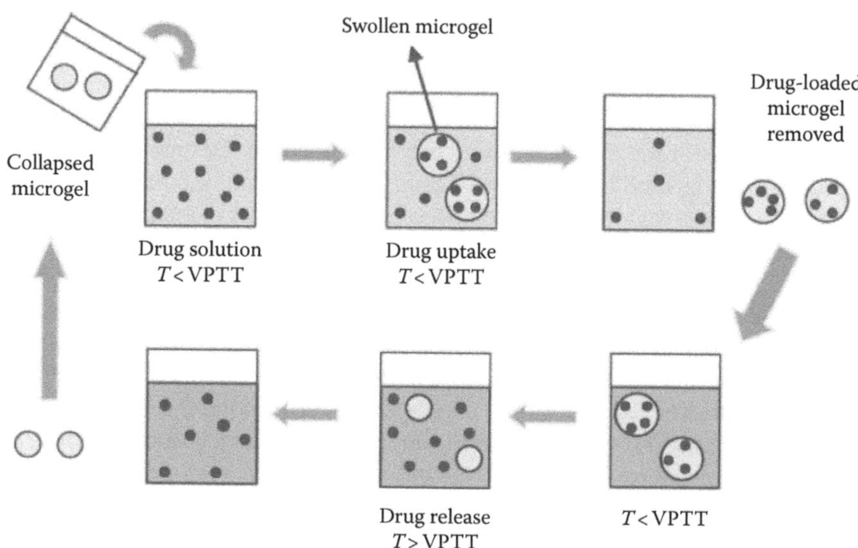

FIGURE 9.38 Illustration of a temperature-sensitive microgel particle-based drug release process. (From Imaz, A. and Forcada, J., *N-Vinylcaprolactam-Based Microgels for Biomedical Applications*, Vol. 48, J. Polym. Sci., Part A: Polym. Chem., 2010, pp. 1173–1181.)

acid (AAc) at 2°C–4°C enhanced the delivery of the loaded caffeine across epidermis in comparison to the saturated solution of caffeine, through 3.5 orders of magnitude. In a study, it has been found that "In solid tumors with hyperthermia therapy designed thermo-responsive and biodegradable linear-dendritic nanoparticles have a big potential for thermally targeted and sustained release of Ceramide C6" [69].

According to another in vitro and in vivo study, the ocular controlled release in a desired quantity was gained by the thermosensible hydrogel. Moreover, the release was without adverse effects and completely tolerated [70,71].

9.6 CONCLUSION

This chapter covers the preparation, characterization, and biomedical application of thermally sensitive particles. Thermosensitive hydrogel is prepared by precipitation polymerization of *N*-alkylacrylamide or *N*-alkylmethacrylamide as a principal water-soluble monomer, a water-soluble cross-linker (for instance, *N*-methylenebisacrylamide), and an initiator (such as Azobis-amidinopropane derivatives, potassium persulfate, or basically any charged initiator). The core–shell latexes are produced by a combination of emulsion and precipitation polymerization, such as the preparation of polystyrene core and poly(*N*-isopropylacrylamide) shell or encapsulation of colloidal seed using alkylacrylamide derivatives. During the elaboration of such stimuli-responsive particles, various aspects should be considered: (1) a water-soluble cross-linker is needed, (2) the polymerization temperature should be higher than the LCST of the corresponding linear polymer, and (3) the production of water-soluble polymer (which can be controlled by monitoring the polymerization conditions). The polymerization mechanism has been clearly discussed and well illustrated, but the nucleation step remains questionable and requires further work.

The colloidal characteristics of *N*-alkylacrylamide or *N*-alkylmethacrylamide-based particles are temperature related. In fact, the swelling ability, charge density and charge distribution, hydrophilic–hydrophobic balance, hydration and dehydration property, particle size, surface polarity, colloidal stability, water content, turbidity, electrokinetic and rheological properties are indiscernibly temperature dependent. Such polymer particles can be used as a stimuli-responsive model for the investigation of colloidal properties and for theoretical studies.

As can be seen from this chapter, the adsorption and desorption of proteins and nucleic acids can be monitored by controlling the key point governing the driving forces involved in the adsorption process.

The adsorption of proteins onto charged thermally sensitive particles is greatly affected by the incubation temperature. Protein adsorption onto highly hydrated thermosensitive particles below the volume phase transition temperature is negligible. However, the affinity and the amount of protein increase together with the temperature reveal the complex adsorption process. Desorption is easily favored by cooling the temperature and controlling the adsorption time. In addition, the adsorption and desorption processes are pH, time, and ionic strength dependent.

The adsorption process of nucleic acids onto such cationic hydrophilic thermally sensitive colloids is governed by the attractive electrostatic interaction as the driving forces. The adsorption of DNA, RNA, and ssDNA are related to the surface charge density and accessible adsorption sites. The desorption of pre-adsorbed nucleic acid molecules onto polymer particles is favored by reducing the attractive electrostatic interactions by altering the pH and the salinity of the medium. Such hydrophilic particles can be used for specific adsorption and the concentration of nucleic acids from any biological sample containing a complex mixture (proteins, lipids, membrane fragments, etc.). In addition, colloidal particles bearing hydrophilic and cross-linked thermally sensitive shells can be used directly in nucleic amplification processes (i.e., PCR) without any inhibition phenomena.

Thermally sensitive polymers under colloidal nano and microgels are also started to be used in drug delivery and the most examined polymers are poly(*N*-alkylacrylamide) derivatives. Unfortunately, such polymers are biocompatible but not biodegradable, and they are mainly used as model rather than for real in vivo applications. Thermally sensitive polymers have lower critical transition temperature close to the physiological value that suggests several possibilities in the biomedical field. Biocompatibility, biodegradability, permeability, and physical characteristics of hydrogels are the properties that made them good candidates as biomaterials for use in many medical applications, including drug delivery.

ACKNOWLEDGMENTS

I would like to gratefully acknowledge all the students (F. Meunier, L. Holt, V. Bourrel, F. Sauzedde, D. Duracher, L. Housni, P. Hazot, G. Zhou, Y. Cuie, T. Taniguchi, T. Leon, and G. Levourch) involved in the elaboration, the characterization, and the use of thermally sensitive colloidal particles for biomedical applications.

REFERENCES

1. Pelton RH, Chibante P. Preparation of aqueous latices with *N*-isopropylacrylamide. *Colloid Surf.* 1986;20:247–256.
2. Heskins M, Guillet JE. Solution properties of poly(*N*-isopropylacrylamide). *J. Macromol. Sci. Chem.* 1968;A2(8):1441–1455.
3. Schild HG. Poly(*N*-isopropylacrylamide): Experiment, theory and application. *Prog. Polym. Sci.* 1992;17:163–249.
4. Meunier F, Elaissari A, Pichot C. Preparation and characterization of cationic poly(*n*-isopropylacrylamide) copolymer latexes. *Polym. Adv. Technol.* 1995;6:489–496.
5. Duracher D, Elaissari A, Pichot C. Preparation of poly(*N*-isopropylmethacrylamide) latexes kinetic studies and characterization. *J. Polym. Sci. A: Polym. Chem.* 1999;37:1823–1837.
6. Hazot P, Delair T, Elaissari A, Pichot C, Chapel JP, Davenas J. Synthesis and characterization of functionalized poly(*N*-ethylmethacrylamide) thermosensitive latex particles. *Macromol. Symp.* 2000;150:291–296.
7. Loos W, Verbrugghe S, Goethals EJ, Du Prez FE, Bakeeva IV, Zubov VP. Thermo-Responsive Organic=Inorganic Hybrid Hydrogels based on poly(N-vinylcaprolactam). *Macromol. Chem. Phys.* 2003;204:98–103.
8. Bronstein L, Kostylev M, Tsvetkova I, Tomaszewski J, Stein B, Makhaeva EE, Khokhlov AR, Okhapkin AR Core–shell nanostructures from single poly(N-vinylcaprolactam), Macromolecules: Stabilization and Visualization. *Langmuir* 2005;21:2652–2655.
9. Van Durme K, Verbrugghe S, Du Prez FE, Van Mele B. Light scattering and micro calorimetry studies on aqueous solutions of thermo-responsive PVCL-g-PEO copolymers, *Polymer* 2003;44:6807–6814.
10. Van Durme K, Verbrugghe S, Du Prez FE, Van Mele B. Influence of poly(ethylene oxide)-grafts on kinetics of LCST-behavior in aqueous poly(N-vinyl caprolactam) solutions and networks studied by Modulated Temperature DSC. *Macromolecules* 2004;37:1054–1061.
11. Mc Phee W, Tam KC, Pelton R. Poly(*N*-isopropylacrylamide)latices prepared with sodium dodecyl sulfate. *J. Colloid Interf. Sci.* 1993;156:24–30.
12. Kawaguchi H, Kawahara M, Yaguchi N, Hoshino F, Ohtsuka Y. Hydrogel microspheres I. Preparation of monodisperse hydrogel microspheres of submicron or micron size. *Polym. J.* 1988;20:903–909.
13. López-León T, Ortega-Vinuesa JL, Bastos-Gonzales D, Elaissari A. Cationic and anionic poly(*N*-isopropylacrylamide) based submicron gel particles: Electrokinetic properties and colloidal stability. *J. Phys. Chem. B* 2006;110:4629–4636.
14. Vihola H, Laukkanen A, Valtola L, Tenhi H, Hirvonen J. *Biomaterials* 2005;26:3055–3064.
15. Wu X, Pelton RH, Hamielec AE, Woods DR, McPhee W. The kinetics of poly(*N*-isopropylacrylamide) microgel latex formation. *Colloid Polym. Sci.* University of Lyon-1, France, 1994;272:467–477.
16. Dainton FS, Tordoff M. The polymerization of acrylamide in aqueous solution part 3. The hydrogen peroxide photosensitized reaction at 258°C. *Trans. Faraday Soc.* 1957;53:499.
17. Currie DJ, Dainton FS, Watt WS. The effect of pH on the polymerization of acrylamide in water. *Polymer* 1965;6:451.
18. Dainton FS, Sisley WD. Polymerization of methacrylamide in aqueous solution part-1-hydrogen-peroxide photosensitized reaction. *Trans. Faraday Soc.* 1963;59:1369.

19. Meunier F. Synthèse et caractérisation de support polymères particulaires hydrophiles à base de *N*-isopropylacrylamide. Elaboration de conjugués particules/ODN et leur utilisation dans le diagnostic médical, Thèse, 1996.
20. Guillermo A, Cohen-Addad JP, Bazil JP, Duracher D, Elaïssari A, Pichot C. Crosslink density of thermosensitive microgel particles investigated by NMR. *J. Polym. Sci. B: Polym. Phys.* 2000;38(6):889–898.
21. Duracher D, Elaïssari A, Mallet F, Pichot C. Preparation of thermosensitive latexes by copolymerization of *N*-isopropylmethcrylamide with a chelating monomer. *Macromol. Symp.* 2000;150:297–303.
22. Duracher D, Sauzedde F, Elaïssari A, Perrin A, Pichot C. Cationic amino-containing *N*-isopropylacrylamide-styrene copolymer latex particles. 1. Particle size and morphology vs. polymerization process. *Colloid Polym. Sci.* 1998;276:219–231.
23. Hoshino F, Fujimoto T, Kawaguchi H, Ohtsuka Y. N-substituted acrylamide-styrene copolymer lattices. II. Polymerization behavior and thermosensitive stability of latices. *Polym. J.* 1987;19(2):241–247.
24. Duracher D, Sauzedde F, Elaïssari A, Pichot C, Nabzar L. Cationic amino-containing *N*-isopropylacrylamide-styrene copolymer latex particles. 2. Characterization and colloidal stability. *Colloid Polym. Sci.* 1998;276:920–929.
25. Hazot P, Chapel JP, Pichot C, Elaissari A, Delair T. Preparation of poly(*N*-ethyl methyl methacrylamide) particles via an emulsion/precipitation process: The role of the crosslinker. *J. Polym. Sci. A: Polym. Chem.* 2002;40:1808–1817.
26. Hazot P, Delair T, Elaissari A, Chapel JP, Pichot C. Functionalization of poly(*N*-ethylmethacrylamide) thermosensitive particles by phenylboronic acid. *Colloid Polym. Sci.* 2002;280:637–646.
27. Castanheira EMS, Martinho JMG, Duracher D, Charreyre MT, Elaïssari A, Pichot C. Study of cationic *N*-isopropylacrylamide-styrene copolymer latex particles using fluorescent probes. *Langmuir* 1999;15(20):6712–6717.
28. Elaissari A, Rodrigue M, Meunier F, Herve C. Hydrophlic magnetic latex for nucleic acid extraction, purification and concentration. *J. Magn. Magn. Mater.* 2001;225:127–133.
29. Elaissari A, Holt L, Meunier F, Voisset C, Pichot C, Mandrand B, Mabilat C. Hydrophilic and cationic latex particles for the specific extraction of nucleic acids. *J. Biomater. Sci. Polym. Edn.* 1999;10:403–420.
30. Kawaguchi H, Fujimoto K, Mizuhara Y. Hydrogel microspheres: III. Temperature-dependent adsorption of proteins on poly-*N*-isopropylacrylamide hydrogel microspheres. *Colloid Polym. Sci.* 1992;270:53–57.
31. Hoffman S, Afrassiabi A, Dong LS. Thermally reversible hydrogels: II. Delivery and selective removal of substances from aqueous solutions. *J. Control. Release* 1986;4:213–222.
32. Dong LD, Hoffman AS. Synthesis and application of thermally reversible heterogels for drug delivery. *J. Control. Release* 1990;13:21–31.
33. Dong LC, Hoffman AS. Thermally reversible hydrogels: III. Immobilization of enzymes for feedback reaction control. *J. Control. Release* 1986;4:223–227.
34. Stayton PS, Shimoboji T, Long C, Chilkoti A, Chen G, Harris JM, Hoffman AS. Control of protein-ligand recognition using a stimuli-responsive polymer. *Nature* 1995;378:472–474.
35. Park TG, Hoffman AS. Immobilization and characterization of b-galactosidase in thermally reversible hydrogel beads. *J. Biomed. Mater. Res.* 1990;24:21–38.
36. Chen G, Hoffman AS. Preparation and properties of thermoreversible, phase-separating enzyme-oligo (*N*-isopropylacrylamide) conjugates. *Bioconjug. Chem.* 1993;4:509–514.
37. Hoffman S. Applications of thermally reversible polymers and hydrogels in therapeutics and diagnostics. *J. Control. Release* 1987;6:297–305.
38. Chilkoti A, Chen G, Stayton PS, Hoffman AS. Site-specific conjugation of a temperature-sensitive polymer to a genetically-engineered protein. *Bioconjug. Chem.* 1994;5:504–507.
39. Duracher D, Elaïssari A, Mallet F, Pichot C. Adsorption of modified HIV-1 capsid p24 protein onto thermosensitive and cationic core-shell poly(styrene)-poly(*N*-isopropylacrylamide) particles. *Langmuir* 2000;13(23):9002–9008.
40. Elaissari A, Ganachaud F, Pichot C. Biorelevant latexes and microgels for the interaction with nucleic acids. *Top. Curr. Chem.* 2003;227:169–193.
41. Elaissari A, Chauvet JP, Halle MA, Decavallas O, Pichot C, Cros P. Effect of charge nature on the adsorption of single-stranded DNA fragments onto latex particles. *J. Colloid Interf. Sci.* 1998;202:2252–2260.
42. de Gennes PG. *Scaling Concept in Polymer Physics*. Cornell University Press, Ithaca, NY, 1979.
43. Elaissari A, Cros P, Pichot C, Laurent V, Mandrand B. Adsorption of oligonucleotides onto negatively and positively charged latex particles. *Colloid Surf.* 1994;83:25–31.
44. Ganachaud F, Elaissari A, Pichot C, Laayoun A, Cros P. Adsorption of single-stranded DNA fragments onto cationic aminated latex particles. *Langmuir* 1997;13:701–707.

45. Charles MH, Charreyre MT, Delair T, Elaissari A, Pichot C. Oligonucleotide-polymer nanoparticle conjugates: Diagnostic applications. *STP Pharm. Sci.* 2001;11(4):251–263.
46. Uguzdogan E, Denkbas EB, Tuncel A. RNA-sensitive *N*-isopropylacrylamide/vinylphenylboronoic acid random copolymer. *Macromol. Biosci.* 2002;2(5):214–222.
47. Çamli ST, Senel S, Tuncel A. Nucleotide isolation by boronic acid functionalized hydrophilic supports. *Colloid Surf. A: Physicochem. Eng. Aspects* 2002;207:127–137.
48. Elmas B, Onur MA, Senel S, Tuncel A. Thermosensitive *N*-isopropylcarylamide-vinylphenyl boronic acid copolymer latex particles for nucleotide isolation. *Colloid Surf. A: Physicochem. Eng. Aspects* 2004;232:253–259.
49. Elmas B, Onur MA, Senel S, Tuncel A. Temperature controlled RNA isolation by *N*-isopropylacrylamide-vinylphenyl boronic acid copolymer latex. *Colloid and Polym. Sci.* 2002;280(12):1137–1146.
50. Aguilar MR, Elvira C, Gallardo A, Vázquez B, Román JS. Smart polymers and their applications as biomaterials. *Top. Tissue Eng.* 2007;3:2.
51. Medeiros SF, Santos AM, Fessi H, Elaissari A. Thermally-sensitive and magnetic poly(*N*-vinylcaprolactam)-based nanogels by inverse miniemulsion polymerization. *Colloid Sci.* 2012;1: 2.
52. Honey Priya J, Rijo J, Anju A, Anoop KR. Smart polymers for the controlled delivery of drugs—A concise overview. *Acta Pharm. Sin. B* 2014;2:120–127.
53. Yong Q, Kinam P. Environment-sensitive hydrogels for drug delivery. *Adv. Drug Deliv. Rev.* 2012;64:49–60.
54. Nguyen KT, West JL. Photopolymerizable hydrogels for tissue engineering applications. *Biomaterials* 2002;23:4307–4314.
55. Sultana F, Manirujjaman M, Imran-Ul-Haque MA, Sharmin S. An overview of nanogel drug delivery system. *Appl. Pharm. Sci.* 2013;3:S95–S105.
56. Gonçalves C, Pereira P, Gama M. Self-assembled hydrogel nanoparticles for drug delivery applications. *Materials* 2010;3:1420–1460.
57. Buenger D, Topuz F, Groll J. Hydrogels in sensing applications. *Prog. Polym. Sci.* 2012;37:1678–1719.
58. Viholaa H, Laukkanenb A, Hirvonena J, Tenhub H. Binding and release of drugs into and from thermo-sensitive poly(*N*-vinyl caprolactam) nanoparticles. *Eur. J. Pharm. Sci.* 2002;16:69–74.
59. Kavanagha CA, Rochevb YA, Gallaghera WM, Dawsonc KA, Keenan AK. Local drug delivery in restenosis injury: Thermoresponsive co-polymers as potential drug delivery systems. *Pharmacol. Ther.* 2004;102:1–15.
60. Kim MR, Jeong JH, Park TG. Swelling induced detachment of chondrocytes using RGD-modified poly(*N*-isopropylacrylamide) hydrogel beads. *Biotechnology* 2002;18:495–500.
61. Kabanov AV, Vinogradov SV. *Nanogels as Pharmaceutical Carriers*, Vol. 4. Springer Link, Springer, New York, 2008, pp. 67–80.
62. Alam MA, Rabbi MA, Miah MAJ, Rahman MM, Rahman MA, Ahmad H. A versatile approach on the preparation of dye-labeled stimuli-responsive composite polymer particles by surface modification. *Colloid Sci. Biotechnol.* 2012;1:225–234.
63. Schmaljohann D. Thermo- and pH-responsive polymers in drug delivery. *Adv. Drug Deliv. Rev.* 2006;58:1655–1670.
64. Ruel-Gariépy E, Leroux J-C. In situ-forming hydrogels—Review of temperature-sensitive systems. *Eur. J. Pharm. Biopharm.* 2004;58:409–426.
65. Ramanan RMK, Chellamuthu P, Tang L, Nguyen KT. Development of a temperature-sensitive composite hydrogel for drug delivery applications. *Biotechnol. Prog.* 2006;22:118–125.
66. Imaz A, Forcada J. *N-Vinylcaprolactam-Based Microgels for Biomedical Applications*, Vol. 48. 2010, pp. 1173–1181.
67. Abu Samah NH, Heard CM. Enhanced in vitro transdermal delivery of caffeine using a temperature and pH-sensitive nanogel, poly(NIPAM-*co*-AAc). *Int. J. Pharm.* 2013;453:630–640.
68. Hoare T, Young S, Lawlor MW, Kohane DS. Thermoresponsive nanogels for prolonged duration local anesthesia. *Acta Biomater.* 2012;8:3596–3605.
69. Stover TC, Kim YS, Lowe TL, Kester M. Thermoresponsive and biodegradable linear-dendritic nanoparticles for targeted and sustained release of a pro-apoptotic drug. *Biomaterials* 2008;29:359–369.
70. Hamcerencu M, Popa M, Costin D, Bucatariu P, Desebrieres J, Riess G. New ophthalmic insert. *Oftalmologia* 2009;53:83–90.
71. Ichikawa H, Fukumori Y. New applications of acrylic polymers for thermosensitive drug release. *STP Pharm. Sci.* 1997;7:529–545.

10 Microemulsions and Their Applications in Drug Delivery

Ziheng Wang and Rajinder Pal

CONTENTS

10.1 INTRODUCTION

Microemulsions are being increasingly used as vehicles for lipophilic drugs. They can be used for parenteral drug delivery systems (Date and Nagarsenker 2010) or could be encapsulated into softgel capsules as a convenient solid dosage form (Gullapalli 2010). In contrast to conventional emulsions, microemulsions are small droplet size (typically between 20 and 200 nm) (Talegaonkar et al. 2008) and exhibit long-term stability (McClements 2012). They have been proven to promote the gastric-intestinal (GI) absorption of lipophilic drugs and consequently enhance the bioavailability of some active pharmaceutical ingredients (API) (Lawrence and Rees 2000). Food and Drug Administration in the United States (U.S. FDA) has classified APIs into four groups based on the Biopharmaceutical Classification System (BCS). APIs in class II (high permeability, low solubility) and class IV (low permeability, low solubility) are ideal candidates for microemulsion-based drug delivery systems due to their poor solubility in the aqueous phase. Table 10.1 gives examples of drugs that have been successfully delivered in microemulsion form.

Microemulsions are isotropic, thermodynamically stable systems. They appear to be transparent (or translucent) (Lawrence and Rees 2000). They can be easily prepared and identified in different ways. For example, when 7 mL of a household liquid detergent, 14 mL of white spirit,

TABLE 10.1
Examples of Drugs Delivered in Microemulsion Form

Drug Name	Usage	References
Paclitaxel	Anticancer	Gao et al. (2003)
Fenofibrate	Antihyperlipidemic	Liang et al. (2006)
Cholesterol ester transfer protein (CETP) inhibitors		Gumkowski et al. (2005)
Atorvastatin		Shen and Zhong (2006)
Fluvastatin		Benameur et al. (2003)
Rapamycin	Immunosuppressive drug	Fricker et al. (2006)
Cyclosporine		Ward and Cotter (1987)
Felodipine	Antihypertensive drug	Von Corswant (2003)
Nifedipine		Rudnic et al. (1999)
Indomethacin	Analgesic drug	Farah and Denis (2000)
Ibuprofen		Bauer et al. (2002)
Naproxen		Mulye (2002)
Tipranavir	Anti-HIV drug	Chen and Gunn (2003)
Progesterone	Hormone	Gao and Morozowich (2003)
Testosterone		Gao and Morozowich (2003)
Fish oil	Nutrition supplement	Mishra et al. (2001)
Acyclovir	Antiviral drug	Burnside et al. (1999)
Melatonin	Immunomodulator	Eugster et al. (1996)

and 4 mL of *n*-pentanol (amyl alcohol) or *n*-butanol are gently mixed, a two-phase system is produced at equilibrium (Makoto 1998) where the upper oil phase of the system exhibiting a strong Tyndall effect is identified as the microemulsion phase (Makoto 1998). Microemulsion-based drugs normally exhibit long shelf-life due to the high stability of microemulsions. Other advantages of using microemulsion as a lipophilic drug carrier include the enhanced bioavailability (due to small droplet size) and the ease of formation (due to low interfacial tension) (Lawrence and Rees 2000; Talegaonkar et al. 2008). The small droplet size of microemulsion also facilitates the permeability of the drug passing through the mucous membrane. Microemulsion-based drugs can be delivered through different routes such as oral delivery as softgel capsules (Kovarik et al. 1994), topical or transdermal delivery as lotions (Gupta et al. 2005), and parenteral delivery as intramuscular and intravenous injections (Von Corswant et al. 1998).

Microemulsions are ternary systems containing oil, water, and surfactant. The terms "oil" and "water" in a microemulsion system normally refer to "oil phase (oil and oil soluble components such as cyclosporine)" and "aqueous phase (water and water soluble components such as sodium chloride)," respectively. The phase behavior of water–oil–surfactant mixtures was extensively studied by Winsor (1948). Based on his experimental observations, Winsor classified equilibrium mixtures of water–oil–surfactant into four systems: (1) type I (Winsor I) system where water continuous or oil-in-water (O/W) type microemulsion coexists with the oil phase. In these systems, the aqueous phase is surfactant-rich; (2) type II (Winsor II) system where oil continuous or water-in-oil (W/O) type microemulsion coexists with the aqueous phase. In these systems, the oil phase is surfactant-rich; (3) type III (Winsor III) system where bicontinuous type microemulsion (also referred to as surfactant-rich middle-phase) coexists with excess oil at the top and excess water at the bottom; and (4) type IV (Winsor IV) system where only a single-phase (microemulsion) exists. The surfactant concentration in type IV microemulsion is generally greater than 30 wt%. Type IV microemulsion could be water continuous, bicontinuous, or oil continuous depending on the chemical composition. The phase behavior of microemulsions is often described as a *fish* diagram shown in Figure 10.1 (Komesvarakul et al. 2006).

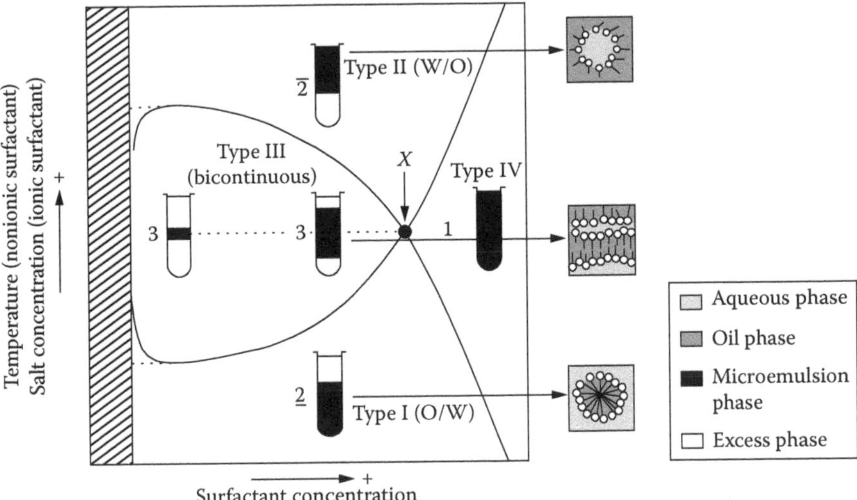

FIGURE 10.1 *Fish* diagram: 1—single-phase region; 2—two-phase region (upper bar: microemulsion phase at the top; lower bar: microemulsion phase at the bottom); and 3—three-phase region. *X*: tri-critical point. This diagram assumes that the density of the aqueous phase is greater than that of the oil phase. (Adapted from Komesvarakul, N. et al., *J. Cosmet. Sci.*, 55, 309, 2006.)

10.1.1 Continuous and Dispersed Phases

The fluid phase of a microemulsion in which oil or water is distributed as droplets is called continuous phase. Oil or water droplets dispersed in the continuous phase are collectively called dispersed phase. In type I microemulsion, the dispersed phase is oil and the continuous phase is water. Type II microemulsions have a reverse arrangement in that the dispersed phase is water and the continuous phase is oil. There is no dispersed phase in type III microemulsion as type III microemulsion is bicontinuous in nature. Likewise the bicontinuous form of type IV microemulsion has no dispersed or continuous phase. The concept of dispersed and continuous phases is important in quantifying the droplet size of microemulsion. The droplet size distribution can be obtained for different types of microemulsions using dynamic light scattering (DLS). However, one should keep in mind that the droplet size distribution is meaningless for type III or type IV bicontinuous microemulsion due to its bicontinuous nature.

10.1.2 Surfactants

Surfactants are amphiphilic molecules composed of both hydrophilic part (e.g., ethylene oxide group) and hydrophobic part (e.g., alkyl chain). There are three types of surfactants: ionic (anionic and cationic), nonionic, and zwitterionic. The hydrophilic part of surfactant bears either negative charge (anionic) or positive charge (cationic) in the case of ionic surfactant, no charge in the case of nonionic surfactant, and both negative and positive charges in the case of zwitterionic surfactant. At a very low concentration of surfactant, the surfactant molecules exist as monomers in the base liquid. The surfactant molecules also adsorb onto the interface between two immiscible phases (oil–water or gas–liquid interfaces). With the increase in surfactant concentration, the surfactant molecules begin to aggregate. As an example, consider the addition of surfactant to water. The surfactant molecules initially disperse in water as monomers and also adsorb on to the air–water interface till they reach surface saturation. Then monomers in water begin to aggregate as clusters with their hydrophobic groups toward the interior of clusters and their hydrophilic groups toward water. These clusters are called *micelles* and the concentration where micelles begin to form is called the *critical micelle concentration* (cmc) of the surfactant (Rosen 2004). The cmc of a surfactant makes it unique from other amphiphilic molecules. For example, the molecules of a co-surfactant are also amphiphilic but co-surfactants generally have no cmc. With continuous addition of surfactant to water above cmc, changes in the microstructure of surfactant/water system occur as shown in Figure 10.2a from spherical micelle solution to cylindrical micelle solution to hexagonal liquid crystal (LC) to cubic liquid crystal and to lamellar liquid crystal. Note that liquid crystal (LC) is a material in semisolid state that has both organized structure (like solid) and disordered structure (like liquid) at a molecular level. The changes in the microstructure summarized in Figure 10.2a are normally accompanied by changes in the viscosity of the system. For example, the viscosity of LC is much higher than that of micellar solutions; hexagonal LC has the highest viscosity among the three types of liquid crystals (Rosen 2004). It should be noted that the lamellar structure could be either flat or curved depending on the surfactant structure (Kik et al. 2005).

When surfactant is added to oil, the surfactant molecules form micelles as usual at surfactant concentrations above cmc but the micelles formed in oil are called *reverse micelles* as they have a reverse arrangement of surfactant molecules as compared with the arrangement of surfactant molecules in micelles formed in aqueous systems. In reverse micelles, the hydrophilic heads of the

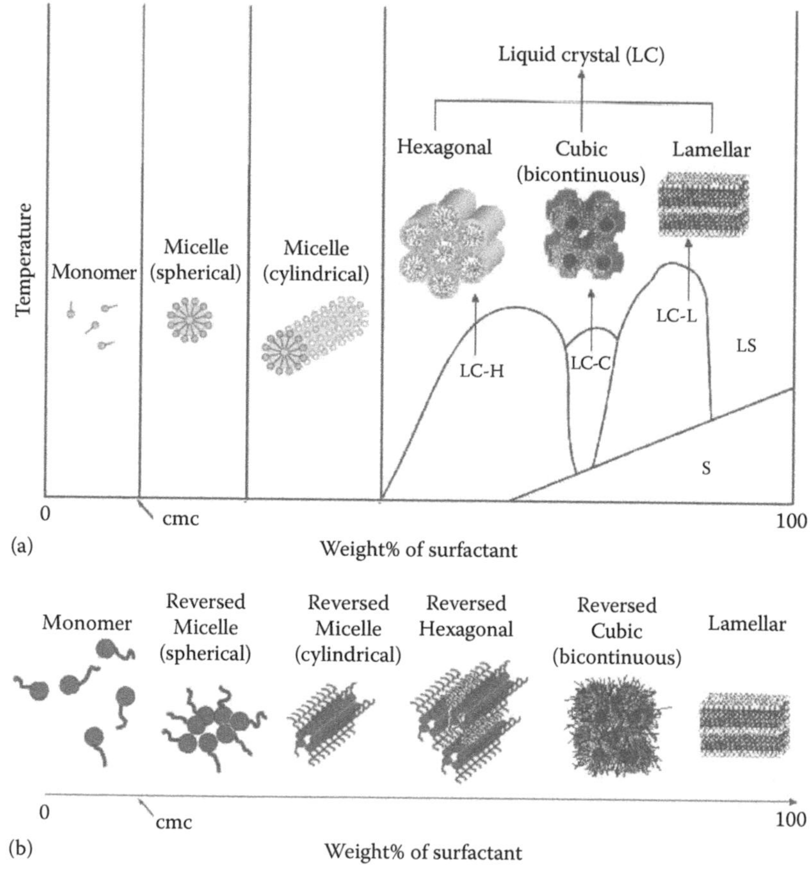

FIGURE 10.2 (a) Phase diagram of surfactant–water system. LC-H, liquid crystal-hexagonal region; LC-C, liquid crystal-cubic region; LC-L, liquid crystal-lamellar region; LS, liquid surfactant region (water in surfactant); and S, solid region. (b) Microstructural changes in surfactant–oil system with the increase in surfactant concentration. (Adapted from Rosen, M.J., *Surfactants and Interfacial Phenomena*, 3rd edn., John Wiley & Sons, Inc., New York, 2004, pp. 1–33, 110–113, 208–234.)

surfactant molecules are inside the micelles and the hydrophobic tails of the surfactant molecules extend away from the core of the micelles to the oil phase. Figure 10.2b shows the microstructural changes that occur upon continuous addition of surfactant to oil (Rosen 2004).

The adsorption of surfactant at the interface alters the interfacial free energy. The interfacial free energy is defined as the minimum work required to create the interface (Rosen 2004). Interfacial tension (γ), which is the interfacial free energy normalized by area, is the minimum work required to create the interface of unit area. Physically, interfacial tension is a measure of the strength of interaction between two phases. The higher the interfacial tension, the weaker is the interaction between the two phases. For example, pure trichloroethylene (TCE) and water are immiscible phases with γ = 39.6 mJ/m^2 at 25°C (Ma et al. 2008). To create interfacial contact area between the two phases, that is, to formulate TCE–water emulsions, energy input exceeding 39.6 mJ/m^2 is required. With the addition of certain amounts of anionic surfactant sodium oleate (SO) and cationic surfactant benzethonium chloride (BC), γ can be lowered to as low as 3.7 ± 0.4 mJ/m^2 (Wang and Acosta 2013). This reduction in interfacial tension occurs because the surfactant partitions between TCE and water phases, and as a result, an increase in interaction between TCE and water phases is observed. Due to a decrease in the interfacial tension, less energy input is required to create new interfacial contact area. It should be noted that γ should be very low (as low as 1 mJ/m^2 or lower) in order to formulate microemulsions (Upadhyaya et al. 2006). Due to a very low interfacial tension, microemulsions are normally formed under gentle agitation with ultralow energy input (e.g., stomach agitation in vivo is sufficient). To achieve ultralow values of interfacial tension, it is often necessary to add a co-surfactant to the system. Thus, microemulsions are usually formulated using a combination of surfactant and co-surfactant.

In the pharmaceutical field, nonionic surfactants are widely used as they are less irritative than ionic surfactants (Mason et al. 2006). Before the design and formulation of any microemulsion-based drug delivery system, it is important to consult the inactive ingredient guide (IIG) published by FDA on their website to check the limits for different surfactants.

10.1.3 Co-Surfactant and Linker

Co-surfactant is a small amphiphilic molecule. It has smaller head and tail groups as compared with a surfactant molecule. The most common types of co-surfactants are C_3–C_6 alcohols, such as *sec*-butanol (Salager et al. 2005). The co-surfactant adsorbs at the oil–water interface and modifies the formulation requirements. For example, it modifies the hydrophilic–lipophilic balance (*HLB*) requirement of the surfactant. Co-surfactants with short chain lengths (C_3–C_5) tend to be more hydrophilic whereas co-surfactants with long chain lengths (C_5 or higher) tend to be more lipophilic. Therefore, a less hydrophilic surfactant is required in the formulation when co-surfactants with short chain lengths (C_3–C_5) are used, whereas a less lipophilic surfactant is required when co-surfactants with long chain lengths (C_5 or higher) are used (Bavière et al. 1981). The co-surfactants also interfere in the surfactant–surfactant interactions by pushing the surfactant molecules apart, disrupting the LC structure, and reducing the viscosity of the mixture (Jones and Dreher 1976). The weight ratio of surfactant to co-surfactant can vary from 1:0.5 (w/w) to 1:3 (w/w), depending on the stability of the system (Kang et al. 2004; Wang and Pal 2014).

As already noted, the addition of co-surfactant to a microemulsion modifies the *HLB* requirement of the surfactant. The addition of co-surfactant also replaces a certain amount of surfactant at the oil–water interface. These effects need to be compensated for in the formulation design in order to obtain a microemulsion that has the same phase behavior as that of one without the presence of a co-surfactant. For example, the contribution of ethanol as a co-surfactant is hydrophilic. Therefore, the surfactant needs to be less hydrophilic, which can be achieved by reducing the number of ethylene oxide groups in the surfactant structure. However, a higher surfactant concentration is required in the formulation since ethanol as a co-surfactant replaces some of the surfactant from oil–water interface (Salager et al. 2005). In practice, *sec*-butanol or a mixture of propanol and butanol (1:1 in weight ratio) is often selected as a co-surfactant in order to disrupt the order of the LC structure while maintaining the original phase behavior (Salager et al. 2005).

With the increase in alkyl chain length of a co-surfactant, the mixture of surfactant and co-surfactant becomes more lipophilic. However, the adsorption of co-surfactant at the interface becomes less significant. This is because a long-chain alcohol has more affinity for the oil phase. Therefore, instead of replacing surfactant at the oil–water interface by pushing the head groups of surfactant molecules apart at oil–water interface, a long chain co-surfactant preferentially stays with the hydrophobic part of the surfactant (e.g., tail group). This type of co-surfactant with long alkyl chain length ($>C_{10}$) is called a lipophilic *linker* (Salager et al. 2005). A linker is different from co-surfactant in that the linker is either hydrophilic or lipophilic enough to co-adsorb at only one side of the oil–water interface whereas co-surfactant interacts with both oil and water phases and replaces the surfactant at the oil–water interface. Due to the co-adsorption effect, microemulsions formulated using linkers may exhibit larger solubilization capacity than the ones without a linker (Acosta et al. 2005). Typical linkers include hexyl glucoside (hydrophilic linker), sorbitan monooleate (lipophilic linker), and long chain ($>C_{10}$) alcohols (lipophilic linker) (Acosta et al. 2005).

10.1.4 Microemulsion and Nanoemulsion: Their Similarities and Differences

Microemulsion is a self-assembling nano-scale emulsion whereas nanoemulsion is a nano-scale emulsion formed under intense mechanical shear (McClements 2012). Microemulsion is an isotropic solution of oil and water, prepared using a high surfactant concentration of around 40 wt% under gentle stirring or shaking. The usage of a large concentration of surfactant ensures ultralow oil–water interfacial tension and spontaneous formation of microemulsion without any mechanical shear. The preparation of nanoemulsions requires very high shear in order to rupture large droplets into nano-scale droplets. The mechanical shear should be intensive enough to overcome the large interfacial tension (McClements 2011).

Thermodynamically, the change in free energy to formulate either a microemulsion or a nanoemulsion from two separate phases (i.e., $\Delta G_{\text{formation}}$) can be expressed as follows (McClements 2012):

$$\Delta G_{\text{formation}} = \Delta G_I + (-T\Delta S) \tag{10.1}$$

where

ΔG_I (J/mol) is the change in interfacial free energy

$-T\Delta S$ (J/mol) is the entropic contribution to free energy of formation

The entropic contribution is due to the change in configuration from two stratified phases to an emulsion with a large number of droplets. The change in the interfacial free energy ΔG_I is given as follows:

$$\Delta G_I = \gamma \Delta A \tag{10.2}$$

where

γ is the interfacial tension

ΔA is the increase in the interfacial contact area between the phases due to the formation of a microemulsion or a nanoemulsion

The change in interfacial free energy ΔG_I is always positive in the formation of a microemulsion or a nanoemulsion as both γ and ΔA are positive; positive ΔA is due to an increase in the interfacial area between oil and water when droplets are formed. The entropy contribution $-T\Delta S$ is always negative as both T and ΔS are positive; positive ΔS is due to an increase in the disorder of the system when droplets are formed. Therefore, $\Delta G_{\text{formation}}$ can be either positive or negative depending on the balance between ΔG_I and $-T\Delta S$. In the case of microemulsions, $\Delta G_{\text{formation}}$ is negative and therefore microemulsions are formed spontaneously and are thermodynamically stable systems. The $\Delta G_{\text{formation}}$ is negative in microemulsions due to an ultralow value of interfacial tension between oil and water. Unlike microemulsions, nanoemulsions are thermodynamically unstable systems as $\Delta G_{\text{formation}}$ is positive due to high interfacial tension between the oil and water phases. Although nanoemulsions are thermodynamically unstable systems, they can be made kinetically stable due to steric stabilization of the droplets.

Kinetically, the rate of separation of nanoemulsion into two separate phases can be described by the Arrhenius equation (Missen et al. 1999):

$$k = Ae^{-E_a/RT} \tag{10.3}$$

where

k is the rate constant
A is the pre-exponential factor
E_a is the activation energy
R is the universal gas constant
T is the absolute temperature (in K)

At constant temperature, the rate of separation of nanoemulsion into separate phases depends on the activation energy.

10.2 CHARACTERIZATION OF MICROEMULSIONS

10.2.1 Types of Microemulsions

As shown in Figure 10.1, four types of microemulsion systems can be formulated. In type I to type III microemulsion systems, two or more phases are present in equilibrium with each other. Only in the case of type IV (Winsor IV) microemulsion, a single phase is present. However, type IV microemulsion could be either water continuous, bicontinuous, or oil continuous. Several techniques could be used to identify different types of microemulsions. For example, the Tyndall effect can be observed in the lower phase of type I, middle phase of type III, and upper phase of type II microemulsion by simply pointing a laser pointer toward the sample as shown in Figure 10.3.

The change in electrical conductivity can be used to differentiate different types of microemulsions. For example, type II or type IV oil continuous microemulsions have a very low electrical conductivity (say less than 1 μS/cm); type III or type IV bicontinuous microemulsions have a medium electrical conductivity (say in between 1 and 10 μS/cm); and type I or type IV water continuous

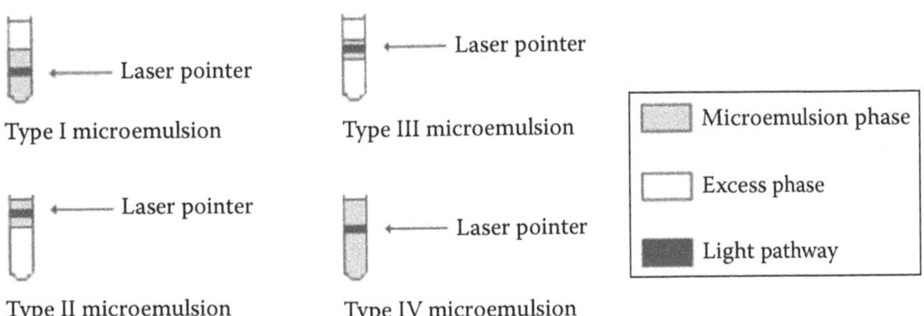

FIGURE 10.3 Tyndall effect in type I, II, III, and IV microemulsions. These observations assume that the density of the oil phase is smaller than that of the water phase.

microemulsions have a high conductivity (say greater than 10 μS/cm) due to water as the continuous phase (Krauel et al. 2005).

The phase transition from type IV oil continuous microemulsion to type IV water continuous microemulsion could be captured by observing changes in the viscosity (Watanabe et al. 2004). For example, the viscosity of type IV oil continuous (W/O) microemulsion rises slowly initially with the increase in water concentration. With further addition of water the viscosity begins to increase sharply. The increase in viscosity is mainly due to the transition from type IV oil continuous (W/O) microemulsion to type IV bicontinuous microemulsion. The viscosity reaches a maximum value at some water concentration. Upon further increase in water concentration, the transition of type IV bicontinuous microemulsion to type IV water continuous (O/W) microemulsion occurs resulting in a sharp decrease in viscosity. The viscosity of type IV water continuous (O/W) microemulsion continues to decrease with further increase in aqueous fraction (Watanabe et al. 2004).

Other methods such as dying (Ho et al. 1996) and cryo-field emission scanning electron microscopy (Cryo-FESEM) (Krauel et al. 2005) could also be used to identify the types of microemulsions.

The droplet size distribution of microemulsions can be determined using the dynamic light scattering (DLS) technique. The average droplet size of about 80–120 nm is an acceptable level for microemulsion-based drug delivery systems (Wang and Pal 2014).

10.2.2 Hydrophilic Lipophilic Difference Method to Prepare Microemulsions

Generally speaking, hydrophilic surfactant is used to formulate water continuous microemulsions and lipophilic surfactant is used to formulate oil continuous microemulsions. The hydrophilicity of surfactant can be measured in terms of the *HLB* (Pasquali et al. 2008). The *HLB* value of a surfactant is defined as follows based on Griffin's method:

$$HLB = 20 \times \frac{M_h}{M} \tag{10.4}$$

where

M_h is the molecular weight of the hydrophilic part of the surfactant
M is the molecular weight of the whole surfactant molecule

According to Equation 10.4, the *HLB* scale ranges from 0 to 20. The surfactant with *HLB* value between 12 and 16 is considered to be suitable for the formulation of O/W (oil in water) microemulsions. Surfactants with *HLB* values between 7 and 11 are more suitable for the preparation of W/O (water in oil) microemulsions. The *HLB* concept is simple to use and the database is available for a large number of surfactants. However, the main disadvantage of the *HLB* concept is that it does not consider the impact of other factors on microemulsion formulation, such as temperature, aqueous phase salinity, oil-chain length, and co-surfactant. For example, water continuous microemulsion could be formulated using Cremophor EL (*HLB* = 13.5) at room temperature (20°C –25°C) (Wang and Pal 2014). However, one may end up with bicontinuous type microemulsion using the same formulation at 37°C in vivo due to the change in temperature. In summary, the impact of factors such as co-surfactant, chemical nature of oil, temperature, salinity of aqueous phase, etc., on microemulsion type are not captured in the *HLB* concept.

To overcome the limitations of the *HLB* method to formulate microemulsions, a different approach called *hydrophilic–lipophilic difference* (HLD) was developed by Salager et al. (1983, 2000). The *HLD* approach captures the impact of various factors on microemulsion type. The *HLD* value for nonionic and ionic surfactants can be calculated as follows (Acosta and Bhakta 2009):

Nonionic surfactants:

$$HLD = b(S) - K \times N_{CO} - \phi(A) + c_T \Delta T + CC \tag{10.5}$$

Ionic surfactants:

$$HLD = \ln(S) - K \times N_{CO} - f(A) - \alpha_T \Delta T + CC \quad (10.6)$$

where

S is the salinity (g/100 mL) of aqueous phase

b is an empirical constant equal to 0.13 for NaCl and 0.1 for $CaCl_2$ (Acosta and Bhakta 2009)

K is a constant that ranges from 0.1 to 0.2 (normally is around 0.17)

N_{CO} is the alkane carbon number (ACN) of oil, which is a measure of hydrophobicity of the oil phase (Ontiveros et al. 2013)

The more hydrophobic (e.g., long chain length) the oil phase, the higher is the N_{CO} value. In the case of non-alkane oils, an equivalent alkane carbon number (EACN) is used (Acosta and Bhakta 2009); $\phi(A)$ or $f(A)$ is a factor that takes into account the influence of co-surfactant. It is related to the partitioning of co-surfactant in the two phases. If there is no partitioning of co-surfactant observed in the formulation (e.g., using *sec*-butanol as noted earlier) or no co-surfactant is present in the system, $\phi(A)$ or $f(A)$ is zero (Salager et al. 2005; Acosta and Bhakta 2009); c_T and α_T are temperature factors, equal to 0.06 and 0.01 K^{-1}, respectively; ΔT is temperature difference from 25°C; and *CC* is called *characteristic curvature* of the surfactant. Like *HLB*, *CC* is also a measure of the hydrophilicity of surfactant. However, the *CC* scale is different from that of *HLB* (Acosta et al. 2008). The smaller the *CC* value, the more hydrophilic is the surfactant (e.g., the *CC* value of sodium oleate is −1.7 whereas oleic acid has a *CC* value of 0) (Acosta et al. 2008).

The *HLD* criteria for the formulation of different types of microemulsions are as follows: for type I microemulsion, $HLD < 0$; for type III microemulsion, $HLD \approx 0$; and for type II microemulsion, $HLD > 0$ (Salager et al. 2000; Mason et al. 2006; Ontiveros et al. 2013). According to Equations 10.5 and 10.6, oil with a long-chain structure (high alkane carbon number) is preferred for the formulation of type I microemulsion, if other factors such as temperature, surfactant, co-surfactant, and electrolyte concentration are kept the same. In the case of nonionic surfactant, the temperature effect is more pronounced than the electrolyte concentration effect. The hydrophilic parts of nonionic surfactants are more sensitive to temperature changes than to electrolyte concentration changes. The breakage of hydrogen bonds at high temperature makes nonionic surfactant more lipophilic (Nilsson and Lindman 1983). On the contrary, ionic surfactant is more sensitive to changes in electrolyte concentration than to temperature changes. A high electrolyte concentration can compress the electrical double layer of the hydrophilic part and make the ionic surfactant more lipophilic (Srinivasan and Blankschtein 2003).

Although the *HLD* method has a limited database (*CC* value is known only for a limited number of surfactants; likewise the N_{CO} value is known for a limited number of oils) as compared with the *HLB* method, this concept can play an important role in the formulation design.

10.2.3 Self-Microemulsifying Drug Delivery System

The self-microemulsifying drug delivery system (SMEDDS) is a very promising drug delivery system for oil-soluble drugs. It is only a pre-mixture of oil-soluble drug, oil, surfactants, and co-surfactants and is able to form a microemulsion spontaneously under gentle agitation in vivo (Kang et al. 2004). During the in vitro tests, temperature is usually maintained at 37°C since this is the actual temperature at which microemulsion would be formed in vivo. No water is loaded in SMEDDS as water comes from the aqueous phase present in vivo. The system with zero water loading can be stored in capsules as reverse micelles. The amount of drug solubilized in the reverse micelles of SMEDDS is a very important factor to evaluate the system. The drug solubilization ability of the system dictates the selection of various ingredients.

Drugs are known to solubilize at the interface of microemulsion droplets or micelles. Research work by Spernath et al. (2003) has demonstrated that reverse micelles have a higher drug solubilization capacity than that of the individual components. They entrapped lycopene (model drug) in the reverse micelles of *R*-(+)-limonene (model oil) and surfactant polysorbate 60 (Tween 60). The drug capacity reached 1500 ppm as compared with 700 ppm in *R*-(+)-limonene alone (Spernath et al. 2002). The reason for the increased capacity of drug solubilization is that the drug can now distribute at the surface of the reverse micelles rather than occupy the core. However, drug solubilization at the surface of reverse micelles is highly dependent on the physical properties of surfactant, co-surfactant, and drug. The interactions between the drug and surfactant, and the hydrophobicity of surfactant are also important. Different components can result in different solubilization capacity of drugs (Narang et al. 2007). It should also be noted that drug solubilization may reduce after oral administration of SMEDDS due to aqueous phase dilution and structural changes in vivo from reverse micelles to O/W microemulsion.

10.3 FORMULATION DEVELOPMENTS

10.3.1 Interaction between the Two Phases

Molecules at the interface between two immiscible liquids have a higher potential energy as compared with the molecules in the bulk phase (Rosen 2004). Figure 10.4 shows a schematic representation of the interface between two phases. The increase in the potential energy of molecules "*a*" per unit area is equal to the interaction energy of molecules in the bulk (γ_{aa}) minus the interaction energy at the interface (γ_{ab}) (Rosen 2004). Likewise, the increase in potential energy of molecules "*b*" is $\gamma_{bb} - \gamma_{ab}$. Therefore, the interfacial free energy per unit area or the interfacial tension (γ_I) can be expressed as follows (Rosen 2004):

$$\gamma_I = (\gamma_{aa} - \gamma_{ab}) + (\gamma_{bb} - \gamma_{ab}) = \gamma_{aa} + \gamma_{bb} - 2\gamma_{ab} \quad (10.7)$$

It should be noted that the interaction energy between the like molecules (i.e., γ_{aa} or γ_{bb}) is always greater than the interaction energy between the unlike molecules (γ_{ab}). When the phase consisting of "*a*" molecules is a gas phase, $\gamma_{aa} \approx 0$ and $\gamma_{ab} \approx 0$, and therefore γ_I is the surface tension of phase "*b*" given as γ_{bb}. Likewise, γ_{aa} is the surface tension of phase "*a*" (Rosen 2004). According to Equation 10.7, the interfacial tension γ_I increases with the decrease in γ_{ab} as the surface tensions γ_{aa}

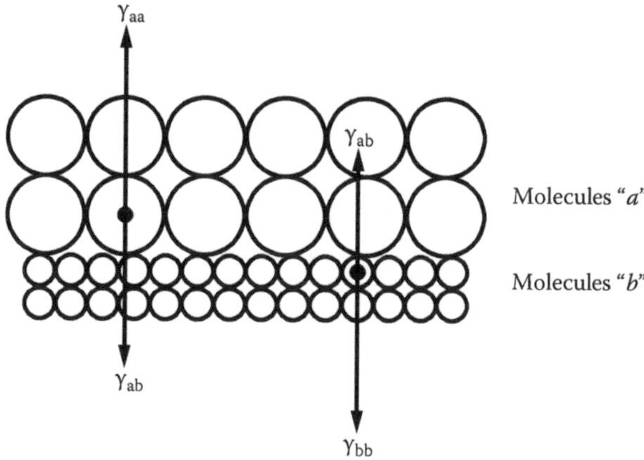

FIGURE 10.4 Interaction between two immiscible liquids. (Adapted from Rosen, M.J., *Surfactants and Interfacial Phenomena*, 3rd edn., John Wiley & Sons, Inc., New York, 2004, pp. 1–33, 110–113, 208–234.)

and γ_{bb} are fixed values. Therefore, interfacial tension (γ_I) is a measure of interaction between the two phases. The higher the γ_I, the weaker is the interaction between the two phases.

To formulate a microemulsion, the interfacial tension (γ_I) should be ultralow (<1 mJ/m^2) (Upadhyaya et al. 2006) and the interaction between the two phases (γ_{ab}) must be high. Therefore, a certain amount of surfactant is needed to increase the interaction between the two phases to a level where a microemulsion is formed spontaneously. This concentration is called critical microemulsion concentration (cμc), which is the minimum surfactant concentration required to formulate a microemulsion (Aveyard et al. 1989).

10.3.2 Phase Scan

The purpose of a phase scan is to determine the temperature (in case of nonionic surfactants) or salinity (in case of ionic surfactants) that can produce a microemulsion of desired type (O/W, W/O, or bicontinuous) and properties (such as solubilization). Figure 10.5 shows a typical phase scan (Rosen 2004). Phase scan is normally run from type I (Winsor I) system to type II (Winsor II) system by increasing the temperature (in case of nonionic surfactant) or salinity (in case of ionic surfactant). As shown in Figure 10.5, oil solubilization increases from sample 1 to 2 with the increase in temperature or salinity. With the increase in temperature/salinity, the surfactant becomes more lipophilic due to the increased dehydration. Consequently, the interaction between oil and water phase (γ_{ab}) increases and the interfacial tension (γ_{OW}) decreases. As surfactant becomes continuously more lipophilic with the increase in temperature/salinity, a B_M phase (bicontinuous or middle phase) begins to separate from W_M phase (water continuous O/W microemulsion phase). At the start of the separation (sample 3), the interfacial tension between oil and B_M phase (γ_{OB}) is still high and the interfacial tension between B_M and water phase (γ_{BW}) is zero (B_M and water are miscible).

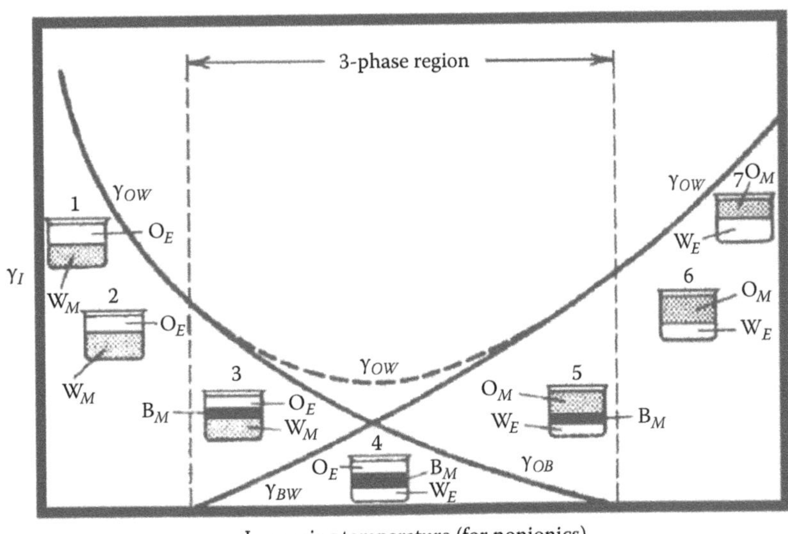

FIGURE 10.5 Phase scan with temperature (for nonionic surfactant) or salinity (for ionic surfactant). O_E, excess oil phase; O_M, microemulsion (W/O) phase; W_E, excess water phase; W_M, microemulsion (O/W) phase; B_M, microemulsion (bicontinuous) phase; γ_{OW}, interfacial tension between oil and water phase; γ_{BW}, interfacial tension between bicontinuous and water phase; and γ_{OB}, interfacial tension between oil and bicontinuous phase. (Adapted from Rosen, M.J., *Surfactants and Interfacial Phenomena*, 3rd edn., John Wiley & Sons, Inc., New York, 2004, pp. 1–33, 110–113, 208–234.)

The apparent interfacial tension between oil and water (γ_{OW}) in the three-phase region, with a middle phase present between oil and water phases, can be calculated as follows (Rosen 2004):

$$\gamma_{OW} = \gamma_{OB} + \gamma_{BW} \tag{10.8}$$

The apparent interfacial tension in the three-phase region is shown as a dashed line in Figure 10.5. When the temperature/salinity is increased to a certain level (sample 4 in Figure 10.5), most of the surfactants from W_M migrate to the middle phase resulting in two excess phases (O_E and W_E) and minimum apparent γ_{OW}. This Winsor III formulation shown as sample 4 in Figure 10.5 is the optimal formulation. The temperature corresponding to the optimal formulation is the phase-inversion temperature (PIT) and the corresponding salinity is the optimum salinity (Rosen 2004).

With further increase in temperature/salinity from sample 4 to 5, the surfactant starts migrating from the middle phase (B_M phase) to the oil phase (O_E phase) and type II microemulsion (O_M phase) begins to form. As surfactant becomes more lipophilic, the middle phase (B_M phase) disappears. Water solubilization decreases and γ_{OW} increases with further increase in temperature/salinity from sample 6 to 7. These changes are due to the transfer of surfactant to the oil phase and the reduction in the interaction between oil and water phases.

As already pointed out, the purpose of the phase scan is to determine the optimal temperature/salinity that can produce microemulsion of desired type and properties. Once the optimal temperature/salinity is established, one can further investigate the effects of composition on phase behavior using the ternary phase diagram.

10.3.3 Pseudo-Ternary Phase Diagram

A pseudo-ternary phase diagram of drug, oil, surfactant, co-surfactant, and water can be very helpful in formulating a suitable composition of self-microemulsifying drug delivery system (Prajapati et al. 2012). In general, three types of phases are encountered in a pseudo-ternary phase diagram: microemulsion (ME), liquid crystal (LC), and coarse emulsion (EM). The microemulsion (ME) region is a single-phase region (Salager et al. 2005) and is the region of main interest in the formulation of SMEDDS. A large microemulsion region can offer more flexibility in the selection of the optimal dosage composition (Wang and Pal 2014). Formulations in this region result in type IV microemulsions (Winsor IV) at equilibrium state and can be identified with their clear and transparent appearance. They also exhibit Tyndall effect. Liquid crystal (LC) could be of three types: hexagonal, cubic, and lamellar LC. Hexagonal and lamellar LCs are anisotropic and exhibit oil streaks or angular and striated textures under crossed polarized microscope (Cistola et al. 1986). Figure 10.6 shows the pictures of liquid crystal samples under crossed polarized microscope. The lamellar LC exhibits oil streaks as shown in Figure 10.6a and the hexagonal LC exhibits angular and striated textures as shown in Figure 10.6b. Cubic LC is an isotropic structure and cannot be observed under crossed polarized microscope. Coarse emulsion (EM) is the traditional thermodynamically unstable emulsion; it appears as milky white during the preparation and ends up into two or three phases at equilibrium (Salager et al. 2005). The droplet size of coarse emulsion can range anywhere from sub-microns to microns (Li et al. 2005). Formulations in the EM region are only kinetically stable. At equilibrium they end up into either two separate phases, or type I microemulsion with excess oil phase, or type II microemulsion with excess water phase, or type III microemulsion with both excess oil and water phase. The boundary lines between the two emulsion regions (ME/EM) are drawn out according to the emulsion appearance and droplet size.

Figure 10.7 shows a ternary diagram without the specification of different phase regions and boundaries. The ternary diagram represents a three-component system (oil, water, and surfactant in the present case). If the surfactant phase is a mixture of surfactant and co-surfactant, the ternary

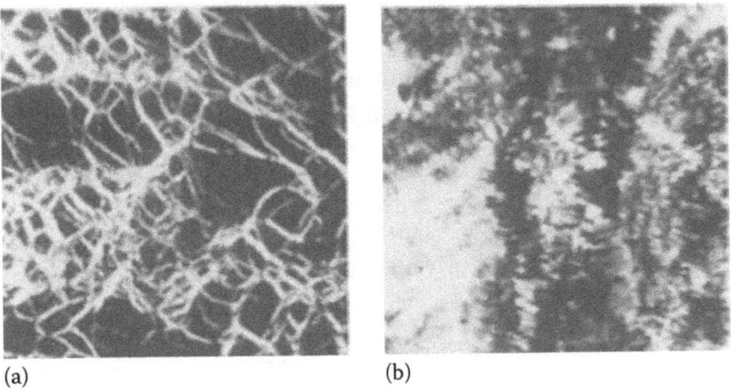

FIGURE 10.6 Liquid crystal structure under crossed polarized microscope. (a) Oil streaks-lamellar LC and (b) angular and striated textures-hexagonal LC. (Adapted from Cistola, D.P. et al., *Biochemistry*, 25(10), 2804, 1986.)

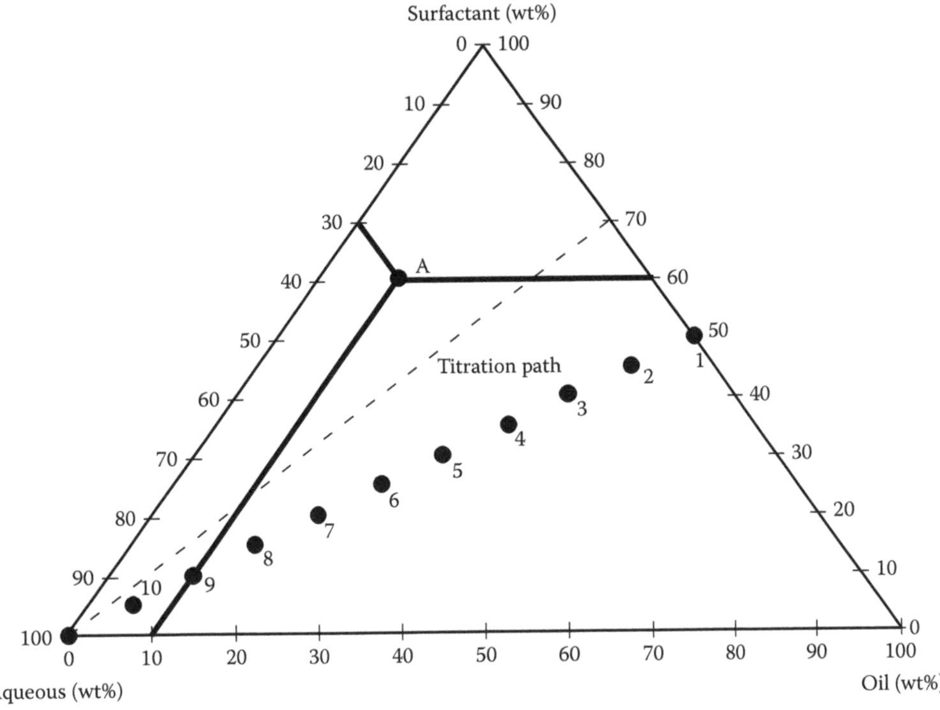

FIGURE 10.7 Ternary diagram. Point A represents 30 wt% of aqueous phase, 10 wt% of oil phase, and 60 wt% of surfactant phase.

diagram is referred to as a *pseudo-ternary* diagram. The ternary diagram can be read following the solid lines shown in the figure. For example, point A corresponds to a composition of 30% water phase, 60% surfactant phase, and 10% oil phase. The phase region to which point A belongs depends on the particle size and the appearance of the sample. In order to mark different phase regions and boundaries on the ternary diagram, a titration technique is employed (Wang and Pal 2014). The titration begins by fixing two components and varying the third component. For example, the dashed line shown in Figure 10.7 is followed with the addition of water. The titration procedure begins with zero loading of aqueous phase and ends up at a point of 100% aqueous phase loading.

TABLE 10.2
An Example of Titrating Water to Surfactant + Oil System (Points Are Shown in Figure 10.7 as Black Dots)

#	Surfactant (wt%)	Oil (wt%)	Water (wt%)
1	50	50	0
2	45	45	10
3	40	40	20
4	35	35	30
5	30	30	40
6	25	25	50
7	20	20	60
8	15	15	70
9	10	10	80
10	5	5	90

The titration procedure is repeated for different ratios of surfactant phase to the oil phase. As an example, Table 10.2 gives the compositions of mixtures for one set of titration experiments represented by black dots shown in Figure 10.7.

Figure 10.8 shows the typical ternary phase diagram with phase regions and boundaries. The phase diagram is developed using the titration method (Prajapati et al. 2012). Three different regions (ME, LC, and EM) are identified on the phase diagram. Titration begins at different surfactant/oil ratios with zero aqueous phase loading. At the start of the titration, type IV oil-continuous microemulsion (W/O) is formed. With continuous addition of water, phase transitions from W/O

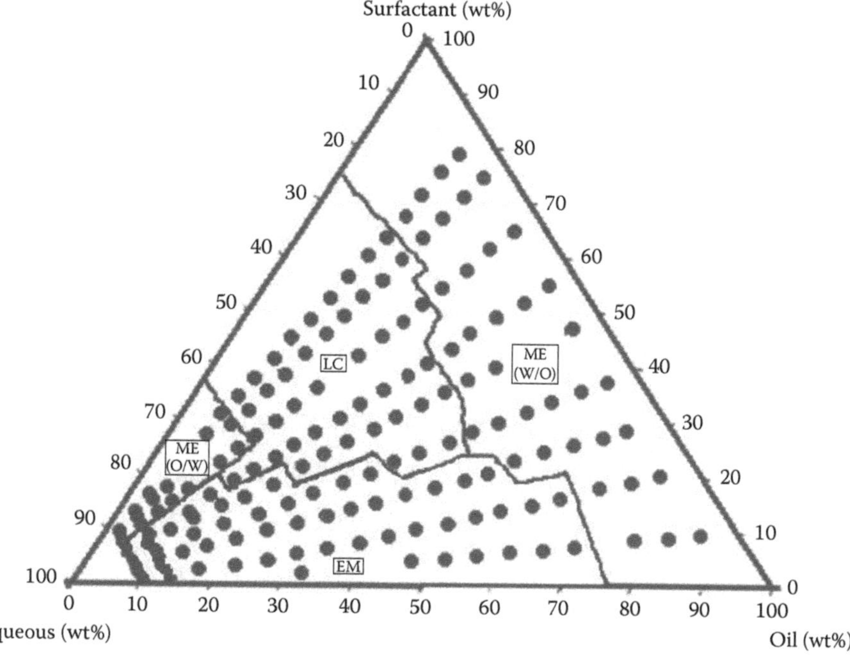

FIGURE 10.8 Ternary phase diagram of water (aqueous phase), PEG-35 castor oil (Cremophor EL, surfactant phase), and mixture of glycerol monocaprylocaprate and caprylic/capric triglycerides (1:3) (oil phase) at 37°C. (Adapted from Prajapati, H.N. et al., *Pharm. Res.*, 29(1), 285, 2012.)

microemulsion to LC and from LC to O/W microemulsion are observed provided that the surfactant/oil ratio is high at the start of the titration (e.g., 80 wt% of surfactant/20 wt% of oil). The continuous addition of water changes the spontaneous curvature of surfactant in oil phase and induces phase inversion from W/O to O/W microemulsion (Fernandez et al. 2004). Note that at a certain water wt%, water and oil phases begin to merge and intertwine together to form LC. This point is called the emulsion inversion point (EIP) and the interfacial tension is minimum at this point (Forgiarini et al. 2001). The formulation ends up with type IV water-continuous microemulsion (O/W) with further addition of water. It should be noted that at low to medium surfactant concentrations (<40 wt% of surfactant at the start of the titration), the formulation may end up in the coarse emulsion (EM) region. Thus, as noted earlier, a certain amount of surfactant (critical microemulsion concentration) is always required to formulate a microemulsion.

10.4 CHALLENGES OF MICROEMULSION-BASED DRUG DELIVERY SYSTEM

10.4.1 Toxicity and Safety of SMEDDS

As a large amount of surfactant is required to form microemulsions, the toxicity of surfactants should be considered in the design and formulation of SMEDDS (Swenson and Curatolo 1992). The presence of a large amount of surfactant can cause irritation or tissue damage as surfactant can disrupt the lipid bilayer of the epithelial cell membrane and interact with the mucosa. For the repeated administration of SMEDDS, a large dosage of surfactant may be given with serious toxicological impact on humans and must be carefully evaluated (Swenson and Curatolo 1992).

Toxicity studies can be divided into two parts: acute oral toxicity and chronic oral toxicity. Swenson et al. (1994) studied the effect of different surfactants on a single pass rat intestinal perfusion system. They uncovered the enhancement ability of drug absorption for different surfactants (Tween 80, bile salts, and sodium dodecyl sulfate) and studied the damage on the intestinal wall resulting from surfactants (Swenson et al. 1994). The studies have shown that the epithelial cells can repair damage upon termination of drug administration. However, long-term effects for repeated drug administration cannot be ignored. A study of chronic oral toxicity is necessary for all the surfactant-containing microemulsion drugs. The study can be executed on a proper animal model using gelatin capsules. The results will reveal relations between the therapeutic effects and the toxicity of a specific surfactant (Constantinides 1995). Extensive research is also needed to reduce the usage of surfactants in drug formulations and maintain the drugs absorption rate at the same time.

10.4.2 Scale-Up and Manufacturing

Compared to the challenge of reducing the drug toxicity, scale-up and manufacturing of SMEDDS is easier. Two important characteristics of SMEDDS, that is, spontaneous formation and thermodynamic stability, are helpful in the scale-up and manufacturing processes. Burskirk et al. (1994) have discussed the general issues related to the SMEDDS manufacturing. Because of the advantages of SMEDDS, the manufacturing process requires only very basic mixing equipment to provide mild agitation to form micelles. The preparation does not require careful in-process control needed in the manufacturing of other drugs (Burskirk et al. 1994). In batch-by-batch manufacturing, the degree of the purity and the chemical instabilities should be monitored carefully. The selection of capsules (soft or hard gelatin capsules), the selection of oil that can maximize drug solubility, and the hygroscopicity of the contents that can either dehydrate or dissolve the gelatin shell are important considerations in the manufacturing of pharmaceuticals (Constantinides 1995). The manufacturing conditions are highly dependent on the nature of the drug. Thus, different drugs should be considered separately to obtain the optimum conditions. Furthermore, the dynamic changes of the drug should be investigated thoroughly before manufacturing the drug.

REFERENCES

Acosta, E.J., A.S. Bhakta. 2009. The HLD-NAC model for mixtures of ionic and non-ionic surfactants. *J Surfactants Deterg* 12(1): 7–19.

Acosta, E.J., T. Nguyen, A. Witthayapanyanon, J.H. Harwell, D.A. Sabatini. 2005. Linker-based bio-compatible microemulsions. *Environ Sci Technol* 39(5): 1275–1282.

Acosta, E.J., J.S. Yuan, A.S. Bhakta. 2008. The characteristic curvature of ionic surfactants. *J Surfactants Deterg* 11: 145–158.

Aveyard, R., B.P. Binks, P.D.I. Fletcher. 1989. Interfacial tensions and aggregate structure in pentaethylene glycol monododecyl ether/oil/water microemulsion systems. *Langmuir* 5: 1210–1217.

Bauer, K., C. Neuber, A. Schmid, K.M. Volker. 2002. Oil in water microemulsion, US Patent 6426078-B1.

Bavière, M., R.S. Schechter, W.H. Wade. 1981. The influence of alcohol on microemulsion composition. *J Colloid Interface Sci.* 81: 266–279.

Benameur, H., V. Jannin, D. Roulot. 2003. Method and formulation for decreasing statin metabolism, US Patent 6652865-B2.

Burnside, B.A., C.E. Mattes, C.M. McGuinness, E.M. Rudnic, G.W. Belendiuk. 1999. Oral acyclovir delivery, US Patent 5883103-A.

Burskirk, G.A., V.P. Shah, D. Adair. 1994. Workshop III report: Scale-up of liquid and semisolid disperse systems. *Eur J Pharm Biopharm* 40: 251–254.

Chen, S., J.A. Gunn. 2003. Oral dosage self-emulsifying formulations of pyranone protease inhibitors, US Patent 6555558-B2.

Cistola, D.P., D. Atkinson, J.A. Hamilton, D.M. Small. 1986. Phase behavior and bilayer properties of fatty acids: Hydrated 1:1 acid-soaps. *Biochemistry* 25(10): 2804–2812.

Constantinides, P.P. 1995. Lipid microemulsions for improving drug dissolution and oral absorption: Physical and biopharmaceutical aspects. *Pharm Res* 12: 1561–1572.

Date, A.A., M.S. Nagarsenker. 2010. Design and evaluation of microemulsions for improved parenteral delivery of propofol. *AAPS PharmSciTech* 9(1): 138–145.

Eugster, C., C.H. Eugster, W. Haldemann, G. Rivara. 1996. Spontaneously dispersible concentrates and aqueous microemulsions with steryl retinates having anti-tumor activity, US Patent 5496813-A.

Farah, N., J. Denis. 2000. Orally administrable composition capable of providing enhanced bioavailability when ingested, US Patent 6054136-A.

Fernandez, P., V. André, J. Rieger, A. Kühnle. 2004. Nano-emulsion formation by emulsion phase inversion. *Colloids Surf A* 251(1): 53–58.

Forgiarini, A., J. Esquena, C. Gonzalez, C. Solans. 2001. Formation of nano-emulsions by low-energy emulsification methods at constant temperature. *Langmuir* 17: 2076–2083.

Fricker, G., B. Haeberlin, A. Meinzer, J. Vonderscher. 2006. Galenical formulations, US Patent 7025975-B2.

Gao, P., W. Morozowich. 2003. Self-emulsifying formulation for lipophilic compounds, US Patent 6531139-B1.

Gao, P., B.D. Rush, W.P. Pfund et al. 2003. Development of a supersaturable SEDDS (S-SEDDS) formulation of paclitaxel with improved oral bioavailability. *J Pharm Sci* 92(12): 2386–2398.

Gullapalli, R.P. 2010. Review-soft gelatin capsules (softgels). *J Pharm Sci* 99: 4107–4148.

Gumkowski, M.J., L. Franco, S.B. Murdande, M.E. Perlman. 2005. Self-emulsifying formulations of cholesteryl ester transfer protein inhibitors, US Patent 6962931-B2.

Gupta, R.R., S.K. Jain, M. Varshney. 2005. AOT water-in-oil microemulsions as a penetration enhancer in transdermal drug delivery of 5-fluorouracil. *Colloids Surf B* 41: 25–32.

Ho, H.O., C.C. Hsiao, M.T. Sheu. 1996. Preparation of microemulsions using polyglycerol fatty acid esters as surfactant for the delivery of protein drugs. *J Pharm Sci* 85(2): 138–143.

Jones, S.C., K.D. Dreher. 1976. Co-surfactant in micellar systems used for tertiary oil recovery. *Soc Petrol Eng J* 16: 161–167.

Kang, B.K., J.S. Lee, S.K. Chon et al. 2004. Development of self-microemulsifying drug delivery systems (SMEDDS) for oral bioavailability enhancement of simvastatin in beagle dogs. *Int J Pharm* 274: 65–73.

Kik, R.A., J.M. Kleijn, F.A.M. Leermakers. 2005. Bending moduli and spontaneous curvature of the monolayer in a surfactant bilayer. *J Phys Chem B* 109: 14251–14256.

Komesvarakul, N., M.D. Sanders, E. Szekeres et al. 2006. Microemulsions of triglyceride-based oils: The effect of co-oil and salinity on phase diagrams. *J Cosmet Sci* 55: 309–325.

Kovarik, J.M., E.A. Mueller, J.B. Van Bree, W. Tetzloff, K. Kutz. 1994. Reduced inter and intra intraindividual variability in cyclosporine pharmacokinetics from a microemulsion formulation. *J Pharm Sci* 83: 444–446.

Krauel, K., N.M. Davies, S. Hook, T. Rades. 2005. Using different structure types of microemulsions for the preparation of poly(alkylcyanoacrylate) nanoparticles by interfacial polymerization. *J Control Release* 106(1–2): 76–87.

Lawrence, M.J., G.D. Rees. 2000. Microemulsion-based media as novel drug delivery systems. *Adv Drug Deliv Rev* 45: 89–121.

Li, P., A. Ghosh, R.F. Wagner, S. Krill, Y.M. Joshi, A.T.M. Serajuddin. 2005. Effect of combined use of nonionic surfactant on formation of oil-in-water microemulsions. *Int J Pharm* 288: 27–34.

Liang, L., A.H. Shojaei, S.A. Ibrahim, B.A. Burnside. 2006. Self-emulsifying formulations of fenofibrate and/or fenofibrate derivatives with improved oral bioavailability and/or reduced food effect, US Patent 7022337-B2.

Ma, H., M. Luo, L.L. Dai. 2008. Influences of surfactant and nanoparticle assembly on effective interfacial tensions. *Phys Chem Chem Phys* 10: 2207–2213.

Makoto, T. (ed.). 1998. *Disperse Systems*. Wiley-VCH, Weinheim, Germany, pp. 244–245, ISBN: 3-527-29458-9.

Mason, T.G., J.N. Wilking, K. Meleson, C.B. Chang, S.M. Graves. 2006. Nanoemulsions: Formation, structure, and physical properties. *J Phys Condens Matter* 18: 635–666.

McClements, D.J. 2011. Edible nanoemulsions: Fabrication, properties, and functional performance. *Soft Matter* 7(6): 2297–2316.

McClements, D.J. 2012. Nanoemulsions versus microemulsions: Terminology, differences, and similarities. *Soft Matter* 8: 1719–1729.

Mishra, A., I. Moussa, Z. Ramtoola, N. Clarke. 2001. Pharmaceutical compositions containing an omega-3 fatty acid oil, US Patent 6284268-B1.

Missen, R.W., C.A. Mims, B.A. Saville. 1999. *Introduction to Chemical Reaction Engineering and Kinetics*. John Wiley & Sons, Inc., New York, p. 44.

Mulye, N. 2002. Self-emulsifying compositions for drugs poorly soluble in water, US Patent 6436430-B1.

Narang, A.S., D. Delmarre, D. Gao. 2007. Stable drug encapsulation in micelles and microemulsions. *Int J Pharm* 345: 9–25.

Nilsson, P.G., B. Lindman. 1983. Water self-diffusion in nonionic surfactant solutions. Hydration and obstruction effects. *J Phys Chem* 87(23): 4756–4761.

Ontiveros, J.F., C. Pierlot, M. Catté et al. 2013. Classification of ester oils according to their equivalent alkane carbon number (EACN) and asymmetry of fish diagrams of C10E4/ester oil/water systems. *J Colloid Interface Sci* 403: 67–76.

Pasquali, R.C., M.P. Taurozzi, C. Bregni. 2008. Some considerations about the hydrophilic-lipophilic balance system. *Int J Pharm* 356(1): 44–51.

Prajapati, H.N., D.M. Dalrymple, A.T. Serajuddin. 2012. A comparative evaluation of mono-, di-and triglyceride of medium chain fatty acids by lipid/surfactant/water phase diagram, solubility determination and dispersion testing for application in pharmaceutical dosage form development. *Pharm Res* 29(1): 285–305.

Rosen, M.J. 2004. *Surfactants and Interfacial Phenomena*, 3rd edn. John Wiley & Sons, Inc., New York, pp. 1–33, 110–113, 208–234.

Rudnic, E., J. McCarty, B. Burnside, C. McGuinness, G. Belenduik. 1999. Emulsified drug delivery systems, US Patent 5952004-A.

Salager, J.L., R.E. Antón, D.A. Sabatini, J.H. Harwell, E.J. Acosta, L.I. Tolosa. 2005. Enhancing solubilization in microemulsions-state of the art and current trends. *J Surfactants Deterg* 8(1): 3–21.

Salager, J.L., N. Marquez, A. Graciaa, J. Lachaise. 2000. Partitioning of ethoxylated octylphenol surfactants in microemulsion-oil-water systems: Influence of temperature and relation between partitioning coefficient and physicochemical formulation. *Langmuir* 16(13): 5534–5539.

Salager, J.L., M. Minana-Perez, M. Perez-Sanchez, C.I. Rojas. 1983. Surfactant-oil-water systems near the affinity inversion part III: The two kinds of emulsion inversion. *J Disp Sci Technol* 4(3): 313–329.

Shen, H., M. Zhong. 2006. Preparation and evaluation of self-microemulsifying drug delivery systems (SMEDDS) containing atorvastatin. *J Pharm Pharmacol* 58(9): 1183–1191.

Spernath, A., A. Yaghmur, A. Aserin, R.E. Hoffman, N. Garti. 2002. Food-grade microemulsions based on nonionic emulsifiers: Media to enhance lycopene solubilization. *J Agric Food Chem* 50: 6917–6922.

Spernath, A., A. Yaghmur, A. Aserin, R.E. Hoffman, N. Garti. 2003. Self-diffusion nuclear magnetic resonance, microstructure transitions, and solubilization capacity of phytosterols and cholesterol in Winsor IV food-grade microemulsions. *J Agric Food Chem* 51: 2359–2364.

Srinivasan, V., D. Blankschtein. 2003. Effect of counterion binding on micellar solution behavior: 2. Prediction of micellar solution properties of ionic surfactant-electrolyte systems. *Langmuir* 19(23): 9946–9961.

Swenson, E.S., W.J. Curatolo. 1992. Means to enhance penetration: (2) Intestinal permeability enhancement for proteins, peptides and other polar drugs: Mechanisms and potential toxicity. *Adv Drug Deliv Rev* 8: 39–92.

Swenson, E.S., W.B. Milisen, W. Curatolo. 1994. Intestinal permeability enhancement: Efficacy, acute local toxicity and reversibility. *Pharm Res* 11: 1132–1142.

Talegaonkar, S., A. Azeem, F.J. Ahmad et al. 2008. Microemulsions: A novel approach to enhanced drug delivery. *Recent Pat Drug Deliv Formul* 2: 238–257.

Upadhyaya, A., E.J. Acosta, J.F. Scamehorn, D.A. Sabatini. 2006. Microemulsion phase behavior of anionic-cationic surfactant mixtures: Effect of tail branching. *J Surfactants Deterg* 9(2): 169–179.

Von Corswant, C. 2003. Microemulsions for use as vehicles for administration of active compounds, US Patent 6602511-B2.

Von Corswant, C., P. Thoren, S. Engstrom. 1998. Triglyceride based microemulsion for intravenous administration of sparingly soluble substances. *J Pharm Sci* 87: 200–208.

Wang, Z., E. Acosta. 2013. Formulation design for target delivery of iron nanoparticles to TCE zones. *J Contam Hydrol* 155: 9–19.

Wang, Z., R. Pal. 2014. Enlargement of nanoemulsion region in pseudo-ternary mixing diagrams for a drug delivery system. *J Surfactants Deterg* 17: 49–58.

Ward, M.V., R. Cotter. 1987. Rapid acting intravenous emulsions of omega-3 fatty acid esters, US Patent 4678808-A.

Watanabe, K., A. Noda, M. Masuda, T. Kimura, K. Komatsu, K. Nakamura. 2004. Bicontinuous microemulsion type cleansing containing silicone oil. II. Characterization of the solution and its application to cleansing agent. *J Oleo Sci* 53(11): 547–555.

Winsor, P.A. 1948. Hydrotropy, solubilisation and related emulsification processes. *Trans Faraday Soc* 44: 376–398.

11 Adhesion and Wetting
A Quantum Mechanics–Based Approach

Costas G. Panayiotou

CONTENTS

11.1 INTRODUCTION

Adhesion and wetting of solid surfaces are major fields in interface science with a remarkable range of technological applications. In spite their importance, their precise determination still remains a challenge due, primarily, to the challenges associated with the practical implementation of an operational definition of the surface tension, γ_S, of solids. The latter is often determined from extrapolations from the melt state and, most commonly, indirectly from the contact angle of liquid drops deposited on the surface and the application of a model/equation for the interfacial tension, γ_{SL}, with the liquid. In general, the work of adhesion of two unit surfaces or the negative of the surface free energy change upon the formation of a unit interface *ij* from the component unit surfaces *i* and *j* is obtained from the Dupre equation [1,2]:

$$W_{ij}^{adh} = \gamma_i + \gamma_j - \gamma_{ij} = -\Delta G_{ij}^{s} \tag{11.1}$$

The work of adhesion of a solid with a liquid, forming a contact angle θ with the solid surface, is given by the classical Young equation [3]:

$$W_{SL}^{adh} = \gamma_L(1 + \cos\theta) \tag{11.2}$$

As seen, a combination of Equations 11.1 and 11.2 does not give the interfacial tension unless we know the surface tension of the solid surface, and vice versa. The use of additional liquids will provide for more equations, but one unknown will always be left undetermined. Thus, it is essential to have an additional model/equation for the interfacial tension or the work of adhesion.

One of the most, if not the most, widely used equations for the interfacial tension is the van Oss–Chaudhury–Good (vOCG) equation [4–6]:

$$\gamma_{SL} = \left(\sqrt{\gamma_S^{LW}} - \sqrt{\gamma_L^{LW}}\right)^2 + 2\left(\sqrt{\gamma_S^A} - \sqrt{\gamma_L^A}\right)\left(\sqrt{\gamma_S^B} - \sqrt{\gamma_L^B}\right) \tag{11.3}$$

or

$$W_{SL}^{adh} = 2\left[\sqrt{\gamma_S^{LW}\gamma_L^{LW}} + \sqrt{\gamma_S^A\gamma_L^B} + \sqrt{\gamma_S^B\gamma_L^A}\right] \tag{11.4}$$

Superscript *LW* in these equations stands for Lifshitz–van der Waals and indicates that in the vOCG model this term encompasses the contributions from the weak London dispersion interactions, as well as the Debye dipole–induced dipole and the Keesom dipole–dipole orientation interactions. Superscripts *A* and *B* stand for acid and base, respectively, and indicate that Lewis acid–base interactions are also considered in the vOCG model but via the asymmetric product of the second term on the rhs of Equation 11.3 and not by a quadratic always positive term like the first term. This model gives the working Equations 11.3 and 11.4 but does not give equations for the separate *LW* and *A* or *B* surface tension components of the compounds in contact. The latter are, in practice, determined by a simultaneous multiparameter fit to experimental contact angle data for various liquid–solid pairs or to experimental interfacial tension data for liquid–liquid interfaces. Through this multifitting process it is attempted to obtain unique values for the surface tension components valid for all kinds of interfaces, including pure liquid–vapor or pure solid–vacuum (or air) interfaces. This results, very often, to very high values of base surface tension components, which are difficult to reconcile with our current understanding of intermolecular interactions. Numerous corrective attempts have been made in the literature [7–9] for reducing the values of these base components but the problem still remains. An often common element to these corrective attempts is the drastic increase of the acid surface tension component of water and the concomitant reduction of the base component, which were equal in the original vOCG model [4–6]. In addition to the preceding problems, there is experimental evidence (and, thus, criticism of the vOCG model for not accounting) for *polarization* of the solid surface by the contacting liquid, that is, for the change of the surface tension components of solids in response to the interaction with (some) contacting liquids [10,11].

This chapter presents a quantum mechanics–based alternative to the vOCG approach whose predictive capacity offers a new way of addressing the previously mentioned controversial interfacial issues. The new approach is built upon the recently introduced molecular descriptors called partial solvation parameters (PSP) [12–16]. The rationale and the working equations of partial solvation parameters are described in this series of recent papers [12–16] where the reader is referred to for the details. A short review is presented in Chapter 3.

Briefly, the PSP approach heavily resides on the quantum mechanics–based COSMO-RS theory of solutions [17–22]. The COSMO model belongs to the class of continuum solvation models (CSM) of quantum mechanics. For the solvation picture, it considers the molecule embedded in a conductor of infinite permittivity that screens perfectly the molecular charges on the surface of its molecular cavity. This molecular cavity is characterized by a volume, V_{cosm}, and a molecular surface area, A_{cosm}. The crucial information is contained in the so-called COSMO file of each compound obtained from quantum chemical calculations at various levels of theory. COSMO files give the detailed surface charge distribution or the σ-profile of each molecule. The σ-profile may be analyzed into its moments of various orders, known as COSMOments, out of which a large number of properties may be calculated, among them the molecular descriptors of Abraham's QSPR/LSER model [23,24].

There are two PSP schemes, the s and the σ-scheme. The two schemes have identical hydrogen bonding parameters but they differ in the way they partition the non-hydrogen-bonding interactions of the molecule. In the s-scheme this partitioning leads to the dispersion, s_d, and polar, s_p PSP, which are equivalent to the more familiar Hansen's dispersion solubility parameter, δ_d, and polar solubility parameter, δ_p, respectively [25]. In the σ-scheme the partitioning leads to the van der Waals, σ_W, and polarity/polarizability, σ_{pz} PSP. The van der Waals PSP is, simply, the weak van der Waals energy density:

$$\sigma_W = \sqrt{\frac{E_{vdW}}{V_{mol}}} \tag{11.5}$$

where
E_{vdW} is the weak van der Waals molar energy
V_{mol} is the molar volume of the compound

Regardless of the partitioning, energy balance implies the following equation:

$$\delta_d^2 + \delta_p^2 = s_d^2 + s_p^2 = \sigma_W^2 + \sigma_{pz}^2 \tag{11.6}$$

Although Equation 11.6 entails a slight recalculation of partial solubility parameters, it is useful in orienting the reader on the physical content of each quantity and in obtaining σ_{pz} from a simple subtraction. The thus calculated σ_{pz} PSP is close to the dielectric PSP obtained from the equation:

$$\sigma_{diel} = \sqrt{\frac{E_{diel}}{V_{mol}}} \tag{11.7}$$

where E_{diel} is the dielectric molar energy of the compound. Both, E_{vdW} and E_{diel}, are directly available from COSMO-RS theory [19], which is the basic source of information for getting PSPs. The name, *van der Waals*, has been retained for the corresponding, σ_W, PSP although it clearly accounts for the atom-specific weak dispersive or London interactions. Similarly, the name, *dispersion* PSP, has been retained for s_d in order to specify the equivalence with the more familiar partial solubility parameter, δ_d. This is not a very accurate terminology, however, since s_d (and δ_d) accounts for the London as well as the Debye induction forces. On the other hand, s_p (and δ_p) account for permanent dipole–dipole (Keesom orientation) interactions and is equal to zero for compounds with zero dipole moment. Thus, σ_{pz} accounts for, both, Keesom and Debye interactions.

In the PSP approach the compounds are divided into two major classes, the homosolvated and the heterosolvated. Heterosolvated are the compounds that can hydrogen bond only with another (heteron, in Greek) different compound, that is, they cannot self-associate. All other compounds are homosolvated. The hydrogen bonding PSPs, σ_{Ga} for the proton donor (electron acceptor or acidic) and σ_{Gb} for the proton acceptor (electron donor or basic) capacity of the compound, are obtained from the corresponding Abraham's LSER descriptors, A and B, which are in turn obtained from the COSMOments of the molecule [19]. For water, $A = 0.676$ and $B = 0.663$ or $A = 1.0198B$. The PSP approach considers perfect neutrality for water and this requires a slight shift of the B scale by ca. 2%. By setting $m\text{-}SUM = A + 1.0198B$, the following defining equations for our hydrogen bonding PSPs are obtained:

$$\sigma_{Ghb}^2 = \sigma_{Ga}^2 + \sigma_{Gb}^2 \tag{11.8}$$

$$\frac{\sigma_{Ga}^2}{\sigma_{Ghb}^2} = 1 - \frac{\sigma_{Gb}^2}{\sigma_{Ghb}^2} = \frac{A}{m\text{-}SUM} \tag{11.9}$$

$$\sigma_{Ghb} = 5.5\sqrt{\frac{m\text{-}SUM}{V_{\text{cosm}}}} \tag{11.10}$$

It is clear from the above description that PSPs may be obtained in a straightforward manner as long as the COSMO file of the molecule is available. COSMO files are already available for thousands of compounds in free or commercial databases. They may also be obtained directly from quantum chemical calculation suites such as the TURBOMOLE [26] or the DMol^3 incorporated in the MaterialStudio suite of Accelrys® and other suites. The PSPs for a number of common solvents are reported in Table 11.1. The PSPs for high polymers cannot be obtained at present from quantum chemical calculations due to their enormous number of conformers and to cpu time limitations. They may be obtained, however, from inverse-gas chromatography (IGC) and other experimental methods [15,16]. It should be made clear that PSPs heavily depend on COSMO-RS model. Any changes in COSMO-RS parameterization or any miscalculations of LSER descriptors in the multilinear regressions with COSMOments will have an impact on PSPs. However, in Reference 14, the PSP approach has been developed into a full activity-coefficient model for concentrated as well as infinitely dilute systems. In this development, the free energy change upon formation of a hydrogen bond between acidic molecule i and basic molecule j is given by the universal equation

$$G_{ab,ij}^{H} = -1.70\sigma_{Ga,i}\sigma_{Gb,j}\sqrt{V_{mol,i}V_{mol,j}} + 2.50 \quad \text{(kJ/mol)} \tag{11.11}$$

in direct analogy with the QSPR/LSER model [23]. Thus, PSPs may be checked now against experimental data on activity coefficients and derived thermodynamic quantities.

With this introduction and the tables with PSPs we may now proceed to the presentation of the new approach to interfacial energies, wetting, and adhesion. This is done in the next section while examples of calculations are shown in Section 11.3. A preliminary version of the new approach is presented in the Appendix of Reference 15.

11.2 MOLECULAR SURFACE TENSION COMPONENTS

The central task and focus in this section is on a one-to-one mapping of PSPs onto the corresponding surface tension components of the molecule. In other words, the molecular descriptors that were designed for properties of bulk phases will now lead to their counterparts for properties of surfaces and interfaces. This is by no means a simple or straightforward task and it will necessarily involve a number of approximations.

For our purpose, we should recall a key distinction between the weak van der Waals or London dispersive intermolecular interactions, reflected by σ_W, and all other intermolecular interactions: Only the first are atom-specific interactions with practically no orientational or directional character. This is crucial when discussing surface properties and interfacial phenomena. A polar molecule, such as ethanol, will exhibit some kind of preferential orientation on the surface and will find itself in a density gradient in order to minimize its surface free energy: Most likely, ethanol molecule will orient its hydroxyl toward the bulk, where there is higher probability to find another hydroxyl and form a hydrogen bond with it. Quantum chemical calculations for molecules able to form inter- and intra-molecular hydrogen bonds, such as ethylene–glycol and glycerol, indicate that their conformers with cooperative intramolecular hydrogen bonds are most likely the most stable ones in the gas phase. In contrast, their more open conformations shown in Figure 11.1 are most likely the most

TABLE 11.1
Partial Solvation Parameters (PSP) of Pure Compounds

Part A: The PSPs for Non-hydrogen-Bonded Compounds

Compound	σ_W (MPa$^{1/2}$)	σ_{pz} (MPa$^{1/2}$)	Compound	σ_W (MPa$^{1/2}$)	σ_{pz} (MPa$^{1/2}$)
n-Pentane	13.65	5.00	Cyclohexane	13.85	8.35
n-Hexane	13.75	6.50	Cycloheptane	13.85	8.55
n-Heptane	13.75	7.00	1-Pentene	13.80	4.35
n-Octane	13.75	7.00	1-Hexene	13.85	5.75
n-Nonane	13.75	7.50	Benzene	14.84	11.43
n-Decane	13.75	7.50	Toluene	14.59	11.09
n-Undecane	13.75	7.50	Ethylbenzene	14.41	10.77
n-Dodecane	13.75	8.00	*n*-Propylbenzene	14.30	10.36
n-Hexadecane	13.75	8.00	*n*-Butylbenzene	14.24	10.17
n-Octadecane	13.75	8.00	*n*-Pentylbenzene	14.17	10.22
n-Eicosane	13.75	8.50	*n*-Hexylbenzene	14.13	10.20
n-Tetracosane	13.75	8.50	*m*-Xylene	14.40	10.89
n-Octacosane	13.75	8.50	*o*-Xylene	14.40	11.45
n-Dotriacontane	13.75	8.50	*p*-Xylene	14.38	10.67
Isobutane	13.17	2.90	Styrene	14.70	12.06
Isopentane	13.37	3.65	Naphthalene	14.43	13.04
2-Methylpentane	13.48	5.10	Carbon tetrachloride	16.97	6.06
3-Methylpentane	13.48	5.80	Methyl chloride	16.92	10.14
2.3-Dimethylpentane	13.37	6.40	Monochlorobenzene	15.51	11.57
2.3.4-Trimethylpentane	13.22	6.95	Benzyl chloride	15.45	12.89
Cyclopentane	14.10	8.05	Carbon disulfide	18.42	8.77

Part B: The PSPs for Hydrogen-Bonded Compounds

Compound	σ_W	σ_{pz}	σ_{Gb}	σ_{Ga}
Homosolvated				
Methanol	15.18	10.62	18.34	15.01
Ethanol	14.65	10.92	15.12	11.28
1-Propanol	14.45	10.96	13.17	9.87
1-Butanol	14.30	11.33	11.91	8.74
1-Pentanol	14.23	10.46	11.19	8.59
1-Hexanol	14.16	9.79	10.51	7.96
1-Heptanol	14.10	10.42	9.95	7.41
1-Octanol	14.07	10.23	9.49	6.96
1-Nonanol	14.03	10.96	9.10	6.58
1-Decanol	13.98	10.82	8.83	6.27
Isopropanol	14.23	10.03	13.39	9.57
2-Butanol	14.23	9.24	11.39	8.43
2-Methyl-1-propanol	14.11	7.76	11.21	9.24
2-Methyl-2-propanol	13.87	8.81	11.82	7.75
2-Pentanol	14.07	10.29	10.61	7.73
3-Pentanol	14.07	10.30	10.27	6.83
Cyclohexanol	14.10	14.04	10.91	7.42
Phenol	14.95	13.93	8.35	11.89
Benzyl alcohol	14.84	15.44	10.63	8.07
m-Cresol	14.73	13.63	8.07	10.73
o-Cresol	14.76	12.21	7.71	10.62

(Continued)

TABLE 11.1 (*Continued*)
Partial Solvation Parameters (PSP) of Pure Compounds

Part B: The PSPs for Hydrogen-Bonded Compounds

Compound	σ_W	σ_{pz}	σ_{Gb}	σ_{Ga}
p-Cresol	14.67	14.18	8.09	10.70
4-Hydroxystyrene	14.03	11.57	7.69	10.67
Ethylene glycol	15.16	20.36	17.51	13.94
1.2-Propylene glycol	14.67	17.76	15.19	11.30
1.3-Propylene glycol	14.65	18.61	16.02	11.62
Glycerol	14.56	24.13	16.59	11.38
Diethylene glycol	14.54	17.36	15.56	7.73
Triethylene glycol	14.46	13.57	15.28	6.70
Tetraethylene glycol	14.37	11.67	15.24	6.05
Formic acid	15.06	8.78	13.90	20.34
Acetic acid	14.27	9.01	13.84	15.55
Propionic acid	14.03	8.38	12.19	13.48
n-Butyric acid	13.90	9.41	11.07	12.09
n-Pentanoic acid	13.86	7.95	10.30	11.05
n-Hexanoic acid	13.81	9.85	9.69	10.25
Methylamine	15.01	4.33	20.55	8.85
Ethylamine	14.24	8.32	17.20	5.70
n-Propylamine	14.07	9.17	15.15	4.35
n-Butylamine	14.05	9.54	13.75	4.00
n-Pentylamine	13.95	8.57	12.15	3.95
n-Hexylamine	13.90	8.36	11.95	3.00
Dimethylamine	14.41	7.48	16.50	5.50
Diethylamine	14.02	6.45	12.55	3.05
DI-*n*-Propylamine	13.93	6.99	10.05	2.75
DI-*n*-Butylamine	13.86	8.27	9.15	2.60
Formamide	15.31	21.85	20.29	18.3
Water	16.26	15.40	28.28	28.28
Hydrogen peroxide	15.80	24.62	17.25	28.70
Heterosolvated				
Acetone	14.10	10.05[1]	10.25	0.00
Methyl ethyl ketone	13.98	8.65	8.65	0.00
2-Pentanone	13.91	8.90	7.95	0.00
Acetaldehyde	14.33	10.25	12.00	0.00
1-Propanal	14.11	10.10	12.50	0.00
1-Butanal	14.00	9.60	11.00	0.00
1-Pentanal	13.93	9.65	8.90	0.00
Ethyl acetate	13.85	8.20	8.15	0.00
n-Propyl acetate	13.80	8.05	8.00	0.00
n-Butyl acetate	13.78	8.45	8.05	0.00
Methyl propionate	13.93	9.25	8.10	0.00
Ethyl propionate	13.79	7.73	7.95	0.00
Propyl propionate	1.01	8.07	7.25[2]	0.00
Methyl isobutyrate	1.01	8.17	8.40[2]	0.00
Butyl isobutyrate	1.02	6.46	8.00[2]	0.00
Dimethyl ether	13.81	2.10	9.80	0.00

(*Continued*)

TABLE 11.1 (*Continued*)
Partial Solvation Parameters (PSP) of Pure Compounds

Part B: The PSPs for Hydrogen-Bonded Compounds

Compound	σ_W	σ_{pz}	σ_{Gb}	σ_{Ga}
Diethyl ether	1.11	3.80	8.70	0.00
DI-*n*-propyl ether	13.76	4.50	6.70	0.00
DI-*n*-butyl ether	1.04	5.25	7.35	0.00
DI-*n*-octyl ether	1.00	7.40	4.60	0.00
Tetrahydrofuran	14.29	8.50	10.25	0.00
Chloroform	17.25	3.20	0.00	7.90
Vinyl chloride	1.00	3.20	0.00	3.55
Trimethylamine	13.80	2.75	11.20	0.00
Triethylamine	13.46	4.15	9.25	0.00
Dimethyl sulfoxide	15.46	9.17	19.52	0.5

stable and preferred ones in the liquid state, not only for entropic reasons but, mainly, because they allow for a more efficient and extensive intermolecular hydrogen bonding network. Across the interface, the molecules will assume various conformations and orientations between those shown in Figure 11.1 that will ultimately lead to surface free energy minimization.

Molecules like ethanol or acetone and the higher member of their homologous series may preferably orient themselves and keep away from the liquid–vapor interface their polar sites. In this way, these molecules (and the majority of polar molecules) will exhibit a polar character on the

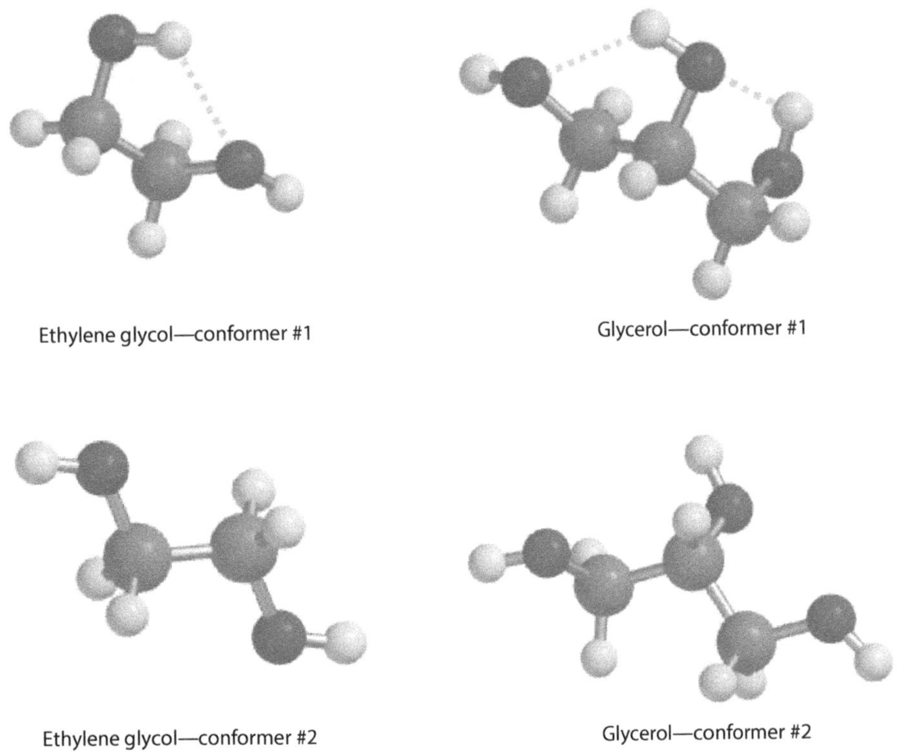

FIGURE 11.1 Conformers of ethylene glycol and glycerol with and without intramolecular hydrogen bonds (shown by dashed lines).

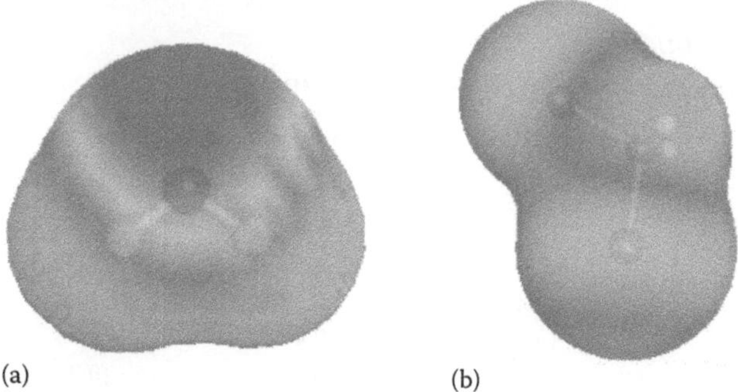

FIGURE 11.2 The COSMO surfaces of (a) water and (b) diiodomethane (DIM). The charge densities vary from highly negative (electron donor or proton acceptor) sites typically marked in deep red over oxygen atom to highly positive (electron acceptor or proton donor) sites marked in deep blue over hydrogen atoms. Intermediate densities are marked with intermediate colors of the visible color spectrum.

interface lower than its corresponding character in the bulk liquid. Such a possibility, however, is not very much available for molecules like water whose COSMO surface is shown in Figure 11.2 with its surface charge density marked in colors. No matter how oriented at the interface, the water molecule will always expose its polar/hydrogen-bonding sites away from the bulk liquid phase. As a consequence, the water molecule is expected to exhibit nearly its full hydrogen bonding capacity at the interface. On the other hand, apolar or less polar molecules, like diiodomethane shown also in Figure 11.2, are relatively more indifferent for specific orientations at interfaces other than those dictated by stereochemical constraints or space filling requirements. Since these two molecules are extensively used in the interface/wetting/adhesion literature, it is worth commenting briefly on their σ-profiles shown in Figure 11.3.

According to COSMO-RS convention, only charge densities in excess of 0.01 e/A^2 in absolute values are capable of participating in hydrogen bonds. As observed in Figure 11.3, the σ-profile of water molecule is nearly symmetric and extended beyond the hydrogen bonding cutoffs on both sides of the profile. This is in contrast to arguments often appearing in the literature for a much more acidic than basic character of water. The nearly symmetric σ-profile of water leads to nearly equal A and B LSER molecular descriptors, as we have seen in the previous section and corroborates our

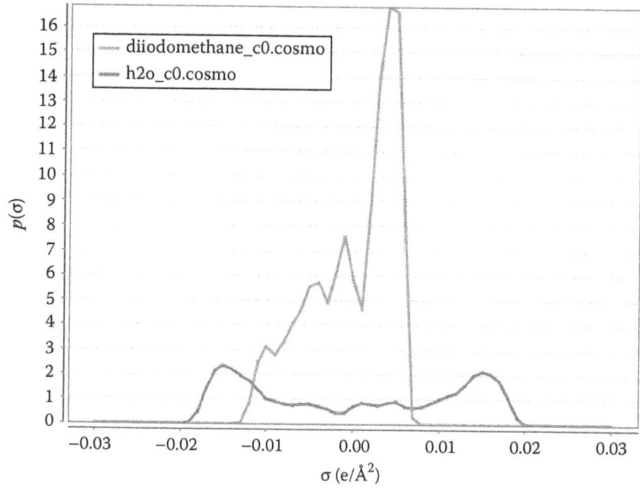

FIGURE 11.3 The sigma profiles of water and diiodomethane. The charge densities refer to the screening charges, that is, to the counter-charges of the real molecular charges.

selection of acidity/basicity scale based on the water neutrality. On the other hand, the σ-profile of the much used diiodomethane is somewhat asymmetric with a minor protrusion into the acidic domain, although its main polarity sites are populating the opposite domain of the profile. In other words, diiodomethane is not an apolar molecule and in no way we may consider it interacting with dispersion forces only.

From the preceding exposition, it becomes clear that preferential orientations and molecular conformations are of central importance at interfaces and ultimately dictate interfacial free energies. This free energy character of interfacial tension makes difficult the mapping of PSPs onto the corresponding surface tension components, γ_W, γ_{pz}, γ_a, and γ_b. We may proceed, however, by confining ourselves, first, to the weakest van der Waals interactions, namely, the London dispersive interactions and the mapping of σ_W into the dispersion or van der Waals surface tension component, γ_W.

From its very definition, σ_W is an energy density (energy per unit volume), while the corresponding surface tension component, γ_W, will be energy per unit area. In a real pure liquid, we could allocate to each molecule an average solvation area and an average solvation volume. The COSMO area, A_{cosm}, and COSMO volume, V_{cosm}, could be considered good measures for this solvation area and the solvation volume, respectively. We could, then, relate the preceding components through the basic defining equations:

$$V_{\text{cosm}}\sigma_W^2 = A_{\text{cosm}}\gamma_W = E_W \quad (11.12)$$

or

$$\gamma_W = \frac{V_{\text{cosm}}}{A_{\text{cosm}}}\sigma_W^2 \quad (11.13)$$

Equation 11.13 is an approximate equation, but it may satisfactorily serve our purpose. Of course, the other components of surface tension cannot be obtained by such a simple equation and, in order to proceed, we should turn to their free energy character. The appropriate equations in this case should resemble Equation 11.11 giving the hydrogen bonding free energy of interacting acidic and basic molecules. Quite generally, this free energy character may simply be expressed as follows:

$$\gamma = \gamma_W + \gamma_{pz} + \gamma_{hb} = \Gamma^h - T\Gamma^s \quad (11.14)$$

The components with superscript h will be referred to as enthalpic or energetic components and those with superscript s as entropic ones. γ_{pz} and γ_{hb} are the surface tension components arising from polarity/refractivity and hydrogen bonding interactions, respectively. Regarding hydrogen bonding quantities of interacting molecules i and j, their relation to PSPs could then be obtained from defining equations of the form (cf. Equations 11.11 and 11.13):

$$\Gamma_{ij}^h = h\sigma_{a,i}\sigma_{b,j}\sqrt{\frac{V_i}{A_i}\frac{V_j}{A_j}} = 2\sqrt{\Gamma_{a,i}^h\Gamma_{b,j}^h} \quad (11.15)$$

and

$$\gamma_{hb,ij} = f\sigma_{Ga,i}\sigma_{Gb,j}\sqrt{\frac{V_i}{A_i}\frac{V_j}{A_j}} + g = 2\sqrt{\gamma_{a,i}\gamma_{b,j}} \quad (11.16)$$

The ratios under the square roots are the solvation volume-to-surface area ratios for each type of interaction. It is not obvious how these ratios could be obtained but we will not need them here. Subscripts *hb*, *a*, and *b* in these equations are referring to the overall hydrogen bonding, to acidic, and to basic components, respectively. For simplicity, in this work, we will consider only one type of acidic and/or basic group per molecule. For pure hydrogen-bonded compounds, then, the defining Equation 11.16 becomes

$$\gamma_{hb} = 2\sqrt{\gamma_a \gamma_b} \tag{11.17}$$

For a molecule with similar acidic and basic character, there is no reason to expect that it will lose this similarity at the interface. For the purposes of this work then and on account of Equation 11.16, we will adopt the following equations:

$$\frac{\gamma_a}{\gamma_b} = \frac{\sigma_{Ga}\sigma_{Ghb}}{\sigma_{Ga}\sigma_{Ghb}} = \frac{\sigma_{Ga}}{\sigma_{Ga}} \tag{11.18}$$

If γ_{hb} were known, Equations 11.17 and 11.18 would form a system of two equations with two unknowns that could be solved for γ_a and γ_b. Even then, however, we would need additional information in order to calculate γ_{pz}. From the preceding discussion, we may expect that, as the molecule at the pure liquid–vapor interface is adopting orientations that will minimize its surface energy, its polar or hydrogen bonding groups are more likely to be oriented toward the bulk liquid phase rather than the interface. As a consequence, this tendency will affect, both, hydrogen bonding and polarity/polarizability components of its surface tension, at least to the extent that high surface charge densities may be counted either as hydrogen bonding sites when interacting with complementary hydrogen bonding sites or as polarity/polarizability sites when interacting with non-hydrogen bonding counterparts [17–21]. In view of this COSMO picture of real solvents and in order to proceed, we will also adopt the following equations:

$$\frac{\gamma_{pz}}{2\gamma_a} = \frac{\sigma_{pz}^2}{u\sigma_{Ga}\sigma_{Ghb}}, \quad \frac{\gamma_{pz}}{2\gamma_b} = \frac{\sigma_{pz}^2}{u\sigma_{Gb}\sigma_{Ghb}} \tag{11.19}$$

or

$$\frac{\gamma_{pz}}{2\sqrt{\gamma_a\gamma_b}} = \frac{\sigma_{pz}^2}{u\sigma_{Ghb}\sqrt{\sigma_{Ga}\sigma_{Gb}}} \tag{11.20}$$

The factor *u* should be, in general, a characteristic property of the molecule and should reflect the peculiarities of the molecular structure and the surface charge distribution that are not accounted for by the PSP terms in the rhs of the equations. This assumption facilitates calculations very much. If the total surface tension, γ, is known, then, Equation 11.14 along with Equations 11.17 through 11.19 form a system of four equations with four unknowns, namely, γ_{pz}, γ_{hb}, γ_a, and γ_b. The solution of this system is facilitated by noting from Equations 11.17 and 11.20 that

$$\gamma_{hb} = 2\sqrt{\gamma_a\gamma_b} = \gamma_{pz} u \frac{\sigma_{Ghb}\sqrt{\sigma_{Ga}\sigma_{Gb}}}{\sigma_{pz}^2} \tag{11.21}$$

Combining with Equation 11.14, we obtain

$$\gamma_{pz} + \gamma_{hb} = \gamma_{pz}\left(1 + u\frac{\sigma_{Ghb}\sqrt{\sigma_{Ga}\sigma_{Gb}}}{\sigma_{pz}^2}\right) = \gamma - \gamma_W \quad \text{or}$$

$$\gamma_{pz} = \frac{\gamma - \gamma_W}{\left(1 + u\left(\sigma_{Ghb}\sqrt{\sigma_{Ga}\sigma_{Gb}}/\sigma_{pz}^2\right)\right)} \tag{11.22}$$

Then, from Equations 11.19 and 11.22 we may obtain γ_a, γ_b, and γ_{hb}. It is worth mentioning that the preceding equations hold for, both, homosolvated and heterosolvated compounds. For the latter, Equation 11.21 gives $\gamma_{hb} = 0$ while Equation 11.22 gives $\gamma_{pz} = \gamma - \gamma_W$. The important point is that the method (Equation 11.19) provides also with γ_a or γ_b components of heterosolvated compounds, for which $\gamma_{hb} = 0$.

The experimental surface tensions of many common liquids are available in the open literature or in critical compilations such as the DIPPR database [27]. From these total surface tensions, one may then obtain the partial surface tensions by using the preceding set of simple equations. Equation 11.13 should be used first in order to obtain γ_W. The value of factor u, however, has not been specified yet.

It is important to point out that for a compound with equal acidic and basic PSPs ($\sigma_{Ga} = \sigma_{Gb}$), the preceding method implies that its corresponding acidic and basic surface tension components will also be equal. This is the case for water in the PSP approach. The COSMO volume and area for water are equal to 0.0256 and 0.4306 nm^2, respectively [19]. Substituting in Equation 11.13, we obtain γ_W = 15.71 mN/m. The DIPPR value for the total surface tension of water at 20°C is 72.8 mN/m. By replacing in Equations 11.19 through 11.22, we obtain the other three surface tension components as functions of factor u. The results are reported in Table 11.2 while in Figure 11.4 is shown the variation of the hydrogen bonding component, γ_α or γ_b, with factor u. In the same figure is also shown that the van Oss [4–6,28] value of 25.5 mN/m for these components corresponds to a value near 2 for the factor u. In order, then, to further simplify our method and have a nearly common reference point with the vOCG approach [4–6,28], we set $u = 2$ and consider it a universal constant applicable to all compounds.

In Table 11.3 are reported the COSMO volumes and areas and the surface tension components for a number of common solvents as calculated by the preceding method. In the last column are reported the total surface tensions as obtained from DIPPR [27]. It should be stressed that the values of the surface tension components in Table 11.3 are predicted ones and not fitted to experimental data other than the overall surface tension of the compound.

TABLE 11.2
The Polar and Hydrogen-Bonding Surface Tension Components of Water for Various Values of the Factor *u* (Equations 11.19 through 11.22)

u	γ_{pz}	γ_a, γ_b	$\gamma_{hb}/\gamma_{total}$
0.75	12.47	22.31	0.61
1	9.90	23.60	0.65
2	**5.42**	**25.84**	**0.71**
3	3.73	26.68	0.73
4	2.84	27.12	0.75

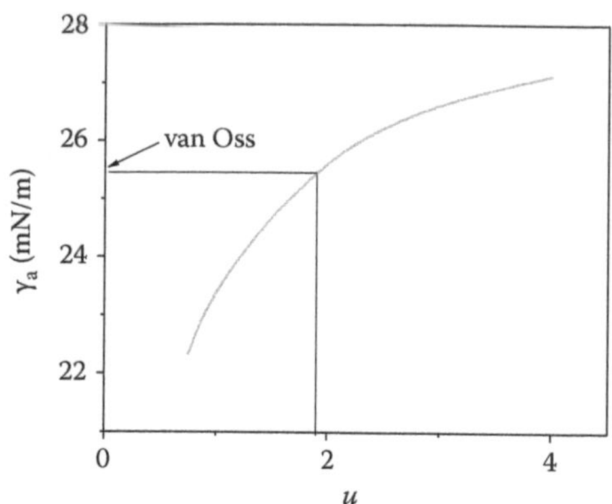

FIGURE 11.4 The HB component of surface tension, γ_a or γ_b, of water as a function of the factor u in the defining Equations 11.19 through 11.22. As shown, the adopted value of 25.5 mN/m by the vOCG approach [4–6,28] corresponds to a value of u near 2.

The major part of the sum of γ_W and γ_{pz}, presumably, corresponds to the widely used Lifshitz–van der Waals surface tension [4–6,28], or

$$\gamma_W + \gamma_{pz} \cong \gamma^{LW} \tag{11.23}$$

The calculations with Equation 11.23 and Table 11.3 are in reasonably good agreement with the reported values for the Lifshitz–van der Waals surface tension component in the literature [4–6,28]. As an example, the γ^{LW} calculated from Equation 11.23 for methanol, ethanol, glycerol, and water are 17.23, 18.03, 31.97, and 21.13 mN/m, respectively, while the corresponding reported values [28] are 18.5, 20.1, 34.0, and 21.8 mN/m, respectively. Having the partial surface tensions of pure solvents, we may now proceed to the next step and propose a method for the surface-tension characterization of polymers and solid surfaces.

The preceding procedure may be applied to polymers and smooth solid surfaces although there are some noticeable difficulties. The problems in this case start with the estimation of γ_W. As already mentioned, it is not possible at present to run quantum chemical calculations for high polymers. The required ratio of COSMO volume-to-surface area of a polymer could be obtained, as an example, from its monomer or its oligomer analogs through the following equation:

$$\left(\frac{V_S}{A_S}\right)_{\text{polymer}} = \left(\frac{V_{\text{cosm}}}{A_{\text{cosm}}}\right)_{\text{monomer}} \left(\frac{r}{q}\right)_{\text{polymer}} \left(\frac{q}{r}\right)_{\text{monomer}} \tag{11.24}$$

q, r being the UNIQUAC/UNIFAC-type parameters [29,30]. The higher the oligomer analog the better is expected to be the approximation with Equation 11.24. When the molecular structures of the solid surface are more complex (branched, cross-linked, dendritic, etc.), Equation 11.24 will be a rather poor approximation. The simplest solution in these cases is the use of one or two experimental data of contact angles with non-hydrogen bonding solvents in order to obtain γ_W and substitute in Equations 11.19 through 11.22 in order to obtain all other surface tension components. However, this requires first an equation for the contact angle or for the work of adhesion or for the interfacial tension. This is done in the next section.

11.2.1 Derivation of the PSP Equation for the Work of Adhesion

The contributions of the various PSP terms in mixing quantities in the case of bulk phases have been discussed in References 14–16. In Reference 14 it was shown that the van der Waals and the

TABLE 11.3
The Partial Surface Tensions of Common Pure Compounds

Compound	V_{cosm} (nm³) [19]	A_{cosm} (nm²) [19]	γ_W (mN/m)	γ_{pz} (mN/m)	γ_b (mN/m)	γ_a (mN/m)	γ (mN/m) [27]
n-Hexane	0.1457	1.569	17.56	0.35	0	0	17.91
n-Heptane	0.1676	1.7689	17.91	1.90	0	0	19.83
n-Octane	0.1895	1.9684	18.21	2.90	0	0	21.13
n-Nonane	0.2113	2.1679	18.46	3.95	0	0	22.40
n-Decane	0.2333	2.3678	18.68	4.70	0	0	23.39
n-Dodecane	0.2768	2.7664	18.97	5.95	0	0	24.93
n-Hexadecane	0.3664	3.5686	19.49	7.65	0	0	27.15
n-Octadecane	0.4089	3.9657	19.49	8.55	0	0	28.01
n-Eicosane	0.4527	4.3648	19.46	9.10	0	0	28.56
n-Docosane	0.4965	4.7636	19.45	9.50	0	0	28.97
n-Tetracosane	0.5403	5.1626	19.42	9.80	0	0	29.29
n-Octacosane	0.6274	5.9555	19.36	10.35	0	0	29.71
Cyclohexane	0.1263	1.3149	18.20	6.45	0	0	24.64
Cycloheptane	0.1498	1.4863	19.25	7.75	0	0	26.99
Benzene	0.1100	1.2137	19.95	8.25	0	0	28.21
Toluene	0.1318	1.4055	19.95	8.00	0	0	27.92
Ethylbenzene	0.1539	1.5860	20.14	8.40	0	0	28.59
n-Propylbenzene	0.1762	1.7865	20.18	8.40	0	0	28.50
n-Butylbenzene	0.1980	1.9860	20.20	8.45	0	0	28.64
Acetaldehyde	0.0643	0.8288	15.93	4.83	6.62	0	20.76
1-Propanal	0.0853	1.0150	16.73	5.23	8.01	0	21.96
1-Butanal	0.1074	1.2153	17.32	7.63	10.02	0	24.95
1-Pentanal	0.1294	1.4194	17.69	7.64	6.50	0	25.33
Acetone	0.0863	1.0268	16.72	6.32	6.57	0	23.04
Methyl ethyl ketone	0.1070	1.2065	17.34	6.62	6.62	0	23.96
Methanol	0.0484	0.6756	16.51	0.72	2.76	2.26	22.22
Ethanol	0.0700	0.8811	17.04	0.99	2.36	1.76	22.10
1-Propanol	0.0918	1.0805	17.04	0.99	2.36	1.76	23.39
1-Butanol	0.1139	1.2814	17.75	1.37	2.47	1.85	24.37
1-Pentanol	0.1355	1.4820	18.18	1.85	2.53	1.86	25.30
1-Hexanol	0.1578	1.6831	18.51	1.92	2.78	2.13	25.90
1-Heptanol	0.1794	1.8818	18.79	2.02	2.92	2.21	26.71
1-Octanol	0.2014	2.0812	18.96	2.62	2.97	2.22	27.10
1-Decanol	0.2449	2.4766	19.15	2.81	3.00	2.20	28.38
2-Pentanol	0.1354	1.4548	18.42	1.55	2.04	1.49	23.45
Cyclohexanol	0.1391	1.4226	19.43	6.35	4.63	3.15	33.42
Ethylene glycol	0.0805	0.9784	18.92	10.82	10.23	8.14	47.99
Glycerol	0.1118	1.2151	19.51	22.46	12.88	8.83	63.30
Tetraethylene glycol	0.2468	2.5551	19.94	8.05	14.77	5.87	46.61
Ethyl acetate	0.1173	1.3316	16.9	6.34	7.33	0.00	23.24
n-Propyl acetate	0.1393	1.5329	17.31	6.55	6.52	0.00	23.86
n-Butyl acetate	0.1611	1.7323	17.65	7.1	5.55	0.00	24.75
Carbon tetrachloride	0.1284	1.3421	26.00	1.00	0	0	27.00
Diiodomethane	0.1164	1.2537	32.71	17.21	0	0	49.92
n-Propylamine	0.0968	1.1245	16.16	1.29	3.67	1.05	21.38
n-Butylamine	0.1186	1.3228	17.2	1.93	4.17	1.21	23.62
Formamide	0.0575	0.7675	17.56	12.74	14.79	13.34	58.40
Dimethyl sulfoxide	0.0988	1.1167	21.14	8.90	40.34	1.03	42.95
Water	0.0256	0.4306	15.71	5.42	25.84	25.84	72.80

polarity/polarizability interactions are contributing with quadratic terms of PSP differences while the hydrogen bonding interactions with logarithmic terms of equilibrium constants, K_{ij}, for the hydrogen bonding between a proton donor (acidic) molecule i and a proton acceptor (basic) molecule j. Assuming analogous contributions at interfaces, Equation 69 of Reference 14 for the residual free energy at infinite dilution per unit volume or the compatibility of components i and j, rewritten here for convenience,

$$\frac{\Delta G_{ij}^{r,\infty}}{V_{mol,i}} = \left\{(\sigma_{Wi} - \sigma_{Wj})^2 + (\sigma_{pzi} - \sigma_{pzj})^2 - \frac{RT}{V_{mol,i}} \ln\left(1 + \frac{a}{r_j} K_{ij}\right)\right\} \quad \text{(69, Reference 14)}$$

could be translated into the following equation for the interfacial tension (or compatibility of surfaces i and j):

$$\gamma_{ij} = \left\{\left(\sqrt{\gamma_{W,i}} - \sqrt{\gamma_{W,j}}\right)^2 + \left(\sqrt{\gamma_{pz,i}} - \sqrt{\gamma_{pz,j}}\right)^2 - \frac{RT}{V_{mol,i}} \ln\left(1 + \frac{a}{r_j} K_{ij}\right)\right\} \quad (11.25)$$

Assuming one interacting site per segment (or, $a/r_j = 1$) and equilibrium constants significantly larger than 1, Equation 11.25 reduces to the following simple equation for cross-associating compounds:

$$\gamma_{ij} = \left\{\left(\sqrt{\gamma_{W,i}} - \sqrt{\gamma_{W,j}}\right)^2 + \left(\sqrt{\gamma_{pz,i}} - \sqrt{\gamma_{pz,j}}\right)^2 - \frac{RT}{V_{mol,i}} \ln(K_{ij})\right\} \quad (11.25a)$$

The corresponding equation for molecules able to self-associate and cross-associate becomes:

$$\gamma_{ij} = \left\{\left(\sqrt{\gamma_{W,i}} - \sqrt{\gamma_{W,j}}\right)^2 + \left(\sqrt{\gamma_{pz,i}} - \sqrt{\gamma_{pz,j}}\right)^2 - \frac{RT}{V_{mol,i}} \ln\left(\frac{K_{ij}}{K_{ii}}\right) - \frac{RT}{V_{mol,j}} \ln\left(\frac{K_{ji}}{K_{jj}}\right)\right\} \quad (11.26)$$

The equilibrium constants K_{ij} are related to hydrogen bonding free energies through the classical equation:

$$-RT \ln K_{ij} = G_{ij}^H \quad (11.27)$$

Substituting from Equation 11.27, Equations 11.25a and 11.26 further simplify to the following forms:

$$\gamma_{ij} = \left\{\left(\sqrt{\gamma_{W,i}} - \sqrt{\gamma_{W,j}}\right)^2 + \left(\sqrt{\gamma_{pz,i}} - \sqrt{\gamma_{pz,j}}\right)^2 + \frac{G_{ij}^H}{V_{mol,i}}\right\} \quad (11.25b)$$

and

$$\gamma_{ij} = \left\{\left(\sqrt{\gamma_{W,i}} - \sqrt{\gamma_{W,j}}\right)^2 + \left(\sqrt{\gamma_{pz,i}} - \sqrt{\gamma_{pz,j}}\right)^2 + \frac{G_{ij}^H - G_{ii}^H}{V_{mol,i}} + \frac{G_{ji}^H - G_{jj}^H}{V_{mol,j}}\right\} \quad (11.26a)$$

The last terms, however, in these equations are energy densities and are more appropriate to PSPs rather than to surface tension components that are energies per unit area. Thus, they also need to be translated into γ terms. This translation is facilitated by the inspection of Equations 11.11 and 11.16 or 11.17, which are the defining equations for the bulk and interface free energy changes upon hydrogen bond formation. Substituting the analogous terms, Equations 11.25b and 11.26a become

$$\gamma_{ij} = \left\{\left(\sqrt{\gamma_{W,i}} - \sqrt{\gamma_{W,j}}\right)^2 + \left(\sqrt{\gamma_{pz,i}} - \sqrt{\gamma_{pz,j}}\right)^2 + 2\sqrt{\gamma_{a,i}\gamma_{b,j}}\right\} \tag{11.25c}$$

and

$$\gamma_{ij} = \left\{\left(\sqrt{\gamma_{W,i}} - \sqrt{\gamma_{W,j}}\right)^2 + \left(\sqrt{\gamma_{pz,i}} - \sqrt{\gamma_{pz,j}}\right)^2 + 2\left(\sqrt{\gamma_{a,i}\gamma_{b,j}} - \sqrt{\gamma_{a,i}\gamma_{b,i}}\right) + 2\left(\sqrt{\gamma_{a,j}\gamma_{b,i}} - \sqrt{\gamma_{a,j}\gamma_{b,j}}\right)\right\} \tag{11.26b}$$

or

$$\gamma_{ij} = \left\{\left(\sqrt{\gamma_{W,i}} - \sqrt{\gamma_{W,j}}\right)^2 + \left(\sqrt{\gamma_{pz,i}} - \sqrt{\gamma_{pz,j}}\right)^2 + 2\left(\sqrt{\gamma_{a,i}} - \sqrt{\gamma_{a,j}}\right)\left(\sqrt{\gamma_{b,i}} - \sqrt{\gamma_{b,j}}\right)\right\} \tag{11.26c}$$

Equation 11.25c is a special case of Equation 11.26c and holds for molecules that cross-associate but do not self-associate (heterosolvated; we avoid the terms monopolar and bipolar used with vOCG model because polar interactions are distinct from hydrogen bonding interactions here). Our discussion will focus then on the general Equation 11.26c. As observed, the last term in this equation is identical to the hydrogen bonding term of the vOCG equation [4–6,28]. The preceding derivation was useful in delineating the assumptions behind this term. Expanding terms in Equation 11.26c and replacing for pure components, we obtain

$$\gamma_{ij} = \gamma_i + \gamma_j - 2\left\{\sqrt{\gamma_{W,i}\gamma_{W,j}} + \sqrt{\gamma_{pz,i}\gamma_{pz,j}} + \sqrt{\gamma_{a,i}\gamma_{b,j}} + \sqrt{\gamma_{a,j}\gamma_{b,i}}\right\} \tag{11.27}$$

which, combined with Dupre equation (11.1), gives for the work of adhesion:

$$W_{ij}^{adh} = 2\left\{\sqrt{\gamma_{W,i}\gamma_{W,j}} + \sqrt{\gamma_{pz,i}\gamma_{pz,j}} + \sqrt{\gamma_{a,i}\gamma_{b,j}} + \sqrt{\gamma_{a,j}\gamma_{b,i}}\right\} \tag{11.28}$$

From Young's Equation 11.2, then, we obtain for the contact angle of a drop of liquid, L, on a solid surface, S, the following working equation:

$$\gamma_L(1+\cos\theta) = 2\left\{\sqrt{\gamma_{W,L}\gamma_{W,S}} + \sqrt{\gamma_{pz,L}\gamma_{pz,S}} + \sqrt{\gamma_{a,L}\gamma_{b,S}} + \sqrt{\gamma_{a,S}\gamma_{b,L}}\right\} \tag{11.29}$$

This is the desired equation, which may be now applied for the characterization of polymer and solid surfaces.

11.3 APPLICATIONS

In this section, we apply Equation 11.29 for the characterization of polymer surfaces on the basis of their PSPs. PSPs for some common polymers [15,16] are reported in Table 11.4. In order to use Equation 11.29 for the prediction of surface tension components of polymers, as we did for liquids in the previous section, we must have first γ_W, which in turn requires A_{cosmo} and V_{cosmo} values. Since

TABLE 11.4
Partial Solvation Parameters of Common Polymers

Polymer	σ_W	σ_{pz}	σ_{Gb}	σ_{Ga}
Polyethylene, HDPE	13.75	9.00	0.0	0.0
Polyethylene, LDPE	13.75	7.75	0.0	0.0
Polypropylene, PP	14.05	5.05	0.0	0.0
Poly(ethyl ethylene), PEE	14.60	5.00	0.0	0.0
Polystyrene, atactic, PS	15.20	7.00	0.0	0.1
Poly(methyl acrylate), PMA	18.00	15.80	9.85 ± 0.40	0.0
Poly(ethyl acrylate), PEA	18.30	11.50	9.0 ± 0.15	0.0
Poly(methyl methacrylate), PMMA	18.50	7.80	7.55 ± 0.70	0.0
Poly(butyl methacrylate), PBMA	17.80	7.50	7.6 ± 0.25	0.0
Poly(hydroxy-ethylacrylate), PHEA	20.80	19.55	4.0 ± 0.20	7.5
Poly(vinyl chloride), PVC	18.50	13.95	0.0	4.2 ± 1.2
Polyepichlorohydrin, PECH	17.30	15.45	7.5	3.0 ± 1.1
Poly(ε-caprolactone), PCL	18.55	9.80	8.25 ± 0.25	0.0
Hyperbranched, Boltorn H_2O	19.95	5.65	10.45 ± 0.15	0.0
Poly(dimethyl siloxane), PDMS	14.50	3.70	5.15 ± 0.15	0.0
Poly(vinyl methyl ether), PVME	16.55	11.25	6.9 ± 0.20	0.0
Polytetrahydrofuran, PTHF	14.85	11.20	9.2	0.0
Poly(ethylene oxide), PEO	16.40	15.70	10.6	0.0

these values cannot be obtained directly for high polymers, we should resort to approximate values from analog oligomers. In Table 11.5 are reported such values for a number of analogs. As observed, there is a non-negligible scatter of values around the average ratio of 0.096. Even for polyethylene, the uncertainty from using data from its analogs may be as high as 5%. From Table 11.3, we observe that γ_W tends to a limiting value of 19.4 mN/m as the size of n-alcane increases. On the other hand, using the two ratios of Table 11.5 and the value of 13.75 MPa$^{1/2}$ for the σ_W PSP of polyethylene

TABLE 11.5
The V_{cosmo}/A_{cosmo} Ratio for Analog Oligomers

Polymer	Analog	A_{cosmo} (nm^2)	V_{cosmo} (nm^3)	V_{cosmo}/A_{cosmo}	γ_W, Pred (mN/m)
PE	Tetradecane	3.1652	0.3211	0.101	19.5 ± 0.5
PE	Hexatriacontane	7.5481	0.8034	0.106	
PP	2,4-Dimethylpentane	1.847	0.190	0.101	20.0 ± 1.0
PP	2,4,6-Trimethylheptane	2.2058	0.2355	0.107	
PVC	Vinyl chloride	0.934	0.076	0.082	28.0 ± 4.0
PVC	2,4-Dichloropentane	1.6962	0.1707	0.101	
PEO	Triethyleneglycol n-propylether	2.6554	0.2561	0.096	26.5 ± 0.5
PEO	Tetraethyleneglycol monomethylether	2.7662	0.2687	0.097	
PEO	Tetraethyleneglycol monobutylether	3.4051	0.3352	0.098	
PS	Styrene	1.5445	0.1476	0.096	22.5 ± 1.0
PS	Ethylbenzene	1.5902	0.1544	0.097	
PVAc	Vinylacetate	1.296	0.1127	0.087	23.0 ± 4.0
PVAc	Ethylacetate	1.3345	0.1175	0.088	
PMMA	Methylmethacrylate	1.4487	0.1342	0.093	33.0 ± 1.0
Average				0.096	

[15,16], we obtain for γ_W the values of 19.1 and 20.0 mN/m. These results lead to an estimated or predicted value of 19.5 ± 0.5 mN/m for the γ_W of polyethylene. Following this procedure, predicted values for γ_W have been obtained for other polymers and are reported in the last column of Table 11.5. The observed large uncertainties for poly(vinyl acetate) (PVAc) and poly(vinyl chloride) (PVC) are due to the large discrepancies between the σ_W of high polymers and of their analog oligomers. Such discrepancies are sometimes unavoidable in view of the fact that the available inverse-gas chromatography (IGC) data are for relatively high temperatures. In these cases of large uncertainties, one may resort to the use of Equation 11.29 with experimental contact angle data for a couple of solvents. Having γ_W, in the one or in the other way, one may follow the procedure of the previous section and predict all other surface tension components of the polymers. Surface tensions of pure polymer surfaces are, however, needed in this procedure.

In their seminal paper, Owens and Wendt [31] have reported surface tension and contact angle data for a number of polymers, which have been widely used as reference data in the literature. These data are reproduced in Table 11.6 for the polymers with available PSPs. In each row of the table there are two sets of data. The upper set uses the experimental contact angle data and calculates the surface tension components of the polymers on the basis of these experimental data. The lower set (in *italics*) gives the predicted surface tension components by the present approach and uses these components in order to predict the contact angles as well. As observed, the calculated surface tension components are in rather satisfactory agreement with the predicted ones. On the other hand, the predicted contact angles with diiodomethane (DIM) are in very good agreement with the experimental ones. As regards contact angles with water, this approach overpredicts them for polystyrene and to a lesser extent for poly(vinyl chloride). The failure for polystyrene may be due to the fact that COSMO-RS and, thus, PSP does not consider the interactions with the pi-electrons of the aromatic ring as hydrogen bonding. This holds true, unless this particular polystyrene surface had acquired extraneous polar groups, which is often the case with such data in the literature. The failure for PVC originates from a rather too high value for σ_W from IGC data, which leads to a high value for γ_W, which, in turn, leads to a low value for γ_{pz} and, thus, a low value for γ_a. The overall picture, however, is rather satisfactory. The important message of the calculations reported in Table 11.6 is that Equation 11.29 along with the surface tension components of solvents from Table 11.3 may be used for a rather reliable estimation of the surface tension components of polymers. In fact, they may be used for the characterization of any smooth solid surface regardless of whether PSP data are available or not.

Before proceeding further or before embarking into calculations of surface tension components for surfaces with unknown PSPs, we should clarify the methodology and set guidelines for the procedure. The first thing we should keep in mind is the potential existence of numerous surfaces

TABLE 11.6
The Surface Tension Components of Polymers

Polymer	γ_{total} (mN/m)	θ, Water	θ, DIM	γ_W	γ_{pz}	γ_a	γ_b
Polyethylene	33.2	104	53	19.7	13.5	0	0
		106	*51*	*19.5*	*13.7*	0	0
Poly(vinyl chloride)	41.5	87	36	24.4	17.1	3.2	0
		92	*35*	*28.0*	*13.5*	*1.2*	0
Poly(methyl methacrylate)	40.2	80	41	32.7	7.5	0	7.0
		81	*40*	*33.0*	*7.2*	0	*6.8*
Polystyrene	42.0	91	35	23.5	18.2	0	1.5
		100	*35*	*22.5*	*19.5*	0	0

Source: Data by Owens, D.K. and Wendt, R.C., *J. Appl. Polym. Sci.*, 13, 1741, 1969.

for the same polymer, depending on intrinsic and extrinsic factors ranging from molecular size distribution, microstructure or end groups, up to specific surface treatment and conditioning. The second thing to keep in mind is that a theoretical approach like PSP can predict unique properties of surfaces if there are such unique properties. If the surface is not clean and not at equilibrium or the probe affects the surface properties, it is difficult to extract unique surface properties. An inspection in the compilation of experimental contact angles on polymers by Lyklema [32] reveals this non-unique property problem. As an example, even for polyethylene, the reported advancing contact angle with water ranges from 28 to 116. Obviously, these polyethylene surfaces are not identical (if the measurements have been conducted properly).

The probe effect on the surface properties is a crucial issue in the study of these interfacial phenomena. It is then worth examining if the current approach can predict contact angles with other solvents on the polymers of Table 11.6. In Table 11.7 are reported such predicted contact angles with four additional solvents.

The results in Table 11.7 deserve several comments. First of all, there are two sets of predictions. Those marked in *italics* are absolute predictions directly from the corresponding PSPs. The other set was obtained from the calculated (modified) surface tension components reported in Table 11.6 in order to reproduce the experimental contact angle data with water and/or DIM. As observed, the predictions with both sets fall, essentially, within the range of measured values. With the expected exception of PVC, the predictions by the two sets are very similar. A similar to PVC discrepancy was expected for polystyrene, on the basis of the data in Table 11.6. As observed in Table 11.7, the absolute predictions for polystyrene are much better than the predictions with the modified surface tension components. This probably indicates that the surface of the polystyrene sample in Table 11.6 may have not been very clean. It is essential to point out that with both sets of predictions, the defining Equations 11.18 through 11.20, which interrelate the polarity/polarizability, acidity, and basicity components, were respected. This is important when a hydrogen bonding solvent is used as a probe of the corresponding character of the solid surface as it cannot attribute arbitrary values in one surface tension component disregarding the others. This in turn guarantees that the hydrogen bonding character of all surfaces, liquid or solid, will be compatible with their bulk-phase hydrogen bonding character and, thus, the information may be interchanged.

TABLE 11.7
Experimental and Predicted Contact Angles on the Polymers of Table 11.6

	θ, Ethylene Glycol		θ, Formamide		θ, Glycerol		θ, DMSO	
Polymer	**Predicted**	**Experim.**	**Predicted**	**Experim.**	**Predicted**	**Experim.**	**Predicted**	**Experim.**
Polyethylene	72	69[a]	85	a77–81	80	a79–94	63	
	72		*85*		*80*	r53–56	*63*	
Polystyrene	52	60[a]	67	a69–88	64	a71–80	44	
	62		*77*	r35–70	*71*		*49*	
PVC	45	45.7[b]	63	a55–66	59	a67–69	<0	Swollen[b]
	52		*69*	56.9[a]	*65*	61.8[a]	*14*	
PMMA	43	36.8[b]	61	a51–64, 41[a]	63	a60–69	43	12[b]
	44		*61*		*63*	58.1[b]	*45*	

Experimental data from the compilation of Lyklema [32] unless otherwise specified. Advancing and receding contact angle data are marked with a and r in front, respectively. Calculations in italics are absolute predictions.

[a] Dann [33].

[b] Della Volpe et al. [34] equilibrium data.

11.4 DISCUSSION AND CONCLUSIONS

The new approach to wetting/adhesion and contact angles described and tested in the previous two sections is based on a drastically different picture of interfacial interactions from that of the vOCG approach [4–6,28]. In the new approach, the surface character of a liquid or a solid at equilibrium is dictated by the solvation character of the compound as expressed by its partial solvation parameters and the requirement for interfacial free energy minimization. This forces the molecules to assume conformations and preferential orientations at interface that will probably keep the molecular sites with high interaction energies toward the bulk phase and away from the interface. As a consequence, the polar or hydrogen bonding character of surfaces of pure compounds is, in general, significantly lower than the corresponding bulk-phase character. We may get a feeling of this difference by calculating the ratio of hydrogen bonding character over the non-hydrogen bonding character in bulk liquid and at the liquid–vapor interface for alkanols. The results are summarized in Table 11.8. As seen, methanol has over five times higher hydrogen bonding character in bulk over the vapor–liquid interface. Even octanol has near two-times higher hydrogen-bonding character in its bulk liquid phase over the liquid–vapor interface. The last column in Table 11.8 indicates the extent of the earlier-mentioned preferential orientations at the interface. Obviously, when these surfaces will come in contact with a polar or hydrogen bonding liquid, these orientations will change and the molecule will exercise its full hydrogen bonding capacity. Depending on the liquid in contact, this may lead to full miscibility.

The case of solid polymer surfaces may be somewhat different depending on whether the polymer is below or above its glass transition temperature and/or melting point. Chain molecules in the glassy or crystalline state are frozen or possessing reduced flexibility for translation and rotation. In order to assume their equilibrium surface conformations, they should be given sufficient time at annealing conditions so that they keep their polar sites with high interaction energies preferentially oriented toward the bulk. When a liquid drop comes in contact with the surface, it will first sense this equilibrium surface character of the polymer, which may be significantly less polar compared to the bulk character. However, if the liquid will remain on the surface for sufficient time, depending on its own polarity, it may act as a local surface plasticizer facilitating main chain or pendant group rotations that may bring on the surface its polar sites. In other words, the liquid may "polarize" the polymer surface [10,11] and reveal (part of) its polar character. Depending on the extent or speed of plasticization, this may take some time. This lag in time may explain the observed contact angle hysteresis and why the advancing angles are always higher than the receding ones. As observed in Table 11.7, the predictions with the modified set of surface tension components are always lower than the absolute predictions of the present approach. The latter reflect the initial equilibrium state

TABLE 11.8
Comparison of Hydrogen Bonding Character in Bulk and in Liquid–Vapor Interface

Alkanol	$B = \dfrac{\sigma_{hb}^2}{\sigma_W^2 + \sigma_{pz}^2}$	$S = \dfrac{\gamma_{hb}}{\gamma_W + \gamma_{pz}}$	B/S
Methanol	1.636	0.290	5.645
Ethanol	1.066	0.226	4.716
Ethanol	1.066	0.226	4.716
1-Propanol	0.823	0.223	3.685
1-Butanol	0.656	0.217	3.025
1-Pentanol	0.638	0.238	2.680
1-Hexanol	0.587	0.244	2.399
1-Heptanol	0.501	0.238	2.104
1-Octanol	0.458	0.234	1.956
1-Decanol	0.375	0.227	1.652

of the pure polymer surface while the former its "polarized" state. Contact angles of absolute predictions are, thus, expected to be closer to advancing contact angles while the ones predicted with the modified surface tension components could be considered to be closer to the receding ones if one could exclude solvent molecules adsorbed on the surface in the receding front.

There are two important points to be made in relation with the vOCG model [4–6,28]. As we have seen, in the present approach there is a direct correspondence of the acidity/basicity surface character of a compound to its corresponding bulk character. In contrast, in the vOCG approach, this correspondence is lost and the basicity character appears disproportionately prevailing. In parallel, the water neutrality hypothesis (equal acidity and basicity) of the original vOCG model is sound and fully supported by quantum chemical calculations and QSPR/LSER data. Any attempts in the literature to drastically alter this neutrality may have improved the fitting performance but they may have obscured the physical basis and the insight in the studied interfacial phenomena.

It should be stressed again that the polymer PSPs used in this work are those obtained from IGC (cf. Table 11.4) at relatively high temperatures. No attempt was made to correct for any temperature dependence. In a refinement of the model, one could certainly account for it. Such an account of temperature dependence could bring σ_W from IGC for high polymers closer to the corresponding one from oligomers.

In conclusion, the present approach to surface tension components has some apparent similarities with the vOCG model [4–6,28] but, in essence, it is a drastically different approach. The key feature is that it provides with a sound basis for the interactions dictating interfacial phenomena, which is fully compatible with the picture we have about intermolecular interactions from quantum chemical calculations. The new approach has not been tested extensively yet, but the tests of this work are indicating a rather satisfactory agreement with experiment. Work is underway in our laboratory toward a more extensive testing of the new approach against experimental data.

ACKNOWLEDGMENTS

The author is grateful to Otto Monsteds Fond for financial support of this work and to Professor G. Kontogeorgis/DTU for valuable discussions.

REFERENCES

1. Dupre, A., *Theorie Mechanique de la Chaleur*, Gauthier, Villars, Paris, France, 1869.
2. Hiemenz, P.C., Rajagopalan, R., *Principles of Colloid and Surface Science*, 3rd edn., Marcel Dekker, New York, 1997.
3. Young, T., *Philos. Trans. R. Soc. Lond.*, 95, 65–87, 1805.
4. van Oss, C.J., Chaudhury, M.K., Good, R.J., *Adv. Colloid Interface Sci.*, 28, 35, 1987.
5. Good, R.J., *J. Adhesion Sci. Technol.*, 6, 1269, 1992.
6. van Oss, C.J., *Interfacial Forces in Aqueous Media*, 2nd edn., CRC Press, Boca Raton, FL, 2006.
7. Della Volpe, C., Siboni, S., *J. Colloid Interface Sci.*, 195, 121, 1997.
8. Della Volpe, C., Siboni, S., *J. Adhesion Sci. Technol.*, 14, 235, 2000.
9. Lee, L.H., *J. Adhesion*, 67, 1, 1998.
10. Yasuda, T., Okuno, T., Yasuda, H., *Langmuir*, 10, 2435, 1994.
11. Carre, A., *J. Adhesion Sci. Technol.*, 21(10), 961, 2007.
12. Panayiotou, C., *Phys. Chem. Chem. Phys.*, 14, 3882, 2012.
13. Panayiotou, C., *J. Chem. Thermodynamics*, 51, 172, 2012.
14. Panayiotou, C., *J. Phys. Chem. B*, 116, 7302, 2012.
15. Panayiotou, C.G., *J. Chromatogr. A*, 1251, 194, 2012.
16. Panayiotou, C., *Polymer*, 54, 1621, 2013.
17. Klamt, A., Jonas, V., Buerger, T., Lohrenz, J.C.W., *J. Phys. Chem.*, 102, 5074, 1998.
18. Klamt, A., *COSMO-RS from Quantum Chemistry to Fluid Phase Thermodynamics and Drug Design*, Elsevier, Amsterdam, the Netherlands, 2005.
19. COSMObase Ver. 1401, COSMOlogic GmbH & Co., K.G., Leverkusen, Germany, 2013.

20. Lin, S.T., Sandler, S.I., *Ind. Eng. Chem. Res.*, 41, 899, 2002.
21. Grensemann, H., Gmehling, J., *Ind. Eng. Chem. Res.*, 44, 1610, 2005.
22. Pye, C.C., Ziegler, T., van Lenthe, E., Louwen, J.N., *Can. J. Chem.*, 87, 790, 2009.
23. Abraham, M.H., *Chem. Soc. Rev.*, 22, 73, 1993.
24. Zissimos, A.M., Abraham, M.H., Klamt, A., Eckert, F., Wood, J., *J. Chem. Inf. Comput. Sci.*, 42, 1320, 2002.
25. Hansen, C.M., *Hansen Solubility Parameters: A User's Handbook*, CRC Press, Boca Raton, FL, 2007.
26. Ahlrichs, R., Baer, M., Haeser, M., Horn, H., Koelmel, C., *Chem. Phys. Lett.*, 162, 165, 1989.
27. Daubert, T.E., Danner, R.P., *Physical and Thermodynamic Properties of Pure Compounds: Data Compilation*, Hemisphere, New York, 2001.
28. van Oss, C.S., Chaudhury, M.K., Good, R.J., *Chem. Rev.*, 88, 927, 1988.
29. Abrams, D.S., Prausnitz, J.M., *AIChE J.*, 21, 116, 1975.
30. Sorensen, J.M., Arlt, W., *Liquid–Liquid Equilibrium Data Collection*, DECHEMA, Chemistry Data Series, Vol. 5, Frankfurt, Germany, 1979.
31. Owens, D.K., Wendt, R.C., *J. Appl. Polym. Sci.*, 13, 1741, 1969.
32. Lyklema, J., *Fundamentals of Interface and Colloid Science*, Vol. III, Academic Press, London, U.K., 2000.
33. Dann, J.R., *J. Colloid Interface Sci.*, 32, 302, 1970.
34. Della Volpe, C., Siboni, S., Maniglio, D., Morra, M., Cassinelli, C., Anderle, M., Speranza, G. et al., *J. Adhesion Sci. Technol.*, 17(11), 1425, 2003.

12 Surface Chemistry of Oil Recovery (Enhanced Oil Recovery)

K.S. Birdi

CONTENTS

12.1 INTRODUCTION

The demand for energy by mankind is the most important issue of this age. As the world's population increases (ca. threefold every 50 years), the demand for energy increases drastically. The energy demand is increasing at about 2% per year. Among energy resources (e.g., oil, gas, coal, atomic energy, hydropower, wind energy, and solar energy), *oil* is the most important at the present stage. Oil supplies about 45% of the world's energy (petroleum) while natural gas (mostly as methane, CH_4) supplies about 20%. At present, worldwide, about 80 million barrels of oil is consumed every day. Most of the oil produced today comes from mature reservoirs. It is estimated that some 10^{12} barrels of oil remain in the reservoirs worldwide after the primary production (1 barrel = 0.159 m^3). Hence, it is a great challenge to scientists to invent methods and procedures to recover these vast amounts of residual oil reserves. There are chemical processes called enhanced oil recovery (EOR) that have been used to recover the residual oil in the past decades (Ahmed, 2001; Bavière, 2007; Birdi, 1999; Carcoana, 1992; Green, 1998; Kenneth, 2012; Lake, 1996; Lake and Walsh, 2008; Lichtner et al., 1996; Mark and Lake, 2003; Rao, 2012; Scheng, 2013; Schramm, 2010; Smith, 1966; Taber, 1968; Taber et al., 1997; Tunio et al., 2011) (Figure 12.1). Further, the EOR methods have developed in various ways as the price of oil has increased over the past decades (from few dollars to around 100 USD/barrel). The oil reservoir production history comprises various phases (Figure 12.1):

- Primary phase
- Secondary phase
- EOR

The conventional oil reservoir is found to be a *complicated* system (under high pressures and temperatures [which depends on the depth and other parameters]) (Figure 12.2) (Alvarado and Manrique, 2010; Bavière, 1991; Birdi, 1999, 2010; Douglas and Tiratsoo, 1975; Green, 1998; Kalfayan, 2008; McCain, 1990; Moritis, 2008; Rao, 2012; Scheng, 2010, 2013; Taber et al., 1997; van Poolen, 1980; Wilhite, 1986):

- Oil phase–solid (porous rock)–pores (size and shape)

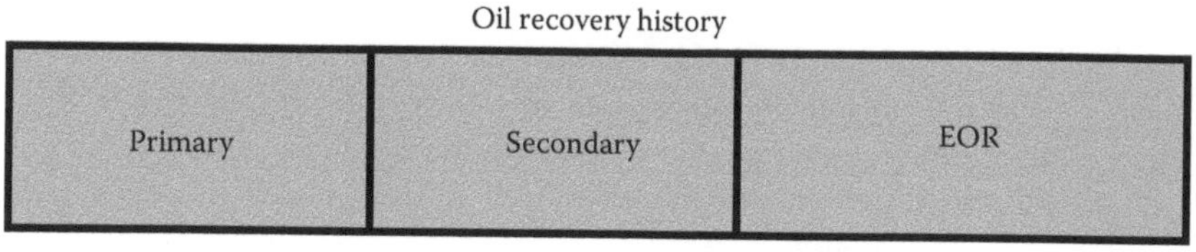

Oil recovery history

Primary	Secondary	EOR

FIGURE 12.1 Oil reservoir (typical) production history (schematic).

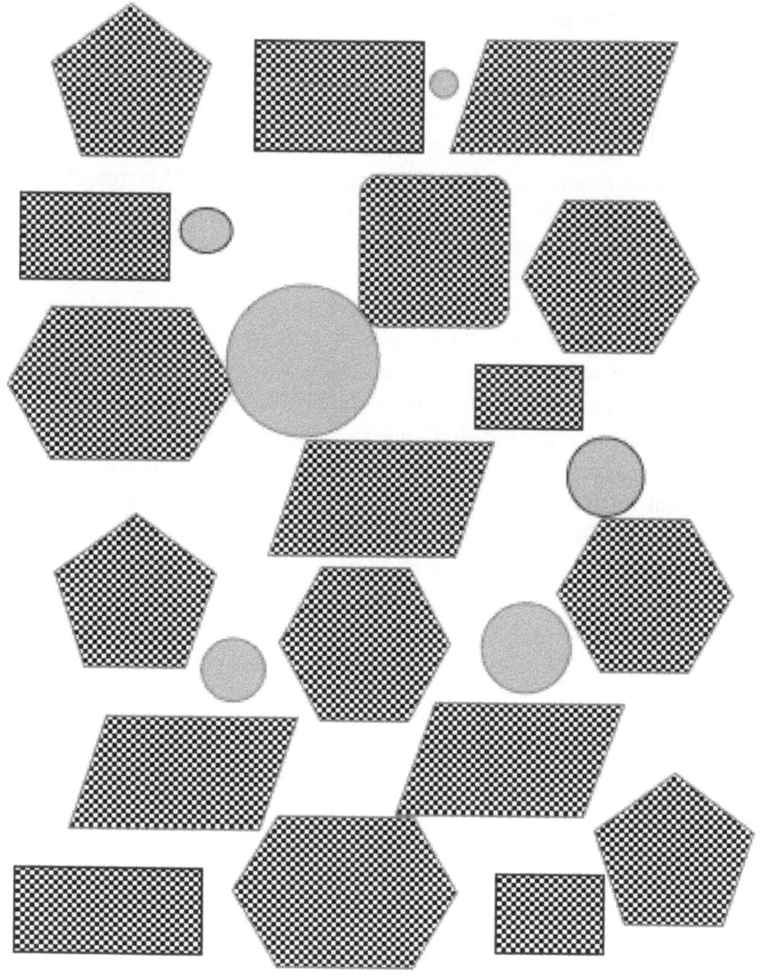

FIGURE 12.2 Oil reservoir rock: oil (circles) in large pores/narrow pores (schematic).

The recovery of oil from reservoir involves the *flow of fluid* through pores under a given temperature and pressure. However, it is known that high pressure alone sometimes does not help in the flow of oil (oil recovery) due to various chemicophysical forces. In these reservoirs, there are various specific *surface forces* (besides other forces), which are important for the oil recovery. In order to recover the vast amounts of residual oil in reservoirs, over the past decade enhanced oil recovery (EOR) methods are being applied. The *conventional* oil reservoirs during the past decades are now being subjected to EOR and with very positive results. The application of surface and colloid chemistry principles has been applied in EOR. In very simple terms, one can further describe the essential *surface forces* involved in EOR:

- Interfacial tension (between oil and solid; oil and water)
- Capillary forces (dependent on porosity and interfacial tension [IFT] between oil and water and pore size and shape)
- Viscosity of oil

Of course, the reservoir characteristics such as pressure and temperature will also determine the EOR procedure to be applied. It is thus seen that the degree of oil recovery will depend on these factors or a combination of these different parameters. The EOR process has thus to resolve these forces in the reservoirs in order to increase the oil recovery.

What Is Oil?

It is of interest to mention briefly about the *origin and composition* of oil as found in the earth reservoirs (Rao, 2012). This is important since the EOR method will be dependent on the composition and chemical characteristics of the oil and the reservoir rock. The composition of oil as found in different parts of the world has been analyzed. Many different trace amounts of molecules have been identified as markers for the source of oil. It is believed that oil (fossil fuel) is formed by natural processes as decomposition of organic matter (such as wood, grass, leaves, etc.). These processes are also known to have taken place over hundreds of millions of years. The composition of fossil fuels is high in carbon. It is thus evident that some millions of years ago, CO_2 from air (at present, CO_2 concentration in air is about 400 ppm [0.04%]) was used to produce plants through photosynthesis (CO_2–water–sunlight: plants, wood, etc.), which after decomposition created coal (solid), oil (liquid petroleum), and natural gas (mostly methane as gas). This also shows that the state (solid–liquid–gas) at which the decomposition ended depends on the physicochemical conditions (i.e., pressure and temperature) in the reservoir.

Decomposition of organic matter: Coal (solid)/oil (liquid)/natural gas (methane)

In the past decades, natural gas and oil have become the most important energy sources. In the United States, gas recovery has increased dramatically in the past decade. This increase arose from the new *unconventional* oil and gas recovery technology applied to tight gas as found in shale gas. Natural gas is recovered by hydraulic fracturing (fracked) technology, which allows natural gas to escape through fractures (Rao, 2012). The International Energy Agency (IEA) estimates show that more than hundreds of trillion cubic meters (tcm) of technically recoverable *unconventional* gas reserves are available worldwide. The United States already produces shale gas, which amounts to about 30% of the domestic demand. Other countries, such as China, Australia, Canada, and the United Kingdom are currently developing similar gas recovery projects. Further, conventional reservoirs are much smaller in volume than the source rocks (such as shale oil and gas). This means that as regards potential oil and gas supply, mankind can recover much larger amounts of oil and gas from shale deposits in the future (estimates indicate that these deposits may be of the order of many decades larger than the conventional reservoirs). Shale oil technology has also developed during the past decades, which now adds a considerable contribution to the world supply (Rao, 2012; van Poolen, 1980).

Due to various physicochemical forces, only about 30% of the total oil in a reservoir (worldwide) has been recovered with the *conventional* techniques. The primary recovery process is based on the reservoir pressure as present initially, and the recovery decreases (and finally stops) as the pressure diminishes. At this stage, mankind has an urgent need to sustain energy supply and therefore the old (i.e., mature) reservoirs with large residual oil reserves left unrecovered require enhanced oil recovery (EOR) methods. In the past decades, a great amount of research has been carried out on EOR processes and the literature shows that this is an important ongoing subject worldwide. This chapter describes the most essential *surface chemistry* aspects about EOR technology. The important surface chemistry parameters are delineated.

General recovery history of oil reservoirs worldwide:

- Primary recovery: Oil production under reservoir pressure.
  - Recovery efficiency: 30%
  - Residual oil: 70%

Following are the distinct stages of oil recovery in the history of most of the reservoirs:

I. In this stage the oil is flowing freely under pressure as found at the well site.
II. As the pressure drops and the flow slows down, additional pressure (by using a gas or water injection) is applied.

As the oil recovery stops after stage II, the reservoir needs to be applied to enhance oil recovery (EOR), if the residual oil is to be recovered. A typical EOR oil recovery in a conventional reservoir shows an increase in oil flow, as shown in Figure 12.3. The data from many different oil reservoirs have been analyzed by using various mathematical models. These models fit the EOR flow pattern very satisfactorily.

Different EOR Processes

- Gas injection
  - Inert gas (nitrogen)
  - Hydrocarbon gas injection
  - Miscible liquid propane
  - CO_2 flooding
- Improved water flooding processes
  - Surfactant/polymer flooding
  - Low IFT flooding
  - Alkaline solutions
  - Polymer flooding
  - Gels
  - Microbial injection
  - Foams

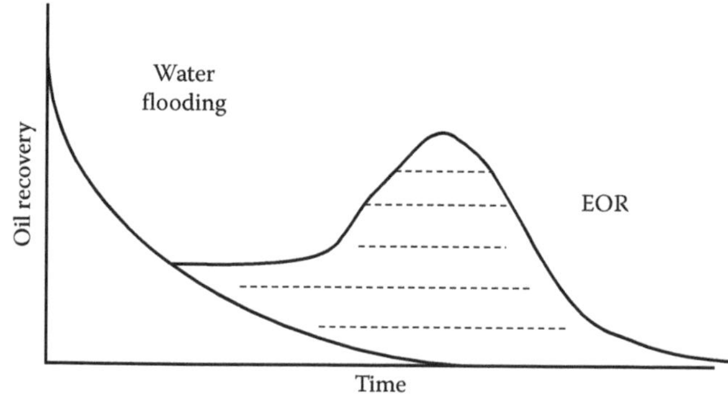

FIGURE 12.3 A typical oil flow pattern after EOR application to a reservoir (schematic).

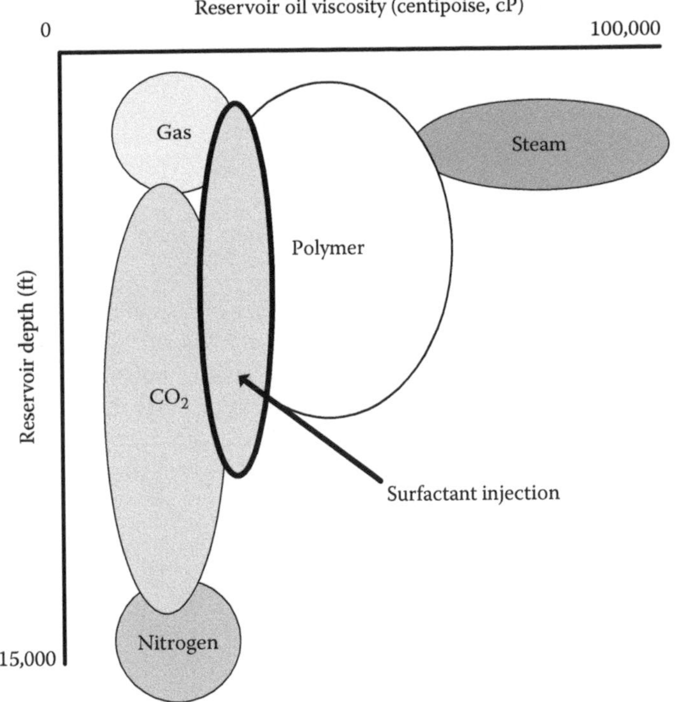

FIGURE 12.4 Correlation between EOR methods and oil reservoir characteristics (schematic).

- Thermal methods
  - In situ combustion
  - Steam and hot water

These different enhanced oil recovery methods become even more involved when the combination of two or more different techniques is used, for example, when CO_2 injection is combined with surfactant–polymer flooding. A simple schematic correlation has been found between the EOR method and the depth of the reservoir and oil viscosity (Figure 12.4). Obviously, in such a complex system no simple correlation can be absolutely valid, but it can be a useful guideline. Field operations have provided evidence for a good relation as depicted in Figure 12.4.

12.1.1 Surface Chemistry Aspects of Enhanced Oil Recovery

The fluid (oil) flow in any reservoir has characteristics that are the same as the flow of fluid through porous rock media. The fluid inside a pore has characteristic properties due to the *shape* of the liquid surface. The shape of a liquid surface in a capillary determines many physical characteristics. A *flat* liquid surface (such as a liquid drop or soap bubble) has different properties than a *curved* surface (Figure 12.5).

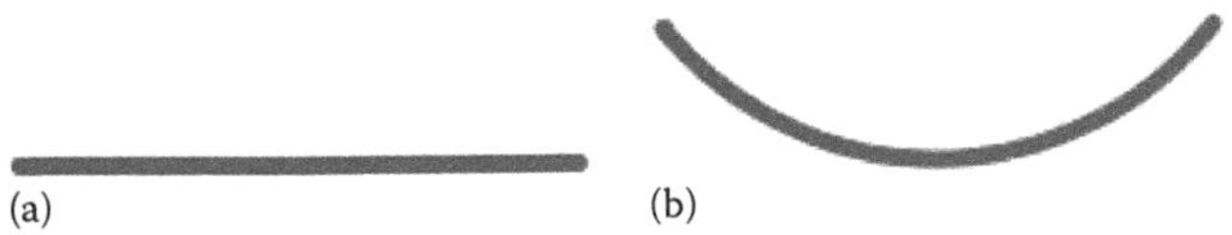

FIGURE 12.5 A flat (a) or a curved (b) liquid surface.

For instance, the pressure inside a soap bubble is higher than outside. The difference in pressure, ΔP_{bubble}, is given as:

$$\Delta P_{bubble} = \frac{2\gamma_{liquid}}{R_{radius}} \tag{12.1}$$

where

γ_{liquid} is the surface tension of the liquid
R_{radius} is the radius of the bubble

It is thus seen that the magnitude of ΔP_{bubble} will be larger in a smaller bubble than inside a larger bubble. Furthermore, the magnitude of ΔP will be zero at a flat surface (since R_{radius} is very large). This also indicates that when two bubbles collide, the smaller bubble will merge into the larger bubble. For example, it is known that the height of meniscus of water in a capillary of smaller radius is higher than in a larger size capillary (Figure 12.6). The magnitude of $\Delta P_{capillary}$ at the surface of fluid is related to the height, $h_{capillary}$, a liquid will rise with a given surface tension, γ_{liquid}:

$$h_{capillary} = \frac{\gamma_{liquid}}{(2R_{radius}\rho_{liquid}\, g_{gravity})} \tag{12.2}$$

where

ρ_{liquid} is the density of the fluid
$g_{gravity}$ is the acceleration of gravity (Birdi, 1999, 2010)

Water: Rise in capillary
$\gamma = 72.8$ mN/m (25°C)
$g_{gravity} = 9.8$ m/s^2
$\rho_{water} = 1000$ kg/m^3

Radius of Capillary	**Magnitude of $h_{capillary}$**
0.015 mm	1 m
1.5 mm	1 cm
14 cm	0.1 mm

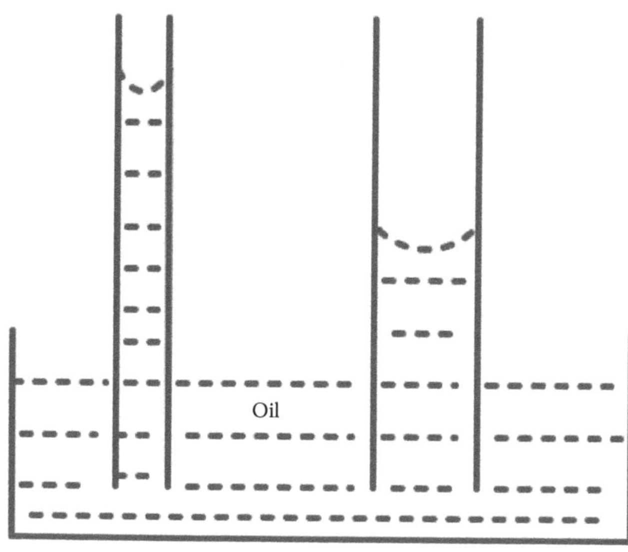

FIGURE 12.6 Rise of a fluid (oil) in large and small capillary (schematic).

These results indicate that oil will be easily produced (which corresponds to the primary oil production) from large size pores than in narrow pores (residual oil will be expected to be present in narrow pores). These data thus indicate that the residual oil is present in very fine pores of the rock. In simple terms, one may describe this process as follows:

- Fluid as found under the surface of the earth is under high pressure and temperature, and will move upward.
- Fluid has to move through fissures or fractures (capillary forces).
- The viscosity of fluid will give resistance to flow.
- Interaction between the fluid and the surface of the rock will determine the process (wetting or nonwetting characteristics).

In the reservoir, besides oil there is also some amount of brine (water with a high content of salt). It is thus obvious that in the system

Oil–Rock–Brine

there are *interfacial forces* involved between the surface tensions of γ_{rock}, γ_{oil}, and γ_{brine}. These surface forces can be analyzed by contact angle measurements (Birdi, 1999, 2002, 2009, 2010). The fact that most of the oil (50%–70%) in a petroleum reservoir remains un-recovered in the ground is due to some main chemicophysical phenomena:

- The interfacial tension of the oil
- Viscosity of oil
- Porosity of the reservoir
- Wetting characteristics of the oil–rock interface

EOR History

- 1930–1960: Gas injection (CO_2)–steam injection
- 1970–1990: Chemical and surfactants
- 2000–2014: Offshore EOR–advanced EOR techniques

The various EOR methods, which have been applied over the past decades, have been regulated by the economics of the oil prices. Mostly, these developments have been regulated by the price increase through time (from ca. a dollar per barrel to over 100 USD/barrel at present stage). The present high price of oil has been the driving factor for the application of more advanced EOR techniques. Furthermore, it must be stressed that these EOR processes are also necessary, considering that about 70% of oil is left behind in the mature oil wells. The EOR methods are dependent on various factors, which are as follows:

- Reservoir rock characteristics
- Reservoir depth
- Oil viscosity

In order to apply the most suitable EOR procedure, one thus needs the knowledge of the various physicochemical data of the reservoir. *Steam injection* reduces the viscosity of oil (due to higher temperature), thus helping the increased production. The addition of polymers gives rise to higher viscosity of the water phase (used for pushing the oil). More recently, there have been some recovery methods where *microbial applications* have been applied in EOR (Banat, 1995; Bryant and Lockhart, 2002; Donaldsen, 1991; Jimoh, 2012). The microbial EOR is complicated and the exact mechanism is also difficult to analyze. However, it is known that microbes produce various chemicals and some of these molecules may exhibit surface activity, which will reduce the oil IFT, and thus help oil recovery.

12.1.1.1 Interfacial Tension in Oil Reservoir

Petroleum, which is generated over millions of years from organic matter that is buried deep in the earth, is usually trapped in underground structures formed by porous rocks. These rocks usually are sandstone or fissured limestone. The petroleum, which is produced in deep-lying source rocks (source reservoir), migrates to these structures and is trapped there, if these structures are sealed at the top. These structures then form what is called a conventional oil field. It is important to note that the source rock will thus be expected to contain many times larger amounts of oil than as one finds in conventional reservoirs.

Formation of oil from organic matter (Source rock, such as shale oil)–movement of oil toward the surface of the earth–trapped in pores and as found in conventional reservoirs

The production of petroleum from a reservoir may be divided into different phases. The *first stage* of oil recovery is where the flow is under the reservoir pressure. Very early in the life of a reservoir, energy must usually be supplied to the porous medium, which bears the crude oil so that it continues to flow to the producing wells. This energy is brought into the reservoir by injection of water or gas (nitrogen, methane, CO_2) (Tunio et al., 2011). With these secondary recovery methods, about 30%–40% of the original oil in place may be recovered, while the rest may be left in the earth.

12.1.1.1.1 Capillary Forces in EOR Process

In any porous solid media, the pore size (*shape and size distribution*) is known to be a very important characteristic. The flow of a fluid through a porous media is primarily dependent on the capillary forces (Figure 12.6). The force holding back the oil in the porous body of the reservoir rock is the interfacial tension between the different phases of oil, water, and gas flowing in the reservoir and the viscosity of the crude oil. In those cases where oil is being displaced by injected water (or other aqueous solutions), the interfacial tension between the oil and water and the wettability of the rock play important roles. In Figure 12.1, the pore space between sand grains is shown (schematic). If the rock is water wet, the floodwater is imbibed into the rock and oil is displaced to producing wells. But, after a short time the situation is such that oil drops also have to displace water in order to allow the oil itself to move in the direction of the producing wells. In these cases, the capillary pressure between the oil and water has to be overcome. This means that oil has to drain water from the reservoir before it can move.

In general, the first step in most EOR processes is to flood the reservoir with water. The pressure gradient in a reservoir is often smaller than the capillary pressure. During this process, therefore, the displacing water bypasses the oil zones (inside small pores). The primary aim of surfactants is to reduce the magnitude of oil–water interfacial tension (IFT). The magnitude of oil–water interfacial tension is related to the chain length of the alkane molecule (Birdi, 1999, 2010).

Interfacial Tension of Alkanes/Water and Alcohol/Water Systems (20°C, 1 atm)

	IFT (mN/m)
Alkane	
n-Hexane	51.0
n-Octane	50.8
Alcohol	
n-Decanol	10.0
n-Octanol	8.5
n-Hexanol	6.8
n-Butanol	1.6

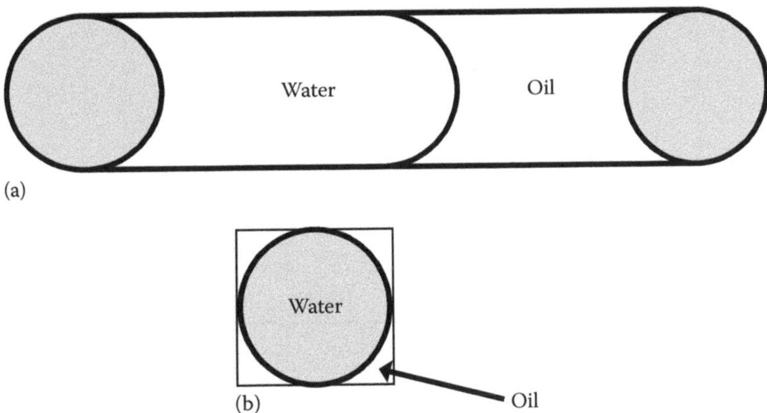

FIGURE 12.7 Water-flooding of oil in a circular shape (a) or square type (end view) pore (b).

These data show that the magnitude of IFT is about 50 mN/m for alkanes and changes very little with chain length. On the other hand, the magnitude of IFT changes to much lower values in the case of alcohol/water systems. The reduction of IFT gives rise to oil recovery from oil as present in the very small pores (with high capillary pressure). The magnitude of IFT of alkane/surfactant solution can be much lower than 1 mN/m (Birdi, 1999, 2002, 2009, 2010), depending on the system.

The high degree of *water-by-pass* is also indicative of that the pores are more nonspherical type. The degree of by-pass is larger in square-type pores than in the circular ones. Since the pores in any reservoir are expected to be nonspherical, one should also expect a large degree of water-by-pass. Actually, this is also observed in real reservoirs and laboratory experiments.

The capillary forces in square tubing have been investigated as a function of size of tubing and surface tension of fluids (Bera et al., 2011; Birdi et al., 1988; Cayias et al., 1976). Of course, one may expect that the pores in reservoirs are a mixture of spherical and nonspherical (square, triangle, pentagon, etc.) shapes. The water-by-pass in rectangular capillary is largely present at the corners (which are absent in circular capillaries) (Figure 12.7). However, this is expected to decrease as the magnitude of IFT decreases (due to the application of suitable surfactant). As can be seen in Figure 12.2, the nonwetting phase (oil) has to pass small throats before it can leave the pore space. It is the diameter of these pore throats that determines the capillary pressure that has to be overcome to displace the nonwetting phase by the wetting phase.

12.1.1.1.2 Capillary Pressure in Square-Shaped Capillary Tubings

The capillary rise of liquids was investigated in square-shaped capillary tubings of different dimensions (300–1000 μm) in a temperature range from 25°C to 35°C (Birdi et al., 1988). The surface tension, γ_{liquid}, data was fitted to the following relation:

$$\gamma_{liquid} = \frac{1}{2}\left(\delta_{liquid}\, g_{gravity}\left(S_{length}\left(\frac{C_{constant}\, H_{rise}}{2} + C_{constant}\, S_{length}\right)\right)\right) \tag{12.3}$$

where

δ_{liquid} is the density
$g_{gravity}$ is the force of gravity
S_{length} is the side length of the square tubing
H_{rise} is the rise of liquid in the tubing
$C_{constant}$ (=1.089) is a capillary constant for square shapes

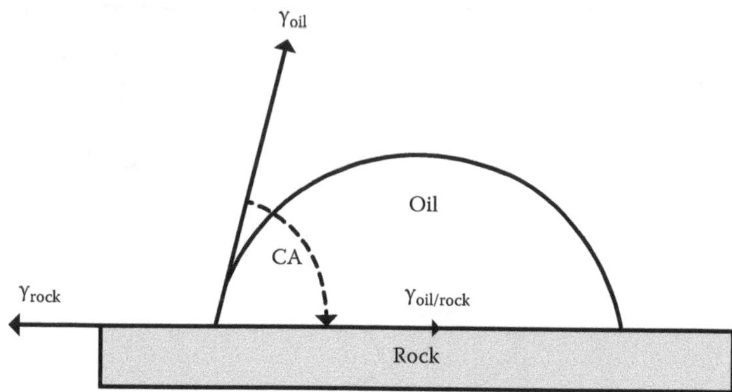

FIGURE 12.8 Wetting characteristics of oil on rock surface as depicted by contact angle (CA), Θ, of surface tension of oil (γ_{oil}), of solid (rock) (γ_{rock}), and of oil/solid (rock) ($\gamma_{oil/rock}$).

The different surface forces, which are present when a drop of oil is placed on a solid (such as reservoir rock) are as follows:

$$\gamma_{rock} = \gamma_{oil} \cos\Theta + \frac{\gamma_{oil}}{\gamma_{rock}} \tag{12.4}$$

where

γ_{rock} is the surface tension of rock
γ_{oil} is the surface tension of oil
$\gamma_{oil/rock}$ is the surface tension of rock–oil interface
contact angle, Θ, is where the three different surface forces reach an equilibrium (Figure 12.8) (Birdi, 2010; Chattoraj and Birdi, 1984)

The contact angle corresponds to the state where the different surface forces are at equilibrium. The magnitude of γ_{rock} can be estimated if the value of contact angle, Θ, and the γ_{oil} is known. The determination of surface tension of solids (such as γ_{rock}) has been extensively studied and its application in real-world systems has been extensively delineated in the current literature (Birdi, 1999, 2010; Chattoraj and Birdi, 1984). If the contact angle is less than 90°, then the liquid is known to wet the solid surface. However, if the contact angle is larger than 90°, then the surface is nonwetting. For example:

Water drop placed on the following solid surfaces exhibits different contact angles:

Glass surface: 30°
Teflon: 105°

This clearly shows that the surface of TEFLON is nonwetting as regards water. Similar considerations will apply when considering the wetting properties of oil–rock systems in any EOR system.

The capillary pressure (i.e., a fluid inside a capillary tubing) is given as:

$$P_{capillary} = \frac{(2\gamma \cos\Theta)}{R_{pore}} \tag{12.5}$$

where

γ_{oil} is the interfacial tension
Θ is the contact angle
R_{pore} is the pore radius

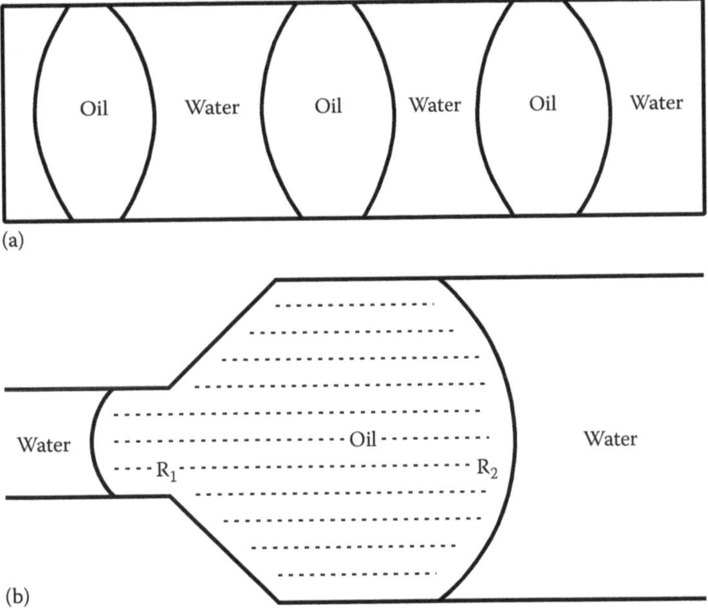

FIGURE 12.9 State of capillary pressure of oil and water in a cylindrical (a: symmetrical and b: asymmetric tubing) (schematic).

If the diameter of the pores were uniform (such as a cylinder), capillary pressure differences would be zero, as shown in Figure 12.9a. However, in an oil drop that is driven by water injection through a pore (with radii R_1 and R_2), the difference in capillary pressure is given as

$$\Delta P_{capillary} = 2\gamma_{oil} \cos\Theta \left(\frac{1}{R_1} - \frac{1}{R_2} \right) \tag{12.6}$$

In a cylindrical pore, $R_1 = R_2$, which means $\Delta P_{capillary} = 0$. The capillary pressure gives rise to water-by-pass, as also observed in real reservoir flooding. It is thus seen that (Figure 12.9b) the quantity capillary pressure counteracts oil recovery. This means that the amount of oil that may be recovered is proportional to the pressure drop applied to the porous medium and inversely proportional to capillary pressure. This can be expressed as follows:

$$1 - S_{capillary} = \left(\frac{\Delta P}{1} \right) \left(\frac{R}{\gamma} \right) \tag{12.7}$$

where

$S_{capillary}$ is the remaining oil or *residual* oil saturation
R is the radius of the pore in the rock (Birdi, 1999)

Combining these relations to permeability gives the following expression:

$$1 - S_{capillary} = \left(\frac{\Delta P}{l\gamma} \right) \left(\frac{k}{\varepsilon} \right)^{0.5} \tag{12.8}$$

where

k is the permeability
ε is the porosity

12.1.1.1.3 *Permeability of Rocks*

The flow of fluids through porous solid material is measured by a method described by Darcy (Ahmed, 2011; Birdi, 1999). The quantity permeability, k_{Darcy}, is defined as the relation between the flow rate of the fluid (V_{flow}), permeability constant, k_{Darcy}, viscosity of the fluid, μ_{fluid}, pressure difference, ΔP, and the thickness, dx_{solid}, of the solid:

$$V_{flow} = \frac{(k_{Darcy}\,\Delta P)}{(\mu_{fluid}\,dx_{solid})} \tag{12.9}$$

A solid with permeability constant of 1 D has the following data:

$V_{flow} = 1\ cm^3/s$
$\mu_{fluid} = 1$ cP (centi poise = 1 mPa s)
$\Delta P = 1$ atm/cm, across a 1 cm^2 area

Typical values of k_{Darcy} for different porous rocks are as follows:

Gravel: 100,000 D
Sand: 1 D
Granite: less than 0.01 micro-Darcy ($\mu D = 10^{-6}$ D)

Laboratory experiments have been carried out to (Birdi, 1991; Taber, 1968) investigate the influence of pressure drop and interfacial tension on residual oil saturation. It was found that the oil recovery from a core of reservoir rock is proportional to the pressure drop across the core and capillary pressure. These experiments were performed on 420 mD Berea-sandstone and 2000 mD sand pack. Though both series of experiments were performed on different materials and in different ways, they showed similar results.

12.1.1.2 Viscosity of Reservoir Oil

It is known that the composition of oil depends on the source. This is found from the fact that the composition of oil from the Middle East is different from that found in Mexico or North Sea or elsewhere. The most significant component is the asphaltene content. Asphaltenes give rise to higher viscosity and other characteristics. As known from fluid flow through porous media, higher viscosity gives rise to a slow movement in the pores. The viscosity of crude oil in a reservoir may vary between about 0.1 and 100,000 mPa s, whereas the viscosity of water lies between 0.5 and 1.2 mPa s, depending on temperature and salinity (Ahmed, 2011; Birdi, 1999).

The dependence of oil recovery (% of oil in place [OIP]) on the viscosity ratio between oil and water is shown in Figure 12.10. It is observed that in the case of a high-viscosity oil, the tendency of the water to bypass oil-filled zones is higher than for the low-viscosity oil. In order to recover more oil from a reservoir, the viscosity of the oil must be reduced or the viscosity of the floodwater increased (by adding polymers).

Viscosity characteristics of oil: Reduction of oil viscosity may be achieved by increasing temperature. This can be done by the injection of hot water or steam into the reservoir. This recovery method is usually applied in the case of heavy oils, for which primary and secondary oil recovery is very low (<10%). The viscosity of these oils is above 100 mPa·s and may reach values of several thousands and more. Application of steam injection gives rise to oil viscosity values of about 10–50 mPa·s, as shown in Figure 12.11.

12.1.2 Surface-Active Substances (Surfactants) Applied in EOR

Surfactants reduce the magnitude of IFT of oil–water interface (Birdi, 1999, 2002, 2009, 2010) from a value of about 40 mN/m to less than 1 mN/m. Surfactant flooding means that surface-active agents

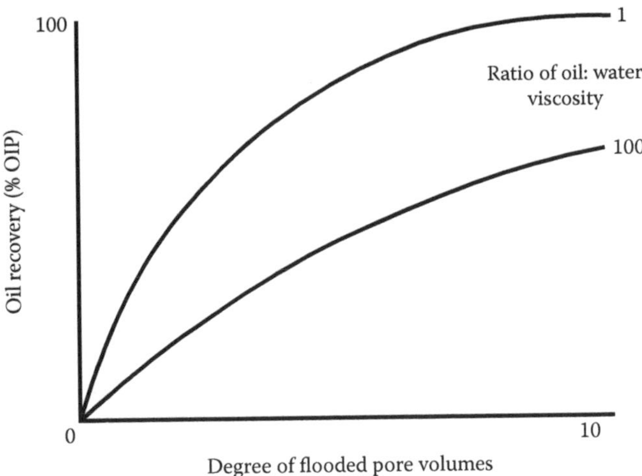

FIGURE 12.10 Variation of oil recovery (% of oil in place [OIP]) versus flooded pore volumes.

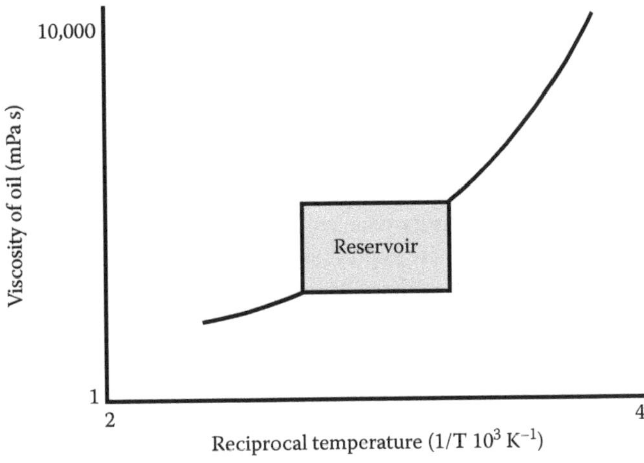

FIGURE 12.11 Variation of viscosity (mPa s) of a typical heavy oil versus temperature (1/T 10^3 K^{-1}).

thus reduce the interfacial tension by a few decades. Oil–water interfacial tension may be reduced by these surfactants from 40 to 10^{-4}–10^{-5} mN/m, depending on the system. The application of surfactants (which gives rise to lower IFT) has been found to be a very important procedure to recover more oil from petroleum reservoirs. There are many other chemicals that exhibit properties similar to those of surfactants (Alvarado and Manrique, 2010; Birdi, 1999; Taber et al., 1997). Surfactants, which have been used in EOR, have to be stable at the oil reservoir conditions (especially high temperature and pressure) (Table 12.1). For instance, arylalkyl sulfonic acids (formed from olefin and SO_3) are found to be stable at high temperatures.

These molecules are also called as amphiphiles, due to their dual nature of properties. The *hydrophobic part*, which is the alkyl chain, dislikes water (hydrophobic) or is insoluble in water, which makes it useful for oil recovery, as this part induces lower IFT. The *hydrophilic part* (like water) allows the surfactant to disperse in water to form large aggregates, micelles (Austad et al., 1994; Birdi, 1999, 2010).

The magnitude of temperatures in oil reservoirs is found to be relatively high (70°C–130°C). The water in these reservoirs is found to contain very high concentrations of electrolytes (mostly NaCl). Further, in some reservoirs one finds besides quartz sand, significant amounts of clay minerals.

TABLE 12.1
Typical Surfactants Used in EOR

Hydrophobic Part	Intermediate Part	Hydrophilic Part	Counter Ions
Iso-octylphenolic	Ethyleneoxide	Carboxylate	Alkaline
Dodecyl-phenolic	Propyleneoxide	Sulfate	Amine
Di-nonylphenolic	Ethylene-propyleneoxide	Sulfonate	Quartz

The surfactant molecule is characterized by its two parts: hydrophobic part and hydrophilic part.

Surfactants have a strong tendency to adsorb rocks in the reservoirs (due to the characteristics of the molecule) (Chattoraj and Birdi, 1984). This is not desirable and therefore one needs to apply surfactants that exhibit low adsorption characteristics. In order to reduce the magnitude of interfacial tension, $\gamma_{oil/water}$, between water and oil, surfactants of various compositions have been tested, and additional surfactants have been developed (Table 12.1). Depending on the oil properties, the salinity of the reservoir brine, and the temperature, the surfactants are tailor-made based on the composition of reservoir fluids.

12.1.3 Polymers and Surfactants in EOR

Flooding petroleum reservoirs with water-soluble polymers and surfactants may be regarded as the most economic tertiary chemical oil recovery method. Polymers for enhanced oil recovery should be water soluble, should develop high viscosity, and should not plug the reservoir during injection (Austad et al., 1994; Birdi, 1999). Further, they should not be degraded by temperature or bacteria and should not precipitate in high-salinity reservoir waters. The water-soluble polyacrylamide can only be used in freshwater systems, whereas polysaccharides such as hydroxyethyl-cellulose or xanthan develop enough viscosity in high-salinity brines. The main problems of using these chemicals in enhanced oil recovery are the stability in the reservoir and the adsorption of the polymers on the reservoir rock. This means that the polymers should develop the desired viscosity throughout the flood process.

As shown in Figure 12.10, in a linear displacement experiment the oil recovery depends on the viscosity ratio between the displaced (oil) and the displacing (water) phase. This is due to capillary pressure, as described earlier, because the pressure drop during flow is proportional to viscosity. In addition, it was found that in a heterogeneous medium, such as sandstone, the sweep efficiency, which represents the portion of the reservoir that is really influenced by the displacing fluid, depends on the viscosity ratio too. It has been found that a quantity defined as mobility ratio, $M_{mobility}$, has been used (Birdi, 1999; Dyes et al., 1954):

$$M_{mobility} = \frac{(k_{water}/\eta_{water})}{(k_{oil}/\eta_{oil})} \tag{12.10}$$

(where k_{water} and k_{oil} are the effective permeabilities to water and oil, respectively, and η_{water} and η_{oil} are the viscosities of water and oil, respectively) influences the areal sweep efficiency. Apart from these two effects, the vertical sweep efficiency determines the performance of a flood. Before the mechanism that improves the vertical sweep efficiency during polymer flooding is discussed, some introductory remarks on the rheology of pseudo-plastic liquids are necessary (Birdi, 1999; Rao, 2012). For water and oil, the viscosity is, in most cases, a constant value. For polymer solutions, this is more or less not the case; the viscosity is a function of the rate of shear strain. In regard to the viscous flow of a liquid, the rate of shear strain is a function of both flow geometry and flow velocity. For flow in porous media, this means that in narrow pores the rate of shear strain is higher than in larger pores, or, in terms of petroleum reservoir engineering, that at the same Darcy velocities the shear rate in less permeable zones is higher than in zones having good permeabilities. Hence, for example, for a Darcy velocity of 0.2 m/day in a sandstone with a porosity of 25% and a permeability of 2000 mD, the rate of shear strain is 8.6 s^{-1} for a polymer solution with a viscosity of 40 mPa^n (n = 0.6). For the same polymer solution at the same Darcy velocity in sandstone with a porosity of 20% and a permeability of 200 mD, the rate of shear strain is 30 s^{-1}. This means that the viscosity of the flowing polymer solution is higher in the sandstone with the high permeability (19.9 mPa·s) than in the sandstone with the low permeability (10.2 mPa·s).

The reservoir oil recovery by various EOR methods has been described for the water (or polymer or surfactant) flooding techniques in EOR (Ahmed, 2011; Alvarado and Manrique, 2010; Birdi, 1999; Taber et al., 1997). Most commonly, the highly permeable zones are invaded by the flood water during secondary operations or natural water drive, and low permeable zones are not flooded so that oil is left in these parts of a reservoir. This means that since the IFT of oil–water (ca. 50 mN/m) is high, the capillary pressure inhibits the water from entering reservoir regions with smaller pores. During polymer flooding (high viscosity), a poor vertical sweep efficiency may be improved because the polymer solution, of course, first follows the paths prepared by water and then, because of its high viscosity, tends to block these parts of the reservoir, so that oil that was previously immobile starts flowing. In the next phase of oil recovery, the surfactant-polymer solution flooding improves the oil recovery due to lower IFT values (less than 1 mN/m) (Ahmed, 2011).

Different oil recovery stages:

- Water flooding:
  - Oil recovery from high permeable zones
- Polymer solution flooding:
  - Oil recovery from low permeable zones
- Surfactant–polymer solution flooding:
  - Oil recovery from very low permeable zones

The pressure gradient in the reservoir and especially in those zones where oil was immobile becomes higher in a polymer flood than it was during the water drive. It has been reported (Sandiford, 1977) that in a system of two parallel flooded cores of different permeabilities, the oil recovery due to polymer flooding is significantly higher than during a flood in one core of uniform permeability.

REFERENCES

Ahmed, T., *Reservoir Engineering Handbook*, 2nd edn., Gulf Professional Publishing, Boston, MA, 2001.

Alvarado, V., Manrique, E., *Energies*, 1529, 3, 2010.

Austad, T., Fjelde, I., Veggeland, K., Taugbol, K., *J. Petrol. Sci.*, 255, 10, 1994.

Banat, I.M., *Biores. Technol.*, 1, 51, 1995.

Baviere, M., *Basic Concepts in Enhanced Oil Recovery Processes*, Society of Chemical Industry, London, U.K., 1991.

Baviere, M., *Basic Concepts in Enhanced Oil Recovery Processes*, Elsevier Applied Science, London, U.K., 2007.

Bera A., Ojha, K., Mandal, A., Kumar, T., *J. Colloid Surf. A*, 114, 383, 2011.

Birdi, K.S., ed., *Handbook of Surface and Colloid Chemistry*, CRC Press, Boca Raton, FL, 1999.

Birdi, K.S., ed., *Handbook of Surface and Colloid Chemistry*, 2nd edn., CRC Press, Boca Raton, FL, 2002.

Birdi, K.S., ed., *Handbook of Surface and Colloid Chemistry*, 3rd edn., CRC Press, Boca Raton, FL, 2009.

Birdi, K.S., *Surface and Colloid Chemistry, Principles and Applications*, CRC Press, Boca Raton, FL, 2010.

Birdi, K.S., Vu, D.T., Noerregaard, A., *Colloid Polym. Sci.*, 470, 266, 1988.

Bryant, S.L., Lockhart, T.P., *Soc. Pet. Eng. Reserv. Eval.*, 365, 5, 2002.

Carcoana, A., *Applied Enhanced Oil Recovery*, Prentice Hall, New York, 1992.

Cayias, J.L., Schechter, R.S., Wade, W.H., *Soc. Petrol. Eng. J.*, 351, 16, 1976.

Chattoraj, D.K., Birdi, K.S., *Adsorption and the Gibbs Surface Excess*, Plenum Press, New York, 1984.

Donaldsen, E.C., ed., *Microbial Enhancement of oil Recovery—Recent Advances*, Development in Petroleum Science, New York, 1991.

Douglas, H., Tiratsoo, E.N., *Introduction to Petroleum Geology*, Scientific Press, Beaconsfield, U.K., 1975.

Dyes, A.B., Caudle, B.H., Ericson, R.A., *Trans. AIME*, 201, 81, 1954.

Green, D.W., *Enhanced Oil Recovery* (*SPE Textbook Series*), Vol. 6, Society of Petroleum Engineers, Richardson, TX, 1998.

Jimoh, I.A., Microbial enhanced oil recovery, PhD thesis, Luma Print, Aalborg University, Esbjerg, Denmark, 2012.

Kalfayan, L., *Production Enhancement with Acid Stimulation*, 2nd edn., Pennwell Corp., 2008.

Kenneth, S.D., *Hubberts Peak: The Impending World Oil Shortage*, Princeton University Press, Princeton, NJ, 2012.

Lake, L.W., *Enhanced Oil Recovery*, Prentice Hall, New York, 1996.

Lake, L.W., Walsh, M.P., Enhanced oil recovery (EOR), Tech. report, Society of Petroleum Engineers, Richardson, TX, 2008.

Lichtner, P.C., Steefel, C.I., Oelekers, E.H., *Reactive Transport in Porous Media*, Mineralogical Society of America, Washington, DC, 1996.

Mark, W., Lake, L.W., *A Generalized Approach to Primary Hydrocarbon Recovery*, Elsevier, Amsterdam, the Netherlands, 2003.

McCain, W.D., *The Properties of Petroleum Fluids*, 2nd edn., Pennwell Publishers, Tulsa, OK, 1990.

Moritis, G., *Oil Gas J.*, 60, 96, 2008.

Rao, V., *Shale Gas*, RTI Press, New York, 2012.

Sandiford, B.B., Enhanced oil and surface chemistry, in: Shah, D.O., Schechter, R.S., eds., *Improved Oil Recovery by Surfactant and Polymer Flooding*, Academic Press, New York, 1977, p. 487.

Scheng, J., *Modern Chemical Enhanced Oil Recovery: Theory and Practice*, Gulf Professional Publishing, New York, 2010.

Scheng, J., *Enhanced Oil Recovery Field Case Studies*, Gulf Professional Publishing, New York, 2013.

Schramm, L.L., *Surfactants: Fundaments and Applications in the Petroleum Industry*, Cambridge University Press, Cambridge, UK, 2010.

Smith, C., *Mechanics of Secondary Oil Recovery*, Reinhold Pub. Corp., New York, 1966.

Taber, J.J., SPE 2098, SPE Reprint Ser. No. 24 (*Surfactant/Polymer Flooding I*), Society of Petroleum Engineers, Richardson, TX, 1968.

Taber, J.J., Martin, F.D., Seright, R.S., *SPE Res. Eng.*, 12, 189, August 1997.

Tunio, S.Q., Tunio, A.H., Ghirano, N.A., El Adawy, Z.M., *Int. J. Appl. Sci. Technol.*, 143, 5, 2011.

van Poolen, H., *Fundamentals of Enhanced Oil Recovery*, Pennwell Corp., Tulsa, OK, 1980.

Wilhite, G.P., *Waterflooding, Textbook Series*, Society of Petroleum Engineers, Richardson, TX, 1986.

13 A Review of Polymer–Surfactant Interactions

Ali A. Mohsenipour and Rajinder Pal

CONTENTS

13.1 INTRODUCTION

The combination of polymer and surfactant additives has been used in a variety of applications such as drug delivery, oil recovery, and cosmetics (Bai et al. 2010; Dan et al. 2009; Harada and Kataoka 2006; Stoll et al. 2010; Villetti et al. 2011; Zhang et al. 2011). In general, polymers are introduced to a surfactant system to control rheology, stability of the system, or to manipulate surface adsorption. The addition of a polymer may help to remove a surfactant from a surface or to improve its adsorption at the surface. In many applications, we need a suitable rheology such as thickening of solution or gelation of solution. The addition of polymer allows the manipulation and control of the solution rheology. The polymer can also speed up the micellization process resulting in a decrease in the free surfactant concentration. This property is exploited in skin formulations in which free surfactant molecules can harm the skin and cause irritation. The combination of polymer and surfactant additives are also being exploited to intensify frictional drag reduction in the turbulent flow of liquids in pipelines (Mohsenipour and Pal 2013).

The majority of the polymer–surfactant interaction studies are focused on polymer–surfactant aggregation at low surfactant concentrations (Goddard and Ananthapadmanabhan 1993; Kwak 1998a,b; Touhami et al. 2001; Trabelsi and Langevin 2007). The polymer–surfactant interaction depends on the types of polymer and surfactant and solution conditions such as pH and temperature (Feitosa et al. 1996a; Jönsson et al. 1998). For example, the interaction between an electrically charged polymer (negative or positive charge) and an oppositely charged ionic surfactant is of electrostatic type, whereas the interaction between a nonionic polymer and a nonionic/ionic surfactant is of hydrophobic type involving hydrophobic parts of the polymer and surfactant (Diamant and Andelman 1999; Goddard and Ananthapadmanabhan 1993; Hansson and Lindman 1996). In spite of the extensive studies conducted on the understanding of polymer–surfactant interactions, the complex behavior of mixed additives in solution has prevented a true understanding of the interaction mechanisms involved.

13.2 BASIC ASPECTS OF POLYMERS AND SURFACTANTS

13.2.1 Polymers

Polymers are large molecules or macromolecules, made up of many small repeating units called monomers. They can be categorized as natural and synthetic polymers. Examples of natural polymers are biopolymers such as DNA and proteins. Examples of synthetic polymers include polyethylene and other plastics that have extensive influence on the daily life of human beings. The number of monomer units in a polymer molecule can vary from few to millions depending on the usage and applications of polymer. The molecular weight has a strong influence on the physical and chemical properties of polymers.

Polymers can be categorized as linear, branched, or cross-linked. In some cases, polymers are synthesized using two or more different types of monomers. Such polymers are called copolymers. If a monomer carries any electric charge, then the resulting polymer will be ionic in nature. Thus, polymers could also be categorized as cationic, anionic, and nonionic polymers.

Polymer solutions exhibit a range of properties depending on the configuration of polymer chains in the solution. The configuration of polymer chains is strongly affected by the interactions between polymer–polymer molecules and polymer–solvent molecules. Polymer chains exist in solution in various formats such as a random coil, an extended configuration, or a helix. The expansion of polymer chains in good solvents cause a significant increase in the solution viscosity. In aqueous solutions, the ionic polymers can produce a highly viscous solution even at a low polymer concentration (Goddard and Ananthapadmanabhan 1993).

Polymers find extensive usage in our daily life and in industrial applications. They are used in solid form as well as in liquid state. In a number of applications they are used as additives to control the rheological properties of liquids. Another important application of polymers is the frictional drag reduction in pipeline turbulent flows. When a small quantity of soluble polymer is added to the pipeline turbulent flow of liquids, a significant reduction in friction is observed. This phenomenon is called drag reduction (DR). Drag reduction by polymeric additives has been used in different applications including pipeline transportation of oil, wastewater treatment, sludge technology, heating and cooling loops, hydraulic and jet machinery, and in biomedical applications (Gasljevic et al. 2001; Harwigsson and Hellsten 1996; Hellsten 2002; Min et al. 2003; Mohsenipour 2013; Mohsenipour and Pal 2013; Ptasinski et al. 2001; Sreenivasan and White 2000; Yang 2009; Zakin et al. 1996).

13.2.2 Surfactants

Surfactants consist of two parts: a hydrophilic head group and a hydrophobic tail group. The hydrophilic head group (an ionizable polar group in case of ionic surfactants) can establish hydrogen bonds. The hydrophobic tail group (a non-polar group) is typically a long chain alkyl group. Surfactants are known for their capability to lower the surface tension of liquids. When a surfactant is dissolved in an aqueous solution, the hydrophobic groups are repelled by water while the hydrophilic groups are attracted to the polar water molecules. This causes the hydrophobic groups to aggregate together in a hydrocarbon phase in order to prevent contact with water. At the same time, the hydrophilic polar groups surrounding the hydrocarbon phase are in contact with the water. Such aggregates of surfactant molecules are called "micelles." Micelles can have different sizes and shapes that are highly dependent on the surfactant molecular structure and system conditions. Examples of different shapes of the micelles are globular/spherical, disk-like, cylindrical or rod like, bilayer spherical (vesicle), and hexagonal. The shape of the micelles can change from one form to another when the solution conditions are changed (Zakin et al. 1998; Zhang 2005; Zhang et al. 2009).

Micelle shapes are determined to a large extent by a packing factor (p) defined as $p = v/a_0 l_c$, where v is hydrodynamic volume of surfactant molecule, l_c is the length of the tail, and a_0 is the head group cross-section area (see Figure 13.1). When p is equal to 1/3 or less, the surfactant will be cone shape and the micelles will be spherical. This is the most commonly encountered micelle shape. For p close to or equal to 1/2, the micelles are of cylindrical shape (rod like). Figure 13.1 shows different micelle shapes and the corresponding p values.

Micelle shapes and sizes are also influenced by the surfactant concentration. With the increase in the surfactant concentration, the micelles undergo a change in shape from spherical to thread-like (Zakin et al. 1998), as shown in Figure 13.2. At high surfactant concentrations, entanglement of thread-like micelles is often observed.

Figure 13.3 shows a simplified phase diagram of an aqueous surfactant system. The Kraft point is the temperature at which the surfactant solubility is equal to the critical micelle concentration (CMC). When the temperature is lower than the Kraft temperature, the surfactant exists in gel or crystal form in the solution. When the temperature is above the Kraft temperature and the surfactant concentration is higher than the CMC, the micelles are formed in the surfactant solution. The micelles are initially spherical in shape. With the increase in surfactant concentration or upon

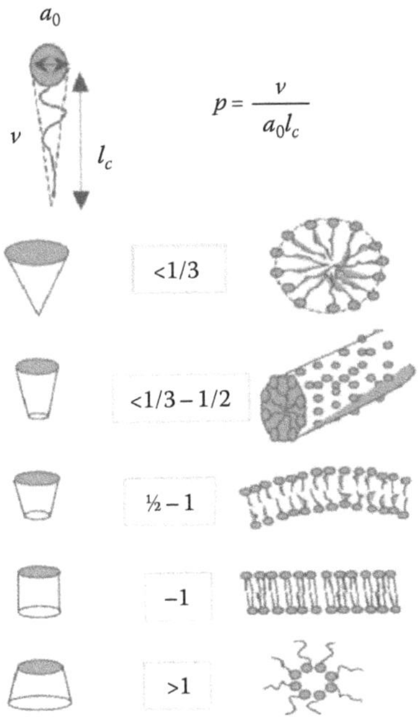

FIGURE 13.1 Different micelle shapes with corresponding packing factor (p) values. (From Zhang, Y., Correlations among surfactant drag reduction, additive chemical structures, rheological properties and microstructures in water and water/co-solvent systems, PhD thesis, The Ohio State University, Columbus, OH, 2005.)

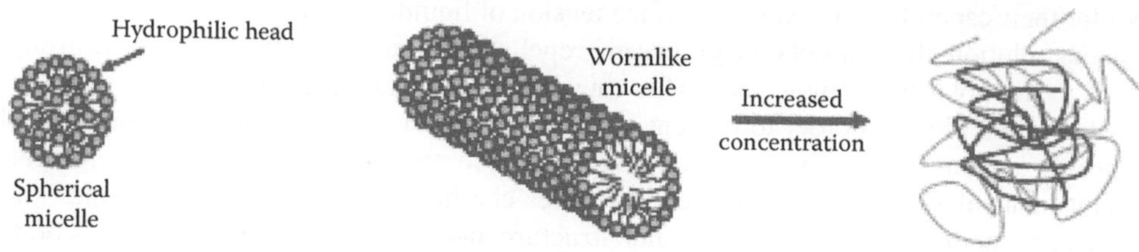

FIGURE 13.2 Spherical and rod-like micelles. (From Rothstein, J.P., *Rheol. Rev.*, 1, 2008.)

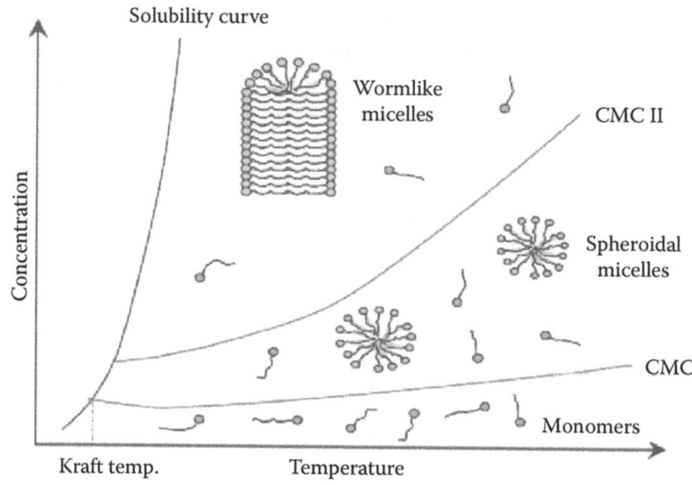

FIGURE 13.3 Simplified phase diagram of aqueous surfactant solution. (From Zhang, Y., Correlations among surfactant drag reduction, additive chemical structures, rheological properties and microstructures in water and water/co-solvent systems, PhD thesis, The Ohio State University, Columbus, OH, 2005.)

addition of counterions in the case of ionic surfactant, the micelle shape changes from spherical to thread-like (Zakin et al. 1998). The concentration at which surfactants form rod-like micelles is sometimes called CMCII. While the CMC is almost independent of temperature, the CMCII increases with the increase in temperature.

Surfactants are categorized into four groups: anionic, cationic, nonionic, and zwitterionic. When the hydrophilic part of a surfactant consists of a negatively charged group like a sulfate, sulfonate, or carboxylate, the surfactant is called anionic. Most of the regular soaps consist of anionic surfactants. Anionic surfactants are sensitive to water hardness, which can force them to precipitate. Cationic surfactants have a positive charge on their hydrophilic part. Quaternary ammonium compounds are the most common cationic surfactants. The cationic surfactants are mostly used as softening, anti-static, soil repellent, and anti-bacterial agents. They are also used as corrosion inhibitors. They are not sensitive to water hardness. These surfactants are expensive compared to anionic surfactants. Nonionic surfactants consist of an uncharged hydrophilic part. This type of surfactants are not capable of being ionized in aqueous solution as their hydrophilic group is of non-dissociable type. Examples of nonionic surfactants are fatty alcohol polyglycosides and alcohol ethoxylates. They are widely used in cleaning detergents. Zwitterionic or amphoteric surfactants consist of both negative and positive charges on their molecules. The solution pH controls the charge on the hydrophilic part of amphoteric surfactants. That is why they act as anionic surfactants in alkalic solution and as cationic surfactants in acidic solution.

Table 13.1 gives examples (names and chemical formulae) of surfactants in different groups (anionic, cationic, nonionic, and zwitterionic). The chemical structures of some of the commonly

TABLE 13.1
Surfactant Classification

Class	Examples	Structures
Anionic	Na stearate	$CH_3(CH_2)_{16}COO^-Na^+$
	Na dodecyl sulfate	$CH_3(CH_2)_{11}SO_4^-Na^+$
	Na dodecyl benzene sulfonate	$CH_3(CH_2)_{11}C_6H_4SO_3^-Na^+$
Cationic	Laurylamine hydrochloride	$CH_3(CH_2)_{11}NH_3^+Cl^-$
	Trimethyl dodecylammonium chloride	$C_{12}H_{25}N^+(CH_3)_3Cl^-$
	Cetyl trimethylammonium bromide	$CH_3(CH_2)_{15}N^+(CH_3)_3Br^-$
Nonionic	Polyoxyethylene alcohol	$C_nH_{2n+1}(OCH_2CH_2)_mOH$
	Alkylphenol ethoxylate	$C_9H_{19}–C_6H_4–(OCH_2CH_2)_nOH$
	Polysorbate 80 $w + x + y + z = 20$ $R=(C_{17}H_{33})COO$	$HO(C_2H_4O)_w$ / $(OC_2H_4)_xOH$ / O / $CH(OC_2H_4)_yOH$ / $CH_2(OC_2H_4)_zR$
	Propylene oxide-modified polymethylsiloxane (EO = ethyleneoxy, PO = propyleneoxy)	$(CH_3)_3SiO((CH_3)_2SiO)_x(CH_3SiO)_ySi(CH_3)_3$ $CH_2CH_2CH_2O(EO)_m(PO)_nH$
Zwitterionic	Dodecyl betaine	$C_{12}H_{25}N^+(CH_3)_2CH_2COO^-$
	Lauramidopropyl betaine	$C_{11}H_{23}CONH(CH_2)_3N^+(CH_3)_2CH_2COO^-$
	Cocoamido-2-hydroxypropyl sulfobetaine	$C_nH_{2n+1}CONH(CH_2)_3N^+(CH_3)_2CH_2CH(OH)CH_2SO_3^-$

Source: Schramm, L.L. et al., *Ann. Rep. Sect. C (Phys. Chem.)*, 99, 3, 2003.

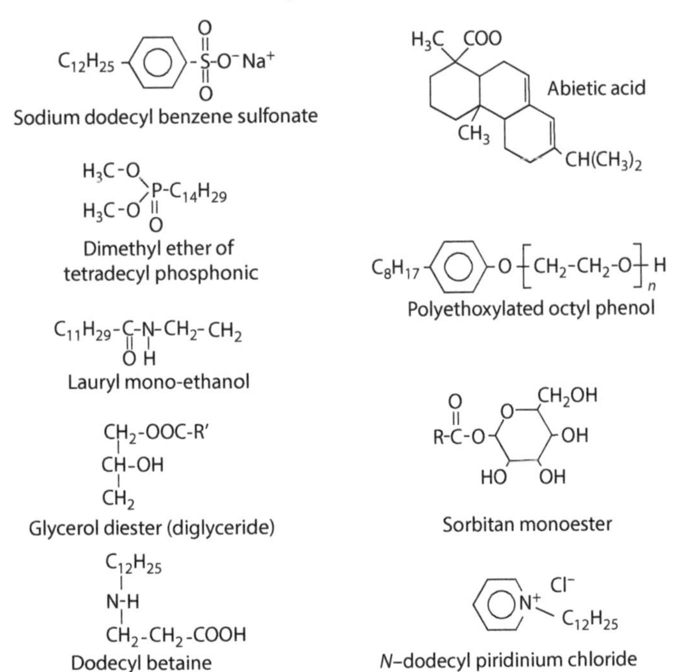

FIGURE 13.4 A few commonly used surfactants. (From Salager, J.-L., Surfactant's types and uses, In: Salager, J.L., ed., *Firep Booket-E300-Attaching Aid in Surfactant Science and Engineering in English*, University of Andes, Merida, Venezuela, 2002, p. 3.)

used surfactants are shown in Figure 13.4. Surfactants are used widely in different industries such as food, paint, and petroleum. A rapid growth in the usage of surfactants is anticipated in the oil field in the years to come. A large potential market exists for surfactants in enhanced oil recovery, but many technical and economic problems still remain that need to be resolved before the surfactants can be used at a large scale (Ahmadi and Shadizadeh 2013; Bachmann et al. 2014; Nelson 1982; Schramm 2000). Another important application of surfactants involves frictional drag reduction in pipeline turbulent flows. Surfactants like polymers are known to cause a significant reduction in friction in the pipeline turbulent flow of liquids. Drag reduction by surfactants did not receive any attention until the work of Mysels (1949). More recent work in this area can be found in Qi and Zakin (2002), Zakin et al. (2006, 2007), Zakin and Lui (1983), and Zhang et al. (2009). Although a good amount of research work has been carried out on surfactants as drag reducers, they have been rarely used at an industrial scale. This area needs more attention in the future work.

Further information about surfactants and their applications can be found in the works of Tadros (2006), Salager (2002), and Schramm et al. (2003).

13.3 INTERACTION OF POLYMERS AND SURFACTANTS

The topic of surfactant and polymer interactions has attracted a lot of attention in the past few decades and many researchers have been involved in this area. The interactions of polymers and surfactants have been studied with different applications in mind such as drug delivery, enhanced oil recovery, and cosmetics (Bai et al. 2010; Dan et al. 2009; Goddard and Ananthapadmanabhan 1993; Harada and Kataoka 2006; Jönsson 1998; Kwak 1998a,b; Stoll et al. 2010; Villetti et al. 2011; Zhang et al. 2011). Researchers have tried to explore and manipulate the interactions between polymer and surfactant to accomplish desirable properties of the solution (Kwak 1998a,b).

The interactions between polymer and surfactant depend on several factors such as the type of polymer, the type of surfactant, and the solution conditions such as pH and temperature (Feitosa et al. 1996a; Jönsson et al. 1998). Some studies have categorized the interactions in two groups: (a) electrostatic interactions and (b) hydrophobic interactions. The first group consists of the interactions of ionic polymers (negative or positive charge) with oppositely charged ionic surfactants. The interaction is electrostatic in nature. The second group involves interactions of nonionic polymer and ionic/nonionic surfactants (Diamant and Andelman 1999; Goddard and Ananthapadmanabhan 1993; Hansson and Lindman 1996). In this second case, the interaction occurs between the hydrophobic parts of polymer and surfactant molecules. Table 13.2 shows possible combinations of polymer and surfactant types relevant in investigating the interactions. The symbol "P" represents polymer, "S" represents surfactant, and the superscripts represent the charge on the species: "0" for neutral, "–" for negative charge, and "+" for positive charge, respectively.

13.3.1 Onset of Self-Assembly in Polymer–Surfactant Systems

When a polymer is present in a surfactant solution, the interaction of polymer and surfactant molecules usually starts at a well-known surfactant concentration called "critical aggregation

TABLE 13.2
Possible Combinations of Polymers and Surfactants for Studying Interactions

Surfactant	Nonionic Polymer	Anionic Polymer	Cationic Polymer
Nonionic	P^0S^0	P^-S^0	P^+S^0
Anionic	P^0S^-	P^-S^-	P^+S^-
Cationic	P^0S^+	P^-S^+	P^+S^+

TABLE 13.3
CMC and CAC Values for Different Polymer/Surfactant Systems

Surfactant	Polymer	*T* (°C)	CMC or CAC (mM)	CAC/CMC	C_2 (mM)
SDS		25	8.0		
SDS	0.1 wt% PEO	25	4.4	0.55	
SDS	0.1 wt% PVP	25	2.1	0.26	
CsDS		30	6.2		
CsDS	0.1 wt% PEO	30	4.2	0.68	9.6
CsDS	0.1 wt% PVP	30	4.1	0.66	8.4
TMADS		25	5.4		
TMADS	0.1 wt% PEO	25	4.6	0.85	8.0
TMADS	1 wt% PEO	25	4.7	0.87	
TMADS	0.1 wt% PVP	25	4.6	0.85	8.6
TEADS		25	3.7		
TEADS	0.1 wt% PEO	25	3.7	1.0	
TEADS		40	3.8		
TEADS	0.1 wt% PEO	40	3.8	1.0	
TP ADS		25	2.2		
TP ADS	1 wt% PEO	25	2.2	1.0	
TP ADS	0.1 wt% PVP	25	2.25	1.01	
TBADS		25	1.15		
TBADS	0.1 wt% PEO	25	1.15	1.0	
TBADS	1 wt% PEO	25	1.15	1.0	

Source: Benrraou, M. et al., *J. Colloid Interf. Sci.*, 267(2), 519, 2003.

concentration (CAC)." In many studies published on the interactions of surfactants and polymers, it is shown that CAC is lower than the CMC (critical micelle concentration) of the surfactant solution alone (Deo et al. 2007; Goddard and Ananthapadmanabhan 1993; Jönsson et al. 1998). In the case of electrostatic interaction between ionic polymer with oppositely charged ionic surfactant, CAC is generally found to be much lower than the CMC of the surfactant. The CAC is close to CMC in the case of nonionic polymer and ionic surfactant. In addition to CAC, there is another critical surfactant concentration, referred to as "polymer saturation point (PSP)," which is relevant in polymer/surfactant systems. While CAC represents the onset of interaction, the polymer saturation point (PSP) reflects the surfactant concentration where the polymer chains become saturated with bound surfactant molecules or micelles (Diamant and Andelman 1999; Goddard and Ananthapadmanabhan 1993; Hansson and Lindman 1996).

Table 13.3 gives the CMC, CAC, and PSP values for a number of polymer/surfactant systems. The variations of CAC with temperature and polymer concentration are also indicated. The ratio CAC/CMC depends on the type of polymer/surfactant combination. For example, the ratio CAC/CMC is 0.26 for SDS/0.1% of PVP solution and 0.55 for SDS/0.1% PEO solution. This shows that even if the type of surfactant is the same, the CAC can be different for different polymers. Thus, the structure of polymer has a strong effect on the CAC.

13.3.2 Available Techniques to Study the Polymer–Surfactant Interactions

There are several analytical techniques available that can be employed to study the interaction between polymers and surfactants. Techniques based on the measurements of surface tension, electrical conductivity, and viscosity are widely used to probe the interactions between the surfactant and polymer molecules. Other techniques such as gel permeation chromatography (GPC), electromotive force (EMF), and isothermal titration calorimetry (ITC) are also used. In this section, a brief discussion of these techniques is presented.

13.3.2.1 Surface Tension Measurements

Figure 13.5 shows the variations of surface tension versus surfactant concentration for pure surfactant system and the mixture of polymer/surfactant where the polymer concentration is constant. In the case of pure surfactant solution, a sharp decrease in the surface tension occurs with the increase in surfactant concentration up to the critical micelle concentration (CMC). For surfactant concentrations higher than the CMC, the surface tension remains constant. In the mixture of polymer and surfactant, the surface tension plot shows two break points. The first point is the CAC point where the interaction between the polymer and the surfactant begins. The second point is the PSP point where the polymer chains become saturated with the surfactant. When the interaction between the polymer and surfactant is weak, CAC and PSP values are close to the CMC of pure surfactant (Mohsenipour 2011).

Figure 13.6 shows surface tension versus surfactant concentration plots for solutions of ionic surfactant SDS and nonionic polymer polyvinyl pyrrolidone (PVP). The addition of polymer changes the surface tension behavior of the surfactant solution. The plots indicate two break points in the present case of nonionic polymer and ionic surfactant. CAC (shown as T_1) represents the beginning of the formation of polymer/surfactant complexes. The CAC value is significantly

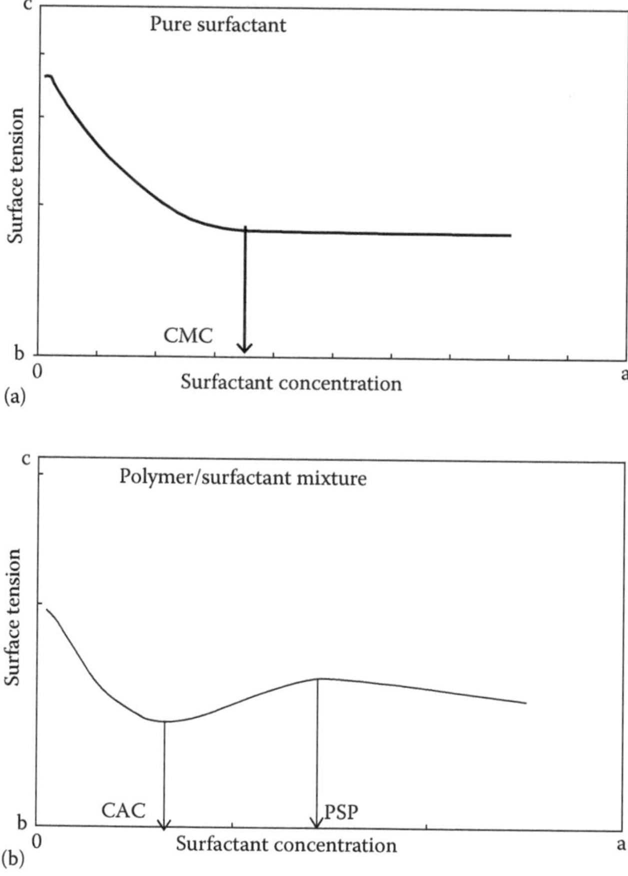

FIGURE 13.5 Schematic plots of surface tension for (a) pure surfactant and (b) mixed surfactant–polymer system. (From Mohsenipour, A.A., turbulent drag reduction by polymers, surfactants and their mixtures in pipeline flow, PhD thesis, University of Waterloo, Waterloo, Ontario, Canada, 2011.)

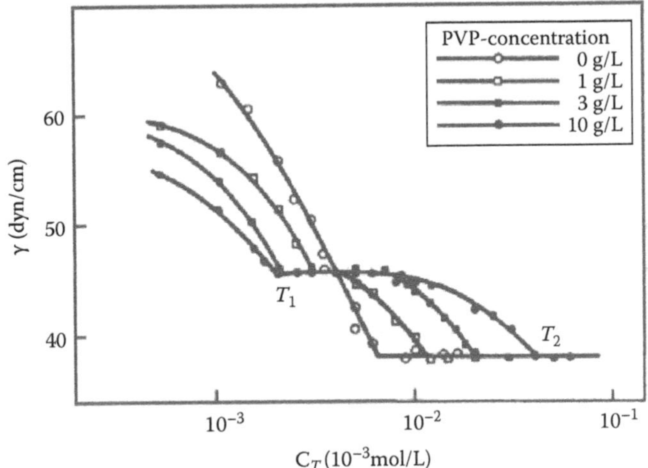

FIGURE 13.6 Surface tension of aqueous solutions of anionic surfactant SDS/PVP versus total SDS concentration. (From Lange, H., *Kolloid Z. Z. Polym.*, 243, 101, 1971.)

smaller than the CMC. At PSP (shown as T_2), the aggregation of polymer/surfactant molecules stops and free surfactant micelles begin to form.

13.3.2.2 Conductivity

The interaction between polymer and surfactant could be investigated by the conductivity measurements. The electrical conductivity measurements are usually used to detect any changes in the solution behavior when an ionic surfactant is added to the aqueous solution. If there occurs any interaction, the solution conductivity is expected to change. Figure 13.7 shows the typical conductivity plots for pure ionic surfactant solution and a mixed polymer/ionic surfactant system.

Curve A shows the trend for pure ionic surfactant solution whereas curve B demonstrates the conductivity trend for a mixture of polymer/ionic surfactant. For pure surfactant solution (curve A), the conductivity is a linear function of the surfactant concentration below the CMC. The ionic surfactant is completely dissociated below the CMC. Above the CMC, micelles are formed. The micellar contribution to conductivity is less than that of the same number of free surfactant molecules.

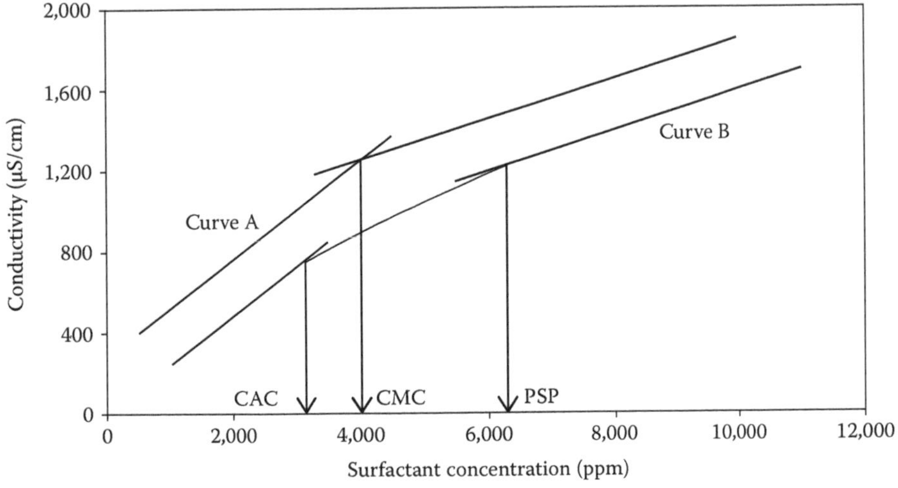

FIGURE 13.7 Schematic conductivity plots for pure surfactant and mixture of surfactant and polymer.

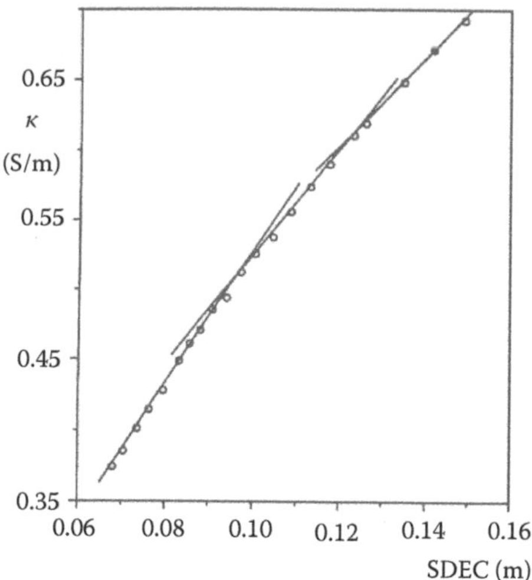

FIGURE 13.8 Specific conductivity of PEO–sodium decyl carboxylate (SDEC) solutions as a function of SDEC concentration at a fixed PEO concentration of 5000 ppm. (From Blokhus, A.M. and Klokk, K., *J. Colloid Interface Sci.*, 230(2), 448, 2000.)

Consequently the slope of the conductivity plot above CMC is lower than that below CMC although the conductivity continues to increase with the increase in the surfactant concentration. For the mixed surfactant and polymer systems, the conductivity plot shows two breakpoints. The first break point is known to occur at CAC, the concentration of surfactant where surfactant monomers begin to associate with the polymer chains. The second break point occurs at PSP (polymer saturation point) where the polymer molecules are saturated with surfactant. Above the PSP, the addition of surfactant results in the formation of free micelles. For systems where the interaction between the surfactant and polymer molecules is weak, the CAC and PSP points are not easily detectable based on the conductivity measurements (Mohsenipour 2011).

Figure 13.8 shows the conductivity data for a system consisting of PEO–sodium decyl carboxylates. Two breaks observed in the conductivity curves are attributed to CAC and PSP.

13.3.2.3 Gel Permeation Chromatography

The principle of gel permeation chromatography is based on the size difference between different components. A pump is used to push the mobile phase through a column consisting of beads as the stationary phase. The mobile phase can flow between the beads and within the pores of the beads. An injection port is used to introduce the test solution into the column. The small size species enter the pores within the beads and have a long residence time. Thus they exit the column slowly. Large size species cannot enter the pores; they have low residence time and exit the column rapidly. At the exit, there are detectors to detect the components as they leave the column. A software is used to control the instrument, and to calculate and display the results.

In the case of polymer/surfactant mixtures, the surfactant solution with a proper concentration is used as the mobile phase. The test solution consists of polymer dissolved in the surfactant solution. The solution is allowed to reach equilibrium before it is injected into the chromatographic system. The polymer/surfactant complex moves through the column faster than the smaller surfactant molecules and is separated from the vacant surfactant solution during the chromatographic run. During this process, the vacant peak, which shows the amount of surfactant that undergoes complexation with polymer, is detected using a refractive index detector (Veggeland and Austad 1993; Veggeland and Nilsson 1995).

Figure 13.9 shows the results of GPC to detect the polymer–surfactant complex of SDS/PEO (Veggeland and Austad 1993). Figure 13.9a shows a sample of GPC analysis of 1,000 ppm PEO

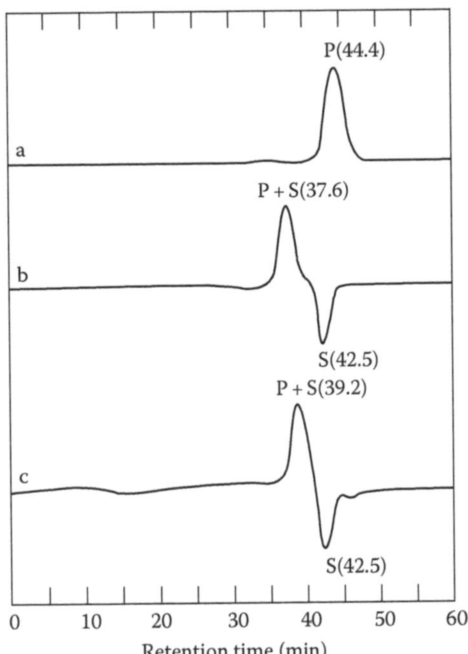

FIGURE 13.9 GPC analysis of SDS/PEO system. The PEO sample was dissolved in the mobile phase; flow rate, 0.5 mL/min; RI detector; injection volume, 50 mL: (a) 1000 ppm PEO (mobile phase—distilled water); (b) 750 ppm PEO (mobile phase—0.017 M SDS); and (c) 750 ppm PEO (mobile phase—0.017 M SDS dissolved in 0.1 wt% NaCl solution). (From Veggeland, K. and Austad, T., *Colloid Surf. A: Physicochem. Eng. Aspects*, 76, 73, 1993.)

(MW = 20,000 g/mol) in pure water using a column of Ultrahydrogel™ 2,000 (13 mm). Figure 13.9b shows the chromatogram of 750 ppm PEO dissolved in 0.017 M SDS. Figure 13.9c gives the chromatogram of 750 ppm PEO dissolved in 0.017 M SDS with 0.1 wt% NaCl. These figures show that although the flow rates are the same, the polymer exits the instrument at different times, that is, the retention time of the peaks is altered. The retention times were 44.4, 37.6, and 39.2 min in distilled water, in the presence of surfactant, and in the presence of both surfactant and salt, respectively.

When SDS is added to PEO, the effective size of the PEO chains increases due to electrostatic repulsion between the negatively charged micellar aggregates bound to the polymer chain. As the size of species changes, the retention time of the species also changes (37.6 min in Figure 13.9a compared to 44.4 min in Figure 13.9b). The addition of salt can cause suppression of electrostatic repulsion, and consequently the size of polymer–surfactant complex and its retention time are affected (see Figure 13.9c). The negative or vacant peaks *S* in the chromatograms shown in Figure 13.9b and c are related to the decrease in the surfactant concentration due to the formation of the polymer–surfactant complexes.

13.3.2.4 Viscosity Measurements

Viscometry is another simple technique that is used widely to detect possible interaction between polymer and surfactant. The relative viscosity (η_r) of a solution could be defined as the ratio of flow time of test solution (t_p) to flow time of water (t_w) through the capillary of a viscometer:

$$\eta_r = \frac{t_p}{t_w} \tag{13.1}$$

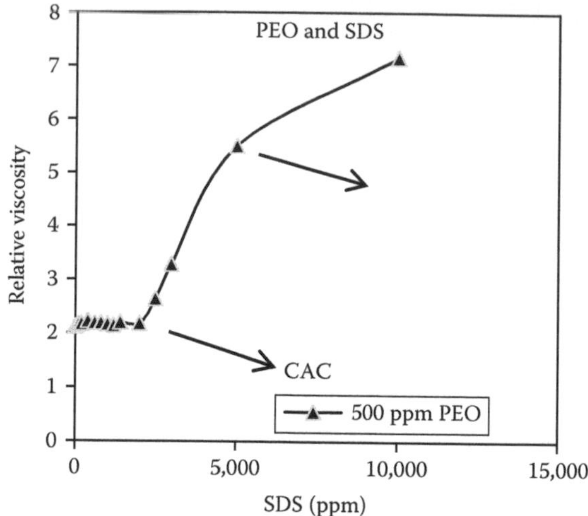

FIGURE 13.10 Relative viscosity of PEO solution as a function of SDS concentration. (From Ghoreishi, S. et al., *Langmuir*, 15(13), 4380, 1999.)

The specific viscosity (η_s) is defined as follows:

$$\eta_s = \frac{t_p - t_w}{t_w} \tag{13.2}$$

To measure the flow time of a solution, the Ubbelohde-type capillary viscometer is used. The flow time of solution is compared with the flow time of base fluid (water). As an example, Figure 13.10 shows the relative viscosity for the PEO/SDS system. When surfactant is added to polymer solution, the hydrodynamic size of polymer is increased due to the attachment surfactant micelles to the backbone of PEO chains. This causes an increase in relative viscosity. The relative viscosity begins to rise at CAC as shown in Figure 13.10.When the surfactant concentration reaches PSP, the relative viscosity begins to flatten.

The Ubbelohde-type capillary viscometer gives viscosity at a low shear rate. To obtain the full viscosity–shear rate profile, a coaxial cylinder viscometer can be employed. The shear rate in a coaxial cylinder viscometer is calculated using the following equations:

$$\text{Newtonian fluids:} \quad \dot{\gamma}_{Ri} = \frac{2S^2}{S^2 - 1}\Omega_0 \tag{13.3}$$

$$\text{Non-Newtonian fluids:} \quad \dot{\gamma}_{Ri} = \frac{2N}{1 - S^{-2N}}\Omega_0 \tag{13.4}$$

where Ω is the angular speed given as:

$$\Omega = \frac{2\pi(\text{rpm})}{60} \tag{13.5}$$

where

S in Equation 13.3 is the ratio of rotor radius to bob radius
N is the slope of ln Ω versus ln(torque) data

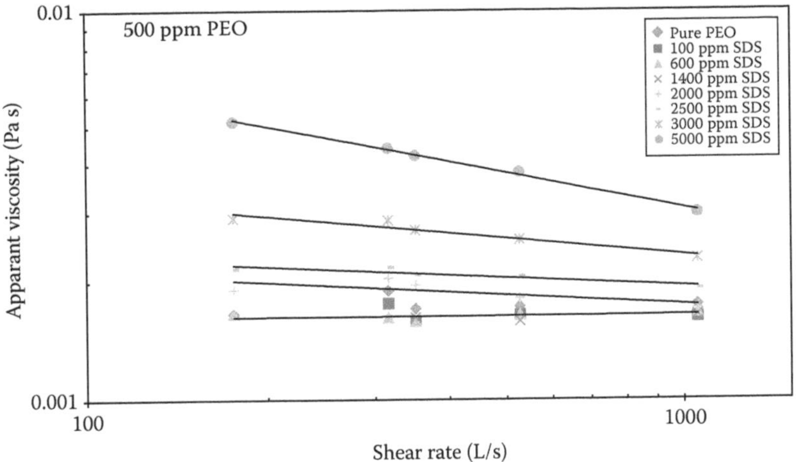

FIGURE 13.11 Apparent viscosity versus shear rate of PEO/SDS mixtures. (From Mohsenipour, A.A., turbulent drag reduction by polymers, surfactants and their mixtures in pipeline flow, PhD thesis, University of Waterloo, Waterloo, Ontario, Canada, 2011.)

The apparent viscosity at any shear rate is given as the ratio of shear stress to shear rate.

Figure 13.11 shows the apparent viscosity versus shear rate plots obtained by a coaxial cylinder viscometer for mixtures of PEO/SDS. The addition of SDS to pure PEO solution causes an increase in apparent viscosity. The solution shows a Newtonian behavior at low SDS concentrations. The solution becomes pseudo plastic shear-thinning at high SDS concentration.

13.3.2.5 Electromotive Force

This method utilizes a surfactant selective electrode to monitor the interactions between the polymer and the surfactant. The EMF of the solution is dependent on the concentration of free surfactant in the solution. The binding of surfactant molecules to the polymer molecules results in a decrease in the free surfactant concentration and hence EMF. A known amount of concentrated surfactant solution is injected into a polymer solution and the EMF value is recorded when the equilibrium is reached. The process is repeated with further injections of surfactant solution. The EMF data are plotted against the surfactant concentration for the solutions with and without the polymer.

Figure 13.12 shows the EMF behavior of SDS/PVI system. In the figure, three regions can be recognized. In the first region (SDS concentration $< T_1$), no interaction occurs between the surfactant and polymer. In the next region, the EMF data begin to deviate from the EMF data of pure surfactant. The SDS concentration where deviation begins is CAC (T_1). With further addition of surfactant to the system, the curve continues to deviate from pure surfactant until the SDS concentration reaches PSP (T_2), where the polymer chains become saturated with surfactant. After this point, the additional surfactant does not go to the polymer molecules and the two curves (mixed system and pure surfactant) merge.

From the EMF curve, it is possible to calculate the concentration of free (monomer) surfactant at any given total surfactant concentration within the interaction zone using the following relation:

$$E = E_0 - \left(\frac{2.303RT}{F}\right)\log(m_1) \tag{13.6}$$

where m_1 is the concentration of free monomer surfactant. The slope $(-2.303RT/F)$ and the intercept E_0 are determined from the linear data in the initial region of the EMF curve, which

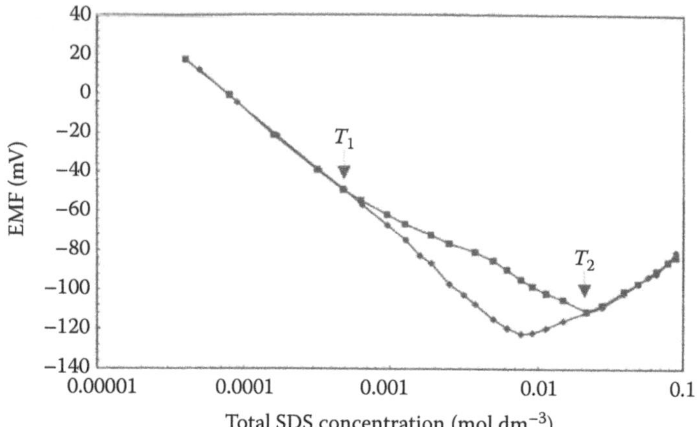

FIGURE 13.12 EMF plot of the SDS electrode (reference Br^-) as a function of the total SDS concentration for the SDS/PVI system in 10^{-4} mol dm^{-3} NaBr: diamond symbol—pure SDS; square symbol—SDS + PVI (0.1% w/v). (From Ghoreishi, S. et al., *Langmuir*, 15(13), 4380, 1999.)

mostly overlaps with the data for pure surfactant. Note that R is the universal gas constant, T the absolute temperature, and F the Faraday constant.

13.3.2.6 Isothermal Titration Calorimetry

When a chemical reaction or physical interaction between different molecular species occurs in a solution, it is accompanied by an enthalpy change. If the process is an endothermic process, then the heat is absorbed from the surroundings. In an exothermic process, heat is released. By measuring the amount of heat absorbed or released, one can determine the degree of interaction between different species. The isothermal titration calorimetry (ITC) is based on this principle of measurement of the heat generated or absorbed upon interaction of different molecules. In the modern ITC instruments, it is possible to measure heat as small as 0.1 mcal (0.4 mJ) and the heat generation rate as small as 0.1 mcal/s, allowing the determination of binding constants, K's, as large as 108–109 mol^{-1} and the precise calculation of reaction rates in the range of 10^{-12} mol/s. The ITC method can be applied to a wide variety of solutions. This method is used to study interactions between protein–small molecule, enzyme–inhibitor, protein–protein, protein–DNA, protein–lipid, protein–carbohydrate, and polymer–surfactant interactions where some other methods may not be applicable. The ITC can be used over a range of biologically relevant conditions (temperature, salt pH, etc.) (Lewis and Murphy 2005). Figure 13.13 shows the major features of the ITC instrument. The microcalorimeter consists of a reference cell and a sample cell. Both are insulated using an adiabatic shield as shown in the figure. The sample cell is filled with the polymer solution to be titrated. The titrant (surfactant solution) is added to the sample cell using the injection syringe. The number of injections is selected and entered as the software data. The injection time and amount of titrant are calculated and controlled by machine. The power differential between the sample and reference cells in maintaining the temperature difference to be zero between the cells is measured, resulting in an instrument output signal.

Figure 13.14 shows the ITC curves for the system consisting of surfactant SDS and polymer PEG. This ITC experiment was done to detect the possible interaction between SDS and two different molecular weight PEGs (MW = 900 and 1450).

The solution of polymer was titrated with 0.2 M SDS. The endothermic peaks observed when SDS is added to PEG-20 and PEG-30 solutions are more pronounced as compared with the peak observed for pure SDS solution. The pronounced endothermic peaks are due to surfactant–polymer complex formation. The titration curves are parallel for SDS concentrations lower than 5.9 mM.

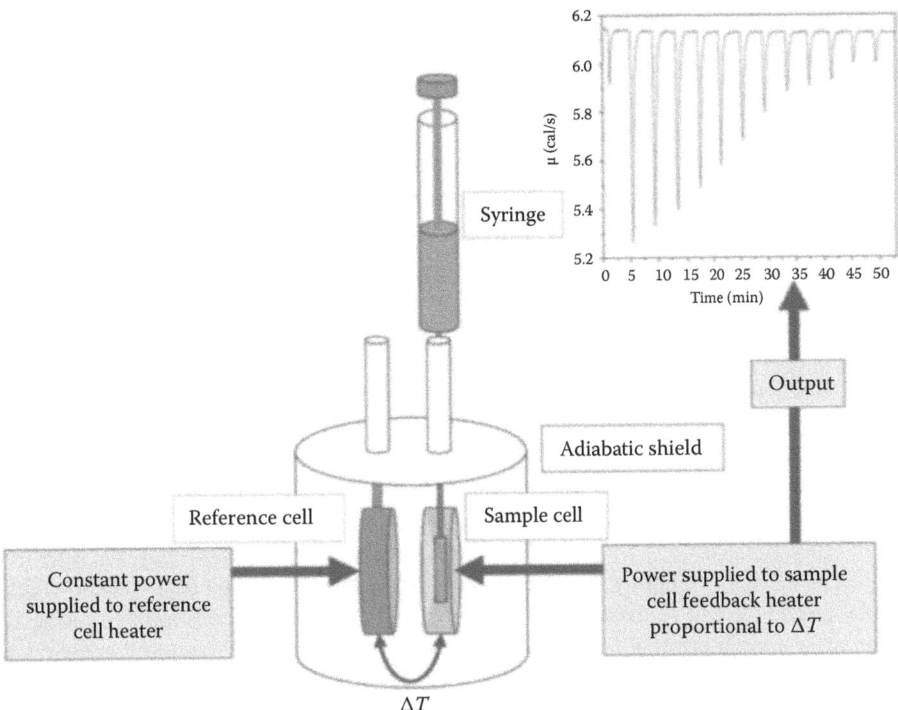

FIGURE 13.13 Schematic diagram of a power compensation ITC. (From Lewis, E.A. and Murphy, K.P., *Isothermal Titration Calorimetry. Protein-Ligand Interactions*, Springer, 2005, pp. 1–15.)

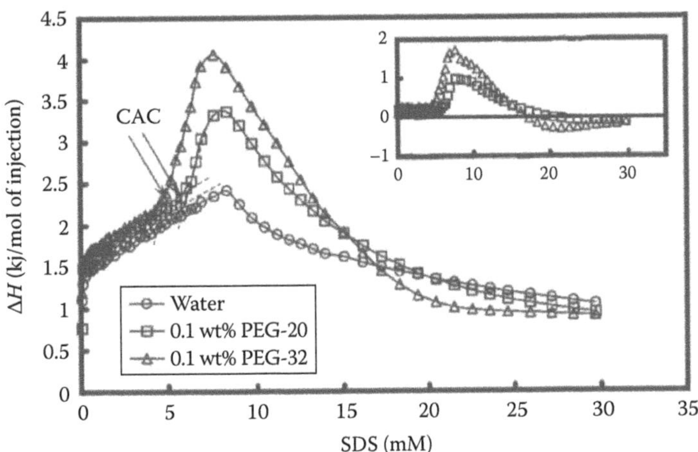

FIGURE 13.14 The ITC curve of 0.2 M SDS titrating into 0.1 wt% PEG-20 □) and 0.1 wt% PEG-32 (Δ) at 298 K. The open circle is the SDS dilution curve in water. The insert is a plot for the difference curves of titrating 0.2 M SDS into PEGs and water. (From Dai, S. and Tam, K., *J. Phys. Chem. B*, 105(44), 10759, 2001.)

Above this point the deviation in the titration curves is observed. This point is CAC. The SDS concentration where the titration curves (pure SDS and SDS/PEG curves) meet each other again is the PSP point. In the present SDS/PEG system, the endothermic reaction is attributed to the dehydration of PEG segments from water phase and the association of these segments with the hydrophobic core of the SDS micelles.

13.3.3 Interactions of Polymer and Surfactant with Opposite Charges

The polymer/surfactant solutions in which the polymer and surfactant carry opposite charges have been investigated in a number of studies. In these solutions, the interaction between the polymer and surfactant is usually strong as the main force of interaction is electrostatic in nature. The addition of ionic surfactant to an oppositely charged polymer solution leads to neutralization of the positive or negative charges of the polymer (electrolyte). The charges on ionic surfactant head groups interact with the charges of ionic polymer molecules. The interaction is fast and occurs even with the addition of a small amount of surfactant to the polymer solution. The interaction of polyelectrolyte and surfactant with opposite charges depends highly on both the polyelectrolyte and surfactant properties (Goddard and Ananthapadmanabhan 1993). The polymer chains may collapse due to charge neutralization resulting in a sharp reduction in the solution viscosity.

When an ionic surfactant is added to an aqueous solution of ionic polymer, the surfactant molecules begin to interact with ionic polymer to form micelles on the backbone of the polymer at critical aggregation concentration (CAC). In general, the surfactant–polymer complex starts to precipitate when the ratio of surfactant to polymer is close to 1. Upon the addition of surfactant to a solution, and hence neutralization of the charges present on the backbone of the polymer chains, the solution tends to become more hydrophobic resulting in the precipitation of the polymer chains (Thalberg et al. 1990). The addition of more ionic surfactant (after the precipitation of complexes) may help the precipitates to re-dissolve. Some studies indicate that the addition of ionic surfactant to the solution after precipitation of complexes cause a charge reversal on the polymer chains resulting in resolubilization of polymer/surfactant aggregates. It seems that the additional surfactant monomers become attached to the complexes and increase the hydrophilicity of these clusters (Goddard and Ananthapadmanabhan 1993).

In the case of oppositely charged polymer and surfactant, it has been generally observed that the critical aggregation concentration (CAC) is an order of magnitude lower than the critical micelle concentration (CMC) of surfactant. As an example, Figure 13.15 shows the comparison of CAC and CMC for a system consisting of oppositely charged surfactant and polymer.

Although the main driving force for the formation of polymer–surfactant complexes is the electrostatic bonding, the hydrophobic interaction of surfactant tail with the polymer also helps to improve the interaction (Kogej and Škerjanc 1999). In the case of hydrophobically modified polyelectrolytes

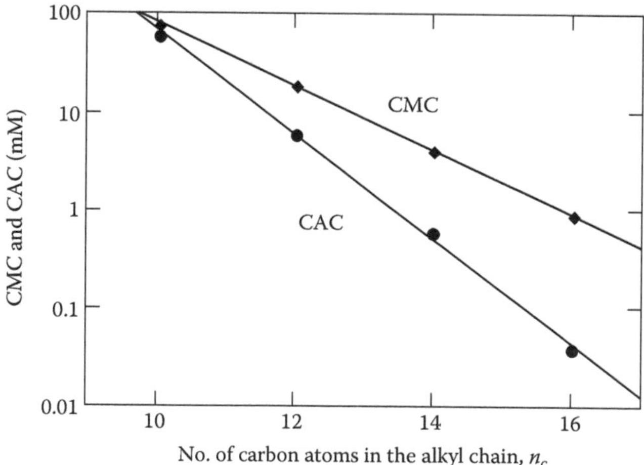

FIGURE 13.15 Relation between CMC and CAC for a system consisting of an ionic surfactant and an oppositely charged polymer. (From Jönsson, B. et al., *Surfactant-Polymer Systems in Surfactant and Polymers in Aqueous Solution*, John Wiley & Sons, New York, 2003, Chapter 13; Thalberg, K. and Lindman, B., *J. Phys. Chem.*, 93(4), 1478, 1989.)

and oppositely charged surfactants, the driving force for the interaction includes both electrostatic and hydrophobic forces. However, the structure of the polymer, especially the distribution of the hydrophobic modifier on the backbone of the polymer chain, has a strong influence on the polymer–surfactant interaction. With microblocky distribution, the interaction is highly dependent on the concentration of surfactant and in this case intermolecular association is improved. In the case of polymers with random distributions, intramolecular associative behavior is increased (Winnik and Regismond 1996). Viscosity measurement is often used to detect the conformational behavior of polymer chains in a surfactant solution. The interaction between oppositely charged polymer and surfactant can have a strong effect on the viscous behavior of polymer/surfactant solutions.

The oppositely charged polymers and surfactants may form a single phase without complex formation, or may form a soluble surfactant–polymer complex, or may result in phase separation (Piculell and Lindman 1992; Wang et al. 1999, 2000). If the interaction leads to liquid–liquid phase separation, the process is called "coacervation." If the interaction leads to a liquid–solid phase separation, then the process is named "precipitation." Many factors affect the interaction of oppositely charged surfactants and polymers that consequently have an effect on the possible phase separation phenomena. Extensive studies have been done on both theoretical and practical aspects of phase separation. It is important to know the conditions under which coacervation occurs in applications dealing with the formulation of cosmetics and pharmaceuticals.

The factors that affect the interaction of oppositely charged surfactants and polymers are the surfactant type, micelles charge density, ionic strength, polymer/surfactant ratio, polymer molecular weight (MW), linear charge density, and temperature. However, the electrostatic factors, such as macromolecular charge densities and ionic strength, are the most significant (Wang et al. 1999).

The subject of interactions between oppositely charged polymers and surfactants can be divided into two categories. One category deals with anionic polymer and cationic surfactant and the other deals with cationic polymer and anionic surfactant. The interactions are similar in principle in both the categories.

13.3.3.1 Interactions of Anionic Polymers with Cationic Surfactants

There are many research studies published on the interaction of anionic polymers with cationic surfactants (Asnacios et al. 1996; Bonnaud et al. 2010; Chandar et al. 1988; Hayakawa and Kwak 1983; Hayakawa et al. 1983; Magny et al. 1994; Malovikova et al. 1984; Mohsenipour et al. 2013a; Sardar and Kamil 2012a,b; Scheuing et al. 2014; Shimizu et al. 1986; Volden et al. 2011; Wang and Tam 2002). Most of these studies are also applicable to the interaction of cationic polymers and anionic surfactants.

Kwak and his team have carried out a good amount of work on the interaction of anionic polymer with cationic surfactant under different conditions, such as added salt concentration, salt type, and temperature (Hayakawa and Kwak 1982, 1983; Hayakawa et al. 1983; Malovikova et al. 1984; Shimizu et al. 1986). For instance, Hayakawa and Kwak (1982) studied the interaction of dodecyltrimethylammonium bromide (DTAB) with two anionic polymers: sodium polystryrenesulfonate (NaPS) and sodium dextran sulfate (NaDxS). In an isothermal condition, they found that the interaction of DTAB with NaPS begins at a lower concentration as compared with the DTAB/NaDxS system due to a higher hydrophobicity of NaDxS.

Although the electrostatic force plays the main role in the interactions of oppositely charged surfactants and polymers, hydrophobicities of polymer and surfactant also play a role. Zana and Benrraou (2000) studied the interaction of two polyelectrolytes PS1 and PS4 (copolymers of disodium maleate and methyl or butyl vinyl ether, respectively), and quaternary ammonium bromide surfactants (3-dodecyldimethyl(alkyl)ammonium bromides and two dimeric surfactants of the polymethylene-α,ω-bis(dodecyldimethylammonium bromide) type). They used the surfactant-binding isotherms method and spectrofluorometry using pyrene as a fluorescent probe to detect the onset of binding. Their results showed that surfactant binding to the polymer is more pronounced when the polymer is more hydrophilic for a given surfactant.

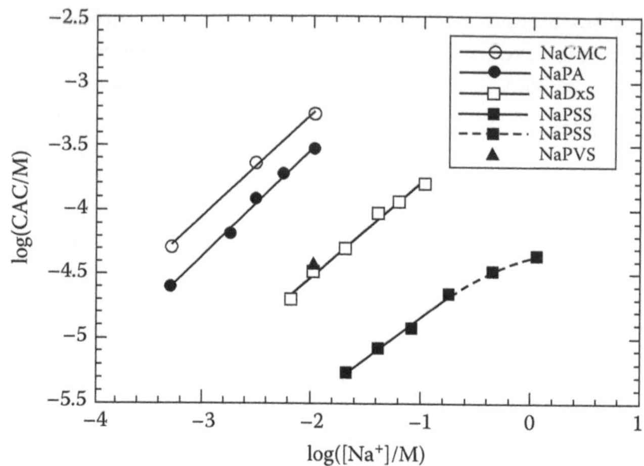

FIGURE 13.16 Effect of salt on the CAC for solutions of $C_{12}TAB$ and different polyelectrolytes. C_P = 0.5 mM. Added salt: NaBr (polymers NaCMC, NaPA, NaPVS), NaCl (polymers NaDxS, NaPSS). (From Hansson, P. and Almgren, M., *J. Phys. Chem.*, 100(21), 9038, 1996.)

Hansson and Almgren (1996) investigated the interaction of alkyltrimethylammonium bromide (CnTAB) with sodium (carboxymethyl)cellulose (NaCMC) in dilute aqueous solutions. They reported that the polyion enhances surfactant aggregation by acting as a counterion cloud. The free energy of formation of this cloud is reflected in the aggregation number and the CAC for the surfactant. They also examined the effect of salt on the CAC (see Figure 13.16). The plots of CAC versus salt concentration are linear on a log–log scale with the following slopes: 0.80 (NaCMC), 0.83 (NaPA), 0.74 (NaDxS), and 0.65 (NaPSS). With the increase in salt concentration, the CAC value increases implying that the polymer–surfactant interaction is delayed with the increase in salt concentration. The increase in salt concentration also results in an increase in the aggregation of surfactant molecules attached to the polyelectrolyte.

By comparing the CAC values for different systems, they concluded that some properties of polyion could have more pronounced effect on interaction and could improve the aggregation of surfactant molecules. They listed the properties as following in order of importance:

Hydrophobicity > Nature of charge carrying group > Flexibility

The effect of linear charge density (ξ) on the interaction has been a part of many studies. These studies have shown that with the decrease in ξ, there is a less effective screening of the charges on the surfactant molecules attached to the backbone of polymer. This results in a larger repulsion and consequently a smaller aggregation number. From a thermodynamic point of view, less work is needed to move the polyelectrolyte from the micelle when the charge density is small. A decrease in the aggregation number with the decrease in ξ is consistent with an increase in CAC.

Almost the same results were obtained by Wallin and Linse (1996a,b) using Monte Carlo simulations to investigate the complexation of a charged micelle with an oppositely charged polymer. They found that the critical micelle concentration (CMC) is greatly reduced in the presence of polyelectrolyte. The most reduction in CMC occurs for the highest linear charge density along with the most flexible polyelectrolyte (110 times smaller as compared with pure surfactant CMC). In the case of lower linear charge density and the more rigid polymer structure, the reduction in CMC was about 10 times. The increase in the linear charge density leads to the stabilization of the system by a decrease in the electrostatic energy and an increase in the entropy due to an increase in the counterions of the macro-ions.

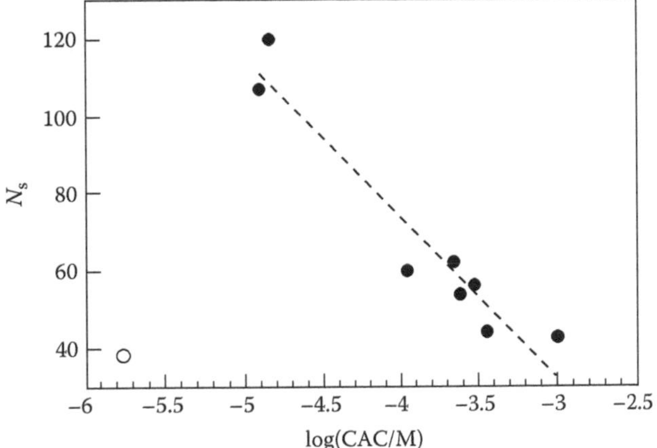

FIGURE 13.17 Relation between aggregation number and CAC for solutions of $C_{12}TAB$ and different polyelectrolytes. (From Hansson, P. and Almgren, M., *J. Phys. Chem.*, 100(21), 9038, 1996.)

Hansson and Almgren (1996) concluded that for almost all of the systems that they investigated, there existed a relation between the aggregation number (N_S) and CAC. They reported that for a fixed salt concentration, a large N_S generally corresponded to a low CAC (see Figure 13.17).

Prajapati (2009) investigated the interaction of CPAM (copolymer of polyacrylamide and sodium polyacrylate) and OTAC-p (octadecyltrimethylammonium chloride) and observed that strong interactions are present between anionic polymer CPAM and cationic surfactant OTAC-p based on the conductivity, viscosity, surface tension, and turbidly measurements. When CPAM is introduced to the surfactant solution, the CMC value is increased due to transfer of free OTAC-p molecules to CPAM chains. The cationic OTAC-p molecules neutralize the charge on the backbone of CPAM chains. At higher CPAM concentrations, parts of the CPAM chains become inaccessible to OTAC-p molecules. The results from the conductivity measurements showed that the CMC value increases from 5700 ppm for pure OTAC-p to 6800 and 7100 ppm in the presence of 500 and 1000 ppm CPAM, respectively (see Figure 13.18).

The viscosity change was also monitored with the increase in surfactant (OTAC-p) concentration. A sharp reduction in viscosity was observed reflecting a strong interaction between the surfactant and the polymer. The electrostatic repulsive forces that cause the extension of polymer chains are neutralized. This leads to the shrinkage of polymer chains resulting in a large reduction in the

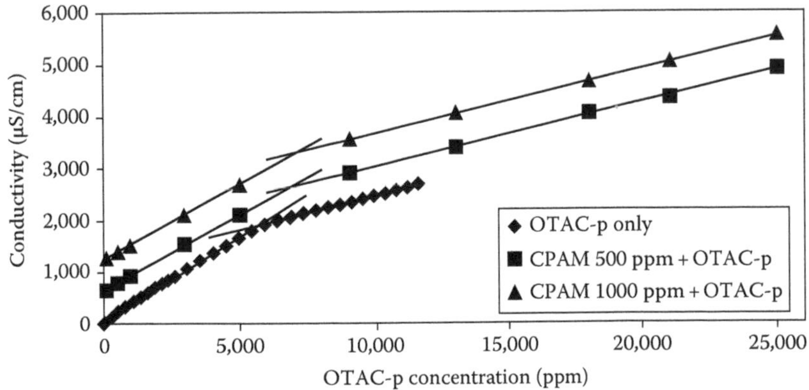

FIGURE 13.18 Conductivity of CPAM/OTAC-p solution in DI water versus OTAC-p concentration. (From Prajapati, K., Interactions between drag reducing polymers and surfactants, MSc thesis, University of Waterloo, Waterloo, Ontario, Canada, 2009.)

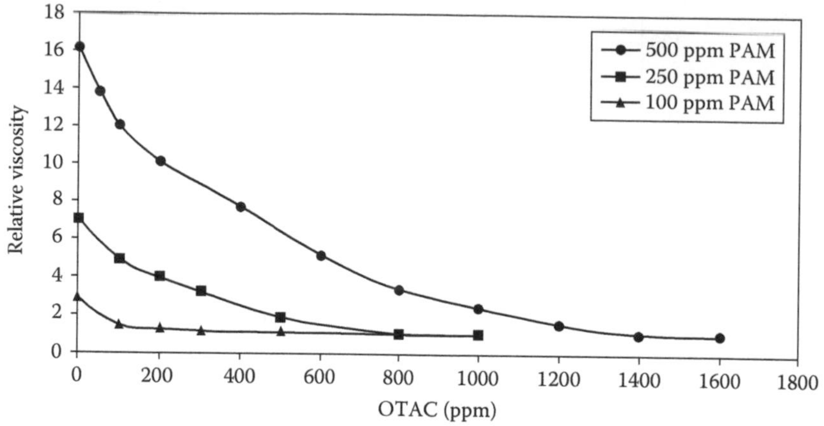

FIGURE 13.19 Relative viscosity for different concentration solutions of CPAM as a function of OTAC concentration. (From Mohsenipour, A.A. et al., *Can. J. Chem. Eng.*, 91(1), 181, 2013.)

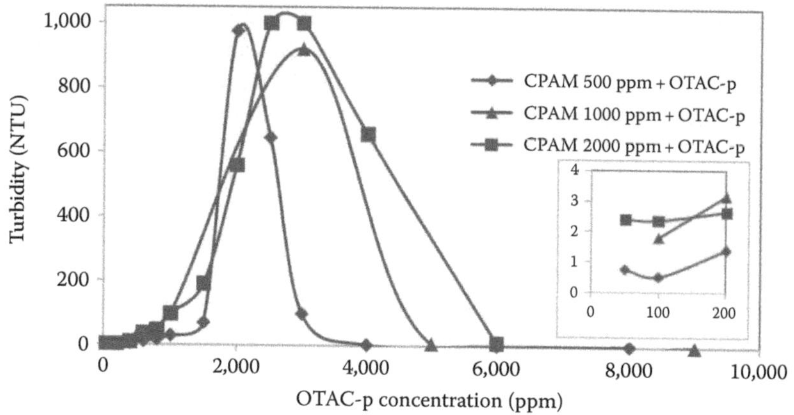

FIGURE 13.20 Turbidity of CPAM/OTAC-p solutions in DI water as a function of OTAC-p concentration. (From Prajapati, K., Interactions between drag reducing polymers and surfactants, MSc thesis, University of Waterloo, Waterloo, Ontario, Canada, 2009.)

relative viscosity. Similar results were reported by Mohsenipour et al. (2013) (see Figure 13.19). The interaction resulted in the insolubility and precipitation of polymer chains. Precipitation was reflected in the turbidity data (see Figure 13.20). The turbidity data of the solutions showed that with the addition of surfactant, the phase separation takes place. The phase separation point is very close to the maximum turbidity shown in Figure 13.20.

Zhang et al. (2013) studied the interaction of long-chain imidazolium ionic liquid (C_{14}mimBr) and polystyrene sulfonate (NaPSS), which is an anionic polyelectrolyte. They used surface tension, isothermal titration microcalorimetry (ITC), dynamic light scatting (DLS), and conductance methods to study the interactions. The surface tension exhibited a complicated behavior with the addition of surfactant to the polymer solution when the surfactant concentration was below the CMC value. The formation of the surfactant/polymer complexes was affected by the concentrations of surfactant and NaPSS. They explained the influence of surfactant on surfactant–polymer interactions mechanistically as shown in Figure 13.21. When cationic surfactant molecules are added to the solution of an anionic polymer, the head groups of surfactant (positive) molecules attach to the negative sites on the backbone of the polymer. This surfactant–polymer complex moves to the surface as more surfactant monomer becomes attached to the backbone of the polymer and the complex becomes more

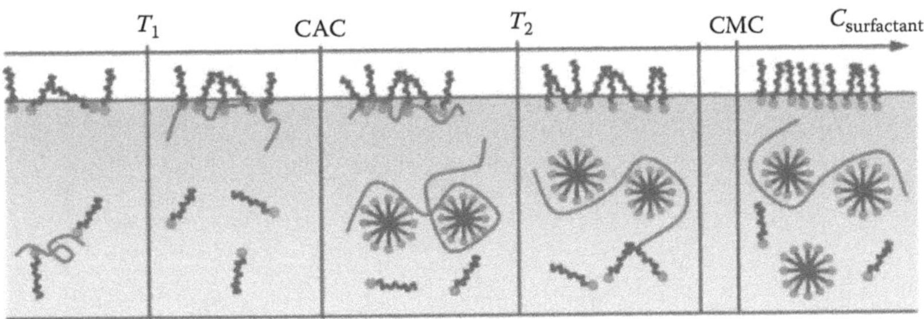

FIGURE 13.21 Precipitation of anionic polymer upon interaction with a cationic surfactant. (From Zhang, Q. et al., *Appl. Surf. Sci.*, 279, 353, 2013.)

hydrophobic. At the CAC point, micelles start to form on the polymer chains. With further increase in the surfactant concentration, the surfactant–polymer complexes present at the surface move to the bulk solution. At some surfactant concentration (indicated as T_2), the polymer chains become saturated with surfactant micelles. This point is the PSP (polymer saturation point). With further addition of surfactant, the surfactant molecules go the surface. When the CMC concentration of surfactant is reached, the surfactant molecules begin to form polymer-free micelles in the solution.

13.3.3.2 Interactions of Cationic Polymers with Anionic Surfactants

The interaction of cationic polymers with anionic surfactants has been investigated in a number of studies (Banerjee et al. 2013; Dan et al. 2010; Dubief et al. 1989; Goddard and Hannan 1976; Goldraich et al. 1997; Han et al. 2012; Li et al. 2012; Mukherjee et al. 2011; Shubin 1994; Winnik et al. 1997). The phase behavior of polymer–surfactant system under interaction and the influences of factors such as micelle surface charge density, polyelectrolyte molecular weight, and polyelectrolyte-to-surfactant ratio have been explored. The interaction behavior of cationic polymer and anionic surfactant is generally found to be similar to the interaction behavior of anionic polymer and cationic surfactant discussed in the preceding section.

Goddard and Hannan (1976) studied the interaction of a water-soluble cationic polymer, substituted cellulose ether, with sodium dodecyl sulfate (SDS) using surface tension, foaming, and electrophoretic mobility measurements. They also measured the force/area to study the effect of polycation in the sub-solution underlying a monolayer spread of sodium docosyl sulfate (SDCS). They reported that the addition of SDS to the polymer solution reduces the solubility and electrophoretic mobility of the polymer. Maximum precipitation was observed close to zero mobility of the polymer. After this point of zero mobility, the addition of more surfactant caused resolubilization of the complex. Surface tension studies revealed that in the presence of the polymer, surface tension of SDS solution is lower compared to SDS solution alone. The presence of polymer in the sub-solution underlying a monolayer spread of sodium docosyl sulfate (SDCS) exerts a strong influence on the force/area characteristics. In another study Goddard (1986a,b) found that the surface tension behavior of a system of polycation and an anionic surfactant is different from the regular surfactant system and other polymer/surfactant mixed systems. They proposed different scenarios with varying concentration of surfactant in a system consisting of cationic cellulosic polymer JR and SDS, as shown in Figure 13.22. At a low concentration of surfactant, the interaction between the surfactant and the polymer is reflected in a pronounced lowering of the surface tension values. Once the polymer is precipitated out and most of polymer and surfactant are out of the solution, the measured values of the surface tension still show low values indicating that the surfactant–polymer complex is a highly surface-active material. The addition of more surfactant increases the amount of polymer-free micelles and reduces the amount of polymer at the surface. The surface tension value finally reaches the same value as that of the pure surfactant system. This mechanistic picture has been supported by the works of other researchers such as Cooke et al. (2000).

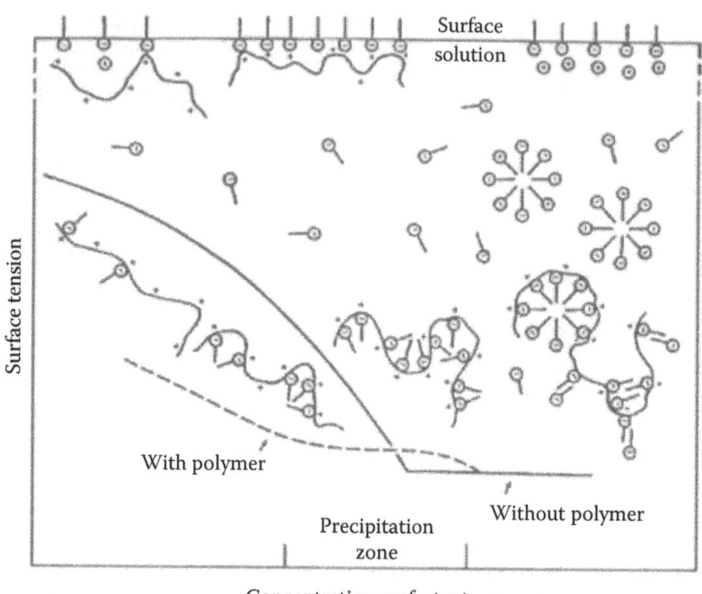

FIGURE 13.22 Conditions in the bulk solution and at the surface of the solution of polycation (fixed concentration) and anionic surfactant. Solid line: hypothetical surface tension curve of surfactant alone; dashed line: surface tension curve for the mixture of surfactant and polycation. (From Goddard, E., *Colloid Surf.*, 19(2), 301, 1986.)

As mentioned earlier, when the polymer is a polyion with charges opposite to that of a surfactant a strong interaction is expected. The starting sign of such interaction is the formation of precipitates. Figure 13.23 shows a simple qualitative phase diagram reflecting the interaction of cationic cellulosic polymer JR* (Union Carbide, Danbury, CT) with triethanolamine lauryl sulfate (TEALS). Goddard (1994) recorded the appearance of the system and developed the diagram shown in Figure 13.23. A clear system is accessible at low and high concentrations of surfactant. This was also observed in the works of Prajapati (2009) and Mohsenipour et al. (2013).

Based on the research work that has been carried out on the interaction of polyelectrolytes and ionic surfactants, the CAC may be looked upon as a solubility parameter. CAC is the solubility threshold when the two components (polymer and surfactant) are mixed. At the CAC, the solution starts becoming turbid and the surfactant–polymer complex begins to precipitate. This process takes place in all the systems containing ionic polymer and an oppositely charged surfactant and is called *colloid titration* (Goddard and Ananthapadmanabhan 1993; La Mesa 2005). In colloid titration, the CAC can be expressed as:

$$\text{CAC} \propto [C_P][C_S]^n \tag{13.7}$$

where the first term indicates the polymer content and the second term surfactant concentration. The exponent n is the number of ions need to reach the stoichiometry. It has been shown that precipitation usually happens at a moderate concentration of surfactant and n is close to unity (La Mesa 2005).

Mukherjee et al. (2011) studied the interaction of cationic polymer poly(diallyldimethylammonium chloride) (PDADMAC) with the anionic surfactants sodium dodecyl sulfate, sodium dodecylbenzenesulfonate, and sodium *N*-dodecanoylsarcosinate using tensiometry, turbidimetry, calorimetry, viscometry, dynamic light scattering (DLS), and scanning electron microscopy (SEM). They found that the morphologies of the surfactant–polymer complexes have different patterns depending on

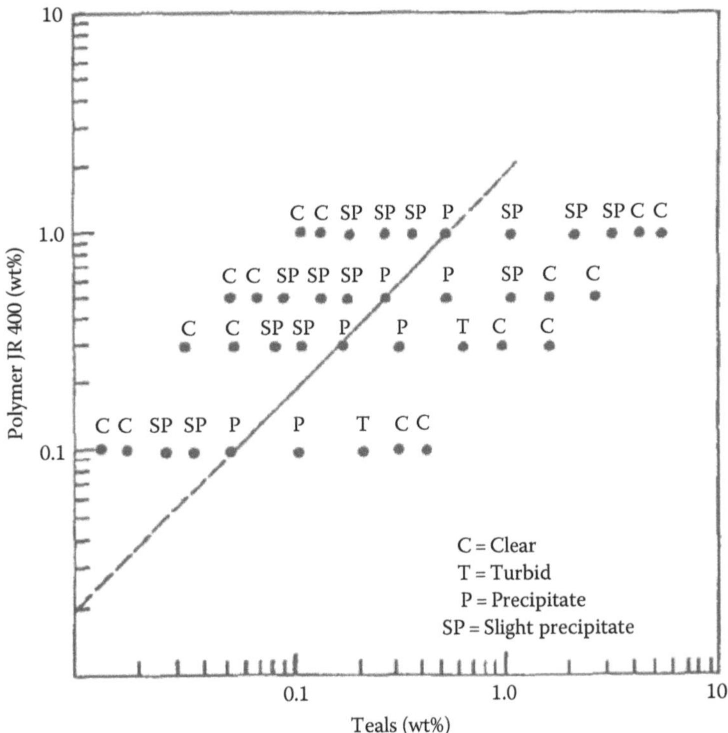

FIGURE 13.23 Solubility diagram of mixed polymer (JR 400) and surfactant (triethanolamine lauryl sulfate) system. (From Goddard, E., *J. Am. Oil Chem. Soc.*, 71(1), 1, 1994.)

the composition and the solvent environment. The type of the surfactant head groups also had a significant influence on the interaction process.

Liu et al. (2014) studied the effects of three sulfonate gemini surfactants with different hydrophobic chain lengths (8, 10, and 12 carbon atoms) on the optical properties of a fluorene-based conjugated cationic polymer poly{[9,9-bis(6′-*N,N,N*-trimethylammonium)hexyl]-fluorene-phenylene}bromide (PFP), which was dissolved in either dimethyl sulfoxide (DMSO)/water solutions (4% v/v), or pure water. They observed a decrease in the photoluminescence (PL) intensity when surfactant and PFP were dissolved in DMSO/water solutions. They also reported a red shift of emission maxima at low surfactant concentrations. With the addition of more surfactant the PL intensity was enhanced. A mechanistic model was presented (see Figure 13.24) to explain their observations. In un-aggregated polymer chains, the surfactant molecules get attached to the backbone of the polymer chains and neutralize the charges present on the polymer chains. The aggregates of polymer chains are formed with surfactant attached to each chain, as shown schematically in Figure 13.24a. For aggregated polymer chains, the surfactant molecules get attached to each chain of the aggregate (see Figure 13.24b). With the addition of more surfactant to the system, the aggregates are separated resulting in charge-neutralized complexes of surfactant and single polymer chains. It would be interesting to explore the surface tension behavior of such systems. Surface tension measurements could bring in more information on the interaction of PFP/surfactants.

13.3.4 Interactions of Uncharged Polymers and Charged Surfactants

The interaction between uncharged polymers and charged surfactants has been studied in a number of research articles. Jones (1967) studied the interaction in polyethylene oxide (PEO)/sodium dodecyl sulfate (SDS) systems. For a fixed amount of polymer concentration, the surfactant concentration

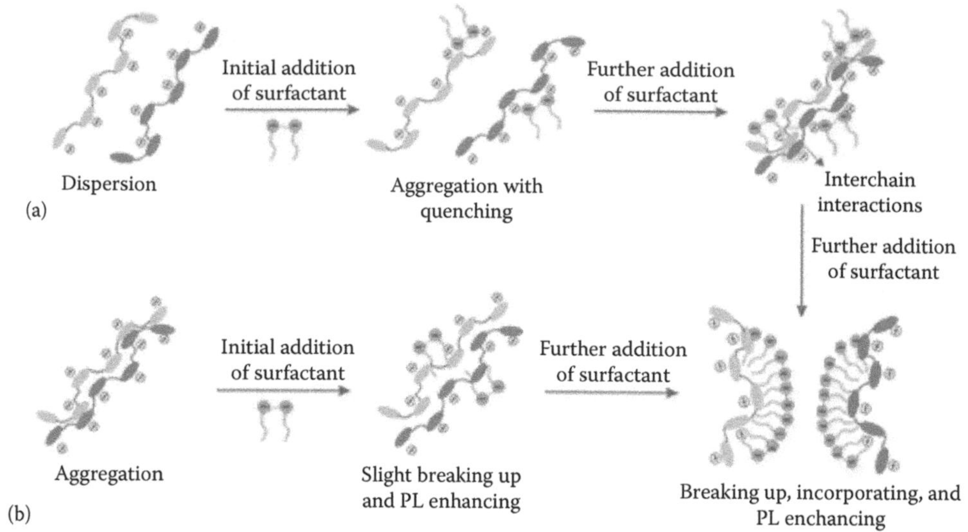

FIGURE 13.24 A mechanistic model describing the interaction of an anionic surfactant with a cationic polymer: (a) 4% DMSO/water solution and (b) aqueous solution. (From Liu, X.-G. et al., *Langmuir*, 30(11), 3001, 2014.)

was varied. He reported two critical concentrations: CAC (critical aggregation concentration) and PSP (polymer saturation point). His work has had a major impact in the area of polymer/surfactant interactions. Jones formalized the important concepts and defined CAC and PSP when surfactant is added to the polymer solution (Jones 1967). A good review on some of the early work on the interaction of nonionic polymer and ionic surfactant is given by Goddard (1986a,b). A considerable effort has been made in understanding the interactions between nonionic polymer and different types of surfactants (Fishman and Elrich 1975; Jiang and Han 2000; Ma and Li 1989). Some of the recent studies are discussed here.

Figure 13.25 shows schematically different regions of interaction between a nonionic polymer and an ionic surfactant. At very low concentrations of surfactant (region I) no significant interaction

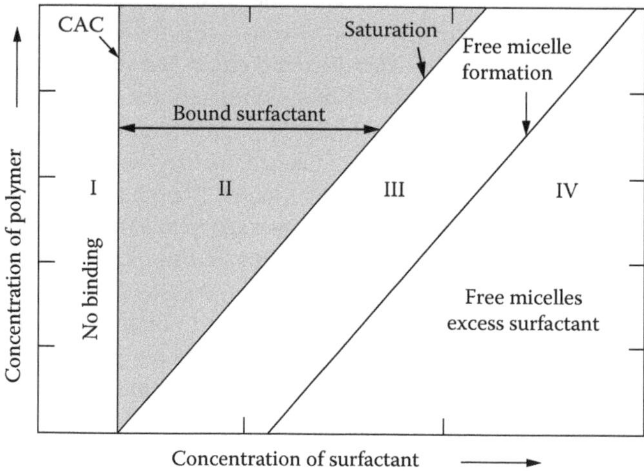

FIGURE 13.25 Different regions of association between nonionic polymer and ionic surfactant. (From Jönsson, B. et al., *Surfactant-Polymer Systems in Surfactant and Polymers in Aqueous Solution*, John Wiley & Sons, New York, 2003, Chapter 13.)

takes place between the polymer and the surfactant. In region II with surfactant concentration ≥ CAC (critical aggregation concentration), the surfactant molecules begin to interact with the polymer molecules resulting in the formation of polymer–surfactant complexes. At a certain concentration of surfactant (corresponding to the end of region II), the polymer molecules become saturated with the surfactant. In region III, the surfactant molecules tend to aggregate and form micelles on the backbone of the polymer molecules. At high surfactant concentrations corresponding to region IV, free micelles (not associated with polymer) are formed.

The small-angle neutron-scattering experiments have proven that the surfactant molecules attach to the polymer chains in the form of micelles. The surfactant micelles on polymer chains are similar to the micelles in the polymer-free surfactant solution. The only difference is the aggregation number, which is less for a surfactant–polymer system compared to the polymer-free solution. However, the radius of gyration of the polymer molecule in a surfactant–polymer system is comparable to that of a "free" macromolecule (Ruckenstein et al. 1987) in the absence of any surfactant.

13.3.4.1 Interaction of Nonionic Polymers with Anionic Surfactants

Many studies have been performed on nonionic polymers and anionic surfactants, which have shown strong interaction between the polymer and surfactant. The studies have shown that anionic surfactants are very effective in binding to the nonionic polymers. The studies support the idea that the size of the anionic head group and its hydrophobicity play a significant role in the overall interaction (Goddard 1986a,b; Hansson and Lindman 1996; Nagarajan 1989).

Some of the factors that have been investigated in the published studies are listed as follows:

1. Chain lengths of polymer and surfactant
2. Polymer hydrophobocity
3. Surfactant headgroup
4. Polymer molecular weight
5. Presence of salt
6. Fraction of micelle ionization

As the interaction of nonionic polymer and ionic surfactant is controlled by the hydrophobic forces between these two components, any factor that increases the hydrophobicity will have a strong effect on the interaction. Polymers and surfactants with long chains and tails are more susceptible to interaction than those with shorter chains or tails. In the case of a polymer, longer chains can promote the hydrophobicity of the polymer. The hydrophobicity of polymer molecules could be increased by the addition of a small number of long alkyl chains to the backbone of the polymer molecules. Such polymers are called "hydrophobically modified (HM)" polymers.

Breuer and Robb (1972) studied the effect of polymer hydrophobicity on the interaction of polymer with surfactant. SDS was used as the anionic surfactant along with several different polymers: polyvinyl acetate (PVAc), polyvinyl alcohol (PVA), polypropylene oxide (PPO), PEO, and methylcellulose (MeC).

They found that the interaction of polymer with SDS was in the following order of strength:

$$\text{PVA} < \text{PEO} < \text{MeC} < \text{PVAc} < \text{PPO}$$

Thus, there occurs an increase in the level of interaction between polymer and surfactant with the increase in hydrophobicity.

The length of the hydrocarbon chain can affect the hydrophobicity of the polymer. As the length increases the hydrophobicity increases. The chain length can also affect some other factors leading to a decrease in the interaction between polymer and surfactant. For example, polysaccharide Dextran has a long and flexible chain but has a little tendency to interact with SDS (Goddard 1986a,b).

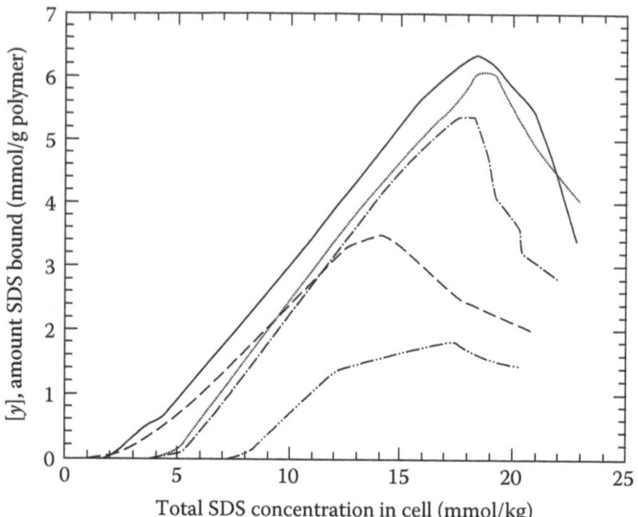

FIGURE 13.26 Surfactant binding isotherms for SDS with nonionic polymers: at 20°C for (—) CST-103; (······) HPMC; (— · —) MC; (— —) HPC, and (— · · —) HEC. (From Singh, S.K. and Nilsson, S., *J. Colloid Interface Sci.*, 213, 152, 1999.)

Figure 13.26 shows another example of the effect of hydrophobicity on the interaction of polymers and surfactants (Singh and Nilsson 1999). The figure shows the binding curves for the cellulose ethers with SDS at 20°C. The polymers investigated are ethyl hydroxyethyl cellulose (EHEC), also called CST-103, hydroxypropyl methyl cellulose (HPMC), hydroxypropyl cellulose (HPC), methyl cellulose (MC), and hydroxyethyl cellulose (HEC). Among them, CST-103 is the most hydrophobic and HEC is the most hydrophilic. The binding curves show a clear trend in the critical aggregation concentration. The CAC decreases with the increase in hydrophobicity. The most hydrophobic polymer (CST-103) has the lowest CAC value. The least hydrophobic polymer (HEC) shows the highest CAC value.

Hydrophobically modified (HM) nonionic polymers exhibit a strong tendency to interact with ionic surfactant. These polymers show a different behavior compared to regular nonionic polymers when they interact with ionic surfactants. The rheology of these fluids has been investigated in great detail. In most of the studies, it is observed that the addition of surfactant causes an increase in viscosity up to a certain concentration and then a reduction in viscosity with further addition of surfactant. Figure 13.27 shows a simple mechanistic model proposed by Winnik and Regismond (1996) for the interaction of ionic surfactant such as SDS, TTAC, and CTAC with a hydrophobically modified polymer such as poly(*N*-isopropylacrylamides) (HM-PNIPAM), a copolymer of NIPAM, and *N*-*n*-alkylacrylamides (*n*-alkyl = decyl, tetradecyl, and octadecyl groups). The increase in viscosity is attributed to the formation of complex on the alkyl substituents, which promotes interlinking and networking of the polymeric chains. At a low surfactant concentration ($C <$ CMC), the surfactant micelles act as a bridge between different chains of polymer. With the increase in surfactant concentration, the amount of polymer chains available to interact with the surfactant molecules is reduced. In this case, the density of micelles on the backbone of polymer chains is increased to a point where the repulsion of polymer chains becomes significant. Consequently, the network of the polymer chains is destroyed resulting in a reduction of viscosity. Experiments with Py-labeled (pyridine) PNIPAM and Py-labeled HM-PNIPAM samples confirmed this conclusion.

In another study, Dualeh and Steiner (1990) reported a strong interaction between C_{12}-grafted hydroxyethylcellulose (HM-HEC) in the presence of SDS. They found that this hydrophobically modified polymer acts differently from the unmodified HEC. The hydrophobic characteristics of HM-HEC was responsible for the strong interaction (Winnik and Regismond 1996).

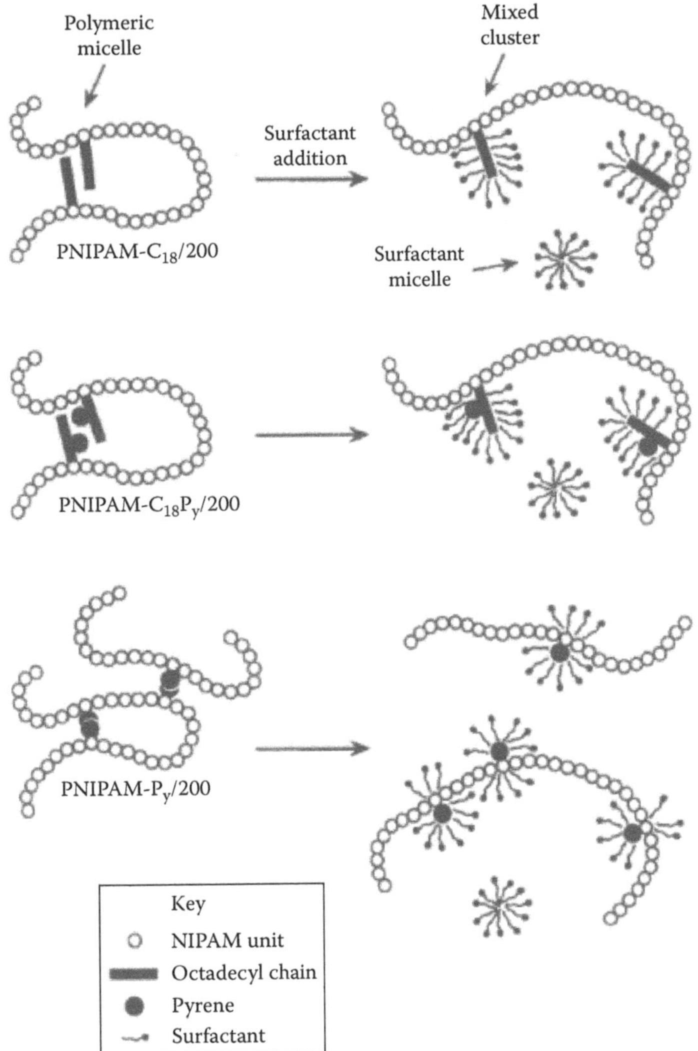

FIGURE 13.27 The interactions between surfactant and HM-PNIPAM, PNIPAM-C_{18}/P_y, and PNIPAM-P. (From Winnik, F.M. and Regismond, S.T., *Colloid Surf. A: Physicochem. Eng. Aspects*, 118(1), 1, 1996.)

The influence of the size of the surfactant head group on the interaction between polymer and surfactant has been investigated in several studies. Blokhus and Klokk (2000) studied the interactions between PEO and the surfactants sodium alkylcarboxylates (octyl, decyl, and dodecyl) and SDS by means of conductivity measurements and gel permeation chromatography (GPC). From the conductivity measurements, the critical aggregation concentration (CAC), the ionization degree, and the binding ratios were determined. The binding ratio was also determined from GPC. The PEO–surfactant interactions were observed in most of the surfactants studied (except sodium octanoate). For the polymer–surfactant complexes, the ionization degree was observed to be about 0.2 higher than the ionization degree for the corresponding aqueous micelles. Further, the binding ratio decreased somewhat with decreasing the chain length of the alkylcarboxylate. The Gibbs free energy analysis showed that the polymer–surfactant interaction decreases with decreasing the chain length of the alkylcarboxylates and is weaker for alkylcarboxylate compared with alkylsulfate of similar chain length (Blokhus and Klokk 2000). Brackman and Engberts (1992) also found out that sodium alkylphosphates interact less with PEO than sodium alkylsulfates of similar chain length.

Benrraou et al. (2003) studied the interaction between nonionic polymers and anionic surfactants. The polymers used were poly(ethylene oxide) and poly(vinylpyrrolidone) and the surfactants used were cesium and tetraalkylammonium (tetramethyl to tetrabutyl ammonium) dodecylsulfates. They used the electrical conductivity method to determine the CAC for different polymer–surfactant combinations. They concluded that the value of the ratio CAC/CMC increases with the increase in the radius of the counterion. Thus, the strength of interaction decreases upon increasing the counterion radius. The sequence in which the ratio CAC/CMC varied is as follows: Na^+ < Cs^+ < tetramethylammonium$^+$ < tetraethylammonium$^+$ = tetrapropylammonium$^+$ = tetrabutylammonium$^+$ = 1.0. The ratio CAC/CMC is unity in the case of the last three large-size ions indicating the absence of any interaction. For small ions such as Na^+, the CAC/CMC ratio is less than unity indicating significant interaction.

Yan and Xiao (2004) carried out an investigation on the effect of anionic surfactant headgroup on interaction between surfactants (alkyl sulfate and alkyl sulfonate) and nonionic polymer (PEO). Their study revealed that $C_{12}SO_3$ and $C_{12}SO_4$ interact differently with PEO. They also concluded that both hydrophobic and electrostatic interactions play important roles in the $C_{12}SO_3$/PEO and $C_{12}SO_4$/PEO interactions. The headgroup of C_nSO_4 has an extra oxygen compared to C_nSO_3, which makes it bigger. Furthermore, the charge on the C_nSO_4 headgroup and $-CH_2$ (first group on alkyl chain of surfactant) are opposite, while the C_nSO_3 headgroup and $-CH_2$ have the same charge. This makes the hydration cosphere of the hydrophobic tail of C_nSO_4 very different from that of the corresponding tail of C_nSO_3. As a result, the hydration effect of C_nSO_3 is stronger than that of C_nSO_4, which means the hydration shell of the headgroup may extend over a larger part of the alkyl chain in C_nSO_3. This effect "shortens" the effective hydrophobic tail of C_nSO_3 as compared with C_nSO_4. As hydrophobicity is the most important factor in the interaction between nonionic polymer and surfactant, the difference in the hydrophobic nature of C_nSO_3 and C_nSO_4 makes the interaction of C_nSO_3–PEO weaker than that of C_nSO_4–PEO.

The effect of polymer molecular weight (MW) on the interaction between nonionic polymer and ionic surfactant has been studied in several papers. Schwuger (1973) investigated the effect of PEO molecular weight on the surface tension of SDS solutions. He found that in the case of low MW (600) PEO, the interaction was weak. The interaction was moderate when the MW of PEO was 1550. However, for PEO molecular weights higher than 4000, the interaction was strong and independent of the molecular weight.

Dai and Tam (2001) studied the interaction of a series of poly(ethylene glycol) (PEG) with anionic surfactant sodium dodecyl sulfate (SDS) by mean of an isothermal titration calorimetry (ITC). Figure 13.28 shows the thermograms for 0.2 M SDS titrating into 0.1 wt% PEG of different molecular weights at 298 K. The CAC is not sensitive to molecular weight of PEG; it decreases only slightly by increasing the molecular weight. This decrease was attributed to a decrease in the hydrophobocity of PEG. They also reported that PSP decreases with increasing PEG molecular weight. It was concluded that the binding between PEG and ionic surfactants (the ratio of EO/SDS) increases with the increase in the polymer molecular weight.

Some investigators have considered the effect of salt on the interaction between nonionic polymer and anionic surfactant. Mandal et al. (2013) investigated the interaction of sodium dodecyl sulfate (SDS) with aqueous polyvinylpyrrolidone (PVP) under the influence of different salts including NaCl, NaBr, NaI, KCl, LiCl, NH_4Cl, Na_2SO_4, and Na_3PO_4. The concentration of SDS was varied and the interactions were detected using techniques such as tensiometry, viscometry, conductometry, and calorimetry. They found that the surfactant SDS binds to the PVP chains. Reverse micelles of surfactant are formed on the backbone of the polymer chains. The presence of salt has a significant influence on the interaction between the surfactant and the polymer.

Minatti and Zanette (1996) reported that the critical aggregation concentration (CAC) and polymer saturation point (PSP) in PEO/SDS mixtures were affected by the presence of salt (NaCl). Masuda et al. (2002) studied the swelling behavior of poly(ethylene oxide) (PEO) gels in aqueous solutions of sodium dodecyl sulfate (SDS). They observed that in the absence of salt, PEO gels start

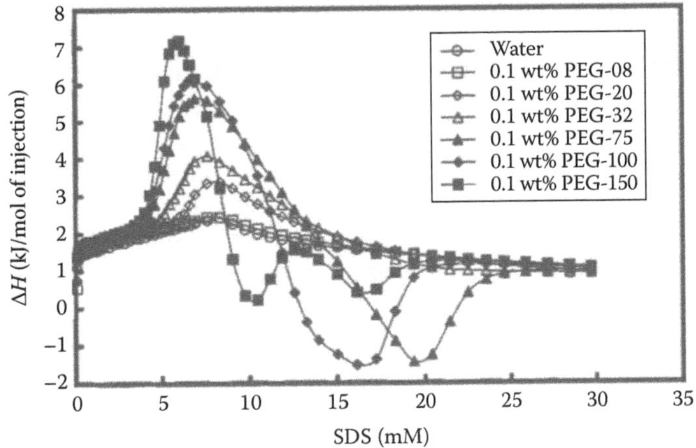

FIGURE 13.28 Thermograms for 0.2 M SDS titrating into 0.1 wt% PEG of different molecular weights at 298 K. The open circle is the SDS dilution curve in water. (From Dai, S. and Tam, K., *J. Phys. Chem. B*, 105(44), 10759, 2001.)

to swell at a surfactant concentration lower than the CMC of SDS. This concentration was in agreement with CAC value reported for the aqueous PEO solution.

Solutions of sodium dodecyl sulfate (SDS) and a variety of nonionic polymers listed in Table 13.4 were studied by Ghoreishi et al. (1999). The EMF method utilizing an SDS membrane selective electrode was applied to investigate the interaction between the polymers and SDS.

They evaluated the CMC, CAC(T_1), and PSP(T_2) from the experimental data. The CAC and PSP and the related information is given in Table 13.5. Almost all the solutions showed a clear interaction. To calculate m_1 (the free monomer surfactant concentration) at any SDS concentration in the interaction zone, they used Equation 13.6 where the slope (~−(2.303RT/F)) and intercept E_0 are derived from the linear data in region I, which also overlaps with the data for pure SDS (see Figure 13.12).

Ghoreishi et al. (1999) reported that at low salt concentrations different polymers show some selectivity toward SDS. A variation in the maximum amount of SDS that could be bound to the polymer was observed from one polymer to another. At high salt concentrations, all the polymers behaved in a similar way.

Mesa and his team (Capalbi and La Mesa 2001; Gasbarrone and La Mesa 2001) studied the interaction of PVP and SDS in a dilute solution. They presented two graphs (see Figure 13.29) showing the phase diagrams of ionic surfactant/nonionic polymer in water systems. The figures show three subregions in which CAC and CMC play the main role. The CMC acts as a "triple" point, where free surfactant, surfactant–polymer complex, and free micelles coexist (Gasbarrone and La Mesa 2001). Referring to Figure 13.29a, free polymer and free surfactant molecules co-exist in the region on the left of the CAC line. In the center portion of the diagram (region in between the CAC and CMC curves), micelles begin to form on the backbone of the polymer molecules. At high surfactant concentrations in a region on the right side of the CMC curve, free micelles co-exist with surfactant/polymer complex.

Chari et al. (1994) reported the effect of SDS on both local polymer chain motions and long range dimensions of the polymer coil. They showed that when a PEO coil is saturated with SDS micelles, it is more swollen compared to free coils when it is in a good solvent. They also revealed that these swollen chains are not fully stretched. Their results showed that the polymer coil at saturation is more like a swollen cage rather than a necklace.

Romani et al. (2005) investigated the effect of addition of different combinations of polyethylene glycol (PEG, MW = 8,000 g/mol) and polyvinylpyrrolidone (PVP, MW = 55,000 g/mol) on

TABLE 13.4
Nonionic Polymers Used in the Research Work of Ghoreishi et al. (1999)

Polymer	Abbreviation	Molecular Weight	Supplier
Methyl cellulose	MC	>50,000	Aqualon
Hydroxypropyl cellulose	HPC	100,000	Aldrich
Ethyl hydroxy ethyl cellulose	EHEC	120,000	Bernol Noble AB, Sweden
Hydrophobically modified EHEC	HM-EHEC	120,000	Bernol Noble AB, Sweden
Hydroxybutylmethyl cellulose	HBMC	120,000	Aldrich
Hydroxypropylmethyl cellulose	HPMC	>50,000	Aqualon
Hydroxyethyl cellulose	HEC	250,000	Aldrich
Poly(vinylpyrrolidone)	PVP	360,000	Sigma
Poly(vinylpyrrolidone)	PVP	40,000	Unilever
Poly(vinylpyrrolidone)	PVP	24,000	Unilever
poly(vinylpyrrolidone)	PVP	15,000	Unilever
Poly(vinylpyrrolidone)	PVP	10,000	Sigma
Poly(vinylpyrridine) nitrogen oxide	PVPy-N-0	200,000	Unilever
Poly(vinylpyrrolidone)/poly(vinylimidazole) copolymer	PVP/PVI	40,000	Unilever
Poly(propylallylamine)	PPAA		Unilever
Poly(vinylimidazole)	PVI		Unilever
Poly(ethylene oxide)	PEO	4,000	BDH
Poly(ethylene oxide)	PEO	6,000	BDH
Poly(propylene oxide)	PPO	1,000	Aldrich
Poly(vinyl methyl ether)	PVME	27,000	Aldrich
Methylvinylimidazole/vinylpyrrolidone copolymer	MVI/VP	40,000	Unilever
N-vinylacryloylpyrrolidine and vinylpyridine dicyanomethylide copolymer	PAPR	10,000	

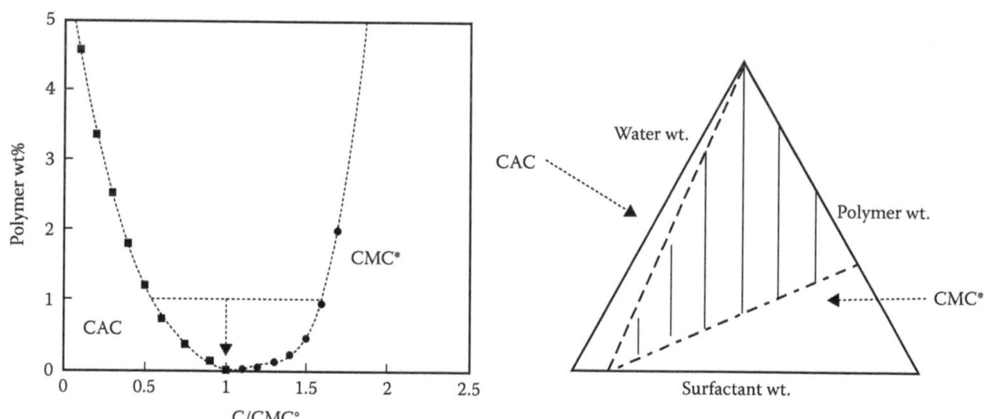

FIGURE 13.29 Phase behavior of a system consisting of nonionic polymer and ionic surfactant. (From Gasbarrone, P. and La Mesa, C., *Colloid Polym. Sci.*, 279(12), 1192, 2001.)

the CMC of surfactant SDS. They measured the electrical conductivity, the zeta potential, and the viscosity. They also used the fluorescence spectroscopy and small-angle X-ray scattering (SAXS). The results showed that the SDS–polymer interaction begins at a low surfactant concentration, and the CAC is dependent on the polymer composition. The average aggregation number varied with surfactant concentration and was highly unstable compared to pure SDS micelles. The zeta

TABLE 13.5
Critical Concentration Associated with Binding in 1×10^4 mol dm^{-3} NaBr

Polymer	$T_1 \times 10^3$ (mol dm^{-3})	$T_2 \times 10^3$ (mol dm^{-3})	$T_f \times 10^3$ (mol dm^{-3})	m_1 at $T_2 \times 10^3$ (mol dm^{-3})	$T_2 - m_1 \times 10^3$ (mol dm^{-3})	$(T_2 - m_1)/C_P \times 10^3$ (mol of bound SDS/g dm^{-3} polymer)	a
MC (0.5%)	0.003	0.025	0.025	0.0069	0.0224	0.00448	N/A
HPC (0.5%)	0.0015	0.023	0.022	0 0033	0.0197	0.00394	6.06
HPC (0.5%)	0.001	0.0232	0.023	0.006	0.0172	0.00344	5.3
EHEC (0.5%) (120,000)	0.002	0.0224	0.02	0.00436	0.0180	0.00361	6.65
HMEHEC (0.5%)	0.002	0.020	0 020	0.0068	0.013	0.0026	4.8
HBMC (0.5%)	0.0035	0.03	0.0103	0.0031	N/A	N/A	N/A
HPMC (0.5%)	0.004	0.022	0.013	0.0032	N/A	N/A	N/A
HEC (0.5%)	0.0024	0.0333	0.0182	0.0030	N/A	N/A	N/A
PVP (1%) (3,600,000)	0.0015	0.045	0042	0.0038	0.0412	0.00412	22.8
PVP (1%) (360,000)	0.002	0.050	0.050	0.0047	0.0453	0.00453	25.1
PVP (0.5%) (10,000)	0.0009	0.048	0.045	0.0035	0.0445	0.0089	5.5
PVP (1%) (24,000)	0.0012	0.055	0.015	0.005	N/A	N/A	N/A
PVP (1%) (15,000)	0.0013	0.052	0.01	0.0045	N/A	N/A	N/A
PVP (1%) (10,000)	0.002	0.05	0.012	0.0047	N/A	N/A	N/A
PV1 (0.1%)	0.0004	0.022	0.014	0.0033	0.0167	0.0167	10.3
PVP$_Y$-NO (0.1%)	<0.00002	0.017	0.014	0.0075	0.0095	0.0095	29.2
PPAA (0.1%)	<0.00002	0.025	0.02	0.004	0.021	0.021	N/A
PVP/PVI (0.1%)	0.0007	0.016	0.015	0.0050	0.011	0.011	6.8
PVME (0.5%)	0.001	0.043	0.043	0.004	0.039	0.0078	3.24
PEO (1%) (4,000)	0.0045	0.038	0.006	0.0033	N/A	N/A	N/A
PEO (1%) (6,000)	0.002	0.038	0.008	0.0035	N/A	N/A	N/A
PAPR (0.5%) (10,000)	0.0025	0.055	0.033	0.00555	N/A	N/A	N/A
PPO (0.5%)	0.001	0.14	0.056	0 00456	N/A	N/A	N/A
MVI/VP (0.5%)	0.0005	0.04	0.038	0.005	0.035	0.007	4.3

Source: Ghoreishi, S. et al., *Langmuir*, 15(13), 4380, 1999.

[a] T_1, SDS the onset of binding; T_2, SDS concentration at which the EMFs of the electrode with and without the polymer merge after binding; m_1, monomer concentration of SDS; T_f, SDS concentration at the maximum in the EMF and hence m_1 data; C_P, concentration of polymer in g dm^{-3}. a = (molar concentration of bound SDS aggregates)/(molar concentration of polymer).

potentials were found to increase linearly with the fraction of PVP at constant SDS concentration. The results of the SAXS indicated that the PVP/PEG/SDS system consists of cylindrically shaped structures with an anisometry ratio of about 3.0.

Meszaros et al. (2005) carried out a thermodynamic analysis of the interaction between 14 different molar mass poly(ethylene oxide) (PEO) and sodium dodecyl sulfate (SDS) based on the measured surfactant-binding isotherms. They measured the surfactant-binding isotherms using the potentiometric method in the presence of 0.1 M inert electrolyte (NaBr). When the molecular weight of PEO was lower than 1000, no PEO/SDS complex formation was detected. In the molecular weight range of $1000 < \text{PEO} < 8000$, the critical aggregation concentration (CAC) and the surfactant aggregation number decreased with the increase in the polymer molecular weight. The polymer concentration had a direct effect on the amount of bound surfactant at saturation. They further observed that when the molecular weight exceeded ~8000, the CAC no longer depended on the polymer molar mass, and the bound amount of the surfactant at saturation became proportional to the mass concentration of the polymer.

Burman et al. (2014) studied the interaction of a variety of anionic surfactants including sodium dodecyl sulfate (SDS), sodium cholate (NaC), sodium deoxycholate (NaDC), and sodium taurodeoxycholate (NaTDC) with a nonionic polymer of hydroxy propyl cellulose (HPC). They applied the microcalorimetric, conductometric, and fluorimetric methods to study the interactions. Using calorimetric and conductometric techniques, they could obtain some of the solution properties such as the critical aggregation concentration (CAC), critical micelle concentration (CMC), polymer saturation concentration (PSP), and the extent of binding of the surfactants with polymer. They concluded that the hydrophobicity and charge density of surfactant have a strong effect on micellization. The fluorescence results showed that increasing the surfactant concentration decreased the micro-polarity.

13.3.4.2 Interactions of Nonionic Polymer with Cationic Surfactant

The early attempts in this area revealed that the nonionic polymer and cationic surfactants do not interact with each other. However, in the presence of some types of ions (such as SCN^- or I^-) as counterions, a weak interaction was observed. The bulkiness of the cationic surfactant head group (compared to anionic surfactants) was the main reason for such observations (Nagarajan 1989; Witte and Engberts 1987). However, recent studies have shown that many nonionic polymers can interact and associate with cationic surfactants. Hydrophobicity has been mentioned as one of the key factors and plays an important role in the interaction between nonionic polymers and cationic surfactant. Typically, polymers with higher hydrophobicity show a better interaction (Anthony and Zana 1996; Thuresson et al. 1995, 1996). Zana et al. (1992) reported that when hydroxyethyl cellulose interacts with hexadecyltrimethylammonium chloride and bromide, the CAC is less than CMC and the aggregation number of the surfactant is also less compared with the polymer-free solution. Brackman et al. (1992) observed no interaction between the nonionic polymers PEO and PVP polymers and the cationic surfactant.

Mya et al. (2000) found that a strong interaction could occur between cationic surfactant hexadecyltrimethylammonium and PEO when the temperature is above 25°C. In this case, the hydrodynamic radius (R_h) of polymer coil increases due to chain expansion. The expansion occurs because of electrostatic repulsions between the bonded micelles on the backbone of the polymer chain. This observation is consistent with the work of Hormnirun et al. (2000).

Muzzalupo et al. (2007) studied the interactions of nonionic homopolymer poly(vinylpyrrolidone) (PVP) and different gemini surfactants at 25°C. They studied the interactions under a wide variety of experimental conditions by changing the amounts of polymer and surfactant. According to their reported data, the cationic gemini surfactants do not interact with nonionic PVP.

The interaction of a nonionic diblock copolymer consisting of ethylene oxide and butylene oxide with cationic surfactant cetyl trimethyl ammonium bromide (CTAB) was investigated by Bibi et al. (2012). They compared their results for copolymer-CTAB system with the results they obtained for the same copolymer and anionic surfactant sodium dodecyl sulfate (SDS). Surface tension, conductivity, and dynamic laser light scattering techniques were employed. Using the surface tension measurements, the critical micelle concentration (CMC), free energy of adsorption (ΔG_{ads}), free energy of micellization (ΔG_m), surface excess concentration (Γ), and minimum area per molecule (A) were calculated. The critical micelle concentration (CMC) and critical aggregation concentration (CAC) at different temperatures were determined by means of the conductivity measurements. The changes in the physicochemical properties of the micellized block copolymer were studied using dynamic laser light scattering. The results showed that the physiochemical properties of diblock copolymer are strongly influenced by the addition of surfactant. Due to the formation of polymer–surfactant complex, the hydrodynamic radius of the diblock copolymer changed significantly. The interaction was observed at a low concentration of surfactant indicating a low CAC. The CAC decreased with the increase in the temperature.

Gemini surfactants are more effective compared to regular surfactants in lowering the surface tension. Sardar and Kamil (2012a,b) studied the interaction between a nonionic polymer, (hydroxypropyl)methyl cellulose (HPMC), and cationic gemini surfactants, bis(hexadecyldimethylammonium) hexane dibromide (16-6-16), bis(hexadecyldimethylammonium)pentane dibromide (16-5-16), and their corresponding monomeric counterpart cetyltrimethylammonium bromide (CTAB). Electrical conductometry, fluorescence, and viscometry methods were used to characterize the interactions. They reported that the gemini surfactants interact strongly with HPMC as compared with conventional monomeric surfactant CTAB. The fluorescence measurements showed that the aggregation number was higher in the case of CTAB as compared with the gemini surfactant. Upon the addition of gemini surfactant to the polymer solution, the viscosity increased profoundly due to the formation of a network between the surfactant micelles and the polymer molecules. The results of this study appear to be in contradiction with the work of Muzzalupo et al. (2007) who did not observe any interaction between cationic gemini surfactants and nonionic polymer PVP.

Peng et al. (2013) studied a thermo-sensitive copolymer based on oligo(ethylene glycol)methacrylates with three types of cationic alkyltrimethylammonium bromide surfactants (RTAB with R equal to C12, C14, and C16): dodecyltrimethylammonium bromide (DoTAB), tetradecyltrimethylammonium bromide (TTAB), and cetyltrimethylammonium bromide (CTAB). They used isothermal titration calorimetry (ITC), surfactant selective electrode (SSE), and dynamic light scattering (DLS) methods to study the interaction. They reported a strong interaction between the polymer and surfactants. The PSP was dependent on the polymer concentration and was independent of the temperature. The binding affinity of different surfactants for polymer varied in the following sequence: CTAB > TTAB > DoTAB.

Mohsenipour and Pal (2013) studied the behavior of polyethylene oxide (PEO) and octadecyltrimethylammonium chloride (OTAC) mixtures and reported that the relative viscosity increased compared with pure PEO and surfactant alone. The increase in viscosity was attributed to the formation of network of polymer molecules and surfactant micelles. Similar patterns were observed by Prajapati (2009) for the same system using a different molecular weight PEO.

13.3.5 Interaction of Charged Polymers and Uncharged Surfactants

The interaction between ionic polymers and nonionic surfactants has not received as much attention as the other cases discussed in the preceding sections. Much of the earlier research in this area has been focused on the interaction of anionic polymeric acids with nonionic surfactants of polyethylene oxide. The polyethylene oxide has shown the ability to form hydrogen bonds with polymeric acid like polycarboxylic acid in water. This causes a reduction in the solution viscosity as the polymer chains tend to shrink. Saito and Taniguchi (1973) studied the interaction of polyacrylic acid with a series of nonionic surfactants $(EO)_nRE$ in which EO is ethylene oxide, R is hydrocarbon group, and E represents ether. They reported that the interaction in this system is a function of the nature of the hydrophobic moiety (R) and the length of the hydrophilic tail (EO).

The rheological behavior of a 1% w/w solution of hydrophobically modified (hydroxypropyl) guar (HMHPG) in water was investigated by Aubry and Moan (1996) in the presence of a nonionic surfactant. The response to steady and oscillatory shear flow, at different surfactant concentrations around the CMC, showed different behaviors below and above the CMC point. Below the CMC, a reinforcement of the intermolecular hydrophobic network occurs due to an increase in the number of intermolecular hydrophobic associations. Above the CMC, the intermolecular hydrophobic network is destroyed.

Smith and McCormick (2001) synthesized a series of terpolymers composed of acrylic acid, methacrylamide, and DiC_6, DiC_8, or DiC_{10} twin-tailed hydrophobic monomers and studied their interactions with different types of surfactants including sodium dodecyl sulfate (SDS) as anionic, cetyl trimethylammonium bromide (CTAB) as cationic, and Triton X-100 as a nonionic surfactant. In the case of terpolymers DiC_6AM and DiC_8AM, the viscosity measurements revealed different behaviors for different surfactants. While a weak interaction was observed with SDS, gelation occurred in the case of CTAB. With nonionic surfactant Triton X-100, hemimicelle formation followed by polymer hydrophobe solubilization was reported. The $DiC_{10}AM$ terpolymer showed similar interaction behavior with CTAB and Triton X-100. Using the fluorescence technique on a dansyl-labeled $DiC_{10}AM$ terpolymer, they concluded that the strength of interaction between the polymer and a surfactant varied according to the following sequence: CTAB > Triton X-100 > SDS.

Zhao and Chen (2006) experimentally investigated the clouding phenomena and phase behaviors of two nonionic surfactants, Triton X-114 and Triton X-100, in the presence of either hydroxyethyl cellulose (HEC) or its hydrophobically modified counterpart (HMHEC). Compared with HEC, HMHEC was found to have a stronger effect in lowering the cloud point temperature of a nonionic surfactant at low concentrations. The difference in the clouding behavior was attributed to different kinds of molecular interactions. Depletion flocculation was the underlying mechanism in the case of HEC, whereas the chain-bridging effect was responsible for the large decrease in the cloud point in the HMHEC system. The analysis of the composition of the macroscopic phases formed was carried out to provide support for associative type phase separation in the case of HMHEC, in contrast to segregative type phase separation observed in the case of HEC. In some mixtures of HMHEC and Triton X-100 at high surfactant concentrations, an interesting three-phase-separation phenomenon was observed.

13.3.6 Interaction of Uncharged Polymers and Uncharged Surfactants

Most of the early studies in this area have reported little or weak interaction between nonionic polymers and nonionic surfactants (Saito 1987).

Feitosa et al. (1996b) studied the interaction between nonionic surfactant $C_{12}E_5$ and a high molar mass ($M = 5.94 \times 10^5$) poly(ethylene oxide) (PEO) in aqueous solution at different temperatures using dynamic light scattering and fluorescence methods. They observed that the surfactant micelles and PEO coil tend to form clusters. The presence of micelles on the polymer chain causes extension of polymer chain. The temperature has a strong effect on the hydrodynamic radius of the complex. The hydrodynamic radius increases with temperature. A similar behavior was found with the increase in the concentrations of surfactant and polymer. At high concentrations of the surfactant, the coil/micellar complexes coexist with free $C_{12}E_5$ micelles in the solution. Fluorescence quenching measurements showed that the average aggregation number (N) of the micelles is smaller in the presence of PEO as compared with no polymer. However, at high concentrations of surfactant, N is larger in the presence of polymers.

Couderc et al. (2001) carried out an investigation of the interactions of different binary systems consisting of nonionic surfactant, hexaethylene glycol mono-n-dodecyl ether ($C_{12}EO_6$), and triblock copolymer "Pluronic" F127 of chemical composition $EO_{97}PO_{69}EO_{97}$. This copolymer is nonionic in which EO represents the ethylene oxide blocks and PO the propylene oxide blocks.

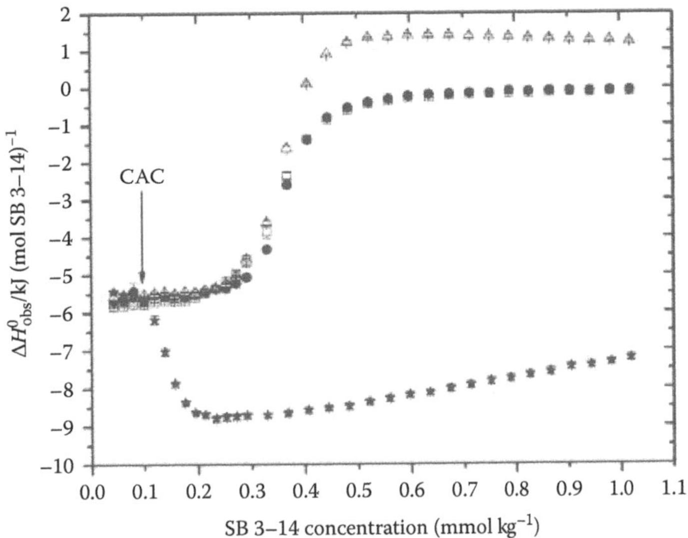

FIGURE 13.30 Calorimetric titration curves for the addition of SB 3–14 (6.0 mmol kg^{-1}) at 15°C to water (□) and to a 0.1 wt% polymeric solution of PEO (●), PPO (Δ), and PAA (★) at pH 3.5. (From Brinatti, C. et al., *Langmuir*, 30(21), 6002, 2014.)

Using isothermal titration calorimetry (ITC) and differential scanning calorimetry (DSC), micellar aggregates of polymer and surfactant were detected in all cases. They also measured the CMC's of the F127-rich micelles.

Nambam and Philip (2012) studied the interaction of a nonionic surfactant (nonylphenolethoxylate [NP_9]) with a triblock copolymer of PEO–PPO–PEO in aqueous solution. The micellar size, zeta potential, and the rheological properties were measured. They also compared the results with the interactions of the same copolymer with an anionic surfactant (sodium dodecyl sulfate [SDS]) and a cationic surfactant (cetyltrimethylammonium bromide [CTAB]). They reported that the addition of NP_9 does not have any effect on the viscoelastic properties. The same result was observed in the case of CTAB whereas a drastic reduction was reported in the case of SDS.

The interaction of a series of zwitterionic surfactants (sulfobetaines) with some nonionic polymers were studied by Brinatti et al. (2014) using an isothermal titration calorimetry (ITC). Polymers were poly(ethylene oxide), PEO, M = 100,000 g/mol, poly(propylene oxide), PPO, M = 1,000 g/mol, and poly(acrylic acid), PAA, M = 2,000 g/mol. The results showed that PAA could induce the formation of micellar aggregates at low surfactant concentrations indicating significant interaction. The ITC curves are presented in Figure 13.30 for the polymers PAA, PEO, and PPO. The curves are similar for PEO and PPO and they follow the same curve as that of the surfactant sulfobetaine in water. The introduction of PEO and PPO in solution changes the endothermic enthalpy value slightly at low surfactant concentration. However, the CMC value is the same. The PEO and PPO curves are different from each other at high surfactant concentrations above the CMC region. As PPO is more hydrophobic compared to PEO, its enthalpy curve lies slightly above the dilution curve of surfactant. The upward shift in the enthalpy curve for PPO relative to the dilution curve is about 1.5 kJ/mol.

13.3.7 Mechanisms of Polymer–Surfactant Associations

In the preceding sections, the focus was on the various characterization techniques and the detection of interactions between different combinations of polymers and surfactants. Based on the literature discussed in the preceding sections, the degrees of interaction between different combinations of polymers and surfactants are summarized in Table 13.6. According to Table 13.6, it seems that the

TABLE 13.6
Table of Surfactant Reactivity with Polymer

Type of Polymer	Degree of Reactivity
P^0 uncharged polymer	$S^- > S^+ \gg S^0$
P^+, polycation	$S^- \gg S^0 \gg S^+$
P^-, polyanion	$S^+ \gg S^0 \gg S^-$

Source: Goddard, E., *J. Am. Oil Chem. Soc.*, 71(1), 1, 1994.
S^-, anionic surfactant; S^+, cationic surfactant, S^0, nonionic surfactant.

most active systems in terms of polymer–surfactant interactions are those consisting of oppositely charged polymer/surfactant system and nonionic polymer/ionic surfactant system.

The interaction of polymers and surfactants is still an ongoing research. The mechanisms and thermodynamics of interactions are not yet fully understood. In this section, we review some of the literature related to the mechanisms of interaction between polymer and surfactant molecules.

The mechanism responsible for the aggregation of surfactant molecules in aqueous solution can be extended to the association of polymer and surfactant. When surfactant is dissolved in water, the headgroups (which are hydrophilic) point toward the water side, and the tails (which are hydrophobic) tend to escape the water phase. To create a hydrophobic environment, the tails form aggregates (micelles). When surfactant molecules are introduced to a polymer solution, the tails of the surfactant molecules are attracted to the hydrophobic sites available on the backbone of polymer chains. Thus, aggregation of surfactant molecules often occurs on the backbone of the polymer chains. This is the main mechanism for interaction and association of polymer and surfactant molecules. The other mechanism for interaction between polymer and surfactant molecules is the electrostatic attraction between oppositely charged polymer and surfactant molecules.

The polymer–surfactant interaction is highly dependent on the type of each component or reactant (polymer and surfactant). The shape and format of complexes formed (if there is any interaction) will be different with different types of components. Nagarajan (2001) has categorized the different types of interactions between polymers and surfactants and has introduced the possible shapes of polymer–surfactant complexes. These formats are presented in Figure 13.31. When both the polymer and the surfactant have the same electrical charge, no interaction is expected. This situation is represented as structure A in Figure 13.31. For the case where the polymer and the surfactant carry opposite charges, the structures shown as B and C in Figure 13.31 are observed. The electrostatic interaction is the main driving force for the formation of such structures. The bonding of the polymer and the surfactant molecules occurs in both structures. However, in the case of C structure, the surfactant molecules lead to the formation of a network between different chains of the macromolecules. Random copolymer or multiblock copolymer with short blocks interact with the surfactant molecules such that the complexes formed are of structural format D. Structures E, F, and G are formed in the case of hydrophobically modified polymers. Many factors can affect the interaction between hydrophobically modified polymers and surfactants. Factors such as the size of the hydrophobic modifier, the grafting density on the backbone of the polymer molecules, and the ratio of polymer to surfactant may change the structure of polymer–surfactant complexes. Structure E is possible at low concentration of surfactant. The individual surfactant molecules associate with the hydrophobic sites on the polymer chains. No conformational change is expected in this case. Increasing the surfactant concentration leads to a change in the structure from E to F where a single cluster or micelle of surfactant molecules may interact with multiple hydrophobic modifier sites present on the backbone of the same polymer chain. This association changes the polymer conformation profoundly. When the surfactant concentration is high, the surfactant molecules form clusters or micelles around each hydrophobic modifier site present on the polymer chain, as shown

A. Polymer molecule does not interact with surfactants for electrostatic or steric reasons. No surfactant is bound to the polymer. For example, the surfactant and the polymer are both anionic or both cationic

B. The polymer and the surfactant are oppositely charged. Single surfactant molecules are bound linearly along the length of the polymer molecules.

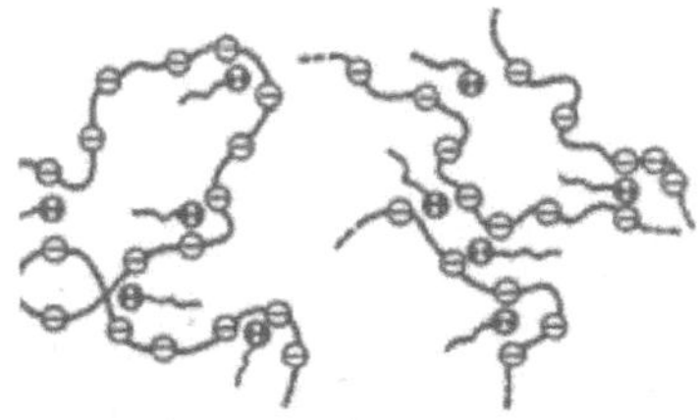

C. The polymer and the surfactant are oppositely charged. A single surfactant molecule binds at multiple sites on a single polymer molecule, giving rise to intra-molecular bridging. Alternatively, it binds to more than one polymer molecule allowing intermolecular bridging.

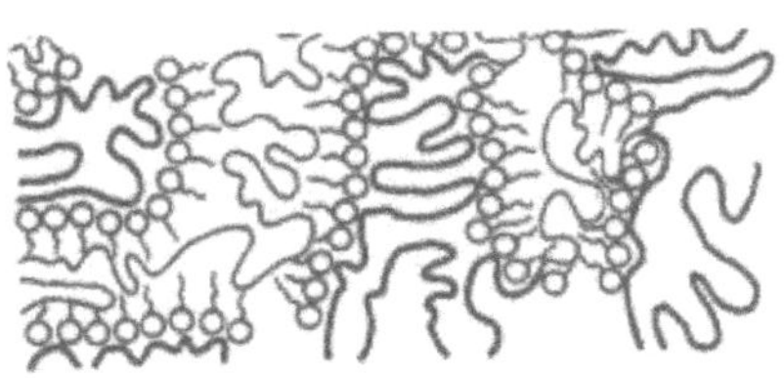

D. The polymer is an uncharged random or multiblock copolymer. The surfactant molecules orient themselves at domain boundaries separating the polymer segments of different polarities.

E. Polymer is hydrophobically modified. Individual surfactant molecules associate with one or more of the hydrophobic modifiers on a single polymer molecule or multiple polymer molecules.

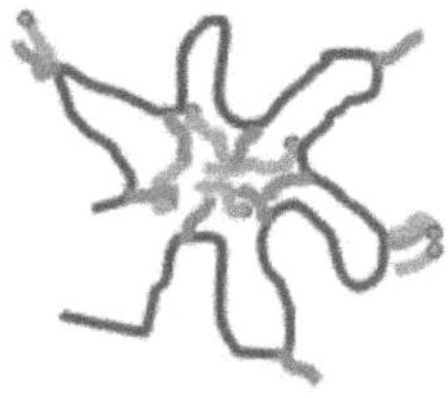

F. Polymer is hydrophobically modified. Clusters of surfactant molecules associate with multiple hydrophobic modifiers on a single polymer molecule.

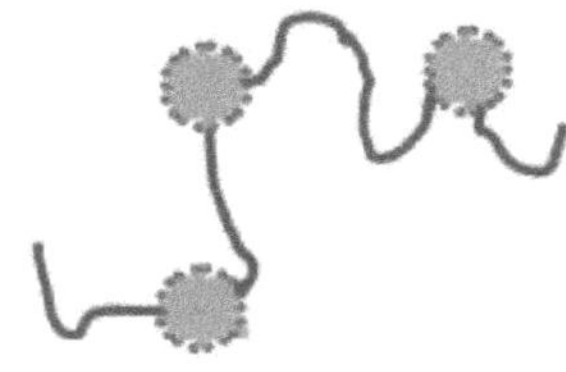

G. Polymer is hydrophobically modified. Clusters of surfactant molecules associate with each of the hydrophobic modifiers on a single polymer molecule.

H. The polymer segments partially penetrate and wrap around the polar head group region of the surfactant micelles. A single polymer molecule can associate with one or more surfactant micelles.

FIGURE 13.31 Different types of polymer-surfactant complexes. (From Nagarajan, R., Polymer-surfactant interactions, *New Horizons: Detergents for the New Millennium Conference Invited Paper*, 2001, Fort Myers, FL.)

in structure G. When a nonionic polymer is interacting with the ionic surfactant, the most probable structure is H shown in Figure 13.31. In this case, the polymer molecule wraps around the surfactant micelles and at the same time some portion of the polymer chain penetrates into the core of surfactant micelles to change the core environment and make it more hydrophobic (which is preferred by the surfactant tails).

The early model proposed for the interaction of nonionic polymer with ionic surfactant by Shirahama et al. (1974), referred to as the "string of beads" model, postulated the formation of surfactant micelles (beads) on the backbone of the polymer chains. Later on two additional models were proposed by Nagarajan (1980) and Ruckenstein et al. (1987). Nagarajan proposed the so-called necklace model. According to this model, the polymer–surfactant complex consists of polymer chain wrapped around one or more surfactant micelles. The polymer segments tend to partially penetrate into the polar head group region of the micelles causing a reduction in the micelle core-water contact area. Structure H of Figure 13.31 describes the necklace model closely. This model suggests that the polymer–surfactant complex brings the micelles in a more stable position from the point of view of free energy. Using this model, one can also explain the increase in the interaction tendency between the polymer and the surfactant when the hydrophobicity of the polymer is increased.

The model by Ruckenstein et al. (1987) describes the formation of polymer–surfactant complex based on the adsorption of micelles on the polymer chains. The presence of the polymer molecules in water changes the micro-environment of the surfactant molecules and the surface free energy between micellar hydrocarbon core and the solvent in the "free space" of the coiled macromolecule. If the surfactant head group is small (such as anionic surfactant head group), the overall surface free energy is decreased in the presence of polymer, and consequently the micellization process is amplified. In such cases, the free aggregation of surfactant molecules starts only after the polymer molecules are saturated with the bound micelles. If the head group is large (such as cationic surfactant head group), the surface free energy is increased and the presence of the polymer has no effect on the micellization of surfactant.

Wallin and Linse (1996a,b) proposed a model for the interaction between a surfactant and an oppositely charged polymer based on Monte Carlo simulation taking into consideration the effects of chain flexibility, linear charge density, and surfactant tail length. They emphasized the importance of electrostatic interaction and polyelectrolyte rigidity. Figure 13.32 shows some of the snapshots produced by Monte Carlo simulation. The figure shows the changes in the structure of the polymer–micelle complex with the increase in polymer rigidity. Clearly the flexibility of the polymer chain can have

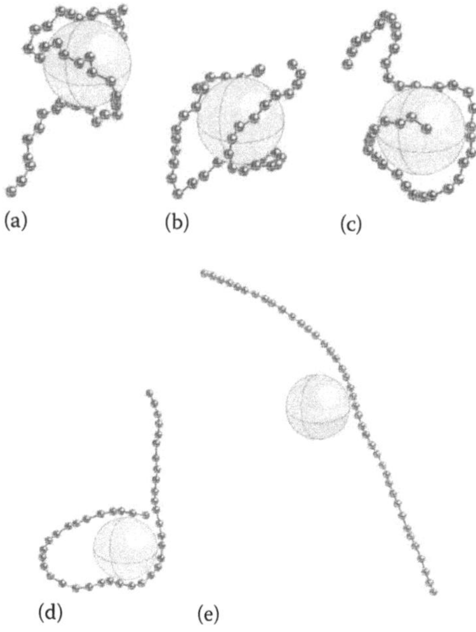

FIGURE 13.32 Snapshots of Monte Carlo simulation for interaction between oppositely charged polymer and surfactant with different angles between consecutive chain segments: (a) 90°, (b) 135°, (c) 150°, (d) 165°, and (e) 175°. (From Wallin, T. and Linse, P., *Langmuir*, 12(2), 305, 1996a.)

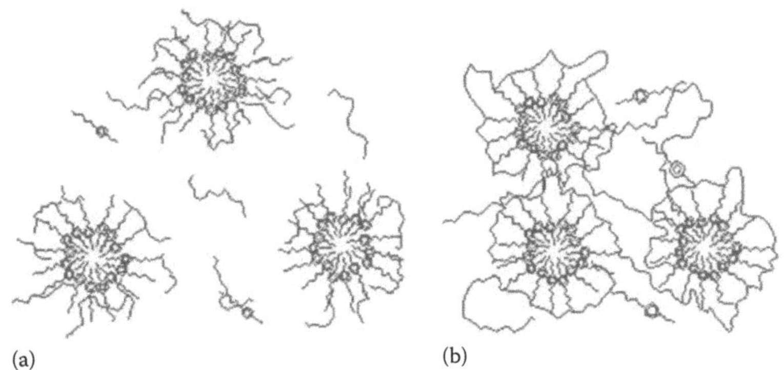

FIGURE 13.33 Schematic illustration of TX-100/PEG complexes formed with low molecular weight (MW < 2000 Da) PEG (a) and high molecular weight (MW > 2000 Da) PEG (b). (From Ge, L. et al., *Polymer*, 48(9), 2681, 2007.)

a great effect on the structure of the complex. When the polymer molecules are flexible, the polymer molecules wrap around the micelles and are completely adsorbed. The rigidity of the polymer molecule makes it difficult for the polymer chain to come close to the micelles and become attached.

Ge et al. (2007) investigated the interaction of nonionic surfactant TX-100 and polyethylene glycol (PEG) using fluorescence resonance energy transfer. Based on their research, they proposed two possible models for the interaction between nonionic polymer PEG and nonionic surfactant TX-100. They concluded that for low molecular weight PEG, the complexes are sphere-like (Figure 13.33a) clusters where the polymer molecule wraps around a single micelle. For high molecular weight PEG, the longer polymer chains mostly tend to form coral-like clusters (Figure 13.33b) where the same polymer molecule is wrapped around multiple micelles.

REFERENCES

Ahmadi, M. A. and S. R. Shadizadeh (2013). Experimental investigation of adsorption of a new nonionic surfactant on carbonate minerals. *Fuel* **104**: 462–467.

Anthony, O. and R. Zana (1996). Interactions between water-soluble polymers and surfactants: Effect of the polymer hydrophobicity. 2. Amphiphilic polyelectrolytes (polysoaps). *Langmuir* **12**(15): 3590–3597.

Asnacios, A., D. Langevin, and J.-F. Argillier (1996). Complexation of cationic surfactant and anionic polymer at the air-water interface. *Macromolecules* **29**(23): 7412–7417.

Aubry, T. and M. Moan (1996). Influence of a nonionic surfactant on the rheology of a hydrophobically associating water soluble polymer. *Journal of Rheology* **40**(3): 441–448 (1978 present).

Bachmann, R. T., A. C. Johnson, and R. G. Edyvean (2014). Biotechnology in the petroleum industry: An overview. *International Biodeterioration & Biodegradation* **86**: 225–237.

Bai, G., M. Nichifor, and M. Bastos (2010). Cationic polyelectrolytes as drug delivery vectors: Calorimetric and fluorescence study of rutin partitioning. *The Journal of Physical Chemistry B* **114**(49): 16236–16243.

Banerjee, R., S. Dutta, S. Pal, and D. Dhara (2013). Spontaneous formation of vesicles by self-assembly of cationic block copolymer in the presence of anionic surfactants and their application in formation of polymer embedded gold nanoparticles. *The Journal of Physical Chemistry B* **117**(13): 3624–3633.

Benrraou, M., B. Bales, and R. Zana (2003). Effect of the nature of the counterion on the interaction between cesium and tetraalkylammonium dodecylsulfates and poly(ethylene oxide) or poly(vinylpyrolidone). *Journal of Colloid and Interface Science* **267**(2): 519–523.

Bibi, I., A. Khan, N. Rehman, S. Pervaiz, K. Mahmood, and M. Siddiq (2012). Characterization of surfactant-diblock copolymer interactions and its thermodynamic studies. *Journal of Dispersion Science and Technology* **33**(6): 792–798.

Blokhus, A. M. and K. Klokk (2000). Interactions between poly(ethylene oxide) and sodium alkylcarboxylates as studied by conductivity and gel permeation chromatography. *Journal of Colloid and Interface Science* **230**(2): 448–451.

Bonnaud, M., J. Weiss, and D. J. McClements (2010). Interaction of a food-grade cationic surfactant (lauric arginate) with food-grade biopolymers (pectin, carrageenan, xanthan, alginate, dextran, and chitosan). *Journal of Agricultural and Food Chemistry* **58**(17): 9770–9777.

Brackman, J. C. and J. B. F. N. Engberts (1992). Effect of surfactant charge on polymer-micelle interactions: *n*-Dodecyldimethylamine oxide. *Langmuir* **8**(2): 424–428.

Breuer, M. and I. Robb (1972). Interactions between macromolecules and detergents. *Chemistry & Industry* **1972**(13): 530–535.

Brinatti, C., L. B. Mello, and W. Loh (2014). Thermodynamic study of the micellization of zwitterionic surfactants and their interaction with polymers in water by isothermal titration calorimetry. *Langmuir* **30**(21): 6002–6010.

Burman, A. D., S. Ghosh, and A. Das (2014). Behavior of cationic and anionic surfactants including bile salts in presence of hydroxy propyl cellulose (HPC). *Journal of Nanofluids* **3**(2): 140–153.

Capalbi, A. and C. La Mesa (2001). Polymer-surfactant interactions. *Journal of Thermal Analysis and Calorimetry* **66**(1): 233–241.

Chandar, P., P. Somasundaran, and N. Turro (1988). Fluorescence probe investigation of anionic polymer-cationic surfactant interactions. *Macromolecules* **21**(4): 950–953.

Chari, K., B. Antalek, M. Lin, and S. Sinha (1994). The viscosity of polymer–surfactant mixtures in water. *The Journal of Chemical Physics* **100**: 5294.

Cooke, D., C. Dong, R. Thomas, A. Howe, E. Simister, and J. Penfold (2000). Interaction between gelatin and sodium dodecyl sulfate at the air/water interface: A neutron reflection study. *Langmuir* **16**(16): 6546–6554.

Couderc, S., Y. Li, D. Bloor, J. Holzwarth, and E. Wyn-Jones (2001). Interaction between the nonionic surfactant hexaethylene glycol mono-*n*-dodecyl ether ($C_{12}EO_6$) and the surface active nonionic ABA block copolymer pluronic F127 ($EO_{97}PO_{69}EO_{97}$) formation of mixed micelles studied using isothermal titration calorimetry and differential scanning calorimetry. *Langmuir* **17**(16): 4818–4824.

Dai, S. and K. Tam (2001). Isothermal titration calorimetry studies of binding interactions between polyethylene glycol and ionic surfactants. *The Journal of Physical Chemistry B* **105**(44): 10759–10763.

Dan, A., S. Ghosh, and S. P. Moulik (2009). Physicochemistry of the interaction between inulin and alkyltrimethylammonium bromides in aqueous medium and the formed coacervates. *The Journal of Physical Chemistry B* **113**(25): 8505–8513.

Dan, A., S. Ghosh, and S. P. Moulik (2010). Interaction of cationic hydroxyethylcellulose (JR400) and cationic hydrophobically modified hydroxyethylcellulose (LM200) with the amino-acid based anionic amphiphile Sodium *N*-Dodecanoyl Sarcosinate (SDDS) in aqueous medium. *Carbohydrate Polymers* **80**(1): 44–52.

Deo, P., N. Deo, P. Somasundaran, A. Moscatelli, S. Jockusch, N. J. Turro, K. Ananthapadmanabhan, and M. F. Ottaviani (2007). Interactions of a hydrophobically modified polymer with oppositely charged surfactants. *Langmuir* **23**(11): 5906–5913.

Diamant, H. and D. Andelman (1999). Onset of self-assembly in polymer-surfactant systems. *Europhysics Letters* **48**: 170.

Dualeh, A. J. and C. A. Steiner (1990). Hydrophobic microphase formation in surfactant solutions containing an amphiphilic graft copolymer. *Macromolecules* **23**(1): 251–255.

Dubief, C., J. F. Grollier, and J. Mondet (1989). Cosmetic compositions containing a cationic polymer and an anionic polymer as thickening agent.

Feitosa, E., W. Brown, and P. Hansson (1996a). Interactions between the non-ionic surfactant C12E5 and poly(ethylene oxide) studied using dynamic light scattering and fluorescence quenching. *Macromolecules* **29**(6): 2169–2178.

Feitosa, E., W. Brown, M. Vasilescu, and M. Swanson-Vethamuthu (1996b). Effect of temperature on the interaction between the nonionic surfactant C12E5 and poly(ethylene oxide) investigated by dynamic light scattering and fluorescence methods. *Macromolecules* **29**(21): 6837–6846.

Fishman, M. and F. Elrich (1975). Interactions of aqueous poly(*N*-vinylpyrrolidone) with sodium dodecyl sulfate. II. Correlation of electric conductance and viscosity measurements with equilibrium dialysis measurements. *The Journal of Physical Chemistry* **79**(25): 2740–2744.

Gasbarrone, P. and C. La Mesa (2001). Interactions of short-chain surfactants with a nonionic polymer. *Colloid and Polymer Science* **279**(12): 1192–1199.

Gasljevic, K., G. Aguilar, and E. Matthys (2001). On two distinct types of drag-reducing fluids, diameter scaling, and turbulent profiles. *Journal of Non-Newtonian Fluid Mechanics* **96**(3): 405–425.

Ge, L., X. Zhang, and R. Guo (2007). Microstructure of Triton X-100/poly(ethylene glycol) complex investigated by fluorescence resonance energy transfer. *Polymer* **48**(9): 2681–2691.

Ghoreishi, S., Y. Li, D. Bloor, J. Warr, and E. Wyn-Jones (1999). Electromotive force studies associated with the binding of sodium dodecyl sulfate to a range of nonionic polymers. *Langmuir* **15**(13): 4380–4387.

Goddard, E. (1986a). Polymer–surfactant interaction part II. Polymer and surfactant of opposite charge. *Colloids and Surfaces* **19**(2): 301–329.

Goddard, E. (1994). Polymer/surfactant interaction—Its relevance to detergent systems. *Journal of the American Oil Chemists' Society* **71**(1): 1–16.

Goddard, E. and R. Hannan (1976). Cationic polymer/anionic surfactant interactions. *Journal of Colloid and Interface Science* **55**(1): 73–79.

Goddard, E. D. (1986b). Polymer–surfactant interaction part I. uncharged water-soluble polymers and charged surfactants. *Colloids and Surfaces* **19**(2): 255–300.

Goddard, E. D. and K. P. Ananthapadmanabhan (1993). *Interactions of Surfactants with Polymers and Proteins*. CRC Press, Boca Raton, FL.

Goldraich, M., J. Schwartz, J. Burns, and Y. Talmon (1997). Microstructures formed in a mixed system of a cationic polymer and an anionic surfactant. *Colloids and Surfaces A: Physicochemical and Engineering Aspects* **125**(2): 231–244.

Han, J., F. Cheng, X. Wang, and Y. Wei (2012). Solution properties and microstructure of cationic cellulose/ sodium dodecyl benzene sulfonate complex system. *Carbohydrate Polymers* **88**(1): 139–145.

Hansson, P. and M. Almgren (1996). Interaction of CnTAB with sodium (carboxymethyl) cellulose: Effect of polyion linear charge density on binding isotherms and surfactant aggregation number. *The Journal of Physical Chemistry* **100**(21): 9038–9046.

Hansson, P. and B. Lindman (1996). Surfactant-polymer interactions. *Current Opinion in Colloid & Interface Science* **1**(5): 604–613.

Harada, A. and K. Kataoka (2006). Supramolecular assemblies of block copolymers in aqueous media as nanocontainers relevant to biological applications. *Progress in Polymer Science* **31**(11): 949–982.

Harwigsson, I. and M. Hellsten (1996). Environmentally acceptable drag-reducing surfactants for district heating and cooling. *Journal of the American Oil Chemists' Society* **73**(7): 921–928.

Hayakawa, K. and J. C. Kwak (1982). Surfactant-polyelectrolyte interactions. 1. Binding of dodecyltrimethylammonium ions by sodium dextransulfate and sodium poly(styrenesulfonate) in aqueous solution in the presence of sodium chloride. *The Journal of Physical Chemistry* **86**(19): 3866–3870.

Hayakawa, K. and J. C. Kwak (1983). Study of surfactant-polyelectrolyte interactions. 2. Effect of multivalent counterions on the binding of dodecyltrimethylammonium ions by sodium dextran sulfate and sodium poly(styrene sulfonate) in aqueous solution. *The Journal of Physical Chemistry* **87**(3): 506–509.

Hayakawa, K., J. P. Santerre, and J. C. Kwak (1983). Study of surfactant-polyelectrolyte interactions. Binding of dodecyl- and tetradecyltrimethylammonium bromide by some carboxylic polyelectrolytes. *Macromolecules* **16**(10): 1642–1645.

Hellsten, M. (2002). Drag-reducing surfactants. *Journal of Surfactants and Detergents* **5**(1): 65–70.

Hormnirun, P., A. Sirivat, and A. Jamieson (2000). Complex formation between hydroxypropylcellulose and hexadecyltrimethylamonium bromide as studied by light scattering and viscometry. *Polymer* **41**(6): 2127–2132.

Jiang, W. and S. Han (2000). Viscosity of nonionic polymer/anionic surfactant complexes in water. *Journal of Colloid and Interface Science* **229**(1): 1–5.

Jones, M. N. (1967). The interaction of sodium dodecyl sulfate with polyethylene oxide. *Journal of Colloid and Interface Science* **23**(1): 36–42.

Jönsson, B. (1998). *Surfactants and Polymers in Aqueous Solution*, John Wiley & Sons, New York, NY.

Jönsson, B., K. Holmberg, B. Kronberg, and B. Lindman (2003). *Surfactant-Polymer Systems in Surfactant and Polymers in Aqueous Solution*. John Wiley & Sons, New York, Chapter 13.

Jönsson, B., B. Lindman, K. Holmberg, and B. Kronberg (1998). *Surfactants and Polymers in Aqueous Solution*. John Wiley & Sons, Chichester, U.K.

Kogej, K. and J. Škerjanc (1999). Fluorescence and conductivity studies of polyelectrolyte-induced aggregation of alkyltrimethylammonium bromides. *Langmuir* **15**(12): 4251–4258.

Kwak, J. C. (1998a). *Polymer-Surfactant Systems*. CRC Press, Boca Raton, FL.

Kwak, J. C. T. (1998b). *Polymer-Surfactant Systems*. Surfactant Science Series, Vol. 77. Marcel Dekker, New York.

La Mesa, C. (2005). Polymer–surfactant and protein–surfactant interactions. *Journal of Colloid and Interface Science* **286**(1): 148–157.

Lange, H. (1971). Interaction between sodium alkyl sulfates and polyvinyl pyrrolidone in aqueous solutions. *Kolloid-Zeitschrift und Zeitschrift für Polymere* **243**: 101–109.

Lewis, E. A. and K. P. Murphy (2005). Isothermal Titration Calorimetry. In: Nienhaus, G.U., ed. *Protein-Ligand Interactions*. Springer, New York, NY, pp. 1–15.

Li, D., M. S. Kelkar, and N. J. Wagner (2012). Phase behavior and molecular thermodynamics of coacervation in oppositely charged polyelectrolyte/surfactant systems: A cationic polymer JR 400 and anionic surfactant SDS mixture. *Langmuir* **28**(28): 10348–10362.

Liu, X.-G., X.-J. Xing, Z.-N. Gao, B.-S. Wang, S.-X. Tai, and H.-W. Tang (2014). influence of three anionic gemini surfactants with different chain lengths on the optical properties of a cationic polyfluorene. *Langmuir* **30**(11): 3001–3009.

Ma, C. and C. Li (1989). Interaction between polyvinylpyrrolidone and sodium dodecyl sulfate at solid/liquid interface. *Journal of Colloid and Interface Science* **131**(2): 485–492.

Magny, B., I. Iliopoulos, R. Zana, and R. Audebert (1994). Mixed micelles formed by cationic surfactants and anionic hydrophobically modified polyelectrolytes. *Langmuir* **10**(9): 3180–3187.

Malovikova, A., K. Hayakawa, and J. C. Kwak (1984). Surfactant-polyelectrolyte interactions. 4. Surfactant chain length dependence of the binding of alkylpyridinium cations to dextran sulfate. *The Journal of Physical Chemistry* **88**(10): 1930–1933.

Mandal, B., S. P. Moulik, and S. Ghosh (2013). Physicochemistry of interaction of polyvinylpyrrolidone (PVP) with sodium dodecyl sulfate (SDS) in salt solution. *Journal of Surfaces and Interfaces of Materials* **1**(1): 77–86.

Masuda, Y., K. Hirabayashi, K. Sakuma, and T. Nakanishi (2002). Swelling of poly (ethylene oxide) gel in aqueous solutions of sodium dodecyl sulfate with added sodium chloride. *Colloid and Polymer Science* **280**(5): 490–494.

Mészáros, R., I. Varga, and T. Gilányi (2005). Effect of polymer molecular weight on the polymer/surfactant interaction. *The Journal of Physical Chemistry B* **109**(28): 13538–13544.

Min, T., J. Yul Yoo, H. Choi, and D. D. Joseph (2003). Drag reduction by polymer additives in a turbulent channel flow. *Journal of Fluid Mechanics* **486**: 213–238.

Minatti, E. and D. Zanette (1996). Salt effects on the interaction of poly(ethylene oxide) and sodium dodecyl sulfate measured by conductivity. *Colloids and Surfaces A: Physicochemical and Engineering Aspects* **113**(3): 237–246.

Mohsenipour, A. A. (2011). Turbulent drag reduction by polymers, surfactants and their mixtures in pipeline flow. PhD thesis, University of Waterloo, Waterloo, Ontario, Canada.

Mohsenipour, A. A. and R. Pal (2013). Drag reduction in turbulent pipeline flow of mixed nonionic polymer and cationic surfactant systems. *The Canadian Journal of Chemical Engineering* **91**(1): 190.

Mohsenipour, A. A., R. Pal, and K. Prajapati (2013). Effect of cationic surfactant addition on the drag reduction behaviour of anionic polymer solutions. *The Canadian Journal of Chemical Engineering* **91**(1): 181–189.

Mukherjee, S., A. Dan, S. C. Bhattacharya, A. K. Panda, and S. P. Moulik (2011). Physicochemistry of interaction between the cationic polymer poly(diallyldimethylammonium chloride) and the anionic surfactants sodium dodecyl sulfate, sodium dodecylbenzenesulfonate, and sodium *N*-dodecanoylsarcosinate in water and isopropyl alcohol–water media. *Langmuir* **27**(9): 5222–5233.

Muzzalupo, R., M. R. Infante, L. Pérez, A. Pinazo, E. F. Marques, M. L. Antonelli, C. Strinati, and C. La Mesa (2007). Interactions between gemini surfactants and polymers: Thermodynamic studies. *Langmuir* **23**(11): 5963–5970.

Mya, K. Y., A. M. Jamieson, and A. Sirivat (2000). Effect of temperature and molecular weight on binding between poly(ethylene oxide) and cationic surfactant in aqueous solutions. *Langmuir* **16**(15): 6131–6135.

Mysels, K. J. (1949). Napalm. Mixture of aluminum disoaps. *Industrial & Engineering Chemistry* **41**(7): 1435–1438.

Nagarajan, R. (1980). Thermodynamics of surfactant-polymer interactions in dilute aqueous solutions. *Chemical Physics Letters* **76**(2): 282–286.

Nagarajan, R. (1989). Association of nonionic polymers with micelles, bilayers, and microemulsions. *The Journal of Chemical Physics* **90**: 1980–1994. doi: 10.1063/1.456041.

Nagarajan, R. (2001). Polymer-surfactant interactions. *New Horizons: Detergents for the New Millennium Conference Invited Paper*, Fort Myers, FL.

Nambam, J. S. and J. Philip (2012). Effects of interaction of ionic and nonionic surfactants on self-assembly of PEO–PPO–PEO triblock copolymer in aqueous solution. *The Journal of Physical Chemistry B* **116**(5): 1499–1507.

Nelson, R. (1982). Application of surfactants in the petroleum industry. *Journal of the American Oil Chemists Society* **59**(10): 823A–826A.

Peng, B., X. Han, H. Liu, and K. C. Tam (2013). Binding of cationic surfactants to a thermo-sensitive copolymer below and above its cloud point. *Journal of Colloid and Interface Science* **412**: 17–23.

Piculell, L. and B. Lindman (1992). Association and segregation in aqueous polymer/polymer, polymer/surfactant, and surfactant/surfactant mixtures: Similarities and differences. *Advances in Colloid and Interface Science* **41**: 149–178.

Prajapati, K. (2009). Interactions between drag reducing polymers and surfactants. MSc thesis, University of Waterloo, Waterloo, Ontario, Canada.

Ptasinski, P., F. Nieuwstadt, B. Van Den Brule, and M. Hulsen (2001). Experiments in turbulent pipe flow with polymer additives at maximum drag reduction. *Flow, Turbulence and Combustion* **66**(2): 159–182.

Qi, Y. and J. L. Zakin (2002). Chemical and rheological characterization of drag-reducing cationic surfactant systems. *Industrial & Engineering Chemistry Research* **41**(25): 6326–6336.

Romani, A. P., M. H. Gehlen, and R. Itri (2005). Surfactant-polymer aggregates formed by sodium dodecyl sulfate, poly(*N*-vinyl-2-pyrrolidone), and poly(ethylene glycol). *Langmuir* **21**(1): 127–133.

Rothstein, J. P. (2008). Strong flows of viscoelastic wormlike micelle solutions. *Rheology Reviews* 1–42.

Ruckenstein, E., G. Huber, and H. Hoffmann (1987). Surfactant aggregation in the presence of polymers. *Langmuir* **3**(3): 382–387.

Saito, S. (1987). Polymer-surfactant interactions. In: Schick., M., ed. *Nonionic Surfactants Physical Chemistry*. Surfactant Science Series, Vol. 23. Marcel Dekker Inc., New York, pp. 881–892.

Saito, S. and T. Taniguchi (1973). Effect of nonionic surfactants on aqueous polyacrylic acid solutions. *Journal of Colloid and Interface Science* **44**(1): 114–120.

Salager, J.-L. (2002). Surfactant's types and uses. In: Salager, J.L., ed. *Firep Booket-E300-Attaching Aid in Surfactant Science and Engineering in English*, University of Andes, Merida, Venezuela, p. 3.

Sardar, N. and M. Kamil (2012a). Interaction between nonionic polymer hydroxypropyl methyl cellulose (HPMC) and cationic gemini/conventional surfactants. *Industrial & Engineering Chemistry Research* **51**(3): 1227–1235.

Sardar, N. and M. Kamil (2012b). Solution behavior of anionic polymer sodium carboxymethylcellulose (NaCMC) in presence of cationic gemini/conventional surfactants. *Colloids and Surfaces A: Physicochemical and Engineering Aspects* **415**: 413–420.

Scheuing, D. R., T. Anderson, W. L. Smith, E. Szekeres, and R. Zhang (2014). Cationic micelles with anionic polymeric counterions compositions thereof. U.S. Patent 8933010.

Schramm, L. L. (2000). *Surfactants: Fundamentals and Applications in the Petroleum Industry*. Cambridge University Press, Cambridge, U.K.

Schramm, L. L., E. N. Stasiuk, and D. G. Marangoni (2003). Surfactants and their applications. *Annual Reports Section C (Physical Chemistry)* **99**: 3–48.

Schwuger, M. (1973). Mechanism of interaction between ionic surfactants and polyglycol ethers in water. *Journal of Colloid and Interface Science* **43**(2): 491–498.

Shimizu, T., M. Seki, and J. C. Kwak (1986). The binding of cationic surfactants by hydrophobic alternating copolymers of maleic acid. *Colloids and Surfaces* **20**(4): 289–301.

Shirahama, K., K. Tsujii, and T. Takagi (1974). Free-boundary electrophoresis of sodium dodecyl sulfate-protein polypeptide complexes with special reference to SDS-polyacrylamide gel electrophoresis. *Journal of Biochemistry* (Tokyo, Japan) **75**(2): 309.

Shubin, V. (1994). Adsorption of cationic polymer onto negatively charged surfaces in the presence of anionic surfactant. *Langmuir* **10**(4): 1093–1100.

Singh, S. K. and S. Nilsson (1999). Thermodynamics of interaction between some cellulose ethers and SDS by titration microcalorimetry: II. Effect of polymer hydrophobicity. *Journal of Colloid and Interface Science* **213**: 152–159.

Smith, G. L. and C. L. McCormick (2001). Water-soluble polymers. 79. Interaction of microblocky twin-tailed acrylamido terpolymers with anionic, cationic, and nonionic surfactants. *Langmuir* **17**(5): 1719–1725.

Sreenivasan, K. R. and C. M. White (2000). The onset of drag reduction by dilute polymer additives, and the maximum drag reduction asymptote. *Journal of Fluid Mechanics* **409**: 149–164.

Stoll, M., H. Al-Shureqi, J. Finol, S. Al-Harthy, S. Oyemade, A. de Kruijf, J. Van Wunnik, F. Arkesteijn, R. Bouwmeester, and M. Faber (2010). Alkaline-surfactant-polymer flood: From the laboratory to the field. *SPE EOR Conference at Oil & Gas West Asia*, Muscat, Oman. doi: 10.2118/129164-MS.

Tadros, T. F. (2006). *Applied Surfactants: Principles and Applications*. John Wiley & Sons, New York.

Thalberg, K. and B. Lindman (1989). Interaction between hyaluronan and cationic surfactants. *The Journal of Physical Chemistry* **93**(4): 1478–1483.

Thalberg, K., B. Lindman, and G. Karlstroem (1990). Phase diagram of a system of cationic surfactant and anionic polyelectrolyte: Tetradecyltrimethylammonium bromide-hyaluronan-water. *Journal of Physical Chemistry* **94**(10): 4289–4295.

Thuresson, K., B. Nystroem, G. Wang, and B. Lindman (1995). Effect of surfactant on structural and thermodynamic properties of aqueous solutions of hydrophobically modified ethyl (hydroxyethyl) cellulose. *Langmuir* **11**(10): 3730–3736.

Thuresson, K., O. Söderman, P. Hansson, and G. Wang (1996). Binding of SDS to ethyl (hydroxyethyl) cellulose. Effect of hydrophobic modification of the polymer. *The Journal of Physical Chemistry* **100**(12): 4909–4918.

Touhami, Y., D. Rana, G. Neale, and V. Hornof (2001). Study of polymer-surfactant interactions via surface tension measurements. *Colloid & Polymer Science* **279**(3): 297–300.

Trabelsi, S. and D. Langevin (2007). Co-adsorption of carboxymethyl-cellulose and cationic surfactants at the air-water interface. *Langmuir* **23**(3): 1248–1252.

Veggeland, K. and T. Austad (1993). An evaluation of gel permeation chromatography in screening surfactant/polymer interaction of commercial products in saline aqueous solutions. *Colloids and Surfaces A: Physicochemical and Engineering Aspects* **76**: 73–80.

Veggeland, K. and S. Nilsson (1995). Polymer-surfactant interactions studied by phase behavior, GPC, and NMR. *Langmuir* **11**(6): 1885–1892.

Villetti, M. A., C. I. D. Bica, I. T. S. Garcia, F. V. Pereira, F. I. Ziembowicz, C. L. Kloster, and C. Giacomelli (2011). Physicochemical properties of methylcellulose and dodecyltrimethylammonium bromide in aqueous medium. *The Journal of Physical Chemistry B* **115**(19): 5868–5876.

Volden, S., J. Genzer, K. Zhu, M.-H. G. Ese, B. Nyström, and W. R. Glomm (2011). Charge-and temperature-dependent interactions between anionic poly(*N*-isopropylacrylamide) polymers in solution and a cationic surfactant at the water/air interface. *Soft Matter* **7**(18): 8498–8507.

Wallin, T. and P. Linse (1996a). Monte Carlo simulations of polyelectrolytes at charged micelles. 1. Effects of chain flexibility. *Langmuir* **12**(2): 305–314.

Wallin, T. and P. Linse (1996b). Monte Carlo simulations of polyelectrolytes at charged micelles. 2. Effects of linear charge density. *The Journal of Physical Chemistry* **100**(45): 17873–17880.

Wang, C. and K. Tam (2002). New insights on the interaction mechanism within oppositely charged polymer/surfactant systems. *Langmuir* **18**(17): 6484–6490.

Wang, Y., K. Kimura, P. L. Dubin, and W. Jaeger (2000). Polyelectrolyte-micelle coacervation: Effects of micelle surface charge density, polymer molecular weight, and polymer/surfactant ratio. *Macromolecules* **33**(9): 3324–3331.

Wang, Y., K. Kimura, Q. Huang, P. L. Dubin, and W. Jaeger (1999). Effects of salt on polyelectrolyte-micelle coacervation. *Macromolecules* **32**(21): 7128–7134.

Winnik, F. M. and S. T. Regismond (1996). Fluorescence methods in the study of the interactions of surfactants with polymers. *Colloids and Surfaces A: Physicochemical and Engineering Aspects* **118**(1): 1–39.

Winnik, F. M., S. T. Regismond, and E. D. Goddard (1997). Interactions of an anionic surfactant with a fluorescent-dye-labeled hydrophobically-modified cationic cellulose ether. *Langmuir* **13**(1): 111–114.

Witte, F. M. and J. B. F. N. Engberts (1987). Perturbation of SDS and CTAB micelles by complexation with poly(ethylene oxide) and poly(propylene oxide). *The Journal of Organic Chemistry* **52**(21): 4767–4772.

Yan, P. and J. X. Xiao (2004). Polymer-surfactant interaction: Differences between alkyl sulfate and alkyl sulfonate. *Colloids and Surfaces A: Physicochemical and Engineering Aspects* **244**(1–3): 39–44.

Yang, S.-Q. (2009). Drag reduction in turbulent flow with polymer additives. *Journal of Fluids Engineering* **131**(5): 051301.

Zakin, J., Y. Zhang, and W. Ge (2007). Drag reduction by surfactant giant micelles. *Surfactant Science Series* **140**: 473.

Zakin, J. L., Y. Zhang, and Y. Qi (2006). Drag reducing agents. In: Lee, S., ed. *Encyclopedia of chemical processing*, Vol. 2. Taylor and Francis, Boca Raton, FL, pp. 767–786.

Zakin, J. L., B. Lu, and H. W. Bewersdorff (1998). Surfactant drag reduction. *Reviews in Chemical Engineering* **14**(4–5): 67.

Zakin, J. L. and H. L. Lui (1983). Variables affecting drag reduction by nonionic surfactant additives. *Chemical Engineering Communications* **23**(1): 77–88.

Zakin, J. L., J. Myska, and Z. Chara (1996). New limiting drag reduction and velocity profile asymptotes for nonpolymeric additives systems. *AIChE Journal* **42**(12): 3544–3546.

Zana, R. and M. Benrraou (2000). Interactions between polyanions and cationic surfactants with two unequal alkyl chains or of the dimeric type. *Journal of Colloid and Interface Science* **226**(2): 286–289.

Zana, R., W. Binana-Limbelé, N. Kamenka, and B. Lindman (1992). Ethyl (hydroxyethyl) cellulose-cationic surfactant interactions: Electrical conductivity, self-diffusion and time-resolved fluorescence quenching investigations. *The Journal of Physical Chemistry* **96**(13): 5461–5465.

Zhang, H., D. Wang, and H. Chen (2009). Experimental study on the effects of shear induced structure in a drag-reducing surfactant solution flow. *Archive of Applied Mechanics* **79**(8): 773–778.

Zhang, Q., W. Kang, D. Sun, J. Liu, and X. Wei (2013). Interaction between cationic surfactant of 1-methyl-3-tetradecylimidazolium bromide and anionic polymer of sodium polystyrene sulfonate. *Applied Surface Science* **279**: 353–359.

Zhang, X., D. Taylor, R. Thomas, and J. Penfold (2011). The role of electrolyte and polyelectrolyte on the adsorption of the anionic surfactant, sodium dodecylbenzenesulfonate, at the air-water interface. *Journal of Colloid and Interface Science* **365**(2): P656–P664.

Zhang, Y. (2005). Correlations among surfactant drag reduction, additive chemical structures, rheological properties and microstructures in water and water/co-solvent systems. PhD thesis, The Ohio State University, Columbus, OH.

Zhao, G. and S. B. Chen (2006). Clouding and phase behavior of nonionic surfactants in hydrophobically modified hydroxyethyl cellulose solutions. *Langmuir* **22**(22): 9129–9134.

Index

G

H

I

Q

R

S

T

U

V

W

Y

Z

Milton Keynes UK
Ingram Content Group UK Ltd.
UKHW051901071024
449327UK00025B/2045

9 780367 575663